An Introduction to Genetic Analysis

An Introduction to Genetic Analysis

FOURTH EDITION

David T. Suzuki

University of British Columbia

Anthony J. F. Griffiths

University of British Columbia

Jeffrey H. Miller

University of California, Los Angeles

Richard C. Lewontin

Harvard University

W. H. FREEMAN AND COMPANY / NEW YORK

Cover illustration by Tomo Narashima

Library of Congress Cataloging-in-Publication Data

An introduction to genetic analysis, 4th ed. / David T. Suzuki
  [et al.]
    p. cm.
  Bibliography, p.
  Includes index.
  ISBN 0-7167-1956-8.   ISBN 0-7167-1996-7 (pbk.)
  1. Genetics.   2. Genetics—Methodology.   I. Suzuki, David T.,
1936–
QH430.I62   1989   88-34252
575.1—dc19   CIP

Printed in the United States of America

4 5 6 7 8 9   KP   9 9 8 7 6 5 4 3 2 1

Contents

Preface

■ The great power of modern genetics and its position of prominence in biological research have grown from a blend of classical and molecular techniques. Each analytical approach has its unique strengths. Classical genetics is unparalleled in its ability to explore uncharted biological terrain; molecular genetics is equally unparalleled in its ability to unravel cellular mechanisms. It would be unthinkable to teach one without the other. By giving both due prominence in this book, we have attempted to present a balanced view of genetics as practiced today.

The partnership of classical and molecular genetics has always presented a teaching problem: the order and manner in which the two partners should be introduced to the student. A common solution is to introduce both classical and molecular sides of each genetic principle concurrently. This approach makes use of the wisdom of hindsight — the teacher or writer addresses the student from the lofty vantage point of current understanding of the diverse parts of genetics and how they interrelate. We have chosen a different approach. It seems to us that students begin much as biologists did at the turn of the century, asking general questions about the laws governing inheritance. Thus the first half of the book is a more or less historical treatment of classical eukaryotic genetics, modified somewhat by the need to organize the material under specific genetic concepts. Some molecular information is provided where necessary, but this aspect is not emphasized. Armed with classical principles, the students can proceed to the second half of the book, which integrates molecular techniques and information into the classical framework. Progression from the general perspective to the specific seems to be a natural one, and makes sense not only in research but in teaching about research.

Geneticists today can be divided into two broad groups: those concerned with the mechanisms of heredity and those using techniques of genetics to probe other fundamental biological processes or phenomena. Both of these approaches are emphasized and recur as themes throughout the text. However, as the title of this book suggests, the main theme is genetic analysis. This emphasis reflects our belief that the best way to understand genetics is by understanding how genetic inferences are made. In any science, findings are important in themselves, but equally important are the modes of inference and the techniques of analysis, because these are the keys to future exploration. Quantitative analysis is particularly important. Many of the abstract ideas in genetics, from independent assortment to the existence of repetitive DNA, have been based on the analysis of quantitative data. The problems at the end of each chapter provide the student with the opportunity to apply these analytical methods to experimental situations.

Once again, this edition contains extensive revisions. These changes were made partly from our continuing attempts to update the contents and improve pedagogy in light of our own teaching experiences and partly from the suggestions of users and reviewes. In recent years, the balance between classical and molecular genetics in research has changed considerably; we have therefore given molecular genetics more emphasis in this edition than in previous ones, in recognition of its increasing application in all areas of the life sciences.

New features have been added and old ones improved. One prominent innovation is the list of key concepts found on the title page of each chapter. These are brief summary statements of the major themes to be found in the chapter, phrased to avoid the use of terms that have not yet been defined. They are intended to act as pedagogical signposts for the material that follows. A second major new feature is the addition of solved problems at the end of each chapter. Though these are often, but not always, representative of the problem sets, they are meant to show how the principles of the chapter can be applied to subsequent analysis. Approximately sixty new problems have been added to the problem sets. Most of these are simpler problems inserted early in the sets, and are designed to give practice and build confidence. As before, the problems are arranged roughly in order of difficulty, with particularly challenging problems marked with asterisks. The selected problem solutions at the back of the book have been assembled and modified by Diane Lavett (SUNY College at Cortland), who is also author of the *Companion,* which contains the complete set of problem solutions and a number of helpful study hints.

Also prominent is the newly added color insert. Although genetics is a somewhat cerebral subject, and is properly taught this way, it is nevertheless true that the subject material of genetics — the organisms themselves — are very often colorful and interesting. Furthermore this richness is a major source of inspiration in research as well as in teaching. We have therefore added a selection of colorful specimens that illustrate a variety of the principles covered in the book. Of course, some topic areas lend themselves naturally to color illustration and these comprise the bulk of the illustrations.

All chapters have been revised, but some contain extensive changes, as follows. In the first chapter on linkage (Chapter 5), the section on mapping has been completely rewritten. In the second linkage chapter (Chapter 6) the section on tetrad analysis has likewise been completely rewritten. The chapter on bacterial and phage genetics (Chapter 10) has been relocated to its original position in the second edition, in order to better accommodate the ensuing discussion on the nature of the gene. The entire central part of the book (broadly, molecular genetics) has undergone extensive revisions. Noteworthy here are the major changes in Chapter 10 and the addition of a new section on DNA manipulation in eukaryotes to Chapter 15. Chapter 16 has been amended to include updated material on eukaryotic genetic control mechanisms. Models of recombination are described more clearly in Chapter 18 and new material concerning mechanisms of transposition is presented in Chapter 19. The material on developmental genetics (Chapter 21 in the Third Edition) has been extended and reorganized into two separate chapters (21 and 22) that reflect the exciting recent developments in research in that area. The two chapters on populations (23 and 24) have also been streamlined.

As before, several special features aid the student. Throughout the text, major conclusions are summarized as Messages. These provide convenient stopping points from which readers may orient themselves within each chapter and may also be used as a convenient way of reviewing the material. Each chapter ends with a concise summary. New terms are set in boldface type; boldface is also used for emphasis. Most boldface words are defined in the Glossary. Suggestions for Further Reading are also found at the end of the book.

Thanks are due to the following people at W. H. Freeman and Company for their considerable support throughout the preparation of this edition: Patrick Fitzgerald, acquisitions editor; Moira Lerner, development editor; Stephen Wagley, project editor; Mary George, copy editor; Mike Suh, designer; Bill Page, illustration coordinator; and Susan Stetzer, production coordinator.

We also extend our thanks to the following reviewers, whose insights and suggestions were most helpful in the revision process: Wyatt W. Anderson, University of Georgia; Anna Berkovitz, Purdue University; Bruce J. Cochrane, University of South Florida; Christopher A. Cullis, Case Western Reserve University; Jeffrey L. Doering, Loyola University of Chicago; James E. Haber, Brandeis University; Robert Holmgren, Northwestern University, Robert Ivarie, University of Georgia; Janet Kurjan, Columbia University; Diane K. Lavett, State University of New York, Cortland; Anthony J. Pelletier, University of Colorado, Boulder; Jeffrey Powell, Yale University; Mark F. Sanders, University of California, Davis; Katherine Spindler, University of

Georgia; Jill Tabor; Laurie Tompkins, Temple University; David A. West, Virginia Polytechnical Institute and State University; and John H. Williamson, Davidson College.

We hope this book will stimulate the reader to do some first-hand experimental genetics, whether as professional scientist, student, or amateur plant or animal breeder. Failing this, we hope to impart some lasting impression of the incisiveness, elegance, and power of genetic analysis.

David T. Suzuki
Anthony J. F. Griffiths
Jeffrey H. Miller
Richard C. Lewontin

An Introduction to Genetic Analysis

Genetics and the Organism

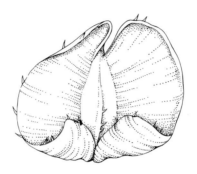

KEY CONCEPTS

Genetics has unified the biological sciences.

■

Genetics may be defined as the study of genes through their
variation.

■

Gene variation contributes to variation in nature.

■

The characteristics of an organism are determined by the
interaction of its unique set of genes with its unique
environment.

■

Genetics is of direct relevance to human affairs.

■ Why study genetics? The answers to this question constitute the major part of this book, but at the outset a summary answer can be given. Although a relatively young discipline, genetics has assumed a position of central importance in the biological sciences, because a knowledge of the structure and function of the genetic material has been found to be essential to an understanding of most aspects of a living organism. In addition to this powerful unifying role, genetics has gained a position of great importance in human affairs. The findings of genetic research have had considerable impact not only in the applied areas of biology, medicine, and agriculture but also in such areas as philosophy, law, and religion. It is a rare newspaper issue nowadays that does not address some aspect of genetics.

The Scope of Genetics

Why has genetics become so important? To answer this question we must first define genetics. The science of **genetics** attempts to understand the properties of the genetic material, *deoxyribonucleic acid,* best known by its abbreviation DNA. Geneticists study the properties of DNA at many levels, ranging from cells to populations. The cells of all organisms, from bacteria to humans, contain one or more sets of a basic DNA complement that is unique to the species. This fundamental complement of DNA is called a **genome.** The genome may be subdivided into **chromosomes,** each of which is a very long single continuous DNA molecule. In its turn, a chromosome can be demarcated along its length into thousands of functional regions, called **genes,** and also into regions of less well understood function. For example, each of the trillions of cells that comprise a human being has 46 chromosomes in two equivalent sets, or genomes, of 23. Each of the 23 chromosomes in a genome is unique; it is matched only by its equivalent partner in the other set and by all other chromosomes of that type in the rest of the members of the species. An average human chromosome represents about a 50-millimeter length of DNA; therefore, the human genome is the equivalent of about one meter of DNA. This amount of DNA can only be packed into a nucleus by very efficient coiling and folding.

The functional units of heredity, the genes, have quite naturally become the focus of geneticists as they try to understand the laws of heredity. However, because genes are relevant to many different biological processes, many other kinds of biologists—from physiologists to ecologists—are also interested in genes, but these scientists would probably not consider themselves to be geneticists. Although it is true that many subdisciplines of biology are separated by "gray areas" today, it is nevertheless an interesting excercise to ask what, if anything, sets genetics apart. To this question, two kinds of answer can be given. First, geneticists tend to concentrate more on the basic properties of the genes themselves. Second, genetics works in a unique way that makes use of naturally occurring or induced gene variation in a population of organisms. In fact, variation is the raw material for genetic studies: if all members of a population were identical, genetic analysis as we know it could not be done. Therefore, it is useful to define genetics as *the study of genes through their variation.* Many examples of the way in which genes are studied through their variation will be found throughout this book, and the topic will be addressed again later in this chapter. The principle is, however, simple: only when a gene is different in some way can the geneticists follow its inheritance and its effects on the organism.

Geneticists study all aspects of genes. The study of the modes of gene transmission from generation to generation is broadly called **transmission genetics,** the study of gene structure and function is called **molecular genetics,** and the study of gene behavior in populations is called **population genetics.** These form the three major subdivisions of the field of genetics, although, as with all categories invented by humans, the subdivisions are to a certain extent arbitrary and there is considerable overlap. It is the knowledge of how genes act and how they are transmitted down through the generations that has unified biology; previously, specific sets of biological phenomena had each been relegated to separate disciplines. An understanding of how genes act is now an essential prerequisite for such biological fields of study as development, cytology, physiology, and morphology. An understanding of gene transmission is a fundamental aspect of areas such as ecology, evolution, and taxonomy. Further unification has resulted from the discovery that the basic chemistry of gene structure and function is very similar across the entire spectrum of life on the earth. These points may seem trite to those who have grown up in the light of current knowledge, but it is important to realize that our modern view of biology and its inter-related parts is a relatively recent phenomenon. Not so long ago, biology was fragmented into many camps that rarely communicated with each other. Today, however, every biologist must be a bit of a geneticist, because the findings and techniques of genetics are being applied and used in all fields. Take physiology, for example. Most aspects of the physiology of a cell, from photosynthesis to microtubule function, are strongly influenced by genes. In medicine, many human diseases, including many cancers, are influenced at the genetic level. Genetics, in fact, provides the modern paradigm for all of biology.

> **Message** Genetics has provided a unifying thread for the previously disparate fields of biology.

What have been some of the success stories of genetics within basic biology? For just about any feature of biological structure or function—such as size, shape, number of parts, color, pattern, behavioral pattern, or biochemical function—that has been looked at in experimental organisms, genes have been found to be involved. Determining the location of these genes on their respective chromosomes is relatively easy. It has been found that the major way in which genes exert their effect is by controlling the myriad chemical reactions that go on inside cells, and thousands of specific genes have each been associated with a specific chemical reaction. The control of gene action (how genes are "turned on" and "turned off") has been intensely studied in many organisms, and the mechanisms in these cases are well understood. Many different genes have been isolated in the test tube, and their particular chemical structures have been determined. Such studies have provided important clues about how genes perform their functions. A gene can be removed from one organism and introduced into another, either for the convenience of propagating large amounts of the gene for later study or to examine its effects in another biological system. Genes can be modified at will to study the effects of these changes on biological processes; such changes can involve a large part of the genes or very localized regions of interest to the experimenter. Most genes have been found to reside rather stably at specific chromosomal locations, but other pieces of DNA have been found to be capable of sudden relocation to new areas. Last but not least, most of these findings have tremendous relevance for evolutionary processes, which of course are concerned with changes in the structure and function of the gene set.

Clearly, the advances in genetics have been truly astounding, particularly over the last three decades. Many recent accomplishments—such as the isolation and characterization of individual genes, which researchers in the 1950s believed could never happen in their lifetimes—have already come to be regarded as routine procedures in current work in genetics.

This chapter presents an overview of the subject of genetics, by way of an orientation to the rest of the book. We shall deal first with generalities about genes, their inheritance, and the ways in which they interact with the environment. Then we shall discover the unique ways in which geneticists identify specific genes and examine the use of these techniques in studying biological phenom-

ena. Finally, we shall consider some of the ways in which genetics has interacted with human society.

Gene Transmission

Genetics embraces two contradictory aspects of nature: offspring resemble their parents, yet they are not identical to their parents. The offspring of lions are lions and never lambs, yet no two lions are identical, even if they come from the same litter. We have no trouble recognizing the differences between sisters, for example, and even "identical" twins are recognized as distinctive individuals by their parents and close friends. But we also can notice subtle similarities between parents and children. As we shall see, **heredity** (the similarity of offspring to parents) and **variation** (the difference between parents and offspring and between the offspring themselves) turn out to be two aspects of the same fundamental mechanism.

Humans necessarily became involved with both heredity and variation when they began to domesticate plants and animals (around 10,000 years ago), because humans had to choose the organisms with advantageous characteristics from among all the organisms at their disposal and then seek to propagate these traits in future generations. References in Egyptian tomb inscriptions and in the Bible to sound breeding practices convince us that conscious concern with genetic phenomena is at least as old as civilization. The farmers and shepherds involved with such concerns could quite deservedly have claimed to be called geneticists. But the formal study of genetics as a coherent and unified theory of heredity and variation is little more than a century old.

Modern genetics as a set of principles and analytic rules began with the work of an Augustinian monk, Gregor Mendel, who worked in a monastery in the middle of the nineteenth century in what is now Brno in Czechoslovakia. Mendel was taken into the monastery by its director, Abbot Knapp, with the express purpose of trying to discover a firm mathematical and physical foundation underlying the practice of plant breeding. Knapp and others in Brno were interested in fruit breeding. They believed that recent advances in mathematics in the physical sciences could establish a model for building a science of variation. Mendel was recommended to Knapp as a good scholar of mathematics and physics, although a rather mediocre student of biology!

Mendel's methods, which he developed in the monastery garden, are still used today (in an extended form) and form an integral part of genetic analysis. (Mendel's work is considered in detail in Chapter 2.) Mendel realized that both the similarities and the differences among parents and their offspring can be explained by a mechanical transmission of discrete hereditary units, which we now call genes, from parent to offspring. We now know that in all organisms—whether bacteria, fungi, animals, or plants—there is a regular passage of hereditary information from parent to offspring by means of the genes. The regularities we observe in heredity and variation are consequences of the regularities of the mechanical lanes of transmission and activity of these genes.

Recall that each gene is a portion of a DNA molecule. In more than one sense, then, DNA is truly the thread of life: not only is a DNA molecule itself a thread-like string of genes, but the DNA handed down from parent to offspring represents a narrow connecting thread between the generations. When we say that a woman has her mother's hair or a man has his father's nose, what we really mean is that the parent has handed on, in egg or sperm, the instructions necessary to direct the synthesis of that specific feature.

Out of these basic considerations emerge two vastly powerful and unique properties of DNA that make it the fundamental molecule of life. The first of these is its ability to serve as a model for the production of replicas of itself, termed **replication.** This property is the key to transmission and forms the basis of transmission genetics. A parental organism transmits a replica of its DNA to the progenitor cell of an individual of the next generation. As this progenitor cell goes through its rounds of division to produce a multicellular organism, each division is accompanied by the production of identical replicas of the DNA of the progenitor cell, which are apportioned into each new cell. Thus, replication is the mechanism through which life persists across the generations in a stable fashion.

The second property of DNA that makes it a fundamental molecule of life is its ability to act as a carrier of information. For example, embodied into the one meter of DNA that constitutes a human chromosome set is the information needed to build a specimen of *Homo sapiens.* The word information means literally "that which is necessary to give form"; this is precisely what the DNA of the genes does. The information is "written" into the sequence of DNA in the form of a molecular code.

Gene and Organism

Precisely how does information become form? At the level of molecules, the answer to this question embraces much of what was defined previously as molecular genetics. Basically, the phenomena and structures of life are produced by an interaction of DNA with the inanimate world, the nonliving environment. The universe naturally tends to disarray; order tends spontaneously to disorder; complex and orderly objects become piles of dust; the reverse does not occur unaided. Yet DNA causes an eddy in this river of chaos; through its interaction with the disorderly components of the universe, the most orderly system that we know about is born: the phenomenon of life. One of the unexpected discoveries arising from the study of DNA function is that the mechanism of converting information into form is virtually identical across all groups of organisms on this planet. We humans share a common genetic chemistry with the entire variety of life forms on the earth—a staggering spectrum, including some 286,000 species of flowering plants, 500,000 species of fungi, and 750,000 species of insects.

A general view of the interaction of DNA with the environment is a necessary prelude to the detailed analyses that are found in the chapters ahead. We must put the gene and the environment into perspective in order to provide a framework on which the details of genetic analysis can be hung.

It is a characteristic of living organisms that they mobilize the components of the world around themselves and convert these components into their own living material, or into artifacts that are extensions of themselves. An acorn becomes an oak tree, using in the process only water, oxygen, carbon dioxide, some inorganic materials from the soil, and light energy.

The seed of an oak tree develops into an oak, while the spore of a moss develops into a moss, although both are growing side by side in the same forest (Figure 1-1).

Figure 1-1. The genes of a moss direct environmental components to be shaped into a moss, whereas the genes of a tree cause a tree to be constructed from the same components. (From Grant Heilman.)

The two plants that result from these developmental processes resemble their parents and differ from each other, even though they have access to the same narrow range of inorganic materials from the environment. The specifications for building living protoplasm from the environmental materials are passed in the form of genes from parent to offspring through the physical materials of the fertilized egg. As a consequence of the information in the genes, the seed of the oak develops into an oak and the moss spore becomes a moss.

What is true for the oak and moss is also true within species. Consider plants of the species *Plectritis congesta*, the sea blush. Two forms of this species are found wherever the plants grow in nature: one form has wingless fruits, and the other has winged fruits (Figure 1-2). These plants will self-pollinate, and we can observe the offspring that result from such "selfs" when these are grown in a greenhouse under uniform conditions. It is commonly observed that the progeny of a winged-fruited plant are all winged-fruited and that the progeny from a wingless-fruited plant all have wingless fruits. Since all the progeny were grown in an identical environment, we can safely conclude that the difference between the original plants must result from the different genes they carry.

The *Plectritis* example involves two inherited forms that can both be considered perfectly normal. Yet the determinative power of genes is equally well demonstrated when a gene becomes abnormal. The human inherited disease sickle-cell anemia provides a good example. In this case, careful study has revealed the chain of events whereby the gene impacts on the organism from the submicroscopic molecular level, through the microscopic level, to the macroscopic anatomical level. The underlying cause of the disease is a variation in hemoglobin, the oxygen-transporting protein molecule found in red blood cells. Normal people have a type of hemoglobin called hemoglobin A, the information for which is encoded in a gene. A minute chemical change at the molecular level in the DNA of this gene results in the production of a slightly changed hemoglobin, termed hemoglobin S. In people possessing only hemoglobin S, the ultimate effect of this small change is severe ill health and usually death. The gene works its effect on the organism through a complex "cascade effect," as summarized in Figure 1-3.

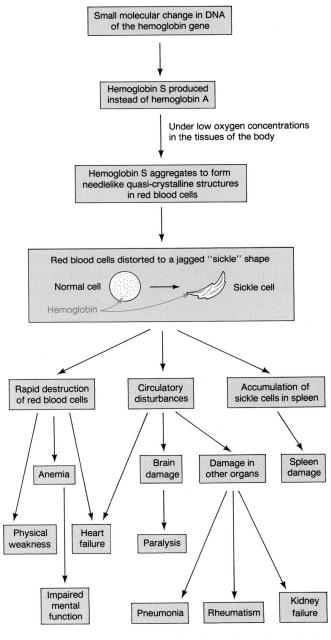

Figure 1-3. Chain of events resulting in sickle-cell anemia in humans.

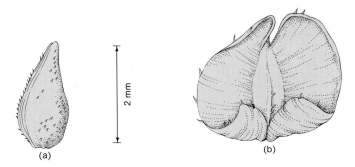

Figure 1-2. The fruits of two different forms of *Plectritis congesta*, the sea blush. (a) Wingless fruits. (b) Winged fruits. Any one plant has either all wingless or all winged fruits. In every other way the plants are identical. The striking difference in the appearance of the fruits is determined by a simple genetic difference.

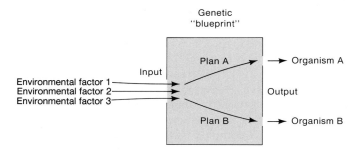

Figure 1-4. A model of determination that emphasizes the role of genes.

Observations like these lead to a model of the interaction of genes and environment like that shown in Figure 1-4. In this view, the genes act as a set of instructions for turning more or less undifferentiated environmental materials into a specific organism, much as blueprints specify what form of house is to be built from basic materials. The same bricks, mortar, wood, and nails can be made into an A-frame or a flat-roofed house, according to different plans. Such a model implies that the genes are really the dominant elements in the determination of organisms; the environment simply supplies the undifferentiated raw materials.

But now consider two monozygotic ("identical") twins, the product of a single fertilized egg that divided and produced two complete individuals with identical genes. Suppose that the twins are born in England but are separated at birth and taken to different countries. If one is raised in China by Chinese-speaking foster parents, she will speak perfect Chinese, while her sister raised in Budapest will speak fluent Hungarian. Each will absorb the cultural values and customs of her environment. Although the twins begin life with identical genetic properties, the different cultural environments in which they live will produce differences between the sisters (and differences from their parents). Obviously, the difference in this case is due to the environment, and genetic effects are of little importance.

This example suggests the model of Figure 1-5, which is the opposite of that shown in Figure 1-4. In the model in Figure 1-5, the genes impinge on the system, giving certain general signals for development, but the environment determines the actual course of change. Imagine a set of specifications for a house that simply calls for "a floor that will support 30 pounds per square foot" or "walls with an insulation factor of 15"; the actual appearance and nature of the structure would be determined by the available building materials.

Our different types of examples — of purely genetic effect versus that of the environment — lead to two very different models. Given a pair of seeds and a uniform growth environment, we would be unable to predict future growth patterns solely from a knowledge of the environment. In any environment we can imagine, if growth occurs at all, the acorn will become an oak and the spore will become a moss. On the other hand, considering the twins, no information about the set of genes they inherit could possibly enable us to predict their ultimate languages and cultures. Two individuals that are *genetically different* may develop differently in the *same environment,* but two *genetically identical* individuals may develop differently in *different environments.*

In general, of course, we deal with organisms that differ in both genes and environment. If we wish to understand and predict the outcome of the development of a living organism, we must first know the genetic constitution that it inherits from its parents. Then we must know the *historical sequence* of environments to which the developing organism is exposed. We emphasize the historical sequence of environments rather than simply the general environment. Every organism has a developmental history from birth to death. What an organism will become in the next moment depends critically both on the environment it encounters during that moment and on its present state. It makes a difference to an organism not only what environments it encounters but in what sequence it encounters them. A fruit fly *(Drosophila)* develops normally at 20°C. If the temperature is briefly raised to 37°C early in the pupal stage of

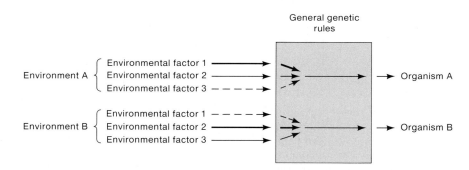

Figure 1-5. A model of determination that emphasizes the role of the environment.

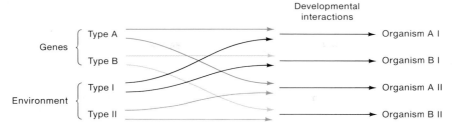

Figure 1-6. A more realistic model of determination that emphasizes the interaction of genes and environment.

development, the adult fly will be missing part of the normal vein pattern on its wings. However, if this "temperature shock" is administered just 24 hours later, the fly develops normally.

Considering the preceding discussion of the general nature of interaction between genes and environment, we see that there is no reason to prefer either of the asymmetrical models of Figures 1-4 and 1-5. Instead, we can create a more general model (Figure 1-6) in which genes and environment are seen as symmetrical factors that jointly determine (by some rules of development) the actual characteristics of the organism.

> **Message** The developmental transformation of an organism from one stage of its life to another is a result of the unique interaction of its genes and its environment at each moment of its life history. Organisms are determined neither by their genes nor by their environment; rather, they are the consequence of the interaction of genes *and* environment.

Genotype and Phenotype

In studying the reaction whereby genes and environment produce an organism, geneticists have developed some useful terms, which are introduced in this section.

A typical organism resembles its parents more than it resembles unrelated individuals. Thus, we often speak as if the individual characteristics themselves are inherited: "He gets his brains from his mother," or "She inherited diabetes from her father." Yet our discussion in the preceding section shows that such statements are invalid. "His brains" and "her diabetes" develop through long sequences of events in the life histories of the affected persons, and both genes and environment play roles in those sequences. In the biological sense, individuals inherit only the molecular structures of the fertilized eggs from which they develop. Individuals inherit their *genes*, not the end products of their individual developmental histories.

To prevent such confusion between genes (which are inherited) and developmental outcomes (which are not), geneticists make a fundamental distinction between the genotype and the phenotype of an organism. Organisms belong to the same **genotype** if they have the same set of genes. Organisms belong to the same **phenotype** if they resemble each other in some manifest way.

Strictly speaking, the genotype describes the complete set of genes inherited by an individual and the phenotype describes *all* aspects of the individual's morphology, physiology, behavior, and ecological relationships. In this sense, no two individuals ever belong to the same phenotype, because there is always some difference (however slight) between them in morphology or physiology. Moreover, except for individuals produced from another organism by asexual reproduction, any two organisms differ at least a little in genotype. In practice, we use the terms genotype and phenotype in a more restricted sense. We deal with some partial phenotype descriptions (say, eye color) and with some subset of the genotype (say, the genes that influence eye pigmentation).

> **Message** When geneticists use the terms phenotype and genotype, they generally mean "partial phenotype" and "partial genotype" with respect to some defined set of traits and genes.

Note one very important difference between genotype and phenotype: the genotype is essentially a fixed character of an individual organism; the genotype remains constant throughout life and is essentially unchanged by environmental effects. Most phenotypes change continually throughout the life of the individual; the direction of that change is a function of the sequence of environments that the individual experiences. Fixity of genotype does not imply fixity of phenotype.

The Norm of Reaction

How can we quantify the relation between the genotype, the environment, and the phenotype? For a particular genotype, we could prepare a table showing the phenotype that would result from the development of that genotype in each possible environment. Such a tabulation of environment-phenotype relationships for a given genotype is called the **norm of reaction** of the genotype. In practice, of course, we can make such a tabulation only for a partial genotype, a partial phenotype, and some particular aspects of the environment. For example, we might specify the eye size of a fruit fly that would result from development at various constant temperatures for several different partial genotypes.

Figure 1-7 represents just such norms of reaction for three partial genotypes in the fruit fly *Drosophila melanogaster*. The graph is a convenient summary of more extensive tabulated data. The size of the fly eye is measured by counting its individual facets, or cells. The vertical axis of the graph shows the number of facets (on a logarithmic scale); the horizontal axis shows the constant temperature at which the flies develop.

Three norms of reaction are shown on the graph. Flies of the *wild-type* genotype (characteristic of flies in natural populations) show somewhat smaller eyes at higher temperatures of development. Flies of the abnormal *ultra-bar* genotype have smaller eyes than *wild-type* flies at any particular temperature of development.

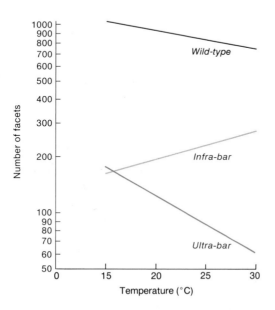

Figure 1-7. Norms of reaction to temperature for three different eye-size genotypes of *Drosophila melanogaster*: *wild-type*, *infra-bar*, and *ultra-bar*. Eye size is measured by the number of facets in the eye. The term "bar" comes from the eye shape produced by fewer facets.

Higher temperatures also have a stronger effect upon eye size in the *ultra-bar* flies (the *ultra-bar* line slopes more steeply). Any fly of the abnormal genotype, *infra-bar*, also has smaller eyes then any *wild-type* fly, but higher temperatures have the opposite effect on flies of this genotype: *infra-bar* flies raised at higher temperatures tend to have larger eyes than those raised at lower temperatures. These norms of reaction indicate that the relationship between genotype and phenotype is complex rather than simple.

Message   A single genotype can produce many different phenotypes, depending on the environment. A single phenotype may be produced by various different genotypes, depending on the environment.

If we know that a fruit fly has the *wild-type* genotype, this information alone does not tell us whether its eye has 800 or 1000 facets. On the other hand, the knowledge that a fruit fly's eye has 170 facets does not tell us whether its genotype is *ultra-bar* or *infra-bar*. We cannot even make a general statement about the effect of temperature on eye size in *Drosophila*, because the effect is opposite in two different genotypes. We see from Figure 1-7 that some genotypes do differ unambiguously in phenotype, no matter what the environment: any *wild-type* fly has larger eyes than any *ultra-bar* or *infra-bar* fly. But other genotypes overlap in phenotypic expression: the eyes of *ultra-bar* flies may be larger or smaller than those of *infra-bar* flies, depending on the temperatures at which the individual developed.

To obtain a norm of reaction like the norms of reaction in Figure 1-7, we must allow different individuals of identical genotypes to develop in many different environments. To carry out such an experiment, we must be able to obtain or produce many fertilized eggs with identical genotypes. For example, to test a human genotype in 10 environments, we would have to obtain genetically identical sibs and raise each individual in a different milieu. Obviously, that is possible neither biologically nor socially. At the present time, we do not know the norm of reaction of any human genotype for any character in any set of environments. Nor is it clear how we can ever acquire such information without the unacceptable manipulation of human individuals.

For a few experimental organisms, special genetic methods make it possible to replicate genotypes and thus to determine norms of reaction. Such studies are particularly easy in plants that can be propagated vegetatively (that is, by cuttings). The pieces cut from a single plant all have the same genotype, so all offspring produced in this

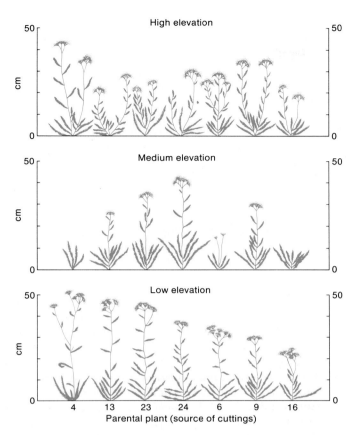

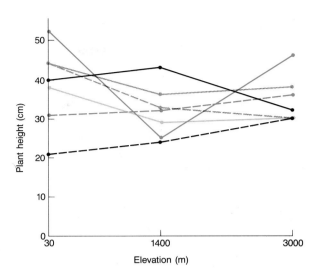

Figure 1-9. Graphic representation of the complete set of results of the type shown in Figure 1-8. Each line represents the norm of reaction of a separate plant. Notice that the norms of reaction cross one another, so that no sharp distinction is apparent.

Figure 1-8. Norms of reaction to elevation for seven different *Achillea* plants (seven different genotypes). A cutting from each plant was grown at low, medium, and high elevations. (Carnegie Institution of Washington.)

way have identical genotypes. Figure 1-8 shows the results of a study using such vegetative offspring of the plant *Achillea*. Many plants were collected, and three cuttings were taken from each plant. One cutting was planted at low elevation (30 meters above sea level), one at medium elevation (1400 meters), and one at high elevation (3050 meters). Figure 1-8 shows the mature individuals that developed from the cuttings of seven collected (parental) plants; the three plants of identical genotypes are aligned vertically in the figure for comparison.

First, we note an average effect of environment: in general, the plants grew poorly at the medium elevation. This is not true for every genotype, however; the cutting of plant 24 grew best at the medium elevation. Second, we note that no genotype is unconditionally superior in growth to all others. Plant 4 showed the best growth at low and high elevations but showed the poorest growth at the medium elevation. Plant 9 showed the second-worst growth at low elevation and the second-best at high elevation. Once again, we see the complex relation-

ship between genotype and phenotype. Figure 1-9 graphs the norms of reaction derived from the results shown in Figure 1-8. Each genotype has a different norm of reaction, and the norms cross one another so that we cannot identify either a "best" genotype or a "best" environment for *Achillea* growth.

We have seen two different patterns of reaction norms. The difference between the *wild-type* and the abnormal eye-size genotypes in *Drosophila* is such that the corresponding phenotypes show a consistent difference, regardless of the environment. Any fruit fly of *wild-type* genotype has larger eyes than any fruit fly of the abnormal genotypes, so we could (imprecisely) speak of "large-eye" and "small-eye" genotypes. In this case, the differences in phenotype between genotypes are much greater than the variation within a genotype for different environments. In the case of *Achillea*, however, the variation for a single genotype in different environments is so great that the norms of reaction cross one another and form no consistent pattern. In this case, it makes no sense to identify a genotype with a particular phenotype except in terms of response to particular environments.

Developmental Noise

Thus far, we have assumed that the phenotype is uniquely determined by the interaction of a specific genotype and a specific environment. But a closer look shows some further unexplained variation. According to Figure 1-7, a *Drosophila* of *wild-type* genotype raised at 16°C has 1000 facets in each eye. In fact, this is only an average value; individual flies studied under these conditions commonly have 980 or 1020 facets. Perhaps these variations are due to slight fluctuations in the local environment or slight differences in genotypes. However, a typical count may show that a fly has, say, 1017 facets in the left eye and 982 in the right eye. In another fly, the left eye has slightly fewer facets than the right eye. Yet the left and right eyes of the same fly are genetically identical. Furthermore, under typical experimental conditions, the fly develops as a larva (a few millimeters long) burrowing in homogeneous artificial food in a laboratory bottle and then completes its development as a pupa (also a few millimeters long) glued vertically to the inside of the glass high above the food surface. Surely the environment does not differ significantly from one side of the fly to the other! But if the two eyes experience the same sequence of environments and are identical genetically, then why is there any phenotypic difference between the left and right eyes?

Differences in shape and size are partly dependent on the process of cell division that turns the zygote into a multicellular organism. Cell division, in turn, is sensitive to molecular events within the cell, and these may have a relatively large random component. For example, the vitamin biotin is essential for growth, but its *average* concentration is only one molecule per cell! Obviously, any process that depends on the presence of this molecule will necessarily be subject to fluctuations in rate resulting from random variations in concentration. But if a cell division is to produce a differentiated eye cell, it must occur within the relatively short developmental period during which the eye is being formed. Thus, we

would expect random variation in such phenotypic characters as the number of eye cells, the number of hairs, the exact shape of small features, and the variations of neurons in a very complex central nervous system — even when the genotype and the environment are precisely fixed. Even in such structures as the very simple nervous systems of nematodes, these random variations do occur.

Message   Random events in development lead to an uncontrollable variation in phenotype; this variation is called **developmental noise.** In some characteristics, such as eye cells in *Drosophila,* developmental noise is a major source of the observed variations in phenotype.

Like noise in a verbal communication, developmental noise adds small random variations to the predictable process of development governed by norms of reaction. Adding developmental noise to our model of phenotype development, we obtain something like Figure 1-10. With a given genotype and environment, there is a range of possible outcomes for each developmental step. The developmental process does contain feedback systems that tend to hold the deviations within certain bounds, so that the range of deviation does not increase indefinitely through the many steps of development. However, this feedback is not perfect. For any given genotype, developing in any given sequence of environments, there remains some uncertainty as to the exact phenotype that will result.

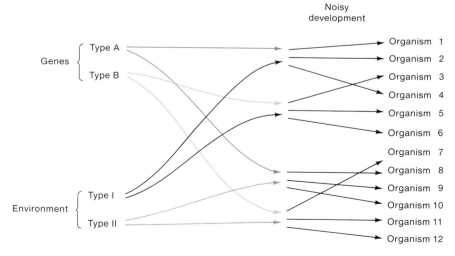

Figure 1-10. A model of phenotypic determination that shows how genes, environment, and developmental noise interact to produce any given phenotype.

Techniques of Genetic Analysis

Our discussion thus far has been based on the wisdom of hindsight. With the wealth of genetic knowledge we now share, we can make generalizations about DNA, genes, phenotypes, and genotypes as though these concepts were self-evident. But obviously this was not always the case; the current wisdom was acquired only after extensive genetic research over the years. Mendel, for example, almost certainly was completely without a conceptual basis for his research at the beginning of his work, but he was able to piece together his genetic principles from the results of his many experiments. This is true of genetic research in general: we start in the unknown, and then ideas and facts emerge out of experimentation. But how does everyday genetics work? We shall explore the answer to this question in much of the rest of this book, but we begin here with an overview of the principles of genetic research.

The process of identification of the specific hereditary components of a biological system is called **genetic dissection.** A geneticist is a type of biologist interested in some aspect of the structure or function of organisms. In the same way that an anatomist probes biological structure and function with a scalpel, the geneticist probes biological structure and function armed with genetic variants, usually abnormal ones. If the geneticist is interested in a biological process X, he or she embarks on a search for genetic variants that affect X. Each variant identifies a separate component of X. In much the same way that a novice auto mechanic can learn a lot about how an internal combustion engine works by pulling out a spark-plug lead, for example, the geneticist "tinkers" with a living system. This approach is tremendously effective in charting the unknown—the invisibility and magnitude of which are often not appreciated by those who have not attempted some kind of research—and represents a truly powerful tool.

> **Message**   Genetic dissection is a powerful way of discovering the components of any biological process.

The use and analysis of hereditary variants embodies the primary research tool of the geneticist. This approach is equally powerful from the molecular level, where it can be used to probe cellular and organismal processes, all the way up to the population level, where it can be used to investigate evolutionary processes. The spectrum of variants that has been obtained and studied effectively by geneticists is staggering; as we have seen, there are variants affecting shape, number, biochemical function, and so on. In fact, the general finding has been that genetic variants can be obtained for virtually any biological structure or process of interest to the investigator.

Recall that a set of variants of any given biological character can fall into either of two broad categories: (1) variants with underlying genotypes that show nonoverlapping norms of reaction, and (2) those with underlying genotypes that show overlapping norms of reaction. The techniques of genetic analysis are different in each case.

The analysis of genotypes with nonoverlapping norms of reaction has produced most of the phenomenal success of geneticists in elucidating the cellular and molecular processes of organisms. The appeal of such systems from the experimenter's viewpoint is that the environment can virtually be ignored, because each genotype produces a discrete, identifiable phenotype. In fact, geneticists have deliberately sought and selected just such variants. Experiments become much easier— actually, they only become possible—when the observed phenotype is an unambiguous indication of a particular genotype. After all, we cannot observe genotypes; we can visibly distinguish only phenotypes.

The availability of such variants made Mendel's experiments possible. He used genotypes that were established horticultural varieties. For example, in one set of variants, one genotype always produces tall pea plants and another always produces dwarf plants. Had Mendel chosen variants with genotypes that have overlapping norms of reaction, he would have obtained complex and patternless results like those in the *Achillea* experiments (Figure 1-8) and could never have identified the simple relationships that became the foundation of genetic understanding. Much of modern molecular genetics is based on research with variants of bacteria that similarly show distinctive characters with genotypes that have nonoverlapping norms of reaction. For instance, modern DNA manipulation technology relies heavily on selection systems based on bacterial genes for drug resistance; the expression of such genes is very clear-cut and reliable.

In this category, the variants are often strikingly different. Many such variants could never survive in nature, simply because they are so developmentally extreme. In other cases, such as the winged and wingless fruits of *Plectritis,* the strikingly different forms do appear regularly in natural populations.

Reflecting the importance of discrete variants in developing our present understanding of the genetic basis of life, most of the rest of this book is devoted to the analysis of this kind of variation. The story begins with Mendel's research and theories, proceeds through classical genetics, and ends with the startling discoveries of molecular biology. It must be emphasized that the entire development of this knowledge critically depended on the availability of phenotypes that have simple relationships to genotypes. Because our study of genetic analysis necessarily puts so much emphasis on such simple trait differences, you may get the impression that they represent the most common relationship between gene and trait. They do not. Geneticists have to pick and choose among cases of genetic variation to find those with a simple correspondence between genotype and phenotype. For example, among the hundreds of classical mutants of *Drosophila* that have been well enough characterized to place them on specific chromosomes, about one-half are too variable in their phenotypic expression to be used in further genetic analysis (so-called Rank 4 and Rank 5 mutants). Only one-quarter (Rank 1 mutants) are ideally suited for the purposes of careful genetic analysis. And this count excludes the large number of mutants that were so unreliable they could not be put on genetic maps in the first place. Environment and developmental age can be quite important to the expression of even the most useful mutants. The widely used mutation *purple* produces an eye color like the normal ruby color in very young flies, which darkens to a distinguishable difference only with age. The mutation *Curly*, one of the most important tools in genetic analysis in *Drosophila*, results in curled wings at 25°C but in normally straight wings at 19°C.

Whereas simple one-to-one relationships between genotype and phenotype dominate the world of experimental genetics, in the natural world such relationships are quite rare. When obtained for organisms taken from natural populations, norms of reaction for size, shape, color, metabolic rate, reproductive rate, and behavior almost always turn out to be like those of *Achillea*. The relationships of genotype to phenotype in nature are almost always one-to-many rather than one-to-one. This is the underlying cause of the rarity of discrete phenotypic classes in natural populations.

Obviously, the analysis of these one-to-many relationships is far more complex. The researcher is confronted with a bewildering range of phenotypes. Special statistical techniques must be used to disentangle the genetic, environmental, and noise components. Chapter 22 deals with such techniques.

There is another problem that is distinct from the purely analytic difficulties. The major discoveries of genetics were founded on work with discrete norms of reaction. Although these discoveries have revolutionized pure and applied biology, great care must be taken in extrapolating such ideas to systems that are based on the complex interactions of gene and environment, especially those found in nature. For example, the difference between yellow-bodied and gray-bodied *Drosophila* in laboratory stocks can be shown to be a simple genetic difference, but it does not follow that skin color in humans obeys similar genetic laws. In fact, even in *Dro-*

sophila, the various intensities of black pigment in flies from natural populations turn out to be a consequence of the interaction of temperature with a complex genetic system.

Message In general, the relationship between genotype and phenotype cannot be extrapolated from one species to another or even between phenotypic traits that seem superficially similar within a species. A genetic analysis must be carried out for each particular case.

Our overview of the scope of genetics would not be complete without a discussion of some of the ways in which genetics affects our everyday lives.

Genetics and Human Affairs

Knowledge about hereditary phenomena has been important to humans for a very long time. Civilization itself became possible when nomadic tribes learned to domesticate plants and animals. Long before biology existed as a scientific discipline, people selected grains with higher yields and greater vigor and animals with better fur or meat. They also puzzled about the inheritance of desirable and undesirable traits in the human population. Despite this long-standing concern with heredity and the practice of selective breeding, it was not until the discovery of Mendel's laws that we were able to elucidate the genetic basis for selection and to place breeding techniques in a scientifically vigorous framework.

Early in this century, a new wheat strain called Marquis was developed in Canada. This high-quality strain is resistant to disease; furthermore, it matures two weeks earlier than other commercially used strains—a very important factor where the growing season is short. At the time of its introduction, the use of Marquis wheat opened up millions of square miles of fertile soil to cultivation in such northerly countries as Canada, Sweden, and the U.S.S.R. Table 1-1 shows how geneticists have bred a wide range of desirable characteristics into another commercial crop, rice. In addition to improving crop varieties, geneticists have learned to alter the genetic systems of insects to reduce their fertility. This technique is providing an important new weapon in the age-old struggle to keep insects out of human crops and habitations.

As with many other areas of science, new knowledge has produced new challenges as well as solutions to some human problems. Breeding successes in recent years led

TABLE 1-1. Development of pest-resistant strains of rice

Strain	Year developed	Diseases					Insects		
		Blast fungus	Bacterial blight	Leaf-streak virus	Grassy stunt virus	Tungro virus	Green leaf-hopper	Brown hopper	Stem borer
IR 8	1966	MR	S	S	S	S	R	S	MS
IR 5	1967	S	S	MS	S	S	R	S	S
IR 20	1969	MR	R	MR	S	R	R	S	MS
IR 22	1969	S	R	MS	S	S	S	S	S
IR 24	1971	S	S	MR	S	MR	R	S	S
IR 26	1973	MR	R	MR	MR	R	R	R	MR

NOTE: The entries describe each strain's susceptibility or resistance to each pest as follows: S = susceptible; MS = moderately susceptible; MR = moderately resistant; R = resistant. (From *Research Highlights*, I.R.R.I. 1973, p. 11.)

to the "Green Revolution" as the scientific answer to the problem of human hunger. Using sophisticated breeding techniques based on new knowledge about genes, geneticists created such high-yield varieties as dwarf wheat (Figure 1-11) and rice (Figure 1-12). Extensive planting of these crops around the world did provide new food supplies, but new problems quickly became apparent. These specialized crops require extensive cultivation and costly fertilizers. Figure 1-12 is a norm-of-reaction curve, a classic example of the variable interaction of genotype with environment. The use of the new high-yield varieties produced a wide range of social and economic problems in the impoverished countries where they were most needed. Furthermore, the spread of monoculture (the extensive reliance on a single plant variety) left vast areas at the mercy of some newly introduced or evolved form of pathogen—say, a plant disease or an insect pest. With the huge population of humans on earth, our dependence on high-yield varieties of crop plants and domestic animals is becoming

Figure 1-11. A specially bred strain of dwarf wheat *(right)* resists crop damage far better than the normal strain *(left)*. (Courtesy of The Rockefeller Foundation.)

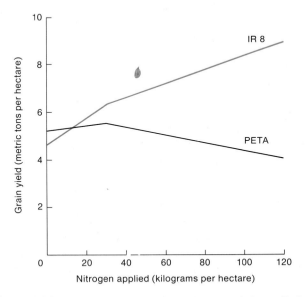

Figure 1-12. The specially bred strain of dwarf rice called IR 8 owes part of its success to its remarkable response to the application of fertilizer. PETA, an older, nondwarf strain, shows a more typical response. (From Peter R. Jennings, "The Amplification of Agricultural Production." Copyright © 1976 by Scientific American, Inc. All rights reserved.)

increasingly obvious. In a very real sense, the stability of human society depends on the ability of geneticists to juggle the inherited traits that shape life forms, keeping the crops a jump ahead of destructive parasites and predators.

In recent years, advances in biotechnology have led to the creation of special genetically engineered strains of bacteria and fungi that carry specific genes from unrelated organisms such as humans. The microbes produce such useful compounds as insulin, human growth hormone, and the antiviral (and possibly anticancer) agent interferon.

But the most exciting and frightening application of genetic knowledge is to the human species itself. Genetic discoveries have had major effects on medicine. We can now diagnose hereditary disease before or soon after birth, and in some cases we can provide secondary treatments. Using family pedigrees, a genetic counselor can give prospective parents the information they need to make intelligent decisions about the risks of genetic disease in their offspring. Such refined techniques as amniocentesis and fetoscopy provide information about possible genetic disease at early stages of pregnancy. A battery of postnatal chemical tests can detect problems in the newborn infant, so that some corrective techniques can be applied immediately to alleviate the effects of many genetic diseases.

Our new ability to recognize genetic disease poses an important moral dilemma. An estimated 5 percent of our population survives with severe physical or mental genetic defects. This percentage probably will increase with extended exposure to various environmental factors—and, paradoxically, with improved medical technology. As geneticist Theodosius Dobzhansky has remarked,

> If we enable the weak and the deformed to live and propagate their kind, we face the prospect of a genetic twilight. But if we let them die or suffer when we can save or help them, we face the certainty of a moral twilight.

Of those patients admitted to pediatric hospitals in North America, 30 percent are estimated to have diseases that can be traced to genetic causes. The financial burden to society is already significant. Are we prepared to shoulder this genetic burden? How much money are we willing to spend to keep the genetically handicapped alive and to enable them to lead as normal a life as possible?

Many other significant issues are raised by the potential applications of genetic knowledge to human beings. Obviously, the human brain is subject to the same rules of genetic determination as the rest of the body. Does this mean that our thoughts and behaviors are extensively determined by inherited predispositions? Or can we view the mind as a clean slate at birth, written on only by individual experience? The nature of inborn constraints on thought and personality and the implications for present sociological problems have fascinated many geneticists and other scientists. Such books as *African Genesis, The Territorial Imperative, On Aggression,* and other popular titles on sociobiology have stimulated widespread public interest. A bitter debate has raged about the differences in intelligence among various racial and social groups. Of course, this topic is not new. It was debated by Lycurgus in Sparta against Plato in Athens, and dreams of producing a pure race of superior humans have motivated many infamous figures in history. As the problem of human overpopulation becomes obvious to almost everyone, there is increased talk of legislated sterilization and the planned selection of human offspring. There is very serious talk, even among some geneticists, about the ability of the human race to take control of its own evolution. Others are frightened by the possibilities for disastrous error or unacceptable sociological consequences.

The sophisticated technology of molecular genetics has given us a wide range of new techniques for shaping our genetic makeup. Even more bizarre procedures loom in the near future. We have moved beyond con-

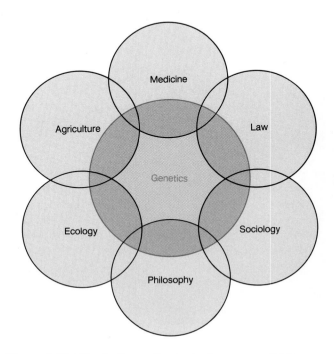

Figure 1-13. The findings of genetics have powerful impacts on many interacting areas of human endeavor.

ventional breeding techniques to the ability to make specific chemical and molecular modifications in the genetic apparatus. In some cases genetic material is added that is completely new to the species being modified. While some scientists emphasize the promised benefits of such research, others raise disturbing questions about possible dangers. Could there be an accidental release from some laboratory of an artificial pathogen that has never existed on this planet before? Such worries have led to calls for a complete moratorium on this type of research—or at least for legislation that requires ruthlessly efficient containment facilities. A number of popular books warn us of a *Genetic Fix,* a *Biological Time Bomb,* the *Genetic Revolution,* a *Fabricated Man,* and the *Biocrats.* Knowledge of genetic mechanisms has made us aware of other new dangers as well. Some geneticists fear that increased exposure to chemical food additives and to the vast array of chemicals in other commercial products may be changing the human genetic makeup in a very undesirable and haphazard way. This type of random genetic change can also be caused by such environmental agents as fallout from nuclear weapons, radioactive contamination from nuclear reactors, and radiation from various X-ray machines. These agents may be contributing to inherited disease, and they almost certainly are contributing to the incidence of cancer, a genetic disease of the somatic ("body") cells.

The study of genetics is relevant not only to the biologist but to any thinking member of today's complex technological society (Figure 1-13). A working knowledge of the principles of genetics is essential for making informed decisions on many scientific, political, and personal levels. Such a working knowledge can come only through an understanding of the way in which genetic inferences are made—that is, from an understanding of genetic analysis, the subject of this book.

Mendelian Analysis

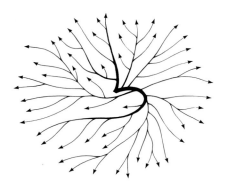

KEY CONCEPTS

The existence of genes can be inferred by observing certain
standard progeny ratios in crosses between hereditary variants.

∎

A discrete character difference is often determined by a
difference in a single gene.

∎

In higher organisms, each type of gene is represented twice
in each cell.

∎

During sex-cell formation, each member of a gene pair
separates into one-half of the sex cells.

∎

During sex-cell formation, different gene pairs are often
observed to behave independently of one another.

■ The gene, the basic functional unit of heredity, is the focal point of the discipline of modern genetics. In all lines of genetic research, it is the gene that provides the common unifying thread to a great diversity of experimentation. Geneticists are concerned with the transmission of genes from generation to generation, with the nature of genes, with the variation in genes, and with the ways in which genes function to dictate the features that constitute any given species.

In this chapter we trace the birth of the gene as a concept. We shall see that genetics is, in one sense, an abstract science: most of its entities began as hypothetical constructs in the minds of geneticists and were later identified in physical form if the reasoning was sound.

The concept of the gene (but not the word) was first proposed in 1865 by Gregor Mendel. Until then, little progress had been made in understanding heredity. The prevailing notion was that the spermatozoon and egg contained a sampling of essences from the various parts of the parental body; at conception, these essences somehow blended to form the pattern for the new individual. This idea of **blending inheritance** evolved to account for the fact that offspring typically show some characteristics that are similar to those of both parents. However, there are some obvious problems associated with this idea, one of which is that offspring are not always an intermediate blend of their parents' characteristics. Attempts to expand and improve this theory led to no better understanding of heredity.

As a result of his research with pea plants, Mendel proposed instead a theory of **particulate inheritance.** According to Mendel's theory, characters are determined by discrete genetic units that are inherited intact down through the generations. This model explained many observations that could not be explained by the idea of blending inheritance. It also proved a very fruitful framework for further progress in understanding the mechanism of heredity

For many reasons, the importance of Mendel's ideas was not recognized until about 1900 (after his death). His written work was then rediscovered by three scientists, after each had independently obtained the same kind of results. Mendel's work constitutes the prototype genetic analysis. He laid down an experimental and logical approach to heredity that is still used today. Let's examine this exemplary work.

Mendel's Experiments

Mendel's studies provide an outstanding example of good scientific technique. Mendel chose research material well suited to the study of the problem at hand,

designed his experiments carefully, collected large amounts of data, and used mathematical analysis to show that the results were consistent with his explanatory hypothesis. The predictions of the hypothesis were then tested in a new round of experimentation. (Some historians of science believe that Mendel may have "fudged" his data to fit his hypothesis, but we shall accept his reports at face value in our discussion here.)

Mendel studied the garden pea *(Pisum sativum)* for two main reasons. First, peas were available through a seed merchant in a wide array of distinct shapes and colors that could be very easily identified and analyzed. Second, peas left to themselves will **self** (self-pollinate) because the male parts (anthers) and female parts (ovaries) of the flower — which produce the pollen and the eggs, respectively — are enclosed in a petal box, or keel (Figure 2-1). The gardener or experimenter can **cross** (cross-pollinate) any two plants at will. The anthers from one plant are clipped off to prevent selfing; pollen from the other plant is then transferred to the receptive area with a paintbrush or on transported anthers (Figure 2-2). Thus, the experimenter can readily choose to self or to cross the pea plants.

Other practical reasons for Mendel's choice of peas were that they are cheap and easy to obtain, take up little space, have a relatively short generation time, and produce many offspring. These considerations are typical of those involved in the choice of organism for any piece of genetic research. This is a crucial decision and is nearly always based on not only scientific criteria but also a good measure of expediency.

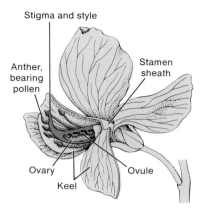

Figure 2-1. Cutaway view of reproductive parts of a pea flower. (After J. B. Hill, H. W. Popp, and A. R. Grove, Jr., *Botany,* McGraw-Hill, 1967.)

Figure 2-2. One technique of artificial cross-pollination, demonstrated with *Mimulus guttatus,* the yellow monkey flower. Anthers from the male parent are applied to the stigma of an emasculated flower, which acts as the female parent.

Plants Differing in One Character

First, Mendel chose several characters to study. It is important to clarify the meaning of character in this sense. Here, a **character** is a specific property of an organism; geneticists use this term as a synonym for characteristic or trait.

For each of the characters he chose, Mendel obtained lines of plants, which he grew for two years to make sure they were pure lines. A **pure line** is a population that breeds true for the particular character being studied; that is, all offspring produced by selfing or crossing within the population show the same form for this character. By ensuring pure lines in his research material, Mendel had made a clever beginning: he had established a basis of identifiable constant behavior, so that any changes observed following deliberate manipulation in his research would be scientifically meaningful; in effect, he had set up a control experiment.

Two of the plant lines Mendel grew proved to breed true for the character of flower color. One line bred true for purple flowers; the other, for white flowers. Any

plant in the purple-flowered line — when selfed or when crossed with others from the same line — produced seeds that all grew into plants with purple flowers. When these plants in turn were selfed or crossed within the line, their progeny also had purple flowers, and so on. The white-flowered line similarly produced only white flowers through all generations. Mendel obtained seven pairs of pure lines for seven characters, with each pair differing in respect to only one character (Figure 2-3).

Each pair of Mendel's plant lines can be said to show a **character difference** — a contrasting difference between two lines of organisms (or between two organisms) with respect to one particular character. The differing lines (or individuals) represent different forms that the character may take: they can be called character forms, character variants, or phenotypes. The useful term phenotype (derived from Greek) literally means "the form that is shown." The term is extensively used in genetics, and we use it in this discussion, even though such words as gene and phenotype were not coined or used by Mendel. We shall describe Mendel's results and hypotheses in terms of more modern genetic language.

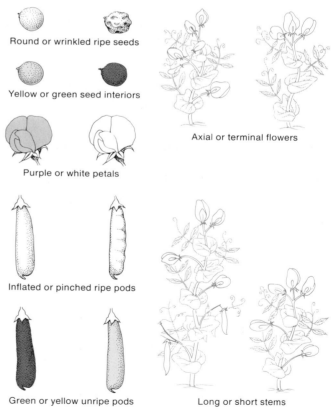

Figure 2-3. The seven character differences studied by Mendel. (After S. Singer and H. Hilgard, *The Biology of People.* Copyright © 1978, W. H. Freeman and Company.)

Figure 2-3 shows seven pea characters, each represented by two contrasting phenotypes. Contrasting phenotypes for a particular character are the starting point for any genetic analysis, by Mendel or by the modern geneticist. This is an illustration of the point made in Chapter 1 that variation is the raw material for any genetic analysis. Of course, the delineation of characters is somewhat arbitrary; there are many different ways to "split up" an organism into characters. For example, consider the following different ways of stating the same character and character difference.

Character	Phenotypes
Flower color	Red versus white
Flower redness	Presence versus absence
Flower whiteness	Absence versus presence

In many cases, the description chosen is a matter of convenience (or chance). Fortunately, the choice does not alter the final conclusions of the analysis, except in the words used.

We turn now to some of Mendel's specific experimental results. In our discussion, we shall follow his analysis of the lines breeding true for flower color.

In one of his early experiments, Mendel used pollen from a white-flowered plant to pollinate a purple-flowered plant. These plants from the pure lines are called the **parental generation** (P). All the plants resulting from this cross had purple flowers (Figure 2-4). This progeny generation is called the **first filial generation** (F_1). (The subsequent generations in such an experiment are called F_2, F_3, and so on.)

Mendel also made a **reciprocal cross.** In most plants, any cross can be made in two ways, depending on which phenotype is used as male (♂) or female (♀). For example, the two crosses

$$\text{phenotype A ♀} \times \text{phenotype B ♂}$$
$$\text{phenotype B ♀} \times \text{phenotype A ♂}$$

are reciprocal crosses. Mendel's reciprocal cross, in which a white flower was pollinated by a purple-flowered plant, produced the same result (all purple flowers) in the F_1 (Figure 2-5). Mendel concluded that it makes no difference which way the cross is made. If one pure-breeding parent is purple-flowered and the other is white-flowered, all plants in the F_1 are purple-flowered. The purple flower color in the F_1 generation is identical to that in the purple-flowered parental plants. In this case, the inheritance obviously is not a simple blending of purple and white colors to produce some intermediate color. To maintain a theory of blending inheritance, we would have to assume that the purple color is somehow

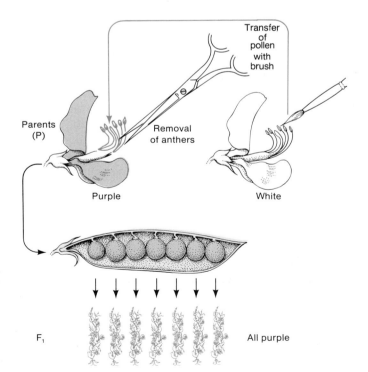

Figure 2-4. Mendel's cross of purple-flowered ♀ × white-flowered ♂.

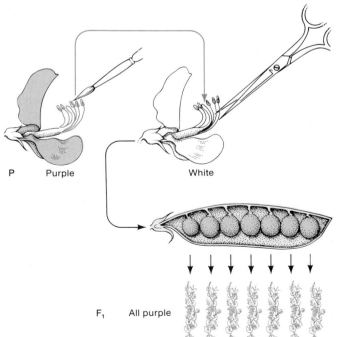

Figure 2-5. Mendel's cross of white-flowered ♀ × purple-flowered ♂.

"stronger" than the white color and completely overwhelms any trace of the white phenotype in the blend.

Next, Mendel selfed the F_1 plants, allowing the pollen of each flower to fall on the stigma within its petal box. He obtained 929 pea seeds from this selfing (the F_2 individuals) and planted them. Interestingly, some of the resulting plants were white-flowered; the white phenotype had reappeared! Mendel then did something that, more than anything else, marks the birth of modern genetics: he *counted* the numbers of plants with each phenotype. This procedure seems obvious to modern biologists after another century of quantitative scientific research but had seldom, if ever, been used in genetic studies before Mendel's work. There were 705 purple-flowered plants and 224 white-flowered plants. Mendel observed that the ratio of 705 : 224 is almost a 3 : 1 ratio (in fact, it is 3.1 : 1).

Mendel repeated this breeding procedure for six other pairs of pea character differences. He found the same 3 : 1 ratio in the F_2 generation for each pair (Table 2-1). By this time, he was undoubtedly beginning to believe in the significance of the 3 : 1 ratio and to seek an explanation for it. The white phenotype is completely absent in the F_1 generation, but it reappears (in its full original form) in one-fourth of the F_2 plants. It is very difficult to devise an explanation of this result in terms of blending inheritance. Even though the F_1 flowers are purple, the plants still must carry the *potential* to produce progeny with white flowers.

Mendel inferred that the F_1 plants receive from their parents the ability to produce both the purple phenotype and the white phenotype and that these abilities are retained and passed on to future generations rather than blended. Why is the white phenotype not expressed in the F_1 plants? Mendel invented the terms **dominant** and **recessive** to describe this phenomenon without explaining the mechanism. In modern terms, the purple phenotype is dominant to the white phenotype and the white phenotype is recessive to the purple phenotype. Thus, the phenotype of the F_1, established by breeding

different pure lines, provides the operational definition of dominance: the phenotype that appears in such F_1 individuals is defined as the dominant phenotype.

Mendel made another important observation when he individually selfed the F_2 plants. In this case, he was working with the character of seed color. Because this character could be observed without growing plants from the peas, much larger numbers of individuals could be counted. (In this species, the color of the seed is a characteristic of the offspring — the seed itself — rather than the parent plant.) Mendel used two pure lines of plants with yellow and green seeds, respectively. In a cross between one plant from each line, he observed that all of the F_1 peas were yellow. Symbolically

$$P \qquad \text{yellow} \times \text{green}$$
$$\downarrow$$
$$F_1 \qquad \text{all yellow}$$

Therefore, in this character, yellow is dominant and green is recessive.

Mendel grew F_1 plants from these yellow F_1 peas and selfed the plants. Of the resulting F_2 peas, $\frac{3}{4}$ were yellow and $\frac{1}{4}$ were green — the 3 : 1 ratio again (see the second line of Table 2-1). He then grew plants from 519 of the yellow F_2 peas and selfed each of these F_2 plants. When the peas appeared (the F_3 generation), Mendel found that 166 of the plants had only yellow peas. The remaining 353 plants bore both yellow and green peas on the same plant. Counting all the peas from these plants, he again obtained a 3 : 1 ratio of yellow to green peas. The green F_2 peas all proved to be pure-breeding green peas; that is, selfing produced only green peas in the F_3 generation.

In summary, all of the F_2 green peas were pure-breeding greens like one of the parents. Of the F_2 yellows, about $\frac{2}{3}$ were like the F_1 yellows (producing yellow and green seeds in a 3 : 1 ratio when selfed), and the remaining $\frac{1}{3}$ were like the pure-breeding yellow parent.

TABLE 2-1. Results of all Mendel's crosses in which parents differed for one character

Parental phenotype	F_1	F_2	F_2 ratio
1. Round × wrinkled seeds	All round	5474 round; 1850 wrinkled	2.96 : 1
2. Yellow × green seeds	All yellow	6022 yellow; 2001 green	3.01 : 1
3. Purple × white petals	All purple	705 purple; 224 white	3.15 : 1
4. Inflated × pinched pods	All inflated	882 inflated; 299 pinched	2.95 : 1
5. Green × yellow pods	All green	428 green; 152 yellow	2.82 : 1
6. Axial × terminal flowers	All axial	651 axial; 207 terminal	3.14 : 1
7. Long × short stems	All long	787 long; 277 short	2.84 : 1

Thus, the study of the F_3 generation revealed that the apparent $3:1$ ratio in the F_2 generation could be more accurately described as a $1:2:1$ ratio.

Further studies showed that these $1:2:1$ ratios existed in all of the apparent $3:1$ ratios that Mendel had observed. Thus, the problem really was to explain the $1:2:1$ ratio.

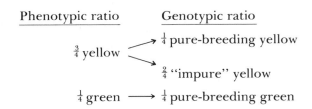

Phenotypic ratio	Genotypic ratio

$\frac{3}{4}$ yellow → $\frac{1}{4}$ pure-breeding yellow

→ $\frac{2}{4}$ "impure" yellow

$\frac{1}{4}$ green → $\frac{1}{4}$ pure-breeding green

Mendel's explanation was a classic example of a model or hypothesis derived from observation that was well suited for testing by further experimentation. Mendel deduced the following explanation of the $1:2:1$ ratio.

1. There are hereditary determinants of a particulate nature. (Mendel saw no blending of phenotypes, so he was forced to draw this conclusion.) We now call these determinants **genes.**

2. Each adult pea plant has two genes—a **gene pair**—in each cell for each character studied. Mendel's reasoning here was obvious: the F_1 plants, for example, must have had one gene, called the **dominant gene,** responsible for the dominant phenotype and one gene, called the **recessive gene,** responsible for the recessive phenotype, which showed up only in later generations.

3. The members of each gene pair segregate (separate) equally into the gametes, or sex cells. The gametes in animals are readily identifiable as eggs and sperm. Plants produce eggs and sperm too, but these forms are less easily identified as such.

4. Consequently, each gamete carries only one member of each gene pair.

5. The union of one gamete from each parent to form the first cell (or zygote) of a new progeny individual is random, occurring without regard to which member of a gene pair is carried.

These points can be illustrated diagrammatically for a general case, using A to represent a dominant gene and a to represent the recessive gene (as Mendel did), much as a mathematician uses symbols to represent abstract entities of various kinds. This is shown in Figure 2-6, which illustrates how these five points explain the $1:2:1$ ratio.

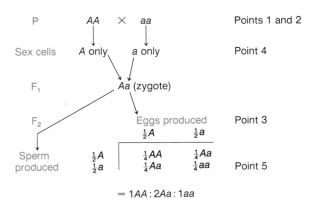

Figure 2-6. Symbolic representation of the P, F_1, and F_2 generations in Mendel's system involving a character difference determined by one gene difference. The five points are those listed in the text.

The whole model made logical sense of the data. However, many beautiful models have been knocked down under test. Mendel's next job was to test his model. He did this by taking (for example) an F_1 yellow and crossing it with a green. A $1:1$ ratio of yellow to green seeds could be predicted in the next generation. If we let Y stand for the dominant gene causing yellow seeds and y stand for the recessive gene causing green seeds, we can diagram Mendel's predictions, as shown in Figure 2-7. In this experiment, he obtained 58 yellow (Yy) and 52 green (yy), a very close approximation to the predicted $1:1$ ratio and confirmation of the equal segregation of Y and y in the F_1 individual. This concept of **equal segregation** has been given formal recognition as Mendel's first law.

Mendel's First Law The two members of a gene pair segregate (separate) from each other into the gametes, so that one-half of the gametes carry one member of the pair and the other one-half of the gametes carry the other member of the gene pair.

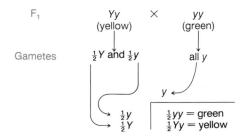

Predicted progeny ratio is 1 yellow : 1 green.

Figure 2-7. Predicted consequences of crossing an F_1 yellow with any green.

Now we need to introduce some more terms. The individuals represented by *Aa* are called **heterozygotes,** or sometimes **hybrids,** whereas the individuals in pure lines are called **homozygotes.** In words such as these, "hetero-" means different and "homo-" means identical. Thus, an *AA* plant is said to be homozygous for the dominant gene, sometimes called **homozygous dominant;** an *aa* plant is homozygous for the recessive gene, or **homozygous recessive.** As we saw in Chapter 1, the designated genetic constitution with respect to the character or characters under study is called the genotype. Thus, *YY* and *Yy*, for example, are different genotypes even though the seeds of both types are of the same phenotype (that is, yellow). You can see that in such a situation the phenotype can be thought of simply as the outward manifestation of the underlying genotype.

Note that the expressions dominant and recessive have been used in conjunction with both the phenotype *and* the gene. This is accepted usage. The dominant phenotype is established in analysis by the appearance of the F_1. Obviously, however, a phenotype (which is merely a description) cannot really exert dominance. Mendel showed that the dominance of one phenotype over another is in fact due to the dominance of one member of a gene pair over the other.

Let's pause to let the significance of this work sink in. What Mendel did was to develop an analytic scheme for the identification of major genes regulating any biological character or function. Starting with two different phenotypes (purple and white) of one character (petal color), he was able to show that the difference was caused by a difference at one gene pair. Let's take the two forms of the petal-color character as an example. Modern geneticists would say that Mendel's analysis had identified a major gene for petal color. What does this mean? It means that in these organisms there is a *kind* of gene that has a profound effect on the color of the petals. This gene can exist in different forms: the dominant form of the gene (represented by *C*) causes purple petals, and the recessive form of the gene (represented by *c*) causes white petals. The forms *C* and *c* are called **alleles** (or alternative forms) of that gene for petal color. They are given the same letter symbol to show that they are forms of the same kind of gene. We could express this another way by saying that there is a kind of gene, called phonetically a "see" gene, with alleles *C* and *c*. Any individual pea plant will always have two "see" genes, forming a gene pair, and the actual members of the gene pair can be either *CC*, *Cc*, or *cc*. Notice that although the members of a gene pair can produce different effects, they obviously both affect the same character.

Students often find the term allele confusing, and the reason is probably that the words allele and gene are used interchangeably in some situations. For example, "dominant gene" and "dominant allele" both refer to the same thing in an interchangeable way. This stems from the fact that the forms (alleles) of any type of gene are of course genes themselves.

The basic route of Mendelian analysis for a single character is summarized in Table 2-2.

Message The existence of genes was originally inferred (and is still inferred today) by observing precise mathematical ratios in the filial generations issuing from two genetically different parental individuals.

Plants Differing in Two Characters

In the experiments described so far we have been concerned with what is sometimes called a **monohybrid cross,** which involves the mating of two pure lines that differ in a single gene that controls a character difference. Now we can ask what happens in a **dihybrid cross,** in which the pure parental lines differ in two genes that control two separate character differences. We can use the same symbolism that Mendel used to indicate the genotype of seed color (*Y* and *y*) and seed shape (*R* and *r*).

A pure-breeding line of *RR yy* plants, on selfing, produces seeds that are round and green. Another pure-breeding line is *rr YY*; on selfing, this line produces wrinkled yellow seeds (*r* is a recessive allele of the seed-shape gene and produces a wrinkled seed; see Figure 2-8).

Figure 2-8. Round (*R*) and wrinkled (*r*) garden peas. (Grant Heilman.)

TABLE 2-2. Summary of the modus operandi for establishing simple Mendelian inheritance

Experimental procedure:	1. Choose pure lines showing a character difference (purple versus white flowers).
	2. Intercross the lines.
	3. Self the F_1 individuals.
Results:	F_1 is all purple; F_2 is ¾ purple and ¼ white.
Inferences:	1. The character difference is controlled by a major gene for flower color.
	2. The dominant allele of this gene causes purple petals; the recessive allele causes white petals.

Symbolic interpretation:

Character	Phenotypes	Genotypes	Alleles	Gene
		CC (homozygous dominant)		
	Purple (dominant)		*C* (dominant)	
Flower color		*Cc* (heterozygous)		Flower-color gene
	White (recessive)		*c* (recessive)	
		cc (homozygous recessive)		

When Mendel crossed plants from these two lines, he obtained round yellow F_1 seeds, as expected. The results in the F_2 are summarized in Figure 2-9. Mendel performed similar experiments using other pairs of characters in many other dihybrid crosses; in each case, he obtained 9 : 3 : 3 : 1 ratios. So, he had another phenomenon to explain, and some more numbers to turn into an idea.

Mendel first checked to see whether the ratio for each gene pair in the dihybrid cross was the same as that for a monohybrid cross. If you look at only the round and wrinkled phenotypes and add up all the seeds falling into these two classes in Figure 2-9, the totals are 315 +

108 = 423 round and 101 + 32 = 133 wrinkled. Hence, the monohybrid 3 : 1 ratio still prevails. Similarly, the ratio for yellow seeds to green seeds is (315 + 101):(108 + 32) = 416:140 ≅ 3:1. From this clue, Mendel concluded that the two systems of heredity are independent. He was mathematically astute enough to realize that the 9 : 3 : 3 : 1 ratio is nothing more than a random combination of two independent 3 : 1 ratios.

This is a convenient point at which to introduce some elementary rules of probability that we will often use throughout this book.

The Rules of Probability

1. *Definition of probability.*

$$\text{Probability} = \frac{\text{The number of times an event is expected to happen}}{\text{The number of opportunities for an event to happen (or the number of trials)}}$$

For example, the probability of rolling a four on a die in a single trial is written

$$p(\text{of a four}) = \tfrac{1}{6}$$

because the die has six sides. If each side is equally likely to turn up, then the average result should be one four for each six trials.

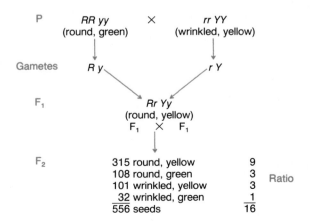

Figure 2-9. The F_2 generation resulting from a dihybrid cross.

2. *The product rule.* The probability of two independent events occurring simultaneously is the product of each of their respective probabilities. For example, with two dice we have independent objects, and

$$p(\text{of two fours}) = \tfrac{1}{6} \times \tfrac{1}{6} = \tfrac{1}{36}$$

3. *The sum rule.* The probability of *either one* of two mutually exclusive events occurring is the sum of their individual probabilities. For example, with two dice

$$p(\text{of two fours } or \text{ two fives}) = \tfrac{1}{36} + \tfrac{1}{36} = \tfrac{1}{18}$$

In the pea example, the composition of the F_2 of a dihybrid cross can be predicted if the mechanism for putting R or r into a gamete is *independent* of the mechanism for putting Y or y into a gamete. The frequency of gamete types can be calculated by determining their probabilities according to the rules just given. Thus, if you pick a gamete at random, the *probability* of picking a certain type of gamete is the same as the frequency of that type of gamete in the population.

We know from Mendel's first law that

$$Y \text{ gametes} = y \text{ gametes} = \tfrac{1}{2}$$

$$R \text{ gametes} = r \text{ gametes} = \tfrac{1}{2}$$

For an $Rr\,Yy$ plant, the probability that a gamete will be R and Y is written $p(RY)$. Similarly, $p(Ry)$ denotes the probability that a gamete will be R and y. Since we have shown that the R/r and Y/y gene pairs act independently, we can use the product rule to calculate that

$$p(RY) = \tfrac{1}{2} \times \tfrac{1}{2} = \tfrac{1}{4}$$

$$p(Ry) = \tfrac{1}{2} \times \tfrac{1}{2} = \tfrac{1}{4}$$

$$p(ry) = \tfrac{1}{2} \times \tfrac{1}{2} = \tfrac{1}{4}$$

$$p(rY) = \tfrac{1}{2} \times \tfrac{1}{2} = \tfrac{1}{4}$$

Thus, we can represent the F_2 generation by a giant grid named (after its inventor) a Punnett square, as shown in Figure 2-10.

The probability of $\tfrac{1}{16}$ shown for each box in the square is derived by further application of the product rule. The constitution of a zygote is the result of two independant entities, the male and female gametes. Thus, for example, the probability (or frequency) of $RR\,YY$ zygotes (combining an RY male gamete with an RY female gamete) will be $\tfrac{1}{4} \times \tfrac{1}{4} = \tfrac{1}{16}$. Grouping all the types that will look the same from Figure 2-9, we find the $9:3:3:1$ ratio (now not so mysterious) in all its beauty.

round, yellow	$\tfrac{9}{16}$ or 9
round, green	$\tfrac{3}{16}$ or 3
wrinkled, yellow	$\tfrac{3}{16}$ or 3
wrinkled, green	$\tfrac{1}{16}$ or 1

The concept of independence of the two systems (round or wrinkled versus yellow or green) is important. This concept of **independent assortment** has been generalized to provide the statement now known as Mendel's second law.

Mendel's Second Law During gamete formation the segregation of one gene pair is independent of other gene pairs

---

A note of warning: we shall see later that the phenomenon of gene linkage is an important exception to Mendel's second law.

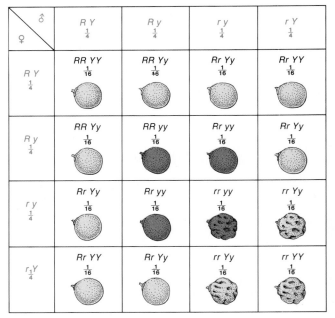

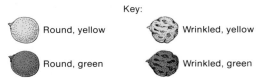

Key:

Round, yellow   Wrinkled, yellow

Round, green   Wrinkled, green

Figure 2-10. Symbolic representation of the genetic and phenotypic constitution of the F_2 generation resulting from parents differing in two characters. (Figure 2-9 shows the P and F_1 generations.)

Note how Mendel's counting led to the discovery of such unexpected regularities as the 9 : 3 : 3 : 1 ratio and how a few simple assumptions (such as equal segregation and independent assortment) can explain this ratio that initially seems so baffling. Although it was unappreciated at the time, Mendel's approach was to provide the key to an understanding of genetic mechanisms.

Of course, Mendel went on to test this second law. For example, he crossed an F_1 dihybrid $Rr\ Yy$ with a double-homozygous recessive strain $rr\ yy$. A cross to a homozygous recessive is now known as a **testcross**. Testcrosses allow the experimenter to focus on the genetic events occurring in one heterozygous individual, because the homozygous recessive tester contributes only recessive alleles to the progeny. We shall see this kind of cross many times in this book. For his testcross, Mendel predicted that the dihybrid $Rr\ Yy$ should produce the gametic types RY, Ry, ry, and rY in equal frequency — that is, as shown along one edge of the Punnett square, in the frequencies $\frac{1}{4}$, $\frac{1}{4}$, $\frac{1}{4}$, and $\frac{1}{4}$. On the other hand, because it is homozygous, the $rr\ yy$ plant should produce only one gamete type (ry), regardless of equal segregation or independent assortment. Thus, the progeny phenotypes should be a direct reflection of the gametic types from the $Rr\ Yy$ parent (because the $r\ y$ contribution from the $rr\ yy$ parent does not alter the phenotype indicated by the other gamete). Mendel predicted a 1 : 1 : 1 : 1 ratio of $Rr\ Yy$, $Rr\ yy$, $rr\ yy$, and $rr\ Yy$ progeny from this testcross, and his prediction was confirmed. He tested the concept of independent assortment intensively on four different gene pairs and found that it applied to every combination.

Of course, the deduction of equal segregation and independent assortment as abstract concepts that explain the observed facts leads immediately to the question of what structures or forces are responsible for generating them. The idea of equal segregation seems to indicate that both alleles of a pair actually exist in some kind of orderly, paired configuration from which they can separate cleanly during gamete formation (Figure 2-11). If any other gene pair behaves independently in the same way, then we have independent assortment (Figure 2-12). But this is all speculation at this stage of our discussion, as it was after the rediscovery of Mendel's work. The actual mechanisms are now known, and we shall discuss them later. (We shall see that it is the chromosomal location of genes that is responsible for their equal segregation and independent assortment.)

The key point to appreciate at this stage is that the ground rules for genetic analysis were established by Mendel. His work made it possible to infer the existence and nature of hereditary particles and mechanisms without ever seeing them. All such theories were based on the analysis of phenotypic frequencies in controlled crosses; this is the experimental approach still used in much of modern genetics.

Methods for Working Problems

We pause here for a few words on the working of problems. The Punnett square is sure and graphic, but it is unwieldy; it is suited only for illustration, not for efficient calculation.

A branch diagram is useful for solving some problems. For example, the 9 : 3 : 3 : 1 phenotypic ratio can be derived by drawing a branch diagram and applying the

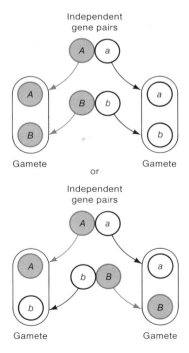

Figure 2-12. Diagrammatic visualization of the segregation of two independent gene pairs into gametes.

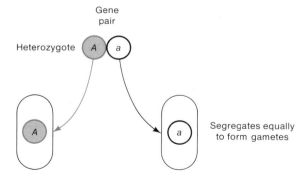

Figure 2-11. Diagrammatic visualization of the equal segregation of one gene pair into gametes.

product rule to determine the frequencies. (Note the use of the convention that $R-$ represents both RR and Rr; that is, either allele can occupy the space indicated by the dash.)

$$\text{F}_1 \ Rr \ Yy \ \text{selfed}$$

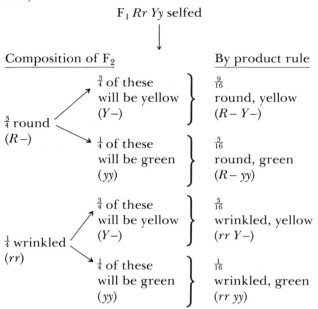

Composition of F$_2$	By product rule

$\frac{3}{4}$ round $(R-)$ → $\frac{3}{4}$ of these will be yellow $(Y-)$ — $\frac{9}{16}$ round, yellow $(R-\ Y-)$

→ $\frac{1}{4}$ of these will be green (yy) — $\frac{3}{16}$ round, green $(R-\ yy)$

$\frac{1}{4}$ wrinkled (rr) → $\frac{3}{4}$ of these will be yellow $(Y-)$ — $\frac{3}{16}$ wrinkled, yellow $(rr\ Y-)$

→ $\frac{1}{4}$ of these will be green (yy) — $\frac{1}{16}$ wrinkled, green $(rr\ yy)$

The branch diagram is, of course, a graphic expression of the product rule. It can be used for phenotypic or genotypic ratios.

The diagram can be extended to a trihybrid ratio (such as $Aa\ Bb\ Cc \times Aa\ Bb\ Cc$) by drawing another set of branches on the end. However, as the number of gene pairs increases, the number of identifiable phenotypes rises startlingly and the number of genotypes climbs even more steeply, as shown in Table 2-3. With such large class numbers, even the branch method becomes unwieldy.

TABLE 2-3. Rise in number of genotypic classes as the power of the number of segregating gene pairs

Number of segregating gene pairs	Number of phenotypic classes	Number of genotypic classes
1	2	3
2	4	9
3	8	27
4	16	81
.	.	.
.	.	.
.	.	.
n	2^n	3^n

In such cases, we must resort to devices based directly on the product and sum rules. For example, what proportion of progeny from the cross

$$Aa\ Bb\ Cc\ Dd\ Ee\ Ff \times Aa\ Bb\ Cc\ Dd\ Ee\ Ff$$

will be $AA\ bb\ Cc\ DD\ ee\ Ff$? The answer is easily obtained if the gene pairs all assort independently, thereby allowing use of the product rule. Thus, $\frac{1}{4}$ of the progeny will be AA, $\frac{1}{4}$ will be bb, $\frac{1}{2}$ will be Cc, $\frac{1}{4}$ will be DD, $\frac{1}{4}$ will be ee, and $\frac{1}{2}$ will be Ff, so we obtain the answer by multiplying these frequencies:

$$p(AA\ bb\ Cc\ DD\ ee\ Ff) = \tfrac{1}{4} \times \tfrac{1}{4} \times \tfrac{1}{2} \times \tfrac{1}{4} \times \tfrac{1}{4} \times \tfrac{1}{2}$$
$$= \tfrac{1}{1024}$$

Let us return now to Mendel's work. When Mendel's results were rediscovered in 1900, his principles were tested in a wide spectrum of eukaryotic organisms (organisms with cells that contain nuclei). The results of these tests showed that Mendelian genetics (sometimes with extensions that we shall discuss in the next three chapters) is universally applicable. Mendelian ratios (such as $3:1$, $1:1$, $9:3:3:1$, and $1:1:1:1$) were extensively reported (Figure 2-13), suggesting that equal segregation and independent assortment are fundamental hereditary processes found throughout nature. Mendel's laws are not merely laws about peas but laws about

Figure 2-13. A $9:3:3:1$ ratio in the phenotype of kernels of corn. Each kernel represents a progeny individual. The progeny result from a self of an individual of genotype $Aa\ Bb$, where A = dark, a = light, B = smooth, and b = wrinkled. Two representative ears of corn are shown.

the genetics of eukaryotic organisms in general. The experimental approach used by Mendel can be extensively applied in plants. However, in some plants and in all animals, the technique of selfing is impossible. This problem can be circumvented by intercrossing identical genotypes. For example, an F_1 animal resulting from the mating of parents from differing pure lines can be mated to its F_1 siblings (brothers or sisters), producing an F_2. The F_1 individuals are identical for the genes in question, so the F_1 cross amounts to a selfing.

Simple Mendelian Genetics in Humans

Other systems present some special problems in the application of Mendelian methodology. One of the most difficult, yet most interesting, is the human species. Obviously, controlled crosses cannot be made, so human geneticists must resort to a scrutiny of matings that have already occurred in the hope that informative matings have been made by chance. The scrutiny of matings is called **pedigree analysis.** A member of a family who first comes to the attention of a geneticist is called the **propositus.** Usually the phenotype of the propositus is

exceptional in some way (for example, the propositus might be a dwarf). The investigator then traces the history of the character shown to be interesting in the propositus back through the history of the family and draws up a family tree or pedigree, using certain standard symbols given in Figure 2-14. (The terms autosomal and sex-linked in the figure will be explained later; they are included to make the list of symbols complete.)

Many human diseases and other exceptional conditions are determined by simple Mendelian recessive alleles. There are certain clues that must be sought in the pedigree. Characteristically, the condition appears in progeny of unaffected parents. Futhermore, two affected individuals cannot have an unaffected child. Quite often such recessive alleles are revealed by consanguineous matings (for example, cousin marriages). This is particularly true of rare conditions where chance matings of heterozygotes are expected to be extremely rare. It has been estimated, for example, that first-cousin marriages account for about 18 to 24-percent of albino children and 27 to 53 percent of children with Tay-Sachs disease; both are rare recessive conditions. Some other examples of disease-causing recessive alleles in humans are those for cystic fibrosis and phenylketonuria (PKU). A typical pedigree for a rare recessive condition is shown in Figure 2-15. Of course, variants that are not

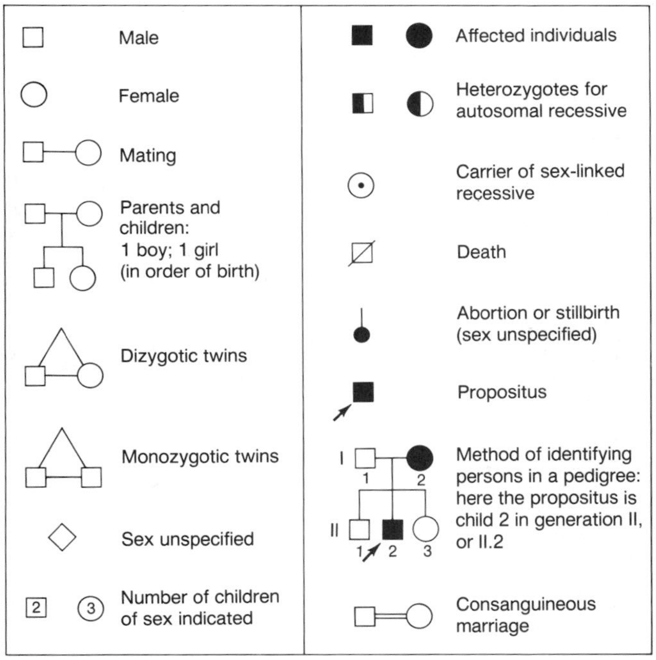

Figure 2-14. Symbols used in human pedigree analysis. (After W. F. Bodmer and L. L. Cavalli-Sforza, *Genetics, Evolution, and Man.* Copyright © 1976 by W. H. Freeman and Company.)

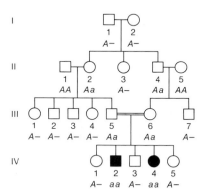

Figure 2-15. Illustrative pedigree, involving an exceptional recessive phenotype determined by the recessive Mendelian allele *a*. Gene symbols normally are not included in pedigree charts, but genotypes are inserted here for reference. Note that individuals II.1 and II.5 marry into the family; they are assumed to be normal because the heritable condition under scrutiny in a pedigree typically is rare. Note also that it is not possible to be certain of the genotype in some individuals with normal phenotype; such individuals are indicated by *A −*.

Figure 2-16. Albinism in an African child. This phenotype is caused by homozygosity for a recessive Mendelian gene. (Courtesy of W. F. Bodmer and L. L. Cavalli-Sforza, *Genetics, Evolution, and Man.* Copyright © 1976 by W. H. Freeman and Company.)

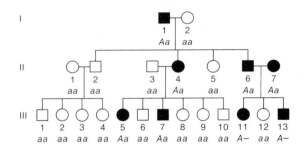

Figure 2-17. Illustrative pedigree involving an exceptional dominant phenotype determined by the dominant Mendelian allele *A*. In this pedigree, most of the genotypes can be deduced.

regarded as diseases also may be caused by recessive alleles. These may be rare, as in albinism (Figure 2-16), or common, as in light eye color (blue or green) in North American populations.

There are also examples of exceptional conditions caused by dominant alleles. (This contrasts with the situation for such conditions as PKU, where the normal condition is attributable to the dominant allele and the recessive allele causes PKU). Once again, there are some simple rules to follow to discern from pedigrees a condition caused by a dominant allele: the condition typically occurs in every generation; unaffected individuals never transmit the condition to their offspring; two affected parents may have unaffected children; and the condition is passed, on average, to one-half of the children of an affected individual. As with recessive alleles acting in a Mendelian manner, both sexes may be equally affected. Achondroplasia (a kind of dwarfism), Huntington's chorea, and brachydactyly (very short fingers) are examples of exceptional conditions in humans caused by dominant alleles. A typical pedigree of a rare dominant condition is shown in Figure 2-17.

Notice that, in this kind of Mendelian analysis, there is again the notion of identification of genes that affect major biological function, this time in humans. Pedigrees for PKU, for example, demonstrate that there is a kind of gene controlling the character we might call "normal PKU function." The two alleles stand for presence and absence. This identification is an important step toward discovery of the precise way in which the

abnormal allele is failing and its possible correction. The medical applications of such genetic analysis obviously are far-reaching. In fact, medical genetics is today a key part of medical training. Simple pedigree analyses have extensive uses, not only in such medical research but also in the day-to-day counseling of prospective parents who fear genetic disease in their children.

Simple Mendelian Genetics in Agriculture

As mentioned in Chapter 1, there has been an interest in plant breeding since prehistoric times. The methods used by Neolithic farmers were probably the same as those used until the discovery of Mendelian genetics. Basically, the approach was to select superior phenotypes from seeds or plants derived from natural populations. Particularly desirable were pure lines of favorable phenotype, because these lines produced constant results over generations of planting. Without the knowledge of Mendelian genetics, how is it possible to develop pure lines? It so happens that self-pollinating plants, such as many crop plants, naturally tend to be homozygous, because the fraction of heterozygotes decreases at each generation of selfing as follows:

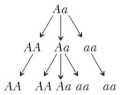

This is illustrated in more detail in Figure 2-18. Thus, pure lines have developed automatically over the years.

However, these less sophisticated breeding practices suffered from a major problem: the breeder was forced to rely on favorable combinations of genes that occurred in nature. With the advent of Mendelian genetics, it

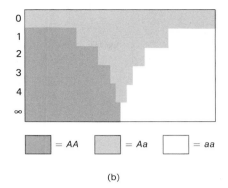

Generation	AA	Aa	aa
0	0	100	0
1	25	50	25
2	37.5	25	37.5
3	43.75	12.5	43.75
4	46.875	6.25	46.875
∞	50	0	50

(a)

(b)

Figure 2-18. Selfing (and inbreeding in general) produces an increasing proportion of homozygous individuals with time. (a) The relative proportions of the three genotypes are shown through several generations, assuming that all individuals in generation 0 are *Aa*. At each generation, *AA* and *aa* breed true, but *Aa* individuals produce *Aa*, *AA*, and *aa* progeny in a $2:1:1$ ratio. (b) A graphic depiction of the process.

became evident that favorable qualities in different lines could be combined through hybridization and subsequent gene reassortment. This procedure forms the basis of modern plant breeding.

The breeding of plants to produce new and improved genotypes works in basically the following way. For naturally self-pollinating plants, such as rice or wheat, two pure lines (each of different favorable genotype) are hybridized by manual cross-pollination and an F_1 is developed. The F_1 is then allowed to self, and its heterozygous gene pairs assort to produce many differ-

ent genotypes, some of which represent desirable new combinations of the parental genes. A small proportion of these new genotypes will be pure-breeding already; in those that are not, several generations of selfing will produce homozygosity of the relevant genes. Figure 2-19 summarizes this method; Figure 2-20 shows an example of the complex series of hybridizations that have taken place in rice.

An example of genetic improvement in a species more familiar to most is the tomato. Anyone who has read a recent seed catalog will be familiar with the abbre-

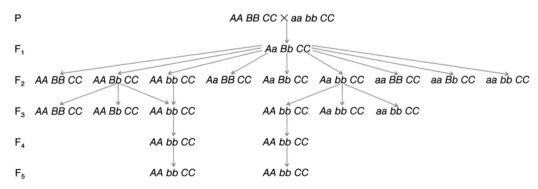

Figure 2-19. The basic technique of plant breeding. A hybrid is generated, and breeding stock is selected in subsequent generations to obtain pure-breeding improved types. For simplicity, we assume here that the parental lines differ in only two genes (*A* and *B*). In this example, the *AA bb CC* genotype is the one desired; that is, tests on the F_2 generation show this to be the most desirable phenotype. A pure-breeding line of this phenotype may be obtained in the F_2. It can also be obtained in the F_3 either from a nonpure parent of similar phenotype *(right)* or from a nonpure parent of different phenotype *(left)*. In typical cases of plant breeding, far larger numbers of genes, individuals, and generations are involved. Furthermore, the breeder usually knows little or nothing about the precise genotypes involved.

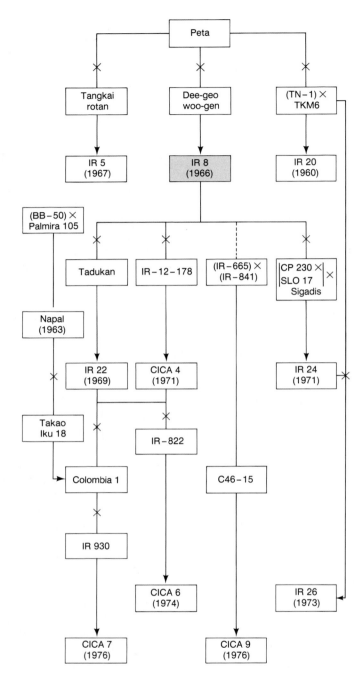

Figure 2-20. The complex pedigree of modern rice varieties. The progenitor of most modern dwarf types was IR 8, selected from a cross between the vigorous Peta (Indonesian) and the dwarf Dee-geo woo-gen (from Taiwan). Most of the other crosses represent progressive improvement of IR 8. The diagram illustrates the extensive plant breeding that produced modern crop varieties. (From Peter R. Jennings, "The Amplification of Agricultural Production." Copyright © 1976 by Scientific American, Inc. All rights reserved.)

viations V, F, and N next to a listed tomato variety. These represent, respectively, resistance to the pathogens *Verticillium, Fusarium,* and nematodes — resistance that has been crossed into domestic tomatoes, typically from wild relatives. Another familiar phenomenon in tomatoes is determinate as opposed to indeterminate growth pattern. Determinate plants are bushier and more compact, and they do not need as much staking (Figure 2-21). Determinate growth is caused by a recessive allele *sp* (self-pruning), which has been crossed into modern varieties. Another useful allele is *u* (uniform ripening); this allele eliminates the green patch or shoulder around the stem on the ripe fruit.

Such examples could be listed for many pages. The point is that simple Mendelian genetics, as described in this chapter, has provided agricultural plant breeding with its rationale and its modern methods. Entire complex genotypes may be constructed from an array of ancestral lines, each showing some desirable feature. When we think of genetic engineering, we think of the genetic techniques of the 1970s and 1980s, but genetic engineering for plant improvement began long ago.

Mendelian genetics also has provided a formal theoretical basis for animal breeding, enabling a greater efficiency than under traditional practices. More recent techniques, such as the use of frozen semen, superovulation, artificial insemination, test-tube fertilization, frozen embryos, and surrogate mothers in livestock species, have enabled breeders to amplify the number of offspring of a specific genotype — a number normally limited by the life span of the animal.

Variants and Genetic Dissection

Genetic analysis, as we have seen, must start with parental differences. Without variants, no genetic analysis is possible. Where do these variants — these raw materials for genetic analysis — come from? This is a question that can be answered in full only in later chapters. Briefly, most of the variants used by Mendel (and by ancestral and modern breeders of plants and animals) arise spontaneously in nature or in the breeders' populations without the deliberate action of geneticists.

Let us distinguish two types of variants, rare and common. Rare variants are generally abnormal. Undoubtedly in a natural setting many of them would be weeded out by natural selection, but they can be kept alive by nurture so that their determinant genes can be studied. On the other hand, for many genes there are two or more common alleles in a population. This results in a condition known as **genetic polymorphism** — the coexistence of several, common genetically determined and distinct variant phenotypes in a population. The

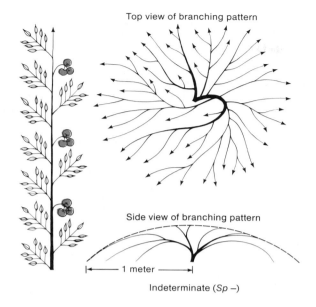

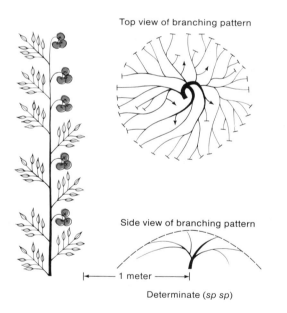

Figure 2-21. Growth characteristics of indeterminate (*Sp–*) and determinate (*sp sp*) tomatoes. Determinate has progressively fewer leaves between inflorescences, and the determinate stems end with a size-arresting inflorescence. Individual stem branches are shown, plus typical branching patterns for mature plants in top and side views. Note the size differences. (The symbol *sp* represents a single gene; many gene symbols involve more than one letter.) (From Charles M. Rick, "The Tomato." Copyright © 1978 by Scientific American, Inc. All rights reserved.)

reason for the existence of polymorphisms is usually not easy to discover, but for the geneticist they are a source of variant alleles for study.

Some of the best examples of naturally occurring variant phenotypes are to be seen in human populations. Cystic fibrosis, for example, is a case of a rare variant phenotype, and the blood groups are cases of genetic polymorphisms.

We have seen that the genetic analysis of variants can identify a gene involved in an important biological process. This is a central aspect of modern genetics; as we saw in Chapter 1, this process is referred to as genetic dissection of biological systems. Mendel was the first genetic surgeon. Using genetic analysis, he was able to identify and distinguish among the several components of the hereditary process in a way as convincing as if he had microdissected those components. The fact that the genes he was using were for pea shape, pea color, and so on, was largely irrelevant. Those genes were being used simply as **genetic markers,** which enabled Mendel to trace the hereditary processes of segregation and assortment. A genetic marker is a variant allele that is used to label a biological structure or process throughout the course of an experiment. It is almost as though Mendel were able to "paint" two alleles two different colors and send them through a cross to see how they behaved! Genetic markers are now routinely used in genetics and in all of biology to study all sorts of processes that the marker genes themselves do not directly affect.

Probably without realizing it, Mendel had also invented another aspect of genetic dissection in which the precise genes used *were* important. In this type of analysis, genetic variants are used in such a way that by studying the variant gene function, we can make inferences about the "normal" operation of the gene and the process it controls. Let's consider the petal-color gene in peas again. This is a major gene affecting an important biological function, which is the color of the petals. Undoubtedly, the success or failure of a plant in nature would largely hinge on its having the right kind of petal color, presumably to attract the appropriate pollinator. So Mendel had pinpointed a gene for a process of central importance to the biology of peas, thus opening the way to extensive further study. Genetic dissection is a versatile tool of modern biological research.

Once a gene has been identified as affecting, say, petal color in peas, we call it a major gene, but what does this really mean? It is major in that it is obviously having a profound effect on the color of the petals. But can we conclude that it is *the* single most important step in the determination of petal color? The answer is no, and the reason may be seen in an analogy. If we were trying to discover how a car engine works, we might pull out various parts and observe the effect on the running of the engine. If a battery cable were disconnected, the engine would stop; this might lead us erroneously to conclude that this cable is *the* most important part of the running of the engine. Other parts are equally necessary, and their removal could also stop or seriously cripple the engine. In a similar way it can be shown (as we will see in Chapter 4) that several genes can be identified, all of which have a major and similar effect of petal coloration.

Mendel's work has withstood the test of time and has provided us with the basic groundwork for all modern genetic study. Yet his work went unrecognized and neglected for 35 years following its publication. Why? There are many possible reasons, but here we shall consider just one. Perhaps it was because biological science at that time could not provide evidence for any real physical units within cells that might correspond to Mendel's genetic particles. Chromosomes had certainly not yet been studied, meiosis had not yet been described, and even the full details of plant life cycles had not been worked out. Without this basic knowledge, it may have seemed that Mendel's ideas were mere numerology.

Message Mendel's work was the prototypical genetic analysis. As such, it is significant for the following reasons.

1. It showed how it is possible to study some central processes of heredity and biology in general through the use of genetic markers.

2. It showed how the biological functions of genes themselves can be elucidated from the study of variant alleles.

3. It had far-reaching ramifications in agriculture and medicine.

In the next chapter, we focus on the physical locations of genes in cells and on the consequences of these locations.

Summary

■ Modern genetics is based on the concept of the gene, the fundamental unit of heredity. In his experiments with the garden pea, Mendel was the first to recognize the existence of genes. For example, by crossing a pure line of purple-flowered pea plants with a pure line of white-flowered pea plants and then selfing the F_1 generation, which was entirely purple, Mendel produced an F_2 generation of purple plants and white plants in a $3:1$ ratio. In crosses such as those of pea plants bearing yellow seeds and pea plants bearing green seeds, he discovered that a $1:2:1$ ratio underlies all $3:1$ ratios. From these precise mathematical ratios Mendel concluded that there are hereditary determinants of a particulate nature (now known as genes). In higher plant and animal cells, genes exist in pairs. The forms of a gene are called alleles. Individual alleles can be either dominant or recessive.

In a cross of heterozygous yellow (Yy) plants with homozygous green (yy) plants, a $1:1$ ratio of yellow to green plants was produced. From this ratio Mendel confirmed his so-called first law, which states that two members of a gene pair segregate from each other during gamete formation into equal numbers of gametes. Thus, each gamete carries only one gene for each gene pair. The union of gametes to form a zygote is random and occurs irrespective of which member of a gene pair is carried.

The foregoing conclusions came from Mendel's work with monohybrid crosses. In dihybrid crosses, Mendel found $9:3:3:1$ ratios in the F_2, which are really two $3:1$ ratios combined at random. From these ratios Mendel inferred that the two gene pairs studied in a dihybrid cross behave independently. This concept has been stated as Mendel's second law.

Although controlled crosses cannot be made in human beings, Mendelian genetics has great significance for humans. Many diseases and other exceptional conditions in humans are determined by simple Mendelian recessive alleles; other exceptional conditions are caused by dominant alleles. In addition, Mendelian genetics is widely used in modern agriculture. By combining favorable qualities from different lines through hybridization and subsequent gene reassortment, plant and animal geneticists are able to produce new lines of superior phenotypes.

Finally, Mendel was responsible for the basic techniques of genetic dissection still in use today. One such technique is the use of genes as genetic markers to trace the hereditary processes of segregation and assortment. The other is the study of genetic variants to discover how genes operate normally.

■   ■   ■

Solved Problems

This section in each chapter contains a few solved problems, which provide some examples of how to approach the problem sets that follow. The purpose of the problem sets is to challenge your understanding of the genetic principles learned in the previous chapter. The best way to demonstrate an understanding of a subject is to be able to use that knowledge in a real or simulated situation. Be forewarned that there is no machine-like way of solving these problems. The three main resources at your disposal are the genetic principles just learned, common sense, and trial and error.

Here is some general advice before beginning. First, it is absolutely essential to read and understand all of the question. Find out exactly what facts are provided, what assumptions have to be made, what clues are given in the question, and what inferences can be made from the available information. Second, be methodical. Staring at the question rarely helps. Restate the information in the question in your own way, preferably using a diagrammatic representation or flowchart to help you think out the problem. Good luck.

1. Consider three yellow round peas, labeled A, B, and C. Each was grown into a plant and crossed to a plant grown from a green wrinkled pea. Exactly 100 peas issuing from each cross were sorted into phenotypic classes as follows:

    A: 51 yellow round
        49 green round
    B: 100 yellow round
    C: 24 yellow round
        26 yellow wrinkled
        25 green round
        25 green wrinkled

    What were the genotypes of A, B, and C? (Use gene symbols of your own choosing; be sure to define each one.)

Solution

Notice that each of the peas is involved in a cross of the type

yellow round $\times$ green wrinkled
$\downarrow$
progeny

so that all the differences among the three progeny populations must be attributable to differences in the underlying genotypes of peas A, B, and C.

You might remember a lot about these analyses from the chapter. This is fine, but let's see how much we can deduce from the data. What about dominance? The key cross for deducing dominance is B. Here, the inheritance pattern is

yellow round $\times$ green wrinkled
$\downarrow$
all yellow round

So yellow and round must be dominant phenotypes, because dominance is literally defined in terms of the phenotype of a hybrid. Now we know that the green wrinkled parent used in each cross must be fully recessive, a very convenient situation because it means that each cross is a testcross, which is generally the most informative type of cross.

Turning to the progeny of A, we see a 1 : 1 ratio for yellow to green. This is a demonstration of Mendel's first law (equal segregation) and shows that for the character of color, the cross must have been heterozygote $\times$ homozygous recessive. Letting Y = yellow and y = green, we have

$$Yy \times yy$$
$$\downarrow$$
$$\tfrac{1}{2} Yy \text{ (yellow)}$$
$$\tfrac{1}{2} yy \text{ (green)}$$

For the character of shape, since all the progeny are round, the cross must have been homozygous dominant $\times$ homozygous recessive. Letting R = round and r = wrinkled, we have

$$RR \times rr$$
$$\downarrow$$
$$Rr \text{ (round)}$$

Combining the two characters, we have

$$Yy\,RR \times yy\,rr$$
$$\downarrow$$
$$\tfrac{1}{2} Yy\,Rr$$
$$\tfrac{1}{2} yy\,Rr$$

Now, cross B becomes crystal clear and must have been

$$YY\,RR \times yy\,rr$$
$$\downarrow$$
$$Yy\,Rr$$

because any heterozygosity in pea B would have given rise to several progeny phenotypes, not just one.

What about C? Here, we see a ratio of 50 yellow : 50 green (1 : 1) and a ratio of 49 round : 51 wrinkled (also 1 : 1). So both gene pairs in pea C, must have been heterozygous, and cross C was

$$Yy\,Rr \times yy\,rr$$
$$Progeny$$

which is a good demonstration of Mendel's second law (independent behavior of different gene pairs).

How would a geneticist have analyzed these crosses? Basically, the same way we just did but with fewer intervening steps. Possibly something like this: "yellow and round dominant; single-gene segregation in A; B homozygous dominant; independent two-gene segregation in C."

2. Phenylketonuria (PKU) is a human hereditary disease that prevents the body from processing the chemical phenylalanine, which is contained in the protein we eat. PKU is manifested in early infancy and, if it remains untreated, generally causes mental retardation. PKU is caused by a simple Mendelian recessive allele.

A couple intends to have children but consults a genetic counselor because the man has a sister with PKU and the woman has a brother with PKU. There are no other known cases in their families. They ask the genetic counselor to determine the probability that their first child will have PKU. What is this probability?

Solution

What can we deduce? If we let the allele for the PKU phenotype be p and the respective normal allele be P, then the sister and brother of the man and woman, respectively, must have been pp. In order to produce these affected individuals, all four grandparents must have been heterozygous normal. The pedigree can be summarized as follows:

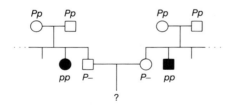

Once these inferences have been made, the problem is reduced to an application of the product rule. The only way the man and woman can have a PKU child is if both of them are heterozygotes (it is obvious that they themselves do not have the disease). Both the grandparental matings are simple Mendelian monohybrid crosses, and both will produce progeny of the following types:

$$\left.\begin{array}{l} \frac{1}{4}\ AA \\ \frac{1}{2}\ Aa \end{array}\right\} \text{Normal } \frac{3}{4}$$
$$\frac{1}{4}\ aa \qquad \text{(PKU) } \frac{1}{4}$$

We know that the man and the woman are normal, so the probability of either being a heterozygote is $\frac{2}{3}$, because within the A – class, $\frac{2}{3}$ are Aa and $\frac{1}{3}$ are AA.

The probability of *both* the man and the woman being heterozygotes is $\frac{2}{3} \times \frac{2}{3} = \frac{4}{9}$. If they are both heterozygous, then one-quarter of their children would have PKU, so the probability that their first child will have PKU is $\frac{1}{4}$, and the total probability of them being heterozygous *and* their first children having PKU is $\frac{4}{9} \times \frac{1}{4} = \frac{4}{36} = \frac{1}{9}$, which is the answer.

Problems

1. What are Mendel's laws?

2. If you had a *Drosophila* that was of phenotype A, what test would you make to determine if it was *AA* or *Aa*?

3. Two black guinea pigs were mated and over several years produced 29 black and 9 white offspring. Explain these results, giving the genotypes of parents and progeny.

4. A woman had a sister who died of Tay-Sachs disease (a rare Mendelian recessive disease). The woman is now worried that her unborn child will also be affected. How would you counsel her?

5. You have three dice: one red (R), one green (G), and one blue (B). When all three dice are rolled at the same time, calculate the probability of the following outcomes:

   a. 6(R) 6(G) 6(B)

   b. 6(R) 5(G) 6(B)

   c. 6(R) 5(G) 4(B)

   d. no sixes at all

   e. two sixes and one five on any dice

   f. three sixes or three fives

   g. the same number on all dice

   h. a different number on all dice

6. **a.** You have three jars containing marbles, as follows:

jar 1	600 red	and 400 white
jar 2	900 blue	and 100 white
jar 3	10 green	and 990 white

   If you blindly select one marble from each jar, calculate the probability of obtaining

   (1) a red, a blue, and a green

   (2) three whites

   (3) a red, a green, and a white

   (4) a red and two whites

   (5) a color and two whites

   (6) at least one white

   *b. In a certain plant, R = red and r = white. You self a red Rr heterozygote with the express purpose of obtaining a white plant for an experiment. What minimum number of seeds do you have to grow to be at least 95 percent certain of obtaining at least one white individual? (HINT: consider your answer to Question 6a(6).)

7. Holstein cattle normally are black and white. A superb black and white bull, Charlie, was purchased by a farmer for $100,000. The progeny sired by Charlie were all normal in appearance. However, certain pairs of his progeny, when interbred, produced red and white progeny at a frequency of about 25 percent. Charlie was soon removed from the stud lists of the Holstein breeders. Explain precisely why, using symbols.

8. Maple syrup urine disease is a rare inborn error of metabolism. It derives its name from the odor of the urine of affected individuals. If untreated, affected children die soon after birth. The disease tends to recur in the same family, but the parents of the affected individuals are always normal. What does this information suggest about the transmission of the disease: is it dominant or recessive? Explain.

9. On a hike into the mountains, you notice a beautiful harebell plant that has white flowers instead of the usual blue. Assuming this effect to be caused by a single gene, outline *precisely* what you would do to find out whether the allele causing white flowers is dominant or recessive to the allele causing blue flowers.

10. Mother and father both find the taste of a chemical called phenylthiourea very bitter. However, two of their four children find the chemical tasteless. Assuming the inability to taste this chemical to be a monogenic trait, is it dominant or recessive? Explain.

11. In humans, the disease galactosemia is inherited as a monogenic recessive trait in a simple Mendelian manner. A woman whose father had galactosemia intends to marry a man whose grandfather was galactosemic. They are worried about having a galactosemic child. What is the probability of this outcome?

12. Huntington's chorea is a rare, fatal disease that usually develops in middle age. It is caused by a dominant allele. A phenotypically normal man in his early twenties learns that his father has developed Huntington's chorea.

   a. What is the probability that he himself will develop the symptoms later on?

   b. What is the probability that his son will develop the symptoms in later life?

13. Achondroplasia is a form of dwarfism inherited as a simple monogenic trait. Two achondroplastic dwarfs marry and have a dwarf child; later, they have a second child who is normal.

   a. Is achondroplasia recessive or dominant? Explain.

   b. What are the genotypes of the two parents in this mating?

   c. What is the probability that their next child will be normal? a dwarf?

14. Suppose that a husband and wife are both heterozygous for a recessive gene for albinism. If they have dizygotic (two-egg) twins, what is the probability that both of the twins will have the same phenotype with respect to pigmentation?

15. The plant blue-eyed Mary grows on Vancouver Island and on the lower mainland of British Columbia. Near Nanaimo, one plant observed in nature had blotched leaves. This plant, which had not yet flowered, was dug up and taken to a laboratory, where it was allowed to self.

Seeds were collected and grown into progeny. One randomly selected (but typical) leaf from each of the progeny is shown:

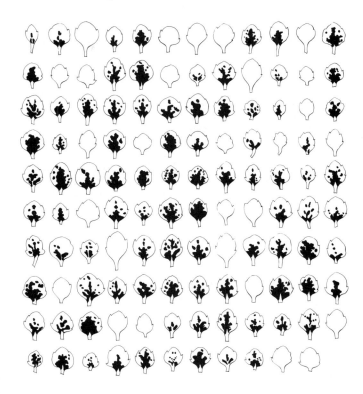

   a. Formulate a concise genetic hypothesis to explain these results. Explain all symbols and show all genotypic classes (and the genotype of the original plant).

   b. How would you test your hypothesis? Be specific.

16. Some species of plants (called heterostylous) produce two forms that differ in their flowers. The following figure shows two flower forms of the Lompoc fiddleneck (*Amsinckia spectabilis*), a heterostylous species.

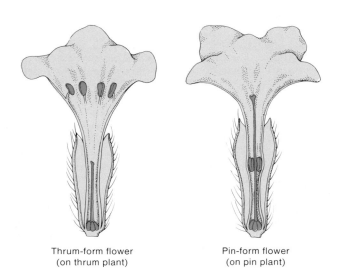

Thrum-form flower   Pin-form flower
(on thrum plant)   (on pin plant)

The following four pollinations are made:

Parents	Progeny
pin plant #1 × pin plant #2	37 pin plants
thrum plant #3 × thrum plant #3	28 thrum plants
thrum plant #3 × pin plant #1	29 thrum plants
thrum plant #4 × pin plant #2	19 pins, 16 thrums

Represent the dominant allele of the heterostyle gene by H and the recessive allele by h.

a. What are the genotypes of (1) pin plant #1, (2) pin plant #2, (3) thrum plant #3, and (4) thrum plant #4?

b. What proportions of pins and thrums would be expected in the following crosses? (1) thrum plant #3 × thrum plant #4, (2) thrum plant #3 × pin plant #2, (3) thrum plant #4 × thrum plant #4.

(Problem 16 courtesy of F. R. Ganders.)

17. Can it ever be proved that an animal is *not* a carrier of a recessive allele (that is, not a heterozygote for a given gene)? Explain.

18. A green seed is planted in a greenhouse and grows into a plant. When the seeds on this plant are examined, they are all yellow! What must have happened?

19. In nature, individual plants of *Plectritis congesta* bear either wingless or winged fruits, as shown in the figure.

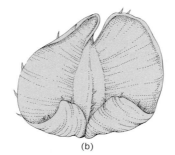

(a)   (b)

Plants were collected from nature before flowering and were crossed or selfed with the results shown in Table 2-4. Interpret these results, and derive the mode of inheritance of these fruit-shaped phenotypes. Use symbols. (NOTE: the progeny marked by asterisks probably have a nongenetic explanation. What do you think it is?)

20. The figure at right shows four human pedigrees. The black symbols represent an abnormal condition (phenotype) inherited in a simple Mendelian manner.

TABLE 2-4.

	Number of progeny plants	
Pollination	Winged	Wingless
Winged (selfed)	91	1*
Winged (selfed)	90	30
Wingless (selfed)	4*	80
Winged × wingless	161	0
Winged × wingless	29	31
Winged × wingless	46	0
Winged × winged	44	0
Winged × winged	24	0

1
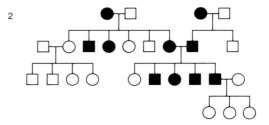

2
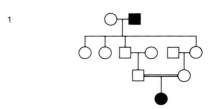

3
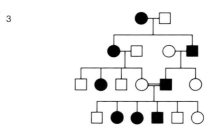

4
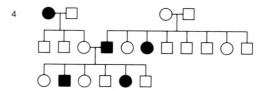

a. For each pedigree, state whether the abnormal condition is dominant or recessive. Try to state the logic behind your answer.

b. In each pedigree, describe the genotypes of as many individuals as possible.

21. When a pea plant of genotype $Aa\ Bb$ produces gametes, what proportion will be $A\ b$? (Assume that the two genes are independent.) Choose the correct answer from the following possible answers: $\frac{3}{4}$, $\frac{1}{2}$, $\frac{9}{16}$, none, or $\frac{1}{4}$.

22. When a fruit fly of genotype $Mm\ Nn\ Oo$ is mated to another fly of identical genotype, what proportion of the progeny flies will be $MM\ nn\ Oo$? (Assume that the three genes are independent.) Choose the correct answer from the following possible answers: $\frac{1}{2}$, $\frac{1}{8}$, $\frac{3}{8}$, $\frac{1}{32}$, or $\frac{1}{64}$.

23. Suppose that you have two lines of plants: one is $AA\ BB$; the other is $aa\ bb$. You cross the two and self the F_1 plants. With respect to these two genes, what is the probability that an F_2 plant will obtain one-half its alleles from one grandparent and one-half from the other? What is the probability that an F_2 plant will obtain all its alleles from a single grandparent?

24. In dogs, dark coat color is dominant over albino and short hair is dominant over long hair. If these effects are caused by two independently assorting genes, write the genotypes of the parents in each of the crosses shown in Table 2-5.

Use the symbols C and c for the dark and albino coat-color alleles and the symbols S and s for the short-hair and long-hair alleles, respectively. Assume homozygosity unless there is evidence otherwise.

(Problem 24 reprinted by permission of Macmillan Publishing Co., Inc., from *Genetics* by M. Strickberger. Copyright © Monroe W. Strickberger, 1968.)

25. In tomatoes, two alleles of one gene determine the character difference of purple versus green stems and two alleles of a separate independent gene determine the character difference of "cut" versus "potato" leaves. Table 2-6 gives the results for five separate matings of tomato plant phenotypes.

a. Determine which alleles are dominant.

b. What are the most probable genotypes for the parents in each cross?

(Problem 25 from A. M. Srb, R. D. Owen, and R. S. Edgar, *General Genetics*, 2d ed. Copyright © 1965 by W. H. Freeman and Company.)

TABLE 2-5.

	Number of progeny			
Parental phenotype	Dark, short	Dark, long	Albino, short	Albino, long
a. Dark, short × dark, short	89	31	29	11
b. Dark, short × dark, long	18	19	0	0
c. Dark, short × albino, short	20	0	21	0
d. Albino, short × albino, short	0	0	28	9
e. Dark, long × dark, long	0	32	0	10
f. Dark, short × dark, short	46	16	0	0
g. Dark, short × dark, long	30	31	9	11

SOURCE: Reprinted by permission of Macmillan Publishing Co., Inc., from *Genetics* by M. Strickberger. Copyright © Monroe W. Strickberger, 1968.

TABLE 2-6.

Mating	Parental phenotype	Number of progeny			
		Purple, cut	Purple, potato	Green, cut	Green, potato
1	Purple, cut × green, cut	321	101	310	107
2	Purple, cut × purple, potato	219	207	64	71
3	Purple, cut × green, cut	722	231	0	0
4	Purple, cut × green, potato	404	0	387	0
5	Purple, potato × green, cut	70	91	86	77

SOURCE: A. M. Srb, R. D. Owen, and R. S. Edgar, *General Genetics*, 2d ed. San Francisco: W. H. Freeman and Co., 1965.

TABLE 2-7.

Mating	Parent #1	Parent #2	Progeny
1	Bowlegs, hairy knees	Bowlegs, hairy knees	$\frac{3}{4}$ bowlegs, hairy knees $\frac{1}{4}$ knock-knees, hairy knees
2	Bowlegs smooth legs	Knock-knees, smooth legs	$\frac{1}{2}$ bowlegs, smooth legs $\frac{1}{2}$ knock-knees, smooth legs
3	Bowlegs, hairy knees	Knock-knees, smooth legs	$\frac{1}{4}$ bowlegs, smooth legs $\frac{1}{4}$ bowlegs, hairy knees $\frac{1}{4}$ knock-knees, hairy knees $\frac{1}{4}$ knock-knees, smooth legs
4	Bowlegs, hairy knees	Bowlegs, hairy knees	$\frac{3}{4}$ bowlegs, hairy knees $\frac{1}{4}$ bowlegs, smooth legs

26. Imagine that a small group of Sasquatches is discovered in the mountains of British Columbia. A study of four matings that occur in the group in the course of several years produces the results shown in Table 2-7.

   a. How many genes are involved in these phenotypes?

   b. Which character differences are controlled by which alleles of these genes?

   c. Which alleles are dominant or recessive?

   d. Only five parents participated in these matings. Give the genotypes of these five individuals.

27. We have dealt mainly with only two pairs of genes, but the same principles hold for more than two pairs at a time. Consider the cross

$$Aa\ Bb\ Cc\ Dd\ Ee \times aa\ Bb\ cc\ Dd\ ee$$

   a. What proportion of progeny will *phenotypically* resemble (1) the first parent, (2) the second parent, (3) either parent, and (4) neither parent?

   b. What proportion of progeny will *genotypically* resemble (1) the first parent, (2) the second parent, (3) either parent, and (4) neither parent?

   Assume independent assortment.

*28. Consider two-child families in which the parents have been identified as carriers of an autosomal recessive gene by virtue of having at least one child with the phenotype. When the children of many such two-child families are totaled, what proportion of children in these failies will show the phenotype? (NOTE: the answer is not 25 percent.)

29. A man is brachydactylous (very short fingers, rare Mendelian dominant), and his wife is not. Both can taste the chemical phenylthiocarbamide (Mendelian dominant, polymorphic), but both their mothers could not.

   a. Give the genotypes of the couple.

   If they have eight children, what is the probability of:

   b. All being brachydactylous

   c. None being brachydactylous

   d. All being tasters

   e. All being nontasters

   f. All being brachydactylous tasters.

   g. None being brachydactylous tasters

   h. At least one being a brachydactylous taster

   i. The first child being a brachydactylous nontaster

   j. The first two children being brachydactylous

   k. Having exactly two brachydactylous children

   l. The first two children being a brachydactylous taster and a nonbrachydactylous nontaster in any order?

Chromosome Theory of Inheritance

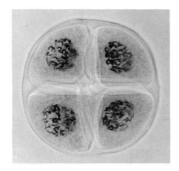

KEY CONCEPTS

Genes are parts of chromosomes.

∎

Mitosis is the nuclear division that results in two daughter nuclei each with genetic material identical to the original nucleus.

∎

Meiosis is the nuclear division by which a reproductive cell with two equivalent chromosome sets divides into four meiotic products, each of which has only one set of chromosomes. Any heterozygosity in the original cell is shuffled to cause variety in the products.

∎

Mendel's laws of equal segregation and independent assortment are based on the separation of members of each chromosome pair and on the independence of different chromosome pairs, both of which occur during meiosis.

∎

Genes on the sex chromosomes show special kinds of phenotypic ratios associated with sex.

■ The beauty of Mendel's analysis is that it is not necessary to know what genes are or how they control phenotypes to analyze the results of crosses and to predict the outcomes of future crosses, according to the laws of equal segregation and independent assortment. All this is possible simply by representing abstract hypothetical factors of inheritance (or genes) by symbols — without any concern about their physical natures or their locations in a cell. Nevertheless, although the validity of Mendelian principles is verified in many different organisms and genotypes, the next question is obvious: what structure (or structures) within cells corresponds to these hypothetical genes?

A major advance in the development of genetics was the notion that the genes, as characterized by Mendel, were associated with specific cellular structures, the chromosomes. This simple concept has become known as the **chromosome theory** of heredity. Although simple, the idea has had profound implications, inextricably uniting the disciplines of genetics and cytology and proving a means of correlating the results of breeding experiments with the behavior of structures that can be actually seen under the microscope. This fusion is still an essential part of genetic analysis today and has important applications in medical genetics, agricultural genetics, and evolutionary genetics.

Mitosis and Meiosis

How did the chromosome theory take shape? Evidence gradually accumulated from a variety of sources. One of the first lines of evidence came from the behavior of chromosomes during nuclear division in cells. The observations leading up to the discovery of the two different types of nuclear division, termed *mitosis* and *meiosis,* were as follows. In the interval between Mendel's research and its rediscovery, many biologists were interested in heredity even though they were unaware of Mendel's results, and they approached the problem in a completely different way. These investigators were interested in the physical nature of the hereditary material. An obvious place to look was in the gametes, because they are the only connecting link between generations. Since egg and sperm differed in size but were believed to contribute equally to the genetic endowment of offspring, the cytoplasm seemed an unlikely seat of the hereditary structures. Nuclei, however, were known to be approximately equal in size in both egg and sperm, so they were considered good candidates for harboring hereditary structures. What was known about the contents of the nuclei? It soon became clear that the most prominent components were the chromosomes, which proved to possess unique properties that set them apart from all other cellular structures. One property that

especially intrigued biologists was the constancy of the number of chromosomes from cell to cell within an organism, from organism to organism within any one species (but different numbers for different species), and from generation to generation within that species. Two questions therefore arose: how is the chromosome number maintained, and why? The first question was answered by observing the behavior of chromosomes under the microscope during mitosis and meiosis; from those observations arose the postulation of the chromosome theory — that chromosomes are the containing structures for genes.

Mitosis is the nuclear division associated with the division of somatic cells (that is, cells of the eukaryotic body that are not destined to become sex cells). This is the kind of division that produces a number of cells from a single progenitor cell, as for example, in the division of a fertilized human egg cell to become a multicellular organism composed of trillions of cells. Each single mitosis is generally associated with a single cell division that produces two genetically identical daughter cells. **Meiosis** is the name given to the nuclear divisions in cells that are undergoing a special kind of division found only in the sexual cycle. A cell that embarks on such divisions is called a **meiocyte.** There are two cell divisions of each meiocyte and two associated meiotic divisions of the nucleus. Hence, each original meiocyte generally produces four cells, which we shall call **products of meiosis.** In humans, meiosis occurs in the gonads, and the products of meiosis are the gametes — sperm (more properly, spermatozoa) and eggs. In flowering plants, meiosis occurs in the anthers and ovaries, and the products of meiosis are **meiospores,** which eventually give rise to gametes. We now turn to the details of these two basic kinds of nuclear division. The following descriptions of the various stages are as general as possible and are applicable to mitosis and meiosis in most organisms in which such divisions occur. Note, however, that the photographic illustrations supplied for each stage (in Figures 3-2 and 3-3) are from one organism (a flowering plant, *Lilium regale*), and a photographic series from any single organism usually cannot show all the details of mitosis and meiosis. Hence, a parallel series of idealized drawings is also included in the illustrations of both processes.

Mitosis

The cell cycle can be divided into several periods: **M, S, G1,** and **G2** (Figure 3-1). Mitosis (M) is usually the shortest period of the cycle, lasting for approximately 5 to 10 percent of the cycle. DNA synthesis occurs in the S period. The G1 and G2 periods stand for the gaps between S and M. Together, G1, S, and G2 constitute **interphase,** the time between mitoses. (Interphase used to be called "resting period"; however, many active cell functions occur in interphase, not the least of which, of course, is DNA synthesis.) The chromosomes are difficult to see during interphase (Figure 3-2a), mainly because they are in an extended state and become intertwined with each other like a tangle of yarn.

The net achievement of mitosis is that each chromosome in the nucleus makes a copy of itself along its length, and then this double structure splits to become two daughter chromosomes, each going to a different daughter nucleus. The result is two daughter nuclei identical to each other and to the nucleus from which they were derived. For the sake of study, scientists divide mitosis into four stages called **prophase, metaphase,**

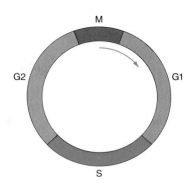

Figure 3-1. Stages of the cell cycle. M = mitosis, S = DNA synthesis, G = gap.

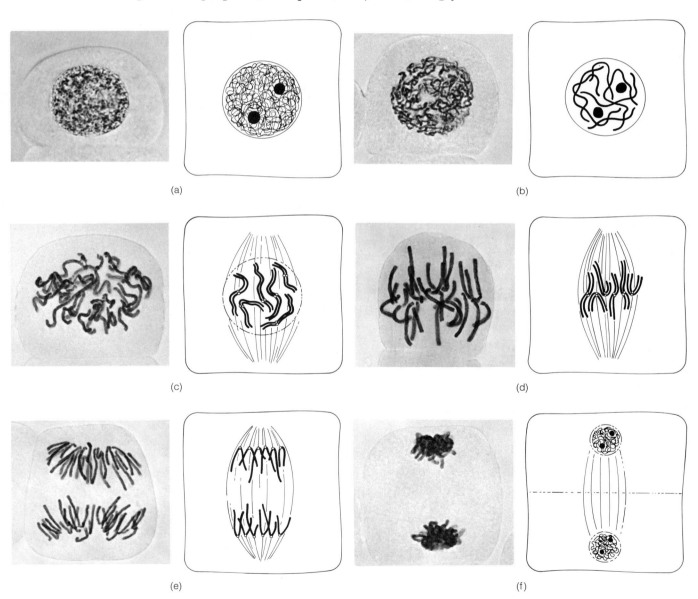

Figure 3-2. Mitosis in root tip cells of *Lilium regale.* (a) Interphase. (b) Early prophase. (c) Late prophase. (d) Metaphase. (e) Anaphase. (f) Telophase. (From J. McLeish and B. Snoad, *Looking at Chromosomes.* Copyright © 1958, St. Martin's, Macmillan.)

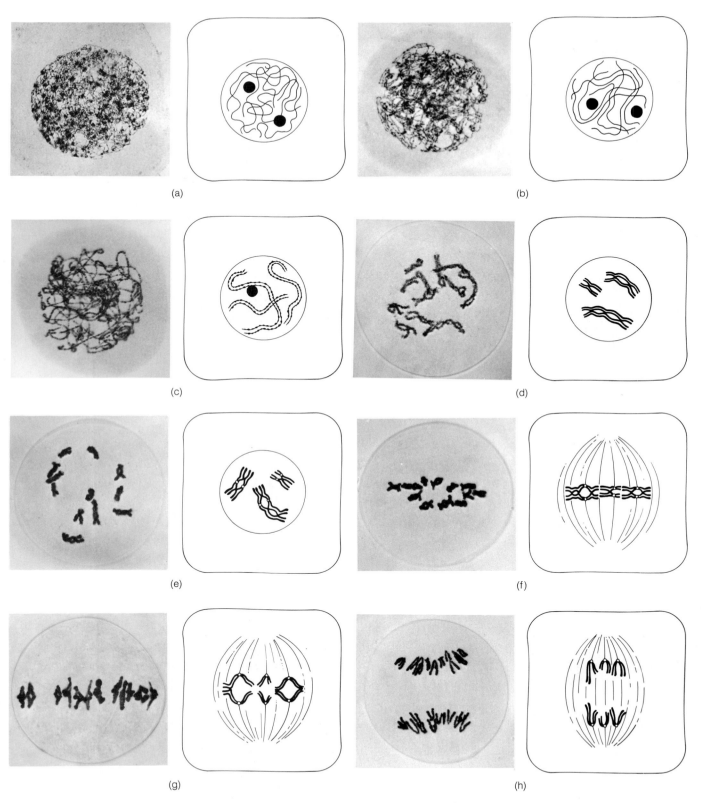

Figure 3-3. Meiosis and pollen formation in *Lilium regale.* (a) Leptotene. (b) Zygotene. (c) Pachytene. (d) Diplotene. (e) Diakinesis. (f) Metaphase I. (g) Early anaphase I. (h) Later anaphase I. (i) Telophase I. (j) Interphase. (k) Prophase II. (l) Metaphase II. (m) Anaphase II. (n) Telophase II. (o) The tetrad. (p) Young pollen grains. Note: For simplicity, multiple chiasmata are drawn as involving only two chromatids; in reality, all four chromatids can be involved. (From J. McLeish and B. Snoad, *Looking at Chromosomes.* Copyright © 1958, St. Martin's, Macmillan.)

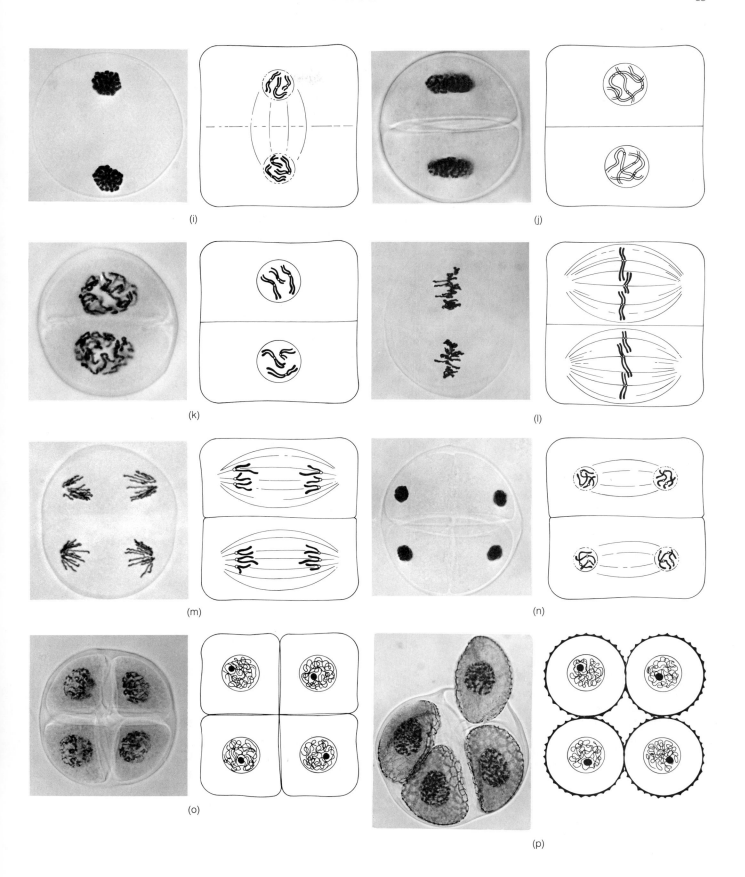

(i)

(j)

(k)

(l)

(m)

(n)

(o)

(p)

anaphase, and **telophase.** It must be stressed, however, that any nuclear division is a dynamic process on which we impose such arbitrary stages only for our own convenience.

Prophase.   The onset of mitosis is heralded by the chromosomes becoming more distinct (Figure 3-2b) as they get progressively shorter through a process of contraction or condensation. This condensation is achieved by the chromosomes being thrown into a series of spirals or coils; the coiling produces structures that are more easily moved around (for the same reason, cotton fibers are packaged commercially on spools). As the chromosomes become visible, they take on a double-stranded appearance, each chromosome being composed of two longitudinal halves called **chromatids** (Figure 3-2c). These "sister" chromatids are joined together at a region called the **centromere.** The **nucleoli** — the large intranuclear spherical structures — disappear at this stage. The nuclear membrane begins to break down, and the nucleoplasm and cytoplasm become one.

Metaphase.   At this stage, the **nuclear spindle** becomes prominent. This is a birdcage-like structure that forms in the nuclear area; it consists of a series of parallel spindle fibers that point to each of two cell poles. The chromosomes move to the equatorial plane of the cell and each becomes attached to spindle fibers at the centromere (Figure 3-2d).

Anaphase.   This stage begins when the pairs of sister chromatids separate, one of a pair moving to each pole (Figure 3-2e). The separation procedure starts with the centromere, which now appears also to have been divided. As each chromatid moves, its two arms appear to trail its centromere; a set of V-shaped structures results, with the points of the V's directed at the poles.

Telophase.   Now a nuclear membrane re-forms around each daughter nucleus, the chromosomes uncoil, and the nucleoli reappear — all of which effectively re-form interphase nuclei (Figure 3-2f). By this time, the spindle has dispersed and the cytoplasm has been divided into two by a new cell membrane.

In each of the resultant daughter cells, the chromosome complement is identical to that of the original cell. Of course, what were referred to as chromatids now take on the role of full-fledged chromosomes in their own right.

Message   Mitosis produces two daughter nuclei that have a chromosomal constitution identical to that of the original nucleus.

Why would organisms devote so much energy to this elaborate mitotic process to ensure the exactly equal distribution of chromosomal material, unless the chromosomes played a key role in directing the development of an organism? From this kind of consideration arose the conclusion that chromosomes very probably are crucial in heredity and development.

Since it was evident that the job of mitosis is to maintain the chromosome number in each nucleus, early investigators noted an apparent conflict in the events of gamete fertilization. They knew that during this process, two nuclei fuse but that the chromosome number nevertheless remains constant. What prevented the doubling of the chromosome number at each generation? This conflict was resolved by the prediction of a special kind of nuclear division that *halved* the chromosome number. This special division, which was eventually discovered in the gamete-producing tissues of plants and animals, was called meiosis.

Meiosis

Meiosis, like mitosis, is preceded by a premeiotic S phase, during which the bulk of DNA synthesis for meiosis occurs. (Some DNA synthesis also occurs during the first prophase of meiosis.) Since meiosis consists of two cell divisions, they are distinguished as meiosis I and meiosis II. The events of meiosis I are quite different from those of meiosis II, and both differ significantly from those of mitosis. Each meiotic division is formally divided into the stages of prophase, metaphase, anaphase, and telophase. Of these, the most complex and lengthy is prophase I, which has its own subdivisions: **leptotene, zygotene, pachytene, diplotene,** and **diakinesis.** Once again, try to imagine those processes as merging dynamically into each other with no clear borders.

Prophase I.   Leptotene.   The chromosomes become visible at this stage as long, thin threads (Figure 3-3a). No longitudinal doubleness is apparent. The process of chromosome contraction continues in leptotene and throughout the entire prophase. One other feature of leptotene is the development of small areas of thickening, called **chromomeres,** along the chromosome, which give it the appearance of a bead necklace.

Zygotene.   This is a time of active pairing (Figure 3-3b) at which it becomes apparent that the chromosome complement of the meiocyte is in fact two complete chromosome sets. Thus, each chromosome has a pairing partner, and the two become progressively paired, or **synapsed,** side by side in zipper fashion. Each pair is called a **homologous pair,** and the two members of a pair are called **homologs.** It should be noted that the occurrence of pairing represents a striking difference from mitosis, in which there is no such process. Furthermore, whereas mitosis can occur in cells with any number of chromosomes, meiosis normally occurs only in cells with two chromosome sets (two **genomes**). Such cells are called **diploid** and are represented symbolically as $2n$, where n is the number of chromosomes in a set. In contrast, cells with only one set (n) are called **haploid.** In higher organisms such as mammals and flowering plants, the cells of the organism are normally diploid and the meiocytes are simply a subpopulation of cells that are set aside to undergo meiosis. In haploid organisms, as we shall see later, a diploid meiocyte has to be constructed as part of the normal reproductive cycle.

How do two homologs find each other in the first place? The probable answer to this is that the ends of the chromosomes, the **telomeres,** are anchored in the nuclear membrane and it is likely that homologous telomeres are close, so that the zippering-up process can begin there. Second, how does the zippering-up work? What is the mechanism whereby two homologs can pair so precisely along their length? Although the mechanism involved is not precisely understood, one important factor is an elaborate structure composed of protein and DNA, called a **synaptonemal complex** (Figure 3-4), that is always found sandwiched between homologs during synapsis.

PACHYTENE.   This stage is characterized by thick threads representing full synapsis (Figure 3-3c). Thus, the number of units in the nucleus is equal to the number n. Nucleoli are often pronounced at this stage. The beadlike thickenings of the chromosomes, called chromomeres, are aligned precisely in the paired homologs, producing a distinctive pattern for each pair.

DIPLOTENE.   Here the DNA synthesis that occurred in the premeiotic S phase becomes manifest as a longitudinal doubleness of each paired homolog (Figure 3-3d). Once again, these units, formed by longitudinal division, are called chromatids. Hence, since each member of a homologous pair produces two sister chromatids, the synapsed structure now consists of a bundle of four homologous chromatids. At diplotene, the pairing

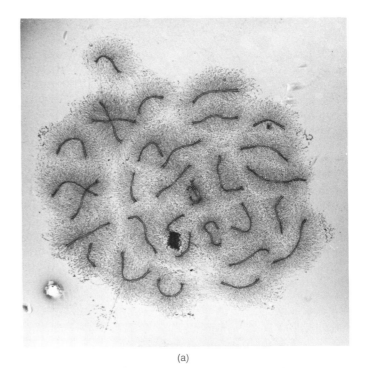

(a)

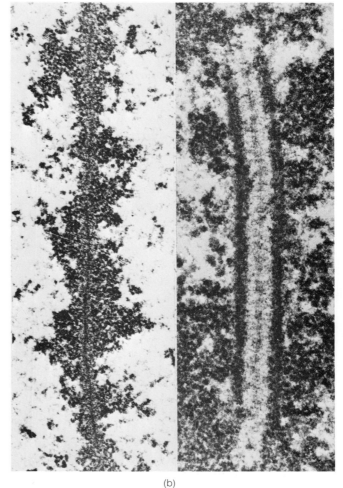

(b)

Figure 3-4. Synaptonemal complexes. (a) In *Hyalophora cecropia*, a silk moth, the normal male chromosome number is 62, giving 31 synaptonemal complexes. In the individual shown here, one chromosome (*center*) is represented three times; such a chromosome is termed trivalent. The DNA is arranged in regular loops around the synaptonemal complex. The black, dense structure is the nucleolus. (b) Regular synaptonemal complex in *Lilium tyrinum*. Note (*right*) the two lateral elements of the synaptonemal complex and also (*left*) an unpaired chromosome, showing a central core corresponding to one of the lateral elements. (Parts a and b courtesy of Peter Moens.)

between homologs becomes less tight; in fact, they appear to repel each other, and as they separate slightly, cross-shaped structures called **chiasmata** (singular, **chiasma**) appear between two nonsister chromatids. One or more chiasmata are found on each chromosome pair. Chiasmata are the visible manifestations of events, called **crossovers,** that occurred earlier, probably during zygotene or pachytene, when there is some DNA synthesis. Crossovers represent one major way in which meiosis differs from mitosis (where they occur only rarely). A crossover is a precise breakage-and-reunion event occurring between two nonsister chromatids. Studies performed on abnormal lines of organisms that undergo crossing-over very inefficiently, or not at all, show severe disruption of the orderly events that partition chromosomes into daughter cells at meiosis. Thus, crossing-over obviously plays a key role in determining the behavior of paired homologs, and the occurrence of at least one crossover per pair is usually essential for proper segregation. Crossovers have another interesting role, which is to promote genetic variation by making new gene combinations, as we shall see in Chapter 5.

DIAKINESIS. This stage (Figure 3-3e) does not differ appreciably from diplotene, except for further chromosome contraction. By this time, the long, filamentous chromosome threads of interphase are replaced by compact units that are far more maneuverable in the movements of the meiotic division.

Metaphase I. By this time, the nuclear membrane and nucleoli have disappeared, and each pair of homologs takes up a position in the equatorial plane (Figure 3-3f). At this stage of meiosis, the centromeres do not divide; this lack of division represents a major difference from mitosis. The two centromeres of a homologous chromosome pair attach to spindle fibers from opposite poles.

Anaphase I. As in mitosis, anaphase begins when chromosomes move directionally to the poles. The members of one homologous pair move to opposite poles (Figures 3-3g and 3-3h).

Telophase I. This telophase (Figure 3-3i) and the ensuing "interphase," called **interkinesis** (Figure 3-3j), are variable aspects of meiosis I. In many organisms, these stages do not exist, no nuclear membrane re-forms, and the cells proceed directly to meiosis II. In other organisms, telophase I and the interkinesis are brief in duration; the chromosomes elongate and become diffuse, and the nuclear membrane re-forms. In any case, there is never DNA synthesis at this time, and the genetic state of the chromosomes does not change. The two nuclei that result from meiosis I are effectively haploid. (Here, haploidy is best demonstrated by counting centromeres,

not chromosomes or chromatids.) The first division is consequently called **reduction division** since it reduces the number of chromosomes by one-half. The second division (at meiosis II) is effectively a mitotic division and resembles mitosis in a haploid cell. It is called **equational division** for this reason.

Prophase II. This stage is characterized by contracted chromosomes showing the haploid number (Figure 3-3k).

Metaphase II. The chromosomes arrange themselves on the equatorial plane at this stage (Figure 3-3l). Here the chromatids often partly dissociate from each other instead of being closely appressed, as in mitosis.

Anaphase II. Centromeres split and chromatids are pulled to opposite poles by the spindle fibers (Figure 3-3m).

Telophase II. The nuclei re-form around the chromosomes at the poles (Figure 3-3n).

The four products of meiosis are shown in Figure 3-3o. Since this series of photographs concerns meiosis in the anthers of a flower, each of the four products of meiosis develops into pollen grains; these are shown in Figure 3-3p. In other organisms, differentiation produces other kinds of structures from the products of meiosis, such as sperm cells in animals.

In summary, meiosis can be seen to involve one doubling of genetic material (the premeiotic S phase) and two cell divisions. Inevitably, this must result in products of meiosis that contain one-half the genetic material of the original meiocyte.

Message Meiosis always occurs in a diploid meiocyte and generally results in four haploid products of meiosis.

A summary of the net events of mitosis and meiosis is shown in Figure 3-5.

The Chromosome Theory of Heredity

The formal statement of the chromosome theory of heredity is usually credited to both Walter Sutton (an American graduate student) and Theodor Boveri (a German biologist). In 1902, these investigators recog-

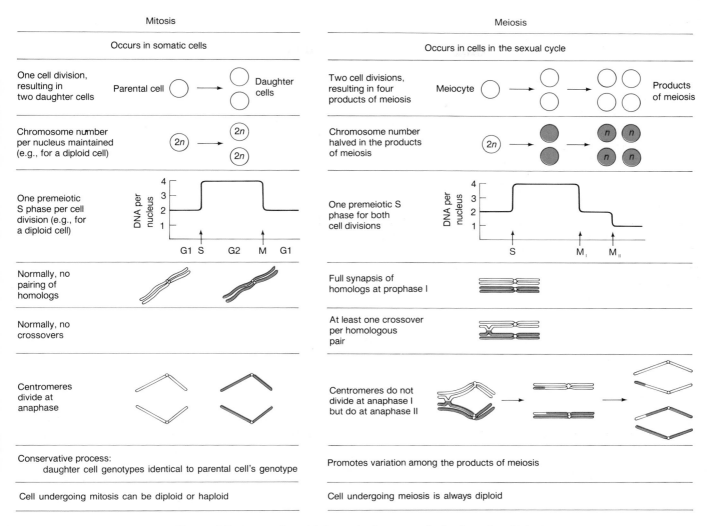

Figure 3-5. Comparison of the main features of mitosis and meiosis.

nized independently that the behavior of Mendel's particles during the production of gametes in peas precisely paralleled the behavior of chromosomes at meiosis: genes are in pairs (so are chromosomes); the members of a gene pair segregate equally into gametes (so do the members of a pair of homologous chromosomes); different gene pairs act independently (so do different chromosome pairs). The similarity is summarized in Figure 3-6.

> **Message**   The parallel behavior of genes and chromosomes led to the suggestion that genes are located on chromosomes.

We saw in Chapter 1 that a major goal of genetics is to explain two apparently conflicting forces of biology: heredity and variation. The two processes of mitosis and meiosis provide a major clue: mitosis is a conservative process that maintains a genetic status quo, whereas meiosis is a process that generates enormous variation, rather like shuffling the gene pack, through independent assortment and (as we shall see later) through crossing-over.

To modern biology students, the chromosome theory may not seem very earthshaking. However, early in the twentieth century, this hypothesis (which potentially united cytology and the infant field of genetics) was a bombshell. Of course, the first response to such an important hypothesis is to try to pick holes in it. For years

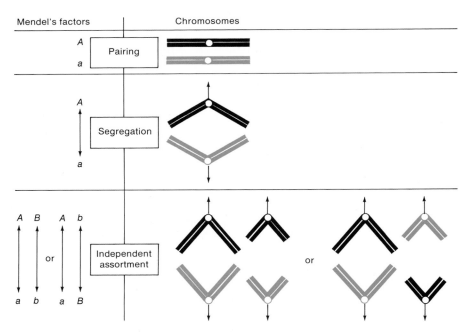

Figure 3-6. Parallels in the behavior of Mendel's hypothetical particles (genes) and chromosomes during meiosis. Black represents one member of a homologous pair; color represents the other member.

after, there was a raging controversy over the validity of what became known as the Sutton-Boveri chromosome theory of heredity.

It is worth considering some of the objections raised to the Sutton-Boveri theory. For example, at the time, chromosomes could not be detected during interphase (between cell divisions). Boveri had to make some very diligent studies of chromosome position before and after interphase in order to argue persuasively that chromosomes retain their physical integrity through interphase, even though they are cytologically invisible at that time. It was also pointed out that in some organisms the chromosomes look alike, so that they might be pairing randomly, whereas Mendel's laws absolutely require the orderly segregation of alleles. However, in species in which chromosomes do differ in size and shape, it was verified that similar chromosomes do occur in pairs and that the homologs pair and segregate during meiosis.

In 1913, Elinor Carothers found an unusual chromosomal situation in a certain species of grasshopper — a situation that permitted a direct test of whether different chromosome pairs do indeed segregate independently. Studying grasshopper testes, she found one chromosome pair (that regularly synapses) which had nonidentical members; this is called a **heteromorphic pair,** and the chromosomes presumably show only partial homology. Furthermore, she found that another chromosome, unrelated to the heteromorphic pair, had

no pairing partner at all. Carothers was able to use these unusual chromosomes as visible cytological markers of the behavior of chromosomes during assortment. By looking at anaphase nuclei, she could count the number of times that each dissimilar chromosome of the heteromorphic pair migrated to the same pole as the chromosome with no pairing partner (Figure 3-7). She found that the two patterns of chromosome behavior occurred with equal frequency. Although these unusual chromosomes obviously are not typical, the results do suggest that nonhomologous chromosomes assort independently.

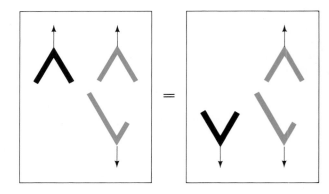

Figure 3-7. The two segregation patterns involving a heteromorphic pair and an unpaired chromosome.

Other investigators argued that all chromosomes appear as stringy structures, so that qualitative differences between them could not be detected. It was suggested that perhaps all chromosomes were just more or less made of the same stuff. It is worth introducing a study out of historical sequence that effectively counters this objection. In 1922, Alfred Blakeslee performed a study on the chromosomes of jimsonweed *(Datura)*, which has 12 chromosome pairs. He obtained 12 different strains, each of which had the normal 12 chromosome pairs plus an extra representative of one pair. Blakeslee showed that each strain was phenotypically distinct from the others (Figure 3-8). This result would

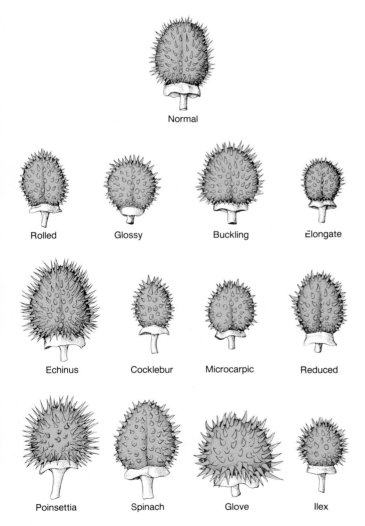

Figure 3-8. Fruits from *Datura* plants, each having one different extra chromosome. Their characteristic appearances show that each chromosome produces a unique effect. (From E. W. Sinnott, L. C. Dunn, and T. Dobzhansky, *Principles of Genetics* 5th ed. McGraw-Hill Book Co.)

not be expected if the nonhomologous chromosomes were all alike.

All these results indicated that the behavior of chromosomes closely parallels that of genes. This of course made the Sutton-Boveri theory attractive, but there was as yet no real proof that genes are located on chromosomes. Further observations, however, did provide such proof.

The Discovery of Sex Linkage

In the crosses discussed thus far, it does not matter which sex of parent is selected from which strain under study. Reciprocal crosses (such as strain A ♀ × strain B ♂ and strain A ♂ × strain B ♀) yield similar progeny. The first exception to this pattern was discovered in 1906 by L. Doncaster and G. H. Raynor. They were studying wing color in the magpie moth *(Abraxas)*, using two different lines: one with light wings; the other with dark wings. If light-winged females are crossed with dark-winged males, all the progeny have dark wings, showing that the allele for light wings is recessive. However, in the reciprocal cross (dark female × light male), all the female progeny have light wings and all the male progeny have dark wings. Thus, this pair of reciprocal crosses does not give similar results, and the wing phenotypes in the second cross are associated with the sex of the moths. Note that the female progeny of this second cross are phenotypically similar to their fathers, as the males are to their mothers. This is sometimes called **crisscross inheritance.** How can we explain these results? Before attempting an explanation, let's consider another example.

William Bateson had been studying the inheritance of feather pattern in chickens. One line had feathers with alternating stripes of dark and light coloring, a phenotype called barred. Another line, nonbarred, had feathers of uniform coloring. In the cross barred male × nonbarred female, all the progeny were barred, showing that the allele for nonbarred is recessive. However, the reciprocal cross (barred female × nonbarred male) gave barred males and nonbarred females. Again, the result is crisscross inheritance. Can we find an explanation for these similar results with moths and with chickens?

An explanation came from the laboratory of Thomas Hunt Morgan, who in 1909 began studying inheritance in a fruit fly *(Drosophila melanogaster)*. Because this organism has played a key role in the study of inheritance, a brief digression about the creature is worthwhile. The life cycle of *Drosophila* is typical of the life cycles of many insects (Figure 3-9).

The flies grow vigorously in the laboratory. In the egg, the early embryonic events lead to the production of a larval stage called the first "instar." Growing rap-

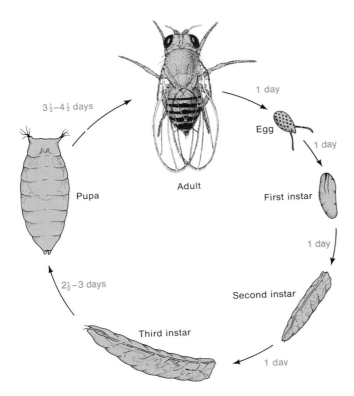

1 day

Egg

1 day

First instar

1 day

Second instar

Third instar

1 day

Pupa

$2\frac{1}{2}$–3 days

Adult

$3\frac{1}{2}$–$4\frac{1}{2}$ days

Figure 3-9. Life cycle of *Drosophila melanogaster,* the common fruit fly.

idly, the larva molts twice, and the third-instar larva then pupates. In the pupa, the larval carcass is replaced by adult structures, and an "imago" (or adult) emerges from the pupal case, ready to mate within 12 to 14 hours. The adult fly is about 2 millimeters in length, so it takes up very little space. The life cycle is very short (12 days at room temperature) in comparison with that of a human, a mouse, or a corn plant; thus, many generations can be reared in a year. Moreover, the flies are extremely prolific: a single female is capable of laying several hundred eggs. Perhaps the beauty of the insect when observed through a microscope added to its early allure. In any case, as we shall see, the choice of *Drosophila* was a very fortunate one for geneticists—and especially for Morgan, whose work earned him a Nobel prize in 1934.

The normal eye color of *Drosophila* is bright red. Early in his studies, Morgan discovered a male with completely white eyes. When he crossed this male with red-eyed females, all the F_1 progeny had red eyes, showing that the allele for white is recessive. Crossing the red-eyed F_1 males and females, Morgan obtained a 3 : 1 ratio of red-eyed to white-eyed flies, but all the white-eyed flies were males. Among the red-eyed flies, the ratio of females to males was 2 : 1. What was going on?

Morgan gathered further data. When he crossed white-eyed males with red-eyed female progeny of the

cross of white males and red females, he obtained red males, red females, white males, and white females in equal numbers. Finally, in a cross of white females and red males (which is the reciprocal of the cross of the original white male with a normal female), all the females were red and all the males were white. This is crisscross inheritance again. However, note that crisscross inheritance was observed in the experiments on chickens and moths when the parental males carried the recessive genes; in the *Drosophila* cross, it is seen when the female parent carries the recessive genes.

Before turning to Morgan's explanation of the *Drosophila* results, we should look at some of the cytological information he was able to use in his interpretations. In 1891, working with males of a species of Hemiptera (the true bugs), H. Henking observed that meiotic nuclei contained 11 pairs of chromosomes and an unpaired element that moved to one of the poles during the first meiotic division. Henking called this unpaired element an "X body"; he interpreted it as a nucleolus, but later studies showed it to be a chromosome. Similar unpaired elements were later found in other species. In 1905, Edmond Wilson noted that females of *Protenor* (another Hemipteran bug) have seven pairs of chromosomes, whereas males have six pairs and an unpaired chromosome, which Wilson called (by analogy) the X chromosome. The females, in fact, have a pair of X chromosomes.

Also in 1905, Nettie Stevens found that males and females of the beetle *Tenebrio* have the same number of chromosomes, but one of the chromosome pairs in males is heteromorphic (of different size). One member of the heteromorphic pair appears identical to the members of a pair in the female; she called this the X chromosome. The other member of the heteromorphic pair is never found in females; she called this the Y chromosome (Figure 3-10). Stevens found a similar situation in *Drosophila melanogaster,* which has four pairs of chromosomes, with

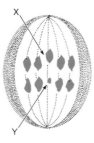

X

Y

Figure 3-10. Segregation of the heteromorphic pair (X and Y) during meiosis in a *Tenebrio* male. The X and Y chromosomes are being pulled to opposite poles during anaphase I. (From A. M. Srb, R. D. Owen, and R. S. Edgar, *General Genetics,* 2d ed. Copyright © 1965 by W. H. Freeman and Company.)

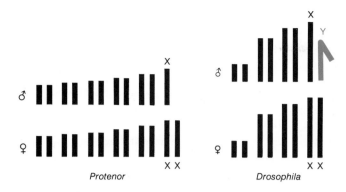

Figure 3-11. Diagrammatic representation of the chromosomal constitutions of males and females in two different insect species.

one of the pairs being heteromorphic in males. Figure 3-11 summarizes these two basic situations. (You may be wondering about the male grasshoppers studied by Carothers that had both a heteromorphic chromosome pair and an unpaired chromosome. This situation is very unusual, and we needn't worry about it at this point.)

With this background information, Morgan constructed an interpretation of his genetic data. First, it appears that the X and Y chromosomes determine the sex of the fly. *Drosophila* females have four chromosome pairs, whereas males have three normal pairs plus a heteromorphic pair. Thus, meiosis in the female produces eggs that each bear one X chromosome. Although the X and Y chromosomes in males are heteromorphic, they seem to synapse and segregate like homologs (Figure 3-12). Thus, meiosis in the male produces two types of sperm, one type bearing an X chromosome and the other bearing a Y chromosome. According to this explanation, union of an egg with an X-bearing sperm pro-

duces an XX (female) zygote and union with a Y-bearing sperm produces an XY (male) zygote. Furthermore, approximately equal numbers of males and females are expected due to the equal segregation of X and Y.

Morgan next turned to the problem of eye color. Assume that the alleles for red or white eye color are present on the X chromosome, with no counterpart on the Y chromosome. Thus, females would have two copies of this gene, whereas males would have only one. This highly unexpected situation proves to fit the data. In the original cross of the white-eyed male with red-eyed females, all F_1 progeny had red eyes, showing that the gene for red eyes is dominant. Therefore, we can represent the two alleles as W (red) and w (white). If we designate the X chromosomes as X^W and X^w to indicate the alleles supposedly carried by them, then we can diagram the two reciprocal crosses as shown in Figure 3-13.

As we can see from the figure, the genetic results of the two reciprocal crosses are completely consistent with the known meiotic behavior of the X and Y chromosomes. This experiment strongly supports the notion of the chromosome location of genes. However, it is only a correlation; it does not constitute a definitive proof of the Sutton-Boveri theory.

Can the same XX and XY chromosome theory be applied to the results of the earlier crosses made with chickens and moths? You will find that it cannot. How-

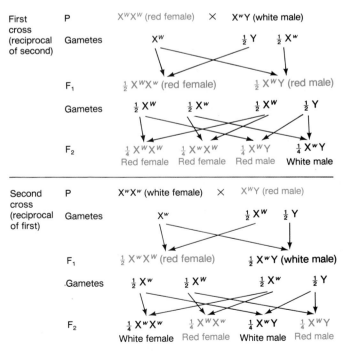

Figure 3-13. Explanation of the different results obtained from reciprocal crosses between red-eyed and white-eyed *Drosophila*.

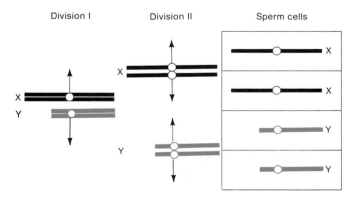

Figure 3-12. Meiotic pairing and the segregation of the X and Y chromosomes into equal numbers of sperm.

Chickens

First cross (reciprocal of second)	P	Z^BZ^B barred males	×	Z^bW nonbarred females
	F_1	Z^BZ^b barred males	⤺⤻	Z^BW barred females
Second cross (reciprocal of first)	P	Z^bZ^b nonbarred males	×	Z^BW barred females
	F_1	Z^BZ^b barred males	⤺⤻	Z^bW nonbarred females

Moths

First cross (reciprocal of second)	P	Z^LZ^L dark males	×	Z^lW light females
	F_1	Z^LZ^l dark males	⤺⤻	Z^LW dark females
Second cross (reciprocal of first)	P	Z^lZ^l light males	×	Z^LW dark females
	F_1	Z^lZ^L dark males	⤺⤻	Z^lW light females

Figure 3-14. Inheritance pattern of genes on the sex chromosomes of two species having the ZW mechanism of sex determinations.

ever, Richard Goldschmidt recognized immediately that these results can be explained with a similar hypothesis: making the simple assumption that the *males* have pairs of identical chromosomes whereas the *females* have a heteromorphic pair. To distinguish this situation from the X-Y situation in *Drosophila,* Morgan suggested that the heteromorphic chromosomes in chickens and moths be called W-Z, with males being ZZ and females being WZ. Thus, if the genes in the chicken and moth crosses are on the Z chromosome, the crosses can be diagrammed as shown in Figure 3-14.

Again, the interpretation is consistent with the genetic data. In this case, cytological data provided a confirmation of the genetic hypothesis. In 1914, J. Seiler verified that both chromosomes are identical in all pairs in male moths whereas females have one heteromorphic pair.

Message The special inheritance pattern of some genes makes it extremely likely that they are borne on the chromosomes associated with sex, which show a parallel pattern of inheritance.

An Aside on Genetic Symbols

In *Drosophila,* a special symbolism for allele designation was introduced to define variant alleles in relation to a "normal" allele. This system is now used by many geneticists and is especially useful in genetic dissection. For a given *Drosophila* character, the allele that is found most frequently in natural populations (or, alternatively, the allele that is found in standard laboratory stocks) is designated as the standard, or **wild-type.** All other alleles are

then non-wild-type. The symbolic designation of a gene is provided by the first non-wild allele found. In Morgan's *Drosophila* experiment, this was white eyes, so the non-wild allele is symbolized by w. The wild-type counterpart allele is conventionally represented by a + superscript, so the normal red-eye allele is written w^+.

In the case of polymorphism, several alleles are common in nature and all could be designated wild-type. Here, appropriate superscripts are used to distinguish alleles. For example, two natural alleles of the alcohol dehydrogenase gene in *Drosophila* are designated Adh^F and Adh^S. (This gene controls the alcohol-metabolizing enzyme alcohol dehydrogenase, and F and S stand for fast and slow movements, respectively, of the enzyme in an electrophoretic gel.)

The wild-type allele is not always dominant over a non-wild-type allele. For the two alleles w and w^+, the use of the lower-case letter indicates that the wild-type is dominant over white (that is, w is recessive to w^+). In another case, the wild-type condition of a fly's wing is straight and flat. A non-wild-type allele causes the wing to be curled. Because this allele is dominant over the wild-type allele, this gene is called Curly; the non-wild-type allele is written Cy, whereas the wild-type allele is written Cy^+. Here, note that the capital letter indicates that Cy is dominant over Cy^+. (Also note that multiple letter symbols represent a single gene, not several genes.)

This kind of symbolism is useful because it helps geneticists to focus on the procedure of genetic dissection. The wild-type allele is defined as the normal functioning situation, and non-wild-type alleles (whether recessive or dominant) can be regarded as abnormal. The abnormal alleles then become probes to investigate how the normal situations work, through an examination of

TABLE 3-1. Symmary of two systems for assigning symbols to genes

Symbolic system	Recessive variant allele, a		Dominant variant allele, A	
	Symbol for wild-type allele	Symbol for variant allele	Symbol for wild-type allele	Symbol for variant allele
Normal/abnormal	a^+ (or $+^a$ or $+$)	a (or a^-)	A^+ (or $+^A$ or $+$)	A (or A^-)
Mendelian	A	a	a	A

the ways in which the normal mechanism can go wrong. Most geneticists who are interested in using genetics to explore biological processes use this kind of symbolism. Note that Mendel's convention (A and a, or B and b) does not define or emphasize normality. In a pea flower, for example, is purple or white normal? However, the Mendelian symbolism is useful for some purposes; it is used extensively in plant and animal breeding. Table 3-1 summarizes the two systems of symbolism.

Proof of the Chromosome Theory

The correlations between the behavior of genes and the behavior of chromosomes made it very likely that genes were parts of chromosomes. But this was not a proof of the chromosome theory, and debate continued. The critical proof of the Sutton-Boveri theory came from one of Morgan's students, Calvin Bridges. Since Bridges's experiments involve complex chromosome behavior that is not fully addressed in this text until Chapter 8, the following description may be omitted for now and returned to later.

We turn now to Bridges's work. Consider a fruit fly cross we have discussed before, now represented in our new symbolism as $X^w X^w$ (white ♀) $\times$ $X^{w+}Y$ (red ♂). We know that the progeny are $X^{w+}X^w$ (red ♀♀) and X^wY (white ♂♂). However, Bridges discovered that rare exceptions occur when the cross is made on a large scale. About one out of every 2000 F_1 progeny is a white-eyed female or a red-eyed male. Because these exceptional progeny resemble their parents of the same sex, the phenotype of the females is said to be **matroclinous** and that of the males is called **patroclinous;** collectively, these individuals are called **primary exceptional progeny.** All the patroclinous males proved to be sterile. However, when Bridges crossed the primary exceptional white-eyed females with normal red-eyed males, 4 percent of the progeny were matroclinous white-eyed females and patroclinous red-eyed males that were fertile. Thus, exceptional offspring were again recovered, but at a higher frequency, and the males were fertile. These exceptional progeny of primary exceptional mothers are called **secondary exceptional offspring** (Figure 3-15). How do we explain the exceptional progeny?

It is obvious that the matroclinous females — which, like all females, have two X chromosomes — must get both of these chromosomes from their mothers because they are homozygous for w. Similarly, patroclinous males must derive their X chromosomes from their fathers because they carry w^+. Bridges hypothesized that rare mishaps occur during meiosis in the female, whereby the paired X chromosomes fail to separate during either the first or second division. This would result in meiotic nuclei containing either two X chromosomes or no X at all. Such a failure to separate is called **nondisjunction;** it produces an XX nucleus and a nullo-X nucleus (containing no X). Fertilization of these two types of nuclei will produce four zygotic classes (Figure 3-16). It is important to note that the line representing a chromosome in these diagrams is in each case not a single chromosome but a pair of daughter chromatids.

If we assume that the XXX and YO classes die, then the two types of exceptional progeny can be expected to

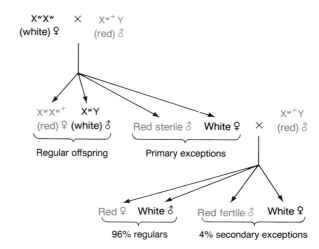

Figure 3-15. *Drosophila* crosses from which primary and secondary exceptional progeny were originally obtained.

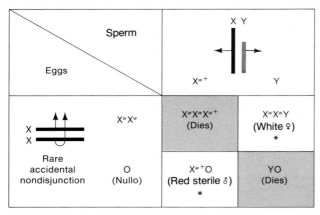

Figure 3-16. Proposed explanation of primary exceptional progeny through nondisjunction of the X chromosomes in the maternal parent.

be X^wX^wY (♀) and $X^{w+}O$ (♂) if their chromosomes are examined. Notice that it is implicit in Bridges's model that the sex of *Drosophila* is determined not by the presence or absence of the Y chromosome but by the number of X chromosomes. In most situations, two X chromosomes produce a female; one X chromosome produces a male.

What about the sterility of the primary exceptional males? This is explicable if we assume that the Y chromosome must be present in order to have male fertility.

How can we explain the secondary exceptional offspring? During meiosis in the XXY females, if the two X chromosomes pair and disjoin most of the time, leaving the Y chromosome unpaired, then we should expect equal numbers of X-bearing and XY-bearing eggs. However, we know that the X and Y chromosomes can pair and segregate, because normal males produce equal

numbers of X-bearing and Y-bearing sperms. To explain the observed results, we must assume that the Y chromosome successfully pairs with an X^w chromosome in approximately 16 percent of the pairings in X^wX^wY females, leaving the other X^w chromosome free to separate to either pole. One-half (8 percent) of these pairings will result in X^w and X^wY eggs, and the other one-half (8 percent) will result in X^wX^w and Y eggs (Figure 3-17).

We can now look at the results of fertilization by equal numbers of X^{w+} and Y sperms (Figure 3-18). We find that one-half of the fertilized X^wX^w and Y eggs will produce $X^wX^wX^{w+}$ and YY zygotes, which we presume die. The other one-half of these fertilized eggs produce the secondary exceptions, X^wX^wY and $X^{w+}Y$. Now we see why the secondary exceptional males are fertile: each of them receives a Y chromosome from the XXY mother.

So far, this is all a model—an intellectual edifice. We have made assumptions about the chromosome location of w and w^+, and we have hypothesized nondisjunction to explain the exceptional progeny. However, if this model is correct, we can now make testable predictions.

1. Cytological study of the primary exceptional progeny (which we have identified through the genetic study) should show that the females are XXY and the males are XO. Bridges confirmed this prediction.

2. Cytological study of the secondary exceptional progeny (which we have identified genetically) should show that the females are XXY and the males are XY. Bridges confirmed this prediction.

3. One-half of the red-eyed daughters of exceptional white-eyed females should be XXY, and one-half should be XX. Bridges confirmed this prediction.

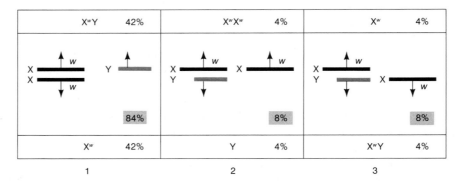

Figure 3-17. Three different segregation patterns in an XXY female fruit fly.

| | | Sperms | |
		X^{w+} (50%)	Y (50%)

| Eggs | X–Y pairing (16%) | X^wX^w (4%) | $X^wX^wX^{w+}$ (dies) (2%) | X^wX^wY (white ♀) (2%) |

The table in Figure 3-18:

Eggs (rows) × Sperms (columns):

X–Y pairing (16%):
- X^wX^w (4%): with X^{w+} → $X^wX^wX^{w+}$ (dies) (2%); with Y → X^wX^wY (white ♀) (2%)
- Y (4%): with X^{w+} → $X^{w+}Y$ (red fertile ♂) (2%); with Y → YY (dies) (2%)
- X^w (4%): with X^{w+} → $X^{w+}X^w$ (red ♀) (2%); with Y → X^wY (white ♂) (2%)
- X^wY (4%): with X^{w+} → $X^wX^{w+}Y$ (red ♀) (2%); with Y → X^wYY (white ♂) (2%)

X–X pairing (84%):
- X^wY (42%): with X^{w+} → $X^wX^{w+}Y$ (red ♀) (21%); with Y → X^wYY (white ♂) (21%)
- X^w (42%): with X^{w+} → X^wX^{w+} (red ♀) (21%); with Y → X^wY (white ♂) (21%)

Secondary exceptional progeny phenotype (4%) (4% die)

"Regular" (expected) progeny phenotypes (92%)

Figure 3-18. The proposed origin of secondary exceptional progeny as a result of specific gamete types produced by the XXY parent.

4. One-half of the white-eyed sons of exceptional white-eyed females should themselves give exceptional progeny, and all of those that do should be XYY. Bridges confirmed this prediction.

Thus, Bridges verified all the testable predictions arising from the assumptions that w and w^+ are indeed on the X chromosome and that an unexplained process of nondisjunction occurs in infrequent cases of meiosis. These confirmations provide unequivocal evidence that genes are associated with chromosomes.

> **Message** When Bridges used the chromosome theory to predict successfully the outcome of certain genetic analyses, the chromosome location of genes was established beyond reasonable doubt.

Sex Chromosomes and Sex Linkage

Humans and all mammals also show an X-Y sex-determining mechanism, with males XY and females XX. Unlike *Drosophila*, however, it is the presence of the Y that determines maleness in humans. This difference is demonstrated by the sexual phenotypes of the abnormal

TABLE 3-2. Chromosome determination of sex in *Drosophila* and humans

Species	Sex chromosomes			
	XX	XY	XXY	XO
Drosophila	♀	♂	♀	♂
Humans	♀	♂	♂	♀

chromosome types XXY and XO (Table 3-2). However, we postpone a full discussion until a later chapter.

Higher plants show a variety of sexual situations. Some species have both male and female sex organs on the same plant, often combined into the same flower — called **hermaphroditic** species (the rose is an example) — or in separate flowers on the same plant — called **monoecious** species (corn is an example). These plant species would be the equivalent of hermaphroditic animals. **Dioecious** species, however, show plants of separate sexes, with female plants bearing flowers containing only ovaries and male plants bearing flowers containing only anthers (Figure 3-19). Some, but not all, dioecious plants have a heteromorphic pair of chromosomes associated with (and almost certainly determining) the sex of the plant. Of the species with heteromorphic sex chro-

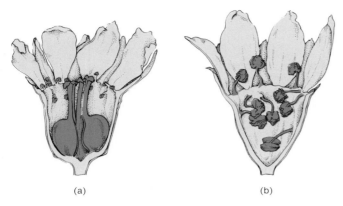

Figure 3-19. Flower forms in *Osmaronia dioica,* a dioecious species. (a) Female flowers. (b) Male flowers.

TABLE 3-3. Sex-chromosome situations in some dioecious plant species

Species	Chromosome number	Sex-chromosome constitution	
		Female	Male
Cannabis sativa (hemp)	20	XX	XY
Humulus lupulus (hop)	20	XX	XY
Rumex angiocarpus (dock)	14	XX	XY
Melandrium album (campion)	22	XX	XY

mosomes, a large proportion have an X-Y system. Critical experiments in a few species suggest a *Drosophila*-like system. Table 3-3 lists some examples of dioecious plants with heteromorphic sex-determining chromosomes. Some dioecious plants have no visibly heteromorphic pair of chromosomes; they may still have sex chromosomes, but not visibly distinguishable types. Other dioecious plants have heteromorphic chromosomes that determine sex, but the constitution of the different sexes is much more complex than the simple X-Y system of the *Drosophila* type. We shall not consider these systems.

The sex chromosomes can be tentatively divided into pairing and differential regions (Figure 3-20). These regions are based on studies of meiosis in males. The pairing regions of the X and Y chromosomes are thought to be homologous. The differential region of each chromosome appears to hold genes that have no counterparts on the other kind of sex chromosome. These genes, whether dominant or recessive, show their effects in the male phenotype. Genes in the differential

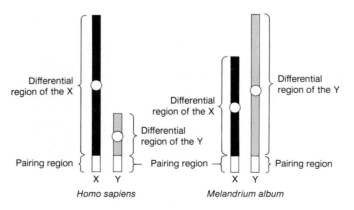

Figure 3-20. Differential and pairing regions of sex chromosomes of humans and of the plant *Melandrium album.* The pairing regions have been located chiefly through cytological examination.

regions are called **hemizygous** ("half-zygous") in the males. Genes in the differential region of the X show an inheritance pattern called **X linkage;** those in the differential region of the Y show **Y linkage.** Genes in the pairing region show what might be called **X-and-Y linkage.** In general, genes on the sex chromosomes show **sex linkage.**

We can introduce some other common terminology here. The sex showing only one kind of sex chromosome (XX♀ or ZZ♂) is called the **homogametic** sex; the other (XY♂ or ZW♀) is called the **heterogamatic** sex. Thus, human and *Drosophila* females (and male birds and moths) are homogametic; that is, they produce only one type of gamete with respect to sex-chromosome constitution. The nonsex chromosomes (what we might call the "regular" chromosomes) are called **autosomes.** Humans have 46 chromosomes per cell: 44 autosomes plus two sex chromosomes. The plant *Melandrium album* has 22 chromosomes per cell: 20 autosomes plus two sex chromosomes.

Genes on the autosomes show the kind of inheritance pattern discovered and studied by Mendel. The genes on the different regions of the sex chromosomes show their own typical patterns of inheritance, as follows.

X-Linked Inheritance

We have already seen an example of X-linked inheritance in *Drosophila:* the inheritance pattern of white eye and its wild-type allele. Of course, eye color is not concerned with sex determination, so we see that not all genes on the sex chromosomes are involved with sexual function. The same is true in humans, where many X-linked genes have been discovered through pedigree analysis hardly any of which could be construed as being connected to sexual function. Just as we earlier listed rules for detecting autosomal inheritance patterns in humans, we can list clues for detecting X-linked genes in human pedigrees. (We shall soon see that Y-linked genes are very rare in humans, so we can usually ignore them.)

Recessive genes showing X-linked inheritance can be detected in human pedigrees through the following clues.

1. Typically, many more males than females show the recessive phenotype. This is because an affected female can result only when both the mother and father bear the gene (for example, $X^A X^a \times X^a Y$), whereas an affected male can result when only the mother carries the gene. If the recessive gene is very rare, almost all observed cases will occur in males.

2. Usually none of the offspring of an affected male will be affected, but all his daughters will carry the gene in masked heterozygous condition, so one-half of their sons will be affected (Figure 3-21).

3. None of the sons of an affected male will inherit the gene, so not only will they be free of the phenotype, but they will also not pass the gene along to their offspring.

Some good examples of X-linked recessive genes in humans are those for hemophilia, red-green color blindness, and Duchenne's muscular dystrophy.

Dominant genes showing X-linked inheritance can be detected in human pedigrees through the following clues.

1. The most important clue here is that affected males pass the condition on to all of their daughters but to none of their sons (Figure 3-22).

2. Females, on the other hand, usually pass the condition on to one-half of their sons and daughters (Figure 3-23).

X-linked dominant conditions are relatively rare. One example is hypophosphatemia (vitamin D-resistant rickets).

Y-Linked Inheritance

In humans, genes on the differential region of the Y chromosome would be inherited only by males, with transmission from father to son. No examples of such inheritance have been confirmed, although hairy ear rims has been suggested as a likely possibility (Figure 3-24). However, in humans (and in other species with similar sex-determination systems), the presence of the Y chromosome and its unique regions determines maleness. Therefore, it seems safe to speculate that "maleness" genes of some kind must exist on the differential

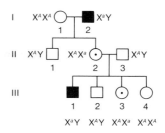

Figure 3-21. Illustrative pedigree showing how X-linked recessives are expressed in males and then carried unexpressed by females in the next generation, to be expressed in their sons. (Note that III.3 and III.4 cannot be distinguished phenotypically.)

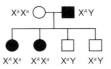

Figure 3-22. Illustrative pedigree showing how X-linked dominants are expressed in all the daughters of affected males.

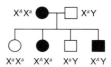

Figure 3-23. Illustrative pedigree showing how females affected by an X-linked dominant condition usually are heterozygous and pass the condition to one-half of their progeny.

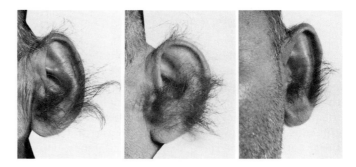

Figure 3-24. Hairy ear rims are thought to be due to a Y-linked gene. (From C. Stern, W. R. Centerwall, and S. S. Sarkar, *The American Journal of Human Genetics* 16, 1964; 467. By permission of Grune & Stratton, Inc.)

region of the Y, and recent studies in humans have identified such an element.

In the fish *Lebistes,* the Y chromosome carries a gene called *maculatus* that determines a pigmented spot at the base of the dorsal fin. This phenotype is passed only from father to son, and females never carry or express the gene, so it seems to be a clear example of Y linkage.

X-and-Y-Linked Inheritance

Are there hereditary patterns that identify genes on homologous regions of the sex chromosomes? There is a gene pair in *Drosophila* that is inherited in such a way that an X-and-Y location is likely. Curt Stern found that a certain non-wild recessive allele in *Drosophila* causes a phenotype of shorter and more slender bristles; he called it bobbed (*b*). If a bobbed $X^b X^b$ female is crossed with a wild-type $X^+ Y^+$ male, all the F_1 progeny are wild-type: $X^+ X^b$ ♀♀ and $X^b Y^+$ ♂♂. The same result, of course, would be expected from an autosomal gene pair. However, the X-and-Y linkage is revealed in the F_2, which shows the following clear, sex-associated pattern.

Sex	Phenotype	Inferred genotype
Males	wild-type	$X^+ Y^+$ and $X^b Y^+$
Females	½ bobbed	$X^b X^b$
	½ wild-type	$X^+ X^b$

It has recently been shown by a combination of genetic and molecular studies that there is a homologous region of the human sex chromosomes in which pairing and crossing-over occur. In fact, it is known that there is an obligatory crossover in this region in every meiosis, so alleles in this vicinity are effectively uncoupled from the unique regions of the X and Y chromosomes and show what is called **pseudoautosomal inheritance.**

The Chromosome Theory in Review

The patterns of inheritance arising from the normal and the nondisjunctional behavior of the sex chromosomes provide satisfying confirmation of the chromosome theory of inheritance, which was originally suggested by the parallel behavior of Mendelian genes and autosomal chromosomes. Now we should pause and state clearly the situation for the regular (autosomal) genes, because these are the genes most commonly encountered. Such a summary is best achieved in a diagram; Figure 3-25 illustrates the passage of a hypothetical cell through meiosis. Two gene pairs are shown on two chromosome pairs.

The hypothetical cell type has four chromosomes: a pair of homologous long chromosomes and a pair of homologous short ones. (Such size differences between pairs are common.) The genotype of the cell is assumed to be *Aa Bb*.

As the diagram shows, two equally frequent kinds of spindle attachments (4a and 4b) result in two basic kinds of segregation patterns of gene pairs. Meiosis then produces four cells of the genotypes shown from each of these segregation patterns. Because segregation patterns 4a and 4b are equally common, the meiotic product cells of genotypes *AB, ab, Ab,* and *aB* are produced in equal freqencies. In other words, each of the four genotypes occurs with frequency ¼. This, of course, is the distribution postulated in Mendel's model and is the one we noted along one edge of the standard Punnett square (see Figure 2-10). We can now understand the exact chromosomal mechanism that produces the Mendelian ratios.

Notice that Mendel's first law (equal segregation) is a direct result of the separation of a pair of homologs (bearing the gene pair under study) into opposite cells at the first division. Notice also that Mendel's second law (independent assortment) results from the independent behavior of separate pairs of homologous chromosomes.

The chromosome theory was important in many ways. At this point in our discussion, we can stress the importance of the theory in centering attention on the role of the cell genotype in determining the organism phenotype. The phenotype of an organism is determined by the phenotypes of all its individual cells. The cell phenotype, in turn, is determined by the alleles present on the chromosomes of the cell. When we say that an organism is (for example) *AA*, we really mean that each cell of the organism is *AA*. This cell genotype determines how the cell functions, thereby controlling the phenotype of the cell and hence the phenotype of the organism.

> **Message** The phenotype of an organism is determined by the kinds of genes it has in its cells and by the forms (alleles) of those genes that are present.

Mendelian Genetics and Sexual Cycles

Thus far, we have been discussing mainly diploid organisms — organisms with two homologous chromosome sets in each cell. As we have seen, diploid is designated $2n$, where n stands for one chromosome set (for example, the pea cell contains two sets of seven chromo-

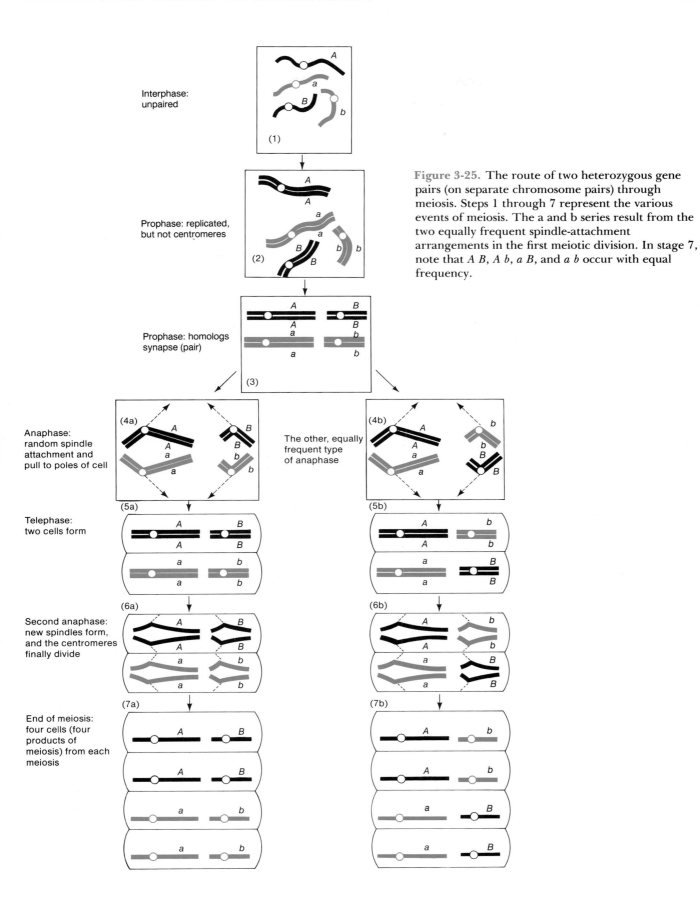

Interphase: unpaired

(1)

Prophase: replicated, but not centromeres

(2)

Prophase: homologs synapse (pair)

(3)

Anaphase: random spindle attachment and pull to poles of cell

(4a)

(4b) The other, equally frequent type of anaphase

Telephase: two cells form

(5a)

(5b)

Second anaphase: new spindles form, and the centromeres finally divide

(6a)

(6b)

End of meiosis: four cells (four products of meiosis) from each meiosis

(7a)

(7b)

Figure 3-25. The route of two heterozygous gene pairs (on separate chromosome pairs) through meiosis. Steps 1 through 7 represent the various events of meiosis. The a and b series result from the two equally frequent spindle-attachment arrangements in the first meiotic division. In stage 7, note that *A B*, *A b*, *a B*, and *a b* occur with equal frequency.

somes, so $2n = 14$). Most of the organisms we encounter in our daily existence are predominantly diploid; these are the so-called higher plants (including flowering plants) and animals (including humans). In fact, evolution seems to have produced a trend toward diploidy (perhaps you can speculate on reasons for this). Nevertheless, a vast proportion (probably the major proportion) of the biomass on the earth is composed of organisms that spend most of their life cycles in a haploid condition, in which each cell has only one set of chromosomes. Important here are the fungi and algae, most of which are predominantly haploid. Bacteria could be considered haploid, but they form a special case because their cells contain no nuclei (they are called prokaryotes), whereas most other life forms (eukaryotes) do have nuclei. (Bacteria are discussed in Chapter 13.) Also important are organisms that spend part of their life cycles as haploid and another part as diploid. Such organisms are said to show **alternation of generations** (a rather imprecise term; "alternation of $2n$ and n" would be more descriptive). All plants, in fact, show alternation of generations. However, higher forms, such as flowering plants, have an inconspicuous haploid plant stage that is dependent on the diploid plant, in which it occurs as a specialized structure. Other forms, such as mosses and ferns, have separate and independent haploid stages.

Do these life cycles show Mendelian genetics? Or is Mendelian inheritance observed only in the higher organisms? The answer is that Mendelian inheritance patterns appear in any organism that has meiosis as part of its life cycle, because Mendelian laws are based on the process of meiosis. All the groups of organisms mentioned, except bacteria, utilize meiosis as part of their cycles. Next, we consider the inheritance patterns shown by less familiar eukaryotes and compare them to more familiar cycles. This is important because it demonstrates the universality of Mendelian genetics. Furthermore, some of the less well-known organisms have been the subject of extensive genetic research, and a knowledge of their life cycles is essential to their genetic analysis. We describe here three major types of cycle, beginning with the more familiar diploid types.

Diploids

Figure 3-26 summarizes the diploid cycle in skeletal form. This is the cycle shown by most animals (including humans). Meiosis occurs in specialized diploid cells, the meiocytes, which are set aside for the purpose but which are part of the diploid adult organism. The products of meiosis are the gametes (eggs or sperm). Fusion of haploid gametes forms a diploid zygote, which (through mitosis) produces a multicellular organism. Mitosis in a diploid proceeds in the fashion outlined in Figure 3-27.

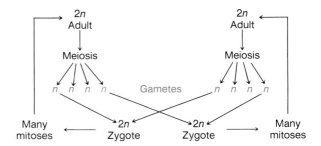

Figure 3-26. The diploid life cycle.

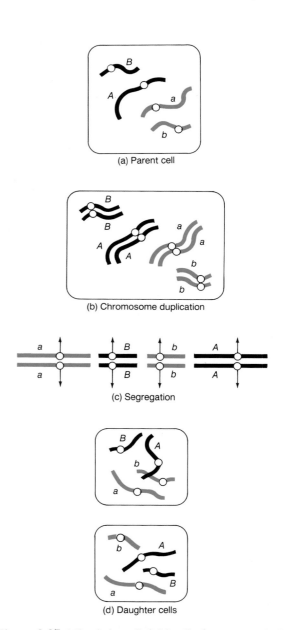

(a) Parent cell

(b) Chromosome duplication

(c) Segregation

(d) Daughter cells

Figure 3-27. Mitosis in a diploid cell of genotype *Aa Bb*.

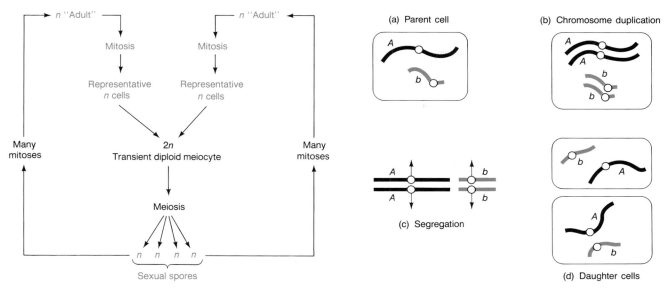

Figure 3-28. The haploid life cycle.

Figure 3-29. Mitosis in a haploid cell of genotype *A b*.

Haploids

Figure 3-28 shows the basic haploid cycle. Here, the "adult" (either multicellular or unicellular) is haploid. How can meiosis possibly occur in a haploid? After all, meiosis involves the pairing of two homologous chromosome sets! The answer is that all haploid organisms that undergo meiosis employ a *transient* diploid stage. In some cases, unicellular haploid adult individuals fuse to form a diploid cell, which then undergoes meiosis. In other cases, specialized (or sometimes representative) haploid cells from different parents fuse to form diploid cells so that meiosis can occur. (These fusing cells are properly called the gametes, so we see that in these cases gametes arise from mitosis.) Meiosis, as usual, produces haploid products of meiosis, which are called **sexual spores.** The sexual spores in some species become new unicellular adults; in other species, they develop through mitosis into a multicellular haploid individual. In haploids, mitosis proceeds as shown in Figure 3-29, which illustrates a cell arbitrarily designated genotype *Ab*. Notice that a cross between two adult haploid organisms involves only one meiosis, whereas a cross between two diploid organisms involves a meiosis in each organism. As we shall see, this simplification makes haploids very attractive for genetic analysis.

Let's consider a cross in a specific haploid. A convenient organism is the pink bread mold *Neurospora*. This fungus is a multicellular haploid in which the cells are joined end to end to form **hyphae,** or threads of cells.

The hyphae grow through the substrate and also send up aerial branches that bear cells known as **asexual spores.** These can detach and disperse to form new colonies; alternatively, they can act as male gametes (Figure 3-30). Another specialized cell (which develops inside a knot of hyphae) can be regarded as the female gamete.

What morphological characters can be studied in such an organism? One is the color of the cells. Variants of the normal pink color can be found — for example, an albino. Figure 3-31 shows a normal culture and an albino culture. Another character is the morphological nature of the culture — perhaps fluffy (normal) versus colonial. We can make a cross by allowing the asexual spores to act as male gametes. A culture cannot self in *Neurospora*, because this fungus has two genetically determined mating types, *A* and *a*. Fertile crosses can occur only if strains of different mating types are paired (in this case, *A* × *a*). Crosses are carried out by adding asexual spores of one culture to another. The nucleus of an asexual spore pairs with a female nucleus. This pair undergoes synchronous mitotic division, and fusions finally occur to generate transient diploid meiocytes. Then meiosis occurs, and sexual spores, called **ascospores,** are formed. These ascospores are black and football-shaped; they are shot out of the knot of hyphae, which is now known as a fruiting body. The ascospores can be isolated, each into a culture tube, where each ascospore will grow into a new culture by mitosis (Figure 3-32).

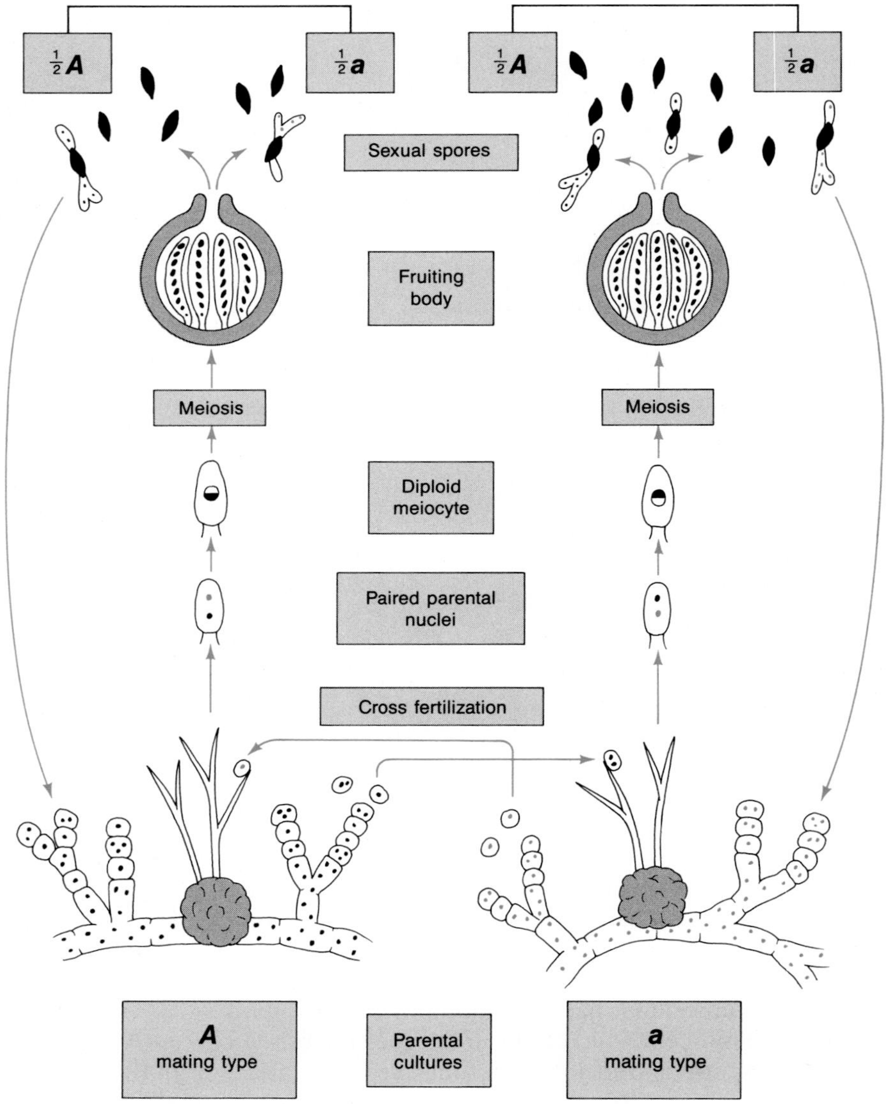

Figure 3-30. Simplified representation of the life cycle of *Neurospora crassa,* the pink bread mold. Self-fertilization is not possible in this species: there are two mating types, determined by the alleles *A* and *a* of one gene. A cross will succeed only if it is *A* × *a*. An asexual spore from the opposite mating type fuses with a receptive hair, and a nucleus travels down the hair to pair with a nucleus in the knot of cells. The *A* and *a* pair, then undergo synchronous mitoses, finally fusing to form diploid meiocytes.

If we cross a fluffy pink culture with a colonial albino culture and isolate and culture 100 ascospores, the resulting cultures (on average) would be

25 fluffy pink cultures

25 colonial albino cultures

25 fluffy albino cultures

25 colonial pink cultures

In total, one-half of the "progeny" are fluffy and one-half are colonial. Thus, these phenotypes are determined by one gene pair that has segregated equally at meiosis. The same is true of the other character: one-half are pink and one-half are albino, so color is determined by a different gene pair. We could represent the four culture types as

$col^+ al^+$ (fluffy pink)

$col\ al$ (colonial albino)

$col^+ al$ (fluffy albino)

$col\ al^+$ (colonial pink)

The 25% : 25% : 25% : 25% ratio is a result of independent assortment, as illustrated in the following branch diagram:

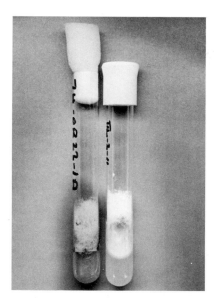

Figure 3-31. A pink wild-type *Neurospora* culture *(left)* and an albino mutant culture lacking the reddish carotenid pigment *(right)*. (Genotypically, the cultures are *al⁺* and *al⁻*, respectively.)

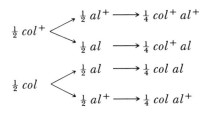

So we see that even in such a lowly organism, Mendel's laws are still in operation.

Alternating Haploid/Diploid

In an organism with alternation of generations, there are two stages to the life cycle: one diploid and one haploid. One stage is usually more prominent than the other. For example, what we all recognize as a fern plant is the diploid stage, but the organism does have a small, independent, photosynthetic haploid stage that is usually much more difficult to spot on the forest floor. In mosses, the green plant is the haploid stage and the brownish stalk that grows up out of this plant is a dependent diploid stage that is effectively parasitic on it. In flowering plants, the main green stage is, of course, diploid. The haploid stages of flowering plants are extremely reduced and dependent on the diploid. These haploids are found in the flower. When meiosis occurs in the anther and ovary meiocytes, the haploid products of meiosis are called **spores.** The spores undergo a few

(a)

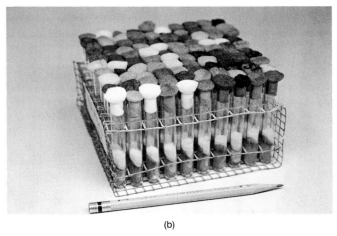

(b)

Figure 3-32. (a) A *Neurospora* cross made in a petri plate. The many small black spheres are fruiting bodies in which meiosis has occurred; the ascospores (sexual spores) were shot as a fine dust into the condensed moisture on the lid (which has been removed and is sitting on the right side of the plate). (b) A rack of progeny cultures, each resulting from one isolated ascospore.

mitotic divisions to produce a small, multicellular haploid stage, as shown in Figure 3-33. In alternation of generations, the diploid stage is called the **sporophyte,** which means sexual-spore-producing plant, and the haploid stage is called **gametophyte,** which means gamete-producing plant. The male gametophyte of seed plants is known as a pollen grain. Figure 3-33 shows that in flowering plants, cells of the gametophyte act as egg or sperm in fertilization. In mosses and ferns, the sperm cells are highly motile and have to travel from one gametophyte to another in a film of water in order to effect fertilization. The generalized cycle of alternation of generations is shown in Figure 3-34.

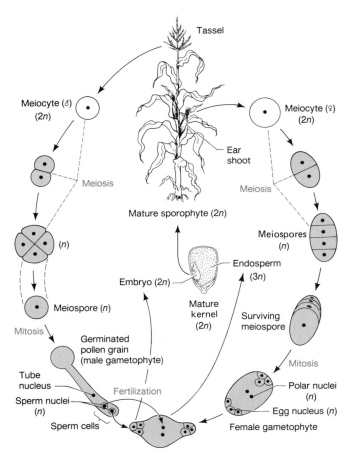

Figure 3-33. Details of the production of the male and female gametophytes and the fertilization process in corn. One sperm cell fuses with the terminal female cell, and the diploid cell thus formed gives rise to the embryo. The other sperm cell fuses with two central female nuclei to give rise to a triploid ($3n$) cell that generates the endosperm tissue surrounding the embryo. The endosperm provides nutrition to the embryo during seed germination.

Little genetic analysis has been done in plants other than flowering plants, but the potential for study is great, so we shall follow a sample cross in a moss. The character to be studied, of course, can pertain to the gametophyte or the sporophyte. Assume that we have a gene pair affecting the "leaves" of the gametophyte, with w causing wavy edges and w^+ causing smooth edges. Also assume that a separate gene pair affects the color of the sporophyte, with r causing reddish coloration and r^+ causing the normal brown coloration. A smooth-leaved gametophyte also bearing an unexpressed allele r is fertilized by transferring onto it male gametes from a wrinkly leaved gametophyte carrying r^+ (Figure 3-35).

A diploid sporophyte develops, actually on the gametophyte, and it is brown (because reddish is recessive). The sexual spores produced from this sporophyte will be

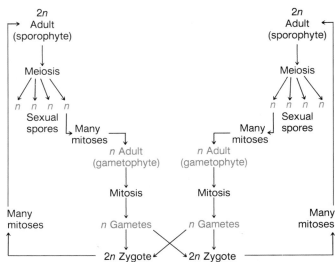

Figure 3-34. The alternation of diploid and haploid stages in the life cycle of plants.

$$25\% \ w^+r^+$$
$$25\% \ w^+r^-$$
$$25\% \ w^-r^+$$
$$25\% \ w^-r^-$$

Of course, only the leaf character will be identifiable in these gametophytes; the r^+ or r^- designation would have to be determined by appropriate intercrosses. (What proportion of reddish sporophytes is expected if crosses are made at random?)

Once again, Mendel's laws rule the inheritance. It is simply a matter of keeping the ploidy levels (the number of chromosome sets) straight in each part of the cycle, and everything is simple Mendelian ratios.

Message   Mendelian laws apply to the products of meiosis in any organism. Their general statement is as follows:

1. At meiosis, the members of a gene pair segregate equally into the haploid products of meiosis.

2. At meiosis, the segregation of one gene pair is independent of the segregation of gene pairs on other chromosome pairs.

Thus, the theory of the chromosome location of genes perfectly explains inheritance patterns. The chromosome theory of inheritance is no longer in doubt; it forms one of the cornerstones of modern biological theory.

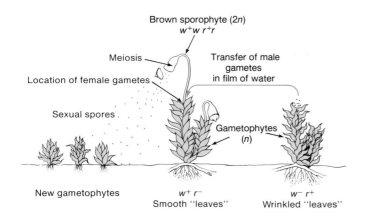

Figure 3-35. Diagrammatic representation of Mendelian genetics in a hypothetical cross in a moss. The w^+ or w alternatives are expressed only in the haploid gametophyte stage, whereas the r^+ or r alternatives are expressed only in the diploid sporophyte stage. In mosses, the diploid stage grows on the haploid stage. Male gametes swim through a water film to get to another plant. There, one male gamete and one female gamete fuse to form a diploid cell that divides at the point of fusion to become the sporophyte.

Summary

■ After the rediscovery of Mendelian principles in 1900, scientists set out to discover what structures within cells correspond to Mendel's hypothetical units of heredity, which we now call genes. Recognizing that the behavior of chromosomes during meiosis parallels the behavior of genes, Walter Sutton and Theodor Boveri suggested that genes were located in or on the chromosomes.

In her experiments with a certain species of grasshopper, Elinor Carothers discovered that nonhomologous chromosomes assort independently. This finding provided further evidence that the behavior of chromosomes closely parallels that of genes. Additional evidence for the validity of the chromosome theory of heredity came from the discovery of crisscross inheritance and the existence of X and Y chromosomes. In his studies of *Drosophila*, Thomas Hunt Morgan showed that the crisscross inheritance of red or white eye color is completely consistent with the meiotic behavior of X and Y chromosomes. Finally, Calvin Bridges's postulation of

nondisjunction (the failure of paired chromosomes to separate) during meiosis enabled him to make testable predictions based on the assumption that the gene for eye color is on the X chromosome and that nondisjunction occurs in infrequent cases of meiosis. The confirmation of these predictions provided unequivocal evidence that genes are located on chromosomes.

We now know which chromosome mechanisms produce Mendelian ratios. Mendel's first law (equal segregation) results from the separation of a pair of homologous chromosomes into opposite cells at the first division. Mendel's second law (independent assortment) is a result of the independent behavior of separate pairs of homologous chromosomes.

Because Mendelian laws are based on meiosis, Mendelian inheritance occurs in any organism with a meiotic stage in its life cycle, including diploid organisms, haploid organisms, and organisms with alternating haploid and diploid generations.

■ ■ ■

Solved Problems

1. The accompanying pedigree concerns a rare human disease.

   a. Deduce the most likely mode of inheritance.

   b. What would be the outcomes of the cousin marriages 1×9, 1×4, 2×3 and 2×8?

   Solution

   a. The most likely mode of inheritance is X-linked dominant. Dominance is revealed by the fact that after the

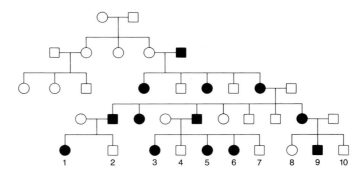

disease is introduced into the pedigree by the male in generation II, every generation shows individuals with the disease phenotype. X-linkage is revealed by the pronounced lack of father to son transmission of the phenotype. If it were autosomal dominant, male-to-male transmission would be common. In theory, autosomal recessive could work but would require too many assumptions. In particular, note the marriages of affected to unaffected people marrying into the family. If the condition were autosomal recessive, the only way these marriages could have affected offspring is if each normal person marrying into the family were a heterozygote; then the matings would be of genotype aa (affected) $\times$ Aa (normal). However, we are told that the disease is rare; in such a case, it is highly unlikely that heterozygotes would be so common. X-linked recessive inheritance is impossible because a mating of an affected woman with a normal man could not produce affected daughters. So we can let A represent the disease-causing allele and a represent normality.

b. 1×9: Number 1 must be heterozygous Aa because she must have obtained a from her normal mother. Number 9 must be AY. Hence, the cross is $Aa\,♀ \times AY\,♂$.

Female gametes | Male gametes | Progeny

$$\frac{1}{2}A \begin{cases} \rightarrow \frac{1}{2}A \longrightarrow \frac{1}{4}AA\,♀ \\ \rightarrow \frac{1}{2}Y \longrightarrow \frac{1}{4}AY\,♂ \end{cases}$$

$$\frac{1}{2}a \begin{cases} \rightarrow \frac{1}{2}A \longrightarrow \frac{1}{4}Aa\,♀ \\ \rightarrow \frac{1}{2}Y \longrightarrow \frac{1}{4}aY\,♂ \end{cases}$$

1×4: Must be $Aa\,♀ \times aY\,♂$.

Female gametes | Male gametes | Progeny

$$\frac{1}{2}A \begin{cases} \rightarrow \frac{1}{2}a \longrightarrow \frac{1}{4}Aa\,♀ \\ \rightarrow \frac{1}{2}Y \longrightarrow \frac{1}{4}AY\,♂ \end{cases}$$

$$\frac{1}{2}a \begin{cases} \rightarrow \frac{1}{2}a \longrightarrow \frac{1}{4}aa\,♀ \\ \rightarrow \frac{1}{2}Y \longrightarrow \frac{1}{4}aY\,♂ \end{cases}$$

2×3: Must be $aY\,♂ \times Aa\,♀$ (same as 1×4).

2×8: Must be $a\,Y♂ \times aa\,♀$ (all progeny normal).

2. Two corn plants are studied; one is Aa, and the other is aa. These two plants are intercrossed in two ways: using Aa as female and aa as male; using aa as female and Aa as male.

a. What endosperm genotypes will be produced in each cross? In what proportions? (Note that the endosperm is $3n$ and is formed by the union of a sperm cell with the two polar nuclei of the female gametophyte.)

b. In an experiment to study the effects of gene "dose," you wish to establish endosperms with known genotypes of aaa, Aaa, AAa, and AAA (carrying 0, 1, 2 and 3 "doses" of A, respectively). What crosses would you make to establish samples of these endosperm genotypes?

Solution

a. In such a question, we have to think about meiosis and mitosis at the same time. The meiospores are produced by meiosis; the nuclei of the male and female gametophytes in higher plants are produced by the mitotic division of the meiospore nucleus. We also need to study the corn life cycle to know what nuclei fuse to form the endosperm.

First cross: $Aa\,♀ \times aa\,♂$

Here, the female meiosis will result in spores of which one-half will be A and one-half will be a. Therefore, an equal number of haploid female gametophytes will be produced. Their nuclei will be either all A or all a, because mitosis reproduces genetically identical genotypes. Likewise, the nuclei in every male gametophyte will all be a. In the corn life cycle, the endosperm is formed from two female nuclei plus one male nucleus, so two endosperm types will be formed as follows.

♀ spore	♀ polar nuclei	♂ sperm	$3n$ endosperm
$\frac{1}{2}A$	A and A	a	$\frac{1}{2}AAa$
$\frac{1}{2}a$	a and a	a	$\frac{1}{2}aaa$

Second cross: $aa\,♀ \times Aa\,♂$

♀ spore	♀ polar nuclei	♂ sperm	$3n$ endosperm
all a	all a and a	$\frac{1}{2}A$	$\frac{1}{2}Aaa$
		$\frac{1}{2}a$	$\frac{1}{2}aaa$

Notice that the phenotype ratio of endosperm characters would still be Mendelian, even though the underlying endosperm genotypes are slightly different. (Of course, none of these problems arise in embryo characters because embryos are diploid.)

b. This kind of experiment has been very useful in studying plant genetics and molecular biology. In answering the question, all we need to realize is that both haploid female polar nuclei contributing to the endosperm are genetically identical. To obtain endosperms, all of which will be aaa, any $aa \times aa$ cross will work. To obtain endosperms, all of which will be Aaa, the cross must be $aa\,♀ \times AA\,♂$. To obtain embryos, all of which will be AAa, the cross must be $AA\,♀ \times aa\,♂$. For AAA, obviously any $AA \times AA$ cross will work. Notice that these endosperm genotypes can be obtained in other crosses, but only in combination with other endosperm genotypes.

Problems

1. What are the major differences between mitosis and meiosis?

2. If mitosis occurs in a cell of genotype *Aa Bb Cc*, where all the gene pairs are on separate chromosome pairs, what will be the genotypes of the resulting cells?

3. In haploid yeast, a cross is made between a purple (*ad⁻*) strain of mating type *a* and a white (*ad⁺*) strain of mating type α. If *ad⁻* and *ad⁺* are one gene pair and *a* and α are an independently inherited gene pair on a separate chromosome pair, what progeny can be predicted? In what proportions?

4. In *Drosophila,* the recessive allele *s* causes small wings and the *s⁺* allele causes normal wings. This gene is known to be X-linked. If a small-winged male is crossed with a normal female, what ratio of normal to small-winged flies can be expected in each sex in the F_1 and F_2? What is the predicted outcome if F_1 females are backcrossed to their father?

5. State where you find mitosis and meiosis occurring in a fern, a moss, a flowering plant, a pine tree, a mushroom, a frog, a butterfly, and a snail.

6. In a colony of flour beetles, which normally are black, a brown individual is found. In this case, the brown color is due to a dominant allele. Invent a symbol for this gene and its wild-type allele.

7. Human cells normally have 46 chromosomes. For each of the following stages, state the number of chromosomes present in a human cell. a. Metaphase of mitosis b. Metaphase 1 of meiosis c. Telophase of mitosis d. Telophase I of meiosis e. Telophase II of meiosis (In your answers, count chromatids as chromosomes.)

8. Four of the following events occur in both meiosis and mitosis, but one occurs only in meiosis. Which one? a. Chromatid formation b. Spindle formation c. Chromosome condensation (shortening and thickening) d. Chromosome movement to poles e. Chromosome pairing

9. a. A cell with 10 pairs of chromosomes undergoes mitosis. How many chromosomes does each of the resulting cells have? (1) 2 pairs (2) 5 (3) 20 pairs (4) 20 (5) 10

   b. A cell with 10 pairs of chromosomes undergoes meiosis. How many chromosomes does each of the resulting cells have? (1) 2 pairs (2) 5 (3) 20 pairs (4) 20 (5) 10

10. Suppose that two interesting *rare* cytological abnormalities are discovered in the karyotype of a human male. There is an extra piece (or satellite) on *one* of the chromosomes of pair 4, and there is an abnormal pattern of staining on *one* of the chromosomes of pair 7. Assuming that all the gametes of this male are equally viable, what proportion of his children will have the same visible karyotype he has? (A karyotype is the total visible chromosome complement.)

11. Suppose that meiosis occurs in the transient diploid stage of the cycle of a haploid organism of chromosome number *n*. What is the probability that an individual haploid resulting from the meiotic division will have a complete parental set of centromeres (that is, a set all from one parent or all from the other parent)?

12. Assuming the sex chromosomes to be identical, name the proportion of all genes you have in common with: a. Your mother b. Your brother

13. In the plant *Rumax hastatulatus,* females normally have the sex-chromosome constitution XX and males normally have the sex-chromosome constitution XYY. Draw diagrams to show how meiosis might proceed to produce the gametic types necessary to maintain a 1:1 sex ratio of males to females.

14. A wild-type female schmoo who is graceful (*G*) is mated to a non-wild-type male who is gruesome (*g*). Their progeny consist solely of graceful males and gruesome females. Interpret these results and give genotypes.

(Problem 14 from E. H. Simon and H. Grossfield, *The Challenge of Genetics,* Addison-Wesley, 1971.)

15. A man with a certain disease marries a normal woman. They have eight children (four boys and four girls); all of the girls have their father's disease, but none of the boys do. What inheritance is suggested? a. Autosomal recessive b. Autosomal dominant c. Y-linked d. X-linked dominant e. X-linked recessive

16. Hypophosphatemia is caused by an X-linked dominant gene in humans. A man with hypophosphatemia marries a normal woman. What proportion of their sons will have hypophosphatemia? a. $\frac{1}{2}$ b. $\frac{1}{4}$ c. $\frac{1}{3}$ d. 1 e. 0

17. A condition known as icthyosis hystrix gravior appeared in a boy in the early eighteenth century. His skin became very thick and formed loose spines that were sloughed off at intervals. When he grew up, this "porcupine man" married and had six sons, all of whom had this condition, and several daughters, all of whom were normal. For four generations, this condition was passed from father to son. From this evidence, what can you postulate about the location of the genes?

18. Duchenne's muscular dystrophy is sex-linked and usually affects only males. Victims of the disease become progressively weaker, starting early in life.

   a. What is the probability that a woman whose brother has Duchenne's disease will have an affected child?

   b. If your mother's brother (your uncle) had Duchenne's disease, what is the probability that you have received the gene?

   c. If your father's brother had the disease, what is the probability that you have received the gene?

19. The following pedigree is concerned with an inherited dental abnormality, amelogenesis imperfecta.

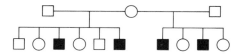

    a. What mode of inheritance *best* accounts for the transmission of this trait?

    b. Write the genotypes of the individual members according to your hypothesis.

20. A sex-linked recessive gene *c* produces a red-green colorblindness in humans. A normal woman whose father was color blind marries a colorblind man.

    a. What genotypes are possible for the mother of the colorblind man?

    b. What are the chances that the first child from this marriage will be a colorblind boy?

    c. Of the girls produced by these parents, what percentage can be expected to be colorblind?

    d. Of all the children (sex unspecified) of these parents, what proportion can be expected to have normal color vision?

21. Male housecats are either black or orange; females are black, tortoise-shell pattern, or orange.

    a. If these colors are governed by a sex-linked gene, how can these observations be explained?

    b. Using appropriate symbols, determine the phenotypes expected in the progeny of a cross between an orange female and a black male.

    c. Repeat part b for the reciprocal of the cross described there.

    d. One-half of the females produced by a certain kind of mating are tortoise-shell, and one-half are black; one-half of the males are orange, and one-half are black. What colors are the parental males and females in this kind of mating?

    e. Another kind of mating produces progeny in the following proportions: $\frac{1}{4}$ orange males, $\frac{1}{4}$ orange females, $\frac{1}{4}$ black males, and $\frac{1}{4}$ tortoise-shell females. What colors are the parental males and females in this kind of mating?

22. A man is heterozygous for one autosomal gene pair *Bb*, and he carries a recessive X-linked gene *d*. What proportion of his sperm will be *bd*? a. 0 b. $\frac{1}{2}$ c. $\frac{1}{8}$ d. $\frac{1}{16}$ e. $\frac{1}{4}$

23. A woman has the rare (hypothetical) disease called quackerlips. She marries a normal man, and all of their sons and none of their daughters are quackerlipped. What is the mode of inheritance of quackerlips? a. Autosomal recessive b. Autosomal dominant c. X-linked recessive d. X-linked dominant

24. The accompanying pedigree concerns a certain rare disease X that is incapacitating but not fatal.

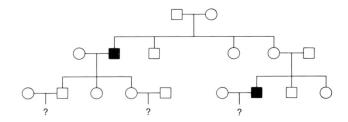

    a. Determine the mode of inheritance of this disease.

    b. Copy the pedigree into your answer book, and write the genotype of each individual according to your proposed mode of inheritance.

    c. If you were this family's doctor, how would you advise the three couples in question about the likelihood of having an affected child?

25. What do you think are the advantages and disadvantages to the organism of having genes organized into chromosomes? Why don't genes float free in the nucleus or the cell? (Try to remember to reconsider this question after finishing the book, and see if your answers have changed.)

26. Assume the pedigree presented here to be straightforward, with no complications such as illegitimacy. Trait W, found in individuals represented by the shaded symbols, is rare in the general population. Which of the following patterns of transmission for W are consistent with this pedigree? Which are excluded? a. Autosomal recessive b. Autosomal dominant c. X-linked recessive d. X-linked dominant e. Y-linked

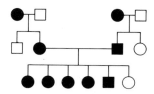

(Problem 26 from A. M. Srb, R. D. Owen, and R. S. Edgar, *General Genetics*, 2d ed. Copyright © 1965 by W. H. Freeman and Company.)

27. A mutant allele in mice causes a bent tail. From the cross results given in Table 3-4, deduce the mode of inheritance of this trait.

    a. Is it recessive or dominant?

    b. Is it autosomal or sex-linked?

    c. What are the genotypes of parents and progeny in all crosses shown in the table?

TABLE 3-4.

Cross	Parents		Progeny	
	Female	Male	Female	Male
1	Normal	Bent	All bent	All normal
2	Bent	Normal	$\frac{1}{2}$ bent, $\frac{1}{2}$ normal	$\frac{1}{2}$ bent, $\frac{1}{2}$ normal
3	Bent	Normal	All bent	All bent
4	Normal	Normal	All normal	All normal
5	Bent	Bent	All bent	All bent
6	Bent	Bent	All bent	$\frac{1}{2}$ bent, $\frac{1}{2}$ normal

28. In the following pedigree, a black dot represents the occurrence of an extra finger and a black square represents the occurrence of an eye disease.

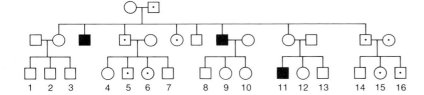

a. What can you tell about the inheritance of an extra finger?

b. What can you tell about the inheritance of the eye disease?

c. What were the genotypes of the original parents?

d. What is the probability that a child of individual 4 will have an extra finger? Will have the eye disease?

e. What is the probability that a child of individual 12 will have an extra finger? Will have the eye disease?

29. In the ovaries of female mammals (including humans), the first meiotic division produces two cells (as normal), one of which becomes a small, nonfunctional cell called a polar body. Consequently, a second meiotic division occurs in only one of the first-division products. Furthermore, one of the products of the second meiotic division also becomes a polar body. Thus, the net result of one meiosis is one ovum (egg) plus two polar bodies.

a. Diagram this system at the cell level and at the chromosome level.

b. Mendelian genetics obviously does not operate in this system: the Mendelian ratios observed in mammals are due solely to segregation in the male. Is this statement true or false? Explain your answer.

30. In ferns, the stage of the life cycle most familiar to us is the diploid sporophyte. Meiosis occurs in specialized cells on the backs of the sporophyte's leaflets. The sexual spores fall to the ground, and each spore can result in a haploid gametophyte. Gametophytes produce gametes by mitosis: these cross-fertilize each other, and a sporophyte develops from each zygote.

A gene p on one chromosome causes pink leaf stalks (which are normally green) in the sporophyte; on another chromosome, the gene h results in hairless stems (which are normally hairy) in the sporophyte. A sporophyte of genotype $+/p\ +/h$ undergoes meiosis, and gametophytes result.

a. What genotypic proportions would be expected in the gametophyte population?

b. Would you expect the p and h genes necessarily to be expressed in the gametophyte?

c. If the gametophytes cross-fertilize each other randomly, what phenotypic and genotypic proportions can be expected in the next sporophyte generation?

31. In *Drosophila*, normal eye color is red but abnormal lines are available with brown eyes. Similarly, wings are normally long, but lines are available with short wings. A female from a pure line with brown eyes and short wings is crossed with a male from a pure normal line. The F_1 con-

sists of normal females and short-winged males. An F_2 is then produced; here *both* sexes of flies show phenotypes as follows:

$\frac{3}{8}$ normal

$\frac{3}{8}$ short wings

$\frac{1}{8}$ brown eyes

$\frac{1}{8}$ brown eyes and short wings

Deduce the inheritance of these phenotypes, using clearly defined genetic symbols of your own invention. State the parental genotypes for both the F_1 and F_2.

32. In the following human pedigree, the black symbols show a rare inherited disease.

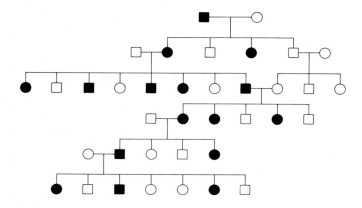

a. Determine the mode of inheritance.

b. Explain your reasoning.

c. According to your hypothesis, if the affected male in the last generation marries a normal woman, what phenotypic proportions can be expected in each sex of progeny?

Extensions to Mendelian Analysis

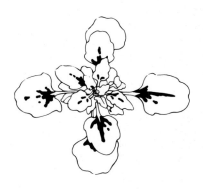

KEY CONCEPTS

A gene can exist in many forms (alleles).

•

The phenotype of the heterozygote is often intermediate, showing that there are several possible kinds of dominance relationships between different alleles.

•

For many genes, allelic forms are known that can kill the organism.

•

Most characters are determined by a specific set of genes that interact with each other and with the environment.

•

Gene interactions are revealed by modified Mendelian ratios.

■ We have seen that Mendel's laws seem to hold across the entire spectrum of eukaryotic organisms — that is, we can identify analogous phenomena that reveal segregation and independent assortment in most eukaryotes. These laws form a base for predicting the outcome of simple crosses. However, it is only a base; the real world of genes and chromosomes is more complex than Mendel's laws suggest, and exceptions and extensions abound. These situations do not invalidate Mendel's laws. Rather, they show that there are more situations than can be explained by segregation and independent assortment of gene pairs and that these situations must be accommodated into the fabric of genetic analysis. This is the challenge we now must meet. Of course, one extension — sex linkage — has already been accommodated. This chapter presents a grab bag of other extensions, based mainly on the complexities of gene expression, and the two following chapters discuss two more major extensions. We shall see that rather than creating a hopeless and bewildering situation, these complexities combine to form a precise and unifying set of principles for the genetic analyst. These principles interlock and support each other in a highly satisfying way that has provided great insight into the mechanics of inheritance.

Variations on Dominance Relations

Dominance is a good place to start. Mendel observed (or at least reported) full dominance (and recessiveness) for all seven gene pairs he studied. He may have been selective in his choice of pea characters to study, because variations on the basic theme of dominance crop up quite often in analysis. These variations center on the phenotype of the heterozygote. Some examples will illustrate this.

In four-o'clock plants, when a pure line with red petals is crossed with a pure line with white petals, the F_1 do *not* have red petals, but pink ones! If an F_2 is produced, the result is

$\frac{1}{4}$ red petals	$1\ C_1C_1$
$\frac{1}{2}$ pink petals	$2\ C_1C_2$
$\frac{1}{4}$ white petals	$1\ C_2C_2$

The occurrence of an intermediate phenotype in the heterozygote introduces the possibility of **incomplete dominance.** The precise position of the heterozygote on some kind of phenotypic "scale" of measurement defines several types of dominance relationships, as summarized in Figure 4-1. In practice, it is often difficult to determine exactly where on the scale the heterozygote is, however.

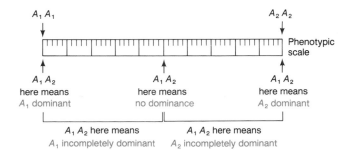

Figure 4-1. Summary of dominance relationships.

The phenotype of the heterozygote is also the key in the phenomenon of **codominance,** in which the heterozygote shows the phenotypes of *both* the homozygotes. In a sense, then, codominance is no dominance at all! A good example is found in the M-N blood-group gene pair in humans. Three blood groups are possible — M, N, and MN — and they are determined by the genotypes L^ML^M, L^NL^N, and L^ML^N, respectively. Blood groups are actually determined by the presence of an immunological antigen on the surface of red blood cells. People of genotype L^ML^N have both antigens. This example shows how the heterozygote can show both phenotypes.

Figure 4-2. Electron micrograph of red blood cells from an individual with sickle-cell anemia. A few rounded cells appear almost normal. (Courtesy of Patricia N. Farnsworth.)

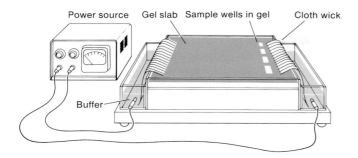

Figure 4-3. Equipment for electrophoresis. Each sample mixture is placed in a well in the gel. The components of the mixture migrate different distances on the gel due to their different electric charges. Several sample mixtures can be tested at the same time (one in each well). The positions of the separated components are later revealed by staining.

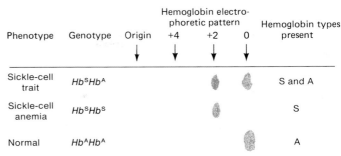

Figure 4-4. Electrophoresis of hemoglobin from a normal individual, an individual with sickle-cell anemia, and an individual with sickle-cell trait. The smudges show the positions to which the hemoglobins migrate on the starch gel. (After D. L. Rucknagel and R. K. Laros, Jr., "Hemoglobinopathics: Genetics and Implications for Studies of Human Reproduction." *Clinical Obstetrics and Gynecology* 12, 1969, 49–75.)

Another interesting example of dominance is found in the human disease sickle-cell anemia. The gene pair concerned affects the oxygen transport molecule hemoglobin, the major constituent of red blood cells. The three genotypes have different phenotypes, as follows:

$Hb^A Hb^A$: Normal. Red blood cells never sickled.

$Hb^S Hb^S$: Severe, often fatal anemia. Red blood cells sickle-shaped.

$Hb^A Hb^S$: No anemia. Red blood cells sickle only under abnormally low oxygen concentrations.

An example of sickle cells is shown in Figure 4-2. In regard to anemia, the Hb^A allele is dominant. In regard to blood cell shape, there is incomplete dominance. Finally, as we shall now see, in regard to hemoglobin there is codominance! The alleles Hb^A and Hb^S actually code for two slightly different forms of hemoglobin that have different chemical charges, and this property can be used in a technique called electrophoresis to study dominance at the molecular level.

In electrophoresis, mixtures of proteins can be separated on the basis of their charges. A small sample of protein (here, hemoglobin from red blood cells) is placed in a small well, cut in a slab of gel. A powerful electric field is applied across the gel, and the hemoglobin moves through the gel according to the degree of electrostatic charge. Figure 4-3 is a sketch of a typical electrophoresis apparatus. The gel is actually a supporting web containing electrolyte solution in its spaces; hence, the hemoglobin can move easily through the field. After an appropriate time, the gel can be stained

for protein, with the results shown in Figure 4-4. The blotches are the stained hemoglobin. We see that homozygous normal people have one type of hemoglobin (A) and anemics have another slow-moving type (S). The heterozygotes have both types, A and S.

Sickle-cell anemia illustrates the somewhat arbitrary nature of the terms incomplete dominance and codominance. The type of dominance depends on the phenotypic level at which the observations are being made — organismal, cellular, or molecular.

Message Complete dominance and recessiveness are not essential aspects of Mendel's laws; those laws deal more with the inheritance patterns and genes than with their nature or function.

Multiple Alleles

Early in the history of genetics, it became clear that it is possible to have more than two forms of one kind of gene. Although only two actual alleles of a gene can exist in a diploid cell (and only one in a haploid cell), the total number of possible different allelic forms that might exist in a population of individuals is often quite large. This situation is called **multiple allelism,** and the set of alleles itself is called an **allelic series.** The concept of allelism is a crucial one in genetics, so we consider several examples. The examples themselves serve also to introduce important areas of genetic investigation.

ABO Blood Group in Humans

The human ABO blood-group alleles afford a modest example of multiple allelism. There are four blood types (or phenotypes) in the ABO system, as shown in Table 4-1. The allelic series includes three major alleles—i, I^A, and I^B—which can be present in any pairwise combination in one individual; thus, only two of the three alleles can be present in any one individual. Note that the series includes cases of both complete dominance and codominance. In this allelic series, the alleles I^A and I^B each determine a unique form of one type of antigen: the allele i determines a failure to produce either form of that type of antigen.

TABLE 4-1. ABO blood groups in humans

Blood phenotype	Genotype
O	ii
A	$I^A I^A$ or $I^A i$
B	$I^B I^B$ or $I^B i$
AB	$I^A I^B$

C Gene in Rabbits

Another example of multiple allelism involves a larger allelic series determining coat color in rabbits. The alleles in this series are C (full color), c^{ch} (chinchilla, a light grayish color), c^h (Himalayan, albino with black extremities), and c (albino). Note the use of a superscript to indicate an allele of a type of gene; this symbolism often is necessary because more than the two symbols C and c are needed to describe multiple alleles. In this series, the alleles are dominant to those listed after them in the order stated (C, c^{ch}, c^h, c); verify this by a careful study of Table 4-2.

TABLE 4-2. C gene in rabbits

Coat-color phenotype	Genotype
Full color	CC or Cc^{ch} or Cc^h or Cc
Chinchilla	$c^{ch}c^{ch}$ or $c^{ch}c^h$ or $c^{ch}c$
Himalayan	$c^h c^h$ or $c^h c$
Albino	cc

Operational Test for Allelism

Now that we have seen two examples of allelic series, it is a good time to pause and consider a question. How do we know that a set of phenotypes is determined by alleles of the same type of gene? In other words, what is the operational test for allelism? For now, the answer is simply the observation of Mendelian single-gene-pair ratios in all combinations crossed. For example, consider three pure-line phenotypes in a hypothetical plant species. Line 1 has round spots on the petals; line 2 has oval spots on the petals; and line 3 has no spots on the petals. Suppose that crosses of the three lines yield the following results:

Cross	F_1	F_2
1 × 2	all round-spotted	$\frac{3}{4}$ round: $\frac{1}{4}$ oval
1 × 3	all round-spotted	$\frac{3}{4}$ round: $\frac{1}{4}$ unspotted
2 × 3	all oval-spotted	$\frac{3}{4}$ oval: $\frac{1}{4}$ unspotted

These results prove that we are dealing with three alleles of a single gene that affects petal spotting because each cross gives a monohybrid Mendelian ratio. We can choose any symbols we wish. Since we don't know which phenotype is the wild-type, we could follow the rabbit system and use S for the round-spotted allele, s^o for the oval-spotted allele, and s for the unspotted allele. Alternatively, we could use S^r for round, S^o for oval, and s for unspotted. The convention provides no firm rules about whether to use capital or small letters, particularly for the alleles in the middle of the series that are dominant to some of their alleles but recessive to others.

What if crosses between pure-breeding variants affecting one character do not produce single-gene Mendelian ratios? This outcome is covered later in the chapter.

Clover Chevrons

Clover is a common name for plants of the genus *Trifolium*. There are many species. Some are native to North America; some grow here as introduced weeds. The red-flowered and white-flowered clovers are familiar to most people. These are different species, but each shows considerable variation among individuals in the curious V or "chevron" pattern on the leaves. Much genetic research has been done with white clover; Figure 4-5 shows that the different chevron forms (and the absence of chevrons) are determined by an allelic series in this species. Furthermore, several types of dominance relations may be seen in the allele combinations shown in the figure.

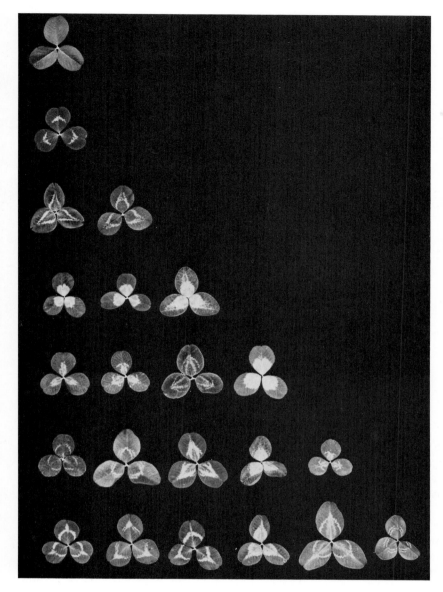

Figure 4-5. Multiple alleles are involved in determining the chevron pattern on the leaves of white clover. The genotype of each plant is shown below. Inspect these data for dominance and codominance. (Courtesy of W. Ellis Davies.)

vv

$V^l V^l$

$V^h V^h$ $V^l V^h$

$V^f V^f$ $V^l V^f$ $V^h V^f$

$V^{ba} V^{ba}$ $V^l V^{ba}$ $V^h V^{ba}$ $V^f V^{ba}$

$V^b V^b$ $V^l V^b$ $V^h V^b$ $V^f V^b$ $V^{ba} V^b$

$V^{by} V^{by}$ $V^l V^{by}$ $V^h V^{by}$ $V^f V^{by}$ $V^{ba} V^{by}$ $V^b V^{by}$

Incompatibility Alleles in Plants

It has been known for millennia that some plants just will not self-pollinate. A single plant may produce both male and female gametes, but no seeds will ever be produced. The same plants, however, will cross with certain other plants, so obviously they are not sterile. This phenomenon is called **self-incompatibility.** Of course, the pea plants used by Mendel were not self-incompatible, for he was able to self his plants with ease. We now know that incompatibility has a genetic basis and that there are several different genetic systems acting in different self-incompatible species. These systems form very nice examples of multiple allelism.

One of the most common systems is found in many monocotyledon and dicotyledon plants, (plants having one and two seed leaves, respectively, in the embryo), including sweet cherries, tobacco, petunias, and evening primroses. In each of these species, one gene determines compatibility-incompatibility relations, and many different allelic forms of the gene are possible in different plants of any one species. Figure 4-6 shows an example in which a fully incompatible reaction, a semicompatible reaction, and a fully compatible reaction are all possible.

If a pollen grain bears an S allele that is also present in the maternal parent, then it will not grow. However, if

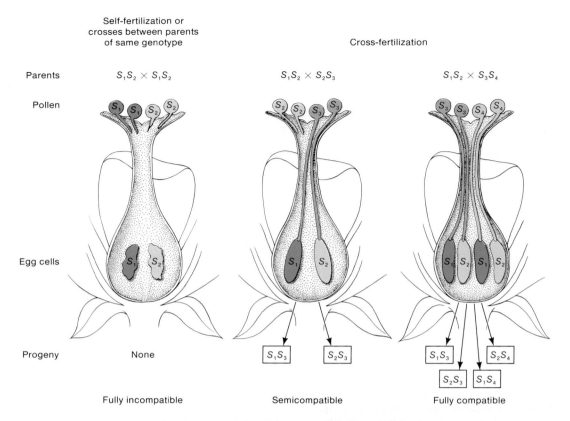

Self-fertilization or
crosses between parents
of same genotype

Cross-fertilization

Figure 4-6. Diagram showing how multiple alleles control incompatibility in certain plants. A pollen tube will not grow if the S allele that it contains is present in the female parent. This diagram shows only four multiple alleles, but many plant incompatibility systems use far larger numbers of alleles. Such systems act to promote the exchange of genes between plants by making selfing impossible and crosses between near relations very unlikely. (From A. M. Srb, R. D. Owen, and R. S. Edgar, *General Genetics*, 2d. ed. Copyright © 1965, by W. H. Freeman and Company.)

that allele is not in the maternal tissue, then the pollen grain produces a pollen tube containing the male nucleus, and this tube effects fertilization. The number of S alleles in a series in one species can be very large (over 50 in the evening primrose and clover), and cases of more than 100 alleles have been reported in some species.

Tissue Incompatibility in Humans

When an organ or a tissue graft is medically necessary in a human, the success or failure of the graft depends on the genotypes of the host and the donor. If the two are mismatched, rejection of the graft eventually occurs, often accompanied by death of the host (Figure 4-7).

The rejection system is based on two important genes called *HLA-A* and *HLA-B*, which determine the immunological acceptability of the introduced tissue. Each gene has a series of allelic forms. There are eight alleles for *HLA-A*, designated *A1*, *A2*, *A3*, *A9*, *A10*, *A11*, *A28*, and *A29*. By coincidence, there are also eight alleles for *HLA-B*, designated *B5*, *B7*, *B8*, *B12*, *B13*, *B14*, *B18*,

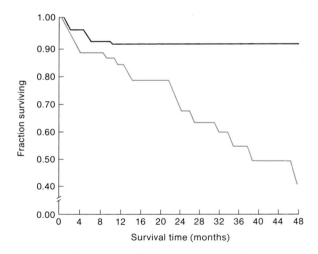

Figure 4-7. The success of kidney transplants in HLA-matched *(black line)* and HLA-mismatched *(colored line)* sibling pairs. HLA refers to genes related to tissue compatibility. (From D. P. Singal et al., *Transplantation* 7, 1965, 246, copyright © 1965 by The Williams & Wilkins Co., Baltimore.)

TABLE 4-3. Success or failure of transplants determined by HLA types

Transplant	Recipient genotype	Donor genotype	Result
1	A1 A2, B5 B5	A1 A1, B5 B7	rejected (due to B7)
2	A2 A3, B7 B12	A1 A2, B7 B7	rejected (due to A1)
3	A1 A2, B7 B5	A1 A2, B7 B7	accepted
4	A2 A3, B7 B5	A3 A3, B5 B5	accepted

and B27. (The two genes do not show independent assortment, but this need not concern us now. For the moment we shall simply use these allelic series as excellent examples of multiple allelism in humans.)

The HLA type of an individual is determined by testing his or her lymphocytes (white blood cells) with a battery of standard antibodies from donors of known HLA type. In a transplant, the recipient's immune system will recognize the graft as "foreign" and reject it if an allele is present in the donor tissue that is not present in the recipient. In the typing test, this rejection reaction is observed as lymphocyte death. The immune system will not reject tissue that lacks some alleles present in the recipient—it only rejects "strange" alleles. Table 4-3 shows examples.

We might wonder what function these HLA genes normally serve; obviously, tissue transplants have not been a normal aspect of human evolution! One possible answer is that some HLA alleles seem to confer resistance or susceptibility to certain specific diseases (Figure 4-8). Thus, these alleles may have played some roles in our evolutionary history in response to some environmental selection pressures. This supposition is supported by the finding that certain ethnic groups and races show a preponderance of specific alleles in their populations. The study of such alleles is important both in deciphering the history of human evolution and in developing modern medical techniques—not only for transplants, but also for the general study of resistance to disease.

Another very interesting possibility emerging from current research is that the HLA genes may be intimately involved in cancer. Presumably, one of the normal functions of the HLA genes is to recognize cancer cells as a kind of "foreign" agent within the body. The development of cancer cells may be a fairly common event in healthy individuals, with these cells being recognized and destroyed before they cause noticeable damage. A cancerous tumor may be the result of a failure of this detection-and-destruction system at some stage. The HLA genes also play an important role in the normal immune response.

> **Message** A gene can exist in several different states or forms—a situation called multiple allelism. The alleles are said to constitute an allelic series, and the members of a series can show any type of dominance to one another.

As a postscript to this section, it is worth noting that several different "kinds" or "species" of cell-surface antigens have been discovered. Each of these kinds is, of course, specified by one kind of gene. The different forms one kind of antigen can take are determined by the multiple alleles of the gene in question.

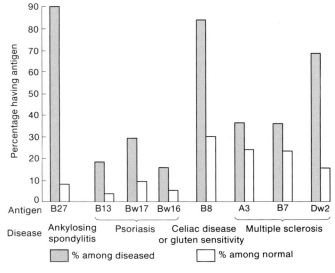

Figure 4-8. Histograms showing the effects of specific HLA alleles on predisposition to certain diseases in humans. For example, the allele B27 is found in 90 percent of individuals with ankylosing spondylitis (a kind of severe rheumatism) but in only 9 percent of normal individuals. Most of the diseases that show such HLA correlations are known to involve the immune system. (From W. F. Bodmer and L. L. Cavalli-Sforza, *Genetics, Evolution, and Man.* Copyright © 1976 by W. H. Freeman and Company. Based on data from H. O. McDevitt and W. F. Bodmer, *Lancet* 1, 1974, 1267.)

Lethal Genes

Normal wild-type mice have coats with a rather dark overall pigmentation. In 1904, Lucien Cuenot studied mice having a lighter coat color called yellow. Mating a yellow mouse to a normal mouse from a pure line, Cuenot observed a 1 : 1 ratio of yellow to normal mice in the progeny. This observation suggests that a single gene determines this aspect of coat color and that an allele for yellow is dominant to an allele for normal color. However, the situation became more confusing when Cuenot crossed yellow mice with one another. The result was always the same, no matter which yellow mice were used:

$$\text{yellow} \times \text{yellow} \begin{cases} \frac{2}{3} \text{ yellow} \\ \frac{1}{3} \text{ normal color} \end{cases}$$

Two features are of interest in these results. First, the 2 : 1 ratio is a departure from Mendelian expectations. Second, Cuenot had failed to detect a homozygous yellow mouse in any generation of breeding.

Cuenot suggested the following explanation for these results. A cross between two heterozygotes would be expected to yield a Mendelian genotype ratio of 1 : 2 : 1. If one of the homozygous classes died before birth, the live births would then show a 2 : 1 ratio of heterozygotes to the surviving homozygotes. In other words, the allele A^Y for yellow is dominant to the normal allele A with respect to its effect on color, and it also acts as a **recessive lethal** allele with respect to a character we could call "survival." Thus, a mouse with the homozygous genotype $A^Y A^Y$ dies before birth and is not ob-

served among the progeny. All surviving yellow mice must be heterozygous $A^Y A$, so a cross between yellow mice will always yield the following results:

$$A^Y A \times A^Y A \longrightarrow \begin{cases} \frac{1}{4} A A \quad \text{normal color} \\ \frac{2}{4} A^Y A \quad \text{yellow} \\ \frac{1}{4} A^Y A^Y \text{ die before birth} \end{cases}$$

The expected Mendelian ratio of 1 : 2 : 1 would be observed among the zygotes, but it is altered to a 2 : 1 ratio of viable progeny due to the lethality of the $A^Y A^Y$ genotype. This hypothesis was confirmed by the removal of uteri from pregnant females of the yellow × yellow cross; one-fourth of the embryos were found to be dead. Figure 4-9 shows a typical litter from a cross between yellow mice.

The A^Y allele produces effects on two characters: coat color and survival. Such genes known to have more than one distinct phenotypic effect are called **pleiotropic** genes. It is entirely possible that both effects of the A^Y pleiotropic allele are the results of the same basic cause, which promotes yellowness of coat in a single dose and death in a double dose.

Lethal genes are quite common, even in humans. Some lethal effects occur in utero, others much later — in infancy, in childhood, or even in adulthood. Only rarely is a lethal gene associated with a distinguishable heterozygous phenotype, as in yellow mice. One other example you may be familiar with is the tail-less condition in Manx cats. The determining gene is homozygous lethal. In general, the only detectable effect of a lethal gene is on survival of the individual. In some cases, a specific abnormality can be pinpointed as the cause of

Figure 4-9. A mouse litter from two parents heterozygous for the yellow coat-color allele, which is lethal in a double dose. The larger mice are the parents. Not all progeny are visible.

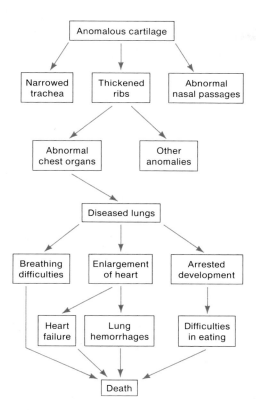

Figure 4-10. Diagram showing how one specific lethal gene causes death in rats. (From I. M. Lerner and W. J. Libby, *Heredity, Evolution, and Society,* 2d ed. Copyright © 1976 by W. H. Freeman and Company; after H. Grüneberg.)

death. For example, a recessive lethal gene in rats may have its pleiotropic effects traced to a basic problem in the nature of the rat's cartilage (Figure 4-10).

The lethality of an allele of a gene is often dependent on the environment in which the organism develops. Whereas certain alleles would be lethal in virtually any environment, others are viable in one environment but lethal in another. For example, remember that many of the phenotypes favored and selected by agricultural breeders would almost certainly be wiped out in nature as a result of competition with the normal members of their population. Modern grain varieties provide good examples; only careful nurturing by the farmer has maintained such phenotypes for our benefit.

In practical genetics, we very commonly encounter situations in which perfect expected Mendelian ratios are consistently skewed in one direction by one allele. For example, in the cross $Aa \times aa$, we predict a progeny phenotypic ratio of 50 percent $A-$ and 50 percent aa, but we might consistently observe a ratio such as 55% : 45% or 60% : 40%. In such a case, the a allele is said to be **subvital,** since a kind of partial lethality is expressed in only some individuals. The partial lethality

may range from 0 to 100 percent, depending on the rest of the genome and on the environment. We shall return to this topic later.

Several Genes Affecting the Same Character

We saw earlier that in a genetic dissection, the identification of a major gene affecting a character does not mean that this is the *only* gene affecting that character. An organism is a highly complex machine in which all functions interact to a greater or lesser degree. At the level of genetic determination, the genes likewise can be regarded as interacting. A gene does not act in isolation; its effects depend not only on its own functions but also on the functions of other genes (and on the environment). In many cases, the complex interactions of major genes are detectable through genetic analysis, and we now look at some examples. Typically, the mode of interaction is revealed by **modified Mendelian ratios.**

Coat Color in Mammals

A character that has been extensively studied at the genetic level is coat color in mammals. These studies have revealed a beautiful set of examples of the interplay between different genes in the determination of one character. The best-studied mammal in this regard, due to its small size and short reproductive cycle, is the mouse. However, the genetic determination of coat color in mice has direct parallels in other mammals, and we shall look at some of these as our discussion proceeds. At least five interacting major genes are involved in determining the coat color of mice: *A, B, C, D,* and *S.*

The A Gene. The wild-type allele *A* produces a phenotype called agouti. Agouti is an overall grayish color with a "brindled" or "mousy" appearance. It is common in many mammals in nature. The effect is produced by a band of yellow on the hair shaft. In the nonagouti phenotype (determined by the allele *a*), the yellow band is absent, so the coat color appears solid (Figure 4-11).

The lethal yellow A^Y allele is another form of this gene. Another form is a^t, which results in a "black-and-tan" effect: a cream-colored belly with dark pigmentation elsewhere. We shall not include these two alleles in the following discussion.

The B Gene. There are two major alleles of the *B* gene. The allele *B* gives the normal agouti color in combination with *A* but gives solid black with *aa*. The genotype $A-bb$ gives a color called cinnamon ("mousy" brown), and *aabb* gives solid brown.

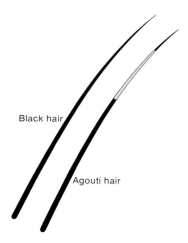

Figure 4-11. Individual hairs from an agouti *(wild-type)* and a black mouse. The yellow band on each hair gives the agouti pattern its "mousy" appearance.

The following sample cross illustrates the inheritance pattern of the *A* and *B* genes:

$$AA\ bb\ \text{(cinnamon)} \times aa\ BB\ \text{(black)}$$

or $AA\ BB$ (agouti) $\times aa\ bb$ (brown)

$$\downarrow$$

F₁   all *Aa Bb* (agouti)

$$\downarrow$$

F₂   9 *A–B–* (agouti)

3 *A–bb* (cinnamon)

3 *aa B–* (black)

1 *aa bb* (brown)

In horses, no *A* agouti gene seems to have survived the generations of breeding, although such a gene does exist in certain wild relatives of the horse. The color we have called brown in mice is called chestnut in horses, and this phenotype also is recessive to black.

The C Gene.   The wild-type allele *C* permits color expression, and the allele *c* prevents color expression. The *cc* constitution is said to be **epistatic** to the other color genes. The word epistatic literally means "standing upon"; in homozygous condition, the *c* allele "stands on" (blots out) the expression of other genes concerned with coat color. The *cc* animals, lacking coat color, are called albino. Albinos are common in many mammalian species, but albinos have also been occasionally reported among birds, snakes, and fish. Epistatic genes produce interesting modified ratios, as seen in the following sample cross (in which we assume that both parents are *aa*):

$$BB\ cc\ \text{(albino)} \times bb\ CC\ \text{(brown)}$$

or $BB\ CC$ (black) $\times bb\ cc$ (albino)

$$\downarrow$$

F₁   all *Bb Cc* (black)

$$\downarrow$$

F₂   9 *B–C–* (black)   9

3 *bb C–* (brown)   3

$$\left. \begin{array}{l} 3\ B\text{–}cc\ \text{(albino)} \\ 1\ bb\ cc\ \text{(albino)} \end{array} \right\} 4$$

A phenotypic ratio of 9 : 3 : 4 is observed. This ratio is the signal for inferring gene interaction of the type called recessive epistasis. In some other organisms, dominant epistasis is observed. (What ratio is produced in dominant epistasis?)

We have already encountered the *c*ʰ (Himalayan) allele in rabbits. It exists also in other mammals, including mice (also called Himalayan) and cats (called Siamese).

It should be pointed out here that the term epistasis is often used in a different way (mainly in population genetics) to describe *any* kind of gene interaction.

The D Gene.   The *D* gene controls the intensity of pigment specified by the other coat-color genes. The genotypes *DD* and *Dd* permit full expression of color in mice, but *dd* "dilutes" the pigment to a milky appearance. Dilute agouti, dilute cinnamon, dilute brown, and dilute black coats all are possible. A gene of this nature is called a **modifier gene.** In the following sample cross, we assume that both parents are *aa CC*:

$$BB\ dd\ \text{(dilute black)} \times bb\ DD\ \text{(brown)}$$

or $BB\ DD$ (black) $\times bb\ dd$ (dilute brown)

$$\downarrow$$

F₁   all *Bb Dd* (black)

$$\downarrow$$

F₂   9 *B–D–* (black)

3 *B–dd* (dilute black)

3 *bb D–* (brown)

1 *bb dd* (dilute brown)

In horses, the *D* allele shows incomplete dominance. Figure 4-12 shows how dilution affects the appearance of chestnut and bay horses.

The S Gene.   The *S* gene controls the presence or absence of spots. The genotype *S–* results in no spots, and

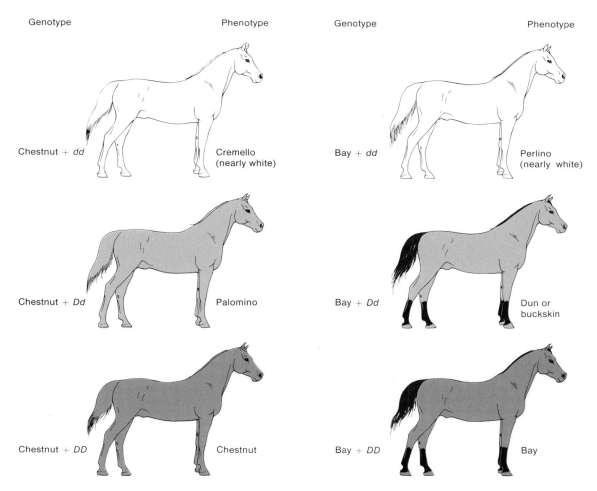

Figure 4-12. The modifying effect of dilution genes on basic chestnut and bay genotypes in horses. Note the incomplete dominance shown by *D*. (From J. W. Evans et al., *The Horse.* Copyright © 1977, by W. H. Freeman and Company.)

ss produces a spotting pattern called piebald in both mice and horses. This pattern can be superimposed on any of the coat colors discussed earlier — with the exception of albino, of course.

By this time, the point of the discussion should be obvious. Normal coat appearance in wild mice is produced by a complex set of interacting genes determining pigment type, pigment distribution in the individual hairs, pigment distribution on the animal's body, and the presence or absence of pigment. Similar situations exist for any character in any organism.

Figure 4-13 illustrates some pigment patterns in mice.

Examples of Gene Interaction in Other Organisms

Some other kinds of gene interaction are best illustrated in other organisms. Peas provide a good example of one

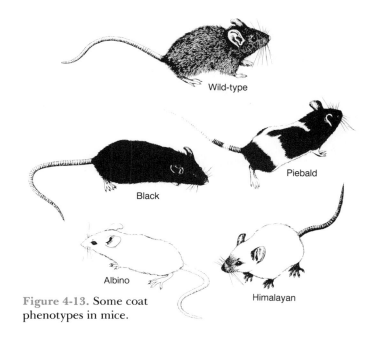

Figure 4-13. Some coat phenotypes in mice.

important situation. When two different, independently obtained pure lines of pea plants that are both white-petaled are crossed, all of the F$_1$ have purple flowers. The F$_2$ shows both purple and white plants in a ratio of 9 : 7. How can these results be explained? By now, you should immediately suspect that the 9 : 7 ratio is a modification of the Mendelian 9 : 3 : 3 : 1 ratio. The explanation is that two different genes in the pea have similar effects on petal color. Let's represent the alleles of these genes by A, a, B, and b:

white strain 1 × white strain 2

AA bb × aa BB

↓

F$_1$ all Aa Bb (purple)

↓

F$_2$ 9 A−B−(purple)   9

3 A−bb (white)

3 aa B−(white)   } 7

1 aa bb (white)

Both gene pairs affect petal color. Whiteness can be produced by a recessive allele of either gene pair, but purpleness is a phenotype produced by a combination of the dominant alleles of *both* gene pairs. This phenomenon is called **complementary gene action**—a term that satisfactorily describes how the two dominant alleles are uniting to produce a specific phenotype: in this example, purple pigment. (Note that we might have mistakenly inferred purpleness to be a specific phenotype of one gene pair if we had only one white strain available for our crosses.)

Another important kind of interaction is **suppression** of one gene by another. This interaction is illustrated by the inheritance of the production of a chemical called malvidin in the plant genus *Primula*. Malvidin production is determined by a single dominant gene K. However, the action of this dominant gene may be suppressed by a nonallelic dominant suppressor D. The following pedigree is informative:

KK dd (malvidin) × kk DD (no malvidin)

↓

F$_1$ all Kk Dd (no malvidin)

↓

F$_2$ 9 K−D−(no malvidin)

3 kk D−(no malvidin)   } 13

1 kk dd (no malvidin)

3 K−dd (malvidin)   3

Recessive suppression of both dominant and recessive genes also is known. The suppressor gene may have its own associated phenotype or (as in the malvidin example) may have no detectable phenotypic effect other than the suppression.

Our final example of gene interaction introduces a concept that we shall develop in a later chapter. It concerns the genes that control fruit shape in the plant called shepherd's purse. Two different lines have fruits of different shapes: one is "round"; the other, "narrow." Are these two phenotypes determined by two alleles of a single gene? A cross between the two lines reveals an F$_1$ with round fruit; this result is consistent with the hypothesis of a single gene pair. However, the F$_2$ shows a 15 : 1 ratio of round to narrow. Again, this ratio immediately suggests a modification of the 9 : 3 : 3 : 1 Mendelian ratio, and it can be explained in terms of two **duplicate genes** (Figure 4-14). Apparently, round fruits are produced as a result of the presence of at least one dominant allele of either gene. The two genes appear to be identical in function. (Contrast this 15 : 1 ratio with the 9 : 7 ratio obtained from complementary genes, where *both* dominant genes are necessary to produce a specific phenotype.)

A summary of the various types of gene interactions producing modified dihybrid Mendelian ratios in diploids is shown in Table 4-4.

Gene interactions also can be detected in haploid organisms. We have already discussed a gene in the fungus *Neurospora* where one allele, al, causes albino asexual spores, compared to the normal pinkish-orange

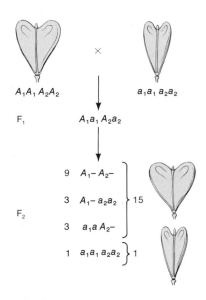

Figure 4-14. Inheritance pattern of duplicate genes controlling fruit shape in shepherd's purse. Either A_1 or A_2 can cause a round fruit.

TABLE 4-4. Modified dihybrid Mendelian ratios produced by gene interaction

Type of gene interaction	$A-B-$	$A-bb$	$aaB-$	$aabb$
Four distinct phenotypes	9	3	3	1
Complementary gene action	9	7		
Dominant suppression by A of dominant gene B	13		3	
Recessive epistasis by aa of B/b genes	9	3	4	
Dominant epistasis by A of B/b genes	12		3	1
Duplicate genes	15			1

color produced by the wild-type allele of the same gene. A cross $al \times al^+$ gives $\frac{1}{2}$ al and $\frac{1}{2}$ al^+ progeny. Another interesting gene, ylo, gives yellow asexual spores, and the cross $ylo \times ylo^+$ gives $\frac{1}{2}$ ylo and $\frac{1}{2}$ ylo^+ progeny. When an al culture is crossed with a ylo culture, the resulting progeny are $\frac{1}{4}$ yellow, $\frac{1}{4}$ wild-type, and $\frac{1}{2}$ albino. How can we explain this result? The answer is a kind of epistasis: al and ylo are forms of separate genes, each of which can affect the normal production of pink pigment. The genotypes in the cross are the following:

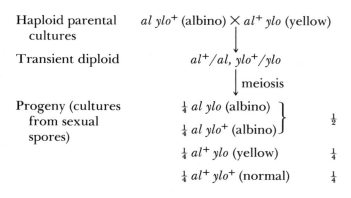

Haploid parental cultures	$al\ ylo^+$ (albino) $\times$ $al^+\ ylo$ (yellow)
Transient diploid	$al^+/al,\ ylo^+/ylo$
	$\downarrow$ meiosis
Progeny (cultures from sexual spores)	$\frac{1}{4}$ $al\ ylo$ (albino) $\Big\}$ $\frac{1}{2}$
	$\frac{1}{4}$ $al\ ylo^+$ (albino)
	$\frac{1}{4}$ $al^+\ ylo$ (yellow) $\quad\frac{1}{4}$
	$\frac{1}{4}$ $al^+\ ylo^+$ (normal) $\quad\frac{1}{4}$

> **Message** Modified Mendelian ratios reveal that a character is determined by the complex interaction of different genes.

Penetrance and Expressivity

The foregoing discussion shows that genes do not act in isolation. A gene does *not* determine a phenotype by acting alone; it does so only in conjunction with other genes and with the environment. Although geneticists do routinely ascribe a particular phenotype to an allele of a gene they have identified, we must remember that this is merely a convenient kind of jargon designed to facilitate genetic analysis. This jargon arises from the ability of geneticists to isolate individual components of a biological process and to study them as part of genetic dissection. Although this logical isolation is an essential aspect of genetics, the message of this chapter is that a gene cannot act by itself.

In the preceding examples, the genetic basis of the dependence of one gene on another has been worked out. In other situations, where the phenotype ascribed to a gene is known to be dependent on other factors but the precise nature of those factors has not been established, the terms *penetrance* and *expressivity* may be very useful in describing the situation.

We have already encountered penetrance in the discussion of lethal alleles. **Penetrance** is defined as the percentage of individuals with a given genotype who exhibit the phenotype associated with that genotype. For example, an organism may be of genotype aa or $A-$ but may not express the phenotype normally associated with its genotype due to the presence of modifiers, epistatic genes, or suppressors in the rest of the genome or a modifying effect of the environment. Penetrance can be used to describe such an effect when the exact cause is not known.

On the other hand, **expressivity** describes the degree or extent to which a given genotype is expressed phenotypically in an individual. Again, the lack of full expression may be due to the rest of the genome or to environmental factors. Figure 4-15 diagrams the distinction between penetrance and expressivity. Obviously, both penetrance and expressivity variation are integral components of the concept of norm of reaction, which was discussed in Chapter 1.

Human pedigree analysis and predictions in genetic counseling can often be thwarted by the phenomena of penetrance and expressivity. For example, if a disease-causing allele is not fully penetrant (as usually is the case), it is difficult to give a clean genetic bill of health to any individual involved as part of a disease pedigree (for example, individual R in Figure 4-16). On the other

hand, pedigree analysis can sometimes identify individuals who do not express but almost certainly do have a disease genotype (for example, individual Q in Figure 4-16).

Further examples of variable expressivity are found in Figure 4-17 and in Problem 15 of Chapter 2.

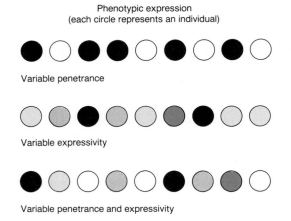

Phenotypic expression
(each circle represents an individual)

Variable penetrance

Variable expressivity

Variable penetrance and expressivity

Figure 4-15. Diagram representing the effects of penetrance and expressivity through a hypothetical character "pigment intensity." In each row, all individuals have the same genotype (say, *PP*), giving them the same "potential to produce pigment." However, the effects of the rest of the genome and of the environment may suppress or modify pigment production in each individual.

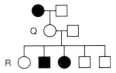

Figure 4-16. Lack of penetrance illustrated by a pedigree for a dominant gene. Individual Q must have the gene (because it was passed on to the progeny), but it was not expressed in the phenotype. An individual such as R cannot be sure that his or her genotype lacks the gene.

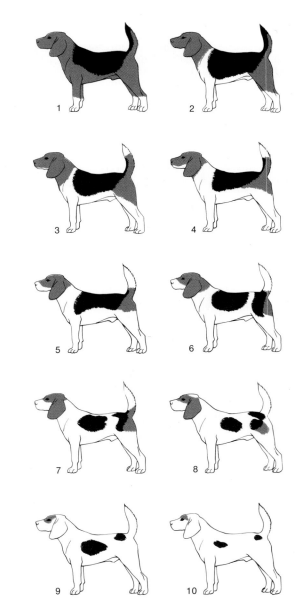

Figure 4-17. Ten grades of piebald spotting in beagles. Each of these dogs has the causative gene S^P; the variation in expressivity is attributed to modifying genes. (Adapted from Clarence C. Little, *The Inheritance of Coat Color in Dogs.* Copyright © 1957 by Cornell University. Used by permission of the publisher, Cornell University Press.)

(From Giorgio Schreiber, *Journal of Heredity* 9, 1930, 403.)

Message   The impact of a gene at the phenotype level depends not only on its dominance relations but also on the conditions of the rest of the genome and the condition of the environment.

In conclusion, notice that the extensions to Mendelian analysis discussed in this chapter are mainly based on the complexities of gene expression. Mendel's principles concerned the inheritance patterns of genes and had little to say about gene expression. When we add the patterns of gene expression to the patterns of gene inheritance, we begin to see the fabric of the modern science of genetics.

The extensions to Mendelian analysis in this chapter and following chapters are not the only ones encountered in routine genetic analysis; they are simply among the most common. The key recognition features for each extension—whether, for example, multiple allelism, incomplete dominance, epistasis, or variable expressivity—occur at the experimental level. The analyst must be constantly on the lookout for these and any other results that might indicate the uniqueness of any given situation. Such signals often lead to the discovery of new phenomena and to the opening up of new research areas.

Summary

■ Although it has been shown that Mendel's laws apply to all eukaryotic organisms, these laws are only a starting point for understanding heredity. The real world of genes and chromosomes is much more complex.

In addition to the full dominance that Mendel observed in his experiments, incomplete dominance and codominance may exist. In incomplete dominance, the phenotype of a heterozygote is intermediate between the phenotypes of the homozygotes. In codominance, the heterozygote shows the phenotypes of both homozygotes.

In his experiments, Mendel reported particles (genes) with two forms. It was later discovered that in fact a gene may have more than two forms. This situation is known as multiple allelism. A member of an allelic series may exhibit any type of dominance relations to the other members of the series. The genes controlling the rejection of incompatible tissues in humans are an example of multiple allelism.

One gene may affect more than one character. Such genes are known as pleiotropic genes. An example is the A^Y allele in mice, which affects both coat color and lethality.

The identification of a major gene affecting a character does not mean that it is the only gene affecting that character; several genes may be interacting to affect a character. A good example of gene interaction is the coat color of mice, which is produced by a complex set of interacting genes that determine pigment type, pigment distribution in the hair, pigment distribution on the animal, and the presence or absence of pigment.

Gene interaction often produces modified Mendelian ratios in the F_2. Some kinds of interaction have specific names, such as complementary gene action, epistasis, suppression, and duplicate gene action. Gene interaction occurs in both diploid and haploid organisms.

Two other important extensions to Mendelian analysis are the concepts of penetrance and expressivity. Penetrance is the percentage of individuals of a specific genotype who express the phenotype associated with that genotype. Expressivity refers to the degree of expression, or severity, of a particular genotype at the phenotypic level.

■ ■ ■

Solved Problems

1. In a species of beetle, the wing covers can be either green, blue, or turquoise. From an interbreeding, mixed, laboratory stock population, individual virgin beetles were selected and mated in specific controlled crosses to determine the inheritance of wing-cover color. The results were as follows:

Cross	Parents	Progeny
1	blue × green	all blue
2	blue × blue	¾ blue : ¼ turquoise

Cross	Parents	Progeny
3	green × green	¾ green : ¼ turqouise
4	blue × turquoise	½ blue : ½ turquoise
5	blue × blue	¾ blue : ¼ green
6	blue × green	½ blue : ½ green
7	blue × green	½ blue : ¼ green ¼ turquoise
8	turquoise × turquoise	all turquoise

a. Deduce the genetic basis of wing-cover color in this species.

b. Write the genotypes of all parents and progeny as completely as possible.

Solution

a. These data seem complex at first, but the inheritance pattern becomes clear if we consider the crosses one at a time. A general principle of solving such problems, as we have seen, is not necessarily to start at the beginning of the data. Look over all the crosses, and try to establish patterns.

One clue that emerges from an overview of the data is that all the ratios are one-gene ratios: there is no evidence of two separate gene pairs being involved at all. How can such variability be explained with a single gene? The obvious answer is variation of that single gene, that is multiple allelism. Perhaps there are three alleles of one gene; let's call the gene w (for wing cover color) and represent the alleles as w^g, w^b, and w^t. Now we have an additional problem, which is to determine the dominance relations of these alleles.

Cross (1) is informative about dominance, because the progeny of a blue × green cross are all blue; hence, blue appears to be dominant to green. This conclusion is supported by cross (5), because the green determinant must have been present in the parental stock to appear in the progeny. Turning to the position of the turquoise determinant, cross (3) is informative because turquoise determinants must have been present unexpressed in the parental stock to show up in the progeny phenotypes. So green must be dominant to turquoise. Hence, we have formed a model in which the dominance is $w^b > w^g > w^t$. Indeed, the inferred position of the w^t allele at the bottom of the dominance series is supported by the results of cross (7), where turquoise shows up in the progeny of a blue × green cross.

b. Now it is just a matter of deducing the specific genotypes. Notice that the question states that the parents are withdrawn from a mixed breeding population, so obviously they could be either homozygous or heterozygous. Of course, a blue strain, for example, can then be homozygous (w^bw^b) or heterozygous (w^bw^g or w^bw^t). Here, a little trial and error and common sense is called for, but by this stage, the question has essentially been answered and all that remains is to "cross the t's and dot the i's." The following genotypes explain the results. A dash indicates either a homozygous allele *or* an allele further down the allelic series.

Cross	Parents		Progeny
1	$w^bw^b \times w^g-$	$\longrightarrow$	w^bw^g or w^b-
2	$w^bw^t \times w^bw^t$	$\longrightarrow$	$\frac{3}{4} w^b- : \frac{1}{4} w^tw^t$
3	$w^gw^t \times w^gw^t$	$\longrightarrow$	$\frac{3}{4} w^g- : \frac{1}{4} w^tw^t$
4	$w^bw^t \times w^tw^t$	$\longrightarrow$	$\frac{1}{2} w^bw^t : \frac{1}{2} w^tw^t$
5	$w^bw^g \times w^bw^g$	$\longrightarrow$	$\frac{3}{4} w^b- : \frac{1}{4} w^gw^g$
6	$w^bw^g \times w^gw^g$	$\longrightarrow$	$\frac{1}{2} w^bw^g : \frac{1}{2} w^gw^g$
7	$w^bw^t \times w^gw^t$	$\longrightarrow$	$\frac{1}{2} w^b- : \frac{1}{4} w^gw^t : \frac{1}{4} w^tw^t$
8	$w^tw^t \times w^tw^t$	$\longrightarrow$	all w^tw^t

2. The leaves of pineapples can be classified into three types: spiny, spiny-tip, and piping (non-spiny). In crosses made between pure strains, the results shown in Table 4-5 appeared.

a. Using alphabetical letters, define gene symbols. Explain these results in terms of the genotypes produced and their ratios.

b. Using the model from part (a), give the phenotype ratios you would expect if you crossed: (1) the F_1 progeny from piping × spiny with the spiny parental stock, and (2) the F_1 progeny of piping × spiny with the F_1 progeny of spiny × spiny-tip.

Solution

a. First, let's look at the F_2 ratios. We have clear 3 : 1 ratios in crosses (1) and (2), indicating single-gene segregations. Cross (3) however, shows a ratio that is almost certainly a 12 : 3 : 1 ratio. How do we know this? Well, there are simply not that many complex ratios in genetics, and trial and error brings us to the 12 : 3 : 1 quite quickly. In the 128 progeny total, the numbers of 96 : 24 : 8 are expected, but the actual numbers fit these expectations remarkably well.

One of the principles of this chapter is that modified Mendelian ratios reveal gene interactions. Cross (3) is certainly a modified dihybrid Mendelian ratio, so it looks as if we are dealing with a two-gene interaction. This seems the most promising place to start; then we can go back to crosses (1) and (2) and try to fit them in later.

Any dihybrid ratio is based on the phenotypic proportions 9 : 3 : 3 : 1. Our observed modification groups them as follows:

TABLE 4-5.

Cross	Parental phenotype	F_1 phenotype	F_2 phenotype
1	spiny-tip × spiny	spiny-tip	99 spiny-tip, 34 spiny
2	piping × spiny-tip	piping	120 piping, 39 spiny-tip
3	piping × spiny	piping	95 piping, 25 spiny-tip, 8 spiny

$$\left.\begin{array}{l} 9\ A-\ B- \\ 3\ A-\ bb \end{array}\right\} \quad 12 \quad \text{piping}$$

$$3\ aa\ B- \qquad 3 \quad \text{spiny-tip}$$

$$1\ aa\ bb \qquad 1 \quad \text{spiny}$$

So without worrying about the name of the type of gene interaction involved (we are not asked to supply this anyway), we can already define our three pineapple leaf phenotypes in terms of the invented gene pairs A/a and B/b:

$$\text{piping} = A- \ (B/b \text{ irrelevant})$$

$$\text{spiny-tip} = aa\ B-$$

$$\text{spiny} = aa\ bb$$

What about the parents of cross (3)? The spiny parent must be $aa\ bb$, and since the B gene is needed to produce F_2 spiny-tip individuals, the piping parent must be $AA\ BB$. (Note that we are *told* that all parents are pure, or homozygous.) The F_1 must therefore be $Aa\ Bb$ (piping).

Without further thought, we can write out cross (1) as follows:

$$aa\ BB \times aa\ bb \longrightarrow aa\ Bb \longrightarrow \tfrac{3}{4}aa\ Bb$$
$$\tfrac{1}{4}aa\ bb$$

Cross (2) can also be written out partially without further thought, using our arbitrary gene symbols:

$$AA-- \times aa\ BB \longrightarrow Aa\ B- \longrightarrow \tfrac{3}{4}A---$$
$$\tfrac{1}{4}aa\ B-$$

We know that this series shows single-gene-pair segregation (the F_2 3:1 ratio), and it seems certain now that the A/a gene pair is involved. But the B allele is needed to produce spiny-tip, so all individuals must be homozygous BB:

$$AA\ BB \times aa\ BB \longrightarrow Aa\ BB \longrightarrow \tfrac{3}{4}A-BB$$
$$\tfrac{1}{4}aa\ BB$$

Notice that the two single-gene segregations in crosses (1) and (2) do not show that the gene pairs are not interacting in such crosses. What is shown is that the two-gene-pair interaction is not *revealed* by these crosses — only by the system in cross (3), which is heterozygous for both gene pairs.

b. Now it is simply a matter of using Mendel's laws to predict cross outcomes:

(1) $Aa\ Bb \times aa\ bb \longrightarrow \left.\begin{array}{l} \tfrac{1}{4}\ Aa\ Bb \\ \tfrac{1}{4}\ Aa\ bb \end{array}\right\} \tfrac{1}{2}$ piping
(independent assortment in a standard $\qquad \tfrac{1}{4}\ aa\ Bb \qquad$ spiny-tip
testcross) $\qquad \tfrac{1}{4}\ aa\ bb \qquad$ spiny

(2) $Aa\ Bb \times aa\ Bb \longrightarrow \tfrac{1}{2}\ Aa \begin{array}{l} \nearrow \tfrac{3}{4}\ B- \longrightarrow \tfrac{3}{8} \\ \searrow \tfrac{1}{4}\ bb \longrightarrow \tfrac{1}{8} \end{array} \left.\right\} \tfrac{1}{2}$ piping

$\qquad\qquad \tfrac{1}{2}\ aa \begin{array}{l} \nearrow \tfrac{3}{4}\ B- \longrightarrow \tfrac{3}{8} \quad \text{spiny-tip} \\ \searrow \tfrac{1}{4}\ bb \longrightarrow \tfrac{1}{8} \quad \text{spiny} \end{array}$

Problems

1. If a man of blood group AB marries a woman of blood group A whose father was of blood group O, what different blood groups can this man and woman expect their children to belong to?

* **2. a.** In Leghorn chickens, C stands for colored feathers and c represents white feathers. At another independently inherited gene, the allele I inhibits color expression, whereas i has no effect on color expression. List all possible allele combinations and their phenotypes.

b. A dihybrid cross between two white Leghorns produces a ratio of 13 white : 3 colored offspring. Explain this result, giving the genotypes of parents and offspring.

3. If Mexican Hairless dogs are interbred, the puppies show a ratio of $\tfrac{2}{3}$ hairless to $\tfrac{1}{3}$ normal; in addition, some deformed dead puppies are also born. Propose an explanation for this outcome, and outline how you would test your idea.

4. In *Clarkia elegans* plants, the allele for white flowers is recessive to the allele for pink flowers. Pollen from a heterozygous pink flower is placed on the pistil of a white flower. What is the expected ratio of phenotypes in the progeny? a. 1 pink : 1 white b. 1 red : 2 pink : 1 white c. all pink d. 3 pink : 1 white e. all white

(Problem 4 courtesy of F. R. Ganders.)

5. "Erminette" fowls have mostly light-colored feathers with an occasional black one, giving a "flecked" appearance. A cross of two erminettes produced a total of 48 progeny, consisting of 22 erminettes, 14 blacks, and 12 pure whites. What genetic basis of the erminette pattern is suggested? How would you test your hypothesis?

6. In a moss, a certain protein is studied through electrophoresis. Two electrophoretic protein forms (P1 and P2) are determined by two alleles ($P1$ and $P2$). In this moss, the protein is found in both sporophyte and gametophyte. A gametophyte showing P1 is crossed with a gametophyte showing P2. What protein forms will be detected on electrophoresis of the sporophyte? When this sporophyte produces spores, what gametophytes will result in what proportions?

7. In the plant incompatibility system described in Figure 4-6, progeny from the cross $S_1S_3 \times S_2S_4$ are intercrossed in all combinations as males and as females. Name the proportion of the crosses for each of the following possible results: a. Fully fertile b. Fully sterile c. Partially fertile

8. In some compatibility systems in plants, pollen compatibility type is determined by the diploid genotype of the male parent. Often in these systems there is dominance of S alleles, not only in the male but also in the female. Assume the dominance series $S_1 > S_2 > S_3 > S_4$. Thus, S_1S_2 is type 1, S_2S_3 is type 2, and so on. Assume that compatibility is possible only between different types. From the cross $S_1S_2 \times S_3S_4$, the progeny are intercrossed in all combinations. What will be the proportion of compatible matings? What progeny will be produced from the compatible matings?

9. In the multiple allele series $C^+ > C^{ch} > C^h$, dominance is from left to right as shown. In a cross of $C^+C^{ch} \times C^{ch}C^h$, what proportion of the progeny will be Himalayan? a. 100 percent b. $\frac{3}{4}$ c. $\frac{1}{2}$ d. $\frac{1}{4}$ e. 0 percent

10. In guinea pigs, black, sepia, cream, and albino are all coat colors. Individual animals showing these colors (not necessarily from pure lines) were intercrossed; the results shown in Table 4-6 appeared.

    a. Deduce the inheritance of these coat colors, using gene symbols of your own choosing. Show all parent and progeny genotypes.

    b. If the black animals in (7) and (8) are crossed, what progeny proportions can you predict using your model?

11. a. The extremities of a mammal show lower temperatures than the rest of its body. What do you suppose this fact might have to do with the action of Himalayan alleles in mice and other animals?

    b. If a small patch of hair is shaved from the back of a Himalayan mouse and the area is kept covered with an ice pack, what color would you expect the hair to be when it grows back in this area? (1) white; (2) agouti; (3) black; (4) cinnamon; (5) yellow.

12. In a maternity ward, four babies become accidentally mixed up. The ABO types of the four babies are known to be O, A, B, and AB. The ABO types of the four sets of parents are determined. Indicate which baby belongs to each set of parents: a. AB × O b. A × O c. A × AB d. O × O

13. The M, N, and MN blood groups are determined by two alleles, L^M and L^N. The Rh$^+$ (rhesus positive) blood group is caused by a dominant allele R of a different gene. In a

court case concerning a paternity dispute, each of two men claimed three children to be his own. The blood groups of the men, the children, and their mother were as follows:

Person	Blood Group		
Husband	O	M	Rh$^+$
Wife's lover	AB	MN	Rh$^-$
Wife	A	N	Rh$^+$
Child 1	O	MN	Rh$^+$
Child 2	A	N	Rh$^+$
Child 3	A	MN	Rh$^-$

From this evidence, can the paternity of the children be established?

14. On a fox ranch in Wisconsin, a mutation arose that gave a "platinum" coat color. The platinum color proved very popular with buyers of fox coats, but the breeders could not develop a pure-breeding platinum strain. Every time two platinums were crossed, some normal foxes appeared in the progeny. For example, in repeated matings of the same pair of platinums, a total of 82 platinums and 38 normals was produced. This result was typical of all other such matings. State a *concise* genetic hypothesis to account for these results.

15. Over a period of several years, Hans Nachtsheim investigated an inherited anomaly of the white blood cells of rabbits. This anomaly, termed the Pelger anomaly, in its usual condition involves an arrest of the typical segmentation of the nuclei of certain white cells. This anomaly does not appear to seriously inconvenience the rabbits.

    a. When rabbits showing the typical Pelger anomaly were mated with rabbits from a true-breeding normal stock, Nachtsheim counted 217 offspring showing the Pelger anomaly and 237 normal progeny.

TABLE 4-6.

		Phenotypes of Progeny			
Cross	Parental phenotypes	Black	Sepia	Cream	Albino
1	black × black	22	0	0	7
2	black × albino	10	9	0	0
3	cream × cream	0	0	34	11
4	sepia × cream	0	24	11	12
5	black × albino	13	0	12	0
6	black × cream	19	20	0	0
7	black × sepia	18	20	0	0
8	black × sepia	14	8	6	0
9	sepia × sepia	0	26	9	0
10	cream × albino	0	0	15	17

What appears to be the genetic basis of the Pelger anomaly?

b. When rabbits with the Pelger anomaly were mated to each other, Nachtsheim found 223 normal progeny, 439 showing the Pelger anomaly, and 39 extremely abnormal progeny. These very abnormal progeny not only had defective white blood cells but also showed severe deformities of the skeletal system; almost all of them died soon after birth. In genetic terms, what do you suppose these extremely defective rabbits represented? Why do you suppose there were only 39 of them?

c. What additional experimental evidence might you collect to support or disprove your answers to part b?

d. In Berlin, about one human in 1000 shows a Pelger anomaly of white blood cells very similar to that described in rabbits. The anomaly is inherited as a simple dominant, but the homozygous type has not been observed in humans. Can you suggest why, if you are permitted an analogy with the condition in rabbits?

e. Again by analogy with rabbits, what genetic situations might be expected among the children of a man and woman who both show the Pelger anomaly?

(Problem 15 is from A. M. Srb, R. D. Owen, and R. S. Edgar, *General Genetics*, 2d ed. Copyright © 1965 by W. H. Freeman and Company.)

16. You have been given a single virgin *Drosophila* female. You notice that the bristles on her thorax are much shorter than normal. You mate her with a normal male (with long bristles) and obtain the following F_1 progeny: $\frac{1}{3}$ short-bristle females, $\frac{1}{3}$ long-bristle females, and $\frac{1}{3}$ long-bristle males. A cross of the F_1 long-bristle females with their brothers gives only long-bristle F_2. A cross of short-bristle females with their brothers gives $\frac{1}{3}$ short-bristle females, $\frac{1}{3}$ long-bristle females, and $\frac{1}{3}$ long-bristle males. Explain these data.

17. In *Drosophila*, a dominant allele H reduces the number of body bristles, giving rise to a "hairless" condition. In the homozygous condition, H is lethal. An independently assorting dominant allele S has no effect on bristle number except in the presence of H, in which case a single dose of S suppresses the hairless phenotype, thus restoring the hairy condition. However, S also is lethal in the homozygous (SS) condition.

a. What ratio of hairy to hairless individuals would you find in the live progeny of a cross between two hairy flies both carrying H in the suppressed condition?

b. If the hairless progeny are backcrossed with a parental fly, what phenotypic ratio would you expect to find in the live progeny?

18. In Labrador retriever dogs, the dominant gene B gives black coat color whereas b gives brown coat color. On another chromosome, the gene E shows dominant epistasis over the B and b genes, resulting in a "golden" coat color, whereas e allows expression of B and b. A breeder wants to determine the genotypes of his three drogs, so he interbreeds them over several years as follows:

Cross	Parents	Progeny
1	dog 1 (golden female) × dog 2 (golden male)	$\frac{6}{8}$ golden: $\frac{1}{8}$ black: $\frac{1}{8}$ brown
2	dog 1 (golden female) × dog 3 (black male)	$\frac{4}{8}$ golden: $\frac{3}{8}$ black: $\frac{1}{8}$ brown

a. What are the genotypes of the three dogs?

b. Show how the observed progeny ratios were produced.

19. In squash plants, white and yellow are fruit colors and disk and sphere are fruit shapes. The cross of white disk × yellow sphere produces an F_1 of all white disk. Selfing the F_1 produces an F_2 of $\frac{9}{16}$ white disk, $\frac{3}{16}$ white sphere, $\frac{3}{16}$ yellow disk, and $\frac{1}{16}$ yellow sphere. Interpret these data, using symbols of your own choosing.

20. Two albinos marry and have four normal children. How is this possible?

21. In *Petunia*, three differently derived pure lines of white-flowered plants were obtained. The following crosses were performed:

Cross	Parents	Progeny
1	line 1 × line 2	F_1 all white
2	line 1 × line 3	F_1 all red*
3	line 2 × line 3	F_1 all white
4	red F_1* × line 1	$\frac{1}{4}$ red : $\frac{3}{4}$ white
5	red F_1* × line 2	$\frac{1}{8}$ red : $\frac{7}{8}$ white
6	red F_1* × line 3	$\frac{1}{2}$ red : $\frac{1}{2}$ white

a. Explain these results, using gene symbols of your own choosing. (Show parent and progeny genotypes in each cross.)

b. If a red F_1 individual from cross 2 is crossed to a white F_1 individual from cross 3, what proportion of progeny will be red?

22. In the Japanese morning glory (*Pharbitis nil*), purple flower color can be produced by dominant alleles of either of two separate gene pairs (A–bb or aaB–). A–B– produces blue petals, and $aabb$ produces scarlet petals. For each of the following cross results, deduce the genotypes of parents and progeny:

Cross	Parents	Progeny
1	blue × scarlet	$\frac{1}{4}$ blue : $\frac{1}{2}$ purple : $\frac{1}{4}$ scarlet
2	purple × purple	$\frac{1}{4}$ blue : $\frac{1}{2}$ purple : $\frac{1}{4}$ scarlet
3	blue × blue	$\frac{3}{4}$ blue : $\frac{1}{4}$ purple
4	blue × purple	$\frac{3}{8}$ blue : $\frac{4}{8}$ purple : $\frac{1}{8}$ scarlet
5	purple × scarlet	$\frac{1}{2}$ purple : $\frac{1}{2}$ scarlet

TABLE 4-7.

Cross	Parental phenotype	F_1 phenotype	F_2 phenotype
1	sun-red × pink	all sun-red	66 sun-red, 20 pink
2	orange × sun-red	all sun-red	998 sun-red, 314 orange
3	orange × pink	all orange	1300 orange, 429 pink
4	orange × scarlet	all yellow	182 yellow, 80 orange, 58 scarlet

23. In corn, pure lines are obtained that have either sun-red, pink, scarlet, or orange kernels when exposed to sunlight (normal kernels remain yellow in sunlight). Table 4-7 gives the results of some crosses between these lines. Analyze the results of each cross, and provide a unifying hypothesis to account for *all* the results. (Explain all symbols you use.)

24. Many kinds of wild animals have the agouti coloring pattern, in which each hair has a yellow band around it (see Figure 4-11).

   a. In black mice and other black animals, the yellow band is not present and the hair is all black. This absence of wild agouti pattern is called nonagouti. When mice of a true-breeding agouti line are crossed with nonagoutis, the F_1 is all agouti and the F_2 has a 3:1 ratio of agoutis to nonagoutis. Diagram this cross, letting A = agouti and a = nonagouti. Show the phenotypes and genotypes of the parents, their gametes, the F_1, their gametes, and the F_2.

   b. Another inherited color deviation in mice substitutes brown for the black color in the wild-type hair. Such brown-agouti mice are called cinnamons. When wild-type mice are crossed with cinnamons, the F_1 is all wild-type and the F_2 has a 3:1 ratio of wild-type to cinnamon. Diagram this cross as in part a, letting B = wild-type black and b = cinnamon brown.

   c. When mice of a true-breeding cinnamon line are crossed with mice of a true-breeding nonagouti (black) line, the F_1 is all wild-type. Use a genetic diagram to explain this result.

   d. In the F_2 of the cross in part c, a fourth color called chocolate appears in addition to the parental cinnamon and nonagouti and the wild-type of the F_1. Chocolate mice have a solid, rich-brown color. What is the genetic constitution of the chocolates?

   e. Assuming that the $A - a$ and $B - b$ allelic pairs assort independently of each other, what would you expect to be the relative frequencies of the four color types in the F_2 described in part d. Diagram the cross of parts c and d, showing phenotypes and genotypes (including gametes).

   f. What phenotypes would be observed in what proportions in the progeny of a backcross of F_1 mice from part c to the cinnamon parent stock? To the nonagouti (black) parent stock? Diagram these backcrosses.

   g. Diagram a testcross for the F_1 of part c. What colors would result, and in what proportions?

   h. Albino (pink-eyed white) mice are homozygous for the recessive member of an allelic pair $C - c$, which assorts independently of the $A - a$ and $B - b$ pairs. Suppose that you have four different highly inbred (and therefore presumably homozygous) albino lines. You cross each of these lines with a true-breeding wild-type line, and you raise a large F_2 progeny from each cross. What genotypes for the albino lines would you deduce from the F_2 phenotypes shown in Table 4-8?

(Problem 24 is adapted from A. M. Srb, R. D. Owen, and R. S. Edgar, *General Genetics*, 2d ed. Copyright © 1965 by W. H. Freeman and Company.)

25. In rats, yellow coat color is determined by an allele A that is not lethal when homozygous. At a separate gene that assorts independently, the allele R produces a black coat. Together, A and R produce a grayish coat, whereas a and r produce a white coat. A gray male is crossed with a yellow female, and the F_1 is $\frac{3}{8}$ yellow, $\frac{3}{8}$ gray, $\frac{1}{8}$ black, and $\frac{1}{8}$ white. Determine the genotypes of the parents.

TABLE 4-8.

	Numbers of F_2 progeny				
F_2 of line	Wild-type	Nonagouti (black)	Cinnamon	Chocolate	Albino
1	87	0	32	0	39
2	62	0	0	0	18
3	96	30	0	0	41
4	287	86	92	29	164

26. Normal *Drosophila melanogaster* have deep red eyes. You are able to establish two homozygous lines: one having bright scarlet eyes, and the other having dark brown eyes. When you cross scarlet-eyed flies with brown-eyed flies, all of the F_1 progeny have deep red eyes. An $F_1 \times F_1$ cross produces the following progeny: 432 red eyes, 158 scarlet eyes, 139 brown eyes, and 52 white eyes. Explain these results; devise your own symbols.

27. In the fowl, the genotype *rr pp* gives single comb, $R-P-$ gives walnut comb, *rr P−* gives pea comb, and $R- pp$ gives rose comb.

 a. What comb types will appear in F_1 and in F_2 in what proportions if single-combed birds are crossed with birds of a true-breeding walnut-combed strain?

 b. What are the genotypes of the parents in a walnut-combed × rose-combed mating from which the progeny are $\frac{3}{8}$ rose-combed, $\frac{3}{8}$ walnut-combed, $\frac{1}{8}$ pea-combed, and $\frac{1}{8}$ single-combed?

 c. What are the genotypes of the parents in a walnut-combed × rose-combed mating from which all the progeny are walnut-combed?

 d. How many genotypes will produce a walnut phenotype? Write them out.

28. F_2 phenotypic ratios of 27 : 37 have been observed. What is their genetic basis? (HINT: You may find it useful to consider trihybrids.)

(Problems 27 and 28 are adapted from A. M. Srb, R. D. Owen, and R. S. Edgar, *General Genetics*, 2d ed. Copyright © 1965 by W. H. Freeman and Company.)

*29. *Answer the parts of this problem in sequence.* You have two homozygous lines of *Drosophila*: one found in Vancouver (line A), and the other found in Los Angeles (line B). Both lines have bright scarlet eyes—a phenotype quite distinct from the deep red eyes of the wild-type.

 a. A cross of line-A males with line-B females produces an F_1 of 200 wild-type males and 198 wild-type females. From this result, what can you say about the inheritance of eye color in the two lines?

 b. A cross of line-B males with line-A females produces an F_1 of 197 scarlet-eyed males and 201 wild-type females. What does this result indicate about the inheritance of eye color?

 c. When you intercross the F_1 progeny from part a, you obtain the following F_2 progeny: 151 wild-type females, 49 scarlet-eyed females, 126 scarlet-eyed males, and 74 wild-type males. Diagram the geno-

types of the parents and of the F_1 offspring. Indicate the expected ratios of F_2 genotypes and phenotypes.

*30. *Answer the parts of this problem in sequence.* In *Drosophila melanogaster*, wild-type eyes are deep red in color. You have obtained two lines of *Drosophila*, line A and line B, in which the eyes are white.

 a. A cross of line-A males with line-B females produces an F_1 of 435 wild-type males and 428 wild-type females. What can you conclude about lines A and B?

 b. A cross of line-B males with line-A females produces an F_1 of 420 white-eyed males and 405 wild-type females. What can you conclude about lines A and B?

 c. Two sets of F_2 progeny are obtained by separately intercrossing the F_1 flies from parts a and b. Two new eye colors appear in the F_2 progeny: bright scarlet and brownish. Table 4-9 summarizes the phenotypes observed in the F_2 generations. Explain these results. Diagram the crosses for all parts of this problem.

31. When true-breeding brown dogs are mated with true-breeding white dogs, all the F_1 pups are white. When $F_1 \times F_1$ crosses are made, the F_2 progeny are 118 white, 32 black, and 10 brown pups. What is the genetic basis for these results?

32. In corn, three dominant alleles, called *A*, *C*, and *R*, must be present to produce colored seeds. Genotypes $A-C-R-$ are colored: all others are colorless. A colored plant is crossed with three tester plants of known genotype. With *aa cc RR*, it produces 50 percent colored seeds; with *aa CC rr*, it produces 25 percent colored; and with *AA cc rr*, it produces 50 percent colored seeds. What is the genotype of the colored plant?

33. In the fungus *Neurospora* (haploid), a mutant gene *td* results in an inability of the fungus to make its own tryptophan, so that growth can occur only if tryptophan is supplied in the medium. The allele *su* assorts independently of *td*; its only known effect is to suppress the td phenotype. Therefore, strains carrying both *td* and *su* do not require tryptophan for growth.

 a. If a *td su* strain is crossed with a genotypically wild-type strain, what genotypes are expected in the progeny, and in what proportions?

 b. What will be the ratio of tryptophan-dependent to tryptophan-independent progeny in the cross of part a?

34. In horses, assume that there are three color-affecting genes that produce the following effects: *WW* is lethal, *Ww* is white, and *ww* allows color; *BB* or *Bb* is black, and *bb* is

TABLE 4-9.

F_2 from cross of part	Wild-type		White-eyed		Scarlet-eyed		Brownish-eyed	
	♂	♀	♂	♀	♂	♀	♂	♀
a	44	88	86	11	14	34	16	29
b	88	90	180	175	28	29	24	26

chestnut; *OO* or *Oo* is solid, and *oo* has white spots on color. A white stallion and a white mare, each heterozygous for all these genes, are mated. What are the expected frequencies of possible phenotypes in the live offspring? What frequencies of phenotypes would be expected if the stallion were *Ww Bb oo?*

35. In mice, a cross is made of *AA BB CC DD SS* × *aa bb cc dd ss.* (These gene symbols are explained in the text of this chapter.) What phenotypes will be produced in the F_2, and in what proportions?

36. Bean anthracnose is a fungus disease affecting beans. Two different bean varieties, A and B, and two different lines of parasitic fungi, α and β, are obtained. Fungus line α produces disease in bean variety A but not in variety B. Fungus line β produces disease in bean variety B but not in variety A. The cross A × B is made, and an F_1 and F_2 are obtained. Both generations are treated with a mixture containing fungi from both lines α and β. None of the F_1 show any disease reaction. In the F_2, the ratio of unaffected to diseased plants is 9 : 7.

   a. Interpret these results genetically.

   b. If the two parasite lines were crossed, how would you expect their progeny to react with bean variety A? With variety B? Make a testable prediction. (For simplicity, assume that the fungus is haploid.)

37. The following pedigree is for a dominant, autosomal gene. How is this pedigree possible, and what can you deduce about the genotype of individual A?

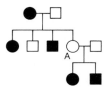

38. Three plants of *Collinsia grandiflora* (blue-eyed Mary) are found in different populations, all having spots on their leaves, as shown in the figure. From the following, select the reason for this variation; then outline the experimental steps that led to your conclusion: a. Different alleles of one gene b. Variable expressivity due to modification by other genes c. Variable expressivity due to environmental effects

39. In foxgloves, the gene *M* causes the production of magenta pigment in the petal and *m* produces no pigment, resulting in white petals with faint yellowish spots. At an unlinked locus, the gene *D* enhances pigment production, resulting in a dark magenta, whereas *d* does not enhance pigmentation, resulting in light magenta. At another unlinked locus, *W* prevents pigment deposition in all parts of the petals except the spots resulting (in the presence of *M*)

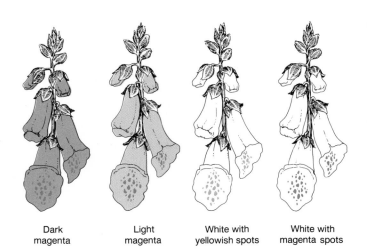

| Dark magenta | Light magenta | White with yellowish spots | White with magenta spots |

in white petals with magenta spots. The allele *w* does not prevent pigment deposition. The four possible phenotypes are summarized in the following diagram.

Consider the following two crosses.

Cross	Parents	Progeny
1	dark magenta × white with yellowish spots	½ dark magenta: ½ light magenta
2	white with yellowish spots × light magenta	½ white with magenta spots: ¼ dark magenta: ¼ light magenta

In each case, give the genotypes of parents and progeny with respect to these three gene pairs.

40. In *Antirrhinum* plants, two pure-breeding, white-flowered lines are intercrossed with the following results:

pure line 1 white × pure line 2 white

↓

F_1 all white

↓

F_2 131 white

29 red

   a. Deduce the inheritance of these phenotypes, using clearly defined gene symbols. State the parental genotypes for both the F_1 and F_2.
   b. Predict the outcome of crosses of the F_1 to each parental line.

41. A phenotype is the result of the interaction of an individual's genotype with the environment. Monozygotic ("identical") twins represent an interesting genetic phenomenon, and the study of such twins sometimes can shed light on the relative contributions of genotype and environment to phenotype. The table on page 93 shows finger-

		Thumb		Forefinger		Middle finger		Ring finger		Little finger	
		Right	Left	Right	Left	Right	Left	Right	Left	Right	Left
Twins	Giorgio										
	Bruno										
Other sibs	Paolo										
	Tullio										
	Laura										
Parents	Father										
	Mother										

print patterns from a family whose members included a pair of monozygotic twins ("sibs," or siblings, is a general term for brothers and sisters). Which of the differences within the family do you think can be attributed to environmental variation (or to developmental noise)? Can you find any differences that must be genetic? (NOTE: monozygotic twins begin as the identical daughter cells from the first mitotic division of a zygote.)

(From Giorgio Schreiber, *Journal of Heredity* 9, 1930, 403.)

*42. In this problem, it is assumed that you know that enzymes are coded by genes—a topic not formally treated until Chapter 12. However, it is a good problem to try here, because it makes a strong connection between gene interactions and cell chemistry.

Assume that in petunias, the normal purple color is a result of a mixture of two pigments, red and blue, which are synthesized in separate biochemical pathways, as shown here:

$$\text{Pathway I} \quad \cdots \longrightarrow \text{white}_1 \xrightarrow{E} \text{blue}$$

$$\text{Pathway II} \quad \cdots \rightarrow \text{white}_2 \xrightarrow{A} \text{yellow} \xrightarrow{B} \text{red}$$

$$\Big\uparrow C$$

$$\text{Pathway III} \quad \cdots \longrightarrow \text{white}_3 \xrightarrow{D} \text{white}_4$$

Red is formed through a yellow intermediate that normally does not reach detectable levels.

A third pathway involving only white compounds normally does not affect the blue and red pathways, but if one of its intermediates (white_3) should build up in concentration, it can be converted to the yellow intermediate of the red pathway.

In the diagram, A to E represent enzymes and their corresponding genes, all of which are unlinked.

Assuming that wild-type alleles are dominant and represent enzyme function and that recessive alleles are mutant and represent lack of enzyme function, state which gene pairs *in dihybrid crosses* will yield

a. 9 purple : 3 green : 4 blue

b. 9 purple : 3 red : 3 blue : 1 white

c. 13 purple : 3 blue

d. 9 purple : 3 red : 3 green : 1 yellow

(NOTE: blue mixed with yellow makes green; assume that no mutations are lethal.)

Linkage I: Basic Eukaryotic Chromosome Mapping

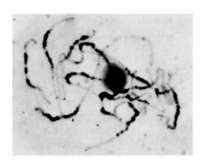

KEY CONCEPTS

Two gene pairs located close together on the same chromosome pair do not show independent assortment at meiosis.

∎

Recombination is the production of genotypes with new combinations of parental alleles.

∎

A pair of homologous chromosomes can exchange parts during a process called crossing-over.

∎

Recombination is caused either by independent assortment or by crossing-over.

∎

Gene loci on a chromosome can be mapped by measuring the frequencies of recombinants produced by crossing-over.

∎

Interlocus map distances based on recombination measurements are roughly additive.

∎

The occurrence of a crossover can influence the occurrence of a second crossover in an adjacent region.

■ We have already established the basic principles of segregation and assortment, and we have correlated them with chromosome behavior during meiosis. Thus, from the cross $Aa\ Bb \times Aa\ Bb$, we expect a $9:3:3:1$ ratio of phenotypes. As we learned from Bridges's study of nondisjunction (page 53), exceptions to simple Mendelian expectations can direct the experimenter's attention to new discoveries. Just such an exception involving a dihybrid cross provided the clue to the important concepts discussed in this chapter.

> **Message** In genetic analysis, exceptions to predicted behavior often are the source of important new insights.

The Discovery of Linkage

In the early years of this century, William Bateson and R. C. Punnett were studying inheritance in the sweet pea. They studied two pairs of genes: one affecting flower color (P, purple, and p, red), and the other affecting the shape of pollen grains (L, long, and l, round). They then made a dihybrid cross, $PP\ LL$ (purple, long) $\times pp\ ll$ (red, round), and selfed the F_1 $Pp\ Ll$ heterozygotes to obtain an F_2. Table 5-1 shows the proportions of each phenotype in the F_2 plants.

The results of this cross are a striking deviation from the expected $9:3:3:1$ ratio. What is going on? This does not look like something that can be explained as a modified Mendelian ratio. Note that two phenotypic classes are larger than expected: the purple, long phenotype; and the red, round phenotype. Bateson and Punnett speculated that the size of these two classes is due to an excess of the two gametic types $P\ L$ and $p\ l$. Because these were the original parental gametic types, the researchers thought that a physical **coupling** between the dominant genes (P and L) and between the recessive

genes (p and l) might have prevented their independent assortment in the F_1. However, the researchers did not know what the nature of this coupling could be.

The explanation of Bateson and Punnett's results had to await the development of *Drosophila* as a genetic tool. After coupling was first described, Thomas Hunt Morgan found a similar deviation from Mendel's second law while studying two autosomal gene pairs in *Drosophila*. One of these gene pairs affects eye color (pr, purple, and pr^+, red), and the other affects wing length (vg, vestigial, and vg^+, normal). Morgan crossed $prpr\ vgvg$ flies with $pr^+pr^+\ vg^+vg^+$ and then testcrossed the doubly heterozygous F_1 females: $pr^+pr\ vg^+vg\ ♀ \times prpr\ vgvg\ ♂$.

The use of the testcross is extremely important. Because one parent (the tester) contributes gametes carrying only recessive alleles, the phenotypes of the offspring represent the gametic contribution of the doubly heterozygous other parent. Hence, the analyst can concentrate on one meiosis and forget about the other. This is in contrast to the situation in an F_1 self, where there are two sets of meiotic divisions to consider: one for the male parental gametes, and one for the female parental gametes. Morgan's results follow; the F_2 individuals are designated in terms of the genes contributed by the F_1 female:

$pr^+\ vg^+$	1339
$pr\ vg$	1195
$pr^+\ vg$	151
$pr\ vg^+$	154
	2839

Obviously, these numbers are a drastic deviation from the Mendelian prediction of a $1:1:1:1$ ratio, and they indicate a coupling of genes. The largest classes are the two gene combinations, $pr^+\ vg^+$ and $pr\ vg$, originally introduced by the parental flies. You can see that the testcross makes the situation much clearer. It directly reveals the gene combinations in the gametic population from one sex in the F_1, thus clearly showing the coupling that could only be inferred from Bateson and Punnett's F_1 self. The testcross also reveals something new: there is approximately a $1:1$ ratio between the two parental types and also between the two nonparental types.

When each parent in a cross was homozygous for one of the recessive genes, an F_1 testcross produced different results:

P $pr^+pr^+\ vgvg \times prpr\ vg^+vg^+$

$\downarrow$

F_1 $pr^+pr\ vg^+vg\ ♀ \times prpr\ vgvg\ ♂$

The following progeny were obtained from the testcross:

TABLE 5-1. Observed and expected phenotypes in the F_2 from the dihybrid cross

Phenotype	Number of individuals observed	Approximate number of individuals expected (from $9:3:3:1$ ratio)
Purple, long ($P-\ L-$)	284	215
Purple, round ($P-\ ll$)	21	71
Red, long ($pp\ L-$)	21	71
Red, round ($pp\ ll$)	55	24
	381	381

$$
\begin{array}{ll}
pr^+ \, vg^+ & 157 \\
pr \, vg & 146 \\
pr^+ \, vg & 965 \\
pr \, vg^+ & \underline{1067} \\
& 2335
\end{array}
$$

Again, these results are not even close to a $1:1:1:1$ Mendelian ratio. Now, however, the largest classes are those that have one dominant gene or the other rather than, as before, two dominants or two recessives. But notice that once again the gene combinations that were originally contributed to the F_1 by the parental flies provide the most frequent classes in the F_2. In the early work on coupling, Bateson and Punnett coined the term **repulsion** to describe this situation, because it seemed to them that in this case the nonallelic dominant genes were "repelled" from each other — the opposite of the situation in coupling, where the dominant genes seemed to "stick together." How can we explain these two phenomena: coupling and repulsion?

Morgan suggested that both pairs of genes being studied in coupling are located *on the same pair of homologous chromosomes*. Thus, when pr and vg are introduced from one parent, they are physically located on the same chromosome, whereas pr^+ and vg^+ are located on the homologous chromosome from the other parent (Figure 5-1). This hypothesis also explains repulsion. In that case, one parental chromosome carries pr and vg^+ and the other carries pr^+ and vg. Repulsion, then, is just another case of coupling: in this case, the coupled genes are one dominant and one nonallelic recessive. This hypothesis explains why gene combinations from P remain together, but how do we explain the existence of nonparental combinations?

Morgan suggested that when homologous chromosomes pair during meiosis, a physical exchange of chromosome parts occasionally occurs during a process

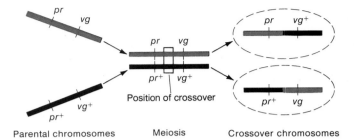

Figure 5-2. Diagrammatic representation of the process of crossing-over during meiosis. Homologous chromosomes exchange parts in the crossover.

called crossing-over. Figure 5-2 illustrates this process, which results in a physical exchange of chromosome segments. The original gene arrangement on the two chromosomes is called the **parental** combination. The two new combinations are called **crossover types, exchange products, intrachromosomal recombinants,** or, simply, **recombinants.**

This hypothesis may seem a bit farfetched. Is there any cytologically observable process that could account for something like crossing-over? We saw in Chapter 3 that during meiosis, when duplicated homologous chromosomes are paired with each other, two nonsister chromatids often appear to cross each other, whereas the other two do not. This is diagrammed in Figure 5-3. Recall that the resulting cross-shaped structure is called a chiasma. To Morgan, the appearance of the chiasmata was a perfect visual corroboration of the concept of crossing-over. (Note that they seem to indicate that crossing-over occurs between chromatids, not between unduplicated chromosomes. We shall return to this point later.) For the present, let's accept Morgan's interpretation that chiasmata are the cytological counterparts of crossovers and leave the proof until Chapter 18. Note that Morgan did not arrive at this interpretation out of nowhere; he was looking for a *physical* explanation for his *genetic* results. His achievement in correlating the results of breeding experiments with cytological phenomena thus serves to emphasize the importance of the chromosome theory as a powerful basis for research.

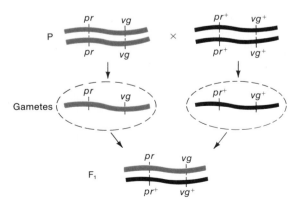

Figure 5-1. Simple inheritance of two gene pairs located on the same chromosome pair.

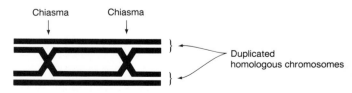

Figure 5-3. Highly diagrammatic representation of chiasmata at meiosis. Each line represents a chromatid of a pair of chromosomes in synapsis during meiosis.

> **Message** Chiasmata are the visible manifestations of crossovers.

Data like those just presented, showing coupling and repulsion in testcrosses and in F_1 selfs, are commonly encountered in genetics. Clearly, results of this kind represent a departure from independent assortment. Such exceptions, in fact, constitute a major addition to Mendel's view of the genetic world.

> **Message** When two gene pairs are located close together on the same chromosome pair, they do not show independent assortment.

The general situation in which gene pairs reside on the same chromosome pair is termed **linkage**. Two gene pairs on the same chromosome pair are said to be linked. It is also proper to refer to the linkage of specific alleles: for example, in one $Aa\ Bb$ individual, A might be linked to b; a would then of necessity be linked to B. These terms graphically allude to the existence of a physical entity linking the genes — that is, the chromosome itself! You may wonder why such genes are referred to as "linked" rather than "coupled"; the answer is that coupling and repulsion have come to indicate two different types of linkage conformation in a double heterozygote, as follows:

$$\text{Coupling conformation} \quad \frac{pr \qquad vg}{pr^+ \qquad vg^+}$$

$$\text{Repulsion conformation} \quad \frac{pr \qquad vg^+}{pr^+ \qquad vg}$$

In other words, coupling refers to the linkage of two dominant or two recessive genes, whereas repulsion indicates the linkage of one dominant and one recessive gene in a double heterozygote. A double heterozygote can be assigned a coupling or a repulsion conformation only from a consideration of the parental genotypes or by testcrossing.

Recombination

We have already introduced the term recombination. This term is widely used in many areas of practical and theoretical genetics, so it is absolutely necessary to be clear about its meaning at this stage. Recombination can occur in a variety of situations in addition to meiosis, but for the present, let's define it in relation to meiosis. To adapt the definition to other situations, we shall simply replace the words "meiotic" and "meiosis" with other appropriate terms.

> **Definition** Meiotic recombination is any meiotic process that generates a haploid product with a genotype that differs from the two haploid genotypes that constituted the meiotic diploid. The product of meiosis so generated is called a **recombinant.**

This definition makes the important point that the detection of recombination is based on a comparison between the *output* (or product) genotypes of meiosis and the *input* genotypes (Figure 5-4). The input genotypes are the two haploid types that combined to make the meiotic diploid (that is, the meiocyte, the diploid cell we are considering at meiosis).

The definition of recombination applies to both haploid and diploid life cycles. In haploids, the situation is identical to that shown in Figure 5-4. The phenotypes of the haploid input and output are used to a determine genotypes directly, because the input and the output in this case are individuals. In diploid cycles, however, the input and the output are gametes! To detect recombination in a diploid cycle, we must have pure-breeding parents, so that their gametic contributions are known. Furthermore, we cannot detect recombinant output gametes directly: we must testcross the meiotic diploid under study to reveal the recombinants produced by

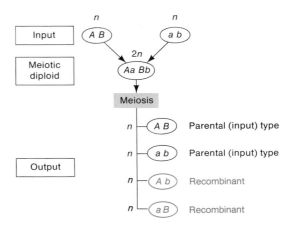

Figure 5-4. Recombination. This diagram shows a meiosis, comparing its input haploid genotypes to its output haploid genotypes.

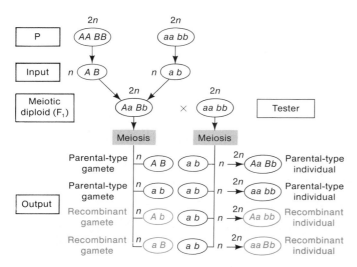

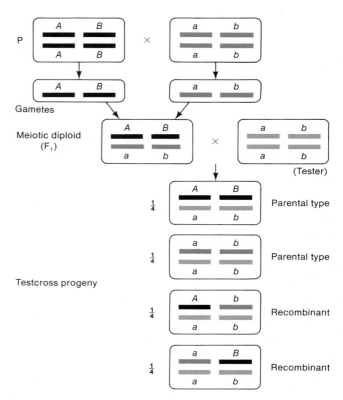

Figure 5-5. An expansion of Figure 5-4 to clarify the process of recombination in diploid organisms. Note that Figure 5-4 is actually a part of this figure. In a diploid meiosis, the recombinant meiotic products are most readily detected by a cross to a recessive tester.

meiosis (Figure 5-5). If a testcross offspring is shown to have been constituted from a recombinant product of meiosis, it too is called a recombinant. Notice again that the testcross allows us to concentrate on *one* meiotic system and avoid ambiguity. In a self of the F_1 in Figure 5-5, for example, an *AA Bb* offspring (which would demonstrate recombination in F_1) cannot be distinguished from *AA BB* without further extensive crosses.

There are two kinds of recombination, and they produce recombinants by completely different methods. But a recombinant is a recombinant, so how can we decide which type of recombination has occurred in any particular progeny? The answer lies in the *frequency* of recombinants, as we shall soon see. First, though, let's consider the two types of recombination: interchromosomal and intrachromosomal.

Interchromosomal Recombination

Interchromosomal recombination is achieved by Mendelian independent assortment. The two recombinant classes always make up 50 percent of the progeny; that is, there are 25 percent of each recombinant type among the progeny (Figure 5-6). If we observe this frequency, we can infer that the gene pairs under study are assorting independently.

The simplest interpretation of these frequencies is that the gene pairs are on separate chromosome pairs. However, as has already been hinted, gene pairs that are far apart on the *same* chromosome pair can act virtually

Figure 5-6. Interchromosomal recombination, which always produces a recombinant frequency of 50 percent. This diagram shows a diploid organism, but we can see the haploid situation by removing the part marked P and the testcross.

independently and still produce the same result. In Chapter 6, we shall see how a large amount of crossing-over effectively unlinks the distantly linked gene pairs.

Intrachromosomal Recombination

Intrachromosomal recombination is produced by crossing-over. Crossing-over occurs between any two nonsister chromatids (we shall show proof of this in Chapter 6). Of course, a crossover does not occur between two loci in all meioses, but when one does occur, one-half of the products of that meiosis will be recombinant, as shown in Figure 5-7. Meioses with no crossover between loci will produce all parental genotypes for these gene pairs.

The sign of intrachromosomal recombination is a recombinant frequency of less than 50 percent. The physical linkage of parental gene combinations prevents the free assortment of gene pairs (Figure 5-8). We saw an example of this situation in Morgan's data (page 95), where the recombinant frequency was $(151 + 154) \div 2839 = 10.7$ percent. This is obviously much less than

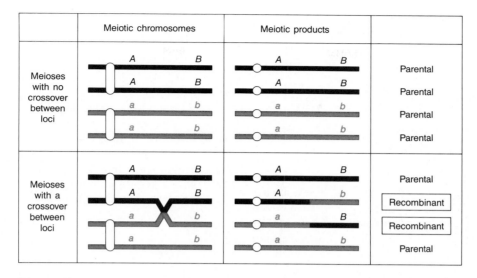

Figure 5-7. Intrachromosomal recombinants arise from that subpopulation of meioses in which crossing-over occurs between the loci under study.

the 50 percent we would expect with independent as-sortment. What about recombinant frequencies greater than 50 percent? The answer is that such frequencies are *never* observed, as we shall prove in Chapter 6.

Linkage Symbolism

Our symbolism for describing crosses becomes cumber-some with the introduction of linkage. We can show the genetic constitution of each chromosome in the *Drosophila* cross as

$$\frac{pr \qquad vg}{pr^+ \qquad vg^+}$$

where each line represents a chromosome; the genes above are on one chromosome, and those below are on the other chromosome. A crossover is represented by placing a × between the two chromosomes, so that

$$\frac{pr \qquad vg}{\times}$$
$$\overline{pr^+ \qquad vg^+}$$

is the same as

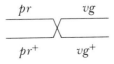

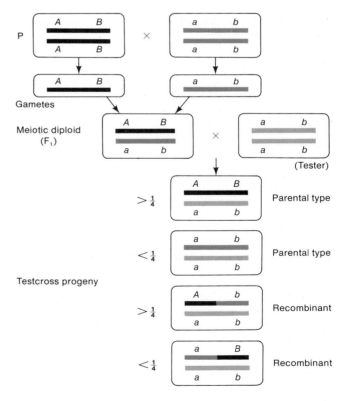

Figure 5-8. Intrachromosomal recombination, which always produces a recombinant frequency equal to or (usually) less than 50 percent. Again, a diploid organism is shown.

We can simplify the genotypic designation of linked genes by drawing a single line, with the genes on each side being on the same chromosome; now our symbol is

$$\frac{pr \qquad vg}{pr^+ \qquad vg^+}$$

But this is still inconvenient for typing and writing, so let's tip the line to give us $pr\,vg/pr^+\,vg^+$, still keeping the genes of one chromosome on one side of the line and those of its homolog on the other. We always designate linked genes on each side in the same order; it is always $a\,b/a\,b$, never $a\,b/b\,a$. That being the case, we can indicate the wild-type allele with a plus sign (+), and $pr\,vg/pr^+\,vg^+$ becomes $pr\,vg/++$. From now on, we shall use this kind of designation unless it creates an ambiguity.

Now if we reconsider the results obtained by Bateson and Punnett, we can easily explain the coupling phenomenon by means of the concept of linkage. Their results are complex because they did not do a testcross. See if you can derive estimated numbers for recombinant and parental types in the gametes.

Linkage of Genes on the X Chromosome

Until now, we have been considering recombination in autosomal genes. If, however, we were to look at sex-linked genes, a testcross would not be necessary. (Don't confuse linkage with sex linkage. Linkage describes the association of genes on the same chromosome, whereas sex linkage describes the association of genes with an X chromosome or other sex-determining chromosome.) Consider what happens when recombination involves the X chromosome. A female, heterozygous for two different sex-linked genes, will produce male progeny hemizygous for those genes, so that the mother's gametic genotype will be the son's phenotype. Let's consider an example, using the following gene pairs: y (yellow body) and y^+ (brown body); w (white eye) and w^+ (red eye). In the following cross, Y represents the Y chromosome; do not confuse it with the y/y^+ gene.

P $y+/y+♀ \times +\,w/Y\,♂$ (that is, $yw^+/yw^+ ♀ \times y^+w/Y ♂$)

F₁ $y+/+\,w ♀ \times y+/Y ♂$ (that is, $yw^+/y^+w ♀ \times yw^+/Y ♂$)

The number of F₂ males in each phenotypic class is

y	w	43
+	w	2146
y	+	2302
+	+	22
		4513

These classes reflect the products of meiosis in the F₁ female. The system is in effect a testcross, because the F₂ males obtain only a Y chromosome from the F₁ $y+/Y$ males; thus, the F₂ males have only one allele (obtained from the mother) for each of the genes we are considering. The recombinant frequency in this case is $(43 + 22) \div 4513 = 1.4$ percent.

Linkage Maps

The frequency of recombinants for the *Drosophila* autosomal genes we studied (pr and vg) was 10.7 percent of the progeny—a much greater frequency than that for the linked genes on the X chromosome. Apparently, the amount of crossing-over between linked gene pairs is not constant. Indeed, there is no reason to expect that crossing-over between different linked gene pairs would occur with the same frequency. As Morgan studied more linked genes, he saw that the proportion of recombinant progeny varied considerably, depending on which linked gene pairs were being studied, and he thought that these variations in crossover frequency might somehow reflect the actual distances separating genes on the chromosomes. Morgan assigned the study of this problem to a student, Alfred Sturtevant, who (like Bridges) became one of the great geneticists. Morgan asked Sturtevant, still an undergraduate at the time, to make some sense of the data on crossing-over between different linked genes. In one night, Sturtevant developed a method for describing relationships between genes that is still used today.

For example, consider a testcross from which we obtain the following results:

$pr\,vg/pr\,vg$	165	Parental (noncrossover) types
$++/pr\,vg$	191	
$pr+/pr\,vg$	23	Recombinant (crossover) types
$+vg/pr\,vg$	21	
	400	

The progeny in this example represent 400 female gametes, of which 44 (11 percent) are recombinant. Sturtevant suggested that we can use this percentage of recombinants as a quantitative index of the linear distance between two gene pairs on a **genetic map,** or **linkage map,** as it is sometimes called.

The basic idea here is quite simple. Imagine two specific gene pairs positioned a certain fixed distance apart (Figure 5-9). Now imagine crossovers occurring randomly along the paired homologs. In some meiotic divisions, a crossover occurs by chance in the chromosomal region between these gene pairs; from these meioses, the recombinants are produced. In other meiotic divisions, no crossover occurs in the intervening region; no recombinants result from these meioses. Sturtevant postulated a rough proportionality: the greater the distance between the linked genes, the greater the chance that a crossover will occur in the intervening region and, hence, the greater the proportion of meioses in which a crossover occurs there. Thus, by measuring the frequency of recombinants, we can obtain a measure of the map distance between the gene pair.

> **Message** One **genetic map unit (m.u.)** is the distance between gene pairs for which one product of meiosis out of 100 is recombinant. Put another way, a **recombinant frequency (RF)** of 0.01 (or 1 percent) is defined as 1 m.u. (A map unit is sometimes referred to as a **centimorgan (cM),** in honor of Thomas Hunt Morgan.)

A direct consequence of this relationship is that if genes A and B are separated by 5 map units (5 m.u.) whereas genes A and C are separated by 3 m.u., then B and C should be either 8 or 2 m.u. apart (Figure 5-10).

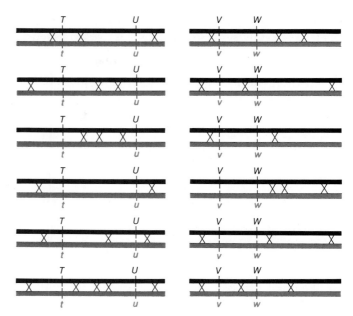

X = Position of crossover

Figure 5-9. Proportionality between chromosome distance and recombinant frequency. Each line represents a chromosome pair (four chromatids) during meiosis, and the crosses represent the positions of the crossovers that are taking place. On the left, two distant gene pairs (T and U) are shown; on the right (a different chromosome) are two close gene pairs (V and W). Crossovers occur at random along the chromosomes, and there are more of them between T and U than between V and W, so the recombinant frequency for T and U will be higher than that for V and W.

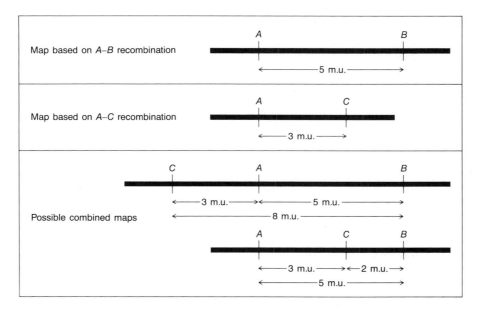

Figure 5-10. The additivity of map distances. If the $A-B$ and $A-C$ distances are calculated, the $B-C$ distance will be either of the two possibilities shown.

Sturtevant found this to be the case. In other words, his analysis strongly suggested that genes are arranged in some linear order.

The place on the map that represents a gene pair is called the **gene locus** (plural, **loci**). The locus of the eye-color gene pair and the locus of the wing-length gene pair, for example, are 11 m.u. apart. They are usually diagrammed this way:

$$\underset{pr}{\vdash}\qquad 11.0\qquad \underset{vg}{\dashv}$$

although they could be diagrammed equally well like this:

$$\underset{pr^+}{\vdash}\qquad 11.0\qquad \underset{vg^+}{\dashv}$$

or like this:

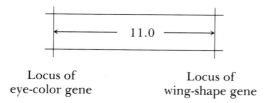

Locus of   Locus of
eye-color gene   wing-shape gene

Usually the locus of this eye-color gene is referred to in shorthand as the "*pr* locus," after the first discovered non-wild allele, but it means the place on the chromosome where any allele of this gene will be found.

Given a genetic distance in map units, we can predict frequencies of progeny in different classes. For example, of the progeny from a testcross of a *pr vg/++* heterozygote, 5½ percent will be *pr +/pr vg* and 5½ percent will be *+vg/pr vg*; of the progeny from a testcross of a *pr+/+vg* heterozygote, 5½ percent will be *pr vg/pr vg* and 5½ percent will be *++/pr vg*.

There is a strong implication that the "distance" on a genetic map is a physical distance along a chromosome, and Morgan and Sturtevant certainly intended to imply just that. But we should realize that the genetic map is another example of an entity constructed from a purely genetic analysis. The genetic map could have been derived without even knowing that chromosomes existed. Furthermore, at this point in our discussion, we cannot say whether the "genetic distances" calculated by means of recombinant frequencies in any way reflect actual physical distances on chromosomes, although cytogenetic analysis of *Drosophila* has shown that genetic distances are, in fact, roughly proportional to chromosome distances. Nevertheless, it must be emphasized that the hypothetical structure (the linkage map) was developed with a very real structure (the chromosome) in mind. In

other words, the chromosome theory provided the framework for the development of linkage mapping.

Three-Point Testcrosses

So far, we have looked at linkage in crosses of double heterozygotes to double recessive testers. The next level of complexity is a cross of a triple heterozygote to a triple recessive tester. This kind of cross, called a **three-point testcross,** is a good illustration of the standard kind of approach used in linkage analysis. We will consider two examples of such crosses here.

First, we will follow through an example from *Drosophila* that involves the non-wild-type alleles *sc* (scute, or loss of certain thoracic bristles), *ec* (echinus, or roughened eye surface), and *vg* (vestigial wing). We can take *sc/sc, ec/ec, vg/vg* triple recessive flies and cross these with wild types to generate triple heterozygotes *sc/+, ec/+, vg/+*. Recombination in these heterozygotes is analyzed by testcrossing heterozygous females with triple recessive tester males. The results of this testcross follow. The progeny are listed as gametic genotypes derived from the heterozygous females. Eight gametic types are possible, and these are seen in the following numbers in a sample of 1008 progeny flies:

sc	ec	vg	235
+	+	+	241
sc	ec	+	243
+	+	vg	233
sc	+	vg	12
+	ec	+	14
sc	+	+	14
+	ec	vg	16
			1008

The systematic way to analyze such crosses is to calculate all possible recombinant frequencies, but it is always worthwhile to inspect the data for obvious patterns before doing this. At first glance, we can note in the above data that there is a considerable deviation from the $1:1:1:1:1:1:1:1$ ratio that is expected if the genes are all unlinked. So let's begin to calculate RF values, taking the loci a pair at a time. Starting with the *sc* and *ec* loci (we essentially ignore the *vg* locus for the time being), we determine which of the gametic genotypes are recombinant for *sc* and *ec*. Since we know that the heterozygotes were established from *sc ec* and *++* gametes, we know that the recombinant products of meiosis must be *sc+* and *+ec*. There are $12 + 14 + 14 + 16 = 56$ of these types; therefore, RF $= (56/1008) \times 100 = 5.5$ m.u. Now we know that these loci must be closely linked on the same chromosome, as follows:

sc ————————— ec
← —— 5.5 m.u. —— →

Now let's look at recombination between the *sc* and the *vg* loci. Once again, the "input" parental genotypes were the double recessive and the wild-type combinations, so we must calculate the frequency of *sc+* and *+vg* progeny types (this time, we ignore the *ec* designations). We see that there are $243 + 233 + 14 + 16 = 506$ recombinants; $506/1008$ is very close to an RF of 50 percent, so the *sc* and *vg* loci are not linked. We can summarize the linkage relationship as follows:

sc ————————— ec   vg
← —— 5.5 m.u. —— →

Now it is obvious that the *ec* and *vg* loci must also be unlinked. Recombination analysis will confirm this. (Try it.)

A second example, using some other loci from *Drosophila,* will introduce some more important genetic concepts. Here the non-wild alleles will be *v* (vermilion eyes), *cv* (crossveinless, or absence of a crossvein on the wing), and *ct* (cut, or snipped wing edges). This time the parental stocks are homozygous double-recessive flies of genotype $+/+$, ct/ct, cv/cv and homozygous single-recessive flies of genotype v/v, $+/+$, $+/+$. From this cross, triply heterozygous progeny of genotype $v/+$, $ct/+$, $cv/+$ are obtained, and females of this genotype are testcrossed to triple recessives of genotype v/v, ct/ct, cv/cv. The gametes representing the eight progeny types from this testcross are shown here, with their numbers out of a total sample of 1448 flies:

v	+	+	580
+	cv	ct	592
v	cv	+	45
+	+	ct	40
v	cv	ct	89
+	+	+	94
v	+	ct	3
+	cv	+	5
			1448

Once again, the standard recombination approach is called for, but we must be careful in our classification of parental and recombinant types. Note that the parental input genotypes for the triple heterozygotes are $+ct\ cv$ and $v++$; we make take this into consideration when we decide what constitutes a recombinant.

Starting with the *v* and *cv* loci, we see that the recombinants are of genotype *v cv* and $++$ and that there are

$45 + 40 + 89 + 94 = 268$ of these. Out of a total of 1448 flies, this results in an RF of 18.5 percent.

For the *v* and *ct* loci, the recombinants are *v ct* and $++$. There are $89 + 94 + 3 + 5 = 191$ of these out of 1448, so that RF = 13.2 percent.

For *ct* and *cv*, the recombinants are $cv+$ and $+ct$. There are $45 + 40 + 3 + 5 = 93$ of these out of 1448, so that RF = 6.4 percent.

Obviously, all of the loci are linked on the same chromosome, because the RF values are all considerably less than 50 percent. Since the *v* and *cv* loci show the largest RF value, they must be the furthest apart; therefore, the *ct* locus must be in between them. A map can be drawn as follows:

v   ct   cv
← —— 13.2 m.u. —— →←— 6.4 m.u.—→

There are several important points to note here. The first is that the gene order we have deduced is different from the order in which the genes were listed in the parental and progeny individuals. Since the point of the exercise was to determine the linkage relationships of these genes, the original listing was of necessity arbitrary; the order simply was not known before the data were analyzed.

Secondly, we have definitely established the position of *ct* between *v* and *cv* and the distances between *ct* and these loci in map units. But we have arbitrarily placed *v* to the left and *cv* to the right: the map could equally well be inverted!

A third point to note is that the two smaller map distances, 13.2 m.u. and 6.4 m.u., add up to 19.6 m.u., which is greater than 18.5 m.u., the distance calculated for *v* and *cv*. Why is this so? The answer to this question lies in the way in which we have analyzed the two rarest classes in our classification of recombination for the *v* and *cv* loci. Now that we have the map, we can see that these two rare classes are in fact double recombinants, arising from two crossovers. An example of the generation of double recombinants from two crossovers is shown in Figure 5-11. However, the *v ct +* and

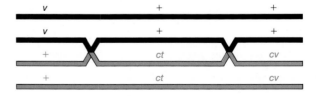

Figure 5-11. An example of a double crossover. Notice that double recombinant chromatids are produced and that double recombinants have the parental allele combination for the outer loci.

$++cv$ genotypes were not counted in our calculation of the RF value for v and cv; after all, with regard to v and cv, they are parental combinations ($v+$ and $+cv$). In the light of our map, however, we see that this led to an underestimate of the distance between the v and cv loci. Not only should we have counted the two rare classes, we should have counted each of them twice because each represents a double recombinant class! Hence, we can correct the value by adding the numbers $45 + 40 + 89 + 94 + 3 + 3 + 5 + 5 = 284$. Out of the total of 1448, this is exactly 19.6 percent, which is identical to the sum of the two component values.

Now that we have had some experience with the data of this cross, we can look back at the progeny listings and see that it is usually possible to deduce gene order by inspection, without a recombinant frequency analysis. It is generally true that the double recombinant classes will be the smallest ones, so that the gene order can be deduced directly. Since only three gene orders are really possible, each representing a different gene in the middle position, only one order should be compatible with the double recombinant classes observed, as shown in Figure 5-12. Only one order gives double recombinants of genotype $v\ ct +$ and $++cv$. Notice in passing that the ability to detect double crossovers depends on having a heterozygous gene pair spanning each crossover; if this cross had not been heterozygous for $ct/+$, we could never have identified the doubles.

Finally, note that linkage maps merely map the loci in relation to each other, using standard map units. We do not know where the loci are on a chromosome—or even which specific chromosome is involved. The linkage map is essentially an abstract construct that can only be correlated with a specific chromosome and with specific chromosome regions by applying special kinds of cytogenetic analyses, as we shall see in Chapter 8.

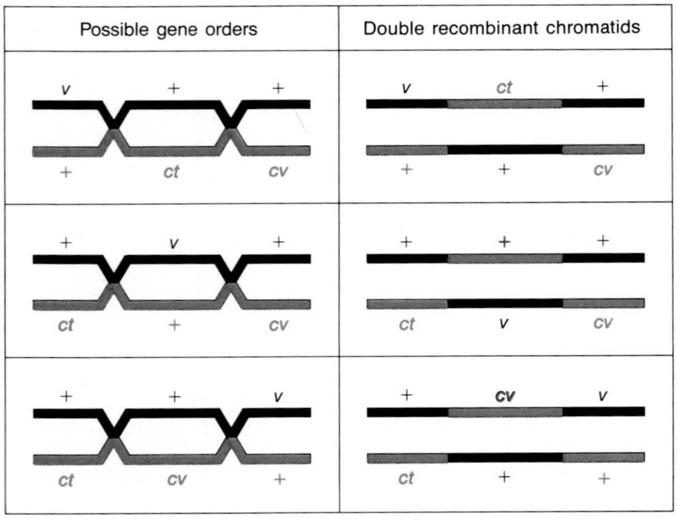

Interference

The detection of the double recombinant classes shows that double crossovers must occur. The existence of double crossovers raises many questions about the nature and distribution of such multiple intrachromosome recombination events. One of the first questions that comes to mind is whether the crossovers that occur in adjacent chromosome regions are independent or whether they interact with each other in some way. For example, we might ask if the occurrence of a crossover in one region decreases the likelihood of the occurrence of a crossover in an adjacent region.

The analysis can be approached in the following way, using data such as these for the $v\ ct\ cv$ region. If the crossovers in the two regions are occurring independently, then according to the product rule (see page 24), double recombinants should be equal in frequency to the product of the recombinant frequencies in the adjacent regions. In the $v - ct - cv$ recombination data, the $v - ct$ RF value is 0.132 and the $ct - cv$ value is 0.064, so double recombinants might be expected at the frequency $0.132 \times 0.064 = 0.0084$ (0.84 percent) if there is independence. In the sample of 1448 flies, $0.0084 \times 1448 = 12$ double recombinants are expected. However, the data show that only 8 were actually observed. This shows us that the two regions are not independent and suggests that the distribution of crossovers favors singles at the expense of doubles. In other words, there is some kind of **interference**—an effect in which the occurrence of a crossover reduces the probability of a crossover in the adjacent region.

Interference is quantified by first calculating a term called the **coefficient of coincidence (c.o.c.)**, which is the ratio of observed to expected double recombinants and then subtracting this value from 1. Hence

$$\text{Interference } (I) = 1 - \text{c.o.c.} =$$
$$1 - \left[\frac{\text{Frequency or number of observed double recombinants}}{\text{Frequency or number of expected double recombinants}} \right]$$

In our example

$$I = 1 - \tfrac{8}{12} = \tfrac{4}{12} = \tfrac{1}{3}, \text{ or } 33\%$$

Possible gene orders			Double recombinant chromatids		
v	$+$	$+$	v	ct	$+$
$+$	ct	cv	$+$	$+$	cv
$+$	v	$+$	$+$	$+$	$+$
ct	$+$	cv	ct	v	cv
$+$	$+$	v	$+$	cv	v
ct	cv	$+$	ct	$+$	$+$

Figure 5-12. Using the parental chromosomes $v++$ and $+ct\ cv$, only three gene orders are possible, and each results in different double recombinant genotypes. Only the first is compatible with the data shown in the text.

In some regions, there are never any observed double recombinants. In these cases, c.o.c. $= 0$, so $I = 1$ and interference is complete. Most of the time, the interference values that are encountered in mapping chromosome loci are between 0 and 1, but in certain special situations, negative interference results from an excess of observed over expected doubles.

Three-point testcrosses and extended versions of them constitute such a basic kind of recombination analysis that it is worth making a step-by-step summary of the analysis, ending with an interference calculation. We shall use numerical values from the cross involving v, ct, and cv.

1. Calculate recombinant frequencies for each pair of gene pairs:

$$v - cv = 18.5\%$$
$$cv - ct = 6.4\%$$
$$ct - v = 13.2\%$$

2. Represent linkage relationships in a linkage map:

$$\begin{array}{ccc} v & ct & cv \\ \hline \end{array}$$
$$\longleftarrow 13.2 \text{ m.u.} \longrightarrow\!\!\times\!\!\longleftarrow 6.4 \text{ m.u.} \longrightarrow$$

3. Determine the double recombinant classes.

4. Calculate the frequency or number of double recombinants to be expected if there is no interference:

Expected frequency $= 0.132 \times 0.064 = 0.0084$

Expected number $= 0.0084 \times 1448 = 12$

5. Calculate interference:

Observed number of double recombinants $= 8$

Expected number of double recombinants $= 12$

$$\therefore I = 1 - \tfrac{8}{12} = \tfrac{4}{12} = 0.33, \text{ or } 33\%$$

Figure 5-13 shows the redistribution of progeny types due to interference, using two adjacent regions each of which is 20 m.u. long. When there is no interference, the expected frequency of double recombinants is $0.2 \times 0.2 = 0.04$. The reason is that recombination in each region is independent of the other, so the product rule can be used. For 10 percent interference, the observed number of double recombinants can be calculated from the formula

$$I = 0.1 = 1 - \frac{\text{observed doubles}}{0.04}$$

The frequency of observed doubles comes to 0.036. The difference of $0.004 (= 0.040 - 0.036)$ obviously is the reduction in the double-recombinant class. But at the same time, there is a reduction in the nonrecombinant class and an increase in both the single-recombinant classes by the same amount. (Note that the map distances in I and II in the figure remain the same.)

You may have noticed that we always used heterozygous females for testcrosses in our crossover studies in *Drosophila*. When $pr\ vg/++$ males are crossed with $pr\ vg/pr\ vg$ females, only $pr\ vg/++$ and $pr\ vg/pr\ vg$ progeny are recovered. This result shows that crossing-over does not occur in *Drosophila* males. However, this absence of crossing-over in one sex is limited to certain species; it is not the case for males of all species (or for the heterogametic sex). In other organisms, crossing-over

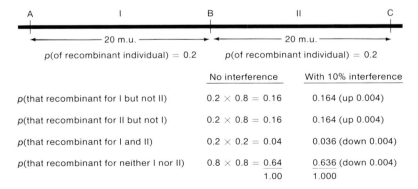

	No interference	With 10% interference
p(that recombinant for I but not II)	$0.2 \times 0.8 = 0.16$	0.164 (up 0.004)
p(that recombinant for II but not I)	$0.2 \times 0.8 = 0.16$	0.164 (up 0.004)
p(that recombinant for I and II)	$0.2 \times 0.2 = 0.04$	0.036 (down 0.004)
p(that recombinant for neither I nor II)	$0.8 \times 0.8 = 0.64$	0.636 (down 0.004)
	1.00	1.000

Figure 5-13. How to calculate the expected frequencies of nonrecombinant, single-recombinant, and double-recombinant types, with and without interference. The loci A and B are separated by distance I (20 m.u.); loci B and C are separated by distance II (20 m.u.).

can occur in XY males or in WZ females. The reason for this sex difference is that *Drosophila* males have an unusual prophase I, with no synaptonemal complexes.

Linkage maps are an essential aspect of the experimental genetic study of any organism. They are the prelude to any serious piece of genetic manipulation. Many organisms have had their genes intensively mapped in this way. The resultant maps represent a tremendous amount of basic applied genetic analysis. Figures 5-14 (page 107) and 5-15 (page 108) show two examples of linkage maps: one from *Drosophila,* and one from the tomato.

The χ^2 Test

We come now to a subject that emerges naturally at this point in the discussion — a subject that is particularly relevant to the detection of linkage. It has been stated that the functional test for the presence or absence of linkage is based on the relative frequencies of the meiotic product types. If there is no linkage, then the four product-cell types *A B*, *a b*, *A b*, and *a B* are produced in a 1 : 1 : 1 : 1 ratio, which fixes the recombinant frequency at 50 percent. If there is linkage, then there is deviation from the 1 : 1 : 1 : 1 ratio and two types (the recombinants) are present in a minority (<50 percent). In practical terms, the question "Is this a 1 : 1 : 1 : 1 ratio?" must be answered in any particular case. An answer of "yes" indicates absence of linkage, and "no" indicates linkage. "Obvious" departures from the 1 : 1 : 1 : 1 ratio present no decision problems, but smaller departures are tricky to handle and require further analytic approaches.

What is the precise problem here? An example will be helpful. A double heterozygote in coupling conformation produces 500 meiotic products, distributed as follows:

$$A\ B \qquad 140$$
$$a\ b \qquad 135$$
$$A\ b \qquad 110$$
$$a\ B \qquad 115$$

By applying the recombinant-frequency test, we find 225 recombinants, or 45 percent. This is admittedly less than 50 percent, but not convincingly so. The skeptic would say, "This is merely a chance deviation from a 1 : 1 : 1 : 1 ratio! If you repeatedly grabbed samples of 500 marbles from a dark sack containing equal numbers of red, blue, yellow, and green marbles, you would get a variation this great fairly often from the 1 : 1 : 1 : 1 ratio." What should we decide? The χ^2 test can help in this kind of predicament.

In general, the χ^2 test tells us how often deviations from expectations will occur purely on the basis of chance. The procedure is as follows:

1. *State a simple hypothesis that gives a precise expectation.* In our example, the best hypothesis is "lack of linkage," which yields an expected 1 : 1 : 1 : 1 ratio. This **null hypothesis** is obviously better than a hypothesis of linkage, which is not precise because (as we have seen) the recombinant frequency can be large or small.

2. *Calculate χ^2.* χ^2 is always calculated from actual numbers — never from percentages, fractions, or decimal fractions. In fact, part of the usefulness of the χ^2 test is that it takes sample size into consideration. The sample is composed of several operational classes, with O representing the observed number in any class and E representing the expected number based on the null hypothesis. The formula for calculating χ^2 is as follows:

$$\chi^2 = \text{total of } \frac{(O - E)^2}{E} \text{ over all classes}$$

In our example, we would set up the calculation as shown in Table 5-2.

3. *Estimate p.* Using χ^2, we estimate the probability p of obtaining the observed results if the null hypothesis is correct. Before this can be done, however, we must compute another item: the number of **degrees of freedom (df)**. In the present context, the number of degrees of freedom can be simply defined as

$$\text{df} = (\text{number of classes} - 1)$$

In our example

$$\text{df} = (4 - 1) = 3$$

We now turn to the table of χ^2 values (Table 5-3), which will give us our p values if we plug in our

TABLE 5-2. Calculation of χ^2

Class	O	E	$(O - E)^2$	$\dfrac{(O - E)^2}{E}$
A B	140	125	225	1.8
a b	135	125	100	0.8
A b	110	125	225	1.8
a B	$\dfrac{115}{500}$	$\dfrac{125}{500}$	100	$\dfrac{0.8}{\chi^2 = \overline{5.2}}$

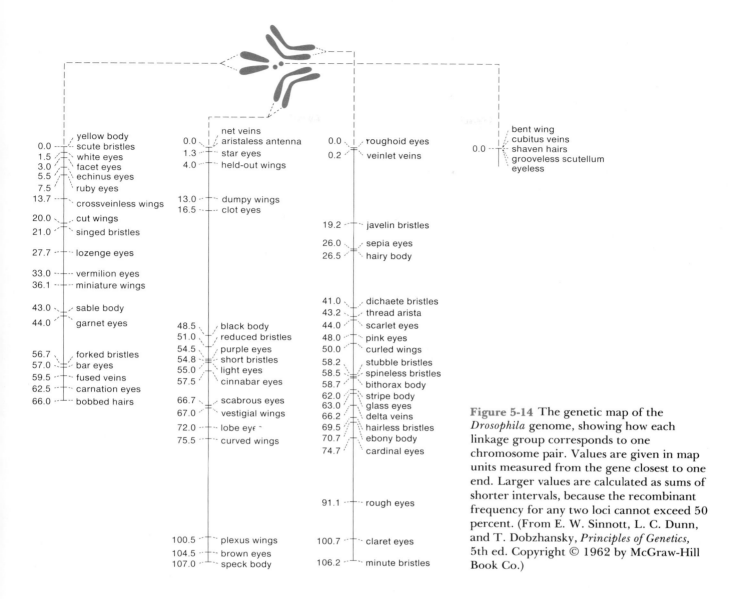

Figure 5-14 The genetic map of the *Drosophila* genome, showing how each linkage group corresponds to one chromosome pair. Values are given in map units measured from the gene closest to one end. Larger values are calculated as sums of shorter intervals, because the recombinant frequency for any two loci cannot exceed 50 percent. (From E. W. Sinnott, L. C. Dunn, and T. Dobzhansky, *Principles of Genetics,* 5th ed. Copyright © 1962 by McGraw-Hill Book Co.)

TABLE 5-3. Critical values of the χ^2 distribution

df	0.995	0.975	0.9	0.5	0.1	0.05	0.025	0.01	0.005	df
1	.000	.000	0.016	0.455	2.706	3.841	5.024	6.635	7.879	1
2	0.010	0.051	0.211	1.386	4.605	5.991	7.378	9.210	10.597	2
3	0.072	0.216	0.584	2.366	6.251	7.815	9.348	11.345	12.838	3
4	0.207	0.484	1.064	3.357	7.779	9.488	11.143	13.277	14.860	4
5	0.412	0.831	1.610	4.351	9.236	11.070	12.832	15.086	16.750	5
6	0.676	1.237	2.204	5.348	10.645	12.592	14.449	16.812	18.548	6
7	0.989	1.690	2.833	6.346	12.017	14.067	16.013	18.475	20.278	7
8	1.344	2.180	3.490	7.344	13.362	15.507	17.535	20.090	21.955	8
9	1.735	2.700	4.168	8.343	14.684	16.919	19.023	21.666	23.589	9
10	2.156	3.247	4.865	9.342	15.987	18.307	20.483	23.209	25.188	10
11	2.603	3.816	5.578	10.341	17.275	19.675	21.920	24.725	26.757	11
12	3.074	4.404	6.304	11.340	18.549	21.026	23.337	26.217	28.300	12
13	3.565	5.009	7.042	12.340	19.812	22.362	24.736	27.688	29.819	13
14	4.075	5.629	7.790	13.339	21.064	23.685	26.119	29.141	31.319	14
15	4.601	6.262	8.547	14.339	22.307	24.996	27.488	30.578	32.801	15

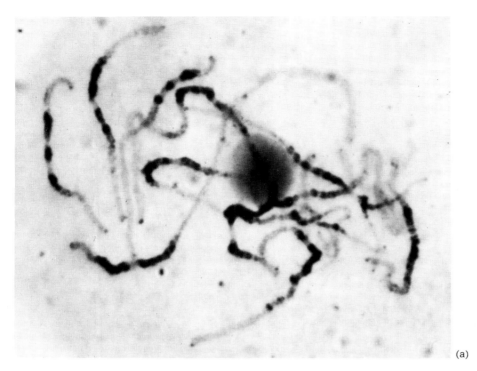

(a)

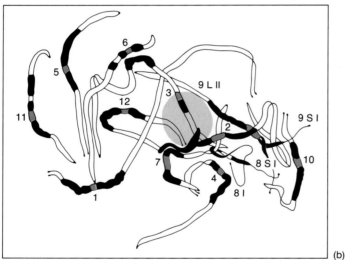

(b)

Figure 5-15 The tomato genome. (a) Photomicrograph of a meiotic pachytene (prophase I) from anthers. (b) Diagram corresponding to part a, identifying the 12 chromosome pairs. Colored centromeres are flanked by densely staining chromosome regions called heterochromatin. Heterochromatin is thought to be genetically inert. (Opposite page, c) Genetic map of the tomato genome. The 12 linkage groups correspond to the 12 chromosome pairs of part b. Centromeres and heterochromatin regions are indicated. (Part a courtesy of Charles M. Rick, Parts b, c from Charles M. Rick, "The Tomato." Copyright © 1978 by Scientific American, Inc. All rights reserved.)

computed χ^2 and df values. Looking along the df = 3 line, we find that our χ^2 value of 5.2 lies between the p values of 0.5 (50 percent) and 0.1 (10 percent). Hence, we can conclude that our p value is just greater than 10 percent (0.1). This means that if our null hypothesis is correct, a deviation *at least* as great as that observed is expected a little more than 10 percent of the time.

4. *Reject or accept the null hypothesis.* How do we know which p value is too low to be acceptable? Scientists in general arbitrarily use the 5 percent level. Thus, any p value less than 5 percent results in rejection of

the null hypothesis, and any p value greater than 5 percent results in acceptance of the null hypothesis. In the case of acceptance, of course, the null hypothesis is not proved, merely possible. In the case of rejection, the hypothesis is not disproved, merely improbable. Now that we have accepted our null hypothesis of "lack of linkage," we must live with it and acknowledge that the skeptic had a point!

In this way, the χ^2 test helps us decide between linkage and absence of linkage. In fact, a major use of the χ^2 test in genetics is in the determination of linkage. But there

(c)

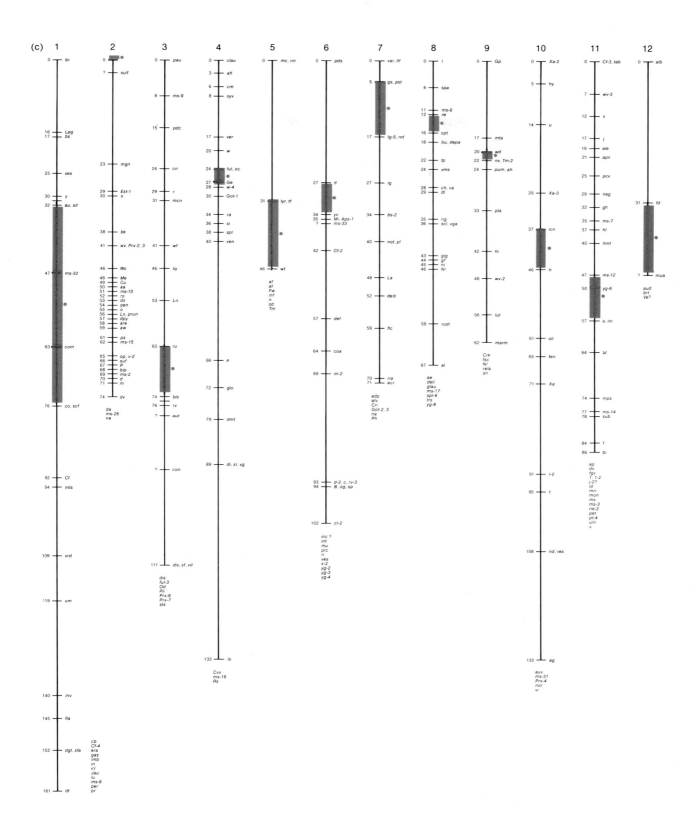

are several other situations in which the χ^2 test is useful, and these again involve testing actual results against simple expectations of a hypothesis.

For example, χ^2 is ideal (as we have seen) for testing deviations from any genetic ratio: $3:1$, $9:3:3:1$, $9:7$, $1:1$, and so on. But there are pitfalls here. Suppose, for example, you believe you have identified a major gene that affects a specific biological function. You testcross a presumed heterozygote Aa to a homozygous recessive individual aa, expecting a $1:1$ phenotypic ratio in the progeny. Out of 200 progeny, 116 are Aa and 84 are aa. Here $\chi^2 = (16^2 \div 100) + (16^2 \div 100) = 5.12$ with 1 df. The p value is 2.5 percent, so you reject the null hypothesis that there is a single gene pair segregating. Actually, however, you must recall that the $1:1$ ratio is expected if there is a single pair of alleles *of equal viability* segregating. With the rejection of the null hypothesis, you must now reject the notion *either* of a single allele pair *or* of the equal viability of alleles (or both). The χ^2 test cannot tell you which portion of a compound null hypothesis to reject, so care must be taken to determine all of the hidden assumptions in a null hypothesis if error is to be avoided in the application of statistical testing.

> **Message** The χ^2 test is used to test experimental results against the expectations derived from a null hypothesis. The test generates a p value that is the probability of obtaining a specific deviation at least as great as the one observed, assuming that the null hypothesis is correct.

Early Thoughts on the Nature of Crossing-Over

The idea that intrachromosomal recombinants were produced by some kind of exchange of material between homologous chromosomes was a compelling one. But experimentation was necessary to test this idea. One of the first steps was to correlate the occurrence of a genetic recombinant with the occurrence of chromosome exchange. Several investigators approached this problem in the same way. In 1931, Harriet Creighton and Barbara McClintock were studying two loci of chromosome 9 of corn: one affecting seed color (C, colored; c colorless), and the other affecting endosperm composition (Wx, waxy; wx, starchy). Furthermore, the chromosome carrying C and Wx was unusual in that it carried a large, densely staining element (called a knob) on the C end and a longer piece of chromosome on the Wx end; thus, the heterozygote was

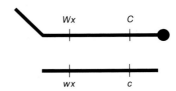

When they separated genetic recombinants from nonrecombinants, Creighton and McClintock found that all the nonrecombinants retained the parental chromosome arrangements, whereas all the recombinants were

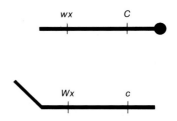

Thus, the correlation between the genetic and cytological events of intrachromosomal recombination was firmly established. The chiasmata seemed to represent the sites of the exchange process, but the final proof of this did not come until 1978.

But what is the mechanism of chromosome exchange in a crossover event? Is it a breakage and reunion process, like splicing sound tapes? Or is it a process that only *appears* to consist of a breakage and a reunion? An idea that favors the latter explanation was proposed in 1928 by John Belling, who studied meiosis in plant chromosomes and observed bumps along the chromosome (the chromomeres), which he thought might correspond to genes (Figure 5-16). Belling visualized the genes as beads strung together with some nongenic linking substance. He reported that during prophase I of meiosis, chromomeres duplicate, so that newly made chromomeres are attached to the originals (Figure 5-17). After duplication, the newly formed chromomeres are fastened together, but because all the chromomeres are tightly juxtaposed, the linking elements could switch from a newly made chromomere on one homologous chromosome to an adjacent one on the other homolog. Belling's model became known as the **copy-choice model** (or **switch model**) of crossing-over. You can see how it can generate a crossover chromatid that would seem to have arisen from a physical breakage and the reunion of chromosomes (Figure 5-18). You can also see that it suggests that the event producing recombinant chromosomes can take place only between newly made chromomeres (and hence newly made chromatids), so that any multiple crossover could involve only two chromatids.

Thus, the copy-choice model predicts that in every meiosis in which multiple crossovers occur (for example,

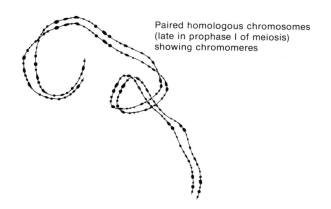

Paired homologous chromosomes
(late in prophase I of meiosis)
showing chromomeres

Figure 5-16. Diagrammatic representation of the chromomeres that are important in Belling's copy-choice model.

Chromomeres

Duplication

Figure 5-17. Division of the chromomeres according to Belling's model.

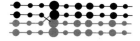

Figure 5-18. The hooking together of the newly synthesized chromomeres according to Belling's model. Joining usually forms parental combinations, but sometimes a switch can occur.

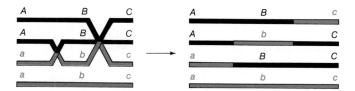

Figure 5-19. One of the several possible types of double-crossover tetrads that are regularly observed, showing how more than two chromatids must be involved. This evidence makes Belling's copy-choice (switch) model very unlikely.

trated in the following diagram, in which the genes are linked in the order shown:

$$ABC \times abc$$

$$\downarrow$$

$$\left. \begin{array}{l} ABc \\ AbC \\ aBC \\ abc \end{array} \right\} \text{ The four products of a single meiotic division}$$

Note that a recombination event has occurred between the first two loci (A/a and B/b) *and* between the second two loci (B/b and C/c) in the same meiosis; a double-exchange event has occurred (and one of the two products, $A b C$, is a double recombinant). *But* more than two chromatids must have been involved; in fact, *three must* have been involved in this case (Figure 5-19). Thus, Belling's suggestion (which would predict that only two chromatids could ever be involved) is most unlikely. (We shall see some *positive* evidence for the model of breakage and reunion in Chapter 18.)

Message The breakage and reunion model of chromosome crossing-over wins by default.

---

double crossovers and even higher multiples), only two chromatids out of the four will ever be involved. This prediction could be tested if only we had some way of recovering all four products from individual meiotic divisions. Each group of four could then be examined for the presence of multiple recombinants. In any group of four that contains at least one multiple recombinant, there must be two accompanying parental types for the copy-choice model to be possible. Luckily, several haploid organisms are suited to the recovery of all four products of a single meiosis. These organisms are some fungi and some unicellular algae. We consider the full analysis of these groups of four (called *tetrad analysis*) at length in Chapter 6. But we need them right now to answer one specific question: What companion genotypes are found in tetrads containing multiple recombinants? Several types of such tetrads are possible, but one interesting situation (often encountered in tetrad analysis) is illus-

The ability to isolate the four products of meiosis in fungi and algae also clears up another mystery, about whether crossing-over occurs at the two-strand (two-chromosome) stage or at the four-strand (four-chromatid) stage. If it occurs at the two-strand stage (before replication), there can never be more than two different products of a given meiosis. If it occurs at the four-strand stage (after replication), up to four different products of meiosis are possible (Figure 5-20). In fact, four different products of a single meiosis are regularly observed, showing that crossing-over occurs at the four-strand stage of meiosis.

Crossing-over is a remarkably precise process. The synapsis and exchange of chromosomes is such that no

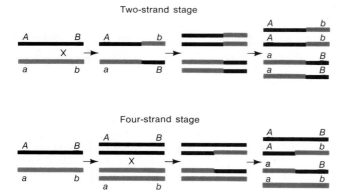

Two-strand stage

Four-strand stage

Figure 5-20. Tetrad analysis provides evidence that enables geneticists to decide whether crossing-over occurs at the two-strand (two-chromosome) or at the four-strand (four-chromatid) stage of meiosis. Because more than two different products of a single meiosis can be seen in some tetrads, crossing-over cannot occur at the two-strand stage.

segments are lost or gained, and four complete chromosomes emerge in a tetrad. At present, however, the actual mechanisms of crossing-over and interference are still somewhat of an enigma, although several attractive models do exist and a lot is known about the sorts of chemical reactions that might ensure precision at the molecular level. We shall return to these points in Chapter 18.

Linkage Mapping by Recombination in Humans

Humans—due partly to their relatively large chromosome number and partly to the lack of suitable pedigrees containing what amount to testcrosses—have revealed very few examples of autosomal linkage through recombination analysis. The X chromosome, however, has been far more amenable to analysis due to hemizygosity in males. Consider the following situation involving the rare recessive genes for defective sugar processing (*g*) and for color blindness (*c*). A doubly affected male (*cg*/Y) marries a normal woman (who is almost certainly *CG/CG*). The daughters of this mating are coupling-conformation heterozygotes. When they marry, they will almost certainly marry normal men (*CG*/Y), and their male children will provide an opportunity for researchers to study the frequency of recombinants (Figure 5-21). Using such pedigrees, the total frequency of recombinants may be estimated. A map based on such techniques is shown in Figure 5-22. Further data on mapping in humans have been provided by less conventional techniques, as we shall see in Chapter 6.

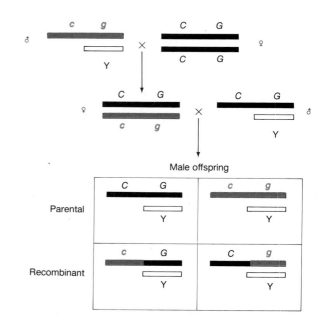

Male offspring

Parental

Recombinant

Figure 5-21. The male children of women heterozygous for two X-linked genes can be used to calculate recombinant frequency. Thus, mapping of the X chromosome is possible by studying certain selected pedigrees.

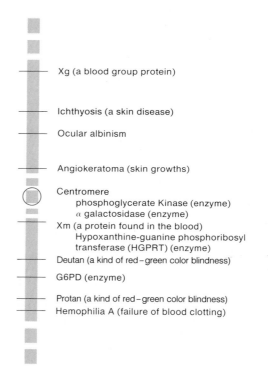

Xg (a blood group protein)

Ichthyosis (a skin disease)

Ocular albinism

Angiokeratoma (skin growths)

Centromere
    phosphoglycerate Kinase (enzyme)
    α galactosidase (enzyme)
Xm (a protein found in the blood)
    Hypoxanthine-guanine phosphoribosyl
    transferase (HGPRT) (enzyme)
Deutan (a kind of red–green color blindness)

G6PD (enzyme)

Protan (a kind of red–green color blindness)
Hemophilia A (failure of blood clotting)

Figure 5-22. A map of the human X chromosome, derived from analysis of recombinant frequencies, as indicated in Figure 5-21. (From W. F. Bodmer and L. L. Cavalli-Sforza, *Genetics, Evolution, and Man.* Copyright © 1976 by W. H. Freeman and Company.)

Linkage studies occupy a large proportion of the routine day-to-day activities of geneticists. When a new variant gene is discovered, one of the first questions to be asked concerning it is "Where does it map?" Not only is this knowledge a necessary component in the engineering and maintenance of genetic stocks for research use, but it is also of fundamental importance in piecing together an overall view of the architecture of the chromosome—and, in fact, of the entire genome. The operational key for the detection of linkage is the same in studies from viruses to humans, and that key is the nonindependence of genes in their transmission from generation to generation. Look for this key in discussion of linkage in following chapters.

Summary

■ In dihybrid crosses of sweet pea plants, William Bateson and R. C. Punnett discovered deviations from the $9:3:3:1$ ratio of phenotypes expected in the F_2 generation. The parental gametic types occurred with much greater frequency than the other two classes. Later, in his studies of two different autosomal gene pairs in *Drosophila*, Thomas Hunt Morgan found a similar deviation from Mendel's law of independent assortment. Morgan postulated that the two gene pairs were located on the same pair of homologous chromosomes. This phenomenon is called linkage.

Linkage explains why the parental gene combinations stay together but not how these nonparental combinations arise. Morgan postulated that during meiosis there may be a physical exchange of chromosome parts by a process now called crossing-over, or intrachromosomal recombination. Thus, there are two types of meiotic recombination. Interchromosomal recombination is achieved by Mendelian independent assortment and results in a recombinant frequency of 50 percent. Intrachromosomal recombination occurs when the physical linkage of the parental gene combinations prevents full assortment of the gene pairs and results in a recombinant frequency of less than 50 percent.

As Morgan studied more linked genes, he discovered considerable variation in recombinant frequency and wondered if this variation reflected the actual distance between genes on a chromosome. Alfred Sturtevant, a student of Morgan's, developed a method of determining the distance between gene pairs on a genetic map, based on the percentage of recombinants. A genetic map is another example of a hypothetical entity based on genetic analysis.

Although the basic test for linkage is deviation from the $1:1:1:1$ ratio of progeny types in a testcross, such a deviation may not be all that obvious. The χ^2 test, which tells how often deviations from expectations will occur purely by chance, can help us to determine whether linkage exists or not. The χ^2 test has other applications in genetics in the testing of observed against expected events.

Several theories about how recombinant chromosomes are generated have been set forth. We now know that crossing-over is the result of a physical breakage and a reunion of chromosome parts and that it occurs at the four-strand stage of meiosis. The actual mechanism of crossing-over, however, remains largely a mystery.

■ ■ ■

Solved Problems

1. In *Drosophila*, the allele *b* gives black body (normally brownish); at a separate gene, the allele *wx* gives waxy wings (normally nonwaxy); and at a third gene, the allele *cn* gives cinnabar eyes (normally red). A female individual heterozygous for these three genes is testcrossed, and 1000 progeny are classified as follows: 5 normal; 6 black, waxy, cinnabar; 69 waxy, cinnabar; 67 black; 382 cinnabar; 379 black, waxy; 48 waxy; and 44 black, cinnabar.

   a. Explain these numbers.

   b. Draw the alleles in their proper positions on the chromosomes of the triple heterozygote.

   c. If it is appropriate according to your explanation, calculate interference.

   d. If two triple heterozygotes of the above type were crossed, what proportion of progeny would be black, waxy? (Remember that there is no crossing-over in *Drosophila* males.)

Solution

a. One of the general pieces of advice given earlier is to be methodical. Here it is a good idea to write out the phenotypes and their inferred genotypes. The cross is a testcross of type

$$+/b \quad +/wx \quad +/cn \times b/b \quad wx/wx \quad cn/cn$$

Notice that there are distinct pairs of progeny classes in terms of frequency. Already, we can guess that the two very common classes of around 380 represent parental chromosomes, the two classes of around 68 represent single crossovers in one region, the two classes of around 45 represent single crossovers in the other region, and the two classes around 5 represent double-crossover derivatives. We can write out the progeny as gametic classes stemming from the female, grouped as follows:

$$
\begin{array}{rccc}
382 & + & + & cn \\
379 & b & wx & + \\
69 & + & wx & cn \\
67 & b & + & + \\
48 & + & wx & + \\
44 & b & + & cn \\
6 & b & wx & cn \\
\underline{5} & + & + & + \\
1000
\end{array}
$$

Writing the classes out this way confirms that the pairs of classes are in fact reciprocal genotypes arising from zero, one, or two crossovers.

At first, because we do not know the parents of the triple heterozygous female, it looks as if we cannot apply the definition of recombination in which gametic genotypes are compared with the two input genotypes that form an individual. But on reflection, the only parental types that make sense in terms of the data presented are $+/+$ $+/+$ cn/cn and b/b wx/wx $+/+$, since these are still the most common classes.

Now we can calculate the recombinant frequencies. For

$$b - wx, \text{ the RF} = \frac{69 + 67 + 48 + 44}{1000} = 22.8\%$$

$$b - cn, \text{ the RF} = \frac{48 + 44 + 6 + 5}{1000} = 10.3\%$$

$$wx - cn, \text{ the RF} = \frac{69 + 67 + 6 + 5}{1000} = 14.7\%$$

The map is therefore

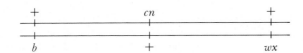

b. The parental chromosomes can now be seen to have been present in the triple heterozygote as

$$
\begin{array}{ccc}
+ & cn & + \\
\hline
b & + & wx
\end{array}
$$

c. The expected number of double recombinants is $0.103 \times 0.147 \times 1000 = 15.141$. The observed number is $6 + 5 = 11$, so interference can be calculated as $I = 1 - 11/15.141 = 1 - 0.726 = 0.274 = 27.4$ percent.

d. Here we are asked to use the newly gained knowledge of the linkage of these genes to predict the outcome of a cross. We are asked for the expected frequency of black, waxy individuals in a self (not a testcross) of the triple heterozygote. Because there is no crossing-over in the male, one-half of his gametes will be $b + wx$ and the other half will be $+ cn +$, but the latter are incapable of contributing to the black, waxy phenotype. In the female, two gametic types will be capable of contributing to black, waxy offspring: $b + wx$ and $b\,cn\,wx$. The contributions of the parents to the black, waxy phenotype are shown here:

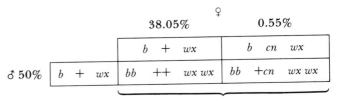

The type $b + wx$ is a nonrecombinant parental chromosome. There is a total of $382 + 379/1000 = 76.1$ percent of these chromosomes, of which 38.05 percent will be $b + wx$. The frequency of $b\,cn\,wx$ will be $11 \div 2/1000 = 0.55$ percent. In total, then, the female will provide $38.05 + 0.55 = 38.6$ percent of gametes capable of forming a black, waxy offspring; one-half of these will fuse with a $b + wx$ gamete from the male, so the frequency of black, waxy offspring will be $38.6/2 = 19.3$ percent.

In summary, the appropriate gametic fusion can be shown this way:

	♀ 38.05%	♀ 0.55%
	b $+$ wx	b cn wx
♂ 50% b $+$ wx	bb $++$ $wx\,wx$	bb $+cn$ $wx\,wx$

Black, waxy offspring

2. The human pedigree at the top of page 115 shows the inheritance of the rare nail-patella syndrome (misshapen nails and kneecaps). Also the ABO blood-group genotype of each individual is shown. Both the loci concerned are autosomal.

a. Indicate whether nail-patella syndrome is dominant or recessive. Give reasons to support your answer.

b. Is there evidence of linkage between the nail-patella gene and the gene for ABO blood type, as judged from this pedigree? Why or why not?

c. If there is evidence of linkage, then draw the alleles on the relevant homologs of the grandparents. If there is no evidence of linkage, draw the configuration of two homologous pairs.

d. According to your model, which descendents represent recombinants?

e. What is the best estimate of RF?

f. If man III-1 is mated to a normal female of blood type O, what is the probability that their first child will be blood type B with nail-patella syndrome?

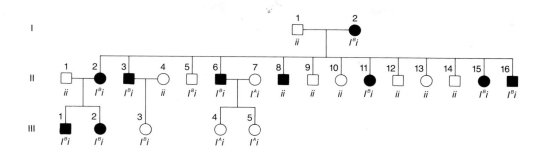

Solution

a. Nail-patella syndrome is most likely dominant. We are told that it is a rare abnormality, so it is most unlikely that the unaffected people marrying into the family carry a presumptive recessive allele for nail-patella syndrome. Let N be the causative allele. Then all people with the syndrome are heterozygotes Nn because all (probably including the grandmother, too) result from a cross to an nn normal individual. Notice that the syndrome occurs in all three successive generations—another indication of dominant inheritance.

b. There is evidence of linkage. Notice that most of the dark symbols—those that carry the N allele—also carry the I^B allele; most likely, these alleles are linked on the same chromosome.

c. The grandparents must be

$$\frac{n \qquad i}{n \qquad i} \times \frac{N \qquad I^B}{n \qquad i}$$

(The grandmother must carry both recessive alleles in order to produce offspring of genotype ii and nn.)

d. Notice that the grandparental mating is equivalent to a testcross, so the recombinants in generation II are

$$\#5:\ n\,I^B/n\,i \quad \text{and} \quad \#8:\ N\,i/n\,i$$

whereas all others are nonrecombinants, being either $N\,I^B/n\,i$ or $n\,i/n\,i$.

e. Notice that the grandparental cross and the first two crosses in generation II are identical and, of course, are all testcrosses. There is a total of 16 progeny; of these, three are recombinant (II-5, II-8, and III-3). This gives a recombinant frequency of $\text{RF} = \frac{3}{16} = 18.8$ percent. (We cannot include the cross of II-6 × II-7, because the progeny cannot be designated as recombinant or not.)

f. (III-1 ♂) $\dfrac{N \qquad I^B}{n \qquad i} \times \dfrac{n \qquad i}{n \qquad i}$ (normal type O ♀)

$$\downarrow$$

Gametes

$81.2\% \begin{cases} N\,I^B & 40.6\% \longleftarrow \text{Nail-patella, blood} \\ n\,i & 40.6\% \qquad\quad \text{type B} \end{cases}$

$18.8\% \begin{cases} N\,i & 9.4\% \\ n\,I^B & 9.4\% \end{cases}$

The two parental classes are always equal, and so are the two recombinant classes. Hence, the probability that the first child will have nail-patella syndrome and be blood type B is 40.6 percent.

Problems

1. A plant of genotype

$$\frac{A \qquad B}{a \qquad b}$$

is testcrossed to

$$\frac{a \qquad b}{a \qquad a}$$

If the two loci are 10 m.u. apart, what proportion of progeny will be $Aa\ Bb$?

2. The two loci A/a and D/d are so tightly linked that no recombination is ever observed. If $AA\ dd$ is crossed to $aa\ DD$, what phenotypes will be seen in the F_2 and in what proportions?

3. The loci R/r and S/s are linked 35 m.u. apart. If a plant of genotype

$$\frac{R \qquad S}{r \qquad s}$$

is selfed, what progeny phenotypes will be seen and in what proportions?

4. The cross $EE\ FF \times ee\ ff$ is made, and the F_1 is then backcrossed to the recessive parent. The following results are observed:

Phenotype	Proportion
$E\ F$	$\frac{2}{6}$
$E\ f$	$\frac{1}{6}$
$e\ F$	$\frac{1}{6}$
$e\ f$	$\frac{2}{6}$

Explain these results.

5. A geneticist studying bloops (an exotic organism found only in textbook problems) has been unable to find any examples of linkage. Suggest some explanations for this situation.

6. A strain of *Neurospora* with the genotype *H I* is crossed with a strain with the genotype *h i*. One-half of the progeny are *H I*, and one-half are *h i*. Explain how this is possible.

7. A female animal with genotype *Aa Bb* is crossed with a double-recessive male *(aa bb)*. Their progeny includes 442 *Aa Bb*, 458 *aa bb*, 46 *Aa bb*, and 54 *aa Bb*. Explain these results.

8. In a haploid organism, the loci *C/c* and *D/d* are linked 8 m.u. apart. In a cross *Cd × cD*, name the proportion of each of the following progeny classes: a. *C D* b. *c d* c. *C d* d. recombinants

9. A fruit fly of genotype *B R/b r* is testcrossed to *b r/b r*. In 84 percent of the meioses, no chiasmata occur between the linked gene pairs; in 16 percent of the meioses, one chiasma occurs between the pairs. What proportion of the progeny will be *Bb rr*? a. 50 percent b. 4 percent c. 84 percent d. 25 percent e. 16 percent

10. An individual heterozygous for four gene pairs *Aa Bb Cc Dd* is testcrossed to *aa bb cc dd*, and 1000 progeny are classified as follows:

a B C D	42
A b c d	43
A B C d	140
a b c D	145
a B c D	6
A b C d	9
A B c d	305
a b C D	310

a. Which gene pairs are linked?

b. If two pure breeding lines were crossed to produce the heterozygous individual, what were their genotypes?

c. Draw a linkage map of the linked genes, showing their order and their distance apart in map units.

d. Calculate an interference value, if appropriate.

11. In the squirting cucumber, *Echballium elaterium*, there are two separate sexes (it is dioecious), determined not by heteromorphic sex chromosomes but by specific genes. It is known that the genes involved are *M* (male fertility), *m* (male sterility), *F* (female sterility), and *f* (female fertility). In populations of this plant, individuals can be male (approximately 50 percent) or female (approximately 50 percent). In addition, a hermaphrodite type is found, but only at a very low frequency. The hermaphrodite has male and female sex organs on the same plant.

a. What must be the full genotype of a male plant? (Indicate linkage relations of the genes.)

b. What must be the full genotype of a female plant? (Indicate linkage relations of the genes.)

c. How does the population maintain an approximately equal proportion of males and females?

d. What is the origin of the rare hermaphrodite?

e. Why are hermaphrodites rare?

12. There is an autosomal gene *N* in humans that causes abnormalities in nails and patellae (kneecaps), called the nail-patella syndrome. In marriages of people with the phenotypes (nail-patella syndrome, blood type A) and (normal nail-patella, blood type O), some children are born with the nail-patella syndrome and blood type A. When marriages between such children (unrelated, of course) take place, their children are of the following types:

66% nail-patella syndrome, blood type A

16% normal nail-patella, blood type O

9% normal nail-patella, blood type A

9% nail-patella syndrome, blood type O

Fully analyze these data.

*13. You obtain two lines of *Drosophila:* one having light yellow eyes, and the other having bright scarlet eyes. (Remember that wild-type *Drosophila* have deep red eyes.) When you cross a yellow female with a scarlet male, you obtain 251 wild-type females and 248 yellow males in the F₁ generation. When you cross F₁ males and females, you obtain the following F₂ phenotypes: 260 wild-type females, 253 yellow females, 77 wild-type males, 179 yellow males, 183 scarlet males, and 80 brown-eyed males (brown is a new phenotype). Explain these results, using diagrams where possible.

14. In the ovaries of higher plants, a haploid nucleus resulting from meiosis undergoes several mitotic divisions without cell division, forming a multinucleate cell called the female gametophyte. (The male nucleus fuses with just one of these nuclei, which acts as the egg nucleus.) In some conifers, the female gametophyte can be quite large — in fact, large enough to be removed and analyzed electrophoretically (Figure 4-3). One tree of lodgepole pine *(Pinus contorta)* is heterozygous for fast and slow "electrophoretic alleles" of the enzymes alcohol dehydrogenase (*ADH^F/ADH^S*) and phosphoglucomutase (*PGM^F/PGM^S*). A sample of 237 female gametophytes is studied, and the following phenotypes are found: 21 *ADH^F PGM^F*, 19 *ADH^S PGM^S*, 95 *ADH^F PGM^S*, and 102 *ADH^S PGM^F*.

a. Explain these results.

b. Can you think of any other uses in genetics for this female-gametophyte system?

15. Using the data obtained by Bateson and Punnett (Table 5-1), calculate the map distance (in m.u.), separating the color and shape genes. (This will require some trial and error.)

16. You have a homozygous *Drosophila* line carrying the autosomal recessive genes *a*, *b*, and *c*, linked in that order. You cross females of this line with males of a homozygous wild-type line. You then cross the F_1 heterozygous males with their heterozygous sisters, and you obtain the following F_2 phenotypes: 1364 +++, 365 $a\,b\,c$, 87 $a\,b+$, 84 ++c, 47 a++, 44 +$b\,c$, 5 a+c, and 4 +b+.

a. What is the recombinant frequency between a and b? Between b and c?

b. What is the coefficient of coincidence?

17. R. A. Emerson crossed two different pure-breeding parental lines of corn and obtained an F_1 that was heterozygous for three recessive genes: *an* (anther), *br* (brachytic), and *f* (fine). He testcrossed the F_1 to a completely homozygous recessive tester and obtained the following progeny phenotypes: 355 anther, 339 brachytic and fine, 88 completely wild-type, 55 anther and brachytic and fine, 21 fine, 17 anther and brachytic, 2 brachytic, and 2 anther and fine.

a. What were the genotypes of the parental lines?

b. Draw a linkage map to illustrate the linkage arrangement of the three genes (include map distances).

c. Calculate the interference value.

18. In corn, the following allelic pairs have been identified in chromosome 3: +/b (plant-color booster versus nonbooster), +/lg (liguled versus liguleless), and +/v (green plant versus virescent). A testcross involving triple recessives and F_1 plants heterozygous for the three gene pairs yields the following progeny phenotypes: 305 +$v\,lg$, 275 b++, 128 b+lg, 112 +v+, 74 ++lg, 66 $b\,v$+, 22 +++, and 18 $b\,v\,lg$. Give the gene sequence on the chromosome, the map distances between genes, and the coefficient of coincidence.

19. A hemophiliac female rabbit with rickets was mated to a male rabbit that lacked a tail. The F_1 females were all wild-type, and the F_1 males had both rickets and hemophilia. The F_1 individuals were intercrossed and produced the following F_2 progeny:

Phenotype	Male	Female
normal	48	485
tail-less	437	0
rickets	4	16
hemophilia	12	14
hemophilia and rickets	439	485
tail-less and hemophilia	2	0
rickets and tail-less	12	0
hemophilia, tail-less, and rickets	46	0

a. Describe the complete linkage arrangements of tail-lessness, rickets, and hemophilia in rabbits.

b. Does interference exist in this system? If so, calculate it.

20. Groodies are useful (but fictional) haploid organisms that are pure genetic tools. A wild-type groody has a fat body, a long tail, and flagella. Non-wild-type lines are known that have thin bodies, or are tail-less, or do not have flagella. Groodies can mate with each other (although they are so shy that we do not know how) and produce recombinants. A wild-type groody is crossed with a thin-bodied groody lacking both tail and flagella. The 1000 baby groodies produced are classified as shown in the following figure. Assign genotypes, and map the three genes.

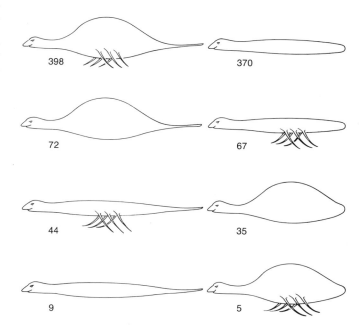

(Problem 16 courtesy of Burton S. Guttman.)

21. Assume that three pairs of alleles are found in *Drosophila:* +/x, +/y, and +/z. As shown by the symbols, each non-wild allele is recessive to its wild-type allele. A cross between females heterozygous at these three loci and wild-type males yields the following progeny phenotypes: 1010 +++ females; 430 $x + z$ males; 441 +y+ males; 39 $x\,y\,z$ males; 32 ++z males; 30 +++ males; 27 $x\,y$+ males; 1 +$y\,z$ male; and 0 x++ males.

a. How were members of the allelic pairs distributed in the members of the appropriate chromosome pair of the heterozygous female parents?

b. What is the sequence of these linked genes in their chromosome?

c. Calculate the map distances between the genes and the coefficient of coincidence.

d. In what chromosome of *Drosophila* are these genes carried?

*22. You have two homozygous lines of *Drosophila*. Line 1 has bright scarlet eyes and a wild-type thorax. Line 2 has dark brown eyes and a humpy thorax. When you cross virgin females of line 2 with males of line 1, you obtain 232 wild-type males and 225 wild-type females in the F_1 generation. You then cross F_1 males with virgin F_1 females and obtain the following F_2 phenotypes: 283 completely wild-type females; 145 completely wild-type males; 139 wild-thorax, scarlet-eyed males, 78 humpy-thorax, brown-eyed females; 40 humpy-thorax, white-eyed males; 39 humpy-thorax, brown-eyed males; 20 wild-thorax, brown-eyed females; 19 humpy-thorax, wild-eyed females; 11 humpy-thorax, scarlet-eyed males; 10 wild-thorax, white-eyed males; 9 wild-thorax, brown-eyed males; and 8 humpy-thorax, wild-eyed males. Explain these results as fully as possible (using symbols wherever you can).

23. The mother of a family with ten children has blood type Rh$^+$. She also has a very rare condition (elliptocytosis, E) that causes red blood cells to be oval rather than round in shape but that produces no adverse clinical effects. The father is Rh$^-$ (lacks the Rh$^+$ antigen) and has normal red cells (e). The children are 1 Rh$^+$ e, 4 Rh$^+$ E, and 5 Rh$^-$ e. Information is available on the mother's parents, who are Rh$^+$ E and Rh$^-$ e. One of the ten children (who is Rh$^+$ E) marries someone who is Rh$^+$ e, and they have an Rh$^+$ E child.

    a. Draw the pedigree of this whole family.

    b. Is the pedigree in agreement with the hypothesis that Rh^+ is dominant and Rh^- is recessive?

    c. What is the mechanism of transmission of elliptocytosis?

    d. Could the genes for E and Rh be on the same chromosome? If so, estimate the map distance between them, and comment on your result.

24. The father of Mr. Spock, first officer of the starship *Enterprise*, came from the planet Vulcan; his mother came from Earth. A Vulcan has pointed ears *(P)*, adrenals absent *(A)*, and a right-sided heart *(R)*. All of these alleles are dominant over normal Earth alleles. These genes are autosomal, and they are linked as shown in this linkage map:

    P/p   A/a   R/r
    |------ 15 m.u. ------|------ 20 m.u. ------|

    If Mr. Spock marries an Earth woman and there is no (genetic) interference, what proportion of their children:

    a. Will show Vulcanian appearance for all three characters?

    b. Will show Earth appearance for all three characters?

    c. Will have Vulcanian ears and heart but Earth adrenals?

    d. Will have Vulcanian ears but Earth heart and adrenals?

    (Problem 24 is from D. Harrison, *Problems in Genetics*, Addison-Wesley, 1970.)

25. In a certain diploid plant, the three loci A/a, B/b, and C/c are linked as follows:

    A/a   B/b   C/c
    |------ 20 m.u. ------|------ 30 m.u. ------|

One plant is available to you (call it the parental plant). It has the constitution $A b c / a B C$.

    a. Assuming no interference, if the plant is selfed, what proportion of the progeny will be of the genotype $a b c / a b c$?

    b. Again assuming no interference, if the parental plant is crossed with the $a b c / a b c$ plant, what genotypic classes will be found in the progeny? What will be their frequencies if there are 1000 progeny?

    c. Repeat part b, this time assuming 20 percent interference between the regions.

26. From several crosses of the general type $AA\ BB \times aa\ bb$, the F_1 individuals of type $Aa\ Bb$ were testcrossed to $aa\ bb$. The results are shown in Table 5-4. In each case, use the χ^2 test to decide if there is evidence of linkage.

TABLE 5-4.

Testcross of F_1 from cross	Number of individuals of genotype			
	$Aa\ Bb$	$aa\ bb$	$Aa\ bb$	$aa\ Bb$
1	310	315	287	288
2	36	38	23	23
3	360	380	230	230
4	74	72	50	44

27. Certain varieties of flax show different resistances to specific races of the fungus called flax rust. For example, the flax variety 77OB is resistant to rust race 24 but susceptible to rust race 22, whereas flax variety Bombay is resistant to rust race 22 and susceptible to rust race 24. When 77OB and Bombay were crossed, the F_1 hybrid was resistant to both rust races. When selfed, it produced an F_2 containing the phenotypic proportions shown in Table 5-5.

TABLE 5-5.

		Rust race 22	
		Resistant	Susceptible
Rust race 24	Resistant	184	63
	Susceptible	58	15

    a. Propose a hypothesis to account for the genetic basis of resistance in flax to these particular rust races. Make a concise statement of the hypothesis, and define any gene symbols you use. Show your proposed genotypes of the 77OB, Bombay, F_1, and F_2 flax plants.

    b. Test your hypothesis, using the χ^2 test. Give the expected values, the value of χ^2 (to two decimal places), and the appropriate probability value. Explain exactly what this value is the probability of. Do you accept or reject your hypothesis on the basis of the χ^2 test?

(Problem 27 is adapted from M. Strickberger, *Genetics,* Macmillan, 1968.)

28. In humans, the genes for colorblindness and hemophilia are both on the X chromosome, with a recombinant frequency of about 10 percent. Linkage of a pathological gene to a relatively harmless one can be used for genetic prognosis. Some of the people from a more extensive pedigree are shown here:

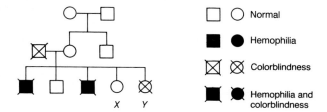

☐ ○	Normal	
■ ●	Hemophilia	
⊠ ⊗	Colorblindness	
▨ ◉	Hemophilia and colorblindness	

What advice could be given to the women *X* and *Y* as to the likelihood of their having sons with hemophilia?

(Problem 28 is adapted from J. F. Crow, *Genetics Notes: An Introduction to Genetics,* Burgess Publishing Co., Minneapolis, 1983.)

29. A geneticist mapping the genes *A*, *B*, *C*, *D*, and *E* makes two three-point testcrosses involving various combinations of these genes:

Series 1		Series 2	
AABBCCDDEE × aabbCCddEE		AABBCCDDEE × aaBBccDDee	
F$_1$ test-crossed to recessive tester		F$_1$ test-crossed to recessive tester	
Progeny		Progeny	
ABCDE	316	ABCDE	243
abCdE	314	aBcDe	237
ABCdE	31	ABcDe	62
abCDE	39	aBCDE	58
AbCdE	130	ABCDe	155
aBCDE	140	aBcDE	165
AbCDE	17	aBCDe	46
aBCdE	13	ABcDE	34
	1000		1000

The geneticist also knew that genes *D* and *E* showed independent assortment.

a. Draw a map of these genes, showing distances in centimorgans wherever possible.

b. Is there any evidence of chromosome interference?

Linkage II: Special Eukaryotic Chromosome Mapping Techniques

KEY CONCEPTS

Double and higher multiple crossovers cause underestimates
in map distances when calculated as recombinant frequencies.

.

This effect of double crossovers can be corrected by using a
mapping formula in random meiotic product analysis or by
directly measuring their frequency in the analysis of meiotic tetrads.

.

Linear meiotic tetrad analysis can be used to introduce
centromeres onto chromosome maps.

.

Tetrad analysis is ideal for studying many aspects of the
segregation, recombination, and assortment of genes at meiosis.

.

Special genetic systems show that crossing-over and gene assortment
can sometimes occur in diploid cells that are undergoing mitosis.

.

Cell lines can be established that are fusion hybrids of human and
rodent cells. Coincident loss or retention of human genetic markers
and chromosomes from these hybrids has provided a
powerful method for localizing human genes to specific chromosomes.

■ This chapter extends the basic treatment of linkage developed in Chapter 5. Three major topics are covered: mapping functions (a more accurate treatment of linkage mapping by recombinant frequency analysis), tetrad analysis, and mitotic genetics. The techniques introduced here are important in many areas of modern genetics, as we shall see in later chapters.

Why should a *second* chapter be devoted to the study of eukaryotic linkage mapping? The simple answer is that the linkage map is such a fundamental tool in the study of heredity that the modern geneticist would be lost without it — much as a traveler might be lost without a road map. Consider the following. One of the most basic concerns of genetics is the structure of the genome. This concern could not be addressed without a knowledge of where specific genes are located. Within this area, there is a wealth of study — not the least part of which is the interaction of genes with other genes in the same vicinity. There are, for example, control genes that profoundly affect genes adjacent to them on the chromosome. Then there is the issue of gene families and clusters of related and/or controlled genes. Furthermore, several molecular techniques of gene cloning require knowledge of map position. All of these areas of study require precise mapping data. A final point: the fantastically complex strains that are an integral part of research in modern biology and biotechnology can be synthesized only through an intricate kind of chromosome mechanics that is almost totally dependent on a detailed knowledge of gene linkage. Mapping analysis is an essential ingredient of modern genetics. For example, mapping the human genome is one of the most active areas of research in medical genetics.

Mapping Functions

In Chapter 5, we defined the genetic map unit (m.u.) as a recombinant frequency (RF) of 1 percent. We have seen that this definition leads to reasonable estimates of map distances. However, when *large* locus-to-locus intervals are being examined, the estimate becomes very imprecise, as in the following example:

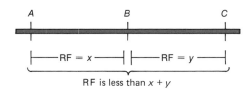

We have seen that we can improve the situation by including the double-recombinant types (twice). However, the point is that map distance (or the "true" distance) between loci is not linearly related to recombinant frequency throughout the range of values possible in mapping experiments. We have the clue provided by double crossovers, but now we must come to grips with the notion that double crossovers are part of a much larger problem of multiple crossovers and their effect on recombinant frequency. It might be pertinent to ask, for example, whether the RF value of x for the $A - B$ interval in the example is an accurate reflection of the true distance between the A and B loci. Perhaps some double crossovers occur in this interval alone; we cannot detect them because we have no markers between A and B, but we should account for them if RF is to reflect physical distance at all well. This section describes a mathematical treatment of the problem that allows for the correction of multiple exchanges without actually detecting them at the phenotypic level.

Any relationship between one variable entity and another is called a function; the relationship between real map distance and recombination frequency is called the **mapping function.**

> **Message**   The relationship between real map distance and recombinant frequency (RF) is not linear. The mapping-function formula provides a closer approximation to the real relationship.

The Poisson Distribution

To calculate the mapping function, we need a mathematical tool that is widely used in genetic analysis because it describes many genetic phenomena well. This tool is called the **Poisson distribution.** A distribution is merely a device that describes the frequencies of various classes of samples. The Poisson distribution describes the frequency of classes of samples containing 0, 1, 2, 3, 4, . . . , i events when the average number of events per sample is small in relation to the total number of times that the event could occur. For example, the *possible* number of tadpoles obtainable in a single dip of a net in a pond is quite large, but most dips yield only one or two or none. The number of dead birds on the side of the highway is potentially very large, but in a sample kilometer the number is usually small. Such samplings are described well by the Poisson distribution.

Let's consider a numerical example. Suppose we randomly distribute 100 one-dollar bills to 100 students in a lecture room, perhaps by scattering them over the class from some vantage point near the ceiling. The average (or mean) number of bills per student is 1.0, but common sense tells us that it is very unlikely that each of the 100 students will capture one bill. We would expect a few lucky students to grab three or four bills each and quite a few students to come up with two bills each. However, we would expect most students to get either one bill or none. The Poisson distribution provides a quantitative prediction of the results.

In this example, the event being considered is the capture of one bill by a student. We want to divide the students into classes, according to the number of events occurring (number of bills captured per student), and find the frequency of each class. Let m represent the mean number of events (here, $m = 1.0$ bill per student). Let i represent the number of events for a particular class (say, $i = 3$ for those students who get three bills each). Let $f(i)$ represent the frequency of the class with i events—that is, the proportion of the 100 students who each capture i bills. The general expression for the Poisson distribution states that

$$f(i) = \frac{e^{-m} m^i}{i!}$$

where e is the base of natural logarithms ($e \cong 2.7$) and $!$ is

the factorial symbol (as examples, $3! = 3 \times 2 \times 1 = 6$ and $4! = 4 \times 3 \times 2 \times 1 = 24$; by definition, $0! = 1$). When computing $f(0)$, recall that any number raised to the power of 0 is defined as 1. Table 6-1 gives values of e^{-m} for m values from 0.000 to 1.000.

In our example, $m = 1.0$. Using Table 6-1, we compute the frequencies of the classes of students who each capture 0, 1, 2, 3, and 4 bills as follows:

$$f(0) = \frac{e^{-1} 1^0}{0!} = \frac{e^{-1}}{1} = 0.368$$

$$f(1) = \frac{e^{-1} 1^1}{1!} = \frac{e^{-1}}{1} = 0.368$$

$$f(2) = \frac{e^{-1} 1^2}{2!} = \frac{e^{-1}}{2 \times 1} = \frac{e^{-1}}{2} = 0.184$$

$$f(3) = \frac{e^{-1} 1^3}{3!} = \frac{e^{-1}}{3 \times 2 \times 1} = \frac{e^{-1}}{6} = 0.061$$

$$f(4) = \frac{e^{-1} 1^4}{4!} = \frac{e^{-1}}{4 \times 3 \times 2 \times 1} = \frac{e^{-1}}{24} = 0.015$$

Figure 6-1 shows a histogram of this distribution. We

TABLE 6-1. Values of e^{-m} for m values of 0 to 1

m	e^{-m}	m	e^{-m}	m	e^{-m}	m	e^{-m}
0.000	1.00000	0.250	0.77880	0.500	0.60653	0.750	0.47237
0.010	0.99005	0.260	0.77105	0.510	0.60050	0.760	0.46767
0.020	0.98020	0.270	0.76338	0.520	0.59452	0.770	0.46301
0.030	0.97045	0.280	0.75578	0.530	0.58860	0.780	0.45841
0.040	0.96079	0.290	0.74826	0.540	0.58275	0.790	0.45384
0.050	0.95123	0.300	0.74082	0.550	0.57695	0.800	0.44933
0.060	0.94176	0.310	0.73345	0.560	0.57121	0.810	0.44486
0.070	0.93239	0.320	0.72615	0.570	0.56553	0.820	0.44043
0.080	0.92312	0.330	0.71892	0.580	0.55990	0.830	0.43605
0.090	0.91393	0.340	0.71177	0.590	0.55433	0.840	0.43171
0.100	0.90484	0.350	0.70469	0.600	0.54881	0.850	0.42741
0.110	0.89583	0.360	0.69768	0.610	0.54335	0.860	0.42316
0.120	0.88692	0.370	0.69073	0.620	0.53794	0.870	0.41895
0.130	0.87810	0.380	0.68386	0.630	0.53259	0.880	0.41478
0.140	0.86936	0.390	0.67706	0.640	0.52729	0.890	0.41066
0.150	0.86071	0.400	0.67032	0.650	0.52205	0.900	0.40657
0.160	0.85214	0.410	0.66365	0.660	0.51685	0.910	0.40252
0.170	0.84366	0.420	0.65705	0.670	0.51171	0.920	0.39852
0.180	0.83527	0.430	0.65051	0.680	0.50662	0.930	0.39455
0.190	0.82696	0.440	0.64404	0.690	0.50158	0.940	0.39063
0.200	0.81873	0.450	0.63763	0.700	0.49659	0.950	0.38674
0.210	0.81058	0.460	0.63128	0.710	0.49164	0.960	0.38289
0.220	0.80252	0.470	0.62500	0.720	0.48675	0.970	0.37908
0.230	0.79453	0.480	0.61878	0.730	0.48191	0.980	0.37531
0.240	0.78663	0.490	0.61263	0.740	0.47711	0.990	0.37158
						1.000	0.36788

SOURCE: F. James Rohlf and Robert R. Sokal, *Statistical Tables*, 2d ed. Copyright © 1981 by W. H. Freeman and Co.

Figure 6-1. Poisson distribution for a mean of 1.0, illustrated in terms of a random distribution of dollar bills to students.

predict that about 37 students will capture no bills, about 37 will capture one bill, about 18 will capture two bills, about 6 will capture three bills, and about 2 will capture four bills. This accounts for all 100 of students; in fact, you can verify that the Poisson distribution yields $f(5) = 0.003$, indicating the likelihood that no student in this sample of 100 will capture five bills.

Similar distributions may be developed for other m values. Some are shown in Figure 6-2 as curves instead of bar histograms.

Derivation of the Mapping Function

The occurrence of crossovers along a chromosome during meiosis also can be described by the Poisson distribution. In any given genetic region, the actual number of crossovers is probably small in relation to the total number of opportunities for such a crossover to occur in that stretch. If we knew the *mean* number of crossovers in the region per meiosis, we could calculate the distribution of meioses with zero, one, two, three, four, and more multiple crossovers. This is unnecessary in the present context

because, as we shall see, the only class we are really interested in is the zero class. The reason for this is that we want to correlate real distances with observable RF values. It turns out that meioses in which there are one, two, three, four, or *any* finite number of crossovers per meiosis all behave similarly in that they produce an RF of 50 percent *among the products of those meioses,* whereas the meiosis with no crossovers produce an RF of 0 percent. Consequently, the determining force in actual RF values is the ratio of class zero to the rest!

The truth of these statements can be illustrated by considering meioses in which zero, one, and two crossovers occur between nonsister chromatids. (Try the three-crossover class yourself.) Figure 6-3 tells us that

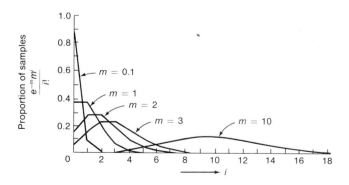

Figure 6-2. Poisson distributions for five different mean values: m is the mean number of events per sample, and i is the actual number of events per sample. (From R. R. Sokal and F. J. Rohlf, *Introduction to Biostatistics.* Copyright © 1973 by W. H. Freeman and Company.)

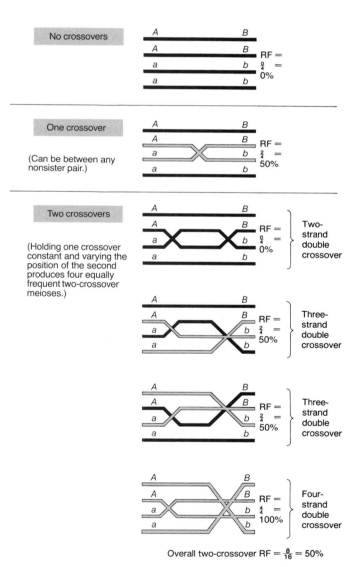

Figure 6-3. Demonstration that the overall RF is 50 percent for meioses in which any nonzero number of crossovers occurs. The color represents a recombinant chromatid.

the only way we can get a recombinant product of meiosis is from a meiosis with at least one crossover in the marked region, and then *always* precisely one-half of the products of such meioses will be recombinant. Note that in the figure we consider only crossovers that occur between nonsister chromatids; **sister-chromatid exchange** is thought to be very rare at meiosis. (If it does occur, it can be shown to have no net effect in most meiotic analyses; see Problem 27 for this chapter.)

At last, we can derive the mapping function. Recombinants will make up one-half of the products of those meioses in which at least one crossover occurs in the region of interest. The proportion of meioses with at least one crossover is 1 minus the fraction with zero crossovers. Hence, $RF = \frac{1}{2}(1 - e^{-m})$, because the zero-class frequency is $e^{-m}m^0/0! = e^{-m}$. When you think about it, m is the best measure we can have of *true* genetic distance — the *actual* average number of crossovers in a region.

If we know an RF value, we can calculate m by solving the equation. When we plot the function as a graph, as in Figure 6-4, several interesting points emerge:

1. No matter how far apart two loci are on a chromosome, we never observe an RF value of greater than 50 percent. Consequently, an RF value of 50 percent would leave us in doubt about whether two loci are linked or are on separate chromosomes. Stated another way, as m gets larger, e^{-m} gets smaller and RF approaches $\frac{1}{2}(1 - 0) = \frac{1}{2} \times 1 = 50$ percent. This may be surprising to you: RF values of 100 percent can never be observed, no matter how far apart the loci are!

2. The function is linear for a certain range corresponding to very small m values (genetic distances). Therefore, RF is a good measure of distance if the distance is small and no multiple exchanges are

likely. In this region, the map unit defined as 1 percent RF has real meaning. Therefore, let's use this region of the curve to define real map units. For small m, such as $m = 0.05$, $e^{-m} = 0.95$ and $RF = \frac{1}{2}(1 - 0.95) = \frac{1}{2}(0.05) = \frac{1}{2} \times m$; for $m = 0.10$, $e^{-m} = 0.90$ and $RF = \frac{1}{2}(1 - 0.90) = \frac{1}{2}(0.10) = \frac{1}{2} \times m$. We see that $RF = m/2$, and this relation defines the dashed line on the ground in Figure 6-4. It allows us to translate m values into real (low-end-of-the-curve-style) map units.

So an m value of 1 is the equivalent of 50 real map units, and we can express the horizontal axis in map units, as indicated on the horizontal scale in Figure 6-4. Here we see that two loci separated by 150 real map units show an RF of only 50 percent. For regions of the graph in which the line is not horizontal, we can use the function to convert the RF into map distance simply by drawing a horizontal line to the curve and dropping a perpendicular to the map-unit axis — a process equivalent to using the equation $RF = \frac{1}{2}(1 - e^{-m})$ to solve for m.

Let's consider a specific numerical example of the use of the mapping function. Suppose we get an RF of 27.5 percent. How many real map units does this represent?

$$0.275 = \tfrac{1}{2}(1 - e^{-m})$$
$$0.55 = 1 - e^{-m}$$

Therefore

$$e^{-m} = 1 - 0.55 = 0.45$$

From e^{-m} tables (or by solving the hard way using logarithms), we find that $m = 0.8$, which is 40 real map units. If we had been happy to accept 27.5 percent RF as meaning 27.5 map units, we would have been considerably underestimating the true distance between the loci.

A note to calculator owners; the mapping function may be rearranged as follows:

$$e^{-m} = 1 - 2\,RF$$
$$-m = \ln(1 - 2\,RF)$$
$$m = -\ln(1 - 2\,RF)$$

This form is more convenient for solution by calculator.

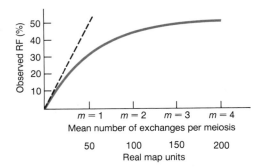

Figure 6-4. The mapping function in graphic form (solid line). The dashed line represents the linear relationship found for small values of m.

Message To estimate map distances most accurately, put RF values through the mapping function. Alternatively, genes may be mapped through the summation of small genetic intervals in which RF has a linear relation to map distance.

A corollary of this statement is that for organisms for which the chromosomes are already well mapped, such as *Drosophila*, a geneticist usually has little need of a map function to locate newly discovered genes. This is because the map is already divided into small, marked regions by the known loci. However, when the process of mapping has just begun in a new organism or when the available genetic markers are sparsely distributed (as in human maps), the corrections provided by the function are needed.

Tetrad Analysis

We have already hinted (in Chapter 5) at the existence of marvelous organisms in which the four products of a single meiosis are recoverable and testable. The group of four is called a **tetrad,** derived from the Greek word for the number four. Tetrad analysis has been possible only in those fungi and single-celled algae in which the products of each meiosis are held together in some way. These organisms are all haploid. There are many advantages to using haploids for genetic analysis; a few are listed here.

1. Because the organisms are haploid, there is no complication of dominance. The nuclear genotype is expressed directly in the phenotype.

2. There is only one meiosis to analyze at a time (refer to the discussion of life cycles in Chapter 3), whereas, in diploids, gametes from two different meiotic events fuse to form the zygote. In diploids, the testcross is an attempt to achieve the same end, but the procedure is technically more laborious and sometimes not possible. In haploids, for example, the products of the cross $++ \times a\ b$ might be

$$
\begin{array}{ll}
++ & 45\% \\
a\ b & 45\% \\
a+ & 5\% \\
+b & 5\%
\end{array}
$$

This result permits a direct calculation of RF as 10 percent, so 10 map units separate the *a* and *b* loci. Note how easy it is to compare product-of-meiosis genotypes with the parental genotypes.

3. Because the organisms are small, fast-growing, and inexpensive to culture, it is possible to produce very large numbers of progeny from a cross. Thus, a good statistical accuracy is possible; also, very rare events occurring at frequencies as low as 10^{-8} are detectable. Furthermore, useful selective techniques to obtain desirable genotypes are easy to apply.

4. In several well-studied species (such as yeasts and *Neurospora*), the structure and behavior of the chromosomes and the gene action are similar to those found in higher organisms. Thus, these simpler forms provide very useful, easily analyzed eukaryotic models.

5. Several haploid types are important in biotechnology. Yeasts and filamentous fungi have long been used industrially, and much selective breeding of these forms has occurred. Now these same forms have become important hosts for novel DNA types created in the test tube through genetic engineering.

6. Last, but not least, there is the possibility of tetrad analysis.

Tetrad analysis itself has proved useful for several reasons:

1. It provides an opportunity to test *directly* some of the assumptions of the chromosome theory of heredity. Thus far in this book, our analyses have been essentially *random-meiotic-product* analyses. In these studies, individuals are examined and inferences are made about the meioses that produced them. This was the basic approach used by Mendel and by most eukaryotic geneticists since. The following example illustrates this kind of inference. A testcross of *Aa* to *aa* produces a 1 : 1 ratio from which the equal segregation of *A* and *a* in a single meiosis is inferred. The use of tetrads provides a far more direct test of this notion because the direct products of a single meiosis are examined. Furthermore, each meiotic product in the tetrad reinforces inferences about the other three products; the data are interlocking, and they increase the confidence with which we can make judgments of meiotic behavior. In research, this is an extremely useful facility.

2. It makes possible the mapping of centromeres as genetic loci, which will be explained in the next section. This seemingly esoteric technique led to the ability to construct new synthetic chromosomes in yeast.

3. It permits examination of the distribution of crossovers between the four chromatids and hence investigation of the possibility of **chromatid interference.**

4. It permits several approaches to studying the mechanism of chromosome exchange (crossing-over). We have already used tetrad analysis in Chapter 5 to deduce the stage at which crossing-over occurs (four strands) and to rule out Belling's copy-choice hypothesis. But perhaps the most significant use is in the analysis of gene conversion (see Chapter 18).

5. It provides a unique approach to the study of abnormal chromosome sets (see Chapters 8 and 9).

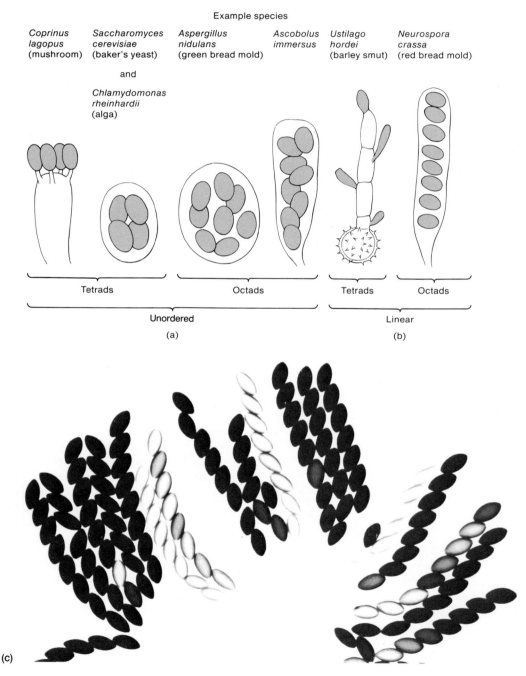

Figure 6-5. Various forms of tetrads and octads found in different organisms. (a) Unordered. (b) Linear. (c) Normally maturing asci of *Neurospora crassa* (Namboori B. Raju, *European Journal of Cell Biology*, 1980).

In summary, tetrad analysis is an ideal way to study meiosis. Meiosis is an extremely important process in eukaryotic biology. It is the reason for the existence of sex and the sexual cycle and is retained in very similar forms by all the major groups of eukaryotic organisms.

> **Message**   Tetrads are ideal structures for studying any aspect of the genetics of meiosis.

In some eukaryotic organisms in which tetrads can be isolated and analyzed, the four products of meiosis develop directly into four sexual spores (see Figure 6-5). In other such organisms, each of the four product-of-meiosis cells undergoes a mitotic division, yielding a group of eight cells called an **octad.** However, an octad is simply a double tetrad composed of four spore pairs. The members of a spore pair are identical, being mitotic daughter cells of one of the four products of meiosis.

The sexual spores, whether four or eight in number, can be found in a variety of arrangements. In some species, the spores are found in a jumbled arrangement called an **unordered tetrad,** shown in Figure 6-5(a). In other species, the spores are arranged in a striking linear arrangement called a **linear tetrad,** shown in Figure 6-5(b). Unordered and linear tetrads each have their

specific uses in genetic analysis. We shall deal with the linear types first.

Linear Tetrad Analysis

How are linear tetrads produced? The key fact is that the spindles of the first and second meiotic divisions and of the post-meiotic mitosis do not overlap. The reason is probably related to the fact that these divisions occur in a tubelike structure, which physically prevents the spindles from overlapping. In any case, the absence of spindle overlap means that the nuclei are laid out in a straight array, and the lineage of each of the eight nuclei of an octad can therefore be traced back through meiosis, as shown in Figure 6-6.

As you might expect and as we shall see, linear tetrads beautifully illustrate the segregation and independent assortment of genes at meiosis. However, linear tetrads are also ideally suited to a special kind of analysis that is not possible in most organisms: **centromere mapping,** or the mapping of centromeres to other gene loci. The centromere is a fascinating region of the chromosome that interacts with the spindle fibers and ensures proper chromosome movement during nuclear division. When this process fails, the daughter cells have abnormal chromosome numbers, which can lead to death or phenotypic abnormality. For example, abnormal chromosome numbers in humans have produced a

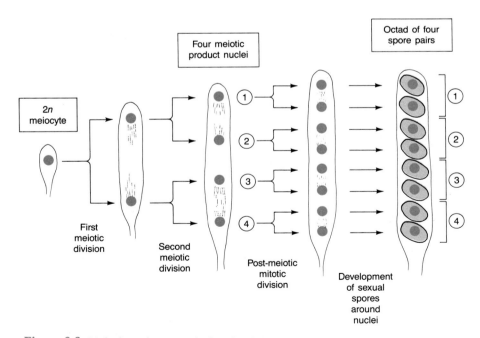

Figure 6-6. Meiosis and post-meiotic mitosis in a linear tetrad. The nuclear spindles do not overlap at any division, so that nuclei never pass each other in the sac. The resulting eight nuclei are laid down in a linear array that can be traced back through the divisions.

major class of genetic diseases. Therefore, the centromere has been the subject of considerable attention in genetics.

How does centromere mapping work? We shall use the fungus *Neurospora crassa* as an example. This fungus belongs to a group called Ascomycetes; in these fungi, a membranous sac encloses the sexual spores; together, the spores and sac are called an ascus (see Figure 6-5 for examples). *Neurospora* produces linear octads. In its simplest form, centromere mapping considers a gene locus and asks how far this locus is from its centromere. We could pick any *Neurospora* locus to illustrate this technique, but we will use the **mating-type locus** because it was one of the first to be used in such an analysis and because it provides some convenient data as well. This locus has two alleles, which are represented as *A* and *a* even though neither is dominant or recessive. These alleles determine the *A* and *a* mating types. Although the mating-type phenomenon is interesting in itself, here we are merely using the locus as a genetic marker to illustrate the analysis.

Centromere mapping is based on the fact that a meiosis in which a crossover occurs in the centromere-to-locus region produces a different allele pattern in the octad than a meiosis in which no crossover occurs in that region. Figures 6-7 and 6-8 show examples of these two patterns. The simpler pattern, shown in Figure 6-7, is produced when there is no crossover. This pattern is typified by one of the alleles occupying one end of the tetrad or octad and the other allele occupying the other end. You can see from the diagram how this pattern is produced: because there is no spindle overlap, the *A*-bearing nuclei and the *a*-bearing nuclei never pass each other in the ascus. Also notice that although the *A* and *a* alleles are together in the diploid meiocyte nucleus, the first meiotic division cleanly segregates the *A* and *a* alleles and these alleles remain separate throughout the second division of meiosis. This gives rise to the term **first-division segregation,** and the allele pattern in the spores is called a **first-division segregation pattern** or **M_I pattern.**

When a crossover occurs in the centromere-to-locus region (see Figure 6-8), *A* and *a* will still be together in the nuclei at the end of the first division of meiosis. Hence, first-division segregation has not occurred. However, the second meiotic division does move the *A* and *a* alleles into separate nuclei, giving rise to the term **second-division segregation.** The allele pattern in the tetrad or octad shows that this has occurred and is therefore called a **second-division segregation pattern** or **M_{II} pattern.** This pattern is typified by any arrangement in which a half tetrad or a half octad contains both alleles.

Now let's look at the following experimental results and interpret them in light of these ideas.

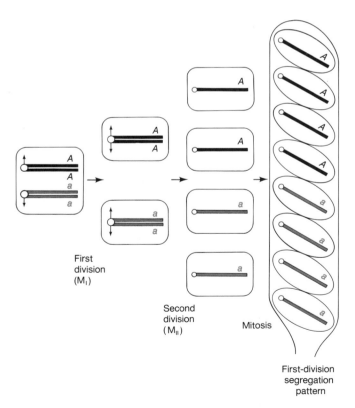

Figure 6-7. Segregation of *A* and *a* into separate nuclei at the first meiotic division. The resultant allele pattern in the octad is called a first-division segregation pattern.

Octads	*A*	*a*	*A*	*a*	*A*	*a*
	A	*a*	*A*	*a*	*A*	*a*
	A	*a*	*a*	*A*	*a*	*A*
	A	*a*	*a*	*A*	*a*	*A*
	a	*A*	*A*	*a*	*a*	*A*
	a	*A*	*A*	*a*	*a*	*A*
	a	*A*	*a*	*A*	*A*	*a*
	a	*A*	*a*	*A*	*A*	*a*
Number	126	132	9	11	10	12
			Total = 300			

The first two octad types show 4 : 4 blocks; these are, by definition, first-division (M_I) segregation patterns. The first octad is simply an upside-down version of the second. Notice that these two types are more or less equal in frequency (126 versus 132). This equality simply reflects random spindle attachment at the first meiotic division, which pulls either *A* "up" and *a* "down" or *a* "up" and *A* "down." We can deduce that 126 +

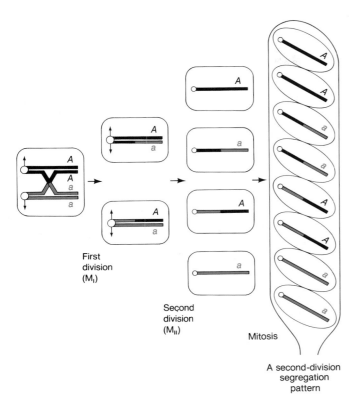

Figure 6-8. Segregation of *A* and *a* into separate nuclei at the second meiotic division due to a crossover. The allele pattern in the octad is called a second-division segregation pattern. (Other second-division patterns are possible.)

132 = 258 meioses out of 300, or 86 percent of meioses, had no crossover in the region between *A/a* and its centromere.

The remaining four ascus types all show half asci with both *A* and *a* present, showing that *A* and *a* segregated not at the first but at the second division of meiosis. What causes these four different variations on the same basic M$_{II}$ theme? Once again, it is the randomness of

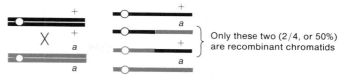

Figure 6-10. Only one-half of the chromatids from a meiosis with a single crossover are recombinant.

spindle attachment, not only at the first but also the second division of meiosis. This random spindle attachment is illustrated in Figure 6-9. The M$_{II}$ (second-division segregation) patterns total 9 + 11 + 10 + 12 = 42, or 14 percent, and show that a crossover occurred between *A/a* and its centromere in 14 percent of meioses.

Well, we have measured the M$_{II}$ pattern frequency at 14 percent. Does this mean that the *A/a* locus is 14 map units from its centromere? The answer is no, but this value can be used to calculate the distance in terms of map units. The 14 percent value is a percent of meioses, and this is not the way map units are defined. Map units are defined in terms of the percentage of recombinant chromosomes issuing from meiosis. Figure 6-10 shows that when a crossover occurs in the region spanned by the centromere and the locus used as a marker, then only one-half of the chromosomes issuing from that meiosis will be recombinant. So to deduce the size of the region in map units, it is necessary to divide the M$_{II}$ pattern frequency by 2. In our example, the distance of the *A/a* locus from its centromere is therefore 14 ÷ 2 = 7 map units.

> **Message**   To calculate the distance of a locus from its centromere in map units, measure the percentage of tetrads showing second-division segregation patterns for that locus and divide by 2.

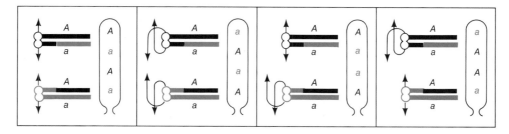

Figure 6-9. Four second-division segregation ascus patterns are equally frequent in linear asci. These patterns are produced by random spindle-to-centromere attachments at the second meiotic division.

The above analysis can be extended to any number of heterozygous gene pairs segregating in a cross. The case of two gene pairs can be discussed as an example. Let's assume that we do not know the linkage relations of two loci a and b with each other or with their respective centromere or centromeres. There are only limited possibilities, however:

1. The loci are on separate chromosomes.

2. The loci are on opposite sides of the centromere on the same chromosome.

3. The loci are on the same side of the centromere (obviously on the same chromosome).

The first two possibilities should both show independence of the M_{II} patterns for both loci. The third possibility is more interesting in that a crossover in the region between the centromere and the proximal locus should produce an M_{II} pattern for both loci in the same ascus (barring rare double crossovers). Hence, if a is proximal to b, for example, we would expect most M_{II} patterns for a also to show M_{II} patterns for b (Figure 6-11). This kind of interdependence of the M_{II} patterns can provide clues

about the linkage relations between unmapped loci and often is revealed in a straightforward centromere mapping study.

> **Message** Loci on the same chromosome arm can show interrelated second-division segregation patterns.

M_{II} frequencies will obviously be greater when a locus is further from the centromere of that chromosome. But the M_{II} frequency never reaches 100 percent; in fact, the theoretical maximum is 67 percent, or $\frac{2}{3}$. The reason for this is that multiple crossovers (especially doubles) become more and more prevalent as the interval becomes larger, and double crossovers can generate M_I patterns as well as M_{II} patterns. For example:

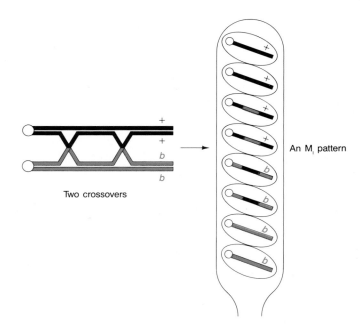

Two crossovers

An M_I pattern

One easy way to calculate the maximum M_{II} frequency is by the following "thought experiment." Consider a heterozygous locus $+/b$ that is so far from its centromere that many crossovers occur in the intervening region. These many crossovers effectively uncouple the locus from its centromere, and the two b alleles and the two $+$ alleles can end up in the tetrad in any arrangement (Figure 6-12). We can simulate this by considering how many different ways there are of dropping four marbles (two b and two $+$) into a test tube (see Figure 6-13). The first marble can be $+$ or b; it makes no differ-

Second-division (M_{II}) segregation pattern for $a/+$ and $b/+$

Figure 6-11. When two loci are in the same chromosome arm, a crossover between the centromere and the most proximal locus produces a second-division (M_{II}) pattern for both loci in the ascus.

Figure 6-12. When a locus is very far from its centromere, we can imagine a large number of crossovers occurring that would effectively unlink the locus from its centromere.

ence. Let's assume that it is +. We then have two b's and one + left. The next marble determines if the pattern will be M_I or M_{II}. One-third of the time, the second marble will be +; thus, the third and fourth marbles must be b, and an M_I pattern is generated. The other two-thirds of the time, the second marble will be b, generating an M_{II} pattern. Therefore, we can see that with this very large (say, infinite) number of crossovers, the $\frac{2}{3}$ frequency of M_{II} patterns can never be exceeded.

> **Message**   The maximum frequency possible for second-division segregation patterns is 66.7 percent ($\frac{2}{3}$).

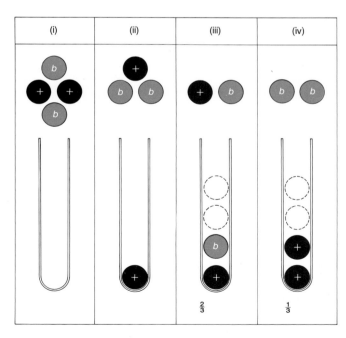

Figure 6-13. Demonstration of the limiting M_I and M_{II} segregation pattern frequencies in linear tetrad analysis. (See text for details.)

This raises an enigma. We feel intuitively that this M_{II} maximum should directly equate with the 50 percent RF maximum observed for very large map distances, but $66.7 \div 2 = 33.3$ percent, which would be 33.3 map units (m.u.). The villain, once again, is the existence of multiple crossovers. We could derive a map function for M_{II} patterns, but in practice, it is simpler to stick to the analysis of smaller intervals and to recognize that the larger M_{II} frequencies will provide increasingly inaccurate estimates of the size of the centromere-to-locus interval.

Unordered Tetrad Analysis

In unordered tetrads or octads, the spores are in no particular sequence, so that centromere mapping studies of the types just discussed are not possible. Even though Mendel's law of equal segregation is demonstrated in every tetrad by a 1 : 1 segregation for any heterozygous locus, the tetrads simply cannot be classified as M_I or M_{II} due to the lack of linearity of the spores in the ascus.

What are some of the uses of unordered tetrad analysis? We have already touched on one. In the course of the routine isolation and analysis of new phenotypes for some specific biological experiments, it is necessary to check on the possible single-gene determination of the phenotype, and unordered tetrad analysis is a fast and convincing way to do this.

In general, most aspects of meiotic genetics can be studied using unordered tetrads. In fact, even in organisms like *Neurospora*, where the normal ascus is linear, it is often faster and more convenient to allow the ascospores to follow the natural process of being shot out of the asci, to collect them in groups of eight on a block of agar, and then to treat them as unordered octads.

Unordered tetrads can be used in routine gene linkage analysis, but generally results are more quickly achieved through the analysis of random meiotic products. However, unordered tetrads provide an improved method of making more accurate measurements, especially of larger distances. Due to the ability to isolate and study the events in individual meioses, it becomes possible to approximate the Poisson distribution of crossovers in a region between two linked loci. Because of this, linkage values can be corrected for the occurrence of double crossovers in a direct way, sidestepping some of the assumptions of the mapping function. We will use this kind of analysis as an example of the use of unordered tetrads.

If we are studying recombination between two loci *a* and *b* in a cross such as *a b* × ++ in an organism like yeast, unordered tetrads can be isolated and the genotypes of the spores can be determined. Three basic ascus types will be revealed:

$$
\text{Spores} \begin{cases} \begin{array}{cc} a & b \\ a & b \\ + & + \\ + & + \end{array} & \begin{array}{cc} a & + \\ a & + \\ + & b \\ + & b \end{array} & \begin{array}{cc} a & b \\ a & + \\ + & b \\ + & + \end{array} \end{cases}
$$

Parental | Nonparental | Tetratype
ditype | ditype | (T)
(PD) | (NPD) |

Remember that these asci are unordered, so even though the first type might look like a case of M_I segregation for both loci, that is not the case: the spores could have been written equally well in any order. These asci merely have been classified according to whether they contain two genotypes (*ditypes*) or four genotypes (*tetratypes*, represented by T). Within the ditype class, both genotypes can be either parental (PD) or nonparental (NPD). What do these classes represent for linked loci such as *a* and *b*? If we assume that individual meioses can have no crossovers (NCO), a single crossover (SCO), or a double crossover (DCO) in the *a* to *b* region, then we can represent the classes of unordered asci that emerge from such meioses as shown in Figure 6-14. Of course, triples and higher numbers of crossovers might occur, but we will assume that these are rare and therefore negligible. Notice that Figure 6-14 is merely an extension of Figure 6-3.

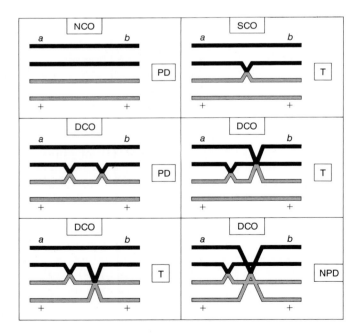

Figure 6-14. The various ways of producing tetratype (T), parental ditype (PD), and nonparental ditype (NPD) tetrads with linked genes. (NCO = noncrossover meioses; SCO = single crossover meioses; DCO = double crossover meioses.)

The key to the analysis is the NPD class, which for linked loci, can only occur from a double crossover involving all four chromatids. Since we are assuming that double crossovers occur randomly between the chromatids, the NPD class should be exactly $\frac{1}{4}$ of the DCO class, and we can estimate that

$$
\text{DCO} = 4\text{NPD}
$$

The single crossover class can also be measured by a similar relationship. Notice that tetratype (T) asci can result from either single crossover or double crossover meioses. But we can estimate the component of the T class that comes from DCO meioses to be 2NPD. Hence, the size of the SCO class can be stated as

$$
\text{SCO} = \text{T} - 2\text{NPD}
$$

Now that we have estimated the sizes of the SCO and DCO classes directly, the noncrossover class can be computed as

$$
\text{NCO} = 1 - (\text{SCO} + \text{DCO})
$$

Thus, we have achieved our estimate of the Poisson distribution of crossovers in this marked region. We can use this estimate to derive a direct estimation of *m*, the mean number of crossovers in this region. The value of *m* is calculated simply by taking the sum of the SCO class plus *twice* the DCO class (because this class contains two crossovers). Hence

$$
\begin{aligned} m &= (\text{T} - 2\text{NPD}) + 2(4\text{NPD}) \\ &= \text{T} + 6\text{NPD} \end{aligned}
$$

In the mapping function section, we saw that to convert an *m* value to map units, it must be multiplied by 50. So

$$
\text{map distance (m.u.)} = 50\,(\text{T} + 6\text{NPD})
$$

Let's assume that in our hypothetical cross of $ab \times ++$, the actual frequencies of the ascus classes are 56 percent PD, 41 percent T, and 3 percent NPD. Using the formula, the map distance between the *a* and *b* loci is

$$
\begin{aligned} 50\,[0.41 &+ (6 \times 0.03)] \\ &= 50\,(0.59) \\ &= 29.5 \text{ m.u.} \end{aligned}
$$

We can compare the accuracy of this method with the result of an analysis of recombinant frequency (RF). First, we measure the RF from unordered asci (even though it would be a waste of time to use unordered tetrads for this purpose). To make this RF calculation, we simply acknowledge that in the three classes PD, NPD and T, the NPD asci are full of recombinant spores, one-half of the T asci are recombinant spores, and the PD asci contain no recombinants. Hence, the RF can be calculated as

$$RF = \tfrac{1}{2}T + NPD$$

In our example

$$RF = 0.205 + 0.03$$
$$= 0.235, \text{ or } 23.5 \text{ m.u.}$$

This is 6 m.u. less than the estimate we obtained using the map-distance formula because we could not correct for double crossovers in the RF analysis.

Message Unordered tetrads can be used in many aspects of meiotic genetics, including the measurement of the distribution of crossovers among different meioses.

Chromatid Interference

Does the occurrence of a crossover between two nonsister chromatids affect the probability of a second crossover between the *same* chromatids? As usual, the best hypothesis to test is a null hypothesis, which is that the second crossover will occur randomly between any pair of nonsister chromatids (Figure 6-15). The randomized second crossover should generate a 1:2:1 ratio of 2-strand:3-strand:4-strand doubles, and any deviation from that ratio can be considered interference between

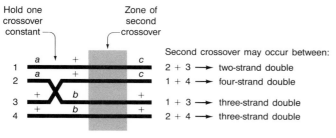

(All four possibilities give separate, distinguishable ascus genotypes.)

Figure 6-15. Double crossovers can produce several distinctive patterns of alleles in the ascus, depending on the strands involved.

chromatids, or chromatid interference. Usually, there is very little or no consistent interference of this sort when the analysis of double-exchange tetrads is performed using fungal asci.

Note that the RF limit of 50 percent applies only when there is no chromatid interference. Positive chromatid interference would favor the production of NPD tetrads in which all the products of meiosis are recombinant, at the expense of PD and T classes. Hence, the RF could rise above 50 percent to a value dependent on the degree of chromatid interference.

Mitotic Segregation and Recombination

We normally think of segregation and recombination as meiotic phenomena, but they do occur (although less frequently) during mitosis. The presence of mitotic segregation and recombination can easily be demonstrated if the genetic system is appropriately chosen.

Mitotic Segregation

In genetics, the term **segregation** is used to describe the separation of two alleles constituting a heterozygote into phenotypically distinguishable individuals or cells. We have seen segregation repeatedly, of course, in our meiotic analyses based on Mendel's first law. However, the alleles of a heterozygote occasionally can be seen to segregate when the heterozygous cell undergoes *mitotic* division. The following example clarifies the way in which this mitotic segregation is detected.

In the 1930s, Calvin Bridges was observing *Drosophila* females that were genotypically $M / +$ (M is a dominant allele that produces a phenotype of slender bristles). Some females had a patch, or **sector,** of wild-type bristles on a body of predominantly M phenotype. Thus, the alleles of the heterozygote showed segregation at the phenotypic level. Bridges concluded that this segregation was the result of mitotic—that is, **somatic**—nondisjunction (Figure 6-16). In heterozygotes of autosomal recessive genes and their wild-type alleles ($+/m$), patches of recessive phenotype on backgrounds of wild-type phenotype again were seen. These patches also can be explained by mitotic nondisjunction.

Other cases of segregation in the somatic tissue of a heterozygote have been found to be due to **mitotic chromosome loss.** Here the chromosome bearing the dominant allele somehow gets left behind when the daughter nuclei reconstitute after mitotic division (Figure 6-17).

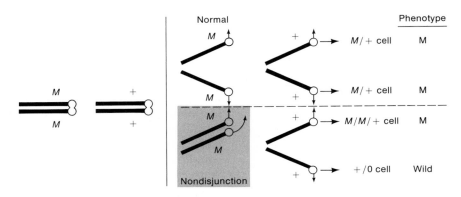

Figure 6-16. Mitotic nondisjunction can lead to phenotypic segregation, as in this example of *M* and +.

Geneticists find two other terms useful in relation to such phenomena. First, **variegation** is the mere existence of different-looking sectors of somatic tissue, whatever the cause. Second, a **mosaic** is an individual composed of tissues of two or more different genotypes, often recognizable by their different phenotypes.

Mitotic Crossing-Over

In 1936, Curt Stern showed that sectors in a mosaic are not always the result of nondisjunction of mitotic chromosome loss. Working with the *Drosophila* sex-linked genes *y* (yellow) and *sn* (singed, representing short, curly bristles), Stern made a cross + *sn*/ + *sn* × *y* + /Y. The female progeny were predominantly wild-type in appearance, as expected. Some females had sectors of yellow tissue or of singed tissue; these could be explained by nondisjunction or chromosome loss. However, some females showed **twin spots.** A twin spot, in this example, is two adjacent sectors—one of yellow tissue, and one of singed tissue—in a background of wild-type tissue (Figure 6-18). Stern reasoned that because the twin sectors of each twin spot were adjacent and occurred too frequently to be chance juxtapositions of single spots, the

twin sectors must be reciprocal products of the same event. That event, he concluded, must have been a crossover between the *sn*/ + locus and the centromere during a mitotic division in which the homologous parental chromosomes had accidentally been in a pairing conformation (Figure 6-19). Figure 6-19 demonstrates that a crossover between the marker loci could also have contributed to the yellow single-spot class.

Twin spots have been observed in other diploid cells, including those of plants. All of these observations can be explained by mitotic crossing-over, which appears to be of quite general (although rare) occurrence.

The definition of mitotic recombination in general is similar to that of meiotic recombination. Compare the following definition with the one on page 97.

> **Definition   Mitotic recombination** is any mitotic process that generates a diploid daughter cell with a different combination of genes from the combination of genes in the diploid parental cell in which mitosis occurred.

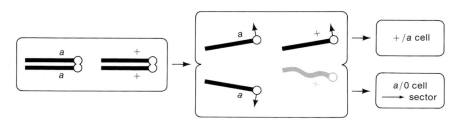

Figure 6-17. Chromosome loss at mitotic division can lead to phenotypic segregation, as in this example of *a* and +.

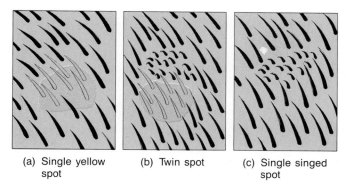

(a) Single yellow spot   (b) Twin spot   (c) Single singed spot

Figure 6-18. Segregation of body-surface phenotypes in *Drosophila* genotype $+ sn/y +$, where *sn* represents singed bristles and *y* represents yellow body. (a) Single yellow spot. (b) Twin spot. (c) Single singed spot.

Fungal Detection Systems

Fungi also are used extensively in the study of mitotic recombination and segregation. It is necessary, however, to generate diploid fungal cells in order to observe these mitotic phenomena, because a haploid cell does not provide the opportunity for two genomes to recombine. Diploids form spontaneously in many fungi. The fungus we will examine is *Aspergillus,* a greenish mold. *Aspergillus* is a highly suitable organism for mitotic analysis for several reasons:

1. The hyphae of the fungus produce long chains of cells called asexual spores. Each of these asexual spores has a single nucleus, and the phenotype of any individual spore is dependent only on the genotype of its own nucleus. This makes certain kinds of selective techniques possible.

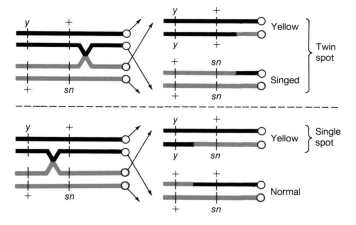

Figure 6-19. A mitotic crossover can lead to phenotypic segregation of the type shown in Figure 6-18.

2. If two haploid cultures are mixed, the hyphae fuse; both nuclear types are then present in a common cytoplasm. This strain is called a **heterokaryon.** Heterokaryons, however, still produce uninucleate asexual spores. Consider the following heterokaryon:

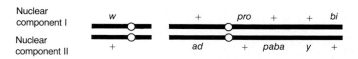

The alleles *ad, pro, paba,* and *bi* are all recessive to their wild-type counterparts, and each confers a requirement for a certain specific chemical supplement to permit growth. The alleles *w* and *y* produce white and yellow asexual spores, respectively, and are also recessive. Thus, the heterokaryon does not require any supplement for growth, but (due to the phenotypic autonomy just described in reason 1) the asexual spores are either yellow or white. Thus, the heterokaryon looks yellowish-white and has a kind of "pepper-and-salt" appearance.

3. In some heterokaryons, green sectors appear. Green is the normal wild-type color of the fungus, and the green coloration in the present example is the result of the spontaneous production of a diploid nucleus, which has propagated to form the sector. The presence of the dominant y^+ and w^+ alleles in the same nucleus produces the wild-type coloration. Diploid asexual spores can be removed from the green sector, and diploid cultures can be prepared for study. Like the heterokaryon, the diploid cells require no growth supplements.

4. When the diploid is fully grown, rare sectors showing either white or yellow asexual spores can be observed. Some of these spores are diploid (recognizable by their large diameters) and some are haploid (small diameters). Two types are particularly suitable for illustrating the phenomena at work: white sectors with white haploid spores, and yellow sectors with yellow diploid spores (Figure 6-20).

Haploid White Sectors. If these sectors are isolated and tested, almost exactly one-half of them prove to have the genotype $w + pro ++ bi$ and one-half show the genotype $w\ ad + paba\ y+$. Thus, the original diploid nucleus has somehow become haploid (a process known as **haploidization**), presumably through the progressive loss of one member of each chromosome pair. By looking at only white haploid spores, we automatically select for the *w*-bearing chromosome. In one-half of the *w* sectors, the $+ pro ++ bi$ chromosome is retained; in the

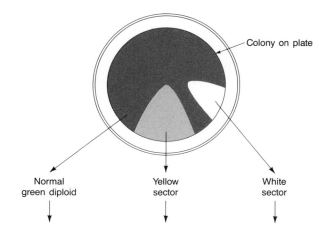

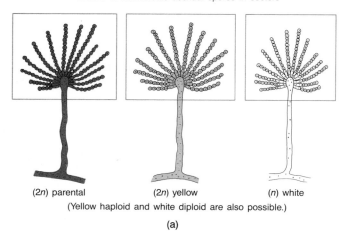

(2n) parental   (2n) yellow   (n) white
(Yellow haploid and white diploid are also possible.)

(a)

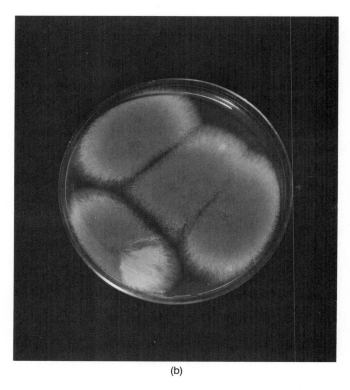

(b)

Figure 6-20. (a) Some sectors showing segregation in an *Aspergillus* diploid of genotype $+/w\ +/y$, where w and y represent white and yellow asexual spores, respectively. Haploids are identified by smaller cell size. (b) Photograph of a white sector in one of several diploid colonies. (Courtesy of Etta Käfer.)

other half, the $ad + paba\ y+$ chromosome is retained (Figure 6-21). In general, the recessive spore-color genes can be used to derive linkage information because haploidization is similar to independent assortment. You can see that this procedure involves selecting the chromosome bearing the spore-color marker and then observing which genes are retained with it and which are independent of it, and in what groupings.

Diploid Yellow Sectors. When a number of these sectors are examined, they usually prove to contain recombinant chromosomes. For example, one sector type was yellow and also required "paba" (*para*-aminobenzoic acid) for growth. Remember that mitotic crossing-over can make heterozygous loci homozygous, so mitotic crossing-over explains this type (Figure 6-22). Notice that we must follow two spindle fibers to each pole in mitotic analysis. Although one chromatid pair would normally *not* lie adjacent to its homologous chromatid pair, this *has* happened (presumably "accidentally"). This yellow diploid arose from a mitotic exchange that obviously took place in the centromere-to-*paba* region; others would arise from mitotic exchanges in the *paba*-

to-y region (these would not require paba for growth). The relative frequencies of these two types would provide a kind of mitotic linkage map for that region. Such mapping can be obtained for several fungi. Expectedly, the gene orders correspond to gene orders in meiotic

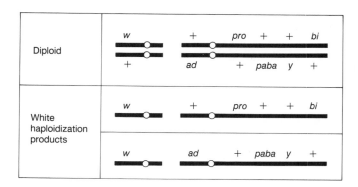

Figure 6-21. Genotype of an *Aspergillus* diploid, and the two white haploidization product genotypes possible for these marked chromosomes. (w = white, y = yellow; *ad, pro, paba,* and *bi* are recessive nutritional markers.)

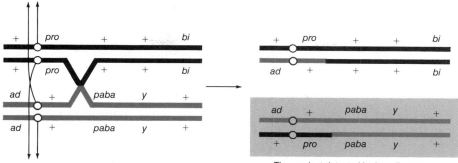

Figure 6-22. A mitotic crossover can produce a diploid yellow sector in the *Aspergillus* diploid shown.

maps, but, unexpectedly, the relative sizes of many of the intervals are very different when meiotic and mitotic maps are compared. Note that every locus from the point of crossing-over to the end of the chromosome arm is made homozygous by a mitotic crossover. This can provide valuable mapping information, and the experiments can be performed much more quickly than a meiotic analysis.

> **Message** Phenotypic sectoring (variegation) in somatic tissue can be due to the mitotic segregation of a heterozygous gene pair (through nondisjunction or through chromosome loss) or to mitotic crossing-over.

In later chapters, we see that there are, in fact, several other genetic ways to produce variegation.

Mapping Human Chromosomes

Human beings do not readily submit themselves to traditional genetic analysis. Until the 1960s, geneticists had to rely on the study of family pedigrees in order to deduce linkage. Most of these pedigree analyses were concerned with the genes on the X chromosome, but such data are hopelessly inadequate from the geneticist's standpoint. Linkage analysis is a more straightforward procedure than is the case with autosomes because the X chromosome in the male is hemizygous and gene combinations can be inferred without a testcross. (See Figures 5-21 and 5-22.) (Some approaches to autosomal mapping are illustrated in Problems 12, 23, and 28 in Chapter 5.) Recently, however, a technique has been developed that has revolutionized the mapping of both sex-linked and autosomal genes in humans. This technique uses human cells grown in culture (Figure 6-23).

There is a virus called Sendai that has a useful property. Normally, a virus has a specific point for attach-

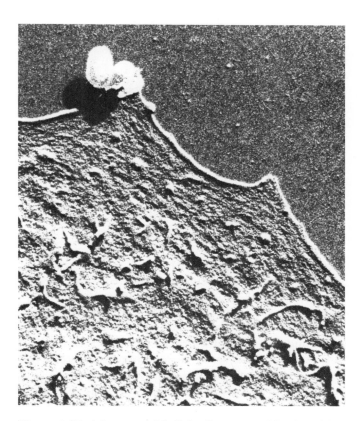

Figure 6-23. A human epithelial cell contrasted in size with two bacterial cells (*Escherichia coli*, from the human intestines). The wavy line is the edge of the human cell, of which only about one-eighth is shown (Courtesy of Jack D. Griffith.)

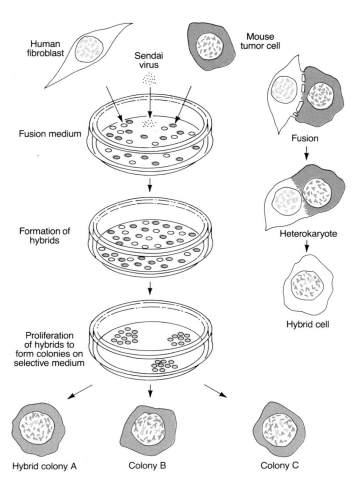

Figure 6-24. Cell-fusion techniques applied to human and mouse cells produce colonies, each of which contains a full mouse genome plus a few human chromosomes (tinted color). A fibroblast is a cell of fibrous connective tissue. (From F. H. Ruddle and R. S. Kucherlapati, "Hybrid Cells and Human Genes." Copyright © 1974 by Scientific American, Inc. All rights reserved.)

ment to and penetration of a host cell. Each Sendai virus has several points of attachment, so that it can simultaneously attach to two different cells if they happen to be close together. A virus, though, is very small in comparison with a cell (similar to the comparison between the earth and the sun), so that the two cells to which the virus is attached are held very close together indeed. In fact, in many cases, the membranes of the two cells fuse together and the two cells become one—a binucleate heterokaryon.

If suspensions of human and mouse cells are mixed together in the presence of Sendai virus (which has been inactivated by ultraviolet light), the virus can mediate fusion of the two kinds of cells (Figure 6-24). Once the cells fuse, the nuclei can fuse to form a uninucleate cell

line. Because the mouse and human chromosomes are very different in number and shape, the hybrid cells can be readily recognized. However, for unknown reasons, the human chromosomes are gradually eliminated at random from the hybrid as the cells divide (perhaps this is analogous to haploidization in *Aspergillus*). This process can be arrested to encourage the formation of a stable partial hybrid in the following way. The mouse cells that are used can be made genetically deficient in some function (usually a nutritional one) so that the function must be supplied by the human genome for growth of cells to occur. This selective technique usually results in the maintenance of hybrid cells that have a complete set of mouse chromosomes and a small number of human chromosomes, which vary in number and type from hybrid to hybrid but which always include the nutritionally sufficient human chromosome.

Luckily, this process can be followed under the microscope because mouse chromosomes can easily be distinguished from human chromosomes. Recently, this procedure has been made a lot easier by the development of stains (such as quinacrine and Giemsa) that reveal a pattern of "banding" within the chromosomes. The size and the position of these bands vary from chromosome to chromosome but the banding patterns are highly specific and constant for each chromosome. Thus, for any hybrid, it is relatively easy to identify the human chromosomes that are present (Figure 6-25). Different hybrid cells are grown separately into lines; eventually a bank of lines is produced that contains, in total, all the human chromosomes.

The mapping technique works as follows. If the human chromosome set contains a genetic marker (such as a gene that controls a specific cell-surface antigen, drug resistance, a nutritional requirement, or a protein variant), then the presence or absence of the genetic marker in each line of hybrid cells can be correlated with the presence or absence of certain human chromosomes in each line (Table 6-2). We can see that in the different hybrid cell lines, genes 1 and 3 are always present or absent together. We can conclude, then, that they are

TABLE 6-2. Comparison of five hybrid lines

		Hybrid cell lines				
		A	B	C	D	E
Human genes	1	+	−	−	+	−
	2	−	+	−	+	−
	3	+	−	−	+	−
	4	+	+	+	−	−
Human chromosomes	1	−	+	−	+	−
	2	+	−	−	+	−
	3	−	−	−	+	+

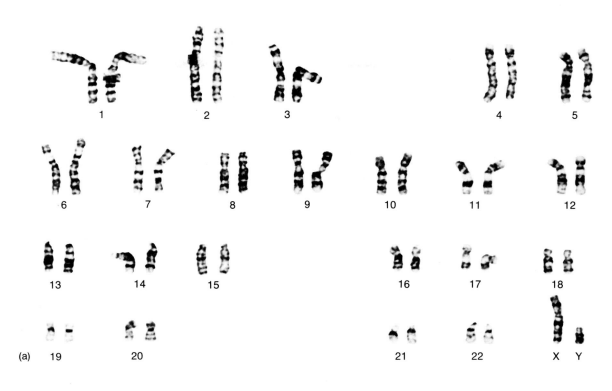

(a)

linked. Furthermore, the presence or absence of genes 1 and 3 is directly correlated with the presence or absence of chromosome 2, so we can assume that these genes are located on chromosome 2. By the same reasoning, gene 2 must be on chromosome 1, but the location of gene 4 cannot be assigned.

Large numbers of human genes have now been localized to specific chromosomes in this way, but of course we cannot derive a linkage map showing the order and distances between genes. Other manipulations are necessary; for example, the loss or gain of variously sized bits of a specific chromosome might be correlated with the presence or absence of genetic markers. Problem 29 at the end of Chapter 8 encourages you to think through the kind of logic involved. The results of this kind of mapping are shown in Figure 6-26.

> **Message**   Mitotic as well as meiotic phenomena can provide information on gene location in fruit flies, fungi, and humans.

In this chapter, we have considered more advanced treatments of transmission genetics, including mapping functions, tetrad analysis, and mitotic genetics. These genetic tools have enabled geneticists to test some of the assumptions of the chromosome theory of heredity and to learn more about how genetic material is passed from one generation to the next.

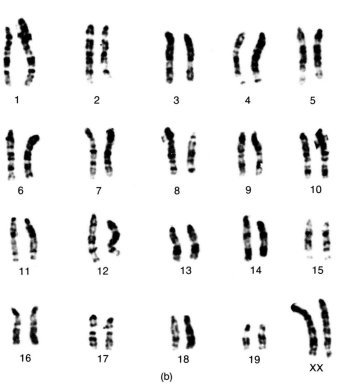

(b)

Figure 6-25. (a) Karyotype of a human male. Note how size, centromere position, and banding pattern (produced by trypsin-Giemsa treatment) can be used to recognize specific human chromosomes. (Photograph by Fred Dill.) (b) Karyotype of a female mouse. (Courtesy of the Jackson Laboratory, Bar Harbor, Me.)

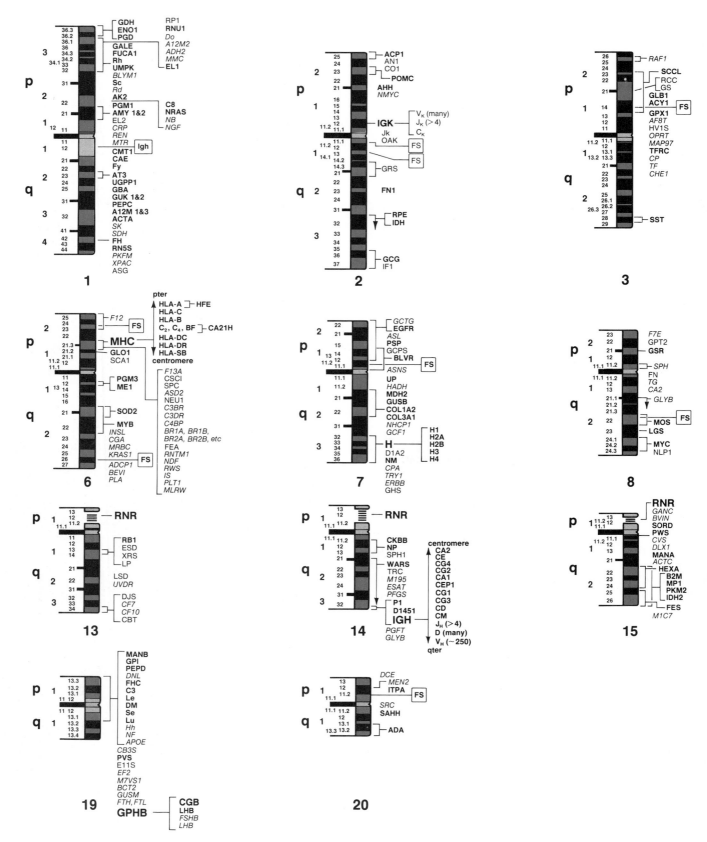

Figure 6-26. Map of human chromosomes, showing banding patterns and the genetic markers that have been assigned to a particular chromosome or to a specific locus on a chromosome. For the 345 autosomal markers, 202 have been assigned by somatic hybridization, 67 by family studies, 14 by both, and 62 by other methods (p and q designate chromosome arms; bold letters designate gene families or clusters; italics represent provisional assignments). (From V. A. McKusick, *Genetic Maps*, Vol. 2. Edited by S. J. O'Brien. Copyright © 1982 by Cold Spring Harbor.)

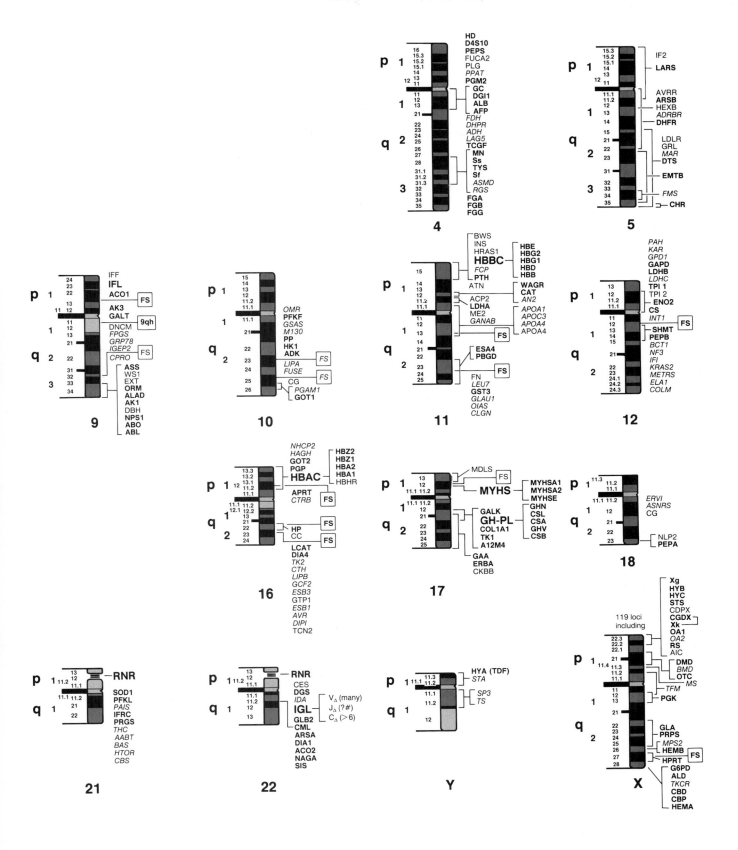

Summary

■ Because multiple crossovers are likely to occur when great distances separate loci, map distance between such loci is not linearly related to recombinant frequency. The ideal relationship between map distance and recombinant frequency is called the mapping function and can be calculated by using the Poisson distribution.

Another useful genetic tool is tetrad analysis, which analyzes the four products of meiosis in those fungi and single-celled algae in which the four products of each meiosis are held together in a kind of bag. Tetrad analysis provides the opportunity to test directly some of the assumptions of the chromosome theory of heredity, to map centromeres as genetic loci, to investigate the possibility of chromatid interference, to examine the mechanisms of chromosome exchange, and to study abnormal chromosome sets.

Tetrads may be linear or unordered. Analysis of linear tetrads is particularly useful because it is possible to map loci in relation to their centromeres and to each other. In crosses involving two linked loci, the asci in a linear or unordered tetrad may be classified as parental ditype (PD), nonparental ditype (NPD), and tetratype

(T). The number of PD asci in relation to the number of NPD asci provides a critical test for linkage. If the number of PD asci is greater than the number of NPD asci, then the recombinant frequency must be less than 50 percent and there must be linkage between the genes. If PD equals NPD, then the proportion of T asci can be used to distinguish independent assortment from loose linkage. Unordered tetrad analysis can be used to correct for multiple crossovers in mapping studies.

Although segregation and recombination are normally thought of as meiotic phenomena, segregation and recombination do occasionally occur during mitosis. Mitotic segregation was first identified in the 1930s, when Calvin Bridges observed patches of M[+] bristles of the body of a female *Drosophila* of predominantly M phenotype. He concluded that the patches were the result of abnormal chromosome segregation at the mitotic level. Around the same time, Curt Stern observed twin spots in *Drosophila* and assumed they must be the reciprocal products of mitotic crossing-over. Fungi are extensively used to study mitotic segregation and recombination.

Humans are generally unsuitable subjects for traditional genetic analysis. However, by using the Sendai virus to fuse human and mice cells, geneticists have been able to locate human genes on specific chromosomes.

■ ■ ■

Solved Problems

1. In *Neurospora,* a cross is made between a haploid strain of genotype $+ad$ and another haploid strain of genotype $nic+$. From this cross, a total of 1000 linear asci are isolated and categorized as follows:

1	2	3	4	5	6	7
+ ad	+ +	+ +	+ ad	+ ad	+ +	+ +
+ ad	+ +	+ +	+ ad	+ ad	+ +	+ +
+ ad	+ +	+ ad	nic ad	nic +	nic ad	nic ad
+ ad	+ +	+ ad	nic ad	nic +	nic ad	nic ad
nic +	nic ad	nic +	+ +	+ ad	+ +	+ ad
nic +	nic ad	nic +	+ +	+ ad	+ +	+ ad
nic +	nic ad	nic ad	nic +	nic +	nic ad	nic +
nic +	nic ad	nic ad	nic +	nic +	nic ad	nic +
808	1	90	5	90	1	5

Map the $+/ad$ and $+/nic$ loci in relation to their centromere(s) and to each other.

Solution

What principles can we draw on to solve this problem? It is a good idea to begin by doing something straightforward, which is to calculate the centromere distances. We do not know if $+/ad$ and $+/nic$ are linked but we do not need to know. The M_{II} pattern frequency for both loci will give us their individual centromere distances, regardless of whether this is the same centromere or not. (We can worry about that later.)

Remember that an M_{II} pattern is anything that is not two blocks of four. Let's start with the centromere distance of $+/nic$. All we have to do is add the ascus types 4, 5, 6 and 7, because these are all M_{II} patterns for the $+/nic$ locus. The total is $5 + 90 + 1 + 5 = 101$, or 10.1 percent. In this chapter, we have seen that to convert this to map units, we must divide by 2, which is 5.05 m.u.

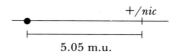

Now we do the same thing for the $+/ad$ locus. Here the M_{II} pattern total is given by asci 3, 5, 6 and 7. The total is $90 + 90 + 1 + 5 = 186$, or 18.6 percent, which is 9.3 m.u.

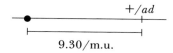

9.30/m.u.

Now we have to put these two together and decide between the following alternatives, all of which are compatible with the above gene-to-centromere distances:

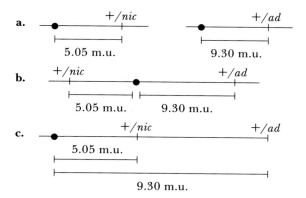

a.

+/nic   +/ad

5.05 m.u.   9.30 m.u.

b.

+/nic   +/ad

5.05 m.u.   9.30 m.u.

c.

+/nic   +/ad

5.05 m.u.

9.30 m.u.

Here a combination of common sense and simple analysis tells us which alternative is correct. First, an inspection of the asci reveals that the most common single type is the one labeled 1, which contains more than 80 percent of all the asci. This type contains only $+ad$ and $nic+$ genotypes, and they are *parental* genotypes. So we know that recombination is going to be quite low and the loci are certainly linked. This rules out alternative 1.

Now consider alternative c; if this were correct, a crossover between the centromere and $+/nic$ would generate not only an M_{II} pattern for that locus but also an M_{II} pattern for $+/ad$, because it is distal. The ascus pattern produced should be

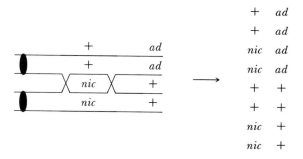

```
                          +    ad
                          +    ad
+    ad                   nic    +
+    ad                   nic    +
nic    +        →         +    ad
nic    +                  +    ad
                          nic    +
                          nic    +
```

Remember that asci 4, 5, 6 and 7 (a total of 101 asci) are M_{II} for $+/nic$; of these, type 5 is the very one we are talking about and contains 90 asci. Therefore, alternative c appears to be correct because ascus type 5 is about 90 percent of the M_{II} asci for $+/nic$. This relationship would not hold if alternative b were correct, because crossovers on either side of the centromere would generate the M_{II} patterns for $+/nic$ and $+/ad$ independently.

Is the map distance from $+/nic$ to $+/ad$ simply $9.30 - 5.05 = 4.25$ m.u.? Close, but not quite. The best way of calculating map distances between gene loci is always by measuring the recombinant frequency (RF). We could go through the asci and count all the recombinant ascospores, but it is simpler to use the formula $RF = \frac{1}{2}T + NPD$. Even though the asci are linear, they can still be scored PD, NPD, and T. The T asci are classes 3, 4 and 7, and the NPD asci are classes 2 and 6. Hence, $RF + [\frac{1}{2}(100) + 2]/1000 = 5.2$ percent or 5.2 m.u., and a better map is

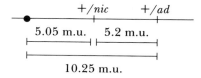

+/nic   +/ad

5.05 m.u.   5.2 m.u.

10.25 m.u.

The reason for the underestimate of the $+/ad$-to-centromere distance calculated from the M_{II} frequency is the occurrence of double crossovers, which can produce an M_I pattern for $+/ad$, as in ascus type 4:

```
                          +    ad
                          +    ad
+    ad                   nic    ad
+    ad                   nic    ad
nic    +        →         +    +
nic    +                  +    +
                          nic    +
                          nic    +
```

2. In *Aspergillus*, the recessive chromosome VI alleles *leu1*, *met5*, *thi3*, *pro2*, and *ad2* confer requirements for leucine, methionine, thiamine, proline, and adenine, respectively, whereas their + alleles confer no such requirements. At the tip of chromosome VI is the locus of a gene with a recessive allele *su* that suppresses *ad2*, so that strains expressing *ad2* and *su* require no adenine. A diploid strain is made by combining the following haploid genotypes, where the loci, with the exception of *su*, are written in no particular order:

su,   leu1,   met5,   thi3,   pro2,   ad2

and

+   +   +   +   +   *ad2*

Diploid asexual spores were spread on a medium containing all supplements except adenine. Most spores did not grow due to the recessiveness of *su*, but a minority did grow into colonies; 100 of these were removed and tested for the phenotypes of the other markers. The following four classes were found (note that these are diploid phenotypes, not haploid genotypes):

1.	su	+	met	+	+	ad	60
2.	su	leu	met	+	+	ad	25
3.	su	+	+	+	+	ad	10
4.	su	leu	met	thi	+	ad	5

a. Explain the production of these four classes and their relative amounts.

b. Why do you think that no colonies expressing *pro2* were recovered?

Solution

a. The experiment concerns the behavior of diploid cells at mitosis, so the four rare classes must be explained by an asexual mechanism. It seems likely that the *su* allele in all classes has been made homozygous *su/su*, because only in that condition can it suppress an *ad2/ad2* homozygote and enable growth to occur without adenine. This then was the basis of the original selection of the 100 colonies. But evidently other alleles have become homozygous too—and in different combinations in different classes.

Chapter 6 demonstrates that mitotic crossing-over can produce homozygosity in any recessive allele that is distal to the crossover. Can we explain the different classes based on the occurrence of mitotic crossovers in different regions of chromosome VI? Inspection of the classes shows that homozygosity can be produced for either *su* alone (class 3), or *su* and *met* (class 1), or *su*, *met* and *leu* (class 2), or *su*, *met*, *leu*, and *thi* (class 4). This virtually dictates to us that the order must be *su – met – leu – thi –* centromere. The homozygous crossovers must then be in the following regions:

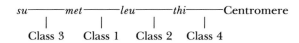

For example, the following crossover is necessary to produce class 2:

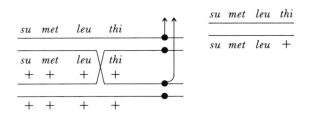

The relative sizes of the classes must reflect the relative sizes of the regions as follows:

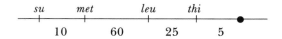

Note that these units are relative proportions and are not the same as meiotic map units.

b. We are told that *pro2* is on chromosome VI, yet it is never made homozygous. Note the importance of the fact that to select for mitotic crossovers, we are making use of an allele *su* which is at the tip of one arm. Recessive alleles in the other arm would not be simultaneously made homozygous by these crossovers, so *pro2* is probably in the other arm. We have no way of knowing where *ad2* is, because it is homozygous from the outset.

Problems

1. In *Neurospora*, the cross + × *al-2* is made. A linear tetrad analysis reveals that the second division segregation frequency is 8 percent.

a. Draw two examples of second-division segregation patterns in this cross.

b. What can the 8 percent value be used to calculate?

2. In the fungal cross *arg-6 al-2* × ++, what will the spore genotypes be in unordered tetrads that are:

a. Parental ditypes? **b.** Tetratypes? **c.** Nonparental ditypes?

3. For a certain chromosomal region, the mean number of crossovers at meiosis is calculated to be two per meiosis. In that region, what proportion of meioses are predicted to have:

a. No crossovers? **b.** One crossover? **c.** Two crossovers?

4. In a *Drosophila* of genotype

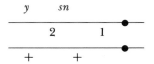

y represents yellow body (+ = brown) and *sn* represents singed hairs (+ = unsinged). What is the detectable outcome if a mitotic crossover occurs during development in:

a. Region 1? **b.** Region 2?

5. In haploid yeast, a cross between *arg⁻ ad⁻ nic⁺ leu⁺* and *arg⁺ ad⁺ nic⁻ leu⁻* produces haploid sexual spores, and 20 of these are isolated at random. The resulting cultures are tested on various media, as shown in Table 6-3 (where + means growth and − means no growth) on page 145.

a. What can you say about the linkage arrangement of these genes?

b. What is the origin of culture 16?

6. Every Friday night, genetics student Jean Allele, exhausted by her studies, goes to the student's union bowling lane to relax. But even there, she is haunted by her genetics studies. The rather modest bowling lane has only four bowling balls: two red, and two blue. These are bowled at the pins and are then collected and returned down the chute in random order, coming to rest at the end stop. Over the evening, Jean notices familiar patterns of the four balls as they come to rest at the stop. Compulsively, she counts the different patterns that occur. What patterns did she see, what were their frequencies, and what is the relevance of this matter to genetics? (This is not a trivial question.)

7. a. Use the mapping function to calculate how many real map units are indicated by a recombinant frequency of 20 percent. Remember that a mean of 1 is equal to 50 real map units.

TABLE 6-3.

Culture	Minimal medium plus			
	Arginine, adenine, nicotin- amide	Arginine, adenine, leucine	Arginine, nicotin- amide, leucine	Adenine, nicotin- amide, leucine
1	+	+	−	−
2	−	−	+	+
3	−	+	−	+
4	+	−	+	−
5	−	−	+	+
6	+	+	−	−
7	+	+	−	−
8	−	−	+	+
9	+	−	+	−
10	−	+	−	+
11	−	+	−	+
12	+	−	+	−
13	+	+	−	−
14	+	−	+	−
15	−	+	−	+
16	+	−	−	−
17	+	+	−	−
18	−	−	+	+
19	+	+	−	−
20	−	+	−	+

b. If you obtain an RF value of 45 percent in one experiment, what can you say about linkage? (The actual figures are 58, 52, 47, and 43 out of 200 progeny.)

*8. Complete Table 6-4 for a situation in *Neurospora* in which the *mean* number of crossovers between the *a* locus ($a/+$) and its centromere is equal to one per meiosis.

a. What is the total M_{II} frequency if the mean is 1?

b. Complete similar tables for means of 0.5, 2.0, and 4.0, and draw a mapping function for the M_{II} frequency (that is, plot the total M_{II} frequency against mean crossover frequency per meiosis).

c. Why does the curve bend downward? How would you correct this?

*9. In a tetrad analysis, the linkage arrangement of the $p/+$ and $q/+$ loci is as follows.

Assume that

In region (i), only two situations can occur at meiosis: no crossover (88 percent of meioses), or a single crossover (12 percent of meioses).

In region (ii), only two situations can occur at meiosis: no crossover (80 percent of meioses), or a single crossover (20 percent of meioses).

There is no interference (in other words, the situation in one region doesn't affect what is going on in the other region).

What proportions of tetrads will be of the following types? a. $M_I M_I$.PD b. $M_I M_I$.NPD c. $M_I M_{II}$.T d. $M_{II} M_I$.T e. $M_{II} M_{II}$.PD f. $M_{II} M_{II}$.NDP g. $M_{II} M_{II}$.T (NOTE: Here the M pattern written first is the one that pertains to the $p/+$ locus.)

10. In *Neurospora*, the cross $+ + + + \times a\ b\ c\ d$ is made (*a, b, c,* and *d* are linked in the order written). Construct crossover diagrams to illustrate how the following unordered (nonlinear) ascus patterns could arise:

```
+ + c +    + b c d    + + c +    + + c +    + b + d
a b c +    + + + d    + b + d    + b + d    + b + d
+ b + d    a b c +    a + c +    a + c +    a + c +
a + + d    a + + +    a b + d    a b + d    a + c +

     + b c d    + b + d    + b + +    + b + d
     a b c +    + + + d    + b + +    a + c +
     + + + d    a b c +    a + c d    + b + d
     a + + +    a + c +    a + c d    a + c +
```

TABLE 6-4.

	Number of exchanges				
	0	1	2	3	4
Probability of this kind of meiosis?					
What proportion of each of these kinds of meiosis will result in an M_{II} pattern for $a/+$?					
What proportion of all asci from this cross will show an M_{II} pattern as a result of each of these kinds of meiosis?					

TABLE 6-5.

	Number of asci of type						
	$a\,b$ $a\,b$ $++$ $++$	$a+$ $a+$ $+b$ $+b$	$a\,b$ $a+$ $++$ $+b$	$a\,b$ $+b$ $++$ $a+$	$a\,b$ $++$ $++$ $a\,b$	$a+$ $+b$ $+b$ $a+$	$a+$ $+b$ $++$ $a\,b$
Cross							
1	34	34	32	0	0	0	0
2	84	1	15	0	0	0	0
3	55	3	40	0	2	0	0
4	71	1	18	1	8	0	1
5	9	6	24	22	8	10	20
6	31	0	1	3	61	0	4
7	95	0	3	2	0	0	0
8	6	7	20	22	12	11	22
9	69	0	10	18	0	1	2
10	16	14	2	60	1	2	5
11	51	49	0	0	0	0	0

11. In *Neurospora*, crosses $a\,b \times ++$ (in which a and b represent different loci in each cross) are made. From each cross, 100 linear asci are analyzed, and the results shown in Table 6-5 are obtained. For each cross, map the genes in relation to each other and to their respective centromere(s).

12. In *Neurospora*, the a locus is 5 m.u. from the centromere on chromosome 1. The b locus is 10 m.u. from the centromere on chromosome 7. From the cross of $a+ \times +b$, determine the frequencies of the following: a. Parental ditype asci b. Nonparental ditype asci c. Tetratype asci d. Recombinant ascospores e. Colonies growing from ascospores grown on minimal medium (inorganic salts, energy source, and vitamins; see Chapter 7) if a and b represent nutritional requirements. (NOTE: Don't bother with mapping-function complications here.)

13. The accompanying figure shows a germinated teliospore of the barley-smut fungus, *Ustilago hordei*, that has just undergone meiosis. Each haploid cell of the promycelium has undergone mitosis to form a sporidium. The four haploid sporidia are numbered in sequence, beginning with the terminal cell. The sporidia can be removed in sequence to establish haploid colonies, which are easy to maintain in the laboratory. Haploid colonies can be subcultured and combined in pairs; when this is done, compatible combinations result in the formation of dikaryons, and incompatible combinations do not. For one such set of haploid cultures, dikaryons were formed by the matings 1×2, 1×4, 3×2, and 3×4, but dikaryons were not formed by the matings 2×4 and 1×3.

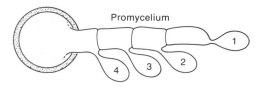

Promycelium

a. What can you conclude about the genetic determination of compatibility?

*b. Dikaryons of *U. hordei* are parasitic on cultivated barley. The four dikaryons listed in the problem, when inoculated into three different barley cultivars (varieties of barley), gave the results shown in Table 6-6. What can you conclude about the genetic determination of virulence (that is, the capacity of dikaryons to incite a severe disease reaction) on these cultivars?

c. Using your own system of symbols, assign genotypes to each of the four haploid cultures.

(Problem 13 courtesy of Clayton Person.)

14. In *Neurospora*, three different crosses are analyzed on the basis of unordered tetrads. In each cross, different linked genes are involved. The results follow:

Cross	Parents	Parental ditypes (%)	Tetratypes (%)	Nonparental ditypes (%)
1	$a+ \times +b$	51	45	4
2	$c+ \times +d$	64	34	2
3	$e+ \times +f$	45	50	5

For each cross, calculate:

a. The frequency of recombinants (RF).

b. The uncorrected map distance, based on RF.

c. The corrected map distance, based on tetrad frequencies.

15. In yeast, a cross is made involving the linked genes *ura3* (uracil requirement) and *lys4* (lysine requirement) as follows:

$$ura3 + \times + \ lys4$$

The 300 *unordered* tetrads are isolated and classified as follows:

$ura3$	$+$	$ura3$	$lys4$	$ura3$	$+$
$+$	$lys4$	$ura3$	$lys4$	$ura3$	$+$
$ura3$	$lys4$	$+$	$+$	$+$	$lys4$
$+$	$+$	$+$	$+$	$+$	$lys4$
138		12		150	

TABLE 6-6.

	Disease on cultivar		
Dikaryon	A	B	C
1×2	None	Severe	None
1×4	Severe	None	None
3×2	Severe	None	None
3×4	Severe	None	Severe

TABLE 6-7.

Segregation ($arg^- : arg^+$)	Frequency (%)
4:0	40
3:1	20
2:2	40

a. What is the recombinant frequency?

b. If it is assumed that either zero, one, or two (never more) crossovers occur between these loci at meiosis, what are the percentages of zero, one, and two crossover meioses?

c. What is the distance between the loci (in map units) *corrected* for the occurrence of double crossovers?

16. In an experiment with haploid yeast, you have two different cultures. Each will grow on minimal medium (see Chapter 7) to which arginine has been added, but neither will grow on minimal medium alone. Using appropriate methods, you induce the two cultures to mate. The diploid cells then undergo meiosis and form unordered (nonlinear) tetrads. You examine a large number of these tetrads and record the data shown in Table 6-7.

a. Using symbols of your own choosing, assign genotypes to the two parental cultures. For each of the three kinds of segregation, assign genotypes to the segregants.

b. Do these data lead to any specific conclusions about the mapping of genes? If so, what are these conclusions?

17. Four histidine loci are known in *Neurospora*. As shown here, each of the four loci is located on a different chromosome.

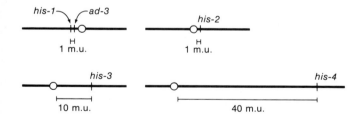

The *his-1* locus is closely linked (1 m.u.) to both *ad-3* and the centromere. The loci *his-2*, *his-3*, and *his-4* are 1, 10, and 40 m.u., respectively, from their centromeres. In your experiment, you begin with an *ad-3* line from which you recover a cell that also requires histidine. Now you wish to determine which of the four histidine loci is involved. You cross the *ad-3 his-?* strain with a wild-type (++) and analyze 10 nonlinear tetrads: two are PD, six are T, and two are NPD. From this result, which of the four *his* loci is most probably the one that changed from *his⁺* to *his*?

(Problem 17 courtesy of Luke deLange.)

18. In the haploid ascomycete fungus *Sordaria,* the ascospores are normally black. Two ascospore-color mutants are isolated. When crossed to a wild-type, mutant 1 gives asci that contain four black spores and four white spores and, mutant 2 gives asci that contain four black spores and four tan spores. When mutants 1 and 2 are intercrossed, some asci contain four black and four white spores, some asci contain four tan and four white spores, and some asci contain four white and two black and two tan spores. Explain these data.

*19. In *Drosophila melanogaster,* two entire X chromosomes can be attached to the same centromere:

The two arms now behave as a single chromosome, called an attached X. Work out the inheritance of sex chromosomes in crosses of attached-X-bearing females with normal males. During meiosis, the duplicated attached-X chromosome segregates from its sister chromatids:

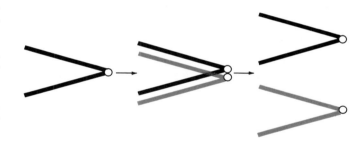

Crossing-over can take place between nonsister chromatids of attached-X chromosomes. Suppose you have an attached-X chromosome of the following genotype:

Diagram all the possible genotypic products and their phenotypes (a) from single crossovers between *a* and *b* and between *b* and the centromere, and (b) from double exchanges between *a* and *b* and between *b* and the centromere. Make sure you consider all possibilities. Describe how such a procedure represents a kind of tetrad analysis.

20. In soybeans, an incompletely dominant gene causes a "yellowish" appearance to the leaves. When yellowish heterozygotes are allowed to mature, some leaves show rare

areas in which there is a patch of dark green (normal) tissue always adjacent to a patch of very pale yellow tissue, all in the yellowish background. Propose an explanation for these rare patches. (Invent symbols and draw diagrams.)

21. In *Aspergillus*, the genes *fw* and *cha* are closely linked on chromosome VIII. The allele *fw* gives fawn-colored conidia (asexual spores), and *cha* gives chartreuse-colored conidia. The normal color of the fungus is dark green. A diploid with *fw* and *cha* heterozygous in repulsion is plated. Most of the diploid colonies are dark green, but a few are bicolored, with one-half fawn and one-half chartreuse. Explain the origin of these bicolored colonies.

22. An *Aspergillus* diploid that is $++++/w\,a\,b\,c$ is haploidized, and many white haploids are scored for *a, b,* and *c*. The results are 25 percent *w a b c*; 25 percent *w+++*; 25 percent *w a + c*; and 25 percent *w + b +*. What linkage relations can you deduce from these frequencies? Sketch your conclusions.

23. In an *Aspergillus* diploid $++/y\,ribo$, the two loci are linked but the order with respect to the centromere is not known. Yellow diploid segregants are obtained; 80 percent of them are ribo$^+$ and 20 percent are ribo-requiring. What is the most likely order?

24. An *Aspergillus* diploid is $+++/pro\,fpa\,paba$, in which *pro* is a recessive allele for proline requirement, *fpa* is a recessive allele for fluorophenylalanine resistance, and *paba* is a recessive allele for *para*-aminobenzoic acid (paba) requirement. By plating asexual spores on fluorophenylalanine, selection for resistant colonies can be made. Of 154 *diploid* resistant colonies, 35 do not require proline or paba, 110 require paba, and 9 require both.

   a. What do these figures indicate?

   b. Sketch your conclusions in the form of a map.

   c. Some resistant colonies (not the ones described) are haploid. What would you predict their genotype to be?

25. Table 6-8 shows the only human chromosomes contained in three colonies of human-mouse hybrid cells. Five enzymes (α, β, γ, δ, and ϵ) are tested in each of the cell colonies, with the following results: α is active only in

colony C; β is active in all three colonies; γ is active only in colonies B and C; δ is active only in colony B; and ϵ shows no activity in any colony. What can you say about the locations of the genes responsible for these enzyme activities?

26. Consider the following set of eight human-mouse clones:

| | Chromosome | | | | | | | | |
Clone	1	2	6	9	12	13	17	21	X
A	+	+	−	*q*	−	*p*	+	+	+
B	+	−	*p*	+	−	+	+	−	−
C	−	+	+	+	*p*	−	+	−	+
D	+	+	−	+	+	−、	*q*	−	+
E	*p*	−	+	−	*q*	−	+	+	*q*
F	−	*p*	−	−	*q*	−	+	+	*p*
G	*q*	+	−	+	+	+	+	−	−
H	+	*q*	+	−	−	*q*	+	−	+

Each clone may carry an intact (numbered) chromosome (+), only its long arm (*q*), or only its short arm (*p*), or it may lack the chromosome (−).

The following human enzymes were tested for their presence (+) or absence (−) in clones A–H:

| | Cell line | | | | | | | |
Enzyme	A	B	C	D	E	F	G	H
Steroid sulfatase	+	−	+	+	−	+	−	+
Phosphoglucomutase-3	−	−	+	−	+	−	−	+
Esterase D	−	+	−	−	−	−	+	+
Phosphofructokinase	+	−	−	−	+	+	−	−
Amylase	+	+	−	+	+	−	−	+
Galactokinase	+	+	+	+	+	+	+	+

Identify the chromosome carrying each enzyme locus. Where possible, identify the chromosome arm.

(Problem 26 from L. A. Snyder, D. Freifelder, and D. L. Hartl, *General Genetics.* Copyright © 1985 by Jones and Bartlett.)

*27. This question will start you thinking about the consequences of sister-chromatid exchange at meiosis. Imagine the diploid cell $A\,B/a\,b$, with its two linked gene pairs A/a and B/b.

   a. Draw a meiosis with a single crossover between the gene pairs. Now insert a single sister-chromatid ex-

TABLE 6-8.

| Hybrid colony | Presence (+) or absence (−) of human chromosome | | | | | | | |
	1	2	3	4	5	6	7	8
A	+	+	+	+	−	−	−	−
B	+	+	−	−	+	+	−	−
C	+	−	+	−	+	−	+	−

change (SCE) involving only one sister pair, also anywhere between the gene pairs. Repeat, but this time put the single SCE outside the gene pairs. Does the SCE make any difference in either case? Explain your answer.

b. Now draw a meiosis with a two-strand double crossover between the gene pairs. Then place a single SCE event (1) between a gene pair and the nearest crossover, (2) between the two crossovers, and (3) outside the gene pairs. Does the SCE make any difference in any case? Explain your answer.

c. Repeat part b for both kinds of three-strand double crossovers and for four-strand double crossovers.

d. Considering that the two-strand, three-strand, and four-strand double crossovers occur in a 1 : 2 : 1 ratio, will single, randomly located SCE events affect the relative frequencies of PD, NPD, and T tetrad types?

Gene Mutation

KEY CONCEPTS

Mutation is the inherent process whereby genes change from one allelic form to another.

■

Mutations in germline cells can be transmitted to progeny, but somatic mutations cannot.

■

Many specialized selective systems make it easier to obtain mutations.

■

Mutagens are agents that increase normally low rates of mutation.

■

Mutation occurs randomly, at any time and in any cell of an organism.

■

Any biological process can be dissected genetically by obtaining mutations that affect that process. (The mutations identify the important components of the process.)

Genetic analysis would not be possible without **variants** — organisms that differ in a particular character. We have considered many examples of analyses that have been performed in organisms that have different phenotypes connected with a particular character. Now we consider the origin of the variants. How, in fact, do genetic variants arise?

The simple answer to this question is that organisms have an inherent tendency to undergo change from one hereditary state to another. This process is called **mutation.** We can recognize two basic levels of mutation:

1. **Gene mutation.** A gene can mutate from one allelic form to another. Such a change occurs at or within a single gene, so it is sometimes called **point mutation.**

2. **Chromosome mutation.** Segments of chromosomes, whole chromosomes, or even entire sets of chromosomes may be involved in genetic change, and this process is collectively called chromosome mutation. Gene mutation is not necessarily involved in such a process; the effects of chromosome mutation are due more to the new arrangements of chromosomes and of the genes they contain.

In this chapter, we explore gene mutation; in Chapters 8 and 9, we consider chromosome mutation.

In any consideration of the subject of change, a fixed reference point, or standard, is necessary. In genetics, that standard is provided by the so-called wild-type. Remember that the wild-type gene may be either a form found in nature or a form commonly used as a standard laboratory stock. Any change away from the standard allele is called **forward mutation;** any change toward the standard allele is called **reverse mutation, reversion,** or **back mutation.** For example

$$
\left.
\begin{array}{l}
a^+ \longrightarrow a \\
D^+ \longrightarrow D
\end{array}
\right\} \text{forward mutation}
$$

$$
\left.
\begin{array}{l}
a \longrightarrow a^+ \\
D \longrightarrow D^+
\end{array}
\right\} \text{reverse mutation}
$$

The non-wild-type form of a gene usually is called a mutation. (To use the same word for the process and the product may sound unnecessarily confusing to you, but in practice little confusion arises!) Thus, we can speak of a dominant mutation (such as D above) or a recessive mutation (such as a). Bear in mind how arbitrary these gene states are; the wild-type of today may have been a mutation in the evolutionary past, and vice versa.

Another useful term is **mutant.** This is, strictly speaking, an adjective and should properly precede a noun. A mutant individual or cell is one whose changed

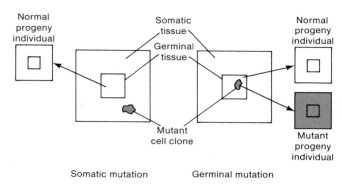

Figure 7-1. Diagrammatic representation of the differing consequences of somatic and germinal mutation.

phenotype is attributable to the possession of a mutation. Sometimes the noun is left unstated; in this case, a mutant always means an individual or cell with a phenotype that shows that it bears a mutation.

One final useful term is **mutation event,** which is the actual occurrence of a mutation.

Somatic Versus Germinal Mutation

Mutations can occur in either somatic or germinal tissue; they are called somatic and germinal mutations, respectively. The two types are diagrammed in Figure 7-1.

Somatic Mutation

A **somatic mutation** — a mutation occurring in developing somatic tissue — can lead to a population of identical mutant cells, all of which have descended from the cell in which the original mutation occurred. A population of identical cells derived asexually from one progenitor cell is called a **clone.** The earlier in development the mutation event occurs, the larger the mutant clone will be (Figure 7-2). Mutant clones sometimes can be identified

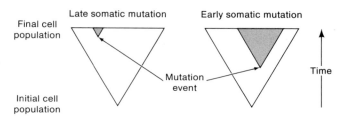

Figure 7-2. Early mutation produces a larger proportion of mutant cells in the growing population than does later mutation.

Figure 7-3. Somatic mutation from red (*R*) to white (*r*) in a rose petal.

by eye if their phenotype contrasts visually with the surrounding wild-type cells (Figure 7-3).

What about transmitting a mutation to progeny? By definition, this is not possible. However, note that if a plant cutting is taken from tissue that includes a mutant clone, the plant that grows from the cutting may develop mutant germinal tissue and transmit the mutant gene to progeny.

The genetic system for the detection of somatic mutation must be able to rule out the possibility that the sector is due to mitotic segregation or recombination. If the individual is a homozygous diploid, such sectoring is almost certainly due to mutation.

Germinal Mutation

Germinal mutation occurs in tissue that ultimately will form sex cells. Then, if these mutant sex cells act in fertilization, the mutation will be passed on to the next generation. Of course, an individual of perfectly normal phenotype and of normal ancestry can harbor undetected mutant sex cells. These mutations can be detected only if they turn up in the zygote. Such a situation is thought to have occurred in Queen Victoria, who was possibly the origin of a germinal mutation to the X-linked recessive mutant allele for hemophilia (failure of the blood to clot). This mutation showed up in some of her male descendants.

At the operational level, the detection of germinal mutation depends on the ability to rule out meiotic segregation and recombination as possible causes.

> **Message** Before any new variant hereditary state can be attributed to mutation, both segregation *and* recombination must be ruled out. This is true for both somatic and germinal mutations.

Mutant Phenotypes

The phenotypic consequences of mutation may be so subtle as to require refined biochemical techniques to detect a difference from the wild-type gene or so severe as to produce gross morphological defects or death. A rough classification follows, based only on the ways in which the mutations are recognized. This is by no means a complete classification.

Morphological Mutations. *Morph* means "form." The mutations in this class affect the outwardly visible properties of an organism, such as shape, color, or size. Albino ascospores in *Neurospora*, curly wings in *Drosophila*, and dwarfism in peas are all considered morphological mutations. Some examples of morphological mutants are shown in Figures 7-4 and 7-5.

Lethal Mutations. Here the new allele is recognized through its lethal effects on the organism. Sometimes a primary cause of death is easy to identify (for example, in

Figure 7-4. A rare morphological mutant in the thimbleberry, *Rubus parviflorus*, which arose spontaneously in nature. *(left)* Three normal wild-type flowers; *(right)* a flower from the mutant. The mutation causes a spectacular increase in the number of petals per flower. Such mutants have been used extensively in horticulture to increase the showiness of flowers.

(a)

(b)

Figure 7-5. Plumage in Japanese quail. (a) Normal and (b) a rare mutant that arose spontaneously in a laboratory population of Japanese quail. The mutation is autosomal recessive that interferes with the normal development of the feathers. (Courtesy of Janet Fulton.)

certain blood abnormalities). But often the gene is recognizable *only* by its effects on mortality.

Conditional Mutations. In this class, a mutant allele expresses the mutant phenotype under a certain condition called the **restrictive condition** but expresses a normal phenotype under another condition called the **permissive condition.** Temperature-conditional mutants have been studied the most frequently. For example, a certain class of mutations in *Drosophila* is known as "dominant heat-sensitive lethal." Heterozygotes in this class (say, H^+/H) are normal at 20°C (the permissive condition) but die if the temperature is raised to 30°C (the restrictive condition).

Many mutant organisms are less vigorous than normal forms. For this reason, conditional mutants are handy in that many of them can be grown under permissive conditions and then shifted to restrictive conditions for study. There are other advantages to conditional mutations, as we shall see in following chapters.

Biochemical Mutations. This class is identified by the loss of or a change in some biochemical function of the cell. This change typically results in an inability to grow and proliferate. In many cases, however, growth of a mutant can be restored by supplementing the growth medium with a specific nutrient. Biochemical mutants have been extensively analyzed in microorganisms. Microorganisms, by and large, are **prototrophic:** they are nutritionally self-sufficient and can exist on a substrate of simple inorganic salts and an energy source; such a growth medium is called a **minimal medium.** Biochemi-

cal mutants, however, often are **auxotrophic:** they must be supplemented with complex nutrients in order to grow. For example, in fungi, a certain class of biochemical mutants will not grow unless specifically supplemented with the important cellular chemical adenine. These auxotrophic mutants are called *ad*, or "adenine-requiring." The method of testing auxotrophs is shown in Figure 7-6.

Resistant Mutations. Here, the mutant cell or organism acquires the ability to grow in the presence of some specific inhibitor, such as cycloheximide, or a pathogen, to which wild-types are susceptible. Such mutants have been extensively used because they are relatively easy to select for, as we shall see.

Table 7-1 illustrates detection of three of these mutation types. The five classes are not mutually exclusive, nor do they cover all mutation types. Nevertheless, they are useful components in the vocabulary of mutation.

The Usefulness of Mutations

Mutation is a biological process that has been occurring as long as there has been life on this planet. As such, it is certainly fascinating and worthy of study. Mutant alleles like the ones mentioned in the previous section obviously are invaluable in the study of the process of mutation itself. In this connection, they are used as genetic markers, or representative genes: their precise function is not particularly important, except as a way to detect them.

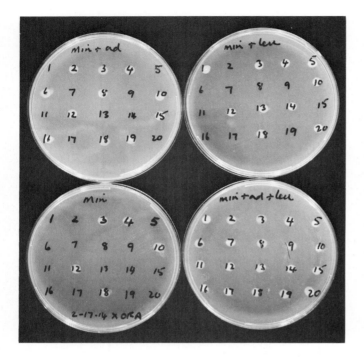

Figure 7-6. Testing for auxotrophy and prototrophy in the fungus *Neurospora crassa*. The process is illustrated here by testing the genotypes of 20 haploid cultures derived from a cross between an adenine-requiring auxotrophic mutant and a leucine-requiring auxotrophic mutant (thus, the cross was $ad^- leu^+ \times ad^+ leu^-$). The cultures are tested by dipping a sterile needle into a culture tube (say, tube 1) and then touching the needle to the surface of each of four different media, where a few cells will adhere. If the cells grow and divide, then a colony is formed; colonies appear as white circles on the photograph. The four plates contain different combinations of adenine and leucine on which different genotypes produce different results. For example, culture 1 obviously requires leucine but not adenine, so its genotype must be $ad^+ leu^-$. Four different genotypes are present in the 20 cultures. Min = the basic "minimal" medium for wild-type *Neurospora,* consisting of simple inorganic salts, an energy (carbon) source, and a nonnutritive gel called agar.

In modern genetics, however, mutant genes have another role in which their precise function *is* important. We have already referred (in Chapter 2) to genetic dissection as an established approach to biological analysis. Mutant genes are like probes, which can be used to disassemble the constituent parts of a biological function and to examine their workings and interrelationships. Thus, it is of considerable interest to a biologist studying a particular function to have as many mutant forms affecting that function as possible. This has led to "mutant hunts" as an important prelude to any genetic dissection in biology. To identify a genetic variant is to identify a component of the biological process. We explore this idea further later in the chapter.

Message Mutations can be used for two purposes: (1) to study the process of mutation itself, and (2) to permit the genetic dissection of a biological function.

Mutation Detection Systems

The tremendous stability and constancy of form of species from generation to generation suggest that mutation must be a rare process. This supposition has been confirmed, creating a problem for the geneticist who is trying to demonstrate mutation.

The prime need is for a detection system — a set of circumstances in which a mutant allele will make its presence known at the phenotypic level. Such a system ensures that any of the rare mutations that might occur will not be missed.

One of the main considerations here is that of dominance. The system must be set up so that recessive mutations will not be masked by a paired dominant normal allele. (Dominant mutations are less of a problem.) As an example, we can use one of the first detection systems ever set up — the one used by Lewis Stadler in the 1920s to study mutation in corn from *C*, expressed phenotypically as a colored kernel, to *c*, expressed as a white kernel. Here we are dealing with the phenotype of the endo-

TABLE 7-1. Operational basis for detection of mutant phenotype in three different types of mutations

Genotype	Conditional mutation (temperature-sensitive)		Auxotrophic mutation		Resistant mutation	
	Low temperature	High temperature	Without supplement	With supplement	Without agent	With agent
Wild-type	Normal	Normal	Growth	Growth	Growth	**No growth**
Mutant	Normal	**Mutant**	**No growth**	Growth	Growth	Growth

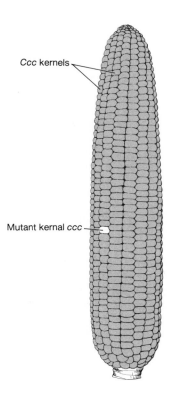

Figure 7-7. Detecting mutants by screening large numbers of progeny (kernels) in a corn cross of *cc* ♀ × *CC* ♂.

sperm of the seed. If you check back to Figure 3-33 (page 64), you will see that this tissue is formed by the fusion of two identical haploid female nuclei with one haploid nucleus from the male pollen cell. Hence, the tissue has three chromosome sets (it is 3*n*). This does not seriously complicate the genetic analysis because dominance generally still works in the presence of two recessive alleles.

Stadler crossed *cc* ♀ × *CC* ♂ and simply examined thousands of individual kernels on the corn ears that resulted from this cross. Each kernel represents a progeny individual. In the absence of mutation, every kernel would be *Ccc* and show the colored phenotype. Therefore, the presence of a white kernel indicates mutation from *C* to *c* in the *CC* parent. Although laborious, this is a very straightforward and reliable method of mutant detection (Figure 7-7).

This basic system can be extended to as many loci as can be conveniently made heterozygous in the same cross. For example

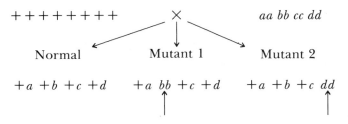

By increasing the number of specific loci under study, the investigator increases the likelihood of detecting a mutation in the experiment. Hence, this type of test is called a **specific locus test** for detecting mutation. It has been used extensively in corn and in mammalian genetics.

In *Tradescantia* plants of the vegetatively propagated strain called 02, there is a simple detection system for somatic mutations. These plants are heterozygous for dominant blue and recessive pink pigmentation alleles of one gene. The pigmentation is expressed in the flower parts: the petals and the stamens. In this plant, the stamens have hairs that are chains of single cells. Millions of single cells can be screened for pink cells representing somatic mutations of the blue allele to the pink allele (Figure 7-8).

In humans, mutation detection is simple in theory but difficult in practice. For example, dominant mutations arising anew in a pedigree should be clearly recognizable. However, there is always the possibility that a person may carry the dominant gene without expressing it; this possibility presents complications in genetic analysis. Thus, the most convincing cases of dominant mutations are shown for conditions known to have high penetrance and expressivity (see page 83). Autosomal recessive mutations, on the other hand, may go unnoticed because they can be transmitted for many generations without being manifested in the phenotype.

Haploids have a great advantage over diploids in mutation studies. The system of detecting haploid mutations is quite straightforward: any newly arising allele will announce its presence unhampered by any dominant partner allele. In fact, the question of dominance or recessiveness need never arise: there is what amounts to a built-in detection facility. In some cases, a direct iden-

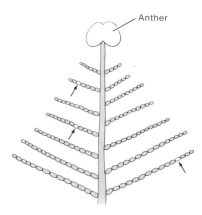

Figure 7-8. *Tradescantia* stamen heterozygous for *P* (blue) and *p* (pink) alleles. In the chains that constitute the lateral hairs, some cells (marked by arrows) are mutant. Shading represents a blue color; lack of shading represents a pink color.

tification of mutants is possible. In *Neurospora,* for example, auxotrophic adenine-requiring mutants have been found to map at several loci. One of these gene loci (*ad-3*) is unique in that auxotrophic mutants accumulate a purple pigment in their cells when grown on a low concentration of adenine. Thus, auxotrophic mutants of this gene may be detected simply by allowing single asexual spores to grow into colonies on a medium with limited adenine. The purple colonies can be identified easily among the normal white colonies.

What about other auxotrophs? Usually, there are no visual pleiotropic effects, as are exhibited with *ad-3*. We will examine the most commonly used detection technique, called replica plating, later in this chapter.

How Common Are Mutations?

If a detection system is available, the investigator can set out to find mutations. One thing will be readily apparent: mutations are in general very rare. This is shown in Table 7-2, which presents some data Stadler collected while working with several corn loci. Mutation studies of this sort are a lot of work! Counting a million of *anything* is no small task. Another feature shown by these data is that different genes seem to generate different frequencies of mutations; a 500-fold range is seen in the corn results. Obviously, one of the prime requisites of mutation analysis is to be able to measure the tendency of different genes to mutate. Two terms are commonly used to quantify mutation:

1. **Mutation rate.** The mutation rate is a number that represents an attempt to measure the probability of a specific kind of mutation event occurring over a specific unit of time. This is obviously getting close to the intrinsic mutation tendency of a gene. Time can be one of several units measured. Instead of an actual time unit, such as hours or days, a unit such as

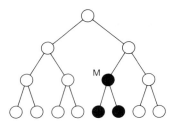

Figure 7-9. A simple cell pedigree, showing a mutation at M.

an organismal generation, cell generation, or cell division is normally used. Consider the lineage of cells in Figure 7-9.

Obviously, only one mutation event (M) has occurred, so the numerator of a mutation rate is established. But what can we use as the denominator? The total opportunity for mutation—the "time" element—may be represented either by the total number of lines in Figure 7-9 (14 total generations) or, alternatively, by the total number of actual cell divisions (7). Either is acceptable, if stated clearly (for example, one mutation per seven cell divisions). (Note that we are dealing with generations—not with generation cycles, of which there are three.)

In practice, mutation rates are not easy to obtain in most organisms. The reason is that the precise number of proliferative steps which generate a population of cells must be very carefully monitored in some way. We shall consider a specific example later in the chapter.

2. **Mutation frequency.** The mutation frequency is the frequency at which a specific kind of mutation (or mutant) is found in a population of cells or individuals. The cell population can be of gametes (in higher organisms) or of asexual spores, or of almost any other cell type. This variable is much easier to measure than the mutation rate. In our example in Figure 7-9, the mutation frequency in the final population of eight cells would be $\frac{2}{8} = 0.25$.

Some mutation rates and frequencies are shown in Table 7-3.

Selective Systems

The rarity of mutations is a problem if the investigator is trying to amass a collection of a specific type for genetic study. Geneticists respond to this problem in two ways. One approach is to use **selective systems** of detection, techniques specially designed to facilitate picking out the desired mutant types from among the rest of the individ-

TABLE 7-2. Forward-mutation frequencies at some specific corn loci

Gene	Number of gametes tested	Number of mutations	Average number of mutations per million gametes
$R \longrightarrow r$	554,786	273	492.0
$I \longrightarrow i$	265,391	28	106.0
$Pr \longrightarrow pr$	647,102	7	11.0
$Su \longrightarrow su$	1,678,736	4	2.4
$Y \longrightarrow y$	1,745,280	4	2.2
$Sh \longrightarrow sh$	2,469,285	3	1.2
$Wx \longrightarrow wx$	1,503,744	0	0.0

TABLE 7-3. Mutation rates or frequencies in various organisms

Organism	Mutation	Value	Units
Bacteriophage T2 (bacterial virus)	Lysis inhibition $r \rightarrow r^+$ Host range $h^+ \rightarrow h$	1×10^{-8} 3×10^{-9}	*Rate:* mutant genes per gene replication
Escherichia coli (bacterium)	Lactose fermentation $lac \rightarrow lac^+$ Histidine requirement $his^- \rightarrow his^+$ $his^+ \rightarrow his^-$	2×10^{-7} 4×10^{-8} 2×10^{-6}	*Rate:* mutant cells per cell division
Chlamydomonas reinhardtii (alga)	Streptomycin sensitivity $str\text{-}s \rightarrow str\text{-}r$	1×10^{-6}	
Neurospora crassa (fungus)	Inositol requirement $inos^- \rightarrow inos^+$ adenine requirement $ad^- \rightarrow ad^+$	8×10^{-8} 4×10^{-8}	*Frequency* per asexual spore
Corn	See Table 7-2		
Drosophila melanogaster (fruit fly)	Eye color $W \rightarrow w$	4×10^{-5}	
Mouse	Dilution $D \rightarrow d$	3×10^{-5}	
Human *to autosomal dominants*	Huntington's chorea Nail–patella syndrome Epiloia (predisposition to type of brain tumor) Multiple polyposis of large intestine Achondroplasia (dwarfism) Neurofibromatosis (predisposition to tumors of nervous system)	0.1×10^{-5} 0.2×10^{-5} $0.4\text{--}0.8 \times 10^{-5}$ $1\text{--}3 \times 10^{-5}$ $4\text{--}12 \times 10^{-5}$ $3\text{--}25 \times 10^{-5}$	*Frequency* per gamete
to X-linked recessives	Hemophilia A Duchenne's muscular dystrophy	$2\text{--}4 \times 10^{-5}$ $4\text{--}10 \times 10^{-5}$	
bone-marrow tissue-culture cells	Normal $\rightarrow$ azaguanine resistance	7×10^{-4}	*Rate:* mutant cells per cell division

SOURCE: R. Sager and F. J. Ryan, *Heredity,* John Wiley, 1961.

uals. The other approach is to try to increase the mutation rate using **mutagens,** agents that have the biological effect of inducing mutations above the background (or spontaneous) rate.

Message   Obtaining rare mutations is facilitated by the use of selective systems and/or mutagens.

Selective systems are many and varied. Their scope is immense and is limited only by the ingenuity of the experimenter. But they all have one thing in common: elevated **resolving power:** a selective system can automatically distinguish between (or resolve) two alternative states—in this case, mutant and nonmutant. In other words, the selective system lets the material, not the experimenter, do the resolving work. We shall see that resolving power is an important aspect of many areas of genetic analysis. It is especially important in that it allows the experimenter to select rare events of any kind. (Remember that rare exceptions are often the key to understanding the normal situation.) Most of the examples of selective systems presented are in microorganisms. This doesn't mean that selective systems are impossible in higher organisms, but merely that selection can be used to much better advantage in microbes. A million spores or bacterial cells are easy to produce; a million mice, or even a million fruit flies, involve a large-scale commitment of money, time, and laboratory space. Microbes appear frequently in the discussion that follows, so a few words on culturing and routine microbial manipulation are appropriate here.

Microbial Techniques

The microbes that we consider in this book are bacteria, fungi, and unicellular algae, all of which can be regarded as haploid. But whereas fungi and algae are eukaryotic (have their chromosomes in a nucleus surrounded by a nuclear membrane), bacteria are **prokaryotic** (their chromosomes are not enclosed in a separate compartment of any kind).

In liquid culture, prokaryotic organisms proliferate as suspensions of individual cells. Each starting cell goes through repeated cell divisions, so that, from each, a series of 2, 4, 8, 16, . . . descendant cells is produced. This exponential growth is limited by the availability of nutrients in the medium, but a dense suspension of millions of cells is soon produced. Such fungi as yeasts follow this growth pattern exactly, but a slightly different situation occurs in the mycelial fungi. Here the descendant cells remain attached as long chains called hyphae; thus, a liquid culture started from asexual spores eventually looks like tapioca pudding, with small fuzzy balls of hyphae in suspension, each originating from one spore.

In solid culture, usually on an agar-gel surface, descendant cells again tend to stay together, so that colonies are produced. One colony develops from each original cell in the suspension that is spread on the surface of the culture medium.

It is usually necessary to know how many cells you have in a culture. A suspension of cells may be counted by using one of several methods:

1. **Microscope counts.** The suspension is placed in a chamber of known depth that is marked off in a grid of known dimensions. This device is known as a hemocytometer. The hemocytometer is placed under the microscope, and the cells are counted directly.

2. **Turbidity.** Bacteria are too small to be counted conveniently under a light microscope, so they often are counted using turbidity measurements. The amount of light passing through suspensions of known density is measured, and a turbidity calibration curve is drawn. Suspensions of unknown density can then be checked off on this calibration curve.

3. **Colony-forming units.** Suspensions of known volume and appropriate dilution are **plated** (spread) on the surface of a nutrient medium solidified with agar (Figure 7-10). Each cell will form a colony, and the colonies can be counted with the unaided eye.

4. **Electronic cell counters.**

Figure 7-11 summarizes some of these methods.

What kinds of phenotypes may be examined in microorganisms? Morphological mutations affecting color, shape, and size of colony are useful but of limited occurrence. Other characters have been far more useful—as examples, auxotrophic mutations (can these cells grow without this specific supplement?), resistance mutations (can these cells grow on this growth inhibitor?), and substrate-utilization mutants (can these cells utilize this sugar as an energy source, as wild-types can?).

Figure 7-10. Plating microbial cells. A cell suspension of appropriate density is being poured over the surface of a plate of medium. Each cell will ultimately produce a visible colony. This is one of the routine techniques used by microbial geneticists and will be referred to several times later in this book.

We now return to selective systems and some examples thereof.

Reversion of Auxotrophs

For the detection of the reversion of auxotrophy to prototrophy, there is a direct selection system. Take an adenine auxotroph, for example. A culture of the auxotrophic mutant is grown on an adenine-containing medium. The cells are then plated on a solid medium containing no adenine. The only cells that can proliferate (grow and divide) on this medium are adenine prototrophs, which must have arisen by reverse mutation in the original culture (Figure 7-12). For most genes (not just those concerned with nutrition), the rate of reversion is generally much lower than the rate of forward mutation. (We shall explore the reason for this later.)

Filtration Enrichment

Filtration enrichment is used in mycelial fungi to select specific auxotrophic mutants. A suspension of (predominantly) prototrophic spores is grown in a liquid culture with no growth supplements. (Recall that such a minimal

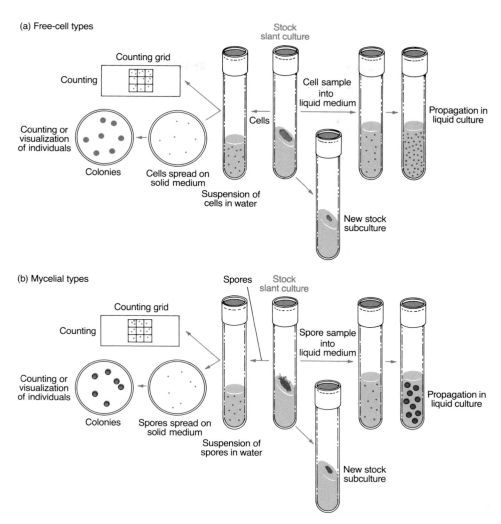

Figure 7-11. Some culture and counting techniques for microorganisms. (a) Free-cell types. (b) Mycelial types.

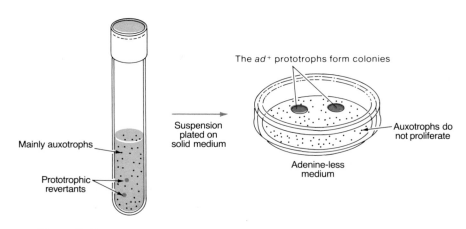

Figure 7-12. Selection for prototrophic revertants of auxotrophic mutants.

medium consists of inorganic salts and an energy source like sugar.) Any auxotrophic mutants that arose in the original culture will not grow in this medium, but prototrophs of course will. The prototrophic colonies may be filtered off using a glass-fiber filter, which allows the auxotrophic cells to pass through. These auxotrophs will be of several different types: some requiring A, some B, and so on. If we are specifically interested in adenine auxotrophs, then we plate the heterogeneous suspension that comes through the filter on a medium supplemented with adenine. Only adenine auxotrophs will respond to this medium and form colonies (Figure 7-13).

Penicillin Enrichment

An analogous technique is available for auxotroph selection in bacteria. Bacteria are in the main highly sensitive to the antibiotic penicillin, but only the proliferating cells are sensitive. If penicillin is added to a suspension of cells in liquid culture, all the prototrophs are killed because they proliferate, but the auxotrophs survive! The penicillin can be removed by washing the cells on a filter.

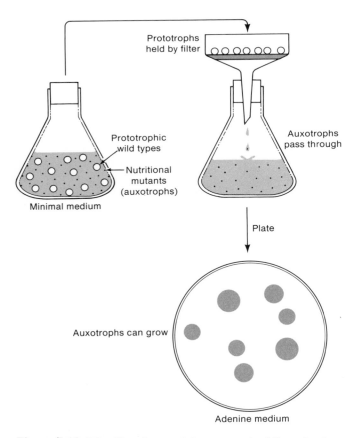

Figure 7-13. The filtration-enrichment method for selecting forward mutations to auxotrophy in filamentous organisms. (This example involves mutation to adenine requirement.)

Then plating on a medium supplemented with a specific chemical will reveal colonies of the auxotrophs that specifically require that compound.

Resistance

Resistance to specific environmental agents not normally tolerated by wild-types is easily demonstrated in microorganisms. We use an example that was important historically in determining the nature of mutation.

Viruses, like most parasites, are highly specific with regard to the hosts they will parasitize. Some viruses are specific to bacteria; these are called either bacterial viruses, or **bacteriophages,** or **phages.** Phages have played a major role in the elucidation of our present level of understanding of molecular genetics. They appear frequently throughout the rest of the book. Although they are introduced properly in the chapter on bacterial genetics, we need them at this point to demonstrate resistance.

The intestinal bacterium *Escherichia coli* is parasitized by many specific phages. One of these, called T1, was used in early bacterial mutation studies. T1 will attack and kill most *E. coli* cells, liberating a host of fresh viruses from each dead cell. T1 is a subcellular lollipop-shaped particle that can be seen only under an electron microscope, but the progress of a phage infection can be followed on a plate. If a plate is spread with large numbers of bacteria (around 10^9) and phages, most of the bacteria are killed. However, T1 phage-resistant bacterial cells survive and produce colonies that can be isolated. These individuals are called T-one resistant (Ton^r) bacterial cells.

During these early studies on the selection of variants, the origin of these Ton^r bacterial mutants was questioned by Salvadore Luria and Max Delbrück (in 1943) in a classic experiment that was highly relevant to the study of mutation in general and, more specifically, to all selection experiments to follow. Although Ton^r individuals are obviously genuine mutants—after all, they represent a stable inherited phenotype—there was doubt about how they originated. Were the Ton^r colonies derived from cells that were genetically Ton^r *before* the exposure to T1 phage? (The Ton^r cells would have originated through random genetic change.) Or did the Ton^r genotype arise in response to the T1 exposure through a kind of physiological adaptation?

How is it possible to distinguish between these two alternatives? Let's assume that the resistant cells result from a random genetic change that can occur at any time during growth. If we initiate a bacterial culture with a small number of cells and the population then increases, random genetic change to resistance to T1 phage could occur in any cell at any time. If we take a large number of small populations of cells and let each one expand into a

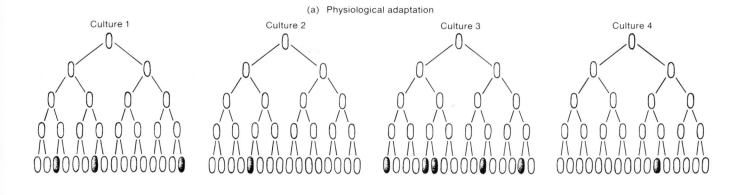

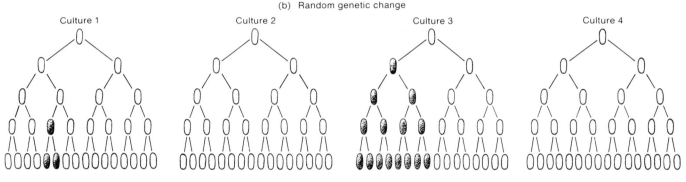

Figure 7-14. Cell pedigrees illustrating the expectations from two contrasting theories about the origin of resistant cells. (a) Physiological adaptation. (b) Random genetic change. (From G. S. Stent and R. Calendar, *Molecular Genetics*, 2d ed. Copyright © 1978 by W. H. Freeman and Company.)

large population that is then exposed to T1 phage, the number of mutants in each population will vary considerably, depending on *when* the change occurred. On the other hand, if each cell has the same probability of becoming resistant through physiological adaptation, then in each population of cells there should be the same general frequency of survivors and little variation of that value (Figure 7-14).

Luria and Delbrück carried out such a test. Into each of 20 culture tubes containing 0.2 milliliters (ml) of medium and into one containing 10 ml, they introduced 10^3 *E. coli* cells per milliliter and incubated them until about 10^8 cells per milliliter were obtained. Each of the 20 0.2-ml cultures was then spread on a plate that had a dense layer of T1 phages. From the 10-ml "bulk" culture, 10 0.2-ml volumes were withdrawn and plated. Many colonies were T1-resistant, as shown in Table 7-4.

A tremendous amount of variation from plate to plate was seen in the individual 0.2-ml cultures but not in the samples from the bulk culture (which represent a kind of control). This situation cannot be explained by physiological adaptation, because all the samples spread had the same approximate number of cells. The simplest explanation is random genetic change, occurring either early (large number of resistant cells), or late (few resistant cells), or not at all (no resistant cells) in the 0.2-ml cultures.

This elegant analysis suggests that the resistant cells are *selected* by the environmental agent (here, phages) rather than produced by it. Can the existence of mutants in a population *before* selection be directly demonstrated? This was done by means of the technique of **replica plating,** developed by Joshua Lederberg and Esther Lederberg in 1952. A sterile piece of velvet placed lightly on the surface of a plate will pick up cells wherever there is a colony (Figure 7-15). (Under a microscope, velvet looks like a series of needles, which explains why lint adheres to it so well and why it picks up colonies of cells.) On touching the velvet to another sterile plate, some of the cells clinging to the velvet are inoculated onto the plate in the same relative positions as the colonies on the original "master" plate. This simple technique allows the investigator to grow cells on a nonselective medium (either with complete nutrients or free of antibiotics or phages) and then to transfer a copy of the colonies to a selective medium (either minimal or

TABLE 7-4. Results of Luria and Delbrück's test

Individual cultures		Bulk culture	
Culture number	Number of T1-resistant colonies	Culture number	Number of T1-resistant colonies
1	1	1	14
2	0		
3	3	2	15
4	0		
5	0	3	13
6	5		
7	0	4	21
8	5		
9	0	5	15
10	6		
11	107	6	14
12	0		
13	0	7	26
14	0		
15	1	8	16
16	0		
17	0	9	20
18	64		
19	0	10	13
20	35		
Mean	$\overline{11.3}$	Mean	$\overline{16.7}$

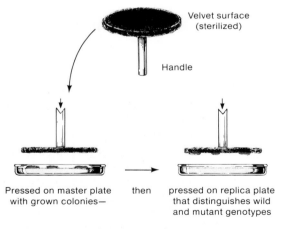

Figure 7-15. Replica-plating methodology. Replica plating is used to identify mutant colonies on a master plate through their behavior on selective replica plates. (From G. S. Stent and R. Calendar, *Molecular Genetics,* 2d ed. Copyright © 1978 by W. H. Freeman and Company.)

Velvet surface (sterilized)

Handle

Pressed on master plate with grown colonies— then pressed on replica plate that distinguishes wild and mutant genotypes

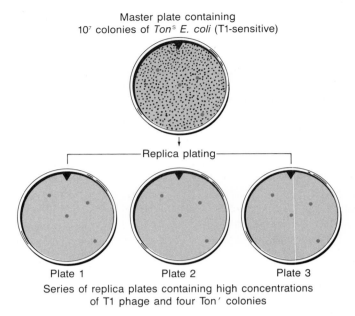

Master plate containing 10^7 colonies of Ton^s *E. coli* (T1-sensitive)

Replica plating

Plate 1   Plate 2   Plate 3

Series of replica plates containing high concentrations of T1 phage and four Ton$'$ colonies

Figure 7-16. Replica plating to demonstrate the presence of mutants before selection. The identical patterns on the replicas show that the resistant colonies are from the master. (From G. S. Stent and R. Calendar, *Molecular Genetics,* 2d ed. Copyright © 1978 by W. H. Freeman and Company.)

with antibiotics or phages), as shown in Figure 7-16. The cells from some of the colonies on the master plate form colonies when transferred to the selective plate; in these cases, cells from those colonies on the master plate can be retested. If they are found to be mutant (in this example, resistant), the investigator has proof that the mutation occurred *before* any selection was applied for the mutant. (It should be noted that there are some situations in which physiological adaptation can occur. The point for now is that mutations do not arise by physiological adaptation.)

> **Message** Mutation is a random process that can occur in any cell at any time.

Incidentally, replica plating has become an important technique of microbial genetics. In general, it is a way of retaining an original set of strains on a master plate while simultaneously subjecting them to a large battery of tests on various kinds of replica plates.

A Way of Calculating Mutation Rate

In passing, let's see how Luria and Delbrück's test, called a **fluctuation test,** provides a way of calculating the mutation rate according to its strict definition. Consider the 20 cultures that were tested for T1 resistance. Here we have a situation that is well described by the Poisson distribution—a great opportunity for mutations to be found, but a large class having no mutations at all in the culture ($\frac{11}{20}$). Look back at Figure 7-9, and convince yourself that the number of cell divisions necessary to produce n cells is n minus the original number in the culture (it would be $8 - 1 = 7$ in the example in Figure 7-9). If n is very large, which it is in the fluctuation-test cultures, and if the original number of cells was relatively very small, then a sufficiently accurate estimate of the number of cell divisions is given by n itself. If the mutation rate per cell division is μ, then each culture tube may be expected to have entertained an average of μn mutation events. The Poisson distribution (see page 121) then tells us that the zero class (no mutants detectable) will equal $e^{-\mu n}$. Because we know that the zero class $= \frac{11}{20} = 0.55$ and that $n = 0.2 \times 10^8$, we can solve the following equation for μ:

$$0.55 = e^{-\mu(0.2 \times 10^8)}$$

We find that $\mu = 3 \times 10^{-8}$ mutation event per cell division. (Note that in the same data, the mutation frequency is $(1 + 3 + 5 + 5 + 6 + 107 + 1 + 64 + 35)/(20 \times 0.2 \times 10^8)$, which equals $227/(4 \times 10^8)$, or 5.7×10^{-7}.)

Microbial-Like Selection Techniques in Cell Culture of Higher Organisms

There are techniques for growing cells of higher organisms (animals and plants) in culture. Cells taken from certain tissues, including cancerous tissue, will proliferate much like microbial cells in culture. Furthermore, many of the techniques used for mutant induction and selection in microbes can also be used on these higher cells, permitting the flourishing of **somatic cell genetics** of higher organisms as an active area of research. The combination of these mutation-selection techniques with sophisticated techniques of cell fusion and hybridization now makes possible a wide variety of in vitro manipulations on higher cells, including human cells.

One extensively worked cell-culture system uses Chinese hamster ovary (CHO) cells. In this system (and in others), many mutant cell types have been isolated. Let's have a look at some examples. In the first place, dominant mutations are common. Dominance is defined in terms of the phenotype observed when the mutant cell line is fused with normal cells. Some mutations that confer resistance to a particular drug, such as ouabain, α-amanitin, methotrexate, or colchicine, are dominant. Dominant mutations are not surprising: we might expect to find *only* dominant mutations because the cells are diploid.

Nevertheless, recessive mutants are common, too! Resistance to lectins, 8-azaguanine, or methotrexate, the auxotrophic requirement for glycine, proline, or lysine, and various temperature-sensitive lethals are all examples of recessive mutants. (Note that certain kinds of mutant phenotypes can be either dominant or recessive in different mutants.)

How is it possible for recessive mutants to show up in diploid CHO cells? No one has the complete answer to this question. However, there is some evidence that although CHO cells are predominantly diploid, they are hemizygous for very small sections of the genome. Thus, recessive mutations would be expressed if they occurred in such regions. But whatever the mechanism of such mutations, their existence provides an invaluable tool for the mammalian-cell somatic geneticist.

Single cells of plants may also be grown in culture. Cells are plated on a synthetic medium containing appropriate combinations of plant hormones. The resulting colonies, called **calluses,** can be subdivided for further growth in culture or they can be transferred to a different medium that promotes differentiation of roots and shoots. Such "plantlets" can then be potted in soil and grown into mature plants. In one application of this technique, selection was imposed for cell resistance to a toxin produced by a fungal plant parasite. Calluses growing on the toxin-containing plates were processed to produce plants that also showed resistance to the fungal toxin.

Mutation Induction

The task of finding mutations in multicellular organisms, compared with that in microorganisms, is tremendously complex. In 1928, Hermann J. Muller devised a method of searching for any lethal mutation on the X chromosome in *Drosophila*. He first constructed a chromosome called ClB, which carries a chromosomal inversion (labeled C for crossover suppressor; see pages 181–183), a lethal (l), and the dominant bar-eye marker (B). *ClB*/Y males die due to hemizygosity for the lethal gene, but the chromosome can be maintained in heterozygous *ClB*/+ females. By mating wild-type males with *ClB*/+ females, Muller could then test for lethal mutations anywhere on the X chromosomes in samples of the male gametes by mating single *ClB*/+ F_1 females with other wild-type males (Figure 7-17). We can see that

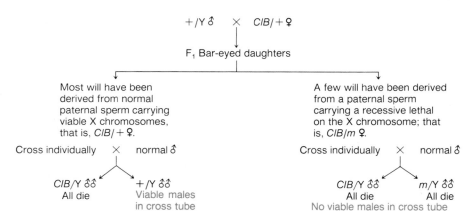

Figure 7-17. The ClB test for mutant detection in *Drosophila*. The symbol *m* represents a recessive lethal mutation anywhere on the X chromosome.

if a mutation has occurred on an X chromosome in one of the original male gametes sampled, then the F_1 female carrying that chromosome will not produce any viable male progeny. Note that only rarely will the new lethal mutation *m* on the X chromosome be an allele of the lethal gene on the ClB chromosome (of course, when it is, the ClB/l female will die). The absence of males in a vial is very easily seen under low-power magnification and readily allows scoring for the presence of lethal mutations on the X chromosome. Muller found their spontaneous frequency to be about 1.5×10^{-3}, still a relatively low value for an entire chromosome with its many different genes.

Muller then asked whether there were any agents that would increase the rate of mutation. Using the ClB test, he scored for sex-linked lethal frequencies after irradiating males with X rays and discovered a striking increase; his results supplied the first experimental evidence of a mutagen — in this case, the X rays. Recall that a mutagen is an agent that causes mutation to occur at higher than spontaneous levels. Mutagens have been invaluable tools not only for studying the process of mutation itself but for increasing the yield of mutants for other genetic studies. It is now known that many kinds of radiation will increase mutations.

Table 7-5 shows the effects of several types of radiation on increasing mutation frequencies in *Drosophila*. To the list in Table 7-5 must be added ultraviolet (UV) radiation, which is also mutagenic. Radiation is often categorized as ionizing or nonionizing, depending on whether or not ions are produced in the tissue through which it passes. X rays and γ (gamma) rays are often-used examples of the former, and UV radiation, of the latter.

The harnessing of nuclear energy for weapons and fuel has become a social issue due to the mutagenic effect of radiation. The pros and cons of the use of nuclear

energy must be weighed by each of us individually, but here let us address a relevant point that is often poorly understood. For any organism, the vast majority of newly formed mutations are deleterious. But, we may ask, if organisms evolved through an advantage conferred by a mutant condition, why aren't many mutations improvements? To answer this question, let's use an analogy often cited in this connection in which a cell is compared to a highly complicated watch that has many delicate parts. This watch evolved through generations of minor changes in the design of the machine. If we expose the cogs and wheels by removing the back casing, close our eyes, and plunge a thick needle (analogous to a mutagen) into the workings, there is a remote possibility that this random hit will improve the efficiency of the watch, but the chances are overwhelmingly high that any change inflicted in this way will damage it. So it is with a cell or an organism. (We consider good molecular

TABLE 7-5. Relative efficiencies of various types of radiation in producing mutations in *Drosophila*

Type of radiation	Sex-linked recessive lethals per 1000 roentgens*	Irradiated male X chromosomes (%)
Visible light (spontaneous)	0.0015	0.15
X rays (25 Mev)	0.0170	1.70
β rays, γ rays, hard X rays	0.0290	2.90
Soft X rays	0.0250	2.50
Neutrons	0.0190	1.90
α rays	0.0084	0.84

* The roentgen (r) is a unit of radiation energy.

reasons for the detrimental nature of mutations in Chapter 12.)

> **Message** Most mutations are deleterious. Evolution is possible because a very few are not.

The kinds of gene mutations considered in this chapter are point mutations. (Recall that such mutations affect one point on the chromosome, the gene in question; a more precise meaning at the molecular level comes later.) It is clear that within a certain range of radiation dosage, point-mutation induction is linear; that is, if we double or halve the radiation level, the number of mutants produced will vary accordingly. From the kind of graph shown in Figure 7-18, it is possible to extrapolate to very low radiation levels and to infer very low frequencies of mutation induction. Because exposure to radiation from X-ray machines, to radioactive fallout from bomb testing, and to contamination from nuclear plants is very low, it is possible to conclude that the effects are negligible. Yet even though

increases in mutation rates might be low, population numbers are large: every year 200 million new gametes form 100 million new babies in the world. In this very large annual "mutation experiment," even low mutation frequencies are potentially translatable into large numbers of mutant babies.

Radiation doses are cumulative. If a population of organisms is repeatedly exposed to radiation, the frequency of mutations induced will be in direct proportion to the *total amount* of radiation absorbed over time. However, there are exceptions to both additive and cumulative effects. For example, if mice given *x* rads (a biological measure of a dose of radiation) in one short burst (called an "acute" dose) are compared with those given the same dose gradually over a protracted period of weeks or months (a "chronic" dose), significantly fewer mutations are found in the chronically exposed group. This has been interpreted to mean that there is some form of repair of radiation-induced genetic damage over time.

Of even greater importance for geneticists was the discovery that certain chemicals also may be mutagenic. The first demonstration of chemical mutagenesis was made in 1947 by Charlotte Auerbach and J. M. Robson, who conducted experiments on mustard gases that were used earlier in gas warfare. (You might be interested to know that their discovery was kept from publication by the British military for several years.) This initial study opened a floodgate to research into the mutagenic effects of a wide variety of chemicals.

The mutagenicity of chemicals is an important phenomenon because, in many cases, the chemical reactions responsible for the mutagenic action of a compound can be determined, thereby providing a clue to the molecular basis of the mutation. Furthermore, many chemicals are much less toxic to an organism than radiation is and yet result in much higher frequencies of mutation. So, as a tool, chemical mutagens have been used to induce much of the wide array of mutants now available for genetic studies. Finally, great controversy now exists over the potential mutagenic effects of a host of molecules in the human environment, ranging from caffeine to pollutants, pesticides, and LSD.

Mutagens in Genetic Analysis

Mutations are very useful. In the same way that we can learn how the engine of a car works by tinkering with its parts one at a time to see what effect each has, we can see how a cell works by altering its parts one at a time by means of induced mutations. The **mutation analysis** is a further aspect of the process we have called genetic dissection of living systems.

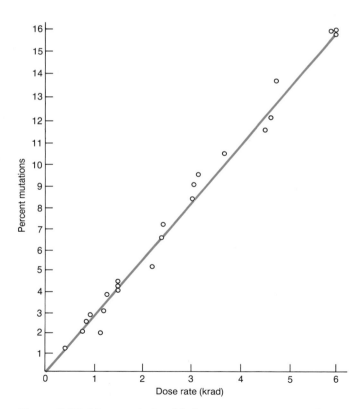

Figure 7-18. Linear relationship between percent mutations (mainly sex-linked recessive lethals) and X-ray dose in *Drosophila melanogaster..*

TABLE 7-6. Forward-mutation frequencies obtained with various mutagens in *Neurospora*

Mutagenic treatment	Exposure time (minutes)	Survival (%)	Number of *ad*-3 mutants per 10^6 survivors
No treatment (spontaneous rate)	—	100	~0.4
Amino purine (1–5 mg/ml)	during growth	100	3
Ethyl methane sulfonate (1%)	90	56	25
Nitrous acid (0.05 M)	160	23	128
X rays (2000 r/min)	18	16	259
Methyl methane sulfonate (20 mM)	300	26	350
UV rays (600 erg/mm² per min)	6	18	375
Nitrosoguanidine (25 μM)	240	65	1500
ICR-170 acridine mustard (5 μg/ml)	480	28	2287

NOTE: The assay measures the frequency of purple *ad-3* colonies among the white colonies produced by the wild-type *ad-3*⁺.

The first stage of any mutational dissection is the **mutant hunt.** Before embarking on a mutant hunt, the investigator usually has some specific biological process in mind. This process must be associated with some recognizable character of an organism. For example, the investigator might be interested in phototropic responses in algae—their tendency to swim toward the light. It would be comparatively easy to establish a selective system to recover mutants defective in this particular response. These mutants would then be tested for their single-gene inheritance, the number of different loci involved in the mutant set, and finally, for specific cellular effects. In this way, the components of the response could be determined unambiguously.

The mutagenic agents are potent tools for the geneticist engaged in mutational dissection. Let's consider a few examples of the great usefulness of mutagens in such research.

First, Table 7-6 shows the relative frequencies of *ad-3* forward mutants in *Neurospora* after various mutagenic treatments. Note that recovery of a single mutant requires the testing of some 2.5 million cells without mutagenic treatment, whereas treatment with ICR-70 produces about one mutant in each 450 surviving cells tested. Table 7-7 provides another example of the great increase in available mutants obtained with an appropriate mutagen. The term **supermutagen** has been introduced to describe some of these highly potent agents.

For a third example, we turn to *Drosophila*. Until the mid-1960s, most researchers working with *Drosophila* used radiation to induce mutations. Doses of 4000

roentgens produce lethal mutations in perhaps 10 to 11 percent of all X chromosomes of irradiated males. At higher doses, however, there is increased infertility in the treated flies. Another problem is that many of the mutations induced by X rays involve chromosome rearrangements, and these can complicate the genetic analysis.

In contrast, the chemical ethyl methanesulfonate (EMS) induces a vastly higher proportion of point mutations. This mutagen is very easily administered simply by placing adult flies on a filter pad saturated with a mixture of sugar and EMS. Simple ingestion of the EMS

TABLE 7-7. The potency of various chemical mutagens for *his*⁻ reversion in *Salmonella*

Mutagen	Revertants/nanomole	Ratio
1,2-epoxybutane	0.006	1
Benzyl chloride	0.02	3
Methyl methane sulfonate	0.63	105
2-naphthylamine	8.5	1400
2-acetylaminofluorene	108	18,000
Aflatoxin B₁	7057	1,200,000
Furylfuramide (AF-2)	20,800	3,500,000

SOURCE: J. McCann and B. N. Ames, in *Advances in Modern Toxicology,* vol. 5. Edited by W. G. Flamm and M. A. Mehlman. Hemisphere Publishing Corp., 1976.

produces very large numbers of mutations. For example, males fed 0.025 molar EMS produce sperm carrying lethals on more than 70 percent of all X chromosomes and on virtually every chromosome 2 and chromosome 3. At these levels of mutation induction, it becomes quite feasible to screen for many mutations at specific loci or for defects with unusual phenotypes.

It is possible to screen mutagenized chromosomes immediately in the F_1 generation if the region of interest is hemizygous. The X chromosome can be tested in this way by crossing EMS-fed males with females carrying an attached-X chromosome and a Y chromosome ($\widehat{X}X/Y$). An attached-X ($\widehat{X}X$) is a compound chromosome formed by the fusing of two separate X chromosomes. It is inherited as a single unit. All F_1 males carry a mutagenized paternal X chromosome, and each fly represents a different treated sperm. Thus, if individual F_1 males are crossed with $\widehat{X}X/Y$ females, each culture will represent a single cloned X. All F_1 zygotes carrying a sex-linked lethal will die, but any newly induced visible mutation will be expressed (Figure 7-19). In this way, it has been possible to select a wide range of behavioral and visible mutants involving a particular region of the genome; this approach is known as **saturating** the region. The F_1 flies can be reared at 22°C, and then clones of each individual X can be established at 22°C, 17°C, and 29°C to permit ready detection of heat-sensitive or cold-sensitive lethals. These temperature-sensitive (ts) lethals are found to represent about 10 to 12 percent of all EMS-induced lethals. We have already examined the utility of such conditional mutants in genetic analysis. Many laboratories now routinely test for ts mutants in any *Drosophila* screening tests. Again, the approach is possible because large numbers of mutants are so easily obtained.

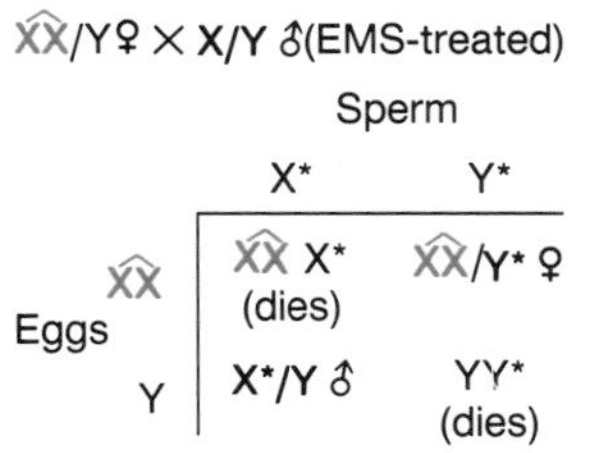

$$\widehat{X}X/Y ♀ \times X/Y ♂ (EMS\text{-treated})$$

Sperm

	X*	Y*
$\widehat{X}X$	$\widehat{X}X$ X* (dies)	$\widehat{X}X/Y*$ ♀
Y	X*/Y ♂	YY* (dies)

Eggs

Figure 7-19. The use of attached-X chromosomes ($\widehat{X}X$) in *Drosophila* to facilitate the recovery of X-linked mutations. Sperm treated with EMS or another mutagen will fertilize eggs containing either the attached-X chromosome or a Y chromosome. The treated X chromosome from the male will show up as the hemizygous X of the sons, revealing phenotypically recessive mutations. (Recall that the ability to carry out this test is dependent on the mechanism of sex determination in *Drosophila*; see Chapters 3 and 21.) The asterisk (*) denotes the chromosome exposed to the mutagens.

The recovery of ts lethals has a side benefit in that it simplifies laboratory procedures. One of the bothersome tasks involved in any large-scale *Drosophila* experiment is the need to separate all females from males within 12 hours after emergence from the pupae to prevent undesired mating. (New males and females do not mate for 12 to 14 hours after emergence.) The tiresome procedure of collecting virgin females can be eliminated by using ts lethals (represented as l^{ts}) to produce unisexual cultures at will. For example, the cross $\widehat{X}X/Y ♀ \times l^{ts}/Y ♂$ produces progeny of both sexes at permissive temperatures. If the culture is shifted to restrictive temperatures, the l^{ts}/Y males die and only the females hatch. Similarly, homozygosis of an l^{ts} in an $\widehat{X}X$ chromosome will produce only wild-type males at restrictive temperatures.

> **Message** The mutagen EMS has revolutionized the genetic versatility of *Drosophila* by providing a potent method for the recovery of a wide range of point mutants.

Mutation Breeding

Mutation is used in other ways than the mutational dissection of biological systems. We saw in Chapter 2 that one way of breeding a better crop plant is to make a hybrid and then to select the desired recombinants from the progeny generations. That approach makes use of the variation naturally found between available stocks or isolates from nature. Another way to generate variability for selection is to treat with a mutagen. In this way, the variability is produced through human intervention.

A variety of procedures may be used. Pollen may be mutagenized and then used in pollination. Dominant mutations will appear in the next generation, and further generations of selfing will reveal recessives. Alternatively, seeds may be mutagenized. A cell in the enclosed embryo of a seed may become mutant, and then it may become part of germinal tissue or somatic tissue. If the mutation is in somatic tissue, any dominant mutations will show up in the plant derived from that seed (Figure 7-20), but this will be the end of the road for such mutations. Germinal mutations will show up in later generations, where they can be selected as appropriate. Figure 7-21 summarizes mutation breeding.

In this chapter, we have seen how gene mutation not only is of biological interest in itself but can also be put to a variety of experimental and practical uses. In the next chapter, we shall see that much the same kind of statement can be made about chromosome mutation.

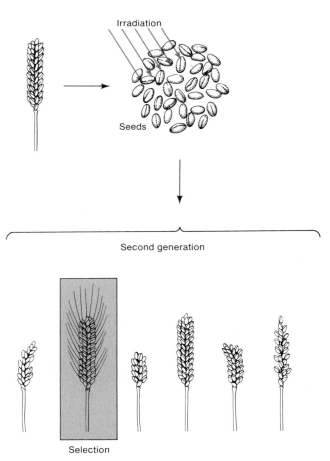

Figure 7-20. Irradiating seeds of *Collinsia grandiflora* with gamma rays to produce mutations in the embryo. (a) The effects of increasing dose (left to right) on survival and vigor. (b) A somatic mutation in a survivor, causing leaf-shape abnormalities. (c) Another somatic mutation causing a mutant sector. Germinal mutations also were recovered in the next generation.

Figure 7-21. Mutation breeding in crops. (Courtesy of Björn Sigurbjörnsson, "Induced Mutations in Plants." Copyright © 1970 by Scientific American, Inc. All rights reserved.)

Summary

■ Gene mutation may occur in somatic cells or germinal cells. Within these two categories, there are several kinds of mutations, including morphological, lethal, conditional, biochemical, and resistant mutations.

Mutations can be used to study the process of mutation itself or to permit genetic dissection of biological functions. In order to carry out such studies, however, it is necessary to have a system for detecting mutant alleles at the phenotypic level. In diploids, dominant mutations should be easily detected; recessive mutations, on the other hand, may never be manifested in the phenotype. For this reason, detection systems are much more straightforward in haploids, where the question of dominance does not arise.

Because spontaneous mutations are rare, geneticists use selective systems or mutagens (or both) to obtain mutations. Selective systems automatically distinguish between mutant and nonmutant states and have been used mainly in microbes. Mutagens are valuable not only in studying the mechanisms of mutation but also in inducing mutations to be used in other genetic studies. In addition, mutagens are frequently used in crop breeding.

■ ■ ■

Solved Problems

1. A certain plant species normally has the purple pigment anthocyanin dispersed throughout the plant. This makes the green parts of the plant look brown (green plus purple) and the parts of the flower that lack chlorophyll (petals, ovary, anthers) look bright purple. The gene A is essential for anthocyanin production, and recessive alleles a result in a lack of anthocyanin. An interesting new allele called a^u arises, which is unstable. The a^u allele reverts back to A at a frequency thousands of times greater than regular (stable) a alleles do.

    a. What phenotype would you expect in plants of genotype (1) $a^u a^u$; (2) $a^u a$; (3) Aa^u?

    b. How can you confirm that the so-called reversion events are true mutations?

Solution

a. (1) We have seen that mutations tend to be very rare; in the normal course of events, they are seldom observed unless they are seriously sought out. However, an allele that is reverting as frequently as a^u is likely to announce its reversion behavior prominently if a proper detection system is available. In a plant containing billions of cells in the somatic tissue, many cells will undoubtedly experience reversion events at some stage of development. As development proceeds, each initial revertant cell will give rise to a clone of revertant cells that should be visible as a purple or brown sector (that is, a spot or blotch). Therefore, a plant of $a^u a^u$ genotype should show white petals with purple blotches of varying size and photosynthetic tissues that are basically green with brown blotches.

(2) Each blotch is derived from a single cell reverting $a^u \rightarrow A$. The allele a^u is expected to be dominant over a, because a is essentially inactive and will not prevent a^u from reverting. Hence, $a^u a$ plants will look the same as $a^u a^u$, but possibly with fewer blotches because there are two chances for reversion in $a^u a^u$ cells.

(3) Since A produces pigment in all cells, Aa^u will be indistinguishable from AA: even though a^u is reverting, the revertant events have no chance to show up —a detection system is lacking. Such plants will be purple-brown throughout.

b. How can we prove that a mutation is a mutation? And why should we bother? In the latter connection, it is worth mentioning that some spotting patterns (as in, for example, Dalmatian dogs) are not caused by mutant clones but rather by special developmental regulatory effects. So spotting does not prove mutation— a skeptic could invoke some other cause.

The key is that genes are hereditary units and mutant genes should be transmitted to subsequent generations of cells or individuals. How could we detect the transmission of A revertant alleles to progeny cells or individuals? Intuitively, we feel that if a few cells could be scraped out of a spot and grown into a plant, that plant would be purple and the reversion hypothesis would be proved. Well, the technology for doing such an experiment is almost (but not quite) available at the time of this writing, but it is a good idea. Is there any other way? If reversion is seen as blotches in the flower, then presumably the germinal tissue (anthers and ovaries) should sometimes be part of a blotch. Therefore, one could collect pollen from a heavily blotched flower and use this to pollinate a plant of genotype aa. If there is some A pollen and if enough progeny plants are examined, then some should be Aa and purple-brown throughout. This would prove the reversion hypothesis because there is nowhere else the A gene could have come from but reversion.

2. In yeast, a mutation experiment is performed on the *tryp4* locus. A *tryp4* mutation confers a requirement for the amino acid tryptophan. A *tryp4* allele named *tryp4-1* was known to be revertable; the experiment was to measure reversion frequency in a population of haploid cells. A culture of mating type α and of genotype *tryp4-1* was grown and 10 million cells were plated on a medium lacking tryptophan; 120 colonies were obtained. The genotypes of these colonies were checked by crossing them to a wild-type culture of mating type a. Based on the results of these crosses, it was found that the colonies were of two types: two-thirds of type 1 and one-third of type 2:

Type 1 α × wild-type $a \longrightarrow$ progeny all tryptophan-independent

Type 2 α × wild-type $a \longrightarrow \frac{3}{4}$ progeny tryptophan-independent

$\frac{1}{4}$ progeny tryptophan-requiring

a. Propose a genetic explanation for the two types.

b. Calculate the frequency of revertants.

Solution

a. The technique of plating on a medium lacking tryptophan should be a selection system for $tryp^+$ revertants, because they do not require tryptophan for growth. When backcrossed to a wild type ($tryp4\text{-}1^+$), all the progeny should be tryptophan-independent. This behavior is exhibited by type 1 colonies, so they represent revertants.

Now what about the type 2 colonies? The fact that some progeny are tryptophan-requiring shows that the $tryp4$ mutation could not have genuinely reverted. Rather, the requirement for tryptophan appears to have been merely masked or suppressed. We have already studied several examples of the suppression of a mutant allele by a new mutation at a separate locus. But even if we had not remembered about these examples, the 3 : 1 ratio in a haploid organism should provide the clue that two independent loci are involved. Let's designate the suppressor mutation as su and its inactive wild-type allele as su^+. Type 2 colonies are then $tryp4\text{-}1\ su$, and the wild-types are $tryp4\text{-}1^+\ su^+$. A cross of these strains produces

$$\frac{1}{2}\,tryp \begin{cases} \frac{1}{2}+ \longrightarrow \frac{1}{4}\text{ requires tryptophan} \\ \frac{1}{2}\,su \end{cases}$$
$$\frac{1}{2}+ \begin{cases} \frac{1}{2}+ \\ \frac{1}{2}\,su \end{cases} \left.\right\} \frac{3}{4}\text{ does not}$$

(Notice that su has no effect on the $tryp^+$ allele.)

b. We are told that two-thirds of the colonies are type 1, or true revertants. Thus, there are $120 \times \frac{2}{3} = 80$ revertants out of a total of 10^7, or a revertant frequency of 8×10^{-6} cells.

Problems

1. In a certain species of flower, the petals are normally blue. The mutation w produces white petals. In a plant of genotype ww, a single reversion event occurs during the development of a petal. What detectable outcome would this reversion produce in the resulting petal?

2. *Penicillium* (a commercially important filamentous fungus) normally can synthesize its own leucine (an amino acid). How would you go about selecting mutants that are leucine-requiring (that cannot synthesize their own leucine)? (NOTE: Like many filamentous fungi, *Penicillium* produces profuse numbers of asexual spores.)

3. How would you select revertants of the yeast allele *pro-1*? (This allele confers an inability to synthesize the amino acid proline, which can be synthesized by wild-type yeast and which is necessary for growth to occur.)

4. More than 10,000 new molecules are synthesized every year. Many of them could be mutagenic. How would you readily screen large numbers of compounds for their mutagenicity?

5. How would you use the replica-plating technique to select arginine-requiring mutants of haploid yeast?

6. Using the filtration-enrichment technique, you do all your filtering with a minimal medium and do your final plating on a complete medium that contains every known nutritional compound. How would you find out what *specific* nutrient is required? After replica plating onto every kind of medium supplement known to science, you still can't identify the nutritional requirement of your new yeast mutant. What could be the reason(s)?

7. *Meiotic (mei)* mutations are known in several organisms. The observable effect of this mutation on the phenotype is to cause a drastic disruption of the meiotic process, often producing only aborted meiotic products. In *Neurospora*, a cross is made between two strains called 1 (mating type A) and 2 (mating type a). The cross is perfectly fertile, with no spore abortion. Seven randomly selected progeny labeled 3 to 9 are intercrossed and crossed to the parental strains. Table 7-8 shows the fertility of the resulting progeny. In the table, F means fully fertile and O means almost sterile (with many aborted white ascospores and with the few normal spores showing reduced recombination throughout the genome). Explain these results in terms of a recessive *mei* mutation in one of the parental strains. Suggest a possible mode of action for the *mei* mutation at the cellular level.

TABLE 7-8.

Strains of mating type a	Strains of mating type A				
	1	3	4	5	6
2	F	F	O	F	O
7	F	F	O	F	O
8	F	F	O	F	O
9	F	F	F	F	F

8. An experiment is initiated to measure the reversion rate of an *ad-3* mutant allele in haploid yeast cells. Each of 100 tubes of a liquid adenine-containing medium are inoculated with a very small number of mutant cells and incubated until there are 10^6 cells per tube. Then the contents of each tube are spread over a separate plate of solid medium containing no adenine. The plates are observed after one day, and colonies are seen on 63 plates. Calculate the reversion rate of this allele per cell division.

9. Suppose that you want to determine whether caffeine induces mutations in higher organisms. Describe how you might do this (include control tests).

10. Certain mealybugs called coccids have a diploid number of 10. In the cells of the males, five of the chromosomes are always seen to be heterochromatic (densely staining) and five are euchromatic (not densely staining). In female cells, all ten chromosomes are always euchromatic. Spencer Brown and Walter Nelson-Rees gave large doses of X radiation to males and females. Interpret the following results:

female (X-rayed) × male (non-X-rayed)

↓

no surviving progeny

female (non-X-rayed) × male (X-rayed)

↓

lots of male progeny

but no female progeny

11. Assume that an albino mutation in a leaf is present in the inner meristem layers but not in the epidermal layer. During development, a few cells of the epidermis migrate into the photosynthetic layers of the leaf. What do you predict will be the appearance of the leaf? (NOTE: the epidermis normally has no chlorophyll.)

12. In corn, a single gene determines presense (Wx) or absence (wx) of amylose in the cell's starch. Cells that have Wx stain blue with iodine; those that have only wx stain red. Design a system for studying the frequency of rare mutations from $Wx \rightarrow wx$ without using acres of plants. (HINT: you might start by thinking of an easily studied cell type.)

13. A new high-yielding strain of wheat suffers from the disadvantage that it tends to "lodge" (fall over) during storms. You have an X-ray source (and a couple of years) to correct this defect. How will you proceed? State which part of the plant (or which stage of its life cycle) you will treat, what results you will look for, which generation you might expect your results in, and so forth.

14. Suppose that you cross a single male mouse from a homozygous wild-type stock with several homozygous black, virgin females. The F₁ consists of 38 wild-type females and males and five black males and females. How can you explain this result?

15. Joe Smith accidentally receives a heavy dose of radiation in the gonadal region. Nine months later, his wife has a daughter, Mary. Mary appears perfectly normal, and she eventually marries a homozygous normal man. The figure below shows their pedigree.

   a. What possible genetic mechanisms could explain these results?

   b. How would you prove which explanation is correct?

16. A man employed for several years in a nuclear power plant becomes the father of a hemophiliac boy — the first occurrence in the extensive family pedigrees of both his own and his wife's ancestry. Another man, also employed for several years in the same plant, has an achondroplastic dwarf child — the first occurrence in his ancestry or in that of his wife. Both men sue their employer for damages. As a geneticist, you are asked to testify in court. What do you say about each situation? (NOTE: hemophilia is an X-linked recessive; achondroplasia is an autosomal dominant.)

17. In a large maternity hospital in Copenhagen, there were 94,075 births. Ten of the infants were achondroplastic dwarfs. (Achondroplasia is an autosomal dominant, showing virtually full penetrance.) Only two of the dwarfs had a dwarf parent. What is the mutation frequency to achondroplasia in gametes? Do you have to worry about reversion rates in this problem? Explain.

18. One of the jobs of the Hiroshima-Nagasaki Atomic Bomb Casualty Commission was to assess the genetic consequences of the blast. One of the first things they studied was the sex ratio in the offspring of the survivors. Why do you suppose they did this?

19. The government wants to build a nuclear reactor near the town of Poadnuck. The townspeople are very upset about the possibility of an accidental release of radioactivity and are putting up a stiff fight to keep it out. They call on you, a geneticist, to inform them about the biological hazards of an accident. What would you say in your speech?

20. The nuclear plant in Problem 19 has been built, and you are now a professor at Poadnuck State College. A radical group manages to infiltrate the plant and blow it up, releasing a large amount of radioactivity into the area. Fortunately, a thunderstorm washes most of the radioisotopes to the ground, preventing widespread contamination. You set out immediately to determine the genetic effects of radiation. How do you go about it? (Remember, you must have controls.)

21. In many organisms, there are examples of unstable recessive alleles that revert to wild-types at very high frequencies. Different unstable alleles often revert in different ways. One gene in corn (C), which produces a reddish pigment in the kernels (seeds), has several unstable alleles. The unstable allele c^{m1} reverts late in the development of the seed and at a very high rate. On the other hand, the unstable allele c^{m2} reverts earlier in the development of the seed and at a lower rate. What phenotype do you predict in plants of genotype:

   a. $C\ c^{m1}$ **e.** $c^{m1}c$ (c is a stable mutant allele)

   b. $C\ c^{m2}$ **f.** $c^{m2}\ c$

   c. $c^{m1}\ c^{m1}$ **g.** $c^{m1}\ c^{m2}$

   d. $c^{m2}\ c^{m2}$

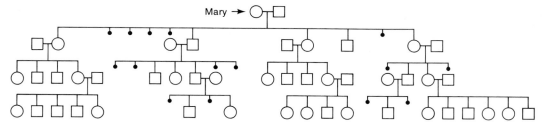

Mary →

• Early miscarriages, sex undetermined

Chromosome Mutation I: Changes in Chromosome Structure

KEY CONCEPTS

Chromosomes have many landmarks that can be used for the detection of abnormal rearrangements.

▪

Due to the strong meiotic pairing affinity of homologous regions, diploids with one normal and one rearranged chromosome set produce pairing structures that have shapes and properties unique to that rearrangement.

▪

A deletion in one chromosome set is generally deleterious as a result of gene imbalance and the unmasking of deleterious alleles in the other chromosome set.

▪

Duplications can lead to gene imbalance but also provide extra material for evolutionary divergence.

▪

Heterozygous inversions show semisterility and reduced recombination in the region spanned by the inversion.

▪

A heterozygous translocation shows semisterility and linkage of genes on the chromosomes involved in the translocation.

■ The gene mutations discussed in Chapter 7 cannot be detected by an examination of chromosomes. A chromosome bearing a gene mutation looks the same as one bearing the normal allele. However, visible changes in the genome do occur regularly; such changes are called **chromosome mutations** or **chromosome aberrations.** In Chapters 8 and 9, we will see that such changes can be detected not only with a microscope but also by standard genetic analysis.

Chromosome mutations are changes in the genome that involve chromosome parts, whole chromosomes, or whole chromosome sets. Thus far, we have treated the eukaryotic chromosome rather naively as merely a string of genes. A great deal is now known about the genetic organization of chromosomes (see Chapter 14), but there still are major gaps in our understanding at the molecular level. Luckily, the simplistic approach is not a major impediment in our understanding of chromosomes at the behavioral level. A large amount of information has been amassed about normal chromosome behavior and about how this behavior can go awry in the creation of chromosome mutations. This information has come from cytological and genetic studies; the union of these two areas of study constitutes the discipline of **cytogenetics.**

Genetics, perhaps more than any other field of biology, makes extensive use of deviations from the norm, and the study of chromosomes is no exception. Although this means of investigation may seem very esoteric to some, the findings have proved to be very important in applied biology — especially in agriculture, animal husbandry, and medicine. Many genetic tricks and devices that geneticists routinely use to build certain genotypes are discussed in this chapter. As you read, keep in mind that although this point is not given the attention it deserves, many of the aberrations we consider here have played important roles in building the theories of evolution and speciation.

The Topography of Chromosomes

Chromosome mutations may be detected either by appropriate genetic tests or, cytologically, simply by viewing the chromosomes under the microscope. Not all chromosome mutations can be detected under the microscope because some are too small or subtle to produce any visible changes. Nevertheless, many cases are detectable in this way, and in this approach the cytogeneticist makes use of a variety of features of chromosome topography that act as landmarks, or chromosome markers. Once the topography of the normal chromosomes is well known, then changes can be identified readily. The following features of chromosomes are commonly used as landmarks in cytogenetic analysis.

Chromosome Size. There can be considerable variation in chromosome size within a genome. In the human genome, for example, there is about a three- to fourfold range in size from chromosome 1 (the biggest) to chromosome 21 (the smallest), as we saw in Figure 6-26. If individual chromosomes are difficult to identify on the basis of size alone, then at least groups of chromosomes of similar size may be defined. A change can be pinpointed as involving, for example, "one of the chromosomes in size group A."

Centromere Position. The region of the centromere usually appears to be pinched; this region is called the **primary constriction.** The position of a centromere defines the ratio between the lengths of the two chromosome arms, and this ratio is a useful characteristic (Figure 8-1). Centromere positions can be categorized as **telocentric** (at one end), **acrocentric** (off center), or **metacentric** (in the middle). In addition to arm ratio, the centromere position determines the shapes of chromosomes during anaphase, ranging from a rod to a J to a V (Figure 8-2). In some organisms, such as the Lepidoptera, centromeres are "diffuse," so that spindle fibers attach all along the chromosome. When such a chromosome is broken, both parts can still migrate to the poles. In contrast, a break in a chromosome with a single centromere results in a fragment that has no centromere and therefore cannot move to the pole. Chromosome segments lacking a centromere are categorized as **acentric** (Figure 8-3).

The molecular structure of the centromeres of a few organisms, such as yeasts, is now known. This is also true

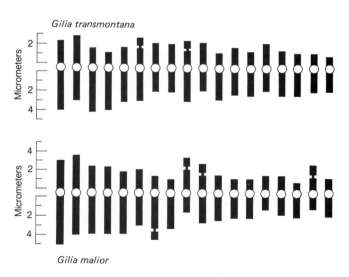

Figure 8-1. Chromosome diagram of the basic chromosome sets of two species of the plant genus *Gilia.* The chromosomes are arranged to show the characteristic arm ratios. Satellites are also shown.

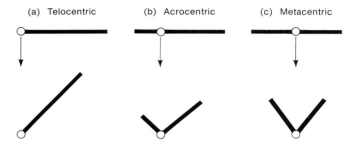

(a) Telocentric   **(b)** Acrocentric   **(c)** Metacentric

Figure 8-2. The classification of chromosomes by the position of the centromere. A telocentric chromosome has its centromere at one end; when the chromosome moves toward one pole of the cell during the anaphase of cellular division, it appears as a simple rod. An acrocentric chromosome has its centromere somewhere between the end and the middle of the chromosome; during anaphase movement, the chromosome appears in the shape of a J. A metacentric chromosome has its centromere in the middle and appears as a V during anaphase.

for the tips of chromosomes, the **telomeres.** Although not morphologically distinct, the telomeres have a unique molecular structure that is crucial to normal chromosome behavior. Such knowledge about molecular structure has permitted a new line of research in yeast in which completely novel chromosomes can be assembled, with a functional centromere, two telomeres, and any other genetic material of interest to the experimenter (see Chapter 15).

Position of Nucleolar Organizers. Different organisms are differently endowed with nucleoli, which range in number from one to many. The nucleoli contain ribosomal RNA, an important component of ribosomes. A

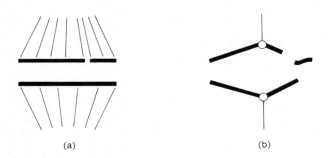

(a)   (b)

Figure 8-3. The spindle-fiber attachment to chromosomes. (a) An example of a relatively rare, diffuse centromere, with many spindle fibers attaching along the length of a single chromosome. A break in the chromosome does not result in the loss of any chromosome material. (b) Spindle-fiber attachment to a single centromere (the more usual form). In this case, a break in the chromosome results in an acentric fragment, which is lost during cellular division.

common situation is to have two nucleoli in a diploid cell. The nucleoli are found next to secondary constrictions of the chromosomes, called **nucleolar organizers,** which have highly specific positions in the chromosome set. Nucleolar organizers contain the genes for ribosomal RNA. Their positions can also be used as landmarks in cytogenetic analysis. A small piece of chromosome distal to the nucleolar organizer is called a **satellite** (Figure 8-1).

Chromomere Patterns. We have already encountered chromomeres in Chapter 5 (pages 110–111). They are visible during certain stages of cell division as beadlike, localized thickenings of the chromosome.

Heterochromatin Patterns. When a chromosome is stained with standard reagents that react with DNA, such as Feulgen stain, distinct regions with different staining characteristics are visually revealed. Densely staining regions are designated **heterochromatin;** poorly staining regions are said to be **euchromatin.** The distinction is thought to reflect the degree of compactness of the DNA in the chromosome. Heterochromatin can be **constitutive** or **facultative.** The constitutive type is a permanent feature of a specific chromosome location and is, in this sense, a hereditary feature; Figure 5-15 shows good examples of constitutive heterochromatin in the tomato. The facultative type of heterochromatin can be either present or absent at any particular chromosomal location. The patterns of heterochromatin and euchromatin along a chromosome constitute good cytogenetic markers. Heterochromatin is also of interest at the level of chromosome function because heterochromatic regions are for the most part genetically inert.

Banding Patterns. Recently developed staining procedures have revealed several intricate chromosome banding systems in a wide range of organisms. Some of these systems are Q bands (produced by quinacrine hydrochloride), G bands (produced by Giemsa stain), and R bands (reversed Giemsa). An example of G bands in human chromosomes is shown in Figure 6-25(a).

A rather specialized kind of banding system, which has been used extensively by cytogeneticists for many years, is found in the so-called giant chromosomes in certain organs of the Diptera. In 1881, E. G. Balbiani recorded peculiar structures in the nuclei of certain secretory cells of two-winged flies. These structures were long and sausage-shaped and were marked by swellings and cross striations. Unfortunately, Balbiani did not recognize them as chromosomes, and his report remained buried in the literature. It was not until 1933 that Theophilus Painter, Ernst Heitz, and H. Bauer rediscovered them and realized these structures are in fact chromosomes.

In secretory tissues, such as the Malpighian tubules, rectum, gut, footpads, and salivary glands of the Diptera, the chromosomes apparently replicate their genetic material many times without actually separating into distinct chromatids. Thus, as the chromosome increases in replicas, it elongates and thickens. *Drosophila melanogaster,* for example, has an *n* number of 4, and only four chromosomes are seen in the cells of such tissues because, for some reason, homologs are tightly paired. Furthermore, all four chromosomes are joined at the **chromocenter,** which represents a coalescence of the heterochromatic areas around the centromeres of all four chromosome pairs. The chromocenter of *Drosophila* salivary-gland chromosomes is shown in Figure 8-4, where L and R stand for arbitrarily assigned left and right arms.

Such giant chromosomes are formed by **endomitosis,** a process in which many chromosome replicas are produced but do not separate. The bundles of multiple replicas are called **polytene chromosomes.** The most commonly examined polytene chromosomes are in the salivary gland nuclei. These nuclei never divide. Along the chromosome length, characteristic stripes called **bands,** which vary in width and morphology, can be observed and identified. In addition, there are regions that may at times appear swollen (**puffs**) or greatly distended (**Balbiani rings**); these are presumed to correspond to regions of genetic activity. Recently, it has been found in *Drosophila* that, in general, each band contains the genetic material of a single gene. (However, the significance of the bands, such as Q and G bands, in human and other chromosomes is not known.) The polytene chromosomes corresponding to specific linkage groups in *Drosophila* have been identified through the use of chromosome aberrations. Such aberrations, as we shall see, have also been useful in the specific localization of genes along the chromosomes.

So much for the cytological properties of *normal* chromosomes. We turn now to the properties of *abnormal* chromosomes. These chromosome aberrations are classified into aberrations of chromosome structure (the subject of the rest of this chapter) and aberrations of chromosome number (the subject of Chapter 9).

First, however, let's consider two important features of chromosome behavior that will be useful in understanding the structural abnormalities. One is that during prophase I of meiosis, homologous regions of chromosomes show a very strong pairing affinity and often go through considerable contortions in order to pair. This property accounts for many of the curious structures seen in cells containing one normal chromosome set plus an aberrant set. Similar pairing contortions between homologs are seen in polytene chromosomes, and equiva-

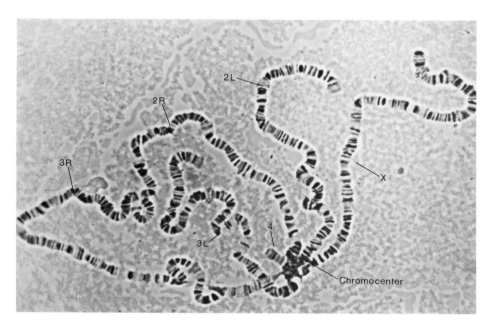

Figure 8-4. *Drosophila* larval salivary chromosomes. Note that the two homologs of each chromosome pair have fused to form a single banded unit. The centromeres of all chromosomes are united at the common chromocenter. Thus, the telocentric X chromosome appears as a single unit, whereas the metacentric second and third chromosomes are seen as left (2L and 3L) and right (2R and 3R) portions. (Photograph by Tom Kaufman.)

lent shapes result. The other important property is that changes in structure usually involve chromosome breakage, and the broken chromosome ends are highly "reactive," showing a strong tendency to join with other broken ends. This property is not exhibited by the telomeres (the regular chromosome ends), however.

Types of Changes in Chromosome Structure

It is possible that a segment of a chromosome might be lost. In order to discuss such events, let's use letters to represent arbitrary chromosome regions, each of which contains many genes. For example, a chromosome described by a break and subsequent rejoining could result in

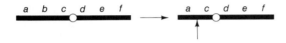

This type of change is called a **deletion** or a **deficiency.** The reciprocal of such a change would be its **duplication:**

We can also conceive of a segment of a chromosome that has rotated 180 degrees and rejoined the chromosome as an **inversion:**

Finally, parts of two nonhomologous chromosomes might be exchanged to produce a **translocation:**

In fact, all of these types of chromosome aberrations do occur, and we will now examine their genetic and cytological properties. Collectively, this class of aberrations is known as **chromosome rearrangements.**

Deletions

Obviously, the occurrence of chromosomal rearrangements results from a break or disruption in the linear continuity of a chromosome. Deletions and duplications

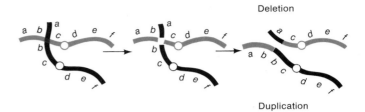

Figure 8-5. One possible way of producing deletions and duplications: by the reunion of broken homologs.

can be produced by the same event if breaks occur simultaneously at different points in two homologs; this can be visualized as occurring when the homologs overlap (Figure 8-5). This does not mean, however, that duplications and deletions are *always* reciprocal products of the same event.

If a deletion is made homozygous (that is, if both homologs have the same deletion), then the combination is generally lethal. This suggests that most regions of the chromosomes are essential for normal viability and that complete elimination of any segment from the genome is deleterious. Even individuals heterozygous for a deletion—those with one normal homolog and one that carries the deletion—may not survive because the genome has been "fine-tuned" during evolution to require a specific balance or ratio of most genes; the presence of the deletion upsets this balance. Nevertheless, in some cases, individuals with relatively large deficiencies can survive if these deletions occur together with normal chromosomes. If meiotic chromosomes in such heterozygotes can be examined, the region of the deletion can be detected by the failure of the corresponding segment on the normal chromosome to pair properly, which results in a **deletion loop** (Figure 8-6). Deletion loops are

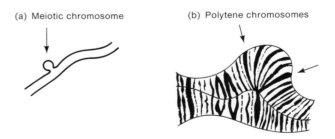

Figure 8-6. Cytogenetic configurations in a deletion heterozygote in *Drosophila*. (a) The "looped-out" portion of the meiotic chromosome is the normal form. The genes in this deletion loop have no alleles with which to pair during synapsis. (b) Since polytene chromosomes in *Drosophila* have specific banding patterns, the missing bands of the deleted chromosome can be observed in the deletion loop of the normal chromosome.

also detected in polytene chromosomes in which the homologs are fused. Thus, deletions can be located on the chromosome by this technique.

Deletions of some chromosome regions produce their own unique phenotypes. A good example is a deletion of one specific small chromosome region of *Drosophila.* When one homolog carries the deletion, the fly shows a unique notch-wing phenotype, so the deletion acts like a dominant mutation in this regard. But because it is a deletion, it is lethal when homozygous and therefore acts as a recessive in regard to its lethal effect.

An effective mutagen for inducing chromosome rearrangements of all kinds is ionizing radiation. This kind of radiation, of which X rays and γ rays are examples, is highly energetic and causes chromosome breaks. The way in which the breaks rejoin determines the kind of rearrangement produced. In the case of deletion, a single break can cause a **terminal deletion** and two breaks can produce an **interstitial deletion** (Figure 8-7).

But what about the genetic properties of deletions in general? Cytological detection of a deletion loop (for example, in the salivary-gland chromosomes of *Drosophila*) is useful, but when cytological approaches are not possible, there are genetic criteria for inferring the presence of a deletion. One is the failure of the chromosome to survive as a homozygote, but this criterion, of course, could also be produced by any lethal gene. Another is the suppression of crossing-over in the region spanning the deficiency, but again this condition could occur with other aberrations and small deficiencies may have only minor effects on crossing-over. The best criterion is that chromosomes with deletions can never revert to a normal condition. Another reliable criterion is the phenotypic expression of a recessive gene on a normal chromosome when the region in which it is located has been deleted from the homolog. Such **pseudodominance** (the

expression of a recessive gene when present in a single dose) also allows the physical localization of that gene by plotting the locations of the deletions that can cause the gene to be pseudodominant. This plotting in turn permits a correlation between the genetic map (based on linkage analysis) and the cytological map (devised by marking the position of deficiency loops in specific cases of pseudodominance). By and large, where this has been done, the maps correspond well—a satisfying cytological endorsement of a purely genetic creation:

$$
\begin{array}{c}
a\ \ b\ \ c\ \ \ \ d\ \ e\ \ \ f \\
\hline
\rule{0pt}{8pt}
\end{array}
\left.\vphantom{\begin{array}{c}a\\b\end{array}}\right\}
$$

$$
\begin{array}{c}
\text{\small +\ \ \ \ \ +\ \ \ +\ \ +} \\
\hline
\end{array}
\left.\vphantom{\begin{array}{c}a\\b\end{array}}\right\}\ \ \text{Phenotype is } + b + + + +
$$

> **Message** Deletion analysis proves that linkage maps are in fact reflections of physical chromosome maps.
>
> Chromosome map ▓▓▓▓▓▓▓▓▓▓▓▓▓▓▓▓▓▓▓▓▓▓▓
> Loci   p q   r   st   u
>
> Linkage map ─────────────────────────────

Conversely, pseudodominance can be used to map a small deletion that cannot be visualized microscopically. Let's consider an X chromosome in *Drosophila* that carries a recessive lethal suspected of being a deletion; we will call this chromosome "X?." We can cross X?-bearing females with males carrying different recessive genes known to reside in that interval. A map of genes in the tip region is

$$
\begin{array}{ccccccccc}
y & dor & br & gt & swa & w & rst & vt \\
\vdash\!\!\!\!+ & \!\!\!\!+ & \!\!\!\!+ & \!\!\!\!+ & \!\!\!\!+ & \!\!\!\!+ & \!\!\!\!+ & \!\!\!\!+ \!-\!-\!-\!-\!-\!-\! \\
\end{array}
$$

0.3   0.3 0.3 0.4   0.2 0.2   0.6

Suppose we obtain all wild-type flies in crosses between X?/X females and males carrying *y, dor, br, gt, rst,* or *vt* but obtain pseudodominance of *swa* and *w* with X? (that is, X?/*swa* is swa and X?/*w* is w). Then we have good genetic evidence for a deletion of the chromosome that includes at least the *swa* and *w* loci but not *gt* or *rst.*

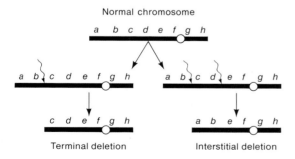

Figure 8-7. Production of terminal and interstitial deletions. Chromosomes can be broken when struck by ionizing radiation (wavy arrows). A terminal deletion is formed when the end piece of a chromosome is lost. An interstitial deletion is formed when two breaks are induced. The acentric fragment *(cd)* is lost; the terminal portion *(ab)* rejoins the main body of the chromosome.

> **Message** Deletions are recognized genetically by (1) lack of revertability, (2) pseudodominance, and (3) recessive lethality, and, cytologically, by (4) deletion loops.

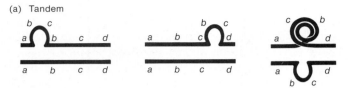

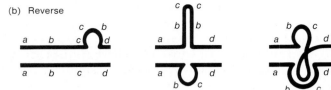

Figure 8-8. Possible pairing configurations in duplication heterozygotes. Duplicated segments may occur (a) in tandem or (b) in reverse order. There are several ways in which the homologs of duplicated heterozygotes may pair, illustrating the high affinity of homologous regions for pairing.

An interesting difference between animals and plants is revealed by deletions. In animals, a male that is heterozygous for a deletion chromosome and a normal one will produce functional sperm carrying each of the two chromosomes in approximately equal numbers. In other words, sperm seem to function to some extent regardless of their genetic content. In diploid plants, on the other hand, the pollen produced by a deletion heterozygote is of two types: functional pollen carrying the normal chromosome, and nonfunctional (or aborted) pollen carrying the deficient homolog. Thus, pollen cells seem to be sensitive to changes in *amount* of chromosome material, and this sensitivity might act to weed out deletions. The situation is somewhat different for polyploid plants, which are far more tolerant of pollen deletions. This tolerance is due to the fact that there are several chromosome sets even in the pollen, and the loss of a segment in one of these sets is less crucial than it would be in a haploid pollen cell. Ovules in either diploid or polyploid plants also are quite tolerant of deletions, presumably due to the nurturing effect of the surrounding maternal tissues.

Duplications

Duplications are very important chromosome changes from the standpoint of evolution because they supply additional genetic material that is potentially capable of assuming new functions. Adjacent duplicated segments may occur in **tandem sequence** with respect to each other (*a bc bc d*) or in **reverse order** (*a bc cb d*). The pairing patterns obtained in these two sequences are different and illustrate the high affinity of homologous regions for pairing. Thus, chromosomes in meiotic nuclei containing a normal chromosome and a homolog with a duplication are seen to pair in the configurations shown in Figure 8-8. Alternatively, duplicated segments may be nonadjacent, either in the same chromosome or in separate chromosomes. Interestingly, once an adjacent tandem duplication arises in a population, homozygosity for such a duplication can result in higher orders of duplication by crossing-over when the chromosomes are **asymmetrically paired** (Figure 8-9).

When a duplication of part of a genome occurs, one of the duplicate regions is free to undergo gene mutation because the necessary basic functions of the region are provided by the other copy. This produces an opportunity for divergence in the function of genes, which could be potentially advantageous in genome evolution. Indeed, in situations in which different gene products with related functions can be compared, such as the globins (to be discussed later), there is good evidence that these products arose as duplicates of each other.

> **Message** Duplications are important chromosome alterations that supply additional genetic material capable of evolving new functions.

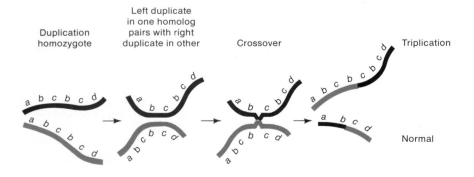

Figure 8-9. Generation of higher orders of duplications by asymmetric pairing and recombination in a duplication homozygote.

Like some deletions, duplications of certain genetic regions may produce specific phenotypes and act like a gene mutation. For example, the dominant mutation *Bar* in *Drosophila* produces a slitlike eye instead of the normal oval one. Cytological study of the polytene chromosomes have shown that *Bar* is in fact a tandem duplication that probably results from an **unequal crossover** (Figure 8-10). Evidence for the asymmetric pairing and crossing-over in *Drosophila* comes from studying homozygous *Bar* females. Occasionally, such females produce offspring with extremely small eyes called "double bar." Each offspring is found to carry three doses of the *Bar* region in tandem (Figure 8-11).

In general, duplications are hard to detect and are rare. However, they are useful tools and can be generated from other aberrations by manipulations that we shall learn later.

Duplications and deficiencies are detectable in human chromosomes. As a matter of fact, some of the best evidence for the unequal crossover origin of tandem duplications (and their reciprocal deletions) comes from studies of the genes that determine the structure of the oxygen-transport molecule, hemoglobin, in humans. The hemoglobin molecule is composed of two different kinds of subunits, or components. Furthermore, there are always two of each kind of subunit, so that there is a

Figure 8-10. Production of *Bar*-eye duplication by nonreciprocal crossing-over. Since this event will occur during meiosis, gametes containing the deletion chromosome will presumably die or produce an inviable zygote. The gamete containing the *Bar* duplication, however, will produce a male offspring with severely reduced eye size (*Bar*) or a female offspring with a slightly reduced eye size (*Bar* heterozygote).

total of four subunits. The kinds of subunits that constitute hemoglobin are different at different stages of development. For example, the fetus has two α subunits and two γ subunits ($\alpha_2\gamma_2$), whereas the adult has two α subunits and two β subunits ($\alpha_2\beta_2$). The structures of these subunits are determined by different genes, some of which are linked and some of which are not. The situation is summarized in Figure 8-12. The linked $\gamma - \sigma - \beta$ group provides the data we need on unequal crossover.

Some thalassemias (a kind of inherited blood disease) have been proved to involve hemoglobin subunits that are part δ and part β (Lepore hemoglobin) or part γ

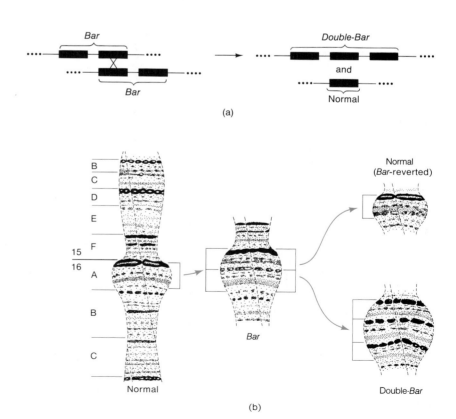

Figure 8-11. Production of double-*Bar* (triplication) and *Bar*-revertant (normal) chromosomes by asymmetric pairing and recombination in a duplication homozygote. (a) Diagrammatic representation. (b) Cytological representation. (Part b is from Bridges, *Science* 83, 1936, 210.)

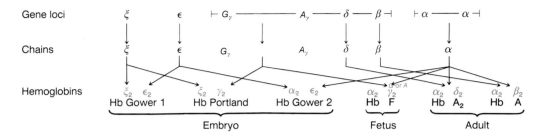

Figure 8-12. Hemoglobin genes in humans. Each Greek letter represents a different gene locus and its type of hemoglobin subunit chain. These loci combine in different ways at different stages in development, as indicated, to yield functioning hemoglobin molecules. (From D. J. Weatherall and J. B. Clegg, "Recent Developments in Molecular Genetics of Human Hemoglobin," *Cell* 16, 1979. Copyright © 1979 by M.I.T. Press.)

and part β (Kenya hemoglobin). The origin of these rare hemoglobin forms can be explained by the unequal crossover models shown in Figure 8-13. Also represented in the diagram are the reciprocal crossover products called anti-Lepore and anti-Kenya. It can be seen that these deletion types lead to the anemic blood disease symptoms.

Multiple tandem and nontandem repeats of DNA are inherent features of normal eukaryotic chromosomes, as we shall see in Chapter 14. However, the origin of these repeats is not clear in most cases. It is also worth noting that some cancers are associated with the production of localized multiple repeats, which appear as homogeneously staining regions.

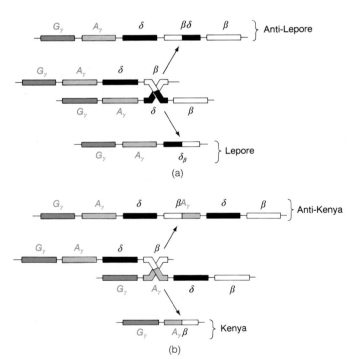

Figure 8-13. Proposed generation of variant human hemoglobin chains by unequal crossing-over in the $\gamma - \delta - \beta$ genetic region: (a) the Lepore variant; (b) the Kenya variant. These deletion types, not the anti-Lepore and the anti-Kenya, result in anemia. (From D. J. Weatherall and J. B. Clegg, "Recent Developments in Molecular Genetics of Human Hemoglobin." *Cell* 16, 1979. Copyright © 1979 by M.I.T. Press.)

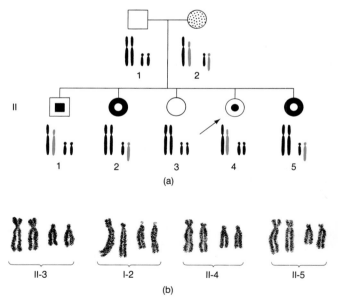

Figure 8-14. Cri du chat syndrome. (a) Pedigree with cases of the syndrome: I. Both parents are phenotypically normal. II. Of the offspring, II-1 and II-4 have cri du chat syndrome, II-2 and II-5 have the reciprocal chromosome abnormality, and II-3 is normal. Symbols represent different karyotypes, and the arrow points to the *proband* or affected individual with whom the study began. (b) Chromosomes 5 and 13 of four persons in the pedigree. (From J. Lejeune, J. Lafourcade, H. Berger, and R. Turpin, *Comptes Rendius*, 1964, 258. *L'Academie des Sciences, Paris.*)

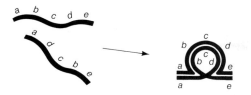

Figure 8-15. Inversion loop in paired homologs of an inversion heterozygote. Tight meiotic pairing produces this cytological configuration.

Visible deletions of chromosome segments in humans are always associated with major incapacities. For example, the deletion of one-half of the short arm of chromosome 5 results in a syndrome characterized by severe mental retardation; this syndrome is called cri du chat (in French, "cry of the cat") syndrome, because of the cat-like, mewing cry of affected infants. Figure 8-14 shows a pedigree involving cri du chat syndrome; examine it now, but don't worry about the details of the production of the deletion until after we have dealt with translocations. Deletions for certain other human chromosomes are always associated with severe incapacity or with intrauterine lethality.

Inversions

Homozygosity for a chromosome carrying an inverted gene sequence will result in a linkage map with a different gene order. In a heterozygote having a chromosome that contains an inversion and one that is normal, there are important genetic and cytological effects. However, because there is no net loss or gain of material, heterozygotes usually are perfectly viable. The location of the inverted segment can be recognized cytologically in the meiotic nuclei of such heterozygotes by the presence of an **inversion loop** in the paired homologs (Figure 8-15).

The location of the centromere relative to the inverted segment determines the genetic behavior of the chromosome. If the centromere is not included in the inversion, then the inversion is said to be **paracentric,** whereas inversions spanning the centromere are **pericentric** (Figure 8-16).

What do inversions do genetically? In a paracentric inversion, heterozygote crossing-over within the inversion loop has the effect of connecting homologous centromeres in a **dicentric bridge,** as well as producing an **acentric** piece of chromosome—one without a centromere (Figure 8-17). Thus, as the chromosomes separate during anaphase I, the disjoining centromeres will remain linked by means of the bridge. This orients the centromeres so that the noncrossover chromatids lie farthest apart. The acentric fragment cannot align itself or move and, consequently, is lost.

Remarkably, in *Drosophila* eggs and in plant megaspores, the dicentric bridge may remain intact long after

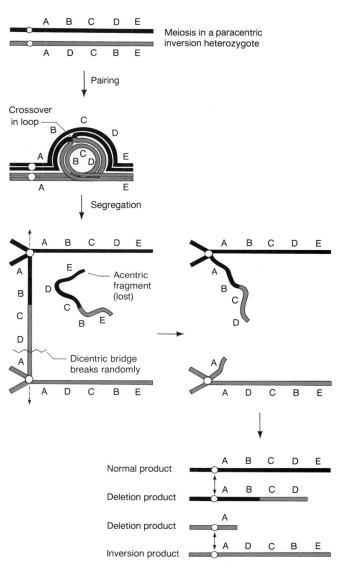

Figure 8-17. Meiotic products resulting from a single crossover within a heterozygous paracentric inversion loop. Crossing-over occurs in the four-strand stage (two identical chromatids connected to each centromere).

Figure 8-16. The location of the centromere relative to the inverted segment. If the centromere is not included in the inversion, then the inversion is paracentric. Inversions that include the centromere are pericentric.

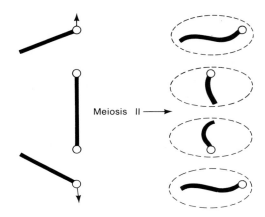

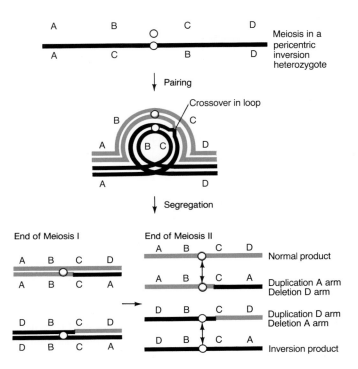

Figure 8-18. In some organisms, such as *Drosophila* and some plants, the dicentric bridge resulting from a single crossover within a heterozygous paracentric inversion loop will not break during anaphase I. As the second meiotic division begins, the noncrossover chromatids are directed to the outermost nuclei. Thus, the two inner nuclei either are linked by the dicentric bridge or contain fragments of the bridge if it breaks.

Figure 8-19. Meiotic products resulting from a meiosis with a single crossover within a heterozygous pericentric inversion loop.

anaphase I; then as the second meiotic division begins, the noncrossover chromatids are directed to the outermost nuclei (Figure 8-18). Thus, the two inner nuclei either are linked by the dicentric bridge or contain fragments of the bridge if it breaks, whereas the outer nuclei contain the noncrossover chromatids. Fertilization of a nucleus carrying the broken bridge should produce defective zygotes that die because they have an unbalanced set of genes. Consequently, in a testcross, the recombinant chromosomes end up in dead zygotes and recombinant frequency is lowered. However, in *Drosophila,* the presence of large inversions does not result in a large increase in zygotic mortality. The inner nuclei linked by the dicentric bridge never participate in fertilization, and only one of the outer nuclei can become the egg nucleus. Thus, we can see that the chromatids participating in a crossover event will be selectively retained in the central nuclei, thereby allowing recovery of the noncrossover chromatids in the egg nuclei. This remarkable process was suggested originally by genetic results and was later confirmed cytologically in *Drosophila.* It has also been shown in plants. (How would you test such a possibility in *Neurospora* using tetrad analysis?) Nevertheless, the genetic consequence of inversion heterozy-

gosity is the same: the selective recovery of noncrossover chromatids from exchange tetrads. In addition, inversion heterozygotes often have mechanical pairing problems in the area of the inversion; this also reduces crossing-over and recombinant frequency in the vicinity.

> **Message**   Although inversion heterozygosity does reduce the number of recombinants recovered, it in fact does so by two mechanisms: by inhibiting the process of chromosome pairing in the vicinity of the inversion, and by selectively eliminating the products of crossovers in the inversion loop.

It is worth noting that paracentric inversions have been useful in studying crossing-over in organisms that have no known genetic markers but that do have chromosomes that can be studied. Dicentric bridges are the con-

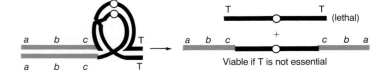

Figure 8-20. Generation of a viable nontandem duplication from a pericentric inversion close to a dispensable chromosome tip.

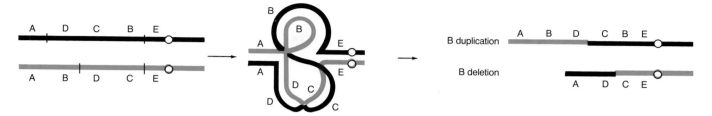

Figure 8-21. Generation of a nontandem duplication from two overlapping inversions.

sequence of crossovers within the inversion loop, so their frequency is related to the amount of crossing-over.

The net genetic effect of a pericentric inversion is the same as that of a paracentric one (crossover products are not recovered)—but for different reasons. In a pericentric inversion, because the centromeres are contained within the inverted region, disjunction of crossover chromosomes occurs in the normal fashion, without the creation of a bridge. However, a crossover within the inversion produces chromatids that contain a duplication and a deficiency for different parts of the chromosome (Figure 8-19). In this case, fertilization of a nucleus carrying a crossover chromosome generally results in its elimination through zygotic mortality caused by an imbalance of genes. Again, the result is the selective recovery of noncrossover chromosomes as viable progeny.

To generate a duplication "on purpose," it is possible to use a pericentric inversion having one breakpoint at the tip (T) of the chromosome (Figure 8-20). A crossover in the loop produces a chromatid type in which the entire left arm is duplicated; if the tip is nonessential, a duplication stock is generated for investigation. Another way to make a duplication (and a deficiency) is to use two paracentric inversions with overlapping breakpoints (Figure 8-21). (These manipulations are possible only in organisms with thoroughly mapped chromosomes for which large sets of standard rearrangements are available.)

We have seen that genetic analysis and meiotic chromosome cytology are both good ways of detecting inversions. As with most rearrangements, there is also the possibility of detection through mitotic chromosome analysis. A key operational feature is to look for new arm ratios (Figure 8-22). Note that the ratio of the long to the short arm has been changed from about 4 to about 1 by the pericentric inversion. (A pericentric inversion was the cause of the chromosome heteromorphism in the grasshoppers observed by Carothers; see page 48.) Paracentric inversions are more difficult to detect in this way,

but they may be observed if banding or other chromosome landmarks are available.

Translocations

Here we consider **reciprocal translocations,** the most common kind of translocation. A segment from one chromosome is exchanged with a segment from another nonhomologous one, so that, in reality, two translocation chromosomes are simultaneously achieved.

The exchange of chromosome parts between nonhomologs establishes new linkage relationships. These new linkages are revealed if the translocated chromosomes are homozygous and, as we shall see, even when they are heterozygous. Furthermore, translocations may drastically alter the size of a chromosome as well as the position of its centromere. For example

Here a large metacentric chromosome is shortened by one-half its length to an acrocentric one, and the small chromosome becomes a large one. Examples in natural populations are known in which chromosome numbers have actually been changed by translocation between acrocentric chromosomes and the subsequent loss of the resulting small chromosome elements (Figure 8-23).

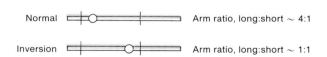

Figure 8-22. The ratio of the lengths of the left and right arms of a chromosome can be changed by a pericentric inversion. Thus, changes in arm ratios can be used to detect pericentric inversions.

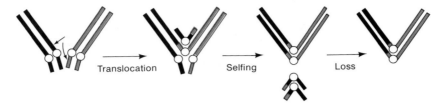

Figure 8-23. Genome restructuring by translocations. Small arrows indicate breakpoints in one homolog of each of two pairs of acrocentric chromosomes. The resulting fusion of the breaks yields one short and one long metacentric chromosome. If, as in plants, self-fertilization (selfing) takes place, an offspring could be formed with only one pair of long and only one pair of short metacentric chromosomes. Under appropriate conditions, the short metacentric chromosome may be lost. Thus, we see a conversion from two acrocentric pairs of chromosomes to one pair of metacentrics.

In heterozygotes having translocated and normal chromosomes, the genetic and cytological effects are important. Again, the pairing affinities of homologous regions dictate a characteristic configuration when all chromosomes are synapsed in meiosis. In Figure 8-24, which illustrates meiosis in a reciprocally translocated heterozygote, the configuration is that of a cross.

Remember, the configuration presented in the figure lies on the metaphase equatorial plate with the spindle fibers perpendicular to the page. Thus, the centromeres would actually migrate up out of the page or down into it. Homologous paired centromeres disjoin, whether or not a translocation is present. Because Mendel's second law still applies to *different paired centromeres,* there are two common patterns of disjunction. The segregation of each of the structurally normal chromosomes with one of the translocated ones (T_1 with N_2 and T_2 with N_1) is called **adjacent-1 segregation.** Both meiotic products are duplicated and deficient for different regions. On the other hand, the two normal chromosomes may segregate together, as do the reciprocal parts of the translocated ones, to produce $T_1 + T_2$ and $N_1 + N_2$ products. This is called **alternate segregation.** There is another event, called **adjacent-2 segregation,** in which homologous centromeres migrate to the same pole, but in general this is a rare occurrence.

Semisterility in Plants.   Once again, in the study of rearrangements, a careful distinction must be made between animals and plants. In animals, the unbalanced products of adjacent-1 segregation in reciprocal translocation heterozygotes usually produce viable gametes. Thus, in *Drosophila,* for example, the gamete population is composed of approximately equal numbers of alternate and adjacent segregation meioses. However, in diploid plants, the T_1N_2 and T_2N_1 gametes normally abort, producing a situation known as semisterility. (Recall that semisterility in plants also can be produced in gametes containing deletions and in the products of crossovers in

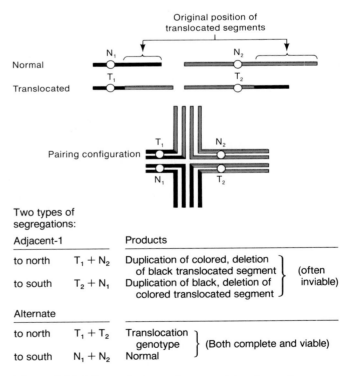

Figure 8-24. The meiotic products resulting from the two most commonly encountered chromosome segregation patterns in a reciprocal translocation heterozygote.

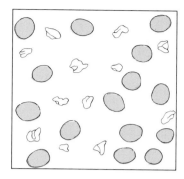

Figure 8-25. Sketch of normal and aborted pollen in a semisterile corn plant. The small, shriveled pollen grains contain chromosomally unbalanced meiotic products of a reciprocal translocation heterozygote. The normal-shaped pollen grains, which contain either the complete translocation genotype or normal chromosomes, are functional in fertilization and development.

inversion loops.) Semisterility is often directly identifiable through the observation of a mixture of shriveled, abnormal pollen grains with normal pollen grains (Figure 8-25).

Even in *Drosophila* and other tolerant animals, the unbalanced gametes (when fertilized by normal gametes) give rise to unbalanced zygotes that tend not to survive unless the imbalance is for very small translocated regions.

Message Translocations, inversions, and deletions produce semisterility by generating unbalanced meiotic products that may themselves be lethal or that may result in lethal zygotes.

Genetically, markers on nonhomologous chromosomes will appear to be linked if these chromosomes are involved in a translocation. Figure 8-26 shows a situation in which a translocation heterozygote has been established by crossing an *aa bb* individual with a translocation bearing the wild-type genes. We shall assume that *a* and *b* are close to the translocation breakpoint. On testcrossing the heterozygote, the only viable progeny are those bearing the parental genotypes, so linkage is seen between loci that were originally on different chromosomes. In fact, if all four arms of the meiotic pairing structure are genetically marked, recombination studies should result in a cross-shaped linkage map. Linkage of genes known to be on separate chromosomes is often a genetic giveaway for the presence of a translocation.

The Importance of Translocations in Human Affairs
Translocations are economically important. In agriculture, the occurrence of translocations in certain crop strains can reduce yields considerably due to the number of unbalanced zygotes that form. On the other hand, translocations are potentially useful: it has been proposed that the high incidence of inviable zygotes could be used to control insect pests by the introduction of translocations into the wild. Thus, 50 percent of the offspring of crosses between insects carrying the translocation and wild-types would die, and $\frac{10}{16}$ of the progeny of crosses between translocation-bearing insects would die.

Translocations occur in humans, usually in association with a normal chromosome set in a translocation heterozygote. Down's syndrome (previously referred to as mongolism) can arise in the progeny of an individual heterozygous for a translocation involving chromosome 21. The heterozygous person is phenotypically normal and is called a carrier. During meiosis, an adjacent-1 segregation will produce gametes carrying duplicated parts of chromosome 21 and possibly a deficiency for some part of the other chromosome involved in the translocation. For unknown reasons, the extra chunk of chromosome 21 is the cause of Down's syndrome. One-

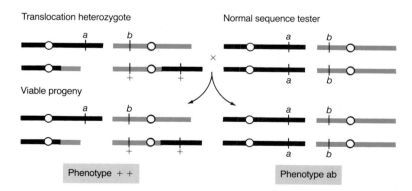

Figure 8-26. Inviability in some translocation progeny produces apparent genetic linkage because only parental types are recovered in the progeny (assuming that *a* and *b* are close to the breakpoint).

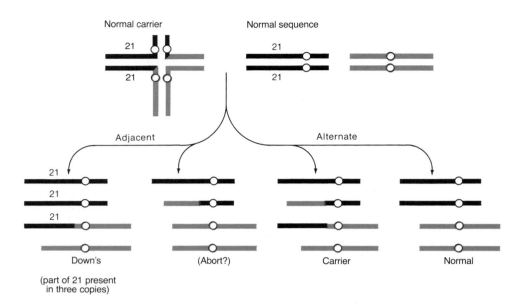

Figure 8-27. Production of the translocation form of Down's syndrome. This form is much rarer than that discussed on page 210. (NOTE: the translocation shown here has been simplified for the purpose of explanation.)

half of the normal children of a carrier will be carriers (Figure 8-27).

Later we shall examine another way in which Down's syndrome is generated. Bear in mind for now that under the mechanism just outlined, there should be a high recurrence rate in the pedigree of the affected family: a translocation heterozygote or carrier can repeatedly produce children with Down's syndrome, and of course the carriers can do the same in subsequent generations. The other method does *not* show this recurrence within a family. The factors responsible for numerous other hereditary disorders have been traced to translocation heterozygosity in the parents (as, for example, in the cri du chat pedigree in Figure 8-14).

The Use of Translocations in Producing Duplications and Deficiencies. Another method of producing duplications and deficiencies for specific chromosome regions makes use of translocations. Let's take *Drosophila* as an example. For reasons that are as yet unclear, heterochromatin near the centromere, is physically very extensive but contains few genes. In fact, for a long time, heterochromatin was considered useless and inert material. In any case, for our purposes, *Drosophila* can tolerate a loss or an excess of heterochromatin with little effect on viability or fertility.

Now let's select two different reciprocal translocations involving the same two chromosomes. Each of them has a breakpoint somewhere in heterochromatin, and each has a euchromatic break on one side or the other of the region we want to be duplicated and deleted (Figure 8-28). It can be seen that if we have a large collection of translocations having one heterochromatic break and euchromatic breaks at many different sites, then duplications and deletions for many parts of the genome can be produced at will for a variety of experimental purposes.

Position-Effect Variegation. In previous chapters, we considered several ways of generating variegation in the somatic cells of a multicellular organism through somatic segregation, somatic crossover, and somatic mutation.

A further cause of variegation associated with translocations is called **position-effect variegation.** The following example is from *Drosophila*.

The locus for white eye color in *Drosophila* is near the tip of the X chromosome. If the tip of a chromosome carrying w^+ is translocated to a heterochromatic region (say, of chromosome 4), then in a heterozygote for the translocation and a normal chromosome 4, with the normal X chromosome carrying w, the eye color is a mosaic of white and red cells. Because the translocation carries the dominant w^+ allele, we would expect the eye to be completely red. How have the white areas come about? We could suppose that when the translocation was formed, the w^+ gene itself was somehow changed to a state that made it more mutable in somatic cells; so the white eye tissue reflects cells in which w^+ has mutated to w.

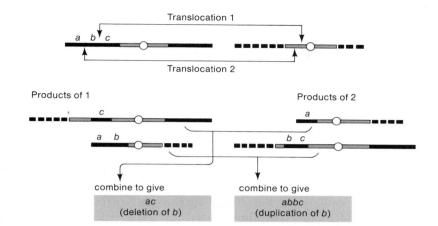

Figure 8-28. Using translocations with one breakpoint in heterochromatin to produce a duplication and a deletion. (Heterochromatin is colored.) If the upper product of translocation 1 is combined with the upper product of translocation 2 by means of an appropriate mating, a deletion of *b* results. If the lower products of the two translocations are combined, the genotype *a b b c* is produced.

In 1972, Burke Judd tested this hypothesis by breeding the *w*⁺ allele out of the translocation and onto a normal X chromosome and by breeding a *w* gene from the normal X chromosome onto the translocation (Figure 8-29). Judd found that when the *w*⁺ gene on the translocation was crossed onto a normal X chromosome and *w* was then inserted into the translocation, the eye color was red; so obviously the *w*⁺ was not defective. When Judd crossed a *w*⁺ gene back onto the translocation, the gene again variegated. Thus, we can conclude that, for some reason, the *w*⁺ gene in the translocation is not expressed in some cells, thereby allowing the expres-

sion of *w*. This kind of variegation is called position-effect variegation because the unstable expression of a gene is a reflection of its position in a rearrangement.

Such position effects on genes are seen in certain cancers that are caused by translocations. For example, most cases of Burkitt's lymphoma, a cancer of certain human antibody-producing cells called B cells, are caused by the relocation of a cancer-causing gene to a position next to a region that normally enhances the production of antibodies (Figure 8-30). The cancer-causing gene (oncogene) is then activated, resulting in cancer.

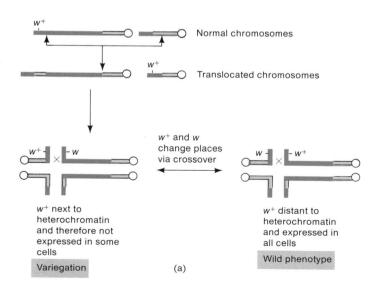

Figure 8-29. Position-effect variegation. (a) The translocation of *w*⁺ to a position next to heterochromatin (gray) causes the *w*⁺ function to fail in some cells, producing position-effect variegation. (b) Appearance of a *Drosophila* eye, showing position-effect variegation. Colored = red eye cells; white = white eye cells. (Part b courtesy of Randy Mottus.)

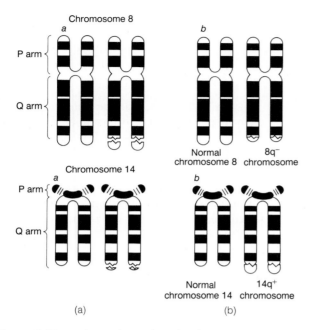

Figure 8-30. Reciprocal translocation between chromosomes 8 and 14 causes most cases of Burkitt's lymphoma. An oncogene on the tip of chromosome 8 becomes relocated next to an antibody gene enhancer region on chromosome 14. (a) Normal situation. (b) Translocation heterozygote.

> **Message**   The expression of a gene can be affected by its position in the genome.

In fungi, a tetrad analysis can be very useful in detecting chromosome aberrations in general. In *Neurospora*, for example, any genome containing a deletion will produce an ascospore that does not ripen to the normal black color, and parents of crosses with high proportions of such "aborted" white ascospores usually contain rearrangements (duplications are generally recovered as black, *viable* ascospores). Specific spore-abortion patterns sometimes identify specific rearrangements. The pattern resulting from a translocation is an equal number of asci having eight black and zero white spores (alternate-segregation meioses) and zero black and eight white spores (adjacent-1-segregation meioses) and some asci having four black and four white spores. The 4 : 4 asci are produced by crossing-over between either centromere and the translocation breakpoint (Figure 8-31).

You may have noticed that little has been said in this chapter about how changes in chromosome structure are brought about. Unequal crossing-over has been considered as a mechanism for the production of tandem duplications, but what about the other rearrangements? Obviously, chromosome breakage is an integral part of most of the mechanisms for most rearrangements. Ionizing radiation, for example, is a potent inducer of chromosome breaks in the laboratory. However, such rearrangements also occur spontaneously in nature. Until quite recently, such spontaneous rearrangements were believed to result from naturally occurring types of radiation, but it is now known that certain genetic factors — special regions in the DNA — contribute in a major way to the instability of chromosomes. These regions can produce chromosome breaks and major rearrangements. A detailed look at this process must wait until Chapter 19, after we have dealt with the relevant molecular genetics.

Summary

■ The morphology of chromosomes provides a way of identifying them. Useful features are chromosome size, centromere position, nucleolar organizer position, and chromomere, heterochromatin, and banding patterns.

Four types of abnormalities of chromosome structure are deletions, duplications, inversions, and translocations. Deletions represent missing sections of chromosomes. If the region removed is essential to life, a homozygous deletion will be lethal. Heterozygous deletions can be nonlethal or lethal and can express recessive genes uncovered by the deletion.

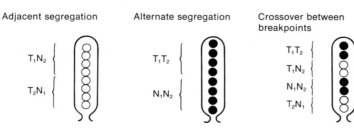

Cross $T_1T_2 \times N_1N_2$

Figure 8-31. Consequences of various meioses in a cross of *Neurospora* heterozygous for a reciprocal translocation. White sexual spores abort; black are viable. T_1 and T_2 represent the respective translocated chromosomes; N_1 and N_2 represent the normal chromosomes. The 4 black – 4 white spores are produced by crossing-over between either centromere and the translocation breakpoint.

Duplications can cause an imbalance in the genetic material, thereby producing a phenotypic effect in the organism. However, there is good evidence from a number of species, including humans, that duplications can lead to an increased variety of gene functions. In other words, duplications can be a source of new material for evolution.

An inversion is caused by a 180° turn of a portion of a chromosome. In the homozygous state, inversions may cause little problem for an organism unless heterochromatin is involved and a position effect is thus exhibited. On the other hand, inversion heterozygotes often have pairing difficulties at meiosis, and an inversion loop may result. Crossing-over within the loop results in inviable products. The crossover products will be different for inversions that are pericentric (span the centromere) and paracentric (do not span the centromere).

A translocation involves the relocation of a chromosome segment to another position in the genome. In the heterozygous state, translocations produce duplication-deletion meiotic products, which can lead to unbalanced zygotes. New gene linkages can be produced by translocations. Both translocation and inversion heterozygotes can have reduced fertility.

Chromosome rearrangements are an important cause of ill health in human populations and are useful in engineering special strains of organisms in pure and applied biology.

■　■　■

Solved Problems

1. A corn plant is obtained that is heterozygous for a reciprocal translocation and therefore is semisterile. This plant is backcrossed to a chromosomally normal-strain homozygous for the recessive gene brachytic (*b*) on chromosome 2. A semisterile F_1 plant is then backcrossed to the homozygous brachytic strain. The progeny obtained show the following phenotypes:

Nonbrachytic		Brachytic	
Semisterile	Fertile	Semisterile	Fertile
334	27	42	279

   a. What ratio would you expect if the chromosome carrying brachytic is not involved in the translocation?

   b. Do you think that chromosome 2 is involved in the translocation? Explain your answer, showing the conformation of the relevant chromosomes of the semisterile F_1 and the reason for the specific numbers obtained.

Solution

   a. Here we should start with the methodical approach and simply restate the data given in the form of a diagram, where

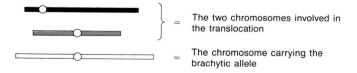

= The two chromosomes involved in the translocation

= The chromosome carrying the brachytic allele

To simplify the diagram, we will not show the chromosomes divided into chromatids, although they would be at this stage of meiosis:

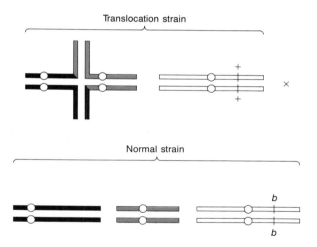

All of the progeny from this cross will be heterozygous for the brachytic chromosome, but what about the chromosomes involved in the translocation? In Chapter 8, we have seen that only alternate-segregation products survive and that one-half of these survivors will be chromosomally normal and one-half will carry the two rearranged chromosomes. The rearranged combination will regenerate a translocation heterozygote when it combines with the chromosomally normal complement from the normal parent. These latter types—the semisterile F_1's—are diagrammed below as part of the backcross to the parental brachytic strain:

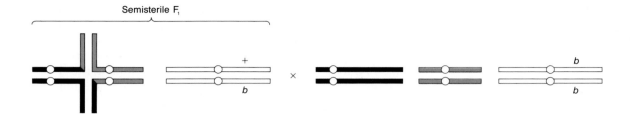

In predicting the progeny from this cross, we can treat the behavior of the translocated chromosomes independently of the pair bearing the brachytic gene. Hence, we can predict that the progeny will be

$\frac{1}{2}$ translocation heterozygotes (semisterile) $\begin{cases} \nearrow \frac{1}{2}\,+b \longrightarrow \frac{1}{4} \text{ semisterile nonbrachytic} \\ \searrow \frac{1}{2}\,bb \longrightarrow \frac{1}{4} \text{ semisterile brachytic} \end{cases}$

$\frac{1}{2}$ normal (fertile) $\begin{cases} \nearrow \frac{1}{2}\,+b \longrightarrow \frac{1}{4} \text{ fertile nonbrachytic} \\ \searrow \frac{1}{2}\,bb \longrightarrow \frac{1}{4} \text{ fertile brachytic} \end{cases}$

This predicted $1:1:1:1$ ratio is quite different from that obtained in the actual cross.

b. Since we have observed a departure from the ratio predicted on the basis of independence, then it seems likely that the brachytic chromosome (2) *is* involved in the translocation. Let's assume that the recessive gene brachytic (*b*) is on the solid black chromosome. But where? For the purpose of the diagram, it doesn't matter where we put it, but it does matter genetically because the position of the *b* affects the ratios in the progeny. If we assume that *b* is located near the tip of the piece that is translocated, we can redraw the pedigree as follows:

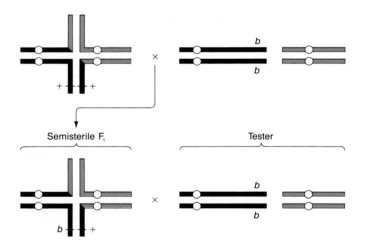

If the chromosomes of the semisterile F_1 segregate in the forms diagrammed here, we could then predict

$\frac{1}{2}$ fertile brachytic

$\frac{1}{2}$ semisterile nonbrachytic

Most progeny are certainly of this type, so we must be on the right track. How are the two minority types produced? Somehow we have to get the b^+ allele onto the normal solid black chromosome and the *b* allele onto the translocated chromosome. This must be achieved by crossing-over between the translocation breakpoint (the center of the cross-shaped structure) and the brachytic locus. To represent this, we must show chromatids, because crossing-over occurs at the chromatid stage:

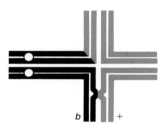

The recombinant chromosomes will produce progeny that are fertile nonbrachytic and semisterile brachytic (69 of these out of a total of 682, or a frequency of about 10 percent). We can see that this frequency is really a measure of the map distance (10 m.u.) of the brachytic locus from the breakpoint. (The same basic result would have been obtained if we had drawn the brachytic locus in the part of the chromosome on the other side of the breakpoint.)

2. A maize geneticist is studying recombination between two genes, *b* and *l*, which are 18 m.u. apart on the right arm of chromosome 12. She is particularly interested in one pure-breeding line (M) isolated from nature. Crosses within line M give the expected RF of 18 percent between *b* and *l*, exactly the same RF obtained when working within the conventional line (P). However, when an appropriately marked stock from M is crossed to an appropriately marked stock from P ($++/++ \times bl/bl$) and the F_1 plants are testcrossed to a bl/bl tester from M, the RF value drops to 2 percent. The same value is obtained when the F_1 is crossed to a bl/bl tester from P. Formulate a model and use it to explain:

a. The RF value within line M.

b. The RF value within line P.

c. The RF value in the $F_1 \times M$ testcross.

d. The RF value in the $F_1 \times P$ testcross.

e. The *origin* of recombinants in the testcrosses.

f. How would you test your model?

Solution

In this problem, as in many pieces of genetic analysis, the data contain an important clue; once the significance of the clue is realized, the details of the experimental results fall rapidly into place. Here the clue is the drastically reduced recombinant frequency. Not very many genetic mechanisms can cause such a reduction. In this chapter, we have studied two major ones: inversions and deletions. In this problem, however, the deletion hypothesis is highly unlikely for two reasons. First, it is clear that the unusual pure line M is viable, and we have seen that large deletions generally are not viable as homozygotes. Second, the recombination frequency is normal within line M, and this would not be true for a deletion. So we are left with the basic hypothesis of a large inversion spanning all or most of the $b-l$ region. Initially, we might consider the possibility that one locus is located inside and one locus is located outside the inversion, but this does not explain the values from crosses within line P (see answer to b).

a.

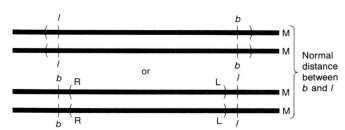

b.

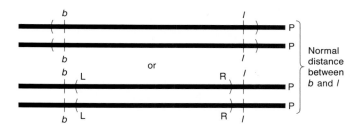

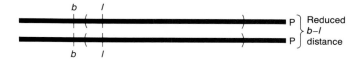

Notice that if only one locus were inside the inversion and the other locus were outside, then a normal map distance would not prevail. For example

c.

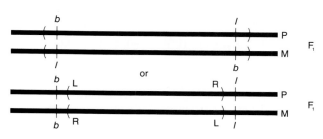

d. Same F_1 as in (c). The tester makes no difference.

e. We have been carrying two basic alternatives. In the first, the recombinants would have to come from double crossovers: one between the loci and one outside the loci but within the inversion. In the second alternative, the recombinants could come from single crossovers in the small part of the $b-l$ region that is not spanned by the inversion.

f. Cytologically, a simple way is to look for inversion loops at meiosis or, if there are chromosome markers such as constrictions or staining bands, to look for inversion in mitotic chromosomes. Genetically, you could map the inverted loci in relation to chromosome markers outside the inversion, as shown here for r and s:

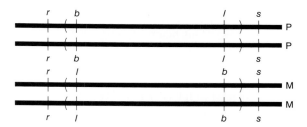

In the normal line P, b would be linked to r; in line M, b would be linked to s.

Problems

1. What features (genetic or cytological) identify and distinguish between: a. deletions b. duplications c. inversions d. reciprocal translocations?

2. The two loci P and Bz are normally 36 m.u. apart on the left arm of a certain plant chromosome. If a paracentric inversion spans about one-quarter of this region but not the loci, what approximate recombinant frequency is predicted between these loci in plants that are: a. heterozygous for the paracentric inversion? b. homozygous for the paracentric inversion?

3. Assume that in *Drosophila,* the following loci are linked in the order a–b–c–d–e–f. A fly of genotype *aa bb cc dd ee ff* is crossed to another fly from a wild-type line. About one-half the progeny is fully wild-type in phenotype, but the others show the recessive phenotype corresponding to *d* and *e*. Propose an explanation for these results.

4. The normal sequence of a certain *Drosophila* chromosome is 123 · 456789, where the dot represents the centromere. Some chromosome aberrations that were isolated have the following structures: a. 123 · 476589 b. 123 · 46789 c. 1654 · 32789 d. 123 · 4566789. Name each type, and draw diagrams to show how each would pair with the normal chromosome.

5. A *Neurospora* heterokaryon is established between nuclei of the genotypes shown in the following diagram in a common cytoplasm:

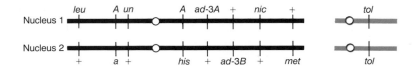

Here *leu, his, ad, nic,* and *met* are all recessive alleles causing specific nutritional requirements for growth. *A* and *a* are the mating-type alleles (one parent must be *A* and the other *a* for a cross to occur). Usually, "*A* plus *a*" heterokaryons are incompatible, but the recessive mutant *tol* suppresses this incompatibility and permits heterokaryotic growth on vegetative medium. The allele *un* is recessive, prevents the fungus from growing at 37°C (it is a temperature-sensitive allele), and cannot be corrected nutritionally. This heterokaryon grows well on a minimal medium, as do most of the cells derived mitotically from it. However, some rare cells show the following traits:

They will not grow on a minimal medium unless it is supplemented with leucine.

When the cells are transferred to a crossing medium, a cross does not occur (they will not self).

They will not grow when moved into a 37°C temperature *even* when supplied with leucine.

When haploid wild-type *a* cells are added to these aberrant cells, a cross occurs, but the addition of *A* does not cause a cross.

From the cross with wild-type *a,* progeny with the genotype of nucleus 1 are recovered, but no alleles from nucleus 2 ever emerge from the cross.

Formulate an explanation for the origin of these strange cells in the original heterokaryon, and account for the observations concerning them.

6. Certain mice called "waltzers" execute bizarre steps in contrast to the normal gait for mice. The difference between normal mice and the waltzers is genetic, with waltzing being a recessive characteristic. W. H. Gates crossed waltzers with homozygous normals and found among several hundred normal progeny a single waltzing mouse (♀). When mated to a waltzing ♂, she produced all waltzing offspring. Mated to a homozygous normal ♂, she produced all normal progeny. Some ♂♂ and ♀♀ of this normal progeny were intercrossed, and there were no waltzing offspring among their progeny. Painter examined the chromosomes of waltzing mice that were derived from some of Gates's crosses and that showed a breeding behavior similar to that of the original, unusual waltzing ♀. He found that these individuals had 40 chromosomes, just as in normal mice or the usual waltzing mice. In the unusual waltzers, however, one member of a chromosome pair was abnormally short. Interpret these observations as completely as possible, both genetically and cytologically.

(Problem 6 is from A. M. Srb, R. D. Owen, and R. S. Edgar, *General Genetics,* 2d ed. Copyright © 1965 by W. H. Freeman and Company.)

7. In *Neurospora crassa,* mutants of the *ad-3B* gene are relatively easy to amass because they have a purple coloration in addition to a requirement for adenine. In haploid cultures, 100 spontaneous *ad-3B* mutants were obtained. Cells from each were then plated on a medium containing no adenine to test for reversion. Even after extensive platings involving a wide array of mutagens, 13 cultures produced no colonies. What is the probable nature of these mutants? Account for both the lack of revertability and the haploid viability of these strains.

8. Six bands in a salivary-gland chromosome of *Drosophila* are shown in the figure below, along with the extent of five deletions (Del1 to Del5):

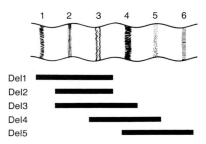

Recessive alleles *a, b, c, d, e,* and *f* are known to be in the region, but their order is unknown. When the deletions are heterozygous with each allele, the following results are obtained:

	a	*b*	*c*	*d*	*e*	*f*
Del1	−	−	−	+	+	+
Del2	−	+	−	+	+	+
Del3	−	+	+	+	−	+
Del4	+	+	−	−	−	+
Del5	+	+	+	−	−	−

In this table, a − means that the deletion is missing the corresponding wild-type allele (the deletion uncovers the recessive) and a + means that the corresponding wild-type allele is still present. Use these data to infer which salivary band corresponds to each gene.

Problem 8 is from D. L. Hartl, D. Friefelder, and L. A. Snyder, *Basic Genetics.* Jones and Bartlett, 1988.

9. In *Drosophila,* five recessive lethal mutations are all shown to map on chromosome 2. Each chromosome bearing the lethal mutation is then paired with a chromosome 2 from a

TABLE 8-1.

Lethal Mutation	Chromosome-2 marker					
	h	i	j	k	l	m
1	M	M	W	W	W	W
2	W	W	W	M	M	W
3	W	W	W	W	W	M
4	W	W	W	M	M	M
5	W	W	W	W	W	W

stock having six recessive mutations (h, i, j, k, l, and m), which are distributed throughout the chromosome in that order. The appearance of the resulting flies is shown in Table 8-1, where M stands for mutant for any particular phenotype and W stands for wild.

a. What is the probable nature of lethal mutations 1 to 4?

b. What can you say about the nature of lethal mutation 5?

10. Two pure lines of corn show different recombinant frequencies in the region from $P1$ to sm on chromosome 6. The normal strain (A) shows an RF of 26 percent, and the abnormal strain (B) shows an RF of 8 percent. The two lines are crossed, producing hybrids that are semisterile.

a. Decide between an inversion and a deletion as the possible cause of the low RF in the abnormal strain. List your reasons.

b. Sketch the approximate relation of the chromosome aberration to the genetic markers.

c. Why are the hybrids semisterile?

11. In the meiosis $P\ Bar\ Q/p\ Bar\ q$, the P/p and Q/q genes represent flanking markers that are very close to the left and right of a homozygous bar-eye mutation. In an appropriate testcross, some normal-eye and some double-bar types are recovered at low frequencies. These show the flanking marker combinations Pq or pQ. Explain the following with diagrams:

a. The origin of the rare normal and double-bar types

b. The association with the flanking marker genotypes

12. In *Neurospora*, a nontandem duplication of the following type grows quite well as a haploid culture:

$$AB \bullet \qquad\qquad ba$$

From such a culture, rare cells of the following constitution are detected:

$$Ab \bullet \qquad\qquad Ba$$

a. What kinds of somatic pairing and crossover could produce such rare types?

b. What other kinds of rare types might you expect to find if you looked hard enough?

13. In *Drosophila*, a pure line is developed, carrying a duplication of the X-chromosome segment that contains the vermilion-eye gene. The stock is

$$\frac{v^+ \qquad v^-}{v^+ \qquad v^-}$$

and has wild-type eye color. Females of this stock are mated to nonduplicated vermilion males:

$$
\begin{array}{c}
X \dfrac{v^+ \qquad v^-}{} \\
X \dfrac{}{v^+ \qquad v^-} \\
\times \\
X \dfrac{}{v^-} \\
Y \underline{}
\end{array}
$$

The male offspring all have wild-type eye color, and the female offspring all have vermilion eyes. Explain why these are surprising results in regard to the theory of dominance. Explain the phenotype of the following: a. Female and male parents b. Female and male progeny

14. In *Neurospora*, the $un3$ locus is just to the left of the centromere on chromosome 1 and always segregates at the first meiotic division. The $ad3$ locus is $10\ m.u.$ to the right of the same centromere.

a. Assuming that only single crossovers or no crossovers occur in the $un3$–$ad3$ region, what linear asci are predicted and in what frequencies in a normal cross of $un3$, $ad3$ × wild-type?

b. Most of the time such crosses behave predictably, but in one case a standard $un3$, $ad3$ strain is crossed to a wild-type isolated from a field of sugar cane in Hawaii. The results follow:

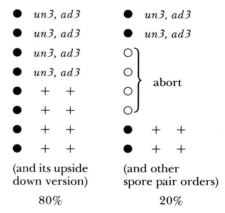

(and its upside down version)   (and other spore pair orders)
80%   20%

Provide an explanation for these results, and state how you could test your idea.

15. *Drosophila* is found all over the world. A linkage study of chromosome 2 involving pure-breeding strains in the Okanagan Valley (Canada) and in Spain gave the following recombinant frequencies (in percent) for the six loci A–F on chromosome 2:

a. Analyze these data with regard to the chromosome arrangement of these genes in each location, and provide an explanation for any difference(s). Draw a map for each set of data, indicating map distances.

Okanagan

	A	B	C	D	E	F
A	0	12	14	23	3	29
B		0	2	35	15	17
C			0	37	17	15
D				0	20	I*
E					0	32
F						0

Spain

	A	B	C	D	E	F
A	0	12	7	30	3	22
B		0	19	18	15	34
C			0	37	4	15
D				0	33	I*
E					0	19
F						0

*I = independent assortment; RF = 50

b. Show what recombinant frequencies might be expected in the five chromosome regions delineated by these genes in Okanagan-Spanish hybrids.

16. An aberrant corn plant gives the following results when testcrossed:

	Interval				
	d–f	f–b	b–x	x–y	y–p
Control RF values	5	18	23	12	6
Aberrant plant RF values	5	2	2	0	6

The locus order is centromere-d–f–b–x–y–p. The aberrant plant is a healthy plant, but it produces far fewer normal ovules and pollen than the control plant.

a. Propose a hypothesis to account for the abnormal recombination and the abnormal fertility.

b. Use diagrams to explain the origin of the recombinants according to your hypothesis.

17. In *Neurospora,* a cross is heterozygous for a paracentric inversion. The breakpoints of the inversion are known to be very close to two loci that recombine with an RF of 10 percent.

a. Use the mapping function RF = $\frac{1}{2}(1 - e^{-m})$ to calculate the *mean* number of exchanges expected in the inversion loop per meiosis.

b. Use this mean frequency to calculate the frequency of meiosis with (1) no, (2) one, and (3) two exchanges in the loop if the Poisson formula is

$$e^{-m}\left(\frac{1}{0!} + \frac{m}{1!} + \frac{m^2}{2!} + \cdots\right).$$

c. Remembering that ascospores bearing deficient chromosome complements do not darken in *Neurospora*, predict how many light and dark ascospores you would find in eight-spored asci resulting from meioses in which there had been (1) no, (2) one, and (3) two crossovers in the loop. (Remember that there are three kinds of double crossovers, so the progeny population of asci from two crossovers could be heterogeneous.)

d. Using your predicted frequencies of zero, one, and two crossover meioses, what *overall* frequencies of the following asci would you find from this cross?

(1) 8 dark : 0 light

(2) 0 dark : 8 light

(3) 4 dark : 4 light

NOTE: your total from part (b) should be less than 100 percent because triple and higher exchanges have been ignored, but it should be close to 100 percent because these events are rare. Simply make your total for this part equal to your total for part (b).

18. A *Drosophila* geneticist has a strain of fruit flies that is true-breeding and wild-type. She crosses this strain with a multiply marked X-chromosome strain carrying the recessive genes y (yellow), cv (crossveinless), v (vermilion), f (forked), and car (carnation), which are equally distributed along the X chromosome from one end to the other. She collects the heterozygous F$_1$ female offspring and mates them with y cv v f B car males (B = bar, or slitlike eye). The geneticist obtains the following classes among the male offspring:

1. y cv v f   car
2. + + + +   +
3. y + + +   car
4. + cv v f   +
5. y cv + f   car
6. + + v +   +
7. y cv + +   car
8. + + v f   +
9. y + + f   car
10. + cv v +   +
11. y cv v f B car

a. Account for the results in classes 1 to 10.

b. How can you account for class 11? (Give two ways.)

c. How would you test your hypotheses?

(Problem 18 courtesy of Tom Kaufman.)

*19. Suppose that you are given a *Drosophila* line from which you can get males or virgin females at any time. The line is homozygous for a second chromosome, which has an inversion to prevent crossing-over, a dominant gene (Cu) for curled wings, and a recessive gene (pr, purple) for dark eyes. The chromosome can be drawn as

You have irradiated sperm in a wild-type male and wish to determine whether recessive lethal mutations have been induced in chromosome 2. How would you go about doing this? (HINT: remember that each sperm carries a *different* irradiated second chromosome.) Indicate the kinds and numbers of flies used in each cross.

20. Predict the chromosome shapes that will be produced at anaphase 1 of meiosis in a reciprocal translocation heterozygote undergoing   a. alternate segregations   b. adjacent-1 segregations.

21. In *Neurospora,* the genes *a* and *b* are on separate chromosomes. In a cross of a standard *a b* strain with a wild-type obtained from nature, the progeny are as follows: *a b*, 45 percent; + +, 45 percent; *a* +, 5 percent; + *b*, 5 percent. Interpret these results, and explain the origin of all the progeny types according to your hypothesis.

22. Curly wings (*Cy*) is a dominant mutation in the second chromosome of *Drosophila*. A *Cy*/+ male is irradiated with X rays and crossed with +/+ females. The *Cy*/+ sons are then mated in single pairs with +/+ females. From one cross, the progeny are

curly males	146
wild-type males	0
curly females	0
wild-type females	163

What abnormality in chromosome structure is the most likely explanation for these results? Use chromosome diagrams of all strains referred to in the question in your explanation. (HINT: remember that crossing-over does not occur in male *Drosophila*.)

23. You discover a *Drosophila* male that is heterozygous for a reciprocal translocation between the second and third chromosomes, each break having occurred near the centromere (which is near the center of each of these chromosomes).

   a. Draw a diagram showing how these chromosomes would synapse at meiosis.

   b. You find that this fly has the recessive genes *bw* (brown eye) and *e* (ebony body) on the nontranslocated second and third chromosomes, respectively, and wild-type alleles on the translocated chromosomes. The fly is mated with a female that has normal chromosomes and is homozygous for *bw* and *e*. What type of offspring would you expect and in what ratio? (HINT: remember that zygotes that have an extra chromosome arm or that are deficient for one chromosome do not survive. There is no crossing-over in *Drosophila* males.)

24. An *insertional* translocation consists of the insertion of a piece from the center of one chromosome into the middle of another (nonhomologous) chromosome. Thus

becomes

How will genomes that are heterozygous for such translocations pair at meiosis? In *Neurospora,* what spore abortion patterns will be produced and in what relative proportions in such translocation heterozygotes? (HINT: remember that duplications survive and have dark spores but deficiencies are light-spored.)

25. In *Neurospora,* the markers *ad-3* and *pan-2* are auxotrophic mutations located on chromosomes 1 and 6, respectively. An unusual *ad-3* line arises in the laboratory, giving the following results:

	Ascospore appearance	RF between *ad-3* and *pan-2*
1. Normal *ad-3* × normal *pan-2*	all black	50%
2. Abnormal *ad-3* × normal *pan-2*	about ½ black and ½ white (inviable)	1%

3. Of the black spores from cross 2, about one-half were completely normal and one-half repeated the same behavior as the original abnormal *ad-3* strain.

Explain all three results with the aid of clearly labeled diagrams. (NOTE: in *Neurospora,* ascospores with extra chromosome material survive and are the normal black color, whereas ascospores lacking any chromosome region are white and inviable.)

26. In corn, the following linkage arrangement holds in normal plants:

P = dark green

p = pale green

S = large ears

s = shrunken ears

An original plant of genotype *Pp Ss* has the expected phenotype (large ears, dark green) but gives unexpected results in crosses as follows:

On selfing, fertility is normal, but the frequency of *pp ss* types is ¼ (not 1/16, as expected).

When crossed to a normal tester of genotype *pp ss*, the F₁ progeny are ½ *Pp Ss* and ½ *pp ss*; fertility is normal.

When an F₁ *Pp Ss* plant is crossed to a normal *pp ss* tester, it proves to be semisterile, but again the progeny are ½ *Pp Ss* and ½ *pp ss*.

Explain these results, showing the full genotypes of the original plant, the tester, and the F_1 individuals. How would you test your hypothesis?

27. A corn plant pr/pr that has standard chromosomes is crossed with a plant that is homozygous for a reciprocal translocation between chromosomes 2 and 5 and for the Pr allele. The F_1 is semisterile and phenotypically Pr (a seed color). A backcross to the parent with standard chromosomes gives 764 semisterile Pr; 145 semisterile pr; 186 normal Pr; and 727 normal pr. What is the map distance of the Pr/pr locus from the translocation point?

28. In *Neurospora,* a reciprocal translocation of the following type is obtained:

The following cross is then made:

Assuming that the small, colored piece of the chromosome involved in the translocation does not carry any essential genes, how would you select products of meiosis that are duplicated for the translocated part of the solid chromosome?

29. Assume that in a study of hybrid cells, three genes in humans are assigned to chromosome 17. These genes (a, b, and c) are concerned with making the compounds a, b, and c—all of which are essential for growth. If $a^- b^- c^-$ mouse cells are fused with $a^+ b^+ c^+$ human cells, assume that you find a hybrid in which the only human component is the right arm of chromosome 17 (17R), translocated by some unknown mechanism to a mouse chromosome. The hybrid can make the compounds a, b, and c. Treatment of cells with adenovirus causes chromosome breaks. Assume that you can isolate 200 lines in which bits of the translocated 17R have been clipped off. These lines are tested for the ability to make a, b, and c, and the following results are obtained:

Number	Can make
0	a only
0	b only
12	c only
0	a and b only
80	b and c only
0	a and c only
60	a, b, and c
48	nothing

a. How would these different types arise?

b. Are a, b, and c all located on the right arm of chromosome 17? If so, draw a map indicating their relative positions.

c. How would quinacrine dyes help you in your analysis? (NOTE: this *kind* of approach has actually been used, although the details of this particular question are largely hypothetical.)

*30. Complex translocations often are found in natural plant populations. The best example is in the evening primrose, *Oenothera* ($2n = 14$). In this genus, the centromeres tend to be positioned more or less in the middle of the chromosomes. Furthermore, chromosome breakage in the production of translocations tends to be close to or at the centromere. The basic haploid chromosome set can be represented as follows, where a dot represents a centromere:

$$1^L \cdot 1^R \quad 2^L \cdot 2^R \quad 3^L \cdot 3^R \quad 4^L \cdot 4^R \quad 5^L \cdot 5^R \quad 6^L \cdot 6^R \quad 7^L \cdot 7^R$$

The species *O. lamarckiana* contains one such basic haploid set plus a haploid set bearing many translocations as follows:

$$1^L \cdot 1^R \quad 3^L \cdot 5^R \quad 2^L \cdot 7^L \quad 6^R \cdot 5^L \quad 4^L \cdot 2^R \quad 7^R \cdot 3^R \quad 6^L \cdot 4^R$$

a. What patterns of chromosome pairing would you expect at meiosis if all homologous regions pair? (Draw a diagram.)

b. In *O. lamarckiana,* the segregation is always alternate. What are the cytological consequences of this?

c. *O. lamarckiana* always contains one "basic" plus one "translocated" chromosome set, as shown, never basic plus basic or translocated plus translocated. Can

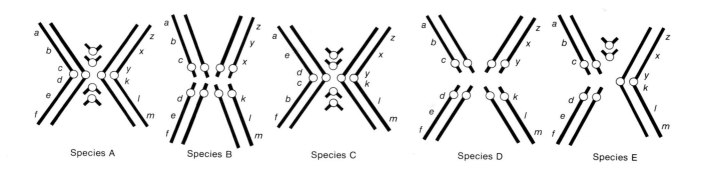

you think of a mechanism whereby the plant might maintain this situation?

d. Can you think of a reason why it is advantageous to the plant to maintain the situation described in (c)?

31. Suppose that you are studying the cytogenetics of five closely related species of *Drosophila*. The figure at the bottom of page 196 shows the gene orders (the letters indicate genes that are identical in all five species) and the chromosome pairs that you find in each species. Show how these species probably evolved from each other, describing the changes occurring at each step. (NOTE: be sure to compare gene order carefully.)

32. Show how the $\frac{10}{16}$ ratio on page 185 was derived.

Chromosome Mutation II: Changes in Number

KEY CONCEPTS

Organisms with multiple chromosome sets (polyploids) are generally larger than normal, but meiotic pairing anomalies in these organisms can produce sterility.

•

An even number of polyploid sets is generally more likely to result in fertility. Then the single-locus segregation ratios are different from those of diploids.

•

Crosses between two different species and the subsequent doubling of the chromosome number in the hybrid produces a special kind of fertile interspecific polyploid.

•

Variants in which a single chromosome has been gained or lost generally arise by nondisjunction (abnormal chromosome segregation at meiosis or mitosis).

•

Such variants tend to be sterile and show the abnormalities attributable to gene imbalance.

•

When fertile, such variants show abnormal gene segregation ratios for the misrepresented chromosome only.

■ Like the structure aberrations treated in Chapter 8, changes in chromosome number occur spontaneously in both natural and laboratory populations of organisms. Alternatively, they may be experimentally induced by applying certain standard mutagenic agents to appropriate cells or tissues. Either standard genetic tests or cytological examination may be used to detect such changes.

Changes in chromosome number are usually classified into changes involving whole chromosome sets and changes involving parts of chromosome sets. Before we consider examples of both classifications, however, we must define some useful terms. The number of chromosomes in a basic set is called the **monoploid number** (x). Organisms with multiples of the monoploid number of chromosomes are called **euploid**. Euploid types that have a greater number of sets than two are called **polyploid**. Thus, $1x$ is **monoploid**, $2x$ is **diploid**, and the polyploid types are $3x$ (**triploid**), $4x$ (**tetraploid**), $5x$ (**pentaploid**), $6x$ (**hexaploid**), and so on. The haploid number (n), which we have already used extensively, refers strictly to the number of chromosomes in gametes. In most animals and many plants that we are familiar with, the haploid number and monoploid number are the same. Hence, n or x (or $2n$ or $2x$) can be used interchangeably. However, in certain plants such as modern wheat, n and x are different. Wheat has 42 chromosomes, but careful study reveals that it is hexaploid, with six rather similar but not identical sets of seven chromosomes. Hence, $6x = 42$ and $x = 7$. However, the gametes of wheat contain 21 chromosomes, so $2n = 42$ and $n = 21$.

Changes that involve parts of a chromosome set result in individuals that are **aneuploid** (not euploid). The addition of chromosomes produces individuals that are **hyperploid**, and subtraction results in **hypoploid** individuals. Changes involving only one or a few chromosomes give rise to the following terms. In an organism that is predominantly diploid, $2n - 1$ is **monosomic**, $2n + 1$ is **trisomic**, $2n - 2$ is **nullisomic** (two homologs are lost), and $2n + 1 + 1$ is a **double trisomic**. In a haploid organism, $n + 1$ is called a **disomic**. It can be seen that these terms refer to the number of copies of a particular type of chromosome that are present in the aneuploid.

Now we can examine the genetic and cytological properties of these variants.

Abnormal Euploidy

Monoploids

In this section, we shall consider monoploidy as an unusual condition. Monoploid individuals can arise spontaneously in natural populations as rare aberrations, but in several forms (such as bees, wasps, and ants), the males are normally monoploid, having been derived from unfertilized eggs.

In the germ cells of a monoploid meiosis cannot occur normally because the chromosomes have no pairing partners. Thus, monoploids are characteristically sterile. (However, meiosis can be bypassed in some monoploid animals, such as male honeybees, which produce gametes essentially by mitotic division.) If meiosis occurs and the single chromosomes segregate randomly, then the probability of all chromosomes going to one pole is $\frac{1}{2}^{x-1}$, where x is the number of chromosomes. This will determine the frequency of viable (whole-set) gametes, obviously a vanishingly small number if x is large.

Monoploids play a major role in modern approaches to plant breeding. Diploidy is an inherent nuisance in the induction and selection of new plant mutations that are favorable and new combinations of genes that are already present. New recessive mutations have to be made homozygous for expression to occur; favorable gene combinations in heterozygotes are broken up by meiosis. Monoploids provide a way around some of these problems. In some plants, monoploids may be artificially derived from the products of meiosis in the plant's anthers. A cell destined to become a pollen grain may instead be induced by cold treatment to grow into an **embryoid**, a small dividing mass of cells. The embryoid may be grown on agar to form a monoploid plantlet, which can then be potted in soil and left to mature (Figure 9-1).

Monoploids may be exploited in several ways. In one, they are first examined for favorable traits or gene combinations, which may arise from heterozygosity already present in the parent or induced in the parent by mutagens. The monoploid can then be subjected to chromosome doubling to achieve a completely homozygous diploid with a normal meiosis, capable of providing seed. How is this achieved? Quite simply, by the application of a compound called **colchicine** to meristematic tissue. Colchicine—an alkaloid drug extracted from the autumn crocus—inhibits the formation of the mitotic spindle, so that cells with two chromosome sets are produced (Figure 9-2). These cells may proliferate to form a sector of diploid tissue that can be identified cytologically.

Another way in which the monoploid may be used is to treat its cells basically like a population of haploid organisms in a mutagenesis-and-selection procedure. The cells are isolated, their walls are removed by enzymatic treatment, and they are treated with mutagen. They are then plated on a selective medium (perhaps a toxic compound normally produced by one of the plant's parasites or an insecticide) to select resistant cells. Resistant plantlets eventually grow into haploid plants,

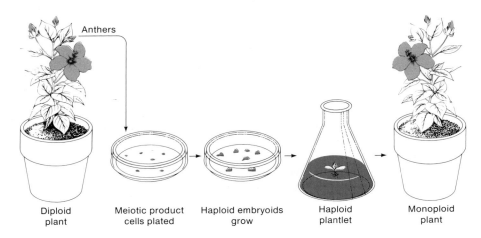

Anthers

Diploid
plant

Meiotic product
cells plated

Haploid embryoids
grow

Haploid
plantlet

Monoploid
plant

Figure 9-1. Generating a monoploid plant by tissue culture. Appropriately treated pollen grains (haploid) can be plated on agar containing certain plant hormones. Under these conditions, haploid embryoids will grow into monoploid plantlets. With another change in plant hormones, these plantlets will grow into mature monoploid plants with roots, stems, leaves, and flowers.

which can then be doubled (using colchicine) into a pure-breeding resistant type (Figure 9-3).

These are potentially powerful techniques that can circumvent the normally slow process of what is basically meiotic plant breeding. The techniques have been successfully applied to several important crop plants, such as soybeans and tobacco. This is, of course, another aspect of somatic-cell genetics in higher organisms.

The anther technique for producing monoploids does not work in all organisms or in all genotypes of an organism. Another useful technique has been developed in barley, an important crop plant. When diploid barley, *Hordeum vulgare,* is pollinated using a diploid wild relative called *Hordeum bulbosum,* fertilization occurs. During the ensuing somatic cell divisions, however, the chromosomes of *H. bulbosum* are preferentially eliminated from the zygote, resulting in a haploid embryo. (The haploidization process appears to be caused by a genetic incompatibility between the chromosomes of the different species.) The resulting haploids can be doubled with colchicine. This approach has led to the rapid production and widespread planting of several new barley varieties, and it is being used successfully in other species too.

> **Message**   Monoploids are used to create new genotypes, which are then doubled to form fertile, homozygous diploid lines.

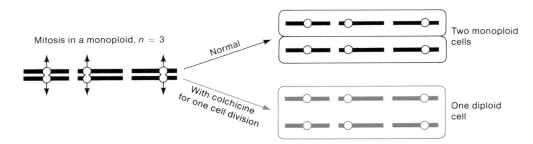

Mitosis in a monoploid, *n* = 3

Normal

With colchicine
for one cell division

Two monoploid
cells

One diploid
cell

Figure 9-2. Using colchicine to generate a diploid from a monoploid. Colchicine added to mitotic cells during metaphase and anaphase disrupts spindle-fiber formation, preventing the separation of chromatids after the centromere is split. A single cell is created that contains pairs of identical chromosomes that are homozygous at all loci.

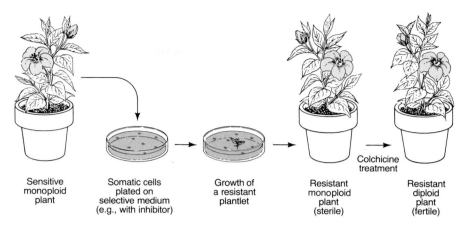

Figure 9-3. Using microbial techniques in plant engineering. The cell walls of haploid cells are removed enzymatically. The cells are then exposed to a mutagen and plated on an agar medium containing a selective agent, such as a toxic compound produced by a plant parasite. Only those cells containing a resistance mutation that allows them to live within the presence of this toxin will grow. After treatment with the appropriate plant hormones, these cells will grow into mature monoploid plants and, with proper colchicine treatment, can be converted into homozygous diploid plants.

Polyploids

Once we are into the realm of polyploids, we must distinguish between **autopolyploids,** which are composed of multiple sets from within one species, and **allopolyploids,** which are composed of sets from different species. Allopolyploids form only between closely related species; however, the different chromosome sets are **homeologous** (only partially homologous)—not fully homologous, as they are in autopolyploids.

Triploids

Triploids are usually autopolyploids. They are constructed from the cross of a $4x$ (tetraploid) and a $2x$ (diploid). The $2x$ and the x gametes unite to form a $3x$ triploid.

Triploids also are characteristically sterile. The problem again involves pairing at meiosis. Although pairing can take place in several ways, it usually occurs between only two chromosomes at a time (Figure 9-4). The net result is always the same, an unbalanced segregation of one of the following types, where the numbers stand for three homologous chromosomes:

$$\frac{1+2}{3} \quad \text{or} \quad \frac{1+3}{2} \quad \text{or} \quad \frac{2+3}{1}$$

This happens for every chromosome threesome, and the probability of obtaining either a $2x$ or an x gamete is

$(\frac{1}{2})^{x-1}$, where x is the number of chromosomes in a set. The others will be unbalanced gametes, having two of one chromosome type, one of another, two of another, and so on; most of these will be nonfunctional. Even if the gametes are functional, the resulting zygotes will be unbalanced. A practical application of the sterility associated with triploidy lies in the production of seedless varieties of watermelons and bananas.

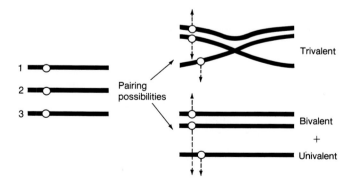

Figure 9-4. Meiotic pairing possibilities in a triploid. (Each chromosome is really two chromatids.) Pairing in meiosis, with the resultant segregation, always occurs between only two of the three homologs. The possibility that one gamete will receive the univalent of all chromosome sets is very small. Consequently, gametes of unbalanced chromosome number are produced, and sterility results.

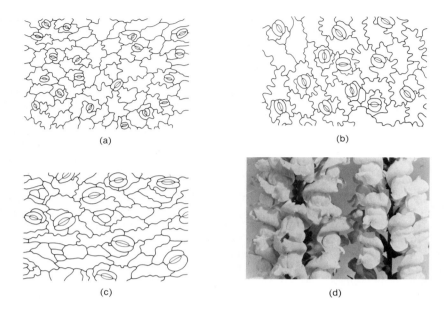

(a)

(b)

(c)

(d)

Figure 9-5. Epidermal leaf cells of tobacco plants, showing an increase in cell size, particularly evident in stomata size, with an increase in autopolyploidy. (a) Diploid. (b) tetraploid. (c) octoploid. (From W. Williams, *Genetic Principles and Plant Breeding*, Blackwell Scientific Publications, Ltd.) (d) Diploid *(right)* and tetraploid *(left)* snapdragons. (W. Atlee Burpee Company.)

Autotetraploids

Autotetraploids occur either naturally, by the spontaneous accidental doubling of a $2x$ genome to a $4x$ genome, or artificially, through the use of colchicine. Autotetraploids are evident in many commercially important crop plants because, as with other polyploids, the larger number of chromosome sets is often associated with the increased size of the plant. This is manifested in increased cell size, fruit size, stomata size, and so on (Figure 9-5).

Because 4 is an even number, autotetraploids can have a regular meiosis, although this is by no means always the case. The crucial factor is how the four chromosomes of one type pair and segregate. There are several possibilities, as shown in Figure 9-6. The two-bivalent and the quadrivalent pairing modes tend to be most

regular in segregation, but even here there is no guarantee of a $2 \leftrightarrow 2$ segregation. If a regular $2 \leftrightarrow 2$ segregation is achieved at each chromosome type, as is the case in some species, then the gametes will be functional and a formal genetic analysis can be developed for such autotetraploids.

Let's consider the genetics of a fertile tetraploid. We can hypothesize an experiment in which colchicine is used to double the chromosomes of an Aa plant into an $AAaa$ autotetraploid, which we will assume shows a $2 \leftrightarrow 2$ segregation. We now have a further worry because polyploids such as tetraploids give different segregation ratios in their progeny, depending on whether or not the locus in question is tightly linked to the centromere. First, we consider a centromeric gene. The three possible pairing and segregation patterns are presented

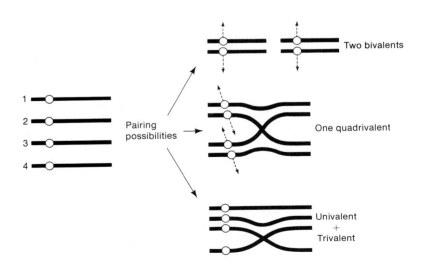

Two bivalents

One quadrivalent

Univalent + Trivalent

Pairing possibilities

Figure 9-6. Meiotic pairing possibilities in tetraploids. (Each chromosome is really two chromatids.) The four chromosomes of one type may pair as two bivalents or as a quadrivalent. Both possibilities can yield functional gametes. However, the four chromosomes may also pair in a univalent–trivalent combination, yielding nonfunctional gametes. Some tetraploids have adopted the first or second method as a routine in meiotic segregation and thus can function normally.

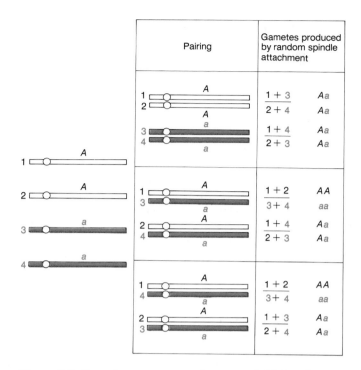

Figure 9-7. Genetic consequences in a tetraploid showing orderly pairing by bivalents. (Each chromosome is really two chromatids.) The locus is assumed to be close to the centromere. Self-fertilization could yield a variety of genotypes, including *aaaa*.

in Figure 9-7; these occur by chance and with equal frequency. As the figure shows, the $2x$ gametes are *Aa*, *AA*, or *aa*, produced in a ratio of $8:2:2$, or $4:1:1$. If such a plant is selfed, the probability of an *aaaa* phenotype in the offspring is obviously $\frac{1}{6} \times \frac{1}{6} = \frac{1}{36}$. In other words, a $35:1$ phenotypic ratio will be observed if A is fully dominant over three *a* alleles.

If, in the same kind of plant, a genetic locus B/b is very far removed from the centromere, crossing-over must be considered. This forces us to think in terms of chromatids instead of chromosomes; there are four *B* chromatids and four *b* chromatids (Figure 9-8). Because the number of crossovers in such a long region will be large, the genes will become effectively unlinked from their original centromeres. The packaging of genes two at a time into gametes is very much like grabbing two balls at random from a bag of eight balls: four of one kind, and four of another. The probability of picking two *b* genes is then

$$\frac{4}{8}\text{(the first one)} \times \frac{3}{7}\text{(the second one)} = \frac{12}{56}$$
$$= \frac{3}{14}$$

So, in a selfing, the probability of a *bbbb* phenotype is $\frac{3}{14} \times \frac{3}{14} = \frac{9}{196} \cong \frac{1}{22}$. Hence, there will be a $21:1$ pheno-

type ratio of $B\text{---}:bbbb$. For genetic loci of intermediate position, intermediate ratios will, of course, result.

Allopolyploids

The "classic" allopolyploid was synthesized by G. Karpechenko in 1928. He wanted to make a fertile hybrid that would have the leaves of the cabbage *(Brassica)* and the roots of the radish *(Raphanus)*. Each of these species has 18 chromosomes, and they are related closely enough to allow intercrossing. A viable hybrid progeny individual was produced from seed. However, this hybrid was functionally sterile because the nine chromosomes from the cabbage parent were different enough from the radish chromosomes that homology was insufficient for normal synapsis and disjunction:

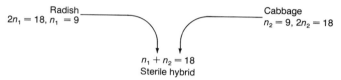

However, one day a few seeds were in fact produced by this (almost!) sterile hybrid. On planting, these seeds produced fertile individuals with 36 chromosomes. All of these individuals were allopolyploids. They had apparently been derived from spontaneous, accidental chromosome doubling to $2n_1 + 2n_2$ in the sterile hybrid, presumably in tissue that eventually became germinal and underwent meiosis. Thus, in $2n_1 + 2n_2$ tissue, there is a pairing partner for each chromosome and balanced gametes of the type $n_1 + n_2$ are produced. These fuse to give $2n_1 + 2n_2$ allopolyploid progeny, which are also

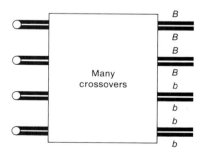

Figure 9-8. Highly diagrammatic representation of a tetraploid meiosis involving a heterozygous locus distant from the centromere. Although tetravalents may not form, the net effect of multiple crossovers in such a long region will be that the genes become effectively unhooked from their original centromeres. Genes are packaged two at a time into gametes, much as two balls may be grabbed at random from a bag containing eight balls: four of one kind, and four of another.

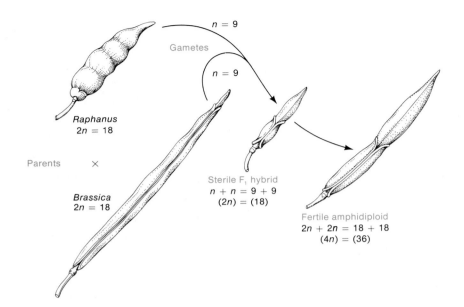

Figure 9-9. The origin of the amphidiploid (*Raphanobrassica*) formed from cabbage (*Brassica*) and radish (*Raphanus*). The production of the fertile amphidiploid in this case from the $2n = 18$ hybrid occurred in an accidental fashion, but one similar to the method of producing tetraploids by using colchicine. (From A. M. Srb, R. D. Owen, and R. S. Edgar, *General Genetics,* 2d ed. Copyright © 1965 by W. H. Freeman and Company. After G. Karpechenko, *Z. Indukt. Abst. Vererb.* 48 1928, 27.)

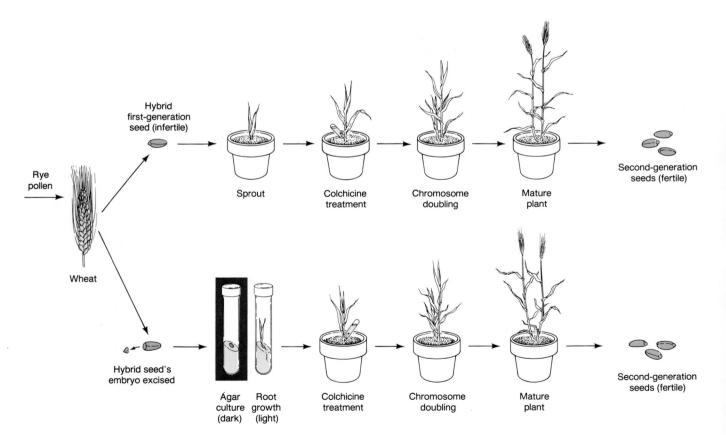

Figure 9-10. Techniques for the production of the amphidiploid *Triticale.* If the hybrid seed does not germinate, then tissue culture *(below)* may be used to obtain a hybrid plant. (From Joseph H. Hulse and David Spurgeon, "Triticale." Copyright © 1974 by Scientific American, Inc. All rights reserved.)

fertile. This kind of allopolyploid is sometimes called an **amphidiploid** (Figure 9-9). (Unfortunately for Karpechenko, his amphidiploid had the roots of a cabbage and the leaves of a radish!)

When the allopolyploid was crossed to either parent species, sterile offspring resulted. In the case of the cross to radish, these offspring were $2n_1 + n_2$, constituted from an $n_1 + n_2$ gamete from the allopolyploid and an n_1 gamete from the radish. Obviously, the n_2 chromosomes had no pairing partners, so sterility resulted. Consequently, Karpechenko had effectively created new species, with no possibility of gene exchange with its parents. He called his new species *Raphanobrassica*.

Today, allopolyploids are routinely synthesized in plant breeding. The goal obviously is to combine some of the worthwhile features of both parental species into one type. This kind of endeavor is very unpredictable, as Karpechenko found out. In fact, only one amphidiploid of potentially widespread use has ever been intentionally produced. This is *Triticale*, an amphidiploid between wheat (*Triticum*, $2n = 6x = 42$) and rye (*Secale*, $2n = 2x = 14$). *Triticale* combines the high yields of wheat with the ruggedness of rye. A massive international *Triticale* testing program is now under way, and many breeders have great hopes for the future of this artificial amphidiploid. Figure 9-10 shows the procedure for synthesizing *Triticale*.

In nature, allopolyploidy seems to have been a major force in speciation of plants. There are many different examples. One particularly satisfying one is shown by the genus *Brassica*, as illustrated in Figure 9-11. Here three different parent species have been hybridized in all possible pair combinations to form new amphidiploid species. This has all taken place in nature, but *Brassica* amphidiploids also have been artificially synthesized (Figure 9-12).

A particularly interesting natural allopolyploid is bread wheat, *Triticum aestivum* ($2n = 6x = 42$). By a study of various wild relatives, it has been possible to reconstruct a probable evolutionary history of bread wheat (Figure 9-13). In a wheat meiosis, there are always 21 pairs of chromosomes. Furthermore, it has been possible to establish that any given chromosome has only one specific pairing partner (homologous pairing) — not five other potential partners (homeologous pairing). The suppression of such homeologous pairing (which would lead to much reduced stability of the species) is maintained by a gene *Ph* on the long arm of chromosome 5 of the **B** set. Thus, *Ph* ensures a diploid-like genetics for this basically hexaploid species. Without *Ph*, bread wheat could probably never have arisen. It is interesting to speculate as to whether Western civilization could have begun or progressed without this species — in other words, without *Ph*.

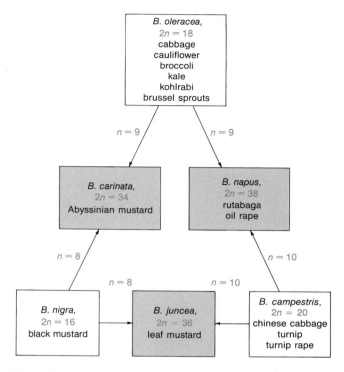

Figure 9-11. A species triangle, showing how amphidiploidy has been important in the production of new species of *Brassica*.

Somatic Allopolyploids from Cell Hybridization

Another innovative approach to plant breeding is to try to make allopolyploid-like hybrids by asexual cell-fusion methods. Theoretically, such a technique would permit the combination of widely differing parental species. The technique does indeed work, but the only allopolyploids that have been produced so far can also be made by the sexual methods we have considered already. In the cell-fusion procedure, cell suspensions of the two parental species are prepared and stripped of their cell walls by special enzyme treatments. The stripped cells are called **protoplasts.** The two protoplast suspensions are then combined with polyethylene glycol, which enhances protoplast fusion. The parental cells and the fused cells proliferate on an agar medium to form colonies (in much the same way as microbes). If these colonies, or calluses, are examined, a fair percentage of them are found to be allopolyploid-like hybrids with chromosome numbers equal to the sum of the parental types. Thus, not only do the protoplast cell membranes fuse to form a kind of heterokaryon, but the nuclei fuse too.

A good example of an allopolyploid-like hybrid is commercial tobacco, *Nicotiana tabacum*, which has 48 chromosomes. This species of tobacco was originally

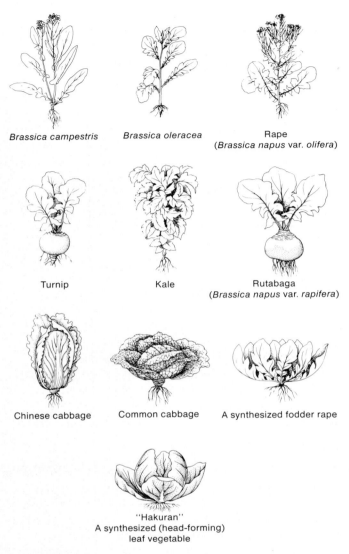

Brassica campestris

Brassica oleracea

Rape
(*Brassica napus* var. *olifera*)

Turnip

Kale

Rutabaga
(*Brassica napus* var. *rapifera*)

Chinese cabbage

Common cabbage

A synthesized fodder rape

"Hakuran"
A synthesized (head-forming)
leaf vegetable

Figure 9-12. Some of the species from the species triangle of *Brassica* and two man-made *Brassica* amphidiploids. (Courtesy of H. Kihara, *Seiken Ziho* 20, 1968.)

found in nature as a spontaneously occurring amphidiploid. The two probable parents are *N. sylvestris* and *N. tomentosiformis,* each of which has 24 chromosomes. A sexual cross between *N. tabacum* and either of these two probable parents gives a 36-chromosome hybrid containing 12 chromosome pairs plus 12 unpaired chromosomes. A cross between *N. sylvestris* and *N. tomentosiformis* yields a 24-chromosome hybrid in which there is no pairing at all. Hence, it appears that part of the *N. tabacum* genome is from *N. sylvestris* and part is from *N. tomentosiformis.* This amphidiploid can be re-created either sexually, by the processes involving colchicine described previously, or somatically by cell fusion. When

cells of the prospective parental species are fused, a 48-chromosome hybrid cell line is produced from which plants identical in behavior to *N. tabacum* may be grown. (Note that in the latter method, colchicine is not required, since the fusion product is already amphidiploid.)

The recovery of somatic hybrids may be enhanced if a selective system is available. For example, two different monoploid lines of *N. tabacum* have light-sensitive yellowish and light-sensitive whitish leaves, respectively. The hybrid calluses (diploid in this example) prove to be green and light-resistant due to complementation between the parental genotypes. The calluses can be grown into plantlets, which then are either grafted onto a mature plant to develop or potted themselves. The protocol for this experiment is illustrated in Figure 9-14.

Message Allopolyploids can be synthesized by crossing related species and doubling the chromosomes of the hybrid or by asexually fusing the cells of different species.

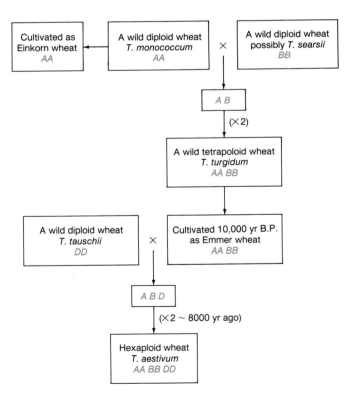

Figure 9-13. Diagram of the proposed evolution of modern hexaploid wheat involving amphidiploid production at several points. *A, B,* and *D* are different chromosome sets.

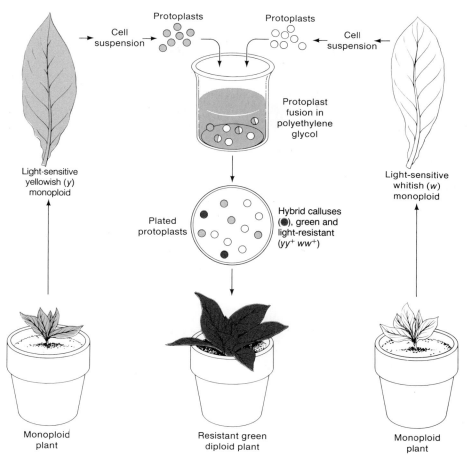

Figure 9-14. Creating a hybrid of two monoploid lines of *Nicotiana tabacum* by cell fusion. One strain has light-sensitive yellowish leaves, and the other has light-sensitive whitish leaves. Protoplasts are produced by enzymatically stripping the cell walls from the leaf cells of each strain. Fusion of the protoplasts can occur, as indicated; those that fuse as hybrids can be grown into calluses that are light resistant as a result of recessiveness of the parental genotypes. The calluses, under the appropriate hormone regime, can be grown into green diploid plants.

Polyploidy in Animals

Notice that all of our examples of polyploidy have been from plants. Polyploid animals do exist, mainly in such lower organisms as flatworms, leeches, and brine shrimp. In some of these organisms, reproduction is **parthenogenetic** (does not involve a normal meiotic sexual cycle). Progeny are produced essentially by the mitotic division of parental cells. Parthenogenetic plants, such as dandelions, also are commonly polyploid. Furthermore, **endopolyploidy** (a polyploidy of a somatic sector) also is common in plants, animals, and tumor cells. The reason for the rarity of polyploid species among higher animals is not known, but the most widely held hypothesis is that their complex sex-determining mechanisms depend on a delicate balance of chromo-

some numbers. In humans, polyploids always abort while still developing in the uterus.

Aneuploidy

Nullisomics (2*n* − 2)

Although nullisomy is a lethal condition in regular diploids, an organism like wheat (which "pretends" to be diploid but is fundamentally hexaploid) can tolerate nullisomy. In fact, all of the possible 21 wheat nullisomics have been produced; these are illustrated in Figure 9-15. Their appearances differ from normal wheat; furthermore, most of them show less vigorous growth.

Figure 9-15. The nullisomics of wheat. Although nullisomics are usually lethal in regular diploids, organisms like wheat, which "pretends" to be diploid but is fundamentally hexaploid, can tolerate nullisomy. Nullisomics, however, are less vigorous growers. (Courtesy of E. R. Sears.)

Monosomics ($2n - 1$)

Monosomic chromosome complements are generally deleterious for two main reasons. First, the balance of chromosomes, carefully put together during evolution, that is necessary to produce a finely tuned cellular homeostasis is grossly disturbed. For example, if a genome consisting of two each of chromosomes a, b, and c becomes monosomic for c (that is, $2a + 2b + 1c$), then the ratio of these chromosomes is changed from $1c:1$ $(a + b)$ to $1c:2 (a + b)$. Second, any deleterious recessive on the single remaining chromosome becomes hemizygous and may be directly expressed phenotypically. (Note that these are the same effects produced by deletions.)

Monosomics, trisomics ($2n + 1$), and other chromosome aneuploids are probably produced by nondisjunction during mitosis or meiosis. In meiosis this can happen at either the first or the second division (Figure 9-16). (Ask yourself whether products of nondisjunction at these two times can be distinguished genetically.) If an $n - 1$ gamete is fertilized by an n gamete, a monosomic ($2n - 1$) zygote is produced. An $n + 1$ and an n gamete yield a trisomic $2n + 1$; an $n + 1$ and an $n + 1$ produce a tetrasomic if the same chromosome is involved or a double trisomic if different chromosomes are involved, and so on.

In *Neurospora* (a haploid), the $n - 1$ meiotic products abort and do not darken like the normal ascospore; so M_I and M_{II} nondisjunctions are detected as asci with $4:4$ and $6:2$ ratios of normal to aborted spores, respectively. (Diagram the chromosome content of the various spores to convince yourself of the relation of the spore pattern to nondisjunction.) What ascus genotypes are produced for loci on the aneuploid chromosomes?

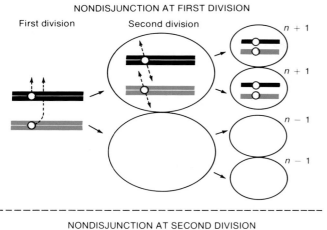

NONDISJUNCTION AT FIRST DIVISION

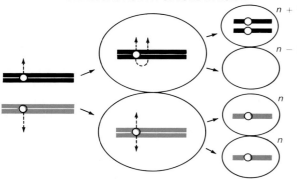

NONDISJUNCTION AT SECOND DIVISION

Figure 9-16. The origin of aneuploid gametes by nondisjunction at either the first or second meiotic division.

In humans, the sex-chromosome monosomic (44 autosomes + 1X) produces a phenotype known as Turner's syndrome. Affected people have a characteristic, easily recognizable phenotype: they are sterile females, are short in stature, and often have a web of skin extending between the neck and shoulders. Their intelligence is near-normal, although some specific cognitive functions are defective. Their frequency is about 1 in 5000 female births. Monosomics for all autosomes die in utero.

If viable, nullisomics and monosomics are useful in locating newly found recessive genes on specific chromosomes in plants. In one such approach, different monosomic lines are obtained, each of which lacks a different chromosome. Homozygotes for the new gene are crossed with each monosomic line, and the progeny of each cross are inspected for expression of the recessive phenotype. The cross in which the phenotype appears identifies its chromosome location. In nullisomics and monosomics, of course, $n - 1$ gametes are produced (see Figure 9-17 for monosomics). In general, these gametes tend to be more viable in a female than in a male parent. The union of these $n - 1$ gametes with n ga-

metes, bearing the new mutation, provides the crucial progeny types for the linkage test.

A similar approach can be used in humans. For example, two people whose vision is normal may produce a daughter who has Turner's syndrome and who is also red-green colorblind. This shows that the allele for red-green colorblindness is recessive, that it is located on the X chromosome of the mother, and that the nondisjunction must occur in the father. (Can you see why?)

Trisomics ($2n + 1$)

In trisomics, trivalents are regularly seen (Figure 9-18; see also Figure 3-4). For genes that are tightly linked to the centromere of a trisomic chromosome set, the random segregations can be represented as shown in Figure 9-19 in a trisomic Aaa. All types occur equally frequently, and a gamete ratio of $1A : 2Aa : 2a : 1aa$ is produced. Trisomics are sometimes recognized by these ratios, which are also useful in locating genes on chromosomes. We have already observed a complete set of trisomic lines in *Datura* (Figure 3-8). Once again, note that chromosome imbalance produces highly chromosome-specific deviations from the normal appearance.

There are several examples of viable trisomics in humans. The combination XXY (1 in 1000 male births) results in Klinefelter's syndrome, producing males with lanky builds who are mentally retarded, and sterile. Another combination, XYY, also occurs in about 1 in 1000 male births. A lot of excitement was aroused when an attempt was made to link the XYY condition with a predisposition toward violence. This is still hotly debated, although it is now clear that an XYY condition in no way guarantees such behavior. Nevertheless, several enterprising lawyers have attempted to use the XYY genotype as grounds for acquittal or compassion in crimes of violence. The XYY males are usually fertile.

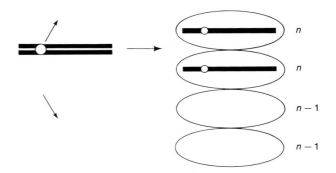

Figure 9-17. Behavior of a monosomic chromosome at meiosis. Two of the resulting gametes contain a normal haploid set of chromosomes (n); the other two contain a set missing the monosomic chromosome of the parent ($n - 1$).

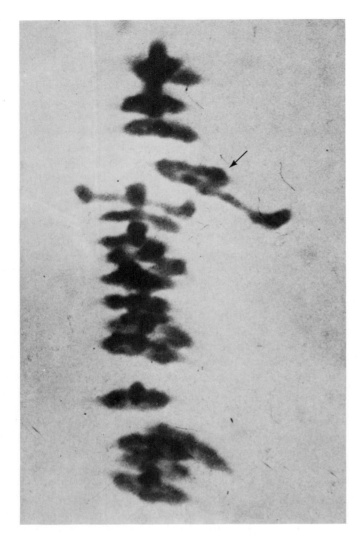

Figure 9-18. Chromosome pairing in a trisomic of wheat. The trivalent chromosome is shown by the arrow. (Courtesy of Clayton Person.)

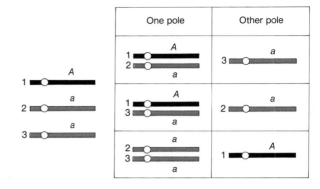

Figure 9-19. Genotypes of the meiotic products of an *Aaa* trisomic. Three segregation patterns are equally likely.

some 21 that must be trisomic to produce this syndrome; these advances offer some hope of a more precise understanding of the chemical nature of and possible therapy for Down's syndrome. In humans, the only other two autosomal trisomics known to survive past birth are individuals with trisomy 13 and trisomy 18. Affected children are even more severely handicapped, both mentally and physically, than in the case of trisomy 21 and rarely survive to 1 year of age.

Chromosome mutation in general plays a prominent role in determining genetic ill health in humans. Figure 9-21 summarizes the surprisingly high levels of various

We have already looked at the generation of Down's syndrome through adjacent segregation in translocation heterozygotes. Down's syndrome also occurs much more commonly as a result of nondisjunction during meiosis. In this form of Down's syndrome, called trisomy 21, there is generally no family history of the phenotype; however, the frequency of this form is dramatically higher among children born to older mothers (Figure 9-20). The overall incidence of this abnormality is about 0.15 percent of all births.

Down's syndrome is a severely incapacitating condition. Affected individuals are mentally retarded, and about one-third die by the age of 10 years. Recent advances in mapping the human genome allow the identification of the precise genes on the long arm of chromo-

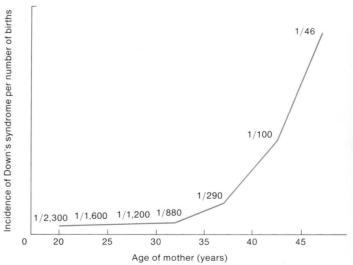

Figure 9-20. Maternal age and the production of Down's syndrome offspring. (From L. S. Penrose and G. F. Smith, *Down's Anomaly*. Copyright © 1966 by Little, Brown and Company.)

CHROMOSOME MUTATION II: CHANGES IN CHROMOSOME NUMBER

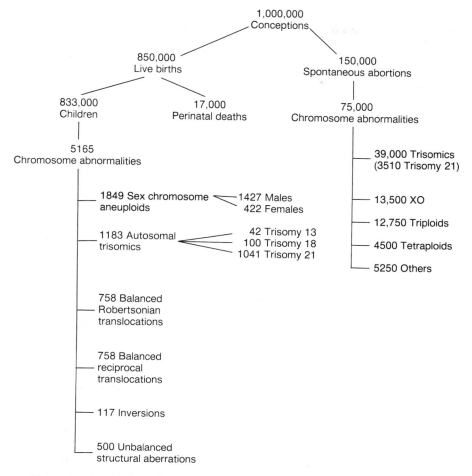

Figure 9-21. The fate of a million implanted human zygotes. (Robertsonian translocations involve fusion or dissociation of centromeres.) (Courtesy of K. Sankaranarayanan, *Mutation Research* 61, 1979.)

chromosome abnormalities at different developmental stages of the human organism. In fact, the incidence of chromosome mutations ranks close to that of gene mutations in human live births (Table 9-1)—a particularly surprising result when we realize that virtually all chromosome mutations arise anew with each generation. In contrast, gene mutations (as we shall see in Chapter 23 owe their level of incidence to a complex interplay of mutation rates and environmental selection that acts over many generations of the history of the human species.

Somatic Aneuploids

Aneuploid cells can arise spontaneously in somatic tissue or in tissue culture. In such cases, the initial result is a genetic mosaic of cell types. Good examples are provided by certain conditions in humans.

TABLE 9-1. Relative incidence of human ill health due to gene mutation and to chromosome mutation

Type of mutation	Percentage of live births
Gene mutation	
Autosomal dominant	0.90
Autosomal recessive	0.25
X-linked	0.05
Total gene mutation	1.20
Chromosome mutation	
Autosomal trisomies (mainly Down's syndrome)	0.14
Other unbalanced autosomal aberrations	0.06
Balanced autosomal aberrations	0.19
Sex chromosomes	
XYY, XXY, and other ♂♂	0.17
XO, XXX, and other ♀♀	0.05
Total chromosome mutation	0.61

Sexual mosaics—people whose bodies are a mixture of male and female tissue—provide the first example. One type of sexual mosaic, XO/XYY, can be explained by postulating an XY zygote in which an early mitotic division involves a nondisjunction of the Y chromosomes, so that both go to one pole:

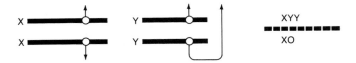

The phenotypic sex of such individuals depends on where the male and female sectors end up in the body. In this case, nondisjunction at a later mitotic division produces a three-way mosaic XY/XO/XYY, which of course contains a clone of normal male cells as well. Other sexual mosaics have different explanations; as examples, XO/XX is probably due to chromosome loss in a female zygote and XX/XY is probably the result of a double fertilization (fused twins).

Somatic aneuploidy and its resulting mosaics are often observed to occur in association with cancer. People suffering from chronic myeloid leukemia (CML), a cancer of the white blood cells, frequently harbor cells containing the so-called Philadelphia chromosome. This chromosome was once thought to represent an aneuploid condition, but it is now known to be a translocation product in which part of the long arm of chromosome 22 attaches to the long arm of chromosome 9. However, CML patients often show aneuploidy in addition to the Philadelphia chromosome. In one study of 67 people with CML, 33 proved to have an extra Philadelphia chromosome and the remainder had various aneuploidies; the most common aneuploidy was trisomy for the long arm of chromosome 17, which was detected in 28 people. Of 58 people with acute myeloid leukemia, 21 were shown to have aneuploidy for chromosome 8; 16 for chromosome 9; and 10 for chromosome 21. In another study of 15 patients with intestinal tumors, 12 had cells with abnormal chromosomes, at least some with trisomy for chromosome 8, 13, 15, 17, or 21. Of course, such studies merely established correlations, and it is not clear whether the abnormalities are best thought of as a cause or as an effect of cancer.

Message   Aneuploids are produced by nondisjunction or some other type of chromosome misdivision at either meiosis or mitosis.

Figure 9-22. (a) Wheat. (b) *Aegilops umbellulata.* Both whole plants and seed heads are shown. (Courtesy of E. R. Sears.)

Chromosome Mechanics in Plant Breeding

Some of the material covered in this chapter may seem somewhat esoteric. The purpose of this section, then, is to provide convincing evidence that the details covered here are of immense significance in the genetic engineering that is so necessary to produce and maintain new crop types in our hungry world. The sole experiment to be described, performed by E. R. Sears in the 1950s, concerns the transfer of a gene for leaf-rust resistance from a wild grass, *Aegilops umbellulata*, to bread wheat, which is highly susceptible to this disease, to offset a major problem in the wheat industry. This is a classic experiment of its kind.

The first problem that Sears encountered was that these two species (Figure 9-22) are not interfertile, so the feat of gene transfer seemed impossible. Sears sidestepped this problem with a **bridging cross,** in which he crossed *A. umbellulata* to a wild relative of bread wheat called emmer, *Triticum dicoccoides*. (Follow the process in Figure 9-23.) *A. umbellulata* is a diploid, $2n = 2x = 14$. We shall call its chromosome sets CC. *T. dicoccoides* is a tetraploid, $2n = 4x = 28$, with sets AA BB. From this

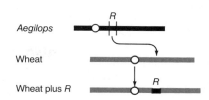

Figure 9-24. Translocation of *Aegilops* R segment to wheat using radiation as a means of breaking the chromosomes.

cross, the resulting sterile hybrid ABC was doubled into a fertile amphidiploid AA BB CC with 42 chromosomes. This amphidiploid was fertile in crosses with wheat (*T. aestivum*), which is represented as $2n = 6x = 42$, AA BB DD. The offspring, AA BB CD, were almost completely sterile due to pairing irregularities between the C and D sets. However, crosses to wheat did produce a few rare seeds, some of which grew into resistant plants. Some of these were almost the desired type, having 43 chromosomes (42 of which were wheat, plus one *Aegilops* chromosome bearing the resistance gene). Thus, in the AA BB CD hybrid, some aberrant form of chromosome assortment had produced a gamete with 22 chromosomes: ABD plus one from the C group.

Unfortunately, the extra chromosome carried just too many undesirable *Aegilops* genes along with the good one, and the plants were weedy and low producers. So the *Aegilops* gene linkage had to be broken. Sears accomplished this by using irradiated pollen from these plants to pollinate wheat. He was looking for translocations of part of the *Aegilops* chromosome tacked onto the wheat chromosomes. These were quite common, but only one turned out to be ideal—a very small, *insertional*, unidirection translocation (Figure 9-24). When bred to homozygosity, the resistant plants were indistinguishable from wheat.

The study of chromosome mutation is obviously of great importance—not only in pure biology, where there are ramifications in many areas (especially in evolution), but also in applied biology. The application of such studies is especially important in agriculture and medicine. Rather curiously, very little is known about the mechanisms of chromosome mutation, especially at the molecular level. More details are needed about the normal chemical architecture of chromosomes. We return to this subject in Chapter 14.

Summary

■ Changes in chromosome number can involve whole chromosome sets, resulting in abnormal euploidy, or parts of sets, resulting in aneuploidy.

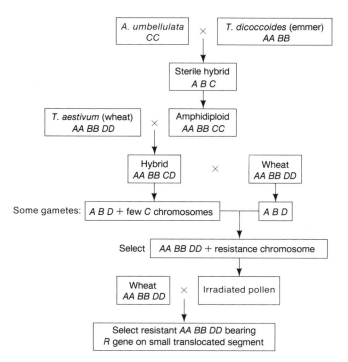

Figure 9-23. Summary of Sears's program for transferring rust resistance from *Aegilops* to wheat. *A, B, C,* and *D* represent chromosome sets of diverse origin. *R* represents the genetic determinant (single gene?) of resistance.

The most common abnormal euploids are polyploids; examples are triploids ($3x$) and tetraploids ($4x$). Odd numbers of sets lead to sterility as a result of unpaired chromosomes at meiosis, whereas even numbers of sets can produce standard (although abnormal) segregation ratios. Allopolyploids (polyploids formed by combining sets from different species) can be made by crossing two related species and then doubling the progeny chromosomes through the use of colchicine or through somatic cell fusion. These techniques have important applications in crop breeding, since allopolyploids are effectively new species. Polyploidy can result in an organism of larger dimensions; this discovery has permitted important advances to be made in horticulture and in crop breeding.

Aneuploids have also been important in the engineering of specific crop genotypes, although aneuploidy per se usually results in an unbalanced genotype with an abnormal phenotype. Examples of aneuploids include monosomic ($2n - 1$) and trisomic ($2n + 1$) chromosome complements. Aneuploid conditions are well studied in humans. Down's syndrome (trisomy 21), Klinefelter's syndrome (XXY), and Turner's syndrome (XO) are well-documented examples. The spontaneous level of aneuploidy in humans is quite high and produces a major portion of genetically based ill health in human populations. Aneuploidy is believed to result in large part from chromosome nondisjunction.

■ ■ ■

Solved Problems

1. In a certain plant, there is a controversy about the type of chromosome pairing seen in autotetraploids formed between two geographical races. It is known that chromosome association is by pairs, but three hypotheses are put forward concerning how this occurs:

   a. Pair formation is random.

   b. Pairs occur only between chromosomes of the same race.

   c. Pairs occur only between chromosomes of different races.

   For a gene A/a, which is closely linked to its centromere, the following cross is made:

$$\text{Race 1}: AAAA \times \text{Race 2}: aaaa$$
$$\downarrow$$
$$\text{Autotetraploid}: AAaa$$

   The autotetraploid is then selfed. What ratio of phenotypes can be expected under each hypothesis of chromosome pairing? Explain your answer.

   Solution

   a. Under random pairing, all possible pairing combinations are equally likely. If we label the four homologous chromosomes 1, 2, 3, and 4, then the equally probable combinations are $1-2/3-4$, $1-3/2-4$, and $1-4/2-3$. This kind of situation in an $AAaa$ tetraploid was examined in the chapter (see page 203). We saw that aa gametes are produced at a frequency of $\frac{1}{6}$ and that the frequency of $aaaa$ progeny when selfing occurs must therefore be $\frac{1}{36}$. All other progeny types will contain at least one A allele. Hence, the expected phenotypic ratio will be 35 $A---$: 1 $aaaa$.

   b. In this alternative, the chromosome pairs will look like

   because both the A chromosomes come from race 1 and both the a chromosomes come from race 2. This diagram shows how being methodical can help in problem solving, because it makes it very clear that the only possible segregation is that of one A and a to each pole. Hence, all gametes will be Aa and, on fertilization, all genotypes will be $AAaa$, and only the A phenotype will be seen.

   c. Here the pairing will look like this:

   Because segregation of the pairs is independent, the gametic population can be represented as follows:

Pair 1	Pair 2	Gametes
$\frac{1}{2}A$	$\frac{1}{2}A$	$\frac{1}{4}AA$
	$\frac{1}{2}a$	$\frac{1}{4}Aa$
$\frac{1}{2}a$	$\frac{1}{2}A$	$\frac{1}{4}Aa$
	$\frac{1}{2}a$	$\frac{1}{4}aa$

Self-fertilization produces *aaaa* progeny at the predicted frequency of $\frac{1}{4} \times \frac{1}{4} = \frac{1}{16}$. Therefore, $\frac{15}{16}$ is *A---* and a 15 : 1 ratio is obtained.

In conclusion, the three ratios are:

a. 35 : 1

b. 1 : 0

c. 15 : 1

2. Due to the small size of the *Drosophila* chromosome 4, monosomics and trisomics for this chromosome are viable but tetrasomics and nullisomics are not. If a fly that is trisomic for chromosome 4 and carries the recessive gene for bent bristles *(b)* on all copies of chromosome 4 is crossed to a phenotypically normal fly that is monosomic for chromosome 4:

a. What genotypes and phenotypes can be expected in the progeny and in what proportions?

b. If trisomics from these progeny are interbred, what phenotypic ratio can be expected in the next generation? (Assume that only one copy of b^+ is needed to produce normal (nonbent) bristles, that unpaired chromosomes pass to either pole at random, and that aneuploid gametes of any kind survive.)

Solution

a. The cross is

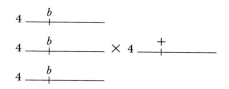

From the bent parent, the gametes will be $\frac{1}{2}bb$ and $\frac{1}{2}b$. From the monosomic parent, the gametes will be $\frac{1}{2}+$ and $\frac{1}{2}0$ (the latter contains no chromosome 4). Hence, the progeny will be $\frac{1}{4}+bb$, $\frac{1}{4}bb$, $\frac{1}{4}+b$ and $\frac{1}{4}b$, which provides a phenotypic ratio of 1 : 1.

b. The trisomics referred to are of genotype $+bb$. If we label these chromosomes 1, 2, and 3, then

$$1 = b$$
$$2 = b$$
$$3 = +$$

and segregation produces

$$1, 2 \; (bb) \; / \; 3 \; (+)$$
$$1, 3 \; (b \; +) \; / \; 2 \; (b)$$
and $$2, 3 \; (b \; +) \; / \; 1 \; (b)$$

Fertilization can be represented as follows:

	$bb(\frac{1}{6})$	$+b(\frac{2}{6})$	$b(\frac{2}{6})$	$+(\frac{1}{6})$
$bb(\frac{1}{6})$	$\times$	$\times$	bbb $\frac{2}{36}$	$\frac{1}{36}$
$+b(\frac{2}{6})$	$\times$	$\times$	$\frac{4}{36}$	$\frac{2}{36}$
$b(\frac{2}{6})$	bbb $\frac{2}{36}$	$\frac{4}{36}$	bb $\frac{4}{36}$	$\frac{2}{36}$
$+(\frac{1}{6})$	$\frac{1}{36}$	$\frac{2}{36}$	$\frac{2}{36}$	$\frac{1}{36}$

The X's represent tetrasomics which, we are told, are not viable. Out of the remaining $\frac{27}{36}$, $\frac{8}{36}$ are *b* phenotype; therefore, a ratio of 19 : 8 is predicted.

Problems

1. Distinguish between Klinefelter's, Down's, and Turner's syndromes in humans.

2. List two ways in which you could make an allotetraploid between two related plant species, both $n = 14$.

3.

a. If nondisjunction of *Neurospora* chromosome 3 occurs at the second division of meiosis, show the content of each of the eight ascospores with regard to chromosome 3.

b. *Neurospora* normally has seven chromosomes. How many chromosomes are present in each of the ascospores in the ascus in (a)?

4.

a. How would you synthesize a pentaploid (5x)?

b. How would you synthesize a triploid (3x) of genotype *Aaa*?

c. You have just obtained a rare recessive mutation *a** in a diploid plant, which Mendelian analysis tells you is Aa*. From this plant, how would you synthesize a tetraploid (4x) of genotype *AAa*a**?

d. How would you synthesize a tetraploid of genotype *Aaaa*?

e. How would you synthesize a plant that is resistant to a chemical herbicide? (Assume that mutation to this trait is very infrequent.)

5. In corn, the part we eat (the kernel) is predominantly triploid tissue called endosperm, which is formed in the following way. The haploid egg-cell nucleus divides to produce several identical nuclei, one of which acts as the gametic nucleus and two of which act as so-called polar nuclei. The pollen-cell nucleus also divides to form several identical haploid nuclei, one of which fuses with the gametic nucleus to produce the embryo and one of which fuses with the two polar nuclei to form the endosperm. What are the constitutions of the endosperm types in the cross *Aa Bb* ♀ × *Aa Bb* ♂? (Assume independent assortment.)

6. Allopolyploids are: a. not fertile at all; b. fertile only among themselves; c. fertile with one parent only;   d. fertile with both parents only;   e. fertile with both parents and themselves.

7. Tetraploid yeast can be created by fusing two diploid cells. These tetraploids undergo meiosis like any other tetraploid and produce four diploid products of meiosis. Assuming that homologous chromosomes synapse randomly in pairs and that there is no crossing-over in the genecentromere interval, what nonlinear tetrads are produced by a tetraploid of genotype *BBbb*? What are the frequencies of the ascus types? (NOTE: this question involves tetrad analysis of tetraploid cells instead of the usual diploid cells.)

8. In a tetraploid *AAaa*, there is no pairing between chromosomes from the same parent. What phenotypic ratio results from selfing? In another tetraploid *BBbb*, pairing only occurs between chromosomes from the same parent. What phenotypic ratio results from selfing in this case? (Assume that one parent carries the dominant allele and that the other carries the recessive allele in each case.)

9. The New World cotton species *Gossypium hirsutum* has a $2n$ chromosome number of 52. The Old World species *G. thurberi* and *G. herbaceum* each have a $2n$ number of 26. Hybrids between these species show the following chromosome pairing arrangements at meiosis:

Hybrid	Pairing Arrangement
G. hirsutum ✕ *G. thurberi*	13 small bivalents + 13 large univalents
G. hirsutum ✕ *G. herbaceum*	13 large bivalents + 13 small univalents
G. thurberi ✕ *G. herbaceum*	13 large univalents + 13 small univalents

Draw diagrams to interpret these observations phylogenetically, clearly indicating the relationships between the species. How would you go about proving that your interpretation is correct?

(Problem 6 is from A. M. Srb, R. D. Owen, and R. S. Edgar, *General Genetics*, 2nd ed. Copyright © 1965 by W. H. Freeman and Company.)

10. An autotetraploid is heterozygous for the two gene loci *FFff* and *GGgg*. Each locus affects a different character and is located on a different set of homologous chromosomes very close to its respective centromere.

   a. What gametic genotypes are produced by this individual, and in what proportions?

   b. If the individual is self-fertilized, what proportion of the progeny will have the genotype *FFFf GGgg*? the genotype *ffff gggg*?

11. Which of the following is *not* caused by meiotic nondisjunction? a. Turner's syndrome b. Down's syndrome c. Klinefelter's   syndrome   d. XYY   syndrome e. Achondroplastic dwarfism

12. A patient with Turner's syndrome is found to be colorblind. Both her mother and father have normal vision. How can this be explained? Does this outcome tell us whether nondisjunction occurred in the father or in the mother? If the colorblindness gene were close to the centromere (it is not, in fact), would the clinical data tell us whether the nondisjunction occurred at the first or at the second meiotic division? Repeat the question for a colorblind patient with Klinefelter's syndrome.

13. Individuals have been found who are colorblind in one eye but not in the other. What would this suggest if these individuals were: a. only or mostly females? b. only or mostly males? (Assume that this is an X-linked recessive trait.)

14. Some men and women with Down's syndrome are able to mate with each other and have offspring, although this is rare. What chromosome constitutions could be expected in the zygotes of such matings, and what would become of these zygotes?

15. When human sperm are treated with quinacrine dihydrochloride, about one-half of the sperm show a fluorescent spot thought to be the Y chromosome. About 1.2 percent of the sperm show two fluorescent spots. The sperm of some industrial workmen who are exposed over a period of about one year to the chemical dibromochloropropane were examined, and the frequency of sperm with two spots was found to be on average 3.8 percent. Propose an explanation for this result, and explain how you would test it.

16. People with Down's syndrome have about a 15-fold higher risk of contracting leukemia. In the progression of the leukemia disease, complex chromosome aneuploidies usually are seen in the cancer cells. Discuss the possible relationship between these two statements.

17. In humans, the only autosomal trisomics that survive until birth are for chromosomes 13, 18, and 21. All three types are severely deformed. If you were a medical geneticist, how would you go about studying aneuploidy for the other chromosomes? Do you think aneuploids for the other chromosomes never occur, or are they very rare?

18. In British Columbia, the mean age of mothers of Down's syndrome babies fell from 34 years to 28 years between 1952 and 1972. What are some possible causes of this population trend, and how could the related hypotheses be tested?

19. Several kinds of sexual mosaics are well documented in humans. Suggest how each of the following examples may have arisen:

   a. XX/XO (that is, there are two cell types in the body, XX and XO)

   b. XX/XXYY

   c. XO/XXX

   d. XX/XY

   e. XO/XX/XXX

20. The discovery of chromosome banding in eukaryotes has greatly improved our ability to distinguish various cytogenetic events. Particularly useful are banding polymorphisms, because the "morphs" can be used as chromosome markers. (These morphs are forms of chromosome that are cytologically distinguishable by virtue of minor variation in band size, position, and so forth.) Let's consider chromosome 21 in humans. Assume that one set of parents are 21^a21^b ♀ × 21^c21^d ♂ where a, b, c, and d represent morphs of a polymorphism for this chromosome. Also assume that fetuses of the following types (where 42A stands for the rest of the autosomes) are produced:

(1) $42A + 21^b21^b21^c + XY$

(2) $42A + 21^a21^b21^d + XX$

(3) $42A + 21^b21^d + XY$

(4) $42A + 21^a21^c21^c + XX$

(5) $42A + 21^a21^b + XY$

(6) $42A + 21^a21^c + XYY$

(7) $(42A + 21^a + XY)(42A + 21^a21^b21^b + XY)$ (mosaic)

(8) $(42A + 21^a21^c + XY)(42A + 21^b21^d + XX)$ (mosaic)

In each case:

a. State the genetic term for the condition.

b. Diagram the event(s) that gave rise to the condition.

c. State in which individual the event(s) took place.

21. In tomatoes, an attempt was made to assign five recessive genes to specific chromosomes using trisomics. Each homozygous mutant ($2n$) was crossed to three trisomics, involving chromosomes 1, 7, and 10. From these crosses, trisomic progeny (less vigorous) were selected. These trisomic progeny were backcrossed to the appropriate homozygous recessive, and *diploid* progeny from these crosses were examined. The results follow, where the ratios are wild-type : mutant:

Trisomic chromosome	Gene				
	d	y	c	h	cot
1	48:55	72:29	56:50	53:54	32:28
7	52:56	52:48	52:51	58:56	81:40
10	45:42	36:33	28:32	96:50	20:17

Which of the genes can be assigned, and to which chromosomes? (Explain your answer fully.)

22. In a *Petunia* plant, the genes *A, B, C*, and *D* are very closely linked. A plant of genotype *a B c D/A b C d* is irradiated with gamma rays and then crossed to *aa bb cc dd*. In the progeny, plants of phenotype *A−B−C−D−* are not rare. What are two possible modes of origin, and which is the more likely?

23. In *Neurospora*, a cross between the multiply marked chromosomes $a+c+e$ and $+b+d+$ produces one product of meiosis that grows on minimal medium (assume that *a, b, c, d*, and *e* are nutritional markers). When the rare colony grows up, some *asexual* spores are $a+c+e$ in genotype, some are $+b+d+$, and the remainder grow on minimal medium. Explain the origin of the rare product of meiosis.

24. In *Sordaria brevicollis* (an ascomycete), the mutations b_1 and b_2 are closely linked complementing markers that result in buff-colored (light brown) ascospores. In the cross $b_1 \times b_2$, if one or both centromeres divide and separate precociously at the first division of meiosis, what patterns of spore colors are produced in those asci? How can these asci be distinguished from normal asci and from asci in which nondisjunction has occurred? (NOTE: ascospores in this fungus are normally black, and hypoploid ascospores are white; assume no crossing-over between b_1 and b_2.)

25. H. Sharat Chandra recovered triploids of mealybugs, *Planococcus* ($x = 5$). He found that all cells in the gonads had 15 chromosomes at the end of meiosis I and variable numbers of chromosomes ranging from 0 to 15 at the end of meiosis II. Interpret these results.

26. In yeast, a diploid is made of genotype

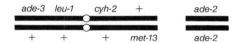

where all mutant alleles are recessive. On a supplemented medium, most colonies appear red (due to an accumulation of red pigment at the block caused by *ade-2* in adenine synthesis). However, a few colonies are one-half red and one-half white. (White is the normal yeast color.) The white sectors are leucine-requiring and cycloheximide-resistant (*cyh* is the resistant allele). The red sectors did not require methionine.

a. Given that *ade-3* is an earlier adenine block than *ade-2*, what mechanism may have given rise to the sectored colonies?

b. How would you test the hypothesis?

27. An *Aspergillus* diploid is

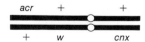

The diploid is green but produces rare white diploid sectors of three different genotypes:

(A) $\dfrac{acr \quad w \quad +}{+ \quad w \quad cnx}$ (B) $\dfrac{+ \quad w \quad +}{+ \quad w \quad cnx}$ (C) $\dfrac{+ \quad w \quad cnx}{+ \quad w \quad cnx}$

Which of these genotypes is most probably due to: a. mitotic crossing-over; b. mutation; c. mitotic nondisjunction? Explain each answer.

28. Design a test system for detecting agents in the human environment that are potentially capable of causing aneuploidy in eukaryotes.

Recombination in Bacteria and Their Viruses

KEY CONCEPTS

The fertility factor (F) permits bacterial cells to transfer
DNA to other cells through the process of conjugation.

•

F can exist in the cytoplasm or can be integrated into the
bacterial chromosome.

•

When F is integrated in the chromosome, chromosome
markers can be transferred during conjugation.

•

During generalized transduction, random chromosome
fragments are incorporated into the heads of certain bacterial
phages and transferred to other cells by infection.

•

During specialized transduction, specific genes near the
phage integration sites on the bacterial chromosome are
mistakenly incorporated into the phage genome and
transferred to other cells by infection.

•

The different methods of gene transfer in bacteria allow
geneticists to make detailed maps of bacterial genes.

■ Thus far, we have dealt almost exclusively with genes that are packed into chromosomes and enclosed within the nuclei of eukaryotic organisms. However, a very large part of the history of genetics and current genetic analysis (particularly molecular genetics) is concerned with prokaryotic organisms, which have no distinct nuclei, and with viruses. Viruses are a problem for biologists to classify. Although they share some of the definitive properties of organisms, many biologists regard viruses as distinct entities that in some sense are not fully alive. They are not cells; they cannot grow or multiply alone. To reproduce, they must parasitize living cells and use their metabolic machinery. Nevertheless, viruses do have hereditary properties that can be subjected to genetic analysis. Genetic analysis of bacteria and their viruses has yielded key insights into the nature of the genetic material, the genetic code, and mutation, and into the structure of the genetic material.

Compared to eukaryotes, prokaryotic organisms and viruses have very simple chromosomes that are not contained within a nuclear membrane. Because they are monoploid, these chromosomes do not undergo meiosis, but they do go through stages analogous to meiosis. The approach to the genetic analysis of recombination in these organisms is surprisingly similar to that for eukaryotes. Although we treat them in separate chapters here, there is no fundamental difference between the two systems.

The prokaryotes are blue-green algae, now classified as "cyanobacteria," and bacteria. The best studied bacterial viruses—those that parasitize bacteria—are called **bacteriophages,** or simply **phages.** Pioneering work with bacteriophages has led to a great deal of recent research on tumor-causing viruses and other kinds of animal and plant viruses that are now becoming better understood at the genetic level.

Working with Microorganisms

How can the geneticist study inheritance in organisms that are too small to be seen without a microscope? Bacteria can be grown in a liquid medium or on a solid surface, such as an agar gel, as long as basic nutritive ingredients are supplied. In a liquid medium, the bacteria divide by binary fission: they multiply geometrically until the nutrients are exhausted or until toxic factors (waste products) accumulate to levels that halt the population growth. A small amount of such a liquid culture can be pipetted onto a petri plate containing an agar medium and spread evenly on the surface with a sterile spreader. Each cell then reproduces by fission. Because the cells are immobilized in the gel, all of the daughter cells remain together in a clump. When this mass reaches more than 10^7 cells, it becomes visible to the naked eye as a **colony.** This process is called **plating** (Figure 10-1). If the initially plated sample contains very few cells, then each distinct colony on the plate will be derived from a single original cell. Members of a colony that share a single genetic ancestor are known as **clones.**

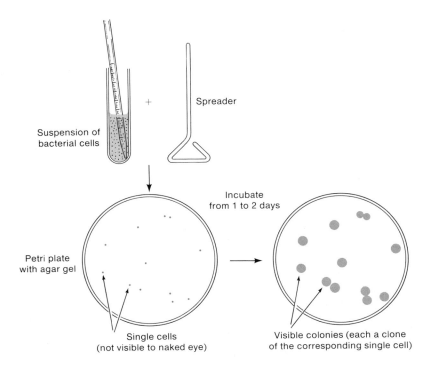

Figure 10-1. Methods of growing bacteria in the laboratory. A few bacterial cells that have been grown in a liquid medium containing nutrients can be spread on an agar medium, also containing the appropriate nutrients. Each of these original cells will divide many times by binary fission and eventually give rise to a colony. All cells in a colony, being derived from a single cell, will have the same genotype and phenotype.

TABLE 10-1. Genotypic symbols used in bacteria

Symbol	Character or phenotype associated with symbol
bio⁻	Requires biotin added as a supplement to minimal medium
arg⁻	Requires arginine added as a supplement to minimal medium
met⁻	Requires methionine added as a supplement to minimal medium
lac⁻	Cannot utilize lactose as a carbon source
gal⁻	Cannot utilize galactose as a carbon source
str^r	Resistant to the antibiotic streptomycin
str^s	Sensitive to the antibiotic streptomycin

For many characters, the phenotype of a clone can be determined readily through visual inspection or simple chemical tests. This phenotype can then be assigned to the original cell of the clone, and the frequencies of various phenotypes in the pipetted sample can be determined. Table 10-1 lists some phenotypes and genetic symbols. Also refer to page 157 for a further review of microbial genetic techniques.

Bacterial Conjugation

The following sections describe the discovery of gene transfer in bacteria and explain different types of gene transfer and their use in bacterial genetics. First, we consider **conjugation,** a process by which certain bacterial cells can transfer DNA to a second cell with which they make contact. Conjugation was first discovered by using strains of the bacterium *Escherichia coli.* Before we examine the historical material, let's summarize what we now know about conjugation in *E. coli,* since it will facilitate our understanding of the early experiments that led to this knowledge.

The ability to transfer DNA by conjugation is dependent on the presence in the cytoplasm of an entity termed the **F** or **fertility factor.** Cells carrying F are termed F⁺; cells without F are termed F⁻. The F factor is a small, circular DNA element which acts like a minichromosome. F contains approximately 100 genes, which give F the capacity to:

1. Replicate its DNA independently of the host chromosome, which allows F to be maintained in a cellular population that is dividing, as illustrated in Figure 10-2(a).

2. Produce **pili**—fibrous, proteinaceous structures that allow cells carrying F to attach to and maintain contact with other cells, as shown in Figure 10-2(b).

3. Transfer a newly synthesized copy of F DNA from a fixed point in the circular F genome to a recipient cell. Figure 10-2(c) depicts the transfer of F to a recipient cell; note how a copy of F always remains behind in the donating cell.

4. Occasionally recombine into the host bacterial chromosome, as diagrammed in Figure 10-2(d). When this occurs, F can also transfer host chromosome markers, since F will transfer any DNA that is linked to it, as shown in Figure 10-2(e). Two consequences of this linked transfer are important:

    a. In any population of cells containing the F factor, F will integrate into the chromosomes of a small fraction of cells, which will therefore exhibit a small but detectable level of transfer of chromosome markers to a second strain. This phenomenon, depicted in Figure 10-2(f), is what led to the initial discovery of gene transfer by conjugation.

    b. It is possible to isolate the specific bacteria in the population that have the F factor integrated in the host chromosome and to cultivate pure strains derived from these cells. In such strains, every cell in the population donates chromosomal markers during F transfer, so that the frequency of transfer for the population is much higher than it is for cells in which the F factor is in the cytoplasm. Therefore, strains with an integrated F factor are termed **high-frequency transfer (Hfr)** strains to distinguish them from normal F⁺ strains, which display a low frequency

Figure 10-2. Some consequences of the presence of the F factor in *E. coli* (a) The F factor is a small, circular chromosome that can replicate independently on the host chromosome and is thus maintained in dividing populations. (b) F confers on the cell the ability to produce pili, which facilitate the cell's attachment of other cells. (c) F can transfer a copy of itself to a second cell during replication. A copy of F remains behind in the donor cell. (d) F can recombine into the host chromosome. (e) When F transfers a copy to a second cell during replication, it also transfers any DNA that is linked to the cell. Here, the integrated F transfers host chromosomal markers during replication. As in (c), a copy of the host chromosome with the integrated F remains behind in the original (donor) cell. (f) In a population of cells with the F factor, a small fraction of cells will have F integrated into the chromosome (see d). This fraction of cells will transfer chromosomal markers (see e). Therefore, when a population of cells carrying F (tube A) is mixed with a population of cells with no F (tube B), a small amount of A → B transfer of chromosomal markers can be detected.

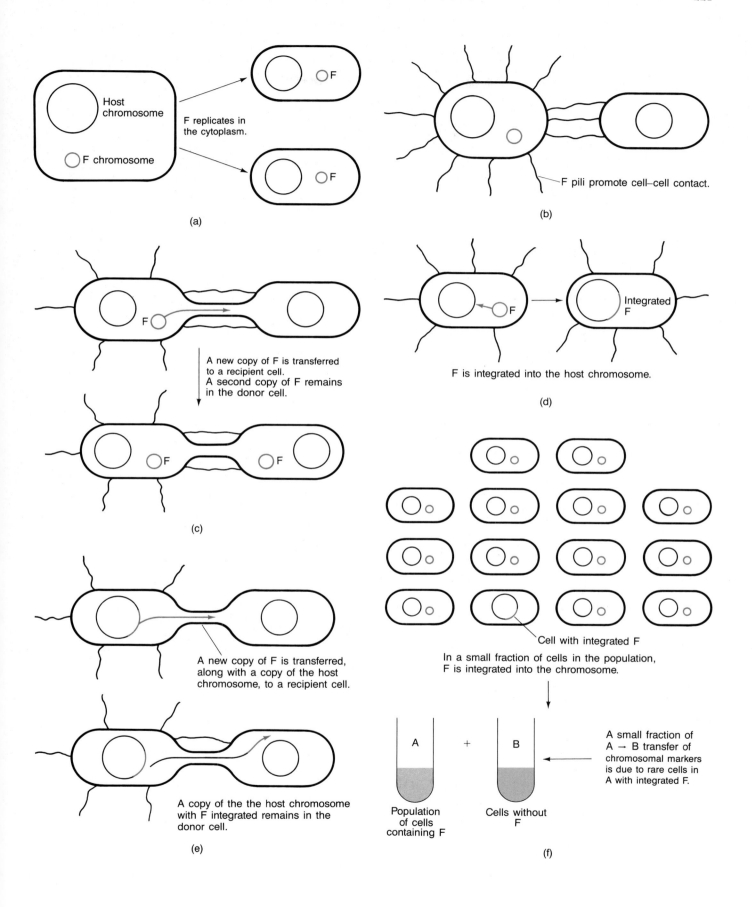

(a) Host chromosome, F chromosome — F replicates in the cytoplasm.

(b) F pili promote cell–cell contact.

(c) A new copy of F is transferred to a recipient cell. A second copy of F remains in the donor cell.

(d) F is integrated into the host chromosome.

(e) A new copy of F is transferred, along with a copy of the host chromosome, to a recipient cell. A copy of the the host chromosome with F integrated remains in the donor cell.

(f) Cell with integrated F. In a small fraction of cells in the population, F is integrated into the chromosome. Population of cells containing F (A) + Cells without F (B). A small fraction of A → B transfer of chromosomal markers is due to rare cells in A with integrated F.

transfer of chromosomal markers. F⁺ strains transfer chromosomal markers only due to the few Hfr cells in the population. Hfr strains have proved to be very useful for genetic mapping, and we consider them in more detail later.

Let's now look at how some of these processes were discovered by pioneers in the field.

The Discovery of Bacterial Gene Transfer

To appreciate fully the reasoning behind early experiments in bacterial genetics, you must try to grasp the level of understanding that existed at the time. It was not known whether bacteria had regular mechanisms for exchanging genetic information, nor even whether they had chromosomes. Indeed, about all that was known was that fairly stable true-breeding lines could be established with different phenotypes (usually involving different nutritional requirements) and that new forms occasionally appeared in these lines, apparently as the result of mutation.

Starting from this lack of knowledge, it is easy to see that experimental questions regarding the very fundamentals of bacterial inheritance needed to be asked. For example, can bacteria exchange hereditary information? (Do they possess any processes similar to sex and recombination?) This question was answered in 1946 by the elegantly simple experimental work of Joshua Lederberg and Edward Tatum, who studied two strains of *E. coli* with different nutritional requirements. Strain A would grow on a "minimum medium" (a medium, designated MM, that contains inorganic salts, a carbon source, and water) only if the medium were supplemented with methionine and biotin; strain B would grow on a minimal medium only if it were supplemented with threonine, leucine, and thiamine. Thus, we can designate strain A as *met⁻ bio⁻ thr⁺ leu⁺ thi⁺* and strain B as *met⁺ bio⁺ thr⁻ leu⁻ thi⁻*. It is perhaps presumptuous even to use gene symbols to designate these phenotypes because it had not been proved at the time that they were determined by single genes. Note that no linkage relationship is intended in writing the symbols in a specific order. The symbolism is merely a convenience.

Experimental Design

In their experiment, Lederberg and Tatum plated bacteria into dishes containing only unsupplemented minimal medium. Some of the dishes were plated with only strain-A bacteria, some, with only strain-B bacteria, and some with a mixture of strain-A and strain-B bacteria, which had been incubated together for several hours in a liquid medium containing all of the supplements (Figure

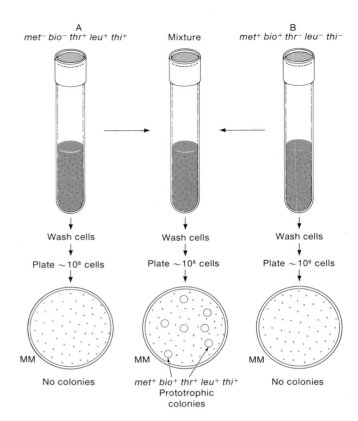

Figure 10-3. Demonstration by Lederberg and Tatum of genetic recombination occurring between bacterial cells. Cells of type A or type B cannot grow on an unsupplemented medium, because both A and B carry mutations and lack the ability to synthesize constituents needed for cell growth. When A and B are mixed for a few hours and then plated, however, a few colonies appear on the agar plate. These colonies derive from single cells in which an exchange of genetic material has occurred; they are therefore capable of synthesizing all of the required constituents of metabolism.

10-3). The plates that received the mixture of strain A and strain B produced growing colonies with a frequency of one in every 10,000,000 cells plated (in scientific notation, 1×10^{-7}), whereas no colonies arose on plates containing either strain A or strain B alone. Because only prototrophic *met⁺ bio⁺ thr⁺ leu⁺ thi⁺* bacteria can grow on a minimal medium, this observation suggests that some form of recombination of genes has occurred between the genomes of the two strains.

Now if you are thinking critically about this experiment (as you should be), you might object: "But what about the possibility that these wild-type colonies were produced by mutation?" The answer is that if the results were due to mutation, then wild-type colonies should have been obtained when the strain-A and strain-B bacteria were plated *by themselves* onto the minimal medium. But they weren't. Therefore, we can conclude that the wild-type colonies were most likely produced by an exchange of genetic material between the two strains.

Requirement for Physical Contact

It could be suggested that the cells of the two strains do not really exchange genes but instead leak substances that the other cells can absorb and use for growing (this won't seem so farfetched later). This possibility of "cross-feeding" was ruled out by Bernard Davis. He constructed a U tube in which the two arms were separated by a fine filter. The pores of the filter were too small to allow bacteria to pass through but large enough to allow the easy passage of the fluid medium and any dissolved substances (Figure 10-4). Strain A was put in one arm; strain B, in the other. After the strains had been incubated for a while, Davis tested the content of each arm to see if cells had been able to grow on a minimal medium, and none were found. In other words, *physical contact* between the two strains was needed for wild-type cells to form. It looked as though some kind of gene transfer was involved, and genetic **recombinants** were indeed produced.

Early Attempts to Identify Linkage in Bacteria

Having seen that bacteria can exchange genetic material, as can higher organisms, Lederberg and Tatum hypothesized that the bacterial genes might be organized into linkage groups, as are the genes of higher organisms. The researchers recognized that two possibilities exist: either the bacterial genes are not organized into linkage groups (in which case something like Mendel's law of independent assortment would apply to all combinations of gene pairs) or, alternatively, the bacterial genes are organized into linkage groups (in which case departures from independent assortment could be detected). Furthermore, they recognized that a simple, two-step extension of their work with strain A and strain B could be used to test for independent assortment.

In step one, they crossed A with B and then plated the cells onto a minimal medium supplemented with biotin. Any surviving colonies *must* be met^+ thr^+ leu^+ thi^+ at these four loci but *could* be either bio^+ or bio^-, because either of these can grow when biotin is supplied. Thus, step one produced many colonies that could be designated as $bio^?$ met^+ thr^+ leu^+ thi^+. In step two, they reasoned that, if the loci are unlinked, then their assortment must be independent, so that about one-half of the $bio^?$ colonies should be bio^+ and the other one-half bio^-. (Why? Because one-half of the "parental" alleles were bio^+ and one-half bio^-.) On the other hand, if the biotin locus is linked to any of the other four loci, then there should be a great departure from the 1 : 1 ratio. Specifically, if the biotin locus is linked to the methionine locus, the researchers expected to see an excess of bio^+ over bio^- alleles (because they selected for the met^+ allele from strain B, and that would pull the linked bio^+ allele along). If the biotin locus is linked to any of the other three loci, they expected to see an excess of bio^- over bio^+ alleles (because the selection for the thr^+ leu^+ thi^+ alleles from strain A would pull the linked bio^- along). To determine how many of the $bio^?$ colonies were + and how many were −, Lederberg and Tatum simply plated samples of them on a minimal medium: if they grew, they were bio^+; if not, they were bio^-. Finally, to generate additional data, they repeated the whole procedure, using each of the other supplements in turn. Their results are shown in Table 10-2.

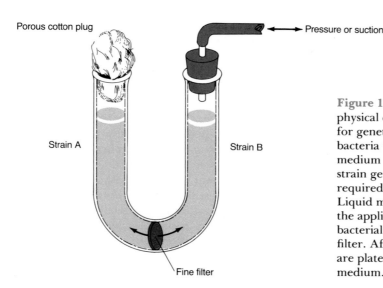

Porous cotton plug

Pressure or suction

Strain A

Strain B

Fine filter

Figure 10-4. Experiment demonstrating that physical contact between bacterial cells is needed for genetic recombination to occur. A strain of bacteria unable to grow on unsupplemented medium is placed in one arm of the U tube. A strain genetically unable to synthesize different required metabolites is placed in the other side. Liquid may be transferred between the arms by the application of pressure or suction, but bacterial cells cannot pass through the center filter. After several hours of incubation, the cells are plated, but no colonies grow on the agar medium.

TABLE 10-2. Allele frequencies observed by Lederberg and Tatum in cross A × B

A: $met^- bio^- thr^+ leu^+ thi^+$

B: $met^+ bio^+ thr^- leu^- thi^-$

Medium supplement	Genotype of surviving colonies	Allele frequencies at the unselected (?) locus	
		+	−
Biotin	$met^+ bio^? thr^+ leu^+ thi^+$	60	10
Threonine	$met^+ bio^+ thr^? leu^+ thi^+$	37	9
Leucine	$met^+ bio^+ thr^+ leu^? thi^+$	51	5
Thiamine	$met^+ bio^+ thr^+ leu^+ thi^?$	8	79

The data clearly show linkage, because the allele frequencies at each of the unselected loci show great departures from a 1 : 1 ratio. By noting whether the departure from 1 : 1 is toward an excess of + alleles or an excess of − alleles, Lederberg and Tatum were able to determine the linkage arrangements among the loci. Let us see how one could determine this from the data in Table 10-2. We have to appreciate, of course, that without knowing the direction of transfer, data of this type can seem confusing. In the case of the biotin locus, there was an excess of + alleles over − alleles. Since we selected for $met^+ thr^+ leu^+ thi^+$, it is clear that the *bio* locus is not linked to the *thr*, *leu*, or *thi* locus, because the strain carrying these three alleles is *bio*⁻. If the biotin locus is linked to any of these loci, then the *bio*⁻ allele would have been pulled along with the + allele of the linked locus. However, in the other parent strain, the *bio*⁺ allele is present together with the *met*⁺ allele. The selection for *met*⁺ has obviously pulled along the *bio*⁺ allele, so that the biotin locus is clearly linked to the methionine locus. To test your understanding of the experiment, try to work out the linkage arrangements among the other loci. In other words, in the second line in Table 10-2, the threonine locus was not selected for. What does the excess of *thr*⁺ alleles over *thr*⁻ alleles among the selected *met*⁺ *bio*⁺ *leu*⁺ *thi*⁺ indicate?

The Discovery of the Fertility (F) Factor

The experiments just discussed represent a complicated way to do a genetic analysis. In fact, Lederberg found it increasingly difficult to interpret his data. In 1953, William Hayes exploited the properties of the antibiotic streptomycin to determine that genetic transfer occurred in one direction in the types of crosses carried out.

Determining the Direction of Transfer

Hayes verified the results of Lederberg and Tatum, using a similar cross:

Strain A   Strain B

$met^- thr^+ leu^+ thi^+ \times met^+ thr^- leu^- thi^-$

However, he treated one of the strains with streptomycin, which prevents cell division and subsequently kills cells but does permit mating to continue for a short period of time. The streptomycin can be washed out after it takes effect on the initial strain, leaving a second strain unexposed to the drug. When Hayes treated strain A with the streptomycin, washed out the streptomycin, mixed in strain B, and then plated the culture on a minimal medium, he obtained the same frequency of colonies he had obtained in the control experiment with both strains untreated. However, when he treated strain B with streptomycin, washed out the streptomycin, mixed in strain A, and then plated the culture on a minimal medium, he obtained no colonies.

This experiment showed that all the recombinants detected in these genetic transfer experiments took place in strain B. Therefore, the production of colonies of strain B — but not of strain A — on the selective minimal medium was required. The obvious interpretation is that the genetic transfer was not reciprocal but occurred only by transfer from strain A to strain B.

> **Message** The transfer of genetic material in *E. coli* is not reciprocal. One cell acts as the **donor,** and the other cell acts as the **recipient.**

In Hayes's experimental strains, the donor could still transmit genes after exposure to streptomycin, but the recipient could not divide to produce colonies if it had been exposed to streptomycin. This kind of unidirectional transfer of genes was originally analogized to a sexual difference, with the donor being termed "male" and the recipient "female." Although the terms "male" and "female" still persist, it should be stressed that this type of gene transfer is not sexual reproduction. In gene transfer, one organism receives genetic information from a donor; this recipient is changed by that information. In sexual reproduction, two organisms donate equally (or nearly so) to the formation of a new organism, but only in exceptional cases is either parent changed.

Loss and Regain of Ability to Transfer

By accident, Hayes discovered a variant of his original strain A (male) that would not produce recombinants on crossing with the B strain (female). Apparently, the A males had lost the ability to transfer genetic material and had changed into females. In his analysis of this sterile variant, Hayes realized that the fertility of *E. coli* could be lost and regained rather easily.

Hayes first isolated a streptomycin-resistant mutant of the sterile A variant. He then mixed the sterile A *str*ʳ cells with the fertile A male *str*ˢ cells. He found that as many as one-third of the A *str*ʳ cells had become fertile, which he judged by their ability to transfer genetic markers to B females. Hayes suggested that maleness, or donor ability, is itself a hereditary state imposed by a **fertility factor (F).** Females lack F and therefore are recipients. Thus, females can be designated F⁻, and males F⁺.

Conjugation

Recombinant genotypes for marker genes are relatively rare in bacterial crosses, but the F factor apparently is transmitted effectively during physical contact, or **conjugation.** Hayes recovered male fertile *str*ʳ cells very soon after mixing F⁻ *str*ʳ cells with F⁺ *str*ˢ cells. There seems to be a kind of "infectious transfer" of the F factor that takes place far more quickly than the regular exchange of genetic markers. The physical nature of the F factor was elucidated much later, but these early experiments clearly showed that it involved some kind of errant particle not closely tied to the genetic markers.

Hfr Strains

An important breakthrough came when Luca Cavalli-Sforza obtained a new kind of male from an F⁺ strain. On crossing with F⁻ females, this new strain produced 1000 times more recombinants for genetic markers than did a normal F⁺ strain. Cavalli-Sforza designated this derivative an **Hfr** strain to indicate a high frequency of recombination. (The derivative later became known as Hfr C to distinguish it from a similar strain, Hfr H, found by Hayes.) In cells resulting from an F⁺ × F⁻ cross, a large proportion of the F⁻ parents are converted to F⁺ by infectious transfer of the fertility particle. However, in the Hfr × F⁻ crosses, virtually none of the F⁻ parents are converted to F⁺ or to Hfr. Thus, infectious transfer of the fertility F does not seem to occur in these crosses, even though the recombination of genetic markers is manyfold more efficient than in F⁺ × F⁻ crosses. Per-

haps at this point you can understand why most geneticists until the late 1950s tried to ignore bacterial genetics!

Determining Linkage from Interrupted-Mating Experiments

The clues all began to fit together in 1957 when Ellie Wollman and François Jacob investigated the pattern of transmission of Hfr genes to F⁻ cells during a cross. They crossed Hfr *str*ˢ *a*⁺ *b*⁺ *c*⁺ *d*⁺ with F⁻ *str*ʳ *a*⁻ *b*⁻ *c*⁻ *d*⁻. At specific time intervals after mixing, they removed samples. Each of these samples was put into a kitchen blender for a few seconds to disrupt the mating cell pairs and then was plated onto a medium containing streptomycin to kill the Hfr donor cells. This is called an **interrupted-mating** procedure. The *str*ʳ cells then were tested for the presence of marker alleles from the donor. Those *str*ʳ cells bearing donor marker alleles must have been involved in conjugation; such cells are called **exconjugants.** Figure 10-5 shows a plot of the results;

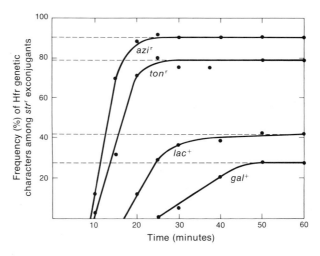

Figure 10-5. Interrupted-mating/conjugation experiments with *E. coli*. F⁻ cells that are *str*ʳ are crossed with Hfr cells that are *str*ˢ. The F⁻ cells produce a number of mutants (indicated by *azi, ton, lac, gal*) that prevent them from carrying out specific metabolic steps. However, the Hfr cells are capable of carrying out all of these steps. At different times after the cells are mixed, samples are withdrawn, disrupted in a blender to break conjugation between cells, and plated on media containing streptomycin. The antibiotic kills the Hfr cells but allows the F⁻ cells to grow and to be tested for their ability to carry out the four metabolic steps. Transfer of the donor allele of each of these steps is obviously dependent on the time that conjugation is allowed to continue. (From E. L. Wollman, F. Jacob, and W. Hayes, *Cold Spring Harbor Symposia on Quantitative Biology* 21, 1956, 141.)

Figure 10-6. Chromosome map from Figure 13-5. A linkage map can be constructed for the *E. coli* chromosome from interrupted-mating studies, using the time at which the donor alleles first appear after mating. The units of distance are given in minutes.

azi[r], *ton*[r], *lac*[+], and *gal*[+] correspond to the *a*[+], *b*[+], *c*[+], and *d*[+] mentioned in our generalized description of the experiment.

The most striking thing about these results is that each donor allele first appears in the F⁻ recipients at a specific time after mating begins. Furthermore, the donor alleles appear in a specific sequence. Finally, the maximal yield of cells containing a specific donor allele is smaller for the donor markers that enter later. Putting all these observations together, we obtain an interpretation of the results.

Message The Hfr chromosome is transferred to the F⁻ cell in a linear fashion, beginning at a specific point (called the origin, or O). The farther a gene is from O, the later it is transferred to the F⁻. For "later" genes, the transfer process most likely will stop before they are transferred.

Wollman and Jacob realized that it would be easy to construct linkage maps from the interrupted-mating results, using as a measure of "distance" the times at which the donor alleles first appear after mating. The units of distance in this case are minutes. Thus, if *b*[+] begins to

enter the F⁻ cell 10 minutes after *a*[+] begins to enter, then *a*[+] and *b*[+] are 10 units apart (Figure 10-6). Like the maps based on crossover frequencies, these linkage maps are purely genetic constructions; at the time, they had no known physical basis.

Chromosome Circularity

When Wollman and Jacob allowed Hfr × F⁻ crosses to continue for as long as two hours before blending, they found that some of the exconjugants were converted into Hfr. In other words, the fertility factor conferring maleness (or donor ability) is eventually transmitted, but at a very low efficiency and apparently as the last element of the linear "chromosome." We now have the following picture:

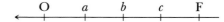

However, when several different Hfr linkage maps were derived by interrupted-mating and "time-of-entry" studies using different separately derived Hfr strains, the maps differed from strain to strain:

Hfr H	O *thr pro lac pur gal his gly thi* F
1	O *thr thi gly his gal pur lac pro* F
2	O *pro thr thi gly his gal pur lac* F
3	O *pur lac pro thr thi gly his gal* F
AB 312	O *thi thr pro lac pur gal his gly* F

At first glance, there seems to be a random reshuffling of genes. However, a pattern does exist; the genes are not thrown together at random in each strain. For example, note that in every case the *his* gene has *gal* on one side and *gly* on the other. Similar statements can be made

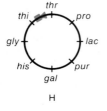 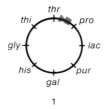 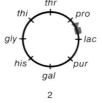  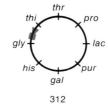

▶■ = F

▶ = origin (first to enter)

■ = terminus (last to enter)

Figure 10-7. Circularity of the *E. coli* chromosome. By using different Hfr strains (H, 1, 2, 3, 312), which have the fertility factor inserted into the chromosome at different points, interrupted-mating experiments indicate that the chromosome is circular. The mobilization points in the various strains are shown.

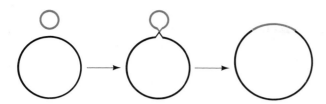

Figure 10-8. Attachment of the fertility (F) factor by crossing-over between two rings.

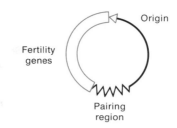

Figure 10-9. The three hypothetical regions of the circular F-factor chromosome.

about each gene, except when it appears at one end or the other of the linkage map. The order in which the genes are transferred is not constant. In two Hfr strains, for example, the *his* gene is transferred before the *gly* gene (*his* is closer to O), but in three strains the *gly* gene is transferred before the *his* gene.

A startling hypothesis was proposed by Alan Campbell to account for these results. Suppose that in an F^+ male, F is a small cytoplasmic element (and therefore easily transferred to an F^- cell on conjugation). If the "chromosome" of the F^+ male is a *ring*, any of the linear Hfr chromosomes could be generated simply by inserting F at some particular place in the ring (Fig. 10-7).

Integration of F

Chromosome circularity was a wildly implausible concept inferred solely from the genetic data; confirmation of its physical reality came only years later. Apparently, the insertion of F determines polarity. The end opposite F becomes the origin. How might we explain F integration? Wollman and Jacob suggested that some kind of crossover event between F and the chromosome might generate the Hfr chromosome. Campbell then came up

with a brilliant extension of that idea. He proposed that F, like the chromosome, is circular. Hence, a crossover between the two rings would produce a single larger ring with F inserted (Figure 10-8).

Now suppose that F consists of three different regions, as shown in Figure 10-9. The bacterial chromosome is pictured as having several regions of pairing homology with the pairing region of F. Then different Hfr chromosomes could easily be generated by crossovers at these different sites. This direct-crossover model of integration, depicted in Figure 10-10, was subsequently confirmed.

Episomes

The fertility factor thus exists in two states: as a free cytoplasmic element F that is easily transferred to F^- recipients and as an integrated part of a circular chromosome that is transmitted only very late in conjugation. The word **episome** was coined for a genetic particle having such a pair of states. A cell containing F in the first state is called an F^+ cell, a cell containing F in the second state is an Hfr cell, and a cell lacking F is an F^- cell.

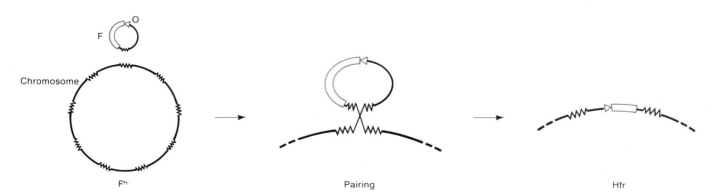

Figure 10-10. Model for the insertion of the F factor into the *E. coli* chromosome. If the F factor is considered to be circular, then an area of the fertility factor may have regions of pairing homology for different regions of the circular *E. coli* chromosome. Crossover between the F factor and the chromosome could thus insert the F factor into various points on the chromosome. (Jagged lines represent possible regions of homology; the arrow indicates the direction of transfer from the origin, or O.)

> **Message**   An **episome** is a genetic factor in bacteria that can exist either as an element in the cytoplasm or as an integrated part of a chromosome. The F factor is an episome.

Infectious elements other than F have been found in *E. coli* and other bacteria. Some are episomes that integrate into the ring chromosome as the F factor does. Others, called **plasmids,** do not integrate into the ring. Episomes and plasmids are of central importance in genetic engineering; we consider them in more detail later in this book.

Mechanics of Transfer

Does an Hfr die after donating its chromosome to an F⁻ cell? The answer is no (unless the culture is treated with streptomycin). There is evidence that the Hfr chromosome replicates shortly before or during conjugation, transferring a single strand to the F⁻ cell, which is then replicated in the F⁻ cell. This ensures a complete chromosome for the donor after mating. We assume that the F⁻ chromosome also is circular, because F⁻ is readily converted into an F⁺ from which an Hfr can be derived.

The picture emerges of a circular Hfr unwinding a copy of its chromosome, which is then transferred in a linear fashion into the F⁻ cell. How is the transfer achieved? Electron-microscope studies show that Hfr and F⁺ cells have **conjugation tubes,** or **F pili** (singular, pilus), protruding from their cell walls. These tubes are hollow, and they almost certainly represent the path of chromosome transfer to the F⁻ cell, which has no pili. Evidently, the pili are fractured during an interrupted-mating experiment, breaking the donor chromosome. Figure 10-11 shows conjugating bacteria, and Figure 10-12 is a diagrammatic representation of the passage of the chromosome.

We can now summarize the various aspects of the conjugational cycle in *E. coli* (Figure 10-13).

Review

Let's review some of the differences between F⁻, F⁺, and Hfr strains, since these designations are often confusing.

1. F⁻ strains do not contain the F factor and cannot transfer DNA by conjugation. They are, however, excellent recipients of DNA transferred from other cells by conjugation.

2. F⁺ cells contain the F factor in the cytoplasm. This element can be transferred in a highly efficient manner to F⁻ cells during conjugation. The presence of the F factor also confers on the host cell the ability to transfer chromosome markers to a recipient strain, although at a low rate compared to the transfer of the F factor itself. Chromosomal markers are transferred because in any population of F⁺ cells, a small fraction of cells (about one in 1000) has F integrated into the bacterial chromosome. Since conjugation experiments are usually carried out by mixing $10^7 - 10^8$ cells of prospective donor and recipient, the population of F⁺ cells contains many different cells derived from independent integrations of F into the

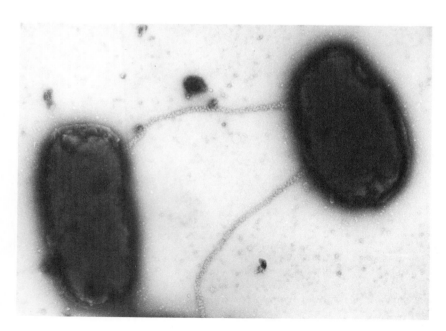

Figure 10-11. Electron micrograph of a cross between two *E. coli* cells ($\times 34{,}300$). The pili of the Hfr cell (not present in the F⁻ cell) have been visualized through the addition of viruses, which attach specifically to them. (Electron micrograph courtesy of David P. Allison, Biology Division, Oak Ridge National Laboratory.)

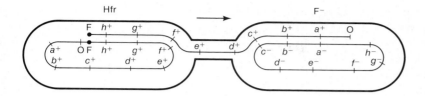

Figure 10-12. Diagrammatic representation of the sequential transfer of genes in a bacterial cross. (F = fertility factor; O = origin)

chromosome. Therefore, chromosomal markers are transferred by different cells in the population, starting at different points on the chromosome. This results in an approximately equal transfer of markers all around the chromosome, although at a low frequency. This type of F⁺-mediated transfer is what Lederberg and Tatum observed when they discovered gene transfer in bacteria. Each of the cells with an integrated F particle can be the source of a new Hfr strain if it is isolated and used to start a clone.

3. Hfr strains are derived from a clone of cells in which a specific integration of F into the bacterial chromosome has occurred. Therefore, all cells in an Hfr population have F integrated into the chromosome at exactly the same point. Hfr cells transfer chromo-

somal markers at a high rate relative to F⁺ cells (where only a fraction of the cells have F integrated into the chromosome). The markers are transferred from a fixed point in a specific order. This contrasts with F⁺ populations, which transfer chromosomal markers in no particular fixed order due to the presence in the population of many different Hfr cells.

Recombination Between Marker Genes After Transfer

Thus far, we have studied only the process of the transfer of genetic information between individuals in a cross. This transfer is inferred from the existence of recombinants produced from the cross. However, before a stable recombinant can be produced, obviously

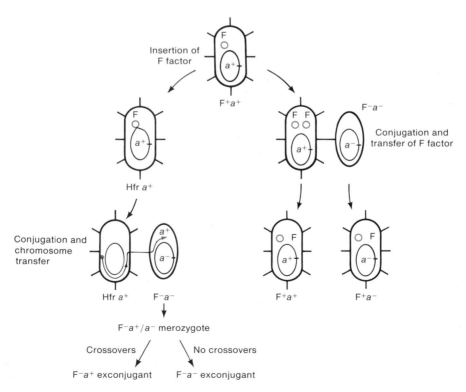

Figure 10-13. Summary of the various events that occur in the conjugational cycle of *E. coli.*

the transferred genes must be integrated or incorporated into the host's genome by an exchange mechanism. We now consider some of the special properties of this exchange event. Genetic exchange in prokaryotes does not take place between two whole genomes (as it does in eukaryotes); rather, it takes place between one *complete* genome, called the F⁻ **endogenote,** and an *incomplete* one, called the donor **exogenote.** What we have in fact is a partial diploid, or **merozygote.** Bacterial genetics is merozygous genetics. Figure 10-14a is a diagram of a merozygote.

It is obvious that a single crossover would not be very useful in generating viable recombinants, because the ring is broken to produce a strange, partially diploid linear chromosome (Figure 10-14b). To keep the ring intact, there must be an even number of crossovers (Figure 10-14c). The fragment produced in such a crossover is only a partial genome; in most cases, it is lost during subsequent cell growth. (We say "in most cases" because there are ways to maintain stable partial diploids which will not be discussed here.) Hence, it is obvious that both reciprocal products of recombination do not survive — only one does. A further unique property of bacterial exchange, then, is that we must forget about reciprocal exchange products in most cases.

Message In the merozygous genetics of bacteria, we generally are concerned with double crossovers and we do not expect reciprocal recombinants.

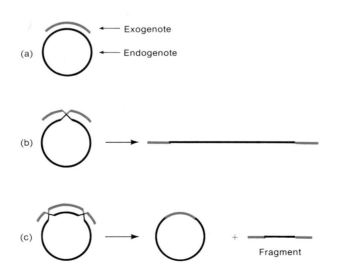

Figure 10-14. Crossover between exogenote and endogenote in a merozygote. (a) The merozygote. (b) A single crossover leads to a partially diploid linear chromosome. (c) An even number of crossovers leads to a ring plus a linear fragment.

The Gradient of Transfer

Only partial diploids exist in the merozygote. Some genes don't even get into the act! To better appreciate this, let's look again at the consequences of gene transfer. Note how a fragment of the donor chromosome appears in the recipient. This is because there is spontaneous breakage of the mating pairs, so that the entire chromosome is rarely transferred. The spontaneous breakage creates a natural **gradient of transfer,** which makes it less and less likely that a recipient cell will receive later and later genetic markers. ("Later" here refers to markers donated late in the order of markers transferred). For example, in a cross of Hfr-donating markers in the order *met–arg–leu,* we would expect a distribution of fragments such as the one represented here:

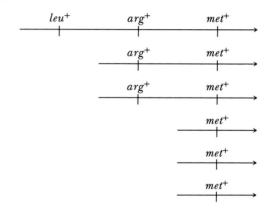

Note that many more fragments contain the *met* locus than the *arg* locus and that the *leu* locus is represented on only one fragment. It is easy to see that the earlier the marker, the greater the chance it will be transferred during conjugation. This bias in fragment size facilitates the ordering of genes (see next section) but greatly complicates the quantitative determination of linkage relationships between genes.

The concept of the gradient of transfer is the same as the one described earlier for interrupted matings, except that here we are allowing the natural disruption of mating pairs to occur instead of interrupting the pairs mechanically.

Ordering Genes with the Gradient of Transfer

We can exploit the natural gradient of transfer to order genetic markers, provided we select for an early marker which enters before the markers we are ordering. Let's see how this works. Suppose that we use an Hfr strain which donates markers in the order *met, arg, aro, his.* In a cross of an Hfr that is *str^s, met$^+$, arg$^+$, aro$^+$, his$^+$* with an F⁻ that is *str^r, met$^-$, arg$^-$, aro$^-$, his$^-$*, recombinants are selected that can grow on a minimal medium without methionine but with arginine, aromatic amino acids, histidine, and streptomycin. Here we are selecting for recombinants in the F⁻ strain that are *met⁺* in a cross in which the *met* locus is transferred as the earliest marker. We can then score for inheritance of the other markers present in the Hfr. A typical result would be

$$met^+ = 100\%$$

$$arg^+ = 60\%$$

$$aro^+ = 20\%$$

$$his^+ = 4\%$$

Note how the frequency of inheritance corresponds to the order of transfer. For this method to work, it is crucial that it be applied only to genetic markers that enter after the selected marker — in this case, after *met*.

Higher Resolution Mapping by Recombinant Frequency in Bacterial Crosses

Interrupted-mating experiments and the natural gradient of transfer are ideal procedures to use to obtain a rough set of gene locations over the entire map. Some other methods are needed to obtain a higher resolution between marker loci that are quite close together. Here we consider one approach to the problem of using the frequency of recombinants to measure linkage. We saw earlier that the first attempts to map in this way were failures. With the wisdom of hindsight, we can now see why they failed, and we can compensate for the problem.

The basic problem was the failure to understand that only fragments of the chromosome are transferred and that the gradient of transfer produces a bias toward the inheritance of early markers. In order to measure linkage and to attach any meaning to a calculated map distance, it is necessary to produce a situation in which every marker has an equal chance at being transferred, so that the recombinant frequencies are dependent only on the distance between the relevant genes.

Suppose that we consider three markers: *met*, *arg*, and *leu*. If the order is *met*, *arg*, *leu*, and *met* is transferred first and *leu* last, then we really want to set up the situation diagrammed below to calculate map distances separating these markers:

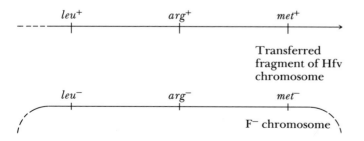

Here we simply have to arrange to select the *latest* marker to enter, which in this case is *leu*. Why? Because if we select for the last marker, then we know that every cell that received fragments containing the last marker also received the earlier markers, namely *arg* and *met*, on the same fragments. We can proceed to calculate map distance in the classical manner, where 1 map unit (m.u.) is equal to the percentage of crossovers in the respective interval on the map. In practice, this is done by calculating, among the total recovered recombinants, the percentage of recombinants produced by crossovers between two markers. Let's look at an example.

Sample Cross

In the cross of the Hfr strain just described (*str^s*, *met⁺*, *arg⁺*, *leu⁺*) with an F⁻ that is *str^r*, *met⁻*, *arg⁻*, *leu⁻*, we would select *leu⁺* recombinants and then examine them for the *arg* and *met* markers. In this case, the *arg* and *met* markers are called the unselected markers. Figure 10-15 depicts the types of crossover events expected. Note how two crossover events are required to incorporate part of the incoming fragment into the F⁻ chromosome. One crossover must be on each side of the selected (*leu*) marker. Thus, in Figure 10-15, one crossover must be on the left side of the *leu* marker and the second must be on the right side. Suppose that the map distance between each marker is 5 m.u. (5 percent recombination). In 5 percent of the total *leu⁺* recombinants, the second crossover occurs between *leu* and *arg* (Figure 10-15a); in another 5 percent of the cases, the second crossover occurs between *leu* and *met*. (Figure 10-15b). We would then expect 90 percent of the selected *leu⁺* recombinants to be *arg⁺ met⁺*, because the second crossover occurs outside the *leu*–*arg*–*met* interval (Figure 10-15c) in 90 percent of the cases. We would also expect 5 percent of the *leu⁺* recombinants to be *arg⁻ met⁻*, resulting from a crossover between *leu* and *arg*, and 5 percent of the *leu⁺* recombinants to be *arg⁺ met⁻*, resulting from a crossover between *arg* and *met*. In reality, then, we are simply determining the percentage of the time that the second crossover occurs in each of the three possible intervals.

In a cross such as the one just described, one class of potential recombinants requires an additional two crossover events (Figure 10-15d). In this case, the *leu⁺*, *arg⁻*, *met⁺* recombinants would require four crossovers instead of two. These recombinants are rarely recovered, because their frequency is sharply reduced compared to the other classes of recombinants.

Deriving Gene Order by Reciprocal Crosses

In many cases, two loci are so close together that it becomes difficult to order them in relation to a third locus. For example, consider three loci linked in this way:

(a) Insertion of late marker only

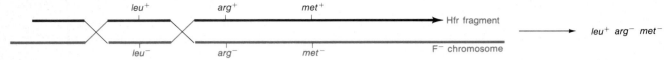

(b) Insertion of late marker and one early marker

(c) Insertion of late and early markers

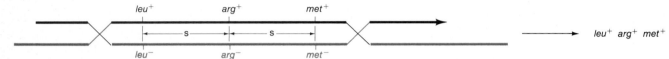

(d) Insertion of late and early markers, but not of market in between

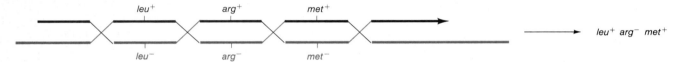

Figure 10-15. Incorporation of a late marker into the F⁻ *E. coli* chromosome. After an Hfr cross, selection is made for the *leu⁺* marker, which is donated late. The early markers (*arg⁺* and *met⁺*) may not be inserted, depending on whether recombination between the Hfr fragment and the F⁻ chromosome occurs at the appropriate place. If the distance between *leu⁺* and *arg⁺* is 5 m.u. and the distance between *arg⁺* and *met⁺* is 5 m.u., then crossovers will occur in each of these intervals 5 percent of the time (a and b), resulting in the *leu⁺ arg⁻ met⁻* (a) and *leu⁺ arg⁺ met⁺* (b) recombinants. Crossovers occurring outside of these intervals can result in the *leu⁺ arg⁺ met⁺* class of recombinants (c). The *leu⁺ arg⁻ met⁺* class of recombinants requires an additional two crossovers (d).

Here *b* and *c* are too closely linked to separate easily. The relative frequency between *a* and *b* under most experimental conditions will be more or less the same as the RF between *a* and *c*. Is the order *a−b−c* or *a−c−b*? One often-used technique is to make a pair of reciprocal crosses, using the same marker genotypes as both donor and recipient. Figure 10-16 shows the crossover events needed to generate prototrophs from the cross *a b c⁺ × a⁺ b⁺ c*, depending on the order of the loci.

Obviously, if the reciprocal crosses give dramatically different frequencies of wild-type survivors on a minimal medium, then we know that the order is *a−c−b*. If we observe no difference in frequencies between the reciprocal crosses, then the order must be *a−b−c*. Once again, this same principle can be used in other bacterial and phage mapping systems.

Infectious Marker-Gene Transfer by Episomes

Now that we understand the F⁻, F⁺, and Hfr states, we can interpret the initially confusing results of recombination work in *E. coli*. Knowing all this, Edward Adelberg began to do recombination experiments with an Hfr strain in 1959. However, the particular Hfr strain he used kept producing F⁺ cells, so the recombination frequencies were not very large. Adelberg called this particular fertility factory F′ to distinguish it from the normal F, for the following reasons:

1. The F′-bearing F⁺ strain reverted back to an Hfr strain quite often.

2. F′ always integrated at the *same place* to give back the original Hfr chromosome. (Remember that randomly selected Hfr derivatives from F⁺ males have origins at many different positions.)

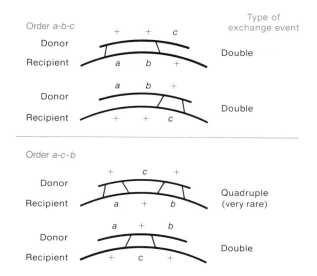

Figure 10-16. Inferring gene order from the relative frequencies of recombinants in reciprocal crosses involving parental genotypes $a\ b\ +\ \times\ +\ +\ c$. If the order is $a-b-c$, then the frequency of $+\ +\ +$ progeny from the reciprocal crosses will be approximately equal. However, if the order is $a-c-b$, then the frequencies will be quite different.

How could these properties of F′ be explained? The answer came from the recovery of a new F′ from an Hfr strain in which the lac^+ locus was near the end of the Hfr chromosome (was transferred very late). Using this Hfr lac^+ strain, François Jacob and Adelberg found an F⁺ derivative that transferred lac^+ to F⁻ lac^- recipients at a very high frequency. Furthermore, the recipients, which became F⁺ lac^+ phenotypically, occasionally produced F⁻ lac^- daughter cells at a frequency of 1×10^{-3}. Thus, the genotype of the recipients appears to be F′ lac^+/lac^-.

Now we have the clue: F′ is a cytoplasmic element that carries a part of the bacterial chromosome. Its origin and reintegration can be visualized as shown in Figure 10-17. This F′ is known as F-lac. Because F-lac^+/lac^- cells are lac^+ in phenotype, we know that lac^+ is dominant over lac^-. As we shall see later, the dominant–recessive relationship between alleles can be a very useful bit of information in interpreting gene function. Partial diploidy, called **merodiploidy,** for specific segments of the genome can be made with an array of F′ derivatives from Hfr strains. The F′ cells can be selected

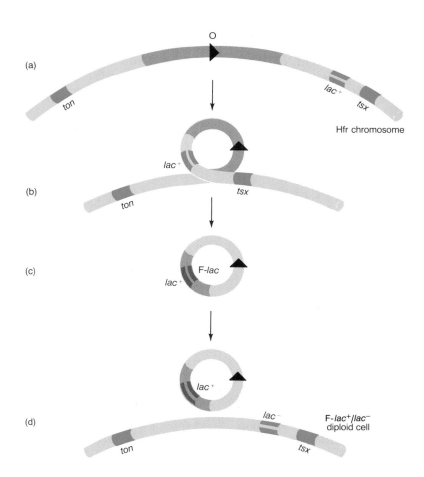

Figure 10-17. Origin and reintegration of the F′ (or F-lac) factor. (a) The F factor is inserted in an Hfr strain between the ton and lac^+ alleles. (b) Abnormal "outlooping" (separation of the F factor) occurs to include lac locus. (c) The F-lac^+ particle (=F′). (d) Constitution of an F-lac^+/lac^- partial diploid produced by the transfer of the F-lac^+ particle to an F⁻ lac^- recipient. (From G. S. Stent and R. Calendar, *Molecular Genetics,* 2d ed. Copyright © 1978 by W. H. Freeman and Company.)

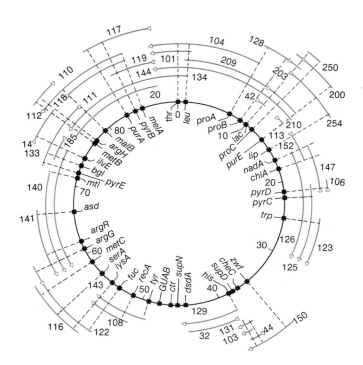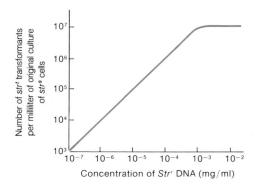

Figure 10-18. Some of the F' factors derived from *E. coli* are shown, together with the regions of the bacterial chromosome that they carry. Markers such as *trp, thr,* and *leu* are indicated on the circular *E. coli* map, which is delineated into 90 minutes (referring to the time of entry of different genes in an Hfr × F⁻ cross, beginning at an arbitrary point). The F' factors transfer markers in a fixed direction, as indicated by the direction of the arrow, and form a fixed point, as shown by the placement of the arrow. (Courtesy of B. Low and J. Falkinham. With permission from *Handbook of Microbiology*. Edited by A. I. Laskin and H. Lechevalier. Chemical Rubber Co., Cleveland, Ohio, 1972.)

by looking for the infectious transfer of normally late genes in a specific Hfr strain.

The use of F' elements to create partial diploids is called **sexduction,** or F'-duction. Some F' strains can carry very large parts (up to one-quarter) of the bacterial chromosome; if appropriate markers are used, the merozygotes generated can be used for recombination studies. Figure 10-18 shows some of the F' factors that have been characterized.

Message During conjugation between an Hfr donor and an F⁻ recipient, the genes of the donor are transmitted linearly, with the inserted fertility factor transferring last.

During conjugation between an F⁺ donor carrying an F' plasmid and an F⁻ recipient, a specific part of the donor genome may be transmitted infectiously to F⁻ cells, via the plasmid. This part was originally adjacent to the F locus in an Hfr strain from which the F⁺ was derived.

Bacterial Transformation

The conversion of one genotype into another by the introduction of exogenous DNA is termed **transformation.** Transformation was discovered in *Streptococcus pneumoniae* in 1928 by Frederick Griffith.

In 1944, Oswald Avery, C. M. McLeod, and M. McCarty demonstrated that the "transforming principle" was DNA. Both results are milestones in the elucidation of the molecular nature of genes. We consider this work in more detail in Chapter 11.

After it was shown that DNA is the agent that determines the polysaccharide character of *S. pneumoniae*, transformation was demonstrated for other genes, such as those for drug resistance (Figure 10-19). The transforming principle, or DNA, is physically incorporated

Figure 10-19. The genetic transfer of streptomycin-resistant (*str*ʳ) to streptomycin-sensitive (*str*ˢ) cells of *E. coli*. The recovery of *str*ʳ transformants among *str*ˢ cells depends on the concentration of *str*ʳ DNA. (From G. S. Stent and R. Calendar, *Molecular Genetics*, 2d ed. Copyright © 1978 by W. H. Freeman and Company.)

into the bacterial chromosome by a physical breakage-and-insertion process analogous to that described for Hfr × F⁻ crosses in Figure 10-14. Thus, if radioactively labeled DNA from an arg^+ bacterial culture is added to unlabeled arg^- cells, the arg^+ transformants (selected by plating on a minimal medium without arginine) can be shown to contain some of the radioactivity. We examine a molecular model of this process in Chapter 18. For now, let's consider transformation simply as a genetic tool.

Linkage Information from Transformation

Transformation has been a very handy tool in several areas of bacterial research. We learn later how it is used in some of the modern techniques of genetic engineering. Here we examine its usefulness in providing linkage information.

When DNA (the bacterial chromosome) is extracted for transformation experiments, some breakage into smaller pieces is inevitable. If the two donor genes are located close together on the chromosome, then there is a greater chance that they will be carried on the same piece of transforming DNA and hence will cause a **double transformation.** Conversely, if genes are widely separated on the chromosome, then they will be carried on separate transforming segments and the frequency of

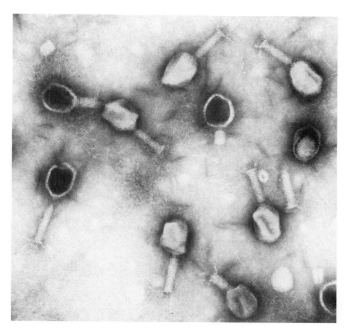

Figure 10-21. Mature particles of the *E. coli* phage T4 (×97,500). (Photograph courtesy of Grant Heilman.)

double transformants will equal the product of the single-transformation frequencies. Thus, it should be possible to test for close linkage by testing for a departure from the product rule.

Unfortunately, the situation is made more complex by several factors — the most important of which is that not all cells in a population of bacteria are competent to be transformed. Because single transformations are expressed as proportions, the success of the product rule obviously depends on the absolute size of these proportions. There are ways of calculating the proportion of competent cells, but we need not detour into that subject now. You can sharpen your skills in transformation analysis in one of the problems at the end of the chapter, which assumes 100 percent competence.

Phage Genetics

Most bacteria are susceptible to attack by bacteriophages, which literally means "eaters of bacteria." A phage consists of a nucleic acid "chromosome" (DNA or RNA) surrounded by a coat of protein molecules. One well-studied set of strains of phage are identified as T1, T2, and so on. Figures 10-20 and 10-21 show the complicated structure of a phage belonging to the class called T-even phages (T2, T4, and so on).

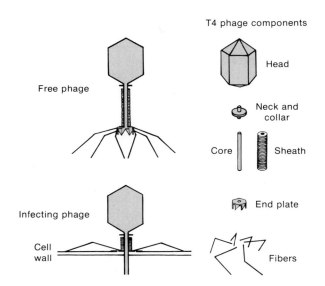

Figure 10-20. Phage T4, shown in its free state and in the process of infecting a cell of *E. coli.* On the right, a phage has been diagrammatically exploded to show its highly ordered structure in three dimensions. (From R. S. Edgar and R. H. Epstein, "The Genetics of a Bacterial Virus." Copyright © 1965 by Scientific American, Inc. All rights reserved.)

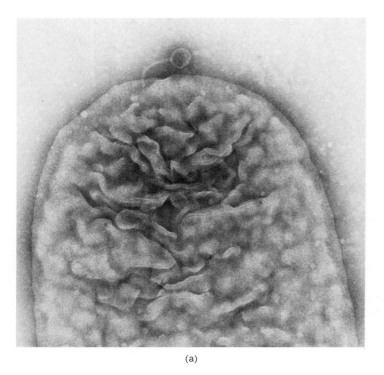

(a)

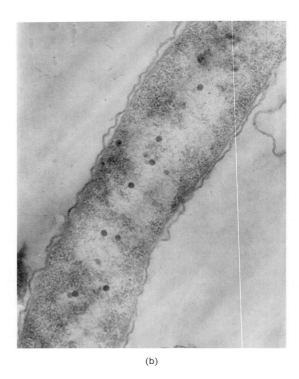

(b)

Figure 10-22. (a) A bacteriophage (called λ) attached to an *E. coli* cell and injecting its genetic material. (b) Progeny particles of phage λ maturing inside an *E. coli* cell. (Photographs courtesy of Jack D. Griffith.)

The Phage Cross

A phage attaches to a bacterium and injects its genetic material into the bacterial cytoplasm (Figure 10-22a). The phage genetic information then takes over the machinery of the bacterial cell by turning off the synthesis of bacterial components and redirecting the bacterial synthetic material to make more phage components (Figure 10-22b). (The use of the word information is interesting in this connection; it literally means "to give form." And of course, that is precisely the role of the genetic material: to provide blueprints for the construction of form. In the present discussion, the form is the elegantly symmetrical structure of the new phages.) Ultimately, many phage descendants are released when the bacterial cell wall breaks open. This breaking-open process is called **lysis.**

But how can we study inheritance in phages when they are so small that they are visible only under the electron microscope? In this case, we cannot produce a visible colony by plating, but we can produce a visible manifestation of an infected bacterium by taking advantage of a phage character involving the effects of phage on bacteria. Let's look at the consequences of a phage infecting a single bacterial cell. Figure 10-23 shows the sequence of events, which leads to the release of progeny phages from the lysed cell. After lysis, progeny phages

infect neighboring bacteria. This is an exponentially explosive phenomenon (an exponential increase in the number of lysed cells). Very soon after starting an experiment of this type (overnight), the effects are visible to the naked eye: a clear area, or **plaque,** is present on the opaque lawn of bacteria on the surface of a plate of solid medium (Figure 10-24). Depending on the phage genome, such plaques can be large or small, fuzzy or sharp, and so forth. Thus, **plaque morphology** is a phage character that can be analyzed.

Another phage phenotype that can be analyzed genetically is **host range.** Certain strains of bacteria are immune to adsorption (attachment) or injection by phages. Phages, in turn, may differ in the spectra of bacterial strains they can infect and lyse.

A phage cross can be illustrated by a cross of T2 phages originally studied by Alfred Hershey. The genotypes of the two parental strains of T2 phage in Hershey's cross were $h^- r^+ \times h^+ r^-$. The alleles are identified by the following characters: h^- can infect two different *E. coli* strains (which we can call strains 1 and 2); h^+ can infect only strain 1; r^- rapidly lyses cells, thereby producing large plaques; and r^+ slowly lyses cells, thus producing small plaques.

In the cross, strain 1 is infected with both parental T2 phage genotypes at a concentration—called **multiplicity of infection,** which is the ratio of phages of

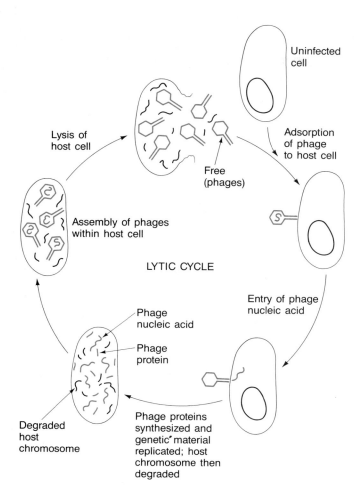

Uninfected cell

Lysis of host cell

Adsorption of phage to host cell

Free (phages)

Assembly of phages within host cell

LYTIC CYCLE

Entry of phage nucleic acid

Phage nucleic acid

Phage protein

Degraded host chromosome

Phage proteins synthesized and genetic material replicated; host chromosome then degraded

Figure 10-23. A generalized bacteriophage lytic cycle. (From J. Darwell, H. Lodish and D. Baltimore, *Molecular Cell Biology.* Copyright © 1986 by W. H. Freeman and Company.)

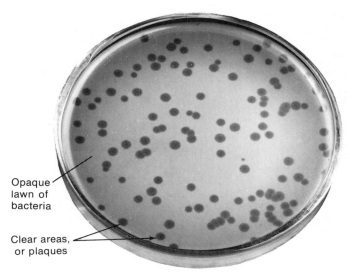

Opaque lawn of bacteria

Clear areas, or plaques

Figure 10-24. The appearance of phage plaques. Individual phages are spread on an agar medium that contains a fully grown "lawn" of *E. coli.* Each phage infects one bacterial cell, producing 100 or more progeny phage that burst the *E. coli* cell and infect neighboring cells. They, in turn, are exploded with progeny, and the process continues until a clear area, or plaque, appears on the opaque lawn of bacterial cells. (From G. S. Stent, *Molecular Biology of Bacterial Viruses.* Copyright © 1963 by W. H. Freeman and Company.)

bacteria — sufficiently high to ensure that a high proportion of cells are simultaneously infected by both phage types. This kind of infection (Figure 10-25) is called a **mixed infection,** or a **double infection.** The phage lysate (the progeny phage) is then analyzed by spreading it onto a bacterial lawn composed of a mixture of strains 1 and 2. Four plaque types are then distinguishable (Figure 10-26 and Table 10-3). These four genotypes can be scored easily as parental ($h^-\ r^+$ and $h^+\ r^-$ Table 10-3) and recombinant, and an RF can be calculated as follows:

$$RF = \frac{(h^+\ r^+) + (h^-\ r^-)}{\text{total plaques}}$$

If we assume that entire phage genomes recombine, then we are not faced with a merozygous situation, as in a

bacterial cross. Presumably, then, single exchanges can occur and produce viable reciprocal products. However, phage crosses are subject to complications that are only mentioned here. First, several **rounds of exchange** can potentially occur within the host. Because hundreds of phages can be released from a single infected cell, each parental infecting phage can be duplicated many times. If recombination does not occur at one specific time in the lytic cycle, then a recombinant produced shortly after infection may undergo further exchange at later times. Second, recombination can occur between genetically similar phages as well as between different types. Thus, $P_1 \times P_1$ and $P_2 \times P_2$ occur in addition to $P_1 \times P_2$. For both of these reasons, recombinants from phage crosses are a consequence of a **population** of events rather than defined, single-step exchange events. Nevertheless, *all other things being equal,* the RF calculation

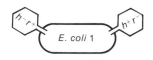

E. coli 1

Figure 10-25. A double infection of *E. coli* by phages.

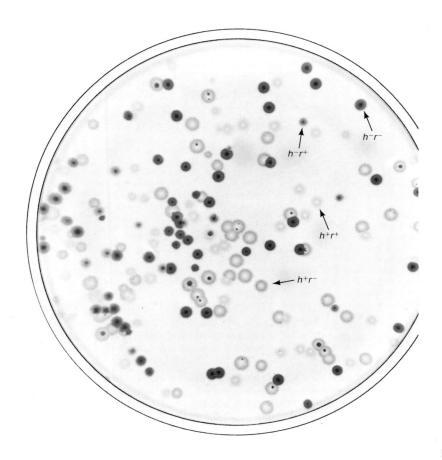

Figure 10-26. Plaque phenotypes produced by progeny of the cross $h^- \ r^+ \times h^+ \ r^-$. Enough phage of each genotype are added to ensure that most bacterial cells are infected with at least one phage of each genotype. After lysis, the progeny phage are collected and added to an appropriate *E. coli* lawn. Four plaque phenotypes can be differentiated, representing two parental types and two recombinants. (From G. S. Stent, *Molecular Biology of Bacterial Viruses*. Copyright © 1963 by W. H. Freeman and Company.)

TABLE 10-3. Progeny-phage plaque types from cross $h^- \ r^+ \times h^+ \ r^-$

Phenotype	Inferred genotype
Clear and small	$h^- \ r^+$
Cloudy and large	$h^+ \ r^-$
Cloudy and small	$h^+ \ r^+$
Clear and large	$h^- \ r^-$

NOTE: clearness is produced by the h^- allele, which allows infection of *both* bacterial strains in the lawn; cloudiness is produced by the h^+ allele, which limits infection to the cells of strain 1.

given does represent a valid index of map distance in phages.

Circularity of the T2 Genetic Map

Hershey obtained several different T2 strains with the rapid-lysis phenotype; he called their genotypes *r1*, *r2*, and so forth (in the order of discovery). Let's indicate three different *r* strains as r_a, r_b, and r_c and make the cross $r_x^- h^+ \times r_x^+ h^-$, where r_x represents one of the *r* genes. Table 10-4 shows the results. We can construct a linkage map in just the same way that Sturtevant constructed the original eukaryotic maps in *Drosophila*. The parental types ($r^- \ h^+$ and $r^+ \ h^-$) occur with the highest frequencies, although not with equal frequency. The two recombinant classes, however, are equally frequent. We can construct linkage maps for each cross (Figure 10-27a). The different recombination values indicate that the loci for the three *r* genes are in different places on the chromosome, so there are four possible linkage maps (Figure 10-27b).

TABLE 10-4. Frequency of progeny-phage types in crosses involving several *r* mutants and an *h* mutant

Cross	Percentage of each genotype			
	$r^- \ h^+$	$r^+ \ h^-$	$r^+ \ h^+$	$r^- \ h^-$
$r_a^- h^+ \times r_a^+ h^-$	34.0	42.0	12.0	12.0
$r_b^- h^+ \times r_b^+ h^-$	32.0	56.0	5.9	6.4
$r_c^- h^+ \times r_c^+ h^-$	39.0	59.0	0.7	0.9

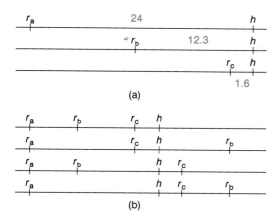

Figure 10-27. (a) Distances between gene pairs for each cross given in Table 10-4. (b) The various possible linkage relationships inferred from the distances in (a).

Can we distinguish among these alternatives? First, let's take only r_b, r_c, and h and ask whether the order is $r_c - h - r_b$ or $h - r_c - r_b$. We can make the cross $r_c^- r_b^+ \times r_c^+ r_b^-$ and compare the RF with the value of 12.3 obtained for the $r_b - h$ interval. From this comparison, we find that h is located between r_c and r_b ($r_c - h - r_b$).

Now we ask whether r_a lies on the side of h next to r_b or on the side next to r_c. The data from crosses of r_a with r_b and r_c do not provide a clear-cut resolution of the answer. After intensive genetic mapping with many different strains of T2, the answer turns out to be that *both* alternative maps are correct. How can both $r_a - r_c - h - r_b$ and $r_c - h - r_b - r_a$ be correct? We have encountered a similar situation before, and the answer is similar. Like the bacterial map, the linkage map of T2 is circular, as shown in Figure 10-28a. The total genetic length of the

T2 linkage map is about 1500 m.u. Figure 10-28b shows a more complex map for another T-even phage.

In the bacterial and phage experiments we have considered, the data are initially confusing. However, once the novelty of the conditions for "crossing" is understood, the analysis of recombination and the construction of linkage maps are fairly straightforward procedures.

Message Recombination between phage chromosomes can be studied by bringing the parental chromosomes together in one host cell through mixed infection. Progeny phages can be examined for parental versus recombinant genotypes.

Lysogeny

In the 1920s, long before *E. coli* became the favorite organism of microbial geneticists, some interesting results were obtained in the study of phage infections of *E. coli*. Some bacterial strains are resistant to infection by certain phages, but these resistant bacteria will cause lysis of nonresistant bacteria when the two bacterial strains are mixed together. The resistant bacteria that induce lysis in other cells are said to be **lysogenic.** Apparently, the lysogenic bacteria somehow "carry" the phages while remaining immune to their lysing action. When nonlysogenic bacteria are infected with phages from a lysogenic strain, a small fraction of the infected cells do not lyse but instead become lysogenic themselves. Initially little attention was paid to this phenomenon after some studies seemed to show that the lysogenic bacteria are simply contaminated with external phages that can be removed by careful purification.

However, in the mid-1940s, André Lwoff examined lysogenic strains of *Bacillus megaterium* and followed the lysogenic behavior of a lysogenic strain through many cell divisions. Carefully observing his culture, he separated each pair of daughter cells immediately after division. One cell was put into a culture; the other was observed until it divided. In this way, Lwoff obtained 19 cultures representing 19 generations (19 consecutive cell divisions). All 19 cultures were lysogenic, but tests of the medium showed no free phage at any time during these divisions, thereby confirming that lysogenic behavior is a character that persists through reproduction in the absence of free phage.

On rare occasions, Lwoff observed spontaneous lysis in one of the cells in his cultures. When the medium was spread on a lawn of nonlysogenic cells after one of these spontaneous lyses, plaques appeared, showing that free phages had been released in the lysis. Lwoff was able to

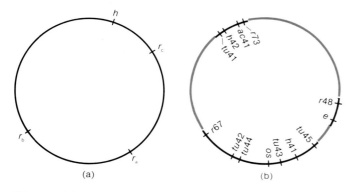

Figure 10-28. Circular maps for T-even phage. (a) Simple linkage map for T2 phage. (b) Map of phage T4, inferred from crosses of nonlethal mutants *r* (rapid lysis), *h* (host range), *ac* (acridine resistance), *tu* (turbid plaques), *os* (resistance to osmotic shock), and *e* (lysis defective).

propose a hypothesis to explain all his observations: each bacterium of the lysogenic strain contains a noninfective factor that is passed from generation to generation, but this factor occasionally gives rise to the production of infective phage (without the presence of free phage in the medium). Lwoff called this factor the **prophage** because it somehow seems to be able to *induce* the formation of a "litter" of infective phage. Later studies showed that a variety of agents, such as ultraviolet light or certain chemicals, can induce lysis in a large fraction of a population of lysogenic bacteria.

We now know exactly how Lwoff's observations occur. A lysogenic bacterium contains a prophage, which somehow protects the cell against additional infection, or **superinfection,** from free phages and which is duplicated and passed on to daughter cells during division. In a small fraction of the lysogenic cells, the prophage is induced to produce infective phage. This process robs the cell of its protection against phage; it lyses and releases infective phage into the medium, thus infecting any nonlysogenic cells present in the culture.

Phages can be categorized into two types. Those for which there are no lysogenic bacteria are called **virulent phages** (resistant bacterial mutants may exist for these phages, but their resistance is not due to lysogeny). Those phages capable of lysogenizing bacteria are called **temperate phages** (resistance to superinfection is an "immunity" conferred by the presence of the prophage). Figure 10-29 diagrams the life cycle of a typical temperate phage.

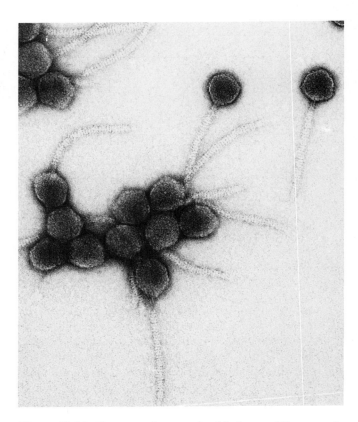

Figure 10-30. Electron micrograph of λ phages. (Photograph courtesy of Jack D. Griffith.)

The Genetic Basis of Lysogeny

What is the nature of the prophage? On induction, the prophage is capable of directing the production of complete mature phage, so all of the phage genome must be present in the prophage. But is the prophage a small particle free in the bacterial cytoplasm, or is it somehow associated with the bacterial genome? Fortuitously, the original strain of *E. coli* used by Lederberg and Tatum (page 222) proved to be lysogenic for a temperate phage called **lambda (λ).** Phage λ has become the most intensively studied and best-characterized phage (Figure 10-30). Crosses between F⁺ and F⁻ cells have yielded interesting results. It turns out that F⁺ × F⁻(λ) crosses yield recombinant lysogenic recipients, whereas the reciprocal cross F⁺(λ) × F⁻ almost never gives lysogenic recombinants.

These results became understandable when Hfr strains were discovered. In the cross Hfr × F⁻(λ), lysogenic F⁻ exconjugants with Hfr genes are readily recovered. However, in the reciprocal cross Hfr(λ) × F⁻, the early genes from the Hfr are recovered among the exconjugants, but recombinants for late markers (those expected to transfer after a certain time in mating) are not recovered. Furthermore, lysogenic exconjugants

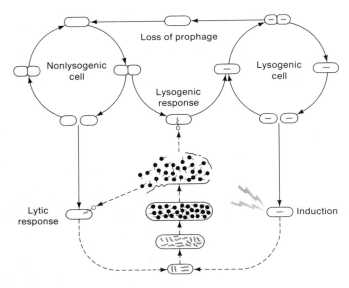

Figure 10-29. Alternative cell cycles associated with a temperate phage and its host. (From A. Lwoff, *Bacteriological Reviews* 17, 1953, 269.)

are almost never recovered from this reciprocal cross. What is the explanation? The observations make sense if the λ prophage is behaving like a bacterial gene locus (that is, as part of the bacterial chromosome). In the cross of a lysogenic Hfr with a nonlysogenic F⁻ recipient, the entry of the λ prophage into the cytoplasm of a nonimmune cell triggers the prophage into a lytic cycle; this is called **zygotic induction.** In interrupted-mating experiments, the λ prophage always enters the F⁻ cell at a specific time, closely linked to the *gal* locus. We can assign the λ prophage to a specific locus next to the *gal* region.

Entry of the λ prophage into an F⁻ cell immediately induces the lytic cycle. But in the cross Hfr(λ) × F⁻(λ), any recombinants are readily recovered (that is, no induction of the prophage occurs). It would seem that the cytoplasm of the F⁻ cell must exist in two different states (depending on whether or not the cell contains a λ prophage), so that contact between an entering prophage and the cytoplasm of a nonimmune cell immediately induces the lytic cycle. Perhaps some cytoplasmic factor specified by the prophage somehow represses the multiplication of the virus. Entry of the prophage into a nonlysogenic environment immediately dilutes this repressing factor, and therefore the virus reproduces. But if the virus specifies the repressing factor, then why doesn't it shut itself off again? Perhaps it does, because a fraction of infected cells do become lysogenic. There may be a race between the λ gene signals for reproduction and those specifying a shutdown. The model of a phage-directed cytoplasmic repressor nicely explains the immunity of the lysogenic bacterial, because any superinfecting phage would immediately encounter a repressor and be inactivated. We discuss this type of model in more detail in Chapter 16.

Prophage Attachment

How is the prophage attached to the bacterial genome? In the days before chromosome circularity was known, two models seemed possible (Figure 10-31).

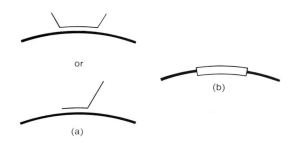

Figure 10-31. Two conceivable models of prophage attachment. (a) Pairing association. (b) Physical incorporation. (Thick line = *E. coli* chromosome; thin line = phage chromosome.)

Figure 10-32. Models of λ affinity for the bacterial chromosome: (a) in the normal phage; (b) in a phage mutant showing deletion.

The clue for a choice between these models came from a mutant strain of λ. A large section of the chromosome was deleted in this strain, and it could not lysogenize although it could reproduce. Perhaps the deleted region (which obviously does not contain the genes for λ production) is a region of homology with the bacterial chromosome (the Campbell model again; see Figure 10-8). Crossing-over between the λ and *E. coli* chromosomes at such a region could incorporate the entire λ genome at a specific point (and as a continuous part) of the *E. coli* chromosome (Figure 10-32).

Can this simple model of lysogeny be tested? The attraction of Campbell's proposal is that it does make testable predictions that λ gives geneticists a chance to test:

1. Physical integration of the prophage should increase the genetic distance between flanking bacterial markers (Figure 10-33). In fact, time-of-entry or recombination distances between the bacterial genes *are* increased by lysogeny.

2. Some deletions of bacterial segments adjacent to the prophage site should delete phage genes (Figure 10-34). Experimental studies do confirm this prediction.

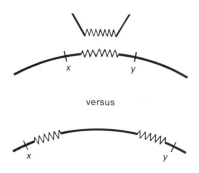

versus

Figure 10-33. The insertion of the λ genome into the *E. coli* chromosome splits the attachment regions, increasing the distance between markers *x* and *y*.

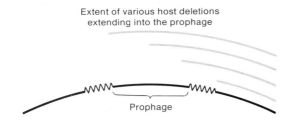

Extent of various host deletions
extending into the prophage

Prophage

Figure 10-34. Host deletions may delete genes from an adjacent prophage.

The phenomenon of lysogeny is a very successful way for a temperate phage to avoid "eating itself out of house and home." Lysogenic cells can perpetuate and carry the phages around. We consider lysogeny and the integration of the λ phage into the host chromosome in more detail in Chapters 16 and 18.

Transduction

Some phages are able to carry bacterial genes from one bacterial cell to another through the process of **transduction.** There are two kinds of transduction: generalized and specialized.

Generalized Transduction

In 1951, Joshua Lederberg and Norton Zinder were testing for recombination in the bacterium *Salmonella typhimurium,* using the techniques that had been successful with *E. coli.* The researchers used two different strains: one was *phe⁻ trp⁻ tyr⁻*, and the other was *met⁻ his⁻*. (We won't worry about the nature of these markers except to note that the mutant alleles confer nutritional requirements.) When either strain was plated on a minimal medium, no wild-type cells were observed. However, after mixing the two strains, wild-type cells occurred at a frequency of about 1 in 10^5. Thus far, the situation seems similar to that for recombination in *E. coli.*

However, in this case, the researchers also recovered recombinants from a U-tube experiment, in which cell contact (conjugation) was prevented by a filter. By varying the size of the pores in the filter separating the two arms, they found that the agent responsible for recombination is about the size of the virus P22, a known temperate phage of *Salmonella.* Further studies supported the suggestion that the vector of recombination is indeed P22. The filterable agent and P22 are identical in properties of size, sensitivity to antiserum, and immunity to hydrolytic enzymes. Thus, Lederberg and Zinder, instead of confirming conjugation in *Salmonella,* had discovered a new type of gene transfer mediated by a virus. They called this process transduction. During the lytic cycle, some virus particles somehow pick up bacterial genes that are then transferred to another host, where the virus inserts its contents. Transduction has subsequently been shown to be quite common among both temperate and virulent phages.

Generation of Transducing Particles

How are transducing phages produced? In 1965, K. Ikeda and J. Tomizawa threw light on this question in some experiments on the temperate *E. coli* phage P1. They found that when a nonlysogenic donor cell is lysed

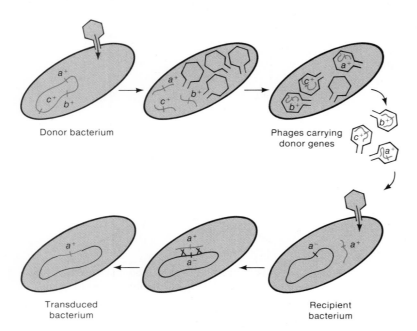

Donor bacterium

Phages carrying donor genes

Transduced bacterium

Recipient bacterium

Figure 10-35. The mechanism of generalized transduction. In reality, only a minority of phage progeny carry donor genes.

by P1, the bacterial chromosome is broken up into small pieces. Occasionally, the forming phage particles mistakenly incorporate a length of *pure* bacterial DNA into a phage head. This is the origin of the transducing phage. A similar process can occur when the prophage of P1 is induced. Because the phage coat proteins determine the phage's ability to attack a cell, such viruses, or transducing particles, can bind to a bacterial cell and inject their contents, which now happen to be donor bacterial genes. When the contents of a transducing phage are injected into a recipient cell, a merodiploid situation is created in which the transduced genes can be incorporated by recombination (Figure 10-35). Because every host marker can be transduced, this type of transduction is termed **generalized transduction.**

Linkage Data from Transduction

From such merozygotes, we can derive linkage information about bacterial genes. Transduction from an $a^+ b^+$ donor to an $a^- b^-$ recipient produces various transductants for a^+ and b^+. We can obtain linkage information from the ratio

$$\frac{\text{single-gene transductants}}{\text{total transductants}} = \frac{(a^+ \ b^-)}{(a^+ \ b^-) + (a^+ b^+)}$$

$$\text{or} \quad \frac{(a^- \ b^+)}{(a^- \ b^+) + (a^+ b^+)}$$

Presumably, the chance of either a^+ or b^+ being included individually in the transducing phage is proportional to the distance between them. If they are close together, they will usually be picked up and transduced by the phage together. Of course, the method gives linkage values only if the genes under test *are* close

TABLE 10-5. Accompanying markers in specific P1 transductions

Experiment	Selected marker(s)	Unselected markers
1	leu^+	50% are azi^r; 2% are thr^+
2	thr^+	3% are leu^+; 0% are azi^r
3	leu^+ and thr^+	0% are azi^r

enough together for *both* to be included in the transducing phage, thus forming an $a^+ b^+$ transductant called a **cotransductant.** Linkages usually are expressed as cotransduction frequencies (Figure 10-36). The closer two genetic markers are, the greater the cotransduction frequency. However, even the failure to find cotransductants provides some linkage information in a negative sense. Thus, transduction joins the battery of modes of genetic transfer in bacteria—along with conjugation, infectious transfer of episomes, and transformation.

The phages P1 and P22 belong to a group that shows generalized transduction (that is, they transfer virtually any gene of the host chromosome). As prophages, P22 probably inserts into the host chromosome and P1 remains free like a large plasmid. But both transduce by faulty headstuffing during lysis.

We can estimate the size of the piece of host chromosome that a phage can pick up from the following type of experiment using P1 phage:

$$\text{donor } leu^+ \ thr^+ \ azi^r \longrightarrow \text{recipient } leu^- \ thr^- \ azi^s$$

We can select for one or more donor markers in the recipient and then (in true merozygous genetics style) look for the presence of the other unselected markers, as outlined in Table 10-5. Experiment 1 in the table tells us

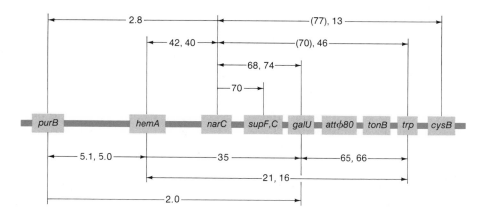

Figure 10-36. Genetic map of the *purB* to *cysB* region of *E. coli* determined by P1 cotransduction. The numbers given are the averages in percent for cotransduction frequencies obtained in several experiments. Where transduction crosses were performed in both directions, the head of each arrow points to the selective marker with the corresponding linkage nearest to each arrow. The values in parentheses are considered unreliable due to interference from the nonselective marker. (Redrawn from J. R. Guest, *Molecular and General Genetics* 105, 1969, 285.)

that *leu* is relatively close to *azi* and distant from *thr*, leaving us with two possibilities:

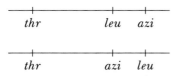

Experiment 2 tells us that *leu* is closer to *thr* than *azi* is, so that the map must be

By selecting for *thr*⁺ and *leu*⁺ in the transducing phage in experiment 3, we see that the transduced piece of genetic material never includes the *azi* locus. If enough markers were studied to produce a more complete linkage map, we could estimate the size of a transduced segment. Such experiments indicate that P1 cotransduction occurs within approximately 1.5 min of the *E. coli* chromosome map (1 min, or minute, equals the length of chromosome transferred by an Hfr in one minute's time at 37°C).

Retrospective

At this point, you might wonder how Lederberg and Zinder were able to detect transduction in their experiment. Recall that they utilized two multiply marked strains. One strain required phenylalanine, tryptophan, and tyrosine for growth; the second required methionine and histidine. Based on what we now know about P22 and P1-mediated generalized transduction, we would not expect to restore either of these strains to the wild-type unless the markers were so closely linked that they could be carried on the same transducing particle. And yet, that is precisely what happened in the experiment. The *his⁻ met⁻* strain was the source of the P22-transducing phage. This strain donated the wild-type markers that replaced the *phe⁻ trp⁻ try⁻* markers. These markers are all part of the same pathway and are very closely linked. (In fact, we now know that the phenylalanine and tyrosine requirements are caused by the same mutation, although the tryptophan requirement is due to a mutation in a different gene.) If Lederberg and Zinder had used a different set of markers, they might never have discovered transduction!

Specialized Transduction

Another class called **specialized** transducing phages carry only restricted parts of the bacterial chromosome. We look now at this process of **specialized transduction.**

Lambda (λ) is a good example of a specialized transducer. Recall that as a prophage, λ always inserts next to the *gal* region of the *E. coli* host chromosome. In transduction experiments, λ can transduce only the *gal* gene and another closely linked gene, *bio*. Let's visualize the mechanism of λ transduction. In Figure 10-37a, we see a schematic representation of the production of a lysogen. The actual recombination between regions of λ and the bacterial chromosome is catalyzed by a specific enzyme system, described more fully in Chapter 18. The phage and bacterial integration regions are not completely identical, so the integration of λ results in two hybrid sites, as shown. We can induce the lytic cycle with ultraviolet light and produce a lysate (progeny phage population). The normal outlooping of the prophage restores the original phage integration site. These phages can integrate normally on subsequent infection of a strain that is not lysogenic for λ (see Figure 10-37a). As seen in Figure 10-37b, very rare abnormal outlooping can result in phage particles that now carry the *gal* genes. These particles are defective in that some genes have been left behind in the host; consequently, they are called *λdgal* (λ-defective *gal*). The *λdgal* particle has a λ body and can infect bacteria, but it is also defective in its integration site. The hybrid integration site left in *λdgal* does not provide a correct substrate for the enzyme that promotes recombination between the phage and bacterial integration sites. Therefore, efficient integration cannot occur in a single infection of *E. coli* by *λdgal*. However, coinfection with a wild-type phage results in efficient integration of the *λdgal* phage. The second phage is often termed a **helper phage** (10-37c). In general, a helper phage provides something that is required for integration into the host chromosome.

In practice, the lysate of λ produced originally (Figure 10-37b) can be used to infect a *gal⁻* recipient culture that is nonlysogenic for λ. These infected cells are plated on a minimal medium, and rare *gal⁺* transductants are the only ones to grow and produce colonies. A small percent of these transductants result from recombination between the *gal* regions of the *λdgal* and the recipient chromosome (Figure 10-37c, ii). The vast majority of the transductants are double lysogens (Figure 10-37c, i). If such a culture is lysed and used as a *gal⁺* donor in transduction, a very high frequency of transduction is obtained. This lysate is called a **high-frequency transduction (HFT)** lysate. HFT lysates contain a significant fraction of specialized *λdgal*-transducing phage, since each lysogen already contained a *λdgal* phage, and induction results in the excision and propagation of both *λdgal* and helper phage. This is in contrast to the original single lysogen (Figure 10-37a), which on induction yielded *λdgal* phage at a very low frequency (Figure 10-37b).

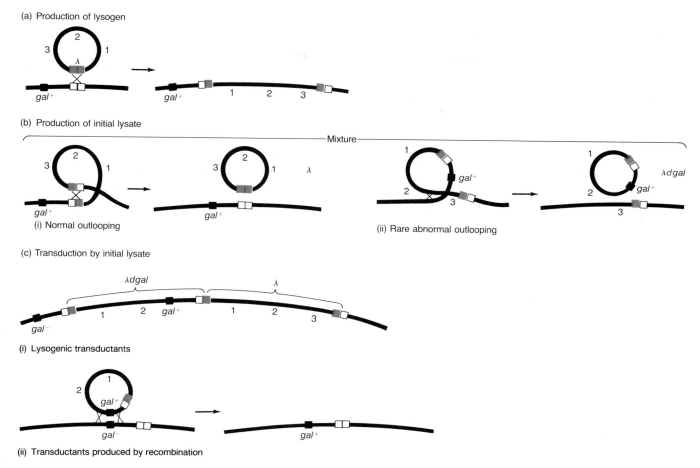

(a) Production of lysogen

(b) Production of initial lysate

Mixture

(i) Normal outlooping

(ii) Rare abnormal outlooping

(c) Transduction by initial lysate

λdgal

λ

(i) Lysogenic transductants

(ii) Transductants produced by recombination

Figure 10-37. Transduction mechanisms in phage λ. (a) The production of a lysogenic bacterium by crossing-over in a specialized region. (b) The lysogenic bacterial culture can produce normal λ or, rarely, an abnormal particle, λdgal, which is the transducing particle. (c) Transduction by the mixed lysate can produce either stable transductants, by crossovers flanking the *gal* gene, or unstable transductants, by the incorporation of λ and λdgal. The open double squares are bacterial integration sites, the colored double squares are λ-phage integration sites, and the pairs of open and colored double squares are hybrid integration sites, partly derived from the bacterial site and partly from the λ site.

Message Transduction occurs when bacteriophage pick up host genes prior to lysis. Generalized transduction is mediated by phage particles that have accidentally incorporated a piece of bacterial chromosome during phage packaging. This can occur in the lytic cycle of some virulent and temperate phages or when lysogenic temperate phages are induced to lysis. Specialized transduction is mediated by temperate phages, because their prophages always are inserted at one specific bacterial locus. In this case, the transducing phage is produced by faulty separation of the prophage from the bacterial chromosome, so that the prophage includes both bacterial and phage genes.

Map of Bacterial Chromosome

Some very detailed chromosome maps for bacteria have been obtained by combining the mapping techniques of interrupted mating, recombination mapping, transformation, and transduction. Today, new genetic markers are typically mapped first into a segment of about 10 to 15 map minutes by using a series of Hfr's that transfer from different points around the chromosome. This allows the selection of markers within the interval to be used for P1 cotransduction. The complexity of the *E. coli* map derived by 1963, which detailed the positions of approximately 100 genes, already illustrated the power and sophistication of genetic analysis at its best (Figure 10-38). After 20 years of further refinement, the 1983

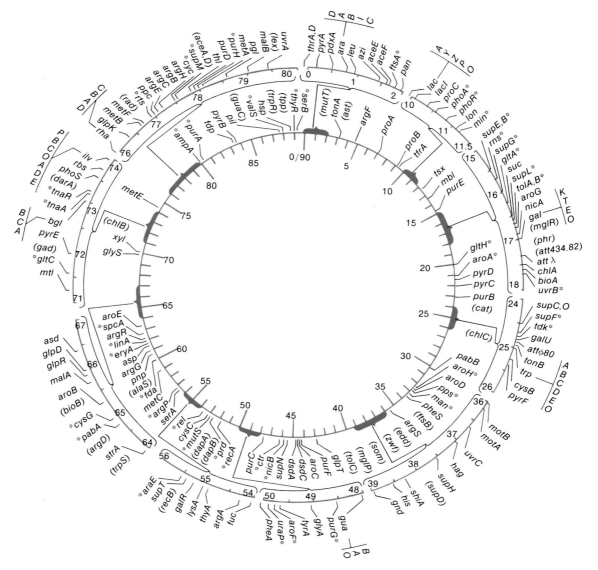

Figure 10-38. The 1963 genetic map of *E. coli*. Units are minutes (timed from an arbitrarily located origin). (From G. S. Stent, *Molecular Biology of Bacterial Viruses*. Copyright © 1963 by W. H. Freeman and Company.)

map, which is adjusted to a scale of 100 min, depicts the positions of more than 1000 genes! A portion of this map is shown in Figure 10-39. More bacterial and phage genetics are discussed in subsequent chapters.

Overview

1. Gene transfer in bacteria can be achieved through conjugation, transformation, and viral transduction.

2. Inheritance of genetic markers via the conjugative transfer of DNA by Hfr strains, transformation of portions of the donor chromosome, and generalized transduction all share one important property. Each

Figure 10-39. Linear scale drawings (pages 247–248) representing the circular map of *E. coli* K12 (Fig. 10-38). The time scale of 100 min, beginning arbitrarily with zero at the *thr* locus, is based on the results of interrupted-conjugation experiments. Parentheses around a gene symbol indicate that the position of the marker is not well known and may have been determined only within 5 to 10 min. An asterisk indicates that a marker has been mapped more precisely but that its position with respect to nearby markers is not known. Arrows beside genes and sets of genes (see Chapter 18) indicate the direction of transcription of these loci. Parentheses around a set of genes indicate that although the direction of transcription of the genes in the set is known, the orientation of the genes on the chromosome is not known. (From B. J. Bachmann, *Microbiological Reviews* 47, 1983, 180–230.)

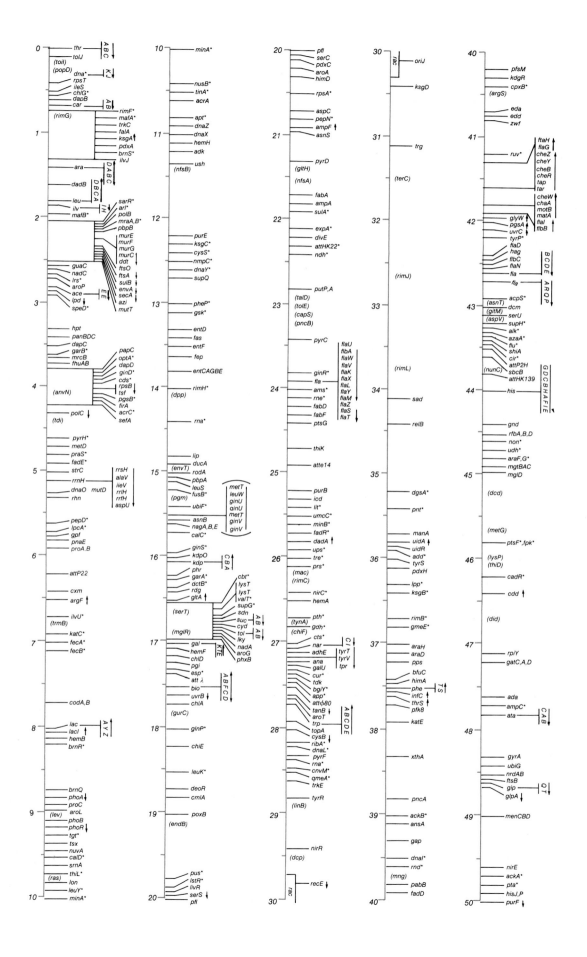

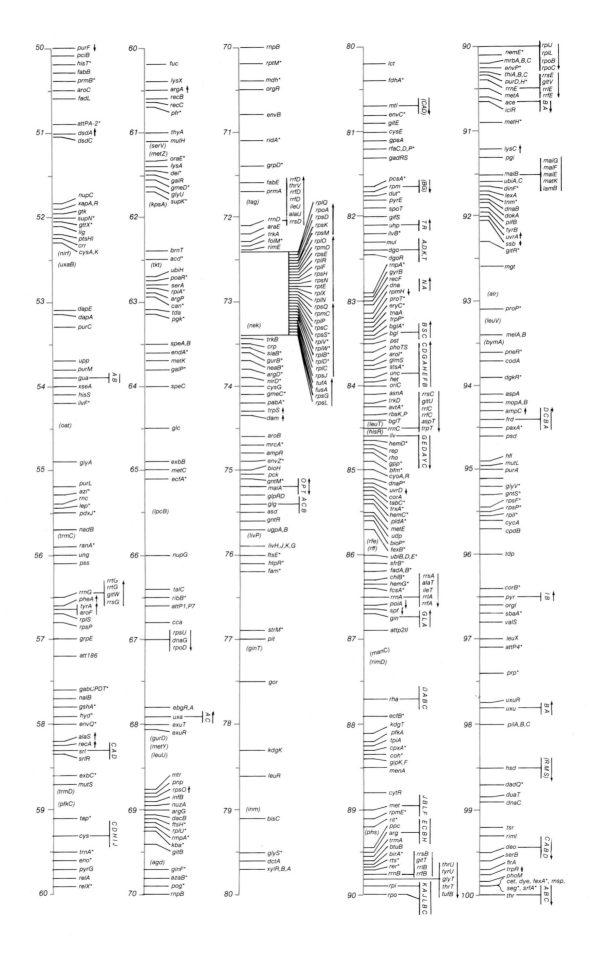

process introduces a DNA fragment into the recipient cell; then a double crossover event must occur if the fragment is to be incorporated into the recipient genome and subsequently inherited. Unincorporated fragments cannot replicate and are diluted out and lost from the population of daughter cells.

3. The conjugative transfer of F′ factors that carry bacterial genes and the specialized transduction of certain genetic markers are similar processes in that a specific and limited set of bacterial genes in each case is efficiently introduced into the recipient cell. Inheritance does not require normal recombination, as in the case of the inheritance of DNA fragments. After the F′ transfer, the F′ factor replicates in the bacterial cytoplasm as a separate entity. The specialized transducing phage DNA is recombined into the bacterial chromosome by a recombination system specific for that phage. In both cases, a partial diploid (merodiploid) results, because each process allows the inheritance of the transferred gene and also of the recipient's counterpart.

Summary

■ Advances in microbial genetics within the past four decades have provided the foundation for recent advances in molecular biology (discussed in the next several chapters). Early in this period, it was discovered that gene transfer and recombination occur between different mating types in several bacteria. In bacteria, however, genetic material is passed in only one direction — from a donor cell (F⁺ or Hfr) to a recipient cell (F⁻). Donor ability is determined by the presence in the cell of a fertility (F) factor acting as an episome.

On occasion, the F factor present in a free state in F⁺ cells can integrate into the *E. coli* chromosome and form an Hfr. When this occurs, gene transfer and subsequent recombination take place. Furthermore, since the F factor can insert at different places on the host chromosome, investigators were able to show that the *E. coli* chromosome is a single circle, or ring. Interruptions of the transfer at different times has provided geneticists with a new method for constructing a linkage map of the single chromosome of *E. coli* and other similar bacteria.

Genetic traits can also be transferred from one bacterial cell to another in the form of purified DNA. This process of transformation in bacterial cells was the first demonstration that DNA is the genetic material. For transformation to occur, DNA must be taken into a recipient cell and recombination between a recipient chromosome and the incorporated DNA then must take place.

Bacteria also have viral diseases, caused by bacteriophages. Phages can affect bacteria in two ways. The phage chromosome may enter the bacterial cell and, using the bacterial metabolic machinery, produce progeny phage that burst the host bacteria. The new phages can then infect other cells. If two phages of different genotypes infect the same host, recombination between their chromosomes can take place during this lytic process. Mapping the genetic loci through these recombinational events has led to the discovery that phage chromosomes can also be circular.

In another infection method, lysogeny, the injected phage lies dormant in the bacterial cell. In many cases, the prophage incorporates into the host chromosome and replicates with it. Either spontaneously or under appropriate stimulation, the dormant phage (prophage) can lyse the bacterial host cell.

Phages can carry bacterial genes from a donor to a recipient. In generalized transduction, pure host DNA is incorporated into the phage head during lysis. In restricted transduction, faulty outlooping of the prophage from a unique chromosome locus results in the inclusion of some host DNA in the phage head.

Figure 10-40 (page 250) summarizes the processes of transduction, transformation, and conjugation.

■ ■ ■

Solved Problems

1. In *E. coli*, four Hfr strains donate the genetic markers shown in the order given:

strain 1: Q W D M T

strain 2: A X P T M

strain 3: B N C A X

strain 4: B Q W D M

All of these Hfr strains are derived from the same F⁺ strain. What is the order of these markers on the circular chromosome of the original F⁺?

Solution

Recall the two-step approach that works well: (1) determine the underlying principle, and (2) draw a diagram. Here the principle is clearly that each Hfr strain donates

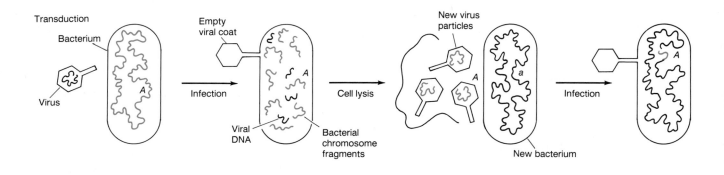

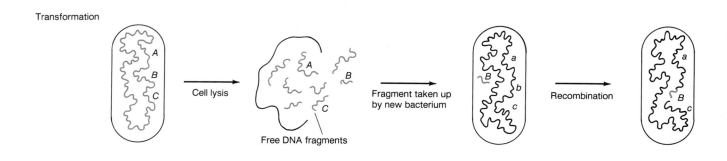

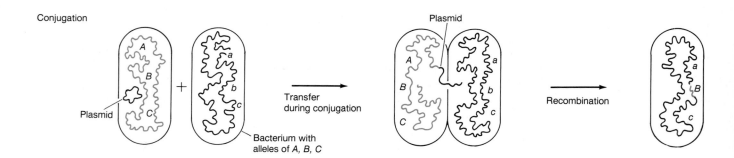

Figure 10-40. Recombination in bacteria requires the introduction into a bacterial cell of an allele obtained from another cell. In transduction, an infecting phage, or bacterial virus, picks up a bacterial–DNA segment carrying allele A and incorporates it, instead of viral DNA, into the virus particle. When such a particle infects another cell, the bacterial–DNA segment recombines with a homologous segment, thereby exchanging allele A for allele a. In transformation, a DNA segment bearing allele B is taken up from the environment by a cell with a chromosome that carries allele b; the alleles are then exchanged by homologous recombination. In conjugation, a plasmid inhabiting one bacterial cell can transfer the bacterium's chromosome (during cell-to-cell contact) to another cell with a chromosome that carries alleles of genes on the transferred chromosome; again, allele B is exchanged for allele b by recombination between homologous DNA segments. (From S. N. Cohen and J. A. Shapiro, "Transposable Genetic Elements." Copyright © 1980 by Scientific American, Inc. All rights reserved.)

genetic markers from a fixed point on the circular chromosome and that the earliest markers are donated with the highest frequency. Since not all markers are donated by each Hfr, only the early markers must be donated for each Hfr. Each strain allows us to draw the following circles:

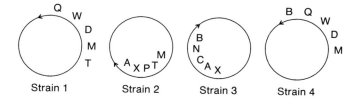

Strain 1   Strain 2   Strain 3   Strain 4

From this information, we can consolidate each circle into one circular linkage map of the order Q W D M T P X A C N B Q.

2. In an Hfr × F⁻ cross, leu^+ enters as the first marker, but the order of the other markers is unknown. If the Hfr is wild-type and the F⁻ is auxotrophic for each marker in question, what is the order of the markers in a cross where leu^+ recombinants are selected if 27 percent are ile^+, 13 percent are mal^+, 82 percent are thr^+, and 1 percent are trp^+?

Solution

Recall that spontaneous breakage creates a natural gradient of transfer, which makes it less and less likely for a recipient to receive later and later markers. Because we have selected for the earliest marker in this cross, the frequency of recombinants is a function of the order of entry for each marker. Therefore, we can immediately determine the order of the genetic markers simply by looking at the percentage of recombinants for any marker among the leu^+ recombinants. Because the inheritance of thr^+ is the highest, this must be the first marker to enter after leu. The complete order is $leu-thr-ile-mal-trp$.

3. A cross is made between an Hfr that is met^+ thr^+ pur^+ and an F⁻ that is met^- thi^- pur^-. Interrupted-mating studies show that met^+ enters the recipient last, so that met^+ recombinants in the F⁻ background are selected on a medium containing supplements that satisfy only the pur and thi requirements. These recombinants are tested for the presence of the thi^+ and pur^+ alleles. The following numbers of individuals are found for each genotype:

1. met^+ thi^+ pur^+ 280
2. met^+ thi^+ pur^- 0
3. met^+ thi^- pur^+ 6
4. met^+ thi^- pur^- 52

a. Why was methionine (met) left out of the selection medium?

b. What is the gene order?

c. What are the map distances in recombination units?

Solution

a. Methionine was left out of the medium to allow selection for met^+ recombinants, because met^+ is the last marker to enter the recipient. This ensures that all of the loci we are considering in the cross will have already entered each recombinant that we analyze.

b. Here it is helpful to diagram the possible gene orders. Since we know that met enters the recipient last, there are only two possible gene orders if the first marker enters on the right: $met-thi-pur$ or $met-pur-thi$. How can we distinguish between these two orders? Fortunately, one of the four possible classes of recombinants requires two additional crossovers. Each possible order predicts a different class that arises by four crossovers rather than two. For instance, if the order were $met-thi-pur$, then met^+ thi^- pur^+ recombinants would be very rare. On the other hand, if the order were $met-pur-thi$, then the four-crossover class would be met^+ pur^- thi^+. From the information given in the table, it is clear that the met^+ pur^- thi^+ class is the four-crossover class and therefore that the gene order $met-pur-thi$ is correct.

c. Refer to the following diagram:

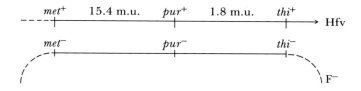

To compute the distance between met and pur, we compute the percentage of met^+ pur^- thi^-, which is $52/338 = 15.4$ m.u. The distance between pur and thi is, similarly, $6/338 = 1.8$ m.u.

4. Compare the mechanism of transfer and inheritance of the lac^+ genes in crosses with Hfr, F⁻, and F' lac^+ strains. How would an F⁻ cell that cannot undergo normal homologous recombination (rec^-) behave in crosses with each of these three strains? Would the cell be able to inherit the lac^+ genes?

Solution

Each of these three strains donates genes by conjugation. In the cases of the Hfr and F⁺ strains, the lac^+ genes on the host chromosome are donated. In the Hfr strain, the F factor is integrated into the chromosome in every cell, so that efficient donation of chromosomal markers can occur, particularly if the marker is near the integration site of F and is donated early. The F⁺ cell population contains a small percentage of cells in which F is integrated into the chromosome. These cells are responsible for the gene transfer displayed by cultures of F⁺ cells. In the cases of Hfr- and F⁺-mediated gene transfer, inheritance requires the incorporation of a transferred fragment by recombination (recall that two crossovers are needed) into the F⁻ chromosome. Therefore, an F⁻ strain that cannot undergo recombination cannot inherit donor chromosomal markers even though they are transferred by Hfr and F⁺ strains. The fragment cannot be incorporated into the chromosome by recombination. Since these fragments do not possess the ability to replicate within the F⁻ cell, they are rapidly diluted out during cell division.

Unlike Hfr and F⁺ cells, the F′ transfer of genes carried on the F′ factor does not require chromosome transfer. In this case, the *lac*⁺ genes are linked to the F′ and are transferred with the F′ at a high efficiency. In the F⁻ cell, no recombination is required, because the F′ *lac*⁺ strain can replicate and be maintained in the dividing F⁻ cell population. Therefore, the *lac*⁺ genes are inherited even in a *rec*⁻ strain.

5. Compare generalized and specialized transduction. How would a *rec*⁻ recipient be affected in its ability to inherit genes via generalized and specialized transduction?

Solution

Generalized transduction involves incorporating chromosomal fragments into phage heads, which then infect recipient strains. Fragments of the chromosome are incorporated randomly into phage heads, so that any marker on the bacterial host chromosome can be transduced to another strain by generalized transduction. On the other hand, specialized transduction involves the integration of the phage at a specific point on the chromosome and the rare incorporation of chromosomal markers near the integration site into the phage genome. Therefore, only those markers that are near the specific integration site of the phage on the host chromosome can be transduced.

Inheritance of markers occurs by different routes in generalized and specialized transduction. A generalized transducing phage injects a fragment of the donor chromosome into the recipient. This fragment must be incorporated into the chromosome by recombination, using the recipient recombination system. Therefore, a *rec*⁻ recipient will not be able to incorporate fragments of DNA and cannot inherit markers by generalized transduction. On the other hand, the major route for the inheritance of markers by specialized transduction involves integration of the specialized transducing particle into the host chromosome at the specific phage integration site. This integration, which sometimes requires an additional wild-type (helper) phage, is mediated by a phage-specific enzyme system that is independent of the normal recombination enzymes. Therefore, a *rec*⁻ recipient can still inherit genetic markers by specialized transduction.

Problems

1. A microbial geneticist isolates a new mutation in *E. coli* and wishes to map its chromosomal location. She uses interrupted-mating experiments with Hfr strains and generalized transduction experiments with phage P1. Explain why each technique, by itself, is insufficient for accurate mapping.

2. In *E. coli*, four Hfr strains donate the markers shown in the order given:

strain 1:   M   Z   X   W   C
strain 2:   L   A   N   C   W
strain 3:   A   L   B   R   U
strain 4:   Z   M   U   R   B

All of these Hfr strains are derived from the same F⁺ strain. What is the order of these markers on the circular chromosome of the original F⁺?

3. Four *E. coli* strains of genotype $a^+ b^-$ are labeled 1, 2, 3, and 4. Four strains of genotype $a^- b^+$ are labeled 5, 6, 7, and 8. The two genotypes are mixed in all possible combinations and (after incubation) are plated to determine the frequency of $a^+ b^+$ recombinants. The following results are obtained, where M = many recombinants, L = low numbers of recombinants, and O = no recombinants.

	1	2	3	4
5	O	M	M	O
6	O	M	M	O
7	L	O	O	M
8	O	L	L	O

On the basis of these results, assign a sex type (either Hfr, F⁺, or F⁻) to each strain.

4. An Hfr strain of genotype $a^+ b^+ c^+ d^- str^s$ is mated with a female strain of genotype $a^- b^- c^- d^+ str^r$. At various times, the culture is shaken violently to separate mating pairs. The cells are then plated on agar of the following three types, where nutrient A allows the growth of a^- cells; nutrient B, of b^- cells; nutrient C, of c^- cells; and nutrient D, of d^- cells (a plus indicates the presence of each nutrient, and a minus indicates its absence):

Agar type	str	A	B	C	D
1	+	+	+	−	+
2	+	−	+	+	+
3	+	+	−	+	+

a. What donor genes are being selected on each pair type of agar?

b. Table 10-6 shows the number of colonies on each type of agar for samples taken at various times after the strains are mixed. Use this information to determine the order of the genes *a*, *b*, and *c*.

c. From each of the 25-minute plates, 100 colonies are picked and transferred to a dish containing agar with all of the nutrients except D. The numbers of colonies that grow on this medium are 89 for the sample from agar type 1, 51 for the sample from agar type 2, and 8 for the sample from agar type 3. Using these data, fit gene *d* into the sequence of *a*, *b*, and *c*.

d. At what sampling time would you expect colonies to first appear on agar containing C and streptomycin but no A or B?

(Problem 4 is from D. Freifelder, *Molecular Biology and Biochemistry*. Copyright © 1978 by W. H. Freeman and Company.)

TABLE 10-6.

Time of sampling (minutes)	Number of colonies on agar of type		
	1	2	3
0	0	0	0
5	0	0	0
7.5	100	0	0
10	200	0	0
12.5	300	0	75
15	400	0	150
17.5	400	50	225
20	400	100	250
25	400	100	250

5. You are given two strains of *E. coli*. The Hfr strain is arg^+ ala^+ glu^+ pro^+ leu^+ T^s; the F^- strain is arg^- ala^- glu^- pro^- leu^- T^r. The markers are all nutritional except T, which determines sensitivity or resistance to phage T1. The order of entry is as given, with arg^+ entering the recipient first and T^s last. You find that the F^- strain dies when exposed to penicillin (pen^s) but the Hfr strain does not (pen^r). How would you locate the locus for *pen* on the bacterial chromosome with respect to *arg, ala, glu, pro,* and *leu?* Formulate your answer in logical, well-explained steps and draw explicit diagrams where possible.

6. A cross is made between Hfr arg^+ bio^+ leu^+ × F^- arg^- bio^- leu^-. Interrupted-mating studies show that arg^+ enters the recipient last, so that arg^+ recombinants are selected on a medium containing *bio* and *leu* only. These recombinants are tested for the presence of bio^+ and leu^+. The following numbers of individuals are found for each genotype:

$$arg^+\ bio^+\ leu^+\quad 320$$
$$arg^+\ bio^+\ leu^-\quad 8$$
$$arg^+\ bio^-\ leu^+\quad 0$$
$$arg^+\ bio^-\ leu^-\quad 48$$

a. What is the gene order?

b. What are the map distances in recombination units?

7. You make the following *E. coli* cross: Hfr Z_1^- ade^+ str^s × F^- Z_2^- ade^- str^r, in which *str* determines resistance or sensitivity to streptomycin, *ade* determines adenine requirement for growth, and Z_1 and Z_2 are two very close sites having Z^- alleles that cause an inability to use lactose as an energy source. After about an hour, the mixture is plated on a medium containing streptomycin, with glucose as the energy source. Many of the ade^+ colonies that grow are found to be capable of using lactose. However, hardly any of the ade^+ colonies from the reciprocal cross Hfr Z_2^- ade^+ str^s × F^- Z_1^- ade^- str^r are found to be capable of using lactose. What is the order of the Z_1 and Z_2 sites in relation to the *ade* locus? (Note that the *str* locus is terminal.)

8. Jacob selected eight closely linked *lac*⁻ mutations (called *lac*-1 through *lac*-8) and then attempted to order the mutations with respect to the outside markers *pro* (proline) and *ade* (adenine) by performing a pair of reciprocal crosses for each pair of *lac* mutants:

Cross A Hfr pro^- $lac\text{-}x$ ade^+ × F^- pro^+ $lac\text{-}y$ ade^-

Cross B Hfr pro^- $lac\text{-}y$ ade^+ × F^- pro^+ $lac\text{-}x$ ade^-

In all cases, prototrophs were selected by plating on a minimal medium with lactose as the only carbon source. Table 10-7 shows the number of colonies in the two crosses for each pair of mutants. Determine the relative order of the mutations.

(Problem 8 is from Burton S. Guttman, *Biological Principles.* Copyright © 1971, W. A. Benjamin, Inc., Menlo Park, California.)

TABLE 10-7.

x	y	Cross A	Cross B	x	y	Cross A	Cross B
1	2	173	27	1	8	226	40
1	3	156	34	2	3	24	187
1	4	46	218	2	8	153	17
1	5	30	197	3	6	20	175
1	6	168	32	4	5	205	17
1	7	37	215	5	7	199	34

9. Linkage maps in an Hfr bacterial strain are calculated in units of minutes (the number of minutes between genes indicates the length of time it takes for the second gene to follow the first after conjugation). In making such maps, microbial geneticists assume that the bacterial chromosome is transferred from Hfr to F^- at a constant rate. Thus, two genes separated by 10 minutes near the origin end are assumed to be the same *physical* distance apart as two genes separated by 10 minutes near the F-attachment end. Suggest a critical experiment to test the validity of this assumption.

10. In the cross Hfr aro^+ arg^+ ery^r str^s × F^- aro^- arg^- ery^s str^r, the markers are transferred in the order given (with aro^+ entering first), but the first three genes are very close together. Exconjugants are plated on a medium containing str (streptomycin, to contraselect Hfr cells), ery (erythromycin), arg (arginine), and aro (aromatic amino acids). The following results are obtained for 300 colonies from these plates isolated and tested for growth on various media: on ery only, 263 strains grow; on ery + arg, 264 strains grow; on ery + aro, 290 strains grow; on ery + arg + aro, 300 strains grow.

a. Draw up a list of genotypes, and indicate the number of individuals in each.

b. Calculate the recombination frequencies.

c.   Calculate the ratio of the size of the *arg*-to-*aro* region to the size of the *ery*-to-*arg* region.

11. A particular Hfr strain normally transmits the *pro*+ marker as the last one during conjugation. In a cross of this strain with an F⁻ strain, some *pro*+ recombinants are recovered early in the mating process. When these *pro*+ cells are mixed with F⁻ cells, the majority of the F⁻ cells are converted to *pro*+ cells that also carry the F factor. Explain these results.

12. F′ strains in *E. coli* are derived from Hfr strains. In some cases, these F′ strains show a high rate of integration back into the bacterial chromosome. Furthermore, the site of integration often is the same site that the sex factor occupied in the original Hfr strain (before production of the F′ strains). Explain these results.

13. You have two *E. coli* strains, F⁻ *str*ʳ *ala*⁻ and Hfr *str*ˢ *ala*⁺, in which the F factor is inserted close to *ala*⁺. Devise a screening test to detect F′ *ala*⁺ sexductants.

14. Five Hfr strains A – E are derived from a single F⁺ strain of *E. coli*. The following chart shows the entry times of the first five markers into an F⁻ strain when each is used in an interrupted conjugation experiment:

A	**B**	**C**	**D**	**E**
mal⁺ (1)	*ade*⁺ (13)	*pro*⁺ (3)	*pro*⁺ (10)	*his*⁺ (7)
*str*ˢ (11)	*his*⁺ (28)	*met*⁺ (29)	*gal*⁺ (16)	*gal*⁺ (17)
ser⁺ (16)	*gal*⁺ (38)	*xyl*⁺ (32)	*his*⁺ (26)	*pro*⁺ (23)
ade⁺ (36)	*pro*⁺ (44)	*mal*⁺ (37)	*ade*⁺ (41)	*met*⁺ (49)
his⁺ (51)	*met*⁺ (70)	*str*ˢ (47)	*ser*⁺ (61)	*xyl*⁺ (52)

a.   Draw a map of the F⁺ strain, indicating the positions of all genes and their distances apart in minutes.

b.   Show the insertion point and orientation of the F plasmid in each Hfr strain.

c.   In using each of these Hfr strains, state which gene you would select to obtain the highest proportion of Hfr exconjugants.

15. *Streptococcus pneumoniae* cells of genotype *str*ˢ *mtl*⁻ are transformed by donor DNA of genotype *str*ʳ *mtl*⁺ and (in a separate experiment) by a mixture of two DNAs with genotypes *str*ʳ *mtl*⁻ and *str*ˢ *mtl*⁺. Table 10-8 shows the results.

a.   What does the first line of the table tell you? Why?

b.   What does the second line of the table tell you? Why?

TABLE 10-8.

	Percentage of cells transformed to		
Transforming DNA	*str*ʳ *mtl*⁻	*str*ˢ *mtl*⁺	*str*ʳ *mtl*⁺
*str*ʳ *mtl*⁺	4.3	0.40	0.17
*str*ʳ *mtl*⁻ + *str*ˢ *mtl*⁺	2.8	0.85	0.0066

TABLE 10-9.

Drug(s) added	Number of colonies	Drug(s) added	Number of colonies
None	10,000	BC	51
A	1156	BD	49
B	1148	CD	786
C	1161	ABC	30
D	1139	ABD	42
AB	46	ACD	630
AC	640	BCD	36
AD	942	ABCD	30

16. A transformation experiment is performed with a donor strain that is resistant to four drugs: A, B, C, and D. The recipient is sensitive to all four drugs. The treated recipient-cell population is divided up and plated on media containing various combinations of the drugs. Table 10-9 shows the results.

a.   One of the genes obviously is quite distant from the other three, which appear to be tightly (closely) linked. Which is the distant gene?

b.   What is the probable order of the three tightly linked genes?

(Problem 16 is from Franklin Stahl, *The Mechanics of Inheritance*, 2d ed. Copyright © 1969, Prentice-Hall, Englewood Cliffs, New Jersey. Reprinted by permission.)

17. In the bacteriophage T4, gene *a* is 1.0 m.u. from gene *b*, which is 0.2 m.u. from gene *c*. The gene order is *a* – *b* – *c*. In a recombination experiment, you recover five double crossovers between *a* and *c* from 100,000 progeny viruses. Is it correct to conclude that interference is negative? Explain your answer.

18. You have infected *E. coli* cells with two strains of T4 virus. One strain is minute (*m*), rapid lysis, (*r*), and turbid (*tu*); the other is wild-type for all three markers. The lytic products of this infection are plated and classified. Of 10,342 plaques, the following numbers are classified as each genotype:

m r tu	3467	*m* + +	520
+ + +	3729	+ *r tu*	474
m r +	853	+ *r* +	172
m + *tu*	162	+ + *tu*	965

a.   Determine the linkage distances between *m* and *r*, between *r* and *tu*, and between *m* and *tu*.

b.   What linkage order would you suggest for the three genes?

c.   What is the coefficient of coincidence in this cross, and what does it signify?

(Problem 18 is reprinted with the permission of Macmillan Publishing Co., Inc., from Monroe W. Strickberger, *Genetics*. Copyright © 1968 by Monroe W. Strickberger.)

19. Using P22 as a generalized transducing phage grown on a $pur^+ pro^+ his^+$ bacterial donor, a recipient strain of genotype $pur^- pro^- his^-$ is infected and incubated. Later, transductants for each donor gene are selected individually. Transductants for pur^+, pro^+, and his^+ are selected in experiments I, II, and III, respectively.

   a. What media are used for these selection experiments?

   b. The transductants are examined for the presence of unselected donor markers, with the following results:

I		II		III	
$pro^- his^-$	87%	$pur^- his^-$	43%	$pur^- pro^-$	21%
$pro^+ his^-$	0%	$pur^+ his^-$	0%	$pur^+ pro^-$	15%
$pro^- his^+$	10%	$pur^- his^+$	55%	$pur^- pro^+$	60%
$pro^+ his^+$	3%	$pur^+ his^+$	2%	$pur^+ pro^+$	4%

   What is the order of the bacterial genes?

   c. Which two genes are closest together?

   d. On the basis of the order you proposed in (c), explain the relative proportions of genotypes observed in experiment II.

   (Problem 19 courtesy of D. Freifelder, *Molecular Biology and Biochemistry*. Copyright © 1978 by W. H. Freeman and Company.)

20. Although most λ-mediated gal^+ transductants are inducible lysogens, a small percentage of these transductants in fact are not lysogens (contain no integrated λ). Control experiments show that these transductants are not produced by mutation. What is the likely origin of these types?

21. An $ade^+ arg^+ cys^+ his^+ leu^+ pro^+$ bacterial strain is known to be lysogenic for a newly discovered phage, but the site of the prophage is not known. The bacterial map is

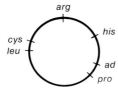

   The lysogenic strain is used as a source of the phage, and the phages are added to a bacterial strain of genotype $ade^- arg^- cys^- his^- leu^- pro^-$. After a short incubation, samples of these bacteria are plated on six different media, with the supplementations indicated in Table 10-10. The table also shows whether or not colonies were observed on the various media.

   a. What genetic process is at work here?

   b. What is the approximate locus of the prophage?

22. You have two strains of λ that can lysogenize *E. coli;* the following figure shows their linkage maps:

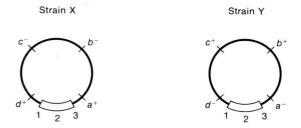

   The segment shown at the bottom of the chromosome, designated 1–2–3, is the region responsible for pairing and crossing-over with the *E. coli* chromosome. (Keep the markers on all your drawings.)

   a. Diagram the way in which strain X is inserted into the *E. coli* chromosome (so that the *E. coli* is lysogenized).

   b. It is possible to superinfect the bacteria that is lysogenic for strain X by using strain Y. A certain percentage of these superinfected bacteria become "doubly" lysogenic (that is, lysogenic for both strains). Diagram how this will occur. (Don't worry about how double lysogens are detected.)

   c. Diagram how the two λ prophages can pair.

   d. It is possible to recover crossover products between the two prophages. Diagram a crossover event and the consequences.

23. You have three strains of *E. coli*. Strain A is F′ $cys^+ trp1/cys^+ trp1$ (that is, both the F′ and the chromosome carry cys^+ and $trp1$, an allele for tryptophan requirement). Strain B is F⁻ $cys^+ trp2$ Z (this strain requires cysteine for growth and carries $trp2$, another allele causing a tryptophan requirement; strain B also is lysogenic for the generalized transducing phage Z). Strain C is F⁻ $cys^+ trp1$ (it is an F⁻ derivative of strain A that has lost the F′).

   a. How would you determine whether $trp1$ and $trp2$ are alleles of the same locus? (Describe the crosses and the results expected.)

TABLE 10-10.

Medium	Nutrient supplementation in medium						Presence of colonies
	ade	arg	cys	his	leu	pro	
1	−	+	+	+	+	+	N
2	+	−	+	+	+	+	N
3	+	+	−	+	+	+	C
4	+	+	+	−	+	+	N
5	+	+	+	+	−	+	C
6	+	+	+	+	+	−	N

NOTE: + indicates the presence of a nutrient supplement; − indicates no supplement present. N indicates no colonies; C indicates colonies present.

b. Suppose that *trp1* and *trp2* are allelic and that the *cys* locus is contransduced with the *trp* locus. Using phage Z to transduce genes from strain C to strain B, how would you determine the genetic order of *cys*, *trp1*, and *trp2*?

24. A generalized transducing phage is used to transduce an $a^- b^- c^- d^- e^-$ recipient strain of *E. coli* with an $a^+ b^+ c^+ d^+ e^+$ donor. The recipient culture is plated on various media with the results shown in Table 10-11. What can you conclude about the linkage and order of the genes?

25. In a generalized transduction system using P1 phage, the donor is $pur^+ nad^+ pdx^-$ and the recipient is $pur^- nad^- pdx^+$. The donor allele pur^+ is initially selected after transduction, and 50 pur^+ transductants are then scored for the other alleles present. The results follow:

Genotype	Number of colonies
$nad^+ pdx^+$	3
$nad^+ pdx^-$	10
$nad^- pdx^+$	24
$nad^- pdx^-$	13
	50

a. What is the cotransduction frequency for *pur* and *nad*?

b. What is the cotransduction frequency for *pur* and *pdx*?

c. Which of the unselected loci is closest to *pur*?

d. Are *nad* and *pdx* on the same side or on opposite sides of *pur*? Explain. (Draw the exchanges needed to produce the various transformant classes under either order to see which requires the minimum number to produce the results obtained.)

26. In a generalized transduction experiment, phages are collected from an *E. coli* donor strain of genotype $cys^+ leu^+ thr^+$ and used to transduce a recipient of genotype $cys^- leu^- thr^-$. Initially, the treated recipient population is plated on a minimal medium supplemented with leucine and threonine. Many colonies are obtained.

a. What are the possible genotypes of these colonies?

b. These colonies are then replica-plated onto three different media: (1) minimal plus threonine only, (2) minimal plus leucine only, and (3) minimal. What genotypes could, in theory, grow on these three media?

c. It is observed that 56 percent of the original colonies grow on (1), 5 percent grow on (2), and no colonies grow on (3). What are the actual genotypes of the colonies on (1), (2), and (3)?

d. Draw a map showing the order of the three genes and which of the two outer genes is closer to the middle gene.

27. In 1965, Jon Beckwith and Ethan Signer devised a method of obtaining specialized transducing phages carrying the *lac* region. In a two-step approach, the researchers first "transposed" the *lac* genes to a new region of the chromosome and then isolated the specialized transducing particles. They noted that the integration site for the temperate phage $\phi80$ (a relative of phage λ), designated *att80*, was located near one of the genes involved in conferring resistance to the virulent phage T1, termed *tonB*:

Beckwith and Signer used an F'*lac* episome that could not replicate at high temperatures in a strain carrying a deletion of the *lac* genes. By forcing the cell to remain *lac*⁺ at high temperatures, the researchers could select strains in which the episome had integrated into the chromosome, thereby allowing the F'*lac* to be maintained at high temperatures. By combining this selection with a simultaneous selection for resistance to T1 phage infection, they found that the only survivors were cells in which the F'*lac* had integrated into the *tonB* locus, as shown in the accompanying figure. Can you see why?

This placed the *lac* region near the integration site for phage $\phi80$. Describe the subsequent steps that the researchers must have followed to isolate the specialized transducing particles of phage $\phi80$ that carried the *lac* region.

TABLE 10-11.

Compounds added to minimal medium	Presence (+) or absence (−) of colonies
C D E	−
B D E	−
B C E	+
B C D	+
A D E	−
A C E	−
A C D	−
A B E	−
A B D	+
A B C	−

NOTE: the allele a^- determines a requirement for A as a nutrient, and so forth.

The Structure of DNA

KEY CONCEPTS

Bacterial cells that express one phenotype can be transformed into cells that express a different phenotype; the transforming agent is DNA.

·

Experiments with labeled T2 phage have established that DNA is the hereditary material.

·

James Watson and Francis Crick showed that the structure of DNA is a double helix, in which each helix is a chain of nucleotides held together by phosphodiester bonds and in which specific hydrogen bonds are formed by pairs of bases.

·

The DNA structure suggests that the fidelity of replication can be ensured if the complementary base of each base is specified by hydrogen bonding.

·

The replication of DNA is semiconservative in that each daughter duplex contains one parental and one newly synthesized strand.

·

Many of the enzymes involved in DNA synthesis in bacteria have been characterized.

■ Until now, we have looked at genes as abstract entities that somehow control hereditary traits. Through purely genetic analysis, we have studied the inheritance of different genes. But what about the physical nature of the gene? This question puzzled scientists for many years until it was realized that genes are composed of deoxyribonucleic acid (abbreviated DNA) and that DNA has a fascinating structure.

The elucidation of the structure of DNA in 1953 by James Watson and Francis Crick was one of the most exciting discoveries in the history of genetics. It paved the way for the understanding of gene action and heredity in molecular terms. Before we see how the solution of DNA structure was achieved, let's review what was known about genes and DNA at the time that Watson and Crick began their historic collaboration:

1. Genes—the hereditary "factors" described by Mendel—were known to be associated with specific character traits, but their physical nature was not understood.

2. The one-gene–one enzyme theory (described more fully in Chapter 12) postulated that genes control the structure of proteins.

3. Genes were known to be carried on chromosomes.

4. The chromosomes were found to consist of DNA and protein.

5. Research by Frederick Griffith and, subsequently, by Oswald Avery and his coworkers pointed to DNA as the genetic material. These experiments, described here, showed that bacterial cells that express one phenotype can be transformed into cells that express a different phenotype and that the transforming agent is DNA.

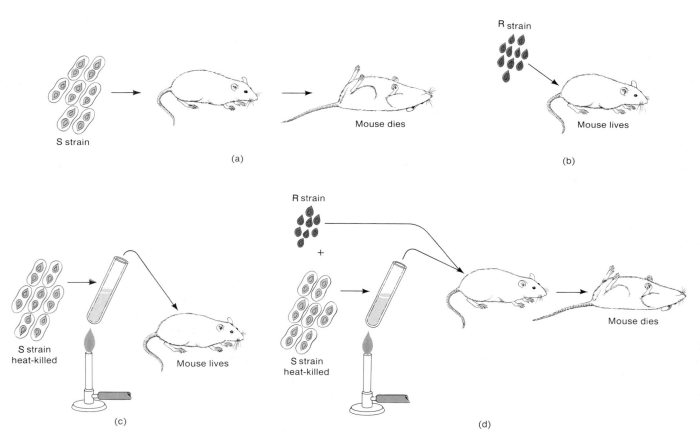

Figure 11-1. The first demonstration of bacterial transformation. (a) Mouse dies after injection with the virulent S strain. (b) Mouse survives after injection with the R strain. (c) Mouse survives after injection with heat-killed S strain. (d) Mouse dies after injection with a mixture of heat-killed S strain and live R strain. The heat-killed S strain somehow transforms the R strain to virulence. Parts (a), (b), and (c) act as control experiments for this demonstration. (From G. S. Stent and R. Calendar, *Molecular Genetics,* 2d ed. Copyright 1978 by W. H. Freeman and Company. After R. Sager and F. J. Ryan, *Cell Heredity,* John Wiley, 1961.)

DNA: The Genetic Material

The Discovery of Transformation

A puzzling observation was made by Frederick Griffith in the course of experiments on the bacterium *Streptococcus pneumoniae* in 1928. This bacterium, which causes pneumonia in humans, is normally lethal in mice. However, different strains of this bacterial species have evolved that differ in virulence (in the ability to cause disease or death). In his experiments, Griffith used two strains that are distinguishable by the appearance of colonies grown in laboratory cultures. In one strain, a normal virulent type, the cells are enclosed in a polysaccharide capsule, giving colonies a smooth appearance; hence, this strain is labeled S. In Griffith's other strain, a mutant nonvirulent type that grows in mice but is not lethal, the polysaccharide coat is absent, giving colonies a rough appearance; this strain is called R.

Griffith killed some virulent cells by boiling them and injected the heat-killed cells into mice. The mice survived, showing that the carcasses of the cells do not cause death. However, mice injected with a mixture of heat-killed virulent cells and live nonvirulent cells died. Furthermore, live cells could be recovered from the dead mice; these cells gave smooth colonies and were virulent on subsequent injection. Somehow, the cell debris of the boiled S cells had converted the live R cells into live S cells. The process is called **transformation.** Griffith's experiment is summarized in Figure 11-1.

This same basic technique was also utilized to determine the nature of the **"transforming principle"** — the agent in the cell debris that is specifically responsible for transformation. In 1944, Oswald Avery, C. M. MacLeod, and M. McCarty separated the classes of molecules found in the debris of the dead S cells and tested them for transforming ability, one at a time. These tests showed first that the polysaccharides themselves do not transform the rough cells. Therefore, the polysaccharide coat, although undoubtedly concerned with the pathogenic reaction, is only the phenotypic expression of virulence. In screening the different groups, Avery and his colleagues found that only one class of molecules, DNA, induced transformation of R cells (Figure 11-2). They deduced that DNA is the agent that determines the polysaccharide character and hence the pathogenic character. Furthermore, it seems that providing R cells with S DNA is tantamount to providing these cells with S genes!

Message The demonstration that DNA is the transforming principle was the first demonstration that genes are composed of DNA.

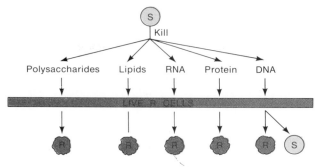

Figure 11-2. Demonstration that DNA is the transforming agent. DNA is the only agent that produces smooth (S) colonies when added to live rough (R) cells.

The Hershey–Chase Experiment

The experiments conducted by Avery and his colleagues were definitive, but many scientists were very reluctant to accept DNA (rather than proteins) as the genetic material. The clincher was provided in 1952 by Alfred Hershey and Martha Chase using the phage (virus)T2. They reasoned that phage infection must involve the introduction into the bacterium of the specific information that dictates viral reproduction. The phage is relatively simple in molecular constitution. Most of its structure is protein, with DNA contained inside the protein sheath of its "head."

Phosphorus is not found in proteins but is an integral part of DNA; conversely, sulfur is present in proteins but never in DNA. Hershey and Chase incorporated the radioisotope of phosphorus (^{32}P) into phage DNA and that of sulfur (^{35}S) into the proteins of a separate phage culture. They then used each phage culture independently to infect *E. coli* with many virus particles per cell. After sufficient time for injection to occur, they sheared the empty phage carcasses (called "ghosts") off the bacterial cells by agitation in a kitchen blender. They used centrifugation to separate the bacterial cells from the phage ghosts and then measured the radioactivity in the two fractions. When the ^{32}P-labeled phages were used, most of the radioactivity ended up inside the bacterial cells, indicating that the phage DNA entered the cells. ^{32}P can also be recovered from phage progeny. When the ^{35}S-labeled phages were used, most of the radioactive material ended up in the phage ghosts, indicating that the phage protein never entered the bacterial cell (Figure 11-3). The conclusion is inescapable: DNA is the hereditary material; the phage proteins are mere structural packaging that is discarded after delivering the vital DNA to the bacterial cell.

Why such reluctance to accept this conclusion? DNA was thought to be a rather simple chemical. How could all the information about the wondrously variable protein structures (with their sequences of the 20 amino

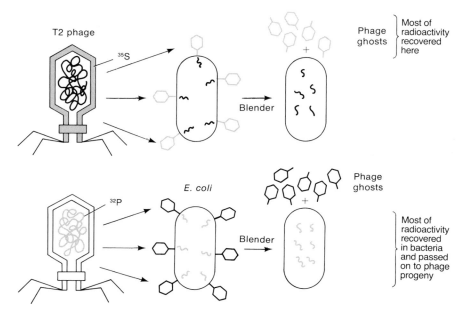

Figure 11-3. The Hershey-Chase experiment, which demonstrated that the genetic material of phage is DNA, not protein. The experiment uses two sets of T2 bacteriophage. In one set, the protein coat is labeled with radioactive sulfur (^{35}S) not found in DNA. In the other set, the DNA is labeled with radioactive phosphorus (^{32}P) not found in protein. Only the ^{32}P is injected into the *E. coli*, indicating that DNA is the agent necessary for the production of new phages.

acids) be stored in such a simple molecule? How could such information be passed on from one generation to the next? Clearly, the genetic material must have both the ability to encode specific information and the capacity to duplicate that information precisely. What kind of structure could allow such complex functions in so simple a molecule?

The Structure of DNA

Although the DNA structure was not known, the basic building blocks of DNA had been known for many years. The basic elements of DNA had been isolated and determined by partly breaking up purified DNA. These studies showed that DNA is composed of only four basic molecules called **nucleotides,** which are identical except that each contains a different nitrogen base. Each nucleotide contains phosphate, sugar (of the deoxyribose type), and one of the four bases (Figure 11-4). In the absence of the phosphate group, the base and the deoxyribose form a **nucleoside** rather than a nucleotide. The four bases are **adenine, guanine, cytosine,** and **thymine.** The full chemical names of the nucleotides are deoxyadenosine 5′-monophosphate (or deoxyadenylate, or dAMP), deoxyguanosine 5′-monophosphate

(or deoxyguanylate or dGMP), deoxycytidine 5′-monophosphate (or deoxycytidylate, or dCMP), and deoxythymidine 5′-monophosphate (or deoxythymidylate, or dTMP). However, it is more convenient just to refer to each nucleotide by the abbreviation of its base (A, G, C, and T, respectively.) Two of the bases, adenine and guanine, are similar in structure and are called **purines.** The other two bases, cytosine and thymine, also are similar and are called **pyrimidines.**

After the central role of DNA in heredity became clear, many scientists set out to determine the exact structure of DNA. How can a molecule with such a limited range of different components possibly store the vast range of information about all the protein primary structures of the living organism? The first to succeed in finding a reasonable DNA structure—Watson and Crick in 1953—worked from two kinds of clues. First, other researchers had amassed a lot of X-ray diffraction data on DNA structure. In such experiments, X rays are fired at DNA fibers, and the scatter of the rays from the fiber is observed by catching them on photographic film, where the X rays produce spots. The angle of scatter represented by each spot on the film gives information about the position of an atom or certain groups of atoms in the DNA molecule. This procedure is not simple to carry out (or to explain), and the interpretation of the

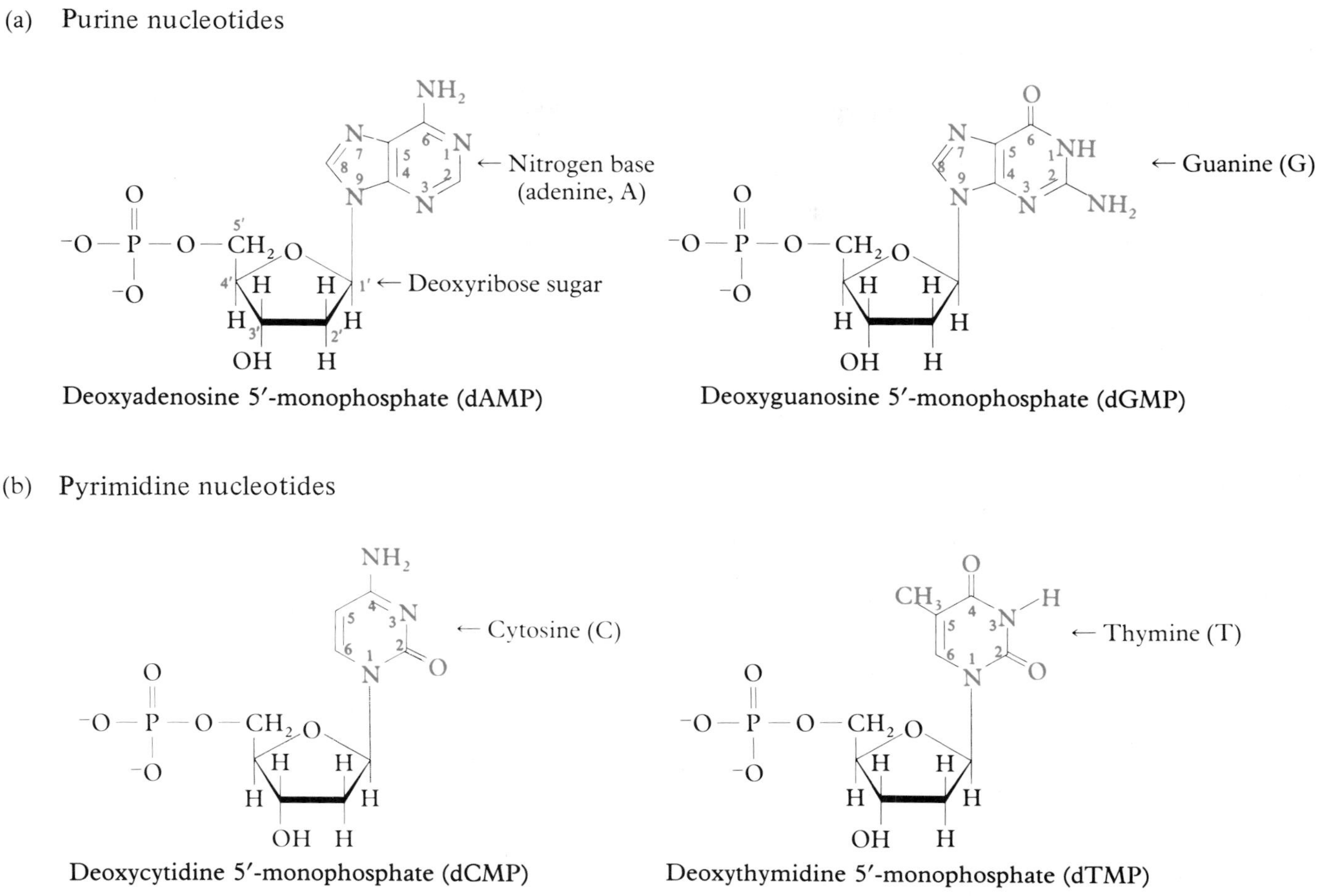

(a) Purine nucleotides

Deoxyadenosine 5′-monophosphate (dAMP)

Deoxyguanosine 5′-monophosphate (dGMP)

(b) Pyrimidine nucleotides

Deoxycytidine 5′-monophosphate (dCMP)

Deoxythymidine 5′-monophosphate (dTMP)

Figure 11-4. Chemical structure of the four nucleotides (two with purine bases and two with pyrimidine bases) that are the fundamental building blocks of DNA. The sugar is called deoxyribose because it is a variation of a common sugar, ribose, that has one more oxygen atom.

spot patterns is very difficult. The available data suggested that DNA is long and skinny and that it has two similar parts that are parallel to one another and run along the length of the molecule. The X-ray data showed the molecule to be helical (spiral-like). Other regularities were present in the spot patterns, but no one had yet thought of a three-dimensional structure that could account for just those spot patterns.

The second set of clues available to Watson and Crick came from work done several years earlier by Erwin Chargaff. Studying a large selection of DNAs from different organisms (see Table 11-1), Chargaff established certain empirical rules about the amounts of each component of DNA:

1. The total amount of pyrimidine nucleotides (T + C) always equals the total amount of purine nucleotides (A + G).

2. The amount of T always equals the amount of A, and the amount of C always equals the amount of G. But the amount of A + T is not necessarily equal to the amount of G + C, as can be seen in the last column of Table 11-1. This ratio varies among different organisms.

The Double Helix

The structure that Watson and Crick derived from these clues is a **double helix**, which looks rather like two interlocked bedsprings. Each bedspring (helix) is a chain of nucleotides held together by **phosphodiester bonds,** in which a phosphate group forms a bridge between —OH groups on two adjacent sugar residues. The two bedsprings (helices) are held together by **hydrogen bonds,** in which two electronegative atoms "share" a

TABLE 11-1. Molar properties of bases (as moles of nitrogenous constituents per 100 g-atoms phosphate in hydrolysate) in DNAs from various sources

Organism	Tissue	Adenine	Thymine	Guanine	Cytosine	$\dfrac{A+T}{G+C}$
Escherichia coli (K12)	—	26.0	23.9	24.9	25.2	1.00
Diplococcus pneumoniae	—	29.8	31.6	20.5	18.0	1.59
Mycobacterium tuberculosis	—	15.1	14.6	34.9	35.4	0.42
Yeast	—	31.3	32.9	18.7	17.1	1.79
Paracentrotus lividus (sea urchin)	Sperm	32.8	32.1	17.7	18.4	1.85
Herring	Sperm	27.8	27.5	22.2	22.6	1.23
Rat	Bone marrow	28.6	28.4	21.4	21.5	1.33
Human	Thymus	30.9	29.4	19.9	19.8	1.52
Human	Liver	30.3	30.3	19.5	19.9	1.53
Human	Sperm	30.7	31.2	19.3	18.8	1.62

SOURCE: E. Chargaff and J. Davidson, eds., *The Nucleic Acids* (New York: Academic Press, 1955).

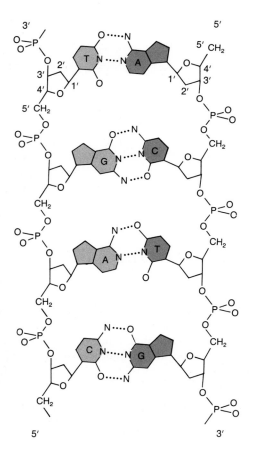

Figure 11-5. The DNA double helix, unrolled to show the sugar–phosphate backbones and base-pair rungs. The backbones run in opposite directions; the 5′ and 3′ ends are named for the orientation of the 5′ and 3′ carbon atoms of the sugar rings. Each base pair has one purine base, adenine (A) or guanine (G), and one pyrimidine base, thymine (T) or cytosine (C), connected by hydrogen bonds *(dotted lines)*. (Figures 11-5, 11-7, 11-11, 11-12, and 11-13 from R. E. Dickerson, "The DNA Helix and How It Is Read." Copyright © 1983 by Scientific American, Inc. All rights reserved.)

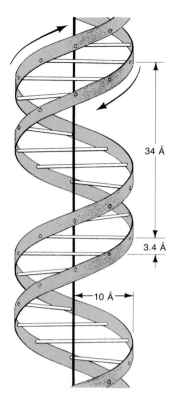

Figure 11-6. A simplified model showing the helical structure of DNA. The sticks represent base pairs, and the ribbons represent the sugar-phosphate backbones of the two antiparallel chains. In fact, the base pairs are more like flat "steps" than the round sticks shown here.

proton, between the bases. The hydrogen bonds are formed by pairs of bases, as can be seen in Figure 11-5, which shows a part of this structure with the helices uncoiled. Each base pair consists of one purine base and one pyrimidine base, paired according to the following rule: G pairs with C, and A pairs with T. In Figure 11-6, a simplified picture of the coiling, each of the base pairs is represented by a "stick" between the "ribbons", or so-called "sugar–phosphate backbones" of the chains. In Figure 11-5, note that the two backbones run in opposite directions; they are said to be **antiparallel,** and (for reasons apparent in the figure) one is called $5' \rightarrow 3'$ and the other $3' \rightarrow 5'$.

The double helix accounted nicely for the X-ray data and also tied in very nicely with Chargaff's data. Studying models they made of the structure, Watson and Crick realized that the observed radius of the double helix (known from the X-ray data) would be explained if a purine base always pairs (by hydrogen bonding) with a pyrimidine base (Figure 11-7). Such pairing would account for the $(A + G) = (T + C)$ regularity observed by

Chargaff, but it would predict four possible pairings: $T \cdots A$, $T \cdots G$, $C \cdots A$, and $C \cdots G$. Chargaff's data, however, indicate that T pairs only with A and C pairs only with G.

Watson and Crick showed that only these two pairings have the necessary complementary "lock-and-key" shapes to permit efficient hydrogen bonding. Hydrogen bonds occur between hydrogen atoms with a small positive charge and acceptor atoms with a small negative charge. For example

Each hydrogen atom in the NH_2 group is slightly positive (δ^+) because the nitrogen atom tends to attract the electrons involved in the N—H bond, thereby leaving the hydrogen atom slightly short of electrons. The oxygen atom has six unbonded electrons in its outer shell, which form an electron cloud around it, making it slightly negative (δ^-). A hydrogen bond forms between one H and the O. Hydrogen bonds are quite weak (only about 3 percent of the strength of a covalent chemical bond), but this weakness (as we shall see) plays an important role in the function of the DNA molecule in heredity. One further important chemical fact: the hydrogen bond is much stronger if the participating atoms are "pointing at each other" in the ideal orientations.

Looking at the hydrogen-bonding potential between the various purine-pyrimidine pairs, we find that only two pairs have the necessary arrangement of δ^+ hydrogen atoms and δ^- acceptor atoms. The A–T pair and the G–C pairs show a beautiful lock-and-key fit

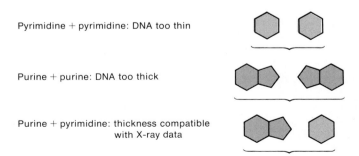

Figure 11-7. The pairing of purines with pyrimidines accounts exactly for the diameter of the DNA double helix determined from X-ray data.

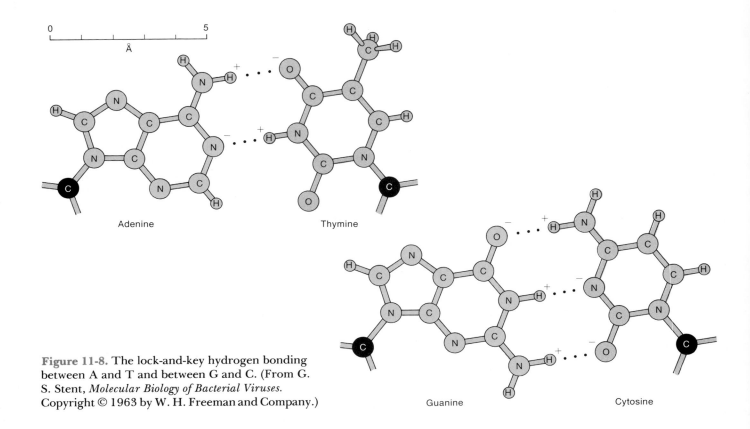

Figure 11-8. The lock-and-key hydrogen bonding between A and T and between G and C. (From G. S. Stent, *Molecular Biology of Bacterial Viruses.* Copyright © 1963 by W. H. Freeman and Company.)

Adenine

Thymine

Guanine

Cytosine

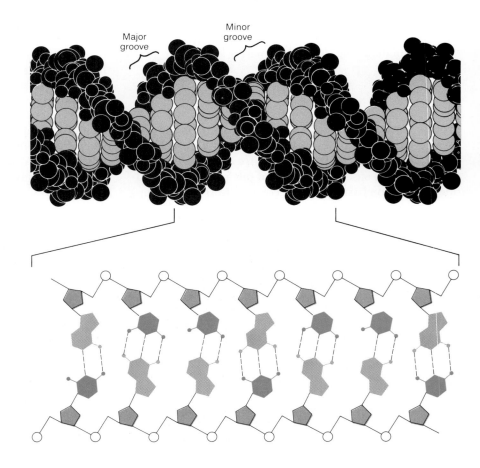

Major groove

Minor groove

Figure 11-9. *(top)* A space-filling model of the DNA double helix. *(bottom)* An unwound representation of a short stretch of nucleotide pairs, showing how A – T and G – C pairing produces the Chargaff ratios. This model is of one of several forms of DNA, termed the B form (see Figure 11-10). (Space-filling model from C. Yanofsky, "Gene Structure and Protein Structure." Copyright © 1967 by Scientific American, Inc. All rights reserved. Unwound structure based on A. Kornberg, "The Synthesis of DNA." Copyright © 1968 by Scientific American, Inc. All rights reserved.)

(Figure 11-8), providing just the proper "width" of the base pair to explain the known radius of the DNA double helix.

Note that the G–C pair has three hydrogen bonds, whereas the A–T pair has only two. We would predict that DNA containing many G–C pairs would be more stable than DNA containing many A–T pairs. In fact, this prediction is confirmed. We now have a neat explanation for the Chargaff's data in terms of DNA structure (Figure 11-9). We also have a structure that is consistent with the X-ray data.

Three-dimensional View of the Double Helix

In three dimensions, the bases actually form rather flat structures (more like steps in a ladder than like the sticks shown in Figure 11-6), and these flat bases partially stack

on top of one another in the twisted structure of the double helix. This stacking of bases adds tremendously to the stability of the molecule by excluding water molecules from the spaces between the base pairs. (This phenomenon is very much like the stabilizing force that you can feel when you squeeze two plates of glass together underwater and then try to separate them.) Subsequently, it was realized that there were two forms of DNA in the fiber analyzed by diffraction. The A form is less hydrated than the B form. Figure 11-10 shows both the A and B forms in a schematic three-dimensional drawing. Note the stacking of bases as represented in this diagram. It is believed that the B form of DNA is the form found most frequently in living cells. The stacking of the base pairs in the double helix results in two grooves in the sugar-phosphate backbones. These are termed the **major** and **minor** grooves and can be readily

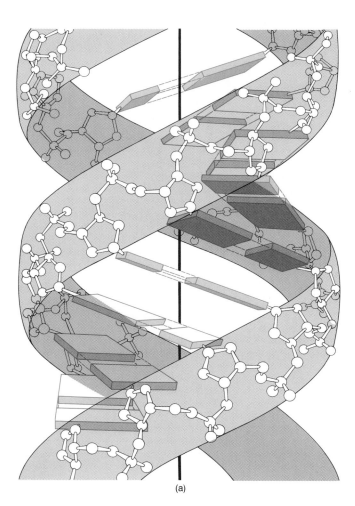

(a)

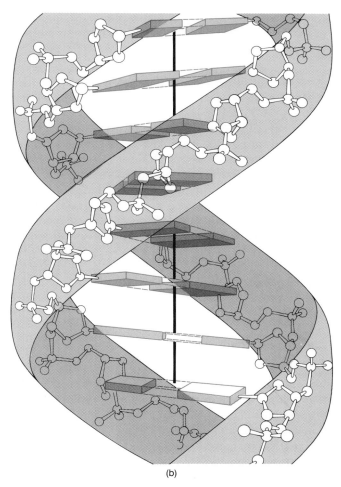

(b)

Figure 11-10. Schematic drawings of the structures of two forms of DNA. (a) A DNA. (b) B DNA. Analyses of DNA fibers by Struther Arnott and others provided the conceptual basis for these drawings. The sugar-phosphate backbones of the double helix are represented as ribbons

and the runglike base pairs connecting them as planks. In A DNA, the base pairs are tilted and are pulled away from the axis of the double helix. In B DNA, on the other hand, the base pairs sit astride the helix axis and are perpendicular to it.

seen in the space-filling (three-dimensional) model in Figure 11-9.

Implications of DNA Structure

Elucidation of the structure of DNA caused a lot of excitement in genetics (and in all areas of biology) for two basic reasons:

1. The structure suggests an obvious way in which the molecule can be **duplicated,** or **replicated,** since each base can specify its complementary base by hydrogen bonding. This essential property of a genetic molecule had been a mystery until this time.

2. The structure suggests that perhaps the *sequence* of nucleotide pairs in DNA is dictating the sequence of amino acids in the protein organized by that gene. In other words, some sort of **genetic code** may write information in DNA as a sequence of nucleotide pairs and then translate it into a different language of amino acid sequences in protein.

This basic information about DNA is now familiar to almost anyone who has read a biology text in elementary or high school, or even magazines and newspapers. It may seem trite and obvious, but try to put yourself back into the scene in 1953 and imagine the excitement! Until then, the evidence that the uninteresting DNA is the genecmlcl ed disappointing and discouraging. But the Watson-Crick structure of DNA suddenly opened up the possibility of explaining two of the biggest "secrets" of life. James Watson has told the story of this discovery (from his own point of view, strongly questioned by others involved) in a fascinating book called *The Double Helix,* which reveals the intricate interplay of personality clashes, clever insights, hard work, and simple luck in such important scientific advances.

Alternate Structures

In addition to the *A* and *B* forms of DNA, a new form has been found in crystals of synthetically prepared DNA that contain alternating purines and pyrimidines on the same strand. This *Z* DNA form has a zigzag-like backbone and generates a left-handed helix, whereas both *A* and *B* DNA form right-handed helices. The perspective drawings in Figures 11-11, 11-12, and 11-13 allow us to compare these three forms, on the basis of precise information obtained from crystal structures of short synthetic DNA sequences called **oligonucleotides.** Whether *Z* DNA occurs naturally in cells is currently under debate.

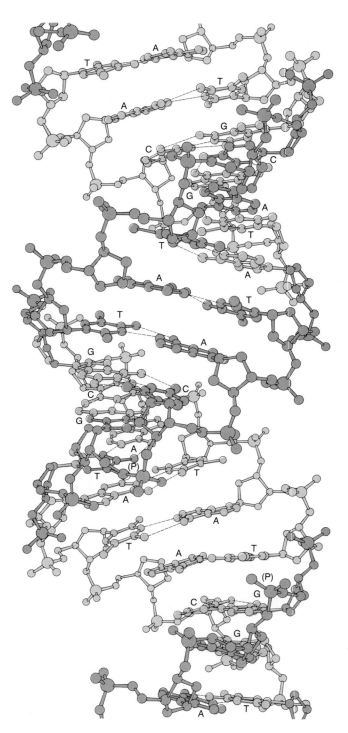

Figure 11-11. The *A*-DNA helix. This perspective drawing was generated from repetition of the central six bases of the octamer GGTATACC. Note how phosphate groups (P) on opposite chains face each other across the major groove.

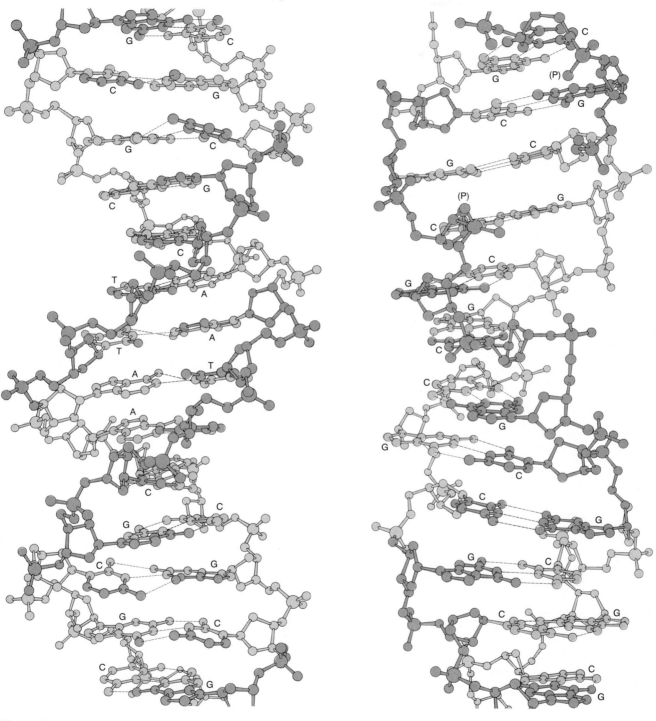

Figure 11-12. The *B*-DNA helix. This perspective drawing was generated from repetition of the central 10 base pairs of the dodecamer CGCGAATTCGCG. Note the large twist relative to *A* DNA and how the twist improves the stacking of the bases along each backbone chain.

Figure 11-13. The *Z*-DNA helix. This drawing depicts the structure as a left-handed helix of alternating guanines and cytosines, generated from the central four base pairs of CGCGCG. Phosphate groups on different chains now face each other across the deep minor groove.

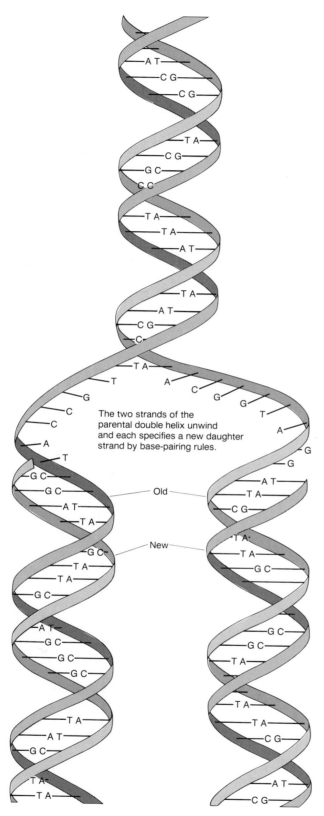

The two strands of the parental double helix unwind and each specifies a new daughter strand by base-pairing rules.

Old

New

Figure 11-14. The model of DNA replication proposed by Watson and Crick is based on the hydrogen-bonding specificity of the base pairs. Complementary strands are shown in different colors.

Replication of DNA

Semiconservative Replication

Figure 11-14 diagrams the possible mechanism for DNA replication proposed by Watson and Crick. Here the sugar–phosphate backbones are represented by lines and the sequence of base pairs is random. Let's imagine that the double helix is like a zipper that unzips starting at one end (the top in this figure). We can see that if this zipper analogy is valid, the unwinding of the two strands will expose single bases of either strand. Because the pairing requirements imposed by the DNA structure are strict, each exposed base will pair only with its complementary base. Due to this base complementarity, each of the two single strands will act as a **template,** or mold, and will begin to reform a double helix identical to the one from which it was unzipped. The newly added nucleotides are assumed to come from a pool of free nucleotides that must be present in the cell.

If this model is correct, then each daughter molecule should contain one parental nucleotide chain (black line in Figure 11-14) and one newly synthesized nucleotide chain (color line). This prediction has been tested in both prokaryotes and eukaryotes. A little thought shows that there are at least three different ways in which a parental DNA molecule might be related to the daughter molecules. These hypothetical modes are called semiconservative (the Watson-Crick model), conservative, and dispersive (Figure 11-15). In **semiconservative** replication, each daughter duplex contains one pa-

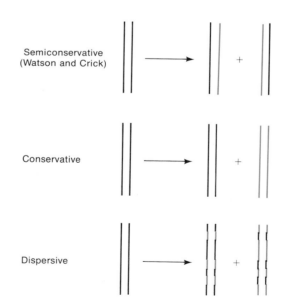

Semiconservative (Watson and Crick)

Conservative

Dispersive

Figure 11-15. Three alternative patterns for DNA replication. The Watson-Crick model would produce the first (semiconservative) pattern. Colored lines represent the newly synthesized strands.

rental and one newly synthesized strand. However, in **conservative** replication, one daughter duplex consists of two newly synthesized strands and the parent duplex is conserved. **Dispersive** replication results in daughter duplexes that consist of strands containing only *segments* of parental DNA and newly synthesized DNA.

The Meselson–Stahl Experiment

In 1958, Matthew Meselson and Franklin Stahl set out to distinguish among these possibilities in an experiment. They grew *E. coli* cells in a medium containing the heavy isotope of nitrogen (^{15}N) rather than the normal light (^{14}N) form. This isotope was inserted into the nitrogen bases, which then were incorporated into newly synthesized DNA strands. After many cell divisions in ^{15}N, the DNA of the cells were well labeled with the heavy isotope. The cells were then removed from the ^{15}N medium and put into a ^{14}N medium; after one and two cell divisions, samples were taken. DNA was extracted from the cells in each of these samples and put into a solution of cesium chloride (CsCl) in an ultracentrifuge.

If cesium chloride is spun in a centrifuge at tremendously high speeds (50,000 rpm) for many hours, the cesium and chloride ions tend to be pushed by centrifugal force toward the bottom of the tube. Ultimately, a **gradient** of Cs^+ and Cl^- ions is established in the tube, with the highest ion concentration at the bottom. Molecules of DNA in the solution also are pushed toward the bottom by centrifugal force. But as they travel down the tube, they encounter the increasing salt concentration, which tends to push them back up due to the buoyancy of DNA (or its tendency to float). Thus, the DNA finally "settles" at some point in the tube where the centrifugal

forces just balance the buoyancy of the molecules in the cesium-chloride gradient. The buoyancy of DNA depends on its density (which in turn reflects the ratio of G–C to A–T base pairs). The presence of the heavier isotope of nitrogen changes the buoyant density of DNA. The DNA extracted from cells grown for several generations on ^{15}N medium can readily be distinguished from the DNA of cells grown on ^{14}N medium by the equilibrium position reached in a cesium-chloride gradient. Such samples are commonly called "heavy" and "light" DNA, respectively.

Meselson and Stahl found that, one generation after the "heavy" cells were moved to ^{14}N medium, the DNA formed a single band of an intermediate density between the densities of the heavy and light controls. After two generations in ^{14}N medium, the DNA formed two bands: one at the intermediate position, and the other at the light position (Figure 11-16). This result would be expected from the semiconservative mode of replication; in fact, the result is compatible *only* with this mode *if* the experiment begins with chromosomes composed of individual double helices (Figure 11-17).

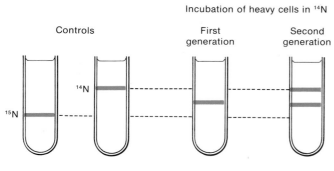

Figure 11-16. Centrifugation of DNA in a cesium chloride (CsCl) gradient. Cultures grown for many generations in ^{15}N and ^{14}N media provide control positions for "heavy" and "light" DNA bands, respectively. When the cells grown in ^{15}N are transferred to a ^{14}N medium, the first generation produces an intermediate DNA band and the second generation produces two bands: one intermediate and one light.

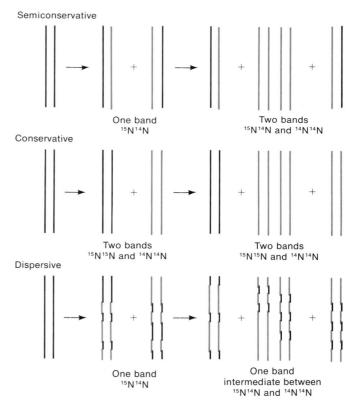

Figure 11-17. Only the semiconservative model of DNA replication predicts results like those shown in Figure 11-16: a single intermediate band in the first generation and one intermediate and one light band in the second generation. (See Figure 11-15 for explanations of symbols.)

Autoradiogaphy

The Meselson–Stahl experiment on *E. coli* was essentially duplicated in 1958 by Herbert Taylor on the chromosomes of bean root-tip cells, using a cytological technique. Taylor put root cells into a solution containing tritiated thymidine ([³H]-thymidine)—the thymine nucleotide labeled with a radioactive hydrogen isotope called tritium. He allowed the cells to undergo mitosis in this solution, so that the [³H]-thymidine could be incorporated into DNA. He then washed the tips and transferred them to a solution containing nonradioactive thymidine. Addition of colchicine to such a preparation inhibits the spindle apparatus so that chromosomes in metaphase fail to separate and sister chromatids remain "tied together" by the centromere.

The cellular location of ³H can be determined by **autoradiography.** As ³H decays, it emits a beta particle (an energetic electron). If a layer of photographic emulsion is spread over a cell that contains ³H, a chemical reaction takes place wherever a beta particle strikes the emulsion. The emulsion can then be developed like a photographic print, so that the emission track of the beta particle appears as a black spot or grain. The cell can also be stained, so that the structure of the cell is visible, to identify the location of the radioactivity. In effect, autoradiography is a process in which radioactive cell structures "take their own pictures."

Figure 11-18 shows the results observed when colchicine is added during the division in [³H]-thymidine or

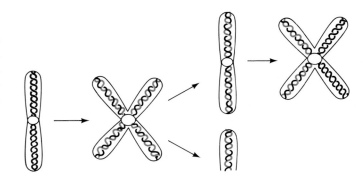

Figure 11-19. An explanation of Figure 11-18 at the DNA level. Colored lines represent radioactive strands.

during the subsequent mitotic division. It is possible to interpret these results by representing each chromatid as a single DNA molecule that replicates semiconservatively (Figure 11-19).

Harlequin Chromosomes

Using a more modern staining technique, it is now possible to visualize the semiconservative replication of chromosomes at mitosis without the aid of autoradiography. In this procedure, the chromosomes are allowed to go through two rounds of replication in bromodeoxyuridine (BUdR). The chromosomes are then stained with fluorescent dye and Giemsa stain; this process produces so-called **harlequin chromosomes** (Figure 11-20). The DNA strands that are newly synthesized in bromodeoxyuridine stain differently from the "original" DNA strands. The basis of this pattern is exactly identical to that of Figure 11-19. (Note, in passing, that harlequin chromosomes are particularly favorable for the detection of sister-chromatid exchange at mitosis; two examples are seen in Figure 11-20.)

Using similar techniques, Taylor showed that chromosome replication at meiosis also is semiconservative. This result drove another nail in the coffin of the copy-choice theory of crossing-over (Chapter 5), which would require *conservative* chromosome replication at meiosis.

Chromosome Structure

Figures 11-18 and 11-19 bring up one of the remaining great unsolved questions of genetics: is a eukaryotic chromosome basically a single DNA molecule surrounded by a protein matrix? Two things strongly suggest that this is, in fact, the case. First, if there were many DNA molecules in the chromosome (whether they were side by side, end to end, or randomly oriented), it would be almost impossible for the chromosome to replicate semiconservatively (with all of the label going into one

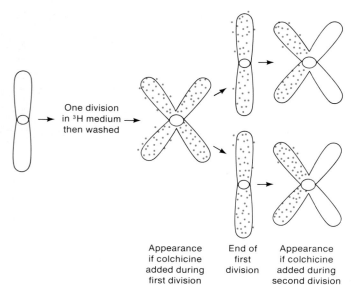

Appearance   End of   Appearance
if colchicine   first   if colchicine
added during   division   added during
first division   second division

Figure 11-18. Diagrammatic representation of the autoradiography of chromosomes from cells grown for one cell division in the presence of the radioactive hydrogen isotope ³H (tritium). Each dot represents the track of a particle of radioactivity.

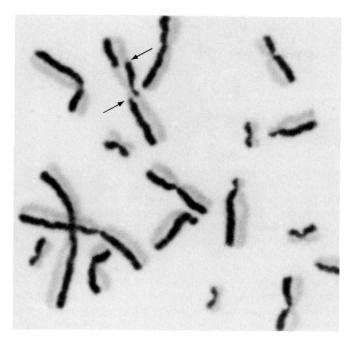

Figure 11-20. Harlequin chromosomes in a Chinese hamster ovary (CHO) cell. The procedure involves letting the chromosomes go through two rounds of replication in the presence of bromodeoxyuridine (BUdR), which replaces thymidine in the newly synthesized DNA. The chromosomes then are stained with a fluorescent dye and Giemsa stain, producing the appearance shown. The DNA strands that are newly replicated in BUdR stain differ from the "original" DNA strands. A chromosome at the top (see arrows) has two sister chromatid exchanges. (Photograph courtesy of Sheldon Wolff and Judy Bodycote.)

chromatid, as in Taylor's results). Look at Figure 11-21, and try to figure out how it could be done. Recent studies on isolated chromosomes and long DNA molecules are consistent with the suggestion that *each chromatid is a single molecule of DNA.* That makes a very long molecule. There is enough DNA in a single human chromosome, for example, to stretch out an inch or two and enough DNA in a single nucleus to stretch one meter. (This raises another interesting problem: how is this long molecule packed into the chromosome to permit easy replication?) The second fact supporting a single-molecule hypothesis is that DNA and genes behave as though they are attached end to end in a single string or thread that we call a linkage group. All genetic linkage data (Chapter 5) tell us that we need nothing more than a single linear array of genes per chromosome to explain the genetic facts.

As just mentioned, there is far too much DNA in a chromosome for it to extend linearly along the chromosome. It must be packed very efficiently into the chromosome. Current thinking (supported by good microscopic evidence) tends toward a process of coiling and

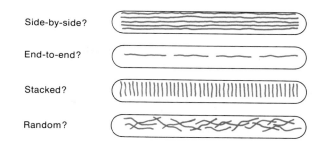

Side-by-side?

End-to-end?

Stacked?

Random?

Figure 11-21. Some theoretical alternative packing arrangements of DNA in a eukaryotic chromosome. Any of these models is hard to reconcile with the data supporting a semiconservative model of DNA replication. The question of the packaging of a long DNA molecule is considered in Chapter 14.

supercoiling of the DNA. Twist a rubber band with your fingers and notice the way it coils; chromosomes may be like this. We return to these questions in Chapter 14.

The Replication Fork

A prediction of the Watson-Crick model of DNA replication is that a fork will be found in the DNA molecule during replication. In 1963, John Cairns tested this prediction by allowing replicating DNA in bacterial cells to incorporate tritiated thymidine. Theoretically, each newly synthesized daughter molecule should then contain one radioactive ("hot") strand and another nonradioactive ("cold") strand. After varying intervals and varying numbers of replication cycles in a "hot" medium, Cairns extracted the DNA from the cells, put it on a slide, and autoradiographed it for examination under the electron microscope. After one replication cycle in [³H]-thymidine, rings of dots appeared in the autoradiograph. Cairns interpreted these rings as shown in Figure 11-22. It is also apparent from this figure that the bacte-

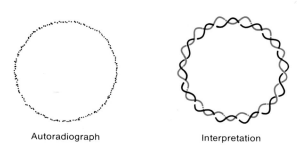

Autoradiograph

Interpretation

Figure 11-22. (left) Autoradiograph of a bacterial chromosome after one replication in tritiated thymidine. According to the semiconservative model of replication, one of the two strands should be radioactive. (right) Interpretation of the autoradiograph.

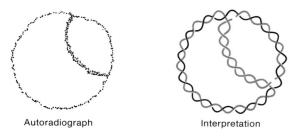

Autoradiograph   Interpretation

Figure 11-23. *(left)* Autoradiograph of a bacterial chromosome during the second round of replication in tritiated thymidine. In this theta θ structure, the newly replicated double helix that crosses the circle could show both strands as radioactive. *(right)* The double thickness of the radioactive tracing on the autoradiogram appears to confirm this.

Autoradiogram

Interpretation

Figure 11-25. A replication pattern in DNA revealed by autoradiography. A cell is briefly exposed to [^{3}H]-thymidine (pulse) and then provided with an excess of nonradioactive (cold) thymidine (chase). DNA is spread on a slide and autoradiographed. In the interpretation shown here, there are several initiation points for replication within one double helix of DNA.

rial chromosome is circular—a fact that also emerged from genetic analysis described earlier (Chapter 10).

During the second replication cycle, the forks predicted by the model were indeed seen. Furthermore, the density of grains in the three segments was such that the interpretation shown in Figure 11-23 could be made. Cairns saw all sizes of these moon-shaped, autoradiographic patterns, corresponding to the progressive movement of the replication zipper, or fork, around the ring. Structures of the sort shown in Figure 11-23 are called theta (θ) structures.

Origin of Replication

In bacteria, replication begins from a fixed **origin** but then proceeds **bidirectionally** (with moving forks at both ends of the replicating piece), as shown in Figure 11-24. In higher cells, replication proceeds from multiple points of origin. Suppose that a eukaryotic cell is briefly exposed to [^{3}H]-thymidine, in a step called a **pulse** exposure, and then is provided an excess of "cold" thymidine, in a step called the **chase;** the DNA is then

extracted, and autoradiographs are made. Figure 11-25 shows the results of such a procedure, with what appear to be distinct, simultaneously replicating regions along the DNA molecule. Replication appears to begin at several different sites on these eukaryotic chromosomes. Similarly, a pulse-and-chase study of DNA replication in polytene (giant) chromosomes of *Drosophila* by autoradiography reveals many replication regions within single chromosome arms (Figure 11-26). As yet there is no firm proof that these regions are indeed different starting points on a single DNA molecule; they could also be interpreted as evidence that the chromosome is made up of many separate DNA molecules. The structure of the eukaryotic chromosome still remains one of the most exciting unresolved problems in genetics (see Chapter 14).

Enzymology of Replication

In the late 1950s, Arthur Kornberg successfully identified and purified an enzyme, termed **DNA polymerase,** that catalyzes the replication reaction:

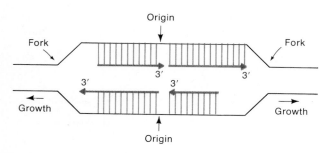

Figure 11-24. Diagrammatic representation of DNA replication proceeding in both directions from the origin. (From A Kornberg, *DNA Synthesis.* Copyright © 1974 by W. H. Freeman and Company.)

$$\text{primer (parental) DNA} + \begin{Bmatrix} \text{dATP} \\ + \\ \text{dGTP} \\ + \\ \text{dCTP} \\ + \\ \text{dTTP} \end{Bmatrix} \xrightarrow{\text{DNA polymerase}} \text{progeny DNA}$$

This reaction works only with the triphosphate forms of the nucleotides (such as deoxyadenosine triphosphate, or dATP). The total amount of DNA at the end of the reaction can be as much as 20 times the amount of origi-

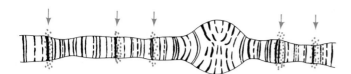

Figure 11-26. Replication pattern in a *Drosophila* chromosome revealed by autoradiography. Several points of replication are seen within a single chromosome, as indicated by the arrows.

nal input DNA, so most of the DNA present at the end must be progeny DNA. Thus, an analysis of this final DNA mixture can be regarded as largely indicative of the nature of the progeny DNA. Figure 11-27 depicts the chain-elongation reaction, or **polymerization** reaction, catalyzed by DNA polymerases.

Although the process of replication appears simple in Figure 11-23, certain complexities arise (refer to Figure 11-28, which summarizes many of the steps in DNA synthesis, as we describe these complexities):

1. The double helix must rotate in the process of replication, because the two strands are intertwined (or interlocked). This rotation is accomplished with the aid of biological catalysts, or **enzymes,** called DNA **topoisomerases,** which convert rings of DNA from one topological form to another. For instance, one topoisomerase, DNA **gyrase,** can induce twisting and coiling of the DNA, called **supercoiling** (Figure 11-29). The supercoiled form may facilitate the un-

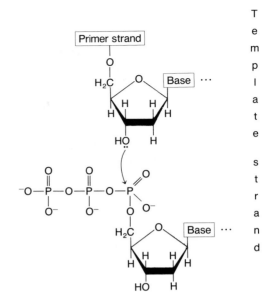

Figure 11-27. Chain-elongation reaction catalyzed by DNA polymerases. (Figures 11-27 and 11-31 from L. Stryer, *Biochemistry,* 3d ed. Copyright © 1988 by W. H. Freeman and Company.)

winding of the helix (Figure 11-30). A **helicase,** the "rep" protein, is probably involved in actually unwinding the helix. The free single-stranded region would be subject to degradation, but it is protected by another protein termed the single-stranded DNA-binding (SSB) protein (Figure 11-28).

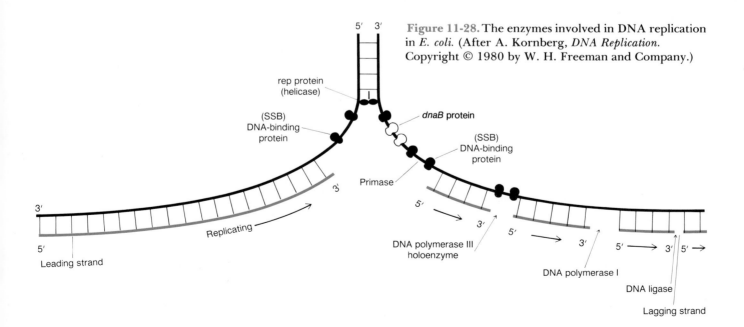

Figure 11-28. The enzymes involved in DNA replication in *E. coli.* (After A. Kornberg, *DNA Replication.* Copyright © 1980 by W. H. Freeman and Company.)

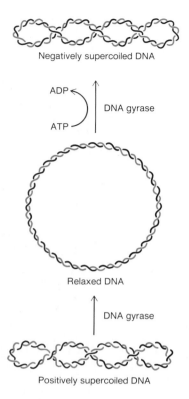

Negatively supercoiled DNA

Relaxed DNA

Positively supercoiled DNA

Figure 11-29. DNA-gyrase-catalyzed supercoiling. Replicating DNA generates "positive" supercoils, depicted at the bottom of the diagram, as a result of rapid rotation of the DNA at the replication fork. DNA gyrase can nick and close phosphodiester bonds, relieving the supercoiling, as shown here (relaxed DNA). Gyrase can also generate supercoils twisted in the opposite direction, termed "negative" supercoils; this arrangement facilitates the unwinding of the helix (see Figure 11-28). (Modified from L. Stryer, *Biochemistry*, 3d ed. Copyright © 1988 by W. H. Freeman and Company.)

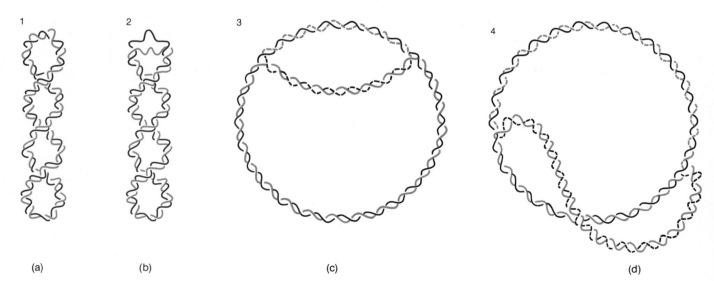

(a)   (b)   (c)   (d)

Figure 11-30. Replication of the information in the sequence of base pairs is necessary for a cell to divide. The replication of a double-stranded DNA ring may depend on the capacity of a topoisomerase to allow the double helix to unwind. One of the common mechanisms for the duplication of a double-stranded ring is the manufacture of a new complementary strand for each original strand. For the new strands to be assembled, the original strands must separate. In bacteria, gyrase negatively supercoils the ring (a). In the negatively supercoilded form, the unwinding of the double helix is made easier. Unwinding is probably required for replication to begin in some organisms (b). Assembly of the new strands is then begun using the original strands as templates (c). In order for replication to proceed, the double helix must be progressively unwound. The unwinding may be accomplished by one or more topoisomerase molecules (d), after which the two double-stranded rings separate; each ring is made up of one original strand and one new strand. (From J. C. Wang, "DNA Topoisomerases." Copyright © 1982 by Scientific American, Inc. All rights reserved.)

2. The simple notion of an unzipping molecule is inadequate. All of the known DNA polymerases synthesize new chains only in the $5' \rightarrow 3'$ direction. (The enzyme first studied by Kornberg, now termed DNA polymerase I, is not the principal DNA replication enzyme, whereas the enzyme termed DNA polymerase III probably is.) It is known that while one strand is synthesized continuously by DNA polymerase III, the other is synthesized in an interrupted fashion. For the latter strand, DNA polymerase III initiates at many different points on the template, leaving gaps that are filled in by DNA polymerase I and then ligated by the enzyme DNA **ligase.**

3. Both DNA polymerase I and DNA polymerase III also possess $3' \rightarrow 5'$ exonuclease activity, which serves as an "editing" or "proofreading" function by removing mismatched bases that were inserted erroneously during polymerization. Figure 11-31 represents the $3' \rightarrow 5'$ exonuclease activity, which requires a free $3'$—OH end that is not base-paired. Cleavage results in the regeneration of a free 3—OH group at the next base. This allows synthesis to continue by the addition of a free nucleotide triphosphate. The hydrolysis of a high-energy phosphate is required for each base that is added. Thus, a polymerase with a $5' \rightarrow 3'$ synthesis and a $3' \rightarrow 5'$ exonuclease editing function preserves the condition necessary for DNA synthesis. If polymerases were $3' \rightarrow 5'$, then the triphosphate would be on the growing strand (consider Figure 11-27) and the free

—OH group would be on the free nucleotide being added. The editing exonuclease activity would be $5' \rightarrow 3'$ and would release the base containing the high-energy phosphate bond, preventing further synthesis! For this reason, $3' \rightarrow 5'$ polymerases have not evolved. The consequence of all polymerases synthesizing in the $5' \rightarrow 3'$ direction is that synthesis on one strand must be discontinuous.

4. DNA polymerase cannot begin a new chain on a single-stranded template without at least a short region of duplex, which serves as a primer. In bacteria, the enzyme **primase,** together with a second protein encoded by *dnaB*, synthesizes an RNA primer for DNA polymerase III.

In Figure 11-28 the simultaneous synthesis of two new DNA chains is shown. On the left side of the diagram, the new strand is synthesized continuously in the $5' \rightarrow 3'$ direction, as the helix is unwound with the aid of the rep protein. Single-stranded regions are stabilized by the SSB protein. On the right side of the diagram, the new strand must be synthesized discontinuously, since all DNA polymerases synthesize in the $5' \rightarrow 3'$ direction. Thus, as new portions of the old template become available, synthesis of fragments of the new strand can begin. This synthesis requires a primase to first synthesize a short RNA stretch that serves as a "primer" onto which DNA polymerase III adds on deoxynucleotides before falling off. This process leaves gaps, as can be seen in the figure, that are filled in by DNA polymerase I, which also removes the short primer at the beginning of the double-stranded fragment that it has now reached. The enzyme DNA ligase then seals the final bond.

DNA and the Gene

We have now learned that DNA is the genetic material and consists of a linear sequence of nucleotide pairs. The obvious conclusion is that the allele maps represent a genetic equivalent of the nucleotide-pair sequences in DNA. We can validate this assumption if we can show that *the genetic maps are congruent with DNA maps.* This step was first taken using some elegant genetic and biochemical manipulations.

The DNA of the λ phage turns out to be a linear stretch of DNA that can be circularized because each $5'$ end of the two strands has an extra terminal extension of 12 bases that is complementary to the other $5'$ end (Figure 11-32). Because they are complementary, these ends can pair to join the DNA into a circle; they are known as "cohesive," or "sticky," ends.

When a linear object such as a DNA molecule is subjected to shear stress (say, by pipetting or stirring), the mechanics of the stress causes breaks to occur, prin-

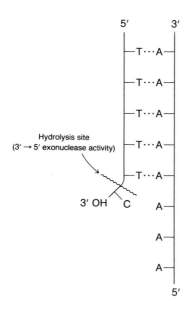

Figure 11-31. $3' \rightarrow 5'$ exonuclease action of DNA polymerase I.

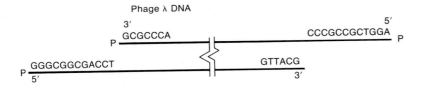

Phage λ DNA

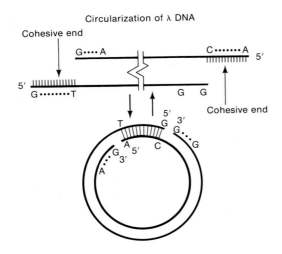

Circularization of λ DNA

Cohesive end

Cohesive end

Figure 11-32. Circularization of DNA. The λ DNA is linear in the phage, but once in a host cell, it circularizes as a prelude to insertion or replication. Circularization is achieved by joining the complementary ("sticky") single-stranded ends. (From A. Kornberg, *DNA Synthesis.* Copyright © 1974 by W. H. Freeman and Company.)

cipally in the middle of the molecule. When DNA from the λ phage is sheared in half, the two halves happen to have different G–C ratios, which means that their buoyant densities differ and they can be separated by centrifugation in CsCl. When the two half-molecules are separated, their genetic content can be assayed by introducing the DNA into bacteria simultaneously on infec-

tion with mutant phage λ. Recombination can occur between the phage DNA and the fragment, as shown in Figure 11-33, in which *a* to *f* are phage genes. The introduced DNA can be incorporated into the normal λ DNA and "rescued" by inducing lysis (the rupture and death of a bacterial cell on the release of phage progeny). You can see that by using different strains, this technique

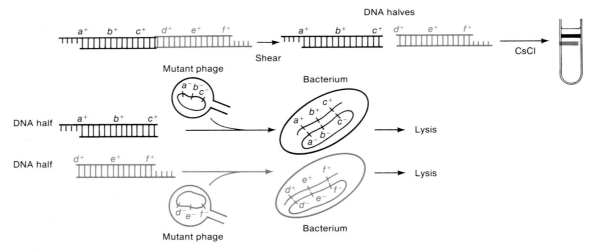

Figure 11-33. When λ DNA is broken, it forms roughly equal halves that happen to have different buoyant densities. Each half can be "rescued" by a multiply mutant phage during simultaneous transformation and infection. From

these experiments, it has been shown that only genes from one specific half of the genetic map could be rescued from one specific half of the DNA. (Note that the bacterial chromosome has been omitted here to simplify the diagram.)

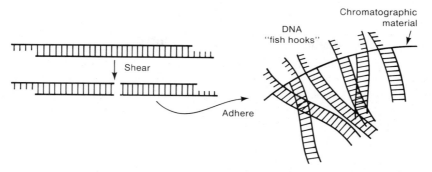

Figure 11-34. One specific sticky end of a DNA molecule can be used as a "fish hook" for the other sticky end plus any genes that may be attached to it. One-half of the λ phage DNA is attached to chromatographic material so that the cohesive or sticky ends are exposed. Any DNA passing over this chromatographic material that has single-stranded base sequences complementary to the "fish hook" cohesive ends will stick. Any other DNA will pass through. Later, the "stuck" DNA can be removed from the chromatographic material by heating it to break the hydrogen bonds.

can lead to the analysis of the genetic marker content of the DNA fractions. This is called a **marker-rescue experiment.** Such experiments demonstrate that one particular half of the DNA molecule carries the information of one particular half of the linkage map.

Dale Kaiser and his associates separated one of the DNA halves carrying a cohesive end and attached the other end to chromatographic material over which DNA could be passed (Figure 11-34). The single-stranded ends dangle free like fish hooks. If the other half of the DNA is sheared into smaller and smaller molecules and then passed over the sticky ends, the complementary sequences will stick. These stuck pieces can be detached easily by raising the temperature to break the

hydrogen bonds in the sticky ends. In this way, a series of fractions (of varying size) of one end of the λ DNA can be separated and tested for content by marker rescue. Kaiser and his associates showed that an unambiguous arrangement of genes on the DNA can be determined and that this sequence is completely congruent with the genetic map (Figure 11-35).

> **Message** We are now justified in concluding that the sequence of bases in the DNA is indeed congruent with the gene map.

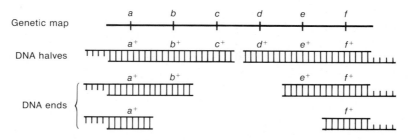

Figure 11-35. As the λ DNA is sheared into progressively smaller and smaller molecules, genes are lost to the fish hooks (Figure 11-32) in the same order as they appear on the genetic map.

Summary

■ Experimental work on the molecular nature of hereditary material has demonstrated conclusively that DNA (not protein, RNA, or some other substance) is indeed the genetic material. Using data supplied by others, Watson and Crick created a double helical model with two DNA strands, wound around each other, running in antiparallel fashion. Specificity of binding the two strands together is based on the fit of adenine (A) to thymine (T) and guanine (G) to cytosine (C). The former pair is held by two hydrogen bonds; the latter, by three.

The Watson-Crick model shows how DNA can be replicated in an orderly fashion—a prime requirement for genetic material. Replication is accomplished semiconservatively. One double helix is replicated into two identical helices, each with identical linear orders of nucleotides; each of the two new double helices is composed of one old and one newly polymerized strand of DNA. This semiconservative replication occurs in both prokaryotes and eukaryotes.

Replication is achieved with the aid of several enzymes, including DNA polymerase, gyrase, and helicase. Replication starts at special regions of the DNA called origins of replication and proceeds down the DNA in both directions. Since DNA polymerase acts only in a $5' \rightarrow 3'$ direction, one of the newly synthesized strands at each replication fork must be synthesized in short segments and then joined using the enzyme ligase. DNA polymerization cannot begin without a short primer, which is also synthesized with special enzymes.

The marker-rescue technique has demonstrated that the sequence of genes on a chromosome corresponds exactly to the linear sequence of DNA and has shown convincingly that the genetic and chemical entities are one and the same thing.

■ ■ ■

Solved Problems

1. If the GC content of a DNA molecule is 56 percent, what are the percentages of the four bases (A, T, G, and C) in this molecule?

   Solution

   If the GC content is 56 percent, then since G = C, the content of G is 28 percent and the content of C is 28 percent. The content of AT is $100 - 56 = 44$ percent. Since A = T, the content of A is 22% and the content of T is 22 percent.

2. Describe the expected pattern of bands in a CsCl gradient for conservative replication in the Meselson-Stahl experiment. Draw a diagram.

   Solution

   Refer to Figure 11-17 for an additional explanation. In conservative replication, if bacteria are grown in the presence of ^{15}N and then shifted to ^{14}N, one strand will be all ^{15}N after the first generation and the other strand will be all ^{14}N, resulting in one heavy band and one light band in the gradient. After the second generation, the ^{15}N DNA will yield one molecule with all ^{15}N and one molecule with all ^{14}N, whereas the ^{14}N DNA will yield only ^{14}N DNA. Thus, only all ^{14}N or all ^{15}N DNA is generated, again, yielding a light band and a heavy band (see figure below).

Problems

1. If thymine makes up 15 percent of the bases in a specific DNA molecule, what percentage of the bases is cytosine?

2. Graph DNA content against time in a cell that undergoes mitosis and then meiosis.

3. *E. coli* chromosomes in which every nitrogen atom is labeled (that is, every nitrogen atom is the heavy isotope ^{15}N instead of the normal isotope ^{14}N) are allowed to replicate in an environment in which all the nitrogen is ^{14}N. Using a solid line to represent a heavy polynucleotide chain and a dashed line for a light chain, sketch the following:

   a. The heavy parental chromosome and the products of the first replication after transfer to a ^{14}N medium, assuming that the chromosome is one DNA double helix and that replication is semiconservative.

   b. Repeat (a), assuming that replication is conservative.

   c. Repeat (a), assuming that the chromosome is in fact two side-by-side double helices, each of which replicates semiconservatively.

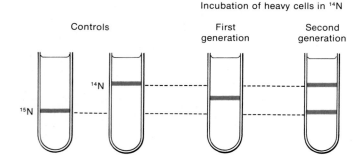

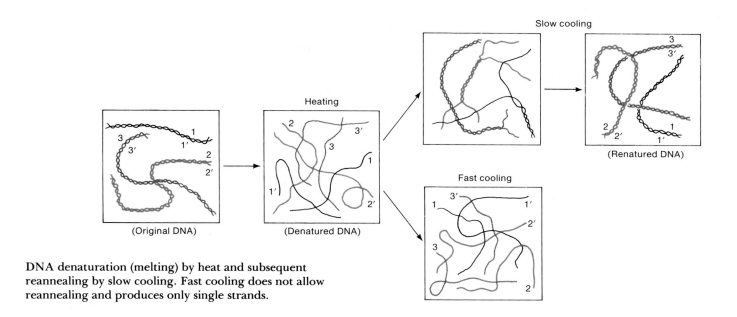

DNA denaturation (melting) by heat and subsequent reannealing by slow cooling. Fast cooling does not allow reannealing and produces only single strands.

d. Repeat (c), assuming that each side-by-side double helix replicates conservatively and that the overall *chromosome* replication is semiconservative.

e. Repeat (d), assuming that the overall chromosome replication is conservative.

f. If the daughter chromosomes from the first division in ^{14}N are spun in a cesium chloride (CsCl) density gradient and a single band is obtained, which of possibilities (a) to (e) can be ruled out? Reconsider the Meselson-Stahl experiment: What does it *prove?*

4. R. Okazaki found that the immediate products of DNA replication in *E. coli* include single-stranded DNA fragments approximately 1000 nucleotides in length after the newly synthesized DNA is extracted and denatured (melted). When he allowed DNA replication to proceed for a longer period of time, he found a lower frequency of these short fragments and long single-stranded DNA chains after extraction and denaturation. Explain how this result might be related to the fact that all known DNA polymerases synthesize DNA only in a $5' \rightarrow 3'$ direction.

5. When plant and animal cells are given pulses of $[^3H]$-thymidine at different times during the cell cycle, heterochromatic regions on the chromosomes are invariably "late replicating." Can you suggest what, if any, biological significance this observation might have?

6. On the planet of Rama, the DNA is of six nucleotide types: A, B, C, D, E, and F. A and B are called marzines, C and D are orsines, and E and F are pirines. The following rules are valid in all Raman DNAs:

Total marzines = total orsines = total pirines

$$A = C = E$$

$$B = D = F$$

a. Prepare a model for the structure of Raman DNA.

b. On Rama, mitosis produces three daughter cells. Bearing this fact in mind, propose a replication pattern for your DNA model.

c. Consider the process of meiosis on Rama. What comments or conclusions can you suggest?

7. If you extract the DNA of the coliphage ϕX174, you will find that its composition is 25 percent A, 33 percent T, 24 percent G, and 18 percent C. Does this make sense in terms of Chargaff's rules? How would you interpret this result? How might such a phage replicate its DNA?

8. The temperature at which a DNA sample denatures can be used to estimate the proportion of its nucleotide pairs that are G–C. What would be the basis for this determination, and what would a high denaturation temperature for a DNA sample indicate?

9. In 1960, Paul Doty and Julius Marmur observed that when DNA is heated to 100°C, all of the hydrogen bonds between the complementary strands are destroyed and the DNA becomes single-stranded (see the figure above). If the solution is cooled slowly, some double-stranded DNA is formed that is biologically normal (for example, it may have transforming ability). Presumably, this **reannealing,** or **renaturation,** process occurs when two single strands happen to collide in such a way that the complementing base sequences can align and reconstitute the original double helix (see figure). This reannealing is very specific and precise, making it a powerful tool because stretches of complementary base sequences in *different* DNAs also will anneal after melting and mixing. Thus, the efficiency of annealing provides a measure of similarity between two different DNAs.

Now suppose that you extract DNA from a small virus, denature it, and allow it to reanneal with DNA taken from other strains that carry either a deletion, an inversion, or a duplication. What would you expect to see on inspection with an electron microscope?

10. DNA extracted from a mammal is heat-denatured and then slowly cooled to allow reannealing. The following graph shows the results obtained. There are two "shoulders" in the curve. The first shoulder indicates the presence of a very rapidly annealing part of the DNA — so rapid, in fact, that it occurs before strand interactions take place.

a. What could this part of the DNA be?

b. The second shoulder is a rapidly reannealing part as well. What does this evidence suggest?

11. Design tests to determine the physical relationship between highly repetitive and unique DNA sequences in chromosomes. (HINT: it is possible to vary the size of DNA molecules by the amount of shearing they are subjected to.)

12. Viruses are known to cause cancer in mice. You have a pure preparation of virus DNA, a pure preparation of DNA from the chromosomes of mouse cancer cells, and pure DNA from the chromosomes of normal mouse cells. Viral DNA will specifically anneal with cancer-cell DNA, but not with normal-cell DNA. Explore the possible genetic significance of this observation, its significance at the molecular level, and its medical significance.

13. Ruth Kavenaugh and Bruno Zimm have devised an elegant technique to measure the maximal length of the longest DNA molecules in solution. They studied DNA samples from the three *Drosophila* karyotypes shown in the figure below. They found the longest molecules in karyotypes (a) and (b) to be of similar length and about twice the length of the longest molecule in (c). Interpret these results.

14. In the harlequin-chromosome technique, you allow *three* rounds of replication in bromodeoxyuridine and then stain the chromosomes. What result do you expect to obtain?

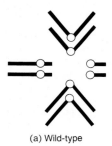

(a) Wild-type

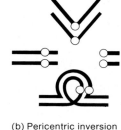

(b) Pericentric inversion

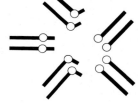

(c) Translocation

The Nature of the Gene

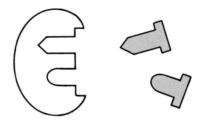

KEY CONCEPTS

The one-gene–one-enzyme hypothesis postulates that genes
control the structure of proteins.

∎

Studies on hemoglobin have demonstrated that a gene
mutation can be connected to an altered amino-acid sequence
in a protein.

∎

The linear sequence of nucleotides in a gene determines the
linear sequence of amino acids in a protein.

∎

A fine structural analysis of the *rII* genes in phage T4 has
showed that the gene consists of a linear array of
subelements, now correlated with nucleotide pairs, that can
mutate and recombine with one another.

∎

A gene can be defined as a unit of function by a
complementation test.

How Genes Work

What is the nature of the gene, and how do genes control phenotypes? For example, how can one allele of a gene produce a wrinkled pea and another produce a round, smooth pea? Today we realize that all reactions of a cell are catalyzed by enzymes, which have a specific three-dimensional configuration that is crucial to their function. We now know that genes specify the structures of proteins, some of which are enzymes, and we can even relate the structure of the genetic material to the structure of proteins. Table 12-1 summarizes our current model of the relationship between genotype and phenotype.

How did we arrive at this point? By following two lines of inquiry:

1. What is the physical structure of the genetic material?

2. How does the genetic material exert its effect? (Or, in detail, how does the structure operate at the level of the gene?)

Chapter 11 recounts the demonstration that DNA is the genetic material and details the unraveling of the structure of DNA. In Chapters 12 and 13, we examine how genes function.

The first clues about the nature of primary gene function came from studies of humans. Early in this century, Archibald Garrod, a physician, noted that several hereditary human defects are produced by recessive mutations. Some of these defects can be traced directly to metabolic defects that affect the basic body chemistry —an observation that has led to the suggestion of "inborn errors" in metabolism. We know now, for example, that phenylketonuria, which is caused by an autosomal recessive allele, results from an inability to convert phenylalanine into tyrosine. Consequently, phenylala-

TABLE 12-1. Model of the relationship of genotype versus phenotype

1. The characteristic features of an organism are determined by the phenotype of its parts, which are in turn determined by the phenotype of the component cells.

2. The phenotype of a cell is determined by its internal chemistry, which is controlled by enzymes that catalyze its metabolic reactions.

3. Enzyme function depends on a specific three-dimensional structure, which in turn depends on a specific linear sequence of amino acids in the protein.

4. The enzymes present in the cell, as well as structural proteins, are determined by the genotype of the cell.

5. Genes specify the linear sequence of amino acids in polypeptides and thus determine phenotypes.

nine accumulates and is spontaneously converted into a toxic compound, phenylpyruvic acid. In a different character, the inability to convert tyrosine into the pigment melanin produces an albino. In any case, Garrod's observations focused attention on metabolic control by genes.

The One-Gene–One-Enzyme Hypothesis

Clarification of the actual function of genes came from research on *Neurospora* by George Beadle and Edward Tatum in the 1940s, who later received a Nobel Prize for their work. Before we describe their actual experiments, let's jump ahead and examine some aspects of **biosynthetic pathways** based on our current understanding. We now know that molecules are synthesized as a series of steps, each one catalyzed by an enzyme. For instance, a biosynthetic pathway might have four steps, where 1 is the starting material and 5 is the final product:

$$\begin{array}{ccccc} A & B & C & D \\ 1 \longrightarrow 2 \longrightarrow 3 \longrightarrow 4 \longrightarrow 5 \end{array}$$

Each step is catalyzed by an enzyme: A, B, C, or D. In turn, each enzyme is specified by a particular gene. We might say that gene a specifies enzyme A, gene b specifies enzyme B, and so on. Therefore, if we inactivate the gene responsible for an enzyme, we eliminate one step in the pathway.

In the following diagram, gene b is eliminated due to a mutation:

$$\begin{array}{cccc} & \times & & \\ \text{—gene } a\text{—} & \text{—gene } b\text{—} & \text{—gene } c\text{—} & \text{—gene } d\text{—} \\ \downarrow & \times & \downarrow & \downarrow \\ A & B & C & D \\ 1 \longrightarrow 2 \longrightarrow 3 \longrightarrow 4 \longrightarrow 5 \end{array}$$

Now we cannot carry out the reaction that converts compound 2 to compound 3. We are blocked at compound 2, and we cannot go further. But what happens if we add back different intermediate compounds? Suppose that, for example, we feed the cell compound 3 or 4. Can either compound synthesize the final product (5)? Yes, either compound 3 or 4 can; subsequent steps are not blocked. What if we add more of compound 1? No, adding compound 1 cannot synthesize product 5 because one of the subsequent steps is blocked.

Let's test our understanding of the concept of biosynthetic pathways by looking at a sample problem that illustrates this principle:

Several mutants that require compound G for growth are isolated. All compounds A–E in the pathway are known, and each compound is tested for its ability to support the growth of each mutant 1–5.

<table>
<thead>
<tr><th></th><th></th><th colspan="6">Compound</th></tr>
<tr><th></th><th></th><th>A</th><th>B</th><th>C</th><th>D</th><th>E</th><th>G</th></tr>
<tr><th></th><th></th><th colspan="6">(+ = growth; − = no growth)</th></tr>
</thead>
<tbody>
<tr><td>*Mutant*</td><td>1</td><td>−</td><td>−</td><td>−</td><td>+</td><td>−</td><td>+</td></tr>
<tr><td></td><td>2</td><td>−</td><td>+</td><td>−</td><td>+</td><td>−</td><td>+</td></tr>
<tr><td></td><td>3</td><td>−</td><td>−</td><td>−</td><td>−</td><td>−</td><td>+</td></tr>
<tr><td></td><td>4</td><td>−</td><td>+</td><td>+</td><td>+</td><td>−</td><td>+</td></tr>
<tr><td></td><td>5</td><td>+</td><td>+</td><td>+</td><td>+</td><td>−</td><td>+</td></tr>
</tbody>
</table>

a. What is the order of compounds A – E and G in the pathway?

How do we approach this type of problem? First, let's find the underlying principle, and then draw a diagram. The main point is that compounds will allow the growth of mutants that are blocked in steps before a compound is made but not of mutants that are blocked in steps after the compound is made. So the compounds that support the growth of the most mutants occur latest in the pathway; compounds that occur earliest support the growth of the least mutants. Therefore, let's ask, which is the latest compound? Or which compound can support the growth of the most mutants? From the table, this is compound G, the final product, followed by D (let's fill in the pathway backwards) and then B, C, A, and E, so that the order is

$$E \longrightarrow A \longrightarrow C \longrightarrow B \longrightarrow D \longrightarrow G$$

b. At which point in the pathway is each mutant blocked?

Clearly, a mutant that is blocked between E and A cannot be satisfied by E but can be satisfied by all the other compounds. Thus, we see that mutant 5 must be blocked in the E – A conversion. We also see that mutant 4 cannot be satisfied by E or A, so it must be blocked in the A – C conversion. By similar logic, we obtain the order 5 – 4 – 2 – 1 – 3, which we can insert into the diagram as follows:

$$\overset{5}{E} \longrightarrow \overset{4}{A} \longrightarrow \overset{2}{C} \longrightarrow \overset{1}{B} \longrightarrow \overset{3}{D} \longrightarrow G$$

Now we can comprehend how Beadle and Tatum first worked out this experiment, using one particular biosynthetic pathway of *Neurospora*.

The Experiments of Beadle and Tatum

Beadle and Tatum analyzed mutants of *Neurospora*. They first irradiated *Neurospora* to produce mutations and then tested cultures from ascospores for interesting mutant phenotypes. They detected numerous **auxotrophs**—strains which cannot grow on a minimal medium unless it is supplemented with specific nutrients. In each case, the mutation that generated the auxotrophic requirement was inherited as a single gene mutation; each gave a 1 : 1 ratio when crossed with a wild-type (recall that *Neurospora* is haploid). Figure 12-1a depicts the procedure Beadle and Tatum used.

One set of mutants required arginine to grow on a minimal medium (Figure 12-1b). These strains provided the focus for much of Beadle and Tatum's further work. First, they found that the mutations mapped into three different locations on separate chromosomes, even though the same supplement (arginine) satisfied the growth requirement for each mutant. Let's call the three loci the *arg-1, arg-2,* and *arg-3* genes. Beadle and Tatum discovered that the auxotrophs for each of the three loci differ in their response to the chemical compounds ornithine and citrulline, which are related to arginine (Figure 12-2).

The *arg-1* mutants grow when supplied with ornithine or citrulline or arginine in addition to the minimal medium. The *arg-2* mutants grow on either arginine or citrulline but not on ornithine. The *arg-3* mutants grow only when arginine is supplied. We can see this more easily by looking at Table 12-2.

It was already known that related compounds are interconverted in cells by biological catalysts called enzymes. Based on the properties of the *arg* mutants, Beadle and Tatum and their colleagues proposed a biochemical model for such conversions in *Neurospora*:

$$\text{precursor} \xrightarrow{\text{enzyme X}} \text{ornithine} \xrightarrow{\text{enzyme Y}}$$
$$\text{citrulline} \xrightarrow{\text{enzyme Z}} \text{arginine}$$

Note how this relationship easily explains the three classes of mutants shown in Table 12-1. The *arg-1* mutants have a defective enzyme X, so they are unable to convert the precursor into ornithine as the first step in producing arginine. However, having normal enzymes

TABLE 12-2. Growth of *arg* mutants in response to supplements

| Mutant | Supplement | | |
	Ornithine	Citrulline	Arginine
	(+ = growth; − = no growth)		
arg-1	+	+	+
arg-2	−	+	+
arg-3	−	−	+

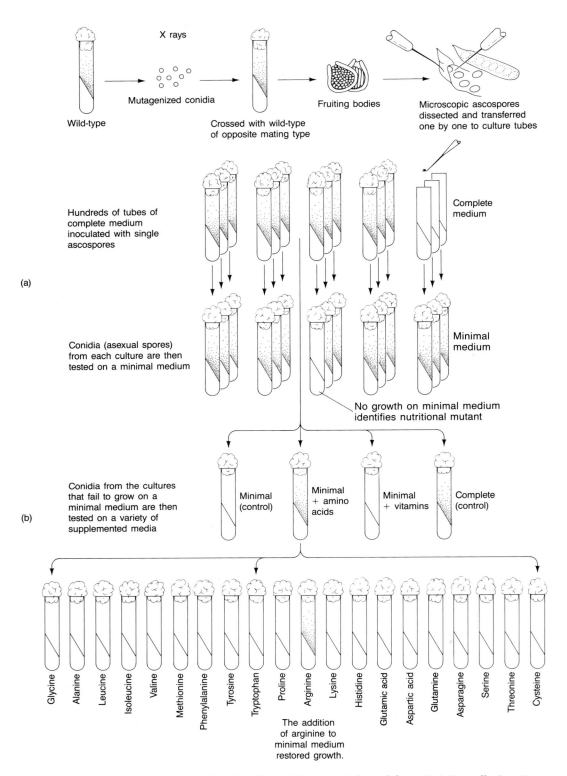

Figure 12-1. The procedure used by Beadle and Tatum. (Adapted from P. J. Russell, *Genetics.* Copyright © 1986 by Little, Brown & Co.)

Y and Z, the *arg-1* mutants are able to produce arginine if supplied with either ornithine or citrulline. The *arg-2* mutants lack enzyme Y, and the *arg-3* mutants lack enzyme Z. Thus, a mutation at a particular gene is assumed to interfere with the production of a single enzyme. The defective enzyme, then, creates a block in some biosynthetic pathway. The block can be circumvented by supplying to the cells any compound that normally comes

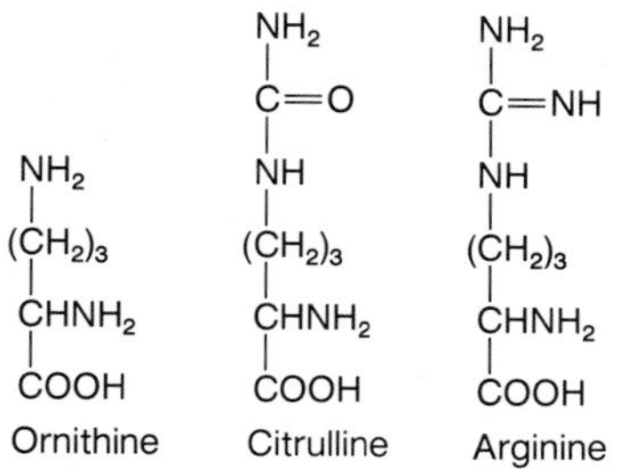

Figure 12-2. Chemical structures of arginine and the related compounds citrulline and ornithine. In the work of Beadle and Tatum, different *arg* auxotrophic mutants of *Neurospora* were found to grow when the medium was supplemented with citrulline or ornithine as an alternative to arginine.

after the block in the pathway. We can now diagram a more complete biochemical model:

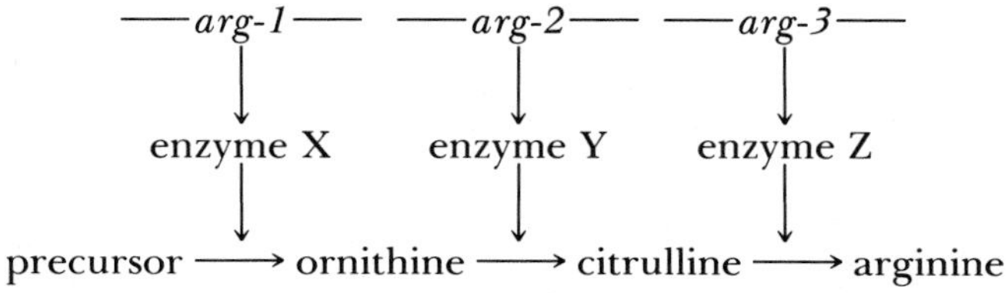

Note that this entire model was inferred from the properties of the mutant classes detected through genetic analysis. Only later were the existence of the biosynthetic pathway and the presence of defective enzymes demonstrated through independent biochemical evidence.

This model, which has become known as the **one-gene–one-enzyme hypothesis,** provided the first exciting insight into the function of genes: genes somehow are responsible for the function of enzymes, and each gene apparently controls one specific enzyme. Other researchers obtained similar results for other biosynthetic pathways, and the hypothesis soon achieved general acceptance. It is one of the great unifying con-

cepts in biology, because it provides a bridge that brings together the concepts and research techniques of genetics and chemistry.

Message Genes control biochemical reactions by controlling the production of enzymes.

We should pause to ponder the significance of this discovery. Let's summarize what it established:

1. **Biochemical reactions in vivo (in the living cell) occur as a series of discrete, stepwise reactions.**

2. **Each reaction is specifically catalyzed by a single enzyme.**

3. **Each enzyme is specified by a single gene.**

Gene–Protein Relationships

The one-gene–one-enzyme hypothesis was an impressive step forward in our understanding of gene function, but just *how* do genes control the functioning of enzymes? Enzymes belong to a general class of molecules called **proteins,** and we must review the basic facts of protein structure in order to follow the next step in the study of gene function.

Protein Structure

In simple terms, a protein is a macromolecule composed of **amino acids** attached end to end in a linear string. The general formula for an amino acid is $H_2N—CHR—COOH$, in which the R group can be anything from a hydrogen atom (as in the amino acid glycine) to a complex ring (as in the amino acid tryptophan). There are 20 common amino acids in living organisms (Table 12-3), each having a different R group.

TABLE 12-3. The 20 amino acids common in living organisms

Amino acid	Three-letter abbreviation	Amino acid	Three-letter abbreviation	Amino acid	Three-letter abbreviation	Amino acid	Three-letter abbreviation
Alanine	Ala	Glutamine	Gln	Leucine	Leu	Serine	Ser
Arginine	Arg	Glutamic acid	Glu	Lysine	Lys	Threonine	Thr
Asparagine	Asn	Glycine	Gly	Methionine	Met	Tryptophan	Trp
Aspartic acid	Asp	Histidine	His	Phenylalanine	Phe	Tyrosine	Tyr
Cysteine	Cys	Isoleucine	Ile	Proline	Pro	Valine	Val

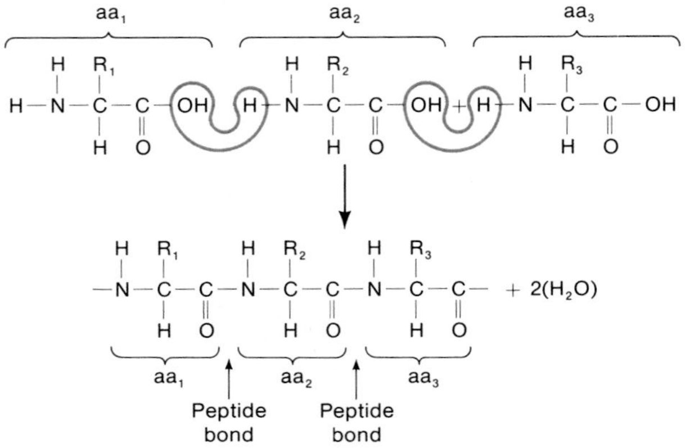

Figure 12-3. Formation of a polypeptide by the removal of water between amino acids to form peptide bonds. R_1, R_2, and R_3 represent side groups that differentiate the amino acids. R can be anything from a hydrogen atom (as in glycine) to a complex ring (as in tryptophan). Each aa indicates an amino acid.

10   20
Met-Glu-Arg-Tyr-Glu-Ser-Leu-Phe-Ala-Gln-Leu-Lys-Glu-Arg-Lys-Glu-Gly-Ala-Phe-Val-

30   40
Pro-Phe-Val-Thr-Leu-Gly-Asp-Pro-Gly-Ile-Glu-Gln-Ser-Leu-Lys-Ile-Ile-Asp-Thr-Leu-

50   60
Ile-Glu-Ala-Gly-Ala-Asp-Ala-Leu-Glu-Leu-Gly-Ile-Pro-Phe-Ser-Asp-Pro-Leu-Ala-Asp-

70   80
Gly-Pro-Thr-Ile-Gln-Asn-Ala-Thr-Leu-Arg-Ala-Phe-Ala-Ala-Gly-Val-Thr-Pro-Ala-Gln-

90   100
Cys-Phe-Glu-Met-Leu-Ala-Leu-Ile-Arg-Gln-Lys-His-Pro-Thr-Ile-Pro-Ile-Gly-Leu-Leu-

110   120
Met-Tyr-Ala-Asn-Leu-Val-Phe-Asn-Lys-Gly-Ile-Asp-Glu-Phe-Tyr-Ala-Gln-Cys-Glu-Lys-

130   140
Val-Gly-Val-Asp-Ser-Val-Leu-Val-Ala-Asp-Val-Pro-Val-Gln-Glu-Ser-Ala-Pro-Phe-Arg-

150   160
Gln-Ala-Ala-Leu-Arg-His-Asn-Val-Ala-Pro-Ile-Phe-Ile-Cys-Pro-Pro-Asn-Ala-Asp-Asp-

170   180
Asp-Leu-Leu-Arg-Gln-Ile-Ala-Ser-Tyr-Gly-Arg-Gly-Tyr-Thr-Tyr-Leu-Leu-Ser-Arg-Ala-

190   200
Gly-Val-Thr-Gly-Ala-Glu-Asn-Arg-Ala-Ala-Leu-Pro-Leu-Asn-His-Leu-Val-Ala-Lys-Leu-

210   220
Lys-Glu-Tyr-Asn-Ala-Ala-Pro-Pro-Leu-Gln-Gly-Phe-Gly-Ile-Ser-Ala-Pro-Asp-Gln-Val-

230   240
Lys-Ala-Ala-Ile-Asp-Ala-Gly-Ala-Ala-Gly-Ala-Ile-Ser-Gly-Ser-Ala-Ile-Val-Lys-Ile-

250   260
Ile-Glu-Gln-His-Asn-Ile-Glu-Pro-Glu-Lys-Met-Leu-Ala-Ala-Leu-Lys-Val-Phe-Val-Gln-

268
Pro-Met-Lys-Ala-Ala-Thr-Arg-Ser-

(a)

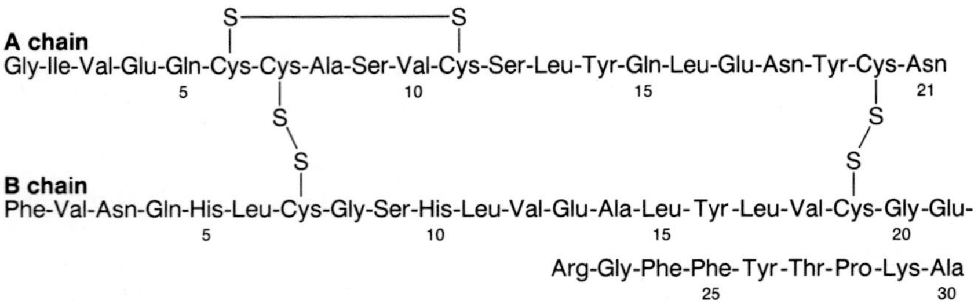

Figure 12-4. Primary sequences of two proteins. (a) The tryptophan synthetase A protein in *E. coli*. (b) Bovine insulin protein. Note that the amino acid cysteine can form unique "sulfur bridges," because it contains sulfur.

(b)

Amino acids are linked together in proteins by covalent (chemical) bonds called **peptide bonds.** A peptide bond is formed through a condensation reaction that involves the removal of a water molecule (Figure 12-3).

Several amino acids linked together by peptide bonds form a molecule called a **polypeptide;** proteins are large polypeptides of the kinds found in living organisms. The linear arrangement of amino acids in a polypeptide chain is called the **primary structure** of the protein. Figure 12-4 shows the primary structures of beef insulin (a hormonal protein) and tryptophan synthetase (an enzyme). Many of the side groups of amino acids attract or repel one another in the protein, resulting in different types of **secondary structure** of the protein. Two of the basic periodic structures are the α-helix and the β-pleated sheet, which are depicted in Figures 12-5 and 12-6, respectively. Protein chains can also formulate a "β turn" (Figure 12-7), in which the direction of the amino acid chain is reversed. The protein chain can be further folded to form a **tertiary structure** (Figure 12-8). In many cases, a number of folded structures can associate to form a **quaternary structure** that is multi-

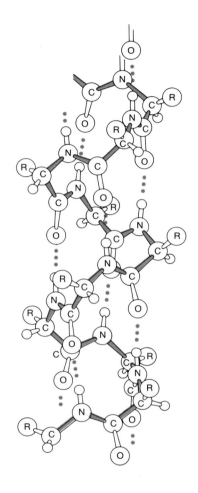

Figure 12-5. The α-helix that is a common basis of secondary protein structure. The backbone of the protein can be seen as the colored lines. Each R is a specific side group on one amino acid. The colored dots are weak, stabilizing hydrogen bonds that maintain the helical shape. (Reprinted from Linus Pauling, *The Nature of the Chemical Bond.* Copyright 1939 and 1940 by Cornell University. Third edition © 1960 by Cornell University. Used by permission of Cornell University Press.)

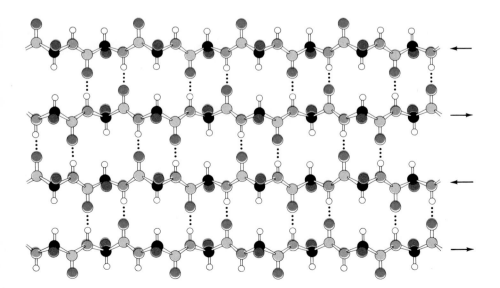

Figure 12-6. Antiparallel β-pleated sheet. Adjacent strands run in opposite directions. Hydrogen bonds between NH and CO groups of adjacent strands stabilize the structure. The side chains (shown in brown) are above and below the plane of the sheet. (From L. Stryer, *Biochemistry,* 3d ed. Copyright © 1983 by W. H. Freeman and Company.)

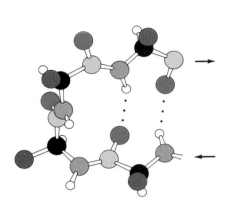

Figure 12-7. Structure of a β-turn. The CO group of residue 1 of the tetrapeptide shown here is hydrogen bonded to the NH group of residue 4, which results in a hairpin turn. (From L. Stryer, *Biochemistry,* 3d ed. Copyright © 1983 by W. H. Freeman and Company.)

Figure 12-9. A model of the hemoglobin molecule (shown as contoured layers to provide a visualization of the three-dimensional shape). The different shadings indicate the different polypeptide chains that combine to form the quaternary structure of the protein. The disks are heme groups—complex structures containing iron. (After M. F. Perutz, "The Hemoglobin Molecule." Copyright © 1964 by Scientific American, Inc. All rights reserved.)

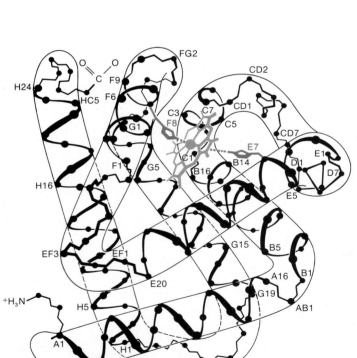

Figure 12-8. Folded tertiary structure of myoglobin, an oxygen-storage protein. Each dot represents an amino acid. The heme group, a cofactor that facilitates the binding of oxygen, is shown in light color. (From L. Stryer, *Biochemistry,* 2d ed. Copyright © 1981 by L. Stryer. Based on R. E. Dickerson, *The Proteins,* 2d ed., vol. 2. Edited by H. Neurath. Copyright © 1964 by Academic Press.)

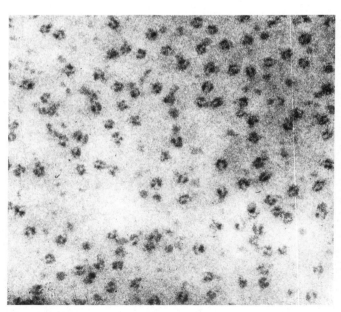

Figure 12-10. Electron micrograph of the enzyme aspartate transcarbamylase. Each small "glob" is an enzyme molecule. Note the quaternary structure: the enzyme is composed of subunits. (Photograph courtesy of Jack D. Griffith.)

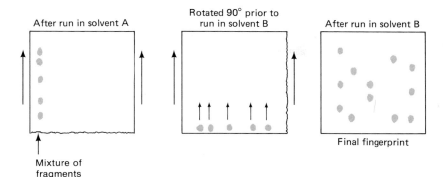

Figure 12-11. Two-dimensional chromatographic fingerprinting of a polypeptide fragment mixture. A protein is digested by a proteolytic enzyme into fragments that are only a few amino acids long. A piece of chromatographic filter paper is then spotted with this mixture and dipped into solvent A. As solvent A ascends the paper, some of the fragments become separated. The paper is then turned 90° and further resolution of the fragments is obtained as solvent B ascends.

meric (composed of several separate polypeptide monomers; see Figures 12-9 and 12-10). Many proteins are basically compact structures; such proteins are called **globular proteins.** Enzymes and antibodies are among the important globular proteins. Other unfolded proteins, called **fibrous proteins,** are important components of such structures as hair and muscle.

In summary, the linear sequence of a protein folds up to yield a unique three-dimensional configuration. This configuration creates specific sites to which substrates bind and at which catalytic reactions occur. The three-dimensional structure of a protein, which is crucial for its function, is determined solely by the primary structure (linear sequence) of amino acids. Therefore, genes can control enzyme function by controlling the primary structure of proteins.

Determining Protein Sequence

Proteins act as biological catalysts, as hormones, and as structural elements in spindle fibers, hair, and muscle, for example. Proteins play a very central role in living systems. If we purify a particular protein, we find that we can specify a particular ratio of the various amino acids for that specific protein. But the protein is not formed by a random hookup of fixed amounts of the various amino acids; it also has a characteristic sequence. For a small polypeptide, the amino-acid sequence can be determined by clipping off one amino acid at a time and identifying it. However, large polypeptides cannot be readily "sequenced" in this way.

Frederick Sanger worked out a brilliant method for deducing the sequence of large polypeptides. There are several different **proteolytic enzymes**—enzymes that can break peptide bonds only between specific amino acids in proteins. A large protein can be broken by such enzymes into a number of smaller fragments, which can be separated according to their migration speeds in a solvent on chromatographic paper. Because the speeds of mobility of different fragments may vary differently

in various solvents, two-dimensional chromatography can be used to enhance the separation of the fragments (Figure 12-11). In this technique, the mixture is separated in one solvent; then the paper is turned 90° and another solvent is used.

When the paper is stained, the polypeptides appear as spots in a characteristic chromatographic pattern called the **fingerprint** of the protein. Each of the spots can be cut out, and the polypeptide fragments can be washed from the paper. Because each spot contains only small polypeptides, their amino-acid sequences can easily be determined. Using different proteolytic enzymes to cleave the protein at different points, we can repeat the experiment to obtain other sets of fragments. The fragments from the different treatments overlap (because the breaks are made in different places with each treatment). The problem of solving the overall sequence then becomes one of fitting together the small-fragment sequences—almost like solving a tricky jigsaw or crossword puzzle (Figure 12-12).

Using this elegant technique, Sanger confirmed that the sequence of amino acids (as well as the amounts of the various amino acids) is specific to a particular protein. In other words, the amino-acid sequence is what makes insulin insulin.

Relationship Between Gene Mutations and Altered Proteins

We now know that the change of just one amino acid is sometimes enough to alter protein function. This was first shown in 1957 by Vernon Ingram who studied the globular protein hemoglobin—the molecule that transports oxygen in red blood cells (see page 73). Hemoglobin is made up of four polypeptide chains. Hemoglobin A (HbA), the protein from normal adults, contains two identical chains of one type called α chains and two identical chains of a second type called β chains. The α chain contains 141 amino acids, and the β chain contains 146. Ingram compared HbA from normal people with

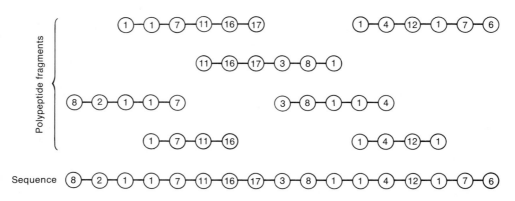

Figure 12-12. Alignment of polypeptide fragments to reconstruct an entire amino-acid sequence. Different proteolytic enzymes can be used on the same protein to form different fingerprints. The amino-acid sequence of each spot can be determined rather easily, and due to the overlap of amino-acid sequences from different spots from different fingerprints, the entire amino-acid sequence of the original protein can be determined. Using this procedure, it took Sanger about six years to determine the sequence of the insulin molecule, a relatively small protein.

HbS from people homozygous for the mutant gene that causes sickle-cell anemia. Using Sanger's technique, Ingram found that the fingerprint of HbS differs from that of HbA in only one spot. Sequencing that spot from the two kinds of hemoglobin, Ingram found that only one amino acid in the fragment differs in the two kinds.

Apparently, of all of the amino acids known to make up a hemoglobin molecule, a substitution of valine for glutamic acid at just one point, position 6 in the β chain, is all that is needed to produce the defective hemoglobin (Figure 12-13). Unless patients with HbS receive medical attention, this single error in an amino acid in one pro-

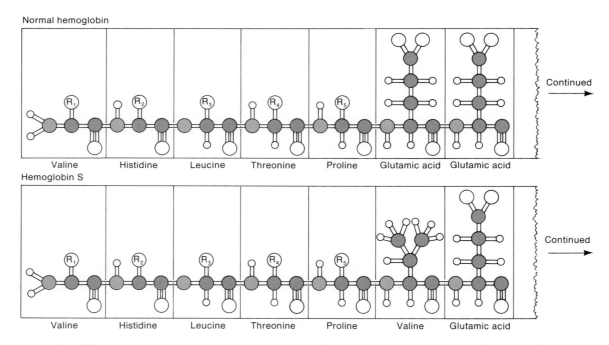

Figure 12-13. The difference at the molecular level between normalcy and sickle-cell disease. Shown are only the first seven amino acids; all the rest not shown are identical.

(From Anthony Cerami and Charles M. Peterson, "Cyanate and Sickle-cell Disease." Copyright © 1975 by Scientific American, Inc. All rights reserved.)

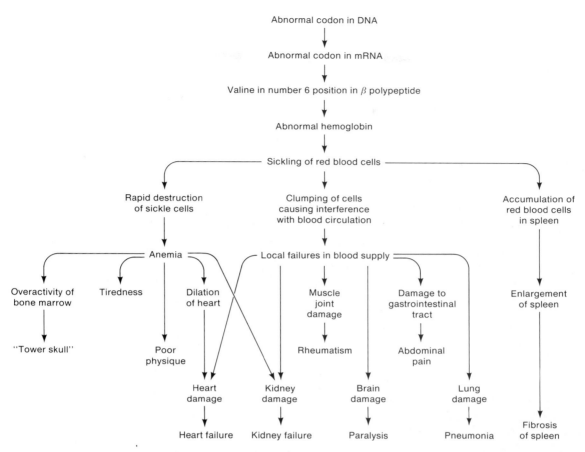

Figure 12-14. The compounded consequences of one amino-acid substitution in hemoglobin to produce sickle-cell anemia. The terms condon and mRNA will be explained later.

tein will hasten their death. Figure 12-14 shows the sequence of events leading to the disease.

Notice what Ingram accomplished. *A gene mutation that is well established through genetic studies has been connected to an altered amino acid sequence in a protein!* Subsequent studies identified numerous changes in hemoglobin, and each is the consequence of a single amino-acid difference. (Figure 12-15 shows a few exam-

ples.) We can conclude that one mutation in a gene corresponds to a change of one amino acid in the sequence of a protein.

Message   Genes determine the primary sequences of amino acids in specific proteins.

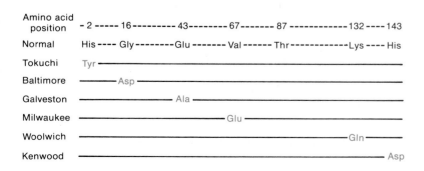

Figure 12-15. A variety of single amino-acid substitutions in human hemoglobin. Amino acids at all residue positions except those indicated are normal. Each type of change causes disease. (Names indicate areas in which cases were first identified.)

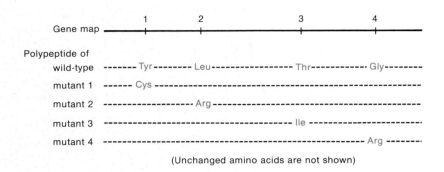

Figure 12-16. Simplified representation of the colinearity of gene mutations. The genetic map of point mutations (determined by recombinational analysis) corresponds linearly to the changed amino acids in the different mutants (determined by fingerprint analysis).

Colinearity of Gene and Protein

The elucidation of the structure of DNA by Watson and Crick led to the realization that the structure of proteins must be encoded in the linear sequence of nucleotides in the DNA. (We see how this genetic code was deciphered in Chapter 13). Following Ingram's demonstration that one mutation alters one amino acid in a protein, a relationship was sought between the linear sequence of mutant sites in a gene and the linear sequence of amino acids in a protein. (It is possible to map mutational sites within a gene due to studies on genetic fine structure to be described in detail later in this chapter).

Charles Yanofsky probed the relationship between altered genes and altered proteins by studying the enzyme tryptophan synthetase in *E. coli*. The enzyme catalyzes the conversion of indole glycerol phosphate into tryptophan. Two genes, *trpA* and *trpB*, control the enzyme. Each gene controls a separate polypeptide; after the A and B polypeptides are produced, they combine to form the active enzyme (a multimeric protein). Yanofsky analyzed mutations in the *trpA* gene which resulted in alterations of the tryptophan synthetase A subunit. He ordered the mutations by P1 transduction (see p. 243)

to produce a detailed gene map, and he also determined the sequence of each respective altered tryptophan synthetase. His results were similar to Ingram's for hemoglobin: each mutant has a defective polypeptide associated with a specific amino-acid substitution at a specific point. However, Yanofsky was able to show an exciting correlation that Ingram was not able to observe due to the limitations of his system. There is an exact match between the sequence of the mutational sites in the gene map of the *A* gene and the location of the corresponding altered amino acids in the A polypeptide chain. The farther apart two mutational sites are in map units, the more amino acids there are between the corresponding substitutions in the polypeptide (Figure 12-16). Thus, Yanofsky demonstrated **colinearity**—the correspondence between the linear sequences of the gene and the polypeptide. Figure 12-17 shows the complete set of data.

Message   The linear sequence of nucleotides in a gene determines the linear sequence of amino acids in a protein.

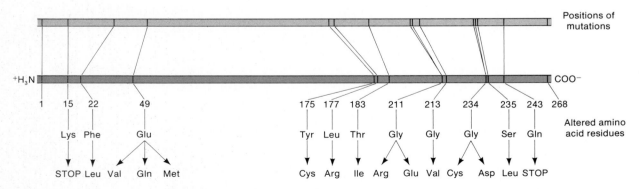

Figure 12-17. Actual colinearity shown in the A protein of tryptophan synthetase from *E. coli*. There is a linear correlation between the mutational sites and the altered amino-acid residues. (Based on C. Yanofsky, "Gene Structure and Protein Structure." Copyright © 1967 by Scientific American. All rights reserved.)

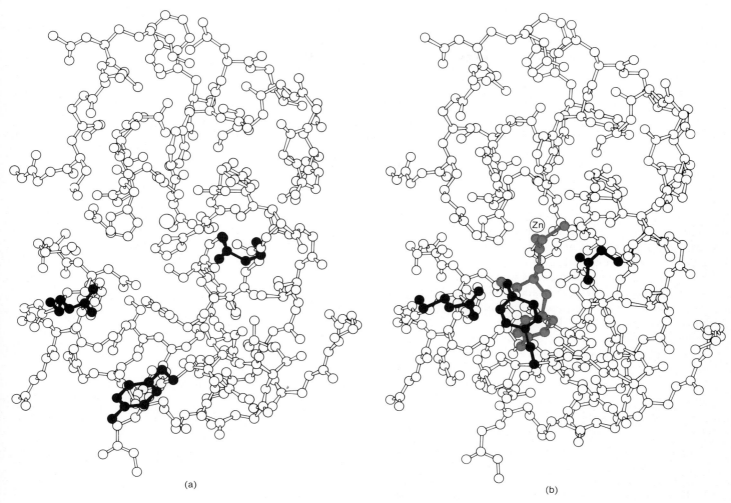

(a)

(b)

Figure 12-18. The active site of a specific enzyme, the digesive enzyme carboxypeptidase. (a) The enzyme without substrate. (b) The enzyme with its substrate (color) in position. Three crucial amino acids (black) have changed positions to engage with the substrate. Carboxypeptidase carves up proteins in the diet. (From W. N. Lipscomb, *Proc. Robert A. Welch Found. Conf. Chem. Res.* 15, 1971, 140–41.)

Protein Function

The genes truly are the master controllers of the cell. They dictate not only cell chemistry through the enzymes encoded by some genes but also biological architecture through the structural proteins encoded by other genes. Furthermore, the blueprints of such important proteins as hormones and hemoglobin are encoded in the structure of genes.

How can a single amino-acid substitution, such as that in sickle-cell hemoglobin (Figure 12-13), have such a profound effect on protein function and the phenotype of an organism? Take enzymes, for example. Enzymes are known to do their job of catalysis by physically grappling with their substrate molecules, twisting or bending the molecules to make or break chemical bonds. Figure 12-18 shows the gastric digestion enzyme car-boxypeptidase in its relaxed position and after grappling with its substrate molecule, glycyltyrosine. The substrate molecule fits into a notch in the enzyme structure; this notch is called the **active site.**

Figure 12-19 diagrams the general concept. (Note that here we have encountered the two basic types of reactions performed by enzymes: the breakdown of substrate into simple products, and the synthesis of a complex product from one or more simpler substrates.)

Much of the globular structure of an enzyme is nonreactive material that simply supports the active site. We might expect amino-acid substitutions throughout most of the structure to have little effect but very specific amino acids to be required for the part of the enzyme molecule that gives the precise shape to the active site.

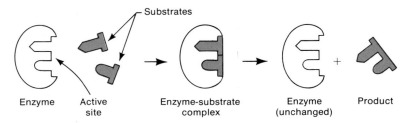

Figure 12-19. Diagrammatic representation of the action of a hypothetical enzyme in putting two substrate molecules together. The "lock-and-key" fit of the substrate into the enzyme's active site is very important in this model.

Hence, the possibility arises that a functional enzyme does not require a *unique* amino-acid sequence for the entire polypeptide. This has been demonstrated to be true in a number of systems in which the effects of different amino-acid substitutions on the catalytic activity of an enzyme have been examined. It is clear that numerous positions in a polypeptide can be filled by several alternative amino acids that are compatible with the enzyme function. At certain other positions in the polypeptide, only the wild-type amino acid will restore activity; in all likelihood, these amino acids form critical parts of the active sites. Some of these critical amino acids are indicated in Figure 12-18 in black.

Message Protein architecture is the key to gene function. A gene mutation typically results in the substitution of a different amino acid into the polypeptide sequence of a protein. The new amino acid may have chemical properties that are incompatible with the proper protein architecture at that particular position; in such a case, the mutation will lead to a nonfunctional protein.

Temperature-sensitive Alleles

Recall that some mutants appear to be wild-type at normal temperatures but can be detected as mutants at high or low temperatures (see page 153). We can now explain such mutations by assuming that the substitution of an amino acid produces a protein that is functional at normal temperatures, called **permissive** temperatures, but distorted and nonfunctional at high or low temperatures, called **restrictive** temperatures (Figure 12-20).

As we have seen, conditional mutations such as temperature-sensitive mutations can be very useful to geneticists. Stocks of the mutant culture can easily be maintained under permissive conditions, and the mutant phenotype can be studied intensively under restrictive conditions. Such mutants can be very useful in the ge-

netic dissection of biological systems. For example, with a temperature-sensitive allele, the time at which a gene is acting can be determined by shifting to a restrictive temperature at various times during development.

Enzymic Explanation of Genetic Ratios and Dominance

When the significance of the gene control of cellular chemistry became clear, a lot of other things fell into place. Many genetic generalizations now could be explained and tied together in a single conceptual model.

Most "classical" (Mendelian) gene-interaction ratios can be explained simply by the one-gene–one-enzyme concept. For example, recall the $9:7$ F_2 dihybrid ratio for flower pigment (page 82):

P   *AA bb* (white) $\times$ *aa BB* (white)

F$_1$ all *Aa Bb*   (purple)

F$_2$ 9   *A — B —*   (purple)

       3   *A — bb*   (white)

       3   *aa B —*   (white)

       1   *aa bb*   (white)

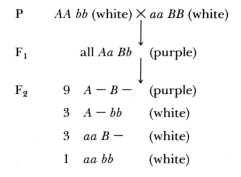

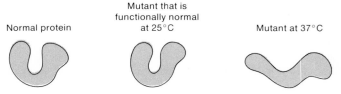

Figure 12-20. Diagram of protein conformational distortion, which is probably the basis for temperature sensitivity in certain mutants. An amino-acid substitution that has no significant effect at normal (permissive) temperatures may cause significant distortion at abnormal (restrictive) temperatures.

We can easily explain this result if we imagine a biosynthetic pathway that leads ultimately to a purple petal pigment in which there are two colorless (white) precursors:

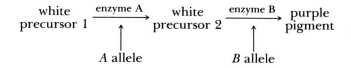

white precursor 1 $\xrightarrow{\text{enzyme A}}$ white precursor 2 $\xrightarrow{\text{enzyme B}}$ purple pigment

A allele B allele

(Try to work out your own models to explain such ratios as 9:3:4 and 13:3.)

The meaning of dominance and recessiveness also becomes a little clearer in light of the biochemical model. In most cases, dominance represents the presence of enzyme function, whereas recessiveness represents the lack of enzyme function. A heterozygote has one dominant allele that can produce the functional enzyme:

precursor X $\xrightarrow[\text{enzyme A}]{}$ product Y

allele A allele a $\longrightarrow$ nonfunctional enzyme (does nothing)

If phenotype Y is due to the presence of product Y and phenotype X is due to the absence of product Y, then it is clear that the heterozygote will show phenotype Y and allele A will be dominant over allele a.

However, this is not the only possible model. We can build in the concept of a threshold. Suppose that phenotype Y is produced only when the concentration of product Y exceeds some threshold level. Further suppose that the homozygote AA produces more enzyme—and, hence, more produce—than the heterozygote Aa. In this case, the phenotype of the heterozygote will depend on the relationship between the threshold and the amount of product Y that is produced by the heterozygote. In Figure 12-21a, the heterozygote A_1A_2 does produce enough product Y to exceed the threshold, so the heterozygote has phenotype Y and therefore A_2 is dominant over A_1. However, the situation could be like the one shown for the alleles B_1 and B_2, where the heterozygote B_1B_2 does not exceed the threshold. In this case (which is less common), B_1 is dominant over B_2 and the dominant phenotype is the one involving a lack of product Y.

Figure 12-21b illustrates a situation in which no threshold exists. The heterozygote has an intermediate phenotype—exactly the situation observed in cases of incomplete dominance.

What determines whether a gene will behave like A, B, or C? Probably many interacting factors are involved: other genes, the chemical nature of the product, and (last but not least) the effect of the environment on that particular cell type. Finally, note that Figure 12-21 shows a simple linear relationship between enzyme concentration and the number of active alleles—but this need not be the case, as we see in Chapter 16.

The recessiveness of the *arg* mutants in *Neurospora* can be demonstrated by making heterokaryons between two *arg* mutants. If an *arg-1* mutant is placed on a minimal medium with an *arg-2* mutant, the two cell types fuse

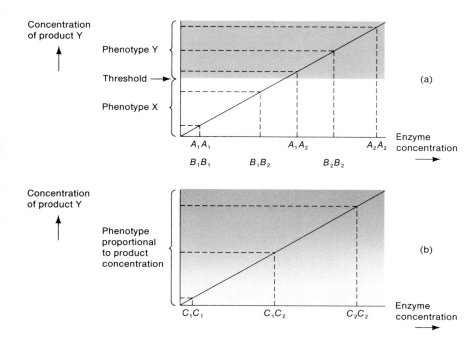

Figure 12-21. Hypothetical curves relating enzyme concentration to the amount of product. Two basic situations are possible. In (a), there is a threshold of the enzyme product above which a sharply contrasting phenotype, Y, is observed. Depending on where the levels in the three possible genotypes occur in relation to this threshold, the heterozygote will show either phenotype X (for example, B_1B_2) or Y (for example, A_1A_2). If the subscript 2 alleles are the active ones, you can see that the active allele (A_2) can be dominant, which is normally the case. Less frequently, the inactive allele (B_1) can be dominant. In (b) there is no threshold, and the heterozygote has an intermediate phenotype. This situation explains incomplete dominance.

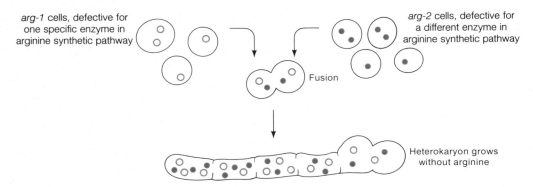

arg-1 cells, defective for one specific enzyme in arginine synthetic pathway

arg-2 cells, defective for a different enzyme in arginine synthetic pathway

Fusion

Heterokaryon grows without arginine

Figure 12-22. Formation of and complementation in a heterokaryon of *Neurospora*. Vegetative cells of this fungus can fuse, allowing the nuclei from the two strains to intermingle within the same cytoplasm. If each cell is blocked at a different point in a pathway, as are *arg-1* and *arg-2*, all functions are present in a heterokaryon and the *Neurospora* will grow; in other words, complementation takes place.

and form a heterokaryon composed of both nuclear types in a common cytoplasm (Figure 12-22). The *arg-1* nuclei produce enzyme for one step, and the *arg-2* nuclei produce enzyme for the other step, so they combine their abilities to produce arginine in the shared cytoplasm.

Genes and Cellular Metabolism: Genetic Diseases

The scale of involvement of the genes in controlling cellular metabolism is staggering. Most of us have been amazed by the charts on laboratory walls showing the myriad interlocking, branched, and circular pathways along which the cell's chemical intermediates are shunted like parts on an assembly line. Bonds are broken, molecules cleaved, molecules united, groups added or removed, and so on. The key fact is that almost every step, represented by an arrow on the metabolic chart, is controlled (mediated) by an enzyme, and each of these enzymes is produced under the direction of a gene that specifies its function. Genes control the enzymes, and the enzymes control the chemical reactions that comprise metabolism. When we study a gene (or a few genes), we are focusing on one tiny corner of this vast network controlled by genes and their enzymes.

Humans provide some startling examples. The list of specific enzyme-associated genetic diseases in Table 12-4 suggests the magnitude of genetic involvement in human disease. Figure 12-23 shows a corner of the metabolic map to illustrate how a set of diseases, some of them common and familiar to us, can stem from the blockage of adjacent steps in biosynthetic pathways.

Genetic Fine Structure

Until the beginning of this chapter, our genetic and cytological analysis led us to regard the chromosome as a linear (one-dimensional) array of genes, strung rather like beads on an unfastened necklace. Indeed, this model is sometimes called the bead theory. According to the bead theory, the existence of a gene as a unit of inheritance is recognized through its mutant alleles. All of these alleles affect a single phenotypic character, all map to one chromosome locus, all give mutant phenotypes when paired, and all show Mendelian ratios when intercrossed. Several points about the bead theory are worth emphasizing:

1. The gene is viewed as a fundamental unit of *structure*, indivisible by crossing-over. Crossing-over occurs between genes (the beads in this model) but never within them.

2. The gene is viewed as a fundamental unit of *change*, or mutation. It changes from one allelic form to another, but there are no smaller components within it that can change.

3. The gene is viewed as the basic unit of *function* (although the precise function of the gene is not specified in this model). Parts of a gene, if they exist, cannot function.

However, the elucidation of the structure of DNA (Chapter 11), together with the realization that the gene consists of a sequence of nucleotides that encode the amino-acid sequence of a protein, are at odds with the view that the gene is the smallest unit of mutation and

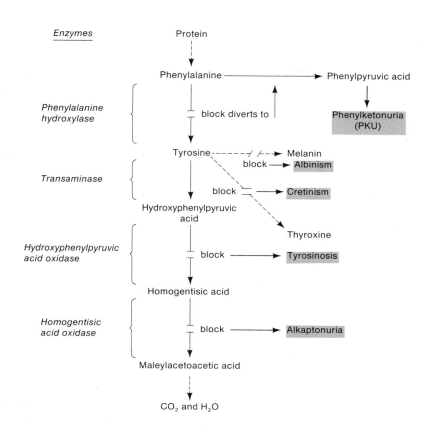

Figure 12-23. One small part of the human metabolic map, showing the consequences of various specific enzyme failures. (Disease phenotypes are shown in colored boxes.) (After I. M. Lerner and W. J. Libby, *Heredity, Evolution, and Society,* 2d ed. Copyright © 1976 by W. H. Freeman and Company.)

recombination. How can these two viewpoints be reconciled? Seymour Benzer's work in the 1950s showed that the bead theory was not correct. Benzer was able to use a genetic system in which extremely small levels of recombination could be detected. He demonstrated that whereas a gene can be defined as a unit of function, a gene can be subdivided into a linear array of sites that are mutable and that can be recombined. The smallest units of mutation and recombination are now known to be correlated with single nucleotide pairs.

TABLE 12-4. Enzymopathies: inherited disorders in which altered activity (usually deficiency) of a specific enzyme has been demonstrated in humans

Condition	Enzyme with deficient activity[1,2]	Condition	Enzyme with deficient activity[1,2]
Acatalasia	Catalase	Angiokeratoma, diffuse (Fabry disease)	Ceramide trihexosidase
Acid phosphatase deficiency	Acid phosphatase	Apnea, drug-induced	Pseudocholinesterase
Adrenal hyperplasia I	20,21-Desmolase*	Argininemia	Arginase
Adrenal hyperplasia II	3-β-Hydroxysteroid dehydrogenase*	Argininosuccinic aciduria	Argininosuccinase
Adrenal hyperplasia III	21-Hydroxylase*	Aspartylglycosaminuria	Specific hydrolase (AADGase)
Adrenal hyperplasia IV	11-β-Hydroxylase*	Ataxia, intermittent	Pyruvate decarboxylase
Adrenal hyperplasia V	17-Hydroxylase*	Carnosinemia	Carnosinase
Albinism	Tyrosinase	Citrullinemia	Arginosuccinic acid synthetase
Aldosterone deficiency	18-OH-Dehydrogenase	Crigler–Najjar syndrome	Glucuronyl transferase
Alkaptonuria	Homogentisic acid oxidase		

(Continued)

TABLE 12-4. *(Continued)*

Condition	Enzyme with deficient activity[1,2]	Condition	Enzyme with deficient activity[1,2]
Cystathioninuria	Cystathionase	Hemolytic anemia	Adenosine triphosphatase
Disaccharide intolerance I	Invertase		Adenylate kinase
Disaccharide intolerance II	Invertase, maltase		Aldolase A
Disaccharide intolerance III	Lactase		Diphosphoglycerate mutase
Ehlers–Danlos syndrome, type V	Lysyl oxidase		Glucose 6-phosphate dehydrogenase
Ehlers–Danlos syndrome, type VI	Collagen lysyl hydroxylase		γ-Glutamylcysteine synthetase
Ehlers–Danlos syndrome, type VII	Procollagen peptidase		Glutathione peroxidase
			Glutathione synthetase
Fanconi panmyelopathy	Exonuclease*		Hexokinase
Farber lipogranulomatosis	Ceramidase		Hexosephosphate isomerase
Formininotransferase deficiency	Formininotransferase*		Phosphoglycerate kinase
			Pyrimidine 5'-nucleotidase
Fructose intolerance	Fructose 1-phosphate aldolase		Pyruvate kinase
			Trisephosphate isomerase
Fructosuria	Hepatic fructokinase	Histidinemia	Histidase
Fucosidosis	α-L-Fucosidase	Homocystinuria I	Cystathionine synthetase
Galactokinase deficiency	Galactokinase	Homocystinuria II	N^5N^{10}-Methylenetetrahydrofolate reductase
Galactose epimerase deficiency	Galactose epimerase	β-Hydroxyisovaleric aciduria and methylcrotonylglysinuria	β-Methylocrotonyl CoA carboxylase*
Galactosemia	Galactose 1-phosphate uridyl transferase		
Gangliosidosis, generalized, type I GM	β-Galactosidase A, B, C	Hydroxyprolinemia	Hydroxyproline oxidase
		Hyperammonemia I	Ornithine transcarbamylase
Gangliosidosis, GM$_1$, type II or juvenile form	β-Galactosidase B, C	Hyperammonemia II	Carbamyl phosphate synthetase
Gangliosidosis, GM(3)	Acetylgalactosaminyl transferase	Hyperglycemia, ketotic form	Propionyl CoA carboxylase*
		Hyperglycinemia, non-ketotic form	Glycine formininotransferase
Gaucher disease	Glucocerebrosidase	Hyperlipoproteinemia, type I	Lipoprotein lipase
Glycogen storage disease I	Glucose 6-phosphatase	Hyperlysinemia	Lysine-ketoglutarate reductase
Glycogen storage disease II	α-1,4-Glucosidase		
Glycogen storage disease III	Amylo-1,6-glucosidase	Hyperprolinemia I	Proline oxidase
Glycogen storage disease IV	Amylo-(1,4–1,6)-transglucosidase	Hyperprolinemia II	δ-1-Pyrroline-5-carboxylate dehydrogenase*
Glycogen storage disease V	Muscle phosphorylase		
Glycogen storage disease VI	Liver phosphorylase*	Hypoglycemia and acidosis	Fructose 1,6-diphosphatase
Glycogen storage disease VII	Muscle phosphofructokinase	Hypophosphatasia	Alkaline phosphatase
Glycogen storage disease VIII	Liver phosphorylase kinase	Immunodeficiency disease	Adenosine deaminase
			Uridine monophosphate kinase
Gout	Hypoxanthine guanine phosphoribosyltransferase PPRP synthetase (increased activity)	Intestinal lactase deficiency (adult)	Lactase
		Isovalericacidemia	Isovaleric acid CoA dehydrogenase
		Ketoacidosis, infantile	Succinyl CoA: 3-ketoacid CoA transferase
		Krabbe disease	A β-Galactosidase
Granulomatous disease	NADPH oxidase	Lactosyl ceramidosis	Lactosyl ceramidase

(Continued)

TABLE 12-4. *(Continued)*

Condition	Enzyme with deficient activity[1,2]	Condition	Enzyme with deficient activity[1,2]
Leigh necrotizing encephalomyelopathy	Pyruvate carboxylase	Oxalosis II (glyceria aciduria)	D-Glycerate dehydrogenase
Lipase deficiency, congenital	Lipase (pancreatic)	Pentosuria	Xylitol dehydrogenase (L-xylulose reductase)
Lysine intolerance	L-Lysine: NAD-oxidoreductase	Phenylketonuria	Phenylalanine hydroxylase
Male pseudohermaphroditism	Testicular 17,20-desmolase Testicular 17-ketosteroid dehydrogenase* α-Reductase*	Porphyria, acute intermittent	Uroporphyrinogen I synthetase
		Porphyria, congenital	Uroporphyrinogen III cosynthetase
Mannosidosis	α-Mannosidase	Pulmonary emphysema and/or cirrhosis	α-1-Antitrypsin
Maple-sugar urine disease	Keto acid decarboxylase	Pyridoxine-dependent infantile convulsions	Glutamic acid decarboxylase
Metachromatic leukodystrophy	Arylsulfatase A (sulfatide sulfatase)	Pyridoxine-responsive anemia	δ-Aminolevulinic acid synthetase*
Methemoglobinemia	NAD-methemoglobin reductase	Pyruvate carboxylase	Pyruvate carboxylase
Methylmalonicaciduria I (B₁₂-unresponsive)	Methylmalonic CoA mutase	Refsum disease	Phytanic acid oxidase
Methylmalonicaciduria II (B₁₂-responsive)	Deoxyadenosyl transferase*	Renal tubular acidosis with deafness	Carbonic anhydrase B
Methylmalonicaciduria III	Methylmalonyl CoA racemase	Richner–Hanhart syndrome	Tyrosine aminotransferase
Mucopolysaccharidosis I	α-L-Iduronidase	Rickets, vitamin-D-dependent	25-Hydroxycholecalciferol*
Mucopolysaccharidosis II	Sulfo-iduronide sulfatase	Sandhoff disease (GM₂-gangliosidosis, type II)	Hexosaminidase A, B
Mucopolysaccharidosis IIIA	Heparan sulfate sulfatase	Sarcosinemia	Sarcosine dehydrogenase*
Mucopolysaccharidosis IIIB	N-Acetyl-α-D-glucosaminidase	Sulfite oxidase deficiency	Sulfite oxidase
Mucopolysaccharidosis IV	6-Sulfatase*	Tay-Sachs disease	Hexosaminidase A
Mucopolysaccharidosis VI	Arylsulfatase B	Thyroid hormonogenesis, defect in, II	Peroxidase*
Mucopolysaccharidosis VII	β-Glucuronidase	Thyroid hormonogenesis, defect in IV	Iodotyrosine dehalogenase (deiodinase)
Myeloperoxidase deficiency with disseminated candidiases	Myeloperoxidase (leukocyte)	Trypsinogen deficiency	Trypsinogen
Niemann–Pick disease	Sphingomyelinase	Tyrosinemia I	Para-hydroxyphenylpyruvate oxidase
Norum disease	Lecithin cholesterol acetyltransferase (LCAT)	Tyrosinemia II	Tyrosine transaminase
Ornithinemia	Ornithine ketoacid aminotransferase	Valinemia	Valine transaminase
		Wolman disease	Acid lipase
Oroticaciduria I	Orotidylic pyrophosphorylase and orotidylic decarboxylase	Xanthinuria	Xanthine oxidase
		Xanthurenic aciduria	Kynureninase
		Xeroderma pigmentosum	Ultraviolet specific endonuclease
Oroticaciduria II	Orotidylic decarboxylase		
Oxalosis I (glycolic aciduria)	2-Oxoglutarate-glyoxylase carboligase	Xylosidase deficiency	Xylosidase

SOURCE: Victor A. McKusick, *Mendelian Inheritance in Man*, 4th ed., Johns Hopkins University Press, 1975.
[1] In some conditions marked * (as well as some not listed), deficiency of a particular enzyme is suspected but has not been proved by direct study of enzymatic activity.
[2] The form of gout due to increased activity of PPRP is the only disorder listed that is due to *increased* enzymatic activity.

Fine–structure Analysis of the Gene

The following material deals with Benzer's classic analysis of the fine structure of the gene. The important concepts in this work are:

1. The life cycle of the bacteriophage.

2. Plaque morphology and the *rII* system of phage T4.

3. The concept of "selection" in genetic crosses with bacteriophages.

4. Deletion mapping.

5. Destruction of the bead theory.

6. Complementation: the difference between complementation and recombination.

The Life Cycle of the Bacteriophage

Benzer's work is much easier to understand if we first review the material covered on pages 235–238. Let's summarize some of the essential points. Recall that bacteriophages are viruses that attack bacteria. The bacteria used most frequently for these studies is the single-cell bacterium *E. coli*. Bacterial viruses usually consist of a protein coat and tail fibers, which allow the virus to attach to the bacterial cell wall. The viral nucleic acid is injected into the cell; this programs the synthesis of proteins that stimulate viral replication and the production of new coat and tail proteins. After the new viral particles assemble, the cell is lysed and the particles are released into the surrounding medium. If the infected cell is on a lawn of bacteria immobilized on the agar surface of a petri plate, then the new virus progeny will infect neighboring cells and their progeny, in turn, will infect other neighboring cells until a small clearing, visible to the naked eye, is produced. This clearing is termed a **plaque**. Figure 10-24 provides an example of phage plaques. Different viruses make different types of plaques. Viruses also differ in the types of cells they infect—a property termed the **host range**. Initially, bacteriophages were characterized according to their host range and given numbers in the "T" series. Thus, we have bacteriophages T1, T2, T3, T4, T5, and so on. Many experiments relating to the nature of the gene and the genetic code have been carried out with phage T4 (Figure 12-24).

The *rII* System

Benzer sought genetic markers that could be used with bacteriophages. When it was realized that the size and shape of a plaque were heritable traits of the virus, a genetic analysis of plaque morphology was initiated.

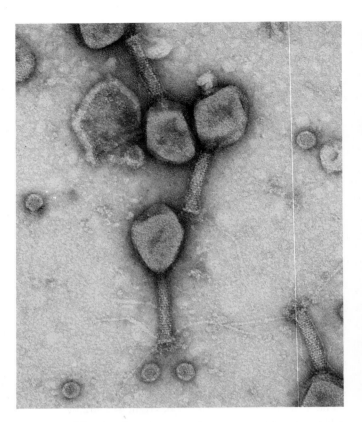

Figure 12-24. Enlargement of the *E. coli* phage T4 showing details of structure: note head, tail, and tail fibers. This was the phage used by Benzer in his experiments on the nature of the *rII* (rapid lysis) gene. (Photograph courtesy of Jack D. Griffith.)

One type of mutant virus (Figure 12-25) produced large plaques that were easy to distinguish from wild-type plaques. The large plaque size resulted from rapid lysis of the bacteria, so the mutants were termed **r** (rapid lysis) **mutants.** Benzer analyzed the *r* mutants genetically, and mapped the respective mutations into two loci: *rI* and *rII*. He then studied the *rII* mutants intensively.

One extraordinary property of *rII* mutants made all of Benzer's work possible: *rII* mutants have a different host range than that of wild-type phages. Two related but different strains of *E. coli*, termed B and K(λ) can be used as different hosts for phage T4. Both bacterial hosts can distinguish *rII* mutants from wild-type phages. *E. coli* B allows both to grow, but different size plaques result: wild-type phages produce small plaques, and *rII* mutants produce large plaques. *E. coli* K, an abbreviation for *E. coli* K(λ), does not permit the growth of *rII* mutants, but it does allow wild-type phages to grow. Figure 12-26 and Table 12-5 show the growth characteristics and plaque morphology of these phages on each host strain. It is essential to know the relationships in Table 12-5 if we are to understand the subsequent work.

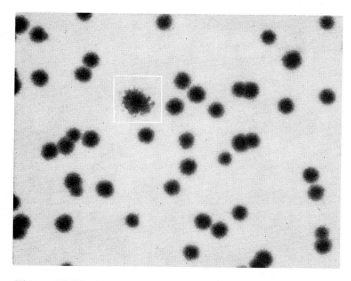

Figure 12-25. A spontaneous mutational event is disclosed by the one mottled plaque *(square)* among dozens of normal plaques produced when standard phage T4 is plated on a layer of colon bacilli of strain B. Each plaque contains some 10 million progeny descended from a single phage particle. The plaque itself represents a region in which cells have been destroyed. Mutants found in abnormal plaques provide the raw material for genetic mapping. (From S. Benzer, "The Structure of the Gene." Copyright © 1962 by Scientific American, Inc. All rights reserved.)

E. coli B is said to be **permissive** for *rII* mutants, because it allows phage growth, whereas *E. coli* K(λ) is said to be **nonpermissive** for *rII* mutants, because it does not allow phage growth.

TABLE 12-5. Plaque phenotypes produced by different combinations of *E. coli* and phage strains

T4 phage strain	*E. coli* strain	
	B	K(λ)
rII	Large, round	No plaques
rII⁺	Small, ragged	Small, ragged

Selection in Genetic Crosses of Bacteriophages

How can we carry out crosses with different phages? In order to do this, two phages are used to infect the same bacterial cell, as described on page 237. If a mixture of phages is used and the ratio of phages to bacteria is high enough, then virtually every bacterium will be infected with at least one phage of each type. Once inside the cell, the DNA from each T4 phage (in this case) have an opportunity to recombine with one another, generating recombinant phages, which are recovered among the progeny. Because both wild-type and *rII* mutants can make plaques and these plaques can be distinguished from one another, we could cross two different *rII* mutants and examine the progeny on *E. coli* B (Figure 12-27, top right photograph), hoping to find small wild-type plaques among the large parental *rII* plaques. If the recombination frequency is high enough to yield several percent or more wild-type plaques, then this method would suffice. However, for recombination that is much less frequent than 1 percent, a lot of work would be involved in generating a map of numerous *rII* mutations.

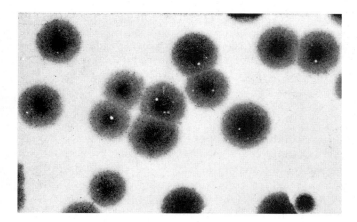

Figure 12-26. Duplicate replatings of a mixed phage population, obtained from a mottled plaque like the one shown in Figure 12-25, give contrasting results, depending on the host. Replated on colon bacilli of strain B *(left),* *r*II mutants produce large plaques. If the same mixed

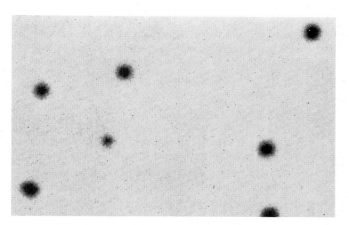

population is plated on strain K *(right),* only the standard phages produce plaques. (From S. Benzer, "The Structure of the Gene." Copyright © 1962 by Scientific American, Inc. All rights reserved.)

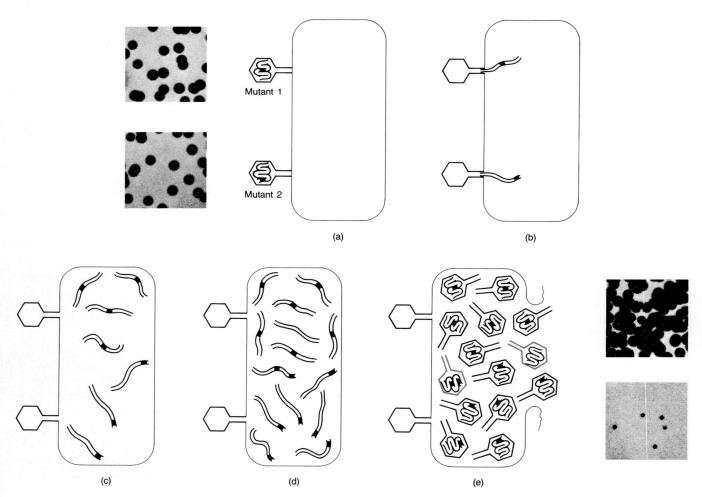

Figure 12-27. The process of recombination permits parts of the DNA of two different phage mutants to be reassembled in a new DNA molecule that may contain both mutations or neither of them (color). Mutants obtained from two different cultures *(photographs at far left)* are introduced into a broth of strain B colon bacilli. Crossing occurs (a) when DNA from each mutant type infects a single bacillus. Most of the DNA replicas are of one type or the other, but occasionally recombination will produce either a double mutant or a standard recombinant containing neither mutation. When the progeny of the cross are plated on strain B *(top photograph at far right)*, all grow successfully, producing many plaques. Plated on strain K, only the standard recombinants are able to grow *(bottom photograph at right)*. A single standard recombinant can be detected among as many as 100 million progeny. (From S. Benzer, "The Structure of the Gene." Copyright © 1962 by Scientific American, Inc. All rights reserved.)

Instead of plating the progeny phages from the cross on *E. coli* B, however, we could plate the progeny on *E. coli* K (Figure 12-27, bottom right photograph), so that only the wild-type recombinant phages could grow. Even if the recombination frequency is very low (say, 0.01 percent), we could detect the recombinant phages easily. Why? Because a typical phage lysate (the phage mixture produced after lysis of the bacteria) from such an infection (whether it involves a cross or not) contains in excess of 10^9 phage per milliliter (ml). If we mix 0.1 ml of such a phage lysate with 0.1 ml of *E. coli* K bacteria, then we would have greater than 10^5 (100,000) wild-type recombinant phages infecting the bacteria at a recombination frequency of 0.01 percent. In practice, increasing dilutions of the phage lysate are used until one dilution yields a countable number of plaques. Now we can see the power of the *rII* – *E. coli* B, K system. In a single milliliter, we can find one recombinant mutant per 10^9 organisms. Contrast this with trying to find one recombinant in 10^9 *Drosophila* or 10^9 mice!

Let's review the sequence of events just described. Samples of two mutant phages are used to coinfect *E. coli* B. Why *E. coli* B? Because *rII* mutants can grow on *E. coli* B, and growth and production of phage progeny is required to provide the opportunity for recombinant phages to appear. The lysate is then used to infect *E. coli* K. In practice, dilutions of the lysate are used in order to calculate the number, or the *titer*, of plaque-forming units (pfu), or active virus particles, in the sample. The number of pfu/ml of wild-type recombinants equals the number of pfu/ml on *E. coli* K, because only the wild-type recombinant phages can grow on *E. coli* K. To calculate the total number of phages in the lysate, regardless of whether they are the parental *rII* or the rare recombinant wild-type, the pfu's are determined on *E. coli* B. Why? Because all of the virus particles can grow. The recombinant frequency can be calculated as the number of pfu on *E. coli* K divided by the number of pfu on *E. coli* B. Can you see why? See Figures 12-27 and 12-28 for diagrammatic representations of these principles.

Finally, in any cross of this type, it is necessary to plate each parental lysate on *E. coli* K to see how many revertants to wild-type there were in the population. Back (reverse) mutations occur at some very low but real frequency. It is important to monitor this frequency and to compare it with the presumed frequency of recombination observed in a cross to be sure that recombination —not simply back reversion of the parental types—has occurred.

In summary, Benzer's use of the *rII* system and two different bacterial hosts provided him with a method for selecting for rare events without having to screen large numbers of plaques.

Message Benzer capitalized on the fantastic resolving power made possible by using selective systems for rare events in phages.

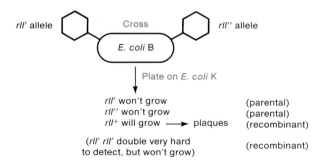

Figure 12-28. Selection of intragenic recombinants at the *rII* locus of phage T4. Mutants within the gene *rII* cannot grow on *E. coli* K. When two different phage carrying different alleles of *rII* infect the same bacterial cell, some progeny phages can grow on *E. coli* K; in other words, some progeny have become *rII*+. This result indicates that recombination has occurred *within* a single gene and not just between genes.

Intragenic Recombination

Benzer started with an initial sample of eight independently derived *rII* mutant strains and set about crossing them in all possible combinations of pairs by the double infection of *E. coli* B and subsequent plating onto a lawn of *E. coli* K (Figures 12-27 and 12-28). The alleles could be mapped unambiguously to the right or the left of each other to give what we now call a **gene map**. In this case, the map units are the frequency of *rII*+ plaques:

Recombination within a gene, called **intragenic recombination,** seems to be the rule rather than the exception. It can virtually always be found at any locus if a suitable selection system is available to detect recombinants. In other words, a mutant allele can be pictured as a length of genetic material (the gene) that has a damaged or non-wild part—a **mutational site**— somewhere. This partial damage is what causes the non-wild phenotype. Different alleles produce different phenotypic effects because they involve damage to different parts (sites) of the wild-type allele.

Thus, an allele a^1 can be represented as

$$+ + + + + * + + + + + + + + + + + + + + + +$$

where the asterisk (*) is the mutant site within an otherwise normal gene (denoted by the sites marked +). A cross between a^1 and another mutant allele a^2, in general, can be represented as

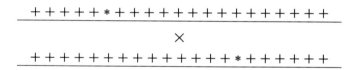

and it is easy to see how

$$+ +$$

could be generated by a simple crossover anywhere between the two mutational sites.

Benzer showed that contrary to the classical view, genes were not indivisible but could be subdivided by recombination. Extending his analysis to hundreds of *rII* alleles, Benzer found that the minimal recombinant frequency in a cross between a pair of different mutant alleles was 0.01 percent, even though his analytical system was capable of detecting recombinant frequencies as low as 0.0001 percent if they occurred. This led to the idea that genes were composed of small units, which Benzer named "recons," and that recombination could occur between but not within "recons." Thus, the "recon," rather than the gene, was the unit of recombination. The gene was assumed to be a completely linear sequence of "recons." Once again, a hypothetical entity (Benzer's "recon") was determined solely through genetic analysis—and once again it was soon to assume physical reality. The elucidation of the structure of DNA (Chapter 11) and subsequent detailed genetic analyses allowed the definition of the single nucleotide pair as the smallest unit of recombination. (The term "recon" is not in current usage.)

> **Message** A gene is composed of units, originally called "recons," that are not divisible by recombination. Each unit consists of a single nucleotide pair.

The occurrence of intragenic recombination permits the construction of detailed gene maps. The relative frequencies of intragenic recombinants in crosses between various mutant alleles reveal the order and relative positions of the mutational sites within a gene. It should be noted that recombination within genes is the same as recombination between genes, except that the scale is different. Figure 12-29 shows a detailed map of the *rII* region obtained by intragenic crosses.

Mutational Sites

Benzer extended his fine-structure analysis to the properties of mutational sites. A useful tool for these studies is **deletion mapping.** The important point to consider here is that deletions are mutations that result from the elimination of segments of DNA. Mutations cannot be converted into wild-types in recombination tests against deletion mutations, because the DNA corresponding to the wild-type region for that particular mutation is no longer present. Therefore, mutations can be ordered against a set of deletions by rapid tests that do not depend on extensive quantitative measurements.

Deletion Mapping

The use of deletion mutants enabled Benzer to locate rapidly new mutational sites in his *rII* gene map. He found some special *rII* mutants that would not give recombinants when crossed with any of several other mutants (which had been shown to be different "recons") but would give recombinants when crossed to still other mutants. Benzer realized that such mutants behaved as if they contained short deletions within the *rII* region, and this model was supported by their lack of reversion.

They could be used for the rapid location of mutational sites in newly obtained mutants. For example, consider the following gene map, showing 12 identifiable mutational sites:

|1|2|3|4|5|6|7|8|9|10|11|12|

One special mutant D_1 fails to give rII^+ recombinants when crossed with mutants carrying altered sites *1, 2, 3, 4, 5, 6, 7,* or *8*; therefore, D_1 behaves as if it carries deletion of sites *1 to 8*:

Another special mutant D_2 fails to give rII^+ recombinants when crossed with *5, 6, 7, 8, 9, 10, 11,* or *12*; therefore, D_2 behaves as if it involves a deletion of sites *5 to 12*:

These overlapping deletions now define three areas of the gene. Let's call them i, ii, and iii:

A new mutant that gives rII^+ recombinants when crossed with D_1 but not when crossed with D_2 must have its mutational site in area iii. One that gives rII^+ recombinants when crossed with D_2 but not with D_1 must have its mutational site in area i. A new mutant that does not give rII^+ recombinants with either D_1 or D_2 must have its mutational site in area ii. For example, assume that a mutant in area iii is crossed with D_1:

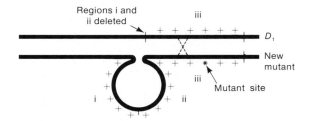

The more deletions there are in the tester set, the more areas can be uniquely designated and the more rapidly new mutational sites can be located (Figure 12-30). Once assigned to a region, a mutation can be mapped against other alleles in the same region to obtain an accurate position. Figure 12-31 shows the complexity of Benzer's actual map.

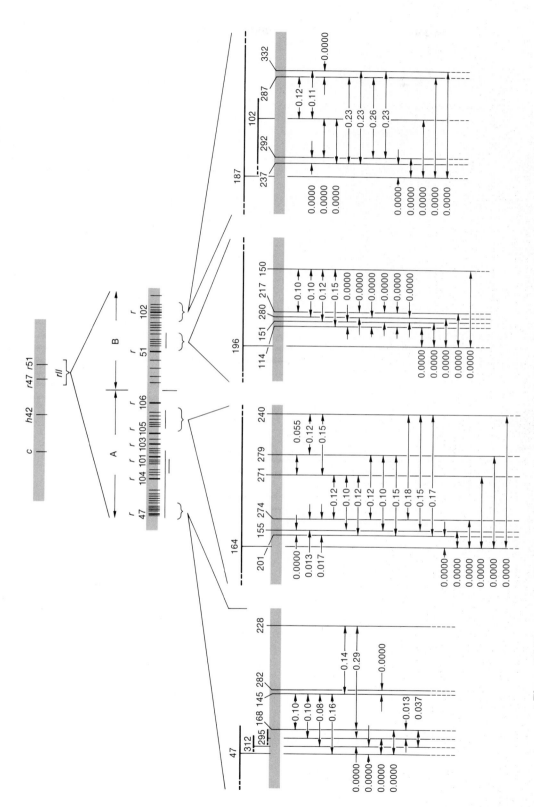

Figure 12-29. Detailed recombination map of the *rII* region of the phage T4 chromosome. The map unit is the percentage of *rII*+ recombinants in crosses between the *rII* mutants. Typical regions are progressively enlarged. Numbers on the map represent mutational sites. Note that the two portions of the *rII* region, A and B, are two different functional units of the region, as described in the text. (After S. Benzer, *Proceedings of the National Academy of Sciences USA* 41, 1955, 344. From G. S. Stent and R. Calendar, *Molecular Genetics*, 2d ed. Copyright © 1978 by W. H. Freeman and Company.)

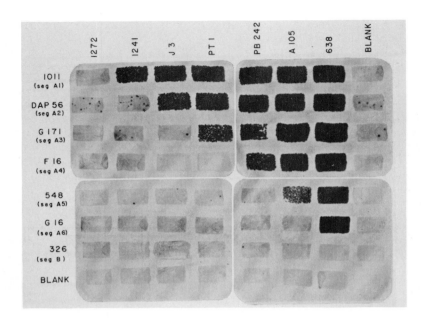

Figure 12-30. Crosses for mapping *rII* mutations. The photograph is a composite of four plates. Each row shows a given mutant tested against the reference deletions of Figure 12-31. Plaques appearing in the blanks are due to revertants present in the mutant stock. The results show each of these mutations to be located in a different segment. (Figures 12-30, 12-32, and 12-33 from S. Benzer, *Proceedings of the National Academy of Sciences USA* 47, 1961, 403–416.)

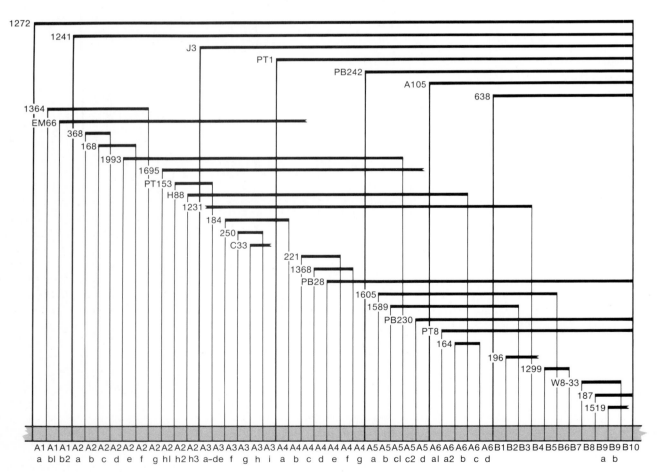

Figure 12-31. Detailed deletion map of the *rII* gene. Each deletion (*horizontal bar*) has an identification number. Along the bottom are the arbitrary identification numbers of the regions defined by the deletions. Note that some deletions extend out of the *rII* gene. (From G. S. Stent and R. Calendar, *Molecular Genetics,* 2d ed. Copyright © 1978 by W. H. Freeman and Company. After S. Benzer, *Proceedings of the National Academy of Sciences USA* 47, 1961, 403.)

Deletions themselves can be intercrossed and mapped just like point mutations. The deleted region is represented by a bar. If no wild-type recombinants are produced in a cross between different deletions, then the bars are shown as overlapping. A typical deletion map might be

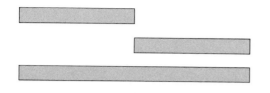

Such deletion maps are useful in delineating regions of the gene to which new point mutations can be assigned.

> **Message** Mutations can be rapidly ordered against a set of deletions.

Analysis of Mutational Sites

The use of deletions enabled Benzer to define the **topology** of the gene (the manner in which the parts are interconnected). The gene is seen as consisting of a linear array of mutable subelements. Benzer's next step was to examine the **topography** of the gene (differences in the properties of the subelements). Operationally, this is determined by asking whether all of the subelements or sites are equally mutable. For this study, it was essential to work with the smallest mutable subelements possible. Instead of multisite mutations (deletions) that exhibited no reversion, Benzer employed revertible mutations, since they probably represented small alterations, or point mutations. Also, mutants with high reversion rates were discarded, because high reversion interferes with recombination tests. Each mutational site was first mapped into one of the short deletion segments and then crossed against all of the other point mutations in the same interval. Benzer concluded that any two revertible mutations that failed to recombine with one another represented recurrences of the same event.

Figure 12-32 shows the distribution of 1612 spontaneous mutations in the *rII* locus. In Benzer's own words, "That the distribution is not random leaps to the eye." This extraordinary nonrandom distribution demonstrates that all sites are not equally mutable. Benzer termed sites that are more mutable than other sites **hotspots.** The most prominent hotspot was represented by over 500 repeated occurrences in the collection of 1612 mutations. By examining a Poisson distribution calculated to fit the number of sites having only one or two occurrences (Figure 12-33), could show that at least

60 sites were truly more mutable than those with only one or two occurrences and also that at least 129 sites were not observed at all (by chance) in this collection, even though they were as mutable as those represented by one or two occurrences. When the analysis was extended to include mutagen-induced mutations, it produced similar results. There were hotspots among the mutations generated by mutagenic agents (see Chapter 17).

From an estimation of the physical size (and thus the number of base pairs) of the *rII* region and the determination of the number of mutational sites, the number of sites was calculated to be approximately one-fifth of the number of nucleotide pairs. In other words, the smallest mutable site was five nucleotide pairs or less. The deciphering of the genetic code (Chapter 13), together with the work (described previously) by Ingram, Yanofsky, and others demonstrating that single amino-acid substitutions resulted from single mutations, allowed Benzer to conclude that a mutation could result from the alteration of a single nucleotide pair. (The direct sequencing of DNA, described in Chapter 15, has since confirmed these conclusions in many examples.)

> **Message** The gene can be divided into a linear array of mutable subelements that correspond to individual nucleotide pairs.

Destruction of the Bead Theory

Now let's review Benzer's work in light of the bead theory. With the aid of deletion mapping, Benzer was able to map an extraordinary number of mutations in the *rII* locus against each other. His experiments have shown that mutations in the same gene can indeed recombine with one another. This result contradicts the ideas of classical genetics, which held that genes on the chromosome could be compared to beads on a necklace and that recombination could occur between genes, but not within genes.

Benzer's analysis of the fine structure of the gene demonstrated that each gene consists of a linear array of subelements or sites that can be altered by mutation and can undergo recombination. This finding also contradicts one tenet of the bead theory, which implies that only the gene as a whole—not parts of the gene—is mutable.

Subsequent work by several investigators identified each genetic site as a base pair in double-stranded DNA. Therefore, Benzer's contribution bridged the gulf between classical genetics and the knowledge of the chemical structure of DNA revealed by Watson and Crick.

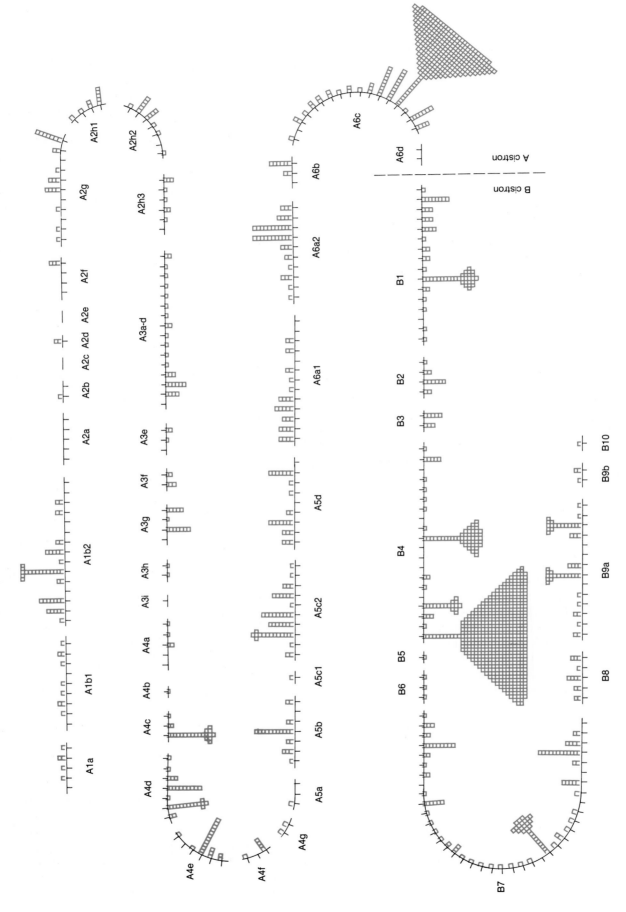

Figure 12-32. The distribution of 1612 spontaneous mutations in the *rII* locus. Each occurrence of an independent mutation is depicted by a square. When mutations are identical, the squares are drawn on top of one another. Many positions are represented by only a single square, whereas others have a large number of squares. These sites are called hotspots. The *rII* locus is subdivided into two regions, *A* and *B*, in this diagram.

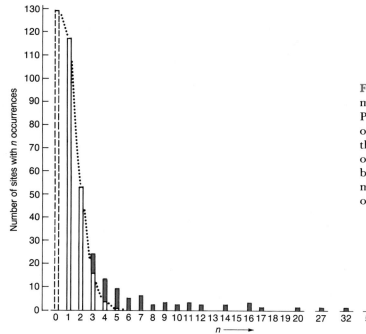

Figure 12-33. Distribution of occurrences of spontaneous mutations at various sites. The dotted curve indicates a Poisson distribution fitted to the numbers of sites having one and two occurrences. It predicts a minimum estimate for the number of sites of comparable mutability that have zero occurrences due to chance (dashed column at $n = 0$). Solid bars indicate the minimum numbers of sites that have mutation rates significantly higher than the one- and two-occurrence class.

According to the bead theory, the Watson-Crick structure made no sense. However, Benzer's demonstration that genes do indeed have fine structures that can be revealed solely by genetic analysis allowed a fusion of the two disciplines and helped to launch the modern era of molecular genetics. Figure 12-34 illustrates the dissection of the *rII* locus and its correspondence with the DNA structure.

Complementation

In another part of his studies, Benzer carried out a series of experiments designed to define the gene in terms of function. Benzer worked out the concept of **complementation.** Traditionally, complementation in bacteriophages has been a difficult idea for students in genetics to master for two reasons. First, it often appears to be a foreign concept, even though so-called complementation for normally haploid organisms simply represents a situation that we have been dealing with routinely in the genetic analysis of diploids. Second, complementation is often confused with recombination. Let's try to deal with each of these situations.

Complementation in Diploids

Suppose that we are studying a phenotype in a diploid cell that requires the active product of two genes. This might be purple flower color in a certain plant that is, say, dependent on two loci: *A* and *B*. In fact, this is pre-

cisely the situation for flower color in peas detailed in Chapter 4. Let's say that mutations in *A* or *B* lead to loss of purple color and result in white flowers and that the wild-type allele is dominant in each case. Thus, *A* is dominant to *a* and *B* is dominant to *b*. The phenotypes resulting from each combination of genes are

Genotype	Color
AABB	purple
AaBB	purple
AABb	purple
AaBb	purple
aaBB	white
aaBb	white
aabb	white
AAbb	white
Aabb	white

Well, we have been doing problems like this all along. What happens when we cross a *AAbb* plant against a *aaBB* plant? Both of these plants, which are white (Can you see why? Recall that both *A* and *B* are required for purple flowers) will yield offspring that are *AaBb*. Since both *A* and *B* are dominant to *a* and *b*, respectively, all progeny from this cross will produce purple flowers.

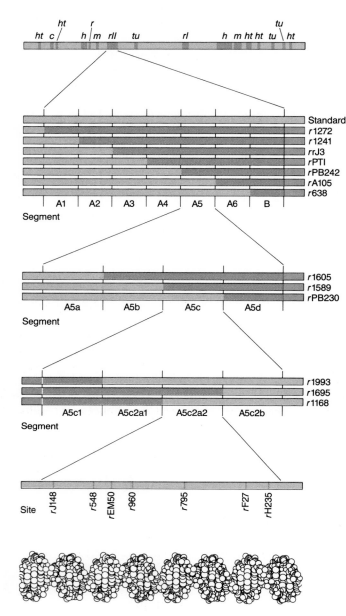

Figure 12-34. Fine-structure analysis of the *rII* locus. This mapping technique localizes the position of a given mutation in progressively smaller segments of the DNA molecule contained in phage T4. The *rII* region represents to start with only a few percent of the entire molecule. The mapping is done by crossing an unknown mutant with reference mutants having deletions (color) of known extent in the *rII* region. Each site represents the smallest mutable unit in the DNA molecule, a single base pair. The molecular segment (*extreme bottom*), estimated to be roughly in proper scale, contains a total of about 40 bp. (From S. Benzer, "The Structure of the Gene." Copyright © 1962 by Scientific American, Inc. All rights reserved.)

Because each parent is homozygous at each locus, we know that all gametes from one parent are *Ab* and that all gametes from the other parent are *aB*. The zygote resulting from the union of these two gametes and all cells descending from this zygote will complete the two-step cellular pathway that allows the production of purple flower pigment because each of the two relevant chromosomes supplies one of the functional genes. We can diagram this as follows, where (1) is the starting compound in the cellular pathway, (2) is an intermediate compound, and (3) is the final product for purple pigment (A and B are hypothetical enzymes produced by genes *A* and *B*):

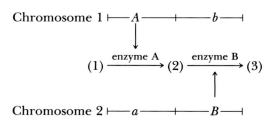

Note that one chromosome supplies enzyme A but not enzyme B and that the second chromosome supplies enzyme B but not enzyme A. Since only one good copy of the gene is required to produce each enzyme in this pathway, the two partially defective chromosomes compensate for their respective deficiencies; that is, they "help one another out," or **complement** one another.

Study the situation just described and diagrammed until you feel that you fully understand it. Now you are ready to tackle complementation in bacteriophage T4, which is very similar to the example given here.

Complementation in Bacteriophage T4

Benzer tested the mutations that he had mapped in the *rII* region to see whether pairwise combinations of mutations would restore the wild-type phenotype. In other words, Benzer looked for complementation in host cells that were temporarily "diploid" for the T4 chromosome. He carried out a mixed infection with different *rII* mutants (Figure 12-35). His criteria for the wild-type phenotype was the ability to lyse *E. coli* K hosts. (Recall that *rII* mutants cannot do this but that wild-type phages can.)

Complementation tests like the one Benzer conducted are carried out in one cycle of infection; they do not involve the multiple cycles of reinfection required for plaque formation. Samples of the two phages to be tested are spread over a strip of host bacteria on a section of a petri plate at a high multiplicity of phages to bacteria to ensure that essentially every bacterium is infected with each phage. After a period of incubation, growth or

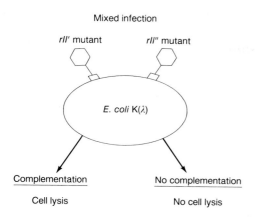

Figure 12-35. Complementation test: A schematic view of *rII* complementation. Two different mutants of *rII* are used to simultaneously infect *E. coli* K (mixed infection). Normally, an *rII* mutant cannot lyse *E. coli* K or generate progeny phages. However, if the two different mutants can complement, then lysis and phage growth will result. If the two *rII* mutants cannot complement one another, then no lysis or phage growth will result.

absence of growth of the bacteria in the strip indicates whether or not the bacteria have lysed as a result of the phage infection. Pairwise tests of many different mutants allowed Benzer to separate the mutations into two groups, labeled *A* and *B*. All mutations in the *A* group complemented those in the *B* group, whereas no mutations in the *A* group complemented any other mutations in the *A* group and no mutations in the *B* group complemented any other mutations in the *B* group. Benzer found that all mutations in group *A* mapped in one-half of the *rII* locus and that all mutations in group *B* mapped in the other half of the *rII* locus (Figure 12-36).

The following diagram depicts a model for the *rII* locus based on Benzer's results:

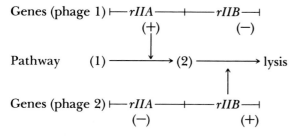

We assume that two genes, termed *rIIA* and *rIIB,* are involved in lysis. We can envision two steps, (1) and (2), each controlled by an enzyme specified by the different *rII* genes. Thus, *rIIA* and *rIIB* would specify enzymes A and B, respectively. Note how this model is similar to the

biosynthetic pathways we studied at the beginning of the chapter. In the diagram, we see two different *rII* mutants, which have defects in either the *rIIA* or *rIIB* region.

Because each phage chromosome can program the synthesis of one of the required functions for the cellular pathway, lysis can occur, which is the wild-type phenotype. Therefore, the two phages can complement one another. Notice how this situation is virtually identical to the example given previously for purple flower color in plants. The only difference is that the diploid cells in the case of the plant persist and remain diploid, whereas the double-infection experiment in T4 phage creates a temporary diploid-like cell for the phage chromosome. During the existence of this "diploid," complementation can occur.

The Cistron

Mutations that fail to complement must affect the same unit of function. Mutations that do complement must affect different functional units. Benzer called this unit of function a **cistron.**

The cistron gets its name from the **cis-trans test,** which is the term Benzer used to describe a complete complementation test arranged with the mutational sites on the same chromosome (cis) or on opposite chromosomes (trans), as shown in Figure 12-37. The cis portion of the test is really a control; the trans portion is the actual test for complementation. In summary, the cis-trans (or complementation) test is performed to determine whether two mutational sites are located within the same functional unit or in different functional units.

Message A cistron is a genetic region within which there is normally no complementation between mutations. The cistron is equivalent to the gene.

Recombination and Complementation

A second source of confusion regarding complementation involves the difference between recombination and complementation. Recombination represents the creation of new combinations of genes through the physical

Figure 12-36. Gene map of *rII*.

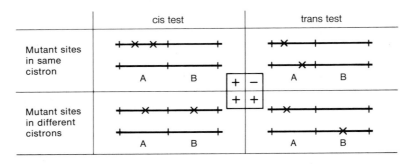

Figure 12-37. The cis-trans test. A cistron is a genetic region within which there is no complementation between mutations. If two different *rII* mutant T4 phages infect the same bacterium (the trans configuration) and no phage growth occurs, then the mutants are said to be in the same cistron *(upper right box)*. However, if complementation does occur (if progeny phages grow), then the mutants are in different functional units or cistrons *(lower right box)*. In the upper right box, only the *B* function can be expressed normally; in the lower right box, both the *A* and *B* functions can be expressed normally. A + represents a successful complementation; a − indicates no complementation; a cross represents a mutational site.

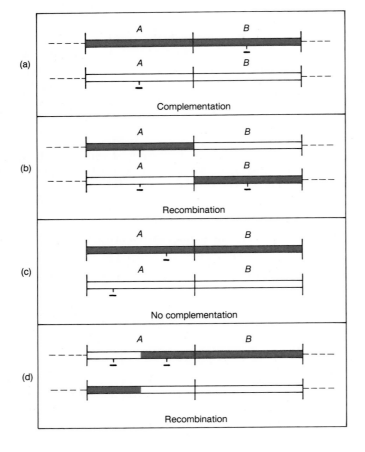

breakage and rejoining of chromosomes. The progeny from a cross in which recombination has occurred have new genotypes that are different from the parental genotypes. Complementation, on the other hand, does not involve any change in the genotypes of individual chromosomes; rather, it represents the mixing of gene products. Complementation occurs during the time that two chromosomes are in the same cell and can each supply a function. Afterward, each respective chromosome remains unaltered. In the case of *rII* mutants, complementation occurs when two different phage chromosomes, with mutations in different *rII* genes, are in the same host cell. However, progeny that result from this complementation carry only the parental genotypes. Figure 12-38 diagrams some of the differences between these two basic processes. The arrangement shown in part (a) of the figure will allow phage growth and lysis

Figure 12-38. A comparison of the genetic consequences of complementation and recombination. (a) The mutant pairs can complement one another, since wild-type gene *A* and gene *B* products can mix in the cytoplasm. (b) If recombination occurs, then a rearrangement of the genomes will take place. (c) The mutant pairs cannot complement because neither mutant can contribute a wild-type gene *A* product. (d) A relatively rare recombination event could result in a wild-type phage.

(complementation) without rearrangement of phage chromosomes, whereas the mutant pair shown in (c) will not. Recombination can occur in each situation, as shown in (b) and (d).

How can we distinguish operationally between complementation and recombination? Fortunately, in the case of *rII* mutants, the incidence of recombination is never more than a few percent (see the map on page 305), and it is only rarely that high. Therefore, lysis will occur in only a very small fraction of the infected cells and will not interfere with the interpretation of the test. To distinguish definitively between these two possibilities, we could conduct an additional genetic analysis; in the case of *rII* mutants, we could analyze the genotypes of the resulting progeny phages (Figure 12-39). The phages resulting from a mixed infection of *E. coli* K(λ) are plated on both *E. coli* B and *E. coli* K(12). All phages grow on B, but only *rII*+ recombinants can form plaques on K. (Here, the multiplicity of infection—the ratio of phages to bacteria—is low, so that each bacterium is infected initially by no more than one phage; complementation cannot result in plaque formation in this situation.) If the progeny phages from the mixed infection of *E. coli* K(λ) result from complementation, then virtually all of the phages will still be *rII* mutants and will not plate on K. However, if recombination is required for phage growth and lysis during the mixed infection, then the analysis of the very low titer of progeny phages will show that about one-half are *rII*+ recombinants and will plate on K (Figure 12-39).

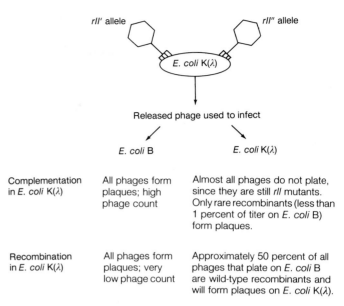

Complementation in *E. coli* K(λ)	All phages form plaques; high phage count	Almost all phages do not plate, since they are still *rII* mutants. Only rare recombinants (less than 1 percent of titer on *E. coli* B) form plaques.
Recombination in *E. coli* K(λ)	All phages form plaques; very low phage count	Approximately 50 percent of all phages that plate on *E. coli* B are wild-type recombinants and will form plaques on *E. coli* K(λ).

Figure 12-39. Analysis of phages resulting from the mixed infection of *E. coli* K(λ). If complementation occurs between two *rII* mutants, then the progeny phages that are released will still be principally *rII* mutants and will fail to plate on *E. coli* K(λ) in single-infection experiments. If recombination but no complementation occurs, then there will be a sharply reduced yield of progeny phages due to the rarity of recombination. However, the few resulting phages will consist of approximately 50 percent wild-type recombinants and will form plaques on *E. coli* K(λ) in single-infection experiments.

Complementation and the Concept of the Gene

Of the different views of what a gene is according to the classical (bead) theory, the one aspect that has held up and seems most essential is the gene as a unit of function. We now see that the gene is equivalent to the cistron, which we can consider to be a unit of function that can be defined experimentally by a cis-trans complementation test. There are occasional exceptions to this operational definition expressed by the cis-trans test; some genes with multimeric proteins and for certain mutations that can affect the expression of more than one gene are examples. However, these exceptions do not upset the basic concept of the gene as a unit of function.

We now know that a cistron is a region of the genetic material that codes for one polypeptide chain. Therefore, the one-gene–one-enzyme hypothesis could be referred to more precisely as the one-cistron–one-polypeptide hypothesis, also emphasizing that cistrons (genes) can code for proteins other than enzymes.

Summary

■ The work of Beadle and Tatum in the 1940s showed that one gene codes for one protein. Further work by Benzer and others illustrated that the gene could be dissected into smaller and smaller pieces. A structural unit of mutation and one of recombination were identified and equated with a single nucleotide pair. A cistron is defined at the phenotypic level as a genetic region within which there is no complementation between mutations. This is the unit that codes for the structure of a single functional polypeptide. The gene is equivalent to the cistron.

The failure of an enzyme to function normally due to a mutation yields a variant phenotype. These variant phenotypes are often the basis of genetic disease in any organism, including humans. In order to understand how abnormal enzymes can cause phenotypic change, we need to understand the structure of proteins. Composed of a specific linear sequence of amino acids connected through peptide bonds, the proteins assume specific three-dimensional shapes as a result of the interaction of the 20 amino acids that in different combinations constitute the polypeptide chain. Different areas of this folded chain are sites for the attachment and interaction of substrates. Furthermore, many functional enzymes and other proteins are built by combining several polypeptide chains in multimeric form.

Specific amino-acid changes can be detected in a protein by the technique of fingerprinting and amino-acid sequencing. Work of this type has demonstrated colinearity between mutational sites on the genetic map and the positions of altered amino acids in a protein.

In the early days of genetics, genes were represented as indivisible beads on a chain. We have now arrived at a far different picture of the gene. Multiple mutational sites exist, and recombination may occur anywhere within a gene. In addition, a closer connection between genotype and phenotype was realized when it was established that one cistron is responsible for the synthesis of one polypeptide.

■ ■ ■

Solved Problems

1. This problem is similar to Problem 14 in the following problem set. Various pairs of *rII* mutants of phage T4 are tested in *E. coli* in both the cis and trans positions. Comparisons are made of "burst size" (the average number of phage particles produced per bacterium). Table 12-6 shows a hypothetical set of results for six different *r* mutants: *rM, rN, rO, rP, rR,* and *rS*. If we assign the mutation *rO*, carried by the *rO* mutant, to the *A* cistron, what are the locations of the other five mutations with respect to *A* and *B* cistrons?

TABLE 12-6. Results of *rII* mutant crosses

Cis genotype	Burst size	Trans genotype	Burst size
rMrN/++	245	*rM* +/+ *rN*	250
rOrP/++	256	*rO* +/+ *rP*	268
rRrS/++	248	*rR* +/+ *rS*	242
rMrO/++	270	*rM* +/+ *rO*	0
rMrP/++	255	*rM* +/+ *rP*	255
rMrR/++	264	*rM* +/+ *rR*	0
rMrS/++	240	*rM* +/+ *rS*	240
rNrO/++	257	*rN* +/+ *rO*	268
rNrP/++	250	*rN* +/+ *rP*	0
rNrR/++	245	*rN* +/+ *rR*	255
rNrS/++	259	*rN* +/+ *rS*	0
rPrR/++	260	*rP* +/ +*rR*	245
rPrS/++	253	*rP* +/+ *rS*	0

Solution

In solving problems like this, first look for the underlying principle and then attempt to draw a diagram to help us work out the solution. The key principle is that in the trans position mutations in the same cistron will not complement and thus will yield no progeny phage (no burst), whereas mutations in different cistrons will complement and yield progeny phages of normal burst size. Since *rO* is in the *A* cistron, we start with the initial diagram

We can now look at Table 12-6, and assign each mutation to a cistron based on whether or not complementation occurs in a mixed infection with *rO* mutants. From the table, it is evident that *N*, *P*, and *S* complement and must be in a different cistron (the *B* cistron), whereas *M*, *O*, and *R* do not complement and must be in the same cistron (the *A* cistron). Therefore, we have the arrangement

We can check this assignment by examining the other crosses to find out if the results are consistent with our answer.

2. The following depletion map shows four deletions (1 – 4) involving the *rIIA* cistron of phage T4:

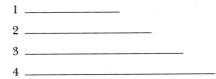

Five point mutations (*a – e*) are tested against these four deletion mutants for their ability to give wild-type (*r⁺*) recombinants; the results are

	a	*b*	*c*	*d*	*e*
1	+	+	+	+	+
2	+	+	+	−	−
3	+	−	+	−	−
4	−	−	+	−	−

What is the order of the point mutations?

Solution

The key principle here is that point mutations can recombine with deletions that do *not* extend past the mutation but cannot recombine to yield wild-type phages with deletions that do extend past the mutation. Looking at the test results given in the problem, any mutation that recombines with deletion 1 must be to the right of the deletion, any mutation that recombines with deletion 2 must be to the right of deletion 2, and so on. Let's look at point mutation *a*. It recombines with deletions 1, 2, and 3 but not with deletion 4. Therefore, it is to the right of deletions 1, 2, and 3 but not to the right of deletion 4 (as drawn in Problem 2). We can then easily place point mutation *a* in the interval between deletions 3 and 4. Point mutation *b* recombines with deletions 1 and 2 and must be to the right of them. It does not recombine with deletions 3 and 4, so it is in the interval between deletion 2 and 3. Point mutation *c* recombines with all the deletions and is to the right of all deletions, even deletion 4. Finally, both point mutations *d* and *e* recombine only with deletion 1 and must therefore be in the interval between deletions 1 and 2. The solution we have just derived can be summarized as

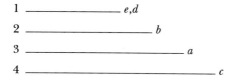

Problem 18 in the following problem set is similar to this sample problem. See if you can apply the reasoning set forth here when you solve Problem 18.

Problems

1. A common weed, Saint-John's-wort, is toxic to albino animals. It also causes blisters on animals that have white areas of fur. Suggest a possible genetic basis for this reaction.

2. In humans, the disease galactosemia causes mental retardation at an early age because lactose in milk cannot be broken down, and this failure affects brain function. How would you provide a secondary cure for galactosemia? Would you expect this phenotype to be dominant or recessive?

3. Amniocentesis is a technique in which a hypodermic needle is inserted through the abdominal wall of a pregnant woman into the amnion, the sac that surrounds the developing embryo, to withdraw a small amount of amniotic fluid. This fluid contains cells that come from the embryo (not from the woman). The cells can be cultured; they will divide and grow to form a population of cells on which enzyme analyses and karyotypic analyses can be performed. Of what use would this technique be to a genetic counselor? Name at least three specific conditions under which amniocentesis might be useful. (NOTE: this technique involves a small but real risk to the health of both the woman and the embryo; take this fact into account in your answer.)

4. Table 12-7 shows the ranges of enzymatic activity (in units we need not worry about) observed for enzymes involved in two recessive metabolic diseases in humans. Similar information is available for many metabolic genetic diseases.

   a. Of what use is such information to a genetic counselor?

   b. Indicate any possible sources of ambiguity in interpreting studies of an individual patient.

   c. Reevaluate the concept of dominance in the light of such data.

5. Two albinos marry and have a normal child. How is this possible? Suggest at least two ways. (This question appeared first in Chapter 4. Reconsider it now in light of biosynthetic pathways.)

6. In humans, PKU (phenylketonuria) is a disease caused by an enzyme inefficiency at step A in the following simplified reaction sequence and AKU (alkaptonuria) is due to an enzyme inefficiency in one of the steps summarized as step B here:

$$\text{phenylalanine} \xrightarrow{A} \text{tyrosine} \xrightarrow{B} CO_2 + H_2O$$

A person with PKU marries a person with AKU. What phenotypes do you expect for their children? a. All normal; b. all having PKU only; c. all having AKU only; d. all having both PKU and AKU; e. some having AKU and some having PKU.

7. Three independently isolated tryptophan-requiring strains of yeast are called trpB, trpD, and trpE. Cell suspensions of each are streaked on a plate supplemented with just enough tryptophan to permit weak growth for a *trp⁻* strain. The streaks are arranged in a triangular pattern so that they do not touch one another. Luxuriant

TABLE 12-7.

| Disease | Enzyme involved | Range of enzyme activity | | |
		Patients	Parents of patients	Normal individuals
Acatalasemia	Catalase	0	1.2–2.7	4.3–6.2
Galactosemia	Gal-1-P uridyl transferase	0–6	9–30	25–40

growth is noted at both ends of the trpE streak and at one end of the trpD streak.

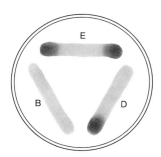

a. Do you think complementation is involved?

b. Briefly explain the patterns of luxuriant growth.

c. In what order in the tryptophan-synthesizing pathway are the enzymic steps defective in trpB, trpD, and trpE?

d. Why was it necessary to add a small amount of tryptophan to the medium in order to demonstrate such a growth pattern?

8. In *Drosophila* pupae, certain structures called imaginal disks can be detected as thickenings of the skin; after metamorphosis, these imaginal disks develop into specific organs of the adult fly. George Beadle and Boris Ephrussi devised a means of transplanting eye imaginal disks from one larva into another larval host. When the host metamorphoses into an adult, the transplant can be found as a colored eye located in its abdomen. The researchers took two strains of flies that were phenotypically identical in terms of bright scarlet eyes: one due to the sex-linked mutant vermilion (*v*); the other due to cinnabar (*cn*) on chromosome 2. If *v* disks are transplanted into *v* hosts or *cn* disks into *cn* hosts, then the transplants develop as mutant scarlet eyes. Transplanted *cn* or *v* disks in wild-type hosts develop wild-type eye colors. A *cn* disk in a *v* host develops a mutant eye color, but a *v* disk in a *cn* host develops wild-type eye color. Explain these results, and outline the experiments you would propose to test your explanation.

9. In *Drosophila*, the autosomal recessive *bw* causes a dark brown eye and the unlinked autosomal recessive *st* causes a bright scarlet eye. A homozygote for both genes has a white eye. Thus, we have the following correspondences between genotypes and phenotypes:

$$+/+ \; +/+ = \text{red eye (wild-type)}$$
$$+/+ \; bw/bw = \text{brown eye}$$
$$st/st \; +/+ = \text{scarlet eye}$$
$$st/st \; bw/bw = \text{white eye}$$

Construct a hypothetical biosynthetic pathway showing how the gene products interact and why the different mutant combinations have different phenotypes.

10. Several mutants are isolated, all of which require compound G for growth. The compounds (A–E) in the biosynthetic pathway are known, and each compound is tested for its ability to support the growth of each mutant (1–5). In the following table, + indicates growth and − indicates no growth:

		A	B	C	D	E	G
	1	−	−	−	+	−	+
	2	−	+	−	+	−	+
Mutant	3	−	−	−	−	−	+
	4	−	+	+	+	−	+
	5	+	+	+	+	−	+

a. What is the order of compounds A–E and G in the pathway?

b. At which point in the pathway is each mutant blocked?

c. Would a heterokaryon composed of double mutants 1,3 + 2,4 grow on a minimal medium? 1,3 + 3,4? 1,2 + 2,4 + 1,4?

11. In *Neurospora* (a haploid), assume that two genes participate in the synthesis of valine. Their mutant alleles are *val-1* and *val-2*, and their wild-type alleles are *val-1*⁺ and *val-2*⁺. These two genes are linked on the same chromosome, and a crossover occurs between them on average in one of every two meioses.

a. In what proportion of meioses are there no crossovers between the genes?

b. Use the map function to determine the recombinant frequency between these two genes.

c. Progeny from the cross *val-1 val-2*⁺ × *val-1*⁺ *val-2* are plated on a medium containing no valine. What proportion of the progeny will grow?

d. The *val-1 val-2*⁺ strains accumulate intermediate compound B, and the *val-1*⁺ *val-2* strains accumulate intermediate compound A. The *val-1 val-2*⁺ strains grow on valine or A, but the *val-1*⁺ *val-2* strains grow only on valine and not on B. Show the pathway order of A and B in relation to valine, and indicate which gene controls each conversion.

12. In a certain plant, the flower petals are normally purple. Two recessive mutations arise in separate plants and are found to be on different chromosomes. Mutation 1 (m_1) gives blue petals when homozygous ($m_1 m_1$). Mutation 2 (m_2) gives red petals when homozygous ($m_2 m_2$). Biochemists working on the synthesis of flower pigments in this species have already described the following pathway:

$$\text{colorless (white) compound} \quad \overset{\text{enzyme } A}{\nearrow} \text{blue pigment}$$
$$\underset{\text{enzyme } B}{\searrow} \text{red pigment}$$

a. Which mutant would you expect to be deficient in enzyme-A activity?

b. A plant has genotype $M_1m_1M_2m_2$. What would you expect its phenotype to be?

c. If the plant in (b) is selfed, what colors of progeny would you expect, and in what proportions?

d. Why are these mutants recessive?

13. In sweet peas, the synthesis of purple anthocyanin pigment in the petals is controlled by two genes, *B* and *D*. The pathway is

$$\text{white intermediate} \xrightarrow[\text{enzyme}]{\text{gene-}B}$$

$$\text{blue intermediate} \xrightarrow[\text{enzyme}]{\text{gene-}D} \text{anthocyanin (purple)}$$

a. What color petals would you expect in a pure-breeding plant unable to catalyze the first reaction?

b. What color petals would you expect in a pure-breeding plant unable to catalyze the second reaction?

c. If the plants in (a) and (b) are crossed, what color petals will the F_1 plants have?

d. What ratio of purple:blue:white plants would you expect in the F_2?

14. Various pairs of *rII* mutants of phage T4 are tested in *E. coli* in both the cis and trans positions. Comparisons are made of the average number of phage particles produced per bacterium (a measure called the "burst size"). Table 12-8 shows a hypothetical set of results for six different *r* mutants: *rU*, *rV*, *rW*, *rX*, *rY*, and *rZ*. If we assign *rV* to the A cistron, what are the locations of the other five *rII* mutations with respect to the A and B cistrons?

TABLE 12-8.

Cis genotype	Burst size	Trans genotype	Burst size
$rU\ rV/++$	250	$rU+/+rV$	258
$rW\ rX/++$	255	$rW+/+rX$	252
$rY\ rZ/++$	245	$rY+/+rZ$	0
$rU\ rW/++$	260	$rU+/+rW$	250
$rU\ rX/++$	270	$rU+/+rX$	0
$rU\ rY/++$	253	$rU+/+rY$	0
$rU\ rZ/++$	250	$rU+/+rZ$	0
$rV\ rW/++$	270	$rV+/+rW$	0
$rV\ rX/++$	263	$rV+/+rX$	270
$rV\ rY/++$	240	$rV+/+rY$	250
$rV\ rZ/++$	274	$rV+/+rZ$	260
$rW\ rY/++$	260	$rW+/+rY$	240
$rW\ rZ/++$	250	$rW+/+rZ$	255

(Problem 14 is from M. Strickberger, *Genetics.* Copyright © 1968 by Monroe W. Strickberger. Reprinted with permission of Macmillan Publishing Co., Inc.)

15. There is evidence that occasionally during meiosis either one or both homologous centromeres will divide and segregate precociously at the first division rather than at the second division (as is the normal situation). In *Neurospora*, *pan2* alleles produce a pale ascospore, aborted ascospores are completely colorless, and normal ascospores are black. In a cross between two complementing alleles *pan2x* × *pan2y*, what ratios of black:pale:colorless would you expect in asci resulting from the precocious division of: a. one centromere? b. both centromeres? (Assume that *pan2* is near the centromere.)

16. *Protozoon mirabilis* is a hypothetical single-celled haploid green alga that orients to light by means of a red eye-spot. By selecting cells that do not move toward the light, 14 white-eye-spot mutants (*eye⁻*) are isolated after mutation. It is possible to fuse haploid cells to make diploid individuals. The 14 *eye⁻* mutants are paired in all combinations, and the color of the eye-spot is scored in each. Table 12-9 shows the results, where + indicates a red eye-spot and − indicates a white eye-spot.

TABLE 12-9.

	1	2	3	4	5	6	7	8	9	10	11	12	13	14
1	−	+	+	+	−	+	+	−	−	+	+	+	+	−
2	+	−	−	−	+	+	+	+	+	+	+	−	+	−
3	+	−	−	−	+	+	+	+	+	+	+	−	+	−
4	+	−	−	−	+	+	+	+	+	+	+	−	+	−
5	−	+	+	+	−	+	+	−	−	+	+	+	+	−
6	+	+	+	+	+	−	−	+	+	−	−	+	−	−
7	+	+	+	+	+	−	−	+	+	−	−	+	−	−
8	−	+	+	+	−	+	+	−	−	+	+	+	+	−
9	−	+	+	+	−	+	+	−	−	+	+	+	+	−
10	+	+	+	+	+	−	−	+	+	−	−	+	−	−
11	+	+	+	+	+	−	−	+	+	−	−	+	−	−
12	+	−	−	−	+	+	+	+	+	+	+	−	+	−
13	+	+	+	+	+	−	−	+	+	−	−	+	−	−
14	−	−	−	−	−	−	−	−	−	−	−	−	−	−

a. Mutant 14 obviously is different from the rest. Why might this be?

b. Excluding mutant 14, how many complementation groups are there, and which mutants are in which group?

c. Three crosses are made, with the results shown in Table 12-10. Explain these genetic ratios with symbols.

d. How many genetic loci are involved altogether, and which of the 14 mutants are at each locus?

e. What is the linkage arrangement of the loci? (Draw a map.)

TABLE 12-10.

Mutants crossed	Number of progeny		
	eye^+	eye^-	Total
1 × 2	31	89	120
2 × 6	5	113	118
1 × 14	0	97	97

17. You have the following map of the *rII* locus:

A cistron   B cistron

r_a r_b r_c r_d r_e r_f

You detect a new mutation r_x, and you find that it does not complement any of the mutants in the *A* or *B* cistron. You find that wild-type recombinants are obtained in crosses with r_a, r_b, r_e, and r_f but not with r_c or r_d. Suggest possible explanations for these results. Describe tests you would use to choose between the explanations.

18. The following map shows four deletions (1 – 4) involving the *rIIA* cistron of phage T4:

1 ———
2 ——————
3   ———————————
4   ————

Five point mutations (*a*–*e*) in *rIIA* are tested against these four deletion mutants for their ability to give r^+ recombinants, with the following results:

	a	b	c	d	e
1	+	+	−	+	+
2	+	+	−	−	−
3	−	−	+	−	+
4	+	−	+	+	+

a. What is the order of the point mutants?

b. Another strain of T4 has a point mutation in the *rIIB* cistron. This strain is mixed in turn with each of the *rIIA* deletion mutants, and the mixtures are used to infect *E. coli* K(λ) at a multiplicity of infection great enough that each host cell will be infected by at least one *rIIA* and one *rIIB* mutant. A normal plaque is formed with deletions 1, 2, and 3, but no plaque

forms with deletion 4. Given that the *B* cistron is to the right of *A*, explain the behavior of deletion 4. Does your explanation affect your answer to (a)?

19. In a phage, a set of deletions is intercrossed in pairwise combinations. The following results are obtained (a + indicates that wild-type recombinants are obtained from that cross):

	1	2	3	4	5
1	−	+	−	+	−
2	+	−	+	+	−
3	−	+	−	−	−
4	+	+	−	−	+
5	−	−	−	+	−

a. Construct a deletion map from this table.

b. The first geneticists to do a deletion-mapping analysis in the mythical schmoo-phage SH4 (which lyses schmoos) came up with this unique set of data:

	1	2	3	4
1	−	−	+	−
2	−	−	−	+
3	+	−	−	−
4	−	+	−	−

Show why this is a unique result by drawing the only deletion map that is compatible with this table. (Don't let your mind be shackled by conventional expectations.)

20. In a haploid eukaryote, four alleles of the *cys-2* gene are obtained. Each allele requires cysteine, and all alleles map to the same locus. The four strains bearing these mutant alleles are crossed to wild-types to obtain a set of eight cultures representing the four mutant alleles in association with each mating type. Then the mutant alleles are intercrossed in all pairwise combinations. The haploid meiotic products from each cross are plated on a medium containing no cysteine. In some crosses, cys^+ prototrophs are observed at low frequencies. The results follow, where the numbers represent the frequencies of cys^+ colonies per 10^4 meiotic products plated:

		Mating type A′			
		1	2	3	4
Mating type A″	1	0	14	2	20
	2	14	0	12	6
	3	2	12	0	18
	4	20	6	18	0

a. Draw a map of the four mutant sites within the *cys-2* gene. Provide a measurement of the relative intersite distances.

b. Do you see any evidence that mutation might be involved in the production of the prototrophs?

21. In *Neurospora*, there is a gene controlling the production of adenine, and mutants in this gene are called *ad-3* mutants. The *his-2* locus is 2.0 m.u. to the left, and the *nic-2* locus is 3.0 m.u. to the right of the *ad-3* locus (*his-2* controls histidine, and *nic-2* controls nicotinamide). Thus, the genetic map is

```
     his-2          ad-3              nic-2
  ───┼──────────────┼─────────────────┼───────
         2 m.u.           3 m.u.
```

Three different *ad-3* auxotrophs are detected: *ad-3^a*, *ad-3^b*, and *ad-3^c*, (Use *a*, *b*, and *c* as labels.) The following crosses are made:

Cross 1: *his-2^+ a nic-2^+* × *his-2 b nic-2*

Cross 2: *his-2^+ a nic-2* × *his-2 c nic-2^+*

Cross 3: *his-2 b nic-2* × *his-2^+ c nic-2^+*

The ascospores are then plated on a minimal medium containing histidine and nicotinamide, and *ad-3^+* prototrophs are picked up. Table 12-11 shows the results obtained. What is the map order of the *ad-3* mutants and the genetic distance between them?

TABLE 12-11.

Genotype of *ad-3^+* recombinants	Number of *ad-3^+* spores picked up		
	Cross 1	Cross 2	Cross 3
his-2 + *nic-2*	0	6	0
his-2^+ + *nic-2^+*	0	0	0
his-2 + *nic-2^+*	15	0	5
his-2^+ + *nic-2*	0	0	0
Total ascospores scored	41,236	38,421	43,600

22. The *Notch* locus of *Drosophila* is assumed to be a single cistron. William Welshons has recovered two classes of *Notch* mutants. Class I mutants show a dominant effect on the wings, bristles, and eyes, and the mutation acts as a recessive lethal. Class II mutations are all nonlethal, and all have recessive mutant phenotypes affecting eyes, bristles, or wings. Heterozygotes for a class I and a class II mutation are viable, but they exhibit the dominant phenotype of the class I mutant and the recessive phenotype of the class II mutant. Construct an explanation, and describe ways to test your model.

23. In a hypothetical diploid organism, squareness of cells is due to a threshold effect such that more than 50 units of "square factor" per cell will produce a square phenotype and less than 50 of these units will produce a round phenotype. Allele *s^f* is a functional gene that actively synthesizes the square factor. Each *s^f* allele contributes 40 units of square factor; thus, *s^f s^f* homozygotes have 80 units and are phenotypically square. A mutant allele (*s^n*) arises; it is nonfunctional, contributing no square factor at all.

a. Which allele will show dominance, *s^f* or *s^n*?

b. Are functional alleles necessarily always dominant? Explain your answer.

c. In a system such as this one, how might a specific allele become changed in evolution so that its phenotype shows recessive inheritance at generation 0 and dominant inheritance at a later generation?

24. Some genes in humans are known to have a multiple (or "pleiotropic") effect on phenotype. Does this constitute an invalidation of the one-gene–one-enzyme hypothesis? Explain.

25. Consider the following three biochemical sequences in *Neurospora* (genes controlling particular reactions through the catalyzing enzymes are indicated for some reactions):

$$\text{glutamate} \xrightarrow{pro\text{-}3} \text{GSA} \longrightarrow \longrightarrow \text{proline}$$

$$\text{precursor X} \xrightarrow{arg\text{-}3} \text{CAP} \xrightarrow[\text{(OTCase)}]{arg\text{-}12} \text{citrulline} \longrightarrow \text{arginine}$$

ornithine

$$\text{precursor Y} \xrightarrow{pyr\text{-}3a} \text{CAP} \xrightarrow[\text{(ATCase)}]{pyr\text{-}3d} \text{ureidosuccinate} \longrightarrow \text{pyrimidine}$$

aspartic acid

(CAP is carbamyl phosphate; note that it occurs in two of these sequences.)

a. The *arg-3* mutants require arginine, and the *pyr-3a* mutants require pyrimidine. Because CAP is found in *both* sequences, what does this observation suggest?

b. The gene *pyr-3* seems to control two consecutive enzymic conversions. How does this fact fit with your answer to (a)? What can you suggest about the probable structure of the *pyr-3* enzyme?

c. The *arg-12* mutants partially suppress *pyr-3a* mutants, and the *pyr-3d* mutants partially suppress *arg-3* mutants. Is this consistent with your answer to (a)?

d. All *pro-3* mutants require proline, but an *arg-12 pro-3* genotype does not require proline; in this case the label added as ornithine ends up in GSA and proline. The conversion of ornithine to GSA is catalyzed by the enzyme ornithine transaminase (OTA). Why do you think *pro-3* single mutants do not have access to ornithine?

e. In rats, enzyme blocks corresponding to *pyr-3a* do *not* lead to pyrimidine requirement. Compare rats and *Neurospora* in this regard.

26. Explain how Benzer was able to calculate the number of sites with zero occurrences, as depicted in Figure 12-33.

27. In *Collinsia parviflora*, the petal color is normally purple. Four recessive mutations are induced, each of which produces white petals and is recessive. The four pure-breeding lines are then intercrossed in the following combinations, with the results indicated:

Mutant crosses	F_1	F_2
1×2	all purple	$\frac{1}{2}$ purple, $\frac{1}{2}$ white
1×3	all purple	$\frac{9}{16}$ purple, $\frac{7}{16}$ white
1×4	all white	all white

a. Explain all these results clearly, using diagrams wherever possible.

b. What F_1 and F_2 do you predict from crosses of 2×3 and 2×4?

28. A biologist is interested in the genetic control of leucine synthesis in the haploid filamentous fungus *Aspergillus*. He treats spores with mutagen and obtains five point mutations ($a-e$), all of which leucine auxotrophs. He first makes heterokaryons between them to check on their functional relationships. He determines the following results, where $+$ indicates that the heterokaryon grew and $-$ indicates that the heterokaryon did not grow on a medium lacking leucine:

	a	b	c	d	e
a	$-$	$+$	$+$	$+$	$-$
b		$-$	$+$	$+$	$+$
c			$-$	$+$	$+$
d				$-$	$+$
e					$-$

The biologist then intercrosses the mutations in all possible combinations. From each cross, he tests 500 ascospore progeny by inoculating them onto a medium lacking leucine. The results follow (the numbers represent the number of leucine prototrophs in the 500 progeny):

	a	b	c	d	e
a	0	125	128	126	0
b		0	124	2*	125
c			0	124	127
d				0	123
e					0

(* found *not* to be due to reversion)

Explain both sets of data genetically.

29. In *Drosophila*, the eye phenotype "star" is caused by recessive mutations (s) mapping to one location on the second chromosome. This region is flanked to the left by the locus A/a and to the right by the locus B/b:

$$\underline{\quad \overset{A/a}{|} \qquad\qquad \overset{star}{|} \qquad\qquad \overset{B/b}{|} \quad\quad}$$

A set of six independently induced *star* mutations is each made homozygous with both *AABB* and *aabb* constitutions and is intercrossed to study complementation at the *star* locus. The results follow, where $+$ indicates wild-eye and s indicates star-eye, both being phenotypes of the F_1.

	AAssBB					
aassbb	1	2	3	4	5	6
1	s	$+$	s	s	$+$	[+]
2		s	$+$	[+]	s	$+$
3			s	s	$+$	$+$
4				s	$+$	$+$
5					s	$+$
6						s

a. How many cistrons are at the *star* location, and which mutational sites are in each cistron?

b. The heterozygotes in brackets are allowed to produce gametes. In both cases, *star*+ recombinant gametes are identified. The gametes are tested for the flanking marker conformation, which is aB in both the 1×6 heterozygote and the 2×4 heterozygote. Order the cistrons in relation to the A/a and B/b loci.

DNA Function

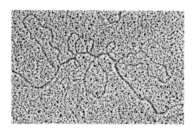

KEY CONCEPTS

DNA is transcribed into an mRNA molecule, which is then
translated during protein synthesis.
▪

Translation requires transfer RNAs and ribosomes.
▪

The genetic code is a nonoverlapping triplet code.
▪

Special sequences signal the initiation and termination of
both transcription and translation.
▪

In eukaryotes, the initial RNA transcript is processed in
several ways to generate the final mRNA.
▪

Many eukaryotic genes contain segments of DNA, termed
introns, which interrupt the normal gene coding sequence.
▪

The primary eukaryotic transcript is spliced in one of a
variety of ways to remove the RNA encoded by the intron
and to yield the final mRNA.

■ The genetic information embodied in DNA can either be copied into more DNA during replication or be translated into protein. These are the processes of information transfer that constitute DNA function. We considered replication in some detail in Chapter 10. Let's now explore the way in which genetic information is turned into protein. Only a part of this story has been revealed through purely genetic analysis; mutants have been useful, of course, but only as tools to shortcut a lot of the biochemical work. However, the kind of reasoning that molecular biologists employ to investigate this aspect of DNA function illustrates an analytical approach also characteristic of work in genetics.

Transcription

Early investigators had good reasons for thinking that information is not transferred from DNA to protein. For one thing, DNA is found in the nucleus (of a eukaryotic cell), whereas protein is known to be synthesized in the cytoplasm. Another nucleic acid, ribonucleic acid (RNA), is necessary as an intermediary.

Early Experiments Suggesting RNA Intermediate

If cells are fed radioactive RNA precursors, then the labeled RNA shows up first of all in the nucleus, indicating that the RNA is synthesized there. In a pulse-chase experiment, a brief "pulse" of labeling is followed by a "chase" of nonlabeled RNA precursors. In samples taken after the chase, the labeled RNA is found in the cytoplasm (Figure 13-1). Apparently, the RNA is synthesized in the nucleus and then moves into the cytoplasm. Thus, it is a good candidate as an information-transfer intermediary between DNA and protein.

In 1957, Elliot Volkin and Lawrence Astrachan made a significant observation. They found that one of the most striking molecular changes when E. coli is infected with the phage T2 is a rapid burst of RNA synthesis. Furthermore, this phage-induced RNA "turns over" rapidly, as shown in the following experiment. The infected bacteria are first pulsed with radioactive uracil (a specific precursor of RNA); the bacteria are then chased with cold uracil. The RNA recovered shortly after the pulse is labeled, but that recovered somewhat longer after the chase is unlabeled, indicating that the RNA has a very short lifetime. Finally, when the nucleotide contents of E. coli and T2 DNA are compared with the nucleotide content of the reduced RNA, the RNA is found to be very similar to the phage DNA.

The tentative conclusion is that RNA is synthesized from DNA and that it passes into the cytoplasm, where it

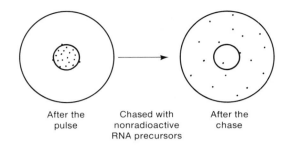

| After the pulse | Chased with nonradioactive RNA precursors | After the chase |

Figure 13-1. RNA synthesized during one short time period is labeled by feeding the cell a brief "pulse" of nonradioactive RNA precursors, followed by a "chase" of radioactive precursors. In an autoradiograph, the labeled RNA appears as dark grains. Apparently, the RNA is synthesized in the nucleus *(small circle)* and then moves out into the cytoplasm.

is somehow used to synthesize protein. We can outline three stages of information transfer (Figure 13-2): **replication** (the synthesis of DNA), **transcription** (the synthesis of an RNA copy of a portion of the DNA), and **translation** (the synthesis of a polypeptide directed by the RNA sequence).

Properties of RNA

Although RNA is a long-chain macromolecule of nucleic acid (as is DNA), it has very different properties. First, RNA is single-stranded, not a double helix. Second, RNA has ribose sugar, rather than deoxyribose, in its nucleotides (hence, its name):

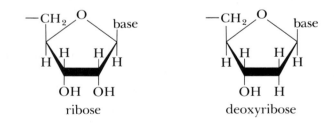

ribose   deoxyribose

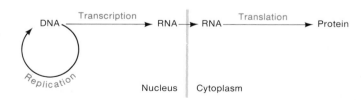

Figure 13-2. The three processes of information transfer: replication, transcription, and translation.

Third, RNA has the pyrimidine base **uracil** (abbreviated U) instead of thymine. However, uracil does form hydrogen bonds with adenine just as thymine does:

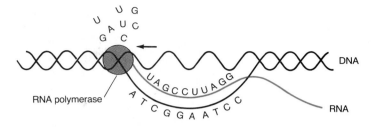

uracil

No one is absolutely sure why RNA has uracil instead of thymine or why it has ribose instead of deoxyribose. The most important aspect of RNA is that it is single-stranded, but otherwise it is very similar in structure to DNA. This suggests that transcription may be based on

Figure 13-3. Synthesis of RNA on a single-stranded DNA template using free nucleotides. The process is catalyzed by RNA polymerase. Uracil (U) pairs with adenine (A).

the complementarity of bases, which is also the key to DNA replication. A transcription enzyme, RNA polymerase, could perform the transcription in a fashion quite similar to replication (Figure 13-3).

In fact, this model of transcription is confirmed cytologically (Figure 13-4). The fact that RNA can be synthe-

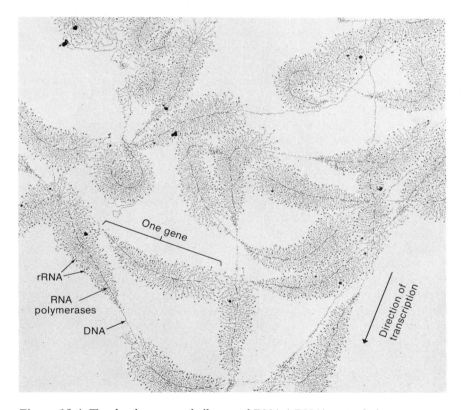

Figure 13-4. Tandemly repeated ribosomal RNA (rRNA) genes being transcribed in the nucleolus of *Triturus viridiscens* (an amphibian). (rRNA is a component of the ribosome, a cellular organelle.) Along each gene, many RNA polymerse molecules are attached and transcribing in one direction. The growing RNA molecules appear as threads extending out from the DNA backbone. The shorter RNA molecules are nearer the beginning of transcription; the longer ones have almost been completed. Hence, the "Christmas tree" appearance. (Photograph O. L. Miller, Jr., and Barbara A. Hamkalo.)

sized with DNA acting as a template is demonstrated by synthesis in vitro of RNA from nucleotides in the presence of DNA, using an extractable RNA polymerase. Whatever source of DNA is used, the RNA synthesized has an (A + U)/(G + C) ratio similar to the (A + T)/(G + C) ratio of the DNA (Table 13-1). This experiment does not indicate whether the RNA is synthesized from both DNA strands or from just one, but it does indicate that the linear frequency of the A-T pairs (in comparison with the G-C pairs) in the DNA is precisely mirrored in the relative abundance of (A + U) in the RNA. (These points are difficult to grasp without drawing some diagrams; Problem 2 at the end of this chapter provides some opportunities to clarify these notions.)

To test the complementarity of DNA with RNA, investigators can apply the specificity and precision of nucleic-acid hybridization. DNA can be denatured and mixed with the RNA formed from it. On slow cooling, some of the RNA strands anneal with complementary DNA to form a DNA-RNA hybrid. The DNA-RNA hybrid differs in density from the DNA-DNA duplex, so its presence can be detected by ultracentrifugation in cesium chloride. Nucleic acids will anneal in this way only if there are stretches of base-sequence complementarity, so the experiment does prove that the RNA transcript is complementary in base sequence to the parent DNA.

Can we determine whether RNA is synthesized from only one or from both of the DNA strands? It seems reasonable that only one strand is used, because transcription of RNA from both strands would produce two complementary RNA strands from the same stretch of DNA and these strands presumably would produce two different kinds of protein (with different amino-acid sequences). In fact, a great deal of chemical evidence confirms that transcription takes place on only one of the DNA strands although not necessarily the same strand throughout the entire chromosome).

The hybridization experiment can be extended to explore this problem. If the two strands of DNA have distinctly different purine : pyrimidine ratios, they can be purified separately because they have different densities in cesium chloride (CsCl). The RNA made from a

TABLE 13-1. Nucleotide ratios in various DNAs and in their transcripts (in vitro)

DNA source	$\dfrac{(A + T)}{(G + C)}$ of DNA	$\dfrac{(A + U)}{(G + C)}$ of RNA synthesized
T2 phage	1.84	1.86
Cow	1.35	1.40
Micrococcus (bacterium)	0.39	0.49

stretch of DNA can be purified and annealed separately to each of the strands to see whether it is complementary to only one. J. Marmur and his colleagues were able to separate the strands of DNA from the *B. subtilis* phage SP8. They denatured the DNA, cooled it rapidly to prevent reannealing of the strands, and then separated the strands in CsCl. They showed that the SP8 RNA hybridizes to only one of the two strands, proving that transcription is **asymmetrical**—that it occurs only on one DNA strand.

RNA is transcribed from a single strand of DNA at a time. However, the same strand is not necessarily transcribed throughout the entire chromosome or through all stages of the life cycle. The RNA produced at different stages in the cycle of a phage hybridizes to different parts of the chromosome, showing the different genes that are activated at each stage. In λ phage, each of the two DNA strands is partially transcribed at a different stage (Figure 13-5). In phage T7, however, the same strand is transcribed for both early-acting and late-acting genes (Figure 13-6).

In most prokaryotes, a single RNA polymerase species transcribes all types of RNA; in higher eukaryotes, three different RNA polymerases transcribe different classes of RNA. We can consider transcription as occurring in three distinct stages: **initiation, elongation,** and **termination.** Let's briefly examine each of these stages, taking bacterial transcription as an example.

nu 1 A W B C *nu* 3 D E FI FII Z U V G T H M L K I J   *b2 int xis red* γ cIII N *rex* cI *cro* cII O P Q S R

Figure 13-5. Map of λ phage DNA, with arrows indicating direction of transcription into mRNA. The upper arrows indicate transcription from one DNA strand, and the lower arrows indicate transcription from the other strand. The DNA is circular during transcription, so the lower two arrows are actually one stretch of transcription. (From G. S. Stent and R. Calendar, *Molecular Genetics,* 2d ed. Copyright © 1978 by W. H. Freeman and Company)

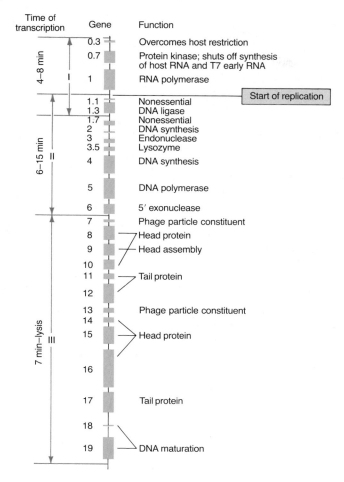

Time of transcription	Gene	Function
4–8 min	0.3	Overcomes host restriction
	0.7	Protein kinase; shuts off synthesis of host RNA and T7 early RNA
I	1	RNA polymerase
6–15 min	1.1	Nonessential
	1.3	DNA ligase
	1.7	Nonessential
	2	DNA synthesis
	3	Endonuclease
	3.5	Lysozyme
II	4	DNA synthesis
	5	DNA polymerase
	6	5′ exonuclease
7 min–lysis	7	Phage particle constituent
	8	Head protein
	9	Head assembly
	10	
	11	Tail protein
	12	
	13	Phage particle constituent
	14	
III	15	Head protein
	16	
	17	Tail protein
	18	
	19	DNA maturation

Start of replication

Figure 13-6. The various parts of the T7-phage genome are transcribed at various times during the phage growth cycle. (After G. S. Stent and R. Calendar, *Molecular Genetics,* 2d ed. Copyright © 1978 by W. H. Freeman and Company)

Initiation

The regions of the DNA that signal initiation of transcription are termed **promoters.** (We consider their role in gene control in Chapter 16). A dissociative subunit of RNA polymerase, the *sigma* factor, allows RNA polymerase to recognize and bind strongly to promoter regions. Figure 13-7 shows the promoter sequences from 12 different transcription initiation points on the *E. coli* genome. The bases are aligned according to homologies, or similar base sequences that appear just prior to the first base transcribed (termed the initiation site in Figure 13-7).

Note in Figure 13-7 how two regions of partial homology appear in virtually each case. These regions have been termed the −35 and −10 regions due to their locations relative to the transcription initiation point. At the bottom of Figure 13-7 an ideal or consensus sequence of a promoter is given. Physical experiments

have confirmed that RNA polymerase makes contact with these two regions when binding to the DNA. The enzyme then unwinds the DNA and begins the synthesis of an RNA molecule.

Elongation

Shortly after initiating transcription, the sigma factor dissociates from the RNA polymerase. The RNA is always synthesized in the 5′–3′ direction (Figures 13-8 and 13-9); in other words, the new chain grows in the direction 5′–3′. This means that the —OH group is the newest end.

Termination

RNA polymerase also recognizes signal for chain termination. In some cases, the recognition is direct; in other cases, the help of an additional protein factor, termed *rho,* is required in order to recognize the termination signals. Many termination sites appear to involve secondary structures of the type shown in Figure 13-10. The exact manner in which these signals bring about chain termination is not yet clear.

Translation

The information-bearing RNA is appropriately called **messenger RNA** (mRNA). It acts as a copy of information from the DNA in the nucleus, sent out to direct protein synthesis in the cytoplasm (rather like a copy of a blueprint sent from the executive office to the production department). However, if you mix mRNA and all 20 amino acids in a test tube and hope to make protein, you will be disappointed. Other components are needed; the discovery of the nature of these components has provided the key to understanding the mechanism of translation.

Application of Sucrose Gradients

Another important experimental tool is involved in this part of our story: **sucrose density-gradient centrifugation.** The CsCl gradient discussed earlier is created by ultracentrifugation of a uniform solution. However, the sucrose gradient is created in a test tube by layering successively lower concentrations of sucrose solution, one on top of the other. The material to be studied is carefully placed on top. When the solution is centrifuged in a machine that allows the test tube to swivel freely, the sedimenting material travels through the gradient at different rates that are related to the sizes and shapes of the molecules. Large molecules migrate farther in a given period of time than smaller molecules do.

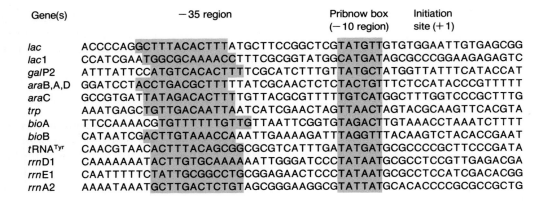

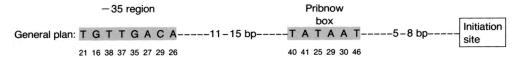

Figure 13-7. Promoter sites have regions of similar sequences, as indicated at the top of the figure by the colored region in the 12 different promoter sequences in *E. coli*. The gene (or genes) governed by each promoter sequence is underlined on the left. Numbering is given in terms of the number of bases before (−) or after (+) the RNA synthesis initiation point. (After J. Darnell, H. Lodish, and D. Baltimore, *Molecular Cell Biology*. Copyright © 1986 by W. H. Freeman and Company.)

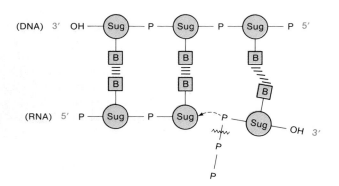

Figure 13-8. Chain elongation during transcription. A nucleoside triphosphate is aligning at the 3′ growing point. Two phosphate groups (P) will be lost as the phosphodiester bond is synthesized. (B = base; Sug = sugar)

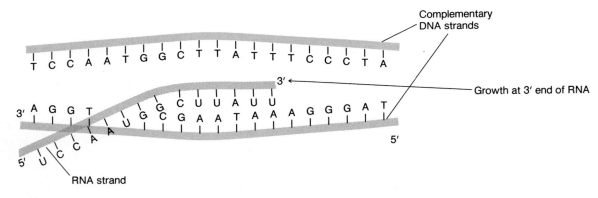

Figure 13-9. An RNA strand is synthesized in the 5′ to 3′ direction from a locally single-stranded region of DNA. (After J. D. Watson, J. Tooze, and D. T. Kurtz, *Recombinant DNA: A Short Course*. Copyright © 1983 by W. H. Freeman and Company.)

```
                UCC
              U     G
              G·C
              A·U—25
              C·G
        15—C·G
              G·C
              C·G   35
         5    C·G    |
         |          |
   5′ UAAUCCCACAG·CAUUUU 3′
```

Figure 13-10. The structure of a termination site for RNA polymerase in bacteria.

The separated molecules can be collected individually by collecting sequential drops from a small opening in the bottom of the tube (Figure 13-11). The time that a fraction takes to move the fixed distance to the tube bottom indicates its position, or sedimentation (S) value, which is a measure of the size of the molecules in the fraction.

Different Classes of RNA Molecules

Using the separatory powers of the sucrose-gradient technique, it is possible to identify several macromole-

cules and macromolecular aggregates in a typical protein-synthesizing system. The main components are **transfer RNA** (tRNA), **ribosomes,** and messenger RNA (mRNA). Transfer RNA is a class of small (4S) RNA molecules of rather similar type and function. In fact, complete nucleotide sequences have been determined for a large number of tRNA molecules from different organisms. They all appear to have some hydrogen-bonded regions and some single-stranded regions, and they form variations of a cloverleaf structure (Figure 13-12).

Ribosomes, on the other hand, are cellular organelles composed of very complex aggregations of ribosomal proteins and **ribosomal RNA** (rRNA) components. In *E. coli,* for instance, at least three separate RNA molecules can be distinguished by size in ribosomes: 23S, 16S, and 5S. The precise function of ribosomes is still the subject of much research; the story of how we have learned what we do know is a fascinating one that lies outside the scope of this book. We do not know precisely how rRNA is bound up with the protein components or how each component functions. Under the electron microscope, a ribosome appears as a "blob." On chemical treatment, the blob splits into two main subblobs (50S

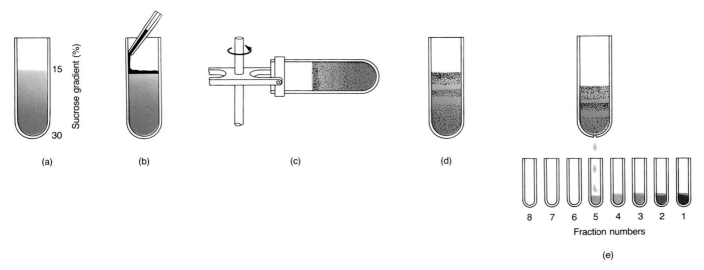

Figure 13-11. The sucrose-gradient technique. (a) A sucrose density gradient is created in a centrifuge tube by layering solutions of differing densities. (b) The sample to be tested is placed on top of the gradient. (c) Centrifugation causes the various components (fractions) of the sample to sediment differentially. (d) The different fractions appear as bands in the centrifuged gradient. (e) The different bands can be

collected separately by collecting samples from the bottom of the tube at fixed time intervals. The S value for the fraction is based on its position in the gradient, which is determined by the time at which it drips from the bottom of the tube. (From A. Rich, "Polyribosomes." Copyright © 1963 by Scientific American, Inc. All rights reserved.)

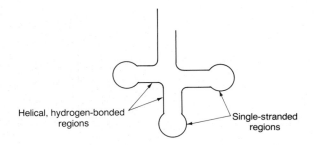

Figure 13-12. The cloverleaf structure of tRNA.

Figure 13-13. The simplified "two-blob" representation of the ribosome.

and 30S in *E. coli*), which we represent by the simplified symbol shown in Figure 13-13.

Both rRNA and tRNA molecules will form RNA–DNA hybrids in vitro, indicating that they are transcribed from the DNA. Figure 13-14 diagrams an exploded view of the components of an *E. coli* ribosome, giving some idea of their relative sizes. Table 13-2 summarizes the main types of RNA that can be found in a typical protein-synthesizing system.

Other components are needed to make protein synthesis work in vitro. These include several enzymes (amino-acyl-tRNA synthetases and peptidyl transferase), several mysterious protein "factors" (probably enzymic), and a chemical source of energy. The energy donor is needed because an orderly structure is being created out of a mess of components—a process that requires energy because the system must lose entropy (a measure of disorder or randomness).

Protein Synthesis

We can regard **protein synthesis** as a chemical reaction, and we shall take this approach at first. Then we shall take a three-dimensional look at the physical interactions of the major components.

In protein synthesis as a chemical reaction:

1. Each amino acid (aa) is attached to a tRNA molecule specific to that amino acid by a high-energy bond derived from GTP. The process is catalyzed by a specific enzyme called a synthetase (the tRNA is said to be "charged" when the amino acid is attached):

$$aa_1 + tRNA_1 + GTP \xrightarrow{synthetase_1} aa_1-tRNA_1 + GDP$$

There is a separate synthetase for each amino acid.

2. The energy of the charged tRNA is converted into a peptide bond linking the amino acid to another one on the ribosome:

$$aa_1-tRNA_1 + aa_2-tRNA_2 \xrightarrow[\substack{\text{on a} \\ \text{ribosome}}]{\text{peptidyl transferase}} \underbrace{aa_1-aa_2}_{\substack{\text{small} \\ \text{polypeptide}}}-tRNA_2 + tRNA_1 \text{ released}$$

3. New amino acids are linked by means of a peptide bond to the growing chain:

$$aa_3-tRNA_3 + aa_1-aa_2-tRNA_2 \longrightarrow \underbrace{aa_1-aa_2-aa_3}_{\substack{\text{larger} \\ \text{polypeptide}}}-tRNA_3 + tRNA_2 \text{ released}$$

4. This process continues until aa_n (the final amino acid) is added. Of course, the whole thing works only in the presence of mRNA, ribosomes, several additional protein factors, enzymes, and inorganic ions.

TABLE 13-2. RNA molecules in *E. coli*

Type	Percentage of cell RNA	Sedimentation coefficient (S)	Molecular weight	Number of nucleotides
Ribosomal RNA (rRNA)	80	23	1.2×10^6	3700
		16	0.55×10^6	1700
		5	3.6×10^4	1700
Transfer RNA (tRNA)	15	4	2.5×10^4	75
Messenger RNA (mRNA)	5		Heterogeneous	

SOURCE: L. Stryer, *Biochemistry*, 2d ed. Copyright © 1981 by W. H. Freeman and Company.

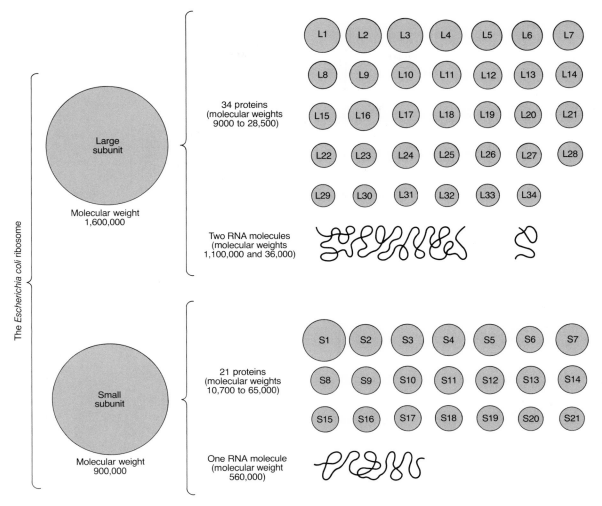

Figure 13-14. The parts of a ribosome of *E. coli*, showing the complexity of what is the single-most important entity in all biological synthesis. The precise roles of the parts are not understood. Peptidyl transferase, the enzymatic activity responsible for the formation of peptide bonds between incoming amino acids and the growing polypeptide chain, appears to be built into the ribosome. (From Donald M. Engelman and Peter B. Moore, "Neutron-scattering Studies of the Ribosome." Copyright © 1976 by Scientific American, Inc. All rights reserved.)

Specificity in Protein Synthesis

To visualize the incredible process of protein synthesis, we must recognize the main interacting sites of the components. We begin with tRNA. Is it the tRNA or the amino acid that recognizes the portion of the mRNA, termed a **codon,** that codes for a specific amino acid? A very convincing experiment has answered this question. In the experiment, cysteinyl tRNA ($tRNA_{Cys}$, the tRNA specific for cysteine) charged with cysteine was treated with nickel hydride, which converted the cysteine (while still bound to $tRNA_{Cys}$) into another amino acid, alanine, without affecting the tRNA:

$$\text{cysteine} - tRNA_{Cys} \xrightarrow{\text{nickel hydride}} \text{alanine} - tRNA_{Cys}$$

Protein synthesized with this hybrid species had alanine wherever we would expect cysteine. Thus, the experiment demonstrated that the amino acids are illiterate; they are inserted at the proper position because the tRNA "adaptors" recognize the mRNA codons and insert their attached amino acids appropriately. We would expect, then, to find some site on the tRNA that recognizes the mRNA codon by complementary base pairing.

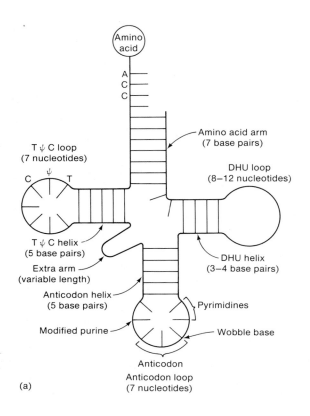

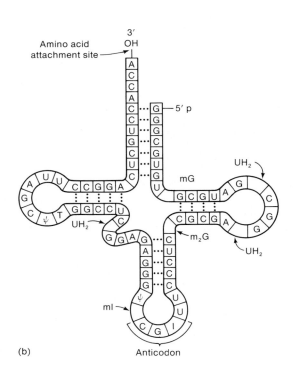

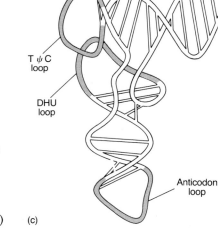

Figure 13-15. The structure of transfer RNA. (a) The functional areas of a generalized tRNA molecule. (b) The specific sequence of yeast alanine tRNA. Arrows indicate several kinds of rare modified bases. (c) The actual three-dimensional structure of yeast phenylalanine tRNA. The symbols ψ, UH₂, mG, m₂G, mI, and DHU are abbreviations for modified bases. (Part a from S. Arnott, "The Structure of Transfer RNA," *Progress in Biophysics and Molecular Biology* 22, 1971, 186; parts b and c from L. Stryer, *Biochemistry,* 2d ed. Copyright 1981 by W. H. Freeman and Company; part c based on a drawing by Dr. Sung-Hou Kim.)

Figure 13-15a shows several functional sites of tRNA molecule. The site that recognizes an mRNA codon is called the **anticodon;** its bases are complementary to the bases of the codon. Another operationally identifiable site is the amino-acid attachment site. The other arms probably assist in binding the tRNA to the ribosome. Figure 13-15b shows a specific tRNA (yeast alanine tRNA). The "flattened" cloverleafs shown in these diagrams are not the normal conformation of tRNA molecules; tRNA normally exists as an L-shaped folded cloverleaf, as shown in Figure 13-15c. These diagrams are supported by very sophisticated chemical analysis of tRNA nucleotide sequences and by X-ray crystallographic data on the overall shape of the molecule. Although tRNA molecules share many structural similarities, each has a unique three-dimensional shape that allows recognition by the correct synthetase. The specificity of charging the tRNAs is crucial to the integrity of protein synthesis.

Where does tRNA come from? If radioactive tRNA is put into a cell nucleus in which the DNA has been partially denatured by heating, the radioactivity appears (by autoradiography) in localized regions of the chromosomes. These regions probably reflect the location of **tRNA genes;** they are regions of DNA that produce tRNA rather than mRNA, which produces a protein. The labeled tRNA hybridizes to these sites because of the complementarity of base sequences between the tRNA and its parent gene. A similar situation holds for rRNA. Thus, we see that even the one-gene–one-polypeptide ideal is not completely valid. Some genes do not code for protein; rather, they specify RNA components of the translational apparatus.

> **Message** Some genes code for proteins; other genes specify RNA (tRNA or rRNA) as their final product.

How does tRNA get its fancy shape? It probably folds up spontaneously into a conformation that produces maximal stability. Transfer RNA contains many "odd" or modified bases (such as pseudouracil, φ) in its nucleotides; these play a direct role in folding and also have been implicated in other tRNA functions.

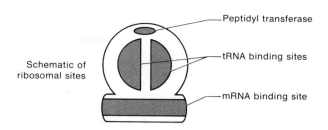

Figure 13-16. A diagrammatic representation of some important functional regions of the ribosome.

Ribosomes

Now we turn to the ribosomes, which have several sites that we can predict from our model of translation, including binding sites for mRNA and tRNA. By adding radioactive tRNA to a solution of ribosomes and measuring how much of it becomes bound to the ribosomes, we could calculate that there are two tRNA binding sites per ribosome. There also is an mRNA binding site, and the enzyme peptidyl transferase is built up into the ribosome (Figure 13-16).

We can now assemble all of the components in a diagram: Figure 13-17 shows an amino acid being added to a growing polypeptide. The ribosome "moves along" the mRNA strand. As each new codon enters the ribo-

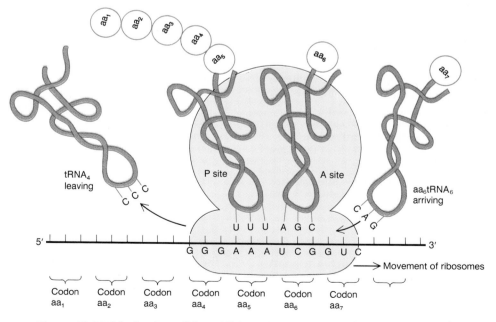

Figure 13-17. Mechanism of the addition of a single amino acid to the growing polypeptide chain during translation of mRNA. Amino acid 6 is being added; amino acids 1 to 5 are already part of polypeptide. The A site is the amino-acid binding site; the P site is the polypeptide binding site.

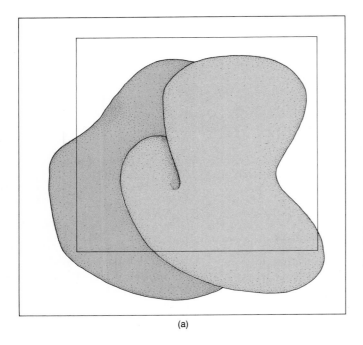

(a)

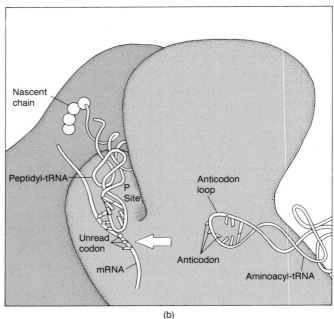

(b)

Figure 13-18. Hypothetical molecular mechanism for the synthesis of a protein by a ribosome is shown in a sequence of drawings. (a) The entire ribosome. The color indicates the large subunit of the ribosome, and the gray depicts the small subunit. The black square indicates the part of the ribosome that appears in each of the following drawings. (b) The events that begin an iteration of the elongation cycle. A peptidyl-tRNA is bound to the small subunit's P site; it bears a nascent protein chain that is shown as entering the large subunit. The anticodon of the peptidyl-tRNA is bound to a codon of the mRNA. The codon was translated in the preceding iteration of the cycle. An aminoacyl-tRNA draws near the small subunit. GTP and the elongation factor EF-Tu, which accompany the tRNA, are not depicted. (c) The arriving tRNA has bound itself to the R site; its anticodon becomes bound to the next unread codon. The anticodon is part of a loop of seven bases. Five of them are stacked on top of one another; these are the five that lie nearest the end of the tRNA designated the 5′ end. (d) Protein synthesis continues. Here the aminoacyl-tRNA has begun to flip into a position that will bind it to the A site. The flip constitutes a test of the strength of the binding between the codon and the anticodon; hence, it improves the likelihood that the arriving amino acid is the one the code requires. (e) The aminoacyl-tRNA has arrived at the A site. The five stacked bases in the anticodon loop are now the ones that lie closest to the 3′ end of the molecule. This change allows the tRNA to flip without tugging on the mRNA. The proper alignment of the tRNAs on the mRNA may be ensured by a bond between bases that form a bridge between the two anticodon loops. The large subunit can now transfer the nascent chain to the amino acid on the aminoacyl-tRNA. (f) The cycle is completed. The naked tRNA is expelled; the A-site tRNA takes its place at the P site, and the mRNA is repositioned for the arrival of the next aminoacyl-tRNA. Neither GTP nor the elongation factor EF-G, both of which act in these final events, is shown. (After J. Lake, "The Ribosome." Copyright © 1981 by Scientific American, Inc. All rights reserved.)

some, conditions are right for the insertion of the appropriate aa–tRNA complex in the amino-acid binding site (A site) and for the formation of a bond between the amino acid in the A site and the polypeptide being held to a tRNA in the polypeptide binding site (P site). When the peptide bond is formed, the tRNA in the P site is liberated, and the new tRNA–polypeptide complex moves into the unoccupied P sites as the ribosome moves farther along the mRNA. Figure 13-18 is a three-dimensional illustration of the formation of the peptide bond on the ribosome.

Much of the model shown in Figure 13-17 was constructed by inference based on genetic and chemical analyses. Its validity was gratifyingly demonstrated recently by electron micrographs of the process of protein synthesis in an *E. coli* gene. You can see in Figure 13-19

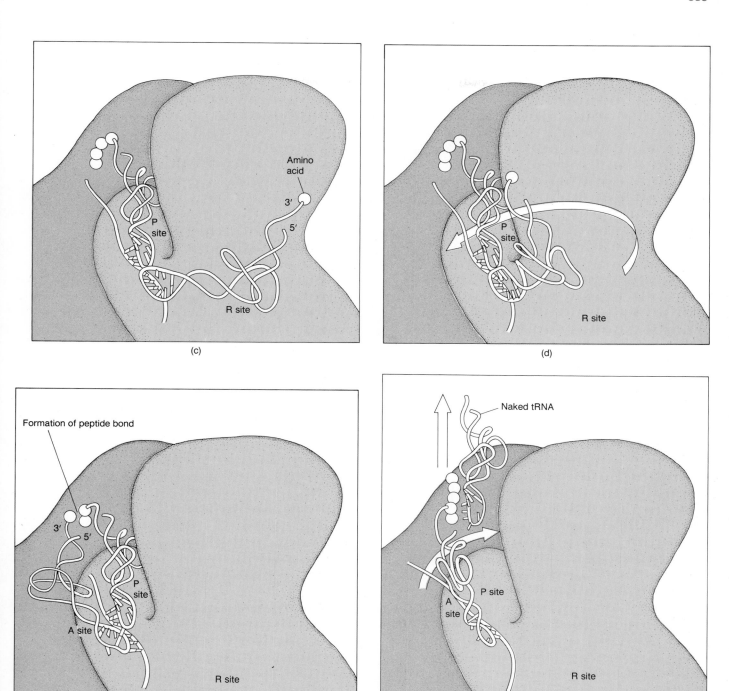

(c)

Amino acid

3′

5′

P site

R site

(d)

P site

R site

(e)

Formation of peptide bond

3′

5′

P site

A site

R site

(f)

Naked tRNA

P site

A site

R site

that protein synthesis in this prokaryotic organism does not wait for the completion of mRNA synthesis. As soon as part of the mRNA is synthesized, ribosomes attach to it and begin translation! Protein synthesis in eukaryotes is different, because mRNA is made in the nucleus and then shipped to the cytoplasm for translation; thus, transcription and translation in the eukaryote are separated in both time and space.

In Figure 13-17, a few imaginary codons have been inserted as illustrations of the process. Of course, this was the general notion in the minds of the investigators at the time also, but the knowledge of *which* specific codons represented *which* specific amino acids had to await another round of sophisticated investigations. This work came to be known as "cracking the genetic code."

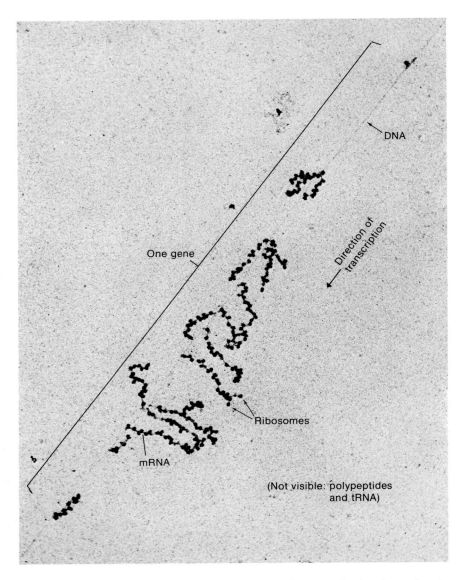

Figure 13-19. A gene of *E. coli* being simultaneously transcribed and translated. (Electron micrograph by O. L. Miller, Jr., and Barbara A. Hamkalo.)

The Genetic Code

If genes are segments of DNA and if DNA is just a string of nucleotide pairs, then how does the sequence of nucleotide pairs dictate the sequence of amino acids in protein? The analogy to a code springs to mind at once. The cracking of the genetic code is the story of the rest of this chapter. The experimentation was sophisticated and swift, and it did not take long for the code to be deciphered once its existence was strongly indicated.

Simple logic tells us that if nucleotide pairs are the *letters* in a code, then a combination of letters could form

words representing different amino acids. We must ask whether the code is overlapping or nonoverlapping. Then we must ask how many letters make up a word, or codon, and which specific codon or codons represent each specific amino acid.

Overlapping Versus Nonoverlapping Codes

Figure 13-20 shows the difference between an overlapping and a nonoverlapping code. In the example, a three-letter or **triplet** code is shown. For the nonoverlapping code, consecutive amino acids are specified

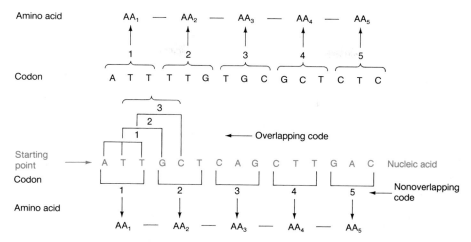

Figure 13-20. The difference between an overlapping and a nonoverlapping code. The case illustrated is for a code with three letters (a triplet code). In a nonoverlapping code, a protein is translated by reading codons that do not share any of the same nucleotides. An overlapping code uses codons that employ some of the same nucleotides as other codons for the translation of a single protein, as shown in the top of the diagram (for the DNA sequence shown at the bottom of the diagram).

by consecutive code words (codons) that do not overlap, as shown in the bottom of the figure. For an overlapping code, consecutive amino acids are encoded by sets of bases (the codons) that overlap, as shown in the top portion of the figure. Thus, for the sequence ATTGCTCAG in a nonoverlapping code, the first three amino acids are encoded by the three triplets ATT, GCT, and CAG, respectively. However, for an overlapping code, the first three amino acids are encoded by the triplets ATT, TTG, and TGC if the overlap is a single base, as shown in the figure.

By 1961, it was already clear that the genetic code was nonoverlapping. The analysis of mutationally altered proteins—in particular, the nitrous acid-generated mutants of tobacco mosaic virus—showed that only a single amino acid changes at one time in one region of the protein. This is predicted by a nonoverlapping code. As you can see from Figure 13-20, an overlapping code predicts that a single base change will alter as many as three amino acids at adjacent positions in the protein.

It should be noted that while the use of an overlapping *code* was ruled out by the analysis of single proteins, nothing precluded the use of alternate reading frames to encode amino acids in two different proteins. In the example here, one protein might be encoded by the series of codons that reads ATT, GCT, CAG, CTT, and so on. A second protein might be encoded by codons that are shifted over by one base and therefore read TTG, CTC, AGC, TTG, and so on. This is an example of storing the information encoding two different proteins in two different reading frames, while still using a ge-

netic code that is read in a nonoverlapping manner during the translation of a specific protein. Some examples of such information storage have been found, as we see in Chapter 15.

Number of Letters in the Code

Reading a DNA molecule from one particular end, one nucleotide pair at a time, only four nucleotide pairs are encountered:

$$-A- \quad -T- \quad -G- \quad \text{and} \quad -C-$$
$$-T- \quad -A- \quad -C- \quad \qquad -G-$$

Thus, if the words are one letter long, then only four words are possible. This cannot be the genetic code because we must have a word for each of the 20 amino acids commonly found in cellular proteins. If the words are two letters long, then $4^2 = 16$ words are possible. For example,

$$-AT- \quad \text{or} \quad -CT- \quad \text{or} \quad -CC-$$
$$-TA- \quad \qquad -GA- \quad \qquad -GG-$$

where the length of the DNA molecule runs across, with one strand shown at the top and the other at the bottom. This vocabulary still is not large enough.

If the words are three letters long, then $4^3 = 64$ words are possible. For example,

$$-ATT- \quad \text{or} \quad -GCG- \quad \text{or} \quad -TGC-$$
$$-TAA- \quad \qquad -CGC- \quad \qquad -ACG-$$

This code provides more than enough words to describe the amino acids. We can conclude that the code word must consist of at least three nucleotide pairs. However, if all words are "triplets," then we have a considerable excess of possible words over the 20 needed to name the common amino acids.

Use of Suppressors to Demonstrate a Triplet Code

Convincing proof that a codon is, in fact, three letters long (and no more than three) came from beautiful genetic experiments first reported in 1961 and later extended by Francis Crick, Sidney Brenner, and their co-workers, who used mutants in the *rII* locus of T4 phage. Mutations causing the rII phenotype (see Chapter 12) were induced using a chemical called proflavin, which was thought to act by the addition or deletion of single nucleotide pairs in DNA. (This assumption is based on experimental evidence not presented here.) The following examples illustrate the action of proflavin:

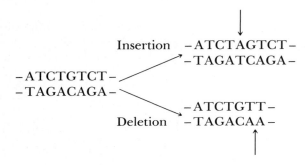

Then, starting with one particular proflavin-induced mutation called FCO, Crick and his colleagues again used proflavin to induce "reversions" that were detected by their wild-type plaques on *E. coli* strain K(λ). Genetic analysis of these plaques revealed that the "revertants" were not identical to true wild-types, thereby suggesting that the back mutation was not an exact reversal of the original forward mutation. In fact, the reversion was found to be caused by the presence of a *second mutation* at a different site from—but in the same cistron as—that of FCO; this second mutation "suppressed" mutant expression of the original FCO. A **suppressor mutation** counteracts or suppresses the effects of another mutation. Some properties of suppressor mutations are:

1. A suppressor mutation is at a different site from that of the mutation it counteracts. Therefore, the original mutation can be recovered by genetic crosses between the wild-type and the "revertant," since the revertant carries both the suppressor and the original mutation.

2. A suppressor mutation may be within the same gene as the mutation it suppresses (internal suppressor), as in the example just given, or it may be in a different gene (external suppressor).

3. Different suppressors may exert their effects in different ways. For instance, some suppressors act at the level of transcription or translation; others alter the physiology of the cell.

The suppressor mutation could be separated from the original forward mutation by recombination. Surprisingly, when this was done, the suppressor was shown to be an *rII* mutation by itself (Figure 13-21).

How can we explain these results? *If the cistron is "read" from both ends, then the two proflavin-induced mutations should still give a mutant phenotype when combined:

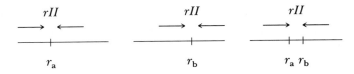

However, if reading is polarized—that is, if the cistron is read from one end only—then the original proflavin-induced addition or deletion could be mutant because it interrupts a normal reading mechanism that establishes the group of bases to be read as words. For example, if each three nucleotide pairs make a word, then the "reading frame" might be established by taking the first three pairs from the end as the first word, the next three pairs as the second word, and so on. In that case, a pro-flavin-induced addition or deletion of a single pair would shift the reading frame from that point on, causing all

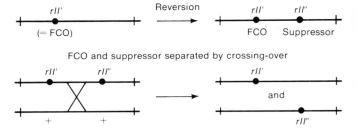

Figure 13-21. The suppressor of an initial *rII* mutation is shown to be an *rII* mutation itself after separation by crossing-over. The original mutant, FCO, was induced by proflavin. Later, when the FCO strain was treated with proflavin again, a revertant was found, which on first appearance seemed to be wild-type. However, it was found that a second mutation within the *rII* region had been induced, and the double mutant *rII'rII"* was shown not to be quite identical to the original wild-type.

following words to be misread. Such a **frame-shift mutation** could reduce most of the genetic message to gibberish. However, the proper reading frame could be restored by a compensatory insertion or deletion somewhere else, leaving only a short stretch of gibberish between the two. Consider the following example in which three-letter English words are used to represent the codons:

THE FAT CAT ATE THE BIG RAT

Delete C: THE FAT↑ATA TET HEB IGR AT

Insert A: THE FAT ATA ATE THE BIG RAT
                              ↑

The insertion suppresses the effect of the deletion by restoring most of the sense of the sentence. By itself, however, the insertion also disrupts the sentence:

THE FAT CAT AAT ETH EBI GRA T

If we assume that the FCO mutant is caused by an addition, then the second (suppressor) mutant would have to be a deletion because, as we have seen, this would restore the reading frame of the gene (a second insertion would not correct the frame). In the following diagrams, we use a hypothetical nucleotide chain to represent DNA for simplicity, realizing that the complementary chain is automatically dictated by the specificity of the base pairing. We also assume that the code words are three letters long and are read in one direction (left to right in our diagrams).

wild-type message

CAT   CAT   CAT   CAT   CAT

rII′ message: distal words changed (x) by frame-shift mutation (words marked ✓ are unaffected)

Addition ────────┐
          CAT   ACA   TCA   TCA   TCA   T
           ✓   x   x   x   x

rII′ rII″ message: few words wrong, but reading frame restored for later words

Deletion ────────────┐
          CAT   ACA   TCT   CAT   CAT
           ✓   x   x   ✓   ✓

The few wrong words in the suppressed genotype could account for the fact that the "revertants" (suppressed

phenotypes) Crick and his associates recovered did not look exactly like the true wild-types phenotypically.

We have assumed here that the original frame-shift mutation was an addition, but the explanation works just as well if we assume that the original FCO mutation is a deletion and the suppressor is an addition. If the FCO is defined as plus, then suppressor mutations are automatically minus; hence, proflavin-induced mutations are also called sign mutations. Experiments have confirmed that a plus cannot suppress a plus and a minus cannot suppress a minus. In other words, two mutations of the same sign never act as suppressors of each other. However, very interestingly, combinations of *three* pluses or *three* minuses have been shown to act together to restore a wild-type phenotype. This observation provided the first experimental confirmation that a word in the genetic code consists of three successive nucleotide pairs, or a triplet. This is because three additions or three deletions within a gene automatically restore the reading frame if the words are triplets. For example,

Deletions

↓   ↓   ↓

CAT   CAT   CAT   CAT   CAT   CAT   CAT

CAT   ACA   TAT   CAT   CAT   CAT

 ✓   x   x   ✓   ✓   ✓

Degeneracy of the Genetic Code

Crick's work also suggested that the genetic code is **degenerate.** That expression is not a moral indictment! It simply means that each of the 64 triplets must have some meaning within the code, so that at least some amino acids must be specified by two or more different triplets. If only 20 triplets are used (with the other 44 being nonsense), then most frame-shift mutations can be expected to produce nonsense words, which presumably would stop the protein-building process. However, if all triplets specify some amino acid, then the changed words would simply result in the insertion of incorrect ("gibberish") amino acids into the protein. Thus, Crick reasoned that many or all amino acids must have several different names in the base-pair code; this hypothesis was later confirmed biochemically.

Proof that the genetic deductions about proflavin were correct came from an analysis of proflavin-induced mutants in a gene with a protein product that could be analyzed. George Streisinger worked with the gene that controls the enzyme lysozyme, which has a known amino-acid sequence. He induced a mutation in the gene with proflavin and selected for proflavin-induced "revertants,'' which were shown genetically to be double mutants (with mutations of opposite sign). When the protein of the double mutant was analyzed, a stretch of "gibberish" amino acids lay between two wild-type ends, just as predicted:

Wild-type
     – Thr – Lys – Ser – Pro – Ser – Leu – Asn – Ala –

"Revertant" type
     – Thr – Lys – Val – His – His – Leu – Met – Ala –

Review

Let's summarize what the work up to this point demonstrates about the genetic code:

1. The code is nonoverlapping.

2. Three bases code for an amino acid. These triplets are termed codons.

3. The code is read from a fixed starting point and continues to the end of the coding sequence. We know this because a single frame-shift mutation anywhere in the coding sequence alters the codon alignment for the rest of the sequence.

4. The code is degenerate in that some amino acids are specified by more than one codon.

Cracking the Code

The actual deciphering of the genetic code — determining the amino acid specified by each triplet — is one of the most exciting genetic breakthroughs of the past two decades. Once the necessary experimental techniques became available, the genetic code was broken in a rush.

The first breakthrough was the discovery of how to make synthetic mRNA. If the nucleotides of RNA are mixed with a special enzyme (polynucleotide phosphorylase), a single-strand RNA is formed in the reaction. No DNA is needed for this synthesis, and so the nucleotides are incorporated at random. The ability to synthesize mRNA offered the exciting prospect of creating specific RNA sequences and then seeing what kinds of amino acids were incorporated when acting as mRNA. The first synthetic messenger obtained, poly-U (· · · – U – U – U – U – · · ·), was made by reacting only uracil nucleotides with the RNA-synthesizing enzyme. In 1961, Marshall Nirenberg and Heinrich Mathaei mixed poly-U with the protein-synthesizing machinery of *E. coli* (ribosomes, aminoacyl tRNA's, a source of chemical energy, several enzymes, and a few other things) in vitro and *observed the formation of a protein!* Of course, the

TABLE 13-3. Expected frequencies of various codons in synthetic mRNA composed of $\frac{3}{4}$ uracil and $\frac{1}{4}$ guanine

Codon	Probability	Ratio
UUU	$p(UUU) = \frac{3}{4} \times \frac{3}{4} \times \frac{3}{4} = \frac{27}{64}$	1.00
UUG	$p(UUG) = \frac{3}{4} \times \frac{3}{4} \times \frac{1}{4} = \frac{9}{64}$	0.33
UGU	$p(UGU) = \frac{3}{4} \times \frac{1}{4} \times \frac{3}{4} = \frac{9}{64}$	0.33
GUU	$p(GUU) = \frac{1}{4} \times \frac{3}{4} \times \frac{3}{4} = \frac{9}{64}$	0.33
UGG	$p(UGG) = \frac{3}{4} \times \frac{1}{4} \times \frac{1}{4} = \frac{3}{64}$	0.11
GGU	$p(GGU) = \frac{1}{4} \times \frac{1}{4} \times \frac{3}{4} = \frac{3}{64}$	0.11
GUG	$p(GUG) = \frac{1}{4} \times \frac{3}{4} \times \frac{1}{4} = \frac{3}{64}$	0.11
GGG	$p(GGG) = \frac{1}{4} \times \frac{1}{4} \times \frac{1}{4} = \frac{1}{64}$	0.03

main excitement centered on the question of the amino-acid sequence of this protein. It proved to be polyphenylalanine — a string of phenylalanine molecules attached to form a polypeptide. Thus, the triplet UUU must code for phenylalanine:

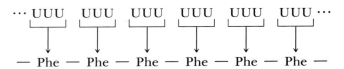

This type of analysis was extended by mixing nucleotides in a known fixed proportion when making synthetic mRNA. In one experiment, the nucleotides uracil and guanine were mixed in a ratio of 3 : 1. When they are incorporated at random into synthetic mRNA, the relative frequency at which each triplet will appear in the sequence can be calculated (Table 13-3). The amino acids produced by this RNA in the protein-synthesizing system in vitro should reflect the same distribution of probabilities. In fact, the ratios of amino acids in the protein produced were those shown in Table 13-5. From this evidence, we can deduce that codons consisting of

TABLE 13-4. Observed frequencies of various amino acids in protein translated from mRNA composed of $\frac{3}{4}$ uracil and $\frac{1}{4}$ guanine

Amino acid	Ratio
Phenylalanine	1.00
Leucine	0.37
Valine	0.36
Cysteine	0.35
Tryptophan	0.14
Glycine	0.12

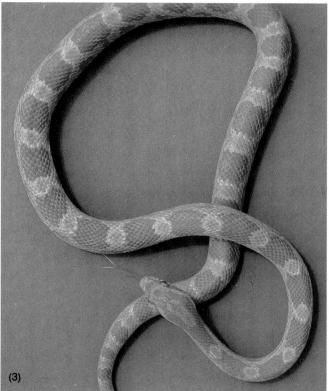

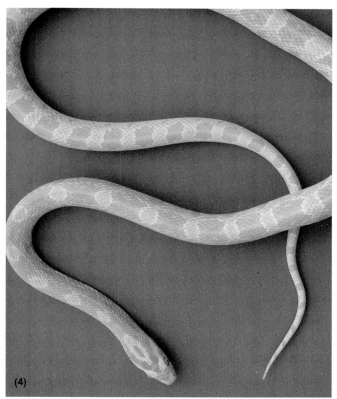

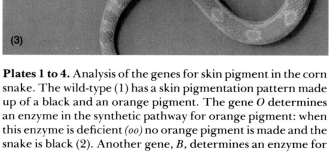

Plates 1 to 4. Analysis of the genes for skin pigment in the corn snake. The wild-type (1) has a skin pigmentation pattern made up of a black and an orange pigment. The gene *O* determines an enzyme in the synthetic pathway for orange pigment: when this enzyme is deficient *(oo)* no orange pigment is made and the snake is black (2). Another gene, *B*, determines an enzyme for black pigment: when this enzyme is deficient *(bb)* the snake is orange (3). When both enzymes are deficient the snake is albino (4). Hence the four homozygous genotypes are *OOBB* (1), *ooBB* (2), *OObb* (3) and *oobb* (4). A cross of 1 × 4 or 2 × 3 would give a dihybrid wild type F$_1$ and a 9 : 3 : 3 : 1 ratio of the four phenotypes in the F$_2$.

(5)

(6)

(7)

Plates 5 to 7. Coat color inheritance in Labrador retrievers. Two alleles of a pigment gene *B* and *b*, determine black (5) or brown (6) respectively. At a separate gene, *E* allows color deposition in the coat, and *ee* prevents deposition resulting in the golden phenotype (7). This is a case of recessive epistasis. Thus the three homozygous genotypes are *BBEE* (5), *bbEE* (6), and *BBee* or *bbee* (7). The dog in 7 is most likely *BBee*—the animal still has the ability to make black pigment, as witnessed by the black nose and lips, but not to deposit this pigment in the hairs. The progeny of a dihybrid cross would produce a 9 : 3 : 4 ration of black : brown : golden.

Plate 8. Pigment phenotypes in foxgloves, determined by three separate genes. *M* codes for an enzyme that synthesizes anthocyanin, the reddish pigment seen in these petals; *mm* produces no pigment and produces the phenotype albino with yellowish spots. *D* is an enhancer of anthocyanin, resulting in a darker pigment; *dd* does not enhance. At the third locus, *ww* allows pigment deposition in petals, but *W* prevents pigment deposition except in the spots, and so results in the white, spotted phenotype. Thus, the four phenotypes from left to right are M – W – – –, mm – – – –, M – ww dd, and M – ww D –.

Plate 9. Pigments in the fungus *Neurospora*. The wild-type *Neurospora* pigment is an orange carotenoid (right). A mutation in a gene for an early enzyme *(al)* in the synthetic pathway for carotenoid results in a pigmentless albino. A mutation in a late enzyme *(ylo)* results in the abnormal accumulation of a yellow carotenoid intermediate. A purine pathway block, *ad-3*, results in the accumulation of a purple pigment, that, mixed with carotenoid gives a brown appearance. The genotypes are, from left to right *ad-3⁺ al ylo⁺*, *ad-3 al⁺ ylo⁺*, *ad-3⁺ al⁺ ylo*, and *ad-3⁺ al⁺ ylo⁺*. The pathways are:

Plate 10. Normal red clover (*Trifolium pratense*, right) has red anthocyanins. A homozygous recessive mutation, *aa*, results in an albino red clover, left, that lacks a normal anthocyanin synthetic enzyme. The white mutant should not be confused with normal white clover (*T. repens*, below), with which it cannot breed because it is a different species.

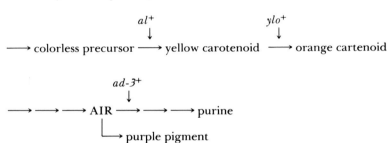

$$\longrightarrow \text{colorless precursor} \xrightarrow{\;al^+\;} \text{yellow carotenoid} \xrightarrow{\;ylo^+\;} \text{orange cartenoid}$$

$$\longrightarrow \longrightarrow \longrightarrow \text{AIR} \xrightarrow{\;ad\text{-}3^+\;} \longrightarrow \longrightarrow \text{purine}$$
$$\qquad\qquad\qquad\quad\searrow \text{purple pigment}$$

Plate 11. Wild type balsam root, *Balsamorhiza sagittata*, normally has flat petals (center). A rolled-petal mutant (at left) typifies the type of sculpturing mutations that have been selected by breeders to increase the diversity of commercial lines of plants such as *Zinnias*. (Photo from R. L. Taylor.)

Plate 12. Wild-type (center) and several horticultural varieties (periphery) of *Potentilla fruticosa*. The varieties have all been derived by selecting for spontaneous mutations that affect size, sculpturing, pigmentation, pattern, etc.

Plate 27. A mosaic of *Euonymus fortunei*, containing a mixture of normal chloroplast DNA (cpDNA) and cpDNA bearing a mutation that affects chlorophyll production. This type of mosaic sometimes throws off sidebranches that are not mixtures, but either normal or wholly albino; intercrosses between these show maternal inheritance.

Plate 28. Chlorophyll production is also affected by nuclear genes, in addition to and in cooperation with the genes of the cpDNA (see Plate 27). A plant *Cc* upon selfing produces 25 percent *cc* plants, deficient in chlorophyll (albino) and hence unable to survive after exhausting their store of food laid down by the maternal plant in the seed.

(29)

(30)

Plates 29 and 30. Quantitative inheritance of bract color in Indian paintbrush (*Castilleja hispida*). Plate 29 shows the two extremes of the range and Plate 30 shows examples from throughout the phenotypic range.

one guanine and two uracils (G + 2U) code for valine, leucine, and cysteine, although we cannot distinguish the specific sequence for each of these amino acids. Similarly, one uracil and two guanines (U + 2G) must code for tryptophan, glycine, and perhaps one other. It looks as though the Watson-Crick model is correct in predicting the importance of the precise sequence (not just the ratios of bases). Many provisional assignments (such as those just outlined for G and U) were soon obtained, primarily by groups working with Nirenberg or with Severo Ochoa.

Specific code words were finally deciphered through two kinds of experiments. The first involved making "mini mRNAs," each only three nucleotides in length. These, of course, are too short to promote translation into protein, but they do stimulate the binding of aminoacyl tRNAs (aa-tRNA's) to ribosomes in a kind of abortive attempt at translation. It is possible to make a specific mini mRNA and determine *which* aminoacyl tRNA it will bind to ribosomes.

For example, the G + 2U problem discussed previously can be resolved by using the following mini mRNAs:

GUU stimulates binding of valyl $tRNA_{Val}$

UUG stimulates binding of leucyl $tRNA_{Leu}$

UGU stimulates binding of cysteinyl $tRNA_{Cys}$

Analogous mini RNA's provided a virtually complete cracking of all the $4^3 = 64$ possible codons.

The second kind of experiment that was useful in cracking the genetic code involved the use of "repeating copolymers." For instance, the copolymer designated $(AGA)_n$, which is a long sequence of · · · AGAAGAA GAAGAAGA · · ·, was used to stimulate polypeptide synthesis in vitro. From the sequence of the resulting polypeptides and the possible triplets that could occur in the respective RNA copolymer, many code words could be verified. (This kind of experiment is detailed in Problem 10 at the end of this chapter. In solving it, you can put yourself in the place of H. Gobind Khorana, who received a Nobel prize for directing the experiments.)

Figure 13-22 gives the genetic code dictionary of 64 words. Inspect this figure carefully, and ponder the miracle of molecular genetics. Such an inspection should reveal several points that require further explanation.

Multiple Codons for a Single Amino Acid

First, note that the number of codons for a single amino acid varies, ranging from one (tryptophan = UGG) to as many as six (serine = UCU or UCC or UCA or UCG or AGU or AGC). Why? The answer is complex but not difficult; it can be divided into two parts:

Second letter

First letter	U	C	A	G	Third letter
U	UUU Phe	UCU	UAU Tyr	UGU Cys	U
	UUC	UCC Ser	UAC	UGC	C
	UUA Leu	UCA	UAA Stop	UGA Stop	A
	UUG	UCG	UAG Stop	UGG Trp	G
C	CUU	CCU	CAU His	CGU	U
	CUC Leu	CCC Pro	CAC	CGC Arg	C
	CUA	CCA	CAA Gln	CGA	A
	CUG	CCG	CAG	CGG	G
A	AUU	ACU	AAU Asn	AGU Ser	U
	AUC Ile	ACC Thr	AAC	AGC	C
	AUA	ACA	AAA Lys	AGA Arg	A
	AUG Met	ACG	AAG	AGG	G
G	GUU	GCU	GAU Asp	GGU	U
	GUC Val	GCC Ala	GAC	GGC Gly	C
	GUA	GCA	GAA Glu	GGA	A
	GUG	GCG	GAG	GGG	G

Figure 13-22. The genetic code.

1. Certain amino acids can be brought to the ribosome by several *alternative* tRNA types (species) having different anticodons, whereas certain other amino acids are brought to the ribosome by only one tRNA.

2. Certain tRNA species can bring their specific amino acids in response to several codons, not just one, through a loose kind of base pairing at one end of the codon and anticodon. This sloppy pairing is called **wobble.**

We had better consider wobble first, and it will lead us into a discussion of the various species of tRNA. Wobble is caused by a third nucleotide of the anticodon (at the 5′ end) that is not quite aligned. This out-of-line nucleotide sometimes can form hydrogen bonds not only with its normal complementary nucleotide in the third position of the codon but also with a different nucleotide in that position. Crick established certain "wobble rules" that dictate which nucleotides can and cannot form new hydrogen-bonded associations through wobble (Table 13-5). In the table, I (inosine) is one of the rare bases found in tRNA, often in the anticodon.

Figure 13-23 shows the possible codons that one tRNA serine species can recognize. As the wobble rules indicate, G can pair with U or with C. Table 13-6 lists all the codons for serine and shows how different tRNA's can service these codons. This is a good example of the effects of wobble on the genetic code.

Sometimes there can be an additional tRNA species that we represent as $tRNA_{Ser_4}$; it has an anticodon identical to any of the three anticodons shown in Table

TABLE 13-5. Codon–anticodon pairings allowed by the wobble rules

5′ end of anticodon	3′ end of codon
G	U or C
C	G only
A	U only
U	A or G
I	U, C, or A

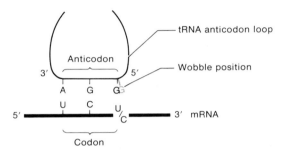

Figure 13-23. In the third site (5′ end) of the anticodon, G can take either of two wobble positions, thus being able to pair with either U or C. This means that a single tRNA species carrying an amino acid (in this case, serine) can recognize two codons—UCU and UCC—in the mRNA.

13-6, but it differs in its nucleotide sequence elsewhere in the tRNA molecule. These four tRNA's are called **isoaccepting tRNAs** because they accept the same amino acid, but they are probably all transcribed from different tRNA genes.

Stop Codons

The second point you may have noticed in Figure 13-22 is that some codons do not specify an amino acid at all.

TABLE 13-6. Different tRNA's that can service codons for serine

Codon	tRNA	Anticodon
UCU UCC	tRNA$_{Ser_1}$	AGG + wobble
UCA UCG	tRNA$_{Ser_2}$	AGU + wobble
AGU AGC	tRNA$_{Ser_3}$	UCG + wobble

These codons are labeled as **stop codons.** They can be regarded as similar to periods or commas punctuating the message encoded in the DNA.

One of the first indications of the existence of stop codons came in 1965 from Brenner's work with the T4 phage. Brenner analyzed certain mutations ($m_1 - m_6$) in a single cistron that controls the head protein of the phage. These mutants had two things in common. First, the head protein of each mutant was a shorter polypeptide chain than that of the wild-type. Second, the presence of a suppressor mutation (*su*) in the host chromosome would cause the phage to develop a head protein of normal (wild-type) chain length despite the presence of the *m* mutation (Figure 13-24).

Brenner examined the ends of the shortened proteins and compared them with wild-type protein, recording for each mutant the next amino acid that *would* have been inserted to continue the wild-type chain. These amino acids for the six mutations were glutamine, lysine, glutamic acid, tyrosine, tryptophan, and serine. There is no immediately obvious pattern to these results, but Brenner brilliantly deduced that certain codons for each of these amino acids are similar in that each of them can mutate to the codon UAG by a single change in a DNA nucleotide pair. He therefore postulated that UAG is a stop (or termination) codon—a signal to the translation mechanism that the protein is now complete.

UAG was the first stop codon deciphered; it is called the **amber codon.** Mutants that are defective due to the presence of an abnormal amber codon are called amber mutants, and their suppressors are amber suppressors. UGA, the **opal codon,** and UAA, the **ocher codon,** are also stop codons and also have suppressors. Stop codons often are called **nonsense codons** because they designate no amino acid, but their sense is real enough (it is unfortunate that this misnomer persists). **Missense mutations,** on the other hand, lead to a change in a single amino acid in the translated polypeptide. Not surprisingly, stop codons do not act as mini mRNA's in binding aa-tRNA to ribosomes in vitro.

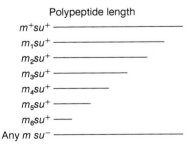

Figure 13-24. Polypeptide-chain lengths of phage-T4 head protein in wild-type *(top)* and various amber mutants *(m)*. An amber suppressor *(su)* leads to phenotypic development of the wild-type chain.

> **Message**   A nonsense mutation changes a codon so that it means "stop" to the translation system. A missense mutation changes a codon so that it stands for a different amino acid.

Nonsense Suppressor Mutations

It is interesting to consider the **nonsense suppressor mutations,** many of which now are known to be mutations in the anticodon loop of specific tRNA's that allow recognition of a nonsense codon in mRNA. Thus, an amino acid is inserted in response to the stop codon, and translation continues past that triplet. Figure 13-25 provides an example. The amber mutation replaces a wild-type codon with a stop codon, which by itself would prematurely cut off the protein at this position. The suppressor mutation in this case produces a $tRNA_{Tyr}$ with an anticodon that recognizes the mutant stop codon. The suppressed mutant contains tyrosine at that position.

Two questions are often asked at this point:

1. What is doing the job of putting tyrosine where tyrosine should be when the gene for tyrosine tRNA becomes a nonsense suppressor? Remember that there typically are several isoaccepting tRNA forms (and hence several $tRNA_{Tyr}$ genes) and that two of these may have the same anticodon loop. The "other" (one or more) wild-type $tRNA_{Tyr}$ genes take over when a suppressor locus is created from the first.

2. What happens to normal stop signals (indicating the ends of normal proteins) when nonsense suppressors are present? Well, normal termination is still the rule. First of all, suppression is not 100 percent efficient. The range of suppression efficiencies varies from less than 1 percent to greater than 50 percent, depending on the suppressor used and on the specific site on which the suppressor acts. Natural termination signals are at the lower end of this scale. Second, the normal end of a protein is often signaled by two different consecutive stop codons, thereby greatly reducing elongation past these sites.

Start Codon

The third point regarding Figure 13-22 is that a single codon (AUG) acts as a **start codon**—rather like the capital letter indicating the start of a sentence. This codon also acts as the codon for the amino acid methio-

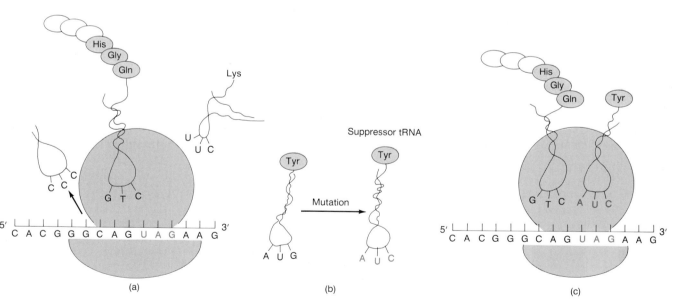

(a)   (b)   (c)

Figure 13-25. (a) A schematic view of translation. Here the translation apparatus cannot go past a nonsense (UAG in this case) codon, because there is no tRNA that can recognize the UAG triplet. This leads to the termination of protein synthesis and to the subsequent release of the polypeptide fragment. This release is promoted by protein factors that are not shown here. (b) The molecular consequences of a mutation that alters the anticodon of a tyrosine tRNA. This tRNA can now read the UAG codon. (c) The suppression of the UAG codon by the altered tRNA, which now permits chain elongation. At the ribosome, the codons of an mRNA molecule base-pair with the anticodons of RNAs, which are charged with amino acids. (From James D. Watson, John Tooze, and David T. Kurtz, *Recombinant DNA: A Short Course.* Copyright © 1983 by W. H. Freeman and Company)

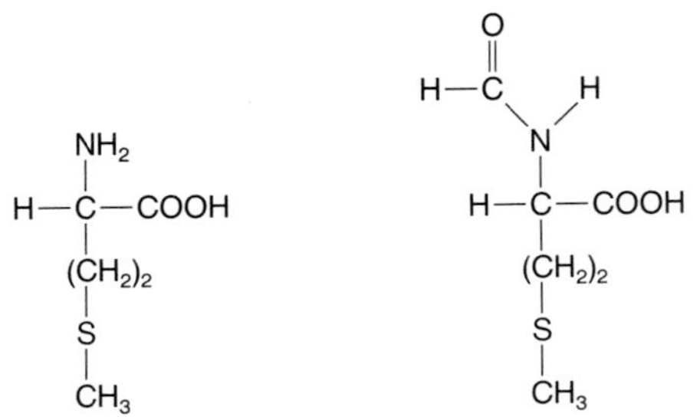

Figure 13-26. The structures of methionine (Met) and N-formylmethionine (N-f-Met). A tRNA bearing N-f-Met can initiate a polypeptide chain but cannot be inserted in a growing chain; a tRNA bearing Met can be inserted in a growing chain but will not initiate a new chain. Both of these tRNA's bear the same anticodon complementing the codon AUG.

nine. In *E. coli* and in most other prokaryotic organisms, the first amino acid in any newly synthesized polypeptide always is N-formylmethionine. It is inserted not by $tRNA_{Met}$, however, but by an initiator tRNA called $tRNA_{N-f-Met}$. This initiator tRNA has the normal methionine anticodon but inserts N-formylmethionine rather than methionine in the polypeptide chain (Figure 13-26). Either the formyl group or a sequence in the tRNA apparently mimics a polypeptide chain; at least the ribosome recognizes some specific signal and shifts one codon to amino-acid position 2. Methionine won't cause this shift; if methionyl $tRNA_{Met}$ inserts, nothing happens. The system then must wait until the Met–$tRNA_{Met}$ diffuses away and an N-f-Met–$tRNA_{N-f-Met}$ chances along to act as an initiator. Similarly, when AUG occurs in the middle of a protein, N-f-Met–$tRNA_{N-f-Met}$ cannot form a peptide bond with the growing chain, so the system must wait for Met–$tRNA_{Met}$. In *E. coli*, GUG and perhaps UUG can serve as initiation codons in some cases. When this occurs, these triplets are recognized by N-f-Met–$tRNA_{N-f-Met}$, and methionine appears as the first amino acid in the chain.

An Unsolved Problem—the Ribosome

The ribosome is an immensely challenging structure. Very little is known about the precise arrangement of the ribosomal protein and rRNA subunits or about how they unite to grapple physically with the mRNA, the tRNA, the amino acid, and the peptide chain. The role of rRNA has always been a mystery. The role of the proteins can be visualized much like a complex enzyme, but the occurrence of what is normally an informational molecule in this structure is baffling. Evidence now suggests that the rRNA is crucial in the binding of various translational components (such as tRNA) to the ribosome. This role must be nonspecific with respect to the tRNA species because a ribosome is basically a nonspecific translational apparatus. Ribosomal RNA is the most abundant RNA in a cell and obviously serves a crucial cell function. It is known to be synthesized in the nucleolus in a repetitive tandem array of rRNA genes from the chromosome region known as the **nucleolar organizer.**

Ribosome Binding Sites

How are the correct initiation codons selected from the many AUG and GUG codons in a mRNA molecule? John Shine and Lynn Dalgarno first noticed that true initiation codons were preceded by sequences that paired well with the 3′ end of 16S rRNA. Figure 13-27 shows some of these sequences. There is a short but variable separation between the Shine-Dalgarno sequence and the initiation codon. Figure 13-28 depicts the base pairing between idealized mRNA and the 16S rRNA that results in ribosome-mRNA complexes leading to protein initiation in the presence of f-Met–tRNA.

Overview of Protein Synthesis

We have followed a highly simplified version of protein synthesis. Several other components or "factors" are required for the system to work. Figure 13-29 summarizes the action of some of them.

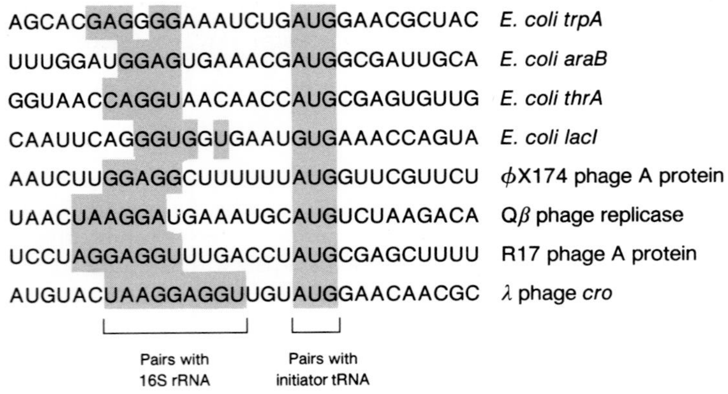

AGCACGAGGGGAAAUCUGAUGGAACGCUAC	*E. coli trpA*
UUUGGAUGGAGUGAAACGAUGGCGAUUGCA	*E. coli araB*
GGUAACCAGGUAACAACCAUGCGAGUGUUG	*E. coli thrA*
CAAUUCAGGGUGGUGAAUGUGAAACCAGUA	*E. coli lacI*
AAUCUUGGAGGCUUUUUUAUGGUUCGUUCU	*φX174 phage A protein*
UAACUAAGGAUGAAAUGCAUGUCUAAGACA	*Qβ phage replicase*
UCCUAGGAGGUUUGACCUAUGCGAGCUUUU	*R17 phage A protein*
AUGUACUAAGGAGGUUGUAUGGAACAACGC	*λ phage cro*

Pairs with 16S rRNA   Pairs with initiator tRNA

Figure 13-27. Ribosomal binding site sequences share certain common features, which are shown in the colored regions. The initiation codon (color) is separated by several bases from a short sequence (color) that is complementary to the 3′-end of 16S rRNA. (After L. Stryer, *Biochemistry*, 3d ed. Copyright © 1988 by W. H. Freeman and Company.)

3' end of 16S rRNA

Figure 13-28. Binding of the Shine-Dalgarno sequence on a mRNA to the 3' end of 16S rRNA (After L. Stryer, *Biochemistry,* 3d ed. Copyright © 1988 by W. H. Freeman and Company.)

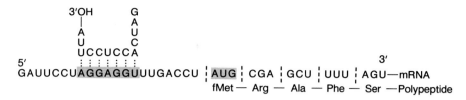

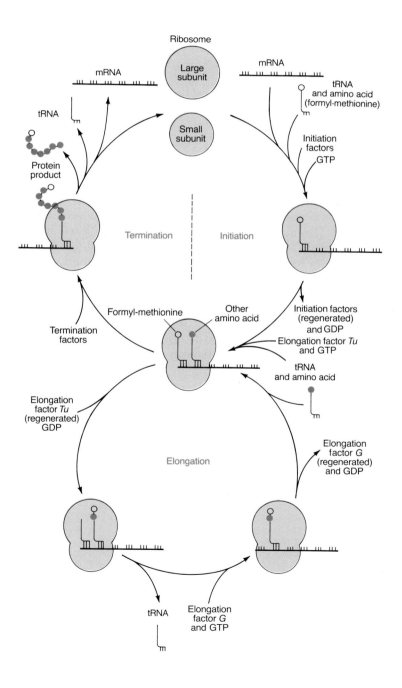

Figure 13-29. The transactions of the ribosome. At initiation, the ribosome recognizes the starting point in a segment of mRNA and binds a molecule of tRNA bearing a single amino acid. In all bacterial proteins, this first amino acid is *N*-formylmethionine. In elongation, a second amino acid is linked to the first one. The ribosome then shifts its position on the mRNA molecule, and the elongation cycle is repeated. When the stop codon is reached, the chain of amino acids folds spontaneously to form a protein. Subsequently, the ribosome splits into its two subunits, which rejoin before a new segment of mRNA is translated. Protein synthesis is facilitated by a number of catalytic proteins (initiation, elongation, and termination factors) and by guanosine triphosphate (GTP), a small molecule that releases energy when it is converted into guanosine diphosphate (GDP). (From Donald M. Engelman and Peter B. Moore, "Neutron-scattering Studies of the Ribosome." Copyright © 1976 by Scientific American, Inc. All rights reserved.)

TABLE 13-7. Mutations in human hemoglobin, in *E. coli* tryptophan synthetase, and in tobacco-mosaic-virus (TMV) coat protein

Protein	Amino acid substitution	Inferred codon change
Hemoglobin	Glu ⟶ Val	GAA ⟶ GUA
Hemoglobin	Glu ⟶ Lys	GAA ⟶ AAA
Hemoglobin	Glu ⟶ Gly	GAA ⟶ GGA
Tryptophan synthetase	Gly ⟶ Arg	GGA ⟶ AGA
Tryptophan synthetase	Gly ⟶ Glu	GGA ⟶ GAA
Tryptophan synthetase	Glu ⟶ Ala	GAA ⟶ GCA
TMV coat protein	Leu ⟶ Phe	CUU ⟶ UUU
TMV coat protein	Glu ⟶ Gly	GAA ⟶ GGA
TMV coat protein	Pro ⟶ Ser	CCC ⟶ UCC

SOURCE: L. Stryer, *Biochemistry*, 2d ed. Copyright © 1981 by W. H. Freeman and Company.

Thus far, our discussion has focused on microbes, but the amazing fact is that the information-transfer, coding, and translation processes are virtually identical in all organisms that have been studied. For example, all of the different single amino-acid substitutions known to occur in human hemoglobin result from single nucleotide-pair substitutions based on the genetic code derived from *E. coli* (Table 13-7). Such observations suggest that the genetic code is shared by all organisms. Furthermore, an information-bearing molecule, such as rabbit red-blood-cell mRNA, which is predominantly hemoglobin gene transcript, will be translated in an alien environment (such as a frog egg) into rabbit hemoglobin (Figure 13-30). Apparently, the translation apparatus is functionally the same in a wide range of different organisms.

Sequencing techniques at the protein, RNA, and DNA levels (see Chapter 15) have verified that the genetic code is universal in all organisms studied to date, ranging from viruses and bacteria to humans. One exception involves mitochondrial DNA (see also Chapter 20). Two codons are translated differently here, due to the properties of tRNA's that are confined to the mitochondrial system. Thus, whereas AUA is normally translated as isoleucine, it is read as methionine in the mitochondria. Also, the mammalian mitochondria translate UGA as tryptophan, although UGA normally specifies a chain-terminating codon. In yeast, the mitochondria translate UGA as tryptophan, as in mammalian mitochondria, but they translate AUA as isoleucine, as

in bacterial systems. As we shall see from genetic engineering experiments in Chapter 15, DNA is DNA no matter what its origin. The nature and message of the DNA represent a universal language of life on earth.

Does this interspecific equivalence of parts in the genetic apparatus indicate a common evolutionary ancestry for all life forms on earth? Or does it simply reflect the fact that this is the only workable biochemical option in the earth environment (biochemical predestination)? Whatever the answer, the wonderful uniformity of the molecular basis of life is firmly established. Minor variations do exist, but they do not detract from the central uniformity of the mechanism we have described.

> **Message** The processes of information storage, replication, transcription, and translation are fundamentally similar in all living systems. In demonstrating this fact, molecular genetics has provided a powerful unifying force in biology. We now know some of the tricks that life uses to achieve persistent order in a randomizing universe.

Eukaryotic RNA

Several aspects of RNA synthesis and processing in eukaryotes are distinctly different from RNA in prokaryotes.

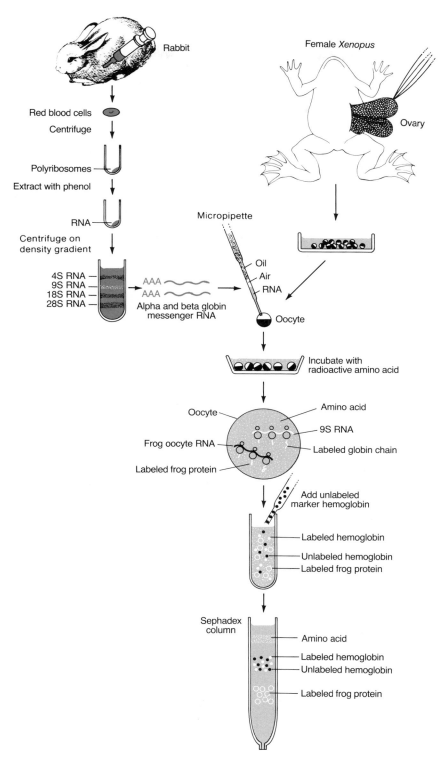

Figure 13-30. Rabbit hemoglobin mRNA is translated into rabbit hemoglobin in a frog *(Xenopus)* egg. This translation occurs with the *Xenopus* translation apparatus. This experiment is one of many that point toward the uniformity of the genetic molecular mechanism in all forms of life.

(Sephadex is a separatory material used in chromatography.) From C. Lane, "Rabbit Hemoglobin from Frog Eggs." Copyright © 1976 by Scientific American, Inc. All rights reserved.)

RNA Synthesis

Whereas a single RNA polymerase species synthesizes all RNA's in prokaryotes, there are three different RNA polymerases in eukaryotes systems:

1. RNA polymerase I synthesizes rRNA.

2. RNA polymerase II synthesizes mRNA. The mRNA molecules are **monocistronic,** whereas many mRNA's are **polycistronic** in prokaryotes.

3. RNA polymerase III synthesizes tRNA's and also small nuclear and cellular RNA molecules.

RNA Processing

The primary RNA transcript produced in the nucleus usually is processed in several ways prior to its transport to the cytoplasm, where it is used to program the translation machinery (Figure 13-31). Figure 13-32 depicts these events in detail. First a "cap" consisting of a 7-methylguanosine residue linked to the 5′ end of the transcript by a triphosphate bond is added during transcription. Then stretches of adenosine residues are added at the 3′ ends. These "poly-A tails" are 150 to 200 residues long. Following these modifications, a crucial "splicing" step removes internal portions of the RNA transcript. The uncovering of this process, and the cor-

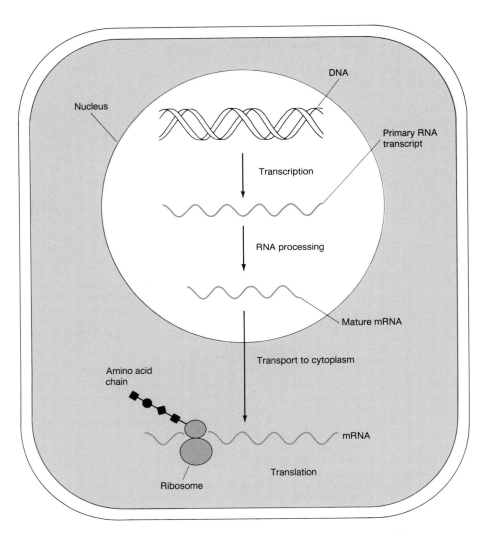

Figure 13-31. Gene expression in eukaryotes. The mRNA is processed in the nucleus prior to transport to the cytoplasm. (Figures 13-31 and 13-32 from J. E. Darnell, Jr., "The Processing of RNA." Copyright © 1983 by Scientific American, Inc. All rights reserved.)

responding realization that genes are "split," with coding regions interrupted by "intervening sequences," constitutes one of the most important discoveries in molecular genetics in the last decade.

Split Genes

Studies of mammalian viral DNA transcripts first suggested a lack of correspondence between the genetic map and specific mRNA molecules. As recombinant-DNA techniques (see Chapter 15) facilitated the physical analysis of eukaryotic genes, it became apparent that primary RNA transcripts were being shortened by the elimination of internal segments before transport into the cytoplasm. In most higher eukaryotes studied, this was found to be true not only for mRNA but also for rRNA—and even for tRNA in some cases.

Figure 13-33 shows the organization of the gene for chicken ovalbumin, a 386-amino-acid polypeptide. The DNA segments that code for the structure of the protein are interrupted by intervening sequences, termed **introns.** In Figure 13-33, these segments are designated with the letters A to G. The primary transcript is processed by a series of "splicing" reactions, much in the same way that a tape-recorded message can be cut and pasted back together. Splicing removes the introns and brings together the coding regions, termed **exons,** to form an mRNA, which now consists of a sequence that is

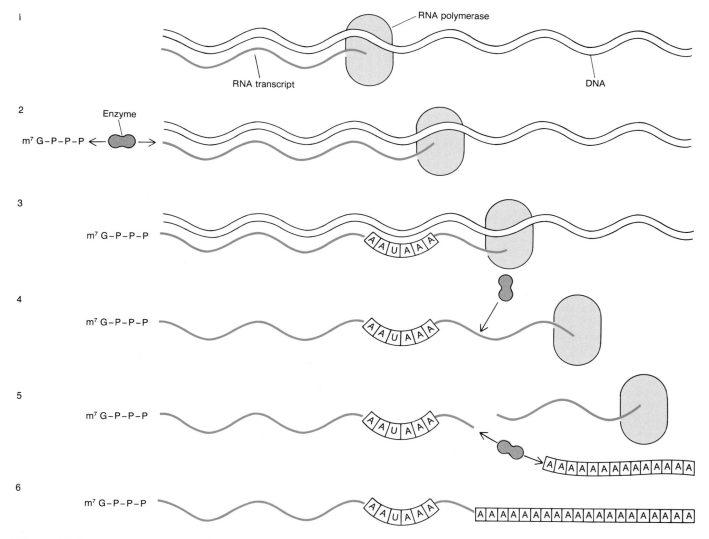

Figure 13-32. A cap consisting of $m^7 G–P–P–P$ and a poly-A tail are added to the RNA transcript in the first stages of mRNA processing. The sequence AAUAAA helps signal a cleavage event about 20 bases downstream. Then, 150 to 200 A residues are added to the 3′ end of the mRNA.

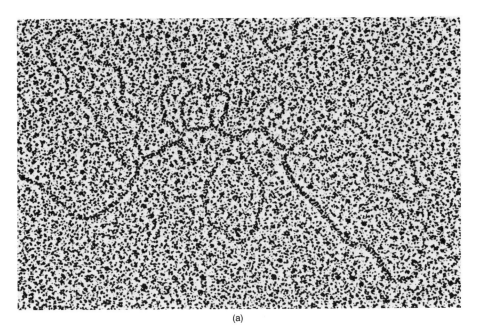

(a)

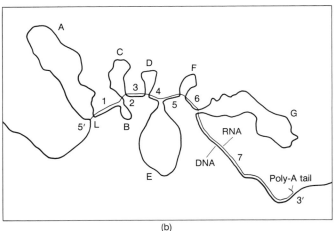

(b)

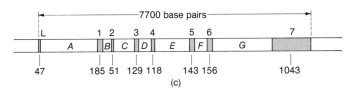

(c)

Figure 13-33. Split-gene organization of the gene for the protein ovalbumin. (a) The electron micrograph and (b) its map show the result of an experiment in which a single strand of the DNA incorporating the gene for the egg-white protein ovalbumin was allowed to hybridize with ovalbumin mRNA, the molecule from which the protein is translated. (c) The looped-out single-stranded segments represent the introns. The schematic representation of the gene shows the seven introns (white) and eight exons (color) and the number of base pairs in each of the exons; the size of the introns ranges from 251 base pairs for (*B*) to about 1600 (for *G*). (Figures 13-33 and 13-34 from P. Chambon, "Split Genes." Copyright © 1981 by Scientific American, Inc. All rights reserved.)

completely colinear with the ovalbumin protein. The exons are numbered 1 to 7 in Figure 13-33. In different genes, introns have been detected that are as large as 2000 base pairs in length. Some genes have as many as 16 introns.

It is clear that splicing occurs after transcription, and in several steps, since RNA transcripts (previously termed "heterogeneous nuclear RNA," or HnRNA) can be isolated that correspond to the entire genetic region (introns + exons), as well as transcripts intermediate in length. In these intermediate-length RNA molecules, certain introns have already been removed, but others are retained. The entire sequence of events for RNA processing and splicing is summarized in Figure 13-34.

There are some interesting examples of the *same* primary transcript generating different mRNAs by using a different splicing route. Figure 13-35 shows how

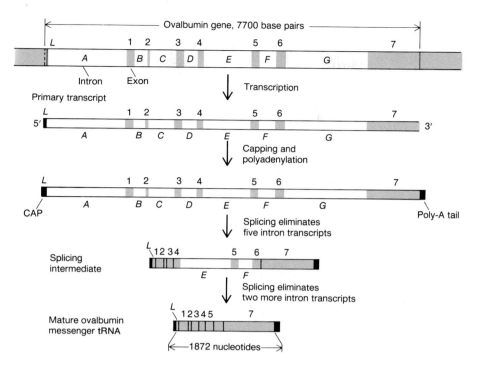

Figure 13-34. Mature mRNA is produced in a number of steps.

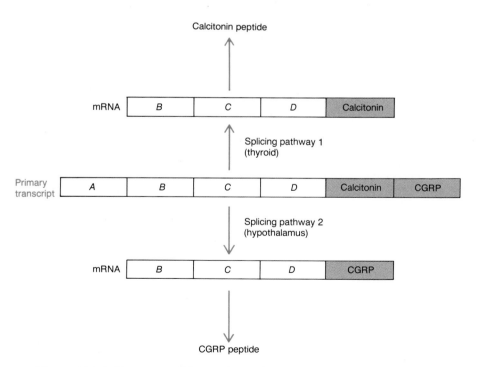

Figure 13-35. Alternate splicing pathways in two different organs generate different translation products from the same primary transcript.

two protein hormones—calcitonin and the related CGRP—are produced from the same transcript. Other examples include mouse amylase and the t-antigen of SV40, each of which is synthesized in different forms by alternate splicing pathways operating on the same primary transcript.

Mechanism of Gene Splicing

There are some sequence homologies at the exon-intron junctures of mRNA's. For instance, there is a $-G-U-$ at the 5' splice site and an $-A-G-$ at the 3' splice site of introns in virtually all mRNAs examined. It is thought that splicing enzymes recognize some common configuration of the mRNA and, perhaps with the help of certain small nuclear ribonucleoprotein particles, or snRNPs, catalyze the cutting and splicing reactions. The snRNP's may help to align the splice sites by hydrogen bonding to the sequences at the exon-intron boundaries (Figure 13-36). One currently favored model employs a lariat-shaped intermediate in the splicing of primary transcripts to yield the final mRNA. Figure 13-37 portrays the **lariat model.** Here the 5' splice site is first cleaved (a) in a reaction that creates a loop (b). The newly formed 3'—OH group then cleaves the 3' splice site (b), releasing the partly looped intron and at the same time joining the two exons (c).

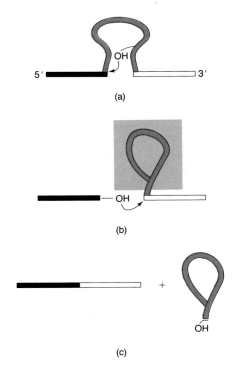

Figure 13-37. Lariat or loop formed by a category of introns as they are removed from their RNA molecules. (a) One of the many 2' hydroxyl groups on the intron attacks the 5' splice site. (b) The subsequent reaction joins the 5' end of the intron not to the 3' end but to a point a short distance away, yielding a branched structure with a loop: the lariat. (c) The branching is accomplished by the formation of a novel 2"–5' phosphodiester bond that enables one adenosine nucleotide to form phosphodiester links with three oher nucleotides rather than with the usual two. Ligation of the exons frees the lariat from the remainder of the RNA. (From T. Cech "RNA as an Enzyme." Copyright © 1986 by Scientific American, Inc. All rights reserved.)

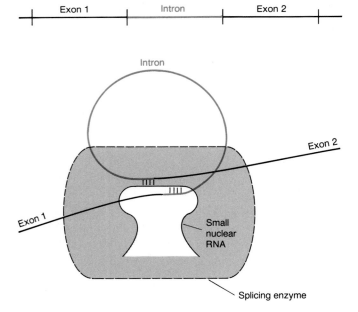

Figure 13-36. Schematic diagram depicting possible participation of small nuclear RNA in splicing reaction.

Self-splicing RNA

One exceptional case of RNA splicing occurs in *Tetrahymena,* where Thomas Cech and his coworkers have demonstrated that the splicing reaction is catalyzed by the RNA molecule itself! Figures 13-38 and 13-39 portray this reaction. This extraordinary finding is the first demonstration that an RNA molecule can serve as a bona fide enzyme in that it can catalyze a specific biological reaction.

Implications of Split Genes

The finding that many eukaryotic genes are interrupted by DNA sequences that are not translated into protein shatters the concept of the gene that we had developed through the end of Chapter 12. Until now we had considered a gene as an uninterrupted sequence of nucleic

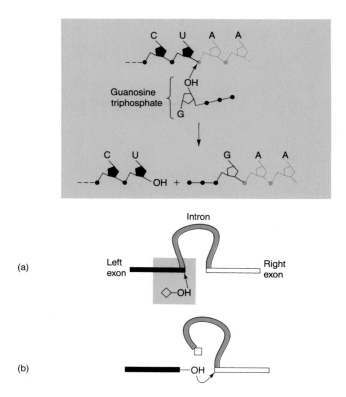

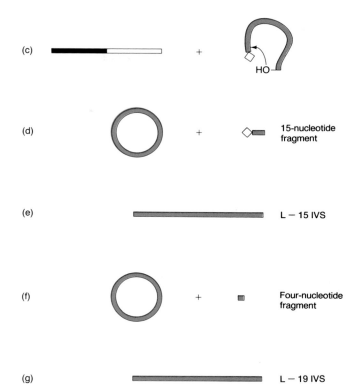

Figure 13-38. Intron removes itself from the *Tetrahymena* rRNA precursor molecule with no assistance from protein enzymes. The cascade of reactions resulting in removal of the intron is unleashed by a free guanosine or guanosine triphosphate molecule *(shown as a diamond)*. (a) A hydroxyl group (OH) attached to the nucleotide "attacks" the phosphate at the 5′ end of the intron. *(inset)* The phosphodiester bond between the intron and the left exon is broken, and a new bond is formed between the guanosine and the intron. (b) This liberates a new hydroxyl group on the end of the left exon, which begins an attack at the 3′ end of the intron (c). The bond there is broken and the exons are ligated, or joined, freeing the intron. (d) A similar reaction enables the intron to form a circle, snipping 15 nucleotides off its end in the process (e). The circle opens into a linear molecule (f) and then closes with the loss of four nucleotides (g). The final, reopened form is known as the L − 19 IVS (Linear Minus 19 Intervening Sequence). (From T. Cech, "RNA as an Enzyme." Copyright © 1986 by Scientific American, Inc. All rights reserved.)

acid coding for a functional macromolecule (RNA or protein). The gene was colinear with the protein it encoded. Clearly, this definition of the gene must now be modified, because it no longer holds in all cases. In prokaryotes and, to a large extent, in lower eukaryotes, genes do represent an uninterrupted coding sequence. For many eukaryotic genes, however, the presence of introns interrupts the coding sequence, and the initial RNA transcript must be processed by splicing reactions in order to generate the finished RNA molecule (either mRNA, tRNA, or rRNA).

Why Split Genes and Splicing?

It is not clear why introns and exons have evolved as such. In some cases, mutations introduced into introns have no noticeable effect on gene expression; in other examples, for example, the removal of a single intron interferes with gene expression. Walter Gilbert has sug-gested that in many cases, exons encode discrete domains of proteins and that the shuffling of exons allows a more rapid evolution of proteins. The introns remain as a vestige of this process, according to Gilbert's explanation.

Protein Processing

Proteins that are secreted from the cell are synthesized with a short leader peptide, called a **signal sequence,** at the amino-terminal (N-terminal) end. This is a stretch of 15 to 25 amino acids, most of which are hydrophobic. The signal sequence allows for transport through the cell membrane; during this process, the signal sequence is cleaved by a peptidase (Figure 13-40). (A similar phenomenon exists for certain bacterial proteins that are secreted.) Moreover, several small peptide hormones, such as corticotropin (ACTH), result from the specific cleavage of a single, large, polypeptide precursor.

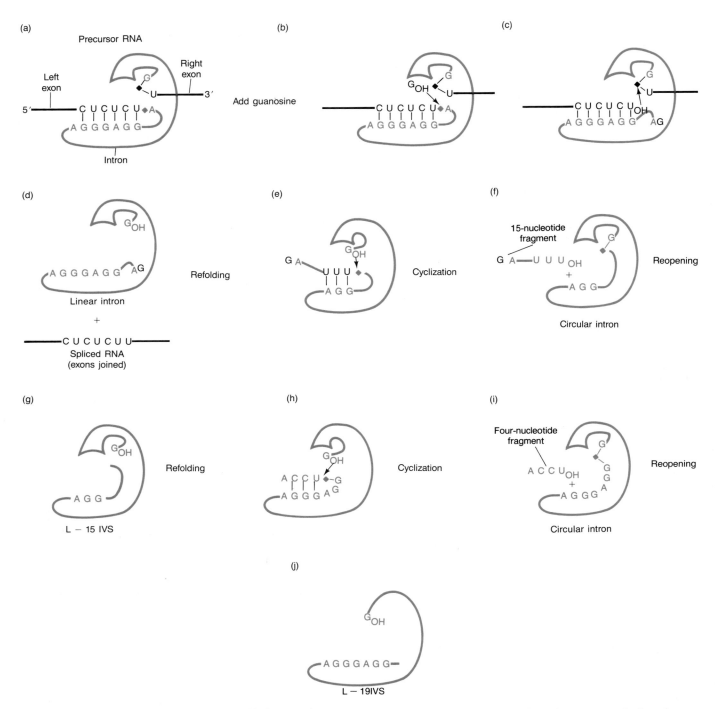

Figure 13-39. Several catalytic strategies aid the *Tetrahymena* intron in its repeated phosphodiester exchanges. (a) The nucleotide sequence *GGAGGG* (such sequences are conventionally given in the 5′–3′ direction) near the end of the intron binds the sequence *CUCUCU* at the end of the left exon. (b) At the same time, the phosphate group at the 5′ end of the intron is activated (*colored diamond*). When guanosine is added, the intron binds it in a position favoring its attack on the exposed, activated phosphodiester bond. (c) Once the bond has been severed, the phosphate at the other end of the intron, which has also been activated (*black diamond*), is attacked by a hydroxyl group attached to the left exon. (d) The exons are thereby joined and the intron is released. (e) The intron then refolds, and the sequence *UUU* near its own 5′ end is positioned for the attack of the guanosine at its 3′ end. (f) The 15 nucleotides lost as the loop closes are those preceding the activated phosphate at the point of attack. After the circle opens (g), it recyclizes (h, i) before opening again (j). (From T. Cech, "RNA as an Enzyme." Copyright © 1986 by Scientific American, Inc. All rights reserved.)

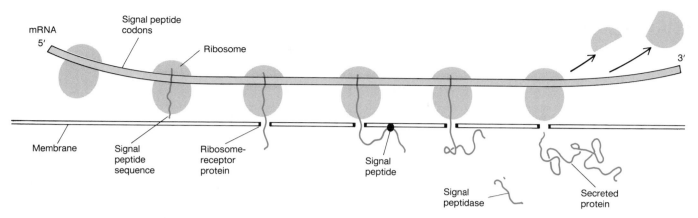

Figure 13-40. Signal sequences. Proteins destined to be secreted from the cell have an amino-terminal sequence that is rich in hydrophobic residues. This "signal sequence" binds to the membrane and draws the remainder of the protein through the lipid bilayer. The signal sequence is cleaved off the protein during this process by an enzyme called signal peptidase. (After J. D. Watson, J. Tooze, and D. T. Kurtz, *Recombinant DNA: A Short Course.* Copyright © 1983 by W. H. Freeman and Company)

Review

This chapter and Chapters 11 and 12 have described the development of the central theory of molecular genetics. A linear sequence of nucleotides in DNA is transcribed into a comparable linear sequence of nucleotides in RNA. This RNA sequence, which is processed in several ways in eukaryotes, then serves as a messenger RNA, which is translated into an amino-acid sequence in protein by a complex translational apparatus. The protein thus made has importance to the organism either as a structural component (such as hair, muscle, or skin protein) or as a regulator of the body chemistry (such as enzymes or hemoglobin). Figure 13-41 summarizes the structural relationship between DNA and protein.

Summary

■ We have discovered in earlier chapters that DNA is the genetic material responsible for directing the synthesis of proteins. The first clue as to how DNA accomplishes this feat came from eukaryotes, when it was shown that RNA is synthesized in the nucleus and then transferred to the cytoplasm. However, most of the details of this transfer of information from DNA to protein were worked out with experiments in bacteria and phages.

RNA is synthesized from only one strand of a double-stranded DNA helix. This transcription is catalyzed by an enzyme, RNA polymerase, and follows rules similar to those followed in replication: A complements with T, G and C, and U (uracil) with A. Ribose is the sugar used in RNA, and uracil replaces thymine. Extraction of RNA from a cell yields three basic varieties: ribosomal, transfer, and messenger RNA. The three sizes of ribosomal RNA (rRNA) combine with an array of specific proteins to form ribosomes that are the machines used for protein synthesis (translation). Transfer RNAs (tRNA) are a group of rather small RNA molecules, each with specificity for a particular amino acid; they carry the amino acids to the ribosome, where they can be attached to a growing polypeptide.

Messenger RNA (mRNA) molecules are of many sizes and base sequences. These are the molecules that contain information for the structure of proteins. The sequence of codons in mRNA determines the sequence of amino acids that will comprise a polypeptide. Each codon is specific for one amino acid, but several different codons may code for the same amino acid; that is, there is redundancy in the genetic code. In addition, there are three codons for which there are no tRNAs; these stop codons terminate the process of translation. Figure 13-42 summarizes the way in which genetic information is turned into protein.

RNA is processed in eukaryotes prior to transport to the cytoplasm. Caps and tails are added, and internal portions of the primary transcript are removed. Many genes are therefore "split" in eukaryotes, and the coding segments of a gene are not colinear with the processed mRNA. Table 13-8 and Figure 13-43 summarize some of the differences between RNA synthesis in prokaryotes and eukaryotes.

The processes of information storage, replication, transcription, and translation are fundamentally similar in all living organisms. In demonstrating this similarity, molecular genetics has provided a powerful unifying force in biology.

■   ■   ■

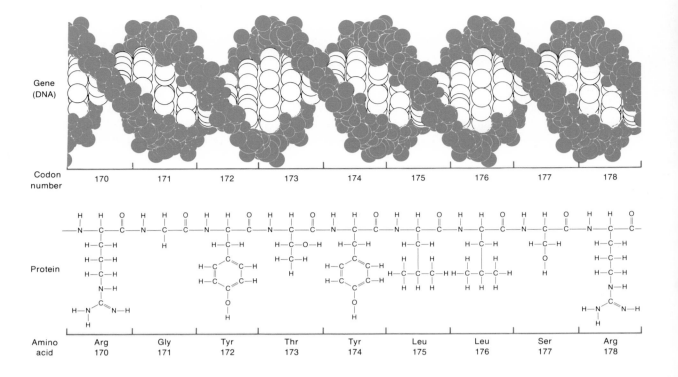

Gene (DNA)									
Codon number	170	171	172	173	174	175	176	177	178

Amino acid	Arg 170	Gly 171	Tyr 172	Thr 173	Tyr 174	Leu 175	Leu 176	Ser 177	Arg 178

Chromosomal DNA   Recognition sequences for RNA polymerase (promoter)

(a)

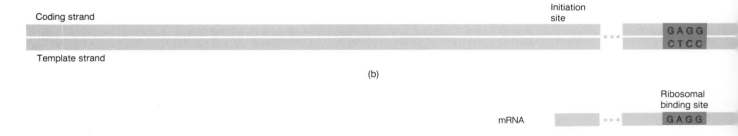

Coding strand

Initiation site

GAGG
CTCC

Template strand

(b)

mRNA

Ribosomal binding site

GAGG

Translation

Protein chain

(c)

Figure 13-42. Genetic information is stored in the double helix of DNA. (a) Each strand of the helix is a chain of nucleotides, each comprising a deoxyribose sugar and a phosphate group, which form the strand's backbone, as well as one of four bases: adenine (A), guanine (G), thymine (T), or cytosine (C). The information is encoded in the sequence of the bases along a strand. The complementarity of the bases (A always pairs with T and G with C) is the basis of the replication of DNA from generation to generation and of its expression (shown here for bacterial DNA) as protein. (b) and (c) Expression begins with the transcription of the DNA base sequence into a strand of mRNA, which corresponds to the coding strand of the DNA except for the fact that uracil (U) replaces thymine. Transcription into RNA and translation into protein are regulated by special sequences (color) in the DNA and the RNA, respectively. The

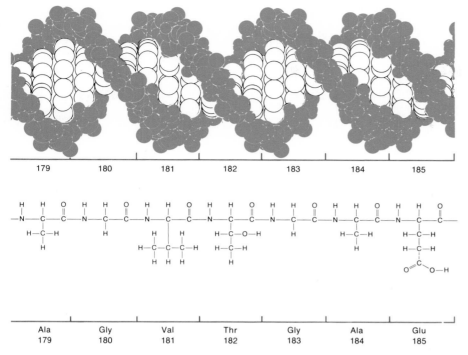

179   180   181   182   183   184   185

Ala	Gly	Val	Thr	Gly	Ala	Glu
179	180	181	182	183	184	185

Figure 13-41. The colinearity of DNA (gene) and polypeptide (protein). The figure shows a stretch of 16 amino acids from the *E. coli* protein called tryptophan synthetase A. (From C. Yanofsky, "Gene Structure and Protein Structure." Copyright © 1967 by Scientific American, Inc. All rights reserved.)

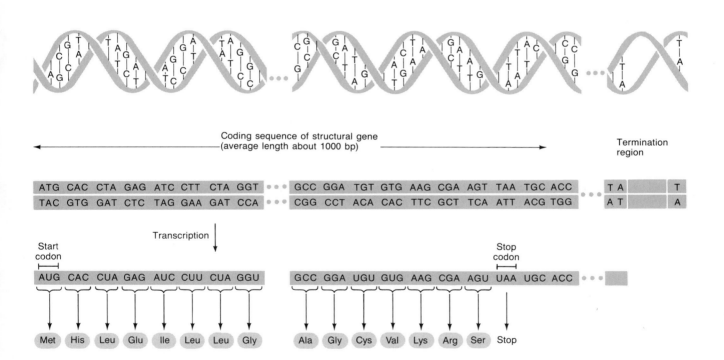

transcribing enzyme, RNA polymerase, binds to a promoter region before a transcription-initiation site; beyond the end of the structural gene, a termination region causes the polymerase to cease transcription. mRNA is translated on cellular organelles called ribosomes; each triplet of bases (codon) encodes a particular amino acid and specifies its incorporation into the growing protein chain. A ribosome binding site on the RNA allows translation to begin at a "start" codon, which is always AUG for the amino acid methionine (Met). Translation proceeds until a "stop" codon is reached (UAA is one of three possibilities), which signals the end of translation and the detachment of the completed protein chain from the ribosome.

Figure 13-43. Protein synthesis involves the same three types of RNA in prokaryotic (a) and eukaryotic (b) cells, but with an important difference: in eukaryotes, the protein-coding sequences of DNA (exons) are often separated by intervening sequences (introns) that must be excised from a primary transcript to make mRNA. In both kinds of cell, tRNA, rRNA, and mRNA are made by the transcription of one strand of the DNA double helix. Three different polymerase enzymes catalyze these reactions in eukaryotes, whereas in prokaryotes there is only a single type of polymerase. In both kinds of cell, the tRNA and rRNA primary transcripts must be processed. The ends of the tRNA transcript are cut and the molecule assumes a looped structure. A single rRNA transcript is cut in several places to form two major types of rRNA, which are then bound to protein molecules to form ribosomal subunits. In prokaryotes, which have no nucleus, the mRNA transcript is generally not processed; ribosomes and tRNA's carrying amino acids begin translating the mRNA into a sequence of amino acids (a protein) as it is being made. In eukaryotes, the nuclear envelope probably facilitates the removal of introns and the splicing of exons from the primary mRNA transcript by protecting it from immediate translation. The mRNA is "read" only after it (like tRNA and the ribosomal subunits) has excited the nucleus through pores in the envelope. Only in the cytoplasm are the mature mRNA, tRNA, and ribosomes united. (From J. E. Darnell, Jr., "RNA." Copyright © 1985 by Scientific American, Inc. All rights reserved.)

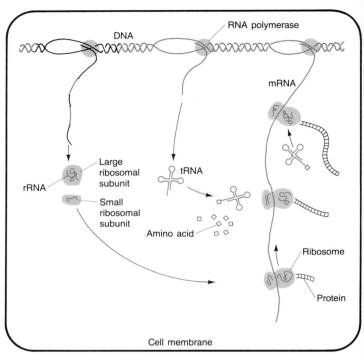

(a) Prokaryote

TABLE 13-8. Differences in gene expression between prokaryotes and eukaryotes

	Prokaryotes		Eukaryotes
1.	All RNA species are synthesized by a single RNA polymerase.	1.	Three different RNA polymerases are responsible for the different classes of RNA molecules.
2.	mRNA is translated during transcription.	2.	mRNA is processed before transport to the cytoplasm, where it is translated. Caps and tails are added, and internal portions of the transcript are removed.
3.	Genes are contiguous segments of DNA that are colinear with the mRNA that is translated into a protein.	3.	Genes are often split. They are not contiguous segments of coding sequences; rather, the coding sequences are interrupted by intervening sequences (introns).
4.	mRNAs are often polycistronic.	4.	mRNAs are monocistronic.

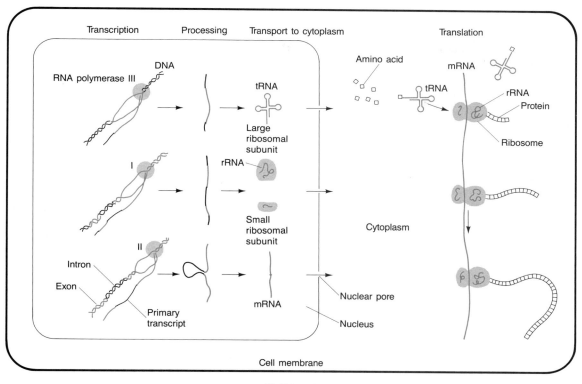

(b) Eukaryote

Solved Problems

1. Using Figure 13-22, show the consequences on subsequent translation of the addition of an – A – codon to the beginning of the following coding sequence:

Ⓐ
↓

– CGA – UCG – GAA – CCA – CGU – GAU – AAG – CAU –
– Arg – Ser – Glu – Pro – Arg – Asp – Lys – His –

Solution

By adding – A – at the beginning of the coding sequence, the reading frame shifts and a different set of amino acids is specified by the sequence, as shown here (note that a set of nonsense codons is encountered, which results in chain termination):

– ACG – AUC – GGA – ACC – ACG – UGA – UAA – GCA
– Thr – Ile – Gly – Thr – Thr – stop – stop –

2. A single nucleotide addition followed by a single nucleotide deletion approximately 20 bp apart in the DNA causes a change in the protein sequence from

– His – Thr – Glu – Asp – Trp – Leu – His – Gln – Asp –

to

– His – Asp – Arg – Gly – Leu – Ala – Thr – Ser – Asp –

Which nucleotide has been added and which nucleotide has been deleted? What are the original and the new mRNA sequences? (HINT: Consult Figure 13-22.)

Solution

We can draw the mRNA sequence for the original protein sequence: (with the inherent ambiguities at this stage)

– His – Thr – Glu – Asp – Trp – Leu – His – Gln – Asp

$$-CA^U_C-AC^U_{\substack{C\\A\\G}}-GA^A_G-GA^U_C-UGG-CU^U_{\substack{C\\A\\G\\UUA\\G}}-CA^U_C-CA^A_G-CA^U_C$$

Because the protein sequence change given to us at the beginning of the problem begins after the first amino acid (His) due to a single nucleotide addition, we can deduce that a – Thr – codon must change to an – Asp – codon. This change must result from the addition of a – G – directly before the – Thr – codon (indicated by a box), which shifts the reading frame, as shown here:

−CA$_C^U$−G̲AC−UGA−$_C^A$GA−UUG−Ⓖ CU−UCA−Ⓤ CA↑−GA$_C^U$−

(under the codons, circled and alternative bases:)
Ⓒ · Ⓖ · C · C
Ⓐ · · U · Ⓐ
G · · · G

(top right of GA codon:) $_G^A$

− His − Asp − Arg − Gly − Leu − Ala − Thr − Ser − Asp −

Also, because a deletion of a nucleotide must restore the final −Asp− codon to the correct reading frame, a −A− or −G− must have been deleted from the end of the original second-last codon, as shown by the arrow. The original protein sequence permits us to draw the mRNA with a number of ambiguities at the first position and, in one case, the third position of the codon. However, the sequence resulting from the frameshift allows us to determine which nucleotide was in the original mRNA. The nucleotide that must have appeared in the original sequence is circled. In only a few cases does the ambiguity remain.

Problems

1. The two strands of phage λ differ from each other in their G−C content. Due to this property, they can be separated in an alkaline cesium chloride gradient (the alkalinity denatures the double helix). When RNA synthesized by phage λ is isolated from infected cells, it is found to form DNA-RNA hybrids with both strands of λ DNA. What does this tell you? Formulate some testable predictions.

2. The data in Table 13-9 represent the base compositions of two double-stranded DNA sources and their RNA products in experiments conducted in vitro.

   a. From these data, can you determine whether the RNA of these species is copied from a single strand or from both strands of the DNA? How? Drawing a diagram will make it easier to solve this problem.

   b. Explain how you can tell whether the RNA itself is single-stranded or double-stranded.

   (Problem 2 is reprinted with permission of Macmillan Publishing Co., Inc., from M. Strickberger, *Genetics.* Copyright © 1968, Monroe W. Strickberger.)

3. Before the true nature of the genetic coding process was fully understood, it was proposed that the message might be read in overlapping triplets. For example, the sequence GCAUC might be read as GCA CAU AUC:

Devise an experimental test of this idea.

4. In protein-synthesizing systems in vitro, the addition of a specific human mRNA to *E. coli* translational apparatus (ribosomes, tRNA, and so forth) stimulates the synthesis of a protein very much like that specified by the mRNA. What does this result show?

5. Which anticodon would you predict for a tRNA species carrying isoleucine? Is there more than one possible answer? If so, state any alternative answers.

6. a. In how many cases in the genetic code would you *fail* to know the amino acid specified by a codon if you know only the first two nucleotides of the codon?

   b. In how many cases would you fail to know the first two nucleotides of the codon if you know which amino acid is specified by it?

7. Deduce what the six wild-type codons may have been in the mutants that led Brenner to infer the nature of the amber codon UAG.

8. If a polyribonucleotide contains equal amounts of randomly positioned adenine and uracil bases, what proportion of its triplets would code for: a. phenylalanine? b. isoleucine? c. leucine? d. tyrosine?

9. You have synthesized three different messenger RNA's with bases incorporated in random sequence in the following ratios: a. 1U:5C; b. 1A:1C:4U; c. 1A: 1C:1G:1U. In a protein-synthesizing system in vitro, indicate the identities and proportions of amino acids that will be incorporated into proteins when each of these mRNAs is tested. (Refer to Figure 13-22.)

10. One of the techniques used to decipher the genetic code was to synthesize polypeptides in vitro, using synthetic mRNA with various repeating base sequences—for example, (AGA)ₙ, which could be written out as AGAA GAAGAAGAAGA. . . . Sometimes the synthesized polypeptide contained just one amino acid (a homopolymer) and sometimes it contained more than one (a heteropolymer), depending on the repeating sequence used. Furthermore, sometimes different polypeptides were made from the same synthetic mRNA, suggesting that the initiation of protein synthesis in the system in vitro does not always start on the end nucleotide of the messenger. For example, from (AGA)ₙ, three polypeptides may have been made: aa₁ homopolymer (abbre-

TABLE 13-9.

Species	DNA base ratio	RNA base ratios	
	$(A + T)/(G + C)$	$(A + U)/(G + C)$	$(A + G)/(U + C)$
Bacillus subtilis	1.36	1.30	1.02
E. coli	1.00	0.98	0.80

viated aa$_1$-aa$_1$), aa$_2$ homopolymer (aa$_2$-aa$_2$), and aa$_3$ homopolymer (aa$_3$-aa$_3$). These probably correspond to the following readings derived by starting at different places in the sequence:

AGA AGA AGA AGA . . .
GAA GAA GAA GAA . . .
AAG AAG AAG AAG . . .

Table 13-10 shows the actual results obtained from the experiment done by Khorana.

a. Why do (GUA)$_n$ and (GAU)$_n$ each code for only two homopolypeptides?

b. Why do (GAUA)$_n$ and (GUAA)$_n$ fail to stimulate synthesis?

c. Assign an amino acid to each triplet in the following list. Bear in mind that there often are several codons for a single amino acid and that the first two letters in a codon usually are the important ones (but that the third letter is occasionally significant). Also remember that some very different-looking codons sometimes code for the same amino acid. Try to carry out this task without consulting Figure 13-22.

AUG	GAU	UUG	AAC
GUG	UUC	UUA	CAA
GUU	CUC	AUC	AGA
GUA	CUU	UAU	GAG
UGU	CUA	UAC	GAA
CAC	UCU	ACU	UAG
ACA	AGU	AAG	UGA

To solve this problem requires both logic and trial and error. Don't be disheartened: Khorana received a Nobel prize for doing it. Good luck!

(Problem 10 is from J. Kuspira and G. W. Walker, *Genetics: Questions and Problems,* McGraw-Hill, 1973.)

11. You are studying a gene in *E. coli* that specifies a protein. A part of its sequence is

– Ala – Pro – Trp – Ser – Glu – Lys – Cys – His –

You recover a series of mutants for this gene that show no enzymatic activity. Isolating the mutant enzyme products, you find the following sequences:

Mutant 1 – Ala – Pro – Trp – Arg – Glu – Lys – Cys – His –

Mutant 2 – Ala – Pro –

Mutant 3 – Ala – Pro – Gly – Val – Lys – Asn – Cys – His –

Mutant 4 – Ala – Pro – Trp – Phe – Phe – Thr – Cys – His –

What is the molecular basis for each mutation? What is the DNA sequence that specifies this part of the protein?

TABLE 13-10.

Synthetic mRNA	Polypeptide(s) synthesized
(UC)$_n$	(Ser – Leu)
(UG)$_n$	(Val – Cys)
(AC)$_n$	(Thr – His)
(AG)$_n$	(Arg – Glu)
(UUC)$_n$	(Ser – Ser) and (Leu – Leu) and (Phe – Phe)
(UUG)$_n$	(Leu – Leu) and (Val – Val) and (Cys – Cys)
(AAG)$_n$	(Arg – Arg) and (Lys – Lys) and (Glu – Glu)
(CAA)$_n$	(Thr – Thr) and (Asn – Asn) and (Gln – Gln)
(UAC)$_n$	(Thr – Thr) and (Leu – Leu) and (Tyr – Tyr)
(AUC)$_n$	(Ile – Ile) and (Ser – Ser) and (His – His)
(GUA)$_n$	(Ser – Ser) and (Val – Val)
(GAU)$_n$	(Asp – Asp) and (Met – Met)
(UAUC)$_n$	(Tyr – Leu – Ser – Ile)
(UUAC)$_n$	(Leu – Leu – Thr – Tyr)
(GAUA)$_n$	None
(GUAA)$_n$	None

NOTE: the order in which the polypeptides or amino acids are listed is not significant except for (UAUC)$_n$ and (UUAC)$_n$.

12. A single nucleotide addition and a single nucleotide deletion approximately 15 sites apart in the DNA cause a protein change in sequence from

– Lys – Ser – Pro – Ser – Leu – Asn – Ala – Ala – Lys –

to

– Lys – Val – His – His – Leu – Met – Ala – Ala – Lys –

a. What are the old and the new mRNA nucleotide sequences? (Use Figure 13-22.)

b. Which nucleotide has been added and which has been deleted?

(Problem 12 is from W. D. Stansfield, *Theory and Problems of Genetics,* McGraw-Hill, 1969.)

13. A mutational event inserts an extra nucleotide pair into DNA. Which of the following do you expect?

a. No protein product at all

b. A protein in which one amino acid is changed

c. A protein in which three amino acids are changed

d. A protein in which two amino acids are changed

e. A protein in which most amino acids following the site of the insertion are changed

14. Suppressors of frame-shift mutations are now known. Propose a mechanism for their action.

15. Use Figure 13-22 to complete the following table. Assume that reading is from left to right and that the columns represent transcriptional and translational alignments.

C					T	G	A			DNA double helix
	C	A		U						mRNA transcribed
						G	C	A		Appropriate tRNA anticodon
		Trp								Amino acids incorporated into protein

TABLE 13-11.

Amino acid at position 68	Frequency among offspring of the cross
Gln	about 0.50
Ser	about 0.50
Arg	4.0×10^{-7}
His	2.0×10^{-7}
Asn	1.0×10^{-11}
Lys	2.0×10^{-7}

16. Consider the gene that specifies the structure of hemoglobin. Arrange the following events in the most likely sequence in which they would occur:

    a. Anemia is observed.

    b. The shape of the oxygen-binding site is altered.

    c. An incorrect codon is transcribed into hemoglobin mRNA.

    d. The ovum (female gamete) receives a high radiation dose.

    e. An incorrect codon is generated in the DNA of the hemoglobin gene.

    f. A mother (an X-ray technician) accidentally steps in front of an operating X-ray generator.

    g. A child dies.

    h. The oxygen-transport capacity of the body is severely impaired.

    i. The tRNA anticodon that lines up is one of a type that brings an unsuitable amino acid.

    j. Nucleotide-pair substitution occurs in the DNA of the gene for hemoglobin.

17. The comparison of physical distances with map distances is facilitated by our knowledge of the genetic code. The code permits us to "translate" numbers of amino acids in a protein into numbers of nucleotides in a corresponding stretch of DNA.

    a. Consider two mutant forms of a particular enzyme in *Neurospora*. The two proteins are known to differ at only two positions in their polypeptide chains. The two positions are separated by 40 other amino acids. Crosses of the two mutants regularly give a recombinant frequency of 1.0×10^{-5} for the two mutations. Approximately how many nucleotides in *Neurospora* lie between a pair of markers that give a recombinant frequency of 10^{-5}?

    b. Consider two other mutant forms of the enzyme. Each has the normal (wild-type) number of amino acids, but these enzymes differ from each other and from the wild-type enzyme at amino acid 68 in the polypeptide chain. The wild-type protein has arginine at position 68, whereas one mutant has gluta-

mine and the other has serine. When mutants 1 and 2 are crossed, six kinds of offspring differing in the amino acid at position 68 are observed. Two of these are parental types in which glutamine or serine is present at position 68, and one has arginine like the wild-type. Three novel types—having lysine, asparagine, or histidine—make up the remainder. Table 13-11 shows the frequencies of these classes in the progeny of the cross. Deduce the nucleotide sequences at codon 68 for each of the six types. What is the frequency of recombinants for crosses in which the markers are at adjacent positions in the DNA?

(Problem 17 is from Franklin W. Stahl, *The Mechanics of Inheritance*, 2d ed., p. 192. Copyright © 1969, Prentice-Hall, Inc., Englewood Cliffs, New Jersey. Reprinted by permission.)

18. An induced cell mutant is isolated from a hamster tissue culture because of its resistance to α-amanitin (a poison derived from a fungus). Electrophoresis shows that the mutant has an altered RNA polymerase; *just one* electrophoretic band is in a position different from that of the wild-type polymerase. The cells are presumed to be diploid. What does this experiment tell you about ways in which to detect recessive mutants in such cells?

19. A double-stranded DNA molecule with the sequence shown at the bottom of the page produces, in vivo, a polypeptide that is five amino acids long:

    a. Which strand of DNA is transcribed, and in which direction?

    b. Label the 5′ and the 3′ ends of each strand.

    c. If an inversion occurs between the second and third triplets from the left and right ends, respectively, and the same strand of DNA is transcribed, how long will the resultant polypeptide be?

    d. Assume that the original molecule is intact and that transcription occurs on the bottom strand from left to right. Give the base sequence, and label the 5′ and 3′ ends of the anticodon that inserts the *fourth* amino acid into the nascent polypeptide. What is this amino acid?

TAC ATG ATC ATT TCA CGG AAT TTC TAG CAT GTA
ATG TAC TAG TAA AGT GCC TTA AAG ATC GTA CAT

The Structure and Function of Chromosomes in Eukaryotes

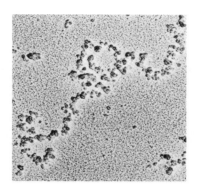

KEY CONCEPTS

A eukaryotic chromosome consists of a single continuous molecule of DNA that is greatly condensed by several orders of folding and wrapping around proteins termed histones.

▪

Chromosome segments can be differentiated by their staining intensity, which is a reflection of the activity of the genetic segments.

▪

Different organisms use different mechanisms to compensate for the differences in gene numbers between sexes for X-linked genes.

▪

In mammalian females, dosage compensation occurs by inactivating one of the X chromosomes in each somatic cell.

■ Unraveling the mysteries of the eukaryotic genome is far more difficult and challenging than understanding simpler prokaryotic chromosomes. The genomes in eukaryotes are much larger and the chromosome structures are more complex than in prokaryotes. In order to get a clearer picture of eukaryotic genetic systems, we consider three aspects of the genome:

1. The complex structure of eukaryotic chromosomes. Is the chromosome simply a structural framework that assembles the genes for proper partition during cell division? Or does the structure of the chromosome play some role in determining the function of the genes within it?

2. The sequence organization of eukaryotic chromosomes.

3. The activity of eukaryotic chromosomes.

Chromosome Structure

DNA Molecules in Chromosomes of Eukaryotes

Let's look at the molecular analysis of chromosomes. By staining chromosomes with dyes that bind to specific macromolecules, we can demonstrate that the visible chromosome contains DNA, RNA, and protein. But how is that material organized during interphase or during cell division? When cells are disrupted mechanically (by squeezing them under high pressure) or osmotically (by exploding them in hypotonic solutions), their contents can be mounted for electron microscopic examination. Such study reveals that the chromosome resembles a mass of spaghetti-like fibrils with diameters of about 230 angstroms (Å), each of which is composed of DNA with associated protein (Figure 14-1). Ernest DuPraw showed that few, if any, ends protrude from the fibrillar mass, as if the DNA is a single long fiber coiled up in a compact mass.

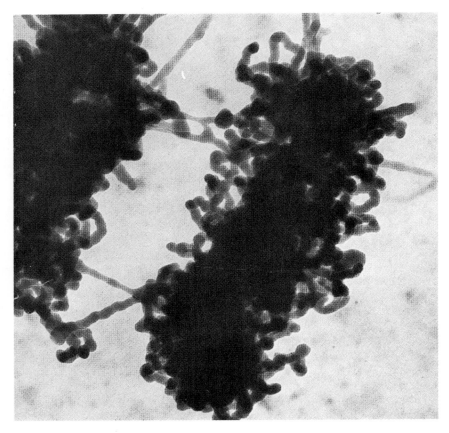

Figure 14-1. Electron micrograph of a metaphase chromatid from an embryonic honeybee cell. It appears to be a tangle of one continuous strand made up of a DNA core complexed with nuclear protein. (From E. J. DuPraw, *Cell and Molecular Biology*, p. 531, Fig. 18-4. Copyright © 1968 by Academic Press, Inc.)

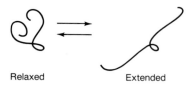

Relaxed   Extended

Figure 14-2. The relaxed and extended states of DNA.

Could it be that each chromosome contains a small number of DNA molecules? In 1973, Ruth Kavenoff and Bruno Zimm resolved this question using a **viscoelastic recoil technique** that essentially measures the size of the largest DNA molecules by their elastic properties in solution. Very simply, if DNA is stretched to an extended state (for example, by spinning a paddle in a DNA solution) and then allowed to recoil toward a random relaxed state (Figure 14-2), the recoil requires a time that is proportional to molecular size. This technique provides a sensitive indicator of the *largest molecules in the solution*, even when they represent a small fraction of the total number of molecules.

Kavenoff and Zimm studied the DNA molecules of *Drosophila melanogaster* (which has four pairs of chromosomes; Figure 14-3). They extracted DNA gently (to avoid shear) from wild-type nuclei and obtained a value of 41×10^9 daltons ($\pm 3 \times 10^9$) from the viscoelastic measurements for the largest DNA molecule in solution.

Using DNA molecules from a mutant having a translocated X-autosome that increases the length of the autosomal portion by one-third (Figure 14-4), Kavenoff and Zimm obtained a viscoelastic value of 58×10^9 daltons for the largest DNA molecule.

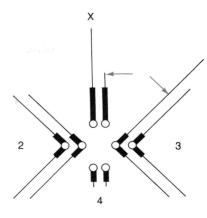

Figure 14-4. Karyotype of a *Drosophila* female heterozygous for an X-autosome translocation. Breakpoints are indicated by arrows. About 60 percent of the mitotic chromosome length of the X is attached to the tip of chromosome 3, thereby increasing the length of chromosome 3 by 37 percent.

Finally, they studied DNA molecules from mutants with a pericentric inversion (Figure 14-5) that increased the length of one chromosome arm without affecting the total DNA content of the chromosome. In this case, they obtained a value of 42×10^9 daltons ($\pm 4 \times 10^9$), not significantly different from the wild-type value. In studies of other *Drosophila* species, Kavenoff and Zimm obtained values for the largest DNA molecules that were proportional to the cytological lengths observed for the largest chromosomes. Therefore, they concluded that each chromosome is composed of a single DNA molecule that extends from one end of the chromosome through the centromere to the other end.

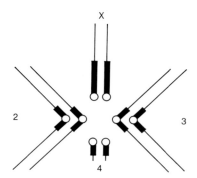

Figure 14-3. The mitotic chromosomes of a *Drosophila melanogaster* female. The X chromosomes are the sex chromosomes; the autosomes are labeled 2, 3, and 4. Open circles represent the centromeres; the thicker and darker regions of the chromosomes represent heterochromatin.

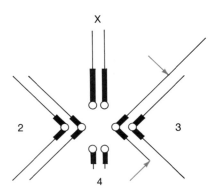

Figure 14-5. Karyotype of a *Drosophila* female heterozygous for a pericentric inversion of chromosome 3. Breakpoints are indicated by arrows. The inversion changes the ratio of the lengths of the two arms of chromosome 3 from 1 : 1 to about 7 : 1 but causes no change in the total length of the chromosome.

DNA Packaging in Chromosomes of Eukaryotes

Each chromosome that becomes visible under the light microscope at division contains one remarkably long molecule of DNA. The typical bacterial or viral chromosome also is a single DNA molecule, but it is a relatively short molecule with only one site at which bidirectional replication begins. The eukaryotic DNA contains multiple sites for initiation of replication.

But how are such long molecules packed into the compact entities we recognize as chromosomes? In terms of end-to-end lengths, the DNA molecule contained in a chromosome may be more than 100,000 times longer than the chromosome itself. How can this compact packaging be achieved in a way that permits orderly replication?

An important insight into the packaging problem came with the recovery from nuclease-digested chromatin of a histone-DNA complex called a **nucleosome**. About 140 base pairs of DNA are wrapped around a core of pairs of four histones (proteins) called H2A, H2B, H3, and H4. Viewed under the electron microscope after special treatment, chromatin appears to be composed of a chain of granules (nucleosomes) about 100 Å in diameter (Figure 14-6). From additional experiments, the DNA appears to be coiled around the protein octamer in two turns or wraps (Figure 14-7). One model that can account for this structure is shown in Figure 14-8.

There is evidence (Figure 14-9) that the nucleosomes in turn are thrown into a coil that has been called a **solenoid structure** (resembling a solenoid coil in an automobile engine). The solenoid structure has a diameter of 200 Å to 300 Å (Figure 14-10). Obviously, the DNA can be compacted extensively by superimposing coils on coiled structures during cell division (Figure 14-11). An incredible cycle of coiling and supercoiling at the level visible in the light microscope appears during the mitosis of gut parasites in termites (Figure 14-12). How are additional layers of coiling superimposed on the nucleosome-solenoid level seen in the interphase cell? That problem remains unsolved.

Message DNA in the eukaryotic chromosome is shortened more than 100,000-fold by superimposing different levels of coiling. At the core of this structure, the DNA is wrapped around a cluster of histone proteins; the fundamental histone-DNA complex thus formed is called the nucleosome.

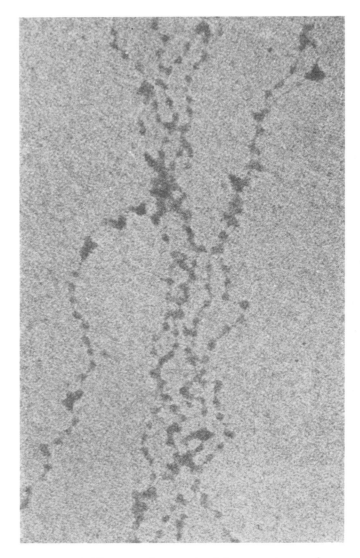

Figure 14-6. Electron micrograph of the central core of chromatin isolated from the eukaryote *Physarum*. The individual granules that form the chains presumably are nucleosomes.

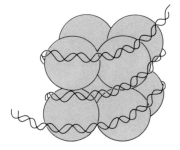

Figure 14-7. A diagrammatic representation of the four parts of histone molecules around which DNA is wound to form a nucleosome.

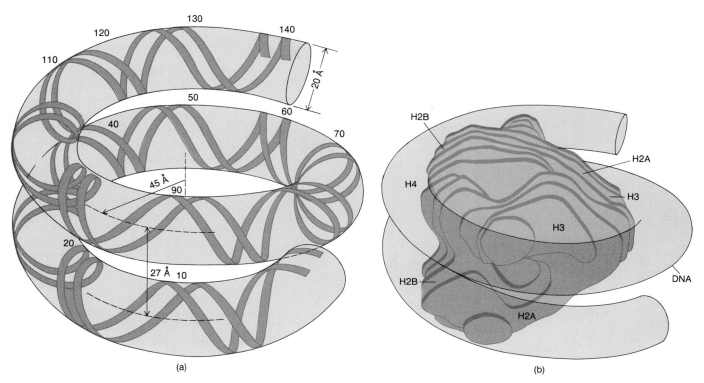

Figure 14-8. (a) The path of DNA around the histone octamer that can account for the bipartite structure of the nucleosome core is a superhelix with an external diameter of 110 Å and a pitch of 27 Å; the turns of the 20 Å-wide DNA helix are nearly in contact. There are about 80 nucleotide pairs of DNA per turn; the nucleosome core, an enzymatically reduced form of the nucleosome consisting of some 140 nucleotide pairs, has about one and three-quarter turns wrapped on it. (b) Model of a nucleosome core made by winding a tube simulating the DNA superhelix on a model of the histone octamer, which was built from a three-dimensional map derived from electron micrographs of the histone octamer. The ridges on the periphery of the octamer form a more or less continuous helical ramp on which a 146-nucleotide-pair length of DNA can be wound. The locations of individual histone molecules (with undefined boundaries at this resolution) are proposed here on the basis of chemical cross-linking data. (Figures 14-8, 14-9, and 14-11 from R. D. Kornberg and A. Klug, "The Nucleosome." Copyright © 1981 by Scientific American, Inc. All rights reserved.)

Inspection of the chromosome with a scanning electron microscope fails to reveal how the tightly coiled strand is held together, but such study does show a mat of fibers that do not appear to have a free end (Figure 14-13). Ulrich Laemmli and his associates developed a method for the gentle removal of histones from chromosomes for inspection under the electron microscope. This inspection reveals that sister chromatids remain paired and that each chromatid has a central structure surrounded by a halo of DNA. The chromosome appears to have a central "scaffold" of protein to which the DNA is anchored. Close inspection reveals the incredible density of DNA packaging (Figure 14-14). The ends of each loop of DNA that extend from the scaffold are attached near the same point, suggesting that the packaging process is regular and precise.

Message   The DNA in the eukaryotic chromosome is organized about a central core from which segments may extend to form loops. The highly condensed metaphase chromosome that is seen under the light microscope results from many different orders of chromatin packing, as summarized in Figure 14-15.

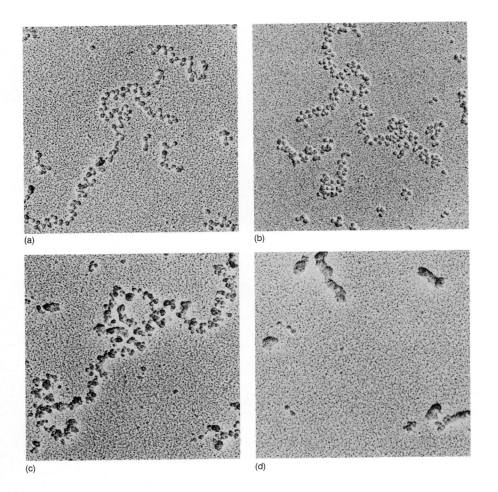

(a)

(b)

(c)

(d)

Figure 14-9. Condensation of chromatin with increasing salt concentration is demonstrated in electron micrographs made by Fritz Thoma and Theo Koller of the Swiss Federal Institute of Technology. At a very low salt concentration, as in (a), chromatin forms a loose fiber about 100 Å thick: nucleosomes connected by short stretches of DNA. At a concentration with an ionic strength closer to that of normal physiological conditions, as in (d), chromatin forms a thick fiber some 250 to 300 Å thick. The origin of this "solenoid" can be deduced by an examination of chromatin at increasing intermediate ionic strengths, as in (b) and (c). It arises from a shallow coiling of the nucleosome filament. The chromatin is enlarged here about 80,000 diameters.

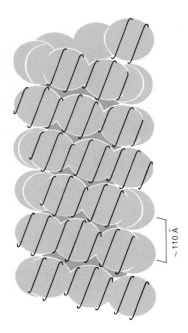

~110 Å

Figure 14-10. A higher level of packing is achieved in DNA when nucleosomes are wound into a solenoid structure. (From J. T. Finch and A. Klug, *Proceedings of the National Academy of Sciences USA* 71, 1976, 1897.)

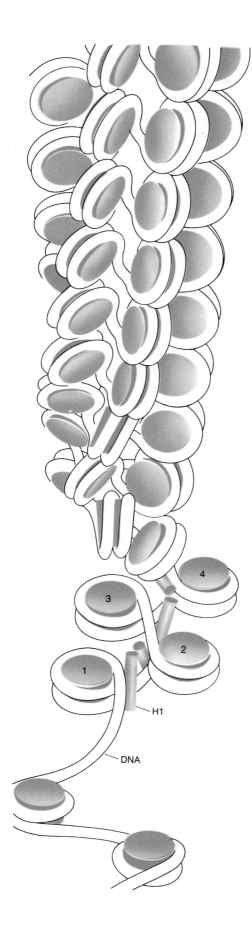

Figure 14-11. Helical superstructures might be formed with increasing salt concentration *(bottom to top)* with the participation of the histone protein H1, as suggested here. The zigzag pattern of nucleosomes (1, 2, 3, 4) closes up and eventually forms a solenoid—a helix with about six nucleosomes per turn. (The helix is probably more irregular than it is in this drawing.) Cross-linking data indicate that H1 molecules on adjacent nucleosomes make contact. Extrapolation from the zigzag form to the solenoid suggests (but does not prove) that the aggregation of H1 at higher ionic strengths gives rise to a helical H1 polymer (not shown) running down the center of the solenoid. In the absence of H1 *(bottom)* no ordered structures are formed. The details of H1 associations are not known at this time; this drawing is meant to indicate only that H1 molecules contract one another and "linker" DNA.

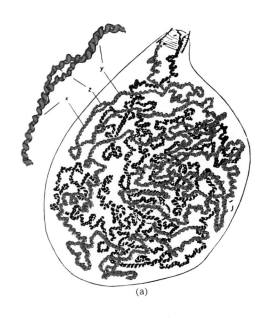

Figure 14-12. Drawings of chromosomes in meiotic prophase in a protozoan, demonstrating different degrees of coiling and supercoiling visible with the light microscope. Two large chromosomes are shown: one colored and the other black; (a) to (d) is a progression. (a) Coiling is seen though duplication becomes apparent. (b) Duplication is well advanced. (c) Supercoiling is beginning. (d) Supercoiling is well advanced. (From L. R. Cleveland, "The Whole Life Cycle of Chromosomes and Their Coiling Systems," *Transactions of the American Philosophical Society* 39, 1949, 1.)

(a)

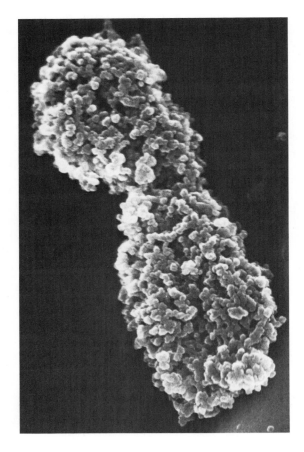

Figure 14-13. A scanning electron micrograph of a mitotic chromosome, showing the compact folding of the strand. (Photo courtesy of Wayne Wray.)

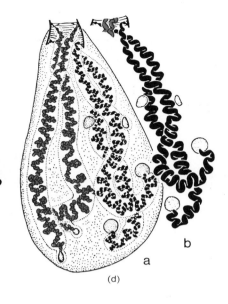

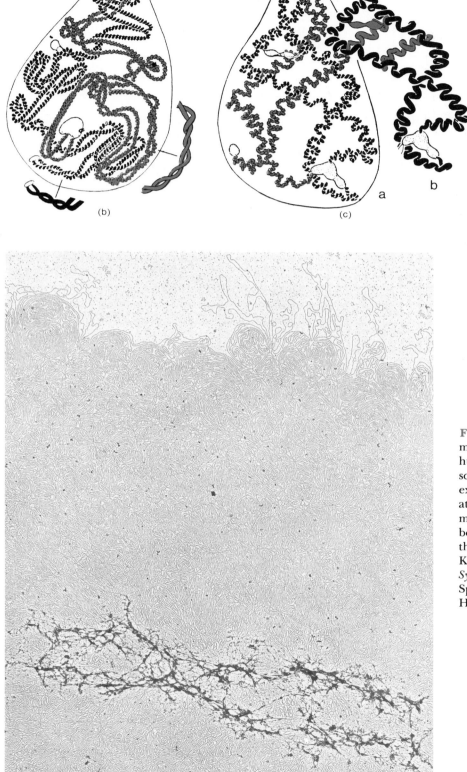

Figure 14-14. Electron micrograph of a metaphase chromosome from a cultured human cell. Note the central core, or scaffold, from which the DNA strands extend outward. No free ends are visible at the outer edge. At even higher magnification, it is clear that each loop begins and ends near the same region of the scaffold. (From W. R. Baumbach and K. W. Adolph, *Cold Spring Harbor Symposium on Quantitative Biology,* Cold Spring Harbor Laboratory, Cold Spring Harbor, New York, 1977.)

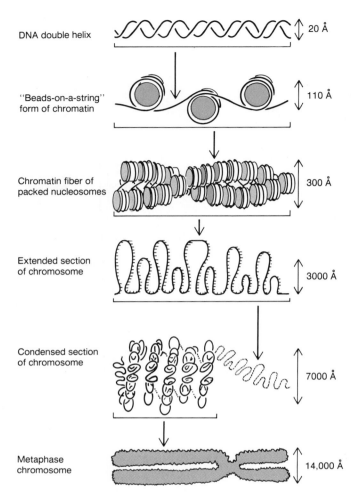

DNA double helix — 20 Å

"Beads-on-a-string" form of chromatin — 110 Å

Chromatin fiber of packed nucleosomes — 300 Å

Extended section of chromosome — 3000 Å

Condensed section of chromosome — 7000 Å

Metaphase chromosome — 14,000 Å

Figure 14-15. Schematic illustration of the many different orders of chromatin packing postulated to give rise to the highly condensed metaphase chromosome. (From B. Alberts et al., *Molecular Biology of the Cell.* Copyright © 1983 by Garland.)

Sequence Organization

Measurements of DNA in different organisms have revealed that—perhaps as expected—eukaryotic genomes consist of much more DNA than prokaryotic or viral genomes. Although, for the most part, higher eukaryotes contain more DNA than lower eukaryotes, there is considerable variation in the amounts of DNA, even among similar species. Certain less complex species have significantly more DNA than other species higher on the evolutionary scale; in fact, amphibians and lilies usually have the highest DNA content per cell of any organism. The lack of correlation between amount of DNA and complexity of an organism is puzzling. Why do eukaryotes have so much extra DNA? Does all of the extra DNA code for protein? Is the extra DNA essential? Scientists are beginning to resolve some of these ques-

tions, although there are still many areas of uncertainty. Estimates of proteins in higher cells suggest that as much as 90 percent of the DNA does *not* code for proteins. Molecular analyses of eukaryotic genomes have revealed several sources of untranslated DNA. These include **repetitive DNA sequences** (Figure 14-16), **intervening sequences** which interrupt the coding sequences of genes, and **spacers** between genes.

Number of Genes in Drosophila

Measurements of DNA show that a haploid nucleus from *Drosophila* contains about 1.5×10^8 base pairs, one genome of *E. coli* contains about 4.6×10^6 base pairs, and a human haploid nucleus contains about 3.1×10^9 base pairs. In *Drosophila*, unique sequences (nonrepetitive DNA) make up about 70 percent of the total nuclear DNA; the remaining 30 percent consists of middle or highly repetitive sequences. A typical *E. coli,* protein is encoded by about 1000 base pairs, so the unique sequences of *Drosophila* could code for around 100,000 such proteins. Looking at the arrangement of DNA in the giant salivary-gland chromosomes, we see that most of the DNA is located in about 5000–6000 densely staining bands—the chromomer. Cytogeneticists have long assumed that each chromomere represents the cytological position of a gene locus. In the haploid nucleus, each chromomere is equivalent to 5000 to 100,000 base pairs, with an average of 30,000. How can this 30-fold excess over the size of a typical *E. coli* gene be explained?

Perhaps the initial assumption of one gene per band is erroneous, and each chromomere in fact represents many genes. Another possibility is that the average *Drosophila* gene is indeed 30 times larger than the average *E. coli* gene, and certainly the existence of large introns (segments that interrupt the coding sequence of the genes) can account for a lot of the excess. Most of the basic metabolic processes exist both in *E. coli* and in multicellular eukaryotes, so the major differences in size of genome could represent codings involved in the regulation of gene activity rather than codings for new gene functions. Much of the eukaryotic DNA may be involved in regulating the gene activity in relation to developmental time, the nature of the tissue, and the extent of activity required.

To distinguish between these alternative explanations, Burke Judd and Thomas Kaufman set out in the late 1960s to detect all of the genes within a limited region of the *Drosophila* X chromosome by inducing and recovering enough mutations to ensure that every locus was represented by at least one mutant. They chose the interval from *zeste (z)* to *white (w),* a span of 10 or 12 bands, for which several overlapping duplications and deficiencies exist with which to map new mutations (Figure 14-17). They selected newly induced mutations in

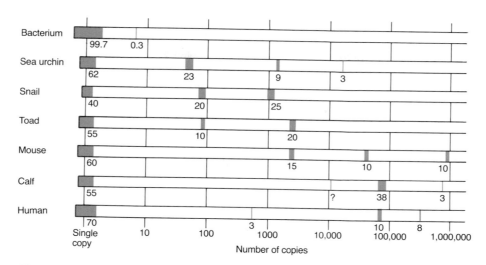

Figure 14-16. Abundance of repeated sequences in the DNA of seven organisms. The width of each band represents the percentage (also given below the band) of the total DNA that appears with a given degree of repetition. (From R. J. Britten and D. E. Kohne, "Repeated Segments of DNA." Copyright © 1970 by Scientific American, Inc. All rights reserved.)

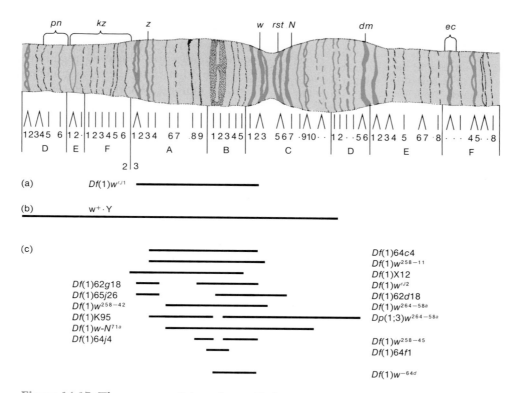

Figure 14-17. The segment of the polytene X chromosome within which the $z-w$ region lies. The known positions of mutations are shown above the chromosome, and the numbering nomenclature is indicated underneath (the $z-w$ region extends from 3A3 to 3C3). The deficiencies and duplications covering this region are indicated at the bottom; the extent of each deficiency or duplication is shown by the horizontal line. Mutations are (a) detected by their mutant phenotype when heterozygous with $Df(1)w^{rJ1}$, (b) maintained in males with $w^+ \cdot Y$, and (c) localized by their phenotypes when heterozygous with different deficiencies. (Figures 14-17, 14-18, and 14-19 from B. Judd, M. Shen, and T. Kaufman, *Genetics* 71, 1972, 139.)

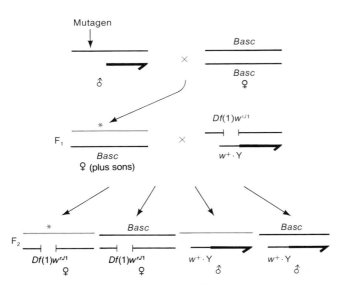

Figure 14-18. The mating scheme used to recover mutations in the $z-w$ region. Wild-type males are treated with a mutagen and then crossed to females carrying Basc, a multiply inverted chromosome marked with B and w^a. Then F_1 females carrying the mutagen-treated chromosome (indicated by an asterisk) are selected and individually mated to males carrying a deficiency on the X chromosome, $Df(1)w^{rJ1}$, and a duplication for this region on the Y chromosome, $w^+ \cdot Y$. A mutation in the $z-w$ region is indicated by the absence (if lethal) of the mutant phenotype of the non-Bar-eyed females. The chromosome is not lost because the mutation is covered by the duplication in non-Basc males.

the $z-w$ interval, using the method for detection and recovery shown in Figure 14-18. Each mutagen-treated X chromosome was initially recovered in an F_1 female and then tested for its survival or phenotype when heterozygous with a $z-w$ deficiency. All mutant chromosomes were "captured" in males carrying a duplication. Each mutant was tested with the duplications and deficiencies to localize it, and all mutations falling within the same segment were tested for complementation and map position by crossing-over. In 1972, Judd and Kaufman reported that 121 point mutations had been assigned to 16 complementation groups that could be mapped by crossing-over (Figure 14-19). The position of each complementation group could be assigned to a single-band, leading to the conclusion that *each gene corresponds to one chromomere.* It is now known that nontranslated spacers exist between genes, large introns are excised from mRNA, and sequences are removed during mRNA maturation. There are also a few cases in which some bands contain several genes (for instance, the chorion protein genes come in pairs). Whether these factors account for the 30-fold excess of DNA per chromomere has not yet been determined.

> **Message** In *Drosophila*, each polytene-chromosome chromomere apparently corresponds to one gene detectable by mutation and complementation. This means that there are 5000 to 6000 genes in *Drosophila*.

Gene Families

Detailed analyses of parts of the eukaryotic chromosome by recombinant DNA techniques (to be described in Chapter 15) have revealed that many genes are present in multiple copies in the genome. Sometimes the gene copies are virtually identical and serve to provide a source of increased gene products, as in the case of *Xenopus* 5S rRNA genes and sea-urchin histone genes (page 470). In other cases, the sequences are related but have diverged during evolution to the point that there are significant differences. Occasionally, chromosome analysis reveals nonfunctional copies of genes, termed **pseudogenes,** that are remnants of the evolutionary process.

Genes that are derived from a common ancestral gene constitute a **gene family.** Often the members of gene families are clustered together on the chromosome; in other cases, they are widely dispersed throughout the genome. An interesting example of a clustered gene family is the globin gene family.

Hemoglobin consists of a tetrameric protein in association with its heme group (recall the analysis of altered hemoglobin molecules in Chapter 12). In adult mammals, the tetramer consists of two α (alpha) and two β (beta) protein chains. However, mammalian embryos have hemoglobin molecules with different constituent chains. Instead of α and β chains, related proteins termed **α-like** and **β-like,** respectively, are found in embryonic hemoglobin. As the embryo develops, yet other α-like and β-like proteins appear to replace those originally present. Finally, after birth, the α and β chains emerge to replace the second set of related polypeptide chains. How the cell programs the production of the different globin proteins during development is a fascinating problem.

Table 14-1 details the specific globin polypeptides in the case of humans. The α-like chain found in embryonic and fetal hemoglobin is termed ζ (zeta). The β-like chain detected in embryonic hemoglobin is the ϵ (epsi-

TABLE 14-1. Globin chains in humans

Globin polypeptides	Embryo	Fetus	Adult
α-like	ζ	α	α
β-like	ϵ	G_γ, A_γ	δ, β

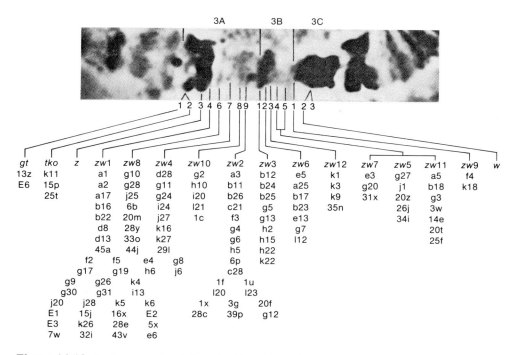

Figure 14-19. A photograph of the 3A – 3C region of the X chromosome of a *Drosophila melanogaster* female. The complementation groups, with the point mutations in each, are shown below the photograph. The cytological position of each complementation group is determined by deletion mapping.

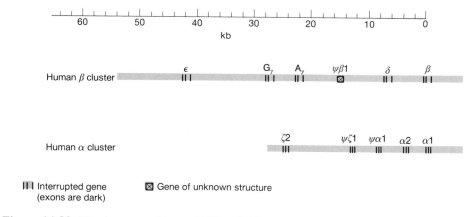

| | Interrupted gene (exons are dark) | ⊠ Gene of unknown structure |

Figure 14-20. The human α-like and β-like globin gene families are each organized into a single cluster that includes functional genes and pseudogenes; the latter are denoted here by ψ (psi). (After B. Lewin, *Genes*. Copyright © 1983 by John Wiley.)

lon) protein, whereas two forms of γ (gamma) — the G_γ and A_γ forms, which differ by only a single amino acid — predominate in fetal hemoglobin. These are replaced by the β chain in adults and, in a small percentage of the molecules, by another β-like chain termed δ (delta).

Figure 14-20 shows the two gene clusters that have been found for the globin genes in humans. The α and β clusters are located on different chromosomes. Here we can see the genetic organization of a 50-kilobase (kb) region of the chromosome, in the case of the β cluster.

Note how the exons (the coding portions of the genes) occupy only a fraction of the region! Some pseudogenes (nonfunctional counterparts) also have been discovered in these clusters; these are indicated in the figure according to the related functional gene.

The Structure and Activity of Chromosomes

Cytological examination allows us to differentiate eukaryotic chromosome segments by their staining intensity. Recall that heterochromatic elements (that is, heterochromatin) are densely stained; these regions generally are assumed to reflect a state of genetic inactivity. Euchromatic regions (that is, euchromatin) are less densely stained and typically less compact than the heterochromatin; these regions are assumed to represent the locations of active genes. The heterochromatic elements can be divided into two classes. First, there are entire chromosomes or specific chromosome segments from a given species that always stain densely; such elements are called **constitutive heterochromatin.** Second, there are chromosomes or chromosome segments from a given species that appear sometimes as euchromatin and sometimes as heterochromatin; heterochromatin of this type is called **facultative heterochromatin.** We consider these two types of heterochromatin separately.

Constitutive Heterochromatin

In many species, heterochromatic chromosomes have very minor biological roles (if they have any role at all). For example, corn may possess small heterochromatic elements called B chromosomes. However, no obvious phenotypic difference can be detected between lines having one, many, or no B chromosomes. Such apparently extraneous chromosomes were long ago named **satellite** or **accessory chromosomes.** They are found in many insect species that infect plants and in some vertebrate species (such as salamanders). Large, metacentric, heterochromatic chromosomes are found in the gonadal (germ) cells of the fly *Sciara,* but these satellite chromosomes do not appear in the somatic cells. What distinguishes satellite chromosomes from the functionally important ones? Why are such apparently useless chromosomes retained? We do not yet know the answers to these questions.

Not all heterochromatic chromosomes appear to be functionally useless. The Y chromosome in *Drosophila* and other organisms is composed of heterochromatin, but it does have a genetic function. In *Drosophila melanogaster,* the Y chromosome carries six distinguishable loci (all of them sites with wild-type alleles necessary for male fertility). However, this chromosome is almost as long as the X chromosome, which carries a large number of genes.

Of even greater interest are the segments of heterochromatin that occur consistently within chromosomes, such as those in Diptera (two-winged flies). In *Drosophila,* for example, these heterochromatic regions contain between one-fourth and one-third of all the chromosomal DNA. All of these regions are found in the **proximal** parts of the chromosomes near the centromere (Figure 14-3). These proximal regions make up a large part of the chromosome complement, but they contain very few genetically detectable loci, which are concentrated in the **distal** (end) parts of the chromosomes. Furthermore, crossing-over occurs rarely (if ever) between loci that closely flank large blocks of this proximal heterochromatin. Since the 1930s, geneticists working with *Drosophila* have assumed that these regions of proximal heterochromatin are genetically inert—virtually devoid of genes that are detectable by mutant alleles.

Does the paucity of detectable mutations in proximal heterochromatin truly reflect the absence of any biological function? The answer is not clear. On the one hand, molecular analyses have revealed that heterochromatin consists mainly of repetitive DNA. In regions of high redundancy, a single point mutation has no noticeable effect. The redundancy makes it difficult to detect active loci if they are present. On the other hand, much of the repetitive DNA may have no function. Also, some heterochromatin is composed of nonrepetitive DNA that probably does not encode proteins.

We saw in Chapter 8 (Figure 8-4) the polytene (giant) chromosomes that form in certain dipteran cells by multiple replication of the DNA strands. In these chromosomes, the heterochromatic segments are replicated to a drastically lesser extent than the euchromatic segments. Hybridization in situ reveals that the satellite DNA's are located in the proximal heterochromatic regions. It is clear that the cytological distinctions between heterochromatin and euchromatin reflect profound functional differences, but the nature of these differences remains a mystery.

Message Heterochromatin is distinguished cytologically from euchromatin by differences in structure and activity.

The Nucleolus Organizer. In early cytological studies, certain discrete structures of the nucleus were found to be unaffected by the stains that dye chromatin; each such structure is called a **nucleolus.** The nucleolus forms at a specific chromosome site called the **nucleolus organizer**

(NO), identified cytologically as a pinched region (secondary constriction) of the chromosome.

In amphibians, a mutation called anucleolate *(an)* is produced by deletion of the NO. The wild-type phenotype of two nucleoli per cell corresponds to the genotype an^+/an^+. The genotype an^+/an has the phenotype of one nucleolus per cell; the *an/an* homozygotes have no nucleoli and die. Thus, genetic studies confirm that the NO is the site of nucleolus production.

What is the function and the structure of the nucleolus? And what is the relationship between the nucleolus and the chromosomal DNA of the NO? Detailed studies reveal structures similar to ribosomes within the nucleolus. In the 1960s, it was suggested that the nucleolus is an assembly point for ribosomes, which then are distributed to daughter cells during mitosis. The NO region then might code for ribosomal RNA's.

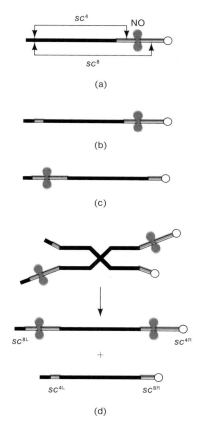

Figure 14-21. The scute-4 and scute-8 inversions. (a) Breakpoints for the two inversions on the *Drosophila* X chromosome. (b) The result of the sc^4 inversion. (c) The result of the sc^8 inversion. (d) A crossover between sc^4 and sc^8 in a heterozygous female yields one crossover product with two nucleolar organizers and another product with no nucleolar organizer. These products are identified as $sc^{8L}sc^{4R}$ and $sc^{4L}sc^{8R}$, respectively. The notation $sc^{8L}sc^{4R}$ indicates that the left part of the chromosome is derived from the sc^8 chromosome and the right part is derived from the sc^4 chromosome.

Ferruccio Ritossa and Sol Spiegelman tested this hypothesis by constructing *Drosophila* strains having different numbers of NO regions per cell. In *Drosophila melanogaster,* the NO is located in the proximal heterochromatin of the X chromosome and in the short arm of the Y chromosome. (In *Drosophila,* the X chromosome is always depicted with the distal region on the left and the proximal region and centromere on the right. We assume this orientation of the chromosome in discussing positions on it.) Several mutant strains of *Drosophila* exist with inversions of the X chromosome in which the left breakpoint is near the scute locus and the right breakpoint is in the proximal region; these inversion mutants are named for the recessive **scute** phenotype they confer. The inversion scute-8 (sc^8) has its right breakpoint between NO and the centromere; the inversion scute-4 (sc^4) has its right breakpoint on the distal side of NO (Figure 14-21). From females heterozygous for sc^4 and sc^8, crossover products can be recovered that carry either two or zero NO regions. Different numbers of NO regions per cell then can be obtained by appropriate genetic combinations (Table 14-2).

By annealing radioactive ribosomal RNA to known amounts of DNA, Ritossa and Spiegelman measured the amount of DNA that hybridized to rRNA. They found a linear relationship between the number of NO regions per cell and the amount of 18S and 28S rRNA that hybridized (Figure 14-22). This result demonstrates that the rRNA loci are located in the NO region. (Subsequently, hybridization in situ has confirmed the NO location of the DNA corresponding to rRNA.) Furthermore, we know the size of the 18S and the 28S rRNA's, and we know the percentage of the total DNA that hybridizes to them. Therefore, we can estimate the number of genes that code for rRNA to be about 200 genes per chromosome. Obviously, such redundancy is one way of ensuring a large amount of rRNA per cell. How-

TABLE 14-2. Genotypes with varying numbers of NO's per cell

Genotype	Number of NO's per cell
$sc^{4L}sc^{8R}/Y\male$	1
$sc^{4L}sc^{8R}/X\female$	1
$X/Y\male$	2
$X/X\female$	2
$sc^{8L}sc^{4R}/X\female$	3
$sc^{8L}sc^{4R}/sc^{8L}sc^{4R}\female$	4

NOTE: here the symbol X indicates a wild-type X chromosome with respect to the scute inversions.

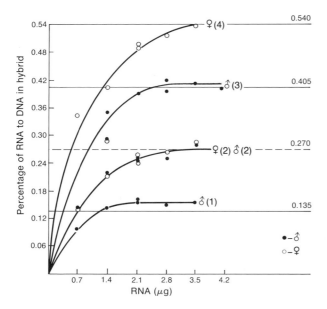

Figure 14-22. The amount of ribosomal RNA (rRNA) that hybridizes to a constant amount of DNA. The plateau is reached when all DNA complementary to the rRNA is hybridized. Each curve is obtained using DNA samples isolated from individuals with a particular number of the nucleolus organizer NO regions per cell (in parentheses). (From F. M. Ritossa and S. Spiegelman, *Proceedings of the National Academy of Sciences USA* 53, 1965, 737.)

ever, this knowledge leads to a new question: how does the cell maintain the genetic equivalence of the multiple copies of each gene? It is likely that unequal crossing-over and gene conversion, described in Chapter 18, serve to maintain gene copy number and to eliminate from the genome point mutations affecting single copies.

Message   The nucleolus organizer (NO) is a region that can be defined cytologically in heterochromatin. It proves to be associated with the DNA segments that code for ribosomal RNA.

Variegated Position Effects.   Some clues about the nature of constitutive heterochromatin come from the study of puzzling **mosaic** phenotypes that sometimes result from chromosome rearrangements. For example, the *white* locus affecting eye color in *Drosophila* is normally found near the left (distal) tip of the X chromosome. The mutant allele w (white eyes) is recessive to the wild-type allele w^+ (red eyes), so that the heterozygote

w^+/w has the wild-type phenotype. Let's use the symbol $R(w^+)$ to represent a chromosome rearrangement (an inversion or translocation) that moves the w^+ allele to a position near the proximal heterochromatin (Figure 14-23a). In some cases, an $R(w^+)/w$ heterozygote shows the expected wild-type phenotype, but many such individuals are **variegated** (with eyes that are mosaics of wild-type and white patches). Some kind of position effect must be involved. Apparently, the heterochromatin causes the inactivation of the wild-type allele in some (but not all) somatic cells.

How far does this effect of the heterochromatin extend into the euchromatin? Let's include a second locus near *white* in our example. The *roughest (rst)* locus affects the surface texture of eye facets. Now we consider the heterozygote $R(w^+ \ rst^+)/wrst$ (Figure 14-23b). Many such heterozygotes do have variegated eyes, but the mosaic patches in the eyes do not exhibit all of the possible phenotypes. Some patches are smooth and red (wild-type), some are roughest and red, and some are roughest and white, but no sectors are ever found to be smooth and white. It would appear that the heterochromatin has a **spreading effect** that moves outward progressively across the adjacent euchromatin. The effect cannot inactivate the w^+ allele without first inactivating the rst^+ allele. The effect can extend quite far from the heterochromatin — to genes that are as much as 60 bands away from the heterochromatin in maps made from the giant (polytene) salivary-gland chromosomes.

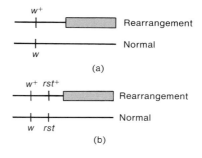

Figure 14-23. Chromosome rearrangements that lead to variegated position effects. (a) In the rearranged chromosome, the allele w^+ is moved to a position near the heterochromatin (colored block). A heterozygote carrying this $R(w^+)$ chromosome and a structurally normal chromosome with the w allele may exhibit the variegated phenotype. (b) The $R(w^+rst^+)$ rearrangement shown at the top can also produce a variegated phenotype when heterozygous with a normal chromosome carrying the *wrst* alleles.

> **Message** Position-effect variegation results when heterochromatin inactivates adjacent euchromatic loci in some somatic cells. This inactivation spreads outward linearly from the heterochromatin through the sequence of genes in the adjacent euchromatin.

In some instances, the mutant tissues in the variegated phenotype appear in patches rather than in a "salt-and-pepper" mixture of individual cells with differing phenotypes. Apparently, all of the cells in a given patch are related by some common event. It is now known that the inactivating effect of the heterochromatin occurs at some early stage in development, so that all daughter cells derived from the affected cell are inactivated.

What effect does the heterochromatin have on the wild-type allele? What determines the stage of development at which this effect is exercised? Why is the effect "permanent" through the divisional cycles of somatic cells, although no fixing of the activity of the allele occurs from generation to generation? These fascinating puzzles about constitutive heterochromatin await molecular explanations.

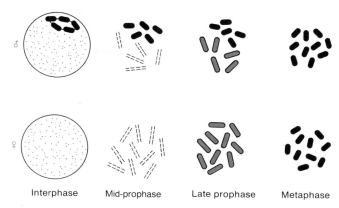

Interphase   Mid-prophase   Late prophase   Metaphase

Figure 14-24. Mitosis in male and female mealybugs (coccids) with a diploid chromosome number of $2n = 10$. In males, five of the chromosomes remain visible during interphase as densely staining heterochromatic elements. During prophase, the other five chromosomes appear less densely stained at first but eventually become heterochromatic. In contrast, none of the chromosomes in the female is visible during interphase; all ten chromosomes in the female behave like the euchromatic set in males. (From S. W. Brown and U. Nur, "Heterochromatic Chromosomes in the Coccids," *Science* 145, 1964, 130–36. Copyright © 1964 by the American Association for the Advancement of Science.)

> **Message** A rearranged chromosome has the potential in each generation to express or not to express its variegating loci. In the somatic cells of an individual, at a specific developmental stage, the potential activity of the loci is determined. After that determination, all daughter cells inherit the loci with their fixed functional states, even if the actual expression of the genetic activity does not occur until days later.

Facultative Heterochromatin

The problems posed by facultative heterochromatin are just as intriguing as those posed by constitutive heterochromatin. We consider here examples drawn from two types of organisms: mealybugs and mammals.

Mealybugs. The insects classified as "true bugs" provide a classic example of the effects of facultative heterochromatin. It has long been known that there is a striking sex difference in chromosome behavior in the scale insects called coccids, or mealybugs. In many coccid species, the diploid ($2n$) number is 10 chromosomes. In females, the chromosomes behave normally, disappearing during interphase and condensing for cell division. In contrast, early in the embryonic development of

males, one chromosome of each pair becomes heterochromatic, and these chromosomes remain visible as a clumped chromocentral mass through interphase (Figure 14-24). After that point in the developmental process, the same chromosomes appear to remain heterochromatic throughout subsequent mitoses.

Chromosome behavior at meiosis also differs between the two sexes. In female coccids, meiosis is normal. However, during spermatogenesis, the first division is equational, with the centromere splitting to allow sister chromatids to separate, and the second division is reductional, with the homologous chromosomes separating (Figure 14-25). This pattern of meiosis in mealybug males is, of course, the opposite of the normal meiotic sequence of divisions. At the second meiotic division in males, the movement of the chromosomes is nonrandom: all of the heterochromatic chromosomes go to one pole, and all of the euchromatic ones go to the other pole. Only the nuclei containing euchromatic elements form functional sperm. What determines which chromosomes become heterochromatic?

In the 1920s, Franz Schrader and Sally Hughes-Schrader described this strange behavior of chromosomes in coccids and suggested that the heterochromatic chromosomes come from the male parent and are genet-

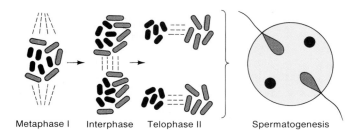

Figure 14-25. Spermatogenesis in the mealybug (coccid). In the male, the first division is equational and the second is reductional—the reverse of the usual meiotic process. In the second meiotic division, all heterochromatic chromosomes go to one pole. Only the euchromatic products form sperm; the heterochromatic products appear as deep-staining residues that slowly degenerate. (From S. W. Brown and U. Nur, "Heterochromatic Chromosomes in the Coccids," *Science* 145, 1964, 130. Copyright © 1964 by the American Association for the Advancement of Science.)

ically inert. Both hypotheses were confirmed in 1957 by Spencer Brown and Walter Nelson-Rees through irradiation studies. In the male offspring of X-irradiated males, radiation-induced chromosome aberrations were found only in the heterochromatic chromosomes. In the male offspring of X-irradiated females, radiation-induced chromosome aberrations were found only in the euchromatic chromosomes. This observation confirmed the paternal origin of the heterochromatic chromosomes.

If the heterochromatic chromosomes are inert, then males are functionally haploid, because only the euchromatic chromosomes are genetically active. Brown and Nelson-Rees applied increasing doses of X rays to male coccids and studied the survival of their offspring. As the dosage was increased, the survival of daughters declined due to the induction of dominant lethal mutations. However, the survival of the sons remained constant, supporting the notion that any mutations induced in the father are not expressed in the active chromosomes of the sons.

This unusual genetic system poses some interesting puzzles. The differing functional states of the chromosomes cannot be determined by assessing the genetic differences between chromosomes. A male's euchromatic chromosomes come from his mother but, in turn, become the heterochromatic chromosomes of his sons. What then controls the heterochromatization of certain chromosomes? Once the male or female origin of chromosomes (and therefore their functional fate in males) is set, each chromosome retains that imprint through subsequent cell generations in the somatic cells of the individual. What event establishes this functional state, and how is it retained through somatic cell divisions? Once

again, a well-documented phenomenon awaits a molecular explanation.

> **Message**   In the male coccid, paternally derived chromosomes remain inactive and heterochromatic in the somatic cells and segregate together into nonfunctional nuclei during spermatogenesis. No molecular explanation is yet known for this pattern of facultative heterochromatin.

Mammals.   Another striking example of facultative heterochromatin is found in mammals. In many of the cells from a mammalian female, the nucleus is characterized by a densely staining heterochromatic element called a **Barr body** (names after its discoverer, Murray Barr). The Barr body contains DNA and is not found in males, so it has been suggested that it may represent an X chromosome. Indeed, the number of Barr bodies is always one less than the number of X chromosomes in the genome (Figure 14-26). Another name commonly applied to the Barr body is **sex chromatin.**

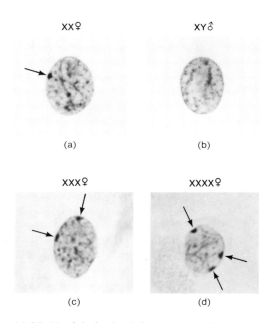

Figure 14-26. Nuclei obtained from cells in the mucous membrane of the human mouth. (a) Nucleus from a female, showing one Barr body (arrow). (b) Nucleus from a male, with no Barr body. (c) Nucleus from an XXX female, showing two Barr bodies. (d) Nucleus from an XXXX female, showing three Barr bodies. (Parts a and b from M. M. Grumbach and M. L. Barr, 14, 1958, 26. Parts c and d courtesy of Dr. M. L. Barr.)

In coccids, the heterochromatic elements are genetically inert. Does the presence of the Barr body indicate that all but one of the female mammal's X chromosomes are inactivated in somatic cells? Such a mechanism would provide an answer to the puzzling problem of how mammals adjust to the presence of twice as many X-linked genes in females as in males, because the X chromosome is known to carry many loci that are necessary for viability. Measurements of enzymes produced by sex-linked genes (glucose 6-phosphate dehydrogenase, for example) show little quantitative difference in enzyme production between the sexes. There must be some **dosage compensation mechanism** to overcome the differences in gene numbers between sexes for X-linked genes, and inactivation of all but one of the female's X chromosomes provides just such a mechanism.

VARIEGATION DUE TO DOSAGE COMPENSATION.   In the late 1950s, Liane Russell obtained a genetic clue to the mode of dosage compensation in mice through studies on mutations after radiation. She irradiated wild-type male mice and mated them to females homozygous for several recessive autosomal coat-color mutations. Any F_1 individual exhibiting a mutant coat color would be presumed to carry a radiation-induced mutant allele of one of the loci. Among the F_1 progeny, Russell recovered several females that exhibited a variegated phenotype, with patches of mutant and wild-type fur. She testcrossed these variegated females and recovered two types of male progeny: completely mutant or completely wild-type. Testcrossing the wild-type male progeny, she obtained completely mutant males and variegated females. Figure 14-27 outlines the crosses and their out-

comes. The last testcross shows that the phenotype is sex-linked in males. What is going on?

Russell and Jean Bangham found that variegation results from a translocation between the chromosome carrying the wild-type color-coat allele and the X chromosome. Apparently, in some cells of females heterozygous for the translocation, the translocated wild-type allele does not function in the production of pigment in fur (so that the cells are mutant). In other cells, the X-autosome part does produce wild-type gene product. In males, on the other hand, the translocated wild-type allele always functions to produce normal fur color in all cells (Figure 14-28). Again it appears that some mechanism is inactivating genes on one X chromosome in the female. The variegation suggests that inactivation occurs during development and that it may involve random selection of one of the X chromosomes, so that different patches show inactivation of different X chromosomes.

X-CHROMOSOME INACTIVATION.   In 1961, Liane Russell and Mary Lyon independently noted that in addition to variegating chromosome rearrangements, many sex-linked point mutations in mice and humans exhibit a variegated phenotype in heterozygous females. They therefore suggested that dosage compensation in mammals may occur through the inactivation of one of the female's two X chromosomes, thereby producing a functional equivalence of X-chromosome genes between males and females.

This hypothesis can be tested in humans by examining cells from females heterozygous for + and − alleles of the X-linked **glucose 6-phosphate dehydrogenase**

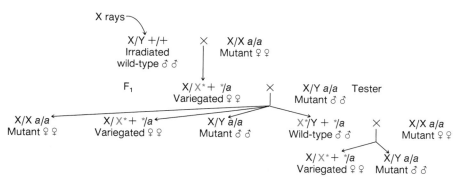

Figure 14-27. Induction of variegating chromosomes in mice. After irradiation, wild-type males are mated to females homozygous for a recessive autosomal marker (*a*). Some F_1 females carrying the irradiated chromosomes (indicated by asterisks) are variegated for the autosomal recessive. Such females are testcrossed, and the offspring show segregation for the mutant and variegating phenotypes in females and for the mutant and wild-type phenotypes in males. Testcrosses of the F_2 wild-type males show that they still transmit the variegating gene, which behaves like a sex-linked locus. These experiments indicate that the radiation must have induced a translocation linking the X chromosome to the autosomal locus *a*.

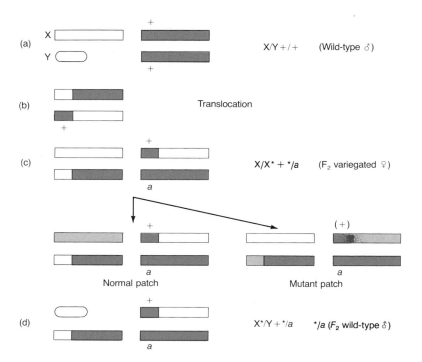

Figure 14-28. A model to explain variegation through an X-autosome translocation. (a) The X/Y+/+ genotype of the parental wild-type males before irradiation. (b) The translocation between X and the autosome induced by irradiation. (c) The X/X* + */a genotype in some F₁ females can produce variegation if one or the other X chromosome is inactivated randomly in various somatic cell lines during development. In cells in which the X chromosome is inactivated (gray), the daughter cells will produce a normal patch because the +/a genotype yields a wild phenotype. In cells in which the X* translocation is inactivated, the + allele may be inactivated due to its proximity to the inactivated X heterochromatin; the daughter cells of this cell will have an a genotype and will exhibit the mutant phenotype. If the inactivation occurs once at some stage of development and then is inherited through all following somatic cell divisions, variegated patches will develop. (d) In the X*/Y + */a males of the F₂ generation, there is no X inactivation, so the +/a genotype leads to a wild phenotype.

(G-6-PD) locus that produce normal and inactive enzymes, respectively. When isolated cells from hair follicles are examined, each cell contains either the normal amount of G-6-PD or no enzyme at all. Also, two alleles that produce electrophoretically distinct enzymes, F and S, can be monitored. When isolated cells, often taken from a skin biopsy, are cloned, each clone contains either the F or the S form but never both forms. These observations tell us a number of things:

1. The G-6-PD locus and, by inference, most or all of the loci on *one* of the female's X chromosomes are inactive.

2. The X chromosome received from either the mother or the father can be the one inactivated.

3. Samples taken from a single hair follicle are mosaic, containing both phenotypes. Therefore, the inacti-

vation occurs at a stage when there are several prospective hair-follicle cells. (It is now estimated that X-chromosome inactivation occurs at about the 20-cell stage.) Figure 14-29 outlines mosaicism observed in hair-follicle cells.

4. Once the inactivation occurs, the state is inherited somatically, because a clone of cells is uniform in the expression of the same allele.

Message   In a mammalian female at some stage in development, either of the X chromosomes in a somatic cell is inactivated by heterochromatization. After this process, which is normally irreversible, all subsequent daughter cells inherit the same inactive X chromosome.

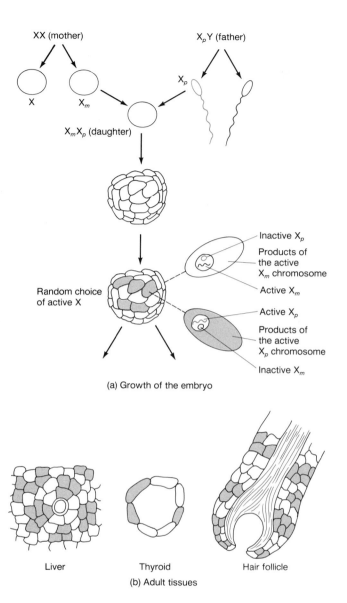

(a) Growth of the embryo

Inactive X_p

Products of the active X_m chromosome

Active X_m

Active X_p

Products of the active X_p chromosome

Inactive X_m

XX (mother)

X_pY (father)

X

X_m

X_p

X_mX_p (daughter)

Random choice of active X

Liver

Thyroid

Hair follicle

(b) Adult tissues

Figure 14-29. (a) Female mosaicism arising from X-chromosome inactivation. A female child has two X chromosomes. This child is the result of the fusion of an egg bearing one of her mother's two X chromosomes with a sperm bearing her father's X chromosome. (A male child has an X and a Y chromosome and results from the fusion of the X-bearing egg with a sperm bearing his father's Y chromosome). Early in the development of the female embryo, when the fertilized egg has multiplied to probably about 20 fetal cells (the precise moment is not known), the process of X inactivation occurs. Each cell selects one of its two X chromosomes for inactivation; the choice is apparently made at random. As a result, some of the cells are left with an active paternal X chromosome (X_p); the others are left with an active maternal chromosome (X_m). Because the choice is never reversed, each female is therefore a mosaic of two kinds of cell. This mosaicism is demonstrated whenever some abnormality (mutation) of one of the two X chromosomes happens to lead to the presence of some detectably abnormal gene product in all the cells in which the abnormal X is active. (b) Demonstration of the female mosaicism in certain adult tissues. (From J. Cairns, *Cancer: Science and Society.* Copyright © 1978 by W. H. Freeman and Company.)

A good illustration of X inactivation can be seen with tortoise-shell cats, which are mosaics with different-colored patches of fur. Because the clonal descendants of cells tend to adhere to one another, the fur in certain mammals (as well as the skin in humans) will be constituted of patches of cells, each derived from a single cell. Due to inactivation, female cats heterozygous for black-coat and yellow-coat traits (C^B/C^Y) will have patched coats (Figure 14-30).

Another very dramatic demonstration of X inactivation is seen in human females who are heterozygous for an abnormality termed X-linked anhidrotic ectodermal dysplasia. The mutant allele causes the absence of sweat glands, and mutant sectors can be detected by the

Figure 14-30. Tortoise-shell cat. An example of a genetic mosaic caused by dosage compensation. Lighter areas are orange; darker areas black. These cats are virtually always female and heterozygous for a pair of X-linked alleles: a dominant black (say, *o*) and a recessive orange (say, *O*). In such cats, orginally X°X° females, one chromosome is inactivated in each cell.

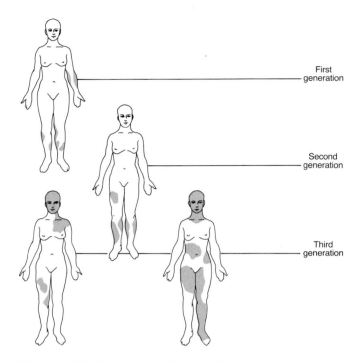

Figure 14-31. Somatic mosaicism in three generations of females heterozygous for sex-linked anhidrotic ectodermal dysplasia. This abnormality in sweat-gland secretion can be demonstrated with a harmless dye. The location of the mutant tissue is determined by chance, but each female does exhibit the characteristic mosaic expression of a single X chromosome.

altered electrical resistance of the skin or by the effects of various sprays. Figure 14-31 shows the phenotypic effects in three generations of women.

SEX DETERMINATION. What activates the dosage compensation in mammalian females? Is it the femaleness (two X chromosomes) or the lack of maleness (absence of a Y chromosome)? To answer this question, we must first understand how sex is determined.

In 1959, William Welshons and William Russell reported a study of a dominant X-linked mutation, *Tabby*

(Ta), that produces a dark coat color in mice. The *Ta/Ta* females have the same dark-furred Ta phenotype as the *Ta/Y* males, but the *Ta/+* females have patches of light and dark fur. Wild-type *+/+* females and *+/Y* males are a uniform light color. Mating *Ta/+* females with *+/Y* males, Welshons and Russell recovered female progeny that were phenotypically similar to *Ta/Ta* females. (You can see that if disjunction is normal, all females should be *Ta/+* or *+/+*.) If sex determination is like that in *Drosophila*, these exceptions probably are *Ta/Ta/Y* (XXY) nondisjunctional females. Welshons and Russell then mated these Ta exceptions with *+/Y* males. The progeny phenotypes were wild-type females, patched females, and Ta males in a 1 : 1 : 1 ratio. If the females were indeed *Ta/Ta/Y*, this cross should have produced the progeny genotypes shown in Table 14-3.

The presence of wild-type females and the absence of Ta females among the progeny obviously show that the assumption of *Ta/Ta/Y* for the parental female is wrong. Welshons suggested that the exceptional Ta female might have only a single X chromosome and no Y; we write the genotype as *Ta/O*. Such a female would produce *Ta* and O eggs, which would result in *Ta/+* and *+/O* female offspring in the cross. Welshons concluded that the Y chromosome in mice determines maleness and its absence determines femaleness. This hypothesis has been confirmed cytologically by the finding that the exceptional Ta females had only a single X chromosome and genetically by the discovery of XXY males.

In humans also, XO and XXY individuals are now recognized. Females who are XO are not completely normal, indicating that some activity of the second X is required at some time or in some tissues during female development. In fact, the short end of the X chromosome is not inactive in Barr bodies. The XO phenotype is called **Turner's syndrome** and includes underdeveloped sex organs, webbing of the neck, and short stature. Cells from females with Turner's syndrome lack the Barr body. An abnormal male genotype, XXY, produces a recognizable phenotype called **Klinefelter's syndrome,** which includes long legs, small testes, and the development of breasts. A Barr body is present in cells

TABLE 14-3. Progeny expected if the exceptional Ta females are XXY

			Sperm	
			+	Y
Eggs	Normal	*Ta*	*Ta/+* ♂	*Ta/Y* ♂
		Ta/Y	Ta/+/Y ♀	Ta/Y/Y ♂
	Nondisjunctional	*Ta/Ta*	?	*Ta/Ta/Y* ♀
		Y	*+/Y* ♂	Y/Y lethal

TABLE 14-4. Correlation between numbers of sex chromosomes and Barr bodies

Sex-chromosome constitution		Number of Barr bodies
Males	Females	
XY, XYY	XO	0
XXY, XXYY, XXYYY	XX	1
XXXY	XXX	2
XXXXY	XXXX	3
XXXXXY	XXXXX	4

from males with Klinefelter's syndrome. The activity of the short end of the X chromosome in the Barr body is responsible for the observed phenotype. Obviously, phenotypic sex is not the factor that determines the formation of a heterochromatic X chromosome (a Barr body). Rather, it is the *number* of X chromosomes that determines whether facultative heterochromatization will occur. This is confirmed through the recognition of individuals who carry several sex chromosomes (many of these individuals are mosaics for different numbers of sex chromosomes), as shown in Table 14-4.

Message X-chromosome inactivation in mammals is determined by the number of X chromosomes per nucleus, not by the phenotypic sex of the individual. Phenotypic sex is determined by the presence or absence of a Y chromosome.

Here is another fascinating phenomenon for which we have no molecular explanation as yet: how does a cell recognize the presence of multiple X chromosomes and ensure that only one X functions?

Dosage Compensation in Drosophila

Heterochromatization inactivates a large genetic segment, but this is a rather crude method of controlling gene expression. Are there modulations of gene expression within a more restricted portion of a chromosome? In fact, dosage compensation was originally defined in regard to the observation in *Drosophila* that some alleles of the *white* locus produce the same eye phenotype in males and females, whereas other alleles produce more eye pigment in females than in males. The former alleles are said to be dosage-compensated; the excess dosage of X-linked genes in the female is not reflected in the phe-

notype. The deletion of the *Notch* gene on the X chromosome produces a mutant phenotype of nicked wings in females heterozygous for the deletion and a normal chromosome, proving that both X chromosomes do function in a normal *Drosophila* female.

In 1965, Ed Grell demonstrated the compensatory capabilities of sex-linked loci, using two loci that affect the enzyme xanthine dehydrogenase (XDH). He constructed strains carrying different numbers of the wild-type alleles of the sex-linked locus maroon-like *(ma-l)* and the chromosome-3 gene rosy *(ry)*. On measuring the relative amounts of enzymatic activity in these *Drosophila* individuals, Grell obtained the results summarized in Table 14-5. Obviously, there is a dosage-compensation for the activity of *ma-l*$^+$, whereas the number of *ry*$^+$ genes is directly reflected in enzyme activity.

In a single cell, is only one allele active or do both alleles function? In *Drosophila*, various alleles of a single gene produce different electrophoretic mobilities of the enzyme 6-phosphogluconate dehydrogenase (6-PGD). Electrophoresis of the enzyme from heterozygous females reveals a hybrid band between the two parental bands. The hybrid band can be explained by assuming simultaneous activity of both alleles, with dimer formation by polypeptides from the two loci. Thus, in *Drosophila* females — unlike mammals — both X chromosomes function in all cells.

Measurement of enzymes controlled by X-linked loci shows that each dosage-compensated locus in a male produces twice as much product as each locus in a female. In different X-autosome translocations, the loci lying in the X portions remain compensated, whereas the autosomal loci are not compensated. These results suggest that no single region or small number of regions control X activity. Rather, each locus or small region apparently has information about the number of gene copies present and is able to control its own activity accordingly. No satisfactory molecular explanation of this method of gene regulation yet exists.

TABLE 14-5. Relative enzymatic activities and numbers of gene copies

Number of gene copies	Relative enzymatic activity for	
	ma-l$^+$	*ry*$^+$
1	1.0	0.5
2	1.0	1.0
3	1.0	1.5

Summary

■ Each eukaryotic chromosome apparently represents a single continuous DNA molecule extending from one end through the centromere to the other end. The molecule of DNA may be 10^5 times as long as the chromosome itself, and it is apparently packaged in a very regular and complex fashion within the chromosome. The packaging begins with the nucleosome—an aggregate of four pairs of histones around which the DNA is coiled twice. Chains of nucleosomes are then coiled, the coils in turn are coiled, and so on.

In eukaryotic chromosomes, differences in gene activity can be recognized by the degree of compaction, which is indicated by staining intensity. Densely staining heterochromatin represents nonactive material. Within a chromosome, heterochromatic regions are typically found at the tips, around the centromere, and around the nucleolus organizer. Such constitutive heterochromatin is generally highly redundant, is relatively devoid of sequences for which RNA transcripts may be found, and contains sequences of satellite DNA.

In some cases, a chromosome may exist in one of two states: functionally active or functionally inactive. Constitutive heterochromatin represents entire chromosomes or specific segments that are predictably heterochromatized and inactive during the cell cycle. In other cases, a chromosome or segment may or may not become heterochromatin; when it does, it is referred to as facultative heterochromatin. Once the functional state of the chromosome is determined, that state is replicated from cell to cell in further reproduction. Early in the development of a mammalian female, one of the two X chromosomes in each cell is inactivated (becoming a heterochromatic Barr body), thereby rendering the cell functionally equivalent to a male cell (which contains only one X chromosome). In contrast, organisms like *Drosophila* compensate for the double dose of sex-linked genes in females (compared to males) by a 50 percent reduction in gene activity on the female X chromosome.

■ ■ ■

Solved Problems

1. A cancer tumor arises from a single abnormal cell. Based on Figure 14-29, draw a diagram which shows the growth of a single family of abnormal cells. Then explain the difference between these cells and the normal surrounding cells.

    Solution

    Whereas the cells in the normal tissue of females are mosaic, as shown in (a) as a mixture of small families of white and shaded cells, tumors are expanding families of cells derived from a single parent cell and are not mosaic. Each cancer is a single family, as shown in color in (b).

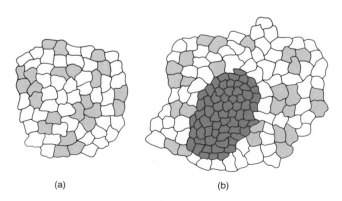

(a)   (b)

(Figure from J. Cairns, *Cancer: Science and Society.* Copyright © 1978 by W. H. Freeman and Company.)

2. Describe how the DNA double helix becomes packed into the metaphase chromosome.

    Solution

    The five stages of packing can be visualized by referring to Figure 14-15. First, about 140 base pairs of DNA are wrapped around a core of histones. The nucleosomes are then thrown into a coil termed a solenoid structure, with a diameter of close to 300 Å. Additional coiling of the coiled solenoid structure results in the chromosome, which is visible in the light microscope.

Problems

1. Which of the following is *not* an advantage resulting from having genes arranged in chromosomes rather than floating free?

    a. Favorable combinations of genes can be inherited more or less as a unit.

    b. Functionally related genes can be controlled in a coordinated fashion.

    c. DNA can pass through the nuclear pore into the cytoplasm.

    d. Orderly separation of genetic information can occur at cell division.

    e. Replication of DNA can occur in orderly fashion.

2   The mealybug, *Planococcus citri*, has a diploid number of 10. The meiotic divisions in *P. citri* are thought to be

reversed—that is, an equational division is followed by a reductional division. In the tetrads formed at meiosis I, it is difficult to distinguish sister chromatids from their paired homologs because they form a tightly paired unit. Sharat Chandra obtained a triploid strain of *P. citri.* What would you expect to observe at the end of meiosis I and II in this strain if: a. the division sequence is conventional? b. the division sequence is reversed?

3. Considering the length of each DNA molecule in a eukaryote and the levels of complexity in its packaged form in a chromosome, the mechanism of pairing between homologous sequences becomes staggering. How could pairing between chromosomes possibly take place so that specific base pairs are properly aligned?

4. In 1917, Alfred Sturtevant noted that crossing-over in any chromosome pair in a *Drosophila* female is increased if the female is heterozygous for an inversion in some different pair. This is called the "interchromosomal effect" on crossing-over. It has since been shown that an inversion of one chromosome always increases crossing-over in the others. Devise a hypothesis to explain this effect. Design an experimental test of your hypothesis.

5. In 1956, R. Alexander Brink noted some aberrant results from crosses with corn involving alleles of the *R* locus, which affects the color pattern in seeds. The alleles studied (and their corresponding phenotypes) were r^r (colorless seed), R^r (dark purple seed), and R^{st} (stippled seed, having irregular spots on a light background). In the cross $R^r r^r \times r^r r^r$, Brink obtained 50 percent dark and 50 percent colorless seeds. In the cross $R^{st} r^r \times r^r r^r$, he obtained 50 percent stippled and 50 percent colorless seeds. In the cross $R^r R^{st} \times r^r r^r$, he obtained 50 percent stippled and 50 percent "weakly colored" seeds. Apparently, in the $R^r R^{st}$ heterozygote, the R^r alleles change to a new and stable allelic state labeled R'. This effect is called **paramutation.** How does paramutation differ from mutation? Devise a hypothesis to explain the results. Design an experimental test of your hypothesis.

6. In embryos of the fungus gnat, *Sciara,* after the seventh cell division, the paternally inherited X chromosomes in all somatic cells migrate through the nuclear membrane and disintegrate in the cytoplasm. Helen Crouse studied the properties of a series of X-autosome translocations with the following X-chromosome breakpoints:

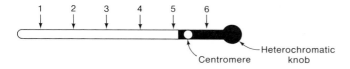

She found that, when males contributed the translocations to the offspring, the translocated portion carrying the X centromere in rearrangements 1 through 5 was lost during embryonic development. However, with translocation 6, the autosomal centromere carrying the right arm of the X was lost. Suggest an interpretation of these results.

7. Erich Wolff has studied the polytene chromosomes of two strains of a European midge, *Phryne cincta.* One strain (Berlin) has no heterochromatic elements on the X chromosome, whereas another strain (alpine) has two heterochromatic sites at which large amounts of DNA are produced and released into a nucleolus-like structure.

   a. Design experiments to study the nature of the DNA in the heterochromatic sites. (NOTE: the strains mate with each other, and fertile F_1 progeny are obtained.)

   b. Wolff also has observed that some *Phryne* strains carry up to seven supernumary chromosomes that are not present in the normal genome. How would you study the DNA in these supernumary chromosomes?

8. When DNA from *Drosophila* is isolated from salivary-gland nuclei and from testes and spun in a CsCl gradient, the following distributions are obtained. What does this result suggest?

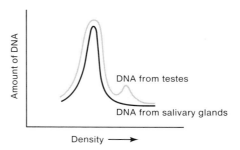

9. Joseph Gall and Mary Lou Pardue recognized that polytene chromosomes are like columns of DNA laid out in proper sequence. They developed a method for denaturing the DNA in these chromosomes without disrupting the spatial organization of the chromosomes. By incubating radioactive RNA on the denatured chromosomes and then removing the unbound RNA, they were able to localize the RNA/DNA hybrids by autoradiography. (This technique is described in more detail in Chapter 15.)

   a. What parameters would affect the resolving power of this technique—that is, the accuracy with which two adjacent sites can be distinguished?

   b. How would you demonstrate that such RNA binding does indeed reflect complementary sequences of DNA?

   c. What do you infer when a certain RNA binds to a certain locus?

10. Highly redundant DNA sequences in *Drosophila* are concentrated in the heterochromatin around all of the centromeres. However, in relation to their DNA content, these same regions have a disproportionately small number of loci that can be detected genetically. For that reason, Theophilus Painter and Hermann J. Müller suggested that this proximal heterochromatin is "inert." Furthermore, very little crossing-over occurs within the heterochromatin. What biological function might be filled by these redundant sequences? Your hypothesis should explain the absence of detectable loci and the nonrandom distribution of the heterochromatin. (NOTE: this is an unsolved problem; no one knows the correct answer, although there are many hypotheses.)

11. You are a molecular biologist on a field trip in the jungles of Surinam, where you discover a purple tree frog with yellow stripes. Struck by its beauty, you capture several dozen males and females and smuggle them back to your laboratory. They breed easily and profusely, and you decide to use them in an investigation of the organization of the ribosomal RNA genes. Design experiments to carry out this investigation without the need for genetic crosses.

12. A class of mutations in *Drosophila* includes loci on all four chromosomes. These mutants are called *Minute*, and they are characterized by similar properties: recessive lethality, a dominant effect on bristles (which become slender and short) and development time (which is delayed by several days), and a Minute phenotype in the presence of a deficiency for any of the loci. There are at least 50 to 60 *Minute* loci. Speculate on the possible biological function of *Minute* genes. How would you study a *Minute* gene from a genetic point of view and from a molecular point of view?

Manipulation of DNA

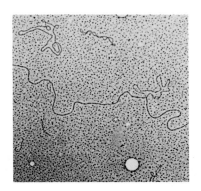

KEY CONCEPTS

The specificity of complementary base-sequence recognition
makes DNA accessible to manipulation.
▪

Restriction enzymes cleave DNA at specific sequences,
facilitating the joining of nonhomologous DNA's from
different sources.
▪

When DNA fragments are inserted into small elements
capable of replication, the segment of DNA can be selectively
amplified in certain hosts.
▪

It is now possible to rapidly determine the nucleotide
sequence of segments of DNA.
▪

Genes can be altered at specific sites by in-vitro mutagenesis
techniques.
▪

Recombinant DNA technology permits the production of
human proteins in bacteria, the early detection of genetic
diseases, and the generation of transgenic animals and plants.

■ We have concentrated on genetic experiments until this point in the text. However, we are now in the midst of a new era in genetics, made possible by the techniques of molecular genetics and the uses of recombinant DNA. Many of the standard methods of genetic analysis have been replaced by biochemical and DNA-sequence analyses that were inconceivable before the early 1970s. In fact, the whole approach to the study of the genetic makeup of higher organisms has been altered by recombinant-DNA techniques. Avenues of study have been opened up to the molecular geneticist at a level of resolution never dreamed possible just a decade before. Let's look at the new arsenal of methods at the disposal of the modern molecular biologist and at the principles on which they are based and see how these techniques are being used to understand heredity, evolution, gene control, and human diseases.

It is interesting to note that many of the operational criteria used in molecular biology are identical to the criteria used by geneticists working at the level of organisms, as shown in Table 15-1.

This chapter presents experiments that do not require genetic crosses but that nevertheless involve similar processes of manipulation. First, we examine the manipulation of DNA before recombinant DNA techniques were available; then we consider more advanced technologies. The chapter is divided into the following topics:

1. The power of base complementarity

2. The discovery of restriction enzymes

3. The formation of recombinant DNA

4. Recombinant DNA methodology

5. Applications of recombinant DNA technology using prokaryotes

6. Recombinant DNA technology in eukaryotes

7. Genetic diseases

8. Recombinant DNA and social responsibility

The Power of Base Complementarity

A significant amount of recombinant DNA technology is based on the recognition properties of complementary base pairs in DNA. Let's examine how base complementarity has played a role in DNA manipulation.

Denaturation of DNA by Heat

The hydrogen bonds that link the complementary base pairs in DNA are disrupted by higher temperatures. Because a G–C pair has three hydrogen bonds, whereas an A–T pair has only two, the **melting point** of a DNA sample (defined as the temperature at which one-half of the DNA sample is no longer hydrogen bonded) is directly proportional to the G–C content (Figure 15-1). DNA with higher G–C content will melt at higher temperatures. Thus, it is possible to separate molecules with different melting points and hence with different G–C contents.

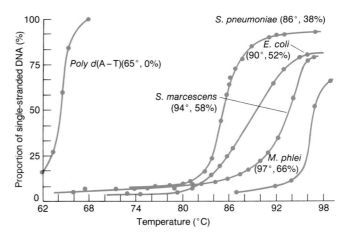

Figure 15-1. Denaturation of DNA from different sources at various temperatures. In parentheses for each sample, the first value is the melting temperature (at which one-half of the DNA molecules have been denatured) and the second value is the percentage of base pairs in the sample that are G–C pairs. There is an obvious relationship between G–C content and melting temperature. (From J. Marmur and P. Doty, *Nature* 183, 1959, 1427.)

TABLE 15-1. Criteria used in the study of molecules and in the study of whole organisms

1. Recognizable differences (comparable to phenotypes) must exist, so that the genetic molecules under study can be distinguished from one another.

2. Techniques must permit the study of individual molecules or groups of molecules (follow the fate of individuals or particular groups) over time.

3. Quantitative measures of the molecules involved (comparable to progeny or frequency counts) must be made.

4. Ways to mix and even to hybridize the molecules in different combinations (equivalent to crosses in organisms) must be devised.

5. Controls for comparison and standardization of results must be established.

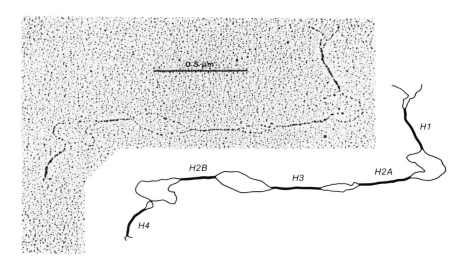

Figure 15-2. Electron micrograph of a partially denatured (at 61°C) DNA molecule from the histone genes of the sea urchin, with a drawing showing the sequence of the histone genes (*H*), separated from one another by the denatured AT-rich spacers. (From R. Portmann and M. Birnstiel, *Nature* 246, 1976, 31.)

Message The relative G–C content of DNA can be inferred from its melting temperature. DNA with a relatively higher content of A–T is denatured (melted) at a lower temperature than is DNA with a higher content of G–C.

Within a DNA molecule, the proportion of G–C versus A–T pairs may vary along the length of the molecule. Partial denaturation can be produced by a hydrogen-bond-breaking agent, such as high temperature, and the denatured regions within the molecule can then be mapped.

The positions and sizes of the melted regions can be reproducibly identified in a population of identical molecules, indicating that such maps do reflect the base composition within the molecules. In 1976, David Wolstenholme used the heat denaturation of AT-rich regions to map the mitochondrial DNA's of different *Drosophila* species.

The histone genes provide another example of differing melting points. Histones are the basic proteins found in association with nuclear DNA, as described in Chapter 14. They are rich in arginine, whose codons are relatively rich in G–C pairs, so the histone genes should have a greater density than other parts of the chromosome. This density differential permits isolation of the histone genes. Furthermore, the histone genes should resist heat denaturation. Indeed, the segments of DNA that code for five histones in sea urchins occur in a single region, with each gene separated from the next by an AT-rich spacer that denatures at lower temperatures (Figure 15-2).

Message Regions within DNA that are relatively rich in A–T pairs denature at lower temperatures than the surrounding regions. Denaturation maps show the effects of these intramolecular differences in melting temperatures.

Reassociation of Complementary Single Strands

In completely denatured DNA, all strands are single. These single strands can reassociate by random collisions that permit matching between complementary sequences. The specific matching of strands after reannealing can be demonstrated by testing the biological activity of reformed duplexes in bacterial transformation.

An important use of reassociation is the physical mapping of deletions or of regions of nucleotide differences in heteroduplexes formed by DNA from different sources. For example, the position and size of a deletion can be determined by annealing DNA from a deletion mutant with DNA from a wild-type, as shown in Figure 15-3. Figure 15-4 shows an electron micrograph of such a heteroduplex. In the same way, hybridization of DNA sequences from sources that are related evolutionarily

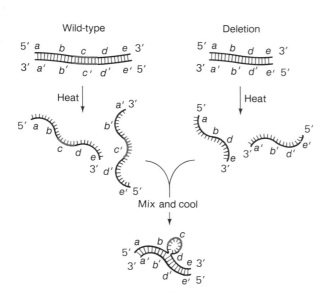

Figure 15-3. Formation of a heteroduplex with deletion and normal DNA.

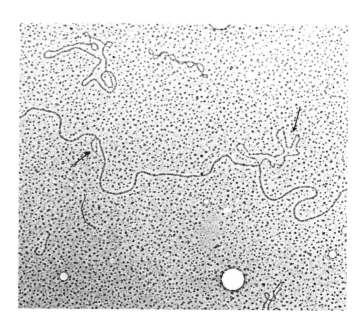

Figure 15-4. A heteroduplex of deletion and wild-type DNA of phage T4. Two different deletion loops are visible (arrows). (From T Homyk.)

provides a visual indication of the extent of divergence in the nucleotide sequences.

Message Strand matching by base complementarity is the property that makes DNA accessible to manipulation. The remarkable specificity of complementary base-sequence recognition provides a powerful tool for actually comparing base sequences in heteroduplexes.

Locating DNA Sequences on Chromosomes

Can the base sequences of a purified nucleic acid be used to locate those sequences of DNA within a chromosome that are complementary to them? In the late 1960s, Mary Lou Pardue and Joseph Gall made use of the giant chromosomes of *Drosophila* salivary glands to answer this question. Because the DNA is duplicated several hundredfold in the polytene chromosome, specific nucleic acids have many potential pairing partners at a complementary site. Suppose that we wish to locate the genes coding for an RNA.

The chromosomes are initially squashed and fixed on slides. Then the DNA is denatured with a mild alkaline treatment that breaks hydrogen bonds without liberating the DNA from the chromosome. The single

strands are prevented from renaturing by treatment with formamide, which combines with free amino groups and inhibits duplex re-formation. Now radioactive RNA is incubated on the chromosomes. After sufficient time for hybridization, the excess label is washed off, single-stranded RNA is removed by treatment with ribonuclease (which breaks single-stranded RNA into nucleotides while leaving untouched any RNA that is double-stranded or paired to DNA), and a photographic emulsion is placed onto the chromosomes to locate DNA-RNA hybrids in an autoradiograph.

In this way, it was possible to locate the DNA that codes for 18S and 28S ribosomal RNA (rRNA) in the chromocenter (Figure 15-5). The locus for 5S rRNA is at 56EF in the right arm of chromosome 2. Several sites for transfer RNA (tRNA) genes have now been identified by localization on chromosomes in situ. RNA made from highly repetitive sequences of DNA (by a procedure we consider later in this chapter) has been used to show the centromere position of this redundant DNA in mammalian chromosomes (Figure 15-6).

Message Chromosome positions of the DNA complementary to specific RNA's can be identified by hybridization to chromosomes in situ.

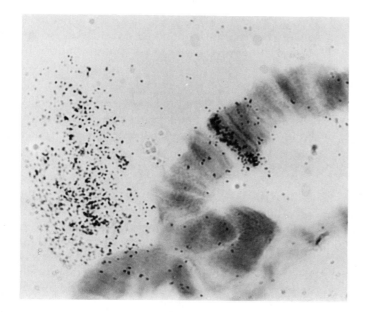

Figure 15-5. Autoradiograph showing ribosomal RNA binding to the chromocenter and 56EF in the right arm of chromosome 2. (From T. Grigliatti et al., *Cold Spring Harbor Symposia on Quantitative Biology* 38, 1974, 461. Cold Spring Harbor Laboratory, Cold Spring Harbor, N.Y.)

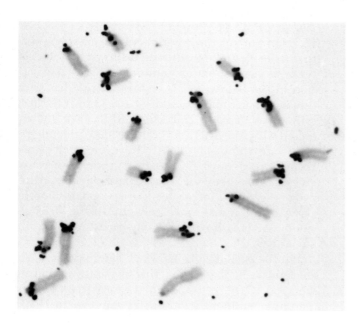

Figure 15-6. Autoradiograph of mouse metaphase chromosomes hybridized in situ with radioactive RNA made from highly redundant mouse DNA. The RNA sequences seem to be present in all chromosomes and to be concentrated in the region immediately adjacent to the centromere. (From M. L. Pardue and J. Gall, *Chromosomes Today,* 3, 1971, 47.)

Isolation of Specific DNA Sequences

In order to probe the properties of DNA in relation to specific functions, we must separate specific classes or segments of DNA from the rest. Several techniques are now available to accomplish this task:

1. One of the earliest techniques was the separation of a rapidly annealing portion of mammalian DNA that represents highly redundant sequences from the more slowly annealing unique segments present only in single copies per genome. This separation is accomplished on nitrocellulose filters or columns of hydroxyapatite. The amount of highly repetitive sequences varies greatly, ranging from 0 percent in many prokaryotes to 30 or 40 percent in some eukaryotes. The highly repetitive sequences involve repetitions of a short basic unit. For example, in the guinea pig, the repeating unit is

2. The highly redundant sequence is a repetition of a short unit, so that the ratio GC to AT in such segments may deviate considerably from the ratio for the bulk of the organism's DNA. Because the equilibrium position of DNA in a cesium chloride (CsCl) gradient is determined by its GC content, DNA fragments with different buoyant densities can be identified as satellites of the main DNA band in the gradient (Figure 15-7). The satellite DNA may be more or less dense than the main band.

These two methods provide DNA fractions with physical properties that distinguish them from the main DNA component. The challenge then is to try to find biological functions for these fractions. But are there ways of purifying DNA sequences with functions that are already known—sequences that therefore are of specific interest? Such methodologies do exist, and they can provide much useful information.

3. A very effective method for separating small plasmids (episomes) from the bacterial chromosomes was developed in 1967 by Jerome Vinograd. The molecule ethidium (Figure 15-8) can insert itself between bases of a DNA molecule if the DNA is sufficiently flexible to untwist a bit, thereby providing space for the ethidium to squeeze in, or intercalate. When DNA has ethidium bound to it, this complex has a density less than that of the pure DNA. Vinograd realized that a bacterial chromosome is so long

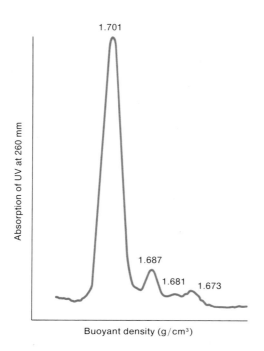

Figure 15-7. Distribution of *Drosophila melanogaster* DNA in a cesium chloride density gradient that separates DNA molecules of differing G–C content. The DNA is prepared in such a manner that it is fragmented into approximately 25-kb segments. The bulk of the DNA appears at 1.701 g/cm³ in the gradient, but there are several satellite bands with a lower buoyant density. (From S. A. Endow, M. L. Polan, and J. G. Gall, *Journal of Molecular Biology* 96, 1975, 670. Copyright © 1975 by Academic Press, Inc., London, Ltd.)

that mechanical stress will break the DNA when it is isolated, thus producing linear DNA molecules that do allow ethidium intercalation. In contrast, smaller circular molecules, such as the fertility factor F, mitochondrial DNA, or chloroplast DNA, remain intact on isolation and have a restricted ability to untwist. Hence, when DNA is isolated from bacterial or eukaryotic cells, the large linear molecules bind much more ethidium than the small circular molecules, which remain more dense. In a CsCl gradient, the circular molecules are readily purified and col-

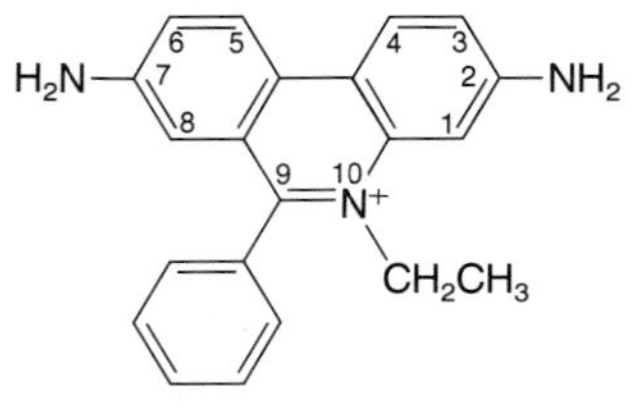

Figure 15-8. The structure of ethidium. This chemical can be used to separate the DNA of *E. coli* from the DNA of its episomes.

lected (Figure 15-9). As we shall see, episomes have played a key role in the manipulation of genes, so the purification of such molecules is an important technique.

4. Another method of separating specific DNA sequences utilizes the ability of some proteins to recognize such sequences. For example, the enzyme RNA polymerase that is responsible for transcription must initiate the process by attachment to specific regions at the beginnings of genes. After a preparation of DNA is exposed to RNA polymerases, the mixture can be digested with a nuclease that breaks down the DNA chains. Those sequences to which the enzyme is bound are sheltered physically from the nucleolytic action, so they can be recovered intact after the nuclease treatment. In principle, any DNA sequence that is bound by a specific protein can be recovered in this manner. This is the principle that permits the purification of the beginning of the *lacZ* gene, which is regulated by the *lac* repressor protein (see Chapter 16). The DNA segment protected by the repressor is 24 base pairs in length.

5. An important tool for the isolation of specific DNA sequences has resulted from a study of tumor-causing viruses. Thus far, we have assumed a unidirectional relationship between DNA, RNA, and protein: DNA directs the synthesis of RNA, which in turn specifies protein production (shown as DNA → RNA → protein). The ability of RNA to function both as the hereditary information and as the molecule to be translated is shown by the existence of viruses that contain only RNA. An exception to the DNA → RNA + protein rule has been discovered in animal tumor viruses known to carry only RNA. Normal cells cultured in vitro and infected with the virus are "transformed" phenotypically into tumor-like cells. However, such transformed cells normally do not produce infective viruses. It has been suggested that the genetic material of the virus is integrated like a prophage into the DNA of the transformed cells. But if the virus genome is RNA, how can it integrate into chromosomal DNA?

In the early 1960s, Howard Temin proposed a mechanism by suggesting that the information in the viral RNA is "transcribed" into DNA that is then inserted into the host cell's genome. In 1970, Temin and David Baltimore independently discovered an enzyme that catalyzes this step; it is called **reverse transcriptase.** The existence of reverse transcriptase requires an amendment of the central dogma to DNA ⇌ RNA → protein and provides a tool for making DNA that complements any RNA that can be purified. Cell biologists have long recog-

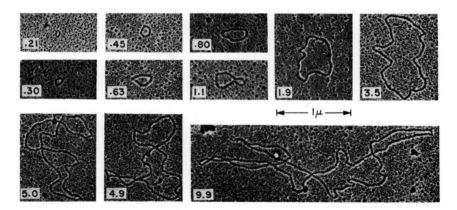

Figure 15-9. Electron micrographs of circular DNA from cultured human cells. The number in each micrograph is the length of the molecule in microns. These may be similar to episomes isolated from bacterial cells. (From R. Radloff, W. Bauer, and J. Vinograd. *Proceedings of the National Academy of Sciences USA* 57, 1967, 1514).

nized that cells specialized to carry out a specific function typically contain large quantities of a single gene product. As examples, the silk-producing cells in the glands of a silkworm produce large quantities of silk protein (fibroin), cells in the mammalian pancreas excrete insulin, red blood cells (erythrocytes) are rich in hemoglobin, cells in chick oviducts make a great deal of ovalbumin, and so on. Isolation of RNA from such specialized cells provides a sample that is enriched for a specific messenger RNA (mRNA). Thus, pure hemoglobin mRNA can readily be prepared from immature red blood cells (reticulocytes). With reverse transcriptase, cDNA (DNA complementary to the mRNA) is formed in vitro from the mRNA. The cDNA can be subjected to sequence analysis, or it can be used to recover complementary sequences from total cellular DNA or from preparations of RNA by the formation of hybrids. When the cDNA is hybridized with RNA, the DNA-RNA hybrids reveal the existence of large RNA transcripts, which are precursors of the active mRNA.

6. The recombinant DNA techniques, described in the following section, allow the isolation of many specific DNA sequences.

> **Message** There are several techniques for isolating specific classes of DNA from a heterogeneous population of molecules. This isolation allows us to focus on a restricted part of the genome, just as geneticists use various techniques to isolate specific loci for genetic analysis.

The Discovery of Restriction Enzymes

Restriction enzymes play a crucial role in the creation and analysis of **recombinant DNA,** or DNA resulting from the fusion of two different DNA segments. Let's review the discovery of these important tools.

The Phenomenon of Host Restriction

We have already seen that a bacterial genotype determines the cell's susceptibility to infection by various phages. Similarly, a phage's genotype determines the range of bacterial cells that it can infect. Clearly, both host and parasite have evolved genetic strategies for coping with each other.

Let's consider the infective capacity of phages grown in bacteria of different genotypes. For example, suppose that a hypothetical phage X is found in two bacterial strains: A and B. We'll label phages from these two sources X-A and X-B, respectively. Suppose that we use X-A to infect strain B and X-B to infect strain A. We find that X-B phages do poorly at infecting A cells, whereas X-A phages infect both strains equally well (Figure 15-10). When the few X-B phages that are recovered from A are used to infect strain A, we find that they are *fully infective!* Perhaps some A-infecting mutants already existed in the X-B population, and selection has now isolated them. However, when the phages are allowed to infect strain B again and then are returned to strain A, we find that they are once again poorly infective in strain A. We have not isolated an A-infecting pure-breeding strain of X-B phages.

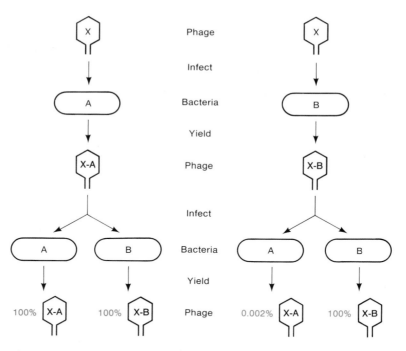

Figure 15-10. Infectivity of hypothetical phage X from two different bacterial strains, A and B. Phages derived from strain B are poorly infective on strain A, whereas phages derived from strain A infect both strains equally well. The percentages indicate the efficiency of plaque formation on each host.

Apparently, the host range of the phage depends on the bacterial strain in which it matured, not on the phage's genotype. In other words, the host genotype seems to modify the phage particle without altering the DNA sequence in the phage.

DNA Modification

But what restricts phage X-B from multiplying in strain A? The mature phage is able to attach to the bacterial cell wall and to inject its DNA into the cell. The factor that prevents it from reproducing is the nucleolytic activity of a host enzyme (nuclease) that breaks the phage DNA into a number of noninfective fragments. The host cell seems to have a defense system that destroys unfamiliar DNA! But what protects the host DNA and the infective X-A phage DNA from this nuclease? The answer to this question reveals the nature of the phage-modification process.

Enzymes exist that modify specific bases at specific sequences *without altering the coding properties of those bases.* For example, in certain bacterial strains, cytosine and adenine are altered by the addition of a methyl group to the forms shown in Figure 15-11. In the 1960s, Werner Arber was able to show that the alteration of

cytosine and adenine by the addition of a methyl group (a process called methylation) protects the DNA from nuclease activity. Arber demonstrated that mutations involving the replacement of a normally methylated base by a base that cannot be methylated results in a loss of protection from the nuclease action. The total number of bases that are methylated is very small, so these particular bases must occupy very specific regions that are attacked by the nucleases. In our example, bacterium A evidently has the capacity to modify DNA, whereas B does not.

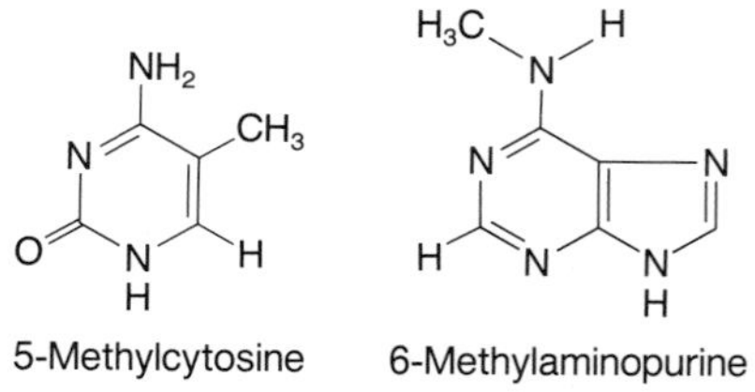

5-Methylcytosine   6-Methylaminopurine

Figure 15-11. Forms of cytosine and adenine that have been modified by the addition of a methyl group. Such methylation does not change the coding capacity of the DNA but does protect the DNA from attack by some highly specific nucleases.

The phenomenon of host restriction is due to the action of nucleases that degrade any DNA not specifically modified for protection from the nucleases of a given host cell.

> **Message** Two interrelated processes are involved in the phenomenon of host restriction: the **restriction** of phage multiplication through the enzymic destruction of invading viral DNA, and the **modification** of phage DNA to render it immune to the effects of restriction enzymes.

Specificity of Restriction Enzymes

The next step in understanding restriction phenomena did not come until 1970, when Hamilton Smith made great progress in his studies of the restriction enzyme from *Haemophilus influenzae*. This enzyme, called HindII, cuts T7 phage DNA into 40 specific fragments. What do the cleavage sites have in common? To find out, Smith took the mixture of fragments and marked the 5' ends by attaching the radioisotope ^{32}P. He then used hydrolyzing enzymes to break the labeled fragments into still smaller pieces (Figure 15-12). He was able to separate the labeled end fragments in small segments only a few nucleotides long, so that he could determine the base sequence of each fragment.

Smith found that the label was always attached to an adenine or a guanine (the purines). Fragments with labeled adenine were always A – A or A – A – C; fragments with labeled guanine were always G – A or G – A – C. Smith suggested that such fragments could be produced if the enzyme cleaves only at the points indicated by the arrows in the following specific sequence (where Py represents a pyrimidine and Pu represents a purine):

$$\downarrow$$
$$5' \quad -G-T-Py-Pu-A-C- \quad 3'$$
$$\bullet$$
$$3' \quad -C-A-Pu-Py-T-G- \quad 5'$$
$$\uparrow$$

Note the rotational symmetry of this sequence: a rotation of 180° around the dot in the center leaves the sequence unchanged. Verify that cleavage at the arrows will produce just the 5'-end fragments that Smith identified: Pu – A – and Pu – A – C –. HindII apparently is very specific in identifying the sequence at which it will cleave. Furthermore, Smith showed that the host-induced methylation that confers protection from HindII

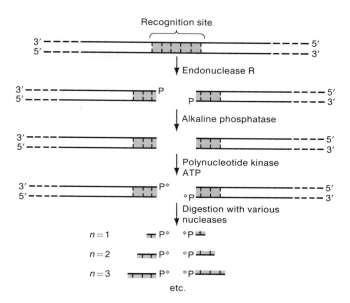

Figure 15-12. Smith's method for identifying the bases at the *HindII* recognition site. The T7 DNA is cleaved by HindII (endonuclease R), which leaves a phosphate on the 5' end of the cleaved strands. This phosphate is removed with alkaline phosphatase, and then polynucleotide kinase is used to catalyze the attachment to the 5' end of a radioactive phosphorus atom in a phosphate group (*P) from labeled ATP. Various nucleases are then used to cleave the fragments into smaller fragments of varying lengths. The labeled fragments are separated, and their base sequences are identified. (From J. T. Kelly, Jr., and H. O. Smith, *Journal of Molecular Biology* 51, 1970, 397. Copyright © 1970 by Academic Press, Inc., London, Ltd.)

activity occurs at this same cleavage site: protection from cleavage by HindII is obtained when the adenines one base away from the cleavage site are methylated (m):

$$5' \quad -G-T-Py-Pu-\overset{m}{A}-C- \quad 3'$$
$$\bullet$$
$$3' \quad -C-A-Pu-Py-T-G- \quad 5'$$
$$\underset{m}{}$$

Studies of other restriction enzymes have yielded similar results. For example, the enzyme EcoRI is produced by a gene on an R plasmid in *E. coli*. EcoRI cleaves the circular DNA of SV40 (a small mammalian virus) at only one site:

$$\downarrow$$
$$5' \quad -G-A-\overset{m}{A}-T-T-C- \quad 3'$$
$$\bullet$$
$$3' \quad -C-T-T-A-A-G- \quad 5'$$
$$\underset{m}{} \quad \uparrow$$

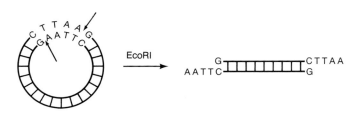

Again the sequence is rotationally symmetrical, but in this case the DNA ring is connected to a linear molecule. The cuts are staggered and therefore produce "sticky ends" (Figure 15-13) like those of linear λ DNA. Once again, methylation of adenines near the cleavage site provides protection against the cleaving action (the methylated bases are indicated in the sequence just displayed, although, of course, *either* methylation *or* cleavage would occur at such a sequence, *not* both). A comparison of the sequences in the two strands for the *EcoRI* site (recall that each is read in the opposite direction) reveals that the sequences are the same. In semantics, a sentence reading the same forward or backward (such as ABLE

Figure 15-13. Cleavage of SV40 DNA by the restriction enzyme EcoRI produces a linear DNA with sticky ends.

TABLE 15-2. Recognition, cleavage, and modification sites for various restriction enzymes

Enzyme	Source organism	Restriction site	Number of cleavage sites in DNA from		
			ϕX174	λ	SV40
EcoRI	*Escherichia coli*	5′ –G–A–A–T–T–C– –C–T–T–A–A–G– 5′	0	5	1
EcoRII	*E. coli*	5′ –G–C–C–T–G–G–C– –C–G–G–A–C–C–G– 5′	2	>35	16
HindII	*Hemophilus influenzae*	5′ –G–T–Py–Pu–A–C– –C–A–Pu–Py–T–G– 5′	13	34	7
HindIII	*H. influenzae*	5′ –A–A–G–C–T–T– –T–T–C–G–A–A– 5′	0	6	6
HaeIII	*H. aegyptius*	5′ –G–G–C–C– –C–C–G–G– 5′	11	>50	19

NOTE: an asterisk (*) is commonly used to indicate methylation sites, rather than the m used here (to avoid confusion with radioactive labeling).

WAS I ERE I SAW ELBA) is called a **palindrome,** and this name is used to describe DNA sequences such as the one in the *EcoRI* site.

These studies have opened an explosively expanding area of research. Dozens of restriction enzymes with different sequence specificities have now been identified, some of which are shown in Table 15-2. Most restriction enzymes recognize specific palindromic sequences. Restriction enzymes are powerful tools for analysis, because they can locate specific base sequences and cut the DNA in a very specific way. Now we are ready to see how we can approach the task of genetic engineering.

> **Message** Restriction enzymes provide a way of cleaving DNA from any source at a specific sequence, thereby producing a heterogeneous population of fragments with identical ends.

Restriction-Enzyme Mapping

The restriction-enzyme target sites can be used as markers for DNA, just as the bubbles of AT-rich melt regions are used. The DNA from a specific source is subjected to successive digestion by different restriction

TABLE 15-2. *(Continued)*

Enzyme	Source organism	Restriction site	Number of cleavage sites in DNA from		
			ϕX174	λ	SV40
HpaII	*H. parainfluenzae*	5′ –C–C–G–G– / –G–G–C–C– 5′	5	>50	1
PstI	*Providencia stuartii*	5′ –C–T–G–C–A–G– / –G–A–C–G–T–C– 5′	1	18	2
SmaI	*Serratia marcescens*	5′ –C–C–C–G–G–G– / –G–G–G–C–C–C– 5′	0	3	0
BamI	*Bacillus amyloliquefaciens*	5′ –G–G–A–T–C–C– / –C–C–T–A–G–G– 5′	0	5	1
BgIII	*B. globiggi*	5′ –A–G–A–T–C–T– / –T–C–T–A–G–A– 5′	0	5	0

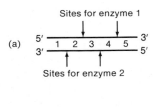

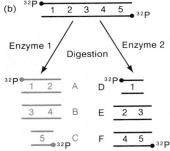

Figure 15-14. (a) A hypothetical DNA with recognition sites for two different restriction enzymes. (b) After labeling of the 5′ ends, digestion of the DNA by enzyme 1 produces fragments A, B, and C, whereas digestion of the DNA by enzyme 2 produces fragments D, E, and F. The fragments can be separated electrophoretically, and the end fragments can be identified by the radioactive labeling.

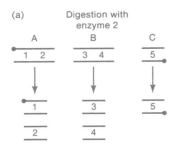

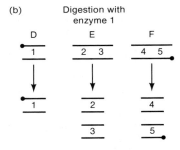

Figure 15-15. (a) Fragments A, B, and C, obtained from the original DNA by digestion with enzyme 1, are now digested with enzyme 2, producing subfragments 1–5. (b) The fragments obtained by digestion with enzyme 2 are now digested with enzyme 1 to produce subfragments 1–5. (c) Comparison of the subfragments (identified by their positions on the electrophoretic gel) obtained from each fragment indicates which fragments must overlap (vertical lines).

enzymes. When the fragments are separated electrophoretically on polyacrylamide gels, the "map" of the restriction sites can be deduced (Figures 15-14, 15-15, and 15-16).

For example, consider a hypothetical DNA molecule having the distribution of restriction sites shown in Figure 15-14a. The 5′ ends can be labeled with ³²P before the DNA is cleaved by each enzyme in separate samples (Figure 15-14b). The resulting fragments are separated on gels. The recovery of three fragments after treatment with each restriction enzyme shows that there are two recognition sites for each enzyme. The absence of radioactivity in fragments B and E shows that these are the center pieces.

The relative positions of the recognition sites can be determined by taking the fragments from one enzyme treatment and treating these fragments with the other enzyme (Figure 15-15). If smaller subfragments are produced by the second treatment, then we know that the original fragments contained sites for the second enzyme. The number of subfragments produced by this double treatment is always one more than the total number of restriction sites (for both enzymes). We now compare the subfragments and determine overlaps. For example, subfragment 2 (Figure 15-15c) is obtained by the further breakdown of both fragments A and E, so we know that A and E must overlap. (The similarity of sub-

fragments in the two experiments is indicated by similar positions on the electrophoretic gel.) Using this kind of reasoning, we can conclude that the overlapping order of fragments in the original DNA must have been DAEBFC.

We know that fragments D and A are both end fragments (due to the radioactive labeling). They share subfragment 1, so we can place subfragment 1 at the end of our map. Fragments A and E share subfragment 2, so we place subfragment 2 next in the map. Proceeding in this way, we can map the subfragments on the original DNA (Figure 15-16). We know that the original treatment with enzyme 2 separated fragments D and E, so we can conclude that a site for enzyme 2 lies between subfragments 1 and 2. Using similar reasoning, we can locate the

Restriction-enzyme sites

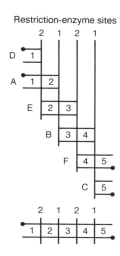

Figure 15-16. Arranging the fragments in the order indicated by their overlapping subfragments, we obtain a map of the restriction-enzyme sites along the DNA *(bottom)*. Compare this map to Figure 15-14a (which, of course, would not be known at the beginning of such an analysis).

other restriction-enzyme sites on the map. The final map at the bottom of Figure 15-16 is, of course, identical to the one shown in Figure 15-14a (which would not have been known in a real experiment). You may find it useful to review this reasoning to be sure you see how the map could be constructed in a real experiment where the order of subfragments is not known initially.

In 1976, Hamilton Smith and Max Birnstiel developed a similar method for locating the restriction-enzyme sites. This method utilizes the ability to distinguish between DNA molecules of differing lengths. Suppose that we have a circular DNA molecule with a single *EcoRI* site, so that the molecule can be converted into a linear rod by treatment with EcoRI (Figure 15-17). Several samples are each partially digested (using concentrations such that around one out of every 50 sites is actually recognized and cleaved) by one of a series of test enzymes, producing different populations of DNA fragments of varying lengths. These fragments can be separated on gels, where they migrate in order of increasing length. By reading across the gels from the bottom up, we can obtain the order of cleavage sites. A map of enzyme sites can now be prepared. Furthermore, the distances that the various fragments migrate in the gels provide a measurement of their sizes, so that the relative distances between cleavage sites can also be determined. Restriction-enzyme maps prepared in this way can be very detailed (Figure 15-18). With such maps, specific segments can be identified and separated for further analysis.

Message The relative order of restriction-enzyme cleavage sites on a DNA molecule can be determined, providing a new kind of chromosome map. Furthermore, the number of nucleotides between any two sites can be estimated from the size of the fragments obtained in the analysis.

The Formation of Recombinant DNA

Let's first look at the larger picture and ask what recombinant DNA is, how it is created, and why it is so important to molecular biology. The complexity of eukaryotic genomes makes a detailed analysis of gene structure and gene regulation extremely difficult. However, the ability to break up chromosomes into small fragments, to isolate and analyze these fragments individually, and even to manipulate them, opens up many areas of study. This is one of the goals of recombinant DNA research: to isolate distinct fragments of DNA and to recombine them into a smaller genome much more amenable to analysis than the original genome from which the DNA fragments came. There are a number of methods for creating such **recombinant DNA molecules,** as we shall see later. Usually, the vehicle into which the DNA fragment has been incorporated is capable of selective replication at some stage, thus allowing the amplification of specific segments of DNA. The increased amounts of DNA of the segment under study greatly facilitate detailed molecular analysis. Bacteria are frequently employed as the host for the maintenance and amplification of recombinant DNA molecules derived from bacteria or higher organisms. In the process by which a specific segment of DNA is selectively amplified in a bacterial cell, termed **cloning** (Figure 15-19), a piece of foreign DNA is inserted into a small plasmid in vitro. The plasmid is then introduced by transformation into a host *E. coli* cell, where it can be maintained by selecting for an antibiotic-resistance gene present on the plasmid.

There are several methods for creating recombinant DNA molecules. Each exploits the properties of different enzymes.

Direct Generation of Sticky Ends

The restriction enzymes described in the preceding sections can cleave DNA from any source at a specific sequence. In the most favorable cases, sticky ends are generated that then anneal to the sticky ends generated in other DNA molecules (the vectors) cleaved by the same restriction enzyme. Recall the cloning procedure depicted schematically in Figure 15-19. Now we can con-

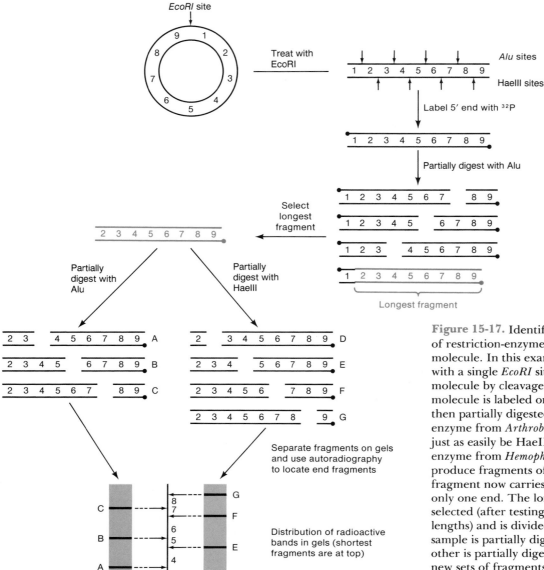

Figure 15-17. Identification of the sequence of restriction-enzyme sites in a circular DNA molecule. In this example, the ring of DNA with a single *EcoRI* site is opened into a linear molecule by cleavage with EcoRI. The molecule is labeled on its 5′ ends with ³²P and then partially digested with Alu (restriction enzyme from *Arthrobacter lutens,* which could just as easily be HaeIII, the restriction enzyme from *Hemophilus eregyptius*) to produce fragments of varying lengths. Each fragment now carries the radioactive label at only one end. The longest fragment is selected (after testing on a gel to compare lengths) and is divided into two samples. One sample is partially digested with Alu, and the other is partially digested with HaeIII. The new sets of fragments are separated on gels. The location of the labeled fragment can be determined by autoradiography of the gels. The sequence of restriction sites can now be read in order as the two gels are compared.

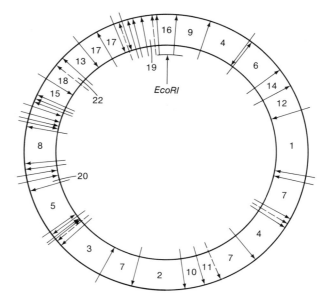

Figure 15-18. Restriction-enzyme cleavage map of SV40 phage DNA. Arrows pointing outward indicate sites for Alu, and arrows pointing inward indicate sites for HaeIII. (From Yang et al., *European Journal of Biochemistry* 61, 1976, 119).

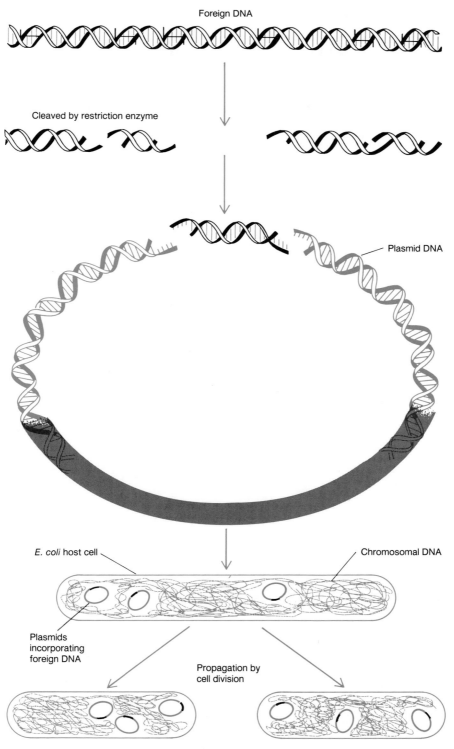

Foreign DNA

Cleaved by restriction enzyme

Plasmid DNA

E. coli host cell

Chromosomal DNA

Plasmids
incorporating
foreign DNA

Propagation by
cell division

Figure 15-19. Cloning recombinant-DNA technique that makes it possible to introduce nucleotide sequences deliberately from the DNA of one strain or species of organism into the DNA of another. The DNA of the "foreign" organism is first fragmented in one of a number of ways. In the example shown here, restriction enzymes (described in the text) are used to fragment the foreign DNA and also to cut the circular plasmid DNA. After the foreign DNA has been recombined with the plasmid DNA, the circular form of the plasmid can be restored, and the structure can be inserted into a suitable host. (From C. Grobstein, "The Recombinant DNA Debate." Copyright © 1977 by Scientific American, Inc. All rights reserved.)

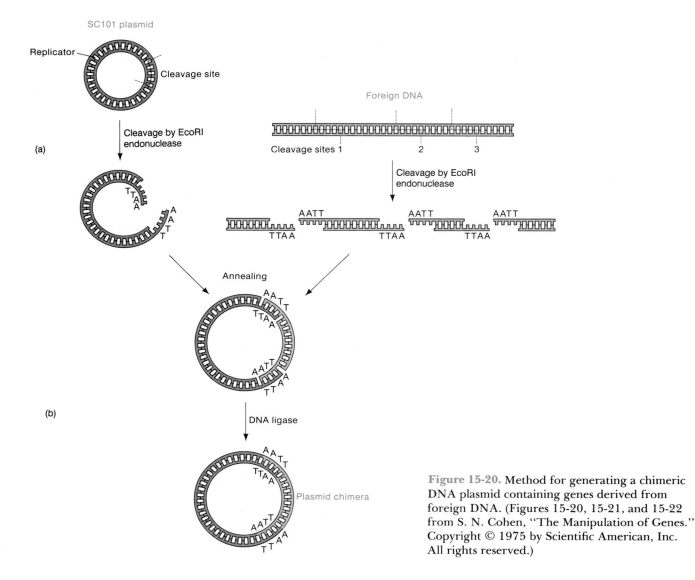

Figure 15-20. Method for generating a chimeric DNA plasmid containing genes derived from foreign DNA. (Figures 15-20, 15-21, and 15-22 from S. N. Cohen, "The Manipulation of Genes." Copyright © 1975 by Scientific American, Inc. All rights reserved.)

sider the formation of the recombinant plasmid in more detail, as shown in Figure 15-20a. In this example, the plasmid called SC101 carries one *EcoRI* restriction site, so digestion with the restriction enzyme EcoRI converts the circular plasmid DNA to a linear molecule with single-stranded sticky ends. DNA from any other source (say, *Drosophila*) can be treated with EcoRI enzyme to produce a population of fragments carrying the same sticky sequence at the fragment ends. When the two populations are mixed, the DNA's from the two sources can combine as duplexes form between their sticky ends. The annealed fragments can then be linked permanently by the enzyme DNA ligase to form a plasmid chimera (Figure 15-20b).

Tailing

The enzyme **terminal transferase** from calf thymus can be used to generate sticky ends on DNA fragments. This enzyme catalyzes the addition of nucleotide "tails" to the 3′ ends of DNA chains. Therefore, if dA (deoxyadenine) residues are added to one DNA fragment (say, the vector) and dT residues are added to the other fragment (in this case, the fragment to be inserted into the vector), then the two fragments can anneal and subsequently be joined by DNA ligase (Figure 15-21). Alternatively, poly-G and poly-C tails can be added to the respective fragments at the 3′ ends.

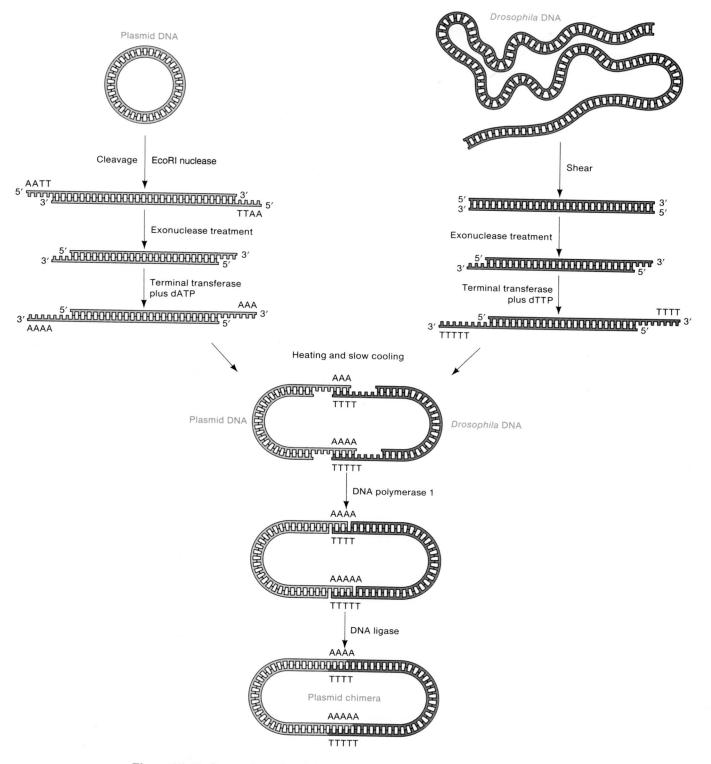

Figure 15-21. Generation of a sticky end by terminal transferase can produce complementary sequences that will allow recombination between sources but will prevent the combination of DNA's from the same source.

Blunt-end Ligation

Under appropriate conditions, T4 DNA ligase can join fragments containing blunt ends. Although less useful for direct cloning, this reaction permits the addition of synthetically prepared DNA duplexes that contain restriction sites. These synthetic DNA **linkers** are very useful, since they allow the creation of restriction sites that are then cleaved with the appropriate enzyme to generate a fragment containing cohesive ends. This procedure facilitates the cloning of fragments containing blunt ends.

Vectors

Cloning vehicles, which are used to enable the replication of cloned fragments, are termed **vectors.** The following general features are common to the most frequently employed vectors:

1. A small, well-characterized molecule

2. An origin of replication within the molecule enabling replication of itself and of the inserted fragment

3. Easy recovery of the hybrid molecule

    There are numerous vectors in current usage.

Plasmids

Plasmids are small, circular DNA molecules that can be introduced into cells by transformation (see Chapter 10), since bacteria like *E. coli* can be made permeable to DNA by treatment with calcium chloride (Figure 15-22). Plasmids can almost be considered minichromosomes that replicate autonomously in the cell. They offer several advantages as vectors, the most important of which is the ability to confer antibiotic resistance to the host cell. This allows direct selection for cells that receive and maintain recombinant DNA plasmids. Different antibiotic-resistance genes can be introduced into plasmids to facilitate these selections. For instance, the commonly used plasmid PBR322 carries the genes that confer resistance to ampicillin and tetracycline. Other commonly used antibiotic-resistance genes are those for chloramphenicol and kanamycin. Plasmids also allow the amplification of cloned DNA; some plasmids are present in 20 to 50 copies during cell growth, and after the arrest of protein synthesis, as many as 1000 copies per cell of a plasmid can be generated.

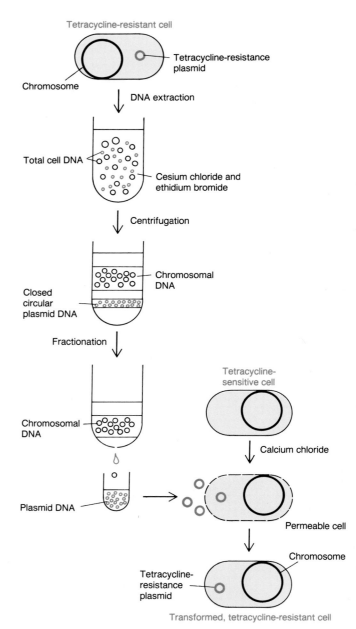

Figure 15-22. Plasmid DNA can be introduced into a bacterial cell by the procedure called transformation. Plasmids carrying genes for resistance to the antibiotic tetracycline *(top left)* are separated from bacterial chromosomal DNA. Because differential binding of ethidium bromide by the two DNA species makes the circular plasmid DNA denser than the chromosomal DNA, the plasmids form a distinct band on centrifugation in a cesium chloride gradient and can be separated *(bottom left)*.

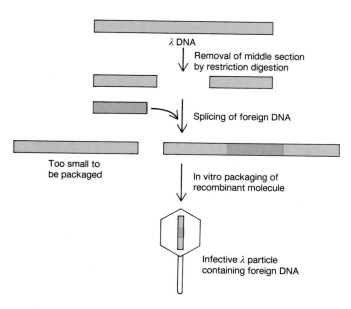

λ DNA

Removal of middle section by restriction digestion

Splicing of foreign DNA

Too small to be packaged

In vitro packaging of recombinant molecule

Infective λ particle containing foreign DNA

Figure 15-23. Use of λ phages as cloning vectors. Altered λ phages containing conveniently situated restriction sites are used to allow selection for a 45-kb packaged molecule containing an insert. (After L. Stryer, *Biochemistry*, 3d ed. Copyright © 1988 by W. H. Freeman and Company.)

Phage λ

Derivatives of phage λ (lambda) are convenient for cloning larger fragments of DNA in the 15 to 20 kilobase (kb; kb = 1000 bases) range. Continuously replicating plasmids that contain large inserts of foreign DNA tend to lose the insert because selection favors smaller plasmids. But phage-λ heads will routinely package a DNA molecule about 45 kb in length. This property can be used to select for inserts of DNA that replace a central portion of the phage genome that is not required for replication of λ in *E. coli* (Figure 15-23).

Single-Stranded Phage

Single-stranded phages can be used as cloning vectors; on infection of *E. coli,* the single infecting strand is converted to a double-stranded replicative form, which can be isolated and used for cloning. The advantages of such phages as cloning vehicles is that on reinfection of suitable hosts, the phage particles that are generated are single stranded and can serve as ready substrates for the rapid dideoxy DNA sequencing technique developed by Fred Sanger (page 411). M13 is the phage most widely used for this purpose.

Expression Vectors

Vectors have been constructed for specific purposes. One set of vectors, termed **expression vectors,** affords the opportunity to express the cloned sequences by fusing them to the appropriate transcription and translation start signals. This allows the foreign DNA to be expressed in the respective host. (We consider aspects of some of these vectors later.) For instance, some vectors have restriction sites located just next to the control region for the *lac* genes, which has been spliced into the vector. This allows expression of the cloned gene by using the *lac* transcription and translation start signals and, in some cases, even allows the expression to be controlled by the *lac* repressor. (See Chapter 16 for a description of gene control by the *lac* repressor.) Similarly, certain plasmids that allow expression in yeast are being used. Some vectors are bifunctional, allowing expression in two different hosts. For example, a certain plasmid contains the origin of replication of the plasmid PBR322 and of the animal virus SV40. These origins allow replication of *E. coli* and some cultured mammalian cell lines, respectively. These plasmids are called **shuttle vectors,** because they can transfer genes back and forth from one type of cell to another.

Recombinant DNA Methodology

Cloning Strategies

Cloning genes involves attaching the desired DNA fragment to the appropriate vector molecule and then propagating the recombinant DNA molecule in a suitable host. We have already examined methods for generating recombinant DNA molecules, as well as different vector systems. However, these techniques must still be used in conjunction with strategies for isolating and recognizing the gene or genetic region to be cloned. As molecular geneticists attempt to analyze very complex eukaryotic genomes, sophisticated methods and strategies become very important.

There are two basic approaches to cloning a specific gene. The first approach involves cloning all the fragments from a restriction digest nonselectively and then screening for the desired gene. The nonselective method for cloning random DNA fragments from a higher organism into bacteria is called **shotgunning.** The entire collection represents a **gene bank,** or **gene library,** which can now be screened for specific genes (Figure 15-24). All of the plasmids or phages comprising a gene library together represent the entire genome of an organism, with each different plasmid or phage carrying a different small fragment from the genome.

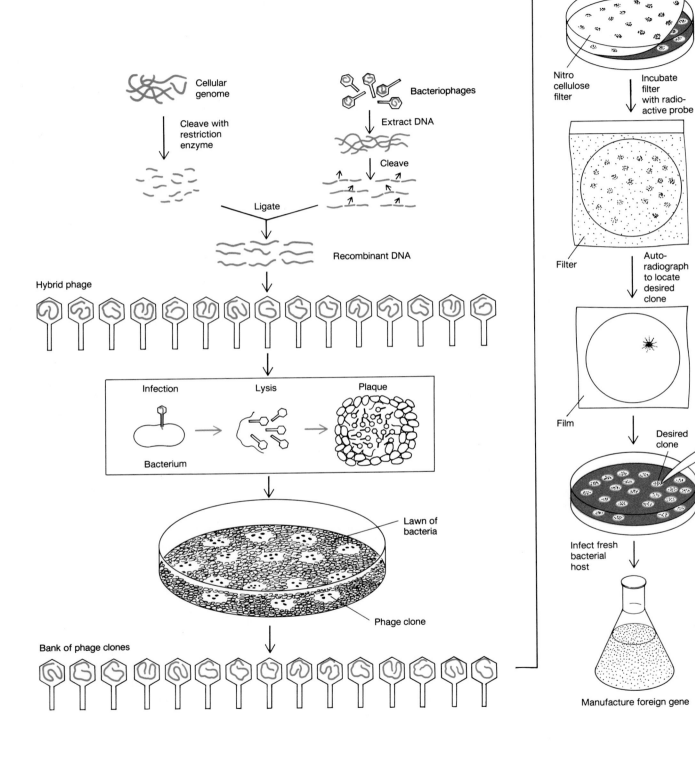

Figure 15-24. (a) A genomic library can be established by cloning genes in λ bacteriophages. When a lawn of bacteria on a Petri plate is infected by a large number of different hybrid phages, each phage in the lawn is inhabited by a single clone of phages descended from the original infecting phage. Each clone carries a different fragment of cellular DNA. The problem now is to identify the clone carrying a particular gene of interest by probing the clones with DNA or RNA known to be related to the desired gene. The plaque pattern is transferred to a nitrocellulose filter and the phage protein is dissolved, leaving the recombinant DNA. (b) The filter is incubated with a radioactivity labeled probe: a DNA copy of the messenger RNA representing the desired gene. The probe hybridizes with any recombinant DNA incorporating a matching DNA sequence, and the position of the clone having the DNA is revealed by autoradiography. Now the desired clone can be selected from the culture medium and transferred to a fresh bacterial host, so that a pure gene can be manufactured. (After R. A. Weinberg, "A Molecular Basis of Cancer," and P. Leder, "The Genetics of Antibody Diversity." Copyright © 1983, 1982 by Scientific American, Inc. All rights reserved.)

How can we determine whether our gene bank is large enough to contain any unique sequence of interest with a high degree of certainty? Louise Clark and John Carbon, who first constructed a gene bank representing the entire *E. coli* genome, have applied the following formula to tabulate the relevant probabilities:

$$P = 1 - (1-f)^N \quad \text{or} \quad N = \frac{\ln(1-P)}{\ln(1-f)}$$

where P equals the probability that a given unique DNA sequence is present in a collection of N transformant colonies and f is the fraction of the total genome used as a source for each fragment. Clearly, f will depend on the size of the fragments and the complexity of the genome. Table 15-3 depicts the results from several genomes at several levels of probability.

This formula applies to cases in which the length x of the DNA sequence being sought is small in relation to the length L of the cloned DNA fragments. Otherwise, a corrected value of f or f' could be used, where $f' = [1 - (x/L)](f)$.

The second approach to cloning involves first using a purified probe for the gene in question, to select the appropriate restriction fragment from the entire collection of fragments, and then cloning that specific fragment.

Detection of Cloned Genes

Southern Blotting. One of the most valuable techniques for identifying cloned genes is **Southern blotting** (Figure 15-25), developed by E. M. Southern. This technique exploits the properties of gel electrophoresis and nitrocellulose filters. Single-stranded but not double-stranded DNA can stick to nitrocellulose. During electrophoresis on agarose gels, DNA fragments from restriction digests will migrate according to their size. They can be transferred to nitrocellulose by a buffer flow, after denaturation, and immobilized on the filters in a pattern that mirrors their positions in the agarose gel. ^{32}P-labeled DNA or RNA probes are then used for hybridization with the affixed DNA on the filters. The probes are denatured to allow annealing with the single-stranded restriction fragments that are anchored on the

TABLE 15-3. Gene bank sizes N needed to contain a particular hybrid plasmid transformant at various probability levels

DNA source	Average size of DNA fragment cloned (dal)	Gene bank size N (number of colonies)		
		$P = 0.90$	$P = 0.95$	$P = 0.99$
E. coli	8.5×10^6	720	940	1440
Yeast	1×10^7	2300	3000	4600
Drosophila	1×10^7	23,000	30,000	46,000

Calculations are based on the formula $P = 1 - (1-f)^N$ and the assumptions that each transformant colony in the bank arises from an independent transformation event and that each hybrid molecule transforms with the same efficiency. (From Louise Clarke and John Carbon, *Cell* 9, 1976, 91–99.)

ple, messenger RNA for the gene in question can be used as a probe, it is difficult to obtain sufficiently pure mRNA. However, the enzyme reverse transcriptase can be used to make a DNA copy of the partially purified mRNA's, as described earlier (page 392), which in turn can program the synthesis of a complementary strand (see Figure 15-26). The resulting DNA duplex is termed **complementary DNA (cDNA).** It is relatively easy to

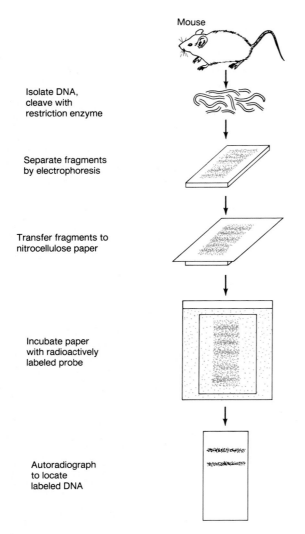

Figure 15-25. The blotting technique developed by E. M. Southern, described in the text. (From P. Leder, "The Genetics of Antibody Diversity." Copyright © 1982 by Scientific American, Inc. All rights reserved.)

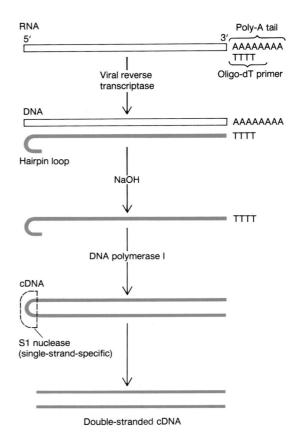

Mouse

Isolate DNA, cleave with restriction enzyme

Separate fragments by electrophoresis

Transfer fragments to nitrocellulose paper

Incubate paper with radioactively labeled probe

Autoradiograph to locate labeled DNA

RNA
5′
3′
Poly-A tail
AAAAAAAA
TTTT
Oligo-dT primer

Viral reverse transcriptase

DNA
AAAAAAAA
TTTT

Hairpin loop

NaOH

TTTT

DNA polymerase I

cDNA

S1 nuclease (single-strand-specific)

Double-stranded cDNA

Figure 15-26. The synthesis of double-stranded cDNA from mRNA. A short oligo-dT chain is hybridized to the poly-A tail of an mRNA strand. The oligo-T segment serves as a primer for the action of reverse transcriptase, which uses the mRNA as a template for the synthesis of a complementary DNA strand. The resulting cDNA ends in a hairpin loop. Once the mRNA strand is degraded by treatment with NaOH, the hairpin loop becomes a primer for DNA polymerase I, which completes the paired DNA strand. The loop is then cleaved by S1 nuclease to produce a double-stranded cDNA molecule. (Figures 15-26 and 15-27 from J. D. Watson, J. Tooze, and D. T. Kurtz, *Recombinant DNA: A Short Course.* Copyright © 1983 by W. H. Freeman and Company.)

gel. Unlabeled single-stranded DNA from an unrelated source is used to saturate the remaining sites on the nitrocellulose to prevent nonspecific binding of the single-stranded probe, which can now bind to the filter only by annealing to complementary DNA fragments. The position of bands that anneal to the probes is revealed by autoradiography.

cDNA. Among the most efficient probes for screening for cloned genes are **cDNA clones.** Although, in princi-

purify individual cDNA fragments. The total population of cDNA fragments can be joined to plasmids so that each plasmid receives only one fragment, and bacterial cells can be transformed so that each cell receives only one plasmid. Clones derived from single cells can be tested for which unique cDNA is present. Amplified DNA is purified and denatured and then used to trap the corresponding mRNA's on a nitrocellulose filter. The mRNA purified in this manner is used to program a cell-free protein synthesis system to allow identification of the protein encoded in the complementary mRNA and thus the gene present in the cloned mRNA. The purified cDNA can itself be used as a probe to screen colonies directly from a gene bank, since replica plating techniques have been developed that employ nitrocellulose filters to detect hybridization of DNA from colonies with specific probes. Then all of the cloned fragments containing parts of the gene sequence used in the probe can be detected. When partial digests are used to generate the restriction fragments that are used to establish the gene bank, there will be fragments of different sizes containing the gene in question and the larger region surrounding the sequences contained in the probe can therefore be analyzed.

Isolation of cDNA.　Some RNA's are easily purified because of their particular size (rRNA, tRNA, 5S RNA), thus facilitating the cloning of the genes encoding these RNA's. Other RNA's are obtainable by virtue of their abundant expression at a specific time or in a particular tissue (for example, the mRNA that encodes the silk protein, fibroin, in the glands of the silkworm). Also, in some cases, the gene product is relatively conserved in evolution, so that isolation of a gene from one organism makes possible its isolation from another, as exemplified by globins, histones, and actin. Sometimes the respective mRNA is rare and difficult to enrich. One technique exploits the possibilities of synthetic oligonucleotides. In cases in which part or all of the amino-acid sequence of the protein encoded by the gene in question has been determined, a series of 15 to 20-base-long oligonucleotides is synthesized, corresponding to each of the possible DNA coding sequences predicted by a stretch of the protein sequence. These oligonucleotides are then used as probes in the manner described above.

It should be stressed that cDNA clones are copies of mature mRNA, whereas genomic clones represent the actual structure of a gene. By analysis of cDNA clones and genomic clones, it is possible to elucidate what portion of a (eukaryotic) gene is transcribed, what parts of the gene are represented in the transcript after processing, and whether any structural rearrangement is associated with the expression of particular genes. One important point to note here is that eukaryotic genes often contain intervening sequences (introns) that do not end up in the mature mRNA. Hence, DNA and genomic DNA can be quite different in size. Chapter 13 describes eukaryotic gene structure and expression in more detail.

Gene Expression.　Screening can also be achieved by testing directly for the production of a protein encoded by the cloned gene. The set of expression vectors described in the preceding section is designed for this purpose. Proteins with immunological activity can be rapidly detected in bacterial colonies by employing replica-plating techniques.

Chromosome Walking

How can we analyze very large segments of DNA from a eukaryotic genome when the segments are much larger than the maximum size of fragments that can be cloned in any individual vector? Figure 15-27 outlines a procedure termed **chromosome walking,** which makes possible the dissection of large regions of DNA. In this method, a cloned portion of eukaryotic DNA is used to screen recombinant DNA clones from the same eukaryotic genome bank for other clones containing neighboring sequences. It is important to note that many gene banks are generated by partial digestion with a restriction enzyme, so that relatively large fragments can be represented. Many of these fragments will contain internal, uncleaved restriction sites, since the digestion was incomplete. Also, some regions of the genome will be present on different-sized fragments.

In the example shown in Figure 15-27, eukaryotic DNA is cut with the restriction enzyme EcoRI and cloned into a λ vector. Recombinant phages containing the "A" gene are identified with an "A" gene probe. The cloned DNA is recut into smaller fragments with a different restriction enzyme, and these pieces are themselves "subcloned" into a plasmid vector. When the subcloned fragment is different from the initial probe ("A" in this case), it can be used to detect other fragments in the original gene bank that contain part of the region carried in the first λ clone. Thus, in Figure 15-27, the subcloned "a" probe is used to identify an overlapping λ recombinant. Subsequent subcloning of different small fragments allows the generation of a series of overlapping clones, as shown in the bottom of Figure 15-27. In this manner, we literally "walk" around the chromosome. One drawback to this method is the requirement that each DNA segment used is not repeated elsewhere in the genome. Therefore, chromosome walking has been used most successfully in *Drosophila*, where repeated sequences are less prevalent than they are in the DNA's of higher cells.

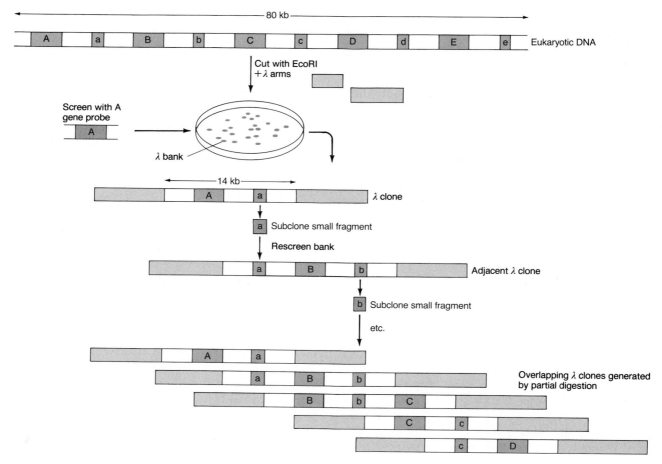

Figure 15-27. Chromosome walking. One recombinant phage obtained from a phage bank made by the partial EcoRI digest of a eukaryotic genome can be used to isolate another recombinant phage containing a neighboring segment of eukaryotic DNA, as described in the text.

> **Message**   Recombinant DNA molecules can be made by joining nonhomologous DNA's from virtually any sources. This technique offers the possibility of bypassing all biological restraints to genetic exchange and mixing—even to the point of permitting genes from widely differing species to combine.

DNA Sequence Determination

The technology for DNA manipulation that we have considered is extensive; combined with a technique for identifying the bases in isolated fragments, it opens the way for genetic engineering. A mere 15 years ago, the ability to determine base sequences easily seemed a long way in the future. Today, however, "DNA sequencing" is performed as routinely as genetic crosses.

We do know what sequence is recognized by each restriction enzyme, so an extensive restriction-enzyme map like the one in Figure 15-18 represents the distribution of known short sequences along the DNA. Ideally, we would like to take each DNA fragment, tag the end with a radioactive label, and then clip it off with an enzyme to identify the nucleotide. However, repeating this process over and over to identify several hundred nucleotides in a typical eukaryotic DNA fragment is far too laborious and time-consuming. As long as this was the only available approach, the sequencing of a small piece of DNA was a major task and as a routine procedure for genetic manipulation was impossible.

The key to DNA sequencing involves starting with a defined fragment of DNA uniquely labeled at one end, then generating a population of molecules that differ in size by one base, and being able to separate these molecules and identify the base in each case. The art of DNA sequencing advanced explosively after 1975, when new techniques were developed that made the task astonishingly simple. The crucial technique involves placing single-stranded DNA on a special gel material (acrylamide or agarose) and subjecting it to an electric current to separate strands on the basis of their lengths. (The use of this technique has already been mentioned in the discussion of restriction-enzyme mapping.) The mobility of a strand is inversely proportional to the logarithm of its length. This technique is so sensitive that fragments differing in length by only a single nucleotide can be separated.

One of the requirements for DNA sequencing is the ability to obtain particular fragments of DNA. There is a strong interdependence of DNA cloning and DNA sequencing technology, since DNA cloning provides large amounts of specific DNA fragments.

A widely used method is the one advanced by Allan Maxam and Walter Gilbert. They label the 3′ ends of DNA with ^{32}P, cleave the DNA and isolate one fragment, and then separate one strand from the other to yield a population of identical strands labeled on one end. They then divide the mixture into four samples, each of which is subjected to a different chemical reagent that destroys one or two specific bases (Figure 15-28). The four reagents destroy (1) only G, (2) A and G, (3) T and C, or (4) only C. The loss of a base makes the sugar-phosphate backbone of the DNA chain more likely to break at that point. The reagent concentration is adjusted so that only about one in 50 of the target bases is destroyed. The procedure is similar to that outlined for restriction-enzyme mapping in Figure 15-19. It results in a mixture of different-sized pieces carrying the ^{32}P label. When these pieces are separated in the different lanes of a gel, they can be arranged in order of length, and the base destroyed at each site can be determined by noting in which lane or lanes the band appears. Thus, the sequence of bases in the strand can quite simply be read from the pattern of bands on the gel. Figure 15-29 shows an actual gel used to determine a portion of a sequence of nucleotides located between the two α-globin genes of human DNA.

> **Message**   It is now possible to carry out rapid determination of nucleotide sequences in DNA. This procedure has become so routine that many hundreds of sequences are known, and the number of known sequences increases daily.

Fred Sanger developed a different sequencing method, with which he and his associates set out to determine the complete nucleotide sequence of φX174 DNA, which codes for nine proteins in this virus. The researchers completed this *tour de force* in 1977. The molecule contains just under 5400 nucleotides! This in-

vestigation was not undertaken solely to achieve the sequencing of a remarkably long DNA molecule; it also sought to resolve an interesting paradox. For the molecular weight of the nine proteins encoded in the φX174 DNA, the number of nucleotides required for the coding can be estimated. That estimate is significantly higher than the number of nucleotides indicated by physical properties of the DNA. It was hoped that knowledge of the complete sequence would lead to an explanation of this puzzle.

Different teams within Sanger's laboratory group worked on the sequencing of different fragments of the φX174 chromosome. The paradox of the "missing nucleotides" was quickly resolved in a surprising way. Within the coding sequence for one protein, a second protein was encoded in a different reading frame! Another pair of such "overlapping" genes was discovered by another team. The *genetic* map of φX174 (Figure 15-30) has thus been completed by the evidence from the molecular study. Figure 15-31 shows the consequences of the overlapping genes D and E. It should be pointed out that the use of two different reading frames to encode two different proteins (each encoded in one reading frame) does *not* constitute an "overlapping" code (see Chapter 13).

> **Message**   A single stretch of nucleotides can be read in more than one way by initiating reading of the sequence in two different reading frames that are offset from each other.

The concept of overlapping genes with offset reading frames had earlier been discarded because it raises problems about the simultaneous evolution of the two proteins. However, once undeniable evidence of such overlapping genes was available for φX174, old observations were reassessed and new studies undertaken; evidence now indicates that overlapping genes exist in other organisms, including *E. coli*. Note that a single point mutation within the region of overlap will produce amino-acid substitutions in two different proteins; this creates an interesting problem for researchers studying the evolutionary effects of various mutations: what happens if a certain mutation is favorable for one protein but unfavorable for the other?

Dideoxy Sequencing

Sanger's original method for sequencing DNA utilizes DNA polymerase to increase DNA chain length. It has been termed the "plus-minus" method. Subsequently, however, Sanger developed a much more powerful method utilizing single-stranded DNA and **dideoxy** nu-

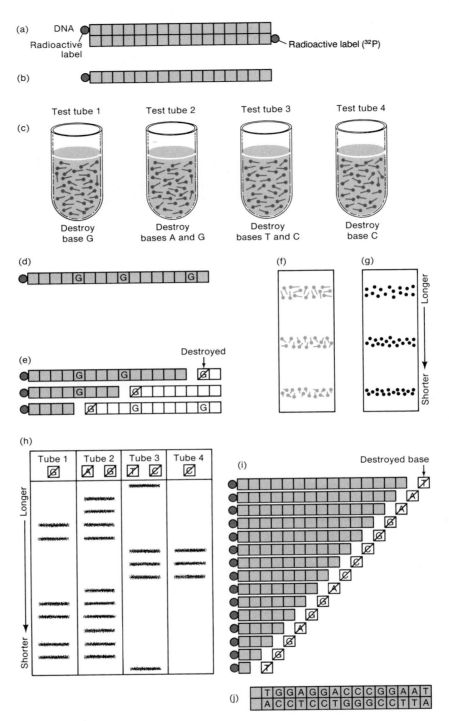

Figure 15-28. Procedure for sequencing DNA, devised by Maxam and Gilbert. (a) The 3′ end of each strand is labeled with ³²P. (b) The strands are separated, and one specific strand is retained. (c) The strands are separated into four equivalent fractions and placed in four test tubes. Each tube is treated with a different reagent that selectively destroys one or two of the four bases. The concentration is adjusted so that only a small proportion of the target bases is attacked, thereby generating a population of fragments of different lengths. (d) A hypothetical example: a strand containing three G bases. (e) After treatment in test tube 1, this strand yields a population of labeled fragments of three different lengths. (f) The fragments are separated on a gel. (g) The bands of labeled fragments are identified by autoradiography. (h) Comparison of the bands produced by labeled fragments from the four different treatments provides a display of successively shorter fragments when read from top to bottom on the gels. (i) The appearance of the band on one or two particular gels indicates which base was destroyed to yield the fragment. (j) The sequence of base pairs in the original DNA can then be inferred. (From W. Gilbert and L. Villa-Komaroff, "Useful Proteins from Recombinant Bacteria." Copyright © 1980 by Scientific American, Inc. All rights reserved.)

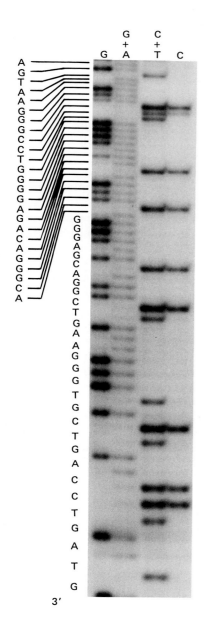

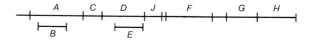

Figure 15-29. Autoradiograph of a Maxam-Gilbert sequencing gel, showing a portion of a sequence of nucleotides located between the two α-globin genes of human DNA. The DNA fragment was labeled at the 3' end and subjected to degradation by methods that cleave at G, C, G and A, C and T (cleavage reactions need not be specific to a *single* base to provide the necessary information). The sequence that can be read from the gel is indicated at the left. (Courtesy of John Hess and C.-K. James Shen, University of California, Davis. From F. J. Ayala and J. A. Kiger, Jr., *Modern Genetics,* 2d ed. Copyright © 1984 by Benjamin/Cummings Publishing Company, 1984.)

Figure 15-30. The genetic map of the virus φX174, including overlapping genes.

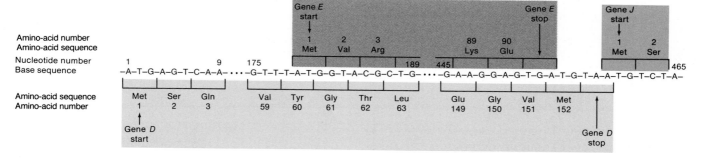

Figure 15-31. The beginnings and ends of genes *D* and *E* (and of their corresponding proteins) in the DNA of φX174. The nucleotide sequence is numbered from the start triplet of gene *D*. The *E* gene begins at nucleotide 179, with its triplet sequence offset from the reading frame of gene *D*. Gene *E* is completely contained within gene *D* and codes for a protein about 60 percent as large as the protein of gene *D*. The final base in the stop triplet for gene *D* is used as the first base of the start triplet for gene *J*.

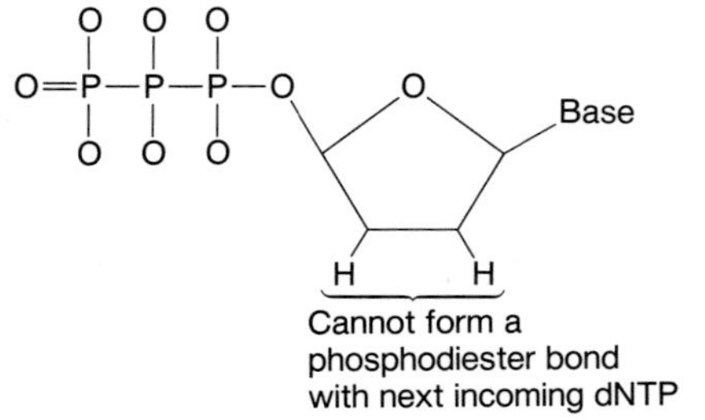

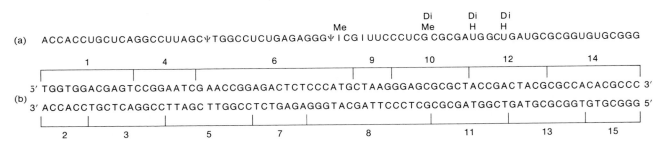

Figure 15-32. The structure of 2′,3′-dideoxy nucleotides, which are employed in the Sanger DNA sequencing method.

cleotides, which lack a hydroxyl group (Figure 15-32). The respective triphosphates can be incorporated into a growing chain, but they terminate synthesis, because they do not permit bond formation with the next nucleotide triphosphate. With four reaction tubes, each containing a small amount of one of the four dideoxy nucleotide triphosphates, the sequence of a DNA molecule can be determined by examining the size of terminated fragments obtained from each reaction on acrylamide gels. With this technique, the entire 48,513 base-pair sequence of bacteriophage λ was determined by Sanger's group.

Due to the ease with which DNA can be sequenced, it now is often convenient to infer the amino-acid sequence of a protein by determining the sequence of codons in the gene that encodes it, although this is not always a simple procedure, as can be seen in the case of split genes in eukaryotes (Chapter 13).

Gene Synthesis

We now have methods to determine the sequence of nucleotides in a DNA segment. Is it possible to synthesize a DNA fragment of some desired sequence from scratch? (We might determine a desired DNA sequence by finding the amino-acid sequence of a desired protein or by determining the nucleotide sequence of an RNA that codes for a desired protein.) Suppose we wish to construct a DNA fragment that has the sequence

T – A – G – C – C – T – C – C – A – G – T – A – A – T
⋮　⋮　⋮　⋮　⋮　⋮　⋮　⋮　⋮　⋮　⋮　⋮　⋮　⋮
A – T – C – G – G – A – G – G – T – C – A – T – T – A

We might set out to construct the bottom strand, with the intention of later synthesizing the complementary strand through normal replication reactions. We add the nuclotides of A and T to an appropriate reaction mix to obtain the A – T dinucleotide, but we must stop the reaction very quickly before A – T – T or A – T – A is formed. Therefore, the yield of the A – T polymer is only a fraction of the amount of A and T supplied. We then add the A – T and C to an appropriate reaction mix, and again we must stop the reaction quickly to avoid forming polymers longer than the desired A – T – C. Thus, the yield of A – T – C represents only a percentage of the A – T. At each step, we have a low yield, which obviously imposes harsh limits on the length of strands that can be constructed in vitro. Such stepwise synthesis of long-chain polynucleotides is impossible.

In the mid-1960s, Gobind Khorana attempted to synthesize the DNA coding for an alanine tRNA molecule (Robert Holley had already determined its sequence). Figure 15-33a shows the sequence of bases in the tRNA, which must be dictated by the DNA sequence shown in Figure 15-33b. Khorana developed a method that bypassed the need to synthesize the entire sequence by the stepwise approach just discussed. He synthesized short polynucleotides of each strand by adding one base at a time. The short fragments were selected to have overlapping complementary sequences. For example, Khorana first synthesized the fragment G – G – T – G – G – A – C – G – A – G – T and then the fragments C – C – A – C – C and T – G – C – T – C – A – G – G – C – C. When

(a)　ACCACCUGCUCAGGCCUUAGCΨTGGCCUCUGAGAGGGΨICGIUUCCCUCG CGCGAUG GCUGAUGCGCGGUGUGCGGG

1　　4　　6　　9　10　12　14

(b)　5′ TGGTGGACGAGTCCGGAATCG AACCGGAGACTCTCCCATGCTAAGGGAGCGCGCTACCGACTACGCGCCACACGCCC 3′
3′ ACCACCTGCTCAGGCCTTAGC TTGGCCTCTGAGAGGGTACGATTCCCTCGCGCGATGGCTGATGCGCGGTGTGCGGG 5′

2　3　5　7　8　11　13　15

Figure 15-33. (a) The base sequence of the major alanine tRNA from yeast. (b) The base sequence of the DNA coding for the alanine tRNA. Numbers indicate the segments Khorana used in synthesizing this DNA molecule. The segments were then annealed and connected with a ligase.

Figure 15-34. Treatment with the enzyme ligase leads to formation of an uninterrupted DNA molecule. In effect, the ligase works its way along the molecule and bonds together any consecutive nucleotides that are not joined in one strand or the other.

the first two fragments are mixed, the complementary sequences anneal to form a double helix with a single-stranded end:

$$G-G-T-G-G-A-C-G-A-G-T$$
$$C-C-A-C-C$$

When the third fragment is added, it anneals to the sticky end and produces a new sticky end:

$$G-G-T-G-G-A-C-G-A-G-T$$
$$C-C-A-C-C \quad T-G-C-T-C-A-G-G-C-C$$

A fourth fragment can then be synthesized to overlap this sticky end and extend beyond. The entire sequence can be "stitched together" in this fashion, obtaining a double helix composed of short fragments. The ends of the fragments then can be linked by the enzyme ligase (Figure 15-34).

As Khorana developed this methodology, it became clear that the gene specifying the tRNA extends beyond the limits indicated by the length of the mature tRNA molecule. He did eventually construct the entire tRNA gene (Figure 15-35). Appended to each end of Khor-

ana's artificial gene was a cohesive terminus of $-T-T-A-A$ complementary to EcoRI-induced ends, so that he could insert the complete gene into λ DNA (Figure 15-36). When the gene was inserted into E. coli by the λ vector, the bacterial cell did synthesize the tRNA. This verified that the start and termination signals inferred to exist did, in fact, work.

Message By construction of short nucleotide sequences which overlap with complementary sequences, large DNA duplexes can be generated.

Solid-phase Methods of DNA Synthesis. The Khorana method has been superseded by easier techniques that rely on the sequential addition of bases to a growing chain fixed to an insoluble resin. One such method employs derivatives of monomers that are protonated

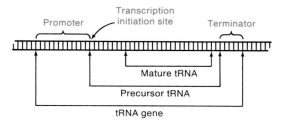

Figure 15-35. The DNA regions included in a complete tRNA gene. The promoter site is the region where the RNA polymerase attaches to begin transcription. Transcription begins at the initiation site and ends at a termination sequence. The precursor tRNA that is transcribed is larger than the active tRNA, which is produced by removing segments from the beginning and end of the precursor molecule.

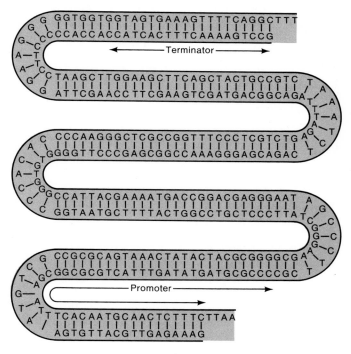

Figure 15-36. The complete sequence of the alanine tRNA gene constructed by Khorana's group. Note the sticky ends that were added to permit insertion in the λ DNA after that DNA was cleaved by EcoRI. (From Graham Chedd, "The Making of a Gene." This first appeared in New Society, London, the weekly review of the social sciences, September 30, 1976. Reprinted with permission of New Science Publications.)

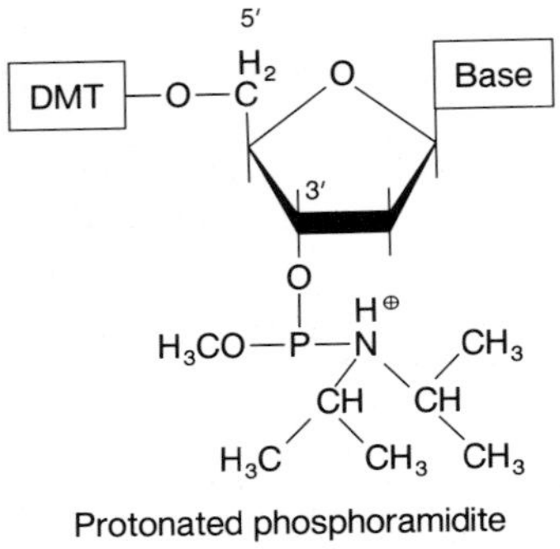

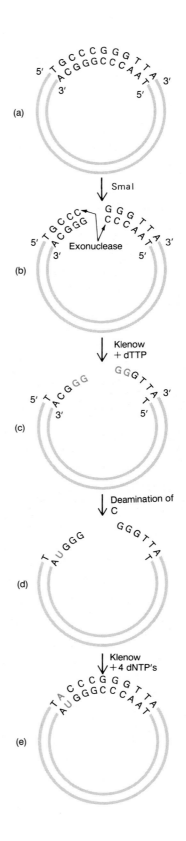

Figure 15-37. Protonated phosphoramidite. The 5′-hydroxyl is blocked by a dimethoxytrityl protecting group (DMT). (From L. Stryer, *Biochemistry*, 3d ed. Copyright © 1988 by W. H. Freeman and Company.)

phosphoramidites (Figure 15-37). In each step, the 5′ oxygen atom of the growing chain reacts with the 3′ phosphorous atom of the added monomer. The 5′ oxygen atom of each monomer is blocked by a protecting group. After each step, the protecting group is removed to allow reactivity with the incoming monomer added in the next step.

The solid-phase synthesis of DNA chains is now automated, and oligonucleotides of lengths up to 60 base pairs are routinely synthesized in several days. Oligonucleotides made in this manner are widely used as probes and in the construction of synthetic genes (described in a following section).

> **Message**   Automated solid-phase methods permit the rapid synthesis of oligonucleotides.

Mutagenesis in vitro

Base substitutions can be introduced into cloned DNA after the generation of short, single-stranded regions, as summarized in Figure 15-38. Single-stranded regions

Figure 15-38. Creation of a substitution mutant through the deamination of cytosine. A restriction enzyme site — in this case, *SmaI* (a,b) — is treated with a modified form of DNA polymerase I, termed the Klenow fragment, in the presence of dATP. (c) 3′ → 5′ digestion of single strands then occurs until an A residue is encountered. (d) The deamination of C residues on exposed single strands is effected by the addition of bisulfite. The molecule is now repaired by the addition of all four dNTP's and the Klenow fragment. (e) The result in this case is the alteration of a G–C base pair to an A–U base pair.

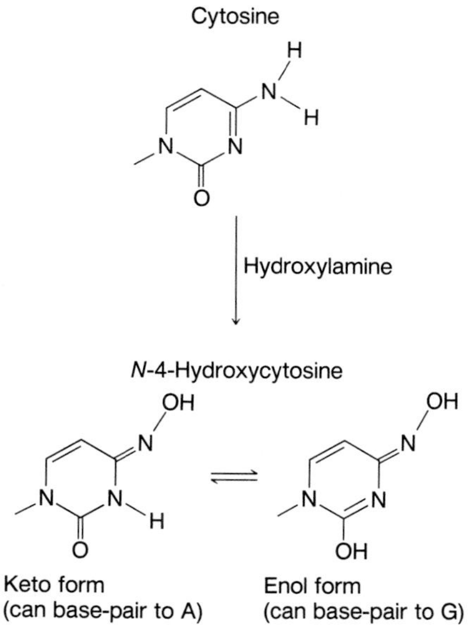

Cytosine

Hydroxylamine

N-4-Hydroxycytosine

Keto form
(can base-pair to A)

Enol form
(can base-pair to G)

Figure 15-39. The formation of N-4-hydroxycytosine with hydroxylamine.

Applications of Recombinant DNA Technology Using Prokaryotes

The essence of recombinant DNA technology is that fragments of DNA from any organism can be inserted into a bacterially based vector molecule. These vectors are amplified in bacteria to provide large amounts of DNA; then the insert can be studied in a variety of ways. Of course, the insert can be from prokaryotic or eukaryotic organisms, and indeed many eukaryotic genes have been intensively studied in this way at the structural and functional levels. Much has been learned about the organization of eukaryotic genes at the structural level by sequencing fragments cloned in bacteria. Much has also been learned at the functional level: eukaryotic genes are sometimes found to be expressed in bacteria in their intact state; in other instances, the eukaryotic sequences have to be manipulated by removing introns or adding bacterial regulatory signals in order to function in a bacterial cell. Furthermore, directed mutagenesis (see below) enables many functional aspects of eukaryotic genes to be dissected in bacterial settings.

In the following two sections, we consider first some examples of genetic engineering in bacteria and then focus on genetic manipulations involving eukaryotic organisms.

Genetic Engineering in Bacteria

This is an exciting period in genetics. For the first time in history, geneticists can program the alteration of the genetic material. The late 1970s and the early 1980s saw the development of techniques that permit the introduction of point mutations, deletions and insertions into segments of cloned DNA, and alterations of specific base pairs. By directing specific mutations into predetermined segments of DNA, we make possible many experiments. Two methods for programming specific alterations of genes are **site-directed mutagenesis** and **gene synthesis**.

Site-Directed Mutagenesis

We have already described methods involving specific or random base substitutions generated in vitro in cloned DNA. However, they all involve short regions surrounding a favorable restriction site. How can we create mutations at specific places that do not happen to be so favorably situated? There is a powerful method for creating specific mutations that circumvents this limitation. This technique employs synthetic oligonucleotides, in which are short (usually 15 to 25 bases in length) DNA segments constructed in vitro (page 415).

are produced either by using limited exonuclease III digestion following restriction enzyme cleavage or by exploiting the exonuclease activity of the *E. coli* DNA polymerase I enzyme. The single-stranded regions allow bisulfite ions to deaminate (cause the loss of an amino group) cytosine to uracil, which leads to C → T transitions, as described in Chapter 14. After treatment with bisulfite, the plasmid is recircularized and the gap is filled by polymerase I and ligase.

Alternatively, the incorporation of base analogs can be employed, which is done by creating short, single-stranded gaps within a duplex (without cleaving both strands). The gap is then repaired using a modified form of DNA polymerase I (see Chapter 10), together with three nucleotide triphosphates and N-4-hydroxycytosine in place of T (actually, dTTP). Both tautomeric forms of N-4-hydroxycytosine (keto and enol) are prevalent (Figure 15-39), allowing pairing with either G or A. The analog is incorporated in place of T but can also pair with G. Therefore, this in vitro method results in T → C transitions.

Less specific changes can be effected at single-stranded gaps by carrying out the polymerization reaction in the absence of one of the four deoxynucleotide triphosphates. At a low rate in vitro, polymerases will add nucleotides at random across from the base complementary to the missing base.

Additional methods for introducing mutations in vitro into cloned DNA segments are covered in the following section.

As a first step, the gene of interest is cloned into a single-stranded phage vector, such as the phage M13. The synthetic oligonucleotide serves as a primer for the in vitro synthesis of the complementary strand of the M13 vector (Figure 15-40). Any desired specific base change can be programmed into the sequence of the

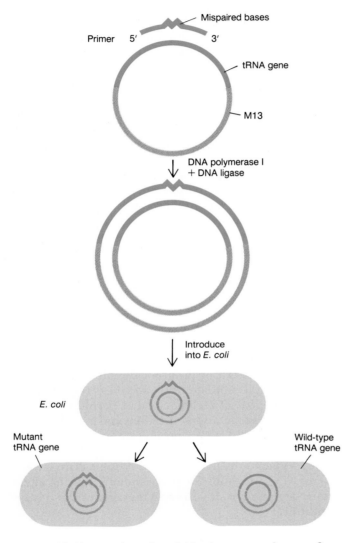

Figure 15-40. Creation of a substitution mutant by use of a synthetic oligonucleotide. The 12- to 15-base oligonucleotide is constructed so that it is complementary to a region of a DNA strand, but with one or two mismatches. When mixed with a clone of the complementary strand, the oligonucleotide will anneal to it even though the match is not exact, as long as the hybridization conditions are not stringent and the mismatches are in the middle of the oligonucleotide segment. The segment then serves as a primer for DNA polymerase I, which synthesizes the remainder of the complementary strand. When the resulting double-stranded molecule is introduced into E. coli, the molecule replicates to re-create either the sequence or the mutant original wild-type sequence.

synthetic primer. Although there will be a mispaired base when the synthetic oligonucleotide hybridizes with the complementary sequence on the M13 vector, one or two mismatched bases can be tolerated when hybridization occurs at a low temperature and a high salt concentration. After DNA synthesis is mediated by DNA polymerase in vitro, the M13 is replicated in E. coli, in which case many of the resulting phages will be the desired mutant.

It should be noted that the synthetic oligonucleotide can be used as a labeled probe to distinguish wild-type from mutant phages by hybridization. The mismatched base will still allow the primer to hybridize with both types of phage at low temperature, but only with the complementary mutant phage at high temperature. This method has been used to create an amber (UAG) suppressor from a human tRNA by converting the anticodon to a sequence that recognizes the UAG triplet instead of the AAA triplet (lysine codon), as depicted in Figures 15-40 and 15-41. The altered tRNA now inserts lysine in response to the UAG codon.

Applications of Gene Synthesis

The Khorana method (page 414) was applied by Herbert Boyer's group to make the gene coding for a small

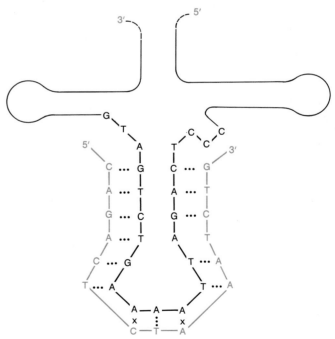

Figure 15-41. Sequences of the antisense strand (the strand that is not transcribed) of the tRNA gene at the anticodon stem and loop, and the synthetic pentadecamer with the two-base-pair mismatch (x) at the anticodon. (After G. F. Temple, et al., Nature 296, 1982, 537.)

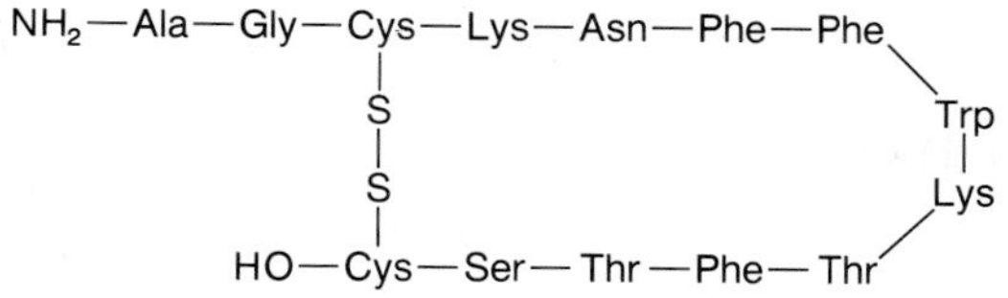

NH₂—Ala—Gly—Cys—Lys—Asn—Phe—Phe

|
S
|
S
|

HO—Cys—Ser—Thr—Phe—Thr

Trp
Lys

Figure 15-42. The amino-acid sequence of the hormone somatostatin.

human growth-regulating hormone, somatostatin. The hormone is a short polypeptide with the sequence shown in Figure 15-42. Boyer's group synthesized the gene using overlapping fragments. Then they added a triplet specifying methionine and an *EcoRI* cleavage site on the amino end; on the other end, they placed two consecutive stop triplets and a *BamHI* site (Figure 15-43). The entire gene was inserted into a plasmid carrying the bacterial gene β-galactosidase, within which there is an *EcoRI* site. The other end of the somatostatin gene hybridized with a *BamHI* site elsewhere in the plasmid. The *E. coli* selected for their possession of the plasmid were found to produce a protein chimera containing part of β-galactosidase fused to somatostatin via a methionine residue. Methionine is cleaved by cyanogen bromide, so the active hormone could be liberated by such treatment (Figure 15-44).

A good example of gene synthesis utilizing synthetic oligonucleotides produced by automated solid-phase synthesis was provided by John Abelson and Jeffrey Miller and their coworkers in 1986 when they constructed artificial tRNA genes in vitro. Specifically, they constructed the phenylalanine tRNA gene, with two changes in the anticodon, so that it now reads the UAG (amber) codon and becomes an amber suppressor tRNA. The gene can be synthesized by combining six oligonucleotides containing the two alterations. These

oligonucleotides are synthesized individually, with short overlaps to allow annealing in vitro. As in the preceding example, specific restriction-enzyme cutting sites are present at each end of the duplex that is generated. Figure 15-45 portrays the steps in the synthesis. After annealing, ligase is added and the gene is inserted into an expression vector that contains a promoter. Because a nonsense suppressor results, the ability to suppress cells containing a nonsense mutation in the *lacZ* gene, which encodes β-galactosidase, is monitored in order to verify the presence of an active suppressor tRNA. The entire experiment can be completed in several days. Compare this with the procedure originally utilized by Khorana to construct a tRNA molecule.

Such techniques, in addition to direct cloning, already have been used to produce recombinant plasmids bearing DNA sequences for human insulin, growth hormone, interferon, and blood-clotting factors. Commercially profitable quantities of such human proteins can be obtained from bacterial cultures.

> **Message** Recombinant DNA technology is sufficiently advanced to allow economically profitable production of human proteins through direct engineering of the genetic content of microorganisms.

Recombinant DNA Technology in Eukaryotes

To the eukaryote biologist, the ultimate interest is the eukaryotic organism itself. Many of the structural and functional questions to be asked about a gene really make sense only in the context of the whole organism. Therefore, it would be of considerable significance and

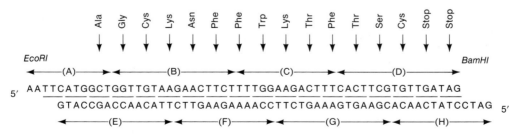

Figure 15-43. The overlapping complementary sequences synthesized to produce the somatostatin gene. A triplet specifying methionine is added to the 5′ end of the somatostatin coding region (inferred from the amino-acid sequence); Adjacent to this, an *EcoRI* restriction sequence is added. A *BamHI* restriction sequence is added at the other

end of the "artificial" gene. (Figures 15-43 and 15-44 from K. Itakura et al., "Expression in *Escherichia coli* of a Chemically Synthesized Gene for the Hormone Somatostatin." *Science* 198, 1977, 1056–1063. Copyright © 1977 by the American Association for the Advancement of Science.)

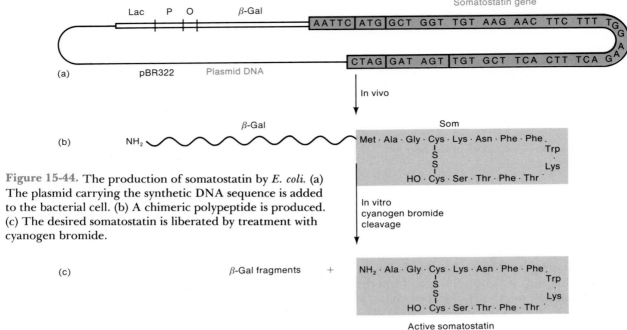

Somatostatin gene

(a) | AATTC | ATG | GCT GGT TGT AAG AAC TTC TTT | T G G A

Lac P O β-Gal

CTAG | GAT AGT | TGT GCT TCA CTT TCA | G A G

pBR322   Plasmid DNA

In vivo

(b)   NH₂ ～～～～～～～～～～

β-Gal   Som

Met · Ala · Gly · Cys · Lys · Asn · Phe · Phe
                    | Trp
                    S
                    | Lys
                    S
                    |
HO · Cys · Ser · Thr · Phe · Thr

Figure 15-44. The production of somatostatin by *E. coli.* (a) The plasmid carrying the synthetic DNA sequence is added to the bacterial cell. (b) A chimeric polypeptide is produced. (c) The desired somatostatin is liberated by treatment with cyanogen bromide.

In vitro
cyanogen bromide
cleavage

(c)   β-Gal fragments   +   NH₂ · Ala · Gly · Cys · Lys · Asn · Phe · Phe
                                        | Trp
                                        S
                                        | Lys
                                        S
                                        |
                            HO · Cys · Ser · Thr · Phe · Thr

Active somatostatin

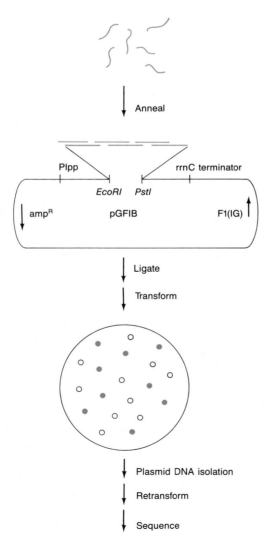

Anneal

Plpp   rrnC terminator

EcoRI   PstI

ampᴿ   pGFIB   F1(IG)

Ligate

Transform

Plasmid DNA isolation

Retransform

Sequence

Figure 15-45. Synthesis of a tRNA suppressor gene. Six oligonucleotides are annealed and ligated into a vector constructed for the expression of the synthetic tRNA gene. The vector contains restriction sites (*EcoRI* and *PstI*) for the insertion of the synthesized segment, as well as a promoter (Plpp) derived from the lipopolysaccharide gene and a transcription terminator (rrnC) derived from ribosomal RNA genes. The plasmid is used to transform cells that carry an amber mutation in the *lacZ* gene. These cells cannot synthesize β-galactosidase due to the chain-terminating amber mutation. However, transformants that synthesis an active suppressor tRNA will be able to produce β-galactosidase (indicated by the colored colonies). These transformants are then chosen for further analysis.

interest to be able to manipulate eukaryotic DNA in the convenience of a bacterial vector and then return the DNA back to the eukaryotic organism from which it came. Furthermore, there is simply no way to detect the expression of some eukaryotic genes in bacterial clones, and the clones in a gene bank need to be tested back in a eukaryotic cell, defective for the gene in question, in order to find the expression of the clone of interest. We shall see in this section that these kinds of reintroductions into eukaryotic cells are now possible, even routine, in several eukaryotes.

In addition to the preceding considerations, the genomes of eukaryotes are much larger and much more complex than, say, those of bacteria. For example, in contrast to the *E. coli* genome of approximately 4000 kilobases (kb), the simple eukaryote *Neurospora crassa* has a haploid genome size of 27,000 kb; in humans, the equivalent figure is 3,000,000 kb. This means that special extensions of recombinant DNA technology must be applied to handle these large genomes, and we explore some of these also in this section.

How to Get DNA into Eukaryotic Cells

It is perhaps a surprise that even bacterial cells will admit DNA through their cell membranes under the appropriate physiological conditions. However, this process now constitutes the basis for bacterial transformation. It is perhaps an even bigger surprise that eukaryotic cells will also take up DNA out of a solution across the cell membrane. Once again, the conditions have to be right. In addition, for DNA to be taken up in plants and fungi, a cell wall must be eroded away to form naked protoplasts. Special enzymes are used to break down the polysaccharides that constitute cell walls; in fungi, one of the enzymes used is extracted from the digestive glands of snails!

Alternatively, DNA can be injected into a cell under the microscope using a microsyringe and a micromanipulator. This permits extra DNA to be inserted into the egg cells of animals from *Drosophila* to humans. There are bizarre techniques, too; one that deserves mention is a specially adapted gun that shoots DNA-coated microscopic tungsten projectiles into a plant cell through the cell wall. The point is, in conclusion, that there are now some standard and some bizarre ways of introducing DNA into the cells of eukaryotic organisms. Of course, once the DNA is inside the cell, it faces numerous possible fates depending on the nature of the DNA itself, its vector sequences, and the idiosyncrasies of the organism in question. We will follow examples from several different organisms to illustrate some of the processes involved. A eukaryotic organism that develops from a cell into which new DNA has been introduced is called a **transgenic organism.** Here we will examine the produc-

tion of several different representative transgenic organisms and some of their unique features.

Transgenic Yeast

The yeast *Saccharomyces cerevisiae* has become the *E. coli* of the eukaryotes. One of the main reasons is that the classical genetics of yeast is extremely well-developed, and the availability of thousands of mutants affecting hundreds of different phenotypes provides a rich source of genetic markers in the development of yeast as a molecular system. Today the blend of classical genetics and molecular biology is indeed a powerful analytical combination in any organism, and the availability of one without the other can be a considerable frustration. In yeast, another important advantage is the occurrence of a circular 6.3 kb natural yeast plasmid, named the **2μ plasmid** after its circumference. This plasmid is transmitted normally to all the products of meiosis either by cell-to-cell contact or in the sexual cycle. The 2μ plasmid forms the basis for several specially engineered, sophisticated yeast vectors.

Yeast Vectors

The simplest **yeast vectors** are essentially bacterial plasmids into which a section of yeast DNA has been inserted (Figure 15-46). When transformed into yeast cells carrying the appropriate mutant allele, these plasmids can be inserted into yeast chromosomes by homologous recombination, involving a single or a double crossover (Figure 15-47). Either the entire plasmid is inserted, or the mutant allele is replaced by the allele on the plasmid. Such integrations can be detected by plating cells on a medium that selects for the allele on the plasmid. Because bacterial plasmids do not replicate in yeast, integration is the only way to generate a stable transgenic phenotype. One of the uses of this type of vector is to select bacterial clones that contain a yeast gene that cannot be detected or selected in bacterial cells: successful transformation of a yeast mutant tells the investigator that the clone has the equivalent wild-type gene.

If the 2μ plasmid is used as the basic vector and other bacterial and yeast segments are spliced into it (Figure 15-46), then a construct is obtained that has several useful properties. First, the 2μ segment confers the ability to replicate autonomously in the yeast cell, and insertion is not necessary for a stable transgenic phenotype. Second, genes can be introduced into yeast, and their effects can be studied in that organism; then the plasmid can be recovered and put back into *E. coli,* provided that a bacterial replication origin and a selectable bacterial marker are on the plasmid. Such **shuttle vectors** are very useful in the routine cloning and manipulation of yeast genes.

With any autonomously replicating plasmid, there is the possibility that a daughter cell will not inherit a copy

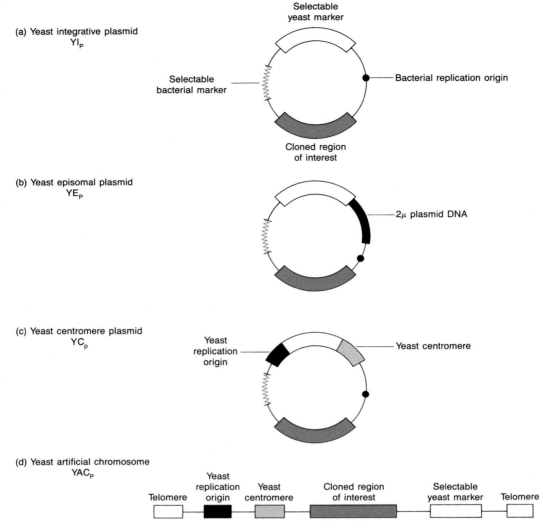

Figure 15-46. Simplified representations of four different kinds of plasmids used in yeast. Each is shown acting as a vector for some genetic region of interest, which has been inserted into the vector. The function of such segments can be studied by transforming a yeast strain of suitable genotype. Selectable markers are needed for the routine detection of the plasmid in bacteria or yeast. Origins of replication are sites needed for the bacterial or yeast replication enzymes to initiate the replication process. (DNA derived from the natural yeast plasmid 2μ has its own origins of replication.)

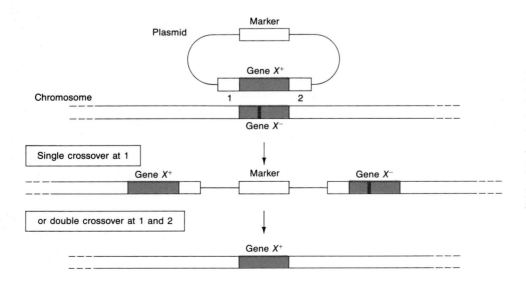

Figure 15-47. Two ways in which a recipient yeast strain bearing a defective gene X^- can be transformed by a plasmid bearing an active allele (gene X^+). The mutant site of gene X^- is represented as a vertical black bar. Single crossovers at position 2 are also possible but are not shown.

because the partitioning of copies to daughter cells is essentially a random process. However, if the section of yeast DNA containing a centromere is added to the plasmid, then the nuclear spindle that ensures the proper segregation of chromosomes will treat the plasmid in somewhat the same way and partition it to daughter cells more efficiently at cell division. The addition of a centromere is one step toward the formation of an artificial chromosome. A further step has been made by linearizing a plasmid containing a centromere and adding the DNA from yeast telomeres to the ends (Figure 15-46d). If this contruct contains yeast replication origins **(autonomous replication sequences, ARS),** then it behaves in many ways like a small yeast chromosome at mitosis and meiosis. For example, when two haploid cells—one bearing a $trp^- ura^+$ artificial chromosome and another bearing a $trp^+ ura^-$ artificial chromosome are brought together to form a diploid, some tetrads will show the clean segregations expected if these two elements are behaving as regular chromosomes. In other words, two ascospores will show $trp^- ura^+$ and the other two will show $trp^+ ura^-$ genotypes.

Applications of Yeast Vectors

One of the great assets of genetic analysis is its incisiveness. Through the analysis of specific blocks in gene function, normal biological processes can be dissected precisely and conclusive inferences can be drawn. Traditionally, the experimenter had to make use of mutations that were produced essentially at random. In vitro mutagenesis (page 416) allows the production of changes at specific places in specific genes. Such an in vitro-mutated gene then is inserted back into the organism to replace the resident wild-type gene and the effects are observed. The yeast-integrative plasmid provides a model of how this can be achieved in a eukaryote. A mutated gene and its flanking regions, carried on an integrating plasmid, provide a region of homology in which a crossover can occur; the entire plasmid then is inserted into the wild-type yeast chromosome at the proper locus (Figure 15-48). Due to the presence of two copies of the gene in question (one mutated and one wild-type), pairing can occur by looping and a crossover at a different site can excise the plasmid—this time bearing the wild-type allele and leaving the mutant allele in the normal chromosome locus.

Selection for Loss of Plasmid

In some cases, the elimination of the plasmid sequence can be selected. For example, if the plasmid selection marker is $lys2^+$ and the recipient yeast chromosome bears $lys2^-$ (Figure 15-48), then insertion produces a

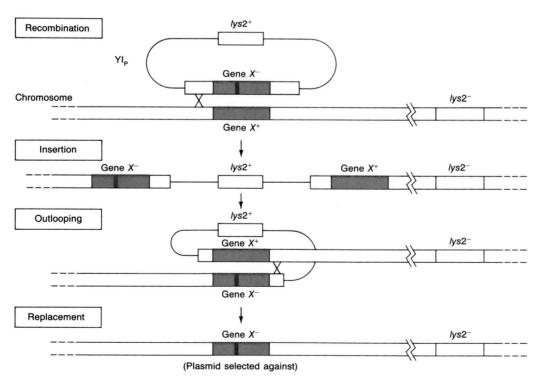

Figure 15-48. A two-step method for replacing an active gene X^+ with a deliberately engineered mutant allele (X^-) for the purpose of observing its effects on phenotype. The $lys2^+$ allele is a yeast marker that can be selected *for* (through the prototrophy that it confers) or *against* (by plating on the chemical α-aminoadipate).

strain with $lys2^+$ and $lys2^-$. The chemical α-aminoadipate permits the growth only of $lys2^-$ strains, so strains that have lost the insert can then be selected by plating on this chemical. About one-half of these strains retain the original wild-type gene of interest (gene X^+); the remainder retain the plasmid-borne mutant allele X. In other cases, the mutant phenotype of the mutant gene X^- can be selected directly by appropriate platings.

Note that YI$_p$ plasmids carry two yeast elements: the gene under investigation X, and the selectable marker. Both can undergo homologous recombination with their respective chromosome loci. The specificity of insertion can be targeted more efficiently by making a restriction cut in the plasmid at the gene in question (gene X). The ends are recombinogenic and direct the plasmid's entry to that specific site (Figure 15-49).

Gene Inactivation

Sometimes, all that is desired is to specifically inactivate a gene of interest. A way of achieving such gene disruptions in one step is actually to insert another gene with a selectable function into the middle of a wild-type allele of the gene of interest, carried in a plasmid. A linear derivative of such a construct will then insert specifically

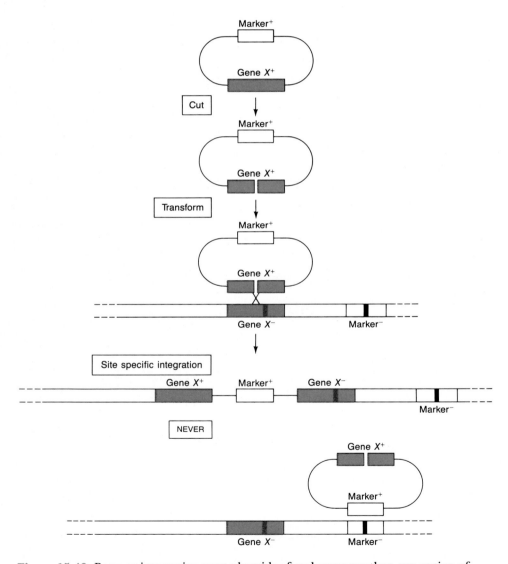

Figure 15-49. Because integrative yeast plasmids often bear more than one region of homology to yeast chromosomes, the site-specificity of integration can be increased by cutting the desired region, thereby producing recombinogenic ends.

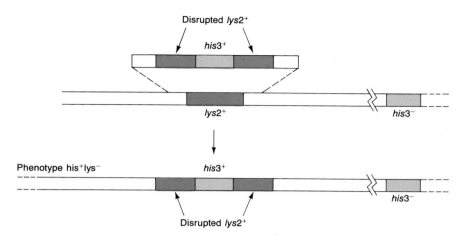

Figure 15-50. A single-step method for replacing a wild-type gene with a disrupted (inactivated) version carried on a linear DNA fragment for the purpose of specifically knocking out gene function. Integrants are selected as *his⁺* in this case.

at the wild-type locus, automatically disrupting it by virtue of the selectable gene inside it.

Thus, if the gene of interest is *lys2* (Figure 15-50), a *his3⁺* gene might be chosen to be inserted into *lys2⁺*, carried in a plasmid. A linear form of this would be used to transform a strain of genotype *lys2⁺ his3⁻*, and *his3⁺* transformants would be selected. Such transformants are found to be also *lys2⁻* by virtue of the replacement of *lys2⁺* by the disrupted *lys2⁺* allele from the plasmid.

Studying Regulation

Centromere plasmids can be used to study the regulatory elements upstream of a gene (Figure 15-51). The relevant coding region and its upstream (5′) region can be spliced into a plasmid, which can be selected by a separate yeast marker such as *URA3*. The upstream region can be manipulated by inducing a series of deletions, which are achieved by cutting the DNA, using a special exonuclease to chew away the DNA in one direction to different extents, and then rejoining it. The experimental objective is then to determine which of these deletions permits the normal functioning of the gene when the plasmid is used to transform a recipient in which the chromosome locus carries a defective mutant allele. The results generally define a specific region that is necessary for normal function and regulation of the gene.

In such regulatory studies, it is often more convenient to use a **reporter gene** instead of the structural gene of interest. Therefore, if the regulation of gene *X* is of interest, the upstream regulatory regions of gene *X* are spliced to the reporter gene. The reporter gene has a phenotype that is easier to monitor than that of gene *X*,

so the normal regulatory signals of gene *X* are expressed through the reporter. A gene that has been extensively used as a reporter is the bacterial *lacZ* gene, which codes for the enzyme β-galactosidase. This enzyme normally breaks down lactose, but it can also break down an analog of lactose, *Xgal* (5-bromo-4-chloro-indolyl-β,D-ga-

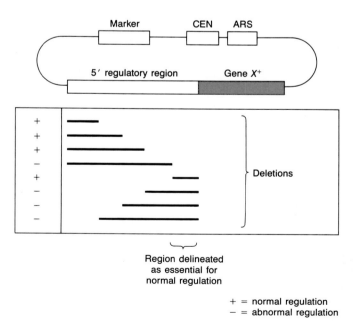

Figure 15-51. The regulation of the yeast gene *X⁺* can be studied by manipulating its regulatory region through deletion analysis in vitro and then transforming the constructs into a yeast strain bearing a defective allele *X⁻*.

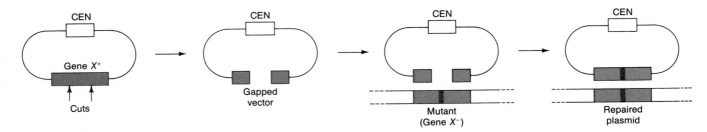

Figure 15-52. A gapped yeast centromeric plasmid is repaired by DNA copied from the homologous chromosome locus. This provides a convenient way of retrieving a mutant sequence of particular interest.

lactoside) very efficiently to yield 5-bromo-4-chloro-in-digo, which is bright blue. The blue color is expressed as a blue yeast colony whenever it is active. Generally, the fusions are constructed in such a way that the upstream regulatory regions of gene X plus a few codons from the structural gene X are fused in the correct reading frame to the region coding for the enzymatically active portion of β-galactosidase. These constructs can be transformed in a nonintegrative vector.

Retrieval

Autonomously replicating vectors can also be used as retrieval agents. If a particularly interesting phenotype is produced by a specific mutant allele, it is useful to be able to easily retrieve that allele and examine its structure and function. This can be achieved by transformation with a gapped centromeric plasmid, which bears a deleted form of the gene of interest. The gap is repaired using information from the in-situ mutant locus. The repaired plasmid is then simply reisolated from the strain and examined at the molecular level (Figure 15-52).

Future Applications

Yeast artificial chromosomes hold great promise as cloning vectors for large sections of mammalian (especially human) DNA. Consider that, for example, the size of the region coding the VIII blood-clotting factor in humans is known to span about 190 kb and that the gene for Duchenne muscular dystrophy spans probably more than 1000 kb! Furthermore, the large size of mammalian genomes in general means that banks or libraries of bacterial vectors are huge. Yeast artificial chromosomes, on the other hand, can carry much longer inserts and are potentially very useful in this regard.

The yeast system is by far the most sophisticated at present. However, the same techniques are being applied to other organisms — especially ones with well-de-fined genetic systems, such as filamentous fungi like *Neurospora* and *Aspergillus*.

Transgenic Plants

Due to their immense economic significance, plants have long been the subject of genetic analysis aimed at developing improved varieties. The advent of recombinant DNA technology has introduced a new dimension to this effort because the genome modifications made possible by this technology are almost limitless. No longer is the breeding confined to selecting variants within the species. DNA can now be introduced from other species of plants, animals, or even bacteria!

The Ti Plasmid

The only vectors routinely used to produce transgenic plants are derived from a soil bacterium called *Agrobacterium tumefaciens*. This bacterium causes what is known as crown gall disease, in which the infected plant produces uncontrolled growths (tumors, or galls), normally at the base (or crown) of the plant. The key to tumor production is a large (200 kb) circular DNA plasmid — the **Ti** (*tumor-inducing*) **plasmid.** When the bacterium infects a plant cell, a part of the Ti plasmid — a region called the T-DNA — is transferred and inserted, apparently more or less at random, into a site on the genome of the host plant (Figure 15-53). The functions required for this transfer are outside the T-DNA on the Ti plasmid. The T-DNA itself carries several interesting functions, including the production of the tumor and the synthesis of compounds called opines. Opines are actually synthesized by the host plant under the direction of the T-DNA. The bacterium then uses the opines for its own purposes, calling on opine-utilizing genes on the Ti plasmid outside of the T-DNA. Two important opines are nopaline and octopine; two separate Ti plasmids are based on them. The structure of Ti is shown in Figure 15-54.

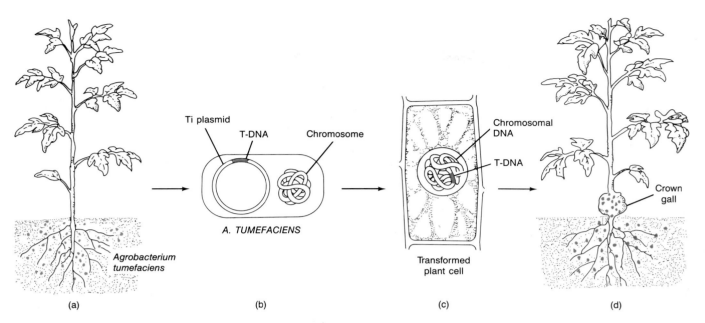

Figure 15-53. In the process of causing crown gall disease, the bacterium *Agrobacterium tumefaciens* inserts a portion of its Ti plasmid—a region called T-DNA—into a chromosome of the host plant.

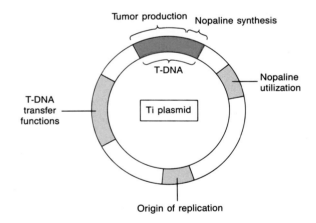

Figure 15-54. Simplified representation of the major regions of the Ti plasmid of *A. tumefaciens*. The T-DNA, when inserted into the chromosomal DNA of the host plant, directs the synthesis of nopaline, which is then utilized by the bacterium for its own purposes. T-DNA also directs the plant cell to divide in an uncontrolled manner, producing a tumor.

Using the Ti Plasmid as a Vector

The natural behavior of the Ti plasmid appears to make it well-suited for the role of a plant vector. If the DNA of interest could be spliced into the T-DNA, then it seems likely that the whole package would be inserted in a stable state into a plant chromosome. This system has indeed been made to work essentially in this way, but with some necessary modifications. Let's follow through a typical protocol.

Ti plasmids are too large to permit easy manipulation, so intermediate vectors must be used to introduce novel DNA into the T region. There are several ways of doing this; one is illustrated in Figure 15-55a. Here the smaller intermediate vector carries a selectable bacterial gene for spectinomycin resistance spc^R as well as a bacterial resistance gene kan^R, engineered for expression in plant cells, and two segments of T-DNA. One T-DNA segment carries the nopaline-synthesis gene plus the right-hand T-DNA border sequence. The second T-DNA segment is a sequence from near the left-hand border and is present to provide a homologous region for recombination. The other major item on the intermediate vector is a convenient cloning site—an engineered segment of DNA with a variety of unique, restriction-cleavage target sequences. The larger plasmid is an engineered Ti plasmid. The entire right-hand region of its T-DNA, including tumor genes and nopaline synthesis genes, have been deleted. This kind of deletion effectively "disarms" the plasmid, rendering it incapable of tumor formation—a "nuisance" aspect of the T-DNA function. Only the Ti plasmid is capable of replication in *Agrobacterium*, so after the intermediate vector is introduced (by conjugation with *E. coli*), plasmid recombinants can be selected by plating on spectinomycin.

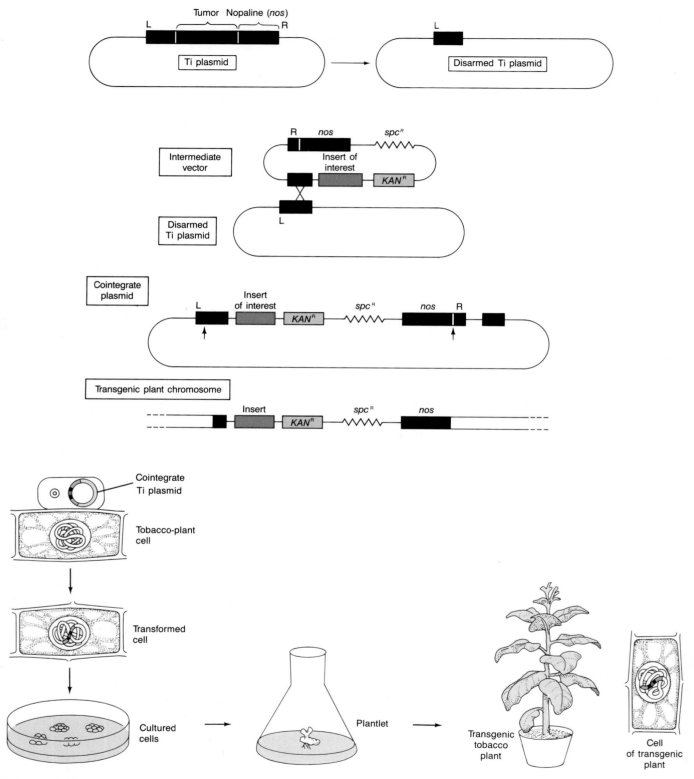

Figure 15-55. (a) To produce transgenic plants, an intermediate vector of manageable size is used to clone the segment of interest. In the method shown here, the intermediate vector is then recombined with a disarmed Ti plasmid to generate a cointegrate structure bearing the insert of interest and a selectable plant kanamycin-resistance marker between the T-DNA borders, which are all the T-DNA that is necessary to promote insertion. (b) The generation of a transgenic plant through the growth of a cell transformed by T-DNA.

Infecting Plant Tissue

Bacteria containing the recombinant double or "cointegrant" plasmid are then used to infect cut segments of plant tissue, such as punched out leaf disks. If bacterial infection of plant cells occurs, anything between the left and right T-DNA border sequences can be inserted into the plant chromosomes. If the leaf disks are placed on a medium containing kanamycin, the only plant cells that will go through cell division are those that have acquired the *kan*^R gene from T-DNA transfer. The growth of such cells results in a clump, or **callus,** which is an indication that transformation has occurred. These calluses can be induced to form shoots and roots and then be transferred to soil, where they develop into transgenic plants (Figure 15-55b). Often only one T-DNA insert is detectable in such plants, where it segregates at meiosis like a regular Mendelian allele (Figure 15-56). The insert can be detected by a T-DNA probe in a Southern hybridization, or by the detection of the chemical nopaline in the transgenic tissue.

Expression of Cloned DNA

What about expressing the DNA cloned into the T-DNA? This can, of course, be any DNA the investigator wants to test in the plant being used. One particularly striking foreign DNA that has been inserted using T-DNA is the gene for the enzyme luciferase, which is isolated from fireflies. The enzyme catalyzes the reaction of a chemical called luciferin with ATP; in this process, light is emitted, which explains why fireflies glow in the dark. A transgenic tobacco plant expressing the luciferase gene will also glow in the dark when watered with a solution of luciferin.

This might seem like a playful experiment, but it has a very important application: the luciferase gene can be used as a reporter gene to study various aspects of gene regulation during development. For example, the upstream regulatory sequences (see Chapter 16) of any

gene of interest can be fused to the luciferase gene and put into a plant via T-DNA. Then the luciferase gene will follow the same developmental pattern as the normally regulated gene does, but the luciferase gene will announce its activity prominently by glowing at various times or in various tissues, depending on the regulatory sequence. An agriculturally important example of inserting foreign DNA via T-DNA is a bacterial gene for resistance to the herbicide glyphosate. This gene confers resistance to the transgenic plant, enabling it to withstand the field application of glyphosate as a weedkiller.

Transgenic Animals

There are several ways of producing transgenic animals. Two major examples are covered elsewhere in this book. The first is the production of transgenic *Drosophila* by the injection of plasmid vectors containing "P elements" into the fly embryo (page 533). The second major example is the production of transgenic mammals by injecting special plasmid vectors into a fertilized egg. In both cases, the extra DNA can find its way into the germline cells, is then passed on to the progeny desired from these cells, and behaves from then on rather like a regular nuclear gene.

Gene Therapy

An interesting experiment in transgenic animals (and other organisms too) is to attempt gene therapy. Here the functions absent in a defective gene of the host are provided via the vector and ultimately expressed in the transgenic animal. The technique has been used in microbes routinely, of course, but is of great relevance in the case of humans in that it offers the hope of correcting hereditary diseases. Gene therapists have solved several hereditary problems in mammals other than humans.

One example is the correction of a growth-hormone deficiency in mice. The recessive mutation *little (lit)* results in dwarf mice. Even though the mouse's growth-hormone gene is present and apparently normal, no messenger RNA (mRNA) is produced. Initially, homozygous *lit/lit* eggs are injected with about 5000 copies of a 5-kilobase linear DNA fragment that contains the rat growth-hormone structural gene *(RGH)* fused to a regulator-promoter sequence from a mouse metallothionine gene *(MP)*. The normal job of metallothionine is to detoxify heavy metals, so that the regulatory sequence is responsive to the presence of heavy metals in the animal. The eggs are then implanted into pseudopregnant mice, and the baby mice are reared. About 1 percent of these babies turn out to be transgenic, showing increased size when heavy metals are administered during development. A representative transgenic mouse is then crossed to a *lit/lit* female, and the ensuing

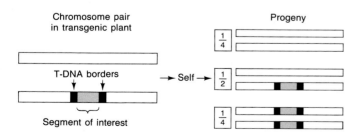

Figure 15-56. T-DNA and any DNA contained within it are inserted into a plant chromosome in the transgenic plant and then transmitted in a Mendelian pattern of inheritance.

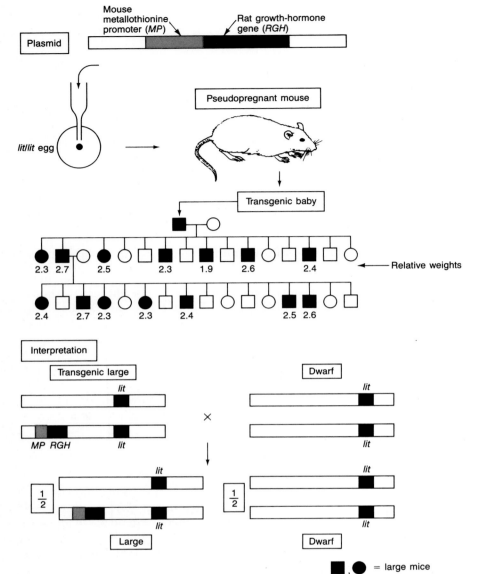

Figure 15-57. The rat-growth hormone gene *(RGH)*, under the control of a mouse promoter region that is responsive to heavy metals, is inserted into a plasmid and used to produce a transgenic mouse. *RGH* compensates for the inherent dwarfism *(lit / lit)* in the mouse. *RGH* is inherited in a Mendelian dominant pattern in the ensuing mouse pedigree.

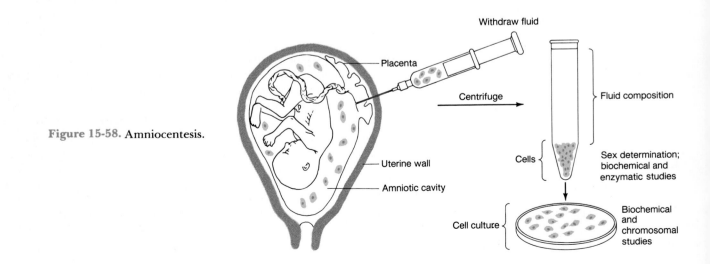

Figure 15-58. Amniocentesis.

pedigree is shown in Figure 15-57. We can see that mice two to three times the weight of their *lit/lit* relatives are produced down through the generations, with the transgenic rat growth-hormone gene acting as a dominant marker, always heterozygous in this pedigree. This kind of technology in mammals, *Drosophila*, and plants is not as controlled as it is in fungi such as yeast. The site of insertion of the introduced DNA in higher eukaryotes can be highly variable, and the DNA is generally not found at the homologous locus. Hence, gene therapy provides not a genuine correction of the original problem but a masking of it.

Regulation

Transgenic *Drosophila* provide us with another example of the use of the bacterial *lacZ* gene as a reporter in the study of gene regulation during development. The *lacZ* gene is fused to the upstream regulatory region of a *Drosophila* heat-shock gene, which is normally activated by high temperatures. This construct is then used to generate transgenic flies. Following heat shock, the flies are killed and bathed in *X-gal* (page 425). The resulting pattern of blue tissues provides information on the major sites of action of the heat-shock gene.

Genetic Diseases

Recessive mutations that follow Mendelian inheritance are responsible for over 500 genetic diseases. Homozygous individuals resulting from marriages involving two carriers of the same recessive trait will be affected by the disease. Screening cells derived from the fetus offers the possibility of predicting genetic defects at an early enough stage to allow the option of abortion to prevent the birth of afflicted individuals. The enzymes or proteins that are altered or missing in a number of genetic diseases are known (refer to the list of "inborn errors of metabolism" in Table 12-2). To detect such genetic defects, fetal cells are taken from the amniotic fluid, separated from other components, and cultured to allow the analysis of chromosomes, proteins, and enzymic reactions, and other biochemical properties. This process, termed **amniocentesis** (Figure 15-58), can already pinpoint a series of known disorders. Table 15-4 lists examples of genetic diseases that can be detected by amniocentesis. Relying on physiological properties or on the presence or absence of enzymic activity in cultured fetal cells limits the screening procedure to those disorders that affect characters or proteins expressed in the cultured cells. The use of recombinant DNA greatly in-

TABLE 15-4. Some common genetic diseases*

Inborn errors of metabolism	Approximate incidence among live births
1. Cystic fibrosis (mutated gene unknown)	1/1600 Caucasians
2. Duchenne muscular dystrophy (mutated gene unknown)	1/3000 boys (X-linked)
3. Gaucher's disease (defective glucocerebrosidase)	1/2500 Ashkenazi Jews; 1/75,000 others
4. Tay-Sachs disease (defective hexosaminidase A)	1/3500 Ashkenazi Jews; 1/35,000 others
5. Essential pentosuria (a benign condition)	1/2000 Ashkenazi Jews; 1/50,000 others
6. Classic hemophilia (defective clotting factor VIII)	1/10,000 boys (X-linked)
7. Phenylketonuria (defective phenylalanine hydroxylase)	1/5000 Celtic Irish; 1/15,000 others
8. Cystinuria (mutated gene unknown)	1/15,000
9. Metachromatic leukodystrophy (defective arylsulfatase A)	1/40,000
10. Galactosemia (defective galactose 1-phosphate uridyl transferase)	1/40,000

Hemoglobinopathies	Approximate incidence among live births
1. Sickle-cell anemia (defective β-globin chain)	1/400 U.S. blacks. In some West African populations, the frequency of heterozygotes is 40%.
2. β-thalassemia (defective β-globin chain)	1/400 among some Mediterranean populations

* Although the vast majority of the over 500 recognized recessive genetic diseases are extremely rare, in combination they represent an enormous burden of human suffering. As is consistent with Mendelian mutations, the incidence of some of these diseases is much higher in certain racial groups than in others.

SOURCE: J. D. Watson, J. Tooze, and D. T. Kurtz, *Recombinant DNA: A Short Course.* Copyright © 1983 by W. H. Freeman and Company.

creases our ability to screen for genetic diseases, how-ever, since we can analyze the DNA directly. In principle, if we could clone out the gene being tested and compare its sequence with that of a cloned normal gene, we could determine whether suspected defects were present. Of course, this would be a very laborious proce-dure, so shortcuts have to be devised to allow more rapid screening. Three useful techniques that have been used for this purpose involve searching for alterations of re-striction sites by the genetic defect itself, probing for altered sequences with synthetic oligonucleotides, and linkage of the mutation in question to altered restriction sites.

Alterations of Restriction Sites

Sickle-cell anemia is an example of a genetic disease that is caused by a well-characterized alteration. Affecting approximately 0.25 percent of U.S. blacks, the disease results from an altered hemoglobin due to the substitu-tion of a valine residue for a glutamic acid residue at position 6 in the β-globin chain (see also Chapter 12). The GAG $\rightarrow$ GTG change eliminates a cleavage site for the restriction enzyme MstII, which cuts the sequence CCTNAGG (where N represents any of the four bases). The change from CCT<u>GAGG</u> to CCT<u>GTGG</u> can thus be recognized by Southern blotting (Figure 15-25) using labeled β-globin cDNA as a probe, since the DNA de-rived from sickle-cell carriers will lack one fragment contained in the DNA from normal individuals (Figure 15-59).

Probing for Altered Sequences

When a genetic disorder can be attributed to a change in a specific nucleotide in all cases, then synthetic-oligonu-cleotide probes can identify that change. The best exam-ple is alpha-antitrypsin deficiency, which leads to a greatly increased probability for developing pulmonary emphysema and results from a single base change at a known position. Using as a probe a synthetic oligonu-cleotide that contains the wild-type sequence in the rele-vant region of the gene, Southern blot analysis (see Fig-ure 15-25) can be employed to determine whether the DNA contains the wild-type or the mutant sequence. At higher temperatures, a complementary sequence will hybridize, whereas a sequence containing even a single mismatched base will not.

Linkage of Altered Restriction Sites

What if a genetic defect itself does not alter a restriction site? Sometimes linkage to a restriction-site alteration can be measured. This strategy is derived from the ob-servation that if a specific genetic region is cloned and

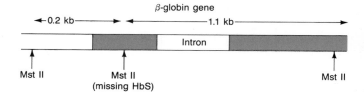

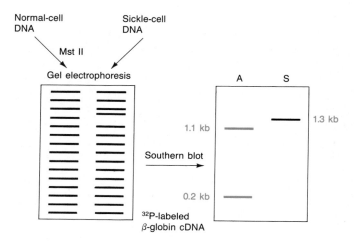

Figure 15-59. Detection of the sickle-cell globin gene by Southern blotting. The base change (A → T) that causes sickle-cell anemia destroys a *MstII* site that is present in the normal β-globin gene. This difference can be detected by Southern blotting. (Modified from J. D. Watson, J. Tooze, and D. T. Kurtz, *Recombinant DNA: A Short Course.* Copyright © 1983 by W. H. Freeman and Company.)

sequenced and this sequence is compared with equiva-lent homologous regions in other individuals, a small percentage of nucleotide differences is seen. One obvi-ous reason for this is the redundancy of the genetic code. But whatever the reason, these differences sometimes create or destroy restriction-enzyme target sequences. In fact, in eukaryotic DNA in general, it is quite easy to find such differences, using a protocol such as the fol-lowing.

When used as a probe, most cloned segments of, say, human DNA are capable of detecting restriction-site variation. So let's begin with a randomly chosen cloned fragment and use it in Southern hybridizations against restriction-enzyme-digested DNA preparations from a sample of people. If we are unlucky, we will see the same pattern on all autoradiograms: either one band or more, depending on whether the specific restriction enzyme used happened to cut into the region spanned by the probe. But eventually it is likely that we will find a variant—a fragment pattern unique from the others

that represents a restriction site with a different location (Figure 15-60).

Such variations in restriction-enzyme sites are almost always neutral; they do not represent coding or other differences that are detectably significant at the functional level. Their real significance is that they represent another kind of chromosome marker, and such

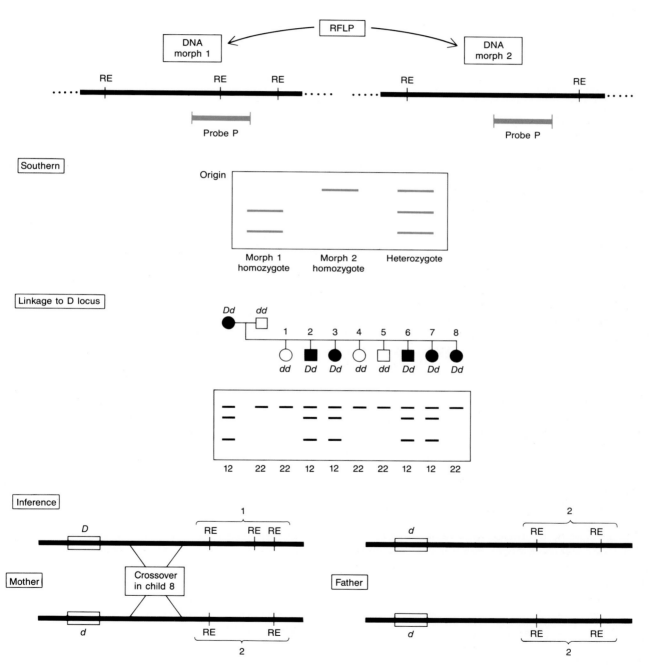

Figure 15-60. The detection and inheritance of a restriction fragment length polymorphism (RFLP). A probe P detects two DNA "morphs" when the DNA is cut by a certain restriction enzyme (RE). The pedigree of the dominant disease phenotype D shows linkage of the D/d locus and to the *RFLP* locus; only child 8 is recombinant.

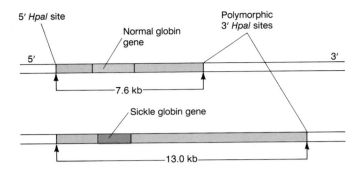

Figure 15-61. *HpaI*-site polymorphism is diagnostic for the sickle β-globin gene in humans.

markers are sorely needed in mapping and manipulating large genomes. The coexistence in a population of two or more phenotypes, generally attributable to the alleles of one gene, is called **polymorphism** (Greek: many forms). The coexistence of two or more restriction fragment patterns revealed, by hybridization, to a particular probe is called **restriction fragment length polymorphism (RFLP)**. In the most useful RFLP's, none of the variants are particularly rare; we can then examine pedigrees for other phenotypes, such as diseases, and look for linkage between the disease locus and the RFLP loci (Figure 15-60). Furthermore, the RFLP loci can also be mapped in relation to each other.

Another example of this method is diagrammed in Figure 15-61, where *HpaI*-site polymorphism is very closely linked to the sickle-cell β-globin gene. Examining linkage to restriction-site changes is simply doing on the DNA level the same type of analysis that has already been applied to fully developed organisms: namely, the rise of an identifiable character to give additional information about the genotype of an individual. For instance, colorblindness is used to yield information about the state of the locus governing the disease hemophilia on the X chromosome (see Problem 28 in Chapter 5). Because the *Cb* (colorblindness) locus is closely linked to the *Hb* (hemophilia) locus, the probability of the presence of the *Hb* alleles in a colorblind individual can be determined, provided the genotypes of the parents are known. The example given in Figure 15-61 employs the same principle. However, instead of determining colorblindness, we are determining the restriction-enzyme cleavage pattern of the DNA.

RFLP's as Map Reference Points

Using RFLP's as map reference points has been so successful that the entire human gene map is now liberally sprinkled with RFLP loci (for an example, see Figure

15-62). These loci are immensely helpful landmarks to use as a guide around the genome. As an example, if a geneticist is interested in cloning and studying a gene that causes a human disease, finding a linked RFLP can be a starting point. If the RFLP is less than about one map unit (1 m.u. = ~1000 kb) away, then the DNA identified by the RFLP probe can be used as a starting point for a chromosome walk (page 409) that ends with a clone of the disease-causing gene.

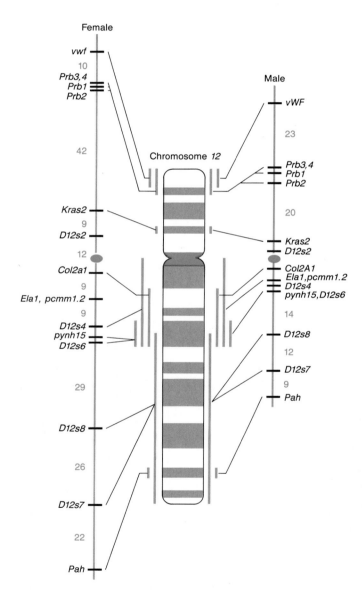

Figure 15-62. The human chromosome 12, showing the location of *RFLP* marker loci that have been detected in various pedigrees. Recombinant frequencies are shown for meiosis in men and in women; note that crossing-over appears to be more frequent in women.

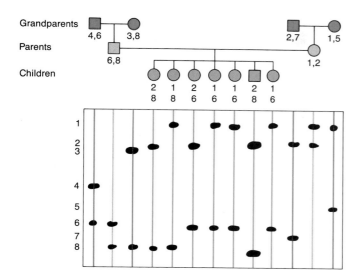

Figure 15-63. A pedigree illustrating the inheritance of alleles of the *RFLP* locus called VNTR. The grandparents display a total of eight RFLP alleles. The Southern blot autoradiogram (*below*) shows the restriction fragments hybridizing to the VNTR probe in individuals from three generations.

In some cases, a mutational event that causes a mutant phenotype, such as a disease, either creates or destroys a restriction site. Then the RFLP becomes useful as a diagnostic test for the presence of the disease allele in heterozygotes or individuals in which the disease has not yet manifested itself. For example, a certain mutation in a hemoglobin gene produces a recessive allele that can cause anemia. This same nucleotide change creates a new restriction site, say for the enzyme HindIII. Therefore, a probe that spans that site will pick up two fragments from the mutant allele and one larger fragment from the wild-type allele. A heterozygous carrier for this disease will show both the two smaller fragments and the larger fragment; that individual can be appropriately counseled concerning having children. Note also that such "direct hits," where the origin of the RFLP is actually in the gene of interest, provide a way of directly locating and sequencing the gene. This is because the probe provides immediate access to this genetic region.

Once the gene has been located and sequenced, a protein sequence can be inferred and a short stretch of the protein can be synthesized. Antibodies to this synthetic protein will usually also bind to the native protein. Thus, labeled antibodies will reveal the distribution of the protein in cells and tissues affected by the disease, which might point the way to a treatment.

We have considered simple two-allele restriction fragment length dimorphism as examples of RFLP's. However, some probes pick up multiple alleles. The family in Figure 15-63 illustrates such a situation.

Pulsed Field Gel Electrophoresis (PFGE)

A technique that is useful in eukaryote mapping, especially in conjunction with restriction enzyme technology, is **pulsed field gel electrophoresis (PFGE)**. In PFGE, instead of applying a single, uniform electric field across a gel, pulses are applied from two separate fields at an angle of about 90°. This allows the separation of much larger DNA fragments than is possible by regular electrophoresis. Apparently, the two fields allow the larger molecules to "snake" more efficiently through the gel, and fragments hundreds of kilobases long will move at a rate proportional to their size. The PFGE technique is so effective that the intact chromosomes of yeast and other fungi can be separated on a gel. Thus, for the first time, cloned genes can be associated with specific chromosomes without any need for meiotic or mitoic recombination analysis. (Figure 15-64).

Mammalian chromosomes are too large to be separated in this way, but special restriction enzymes come to the rescue. These enzymes—the so-called "rare cutters," such as NotI, PvuI or MluI—cleave mammalian DNA into fragments of several hundred to 2000 kb, which then can be separated by PFGE. The human genome measures about 3 million kb, and there are 23 unique chromosomes in the haploid set. Thus, the average chromosome measures about 100,000 kb. The fragments generated by the rare-cutting enzymes are man-

Figure 15-64. Pulsed-field gel electrophoresis of uncut yeast chromosomal DNA. Sixteen bands are resolved. Since there are 22 chromosomes in yeast, some have obviously comigrated in the gel. (All lanes were loaded identically.) (Source: Bio-Rad Laboratories.)

ageable and significant portions of a chromosome. Standard restriction mapping techniques (page 559) can be applied to these large fragments and long-range maps can be produced. Note that such a long-range map would represent invariant restriction sites, rather than the variable sites produced by RFLP mapping. If probes of known chromosome location are shown to hybridize to specific PFGE fragments, then a chromosomal link-up can ultimately be achieved.

Recombinant DNA and Social Responsibility

Recombinant DNA techniques have revolutionized biology with their revelations about gene and chromosome organization, and they promise enormous potential benefits for humanity. However, as these techniques began to be exploited in the early 1970s, some scientists (and some other people) began to express concern about the possible hazards of manipulating gene segments. For example, SV40 is a mammalian virus known to cause cancer in monkeys, and E. coli is a bacterium that normally lives in the human digestive tract. When DNA from SV40 is inserted into E. coli, is it possible that a carcinogenic bacterium might be produced, escape, and thrive as a parasite in humans? Others have wondered whether the combination of genes from eukaryotes with prokaryotic cells might generate new types of pathogenic organisms against which humans would have no natural defenses. Of course, the question of whether such fears are reasonable could be settled definitely only by carrying out the experiments to explore the possible results.

In an unprecedented step, 11 eminent molecular biologists published a letter in 1974 pointing out some of their concerns about the potential biohazards of work with recombinant DNA. They called for the development of guidelines to regulate such research. They asked scientists to observe a moratorium on certain kinds of experiments deemed particularly hazardous (cloning genes for toxins or cancer-causing agents). For the first time in history, a group of scientists publicly declared certain areas of scientific inquiry to be "off limits" and called for possible restrictions on such research.

The call for a moratorium attracted widespread public notice and caused many people to conclude that recombinant DNA research is dangerous. Under considerable public pressure, the National Institutes of Health (NIH) in the United States set out to establish categories of biohazards and guidelines for conducting experiments in each category. Eventually (on June 23,

1976), the NIH announced categories for experiments based on their potential hazards and defined four categories of physical conditions to contain the experiments. These conditions range from the P1 requirements of standard sterile techniques and common-sense precautions to the P4 facilities exercising the most extreme precautions against the escape of any organisms. Standards were set as well for biological restrictions that would minimize the chances of escape. For example, special strains of genetically enfeebled E. coli were constructed to use as recipients of recombinant DNA. These bacteria cannot survive except under special laboratory conditions.

Over the decade of the 1970s, evidence accumulated to indicate that the biohazards of recombinant DNA research were not as serious as some had feared. Meanwhile, increasing impatience built up in the scientific community about the delays in scientific progress and the regulation of scientific research by those with no training in the field. The NIH guidelines have now been relaxed significantly, and the use of recombinant DNA techniques has become routine laboratory practice.

Nonetheless, the turmoil about recombinant DNA did raise important social issues. For example, what is the social responsibility of scientists who are developing powerful new technologies? Should the people who are doing the experiments be the ones who set the guidelines? At what point should the public have an input? Who should be legally liable for any accidental damage that results from scientific research? Should limits be placed on the freedom of scientists to design and conduct research projects? Should a scientist attempt to foresee the possible adverse effects from the future use of discoveries and refuse to advance knowledge in certain directions that might have unfortunate applications?

Other kinds of questions are raised by the controversy over recombinant DNA. Can we predict with confidence the properties of an organism that has been modified by inserting DNA from a totally unrelated source? Can there be deleterious effects that will not be detected until large populations have been exposed for years (as was the case with oral contraceptives)? In the long run, will increasing sophistication in DNA manipulation inevitably lead to the genetic manipulation of human beings? If so, who will decide the conditions?

Although much of the worry about dangers of recombinant DNA research has been laid to rest, the issue has served to raise far more profound questions about the relationship between science and society. These questions have not been answered satisfactorily. They are likely to persist and become even more important in the coming years, as we confront the issues raised by prenatal diagnosis, forensic applications, and the possibility of gene therapy.

Summary

■ The highly specific pairing of complementary bases (to form A–T and G–C pairs) permits the experimental manipulation of DNA. A DNA strand can recognize a complementary strand and anneal with it to renature a double helix or to produce DNA-RNA duplexes.

Because even a short polynucleotide has a sequence that is relatively unlikely to occur by chance, base pairing provides a highly specific method of matching strands. (If all four bases have an equal probability of occurring at any position, then a specific sequence n bases in length should occur with a frequency of $\frac{1}{4}^n$. For example, a particular sequence 5 bases long should occur by chance only about once in 1000 such segments.)

By annealing single strands into double helices, it is possible to identify the sites of chromosomal DNA that are complementary to RNA, to compare DNA's from different sources, and to isolate the DNA that codes for specific RNA's. Studies using DNA–DNA and DNA–RNA hybrids have revealed the existence of eukaryotes of highly redundant short sequences of DNA and of large sections of DNA that do not code for amino acids but insert into genes between regions that do code for amino acids.

Synthetic genes have been constructed by producing short polynucleotides with regions of overlapping complementarity. In this way, a longer duplex can be formed by annealing several short overlapping strands and then using ligase to seal the gaps between adjacent ends.

Restriction enzymes recognize specific nucleotide sequences and cleave the DNA molecule at such sites; they provide a powerful tool for fragmenting DNA in a controlled fashion. Coupled with electrophoretic gels that permit the separation of strands varying in length (by as little as a single base), restriction enzymes have made genetic engineering simple. Large DNA molecules can be cut into small fragments; the fragments can be separated, and their base sequences can be determined. Such a study determined the entire base sequence for the DNA of the phage ϕX174 and revealed the surprising fact that some genes are contained within other genes (but use different reading frames).

Recombinant DNA is produced by linking DNA fragments with sticky ends. Two molecules are prepared with complementary single-stranded ends, and these ends are then annealed. A gene from a eukaryote can be isolated or constructed and then inserted into the DNA of a bacterial plasmid. This recombinant DNA can then be inserted into bacteria, where the recombinant plasmid can persist as a self-replicating cytoplasmic entity. DNA coding for human proteins has been constructed or isolated and inserted into bacteria, where the human proteins are produced in significant quantities.

Recombinant DNA technology provides powerful insights into the structure and regulation of genes. It also promises a way to produce modified organisms that will greatly benefit humankind. However, like any powerful new technique, genetic engineering also involves potential hazards for society that must be assessed carefully.

■ ■ ■

Solved Problems

1. The restriction enzyme HindIII cuts DNA at the sequence AAGCTT, and the restriction enzyme HpaII cuts DNA at the sequence CCGG. On average, how frequently will each enzyme cut double-stranded DNA? (In other words, what is the average spacing between restriction sites?)

Solution

We need only to consider one strand of DNA, because both sequences will be present on the opposite strand at the same site due to the symmetry of the sequences:

$$5'-AAGCTT-3' \quad \text{and} \quad 5'-CCGG-3'$$
$$3'-TTCGAA-5' \quad\quad\quad\quad 3'-GGCC-5'$$

The frequency of the six-base-long HindIII sequence is $1/4^6 = \frac{1}{4096}$, since there are four possibilities at each of the six positions. Therefore, the average spacing between *HindIII* sites is approximately 4 kb. For HpaII, the frequency of the four-base-long sequence is $1/4^4$, or $\frac{1}{256}$. The average spacing between *HpaII* sites is approximately 0.25 kb.

2. Suppose that you want to clone the gene from organism X that encodes a certain tRNA. You possess the purified tRNA—an *E. coli* plasmid that contains a single *EcoRI* cutting site and also confers resistance to ampicillin. How can you clone the gene of interest?

Solution

You can use the tRNA itself to probe for the DNA containing the gene. One method is to digest the DNA from organism X with EcoRI and then to insert this DNA into the plasmid, which you also have cut with EcoRI. Amp^R colonies resulting from a transformation with the relegated plasmid containing the inserts then can be tested against the probe, using applications of Southern hybridization. You can examine those colonies hybridizing with the probe further to verify how much of the tRNA sequence they contain. Alternatively, you can run EcoRI-digested DNA from organism X on a gel and then identify the correct fragment by probing with the tRNA. This fragment can be cut out of the gel and used as a source of highly enriched DNA to clone into the plasmid cut with EcoRI.

Problems

1. The restriction enzyme EcoRI cuts DNA at the sequence GAATTC, and the enzyme HaeIII cuts DNA at the sequence GGCC. On average, how frequently will each enzyme cut double-stranded DNA? (In other words, what is the average spacing between restriction sites?)

2. DNA molecules from three different sources are heat-denatured. The $G-C : A-T$ ratios for the three samples are 1.0, 0.88, and 1.2, respectively. Which sample will melt at the lowest temperature? At the highest? Why?

3. From Table 15-2, determine whether the EcoRI enzyme or the SmaI enzyme would be more useful for cloning. Explain your answer.

4. The bacteriophage ϕX174 has in its head a single strand of DNA as its genetic material. On infection of a bacterial cell, the phage forms a complementary strand on the infective strand to yield a double-stranded replicative form (RF). Design an experiment using ϕX174 to determine whether or not transcription occurs on both strands of the RF double helix.

5. After the irradiation of wild-type T4 phages, an *rII* mutation is recovered that fails to complement with mutants in either the A or the B cistron. DNA's from the mutant and the wild-type strains are mixed, heat-denatured, and cooled slowly to allow annealing. What hybrid molecules will be seen if the *rII* mutation is: a. A double point mutation? b. A deletion? c. An inversion? d. A transposition? e. A tandem duplication?

6. Suppose that the actual function of the *rII* locus is not known. How would you go about determining its primary gene products? (Assume that some techniques not presently available may become available in the future.)

7. Noboru Sueoka showed that some species of crabs contain DNA, of which 30 percent is dAT (a polymer of alternating sequences of adenine and thymine). Suppose that you want to study the cell biology of crab dAT. How would you show: a. Where it is located in the cell? b. If it is nuclear, in which chromosomes it is located? c. Whether there are other DNA sequences linked to the dAT?

8. In 1975, Norman Davidson and his colleagues isolated *Drosophila* DNA and sheared it into pieces. The DNA was denatured into single strands and allowed to renature for a very short period. The DNA was then filtered through a hydroxyapatite column, which retains double-helical DNA while allowing single strands to pass through. About once in every 40,000 to 80,000 kilobases, the researchers recovered a double-helical structure ranging in size from very short to more than 15 kb. In many cases the structure had the following appearance:

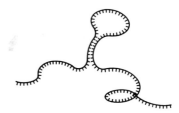

The researchers found approximately 2000 to 4000 such structures per genome. What is the explanation of these structures?

9. In 1973, Eric Davidson and his associates took DNA from *Xenopus* and broke it into pieces a few kilobases in length. The DNA was then denatured and allowed to anneal under conditions in which only redundant sequences anneal. The double-stranded sequences were retained on hydroxyapatite columns and then inspected under an electron microscope. The researchers obtained duplexes that had the following appearance:

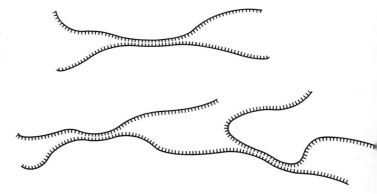

On average, the duplexes were 0.3 kb in length, and the separation between duplexes was approximately 0.8 kb. Interpret these results.

10. You have a purified DNA molecule, and you wish to map restriction-enzyme sites along its length. After digestion with EcoRI, you obtain four fragments: 1, 2, 3, and 4. After digestion of each of these fragments with HindII, you find that fragment 3 yields two subfragments (3_1 and 3_2) and that fragment 2 yields three (2_1, 2_2, and 2_3). After digestion of the entire DNA molecule with HindII, you recover four pieces: A, B, C, and D. When these pieces are treated with EcoRI, piece D yields fragments 1 and 3_1, A yields 3_2 and 2_1, and B yields 2_3 and 4. The C piece is identical to 2_2. Draw a restriction map of this DNA.

11. After treating *Drosophila* DNA with a restriction enzyme, the fragments are attached to plasmids and selected as clones in *E. coli*. Using this "shotgun" technique, David Hogness has recovered every DNA sequence of *Drosophila* in a cloned line.

   a. How would you identify the clone that contains DNA from a particular chromosomal region of interest to you?

   b. How would you identify a clone coding for a specific tRNA?

12. You have isolated and cloned a segment of DNA that is known to be a unique sequence in the genome. It maps near the tip of the X chromosome and is about 10 kb in length. You label the 5′ ends with ^{32}P and cleave the molecule with EcoRI. You obtain two fragments: one is 8.5 kb long; the other is 1.5 kb. You separate the 8.5 kb fragment into two fractions, partially digesting one with HaeIII and the other with HindII. You then separate each sample on an agarose gel. You obtain the following results by autoradiography:

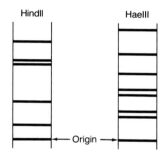

Draw a restriction-enzyme map of the complete 10-kb molecule.

13. As shown at the top of the autoradiographs at the top of the right column the Maxam-Gilbert technique has been used for sequencing each of two DNA fragments from the *a* mating-type locus in yeast. The bases above each gel are the ones attacked by the reagent. What are the base sequences of these two DNA fragments?

14. Suppose that a specific tRNA is encoded by six tandemly duplicated tRNA genes. You don't know what phenotype a tRNA mutation will yield. Design experiments to determine whether there are spacers between the tRNA genes and how the cluster of genes is transcribed. How could you detect mutations in these genes?

15. What results would you expect when adenine is substituted for guanine at nucleotide position 181 of the *D* gene in φX174? (See Figure 15-30)

16. Design an experiment to allow the purification of DNA sequences in the Y chromosome.

17. Calculate the average distances (in nucleotide pairs) between the restriction sites in organism X for the following restriction enzymes:

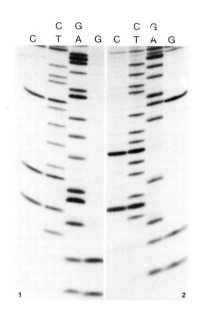

AluI	5′	AGCT	3′
	3′	TCGA	5′
EcoRI	5′	GAATTC	3′
	3′	CTTAGG	5′
Acyl	5′	G Pu CG Py C	3′
	3′	C Py GC Pu G	5′

(NOTE: Py = any pyrimidine; Pu = any purine)

18. Genes *A* and *B*, which map on yeast chromosome 4, are used as genetic markers in a study of two different haploid populations of yeast. The two populations express different allelic forms of the genes: in population 1, gene *A* gives *A1* and gene *B* gives *B1*; in population 2, gene *A* gives *A2* and gene *B* gives *B2*. These alleles are distinguished by the HindIII restriction map of the DNA in the region of the genes:

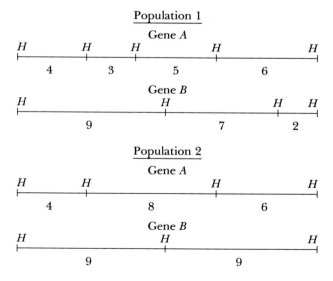

The tetrad products are examined, and the DNA fragments corresponding to genes *A* and *B*, respectively, are given for each type:

Spore type	Frequency	DNA (HindIII) fragments)	
		Gene *A*	Gene *B*
1	15%	4, 3, 5, 6	9, 9
2	15%	4, 8, 6	9, 7, 2
3	35%	4, 3, 5, 6	9, 7, 2
4	35%	4, 8, 6	9, 9

a. What are the allelic forms of the tetrad products?

b. How did they arise?

c. Draw the appropriate linkage map.

(Problem 18 courtesy of Joan McPherson.)

19. a. A fragment of mouse DNa with EcoRI sticky ends carries the gene *M*. This DNA fragment, which is 8 kb long, is inserted into the bacterial plasmid pBR322 at the *EcoRI* site. The recombinant plasmid is cleaved with three different restriction enzymes. The patterns of ethidium-bromide fragments, following electrophoresis on agarose gels, are shown in the accompanying figure.

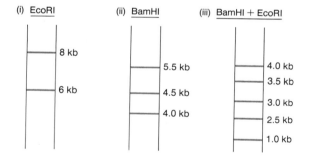

A Southern blot is prepared from gel (iii). Which fragments will hybridize to a probe (^{32}P) of pBR plasmid DNA?

b. Gene *X* is carried on a plasmid consisting of 5300 nucleotide pairs (5300 bp, or base pairs). Cleavage of the plasmid with the restriction enzyme BamHI gives fragments 1, 2, and 3, as indicated in the diagram at the top of the right column (B = *BamHI* restriction site). Tandem copies of gene *X* are contained within a single BamHI fragment. If gene *X* encodes a protein X of 400 amino acids, indicate the approximate positions and orientations of the gene-*X* copies.

(Problem 19 courtesy of Joan McPherson.)

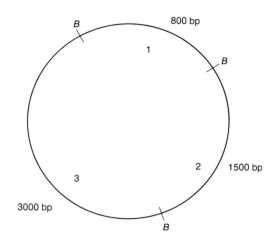

20. A linear fragment of DNA is cleaved with the individual restriction enzymes HindIII and SmaI, and then with a combination of the two enzymes. The fragments obtained are

HindIII	2.5 kb, 5.0 kb
SmaI	2.0 kb, 5.5 kb
HindIII and SmaI	2.5 kb, 3.0 kb, 2.0 kb

a. Draw the restriction map.

b. The mixture of fragments produced by the combined enzymes is cleaved with the enzyme EcoRI, resulting in the loss of the 3-kb fragment (band stained with ethidium bromide on an agarose gel) and the appearance of a band stained with ethidium bromide representing a 1.5-kb fragment. Mark the *EcoRI* cleavage site on the restriction map.

(Problem 20 courtesy of Joan McPherson.)

21. A viral DNA fragment carrying a specific gene, *V*, is transfected (introduced into the cell by transformation) into a muscle-cell culture. Following incubation with ^{32}P-labeled ribonucleotides, the virus-encoded RNA product is isolated at two timed intervals. The radiolabeled viral RNA is treated as follows: First, it is hybridized to a specific cDNA previously constructed from viral-gene-*V* mature mRNA. Second, the hybrid is treated with RNAase. Finally, the hybrid is denatured and electrophoresed on a gel, which is then subjected to autoradiography. The following results suggest that the pathologic nature of the virus is time-related (the number of nucleotides is indicated on the bands observed):

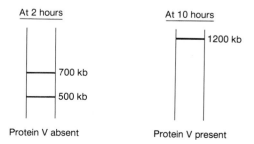

a. What is the size of the mature mRNA for gene *V*?

b. Draw a diagram of each hybride, and indicate what the illustrated bands represent.

c. Why is protein V not produced until after 2 hours?

(Problem 21 courtesy of Joan McPherson.)

22. A yeast plasmid carrying the yeast *leu2⁺* gene is used to transform nonrevertible haploid *leu2⁻* yeast cells. Several *leu⁺*-transformed colonies appear on a leucineless medium. Thus, *leu2⁺* DNA presumably has entered the recipient cells, but now you have to decide what has happened to it inside these cells. Crosses of transformants to *leu2⁻* testers reveal that there are three types of transformants A, B, and C, reflecting three different fates of the *leu2⁺* in the transformantion. The results are

Type A × *leu2⁻* ⟶ ½ *leu⁻*
  ½ *leu⁺*, × standard *leu2⁺* ⟶ ¾ *leu⁺*
      ¼ *leu⁻*

Type B × *leu2⁻* ⟶ ½ *leu⁻*
  ½ *leu⁺*, × standard *leu2⁺* ⟶ 100% *leu⁺*
      0% *leu⁻*

Type C × *leu2⁻* ⟶ 100% *leu⁺*

What three different fates of the *leu2⁺* DNA do these results suggest? Be sure to explain *all* the results according to your hypotheses. Use diagrams if possible.

23. The gene for β-tubulin has been cloned from *Neurospora* and is available, carried in a λ vector. List a step-by-step procedure for cloning the same gene from the related fungus *Podospora*, using as the cloning vector the p^BR *E. coli* plasmid shown here, where kan = kanamycin and tet = tetracycline:

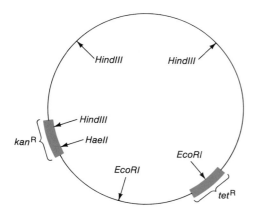

24. A linear phage chromosome is labeled at both ends with ³²P and digested with restriction enzymes. EcoRI produces fragments of sizes 2.9, 4.5, 6.2, 7.4, and 8.0 kb. An autoradiogram developed from a Southern blot of this

digest shows radioactivity associated with the 6.2 and 8.0 fragments. BamHI cleaves the same molecule into fragments of sizes 6.0, 10.1, and 12.9; the label is found to be associated with the 6.0 and 10.1 fragments. When EcoRI and BamHI are used together, fragments of sizes 1.0, 2.0, 2.9, 3.5, 6.0, 6.2, and 7.4 kb are produced.

a. Draw a restriction-enzyme target site map of this molecule, showing relative positions and distances apart.

b. A radioactive probe made from a cloned phage gene *X* is added to Southern blots of single-enzyme digests that have used nonradioactive phage DNA. The autoradiograms show hybridization associated with the 4.5, 10.1, and 12.9 fragments. Draw in the approximate location of gene *X* on the restriction map.

25. Huntington's disease (HD) is a lethal neurodegenerative disorder that exhibits autosomal-dominant inheritance. Because the onset of symptoms is usually not until the third, fourth, or fifth decade of life, patients with HD usually have already had their children and some of them inherit the disease. There has been little hope of a reliable pre-onset diagnosis until recently, when a team of scientists searched for and found a cloned probe (called G8) that revealed a DNA polymorphism (actually, a tetramorphism) relevant to HD. The probe and its four hybridizing DNA types are shown here; the vertical lines represent *HindIII* cutting sites:

					Extent of homology of G8 probe
17.5	3.7	1.2	2.3	8.4	DNA A
17.5	4.9		2.3	8.4	DNA B
15.0	3.7	1.2	2.3	8.4	DNA C
15.0	4.9		2.3	8.4	DNA D

a. Draw the Southern blots expected from the cells of people who are homozygous (AA, BB, CC, and DD) and all who are heterozygous (AB, AC, and so on). Are they all different?

b. What do the DNA differences result from in terms of restriction sites? Do you think they are probably trivial or potentially adaptive? Explain.

c. When human-mouse cell lines were studied, the G8 probe bound only to DNA containing human chromosome 4. What does this tell you?

d. Two families showing HD—one from Venezuela, and one from the United States—are checked to determine their G8 hybridizing DNA type. The results were as follows, where solid black symbols indicate HD and slashes indicate family members who were dead in 1983:

Venezuela:

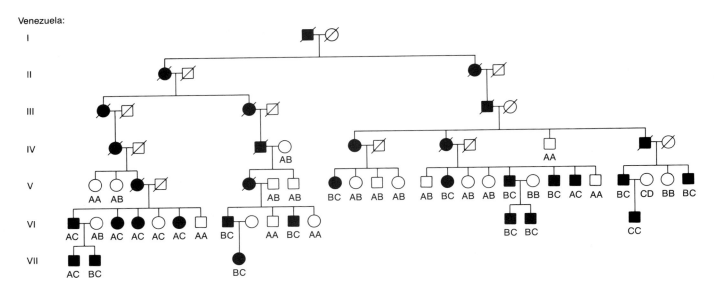

United States:

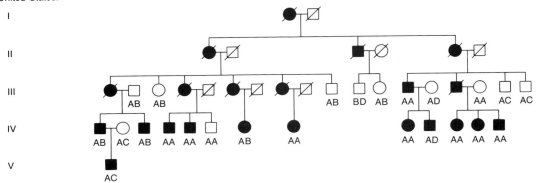

What linkage associations do you see, and what do they tell you?

e. If a 20-year-old member of the Venezuelan family needs genetic counseling, what test would you devise and what advice would you give for each outcome? Repeat this for the U.S. family.

f. How might these data be helpful in finding the primary defect of HD?

g. Could these results be useful in counseling other HD families? Explain.

h. Are there any exceptional individuals in the pedigrees. If so, account for them.

26. DNA studies are performed on a large family that shows a certain autosomal-dominant disease of late onset (approximately 40 years of age). A DNA sample from each family member is digested with the restriction enzyme Taq 1 and run on an electrophoretic gel. A Southern blot is then performed, using a radioactive probe consisting of a portion of human DNA cloned in a bacterial plasmid. The autoradiogram is shown below, aligned with the family pedigree:

Pedigree:

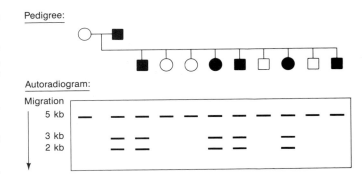

Autoradiogram:

a. Analyze fully the relationship between the DNA variation, the probe DNA, and the gene for the disease. Draw the relevant chromosome regions.

b. How do you explain the last son?

c. Of what use would these results be in counseling people from this family who subsequently married?

Control of Gene Expression

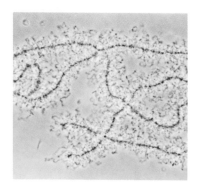

KEY CONCEPTS

Gene regulation is most often mediated by proteins that react to environmental signals by raising or lowering the transcription rates of specific genes.

∎

Negative control in prokaryotes is exemplified by the *lac* system, in which a repressor protein blocks transcription by binding to a site on the DNA termed the operator.

∎

Positive control in prokaryotes occurs when protein factors are required to activate transcription.

∎

Many regulatory proteins have common structural features.

∎

Transcriptional control of eukaryotes is also mediated by trans-acting protein factors, which bind to specific regulatory sequences.

∎

Additional regulatory sites on the DNA, termed enhancers, modulate gene expression in eukaryotes by interacting with specific regulatory proteins.

■ Until now, we have discussed what genes are and how genetic change occurs and is inherited. Also, we saw in Chapter 13 how genes are expressed by being transcribed into RNA molecules, many of which are translated into proteins. But how does the cell *regulate* the expression of all its genes? We can see that control of gene expression is crucial to an organism. In higher cells, specific cell types have differentiated to the point that they are highly specialized. An eye cell in humans synthesizes the proteins important for eye color but does not produce the detoxification enzymes that are synthesized in liver cells. Each cell type has arranged to **repress** (turn off) different groups of genes.

Bacteria also have a need to repress genes. Enzymes involved in sugar metabolism provide an example. Metabolic enzymes are required to break down different carbon sources to yield energy. However, there are many different types of compounds that bacteria could use as carbon sources, including sugars such as lactose, glucose, maltose, rhamnose, raffinose, melibiose, galactose, and xylose. Several enzymes allow each of these compounds to enter the cell and to catalyze different steps in sugar breakdown. If a cell were to synthesize simultaneously all of the enzymes it might possibly need, it would cost the cell much more energy to produce the enzymes than it could ever derive from breaking down any of the prospective carbon sources. Therefore, the cell has devised mechanisms to repress all of the genes encoding enzymes that are not needed and to activate those genes at a time

when the enzymes are needed! Clearly, to do this, two requirements must be met:

1. A method must be found for turning off (or on) each specific gene or group of genes.

2. The cell must be able to recognize situations in which it should activate a specific gene or group of genes.

Basic Control Circuits

Let's take several examples and analyze them in detail. The first system we will focus on is concerned with lactose metabolism in *Escherichia coli*. We have now learned a lot about how this system works. Figure 16-1 shows a physical model for the control of the lactose enzymes. Let's review the current model for gene control, and then consider the studies that have led to the elucidation of the model.

The metabolism of lactose requires two enzymes: a permease to transport lactose into the cell, and β-galactosidase to cleave the lactose molecule to yield glucose and galactose. Permease and β-galactosidase are encoded by two contiguous genes, Z and Y, respectively. A third gene, the A gene, encodes an additional enzyme, termed transacetylase, but this enzyme is not required for lactose metabolism and we will not concentrate on it for now. All three genes are transcribed into a single or

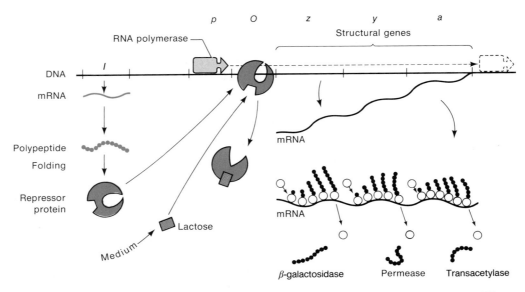

Figure 16-1. Regulation of the *lactose* operon. The *I* gene continually makes repressor. The repressor binds to the *O* (operator) region, blocking the RNA polymerase bound to *P* from transcribing the adjacent structural genes. When lactose is present, it binds to the repressor and changes its shape so that the repressor no longer binds to *O*. The RNA polymerase is then able to transcribe the *Z*, *Y*, and *A* structural genes, so the three enzymes are produced

polycistronic messenger RNA (mRNA) molecule. Thus, it can be seen that by regulating the production of this mRNA, the regulation of the synthesis of all three enzymes can be coordinated. A fourth gene, the *I* gene, which maps near but not directly adjacent to the *Z*, *Y*, and *A* genes, encodes a **repressor** protein, so named because it can block the expression of the *Z*, *Y*, and *A* genes. The repressor binds to a region of DNA near the beginning of the *Z* gene and near the point at which transcription of the polycistronic mRNA begins. The site on the DNA to which the repressor binds is termed the **operator.** One necessary property of the repressor is that it be able to recognize a specific short sequence of DNA — namely, a specific operator. This ensures that the repressor will only bind to the site on the DNA near the genes that it is controlling and not to other random sites all over the chromosome. By binding to the operator, the repressor prevents the initiation of transcription by RNA polymerase. Normally, RNA polymerase binds to specific regions of the DNA at the beginning of genes or groups of genes, termed **promoters** (see Chapter 13), so that it can initiate transcription at the proper starting points. The *POZYA* segment shown in Figure 16-1 constitutes an **operon,** which is a genetic unit of coordinate expression.

The *lac* repressor is a molecule with two recognition sites that can recognize the specific operator sequence for the *lac* operon. It can also recognize lactose and certain analogs of lactose. When the repressor binds to lactose derivatives, it undergoes a conformational change; this slight alteration in shape changes the operator binding site so that the repressor loses affinity for the operator. Thus, in response to binding lactose derivatives, the repressor falls off the DNA. This satisfies the second requirement for such a control system — the ability to recognize conditions under which it is worthwhile to activate expression of the *lac* genes. The relief

of repression for systems such as *lac* is termed **induction;** derivatives of lactose that inactivate the repressor and lead to expression of the *lac* genes are termed **inducers.**

Other bacterial systems operate by using protein "activator" molecules, which must bind to DNA in order to activate transcription. Still additional mechanisms of control require proteins that allow the continuation of transcription in response to intracellular signals. Before we examine some of these control circuits in detail, let's review the classic work that initially described bacterial control systems, for these studies represent landmarks in the use of genetic analysis.

Discovery of the *lac* System: Negative Control

The first major breakthrough in understanding gene control came in the 1950s with the detailed genetic analysis, by François Jacob and Jacques Monod, of the enzymes concerned with lactose metabolism in *E. coli* and of phage-λ immunity. Jacob and Monod used the lactose-metabolism system of *E. coli* (see Figure 16-2) to attack the problem of enzyme **adaptation**— that is, the appearance of a specific enzyme only in the presence of its substrates. This phenomenon had been observed in bacteria for many years. How could a cell possibly "know" precisely which enzymes to synthesize? How could a particular substrate induce the appearance of a specific enzyme?

For the *lac* system, such an "induction" phenomenon could be illustrated by the fact that in the presence of certain galactosides termed "inducers," cells produce over 1000 times more of the enzyme β-galactosidase, which cleaves β-galactosides, than they produce when grown in the absence of such sugars. What role did the inducer play in the induction phenomenon? One idea

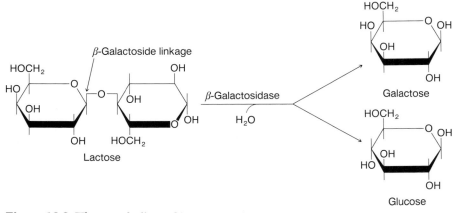

Figure 16-2. The metabolism of lactose. The enzyme β-galactosidase catalyzes a reaction in which water is added to the β-galactoside linkage to break lactose into separate molecules of galactose and glucose. The enzyme lactose permease is required to transport lactose into the cell.

was that the inducer was simply activating a pre-β-galactosidase intermediate that had accumulated in the cell. However, when Jacob and Monod followed the fate of radioactivity labeled amino acids added to growing cells either before or after the addition of an inducer, they could show that induction represented the synthesis of new enzyme molecules. Kinetic studies established that these molecules could be detected as early as three minutes after addition of an inducer! Also, withdrawal of the inducer brought about an abrupt halt in the synthesis of the new enzyme. Therefore, it became clear that the cell possesses a mechanism for turning on and off gene expression in response to environmental signals. What could this mechanism be?

A detailed examination of the substrate specificity of induction provided further important clues about the nature of the induction process. Jacob and Monod wanted to determine the correlation between the catalytic center on the enzyme being induced and the molecular structure of the inducer. At that time, the notion was prevalent that the inducer (the small β-galactoside molecule) could instruct the formation of the catalytic center, in some way serving as a mold for the active site of the enzyme. It was found that although only galactosides serve as inducers, there is no correlation between the inducing capacity of a compound and its affinity of β-galactosidase. Figure 16-3 shows some of the researchers actual data. Note for instance, how phenylethylgalactoside has a very high affinity for β-galactosidase and yet is a poor inducer. Some strong inducers, such as the synthetically prepared isopropyl-β,D-thiogalactoside (IPTG; see Figure 16-4), are not even substrates for the enzyme. The different stereospecificities for inducing β-galactosidase and for binding to β-galactosidase are clues that the element involved in controlling β-galactosidase synthesis is distinct from β-galactosidase itself.

Compound		Concentrations	β-Galactosidase			Galactoside-transacetylase	
			Induction value	V	$1/K_m$	Induction value	V/K_m
β-D-thiogalactosides	(isopropyl)	10^{-4} M	100	0	140	100	80
	(methyl)	10^{-4} M	78	0	7	74	30
		10^{-5} M	7.5	—	—	10	—
	(phenyl)	10^{-3} M	<0.1	0	100	<1	100
	(phenylethyl)	10^{-3} M	5	0	10,000	3	—
β-D-galactosides	(lactose)	10^{-3} M	17	30	14	12	35
	(phenyl)	10^{-3} M	15	100	100	11	—
α-D-galactoside 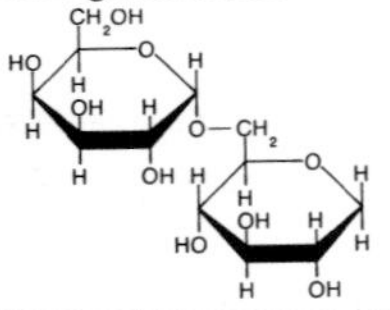	(melibiose)	10^{-3} M	35	0	<0.1	37	<1
β-D-glucoside 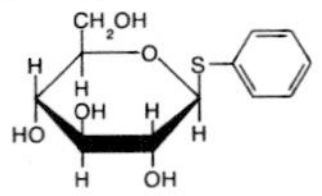	(phenyl)	10^{-3} M	<0.1	0	0	<1	50
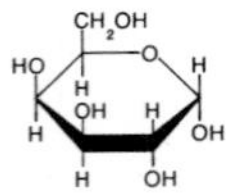	(galactose)	10^{-3} M	<0.1	—	4	<1	<1
Methyl-βD-thiogalactoside (10^{-4} M) phenyl-β-D-thigalactoside (10^{-3} M)			52	—	—	63	—

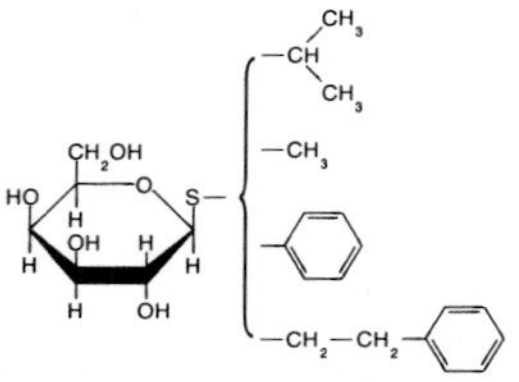

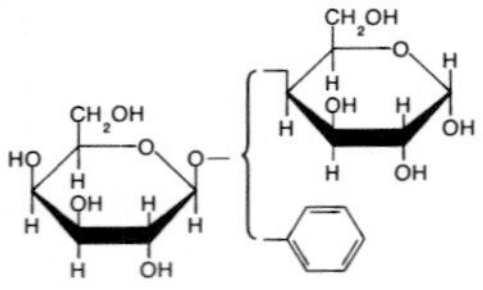

Figure 16-3. Induction of β-galactosidase and galactoside transacetylase by various galactosides. The data are taken directly from the original paper in 1961 by Jacob and Monod. The induction values represent the specific activities of each enzyme, given as the percentage of the maximum induced value. The V column refers to the maximal substrate activity of each compound with respect to β-galactosidase. The $1/K_m$ column expresses the affinity of each compound with respect to β-galactosidase. Both values are given as percentages of these values observed with the reference compound phenylgalactoside. For transacetylase, only relative values of V/K_m are given, because the affinity of this enzyme prevents independent determination of constants. (From F. Jacob and J. Monod, *Journal of Molecular Biology* 3, 1961, 318.)

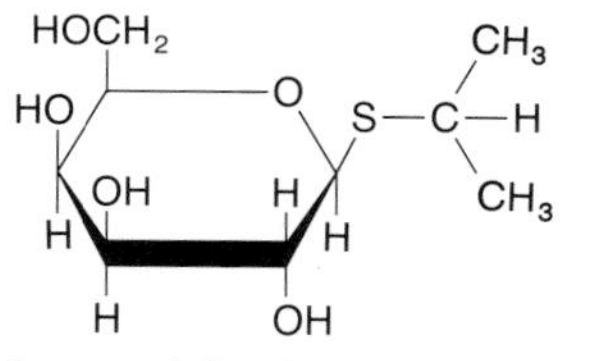

isopropyl-β,D-thiogalactoside
(IPTG)

Figure 16-4. Structure of the inducer of the *lac* operon, IPTG. The β,D-thiogalactoside linkage is not cleaved by β-galactosidase, allowing manipulation of the intracellular concentration of this inducer.

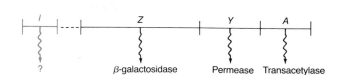

Figure 16-5. The *I* locus: the region controlling the inducibility of the *lac* enzymes.

Genes Controlled Together

Jacob and Monod induced the enzyme permease, which is required to transport lactose into the cell, together with β-galactosidase. The analysis of mutants indicated that each enzyme was encoded by a different gene. The enzyme transacetylase (with a dispensable and as-yet unknown function), also was characterized and later shown to be encoded by a separate gene, although it was induced together with β-galactosidase and permease. Therefore, Jacob and Monod could identify three **coordinately controlled genes:** the *Z* gene encoding β-galactosidase, the *Y* gene encoding permease, and the *A* gene encoding transacetylase. Mapping defined the *Z*, *Y*, and *A* genes as being closely linked on the chromosome. Later studies of these and other coordinately controlled genes led to the realization that in many cases a polycistronic mRNA molecule is produced by a contiguous set of genes. Transcription of this mRNA and its translation into protein proceeded in the same direction (see also Chapter 13). This enables us to understand the basis of an additional class of mutation referred to as **polar mutations.** Polar mutations not only affect the gene within which they map but also reduce or eliminate the expression of all genes that are farther down the line, or "distal" to the gene containing the mutation. Polar mutations exert their effects by interfering with either continued transcription or translation of the polycistronic mRNA. The first type of polar mutation characterized was the chain-terminating nonsense mutation located early in the *Z* gene, which lowers the expression of permease and transacetylase. Other polar mutations turned out to result from the insertion of DNA into the middle of genes (see Chapter 19).

The *I* Gene

Further genetic analysis shed more light on the control circuit. Jacob and Monod characterized a new class of mutant, which synthesized all three enzymes at full levels, even in the absence of an inducer. For the first time, a mutant had been detected that is altered not in the *activity* of an enzyme but in the *control* of enzyme production. These **constitutive** (always expressed in an unregulated fashion) **mutants** were found to have mutations mapping close to but distinct from the *Z*, *Y*, and *A* genes, permitting the definition of the *I* locus as the region controlling the inducibility of the *lac* enzymes. I^+ cells synthesize full levels of the *lac* enzymes only in the presence of an inducer, whereas I^- cells synthesize full levels in the presence or absence of an inducer. Figure 16-5 depicts the *lac* region defined by these experiments.

The Repressor

The discovery of F' factors (see Chapter 10) carrying the *lac* region allows complementation tests that established *I*, *Z*, and *Y* as independent cistrons. These experiments showed that I^+ is dominant over I^- in the trans position. Table 16-1 shows these results. (Recall the complementation tests performed by Seymour Benzer, Chapter 12) in which complementation occurring in the trans position implies the action of a diffusible product.) Therefore, Jacob and Monod formulated the hypothesis that the *I* gene determines the synthesis of a "repressor" molecule that blocks activation of the *lac* gene.

A piece of evidence in support of the repressor model was the characterization of I^s mutations. Although mapping within the *I* gene, these mutations prevented induction of the *lac* enzymes by lactose or by the synthetic inducer IPTG (Figure 16-4). Moreover, they were dominant in trans to both an I^+ and an I^- allele (Table 16-2). The I^s mutation eliminates response to an inducer, presumably by altering the stereospecific binding site and destroying inducer binding. Therefore, even in the presence of IPTG, these molecules can still block *lac* enzyme synthesis. This would also explain their dominance, since the I^s repressor would be unaffected by the wild-type repressor that was inactivated by the inducer. The I^s mutations clearly pointed to a direct interaction between the *I* gene product and the inducer.

TABLE 16-1. Synthesis of β-galactosidase and permease in haploid and heterozygous diploid strains

Strain	Genotype	β-galactosidase		Permease	
		Noninduced	Induced	Noninduced	Induced
1	$I^+Z^+Y^+$	−	+	−	+
2	$I^-Z^+Y^+$	+	+	+	+
3	$I^+Z^-Y^+/FI^-Z^+Y^+$	−	+	−	+
4	$I^-Z^-Y^+/FI^+Z^+Y^-$	−	+	−	+
5	$I^-Z^-Y^+/FI^-Z^+Y^+$	+	+	+	+
6	$\nabla(I, Z, Y)/FI^-Z^+Y^+$	+	+	+	+

NOTE: bacteria were grown in glycerol as a carbon source and induced by *IPTG*. The presence of the maximal level of the enzyme is indicated by +; the absence or very low level of an enzyme is indicated by −; ∇ indicates deletion.

The Operator and the Operon

The specificity of the interaction of the repressor with the *lac* system, which results in turning off enzyme synthesis, suggests a stereospecific complex with an element that Jacob and Monod termed the "operator." The operator was postulated to be a region of DNA near the beginning of the set of genes it controlled. The researchers sought mutations in this recognition element that would allow synthesis of the *lac* enzymes even in the presence of an active repressor. These mutations should be dominant in the cis position. Whereas trans dominance reflects a diffusible product, cis dominance reflects the action of an element that affects only the genes directly adjacent to it. No diffusible product is altered by the mutation. By selecting for constitutivity (unrepressed synthesis) in cells with two copies of the *lac* re-

TABLE 16-2. Synthesis of β-galactosidase and permease by the wild-type and by strains carrying different alleles of the *I* gene

Genotype	Inducer	β-galactosidase	Permease
$I^+Z^+Y^+$	None	−	−
	IPTG	+	+
$I^sZ^+Y^+$	None	−	−
	IPTG	−	−
$I^sZ^+Y^+/FI^+$	None	−	−
	IPTG	−	−
$I^sZ^+Y^+/FI^-$	None	−	−
	IPTG	−	−

NOTE: bacteria were grown in glycerol with and without the inducer IPTG. Presence of the indicated enzyme is represented by +; absence or low levels, by −.

gion, such mutants were found and labeled O^c, for **operator constitutive**. As Table 16-3 indicates, strains carrying these mutations are capable of synthesizing maximal amounts of enzyme in the presence of IPTG, and such mutant strains synthesize 10 to 20 percent of these levels in the absence of an inducer. The O^c mutations are indeed dominant in the cis position, as shown in Table 16-3. Mapping experiments have pinpointed the operator locus between *I* and *Z*.

The *OZYA* segment constitutes a genetic unit of coordinate expression that Jacob and Monod termed the "operon." Figure 16-6 depicts a simplified operon model for the *lac* system. The *lac* operon is said to be under the **negative control** of the *lac* repressor, since the repressor normally blocks expression of the *lac* enzymes in the absence of an inducer.

Let's review the model in Figure 16-6, as postulated by Jacob and Monod. The Z and Y genes code for the structure of two enzymes required for the metabolism of the sugar lactose, β-galactosidase and permease, respectively. The A gene codes for transacetylase. All three genes are linked together on the chromosome. Their transcription into a polycistronic (single) mRNA provides the basis for coordinate control at the level of mRNA synthesis. The synthesis of the polycistronic *lac* mRNA can be blocked by the action of a repressor protein molecule, which binds to an operator region near the start point for transcription. The repressor is the product of the *I* gene. Therefore, mutations in the *I* gene that prevent the synthesis of a functional repressor result in unrepressed or constitutive synthesis of the *lac* enzymes. Repression can also be overcome by certain galactosides, termed inducers, which inactivate the repressor by binding to it and altering its affinity for the operator. In this manner, the inducer can pull the repressor off the DNA.

TABLE 16-3. Synthesis of β-galactosidase and permease by haploid and heterozygous diploid operator mutants

Genotype	β-galactosidase		Permease	
	Noninduced	Induced	Noninduced	Induced
$O^+Z^+Y^+$	−	+	−	+
$O^+Z^+Y^+/FO^+Z^-Y^+$	−	+	−	+
$O^cZ^+Y^+$	+	+	+	+
$O^+Z^+Y^-/FO^cZ^+Y^+$	+	+	+	+
$O^+Z^+Y^+/FO^cZ^-Y^+$	−	+	+	+
$O^+Z^-Y^+/FO^cZ^+Y^-$	+	+	−	+
$I^sO^+Z^+Y^+/FI^+O^cZ^+Y^+$	+	+	+	+

NOTE: bacteria were grown in glycerol with and without the inducer IPTG. The presence and absence of enzyme are indicated by + and −, respectively.

Allostery

The *lac* repressor is a protein with two different binding sites. One site recognizes the inducer molecule; the second site recognizes the *lac* operator sequence on the DNA. Interaction of the repressor with the inducer lowers the affinity of the repressor for the operator. This change in affinity for operator, in response to the binding of an inducer at a distant site, is mediated by a conformational change in the repressor protein. In one conformation, the repressor binds operator well; in a second conformation, it does not. Proteins that function this way are termed **allosteric** proteins. Allosteric transitions, the change from one conformation to another, occur in many different proteins.

The *lac* Promoter

Genetic experiments have suggested that an element essential for *lac* transcription is located between *I* and *O* in the operon model for the *lac* system. This element, termed the "promoter" (*P*), is postulated to serve as an initiation site for transcription. Promoter mutations affect the transcription of all of the genes in the operon in a similar manner. Promoter mutations are cis-dominant, as would be expected for a site on the DNA that serves as a recognition element for transcription initiation, since each promoter governs transcription only for those genes in the operon adjacent to it on the *same* DNA molecule. As outlined in Chapter 13, in vitro experiments have demonstrated that RNA polymerase binds

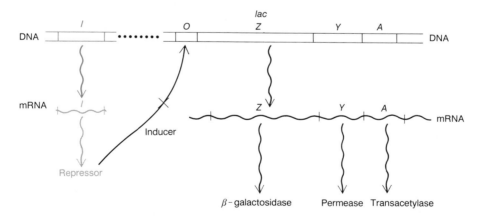

Figure 16-6. A simplified *lac* operon model. The three genes *Z, Y,* and *A* are coordinately expressed. The product of the *I* gene, the repressor, blocks the expression of the *Z, Y,* and *A* genes by interacting with the operator (*O*). The inducer can inactivate the repressor, thereby preventing interaction with the operator. When this happens, the operon is fully expressed. Mutations in *I* or *O* can also result in expression of the three *lac* enzymes, even in the absence of an inducer.

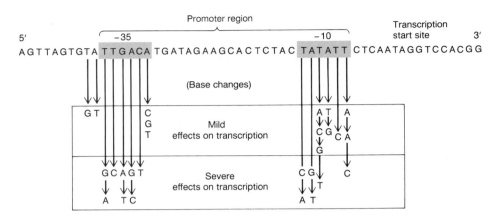

Figure 16-7. Specific DNA sequences are important for the efficient transcription of *E. coli* genes by RNA polymerase. The boxed sequences at approximately −35 and −10 are highly conserved in all *E. coli* promoters. Point mutations in these regions have noticeable effects on transcription efficiency. (From J. D. Watson, J. Tooze, and D. T. Kurtz, *Recombinant DNA: A Short Course.* Copyright © 1983 by W. H. Freeman and Company.)

to the promoter region and that repressor binding to the operator can block RNA polymerase from binding to the promoter. Mutant analysis, physical experiments, and comparisons with other promoters have identified two binding regions for RNA polymerase in a typical prokaryotic promoter. Figure 16-7 summarizes this body of information (see also Figure 13-7).

Message The *lac* operon is a cluster of structural genes that specify enzymes involved in lactose metabolism. These genes are controlled by the coordinated actions of cis-dominant promoter and operator regions. The activity of these regions is, in turn, determined by a repressor molecule specified by a separate regulator gene. Figure 16-1 integrates all this information into a single picture.

Review of Mutations in the *lac* System

Let's review some of the effects of specific mutations in the *lac* system. Clearly, mutations in the *Z* and *Y* genes will result in the Lac⁻ phenotype, since they lead to inactive β-galactosidase or permease. These mutations are recessive in a classic diploid test (here we create stable diploids for the *lac* region by using an F′ *lac* episome), because only one good *Z* gene and one good *Y* gene are needed to produce the Lac⁺ phenotype.

Mutations in the *I* gene that lead to loss of repressor function, termed *I*⁻, do not affect the Lac⁺ phenotype, but they do result in strains that now synthesize the *lac* enzymes at the full rate whether or not an inducer is present. In general, these mutations also are recessive in diploid tests (Figure 16-8), because one functional *I* gene is all that is needed to produce a repressor that can bind to both operators in a diploid.

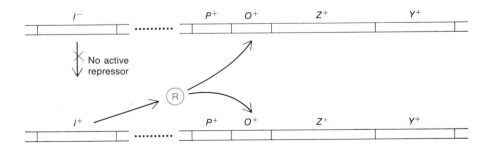

Figure 16-8. *I*⁻ mutations are recessive to wild-type. Although no active repressor is synthesized from the *I*⁻ gene, the wild-type (*I*⁺) gene provides a functional repressor that binds to both operators in a diploid cell and blocks *lac* operon expression (in the absence of an inducer).

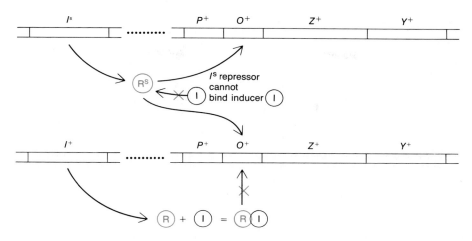

Figure 16-9. I^s mutations are dominant to wild-type. Even though the wild-type repressor is inactivated by an inducer, the I^s repressor is not and can bind to both operators in a diploid cell. Therefore, no enzyme is produced in an I^s–I^+ diploid, even in the presence of an inducer.

Different mutations in the *I* gene have been characterized that alter the inducer binding site so that the repressor no longer binds to the inducer, although it still recognizes the operator. Cells carrying one copy of an *I* gene with an I^s mutation will be Lac⁻, because the altered repressor will always bind to the operator, even in the presence of an inducer, and block synthesis of the *lac* enzymes (Figure 16-9).

Also, mutations in the operator have been found that impair repressor recognition of this short DNA segment. Such mutations allow moderate synthesis of the Lac enzymes, even in the presence of active repressor molecules. (Single mutations do not fully inactivate the operator.) Operator mutations are termed cis-dominant, because they are dominant only for genes directly linked to them on the same chromosome. Thus, if an altered operator is in the same cell with a second chro-

mosome that contains a wild-type operator, the repressor will recognize the wild-type operator and repress the genes linked to it, but it will not recognize the altered operator. Therefore, the genes linked to the altered operator are expressed even in the absence of inducer. Operator mutations are traditionally termed O^c mutations, where c stands for "constitutive," or always active. Figure 16-10 depicts the consequences of O^c mutations.

Promoter mutations, simply termed P^-, prevent RNA polymerase from recognizing the promoter and initiating transcription. These mutations are also cis-dominant.

We can see that the model in Figure 16-1 accounts for the observed phenotypes of different mutants and also for the dominant or recessive nature of each specific type of mutation (see also Figures 16-8, 16-9, and 16-10).

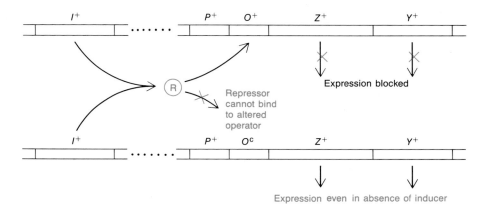

Figure 16-10. O^c mutations are dominant in the cis-position. Because a repressor cannot bind to O^c operators, genes linked to an O^c operator are expressed in the absence of an inducer. However, genes linked to an O^+ operator are still subject to repression.

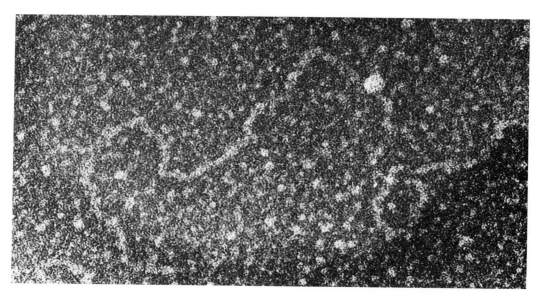

Figure 16-11. The *lac* repressor (large whitish sphere) bound to DNA at the promoter region of the operon. (Photograph courtesy of Jack D. Griffith.)

Characterization of the *lac* Repressor and the *lac* Operator

Several genetic experiments have argued strongly that the repressor is a protein—the most compelling of which was the discovery of suppressible nonsense mutations in the *I* gene, since the resulting nonsense codons exert their effect by provoking termination of the polypeptide chain during translation. The decisive experiment, however, was provided by Walter Gilbert and Benno Müller-Hill, who in 1966 isolated and purified the repressor by monitoring the binding of the radioactively labeled inducer IPTG. They demonstrated that the repressor is a protein consisting of four identical subunits, each with a molecular weight of approximately 38,000 daltons. Each molecule contains four IPTG-binding sites. (A more detailed description of the repressor is given later in the chapter.) In vitro, repressor binds to DNA containing the operator (see Figure 16-11) and comes off the DNA in the presence of IPTG. Gilbert and his coworkers have shown that the repressor can protect specific bases in the operator from chemical reagents. These experiments provide crucial proofs of the mechanism of repressor action formulated by Jacob and Monod.

Gilbert used the enzyme DNase to break apart the DNA bound to the repressor. He was able to recover short DNA strands that had been shielded from the enzyme activity by the repressor molecule and that presumably represented the operator sequence. This se-

quence was determined, and each operator mutation was shown to involve a change in the sequence (Figure 16-12). These results confirm the identity of the operator locus as a specific sequence of 17 to 25 nucleotides situated just before the structural *Z* gene. They also show the incredible specificity of repressor-operator recognition, which is disrupted by a single base substitution. When the sequence of bases in the *lac* mRNA (transcribed from the *lac* operon) was determined, the first 21 bases on the 5′ initiation end proved to be complementary to the operator sequence Gilbert had determined.

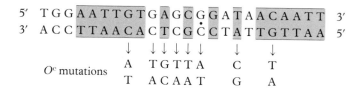

Figure 16-12. The DNA base sequence of the lactose operator and the base changes associated with eight *O*ᶜ mutations. Regions of twofold rotational symmetry are indicated by horizontal lines above and below the symmetric base pairs and by a dot at their axis of symmetry. (From W. Gilbert, A. Maxam, and A. Mirzabekov, in N. O. Kjeldgaard and O. Mallǿe, eds., *Control of Ribosome Synthesis.* Academic Press, © 1976. Used by permission of Munksgaard International Publishers Ltd., Copenhagen.)

Catabolite Repression of the *lac* Operon: Positive Control

There is an additional control system superimposed on the repressor-operator system. This system exists because cells have specific enzymes that favor glucose uptake and metabolism. If both lactose *and* glucose are present, synthesis of β-galactosidase is not induced until all of the glucose has been utilized. Thus, the cell conserves its metabolic machinery by utilizing any existing glucose before going through the steps of creating new machinery to exploit the lactose. The operon model outlined previously will not account for the suppression of induction by glucose, so we must modify it.

Studies indicate that in fact some catabolic breakdown product of glucose (no exact identity is yet known) prevents activation of the *lac* operon by lactose, so this effect was originally called **catabolite repression.** The effect of the glucose catabolite is exerted on an important cellular constituent called cyclic adenosine monophosphate (cAMP). When glucose is present in high concentrations, the cAMP concentration is low; as the glucose concentration decreases, the concentration of cAMP increases correspondingly. The high concentration of cAMP is necessary for activation of the *lac* operon. Mutants that cannot convert ATP to cAMP cannot be induced to produce β-galactosidase because the concentration of cAMP is not great enough to activate the *lac* operon. In addition, there are other mutants that do make cAMP but cannot activate the *lac* enzymes because they lack yet another protein, called CAP (catabolite activator protein), made by the *crp* gene. The CAP protein forms a complex with cAMP, and it is this complex that activates the *lac* operon (Figure 16-13).

How does catabolite repression fit into our model for the structure and regulation of the *lac* operon? Recall

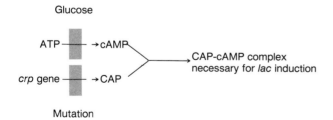

Figure 16-13. Catabolite control of the *lac* operon. The operon is inducible by lactose to the maximal levels when cAMP and CAP form a complex. The *lac* operon cannot be expressed at full levels if formation of cAMP is blocked by excess glucose or if formation of CAP is blocked by mutation of the *crp* gene. (CAP = catabolic activator protein; cAMP = cyclic adenosine monophosphate; *crp* = structural gene responsible for synthesizing CAP.)

$$\overline{\mathrm{GTGAGT}}\ \overline{\mathrm{TAGCTCAC}}$$
$$\bullet$$
$$\underline{\mathrm{CACTCAATCGAGTG}}$$

Figure 16-14. The DNA base sequence to which the CAP–cAMP complex binds. Regions of twofold rotational symmetry are indicated by horizontal lines above and below the symmetric base pairs and by a dot at their axis of symmetry.

the technique that Gilbert used to identify the operator base sequence. In a similar experiment, the CAP-cAMP complex was added to DNA, and the DNA then was subjected to digestion by the enzyme DNase. The surviving strands are presumably those shielded from digestion by an attached CAP-cAMP complex; the sequence of these strands (Figure 16-14) clearly is different from the operator sequence (Figure 16-12), but it also has a rotational twofold symmetry.

The entire *lac* operon can be inserted into λ phage in such a way that the initiation of transcription is prompted by the phage gene adjacent to *lac*. In this case, the transcribed product carries a complementary copy of the base sequence from the *lac* control regions (sequences not transcribed in the *lac* mRNA). We already know the amino-acid sequences for the repressor and β-galactosidase, so these sequences can be identified, and the remaining sequences can be assigned to the control regions (Figure 16-15). We can also fit the known repressor, CAP-cAMP, and RNA polymerase binding sites into the detailed model.

The knowledge about the *lac* operon provides insight into the elegance of gene regulation. The *E. coli* cell normally processes glucose as a source of energy and carbon. It possesses an "emergency" capability to process lactose, but it does not waste energy or materials in preparing the mechanism for that processing as long as glucose is available, even if lactose is also present. This control is accomplished because glucose-breakdown product inhibits formation of the CAP-cAMP complex required for the RNA polymerase to attach at the *lac* promoter site. Even when there is a shortage of glucose catabolites and CAP-cAMP forms, the mechanism for lactose metabolism will be created only if lactose is present. This control is accomplished because lactose must bind to the repressor protein to remove it from the operator site and permit transcription of the *lac* operon. Thus, the cell conserves its energy and resources by producing the lactose-metabolizing enzymes only when they are both needed and useful.

Whereas inducer-repressor control of the *lac* operon is an example of **negative control,** the CAP-cAMP system is an example of **positive control,** because its expression requires the presence of an activating

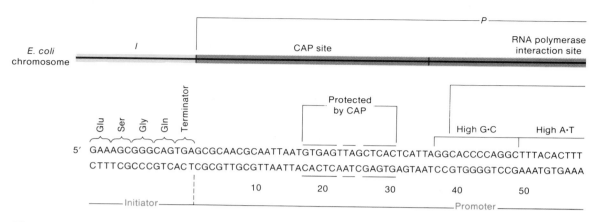

Figure 16-15. The base sequence and the genetic boundaries of the control region of the *lac* operon, with partial sequences for the structural genes. (After R. C. Dickson, J. Abelson, W. M. Barnes, and W. S. Reznikoff, "Genetic Regulation: The *Lac* Control Region." *Science* 187, 1975, 27. Copyright © 1975 by the American Association for the Advancement of Science.)

signal—in this case, the interaction of the CAP-cAMP complex with the CAP region. Figure 16-16 distinguishes between these two basic types of control systems.

> **Message**   The *lac* operon has an added level of control so that the operon remains inactive in the presence of glucose even if lactose is also present. High concentrations of glucose catabolites produce low concentrations of cyclic adenosine monophosphate cAMP, which must form a complex with CAP to permit the induction of the *lac* operon.

By using different combinations of controlling elements, bacteria have evolved numerous strategies for regulating gene expression. Some examples follow.

Dual Positive and Negative Control: The Arabinose Operon

The metabolism of the sugar arabinose is catalyzed by three enzymes encoded by the *araB*, *araA*, and *araD* genes. Figure 16-17 depicts the control circuits for this

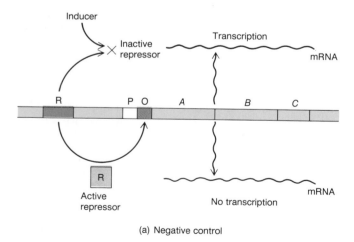

(a) Negative control

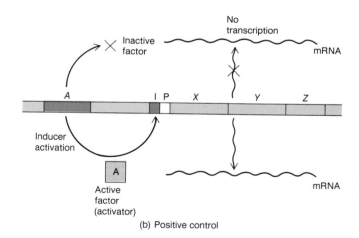

(b) Positive control

Figure 16-16. Comparison of positive and negative control. The basic aspects of negative and positive control are depicted. (a) In negative control, an active repressor (encoded by the *R* gene in the example shown here) blocks gene expression of the *A,B,C* operon by binding to an operator site (*O*). An inactive repressor allows gene expression. The repressor can be inactivated either by an inducer or by mutation. (b) In positive control, an active factor is required for gene expression, as shown for the *X,Y,Z* operon here. Small molecules can convert an inactive factor into an active one, as in the case of cyclic AMP and the CAP protein. An inactive positive control factor results in no gene expression. The activator binds to the control region of the operon, termed *I* in this case. (The positions of both *O* and *I* with respect to the promoter, *P*, in the two examples are arbitrarily drawn.)

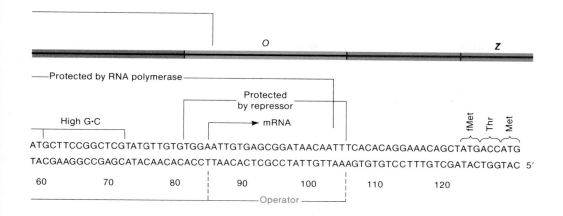

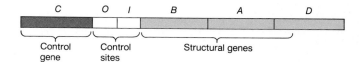

Figure 16-17. Map of the *ara* region. The *DAB* genes together with the *I* and *O* sites constitute the *ara* operon.

operon. Expression is activated at the adjacent **initiator** region, *araI*. Within this region, the product of the *araC* gene, when bound to arabinose, can activate transcription, perhaps by directly affecting RNA polymerase binding in the *araI* region. This represents positive control, since the product of the regulatory gene (*araC*) must be active in order for the operon to be expressed. An additional positive control is mediated by the same CAP-cAMP system that regulates *lac* expression.

In the presence of arabinose, the binding of the C product to the initiator region and the CAP protein are required to allow RNA polymerase to bind to the promoter for the *araB, araA,* and *araD* genes (Figure 16-18a). In the absence of arabinose, the *araC* product assumes a different conformation and actually represses the *ara* operon by binding both to *araI* and an operator

region, *araO*, thereby forming a loop (Figure 16-18b) that prevents transcription. Therefore, the *araC* protein has two conformations that promote two opposing functions at two alternative binding sites. The conformation is dependent on whether the inducer, arabinose, is bound to the protein.

Figure 16-18. Dual control of the *ara* operon. (a) In the presence of arabinose, the *aracC* protein binds to the *araI* region and, when bound to cyclic AMP, the CAP protein binds to a site adjacent to *araI*. This stimulates the transcription of the *araB, araA,* and *araD* genes. (b) In the absence of arabinose, the *araC* protein binds to both the *araI* and *araO* regions, forming a DNA loop. This prevents transcription of the *araB–A–D* operon.

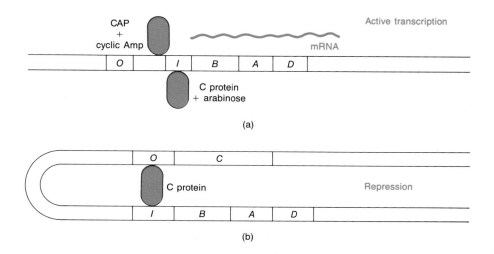

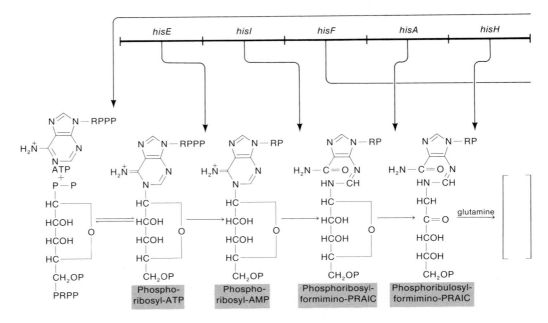

Figure 16-19. The *histidine (his)* gene cluster and the metabolic pathway that it controls. Note that the sequence of genes in the cluster generally corresponds to the sequence of steps that each gene catalyzes in the pathway for the synthesis of histidine. The fact that the final gene in the sequence *(hisG)* catalyzes the first step in the reaction sequence is probably not a coincidence; this pattern is commonly found.

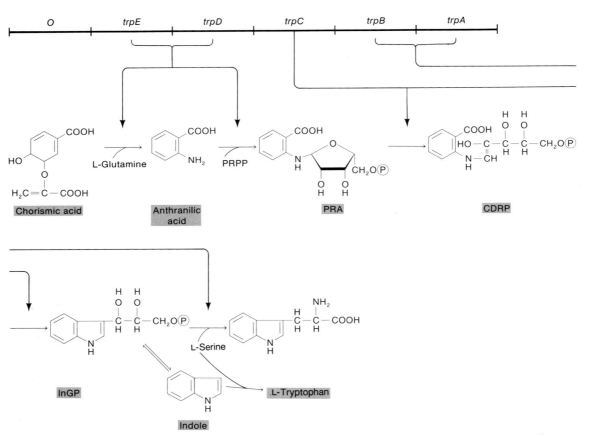

Figure 16-20. The genetic sequence of cistrons in the *trp* operon of *E. coli* and the sequence of reactions catalyzed by the enzyme products of the *trp* structural genes. The products of genes *trpD* and *trpE* form a complex that catalyzes specific steps, as do the products of genes *trpB* and *trpA*. (After S. Tanemura and R. H. Bauerle, *Genetics* 95, 1980, 545–559.)

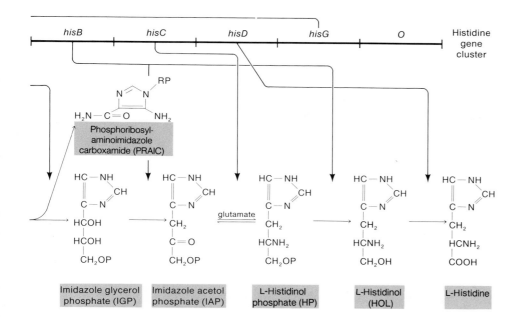

The figure shows the histidine gene cluster (hisB, hisC, hisD, hisG, O) mapped to the metabolic pathway with intermediates: Phosphoribosyl-aminoimidazole carboxamide (PRAIC), Imidazole glycerol phosphate (IGP), Imidazole acetol phosphate (IAP), L-Histidinol phosphate (HP), L-Histidinol (HOL), and L-Histidine.

Metabolic Pathways

Coordinate control of genes in bacteria is widespread. In the early 1960s, when Milislav Demerec studied the distribution of loci affecting a common biosynthetic pathway, he found that the genes controlling steps in the synthesis of the amino acid tryptophan in *Salmonella typhimurium* are clustered together in a restricted part of the genome. In 1964, Demerec then looked at the distribution of genes involved in a number of different metabolic pathways. Analyzing auxotrophic mutations representing 87 different cistrons, he found that 63 could be located in 17 functionally similar clusters. A **cluster** is defined as two or more loci that control related functions, where the loci are carried on a single transducing fragment and are not separated by an unrelated gene.

Furthermore, in cases where the sequence of catalytic activity is known, there is a remarkable congruence between the sequence of genes on the chromosome and the sequence in which their products act in the metabolic pathway. This congruence is strikingly illustrated by the histidine cluster in *Salmonella,* extensively studied in the early 1960s by Philip Hartman and Bruce Ames (Figure 16-19), and by the tryptophan cluster in *E. coli* (Figure 16-20), characterized during the same period by Charles Yanofsky.

> **Message** Genes involved in the same metabolic pathway are frequently tightly clustered on prokaryotic chromosomes, often in the same sequence as the reactions that they control. Furthermore, the genes within a cluster often are expressed at the same time.

The *Tryptophan* Genes: Negative Control with Superimposed Attenuation

The *lac* operon is an example of an inducible system in the sense that the synthesis of an enzyme is induced by the presence of its substrate. Repressible systems also exist, in which an excess of product leads to a shutdown of the production of the enzymes involved in synthesizing that product. Such a control system has been identified for a cluster of genes controlling enzymes in the pathway for tryptophan production. Synthesis of tryptophan is shut off when there is an excess of tryptophan in the medium. Jacob and Monod suggested that the cluster of five *trp* cistrons in *E. coli* forms another operon, differing from the *lac* operon in that the tryptophan repressor will bind to the *trp* operator only when it *is* bound to tryptophan (Figure 16-20). (Recall that the *lac* repressor binds to the operator except when it is bound to lactose.) A second control pathway also modulates tryptophan biosynthesis at the level of enzyme activity. This is termed **feedback inhibition.** Here, the first enzyme in the pathway, encoded by the *trpE* and *trpD* genes, is inhibited by tryptophan itself.

As with the *lac* operon, further analysis of the *trp* operon has revealed yet another level of control superimposed on the basic repressor-operator mechanism. In studying constitutive mutant strains (carrying a mutation in *trpR*, the repressor locus) that continue to produce *trp* mRNA in the presence of tryptophan, Yanofsky found that removal of tryptophan from the medium leads to almost a tenfold increase in *trp* mRNA production in these strains. Furthermore, Yanofsky isolated a

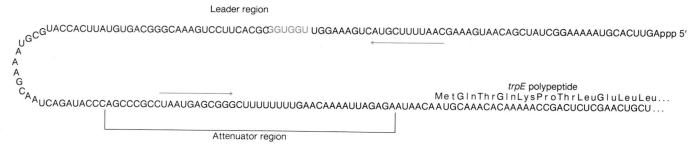

Leader region

```
            U GC G UACCACUUAUGUGACGGGCAAAGUCCUUCACGCGGUGGU UGGAAAGUCAUGCUUUUAACGAAAGUAACAGCUAUCGGAAAAAUGCACUUGAppp 5'
          A                                                            ←
        A
        A
        G
        C AA                                                                           trpE polypeptide
          A UCAGAUACCCAGCCCGCCUAAUGAGCGGGCUUUUUUUUGAACAAAAUUAGAGAAUAACA AUGCAAACACAAAAACCGACUCUCGAACUGCU...
                                                                      MetGlnThrGlnLysProThrLeuGluLeuLeu...
```
Attenuator region

Figure 16-21. The leader sequence and attenuator sequence of the *trp* operon, with the beginning of the *trpE* structural sequence (showing the amino-acid sequence of the *trpE* polypeptide). (From G. S. Stent and R. Calendar, *Molecular Genetics,* 2d ed. Copyright © 1978 by W. H. Freeman and Company. Based on unpublished data provided by Charles Yanofsky.)

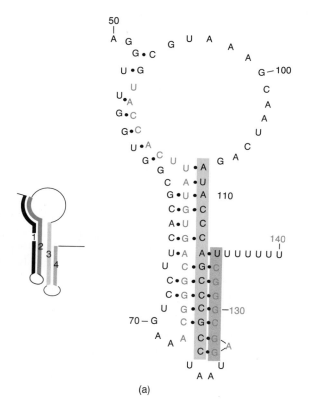

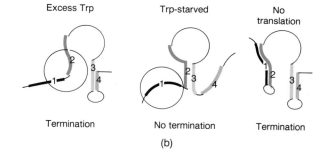

(a)   (b)

Figure 16-22. (a) Proposed secondary structures in *E. coli* terminated *trp* leader RNA. Four regions can base-pair to form three stem-and-loop structures. (b) Model for attenuation in the *E. coli trp* operon. Under conditions of excess tryptophan, the ribosome (closed circles) translating the newly transcribed leader RNA will synthesize the complete leader peptide. During this synthesis, the ribosome will mask regions 1 and 2 of the RNA and prevent the formation of stem-and-loop 1·2 or 2·3. Stem-and-loop 3·4 will be free to form and signal the RNA polymerase molecule (not shown), transcribing the leader region to terminate transcription. Under conditions of tryptophan starvation, charged tRNA^Trp will be limiting and the ribosome will stall at the adjacent *trp* codons in the leader peptide-coding region. Because only region 1 is masked, stem-and-loop 2·3 will be free to form as regions 2 and 3 are synthesized. Formation of stem-and-loop 2·3 will exclude the formation of stem-and-loop 3·4, which is required as the signal for transcription termination. (From D. L. Oxender, G. Zurawski and C. Yanofsky, *Proceedings of the National Academy of Sciences USA* 76, 1949, 5524.)

Met - Lys - Ala - Ile - Phe - Val - Leu - Lys - Gly - Trp - Trp - Arg - Thr - Ser -

〜〜AUG AAA GCA AUU UUC GUA CUG AAA GGU UGG UGG CGC ACU UCC UGA〜〜

Figure 16-23. Translated portion of the *trp* leader region, shown with the corresponding sequence of the leader mRNA.

constitutive mutant strain that produces *trp* mRNA at the maximal level, even in the presence of tryptophan, and he showed that this mutation has a deletion located between the operator and the *trpE* cistron.

Yanofsky was able to isolate the polycistronic *trp* operon mRNA. On sequencing it, he found a long sequence, termed the **leader sequence**, of 160 bases at the 5′ end before the first triplet in the *trpE* gene. The deletion mutant that always produces *trp* mRNA at maximal levels has a deletion extending from base 130 to base 160 (Figure 16-21). Yanofsky called this the **attenuator** region, because its presence apparently leads to a reduction in the rate of mRNA transcription when tryptophan is present. But what is the role of the leader sequence in bases 1 to 130? A surprising observation provides the key to solving this problem.

When studying the mRNA's transcribed from the *trp* operon, Yanofsky discovered that the original constitutive mutant strains continue to produce the first 141 bases of the mRNA at the maximal rate, even in the presence of tryptophan. In other words, the base 141 to 160 segment apparently normally acts as a chain terminator to halt transcription of about nine in every 10 mRNA's if tryptophan is present. When tryptophan is absent, every transcription is carried through this attenuator region and on to completion. In the deletion mutants, the attenuator region is missing, so transcription is carried through in every case, regardless of the presence or absence of tryptophan.

What causes the interference with termination at the attenuator in the absence of tryptophan? Figure 16-22 presents a model based on alternate secondary structures formed by the mRNA in the leader region. The model proposes that one of the two conformations favors transcription termination and that the other favors elongation. Translation of part of the leader sequence would promote the conformation that favors termination.

It is known that the leader sequence can be translated to yield a short peptide of 11 amino acids. There are two tryptophan codons in the translated stretch of the leader mRNA (Figure 16-23). When excess tryptophan is present, there is a sufficient supply of charged tRNA to allow efficient translation through the relevant portion of the leader mRNA, which results in transcription termination. However, in the absence of tryptophan, charge *trp* tRNA is limiting and translation stalls in the leader, allowing a secondary structure of the mRNA to form that is favorable to the continuation of transcription. In this manner, an additional 10-fold range of tryptophan biosynthetic enzymes is superimposed on the normal range that is achieved by repressor-operator interaction. The analysis of numerous point mutations in the *trp* leader sequence that favor or disfavor the respective secondary structures lends strong support to Yanofsky's attenuation model.

Several operons for enzymes in biosynthetic pathways have attenuation controls similar to the one described for tryptophan (Figure 16-24). For instance, the leader region of the *his* operon, which encodes the enzymes of the histidine biosynthetic pathway, contains a translated region with seven consecutive histidine codons. Mutations at outside loci that result in lowering levels of normal charged *his* tRNA produce partially

(a) Met - Lys - His - Ile - Pro - Phe - Phe - Phe - Ala - Phe - Phe - Phe - Thr - Phe - Pro - Stop

5′ AUG AAA CAC AUA CCG UUU UUC UUC GCA UUC UUU UUU ACC UCC CCC UGA 3′

(b) Met - Thr - Arg - Val - Gln - Phe - Lys - His - His - His - His - His - His - His - Pro - Asp -

5′ AUG ACA CGC GUU CAA UUU AAA CAC CAC CAU CAU CAC CAU CAU CCU GAC 3′

Figure 16-24. Amino-acid sequence of the leader peptide and base sequence of the corresponding portion of mRNA from (a) the phenylalanine operon and (b) the histidine operon. Note that seven of the 15 residues in the phenylalanine operon leader are phenylalanine and that seven consecutive residues from the histidine operon leader peptide are histidine. (From L. Stryer, *Biochemistry*. Copyright © 1981 by W. H. Freeman and Company.)

constitutive levels of the enzymes encoded by the *his* operon.

Message The *trp* operon is regulated by a negative repressor–operator control system that represses the synthesis of tryptophan enzymes when tryptophan is present in the medium. A second level of control involves an attenuator region where termination of transcription is induced by the presence of tryptophan.

The λ Phage: A Complex of Operons

At the time they proposed the operon model, Jacob and Monod suggested that the genetic activity of temperate phages might be controlled by a system analogous to the *lac* operon. In the lysogenic state, the prophage genome is inactive (repressed). In the lytic phase, the phage genes for reproduction are active (induced). Since Jacob and Monod proposed the idea of operon control for phages, the genetic system of the λ phage has become one of the better understood control systems. This phage does indeed have an operon-type system controlling its two functional states. By now, you should not be surprised to learn that this system has proved to be more complex than initially suggested.

Alan Campbell induced and mapped many conditionally lethal mutations in the λ phage (Figure 16-25), showing that there is clear evidence for the clustering of genes with related functions. Furthermore, mutations in the *N*, *O*, and *P* genes prevent most of the genome from being expressed after phage infection, with only those loci lying between *N* and *O* being active. We shall soon see the significance of this observation.

When normal bacteria are infected by wild-type λ phage particles, two possible sequences may follow: (1) a phage may be integrated into the bacterial chromosome as an inert prophage (thus lysogenizing the bacterial cell), or (2) the phage may produce the necessary enzymes to guide the production of the products needed for phage maturation and cell lysis. When wild-type phage particles are placed on a lawn of sensitive bacteria, clearings (plaques) appear where bacterial cells are infected and lysed, but these plaques are turbid because lysogenized bacteria (which are resistant to phage superinfection) grow within the plaques.

Mutant phages that form clear plaques can be selected as a source of phages that are unable to lysogenize cells. Such *clear (c)* mutants prove to be analogous to *I* and *O* mutants in *E. coli*. For example, conditional mutants for a site called *cI* are unable to establish a lysogenic state under restrictive conditions but also fail to induce lysis in a cell that has been lysogenized by a wild-type prophage. Apparently, the *cI* mutation produces a defective repressor in the phage control system.

Virulent mutants have been isolated that do not lysogenize cells but that do grow in a lysogenized cell, thus providing defects that are insensitive to the λ repressor. Genetic mapping has revealed *two* operators, designated O_L and O_R, located on the left and right of *cI*, respectively. Furthermore, each operator has a promoter, termed P_L and P_R. Figure 16-26 shows a simplification of the genetic map of the phage control regions. Because transcription extends from the *O* regions away from *cI*, the transcripts must be read from different strands of the DNA.

Mark Ptashne was able to purify the λ repressor. He showed that each λ genome binds six repressor mole-

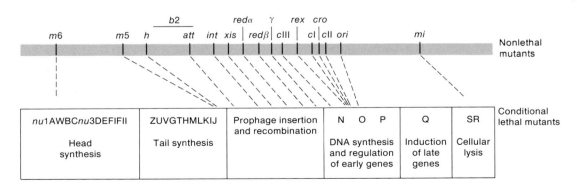

Figure 16-25. The genetic map of the λ phage. The positions of nonlethal and conditionally lethal mutations are indicated, and the characteristic clusters of genes with related functions are shown. (From A. Campbell, *The Episomes,* Harper & Row, 1969.)

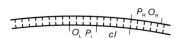

Figure 16-26. Genetic map of the control elements of λ reproduction on the double-stranded phage DNA. Each symbol is placed next to the strand that is transcribed for the corresponding gene.

> **Message** The regulation of the lytic and lysogenic states of the λ phage provides a model for interacting control systems that may be useful for interpreting gene regulation in eukaryotes.

cules, three at each operator. Furthermore, the repressor-binding sites of O_R overlap the promoter for the cI gene itself (Figure 16-27). Different affinities of the repressor for each of the three operator sites allow the repressor to regulate its own synthesis. When there is an excess of repressor, binding to the low-affinity third O_R site occurs, reducing further repressor synthesis. This keeps repressor concentrations from accumulating to the point at which eventual induction would be difficult.

When a phage infects a cell, there is no repressor present. A host-cell enzyme (RNA polymerase) initiates transcription, beginning at O_L and extending through the N gene and beginning at O_R and extending through the cro gene (Figure 16-27).

Now we see why the N gene is important: its product interacts with the RNA polymerase to modify it in such a way that transcription will not terminate in an attenuator region at the ends of N and cro. Instead, transcription proceeds through to genes that are involved in phage DNA replication, recombination, and lysogeny. (The name of the cro gene is derived from the expression "*control of repressor and other things*.") The cro protein acts to inhibit production of the repressor by binding to a promoter region adjacent to the cI locus. Thus, the cro product appears to be a repressor of a repressor, and it must act during the lytic phase of λ growth. On the other hand, the λ repressor binds to O_R, shutting off transcription of the cro gene. Thus, cro and cI activity must be mutually exclusive, with cro activity required for lysis and cI activity required for lysogeny.

The complexity of the λ cycle only increases with study. The description given here is an abbreviated version of a well-characterized process that provides further modifications of the operon model.

Multioperon Repression

Repressors can simultaneously control several or even a large number of operons. For instance, the trp repressor simultaneously regulates the *trp* operon, the *aroH* gene, and the *trp* repressor gene itself. The three respective operators show important sequence homologies. Another multicomponent system subject to the same repressor is the set of functions that constitutes the SOS repair pathway (discussed in conjunction with mutagenesis in Chapter 17). A repressor, encoded by the *lexA* gene, reduces expression of numerous genes involved in DNA repair, including the *recA, uvrA, uvrB,* and *umuC* genes (see Figure 16-28; see also page 490). The LexA repressor also regulates its own synthesis. Some of these LexA-repressed genes still maintain a residual level of expression in the presence of the LexA repressor. In response to an inducing signal, somehow triggered by damaged DNA that blocks replication, a protease activity of the protein encoded by *recA* is activated. This results in the cleavage of the LexA repressor, allowing high levels of synthesis of the SOS functions. UV light and other agents that damage DNA induce the SOS functions. Interestingly, the RecA protease also cleaves the λ repressor (the product of the λcI gene), as well as the repressors of several other phages related to λ. This is the physical basis for the classical UV-light induction of λ lysogens.

Structure of Regulatory Proteins

Protein-sequence analyses and structural comparisons indicate that a number of DNA-binding regulatory proteins share important features. All consist of a DNA-binding domain, located at the amino-terminal end of the protein, which protrudes from the main "core" of the protein. In certain cases, the core protein contains

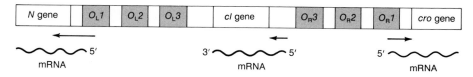

Figure 16-27. The cI control region for λ. (After L. Stryer, *Biochemistry,* 3d ed. Copyright © 1988 by W. H. Freeman and Company.)

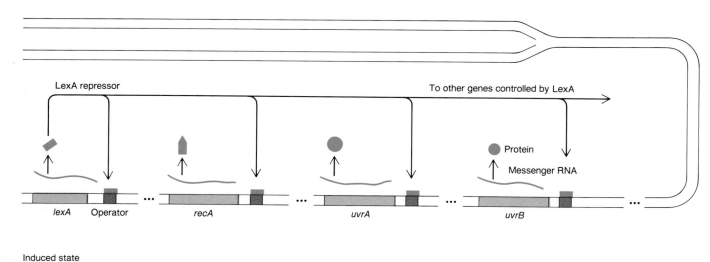

Induced state

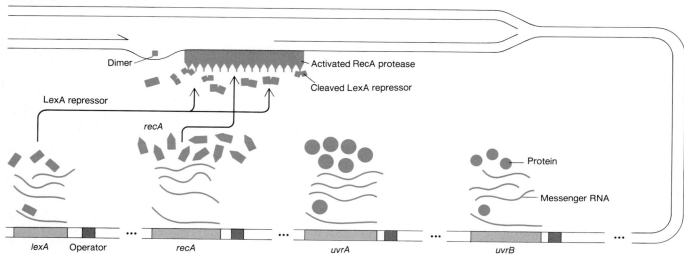

Figure 16-28. Regulatory system based on LexA and RecA. This system is quiescent during normal growth in the absence of damage to DNA *(top)*. The LexA repressor binds to the operators of *lexA, recA, uvrA, uvrB,* and some other genes, keeping the synthesis of messenger RNA and protein at the low level characteristic of uninduced cells. Damage to DNA sufficient to produce a postreplication gap activates the *SOS* response *(bottom)*. The RecA protein binds to the single-stranded DNA opposite gap; its protein-cleaving activity is thereby activated and the LexA repressor is cleaved. In the absence of a functional repressor, the LexA-controlled genes are switched on and protein is synthesized at an increased rate. (From P. Howard-Flanders, "Inducible Repair of DNA." Copyright © 1981 by Scientific American, Inc. All rights reserved.)

the inducer-binding site. This arrangement (Figure 16-29) holds for the Lac, λcI, and λcro repressors, as well as for the CAP protein.

It has been postulated that protruding α-(alpha) helices fit into the major groove of the DNA. The three-dimensional structure of the λCro repressor suggests how such an operator-repressor complex might look

(Figure 16-30). A favored detailed model is shown in Figure 16-31. Here two α-helices from the repressor protein interact with the two consecutive major grooves of the DNA of the operator site. The helices are connected by a turn in the protein secondary structure. This **helix-turn-helix motif,** (Figure 16-32) is common to many regulatory proteins.

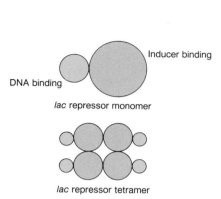

Inducer binding

DNA binding

lac repressor monomer

lac repressor tetramer

Figure 16-29. Schematic diagram showing the arrangement of domains in the *lac* repressor. All mutations affecting DNA and operator binding result in alterations in the amino-terminal end of the protein, whereas mutants defective in inducer binding or aggregation result in alterations in the remaining portion of the protein.

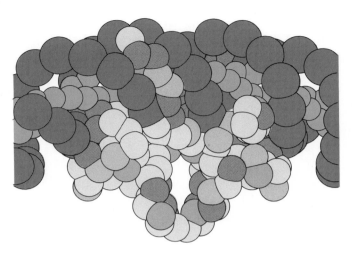

Figure 16-30. Schematic drawing of the proposed sequence-specific complex of Cro with DNA. For the stylized DNA (dark and medium gray), the larger circles indicate the positions of the phosphate groups and the smaller circles follow the bottom of the major and minor grooves. For the Cro dimer (color), one circle is drawn for each amino acid. (From D. H. Ohlendorf et al. *Nature* 298, 1982, 718–723.)

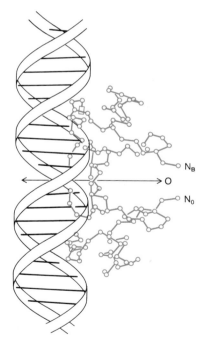

Figure 16-31. Presumed interaction of the Cro repressor with DNA. Two monomers of Cro interact with the DNA. The respective amino termini of the two Cro molecules are labeled N_0 and N_B. A pair of twofold-related α-helices occupy successive major grooves of the DNA. (From W. F. Anderson et al. *Nature* 290, 1981, 754–58.)

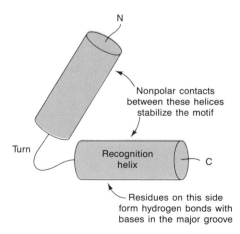

N

Nonpolar contacts between these helices stabilize the motif

Turn

Recognition helix

C

Residues on this side form hydrogen bonds with bases in the major groove

Figure 16-32. Helix-turn-helix motif of DNA binding proteins. These dimeric proteins contain two such units, separated by 34Å — the pitch of the DNA helix. (From L. Stryer, *Biochemistry,* 3d ed. Copyright © 1988 by W. H. Freeman and Company.)

Moreover, the striking partial-sequence homologies of the DNA-binding domains of several regulatory proteins (Figure 16-33) suggest that they bind to other respective operator sites in a similar fashion. It is hoped that the determination of the three-dimensional structure of a repressor-operator complex by X-ray crystallography will allow the elucidation of the rules for DNA sequence recognition by regulatory proteins.

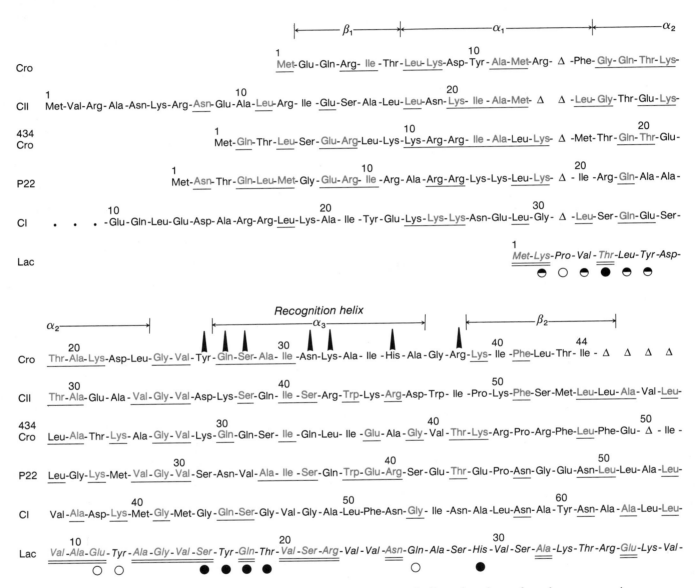

Figure 16-33. Comparison of the amino-terminal amino-acid sequence of the *lac* repressor with five other DNA-binding proteins. Residues that are homologous within the five DNA-binding proteins are indicated by color and a single underline. Residues of the *lac* repressor that are common to one or more of the five proteins are indicated by color and a double underline. α and β show the locations of the α-helices and β-sheet strands of the *cro* protein. The residues of *cro* that are presumed from model building to interact with DNA are capped by an arrowhead. Δ indicates an assumed deletion. The solid circles underneath the *lac* sequence indicate locations where known mutations dramatically reduce DNA-binding ability but do not interfere with inducer binding (do not simply destablize the whole protein). The half-filled circles indicate locations at which amino-acid substitutions may reduce DNA-binding affinity or at which the reduction in binding is weak. Open circles indicate sites at which substitution does not interfere with DNA binding. (Data compiled by B. W. Matthews et al., *Proceedings of the National Academy of Sciences USA* 79, 1982, 1428–1432.)

Transcription: Gene Regulation in Eukaryotes

There are several levels of control of gene expression in eukaryotes. We have already seen in Chapter 14 that chromosome structure can profoundly influence gene activity, as exemplified by the inactivation of the X chromosome (see page 380). Processing of the primary gene transcript (Chapter 13) provides an example of control at a different post-transcriptional level. For instance, in the case of the calcitonin peptide in the thyroid and the CGRP peptide in the hypothalmus (Figure 13-35), alternate splicing pathways in two different organs generate different translation products from the same primary transcript.

What about control of the rate of transcription? What is known about the control signals for eukaryotic genes and about their specific regulation at the level of transcription? There are some similarities between transcriptional control circuits in prokaryotes and eukaryotes. In bacterial systems, positive and negative regulators serve as trans-acting factors that operate on cis-dominant regulatory sequences, such as operators, initiators, and binding sites within promoters. Trans-acting factors and cis-acting regulatory sequences have also been characterized in diverse eukaryotic systems. Moreover, there is a remarkable conversion of control mechanisms from one eukaryotic species to another.

Control of Transcription

As we learned in Chapter 13, there are three different RNA (ribonucleic acid) polymerases in eukaryotic systems. All mRNA molecules are synthesized by RNA polymerase II. In order to achieve maximal rates of transcription, RNA polymerase II requires two basic cis-acting control sequences: promoters, and additional sequence elements, termed **enhancers,** in mammalian cells and **upstream activating sequences (UAS's)** in yeast. Each of these elements is recognized by transacting factors, which serve as positive control elements. Properly functioning promoters are required for tran-

scription initiation; enhancers serve to maximize the rate of transcription for promoters. Enhancers and promoters were originally distinguished by the distance from the point of initiation at which they operate. Promoters function near the initiation site, whereas enhancers can usually function at great distances from this point. More recent work has shown that the internal structure of these elements is more similar than originally imagined.

Promoters

In the same manner that a comparison of promoter sequences in bacteria (Figure 13-7) led to an understanding of important elements in prokaryotic promoters, a compilation of DNA sequences upstream of mRNA start sites in eukaryotes has revealed conserved sequence elements. Figure 16-34 gives a schematic view of sequence elements that have been identified as part of eukaryotic promoters. The **TATA** box is involved in directing RNA polymerase to begin transcribing approximately 30 base pairs (bp) downstream in mammalian systems and 60 to 120 bp downstream in yeast. The TATA box works most efficiently together with two upstream sequences located approximately 40 bp and 110 bp, respectively, from the start of transcription. The **CCAAT** box serves as one of these sequences, and a GC-rich segment often serves as the other sequence (Figure 16-34). Site-directed mutational analysis has demonstrated that altering bases in the TATA box or in the upstream sequences lowers in vivo transcription rates. In one version of this experiment by Richard Myers, Kit Tilly, and Tom Maniatis (Figure 16-35), a "saturation mutagenesis" in vivo alters, in series, each of the bases in the promoter for the β-globin gene. The effect on transcription rates is measured for changes at each base in the promoter sequence. It can be seen that base changes outside the TATA box and the upstream sequences have no effect on levels of transcription, whereas an alteration in either of these elements severely lowers transcription rates.

Unlike prokaryotic promoters, eukaryotic promoters do not provide sufficient recognition signals for

Figure 16-34. Promoter in higher eukaryotes. The TATA box is located approximately 30 base pairs (bp) from the mRNA start site. Usually, two or more upstream elements are found 40 to 110 bp upstream of the mRNA start site. The CCAAT box and the GC box are shown here. Other upstream elements include the sequences GCCACACCC and ATGCAAAT.

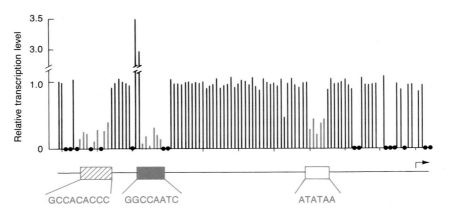

Figure 16-35. Consequences of point mutations in the promoter for the β-globin gene. Point mutations through the promoter region were analyzed for their effects on transcription rates. The height of each line represents the transcription level relative to a wild-type promoter that results from promoters with base changes at that point. A level of 1.0 means that the rates are equal to the wild-type rate; reductions in transcription rates yield levels less than 1.0. Almost every nucleotide throughout the promoter was tested, except for the points shown with black dots. The diagram below the bar graph shows the position of the TATA box and two upstream elements of the promoter. Only the base substitutions that lie within the three promoter elements change the level of transcription. (Figures 16-35 and 16-36 from T. Maniatis, S. Goodbourn, and J. A. Fischer, *Science* 236, 1987, 1237–1245.)

RNA polymerase to initiate transcription in vivo. The TATA box and the upstream sequences must each be recognized by regulatory proteins which bind to these sites and activate transcription, perhaps by enabling RNA polymerase to bind and initiate properly. We will describe some of these trans-acting factors shortly.

Enhancers

Enhancers are cis-acting sequences that can greatly increase transcription rates from promoters on the same DNA molecule. A unique aspect of these elements is their ability to exert their effects at distances of up to several thousand base pairs. Also, they can function in

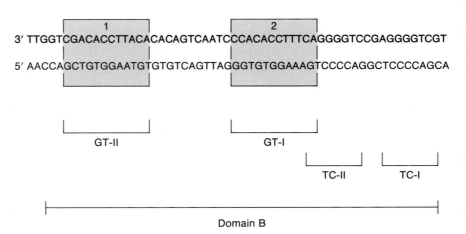

Figure 16-36. Organization of the SV40 enhancer. Boxed sequences 1 to 5 indicate the sequences that are required for maximum levels of enhancer activity. The brackets below each sequence indicate the location of sequence motifs that are repeated within the enhancer. Functional domains A and B are indicated at the bottom of the figure.

either orientation, operating either upstream or downstream from the promoter they are enhancing. Enhancer elements are intricately structured. Figure 16-36 shows the DNA sequence of the SV40 enhancer. There are two basic domains in the enhancer, which, together, contain five sequence elements required for maximal enhancement of transcription. Different sequence motifs serve as recognition sites for specific trans-acting factors. Figure 16-37 depicts the organization of the SV40 controlling elements. The enhancer increases the rates of transcription of both early and late mRNA. Sequences at the end of the enhancer serve as upstream sequence elements for the late mRNA promoter.

In some aspects, enhancers appear to be similar to promoters. Namely, they are organized as a series of cis-acting sequences that are recognized by trans-acting factors. However, enhancers can act at longer distances than promoter elements can.

Trans-acting Factors

Several proteins that recognize the CCAAT box have been identified from mammalian cells. These factors can distinguish between subunits of CCAAT elements. GC boxes are recognized by additional factors. For instance, the Spl protein recognizes the upstream GC boxes in the SV 40 promoter (Figure 16-37). Factors that bind to TATA box sequences have been identified in *Drosophila* and yeast.

Two examples of transacting factors in yeast that operate on the enhancer-like upstream activating sequences (UAS's) are the GCN4 and the GAL4 proteins. GCN4 activates the transcription of many genes involved in amino-acid biosynthetic pathways. In response to amino-acid starvation, GCN4 protein levels rise and, in turn, increase the levels of the amino-acid biosynthetic genes. The UAS's recognized by GCN4 contain the principal recognition sequence element ATGACT CAT. GAL4 is involved in activating UAS's for the promoters of genes involved in galactose metabolism. GAL4 binds to four sites with similar 17-base-pair sequences within these promoters.

Steroid Hormones

In a number of cases, the transcription of specific genes is activated by hormones. These molecules are usually synthesized in one type of cell and then transported to the cells on which they act. Protein hormones, such as insulin, principally act by binding to cell surface receptors and causing a signal to be transmitted across the plasma membrane that stimulates changes inside the cell. In some case, the proteins are transported into the cells themselves. Small hormones and, in particular, steroid hormones enter cells by virtue of their lipid-solubility and bind to receptors in the nucleus. In this respect, they are formally analogous to inducers in certain bacterial systems.

One example is the female sex hormone estrogen. In chicken oviducts, the egg-white protein ovalbumin is specifically synthesized in response to estrogen due to an increased transcription of the ovalbumin gene. The es-

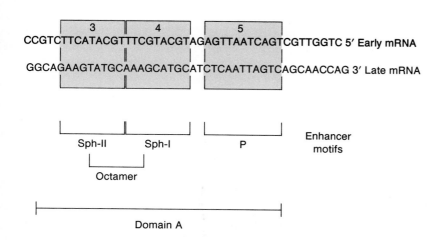

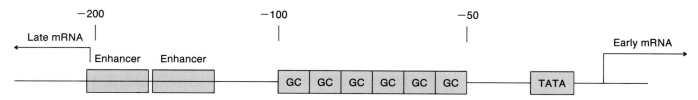

Figure 16-37. SV40 transcriptional control elements. The TATA box, the upstream elements (the GC boxes), and the two enhancer elements are required for maximal rates of early RNA transcription. The enhancer elements also stimulate transcription of late mRNA's. (Modified from W. S. Dynan and R. Tjian, *Nature* 316, 1985, 774–777.)

trogen molecule activates transcription by binding to the DNA at an enhancer site, together with a protein receptor molecule that first recognizes the estrogen molecule in the cytoplasm and transports it to the nucleus. The steroid-receptor activation of genes has been well studied in the mouse mammary tumor virus (MMTV) system. This virus replicates very well in cells with high levels of glucocorticoid steroids. Figure 16-38 diagrams the action of these steroids on MMTV transcription.

Structure of Trans-acting Proteins and Mechanism of Action

Several interesting aspects of trans-acting factors have emerged from recent studies. It is clear that many of these proteins have two separate domains: one of which recognizes cis-acting DNA sequences, and the second of which activates transcription. This can be seen in the GAL4 protein, which consists of 881 amino acids. A fragment of the protein containing only the first 147 amino acids can bind to DNA at specific UAS sites but cannot activate transcription. However, fragments containing later sequences in the protein cannot bind to DNA but can activate transcription when attached to fragments that bind DNA. Also, when the 114 carboxyl-terminal residues of GAL4 are fused to the DNA-binding segment of the human estrogen receptor, this hybrid protein can activate transcription at the binding site normally recognized by the estrogen receptor. This experiment shows not only that several different proteins have similar domain structures but also that the mechanisms of activation are similar in diverse systems. In a related experiment, a hybrid was constructed between two different steroid hormone-receptor molecules: the glucocorticoid receptor and the progesterone receptor. Here the DNA-binding region of the glucocorticoid receptor

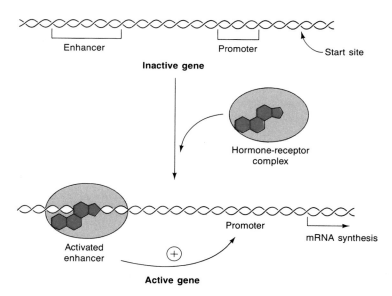

Figure 16-38. The action of steroid hormones at enhancer sequences. A glucocorticoid hormone binds to a soluble receptor protein. This complex, in turn, binds to enhancer sequences, such as the one found in the DNA of the mouse mammary tumor virus (MMTV), and enables them to stimulate the transcription of hormone-responsive genes. (From L. Stryer, *Biochemistry,* 3d ed. Copyright © 1988 by W. H. Freeman and Company.)

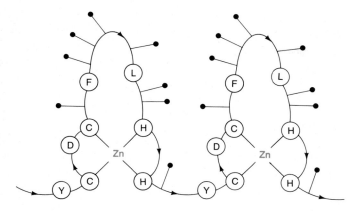

Figure 16-39. Folding scheme for a linear arrangement of repeated domains, each centered on a tetrahedral arrangement of zinc ligands. Encircled residues are conserved amino acids, including the Cys (C) and His (H) zinc ligands. Solid black circles mark the most probable DNA-binding side chains. Here the zinc ion brings the ends of each unit together, leaving the central 12 residues to form a potential DNA-binding loop or "finger." (From J. Miller, A. D. McLachlan, and A. Klug, *The EMBO Journal* 4, 1985, 1609–1614.)

is substituted for the progesterone-receptor binding region, but the activating portion of the progesterone receptor is left intact. The fused protein now responds by binding to glucocorticoid regulatory sequences and stimulating the transcription of the cis-linked genes — but only in response to progesterone.

A second aspect of trans-acting factors that is emerging is a structural motif arising from a cysteine-rich region in the DNA-binding domain that complexes zinc. The protrusions in the protein structure resemble fingers (Figure 16-39); for this reason, the motif has been termed "zinc fingers." One set of trans-acting regulatory proteins has a helix-turn-helix motif similar to that found in many bacterial regulatory proteins (Figure 16-32). Many of these proteins are involved in *Drosophila* development, particularly in determining the spatial arrangement of body segments (Chapter 21). The genes that encode these proteins, called homeotic proteins, contain a similar sequence in the region that encodes the DNA-binding domain. This sequence, named a homeotic box, encodes the recognition helix of the helix-turn-helix segment for these regulatory proteins.

The most reasonable model for the action of trans-acting factors at distances from the genes they activate involves some type of DNA looping reminiscent of the model proposed for the bacterial *ara* operon (Figure 16-18). Figure 16-40 outlines two variations of such a model.

Control of Ubiquitous Molecules in Eukaryotic Cells

Several classes of molecules are found in abundance in every eukaryotic cell: histones, the translational apparatus, membrane components, and so on. Several strategies have evolved to maintain a sufficient amount of these products in the eukaryotic cell. These strategies include continuous and repeated transcription throughout the cell cycle, repetition of genes, and extrachromosomal amplification of specific gene sequences.

Genetic Redundancy

Genes coding for rRNA and tRNA have been extensively analyzed because the availability of purified gene products provides a probe with which to recover com-

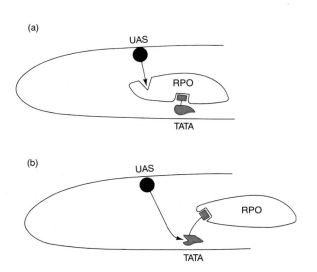

Figure 16-40. Two possible mechanisms of transcriptional activation in eukaryotes at enhancer (or UAS) sites. (a) An activator bound at a UAS (solid black circle) and a TATA box factor bound at a TATA box (color figures) both contact RNA polymerase (RPO). Activation regions of the activator and the TATA box factor are drawn as a triangle and a rectangle, respectively. The DNA between the UAS and TATA box has been looped. (b) The activator contacts the TATA box factor, which in turn contacts the RNA polymerase. (From L. Guarente, *Cell* 52, 1988, 303–305.)

plementary DNA. Simple DNA-RNA hybridization shows that each *Xenopus* chromosome carrying a nucleolus organizer has 450 copies of the DNA coding for 18S and 28S rRNA. In contrast, there are 20,000 copies in each nucleus of the genes coding for 5S rRNA, and these genes are not located in the NO region. Donald Brown and his collaborators during the 1970s analyzed the NO DNA extensively and showed that it carries tandem duplications of a large (40S) transcription unit separated from its neighbors by nontranscribed spacers. The initial transcript is then processed to release the smaller rRNA's finally found in the ribosome.

As we saw in Chapter 15, Max Birnstiel and his associates have shown that the five histone genes in sea urchins also are found as a tightly linked unit (Figure 15-2). There are several hundred copies of a repeating unit with the genes in the sequence *H4–H2B–H3–H2A–H1*. Transcription begins at *H4* and ends at *H1*; each gene is transcribed as a single message. As we have seen in earlier chapters, tandem duplications can pair asymmetrically between homologous chromosomes or within the same chromosome; exchanges then increase or reduce the number of copies. It remains to be seen how (or whether) the number of tandem repeats is kept constant in a particular species.

Gene Amplification

Specific gene amplification has been studied extensively in the case of the rRNA genes in amphibian oocytes. As noted, each chromosome of *Xenopus* carries about 450 copies of the DNA coding for 18S and 28S rRNA. How-

ever, the oocyte contains up to 1000 times this number of copies of these genes. The oocyte is a very specialized cell that is loaded with the nutritive material needed to maintain the embryo until the tadpole stage without any ingestion of food. The cytoplasm also is prepared with the translational apparatus needed to carry out the complex program of differentiation in the many cells of the developing embryo.

During oogenesis, the maturing oocyte increases greatly in size as material is poured in from adjacent nurse cells. (Amazingly, so many ribosomes are present in the egg at fertilization that *an/an* homozygotes carrying no genes for 18S and 28S rRNA will develop and differentiate to the twitching tadpole stage before death.) The oocyte nucleus also contributes material as the cell proceeds to the first meiotic prophase, where further development is arrested. The amphibian meiotic chromosomes are enormous, with numerous lateral loops representing regions of intense genetic activity (Figure 16-41). In the oocyte nucleus, there are hundreds of extrachromosomal nucleoli of varying size. Each nucleolus contains a different-sized ring of rDNA that has been replicated and released from the chromosome. The steps involved in this process of DNA amplification are completely unknown. The DNA rings actively produce rRNA that is assembled into ribosomes, which are stored in the nucleolus until their release during meiosis.

Other examples of genetic amplification are known. Specific puffs of the polytene chromosomes in dipterans are found to produce excess DNA rather than mRNA. Also, George Rudkin showed in the early 1960s that the polytene chromosomes from *Drosophila* salivary glands

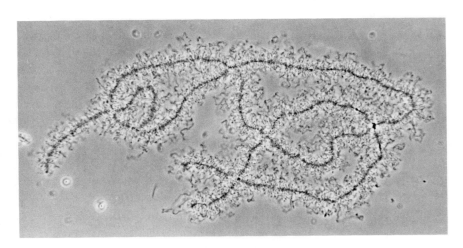

Figure 16-41. The lampbrush chromosomes of an amphibian oocyte. (Photograph courtesy of J. Gall.)

themselves may represent specific amplifications of euchromatic DNA about a thousandfold, whereas the proximal heterochromatic regions may be replicated only a few times. Thus, in addition to the mechanism for replicating chromosomal DNA normally for cell division, there exists a means for amplifying specific loci and not others.

> **Message** Cellular mechanisms exist to ensure an adequate supply of the gene products that are vital to all cells. These mechanisms include increasing the number of gene copies per chromosome (redundancy) and increasing the number of gene copies in the cell (amplification).

Summary

■ The operon model explains how prokaryotic genes are controlled through a mechanism that coordinates the activity of a number of related genes. In negative control, the initiation of transcription is controlled at the operator by a repressor with binding affinities to the operator that may be altered by inducer molecules. Inactivation of the repressor—the negative control element—is required for active transcription. In positive control, transcription initiation requires the activation of a factor. Sometimes one control system is superimposed on another. For instance, superimposed on the repressor–operator system for the *lac* operon is the cAMP–CAP positive control system. Modulation of transcription by an attenuator sequence supplements the repressor–operator control in the *trp* operon. In phage λ, multiple binding sites for repressors, as well as repressors of repressors, add additional levels of control.

A major problem in transcriptional control in eukaryotes is understanding how thousands of promoters can be regulated to yield desired levels of mRNA. It is now clear that promoters are governed both by the number and type of cis-acting control elements and by the action of regulatory proteins that recognize these elements. Also modulating the levels of transcription are enhancers—cis-acting sequences which, when recognized by functional regulatory proteins, can greatly increase the activity of promoters. In some aspects, transcriptional control in eukaryotes is similar to that found in prokaryotes: namely, transacting factors recognize cis-acting (or cis-dominant) sites in both cases. The mechanism of action may even be similar in many cases, as evidenced by the looping out of DNA in the *ara* system in bacteria and by enhancer action in eukaryotes. Also, structural motifs of regulatory proteins share a common thread in the case of the helix-turn-helix motif in bacterial regulatory proteins and in the homeotic proteins in *Drosophila*. However, other structural motifs appear to be unique to classes in eukaryotic regulatory proteins, as exemplified by the zinc-finger motif for many trans-acting factors.

■ ■ ■

Solved Problems

This set of problems, which are similar to Problem 5 at the end of this chapter, are designed to test our understanding of the operon model. Here we are given several diploids and are asked to determine whether Z and Y gene products are made in the presence or absence of an inducer.

1. $$\frac{I^- P^- O^c Z^+ Y^+}{I^+ P^+ O^+ Z^- Y^-}$$

Solution

One way to approach these problems is first to consider each chromosome separately and then to construct a diagram. Figure 16-42 diagrams this diploid. The first chromosome is P^-, so no *lac* enzyme can be synthesized off it. The second chromosome (O^c) can be transcribed, and this transcription is repressible. However, the structural genes linked to the good promoter are defective; thus, no active Z product or Y product can be produced. The symbols to add to Table 16-5 in Problem 5 are − − − −.

2. $$\frac{I^+ P^- O^+ Z^+ Y^+}{I^- P^+ O^+ Z^+ Y^-}$$

Solution

Figure 16-43 diagrams this diploid. The first chromosome is P^-, so no enzyme can be synthesized off it. The second chromosome is O^+, so transcription will be repressed by the repressor supplied from the first chromosome, which can act *in trans* through the cytoplasm. However, only the Z gene from this chromosome is intact. Therefore, in the absence of an inducer, no enzyme will be made; in the presence of an inducer, only the Z gene product β-galactosidase will be produced. The symbols to add to the table in Problem 5 are − + − −.

3. $$\frac{I^+ P^+ O^c Z^- Y^+}{I^+ P^- O^+ Z^+ Y^-}$$

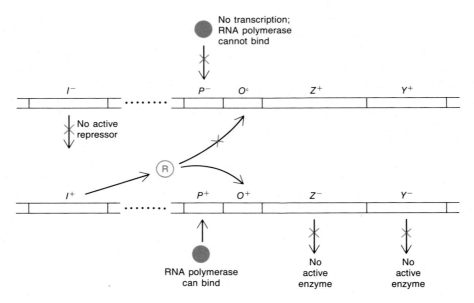

Figure 16-42.

Solution

Because the second chromosome is P^-, we need only to consider the first chromosome, as shown in Figure 16-44. This chromosome is O^c, so that enzyme is made in the absence of an inducer, although only the Y gene is active. The entries in the table in Problem 5 are $--++$.

4. $\dfrac{I^s P^+ O^+ Z^+ Y^-}{I^- P^+ O^c Z^- Y^+}$

Solution

Figure 16-45 diagrams this diploid. In the presence of an I^s repressor, all wild-type operators will be shut off, both with and without an inducer. Therefore, the first chromosome will be unable to produce any enzyme. However, the second chromosome has an altered (O^c) operator and can produce enzyme in both the absence and presence of an inducer. Only the Y gene is active on this chromosome, so the entries in the table in Problem 5 are $--++$.

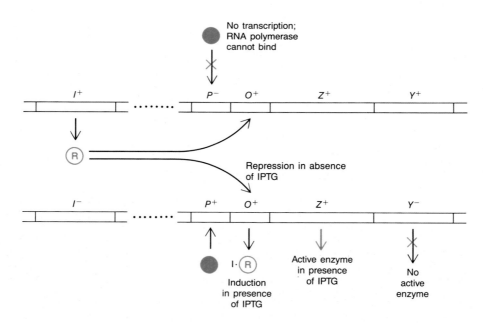

Figure 16-43.

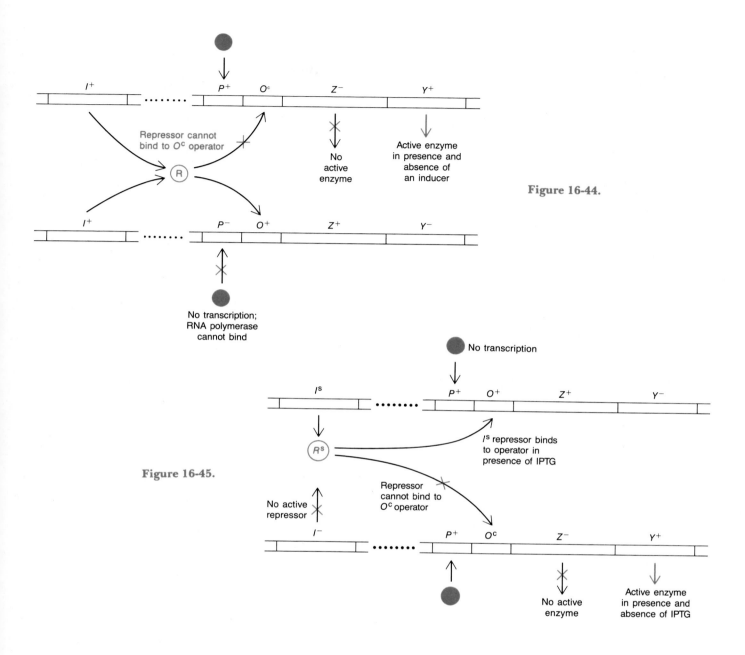

Figure 16-44.

Figure 16-45.

Problems

1. Explain why I^- mutations in the *lac* system are normally recessive to I^+ mutations and why I^+ mutations are recessive to I^s mutations.

2. What do we mean when we say that O^c mutations in the *lac* system are cis-dominant?

3. The genes shown in Table 16-4 are from the *lac* operon system of *E. coli*. The symbols *a*, *b*, and *c* represent the repressor *(I)* gene, the operator *(O)* region, and the structural gene *(Z)* for β-galactosidase, although not necessarily in that order. Furthermore, the order in which the symbols are written in the genotypes is not necessarily the actual sequence in the *lac* operon.

TABLE 16-4.

	Activity (+) or inactivity (−) of z gene	
Genotype	Inducer absent	Inducer present
$a^-b^+c^+$	+	+
$a^+b^+c^-$	+	+
$a^+b^-c^-$	−	−
$a^+b^-c^+/a^-b^+c^-$	+	+
$a^+b^+c^+/a^-b^-c^-$	−	+
$a^+b^+c^-/a^-b^-c^+$	−	+
$a^-b^+c^+/a^+b^-c^-$	+	+

a. State which symbol (*a*, *b*, or *c*) represents each of the *lac* genes *I*, *O*, and *Z*.

b. In Table 16-4, a — on a gene symbol merely indicates a mutant, but you know there are some mutant behaviors in this system that are given special mutant designations. Use the conventional gene symbols for the *lac* operon to designate each genotype in the table.

(Problem 1 is from J. Kuspira and G. W. Walker, *Genetics: Questions and Problems.* Copyright © 1973 by McGraw-Hill.)

4. The map of the *lac* operon is

$$I \quad P\,O\,Z\,Y$$

The promoter (*P*) region is the start site of transcription through the binding of the RNA polymerase molecule before actual mRNA production. Mutationally altered promoters (*P⁻*) apparently cannot bind the RNA polymerase molecule. Certain predictions can be made about the effect of *P⁻* mutations. Use your predictions and your knowledge of the lactose system to complete Table 16-5. Insert a + where an enzyme is produced and a — where no enzyme is produced.

5. In a haploid eukaryotic organism, you are studying two enzymes that perform sequential conversions of a nutrient A supplied in the medium:

$$A \xrightarrow[E_1]{} B \xrightarrow[E_2]{} C$$

Treatment of cells with mutagen produces three different mutant types with respect to these functions. Mutants of type 1 show no E_1 function; all type 1 mutations map to a single locus on linkage group II. Mutations of type 2 show no E_2 function; all type 2 mutations map to a single locus on linkage group VIII. Mutants of type 3 show no E_1 or E_2 function; all types 3 mutants map to a single locus on linkage group I.

a. Compare this system with the *lac* operon of *E. coli*, pointing out the similarities and the differences. (Be

sure to account for each mutant type at the molecular level.)

b. If you were to intensify the mutant hunt, would you expect to find any other mutant types on the basis of your model? Explain.

6. In *Neurospora*, all mutants affecting the enzymes carbamyl phosphate synthetase and aspartate transcarbamylase map at the *pyr-3* locus. If you induce *pyr-3* mutations by ICR-170 (a chemical mutagen), you find that either both enzyme functions are lacking or only the transcarbamylase function is lacking; in no case is the synthetase activity lacking when the transcarbamylase activity is present. (ICR-170 is assumed to induce frame-shifts.) Interpret these results in terms of a possible operon.

7. In 1972, Suzanne Bourgeois and Alan Jobe showed that a derivative of lactose, allolactose, is the true natural inducer of the *lac* operon, rather than lactose itself. Lactose is converted to allolactose by the enzyme β-galactosidase. How does this result explain the early finding that many *Z⁻* mutations, which are not polar, still do not allow the induction of *lac* permease and transacetylase by lactose?

8. Certain *lacI* mutations eliminate operator binding by the *lac* repressor but do not affect the aggregation of subunits to make a tetramer, the active form of the repressor. These mutations are partially dominant to wild-type. Can you explain the partially *I⁻* phenotype of the *I⁻/I⁺* heterodiploids?

9. Explain the fundamental differences between negative control and positive control.

10. Mutants that are *lacY⁻* retain the capacity to synthesize β-galactosidase. However, even though the *lacI* gene is still intact, β-galactosidase can no longer be induced by adding lactose to the medium. How can you explain this?

11. What analogies can you draw between transcriptional trans-acting factors that activate gene expression in eukaryotes and prokaryotes? Give an example.

12. Compare the arrangement of cis-acting sites in the control regions of eukaryotes and prokaryotes.

TABLE 16-5.

Part	Genotype	β-galactosidase		Permease	
		No lactose	Lactose	No lactose	Lactose
Example	$I^+P^+O^+Z^+Y^+/I^+P^+O^+Z^+Y^+$	—	+	—	+
(a)	$I^-P^+O^cZ^+Y^-/I^+P^+O^+Z^-Y^+$				
(b)	$I^+P^-O^cZ^-Y^+/I^-P^+O^cZ^+Y^-$				
(c)	$I^sP^+O^+Z^+Y^-/I^+P^+O^+Z^-Y^+$				
(d)	$I^sP^+O^+Z^+Y^+/I^-P^+O^+Z^+Y^+$				
(e)	$I^-P^+O^cZ^+Y^-/I^-P^+O^+Z^-Y^+$				
(f)	$I^-P^-O^+Z^+Y^+/I^+P^+O^cZ^+Y^-$				
(g)	$I^+P^+O^+Z^-Y^+/I^-P^+O^+Z^+Y^-$				

Mechanisms of Genetic Change I: Gene Mutation

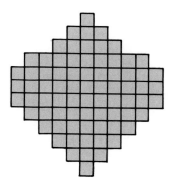

KEY CONCEPTS

Mutations can occur spontaneously due to several different
mechanisms, including errors of DNA replication and
spontaneous damage to the DNA.

∙

Mutagens are agents that increase the frequency of
mutagenesis, usually by altering the DNA.

∙

Biological repair systems eliminate many potentially
mutagenic alterations in the DNA.

∙

Cells lacking certain repair systems have higher than normal
mutation rates.

∙

Potentially mutagenic and carcinogenic compounds can be
detected easily by mutagenesis tests with bacterial systems.

■Genetic change can result from a number of processes. Consider an individual organism that represents a variant from an established control population. What mechanism could have produced the change that created this variant individual? The following mechanisms provide four possibilities:

1. Gene mutation

2. Recombination

3. Transposable genetic elements

4. Chromosomal rearrangements

In previous chapters, we have considered genetic changes merely as useful and interesting phenomena; we did not probe deeply into how such changes come about. Let's now examine, at the molecular levels, the processes that lead to each type of genetic change. This chapter explores the mechanisms of gene mutation; Chapter 18 describes the events leading to recombination; and Chapter 19 introduces the concept of transposable genetic elements, which can effect genetic change by moving from one chromosome location to another. Although some chromosomal rearrangements can result from recombination and from transposable elements, many rearrangement events occur by mechanisms that are not yet understood.

To understand mechanisms of gene mutation requires analysis at the level of DNA and protein molecules. Preceding chapters have described models for

TABLE 17-1. Summary of changes at the molecular level in gene mutations

Type of mutation	Result and example(s)
Forward mutations Single nucleotide-pair (base-pair) substitutions	
At DNA level	
Transition	Purine replaced by a different purine, or pyrimidine replaced by a different pyrimidine: AT $\longrightarrow$ GC GC $\longrightarrow$ AT CG $\longrightarrow$ TA TA $\longrightarrow$ CG
Transversion	Purine replaced by a pyrimidine, or pyrimidine replaced by a purine: AT $\longrightarrow$ CG AT $\longrightarrow$ TA GC $\longrightarrow$ TA GC $\longrightarrow$ CG TA $\longrightarrow$ GC TA $\longrightarrow$ AT CG $\longrightarrow$ AT CG $\longrightarrow$ GC
At protein level	
Silent mutation	Triplet codes for same amino acid: AGG $\longrightarrow$ CGG both code for Arg
Neutral mutation	Triplet codes for different but functionally equivalent amino acid: AAA $\longrightarrow$ AGA changing basic Lys to basic Arg (at many positions, will not alter protein function)
Missense mutation	Triplet codes for a different and nonfunctional amino acid.
Nonsense mutation	Triplet codes for chain termination: CAG $\longrightarrow$ UAG changing from a codon for Gln to an amber termination codon
Single nucleotide-pair addition or deletion: frame-shift mutation	Any addition or deletion of base pairs that is not a multiple of 3 results in a frame-shift in DNA segments that code for proteins.
Intragenic addition or deletion of several to many nucleotide pairs	
Reverse mutations	
Exact reversion	AAA (Lys) $\xrightarrow{\text{forward}}$ GAA (Glu) $\xrightarrow{\text{reverse}}$ AAA (Lys) wild-type mutant wild-type

DNA and protein structure and have discussed the nature of mutations that alter these structures. Table 17-1 draws together this information to provide an explanation of the nature of gene mutation at the molecular level.

Technical advances in the mid-1970s ushered in an exciting new era in molecular genetics, permitting the first direct determination of the sequence of large segments of DNA and also of the sequence changes resulting from mutations. This has greatly increased our understanding of the pathways that lead to mutagenesis and has even helped to unravel the mysteries of mutational hotspots (see page 478). Much work until now on the molecular basis of mutation has been carried out in single-cell bacteria and their viruses. Let's review some

of the recent findings of these studies. We shall also consider biological repair mechanisms, since repair systems play a key role in mutagenesis.

The Molecular Basis of Gene Mutations

Gene mutations can arise spontaneously or they can be induced. **Spontaneous mutations** occur in all cells. **Induced mutations** are produced when an organism is exposed to a mutagenic agent, or mutagen; such mutations typically occur at much higher frequencies than spontaneous mutations do.

TABLE 17-1. (continued)

Type of mutation	Result and example(s)
Equivalent reversion	UCC (Ser) $\xrightarrow{\text{forward}}$ UGC (Cys) $\xrightarrow{\text{reverse}}$ AGC (Ser) wild-type mutant wild-type
	CGC (Arg, basic) $\xrightarrow{\text{forward}}$ CCC (Pro, not basic) $\xrightarrow{\text{reverse}}$ CAC (His, basic) wild-type mutant pseudo-wild-type
Intregenic Suppressor mutations	
Frame-shift of opposite sign at second site within gene	CAT CAT CAT CAT CAT CAT (+) (−) CAT XCA TAT CAT CAT CAT ✓ ✗ ✗ ✓ ✓ ✓
Second-site missense mutation	Still not fully understood at the level of protein function; explained in terms of a second distortion that restores a more or less wild-type protein conformation after a primary distortion.
Extragenic suppressor mutations	
Nonsense suppressors	A gene (for example, for tyrosine tRNA) undergoes a mutational event in its anticodon region that enables it to recognize and align with a mutant nonsense codon (say, amber UAG) to insert an amino acid (here, tyrosine) and permit completion of the translation.
Missense suppressors	A heterogeneous set of mutations with molecular mechanisms that are not fully understood. One missense suppressor in *E. coli* is an abnormal tRNA that carries glycine but inserts it in response to arginine codons. Although *normal* arginine codons also are mistranslated, the observed mutations are not lethal, probably due to the low efficiency of abnormal substitution.
Frame-shift suppressors	Very few examples have been found; in one, a four-nucleotide anticodon in a single tRNA can "read" a four-letter codon caused by a single nucleotide-pair insertion.
Physiological suppressors	A defect in one chemical pathway is circumvented by another mutation (for example, one that opens up another chemical pathway to the same result, or one that permits more efficient transport of a compound produced in small quantities due to the original mutation). Thus, these mutations act as one form of missense suppressors (a very heterogeneous group).

Spontaneous Mutations

Spontaneous mutations arise from a variety of sources, including errors in DNA replication, spontaneous lesions, and even transposable genetic elements. The first two are discussed below; the third is examined in Chapter 19.

Errors in DNA Replication. An error in DNA replication can occur when an illegitimate nucleotide pair (say, A–C) forms during DNA synthesis, leading to a base substitution. Each of the bases in DNA can appear in one of two forms, called **tautomers,** which are in equilibrium. The **keto** form of each base is present in normal DNA, whereas the rarer **imino** and **enol** forms of the bases are not (see Figure 17-1). The ability of the wrong tautomer of one of the standard bases to mispair and cause mutations during DNA replication was first noted by Watson and Crick when they formulated their model for the structure of DNA (Chapter 11). Figure 17-2 demonstrates some possible mispairs resulting from changes of one tautomer to another, termed **tautomeric shifts.** Such mispairs lead to **transition mutations** (Figure 17-3), in which a purine is substituted for a purine or a pyrimidine for a pyrimidine (Table 17-1). The bacterial DNA polymerase III (Chapter 11) has an editing capacity that recognizes such mismatches and excises them, thus greatly reducing the observed mutations. (Another repair system described later in this chapter corrects many of the mismatched bases that escape correction by the polymerase checking function.)

TRANSVERSIONS. **Transversion mutations,** in which a pyrimidine is substituted for a purine, or vice versa, cannot be generated by the mismatches depicted in Figure 17-2. Can you see why? With bases in the DNA in the normal orientation, the creation of a transversion by a replication error would require, at some point during replication, the mispairing of a purine with a purine

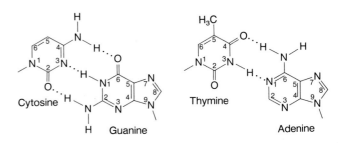

Figure 17-1. Base pairs in DNA. The normal Watson-Crick base pairs are shown.

or a pyrimidine with a pyrimidine. However, the dimensions of the DNA double helix forbid such mispairing on steric grounds. (Refer to some of the figures in Chapter 11 to verify this.) Although mispairing schemes using other rare forms of different bases have been proposed, it is presently unclear how the observed spontaneous transversions are generated.

FRAME-SHIFT MUTATIONS. Replication errors can also lead to **frame-shift mutations.** Recall from Chapter 13 that such mutations result in greatly altered proteins.

In the mid-1960s, George Streisinger and his coworkers deduced the nucleotide sequence surrounding different sites of frame-shift mutations in the lysozyme gene of phage T4. They found that these mutations often occurred at repeated sequences and formulated a model to account for frame-shifts during DNA synthesis. In the Streisinger model (Figure 17-4), frame-shifts arise when loops in single-stranded regions are stabilized by the "slipped mispairing" of repeated sequences. With the advent of DNA sequencing in the mid 1970s, such models could be tested directly. Jeffrey Miller and his coworkers examined mutational **hotspots** in the *lacI* gene of *E. coli.* (Recall from Chapter 12 that Benzer

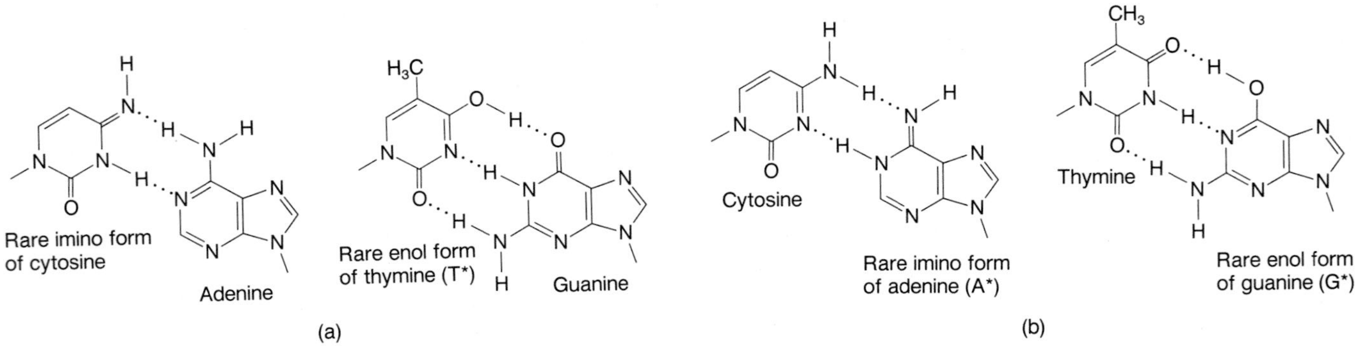

Figure 17-2. Mismatched bases. (a) Mispairs resulting from rare tautomeric forms of the pyrimidines; (b) mispairs

resulting from rare tautomeric forms of the purines are shown in (b).

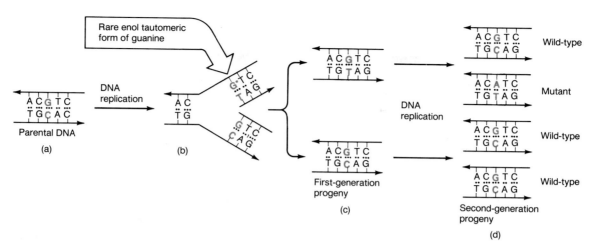

Figure 17-3. Mutation via tautomeric shifts in the bases of DNA. (a) In the example diagrammed, a guanine undergoes a tautomeric shift to its rare enol form (G*) at the time of replication. (b) In its enol form, it pairs with thymine. (c) and (d) During the next replication, the guanine shifts back to its more stable keto form. The thymine incorporated opposite the enol form of guanine, seen in (b), directs the incorporation of adenine during the subsequent replication, shown in (c) and (d). The net result is a GC → AT mutation. If a guanine undergoes a tautomeric shift from the common keto form to the rare enol form at the time of incorporation (as a nucleoside triphosphate, rather than in the template strand diagrammed here), it will be incorporated opposite thymine in the template strand and cause an AT → GC mutation. (From E. J. Gardner and D. P. Snustad, *Principles of Genetics*, 5th ed. Copyright © 1984 by John Wiley.)

demonstrated the existence of such hotspots, or sites in a gene that are much more mutable than other sites, in 1961.) The *lacI* work showed that certain hotspots result from repeated sequences, just as predicted by the Streisinger model. Figure 17-5 depicts the distribution of spontaneous mutations in the *lacI* gene. Compare this with the distribution in the *rII* genes of T4 seen by Benzer (page 308). Note how one or two mutational sites dominate the distribution in both cases. In *lacI*, a four-base-pair sequence repeated three times in tandem is the

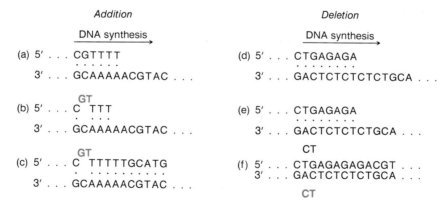

Figure 17-4. A simplified version of the Streisinger model for frame-shift formation. (a) to (c) During DNA synthesis, the newly synthesized strand slips, looping out several base pairs. This loop is stabilized by the pairing afforded by the repetitive-sequence unit (the single A bases, in this case). An addition will result at the next round of replication. (d) to (f) If instead of the newly synthesized strand, the template strand slips, then a deletion results. Here the repeating unit is a C−T dinucleotide. After slippage, a deletion of two base pairs would result at the next round of replication.

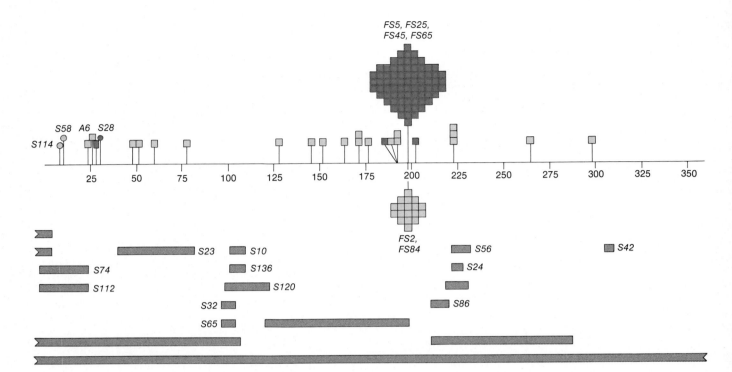

Figure 17-5. The distribution of 140 spontaneous mutations in *lacI*. Each occurrence of a point mutation is indicated by a box. Colored boxes depict fast-reverting mutations. Deletions are represented below the *I* map, which is given in terms of the amino-acid number in the corresponding *I*-encoded *lac* repressor. Allele numbers refer to mutations that have been analyzed at the DNA sequence level. The mutations *S114* and *S58* result from the insertion of transposable elements (see Chapter 19). The encircled *S28* refers to a duplication of 88 base pairs. (From P. J. Farabaugh, U. Schmeissner, M. Hofer, and J. H. Miller, *Journal of Molecular Biology* 126, 1978, 847–857.)

cause of the hotspots (for simplicity, only one strand of the double strand of DNA is indicated):

$$-\text{GTCTGG CTGG CTGG CTGG C}$$

$$FS5,\ FS25,\ FS45,\ FS65$$

wild-type   5′– GTCTGG CTGG CTGG C – 3′

$$FS2 \quad\quad FS84$$

$$\text{GTCTGG CTGG C}$$

The major hotspot, represented here by the mutations *FS5*, *FS25*, *FS45*, and *FS65*, results from the addition of one extra set of the four bases CTGG to one strand of the DNA. This hotspot reverts at a high rate, losing the extra set of four bases. The minor hotspot, represented here by the mutations *FS2* and *FS84*, results from the loss of one set of the four bases CTGG. This mutant does not readily regain the lost set of four base pairs. How can the Streisinger model explain these observations? Remember that the model predicts that the frequency of a particular frame-shift is dependent on the number of base pairs that can be formed during the slipped mis-pairing of repeated sequences. The wild-type sequence shown for the *lacI* gene can slip out one CTGG sequence and stabilize this by forming nine base pairs. (Can you work this out by applying the model in Figure 17-4 to the sequence shown for *lacI*?) Whether a deletion or an addition is generated depends on whether the slippage occurs on the template or on the newly synthesized strand, respectively. In a similar fashion, the addition mutant can slip out one CTGG sequence and stabilize this with 13 base pairs (verify this for the *FS5* sequence shown for *lacI*), which explains the rapid reversion of mutations such as *FS5*. However, there are only five base pairs available to stabilize a slipped-out CTGG in the deletion mutant, accounting for the infrequent reversion of mutations such as *FS2* in the sequence shown for *lacI*.

DELETIONS. Large deletions (more than a few base pairs) represent a sizable fraction of spontaneous mutations, as you can visualize from Figure 17-5. Deletions of up to several thousand base pairs in size have been studied extensively at the DNA sequence level. The major-

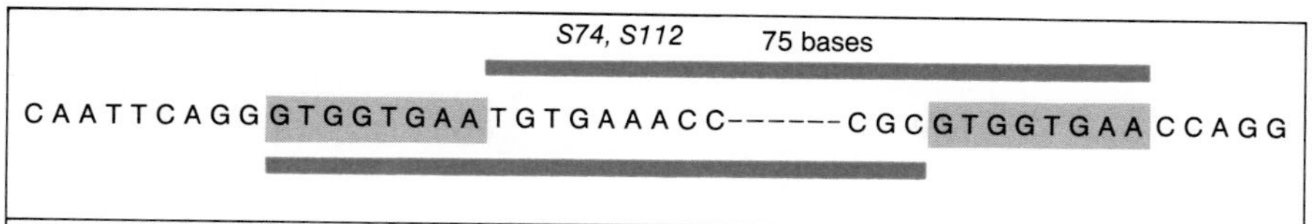

| | S74, S112 | 75 bases | |

C A A T T C A G G GTGGTGAA T G T G A A A C C ------- C G C GTGGTGAA C C A G G

Site (no. of b.p.)	Sequence repeat	Bases deleted	Occurrences	
20 to 95	G T G G T G A A	75	2	S74, S112
146 to 269	G C G G C G A T	123	1	S23
331 to 351	A A G C G G C G	20	2	S10, S136
316 to 338	G T C G A	22	2	S32, S65
694 to 707	C A	13	1	S24
694 to 719	C A	25	1	S56
943 to 956	G	13	1	S42
322 to 393	None	71	1	S120
658 to 685	None	27	1	S86

Figure 17-6. Deletions in *lacI*. Deletions occur as indicated by the colored bars at the top of the figure. One of the sequence repeats and all of the intervening DNA is deleted, leaving one copy of the repeated sequence. All mutations were analyzed by direct DNA sequence determination. (From P. J. Farabaugh, U. Schmeissner, M. Hofer, and J. H. Miller, *Journal of Molecular Biology* 126, 1978, 847–857.)

ity, although not all, of the deletions occur at short repeated sequences. Figure 17-6 shows the results for the first 12 deletions analyzed at the DNA sequence level, presented by Miller and his coworkers in 1978. Further studies have shown that hotspots for deletions involve the longest sequences that are repeated.

Several mechanisms could account for deletion formation. Deletions may be generated as replication errors. For example, an extension of the Streisinger model of slipped mispairing (Figure 17-4) could explain why deletions predominate at short repeated sequences. Alternatively, recombinational mechanisms (to be described in Chapter 18) employed by one or a number of cellular enzyme systems that recognize short-sequence repeats could generate deletions.

Spontaneous Lesions.   In addition to replication errors, **spontaneous lesions,** naturally occurring damage to the DNA, can also generate mutations. Two of the most frequent spontaneous lesions result from depurination and deamination. **Depurination,** the more common of the two, involves the interruption of the glycosidic bond between the base and deoxyribose and the subsequent loss of a guanine or an adenine residue from the DNA (Figure 17-7). A mammalian cell spontaneously loses about 10,000 purines from its DNA during a 20-hour generation period at 37°C. If these lesions were to persist, they would result in significant genetic damage because, during replication, the resulting **apurinic sites** cannot specify a base complementary to the original purine. However, as we see later in the chapter, efficient repair systems remove apurinic sites.

Under certain conditions (to be described later), a base can be inserted across from an apurinic site; this will frequently result in a mutation.

The **deamination** of cytosine yields uracil (Figure 17-8a). Unrepaired uracil residues will pair with adenine during replication, resulting in the conversion of a G–C pair to an A–T pair (a transition). One of the repair enzymes in the cell, uracil-DNA glycosylase, recognizes uracil residues in the DNA and excises them, leaving a gap that is subsequently filled in (a process to be described later in the chapter). An exciting discovery in 1978 revealed that the specificity of this repair enzyme is the cause of one type of mutational hotspot! DNA sequence analysis of GC → AT transition hotspots in the *lacI* gene has shown that 5-methylcytosine residues occur at the position of each hotspot. (Certain bases in prokaryotes and eukaryotes are methylated; see page 394.) Some of the actual data from this study are shown in Figure 17-9. The height of each bar on the graph represents the frequency of mutations at each of a number of sites. It can be seen that the position of 5-methylcytosine residues correlates nicely with the most mutable sites. This is because the deamination of 5-methylcytosine (Figure 17-8b) generates thymine (5-methyluracil), which is not recognized by the enzyme uracil-DNA glycosylase and thus is not repaired. Therefore, C → T transitions generated by deamination are detected more frequently at 5-methylcytosine sites, which occur both in bacteria and in higher cells.

The insertion of certain ''transposable elements'' into genes represents an additional source of spontaneous mutations, as we see in Chapter 19.

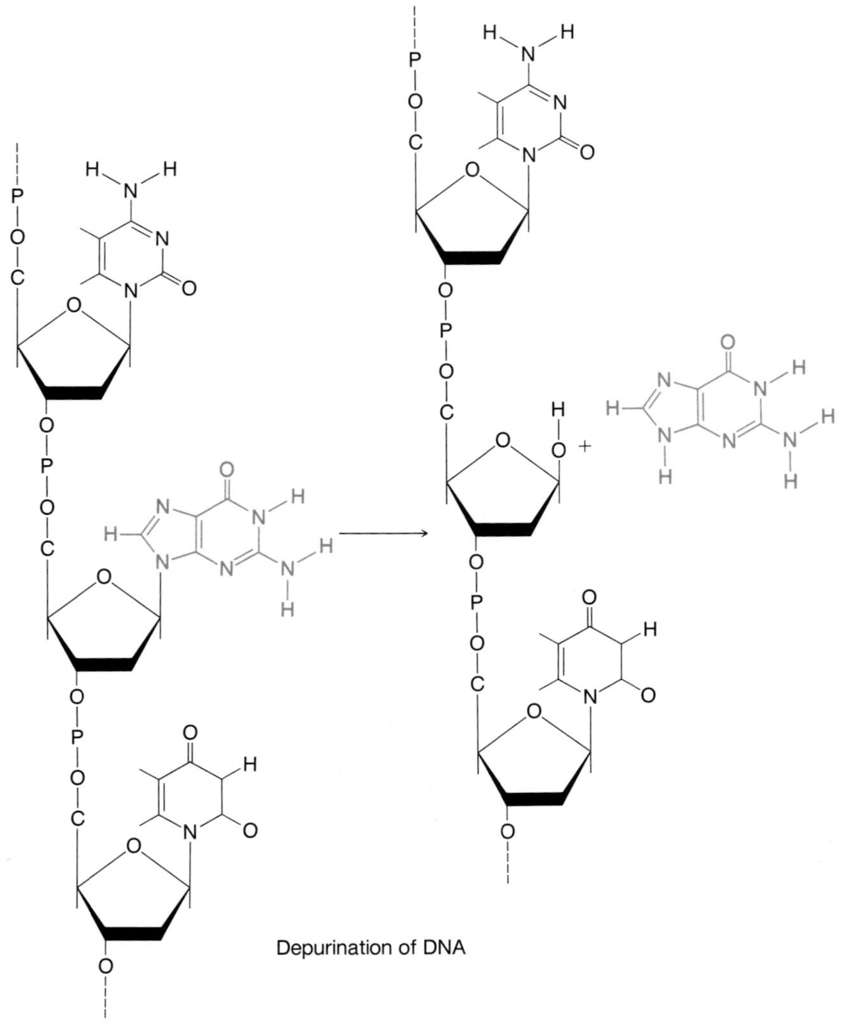

Figure 17-7. The loss of a purine residue from a single strand of DNA. The sugar-phosphate backbone is left intact.

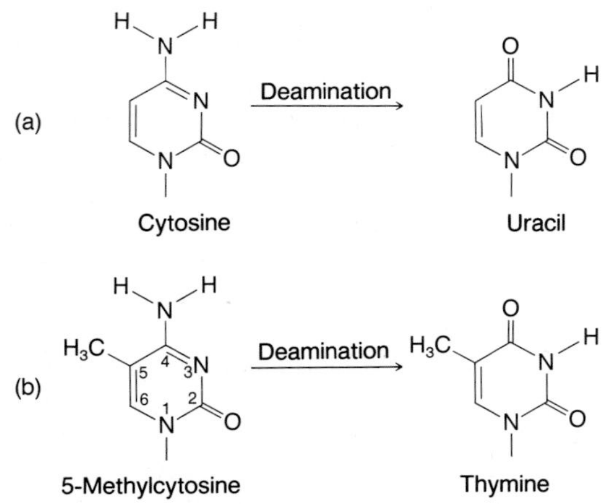

Depurination of DNA

(a)

Cytosine → Uracil

Deamination

(b)

5-Methylcytosine → Thymine

Deamination

Figure 17-8. Deamination of (a) cytosine and (b) 5-methylcytosine.

Message Spontaneous mutations can be generated by different processes. Replication errors and spontaneous lesions generate most of the base-substitution and frame-shift mutations. Replication errors may also cause deletions that occur in the absence of mutagenic treatment.

Induced Mutagenesis

Induced mutations have played an essential role in genetic analysis. When we observe the distribution of mutations induced by different mutagens, we see a distinct specificity that is characteristic of each mutagen. Such **mutational specificity** was first noted in the *rII* system by Benzer in 1961. Specificity arises from a "prefer-

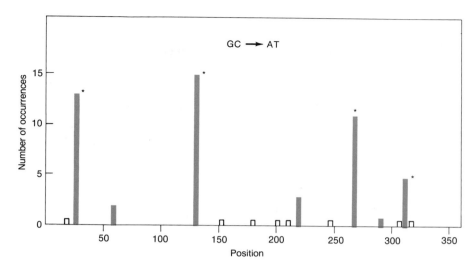

Figure 17-9. 5-methylcytosine hotspots in *E. coli.* Nonsense mutations occurring at 15 different sites and in *lacI* were scored. All result from the GC → AT transition. The asterisks (*) mark the position of 5-methylcytosines. Open bars depict sites at which the GC → AT change could be detected but at which no mutations occurred in this particular collection. It can be seen that 5-methylcytosine residues are hotspots for the GC → AT transition. Of 50 independently occurring mutations, 44 were at the four 5-methylcytosine sites and only six were at the 11 unmethylated cytosines. (From C. Coulondre et al., *Nature* 274, 1978, 775–780.)

ence" both for certain *types* of mutations (for example, GC → AT transitions) and for certain mutational *sites* (hotspots). Due to the recent advances that permit the determination of DNA sequences, we can now visualize mutational specificity at the molecular level. Figure 17-10 shows the mutational specificity in *lacI* for three mutagens described later: ethyl methanesulfonate (EMS), ultraviolet (UV) light, and aflatoxin B_1 (AFB$_1$). In this diagram, the distribution of base-substitution mutations that create chain-terminating UAG codons is seen. Figure 17-10 is similar to Figure 12-31, which shows the distribution of mutations in *rII*, except that the specific sequence changes are known for each site, allowing the diagram to be broken down into each category of substitution.

Figure 17-10 reveals the two components of mutational specificity. First, each mutagen shown favors a specific category of substitution. For example, EMS favors GC → AT transitions, whereas AFB$_1$ favors GC → TA transversions. This is related to the different mechanisms of mutagenesis. Second, even within the same category, there are large differences in mutation rate. This can be seen best with UV light for the GC → AT changes. Some aspect of the surrounding DNA sequence must cause these differences. In some cases, the cause of mutational hotspots can be determined by DNA sequence studies, as previously described for 5-methylcytosine residues and for certain frame-shift sites (Fig-

ures 17-5 and 17-9). In many examples of mutagen-induced hotspots, however, the precise reason for the high mutability of specific sites is still unknown.

Mutagens induce mutations by at least three different mechanisms. They can either replace a base in the DNA, alter a base so that it specifically mispairs with another base, or damage a base so that it can no longer pair with any base under normal conditions.

Incorporation of Base Analogs. Some chemical compounds are sufficiently similar to the normal nitrogen bases of DNA that they occasionally are incorporated into DNA in place of normal bases; such compounds are called **base analogs.** Once in place, these bases have pairing properties unlike those of the bases they have replaced; thus, they can produce mutations by causing incorrect nucleotides to be inserted opposite them during replication. The original base analog exists in only a single strand, but it can cause a nucleotide-pair substitution that is replicated in all DNA copies descended from the original strand.

For example, 5-bromouracil (5-BU) is an analog of thymine that has bromine at the C-5 position in place of the CH$_3$ group found in thymine. This change does not involve the atoms that take place in hydrogen bonding during base pairing, but the presence of the bromine significantly alters the distribution of electrons in the base. The normal structure (the keto form) of 5-BU

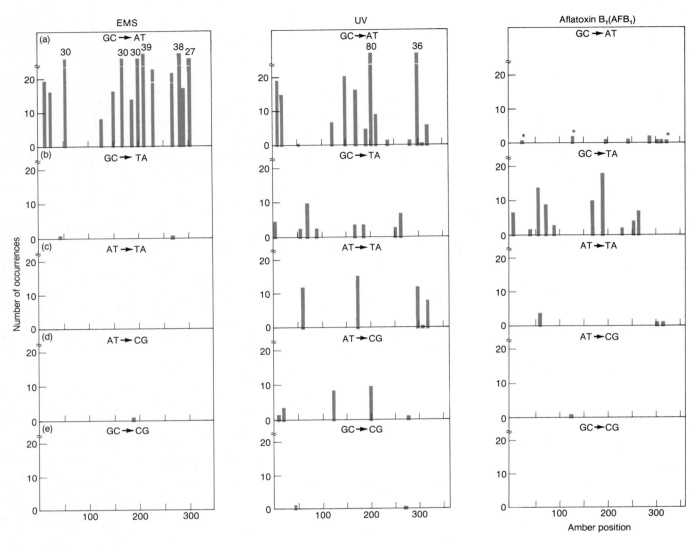

Figure 17-10. Specificity of mutagens. The distribution of mutations among 36 sites in the *lacI* gene is shown for three mutagens: EMS, UV light, and aflatoxin B₁. The height of each bar represents the number of occurrences of mutations at the respective site. Some hotspots are shown offscale, with the number of occurrences indicated directly alongside the respective peak. For instance, in the UV-generated collection, one site resulting from a GC → AT transition is represented by 80 occurrences. Each mutational site represented in the figure generates an amber (UAG) codon in the corresponding mRNA. The mutations are arranged according to the type of base substitution involved. Asterisks mark the position of 5-methylcytosines. (Redrawn from C. Coulondre and J. H. Miller, *Journal of Molecular Biology* 117, 1977, 577–606; and P. L. Foster et al., *Proceedings of the National Academy of Sciences USA,* 80, 1983, 2695–2698.)

undergoes a relatively frequent, spontaneous change to the enol form (Figure 17-11). (We have encountered such tautomeric shifts before; recall Figures 17-2 and 17-3.) The enol form has hydrogen-bonding properties almost identical to those of cytosine! Thus, the nature of the pair formed during replication will depend on the form of 5-BU at the moment of pairing (Figure 17-12). 5-BU causes transitions almost exclusively, as predicted in Figures 17-11 and 17-12.

Another widely used analog is 2-aminopurine (2-AP), which is an analog of adenine that can pair with thymine but can also mispair with cytosine. Therefore, when 2-AP is incorporated into DNA as adenine, it can generate AT → GC transitions by mispairing with cytosine during subsequent replications. Or, if 2-AP is incorporated by mispairing with cytosine, then GC → AT transitions will result. Genetic studies have shown that 2-AP is very specific for transitions.

Specific Mispairing. Some mutagens are not incorporated into the DNA but instead alter a base, causing specific mispairing. Certain **alkylating agents,** such as

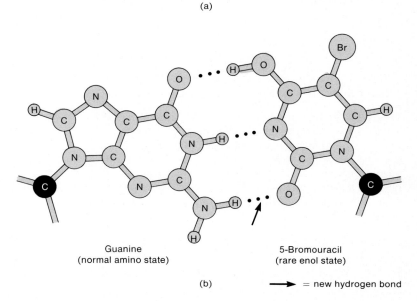

Figure 17-11. The alternative pairing possibilities for 5-bromouracil (5-BU). 5-BU is an analog of thymine that can be mistakenly incorporated into DNA as a base. It has a bromine atom in place of the methyl group. (a) In its normal keto state, 5-BU mimics the pairing behavior of the thymine it replaces, pairing with adenine. (b) The presence of the bromine atom, however, causes a relatively frequent redistribution of electrons, so that 5-BU can spend part of its existence in the rare enol form. In this state, it pairs with guanine, mimicking the behavior of cytosine and thus inducing mutations during replication.

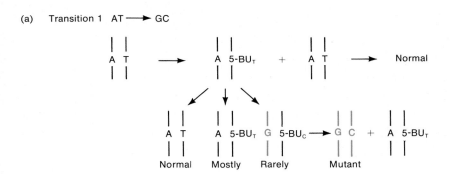

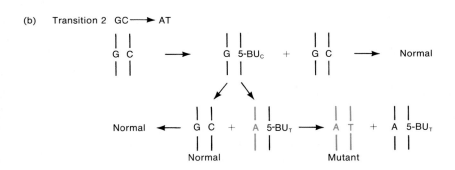

Figure 17-12. The mechanism of 5-BU mutagenesis. (a) In its normal keto state, 5-BU pairs like thymine (5-BU$_T$). (b) In its enol state, 5-BU pairs like cytosine (5-BU$_C$). In (a), 5-BU is incorporated across from adenine and subsequently mispairs with guanine, resulting in AT → GC transitions. In (b), 5-BU is misincorporated across from guanine and subsequently pairs with adenine, resulting in GC → AT transitions.

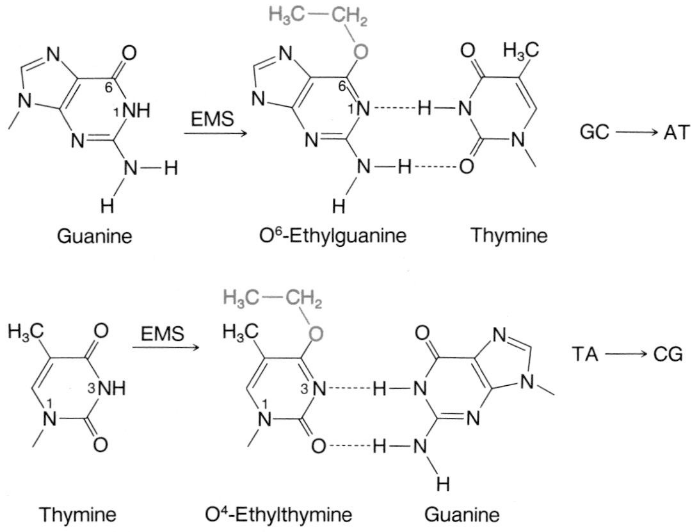

Guanine   O⁶-Ethylguanine   Thymine   GC ⟶ AT

Thymine   O⁴-Ethylthymine   Guanine   TA ⟶ CG

Figure 17-13. Alkylation-induced specific mispairing. The alkylation (in this case, ethylation generated by EMS) of the O-6 position of guanine and also the O-4 position of thymine can lead to direct mispairing with thymine and guanine, respectively, as shown here. In bacteria, where mutations have been analyzed in great detail, the principal mutations detected are GC → AT transitions, indicating that the O-6 alkylation of guanine is most relevant to mutagenesis.

ethyl methanesulfonate (EMS) and the widely used nitrosoguanidine (NG), operate via this pathway:

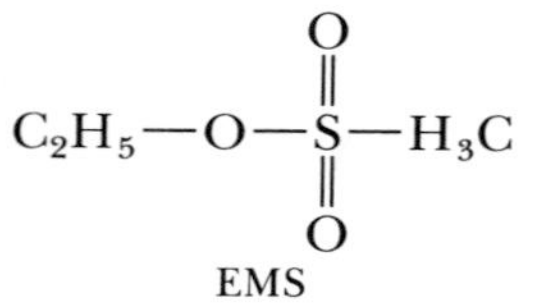

EMS 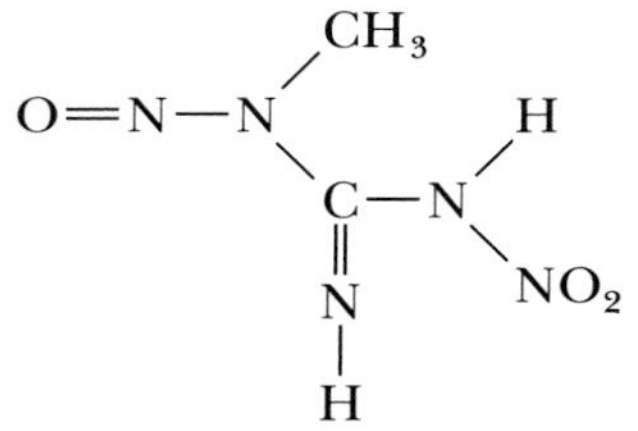   NG

Although such agents add alkyl groups (an ethyl group in the case of EMS and a methyl group in the case

of NG) to many positions on all four bases, mutagenicity is best correlated with an addition to the oxygen at the 6 position of guanine to create O-6-alkylguanine. This leads to direct mispairing with thymine, as shown in Figure 17-13, and would result in GC → AT transitions at the next round of replication. As expected, determinations of mutagenic specificity for EMS and NG show a strong preference for GC → AT transitions. (Verify this by consulting the data for EMS shown in Figure 17-10.)

Hydroxylamine (HA) is a specific inducer of GC → AT transitions. This specific effect is observed particularly in phage and *Neurospora;* the effects of HA are less specific in *E. coli*. Its structure is

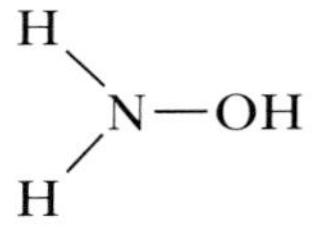

The relative specificity of HA is very probably due to the fact that it preferentially hydroxylates the amino nitrogen at cytosine C-4, creating *N*-4-hydroxycytosine, which can bind like thymine (Figure 17-14). *N*-4-hydroxycytosine prepared in vitro has the same ability to pair with thymine and cause mutations (see Chapter 15), which strongly supports the proposed mechanism.

Other examples of mutagens that alter bases specifically are those that deaminate cytosine to uracil. For instance, bisulfite ions convert cytosine to uracil, as does nitrous acid (NA). The uracil residue, if unrepaired (as shown earlier in Figure 17-8), will pair with adenine instead of guanine, generating a C → T transition. Nitrous acid also deaminates adenine to generate hypoxanthine, which can form A – C mispairs (Figure 17-15).

The **intercalating agents** form another important class of DNA modifiers. This group of compounds includes proflavin, acridine orange, and a class of chemicals termed ICR compounds (Figure 17-16a). These agents are planar molecules, which mimic base pairs and are able to slip themselves in between the stacked nitrogen bases at the core of the DNA double helix through a process called **intercalation** (Figure 17-16b). In this in-

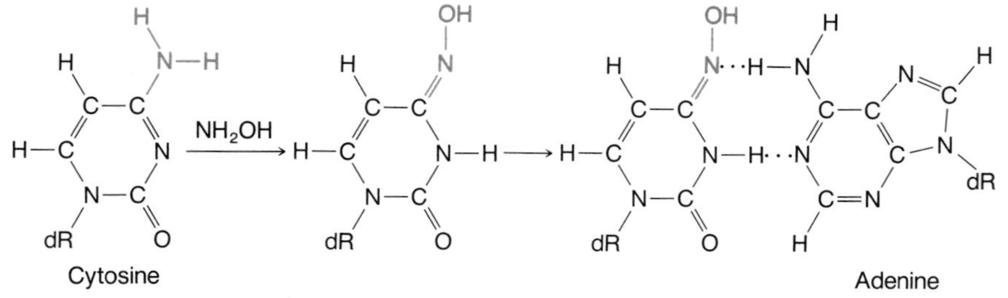

Cytosine   Adenine

Figure 17-14. A possible explanation for the GC → AT specificity of HA in some organisms. Cytosine is modified to pair like thymine, resulting in a GC → AT transition.

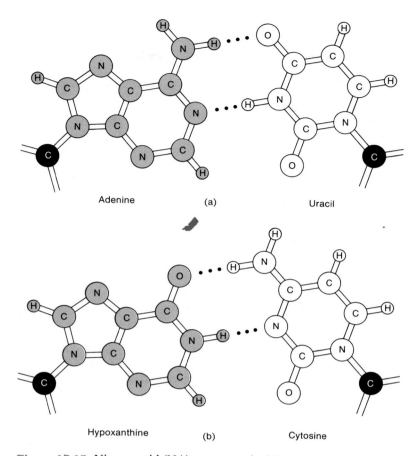

Figure 17-15. Nitrous acid (NA) mutagenesis. (a) NA deaminates cytosine to form uracil, which bonds like thymine. (b) NA deaminates adenine to form hypoxanthine, which bonds like guanine. These altered bonding patterns can lead to mutations. For example, AT may become GC, or GC may become AT. (From E. Freese, in *Structure and Function of Genetic Elements,* Brookhaven Symposia in Biology, No. 12, Brookhaven National Laboratory, Upton, N.Y., 1959.)

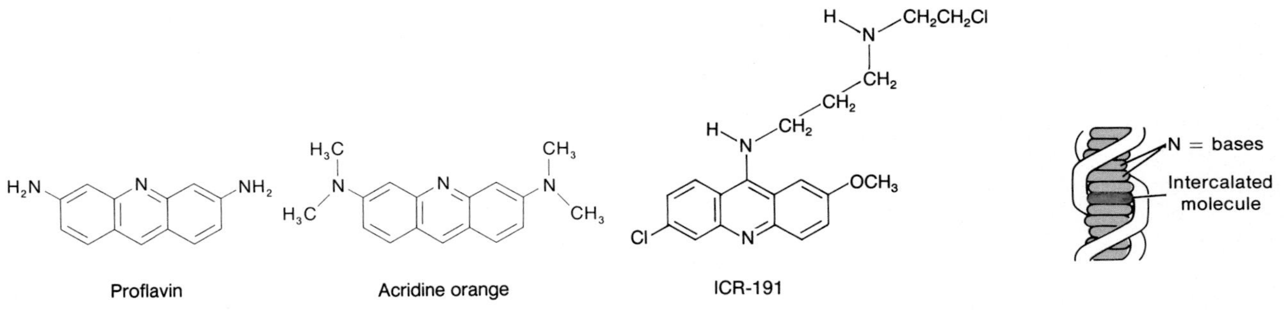

Figure 17-16. Intercalating agents. (a) Structures of the common agents proflavin, acridine orange, and ICR-191. (b) An intercalating agent slips between the stack of bases at the center of the DNA molecule. This occurrence can lead to single-nucleotide-pair insertions and deletions. (From L. S. Lerman, *Proceedings of the National Academy of Sciences USA* 49, 1963, 94.)

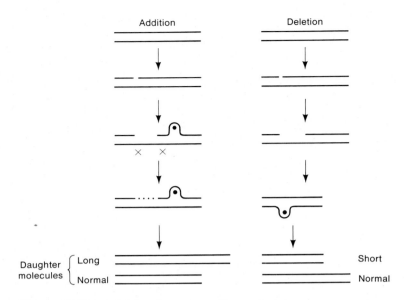

Figure 17-17. Model for the action of intercalating agents to cause short deletions or insertions. Here we assume that the agents are active only during DNA processing (repair or recombination), when they insert into a single-stranded loop and stabilize it. (See also Figures 17-4 and 17-16.) For an addition to occur, the loop can form only if there is a short, repeat length of complementary sequence (indicated by ×).

tercalated position, the agent can cause single-nucleotide-pair insertions or deletions. Intercalating agents may also stack between bases in single-stranded DNA; in so doing, they may stabilize bases that are looped out during frame-shift formation, as depicted in the Streisinger model (Figure 17-4). A model for the actions of such agents is shown in Figure 17-17.

Replication Bypass: The SOS System. A large number of mutagens require the operation of host enzyme systems in bacteria (and presumably in mammalian cells as well) in order to generate viable mutants. These mutagens damage one or more bases so that specific pairing is no longer possible. The damaged bases, if left unrepaired, block replication because DNA synthesis will not proceed past a base that cannot specify its complementary base by hydrogen bonding. This is for a good reason.

The insertion of bases across from noncoding lesions would lead to frequent mutations. The **bypass** of such replication blocks requires the activation of a special system, termed the **SOS system.** Therefore, mutagens that generate noncoding lesions are dependent on the SOS system for their action. The category of SOS-dependent mutagens is important, since it includes most carcinogens, such as ultraviolet light and aflatoxin B_1 (to be examined later; see Figure 17-30), as well as benzo(a) pyrene. The name SOS is derived from the idea that this system is induced as an emergency response to prevent cell death in the presence of significant DNA damage. (This "induction" is really the activation of gene expression, described more fully in Chapter 15.) SOS induction is a last resort, allowing the cell to trade a certain level of mutagenesis for ultimate survival. Figure 17-18 depicts the bypass of a noncoding lesion.

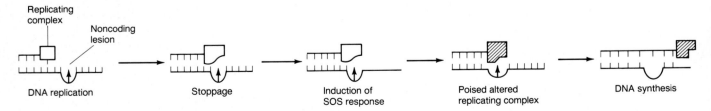

Figure 17-18. The SOS bypass system. This highly schematic diagram represents the stoppage of DNA replication in response to a noncoding lesion. The SOS system relieves this blockage, perhaps by altering the replicating complex.

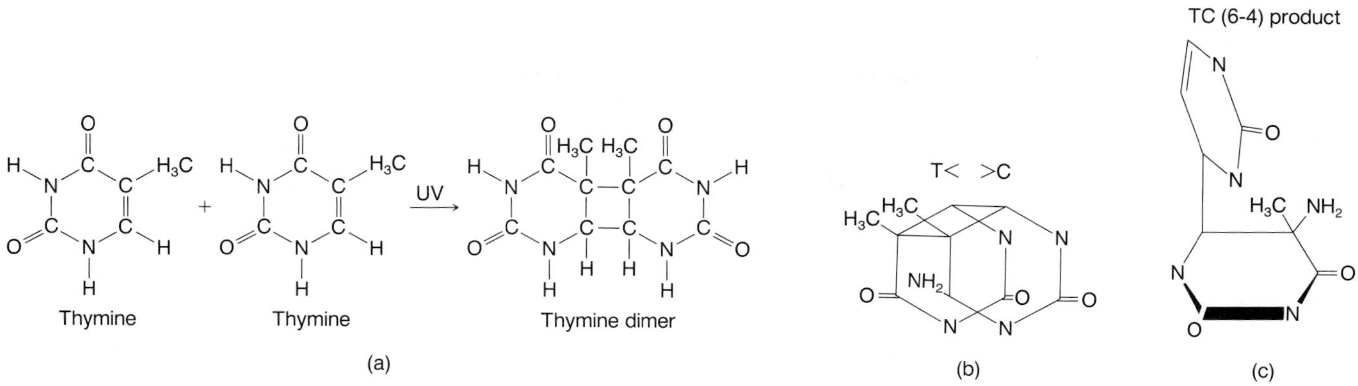

Figure 17-19. UV photoproducts: (a) cyclobutane-pyrimidine photodimer of T and T; (b) three-dimensional view of photodimer of T and C; (c) the 6-4 photoproduct of T and C.

How is the SOS system involved in the recovery of mutants induced by certain mutagens? One hypothesis is that induction of the SOS system lowers the fidelity of DNA replication so much (to permit the bypass of non-coding lesions) that many replication errors occur, even for undamaged DNA. This hypothesis could be tested, since it predicts that most mutations generated by different SOS-dependent mutagens would be similar, rather than specific to each mutagen. Most mutations would result from the action of the SOS system itself on undamaged DNA. The mutagen would then play the indirect role of inducing the SOS system. Studies of mutational specificity, however, have shown that this is not the case. Instead, a series of different SOS-dependent mutagens have markedly different specificities. You can see this by comparing the specificities of ultraviolet (UV) light and aflatoxin B_1 in Figure 17-10. Each mutagen induces a unique distribution of mutations. Therefore, the mutations must be generated in response to specific damaged base pairs. The type of lesion differs in many cases. Some of the most widely studied lesions include UV photoproducts, apurinic sites, and bulky chemical additions on specific bases.

Exactly how the SOS bypass system functions is not clear, although it is known to be dependent on at least two genes: recA, which is also involved in general recombination (described in Chapter 18), and umuC. (Chapter 16 describes the control of the induction of the SOS system.)

Ultraviolet (UV) light generates a number of photoproducts in DNA. Two different lesions occurring at adjacent pyrimidine residues have been most strongly correlated with mutagenesis. The cyclobutane-pyrimidine photodimer and the 6-4 photoproduct (Figure 17-19) interfere with normal base pairing. Induction of the SOS system is required for mutagenesis. Insertion of bases across from UV photoproducts leads most fre-

quently to transition mutations, but other base substitutions (transversions) and frame-shifts are also stimulated by UV light, as are duplications and deletions. The mutagenic specificity of UV light is illustrated in Figure 17-10.

Aflatoxin B_1 (AFB_1) is a powerful carcinogen, which generates apurinic sites following the formation of an additional product at the N-7 position of guanine (Figure 17-20). Studies with apurinic sites generated in vitro have demonstrated a requirement for the SOS system and have also shown that bypass of these sites leads to the preferential insertion of an adenine across from an apurinic site. This predicts that agents that cause depurination at guanine residues should induce preferentially $GC \rightarrow TA$ transversions. Can you see why the insertion of an adenine across from an apurinic site derived from a guanine would generate this substitution at the next round of replication? Figure 17-10 shows the genetic analysis of many base substitutions induced by AFB_1. You can verify that most of the substitutions are indeed $GC \rightarrow TA$ transversions.

AFB_1 is a member of a class of chemical carcinogens that generate bulky addition products by binding cova-

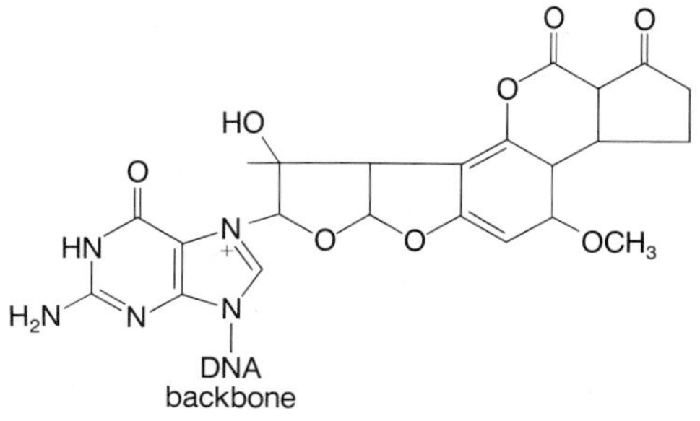

Figure 17-20. The binding of metabolically activated aflatoxin B_1 to DNA.

lently to DNA. Other examples include the diolepoxides of benzo(a)pyrene, a compound produced by internal combustion engines. For many different compounds it is not yet clear *which* DNA addition products play the principal role in mutagenesis. In some cases, the mutagenic specificity suggests that depurination may represent an intermediate step in mutagenesis; in others, the question of which mechanism is operating is completely open.

Message   Mutagens induce mutations by a variety of mechanisms. The active participation of an enzyme system, termed the SOS system, is required to convert some DNA damage into mutations that are recovered in viable cells.

Biological Repair Mechanisms

As we have seen in the previous discussion, there are many potential threats to the fidelity of DNA replication. Not only is there an inherent error rate for the replication of DNA, but there are also spontaneous lesions that can provoke additional errors. Moreover, agents in the environment (mutagens) can damage DNA and greatly increase the mutation rate.

Living cells have evolved a series of enzymatic systems that repair DNA damage in a variety of ways. Some enzymatic systems neutralize potentially damaging compounds before they even react with DNA. One example of such a **detoxification** system involves the enzyme **superoxide dismutase,** which catalyzes the conversion of superoxide radicals (produced during oxidative damage to DNA) to hydrogen peroxide. The enzyme catalase, in turn, converts the hydrogen peroxide to water.

Let's first examine some of the characterized repair pathways, and then consider how the cell integrates these systems into an overall strategy for repair. We can divide repair pathways into several categories.

Direct Reversal of Damage

The most straightforward way to repair a lesion is to remove it directly, thereby regenerating the normal base. Removal is not always possible, since some types of damage are essentially irreversible. In a few cases, however, lesions can be directly removed. One case involves photodamage to DNA caused by ultraviolet light (UV). UV creates a number of photoproducts on the DNA. Some of the lesions involved in mutagenesis occur at neighboring pyrimidines on the same strand of DNA. These UV photodimers are principally of two types (Figure 17-19). The cyclobutane-pyrimidine photodimer

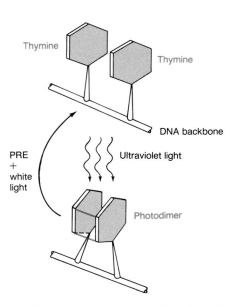

Figure 17-21. Illustration of thymine photodimerization between adjacent thymine bases in a single strand of DNA. (PRE stands for the photoreactivating enzyme, which splits the photodimer.) (After J. D. Watson, *Molecular Biology of the Gene,* 3d ed. Copyright © 1976 by W. A. Benjamin).

can be repaired by a **photoreactivating enzyme (PRE),** whereas the 6-4 photoproduct cannot. PRE operates by binding to the photodimer and splitting it, in the presence of certain wavelengths of visible light, to generate the original bases (Figure 17-21). This enzyme cannot operate in the dark, so other repair pathways are required to remove UV damage.

A second enzyme involved in the direct removal of lesions, **alkyltransferase,** removes certain alkyl groups added to the 0 to 6 positions of guanine (Figure 17-13) by such agents as NG and EMS. The methyltransferase from *E. coli* has been well studied. This enzyme transfers the methyl group from O-6-guanine to a cysteine residue on the protein (Figure 17-22). When this happens, the enzyme is inactivated, so this repair system can be saturated if the level of alkylation is high enough.

Excision Pathways

Excision Repair.   There are several pathways for excising altered bases, together with a stretch of neighboring bases. One general pathway is encoded by three genes in *E. coli* termed *uvrA, uvrB,* and *uvrC.* This system recognizes any lesion that creates a significant distortion of the DNA double helix. An endonucleolytic cut is made several base pairs away on either side of the damaged base, and a 12-base-long segment of single stranded DNA is removed. The short gap is then filled in by repair synthe-

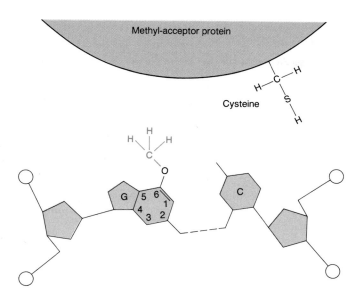

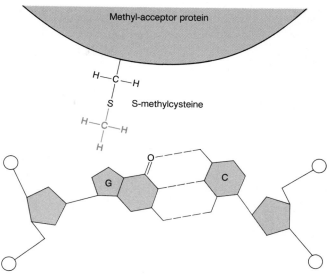

Figure 17-22. Methylation of one site in a guanine by the chemical agent nitrosoguanidine (NG) is repaired by a novel process. The NG adds a methyl group (CH_3) at various sites in the DNA, including an oxygen atom at position 6 of guanine *(left)*, disrupting the hydrogen bonding of guanine to a cytosine. Here the repair is accomplished by a protein that is synthesized in the cell during a period of adaptation when the cell is exposed to a low level of NG. A cysteine (one of the 20 amino acids) on the protein acts as a methyl acceptor: it binds the CH_3 group, thereby restoring the guanine to its original state *(right)*. (From P. Howard-Flanders, "Inducible Repair of DNA." Copyright © 1981 by Scientific American, Inc. All rights reserved.)

sis (probably mediated by DNA polymerase I) and DNA ligase (Figure 17-23). Many types of lesions are removed by means of this general **excision repair,** including the principle UV photoproducts (Figure 17-19) as well as the bulky chemical additions resulting from the binding of compounds such as AFB_1 and epoxides of benzo(a) pyrene. This is a crucial repair pathway in humans also. A rare inherited disorder, *xeroderma pigmentosum* (XP), results from a deficiency in one of the excision-repair enzymes. People suffering from this disorder are extremely prone to UV light-induced skin cancers (Figure 17-24) as a result of exposure to sunlight. The difference in UV photosensitivity between normal and diseased cells is evident from the survival curves in Figure 17-25. Most people with XP die of skin cancer before they reach the age of 30, so we can see how vital a role the excision repair pathway plays.

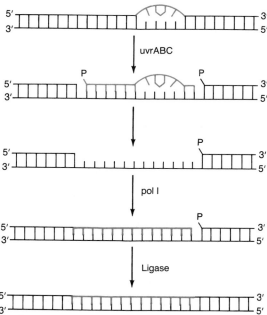

Figure 17-23. A model for nucleotide-excision repair in *E. coli*. Here uvrABC nuclease hydrolyzes the eighth phosphodiester bond 5′ and the fourth phosphodiester bond 3′ to the pyrimidine photodimer, producing a 12-nucleotide-long single-stranded DNA fragment and 3′-OH and 5′-phosphoryl termini. The oligonucleotide carrying the damage is removed, and the resulting gap is filled by DNA polymerase I (pol I) and sealed by DNA ligase. This diagram shows the removal of a 12-nucleotide fragment. The uvrABC nuclease also generates 13 base-long oligonucleotides. (From A Sancar and W. D. Rupp, *Cell* 33, 1983, 249–260.)

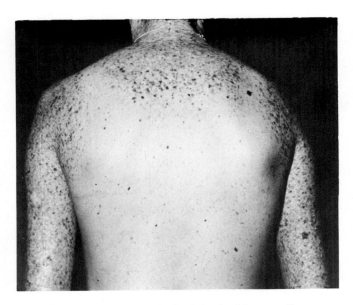

Figure 17-24. Skin cancer in the inherited human disease *xeroderma pigmentosum.* This recessive hereditary disease is caused by a deficiency in one of the excision-repair enzymes, which leads to the formation of skin cancers on exposure of the skin to the UV rays in sunlight. (Photograph courtesy of Dirk Bootsma, Erasmus University, Rotterdam.)

Specific Excision Pathways.   Certain lesions are too subtle to cause a distortion large enough to be recognized by the *uvrA,B,C*-encoded general excision repair system. Therefore, additional excision pathways are necessary.

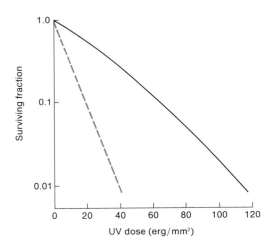

Figure 17-25. Survival curves for human skin cells in tissue culture after exposure to various dosages of UV radiation. The solid line shows the curve for normal cells; the dashed line shows the survival curve for cells from an individual with *xeroderma pigmentosum.* (After J. E. Cleaver, *Advances in Radiation Biology,* vol. 4, 1974.)

AP ENDONUCLEASE REPAIR PATHWAY.   All cells have endonucleases that attack the sites left after the spontaneous loss of single purine or pyrimidine residues. For convenience, the apurinic and apyrimidinic sites are termed **AP sites** because they are biochemically equivalent (see Figure 17-7). The **AP endonucleases** are vital to the cell, since, as noted earlier, spontaneous depurination is a relatively frequent event. These enzymes introduce chain breaks by cleaving the phosphodiester bonds at AP sites. This initiates an excision-repair process mediated by an exonuclease, DNA polymerase I, and DNA ligase (Figure 17-26). (The last two steps are identical to the corresponding steps in the excision repair process shown in Figure 17-23.)

Due to the efficiency of the AP endonuclease repair pathway, it can be the final step of other repair pathways. Thus, if damaged base pairs can be excised, leaving an AP site, the AP endonucleases can complete the restoration to wild-type. The enzymes that excise specific damaged bases are called DNA glycosylases.

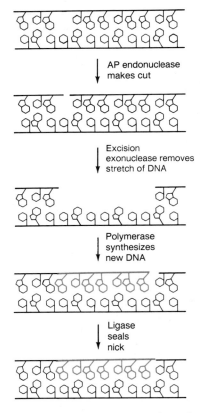

Figure 17-26. Repair of AP sites. AP endonucleases recognize AP sites and cut the phosphodiester bond. A stretch of DNA is removed by an exonuclease, and the resulting gap is filled in by DNA polymerase I and DNA ligase. (After B. Lewin, *Genes.* Copyright © 1983 by John Wiley.)

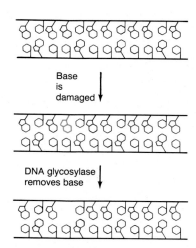

Figure 17-27. Action of DNA glycosylases. Glycosylases remove altered bases and leave an AP site. The AP site is subsequently excised by the AP endonucleases diagrammed in Figure 17-26. (After B. Lewin, *Genes.* Copyright © 1983 by John Wiley.)

DNA GLYCOSYLASE REPAIR PATHWAY. **DNA glycosylases** do not cleave phosphodiester bonds but instead cleave *N*-glycosidic (base-sugar) bonds, liberating the altered base and generating an AP site (Figure 17-27). The AP site resulting from glycosylase action is subsequently repaired by the pathway shown in Figure 17-26. Thus, AP endonucleases cleave the phosphodiester bond, and exonuclease excision is followed by repair mediated by DNA polymerase and ligase. Uracil DNA glycosylase removes uracil from DNA. Uracil residues result from the deamination of cytosine (Figure 17-8) and can lead to a C → T transition if unrepaired. It is possible that the pairing partner of adenine in DNA is thymine (5-methyluracil), rather than uracil, in order to allow recognition and excision of the uracil residues that result from the spontaneous deamination of cytosine. If uracil were a normal constituent of DNA, such repair would not be possible. There is also a glycosylase that recognizes and excises hypoxanthine, the deamination product of adenine.

Different glycosylases have been described that remove alkylated bases (such as 3-methyladenine, 3-methylguanine, and 7-methylguanine), ring-opened purines, and, in some organisms, UV photodimers. New glycosylases are still being discovered.

Post-replication Repair

Some repair pathways are capable of recognizing errors even after DNA replication has already occurred. One of the systems, termed the **mismatch repair system,** can detect mismatches that occur during DNA replication. Suppose you were to design an enzyme system that could repair replication errors. What would this system be required to do? The system would have to have at least three properties:

1. The ability to recognize mismatched base pairs.

2. The ability to excise a mismatch base and carry out repair synthesis.

3. The ability to determine which of the two mismatched bases is the incorrect one.

The third point is the crucial property of such a system. Unless it is capable of discriminating between the correct and the incorrect bases, then the mismatch repair system could not determine base to excise. If, for example, a G – T mismatch occurs as a replication error, how can the system determine whether G or T is incorrect? Both are normal bases in DNA. The mismatch repair system takes advantage of the postreplication modification of certain bases to recognize the old, template strand from the newly synthesized strand. The relevant modification enzyme in this case is adenine methylase, which modifies the sequence

$$5'-G-A-T-C-3'$$
$$3'-C-T-A-G-5'$$

on each strand to create 6-methyladenine. Methylating the 6-position of adenine does not affect base pairing, but it does provide a convenient tag that can be detected by other enzyme systems. Figure 17-28 shows the replication fork. Note that only the old strand is methylated

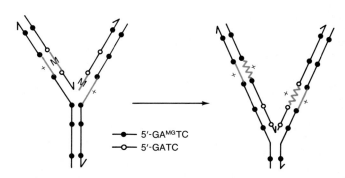

Figure 17-28. Mismatch correction in replicational heteroduplexes: sequence conservation by strand-directed mismatch repair. Mismatched base pairs (+/M) arising as replicational errors (mutations M) are corrected by excision and resynthesis (-\/\/-). Only the newly synthesized strand segments proximal to the replication fork contain unmethylated GATC sequences (-○-) and therefore are subject to efficient mismatch repair. (A^{mg} = adenine with a methyl group at the G position.) (From F. Bourguignon-Van Horen et al., *Biochimie* 64, 1982, 559–564.)

at GATC sequences right after replication. The adenine methylase takes several minutes to recognize and modify the newly synthesized GATC stretches. During that interval, the mismatch repair system can operate because it can now distinguish the old strand from new ones by the methylation patterns. Replication errors produce mismatches on the newly synthesized strand, so it is the bases on this strand that are excised.

A second mismatch repair system has recently been discovered in bacteria. This system removes G – A mispairs. It is not yet clear how this system discriminates between the old and the new strand.

Recombinational Repair. The *recA* gene, which is involved in SOS bypass (Figure 17-18), is also involved in postreplication repair. The DNA replication system stalls at a UV photodimer or other blocking lesion and then restarts past the block, leaving a single-stranded gap. In **recombinational repair,** this gap is patched by DNA cut from the sister molecule (Figure 17-29a). This process seems to lead to few errors. SOS bypass is, however, highly mutagenic, as described earlier. Here the replication system continues past the lesion (Figures 17-18 and 17-29b), accepting noncomplementary nucleotides for new strand synthesis.

Strategy for Repair

We can now assess the overall repair-system strategy used by the cell. It would be convenient if enzymes could be used directly to reverse specific lesions or excise particular altered bases. However, it would cost too much energy to encode and synthesize an enzyme to repair each possible type of cell damage. Therefore, a general excision repair system is used to remove any type of damaged base that causes a recognizable distortion in the double helix. When lesions are too subtle to cause such a distortion, specific excision systems, glycosylases, or removal systems are designed. To eliminate replication errors, a postreplication mismatch repair system operates; finally, postreplication systems eliminate gaps across from blocking lesions that have escaped the other repair systems.

Message Repair enzymes play a crucial role in reducing genetic damage in living cells. Many different repair pathways have been characterized. (Table 17-2 summarizes these pathways.)

Mutators

As the preceding description of repair processes indicates, normal cells are programmed for error avoidance. The repair processes are so efficient that the observed base-substitution rate is as low as 10^{-10} to 10^{-9} per base pair per cell per generation. However, mutant strains have been detected that have increased spontaneous mutation rates. Such strains are termed **mutators.** In

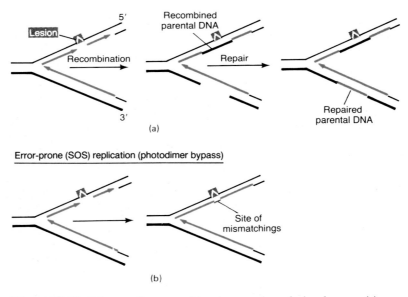

Figure 17-29. Schemes for recombination repair or lesion bypass: (a) postreplication recombination repair; (b) SOS bypass. (From A. Kornberg, *DNA Replication.* Copyright © 1980 by W. H. Freeman and Company.)

TABLE 17-2. Repair systems in *E. coli*

General mode of operation	Example	Type of lesion repaired	Mechanism
Detoxification	Superoxide dismutase	Prevents formation of lesion	Convert peroxides to hydrogen peroxide, which is neutralized by catalase
Direct removal of lesions	Alkyltransferases	O-6-alkylguanine	Transfer alkyl group from O-6-guanine to cysteine residue on transferase
	Photoreactivating enzyme (PRE)	UV photodimiers	Splits dimers in the presence of white light
General excision	*uvrABC*-encoded exonuclease system	Lesions causing distortions in double helix, such as UV photoproducts, and bulky chemical additions	Makes endonucleolytic cut on either side of lesion; resulting gap is repaired by DNA polymerase I and DNA ligase
Specific excision	AP endonuclease	AP sites	Makes endonucleolytic cut; exonuclease creates gap, which is repaired by DNA polymerase I and DNA ligase
	DNA glycosylases	Certain altered bases, such as deaminated bases (uracil, hypoxanthine), certain methylated bases, ring-opened purines, and other modified bases	Removes base, creating AP site, which is repaired by AP endonucleases
Postreplication	Mismatch-repair system	Replication errors resulting in base-pair mismatches	This system recognizes which strand is correct by detecting methylated A residues that are part of the 5′– GATC–3′ sequence, and then excises bases from the newly synthesized strand when a mismatch is detected
	G–A mismatch-repair system	G–A mispairs	Excision and resynthesis; strand discrimination unclear
	Recombinational repair	Lesions that block replication and result in single-stranded gaps	Recombinational exchange
	SOS system	Lesions that block replication	Allow replication bypass of blocking lesion, resulting in frequent mutations across from lesion

many cases, the mutator phenotype is due to a defective repair system. In *E. coli,* the mutator loci *mutH, mutL, mutU,* and *mutS* affect components of the postreplication mismatch repair system, as does the *dam* locus, which specifies the enzyme deoxyadenosine methylase. Dam⁻ strains cannot methylate adenines at GATC sequences, and the mismatch repair system can no longer discriminate between the template and the newly synthesized strands. This leads to a higher spontaneous mutation rate. Figure 17-30 depicts how the mismatch repair system might operate and how the mutator loci affect this system.

The *mutY* locus results in G–C → T–A transversions and affects the recently discovered mismatch system that operates only on G–A mispairs. Ung⁻ mutants are missing the enzyme uracil DNA glycosylase. These mutants cannot repair cytosine deaminations and have elevated levels of C → T transitions. The *mutD* locus is responsible for a very high rate of mutagenesis (at least three orders of magnitude higher than normal). Mutations at this locus affect the proofreading function of DNA polymerase III.

Not all mutators are well understood. For instance, the *mutT* locus raises the level of only AT → CG transversions for reasons that are still not clear.

Reversion Analysis

In this chapter, we have considered the different pathways leading to mutagenesis and have observed that mutagenic processes are often very specific. We can now begin to see how reversion analysis can tell us something

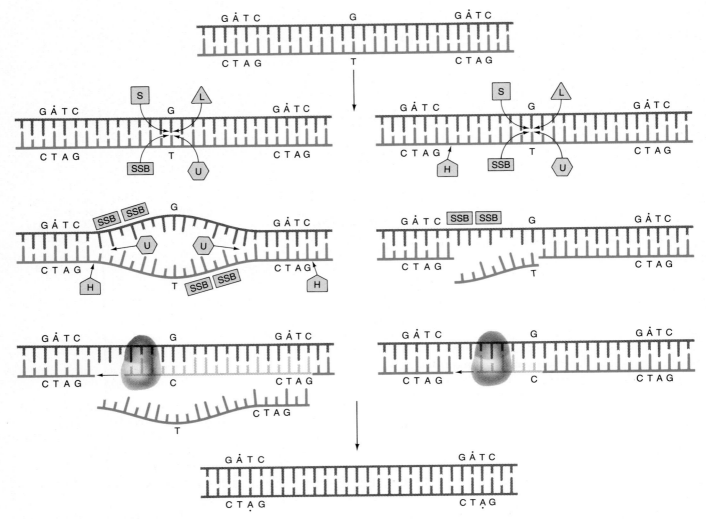

Figure 17-30. Mismatch repair corrects errors after a DNA strand has been synthesized. Two different models have been proposed for the mechanism of repair. In both models, proteins called MutL and MutS interact with the mismatch site (G–T), and a protein called MutH cleaves the newly synthesized strand. The repair apparatus distinguishes the parental strand from the new one by means of the methyl group (black dots) within the parental GATC sequences. The strands surrounding the mismatch are separated with the help of a protein called MutU and are stabilized by a single-stranded binding protein (*SSB*). The main difference between the two models has to do with where the strand containing the incorrect nucleotide is cleaved: at two flanking GATC sequences (*left*) or at one GATC sequence and the mismatch itself (*right*). In each case polymerases synthesize a new segment in place of the excised one, and the corrected strand is eventually methylated like the parental copy. (From M. Radman and R. Wagner "The High Fidelity of DNA Duplication." Copyright © 1988 by Scientific American, Inc. All rights reserved.)

about the nature of a mutation or the action of a mutagen. For example, if a mutation cannot be reverted by action of the mutagen that induced it, then the mutagen must have some relatively specific unilateral action. In the case of a mutation induced by hydroxylamine (HA), it would be reasonable to expect that the original mutation is GC → AT, which of course, cannot be reverted by another specific GC → AT event. Similarly, muta-

tions that can be reverted by proflavin are in all likelihood frame-shift mutations; those induced by nitrous acid (NA) mutations (which are transitions) should not be revertible by proflavin.

Transversions are not as easy to detect by reversion, but they are known definitely to be common among spontaneous mutations, as shown by studies of DNA and protein sequencing. Thus, in the reversion test, if a mu-

TABLE 17-3. Different types of point mutations theoretically distinguishable by their reversion behaviors in response to a battery of specific mutagens

Mutation	Reversion mutagen			
	NA	HA or EMS	Proflavin	Spontaneous
Transition (GC → AT)	+	−	−	+
Transition (AT → GC)	+	+	−	+
Transversion	−	−	−	+
Frame-shift	−	−	+	+

NOTE: a + indicates a measurable rate of reversion due to a given mutagen. NA = nitrous acid; HA = hydroxylamine; EMS = ethyl methanesulfonate.

tation does revert spontaneously but does not revert in response to a transition mutagen or a frame-shift mutagen, then, by elimination, it is probably a transversion. Note that the kinds of logic employed in the reversion test rely heavily on the assumption that the reversion events are not due to suppressors or transposable elements; either of these would make inference from reversion more difficult. Table 17-3 summarizes some reversion expectations based on simple assumptions.

The system outlined in Table 17-3 is intended merely to illustrate the kinds of inferences possible from reversion analysis. Recall that mutagen specificities depend on the organism, the genotype, the gene studied, and perhaps even the *region* of the gene studied.

The Ames Test

There is increasing awareness of a correlation between mutagenicity and carcinogenicity. One study showed that 157 of 175 known carcinogens (approximately 90 percent) are also mutagens. The somatic mutation theory of cancer holds that these agents cause cancer by inducing the mutation of somatic cells. This means that mutagenesis is of great relevance to our society. We are faced not only with the genetic time bomb of germinal mutation—with its potential for increasing inherited disease over the long term—but also with the somatic genetic disease of cancer with its overwhelming immediacy. The modern environment exposes each individual to a wide variety of chemicals in drugs, cosmetics, food preservatives, pesticides, compounds used in industry, pollutants, and so on. Many of these compounds have been shown to be carcinogenic and mutagenic (Figure 17-31). Examples include the food preservative AF-2, the food fumigant ethylene dibromide, the antischistosome drug hycanthone, several hair-dye additives, and the industrial compound vinyl chloride; all are potent, and some have subsequently been subjected to govern-

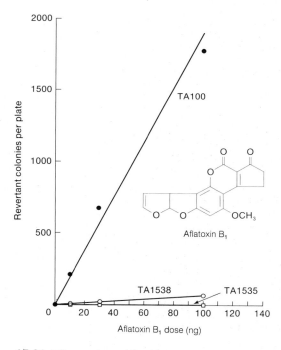

Figure 17-31. The mutagenicity in the Ames test of aflatoxin B_1, which is also a potent carcinogen. TA100, TA1538, and TA1535 are strains of *Salmonella* bearing different *his* auxotrophic mutations. The *Salmonella* strain TA100 is a *his* mutant strain that is highly sensitive to reversion through base-pair substitution. The strains TA1535 and TA1538 are *his* mutant strains that are sensitive to reversion through frame-shift mutation. The results of this test show that aflatoxin B_1 is a potent mutagen that causes base-pair substitutions but not frame-shifts. (From J. McCann and B. N. Ames, in *Advances in Modern Toxicology*, Vol. 5. Edited by W. G. Flamm and M. A. Mehlman. Copyright © by Hemisphere Publishing Corp.)

ment control. However, hundreds of new chemicals and products appear on the market each week. How can such vast numbers of new agents be tested for carcinogenicity before much of the population has been exposed to them?

Many test systems have been devised to screen for carcinogenicity. These are time-consuming tests, typically involving laborious research with small mammals. More rapid tests do exist that make use of microbes (such as fungi or bacteria) and test for mutagenicity rather than carcinogenicity. The most widely used test was developed in the 1970s by Bruce Ames, who worked with *Salmonella typhimurium*. This **Ames test** uses two auxotrophic histidine mutations, which revert by different molecular mechanisms. Further properties were genetically engineered into these strains to make them suitable for mutagen detection. First, they carry a mutation that inactivates the excision-repair system. Second, they carry a mutation that eliminates the protective lipopolysaccharide coating of wild-type *Salmonella*, so that any escaping bacteria will be unable to survive in such natural (but chemically hostile) environments as sewers or intestines.

Bacteria are evolutionarily a long way removed from humans. Can the results of such a test on bacteria have any real significance in detecting chemicals that are dangerous for humans? First, we have seen that the genetic and chemical nature of DNA is identical in all organisms, so that a compound acting as a mutagen in one organism is likely to have some mutagenic effects in other organisms. Second, Ames devised a way to simulate the human metabolism in the bacterial system. Much of the important processing of ingested chemicals in mammals occurs in the liver, where externally derived compounds normally are detoxified or broken down. In some cases, the action of liver enzymes can create a toxic or mutagenic compound from a substance that was not originally dangerous (Figure 17-32). Ames set out to incorporate the mammalian liver enzymes in his bacterial test system.

Rat livers are normally used for this purpose. First, the liver enzymes are mobilized by injection of Arochlor, a polychlorinated biphenyl (PCB), into the rats. The rats are then killed, and their livers are homogenized and centrifuged to remove cell debris. The supernatant, called S9 mix, contains the solubilized batteries of rat liver enzymes. The S9 mix is added to a suspension of auxotrophic bacteria in a solution of the suspected carcinogen being tested. The solution is then plated on a medium containing no histidine, and later the plates are scored for colonies of revertants (Figure 17-33).

The Ames test has detected potential carcinogens among a wide variety of heterogeneous types of chemicals. Of course, chemicals detected by this test can be regarded not only as potential carcinogens (sources of

somatic mutations) but also as possible causes of mutations in germinal cells. Because the test system is so simple and inexpensive, many laboratories throughout the world now routinely test large numbers of potentially hazardous compounds for mutagenicity and potential carcinogenicity.

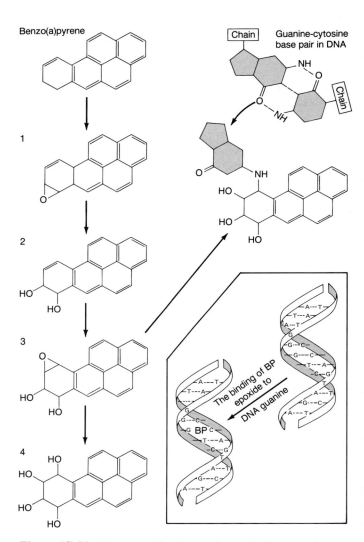

Figure 17-32. The detoxification and metabolic activation of benzo(a)pyrene. The detoxification of the carcinogen benzo(a)pyrene (BP) goes through several steps, as it is made more water-soluble prior to excretion. One of the intermediates in this process (3) is capable of reacting with guanine in DNA *(upper right)*. This leads to a distortion of the DNA molecule and mutations, as described in the text. Benzo(a)pyrene is therefore a mutagen for any cell that has the enzymes that produce this intermediate. (After I. B. Weinstein et al., *Science* 193, 1976, 592–595; From J. Cairns, *Cancer: Science and Society*. Copyright © 1978 by W. H. Freeman and Company.)

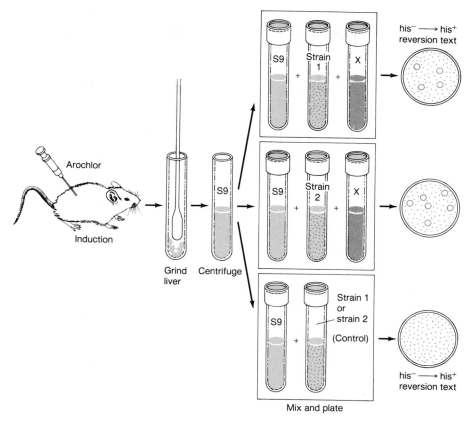

Figure 17-33. Summary of the procedure used for the Ames test. First rat-liver enzymes are mobilized by injecting the animals with Arochlor. (Enzymes from the liver are used because it carries out the processes of detoxifying and toxifying body chemicals.) The rat liver is then homogenized, and the supernatant of solubilized rat-liver enzymes (S9) is added to a suspension of auxotrophic bacteria in a solution of the potential carcinogen (X). This mixture is plated on a medium containing no histidine and revertants of mutant strains 1 and 2 are looked for. A control experiment containing no potential carcinogen is always run simultaneously. The presence of revertants indicates that the chemical is a mutagen and possibly a carcinogen as well.

Summary

■ Gene mutations can arise through many different processes. Spontaneous mutations can result from replication errors or from spontaneous lesions, such as those generated by deamination or depurination. (Recombination and transposable elements can also result in altered genes, as described in Chapters 18 and 19.) Mutagens can increase the frequency of mutations. Some of these agents act by mimicking a base but then pairing differently during DNA replication. Others alter bases in the DNA and convert them to derivatives that pair differently. A third class of mutagens, which includes most carcinogens, damages DNA in such a way that replication is blocked. The activation of an enzymatic pathway, termed the SOS system, is required to replicate past the blocking lesions. This results in mutations that appear mainly across from the blocking lesion. Repair enzymes present in living cells greatly minimize genetic damage, thus averting many mutations. Mutant organisms lacking certain repair enzymes have higher than normal mutation rates.

Our knowledge of the molecular basis of mutation can be exploited for useful purposes. One example is the Ames test, which utilizes mutant bacterial strains to test compounds in the environment for mutagenic activity. Due to the correlation between mutagenicity and carcinogenicity, the identification of potential carcinogens in the environment can be achieved by this rapid assay.

■ ■ ■

Solved Problems

1. Based on the specificity given in the test for aflatoxin B_1 and EMS, describe whether each mutagen would be able to revert amber (UAG) and ochre (UAA) codons back to wild-type.

Solution

EMS induces primarily $G-C \rightarrow A-T$ transitions. UAG codons could not be reverted back to wild-type, since only the UAG $\rightarrow$ UAA change would be stimulated by EMS and that generates a nonsense (ochre) codon. UAA codons would not be acted on by EMS. Aflatoxin B_1 induces primarily $G-C \rightarrow T-A$ transversions. Only the third position of UAG codons would be acted on, resulting in a UAG $\rightarrow$ UAU change (on the mRNA level) that produces tyrosine. Therefore, if tyrosine were an acceptable amino acid at the corresponding site in the protein, aflatoxin B_1 could revert UAG codons. Aflatoxin B_1 would not revert UAA codons because no $G-C$ base pairs appear at the corresponding position in the DNA.

2. A mutant of *E. coli* is highly resistant to mutagenesis by a variety of agents, including ultraviolet light, aflatoxin B_1, and benzo(a)pyrene. Explain one possible cause of this mutant phenotype.

Solution

The mutant might lack the SOS system and perhaps carry a defect in the *umuC* gene. Such strains would not be able to bypass replication-blocking lesions of the type caused by the three mutagens listed. Without the processing of premutational lesions, mutations would not be recovered in viable cells.

Problems

1. Differentiate between the following pairs:

   a. Transitions and transversions

   b. Silent and neutral mutations

   c. Missense and nonsense mutations

   d. Frame-shift and nonsense mutations

2. Why are frame-shift mutations more likely than missense mutations to result in proteins that lack normal function?

3. Describe the Streisinger model for frame-shift formation. Show how this model can explain mutational hotspots in the *lacI* gene of *E. coli*.

4. Diagram two different mechanisms for deletion formation. How do DNA sequencing experiments suggest these possibilities?

5. Describe two spontaneous lesions that can lead to mutations.

6. Compare the mechanism of action of 5-bromouracil (5-BU) with ethylmethanesulfonate (EMS) in causing mutations. Explain the specificity of mutagenesis for each agent in light of the proposed mechanism.

7. Compare the two different systems involved in the repair of AP sites and in the removal of bulky chemical adducts.

8. Describe the repair systems that operate after replication.

9. Normal ("tight") auxotrophic mutants will not grow at all in the absence of the appropriate supplement to the medium. However, in mutant hunts for auxotrophic mutants, it is common to find some mutants (called "leaky") that grow very slowly in the absence of the appropriate supplement but normally in the presence of the supplement. Propose an explanation for the molecular nature of the leaky mutants.

10. Strain A of *Nuerospora* contains an *ad-3* mutation that reverts spontaneously at a rate of 10^{-6}. Strain A is crossed with a newly acquired wild-type isolate, and *ad-3* strains are recovered from the progeny. When 28 different *ad-3* progeny strains are examined, 13 lines are found to revert at the rate of 10^{-6}, but the remaining 15 lines revert at the rate of 10^{-3}. Formulate a hypothesis to account for these findings, and outline an experimental program to test your hypothesis.

11. a. Why is it not possible to induce nonsense mutations (represented at the mRNA level by the triplets UAG, UAA, and UGA) by treating wild-type strains with mutagens that cause only AT $\rightarrow$ GC or TA $\rightarrow$ CG transitions in DNA?

   b. Hydroxylamine (HA) causes only GC $\rightarrow$ AT or CG $\rightarrow$ TA transitions in DNA. Will HA produce nonsense mutations in wild-type strains?

   c. Will HA treatment revert nonsense mutations?

12. Several auxotrophic point mutants in *Neurospora* are treated with various agents to see if reversion will occur. Table 17-4 shows the results (+ indicates reversion).

TABLE 17-4.

Mutant	Mutagen			
	5-BU	HA	Proflavin	Spontaneous
1	—	—	—	—
2	—	—	+	+
3	+	—	—	+
4	—	—	—	+
5	+	+	—	+

   a. For each of the five mutants, describe the nature of the original mutation event (not the reversion) at the molecular level. Be as specific as possible.

   b. For each of the five mutants, name a possible mutagen that could have caused the original mutation event. (Spontaneous mutation is not an acceptable answer.)

c. In the reversion experiment for mutant 5, a particularly interesting prototrophic derivative is obtained. When this type is crossed to a standard wild-type strain, the progeny consists of 90 percent prototrophs and 10 percent auxotrophs. Provide a full explanation for these results, including a precise reason for the frequencies observed.

13. You are using nitrous acid to "revert" mutant *nic-2* alleles in *Neurospora*. You treat cells, plate them on a medium without nicotinamide, and look for prototrophic colonies. You obtain the following results for two mutant alleles. Explain these results at the molecular level, and indicate how you would test your hypotheses.

a. With *nic-2* allele 1, you obtain no prototrophs at all.

b. With *nic-2* allele 2, you obtain three prototrophic colonies, and you cross each separately with a wild-type strain. From the cross prototroph A × wild-type, you obtain 100 progeny, all of which are prototrophic. From the cross prototroph B × wild-type, you obtain 100 progeny, of which 78 are prototrophic and 22 are nicotinamide-requiring. From the cross prototroph C × wild-type, you obtain 1000 progeny, of which 996 are prototrophic and 4 are nicotinamide-requiring.

14. Devise imaginative screening procedures for detecting the following:

a. Nerve mutants in *Drosophila*

b. Mutants lacking flagella in a haploid unicellular alga

c. Supercolossal-sized mutants in bacteria

d. Mutants that overproduce the black compound melanine in normally white haploid fungus cultures

e. Individual humans (in large populations) whose eyes polarize incoming light

f. Negatively phototrophic *Drosophila* or unicellular algae

g. UV-sensitive mutants in haploid yeast

Mechanisms of Genetic Change II: Recombination

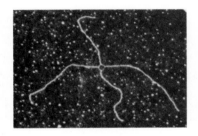

KEY CONCEPTS

Recombination occurs at regions of homology between chromosomes through the breakage and reunion of DNA molecules.

·

Models for recombination, such as the Holliday Model, involve the creation of a heteroduplex branch, or crossbridge, that can migrate and the subsequent splicing of the intermediate structure to yield different types of recombinant DNA molecules.

·

The Holliday Model can be applied to explain genetic crosses.

·

Many of the enzymes involved in recombination in bacteria have been identified.

·

Specific recombination systems catalyze recombination events only between certain sequences.

A normal crossover (a reciprocal intrachromosomal recombination event) is an extraordinary process. Somehow the genetic material from one parental chromosome and the genetic material from the other parental chromosome are "cut up and pasted together" during each meiosis, and this is done with complete reciprocity. In other words, neither chromosome gains or loses any genes in the process. In fact, it is probably correct to say that neither chromosome gains or loses even one nucleotide in the exchange. How is this remarkable precision attained? We do not know for sure, but many interesting phenomena provide important clues about the nature of the answer.

General Homologous Recombination

Some recombination events occur only at specific sequences. We consider these special events at the end of the chapter. Here we focus on the process that results in recombination at any large region of homology between chromosomes.

The Breakage and Reunion of DNA Molecules

Throughout our analysis of linkage, we have implicitly assumed that crossing-over occurs by some process of breakage and reunion of chromatids. The evidence against the copy-choice hypothesis (Chapter 5) provides good *indirect* evidence in favor of breakage and reunion. Furthermore, there is good genetic and cytological evidence that crossing-over occurs during the prophase stage of meiosis, rather than during interphase when chromosomal DNA is replicating. A small amount of DNA synthesis does occur during prophase, but certainly chromosome replication is not associated with crossing-over. One of the first direct proofs that chromosomes (albeit viral chromosomes) can break and rejoin came from experiments on λ phage done in 1961 by Matthew Meselson and Jean Weigle.

Meselson and Weigle simultaneously infected *E. coli* with two strains of λ. One strain, which had the genetic markers c and mi at one end of the chromosome, was "heavy" because the phages were produced from cells grown in heavy isotopes of carbon (^{13}C) and nitrogen (^{15}N). The other strain was $++$ for the markers and had "light" DNA because it was harvested from cells grown on the normal light isotopes ^{12}C and ^{14}N. The two DNA's (chromosomes) can be represented as shown in Figure 18-1a. The multiply infected cells were then incubated in a light medium until they lysed.

The progeny phages released from the cells were spun in a cesium chloride density gradient. A wide band was obtained, indicating that the viral DNA's ranged in density from the heavy parental value to the light parental value, with a great many intermediate densities (Figure 18-1b). Interestingly, some recombinant phages were recovered with density values very close to the heavy parental value. They were of genotype $c+$, and they must have arisen through an exchange event between the two markers (Figure 18-1c). The heavy density of the chromosome would be expected because only the small tip of the chromosome carrying the mi^+ allele would come from the "light" parental chromosome.

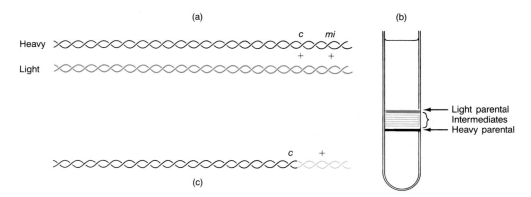

Figure 18-1. Evidence for chromosome breakage and reunion in λ phages. (a) The chromosomes of the two λ strains used to multiply infect *E. coli*. (b) Bands produced when progeny phages are spun in a cesium chloride density gradient. The fact that intermediate densities are obtained indicate a range of chromosome compositions with partly light and partly heavy components. (c) The chromosome of the heavy $c+$ progeny resulting from crossover between the two markers. The density of this crossover product confirms that the crossover involved a physical breakage and reunion of the DNA.

When heavy + + phages were crossed with light *c mi*, the heavy recombinants were found to be + *mi* and the light recombinants were found to be *c* +, as expected. These results can be explained in only one way: the recombination event must have occurred through the physical breakage and reunion of DNA. Of course, we have to be careful about extrapolating from viral to eukaryotic chromosomes. However, this evidence shows that the breakage and reunion of DNA strands is a chemical possibility.

Chiasmata: The Crossover Points

In Chapter 5, we made the simple assumption that chiasmata are the actual sites of crossovers. Mapping analysis indirectly supports this idea: since an average of one crossover per meiosis produces 50 genetic map units, there should be correlation between the size of the genetic map of a chromosome and the observed mean number of chiasmata per meiosis. This correlation has been made in well-mapped organisms.

However, the harlequin chromosome-staining technique (see Chapter 11) has made it possible to test the idea directly. In 1978, C. Tease and G. H. Jones prepared harlequin chromosomes in meioses of the locust. Remember that the harlequin technique produces sister chromatids: one dark, and the other light. When a crossover occurs, it can involve two dark, two light, or dark and light nonsister chromatids. This last situation is crucial because mixed (part dark and part light) crossover chromatids are produced. Tease and Jones found that the dark-light transition occurs right at the chiasma — proving beyond reasonable doubt that these are the crossover sites and settling a question that had been unresolved since the early part of the century (Figure 18-2).

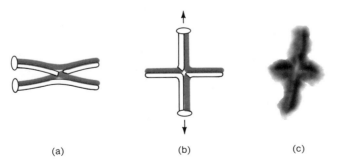

(a)   (b)   (c)

Figure 18-2. Crossing-over between dark- and light-stained nonsister chromatids in a meiosis in the locust. (a) Representation of the chiasma. (b) The best stage for observing is when the centromeres have pulled apart slightly, forming a cross-shaped structure with the chiasma at the center. (c) Photograph of the stage shown in (b). (Photo courtesy of C. Tease and G. H. Jones, *Chromosoma* 69, 1978, 163–178.)

The Holliday Model

Much of the available information on the mechanism of intrachromosomal recombination, especially at the chemical level, has come from studies of bacteria and phages. In addition, we consider a different kind of information based on the genetic analysis of eukaryotes. This is a study of what can be called the genetics of genetics! Several clues about recombination mechanisms have emerged in studies of eukaryotes, resulting in the construction and refinement of models of recombination. Let's examine our current model of recombination and see how it explains experimental results from different genetic systems. Then we can consider some of the actual enzymes involved in the recombination process.

Any model for recombination must explain the basic sequence of events depicted in Figure 18-3. The most plausible recombination model was originally formulated by Robin Holliday and has subsequently been modified by others. The **Holliday model** is typical of the various recombination models that have been proposed both for eukaryotes and prokaryotes. The concept of

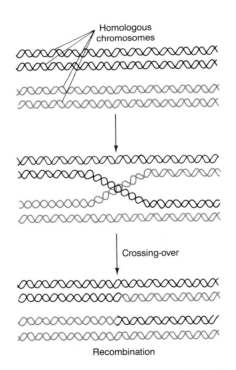

Homologous chromosomes

Crossing-over

Recombination

Figure 18-3. The molecular event of recombination may be schematically represented by two double-stranded molecules breaking and rejoining. (From J. Darnell, H. Lodish, and D. Baltimore, *Molecular Cell Biology.* Copyright © 1986 by Scientific American Books.)

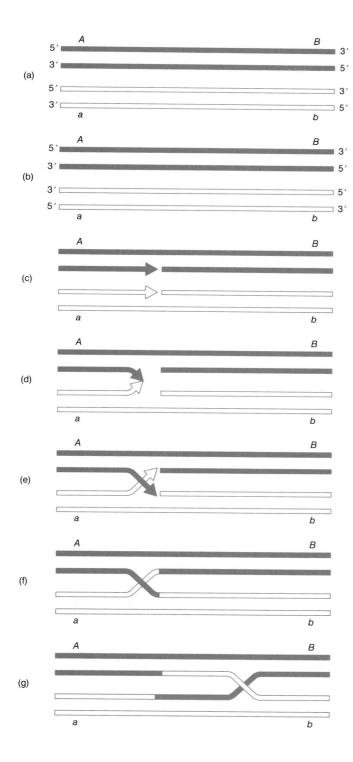

Figure 18-4. A prototype mechanism for genetic recombination. (a). Two homologous double helices are shown. (b) The helices are rotated and aligned so that the bottom strand of the first helix has the same polarity as the top strand of the second helix. (c) The two + or two − strands are cut. (d) The free ends leave the complementary strands to which they had been hydrogen-bonded. (e) The free ends become associated with the complementary strands in the homologous double helix. (f) Ligation creates partially heteroduplex double helices. This is the Holliday model. (g) Migration of the branch point occurs by continued strand transfer by the two polynucleotide chains involved in the crossover. (Figures 18-4 and 18-6 from H. Potter and D. Dressler, *Cold Spring Harbor Symposium on Quantitative Biology* 43, 1979, 970. Cold Spring Harbor Laboratory, Cold Spring Harbor, New York.)

diate structure in one of two ways to yield different types of recombinant molecules. Let's work through the Holliday model in Figure 18-4.

Enzymatic Cleavage and the Creation of Heteroduplex DNA

Looking at 18-4a, we can see that two homologous double helices are aligned, although note that in 18-4b they have been rotated so that the bottom strand of the first helix has the same polarity as the top strand of the second helix ($3' - 5'$, in this case). Then a nuclease cleaves the two strands that have the same polarity (Figure 18-4c). The free ends leave their original complementary strands (Figure 18-4d) and undergo hydrogen bonding with the complementary strands in the homologous dou-

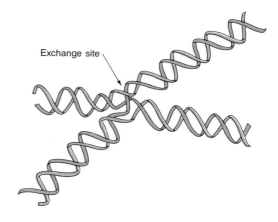

Figure 18-5. Branch migration, the movement of the crossover point between DNA complexes. (After T. Broker, *Journal of Molecular Biology* 81, 1973, 1; from J. D. Watson et al., *Molecular Biology of the Gene,* 4th ed. Copyright © 1987 by Benjamin Cummings.)

hybrid or **heteroduplex** DNA put forth in this model provides a useful way of expressing the present state of knowledge about crossing-over. Figures 18-4, 18-5, and 18-6 depict the Holliday model in its entirety. It is based on the creation of a branch, or **crossbridge,** its migration along the two heteroduplex strands, termed **branch migration,** and the subsequent splicing of the interme-

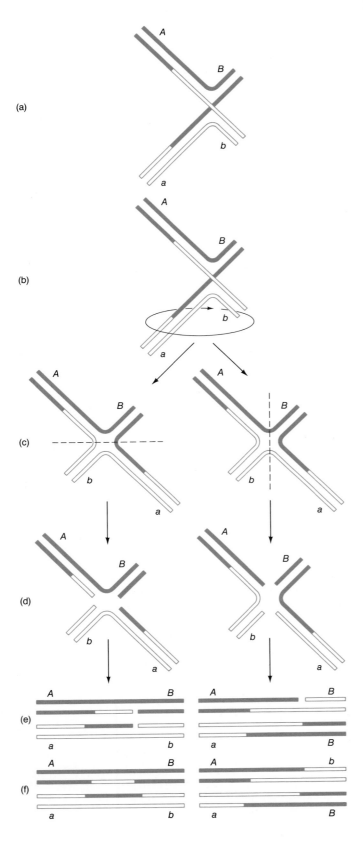

ble helix (Figure 18-4e). Ligation produces the structure shown in Figure 18-4f. This partially heteroduplex double helix is a crucial intermediate in recombination. It has been termed the **Holliday structure.**

Branch Migration

The Holliday structure creates a crossbridge, or branch, that can move, or migrate, along the heteroduplex (Figures 18-4f and 18-4g). This phenomenon of branch migration is a distinctive property of the Holliday structure. Figure 18-5 portrays a more realistic view of this structure as it might appear during branch migration.

Resolution of Holliday Structure

Figure 18-6 demonstrates, in schematic form, how the Holliday structure can be converted to the recombinant structures with which we are familiar. In 18-6a, we can see the structure that we arrived at in Figure 18-4g drawn out in an extended form. Compare 18-4g and 18-6a until you are convinced that these two structures are indeed equivalent. If we rotate the bottom portion of this structure, as shown in Figure 18-6b, then we can generate the form depicted in Figure 18-6c. This last form can be converted back to two uncorrected double helices by enzymatically cleaving only two strands. As indicated in 18-6c, cleavage can occur in either of two ways, each of which generates a different product (Figure 18-6d). These cleaved structures can be viewed more simply (Figure 18-6e). Repair synthesis produces the final recombinant molecules (Figure 18-6f). Note the two different types of recombinants.

Application of the Holliday Model to Genetic Crosses

Let's examine how the model shown in Figures 18-4, 18-5, and 18-6 relate to a genetic cross. We can set up a hypothetical cross of $+ \times m$, in which the $+$ site corresponds to a G–C nucleotide pair and the m site corresponds, after a transition mutation, to an A–T nucleotide pair (Figure 18-7). A hypothetical flanking marker is also indicated at the right end of the molecule. The four chromatids are represented as four DNA double

Figure 18-6. (a) The Holliday model shown in an extended form. (b) The rotation of the structure shown in (a) can yield the form depicted in (c). Resolution of the structure shown in (c) can proceed in two ways, depending on the points of enzymatic cleavage, yielding the structures shown in (d), which can be depicted as shown in (e), and repaired to the forms shown in (f).

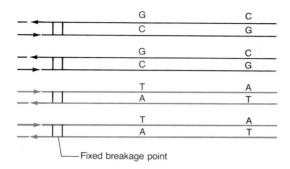

Figure 18-7. In this and the following figures, each double helix represents a meiotic chromatid in a cross $+ \times m$. Black lines represent one parent, and colored lines represent the other. This figure illustrates the normal synapsis arrangement of the four chromatids during prophase of the first meiotic division. Arrows represent the direction of the antiparallel DNA strands. A fixed breakage point, perhaps a recognition site for an endonuclease enzyme, is also represented.

helices. The arrows indicate the directions of the DNA strands ($5' \rightarrow 3'$). Also shown is a hypothetical fixed breakage point (perhaps a recognition site for an endonuclease enzyme; its significance will become clear). The fixed breakage points of two nonsister molecules now become the site of phosphodiester-bond breakage (Figure 18-8), and the broken strands unravel away from the fixed breakage point for some distance. The unraveled strands have the same directionality (arrows), so they can rejoin at each other's breakage points through the action of the enzyme ligase (Figure 18-9). We now have two sets of heteroduplex DNA (the double strands that include both solid and shaded parts) containing the two illegitimate purine-pyrimidine pairs G–T and C–A. (We shall return to these later.) Note that the structure generated in Figure 18-9 can be rotated to generate a planar molecule, as depicted in Figure 18-6c.

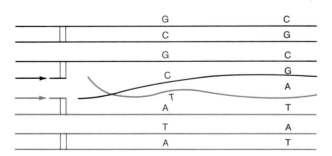

Figure 18-8. The breakage and unwinding of equivalent strands. The broken strands unravel away from the fixed breakage point for some distance and can rejoin at each other's breakage point by means of a ligase enzyme.

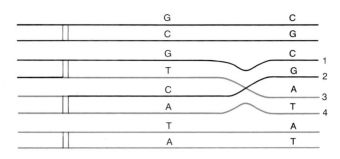

Figure 18-9. Formation of a half-chromatid chiasma and two stretches of heteroduplex DNA extending between the breakpoint and the half-chromatid chiasma. The heteroduplex DNA contains illegitimate base pairs (in this example, GT and CA).

The tangle at the exchange point is now resolved by "nicking." We can assume that strands 2 and 3 break and rejoin in one-half of the cases (Figure 18-10a) and that strands 1 and 4 break and rejoin in the other one-half of the cases (Figure 18-10b). Thus, we see that the result can be either a crossover or a noncrossover situation (Figures 18-6e and f) with respect to flanking

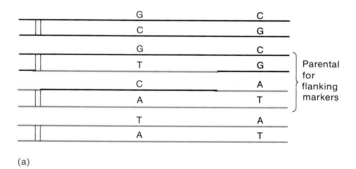

(a)

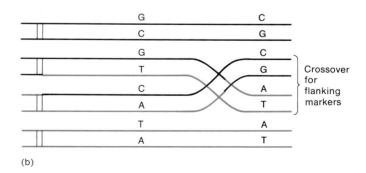

(b)

Figure 18-10. Resolution of the half-chromatid chiasma. (a) In one-half of the cases, resolution will yield a parental conformation for flanking markers. (b) In the other one-half of the cases, resolution will yield a recombinant conformation (crossover) for flanking markers.

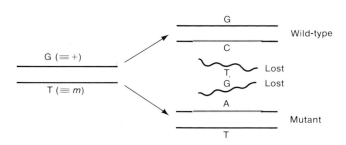

Figure 18-11. Correction of mispaired nucleotides in heteroduplex DNA to either wild-type or mutant pairs. Heteroduplex DNA containing mismatched base pairs is unstable. The mispaired nucleotides can produce a distortion in the DNA at these points that will be recognized by a repair system much like the mismatch repair system described in Chapter 17.

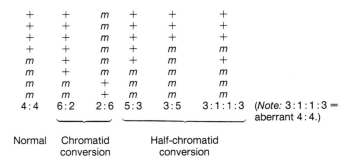

Figure 18-12. Rare aberrant allele ratios observed in a cross of type $+ \times m$ in fungi. (Ascus genotypes are represented here.) When the Mendelian ratio of 4:4 is not obtained, some of the alleles in the cross have been converted to the opposite allele. In some asci, it appears that the entire chromatid has been converted (6:2 or 2:6 ratios). In others, it appears that only half-chromatids have been converted (5:3, 3:5, or 3:1:1:3 ratios).

markers, but that two regions of heteroduplex DNA do exist in either case.

The heteroduplex DNA, which contains mismatched base pairs, is unstable. The mismatched nucleotides can produce a distortion in the DNA at these points, which then is recognized by a repair system, much as the mismatch repair system recognizes and removes mismatched base pairs that occur during replication. For example, the G–T pair can be repaired in one of two ways (Figure 18-11). Thus, the correction of heteroduplex DNA by such a system can result in either a wild-type or a mutant allele in a chromatid. Failure to correct the mismatched nucleotides leads to a heteroduplex chromatid that (on replication) will generate two different daughter chromatids.

The application of the Holliday model to this type of cross and the generation of mismatched alleles nicely explain several genetic results. Let's see what these are.

1: Chromatid Conversion and Half-chromatid Conversion. Much information about recombination in eukaryotes has come from studies of fungi. There is one good reason for this: the ascus. All products of a single meiosis can be recovered and examined, so that records of each meiosis can be kept quite precisely in terms of the total genetic information being lost or gained. Because we can recover all four products of a specific meiosis, we can make inferences about events occurring during that meiosis with a high degree of confidence. The ascus represents a tight system of internally self-consistent controls.

Non-Mendelian allele ratios are detectable in some asci. Mendel would have predicted 4:4 segregations for all monohybrid crosses, but other ratios are obtained

very rarely (in 0.1 to 1 percent of all asci, depending on the fungus species). Figure 18-12 gives the most common aberrant ratios obtained. It appears as though some genes in the cross have been "converted" to the opposite allele (Figure 18-13). The process therefore has become known as **gene conversion;** it can occur only where there is heterozygosity for two different alleles of a gene. In some asci (with a 6:2 or 2:6 ratio), the entire chromatid in meiosis seems to have converted; this process is called **chromatid conversion.** In other asci (with a 5:3 or 3:5 ratio), only one-half of the chromatid seems to have converted, so the process is called **half-chromatid**

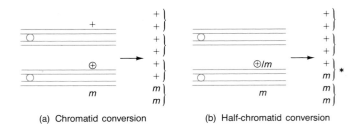

(a) Chromatid conversion   (b) Half-chromatid conversion

Figure 18-13. Gene conversions are inferred from the patterns of alleles observed in asci. (a) In a chromatid conversion, the allele on one chromatid seems somehow to have been converted to an allele like those on the other chromatid pair. The converted allele is shown by the symbol ⊕. One spore pair is of the opposite genotype from that expected in Mendelian segregation. (b) In a half-chromatid conversion, one spore pair (*) has nonidentical alleles. Somehow, one chromatid seems to be "half-converted," giving rise to one spore of the original genotype and one spore converted to the other allele.

conversion. In half-chromatid conversions, different members of a spore pair have different genotypes. Recall that each spore pair is produced by mitosis from a single product of meiosis. Half-chromatid conversion implies that two strands of one double helix carry information for two different alleles at the conclusion of meiosis. Following the next mitotic division to form a spore pair, the two strands segregate to separate nuclei.

It should be emphasized that alleles that are heterozygous at other loci in the same cross typically segregate 4:4, so the aberrant ratios are not the results of accidents of isolation. The process of conversion cannot be mutation, because the process is directional: the allele that is converted always changes to the other specific allele involved in the cross. This specificity has been confirmed in molecular studies on the gene products of converted alleles.

EXPLANATION. Aberrant ratios are explained as a consequence of the generation and subsequent mismatch repair of regions of heteroduplex DNA in the process of crossing-over (Figure 18-11). The various observed aberrant ratios can be produced by the events listed in Table 18-1, which deals only with correction (and hence conversion) from the heteroduplex $m/+$ to $+$. The symbols m and $+$ are used for simplicity instead of the nucleotides used in Figures 18-7 through 18-11. Similar ratios can be developed for the correction of heteroduplex DNA to m.

2: Polarity. In genes for which accurate allele maps are available, we can compare the conversion frequencies of

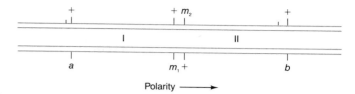

Figure 18-14. Diagram of chromatids involved in a cross. The arrow indicates the polarity of gene conversion in the m locus, pointing toward the end with lower conversion frequency.

alleles at various positions within the gene. In almost every case, the sites closer to one end show higher frequencies than do the sites farther away from that end. In other words, there is a gradient, or **polarity,** of conversion frequencies along the gene (Figure 18-14).

EXPLANATION. Polarity is explained because a heterozygous site will be included more often in heteroduplex DNA if it is located nearer the fixed breakage point (Figure 18-8).

3: Conversion and Crossing-over. In heteroallelic crosses where the locus under study is closely flanked by other genetically marked loci, the conversion event is very often (about 50 percent of the time) accompanied by an exchange in one of the flanking regions. This exchange nearly always occurs on the side nearest the allele that has converted. Furthermore, it almost always involves the chromatid in which conversion has occurred.

TABLE 18-1. Correction of one or both heteroduplex DNA molecules can explain a variety of aberrant allele ratios in an octad

DNA strands at start of meiosis	Strands at heteroduplex DNA stage	No correction	One heteroduplex DNA corrected to +	Both heteroduplex DNA's corrected to +
$-m-$	$-m-$			
$-m-$	$-m-$	} 3 m	} 3 m	} 2 m
$-m-$	$-m-$			
$-m-$	$-+-$	} 1 $+$		
$-+-$	$-m-$	} 1 m	} 5 $+$	} 6 $+$
$-+-$	$-+-$			
$-+-$	$-+-$	} 3 $+$		
$-+-$	$-+-$			
	Aberrant 4:4		5:3	6:2

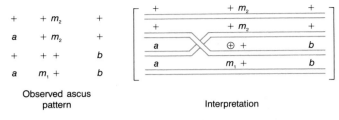

Observed ascus pattern — Interpretation

Figure 18-15. A specific ascus pattern can be explained by both crossover and a chromatid conversion. In this case, a conversion of $m_1 \rightarrow +$ is accomplished by a crossover in the region between a and m_1.

For example, consider the chromatids diagrammed in Figure 18-14. Suppose the polarity is such that alleles toward the left end of the chromatid convert more often than those toward the right end. The cross diagrammed here is between $a^+m_2b^+$ and am_1b, where m_1 and m_2 are different alleles of the m locus and a and b represent closely linked flanking markers. If we look at asci in which conversion has occurred at the m_1 site (the most frequent kind of conversion in this locus), we find that one-half of these asci also will have a crossover in region I and one-half will have no crossover. In the smaller number of asci showing gene conversion at the m_2 site, one-half also will have a crossover in region II and one-half will have no crossover. Such events are detected in ascus genotypes like the one shown in Figure 18-15, which can be interpreted as a conversion of $m_1 \rightarrow +$, accompanied by a crossover in region I. If this is a nutritional locus (say, for arginine requirement), see if you can figure out how the genotypes $m_1 +$, $+ m_2$, and $m_1 m_2$ can be distinguished from each other.

EXPLANATION. The association of crossing-over with about one-half of the cases of gene conversion is explained by the resolution of the exchange point in two equally likely ways. The ascus used to illustrate clue 3 (Figure 18-15) can be explained as shown in Figure 18-16. (Any conversion of the m_2 allele must be ex-

plained by assuming that heteroduplex DNA sometimes "filters in" from some distant breakage point to the right.)

4: Co-conversion. In some asci, a single conversion event seems to include several sites at once. In a heteroallelic cross, this event is called a **co-conversion** (Figure 18-17). The frequency of co-conversion increases as the distance between alleles decreases.

EXPLANATION. Co-conversion is explained by the location of both sites in the region of heteroduplex DNA and by the excision of both sites in the same excision-repair act. This double excision obviously converts both sites to the same parental type.

> **Message** The phenomenon of gene conversion provides clues that lead to a heteroduplex DNA model to explain the mechanism of crossing-over. Mendelian (1 : 1) allele ratios are normally observed in crosses because it is only rarely that a heterozygous locus is the precise point of chromosome exchange. Remember that asci showing gene conversion at a heterozygous locus are relatively rare (on the order of 1 percent).

Visualization of Recombination Intermediates

The Holliday model does formally account for many observed features of recombination. Several of the individual steps that comprise the Holliday model have been demonstrated to occur in vivo or in vitro, such as nicking, strand displacement, branch migration, repair synthesis, and ligation. Recently, H. Potter and D. Dressler have shown that DNA intermediates of the type predicted by the Holliday model can be found in recombin-

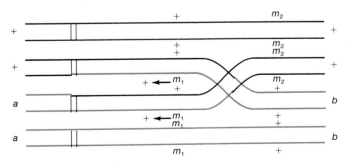

Figure 18-16. An explanation of the ascus in Figure 18-15 in terms of the heteroduplex DNA model. Arrows represent the direction of correction of the heteroduplex DNA.

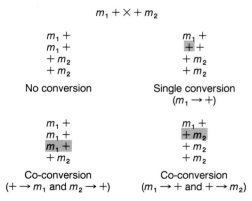

Figure 18-17. Sample ascus patterns obtained from a single conversion and co-conversions.

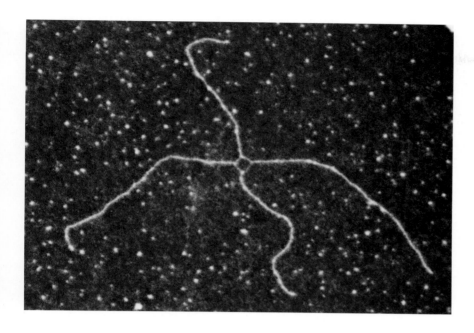

Figure 18-18. Recombination intermediate from recombination between plasmids. Shown are four double helical arms and a single-stranded diamond uniting them. This is the same kind of intermediate as that postulated in the heteroduplex DNA model for recombination in eukaryotes and is formally equivalent to the structure shown in Figure 18-6c. (Photograph courtesy of H. Potter and D. Dressler.)

ing phages or plasmids. Figure 18-18 shows an electron micrograph of a recombinant molecule. It is formally equivalent to the central pair of DNA double helices shown in Figure 18-9, with two arms rotated to produce the central single-stranded "square" (see also Figure 18-6c).

Enzymatic Mechanism of Recombination

Recombination itself is a biological process and, as such, is amenable to analysis by genetic dissection. By isolating mutants defective in some stage of recombination, much light has been shed on the enzymology of recombination. In *E. coli* the products of three genes involved in general recombination — the *recA, recB,* and *recC* genes

— have been well characterized, as has the single-stranded DNA-binding (SSB) protein. Mutants deficient in any of these proteins have reduced levels of recombination.

An initial step in recombination is probably the nicking and unwinding of a DNA duplex by a protein complex encoded by *recB* and *recC*. The SSB protein, which also is involved in DNA replication (Chapter 11), can bind to and stabilize the single strands that are generated. The *recA*-encoded protein, which also plays a role in the induction of the SOS repair system (see Chapters 16 and 17), can bind to single strands and catalyze their invasion of a duplex and subsequent displacement of the corresponding strand from the duplex. These events are depicted in Figure 18-19. The branch migra-

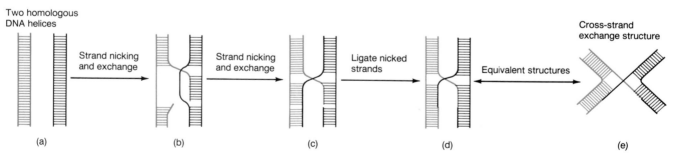

Figure 18-19. A schematic view of some of the steps in recombination. (a) The pairing of two homologous duplexes. (b) A nick is made by the RecB,C nuclease. The helix is partially unwound, and the single-stranded region is extended and stabilized by the SSB protein. The RecA protein catalyzes the invasion by the single strand of the duplex. The SSB protein aids in keeping the single strand free. The arrow points to the position of the next nick. (c) After nicking by the RecB,C nuclease, the free single strand from the second duplex can anneal with the first duplex. (d) RNA ligase can seal this structure. (e) After a 180° rotation, this structure is the one depicted in Figures 18-6c and 18-18. (After B. Alberts et al., *Molecular Biology of the Cell.* Copyright © 1983 by Garland Publishing.)

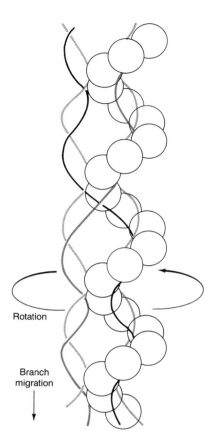

Figure 18-20. Model for branch migration. A recA protein containing one DNA molecule rotates a second DNA molecule joined to the first by a crossover junction. Branch migration is powered by the hydrolysis of ATP by recA. [After M. M. Cox and I. R. Lehman, "Enzymes of General Recombination." *Annual Review of Biochemistry* 56, 252. Copyright © 1987 by Annual Reviews, Inc. All rights reserved. From L. Stryer, *Biochemistry. 3rd ed.* Copyright © 1988 by W. H. Freeman & Company.)

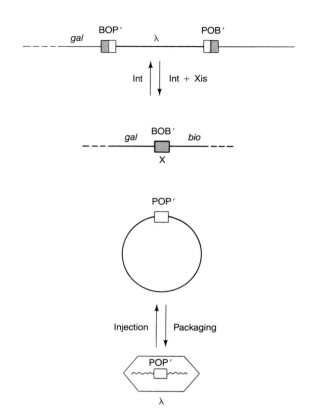

Figure 18-21. λ integration and excision. The λ and *E. coli* chromosomes contain attachment sites that are recognized by the λ integration and excision enzymes. These sites share a short region of homology, indicated by O, but also contain regions flanking the homology that are different. The flanking bacterial sequences are designated B and B′, and the flanking phage sequences are designated P and P′. Thus, the bacterial attachment site is designated BOB′, and the phage attachment site is designated POP′.

tion of the structure produced in Figure 18-19d is driven by the hydrolysis of bound ATP. We can visualize this physically in Figure 18-20.

Site-specific Recombination

Phage Integration

The bacteriophage λ (lambda) integrates and excises via a pathway utilizing enzymes that recognize specific sites on the λ and on the *E. coli* chromosome and catalyze recombination events between them. These attachment, or *att,* sites contain short regions of homology (Figure 18-21). The homologous region of 15 base pairs is only part of each *att* site. The integration reaction is catalyzed

by the λ-encoded **Int** protein, in cooperation with a host protein, while the excision reaction requires these two proteins plus the λ-encoded **Xis** enzyme.

Control of Gene Expression

Several genomes use site-specific recombination systems to control surface antigens or host-range phenotypes. One of the best-characterized systems involves phase variation in the bacterium *Salmonella.* The surface flagella can be one of two types in these strains, termed H1 or H2. The conversion frequency of one type to another is 10^{-3}. This high rate of transition from one "phase" to another is caused by the unique arrangement of genes controlling the *H1* and *H2* loci (Figure 18-22). When the *H2* gene is actively expressed, a repressor protein is also

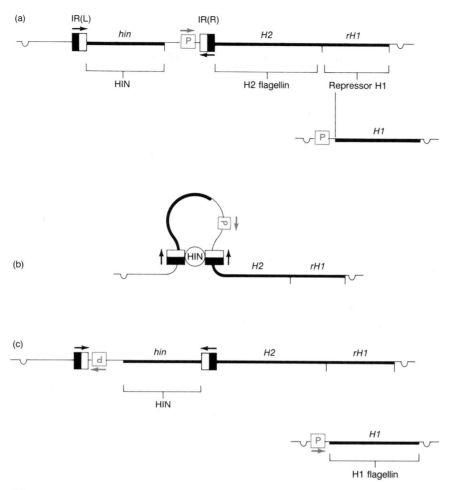

Figure 18-22. Genetic rearrangement controls flagellar phase transition. Inversion of a region of DNA of approximately 1000 base pairs adjacent to the *H2* operon alternately couples and uncouples a promoter element (b). When the promoter is coupled to the gene set *H2*, flagellin is synthesized as well as the product of the *rhI* gene that repressed *H1* expression (a: *H2* on, *H1* off). When the promoter is uncoupled from the *H2* gene set, no *rhI* gene product is synthesized and the *H1* gene is expressed (c: *H2* off, *H1* on). IR(L) and IR(R) describe the cis-acting sites in inverted repeat configurations in which a reciprocal recombination event resulting in inversion takes place. The *hin* gene product is encoded by a sequence within the inversion region and is required for inversion. The fate of DNA sequences rearranged by inversion can be followed by referring to the black areas of the IR boxes. Note that IR(L) and IR(R) are defined for the *H2* (on) position. (After M. Silverman and M. Simon, *Mobile Genetic Elements,* J. A. Shapiro (ed.). Copyright © 1983 by Academic Press.)

synthesized, which prevents transcription of the *H1* gene (Figure 18-22a). Failure to transcribe the *H2* gene and the *H1* repressor gene results in failure to repress *H1* gene expression (Figure 18-22c). The orientation of the intervening genetic segment determines which of the two systems is transcribed, since it contains the transcription-initiation, or "promoter," sites that activate the respective gene systems. The site-specific recombination event (Figure 18-22b) is catalyzed by the Hin

protein encoded by the invertible segment. The Hin protein acts on two 14-base-pair homologous segments in opposite orientation ("inverted repeats"). Recombination at inverted repeats yields an inversion.

Phage *mu* (see Chapter 19) and P1 (Chapter 10) have similar site-specific recombination systems that control host range by allowing a switch from one surface protein to another. The respective *gin* and *cin* genes for *mu* and P1 are closely related to the *hin* gene of *Salmonella*.

Summary

■ We are now beginning to understand the molecular processes behind recombination, which produces new gene combinations by exchanging homologous chromosomes. Both genetics and physical evidence have led to the use of a heteroduplex DNA model to explain the mechanism of crossing-over. The process of recombination itself is under genetic control: there are genes that affect the efficiency of recombination. Several enzymes, including those encoded by the *recA*, *recB*, and *recC* genes, have been implicated in the recombination process in bacteria such as *E. coli*.

■ ■ ■

Solved Problems

1. In *Neurospora* an *ad3* double mutant consisted of two mutant sites within the *ad3* gene, site 1 on the left and site 2 on the right. This was crossed to wild type, using parental stocks heterozygous for two closely linked flanking loci *A* and *B*, as follows, where + represents wild type sequence at the mutant positions:

$$A\ 12\ B \times a + + b$$

Most asci were of the expected type showing regular Mendelian segregations, but there were also some unexpected types, of which several examples are represented here as I through III.

A 12 B	A 12 B	A 12 B
A 12 B	A 12 B	A 12 B
A 12 b	A 12 B	A 12 B
A + + b	A 12 B	A + + B
a 12 B	a + + b	a + 2 b
a + + B	A 12 b	a + + b
a + + b	a + + b	a + + b
a + + b	a + + b	a + + b
I	II	III

Explain the likely origin of the rare types I through III according to molecular recombination models.

Solution

Ascus type I has two nonidentical sister spore pairs. Since the members of a spore pair are derived from a post-meiotic mitosis (see Chapter 6), this proves that the meiotic products in these cases must have contained both 12 and ++ information; in other words, they must have contained heteroduplex DNA. Notice also that these spore pairs are recombinant for *A* and *B*. Therefore it is likely that a crossover occurred between the *A* and *B* loci, that the heteroduplex DNA that constituted the crossover spanned both the *ad3* mutant sites 1 and 2, and that there was no correction of the heteroduplex at those sites.

Type II shows a 5 : 3 ratio of 12 doubles to ++. Here again the heteroduplexes must have spanned both sites, (in a non-crossover configuration) but this time there was correction of a ++/12 heteroduplex to 12/12, presumably by excision and repair of the ++ information (co or double-conversion).

Type III reveals another noncrossover heteroduplex configuration, and this time correction occurred only at site 1, and this is revealed as a 5 : 3 ratio for site 1, with $1 \rightarrow +$ conversion in ascospore 5.

2. In fungal crosses of the following general type

$$
\begin{array}{ccc}
M & \boxed{+\ 2} & N \\
\hline
\times & & \\
\hline
m & \boxed{1\ +} & n
\end{array}
$$

where 1 and 2 are mutant sites of a nutritional gene and *M* and *N* are flanking loci, it is possible to select rare ++ prototrophic recombinants by plating on minimal medium. These prototrophs are then examined for the alleles of the flanking loci. As might be expected, the combination

$$M + + n$$

is commonly encountered, but so, somewhat surprisingly, are

$$M + + N$$

and

$$m + + n$$

a. Explain the origin of these two genotypes in terms of molecular recombination models.

b. If $M + + N$ were more common than $m + + n$, what would that mean?

Solution

a. It is likely that $M + + N$ arose from a noncrossover heteroduplex that spanned site 2, followed by a correction of $2 \rightarrow +$.

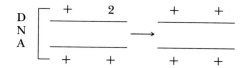

Similarly, $m + + n$ would be explained by a heteroduplex spanning site 1, and corrected $1 \rightarrow +$.

b. Since $M + + N$ arises from gene conversion at the right-hand site, it is likely that heteroduplex DNA is formed more commonly from the right than from the left, possibly because of a closer fixed break point. (Note: prototrophs may also be formed by single site correction in a heteroduplex spanning both sites, but then the inequality would require another explanation.)

Problems

1. What is gene conversion? How can you distinguish gene conversion from mutation?

2. Which of the following linear asci shows gene conversion at the *arg2* locus?

+	+	+	+	+	+
+	+	+	+	+	+
+	arg	+	+	arg	arg
+	arg	arg	arg	arg	arg
arg	arg	arg	+	+	arg
arg	arg	arg	arg	+	arg
arg	+	arg	arg	arg	arg
arg	+	arg	arg	arg	arg
1	2	3	4	5	6

3. At the light spore locus of *Ascobolus*, the 1′ mutant site is in the left portion of the gene and the 1″ mutant site is more to the right. When crosses are made between 1′- and 1″-bearing strains

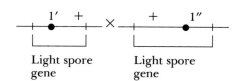

Asci with 6 light and 2 black spores can be selected visually. These are shown to be caused by gene conversion mostly of the following type

+	1″
+	1″
+	1″
+	1″
+	+
+	+
1′	+
1′	+

(+ + rows braced as "black")

but very rarely

+	1″
+	1″
+	+
+	+
1′	+
1′	+
1′	+
1′	+

What can account for this irregularity?

***4.** It has been proposed that the "fixed break point" of the heteroduplex recombination model might correspond to a promoter sequence. The following data relate to this idea. In the fungus *Podospora*, mutants were available in adjacent spore color cistrons 1 through 4, and a cross was made as follows:

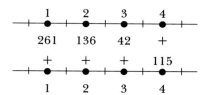

(261, 136, 42, and 115 are merely names for mutant sites.) Many asci were obtained showing gene conversion, but one that was relevant to the above suggestion was as follows:

Spore 1	261	136	42	+
2	+	+	+	+
3	+	+	+	115
4	+	+	+	115

Interpret this ascus in relation to the promoter idea.

5. Many mutagens increase the frequency of sister-chromatid exchange. Discuss possible explanations for this observation.

6. Mutations in locus *46* of the Ascomycete fungus *Ascobolus* produce light-colored ascospores (let's call them *a* mutants). In the following crosses between different *a* mutants, asci are observed for the appearance of dark wild-type spores. In each cross, all such asci were of the genotypes indicated:

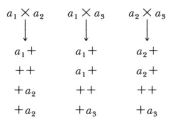

$$a_1 \times a_2 \qquad a_1 \times a_3 \qquad a_2 \times a_3$$

$a_1 +$	$a_1 +$	$a_2 +$
$+ +$	$a_1 +$	$a_2 +$
$+ a_2$	$+ +$	$+ +$
$+ a_2$	$+ a_3$	$+ a_3$

Interpret these results in light of the models discussed in this chapter.

7. In the cross $Am_1m_2B \times am_3b$, the order of the mutant sites m_1, m_2, and m_3 is unknown in relation to each other and to A/a and B/b. One nonlinear conversion ascus is obtained:

$$Am_1m_2B$$
$$Am_1 \quad b$$
$$am_1m_2 B$$
$$am_3 \quad b$$

Interpret this result in the light of the heteroduplex DNA theory, and derive as much information as possible about the order of the sites.

8. In the cross $a_1 \times a_2$ (alleles of one locus), the following acus is obtained:

$$a_1 +$$
$$a_1 a_2$$
$$a_1 a_2$$
$$+ a_2$$

Deduce what events may have produced this ascus (at the molecular level).

9. G. Leblon and J.-L. Rossignol have made the following observations in *Ascobolus*. Single nucleotide-pair insertion or deletion mutations show gene conversions of the $6:2$ or $2:6$ type and only rarely of the $5:3$, $3:5$, or $3:1:1:3$ type. Base-pair transition mutations show gene conversion of the $3:5$, $5:3$, or $3:1:1:3$ type, and only rarely of the $6:2$ or $2:6$ type.

   a. In terms of the hybrid DNA model, propose an explanation for these observations.

   b. Leblon and Rossignol also have shown that there are far less $6:2$ than $2:6$ conversions for insertions and far more $6:2$ than $2:6$ conversions for deletions (where the ratios are $+:m$). Explain these results in terms of heteroduplex DNA. (You might also think about the excision of thymine photodimers.)

   c. Finally, the researchers have shown that when a frame-shift mutation is combined in a meiosis with a transition mutation at the same locus in a cis-configuration, the asci showing *joint* conversion are all $6:2$ or $2:6$ for *both* sites (that is, the frame-shift conversion pattern seems to have "imposed its will" on the transition site). Propose an explanation for this.

10. At the *grey* locus in the Ascomycete fungus *Sordaria*, the cross $+ \times g_1$ is made. In this cross, heteroduplex DNA sometimes extends across the site of heterozygosity and two heteroduplex DNA molecules are formed (as discussed in this chapter). However, correction of heteroduplex DNA is not 100 percent efficient. In fact, 30 percent of all heteroduplex DNA is not corrected at all, whereas 50 percent is corrected to $+$ and 20 percent is corrected to g_1. What proportion of aberrant-ratio asci will be: a. $6:2$? b. $2:6$? c. $3:1:1:3$? d. $5:3$? e. $3:5$?

11. Noreen Murray crossed α and β, two alleles of the *me-2* locus in *Neurospora*. Included in the cross were two markers, *trp* and *pan*, which each flank *me-2* at a distance of 5 m.u. The ascospores were plated onto a medium containing tryptophan and pantothenate but no methionine. The methionine prototrophs that grew were isolated and scored for the flanking markers, yielding the results shown in Table 18-2. Interpret these results in light of the models presented in this chapter. Be sure to account for the asymmetries in the classes.

TABLE 18-2.

Cross	Genotype of *me-2*$^+$ prototrophs			
	trp $+$	$+$ *pan*	*trp pan*	$+ +$
$trp\alpha+ \times +\beta pan$	26	59	16	56
$trp\beta+ \times +\alpha pan$	84	23	87	15

12. In *Neuorspora*, the cross $Ax \times ay$ is made, in which x and y are alleles of the *his-1* locus and A and a are mating-type alleles. The recombinant frequency between the *his-1* alleles is measured by the prototroph frequency when ascospores are plated on a medium lacking histidine; the recombinant frequency is measured as 10^{-5}. Progeny of parental genotype are backcrossed to the parents, with the following results. All *ay* progeny backcrossed to the *Ax* parent show prototroph frequencies of 10^{-5}. When *Ax* progeny are backcrossed to the *ay* parent, two prototroph frequencies are obtained: one-half of the crosses show 10^{-5}, but the other one-half show the much higher frequency of 10^{-2}. Propose an explanation for these results, and describe a research program to test your hypothesis. (NOTE: intragenic recombination is a *meiotic* function that occurs in a diploid cell. Thus, even though this is a haploid organism, dominance and recessiveness could be involved in this question.)

Mechanisms of Genetic Change III: Transposable Genetic Elements

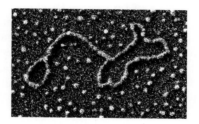

KEY CONCEPTS

A series of genetic elements can occasionally move or transpose from one position on the chromosome to another position on the same chromosome or on a different chromosome.

∎

In bacteria, insertion sequences, transposons, and phage *mu* are examples of transposable genetic elements.

∎

Transposable elements can mediate chromosomal rearrangements.

∎

In higher cells, transposable elements have been extensively characterized in yeast, *Drosophila,* and *maize* and in mammalian systems.

∎

In eukaryotes, some transposable elements utilize an RNA intermediate during transposition, whereas in prokaryotes transposition occurs exclusively at the DNA level.

■ Genetic studies of maize, beginning in the late 1930s, have yielded results that greatly upset the classical genetic picture of genes residing only at fixed loci on the main chromosome. The research literature (see page 537) began to carry reports suggesting the existence of genetic elements of the main chromosomes that can somehow mobilize themselves and move from one location to another. These findings were viewed with skepticism for many years, but it is now clear that we must find a place for such mobile elements in the genetic scheme of things.

A variety of colorful names (some of which help to describe their respective properties) have been applied to these genetic elements: controlling elements, cassettes, jumping genes, roving genes, mobile genes, mobile genetic elements, and transposons. We choose the term **transposable genetic elements,** which is formally most correct and embraces the entire family of types. The term **transposition** has long been used in genetics to describe transfer of chromosomal segments from one position to another in major structural rearrangements. In the present context, what is being transposed seems to be a gene, or a small number of linked genes, or a gene-sized fragment. Any genetic entity of this size can be called a **genetic element.**

Transposable genetic elements seem to be able to move to new positions within the same chromosome or even to move to a different chromosome. The normal genetic role of these elements is not known with certainty. They have been detected genetically through the abnormalities they produce in the activities and structures of the genes near the sites to which they move. A variety of physical techniques have been used to detect them as well. Transposable genetic elements have been found in phages, bacteria, fungi, higher plants, viruses, and insects.

Although transposable elements were first detected in eukaryotes, the molecular nature of transposable genetic elements was first understood in bacteria and phages. Therefore, let's begin with the information derived from the original studies in these prokaryotes.

Transposable Genetic Elements in Prokaryotes

Several types of transposable genetic elements were originally distinguished in bacteria, although geneticists now consider these elements to be members of the same general category of transposable elements found in all organisms.

Insertion Sequences

Insertion sequences, or **insertion-sequence (IS) elements,** are now known to be segments of DNA that

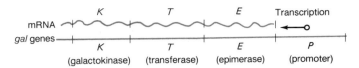

Figure 19-1. The *gal* genes of *E. coli.* Galactokinase, transferase, and epimerase are structural genes of the galactose gene set. The promoter controls their transcription into a single mRNA molecule.

can move from one position on the chromosome to a different position on the same chromosome or on a different chromosome. When IS elements appear in the middle of genes, they interrupt the coding sequence and inactivate the expression of that gene. Due to their size and also because some of them contain transcription and translation termination signals, IS elements can also block the expression of other genes in the same operon if those genes are "downstream" from the promoter of the operon. This property was instrumental in their discovery, as we see in the following section.

IS elements were first found in *E. coli* in the *gal* operon—a set of three genes involved in the metabolism of the sugar galactose. This operon consists of three adjacent genes—*E* (epimerase), *T* (transferase), and *K* (kinase)—which are transcribed together from the same promoter (Figure 19-1).

Polar Mutations

The IS elements were found in mutants that had been selected by their deficiency in kinase activity. These mutations were found to map at various sites throughout the operon. In any particular mutant, the mutational site might map somewhere in the kinase gene itself or in the transferase or epimerase genes. Mutations in the transferase gene that have reduced kinase expression do not reduce epimerase expression. Each mutation obviously affects all structural gene functions transcriptionally downstream (but not upstream) in the operon. This directionality, or **polarity,** had been noted before for certain mutations, which are aptly termed **polar mutations.** For instance, certain chain-terminating nonsense mutations have polar effects. However, this particular kind of polar mutation was new. The mutants were observed to be capable of spontaneous reversion to wild-type (showing that the mutation was not a deletion), but the reversion rate was not increased by any mutagen. Thus, they did not seem to be nonsense mutations, frame-shifts, or any form of point mutation either! Then what were they? In the late 1960s, several experiments demonstrated that these IS elements were insertions of segments of DNA.

Physical Demonstration of DNA Insertion

Recall that the λ phage inserts next to the *gal* operon and that it is a simple matter to obtain λ*dgal* phage particles that have picked up the *gal* region (page 244). When the polar mutations in *gal* are incorporated into λ*dgal* phages and the buoyant density in a cesium chloride (CsCl) gradient of the phages is compared with that of the normal λ*dgal* phages, it is evident that the DNA carrying the polar mutation is longer than the wild-type DNA! This experiment, carried out by Elke Jordon, Heinz Saedler, and Peter Starlinger and, independently, by James Shapiro clearly demonstrates that the mutations are caused by the insertion of a significant amount of DNA into the *gal* operon. Figure 19-2 depicts this experiment in more detail.

Direct Visualization of Inserted DNA

By hybridizing denatured λ*dgal* DNA containing the insertion mutation to denatured wild-type λ*dgal* DNA, the extra piece of DNA actually can be located under the electron microscope. In these experiments, some of the DNA molecules that form in the mixture are not the parent duplexes but heteroduplexes between one mutant and one wild-type strand. When point mutations are analyzed, the heteroduplexes are indistinguishable from the parental DNA molecules. However, in the case of the DNA involving the polar mutations, each heteroduplex shows a single-stranded buckle, or loop (Figure 19-3). This single-stranded buckle confirms the presence of an inserted sequence in the mutated DNA that has no complementary sequence in the wild-type DNA. The length of this single-stranded loop can be calibrated by including standardized marker DNA in the preparation. It proves to be approximately 800 nucleotides in length.

Identification of Discrete IS Elements

Are the segments of DNA that insert into genes merely random DNA fragments or are they distinct genetic entities? Hybridization experiments, which again exploit the properties of the λ*dgal* phage, show that many different insertion mutations are caused by a small set of

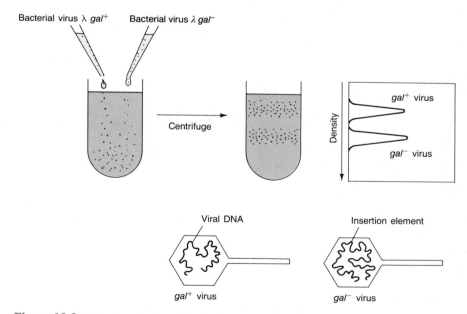

Figure 19-2. Mutation by insertion is demonstrated with phage λ particles carrying the bacterial gene for galactose utilization (*gal*⁺) and carrying the mutant gene *gal*⁻. The viruses are centrifuged in a cesium chloride (CsCl) solution. The *gal*⁻ particles are found to be the denser. Because the virus particles all have the same volume and their outer shells all have the same mass, the increased density of *gal*⁻ particles shows that they must contain a larger DNA molecule: *gal*⁻ mutation was caused by the insertion of DNA. (Figures 19-2, 19-11, 19-12, and 19-22 from S. M. Cohen and J. A. Shapiro. "Transposable Genetic Elements." Copyright © 1980 by Scientific American, Inc. All rights reserved.)

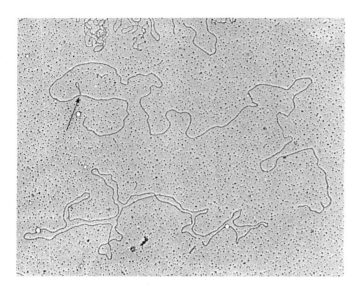

Figure 19-3. Electron micrograph of a λdgal⁺/λdgalᵐ DNA heteroduplex. The single-stranded loop (arrow) is caused by the presence of an insertion sequence in λdgalᵐ. (From A. Ahmed and D. Scraba "The Nature of the *gal3* Mutation of *Escherichia coli,*" *Molecular and General Genetics* 136, 1975, 233.)

insertion sequences. The λdgal phages are isolated from the polar mutants, and their DNA is used to synthesize radioactive RNA in vitro. Certain fragments of this RNA are found to hybridize with the mutant DNA but not with wild-type DNA, reflecting the fact that the mutant contains an extra piece of DNA. These particular RNA fragments also hybridize to DNA from other polar

mutants, showing that the same bit of DNA is inserted in different places in the different polar mutants.

Based on their patterns of cross-hybridization, the insertion mutants are placed into categories. The first sequence, the 800-bp base-pair segment identified in *gal,* is termed IS1. A second sequence, termed IS2, is 1350 bp long. Table 19-1 lists some of the insertion sequences and their sizes.

We now know that the genome of the standard wild-type *E. coli* is rich in IS elements: it contains eight copies of IS1, five copies of IS2, and copies of other less well-studied IS types as well. It should be emphasized that the sudden appearance of an insertion sequence at any given locus under study means that these elements are truly mobile, with a capability for transposition throughout the genome. Presumably, they produce a mutation or some other detectable alteration of normal cell function only when they happen to end up in an "abnormal" position, such as the middle of a structural gene. Insertion sequences also are commonly observed in the F factor. Figure 19-4 shows an example in an F *lac* plasmid.

Message The bacterial genome contains segments of DNA, termed IS elements, which can move from one position on the chromosome to a different position on the same chromosome or on a different chromosome.

TABLE 19-1. Prokaryotic insertion elements

Insertion sequence	(1) Normal occurrence in *E. coli*	(2) Length (bp)	(3) Inverted repeat (bp)
IS1	5–8 copies on chromosome	768	18/23
IS2	5 on chromosome; 1 on F	1327	32/41
IS3	5 on chromosome; 2 on F	1400	32/38
IS4	1 or 2 copies on chromosome	1400	16/18
IS5	Unknown	1250	short
γ–δ	One on F; one or more on chromosome	5700	35
pSC101 segment	On plasmid pSC101	200	30/36

SOURCE: Tables 17-1 and 17-3 from M. P. Calos and J. H. Miller, *Cell* 20, 1980, 579–95.

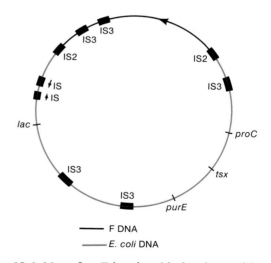

Figure 19-4. Map of an F *lac* plasmid, showing positions of IS elements; *proC, tsx, purE,* and *lac* are bacterial markers. (From H. Ohtsubo and E. Ohtsubo, in *DNA Insertion Elements, Plasmids, and Episomes.* Edited by A. I. Bukhari, J. A. Shapiro, and S. L. Adhya. New York, Cold Spring Harbor Laboratory, 1977.)

Figure 19-5. Diagrammatic representation of the appearance of a heteroduplex DNA molecule formed by annealing corresponding strands of $\lambda dgal^m$ DNA from two specific mutants caused by insertion sequences. This unexpected hybridization between corresponding strands (normally having the same rather than complementary base sequences) can be explained by the model shown in Figure 19-6.

Orientation of IS Elements

Due to the base sequence, the two strands of $\lambda dgal^+$ DNA happen to have different buoyant densities. After DNA denaturation, they can be recovered separately in the ultracentrifuge. In some cases, the *same* strands from two different IS1 mutants form an unexpected hybrid with each other. Under the electron microscope, these hybrids have a peculiar appearance: each is a double-stranded region with four single-stranded tails (Figure 19-5). This observation is explained by assuming that the IS1 elements are inserted in opposite directions in the two mutants (Figure 19-6).

Transposons

A frightening ability of pathogenic bacteria was discovered in Japanese hospitals in the 1950s. Bacteria dysentery is caused by bacteria of the genus *Shigella*. This bacterium proved sensitive to a wide array of antibiotics that were used originally to control the disease. In the Japanese hospitals, however, *Shigella* isolated from patients with dysentery proved to be simultaneously resistant to many of these drugs, including penicillin,

TABLE 19-2. Genetic determinants borne by plasmids

Characteristic	Plasmid examples
Fertility	F, R1, Col
Bacteriocin production	Col E1
Heavy-metal resistance	R6
Enterotoxin production	Ent
Metabolism of camphor	Cam
Tumorigenicity in plants	T1 (in *Agrobacterium tumefaciens*)

tetracycline, sulfanilamide, streptomycin, and chloramphenicol. This multiple drug-resistance phenotype was inherited as a single genetic package, and it could be transmitted in an infectious manner — not only to other sensitive *Shigella* strains, but also to other related species of bacteria! This talent is an extraordinarily useful one for the pathogenic bacterium, and its implications for medical science were terrifying. From the point of view of the geneticist, however, the situation is very interesting. The vector carrying these resistances from one cell to another proved to be a plasmid similar to the F factor. These **R plasmids** (for *r*esistance) are transferred rapidly on cell conjugation, much like the F particle in *E. coli* (Chapter 10).

In fact, these R plasmids proved to be just the first of many similar F-like plasmids to be discovered. These plasmids have been found to carry many different kinds of genes in bacteria. Table 19-2 shows just some of the characteristics that can be borne by plasmids. What is the mode of action of these plasmids? How do they acquire their new genetic abilities? How do they carry them from cell to cell? Some answers to these questions can now be given.

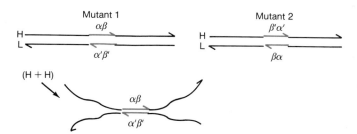

Figure 19-6. A model to explain the heteroduplex DNA structure shown in Figure 19-5. IS1 is represented by the colored lines; α and β represent the ends of the IS1 molecule; H and L represent the strands of $\lambda dgal$ DNA with high and low buoyant densities, respectively. The hybrid can be explained by assuming that the IS1 sequence is inserted in opposite directions in the two mutants.

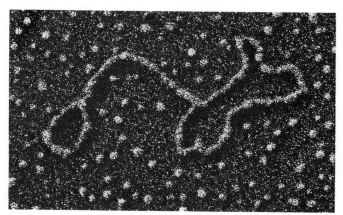

Figure 19-7. Peculiar structure formed when denatured plasmid DNA is reannealed. The double-stranded IR region separates a large circular loop from the small "lollipop" loop. (Photograph courtesy of S. N. Cohen.)

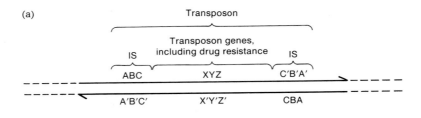

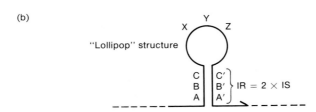

Figure 19-8. An explanation at the nucleotide level for the "lollipop" structure seen in Figure 19-7. The structure is called a transposon. (a) The transposon in its double-stranded form before denaturing. Note the presence of oppositely oriented copies of an insertion sequence (IS). (b) The lollipop structure formed by intrastrand annealing after denaturing. The two IS regions anneal to form the IR of the transposon; the transposon genes are carried in the lollipop loop.

Physical Structure of Transposons

If the DNA of a plasmid conferring drug resistance (carrying the genes for kanamycin resistance, for example) is denatured to single-stranded forms and then allowed to renature slowly, some of the strands form an unusual shape under the electron microscope: a large, circular DNA ring is attached to a "lollipop"-shaped structure (Figure 19-7). The "stick" of the lollipop is double-stranded DNA, which has formed through the annealing of two **inverted repeat (IR) sequences** in the plasmid (Figure 19-8). Subsequent studies have shown that the IR sequences are a pair of IS elements in many cases. For

instance, IS10 is present at the ends of the region carrying the genes for tetracycline resistance (Figure 19-9). In some cases, however, the IR sequences are much smaller.

The genes for drug resistance or other genetic abilities carried by the plasmid are located between the IR sequences on the "lollipop head." The IR sequences together with their contained genes have been collectively called a **transposon (Tn)**. The remainder of the plasmid, bearing the genes coding for resistance-transfer functions (RTF), is called the **RTF region** (Figure 19-10). Table 19-3 lists some of the known transposons.

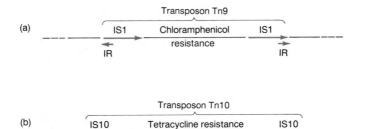

Figure 19-9. Two different transposons having different IR regions and carrying different drug-resistance genes. Tn9 has a short IR region, because the two IS1 elements are in the same orientation and each element has a short inverted repeat. Tn10 has a large IR region because the two IS10 components have opposite orientations.

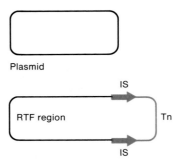

R plasmid bearing a transposon

Figure 19-10. The insertion of a transposon (Tn) into a plasmid. RTF represents the resistance-transfer functional genes of the plasmid. Tn includes both the IS elements and the drug-resistance genes.

TABLE 19-3. Transposons

Transposon	(1) Marker	(2) Length (bp)	(3) Inverted repeat
Tn1	Ampicillin	4957	38
Tn2	Ampicillin		
Tn3	Ampicillin		
Tn4	Ampicillin, streptomycin, sulfanilamide	20,500	Short
Tn5	Kanamycin	5400	1500
Tn6	Kanamycin	4200	Not detectable with electron microscopy
Tn7	Trimethoprim, streptomycin	14,000	Not detectable with electron microscopy
Tn9	Chloramphenicol	2638	18/23
Tn10	Tetracycline	9300	1400
Tn204	Chloramphenicol, fusidic acid	2457	18/23
Tn402	Trimethoprim	7500	Not detectable with electron microscopy
Tn501	Mercuric ions	7800	38
Tn551	Erythromycin	5200	35
Tn554	Erythromycin, spectinomycin	6200	Not determined
Tn732	Gentamicin, tobramycin	11,000	Not determined
Tn903	Kanamycin	3100	1050
Tn917	Erythromycin	5100	Short
Tn951	lac	16,600	Short
Tn1681	Heat-stable enterotoxin	2088	768 (IS1)
Tn1721	Tetracycline	10,900	Short

Movement of Transposons

A transposon can jump from one plasmid to another plasmid (Figure 19-11) or from a plasmid to a bacterial chromosome. Let's consider an actual experiment documenting this transposition.

A transposon (Tn3) containing an ampicillin-resistance gene (Ap^R) is carried in a large *E. coli* plasmid R64-1. This Tn3 is then transferred to a small plasmid RSF1010, which carries the genes for sulfanil amide resistance (Su^R) and streptomycin resistance (Sm^R) in another transposon (Tn4):

R64-1 (Ap^R in Tn3) $\longrightarrow$ RSF1010 ($Su^R Sm^R$ in Tn4)

The following procedure is used to carry out this transposition. First, R64-1 DNA is isolated and added as transforming DNA to a strain carrying RSF1010 that had been treated with $CaCl_2$ (treatment with $CaCl_2$ enhances the probability of uptake of the donor DNA). The Ap^R transformants are selected by plating on an ampicillin medium. These transformants prove to be of several genotypes:

$$Ap^R Su^R Sm^R \quad Ap^R Su^R Sm^S \quad Ap^R Su^S Sm^R \quad Ap^R Su^S Sm^S$$

Apparently, the Ap^R transposon, once it enters the recipient cell, can insert itself into the recipient's genome either outside the $Su^R Sm^R$ transposon or within it. If it inserts within the recipient's original Tn4 transposon, then it knocks out either one or both of the resistance functions originally possessed, depending on the precise point of insertion. New lollipop structures, corresponding to the acquisition of the new transposon, are observed in the transformed genotypes. Figure 19-12 shows a composite diagram of an R plasmid, indicating the various places at which transposons can be located, including the transposons just described. Techniques similar to those used in this experiment can also be used to demonstrate transposition from plasmid to bacterial DNA.

Apparently, the IS regions are genetic elements that can mobilize themselves and can carry with them genes that confer various traits. Thus, multiple drug-resistant plasmids are generated. As can be seen in Table 19-1, even a single IS element has a pair of short terminal repeats 20–50 bp long.

Message Transposons were originally detected as mobile genetic elements that confer drug resistance. Many of these elements consist of recognizable IS elements flanking a gene that encodes drug resistance. IS elements and transposons are now grouped together under the single term "transposable elements."

Phage *Mu*

Phage *Mu* is a normal-appearing phage. We consider it here because, although it is a true virus, it has many features in common with IS elements. The DNA double helix of this phage is 36,000 nucleotides long—much larger than an IS element. However, it does appear to be able to insert itself anywhere in a bacterial or plasmid genome in either orientation. Once inserted, it causes mutation at the locus of insertion—again like an IS element. (The phage was named for this ability: *mu* stands for *mutator*.) Normally, these mutations cannot be re-

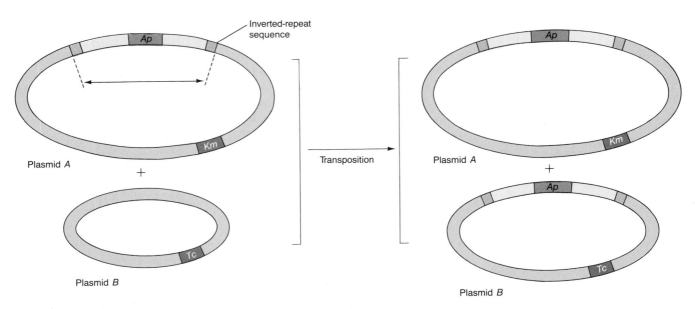

Figure 19-11. Transposition of the transposon Tn3, which carries a gene conferring resistance to the antibiotic ampicillin *(Ap).* It is shown as originally being part of plasmid A, which also includes a gene for resistance to kanamycin *(Km).* A plasmid B, which confers resistance to tetracycline *(Tc),* acquires a copy of the transposon. The new plasmid B confers resistance to ampicillin and tetracycline.

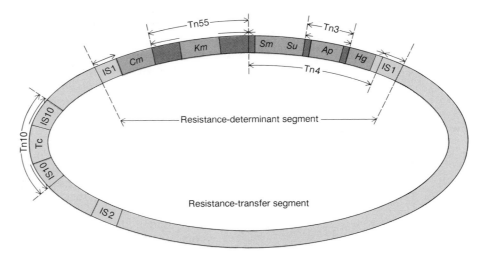

Figure 19-12. The role of transposable elements in the evolution of antibiotic-resistance plasmids is illustrated by a schematic map of a plasmid carrying many resistance genes. The plasmid appears to have been formed by the joining of a resistance-determinant segment and a resistance-transfer segment; there are insertion elements (IS1) at the junctions, where the two segments sometimes dissociate reversibly. Genes encoding resistance to the antibiotics chloramphenicol *(Cm),* kanamycin *(Km),* streptomycin *(Sm),* sulfonamide *(Su),* and ampicillin *(Ap)* and to mercury *(Hg)* are clustered on the resistance-determinant segment, which consists of multiple transposable elements; inverted-repeat termini are designated by arrows pointing outward from the element. A transposon encoding resistance to tetracycline *(Tc)* is on the resistance-transfer segment. Transposon Tn3 is within Tn4. Each transposon can be transferred independently. (Figures 19-12 and 19-13 from S. M. Cohen and J. A. Shapiro, "Transposable Genetic Elements." Copyright © 1980 by Scientific American, Inc. All rights reserved.)

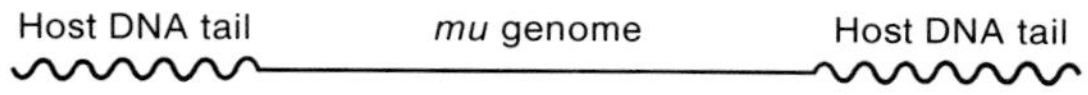

Host DNA tail   *mu* genome   Host DNA tail

Figure 19-13. The DNA of a free *mu* phage has tails derived from its previous host.

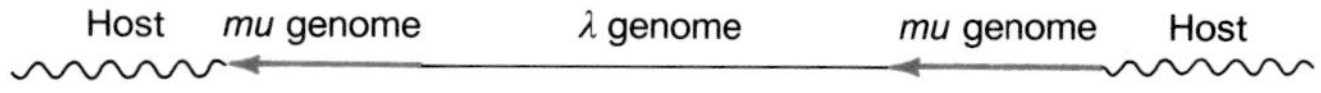

Host   *mu* genome   λ genome   *mu* genome   Host

Figure 19-14. Phage *mu* can mediate the insertion of phage λ into the bacterial chromosome, resulting in a structure like the one shown here.

verted, but reversion can be produced by certain kinds of genetic manipulation. When this reversion is produced, the phages that can be recovered show no deletion, proving that excision is exact and that the insertion of the phage therefore does not involve any loss of phage material either.

Each mature phage particle has a piece of flanking hose DNA on each end (Figure 19-13). However, this DNA is not inserted anew into the next host. Its function is unclear. Phage *mu* also has an IR sequence, but neither of the repeated elements is terminal.

Mu can also act like a genetic snap fastener, mobilizing any kind of DNA and transposing it anywhere in the genome. For example, it can perform this trick on another phage (such as λ) or on the F factor. In such situations, the inserted DNA is flanked by two *mu* genomes (Figure 19-14). It can also transfer bacterial markers

onto a plasmid; here again, the transferred region is flanked by a pair of *mu* genomes (Figure 19-15). Finally, the phage *mu* can mediate various kinds of structural chromosome rearrangements (Figure 19-16).

Mechanism of Transposition

Several different mechanisms of transposition are employed by prokaryotic transposable elements. And, as we shall see later, eukaryotic elements exhibit still additional mechanisms of transposition. The elucidation of different transposition pathways has involved some interesting examples of genetic analysis.

In *E. coli*, we can identify **replicative** and **conservative** (nonreplicative) modes of transposition. In the replicative pathway, a new copy of the transposable element

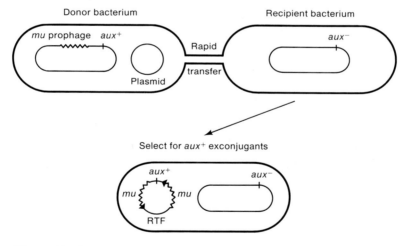

Figure 19-15. Phage *mu* can mediate the transposition of a bacterial gene into a plasmid. The selection procedure for detecting the transposition is indicated here. An auxotrophic mutant gene is indicated by *aux⁻*.

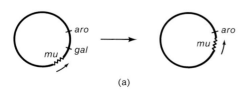

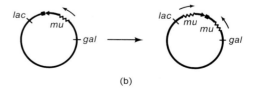

Figure 19-16. Phage *mu* can cause the deletion or inversion of adjacent bacterial segments. (a) The *gal* region is deleted

by transposition of phage *mu*. (b) The F-factor region of an *Hfr* strain is inverted by transposition of phage *mu*.

is generated during the transposition event. The results of the transposition are that one copy appears at the new site and one copy remains at the old site. In the conservative pathway, no replication occurs. Instead, the element is excised from the chromosome or plasmid and is integrated into the new site.

Replicative Transposition

Elisabeth Ljungquist and Ahmad Bukhari first demonstrated that when transposition occurs from one locus to a second locus in certain transposons, a copy of the transposable element is left behind at the first locus. The researchers used restriction-enzyme digestion patterns to show that when phage *mu* transposes, segments of *mu* phage remain at the original location in the *E. coli* chromosome and newly synthesized DNA appears at new locations in the chromosome. Therefore, in these cases the transposable elements do not strictly "jump" from one location to another by excising themselves from the chromosome and migrating through the cytoplasm; instead, they "replicate" into a new location, leaving one copy behind.

An analysis of transposon mutants by Fred Heffron and his coworkers has revealed another interesting fact about the mechanism of transposition. The researchers studied the transposon Tn3 (Figure 19-17). They grouped the mutations that prevent transposition into two categories. A trans-recessive class maps in the gene that encodes the transposase enzyme, which is involved in transposition. A second class of cis-dominant mutations results in the buildup of an intermediate in the transposition process. Figure 19-18 diagrams the transposition pathway involved in Tn3 transposition from one plasmid to another. The intermediate is actually a double plasmid, with both donor and recipient plasmid being fused together. The combined circle resulting from the fusion of two circular elements is termed a

cointegrate. Apparently, the mutations in this second class delete a region on the transposon at which a recombination event takes place that resolves cointegrates into two smaller circles. This region, called the **internal resolution site (IRS),** appears in Figure 19-17.

The finding of a cointegrate structure as an intermediate in transposition has helped to establish a replicative mode of transposition for certain elements. In Figure 19-18, note how the transposable element is duplicated during the fusion event and how the recombination event that resolves the cointegrate into two smaller circles leaves one copy of the transposable element in each plasmid.

Conservative Transposition

Some transposons, such as Tn10, excise from the chromosome and integrate into the target DNA. In these cases, DNA replication of the element does not occur, and the element is lost from the site of the original chromosome. Nancy Kleckner and her coworkers demon-

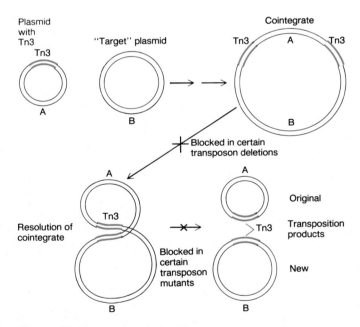

Figure 19-18. Transposition of Tn3 takes place via a cointegrate intermediate. Cointegrates in Tn3 transposition are observed for some internal deletions in the transposon. The correct explanation for this observation is that the cointegrates are intermediates in Tn3 transposition and that their resolution is blocked because the internal deletion has removed an internal resolution site (IRS), where recombination occurs. (From F. Heffron, in *Mobile Genetic Elements,* pp. 223–260. Edited by J. A. Shapiro. Copyright © 1983 by Academic Press.)

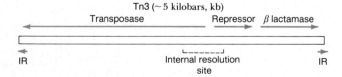

Figure 19-17. The structure of Tn3. Tn3 contains 4957 base pairs and codes for three polypeptides: the transposase is required for transposition, the repressor is a protein that regulates (see Chapter 16) the transposase gene, and *β* lactamase confers ampicillin resistance. Tn3 is flanked by inverted repeats (IR) of 38 base pairs and contains a site within the area designated as the internal resolution site, which is necessary for the resolution of Tn3 cointegrates.

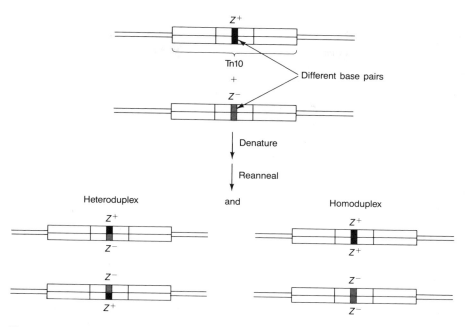

Figure 19-19. Generation of Heteroduplex and Homoduplex Tn10 elements. The denaturation and reannealing of a mixture of two parental λ phages carrying Tn10 elements that differ only at three single bases in the transposon yields a mixture of heteroduplex and homoduplex products. The base difference in the Z gene allows the ultimate determination of the heteroduplex or homoduplex nature of a cell that has received the Z gene via tranposition. (Adapted from J. Bender and N. Kleckner, *Cell* 45, 1986, 801–815.)

strated this by constructing heteroduplexes of λTn10 derivatives containing the *lac* region of *E. coli*. The researchers used DNA from Tn10-*lacZ*+ and Tn10-*lac*− derivatives. The heteroduplexes, therefore, contain one strand with the wild-type *lac* region and a second strand with the mutated (*Z*−) *lac* region. Figure 19-19 diagrams this part of the experiment. The heteroduplex DNA is used to infect cells that have no *lac* genes, and transpositions of the Tet^r Tn10 are selected. Different types of colonies arise from the transposition of a heteroduplex *Z*−/*Z*+ carrying transposon (Figure 19-20). If replication occurs (the replicative mode of transposition), then all colonies are either completely Lac+ or completely Lac−, because the replication will convert the heteroduplex DNA into two homoduplex daughter molecules. However, if the transposition is conservative and does not involve replication, then each colony arises from a *lacZ*+/*lacZ*− heteroduplex. Such colonies are partially Lac+ and partially Lac−. By using media that stains Lac+ and Lac− cells different colors, the Lac+ and Lac− sectors in colonies can be observed.

Therefore, the determination of whether replicative or conservative transposition occurs for Tn10 can be made by observing whether different colored sectors

occur within the same colony resulting from the transposition. Sectored colonies are observed in a majority of cases (Figure 19-21). Thus, Tn10—and perhaps other transposable elements in *E. coli*— transpose by excising themselves from the donor DNA and integrating directly into the recipient DNA.

Molecular Consequences of Transposition

The molecular consequences of transposition reveal an additional piece of evidence concerning the mechanism of transposition: on integration into a new target site, transposable elements generate a repeated sequence of the target DNA. Figure 19-22 depicts the integration of IS1 into a gene. In the example shown, the integration event results in the repetition of a 9-bp target sequence. Analysis of many integration events reveals that the repeated sequence does not result from reciprocal site-specific recombination (as is the case in λ phage integration; see page 512); rather, it is generated during the process of integration itself. The number of base pairs is a characteristic of each element. In bacteria, 9-bp and 5-bp repeats are most common.

The preceding observations have been incorporated into somewhat complicated models of transposi-

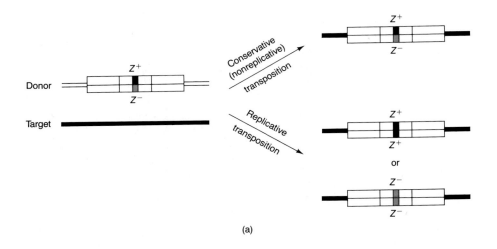

(a)

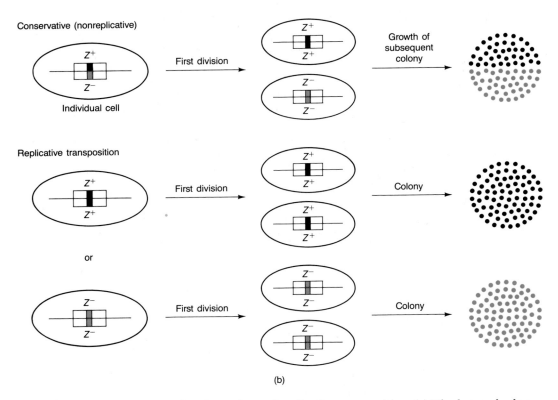

(b)

Figure 19-20. Consequences of conservative and replicative transposition. (a) The heteroduplex or homoduplex nature of DNA (see Figure 19-19) is transposed into a target gene. If the starting DNA is heteroduplex, then the resulting DNA will still be heteroduplex only in a conservative or nonreplicative pathway. (b) Because the heteroduplex results in a transposed cell that maintains the heteroduplex nature of the DNA during conservative transposition, colonies will arise that are part Z^+ and part Z^-. However, in a replicative pathway, transposition results in individual cells that are either all Z^+ or all Z^-, and all the colonies will either be Z^+ or Z^-.

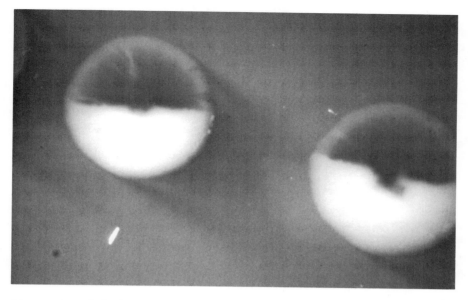

Figure 19-21. Colonies resulting from the experiment illustrated in Figures 19-19 and 19-20. A dye is used to stain the colonies blue. One-half of this colony is Z^+ (dark area), and the other colony is Z^- (white). (Photo courtesy of N. Kleckner.)

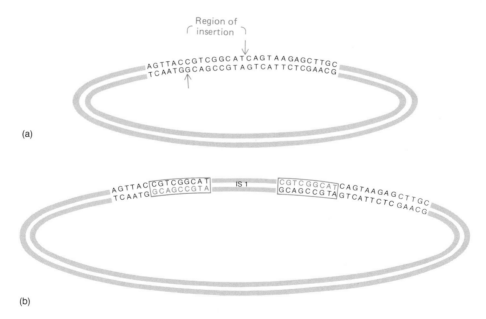

Figure 19-22. Duplication of a short sequence of nucleotides in the recipient DNA is associated with the insertion of a transposable element; the two copies bracket the inserted element. Here the duplication that attends the insertion of IS1 is illustrated in a way that indicates how the duplication may come about. IS1 insertion causes a nine-nucleotide duplication. If the two strands of the recipient DNA are cleaved (arrows) at staggered sites that are nine nucleotides apart, as shown in (a), then the subsequent filling in of single strands on each side of the newly inserted element, seen in (b), with the right complementary nucleotides could account for the duplicated sequences (colored boxes).

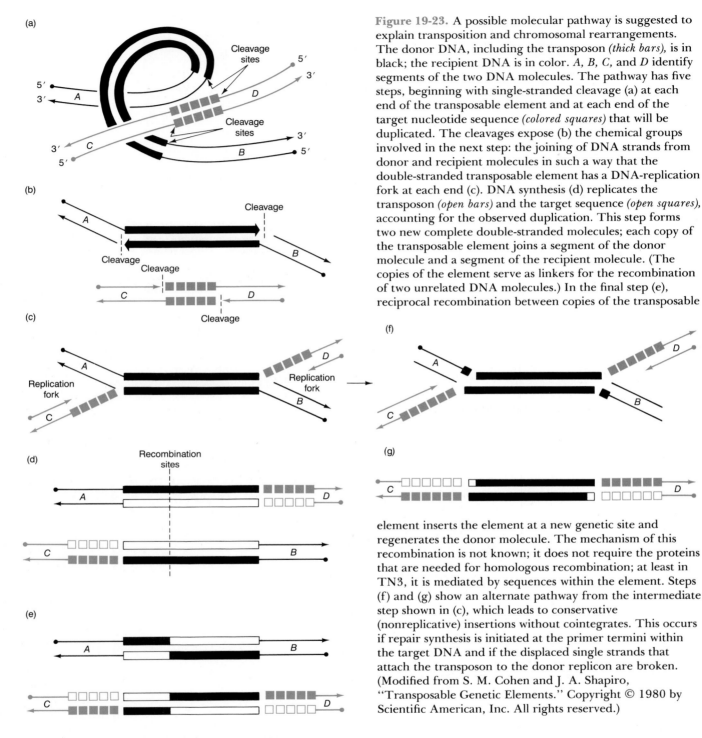

Figure 19-23. A possible molecular pathway is suggested to explain transposition and chromosomal rearrangements. The donor DNA, including the transposon *(thick bars),* is in black; the recipient DNA is in color. *A, B, C,* and *D* identify segments of the two DNA molecules. The pathway has five steps, beginning with single-stranded cleavage (a) at each end of the transposable element and at each end of the target nucleotide sequence *(colored squares)* that will be duplicated. The cleavages expose (b) the chemical groups involved in the next step: the joining of DNA strands from donor and recipient molecules in such a way that the double-stranded transposable element has a DNA-replication fork at each end (c). DNA synthesis (d) replicates the transposon *(open bars)* and the target sequence *(open squares),* accounting for the observed duplication. This step forms two new complete double-stranded molecules; each copy of the transposable element joins a segment of the donor molecule and a segment of the recipient molecule. (The copies of the element serve as linkers for the recombination of two unrelated DNA molecules.) In the final step (e), reciprocal recombination between copies of the transposable element inserts the element at a new genetic site and regenerates the donor molecule. The mechanism of this recombination is not known; it does not require the proteins that are needed for homologous recombination; at least in TN3, it is mediated by sequences within the element. Steps (f) and (g) show an alternate pathway from the intermediate step shown in (c), which leads to conservative (nonreplicative) insertions without cointegrates. This occurs if repair synthesis is initiated at the primer termini within the target DNA and if the displaced single strands that attach the transposon to the donor replicon are broken. (Modified from S. M. Cohen and J. A. Shapiro, "Transposable Genetic Elements." Copyright © 1980 by Scientific American, Inc. All rights reserved.)

tion. Most models postulate that staggered cleavages are made at the target site and at the ends of the transposable element by a transposase enzyme that is encoded by the element. One end of the transposable element is then attached by a single strand to each protruding end of the staggered cut. Subsequent steps depend on which mode of transposition occurs. Figure 19-23 depicts this model of transposition.

Message In prokaryotes, transposition occurs by at least two different pathways. Some transposable elements can replicate a copy of the element into a target site, leaving one copy behind at the original site. In other cases, transposition involves the direct excision of the element and its reinsertion into a new site.

Rearrangements Mediated by Transposable Elements

Transposable elements generate a high incidence of deletions in their vicinity. These deletions emanate from one end of the element into the surrounding DNA (Figure 19-24). Such events, as well as element-induced inversions, can be viewed as aberrant transposition events. Transposons also give rise to readily detectable deletions in which part of the element is deleted together with varying lengths of the surrounding DNA. This process of **imprecise excision** is now recognized as deletions or inversions emanating from the internal ends of the IR segments of the transposon. The process of **precise excision** — the loss of the transposable element and the restoration of the gene that was disrupted by the insertion — also occurs, although at very low rates compared with the frequencies of the events just described.

Message Some DNA sequences in bacteria and phage act as mobile genetic elements. They are capable of joining different pieces of DNA and thus are capable of splicing DNA fragments into or out of the middle of a DNA molecule. Some naturally occurring mobile or transposable elements carry antibiotic-resistance genes.

Review

Let's examine what we have learned up to this point about prokaryotic transposable elements:

1. There are several different types of transposable elements, including insertion sequences (IS1, IS2, . . . and so on), transposons (Tn1, Tn2, . . . and so on), and phage *mu*.

2. Two copies of a transposable element can act in concert to transpose the DNA segments in between them. Some of the antibiotic-resistance conferring transposons are obviously formed in this manner, with two insertion sequences flanking the genes for antibiotic resistance.

3. Most of the transposable elements have recognizable inverted repeat (IR) structures, some of which can be observed under the electron microscope after denaturation and renaturation.

4. Transposable elements are found in bacterial chromosomes, as well as in plasmids.

5. After insertion into a new site on the DNA, transposable elements generate a short repeated sequence, such as 9 base pairs (bp) or 5 bp.

6. Although the detailed mechanism of transposition is not understood, two different pathways for transposition have been identified. In some cases, transposition occurs by replicating a new copy of the element into the target site, leaving one copy behind at the original site. In other cases, transposition involves the excision of the element from the original site and its reintegration into a new site. These two modes of transposition are called replicative and conservative, respectively.

Transposable Genetic Elements in Eukaryotes

Transposable elements have been found in eukaryotes and have close similarities to those observed in bacteria. Several examples follow.

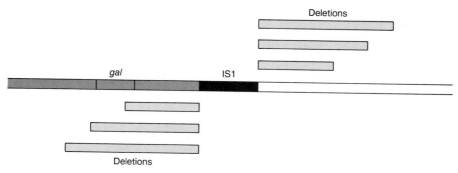

Figure 19-24. Deletion formation mediated by a transposable element. In this example, the transposable element IS1 is shown at a point in the *E. coli* chromosome near the *gal* genes. Deletions can be generated from each end of the IS1 element, extending into the neighboring DNA sequences. In cases where the deletions extend into the *gal* regions, they can be detected as a result of the Gal⁻ phenotype.

Figure 19-25. The structure of a yeast transposable element. The Ty1 sequence occurs approximately 35 times in the yeast genome. It contains two copies of delta (δ) sequence in direct orientation at each end. Delta occurs approximately 100 times in the yeast genome.

Ty Elements in Yeast

Figure 19-25 shows the structure of one of the **Ty elements** in yeast: the Ty1 sequence, which is present in approximately 35 copies in the yeast genome. The 38-bp-long termini (terminal sequences), called δ (delta) sequences, are present in about 100 copies of the genome. Yeast δ sequences, as well as Ty elements, show significant sequence divergence. The terminal δ sequences are present in direct-repeat, orientation, in contrast to transposable elements in bacteria, which carry inverted repeat (IR) sequences. However, like prokaryotic transposons, Ty elements generate a repeated sequence of target DNA (in this case, 5 bp) during transpo-

sition. Also, Ty elements cause mutations by insertion into different genes in the yeast chromosome. It is now known that Ty elements transpose through an RNA intermediate (see the following section on retroviruses).

Transposable Elements in Drosophila

It is now estimated that many spontaneous mutations and chromosomal rearrangements in *Drosophila* are caused by transposable elements. As much as 10 percent of the *Drosophila* chromosome may be composed of families of dispersed, repetitive DNA sequences that move as discrete elements! Three types of transposable elements have been characterized: the *copia*-like elements, the fold-back *(FB)* elements, and the *P* elements. Their structures are summarized in Figure 19-26.

Copia-like Elements. The **copia-like elements** comprise at least seven families, ranging in size from 5 to 8.5 kilobases (kb). Members of each family appear at 10–100 positions in the *Drosophila* chromosome. Each member carries a long, direct terminal repeat and a short, imperfect inverted repeat (Figure 19-26) and is structurally similar to a yeast Ty element (Figure 19-25).

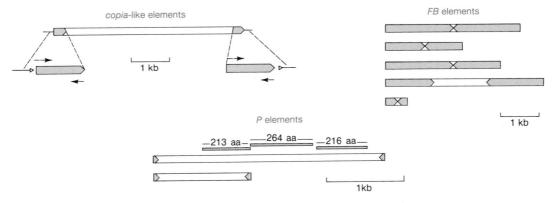

Figure 19-26. Summary of the structures of three classes of *Drosophila* transposable elements. The *copia*-like elements carry long direct terminal repeats. Each repeat makes up about 5 percent of the length of the element. These repeats are shown on an expanded scale below the element to illustrate the presence of short, imperfect inverted repeats at the ends of each long direct repeat ($\longrightarrow$) and the presence of a few base pairs of duplicate target sequence ($\triangleright$) flanking the element that were present in one copy before insertion. The different genomic copies of the family elements are very similar in structure to one another.

The *FB* elements comprise a family of heterogeneous but cross-homologous sequences, ranging in size from a few hundred base pairs to several kilobases. Each *FB* element carries long terminal inverted repeats. In some cases, the entire element consists of these inverted repeats. In other cases, a central sequence is located between the inverted repeats. The inverted repeat sequences themselves are

internally repetitious, having a substructure made up primarily of 31-bp tandem repeats. The number of these 31-bp tandem repeats can differ not only between *FB* elements but also between the termini of a single *FB* element.

The *P* elements have a very different structure from that of the *copia*-like and *FB* elements. *P* elements carry perfect terminal inverted repeats of 31 bp. A fraction of the *P* elements (about one-third in the one strain examined) are very similar in sequence to one another and are 2.9 kb in length. The remainder of the *P* elements are more heterogeneous, but all appear to have structures that are consistent with their having been derived from the 2.9-kb element by one or more internal deletions. DNA sequences analysis of the 2.9-kb element reveals three long, open translational reading frames, which are indicated. (From G. Robin, in *Mobile Genetic Elements*, pp. 329–361. Edited by J. A. Shapiro. Copyright © 1983 by Academic Press.)

Also, *copia*-like elements repeat a characteristic number of base pairs of *Drosophila* DNA on insertion. Certain classic *Drosophila* mutations result from the insertion of *copia*-like and other elements. For example, the white-apricot (w^a) mutation for eye color is caused by the insertion of an element from the *copia* family (from which these elements derive their name) into the white locus. Some *copia*-like families have interesting properties. For instance, all of the insertion mutations detected so far that result from the *gypsy* family of *copia*-like elements are suppressible by a specific allele at one particular outside locus. In other words, the phenotypes resulting from the *gypsy* insertions are affected by unlinked genes. The mechanism of this effect is unknown.

FB Elements.   The **FB elements** range in size from a few hundred to a few thousand base pairs. These elements have sequence homologies, but different elements have sequence differences also. Each carries long inverted repeats at its termini (see Figure 19-26). Sometimes the entire element consists of inverted repeats, but a central sequence separates the inverted repeats in other elements. In either case, *FB* elements can literally fold back on themselves due to the inverted repeats (hence, their name). Several unstable mutations in *Drosophila* have been shown to be caused by the insertion of *FB* elements. Mutations can result either from the interruption of a gene-coding sequence by *FB*-element insertion or from the effects on gene expression due to *FB*-element insertion in or near a control region. The properties of some of the *FB*-insertion mutations suggest that *FB* elements can excise themselves from the genome and promote chromosomal rearrangements at high frequencies.

P Elements.   Of all the transposable elements in *Drosophila*, the most intriguing and useful to the geneticist are the **P elements**. These elements were discovered as a result of studying **hybrid dysgenesis** — a phenomenon that occurs when females from laboratory strains of *Drosophila melanogaster* are mated to males derived from natural populations. In such crosses, the laboratory stocks are said to possess a **M cytotype** (cell type) and the natural stocks are said to possess a **P cytotype**. In a cross of M♀ × P♂, the progeny show a range of surprising phenotypes that are manifested in the germline, including sterility, a high mutation rate, and a high frequency of chromosomal aberations and nondisjunction. These hybrid progeny are termed **dysgenic**, or biologically deficient (hence, the expression "hybrid dysgenesis"). Interestingly, the reciprocal cross P♀ × M♂ produces no dysgenic offspring. An important observation is that a large proportion of the dysgenically induced mutations are unstable — that is, they revert to wild-type or to other mutant alleles at very high frequencies. This insta-

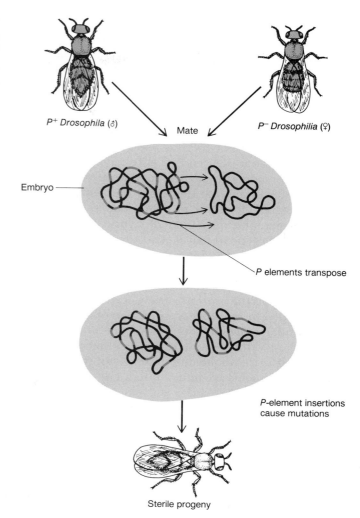

Figure 19-27. The phenomenon known as hybrid dysgenesis results from the mobilization of DNA sequences called *P* elements in *Drosophila* embryos. When a sperm from a *P*-carrying strain fertilizes an egg from a non-*P*-carrying strain, the *P* elements transpose throughout the genome, usually disrupting vital genes. (After J. D. Watson, J. Tooze, and D. T. Kurtz, *Recombinant DNA: A Short Course.* Copyright © 1983 by W. H. Freeman and Company.)

bility is generally restricted to the germline of individuals possessing an M cytotype.

These findings led to the hypothesis that the mutations are caused by the insertion of foreign DNA into specific genes, thereby rendering them inactive. According to this hypothesis, reversion usually would result from the spontaneous excision of these inserted sequences. This hypothesis has been critically tested by isolating dysgenically derived unstable mutants at the eye color locus *white*. A plasmid constructed to carry the white locus was used as a probe to recover the dysgenesis-mutated *white* genes. (This type of experiment is explained in Chapter 15.) The majority of the mutations were found to be caused by the insertion of a genetic element into the middle of the *white*[+] gene. The element, called the *P* element, was found to be present in 30 to 50 copies per genome in P strains but to be completely absent in M strains. The *P* elements vary in size, ranging from 0.5 to 2.9 kb in length (this size difference reflects partially deleted elements derived from a single complete *P* element), but there is always a 31-bp perfect inverted repeat at their ends. There can be as many as three open reading frames in the central area of the *P* element, suggesting that the largest elements have the coding potential for three protein products.

The current explanation of hybrid dysgenesis is based on the proposal that *P* elements encode both the transposase product and P-repressor products. According to this model, which is depicted in Figure 19-27, the transposase is responsible for mobilization of the *P* elements, whereas the repressor prevents transposase production, thereby blocking transposition of the element. (We consider repressor proteins in general in Chapter 16.) In the P cytotype, the high copy number of *P* elements leads to the abundant production of repressor, so that the *P* elements are immobilized. For some reason,

most laboratory strains have no *P* elements; consequently, there is no repressor in the cytoplasm. In hybrids from the cross M♀ × P♂, the *P* elements are in a repressor-free environment and can transpose throughout the genome, causing a variety of damage expressed as the various manifestations of hybrid dysgenesis.

Applications of *P* Element Transposition

Quite apart from their interest as a genetic phenomenon, the *P* elements have become major tools of the modern *Drosophila* geneticist. Two main analytical techniques are possible. In one technique, *P* elements are used to isolate any *Drosophila* gene of interest. First, the investigator simply looks for mutants of that gene in progeny of dysgenic crosses. Then a vector is constructed in which a *P* element is inserted. This vector is used as a probe to identify and isolate DNA segments containing *P* elements (see Chapter 15 for more details of these types of experiments); in a subset of these segments *P* elements are found inserted into the gene of interest. These genes can then be cloned (Chapter 15) and studied.

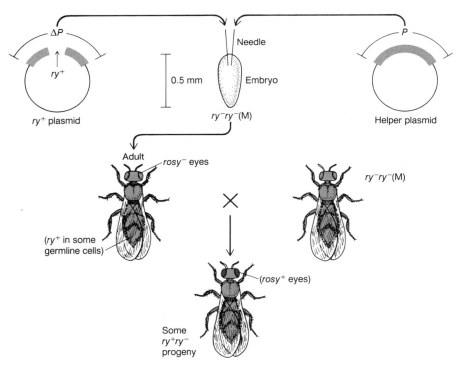

Figure 19-28. *P*-element-mediated gene transfer in *Drosophila*. The *rosy*[+] (*ry*[+]) eye-color gene is inserted into a deleted *P* element (Δ*P*) carried on a bacterial plasmid. At the same time, a helper plasmid bearing an intact *P* element is used. Both are injected into a *ry*[−] embryo, where *ry*[+] transposes with the Δ*P* element into the chromosomes of the germline cells.

The second major analytical technique stems from the discovery by Gerald Rubin and Allan Spradling that *P*-element DNA can be used as an effective vehicle for transferring donor genes into the germline of a recipient fly. Rubin and Spradling devised the following experimental procedure (Figure 19-28). The recipient genotype is homozygous for the *rosy* (*ry⁻*) mutation, which confers a characteristic eye color and is of M cytotype. From this strain, embryos are collected at the completion of about nine nuclear divisions. At this stage, the embryo is one multinucleate cell, and the nuclei destined to form the germ cells are clustered at one end. Two types of DNA are injected into embryos of this type. The first is a bacterial plasmid carrying a deleted *P* element into which the *ry⁺* gene has been spliced. This deleted element is not able to transpose due to the deletion, so a helper plasmid bearing a complete element is also injected. Flies developing from these embryos are phenotypically still *rosy* mutants, but their offspring contain a large proportion of *ry⁺* individuals. These *ry⁺* descendants show Mendelian inheritance of the newly acquired *ry⁺* gene, suggesting that it is located on a chromosome. This has been confirmed by in situ hybridization, which shows that the *ry⁺* gene, together with the deleted *P* element, has been inserted into one of several distinct chromosome locations. None appears exactly at the normal locus of the *rosy* gene. These new *ry⁺* genes are found to be inherited in a stable fashion.

Message   *P* elements in *Drosophila* are a type of transposon that causes hybrid dysgenesis. They are very useful in two ways to the genetic analyst. First, they can be used through transposon mutagenesis to recover selectively any gene with a recognizable mutant phenotype. Marking a specific gene by transposon mutagenesis is termed **transposon tagging**. Second, *P* elements in *Drosophila* can be used as efficient vehicles for the transfer of specific genes to given recipient genotypes.

Retroviruses

Retroviruses are single-stranded animal viruses that employ a double-stranded DNA intermediate for replication. The RNA is copied into DNA by the enzyme reverse transcriptase. The life cycle of a typical retrovirus is shown in Figure 19-29. Some retroviruses, such as mouse mammary tumor virus (MMTV) and Rous sar-

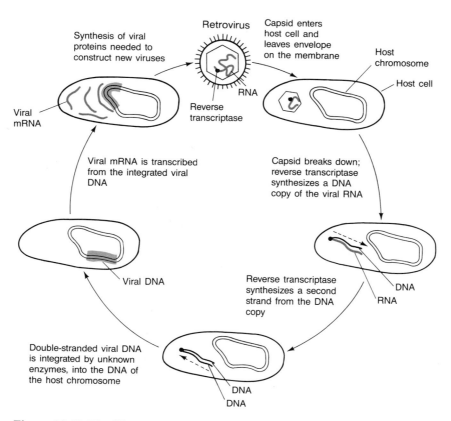

Figure 19-29. The life cycle of a retrovirus. RNA is shown in color; DNA, in black.

coma virus (RSV), are responsible for the induction of cancerous tumors. When integrated into host chromosomes as double-stranded DNA, these retroviruses are termed **proviruses.** Proviruses, like *mu* phage in bacteria, can be considered transposable elements, since they can, in effect, transpose from one location to another.

Retroviruses have structural similarities to some transposable elements from bacteria and other organisms. In particular, the ends of the proviruses have long terminal repeats (LTR's) reminiscent of the sequences of the Ty1 elements in yeast and the long terminal repeats of the *copia*-like elements in *Drosophila*. Also, integration results in the duplication of a short target se-

quence in the host chromosome. For example, in the case of the mouse retrovirus shown in Figure 19-30, 4-bp sequence is duplicated on each side of the integrated provirus. (Note how the staggered cuts and the duplication of the short segment of the host DNA are similar to those shown in Figure 19-22 for the integration of bacterial transposable elements.) On the other hand, the similarities between retroviruses and Ty1 elements in yeast and *copia*-like elements in *Drosophila* suggest that the Ty1 and *copia*-like elements might also be integrated forms of retrovirus-like elements. This has been nicely confirmed in 1985 by Jef Boeke and Gerald Fink and their coworkers in the following experiments.

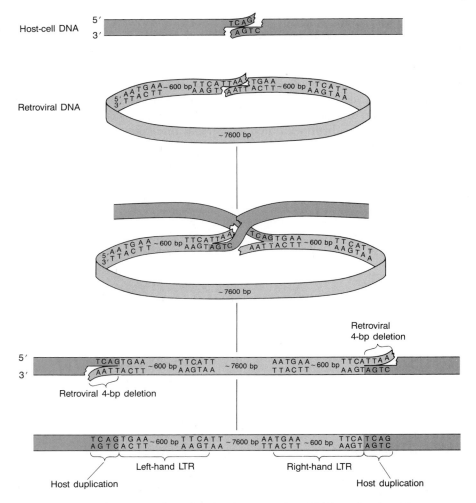

Figure 19-30. The integration of a circular retrovirus DNA molecule with two tandem LTR's. The circular DNA is shown integrating into host-cell DNA at a site at which the sequence TCAG happens to occur. The integration occurs by staggered cleavages of both the cellular DNA and the viral DNA. The illustrated case is a mouse retrovirus that makes a 4 bp repeat of cellular DNA at the site of integration; for other retroviruses, the repeat is 4 to 6 bp long. (From C. Shoemaker and D. Baltimore. *Proceedings of the National Academy of Sciences* 77, 1980, 3932. After J. Darnell et al. *Molecular Cell Biology.* Copyright © 1986 by Scientific American Books, New York.)

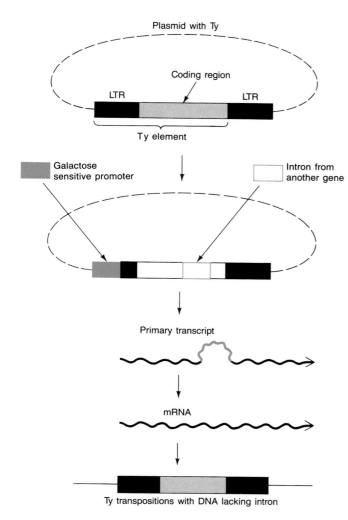

Figure 19-31. Demonstration of transposition through an RNA intermediate. A Ty element is altered by adding a promoter which can be activated by the addition of galactose. Activation of the promoter will increase transcription through the Ty element. Also, an intron from another gene is inserted into the Ty element. Because the final product of transposition contains no intron, the intron must have been spliced out from an RNA transcript (see Chapter 13). This must have occurred as shown here, where the primary transcript contains the intron but the final processed mRNA does not. This RNA is then copied by reverse transcriptase and integrated into the chromosomal DNA. (Modified from J. Darnell, H. Lodish, and D. Baltimore, *Molecular Cell Biology*, p. 450. Copyright © 1986 by Scientific American Books, New York.)

Transposition Via an RNA Intermediate

Figure 19-31 diagrams the experimental design used by Boeke and Fink and their colleagues to alter a yeast Ty element, cloned on a plasmid, in two different ways. First, a promoter inserted near the end of the element can be activated by the addition of galactose to the medium. The use of this galactose-sensitive promoter allows the manipulation of the expression of Ty RNA. In the presence of galactose, more transcription of Ty RNA occurs. Second, an intron (page 347) from another yeast gene is introduced into the Ty transposon coding region.

The addition of galactose greatly increases the frequency of transposition of the altered Ty element. This suggests the involvement of RNA, because galactose-stimulated transcription begins at the galactose-sensitive promoter and continues through the element (Figure 19-31). The key experimental result, however, is the fate of the transposed Ty DNA. When the researchers examined the Ty DNA resulting from transpositions, they found that the intron had been removed! Because introns are excised only during RNA processing (see Chapter 13), the transposed Ty DNA must have been copied from an RNA intermediate transcribed from the original Ty element and then processed by RNA-RNA splicing. The DNA copy of the spliced mRNA is then integrated into the yeast chromosome.

Controlling Elements in Maize

In 1938, Marcus Rhoades analyzed an ear of Mexican black corn. The ear came from a selfing of a pure-breeding pigmented genotype, but it showed a surprising modified Mendelian dihybrid segregation ratio of 12 : 3 : 1 among pigmented, dotted, and colorless kernels. Analysis showed that two events had occurred at unlinked loci. At one locus, a pigment gene A_1 had mutated to a_1; at another locus, a dominant allele Dt (*Dotted*) had appeared. The effect of Dt was to produce pigmented dots in the otherwise colorless phenotype of a_1a_1 (Figure 19-32). Thus, the original line was very probably A_1A_1dtdt, and the mutations generated an A_1a_1Dtdt plant, which on selfing gave the observed ratio of progeny.

But what was causing the dotted phenotype? A reverse mutation of $a_1 \rightarrow A_1$ in somatic cells would be an obvious possibility, but the large numbers of dots in the *Dotted* kernels would require extremely high reversion rates. Using special stocks, Rhoades was able to find anthers in the flowers of a_1a_1Dt- plants that showed patches of pigment (Figure 19-33). He reasoned that these anthers might contain pollen grains bearing the reverted pigment genotype, and he used the pollen from these anthers to fertilize a_1a_1 tested females. Sure enough, some of the progeny were completely pigmented, showing that each dot in the parental plants was in fact the phenotypic manifestation of a genetic reversion event. Thus, a_1 is one of the first known examples of an **unstable mutant allele**—an allele for which reverse mutation occurs at a very high rate. However, the allelic

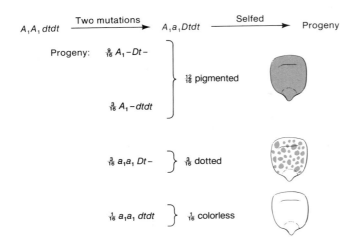

$A_1A_1 dtdt \xrightarrow{\text{Two mutations}} A_1a_1Dtdt \xrightarrow{\text{Selfed}} \text{Progeny}$

Progeny:

$\frac{9}{16} A_1 - Dt -$

$\frac{3}{16} A_1 - dtdt$

$\left.\begin{array}{c}\end{array}\right\} \frac{12}{16}$ pigmented

$\frac{3}{16} a_1a_1, Dt - \left.\begin{array}{c}\end{array}\right\} \frac{3}{16}$ dotted

$\frac{1}{16} a_1a_1, dtdt \left.\begin{array}{c}\end{array}\right\} \frac{1}{16}$ colorless

Figure 19-32. A formal genetic explanation of the appearance of the dotted phenotype in corn. The A_1a_1Dtdt genotype is created by simultaneous mutations of $A_1 \rightarrow a_1$ and $dt \rightarrow Dt$. Upon selfing, this genotype yields the observed $12:3:1$ ratio of kernel phenotypes.

instability is dependent on the presence of the unlinked Dt gene. Once the reverse mutations occur, they are stable; the Dt gene can be crossed out of the line with no loss of the A_1 character.

McClintock's Experiments: The *Ds* Element

In the 1950s, Barbara McClintock demonstrated an analogous situation in another study of corn. She found a genetic factor Ds that causes a high tendency toward

chromosome breakage at the location at which it appears. These breaks can be located either cytologically (Figure 19-34a) or by the uncovering of recessive genes (Figure 19-34b). As you will appreciate, this action of Ds is another kind of instability. Once again, this instability proves to be dependent on the presence of an unlinked gene *Ac (Activator)*, in the same way that the instability of a_1 is dependent on Dt.

McClintock tried to map Ac, but she found it impossible to map! In some plants, it mapped to one position; in other plants of the same line, it mapped to different positions. As if this were not enough of a curiosity, the Ds locus (Figure 19-34) itself was constantly changing position on the chromosome arm, as indicated by the differing phenotypes of the variegated sections (as different recessive gene combinations were uncovered in a system such as the one illustrated in Figure 19-34b).

The wanderings of the Ds element take on new meaning for us in the context of this chapter when we consider the results of the following cross:

$$\male CCDsDsAcAc^+ \times ccDs^+Ds^+Ac^+Ac^+ \female$$

Here C allows color expression and Ds^+ and Ac^+ indicate the lack of the element. Most of the kernels from this

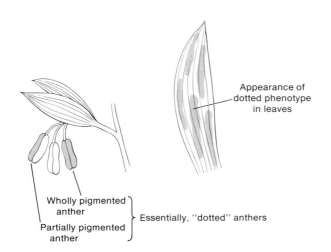

Figure 19-33. Rhoades used special stocks of $a_1a_1 Dt-$ corn plants carrying certain genes that allow pigmented sectors to be detected in tissue other than the kernels. (After M. M. Rhoades, *Genetics* 23:382, 1938.)

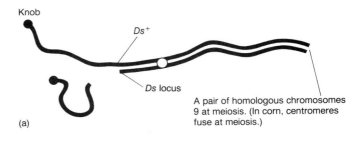

(a)

A pair of homologous chromosomes 9 at meiosis. (In corn, centromeres fuse at meiosis.)

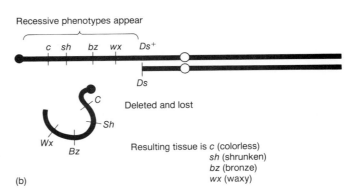

(b)

Figure 19-34. Detection of chromosomal breakage (instability) due to action of the Ds element in corn. (a) Cytological detection of the breakage. (Ds^+ indicates a lack of Ds.) (b) Genetic detection of the breakage.

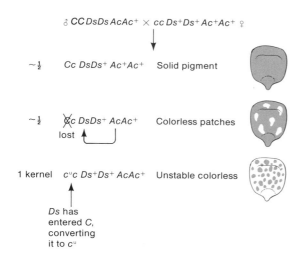

Figure 19-35. Results that indicate the transposition of *Ds* into the *C* gene in corn. (*C* allows color expression; *c* does not. *Ac* = activator.) The action of *Ds* is dependent on the presence of the unlinked gene, *Ac*.

cross were of the expected types (Figure 19-35), but one exceptional kernel was very interesting. In this individual, the *Ds* element seems to have wandered into the middle of the *C* gene and inactivated it, producing the colorless phenotype. Although the presence of *Ac* is still necessary to cause this instability, it now seems to take on the function of rendering the *c* mutation unstable (c^u), and patches of revertant color appear in the kernel. When *Ac* is crossed out of this line, the c^u becomes a stable mutant.

The analogy of this system with the $a_1 Dt$ system is obvious. Perhaps the earlier situation also is due to the insertion of a *Ds*-like element into the A_1 gene. It is natural to ask whether a_1 will respond to *Ac* or c^u will respond to *Dt*. The answer is no; some kind of specificity prevents this cross-activation of mutational instability.

The *wx (waxy)* Locus

The *Ds* element can wander not only into the middle of the *C* gene but also into other genes, rendering them unstable mutants dependent on *Ac*. One such locus, *wx (waxy)*, has been the subject of an intense study on the effects of the *Ds* element. Oliver Nelson has paired many unstable *waxy* alleles in the absence of the *Ac* mutation. In such wx^{m-1}/wx^{m-2} heterozygotes, he has looked for rare wild-type *Wx* recombinants by staining the pollen with KI-I_2 reagent, which stains *Wx* pollen black and *wx* pollen red. By counting the frequency of *Wx* pollen grains in each kind of heterozygote, Nelson has been able to do fine-structure recombination mapping of the *waxy* gene. He has showed that the different "mutable

waxy" mutant alleles are in fact due to the insertion of the *Ds* element in different positions in the gene. In a subsequent experiment, he allowed *Wx*-bearing pollen to fertilize *wx* plants and produce rare *waxy* kernels, which could be detected by staining sliced-off slivers. The *waxy* kernels were then raised into plants, and Nelson was able to show that the *Wx* pollen grains arose from chromosome exchange, which also involved the exchange of flanking markers.

General Characteristics of Controlling Elements

Several systems like $a_1 Dt$ and *DsAc* have now been found in corn. Each shows similar action, having a **target gene** that is inactivated, presumably by the insertion of some **receptor element** into it, and a distant **regulator gene** that maintains the mutational instability of the locus, presumably through its ability to "unhook" the receptor element from the target locus and return the locus to normal function. The receptor and the regulator are termed **controlling elements.**

In the examples discussed so far, the unstable allele is said to be **nonautonomous:** it can revert only in the presence of the regulator. Sometimes, however, a system such as the *Ac−Ds* system can produce an unstable allele that is **autonomous.** Such mutants are recognizable because they show Mendelian ratios (such as 3 : 1 for pigmented to dotted) that apparently are independent of any other element. In fact, such alleles appear to be caused by the insertion of *Ac* itself into the target gene. An allele of this type can subsequently be transformed into a nonautonomous allele. In such cases, the nonautonomy seems to result from the spontaneous generation of a *Ds* element from the inserted *Ac* element. In other words, *Ds* is in all likelihood a partially incomplete version of *Ac* itself.

Figure 19-36 summarizes the overall behavior of the controlling elements in maize as inferred from genetic data. Note that the mutation events that gave rise to Rhoades' original ratio are nicely explained by this mode: a nonautonomous *Ds*-like element was generated from an *Ac*-like progenitor; transposition of the *Ds*-like element into the A_1 gene produced an inactive but mutationally unstable allele a_1.

Molecular Analysis of Controlling Elements

Molecular studies in the last few years on *Ac, Ds,* and on other controlling elements in maize have confirmed McClintock's genetic model in a very satisfying way. One such study focused on *Ac*- or *Ds*-containing unstable alleles on the *Waxy* locus. First, the wild-type *Wx* gene is cloned so that it can be used as a probe to retrieve the unstable alleles. *Waxy* mRNA is identified as the endosperm mRNA which, on translation, produces a protein

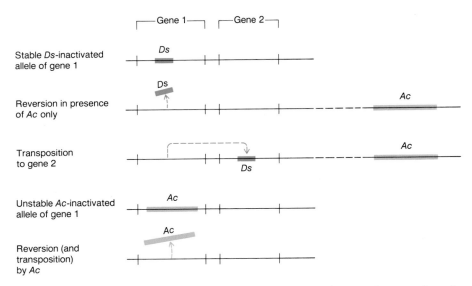

Figure 19-36. Summary of the main effects of controlling elements in corn. *Ac* and *Ds* are used as examples, acting on two hypothetical genes 1 and 2.

which is almost certainly the Waxy structural protein (as judged from the absence of this protein in *wx* mutants). A cDNA clone made from this mRNA is used to fish out, from total genomic DNA, a *Wx* gene, a *wx* gene containing either *Ds* or *Ac* inserted, and *Wx* revertant alleles derived from unstable alleles. In all cases, the DNA is

sequenced, and some of the results are summarized in Figure 19-37. An *Ac* element is about 4500 bp long and has 11-bp imperfect inverted terminal repeats. It has two open reading frames, one of which seems like a good candidate for a transposase. *Ds* elements prove to be deleted *Ac* elements, and the deletion can be small within

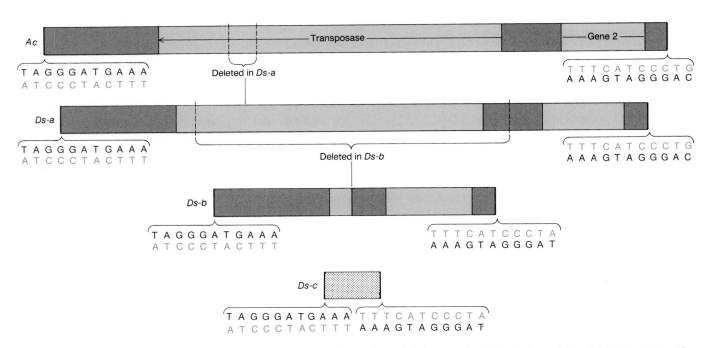

Figure 19-37. The structure of the *Ac* element of maize and several *Ds* elements. (From N. V. Federoff, "Transposable Genetic Elements in Maize." Copyright © 1984 by Scientific American, Inc. All rights reserved.)

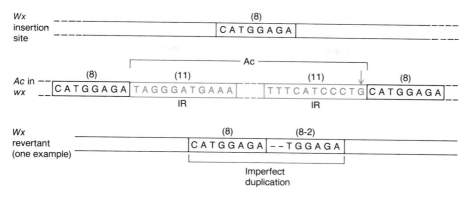

Figure 19-38. Target-insertion-site duplication produced by the transposon *Ac* at the *Wx* gene in maize.

the transposase, as in *Ds-a*, or large, as in *Ds-b* and *Ds-c* (Figure 19-37).

Flanking the inserted transposons are found 8-bp-long duplications of the *Wx* gene DNA. Commonly, the duplication is imperfect. In the revertants, the 8-bp duplication remains, although often with a slight modification that would seem to be necessary to retain the proper reading frame of the gene. These duplications are shown in Figure 19-38.

> **Message**   Controlling elements in maize can inactivate a gene in which they reside, cause chromosome breaks, and transpose to new locations within the genome. Complete elements can perform these functions unaided; other forms with partial deletions can transpose only with the help of a complete element elsewhere in the genome.

Review

Let's examine some of the essential points about eukaryotic transposable elements:

1. Transposable elements exist in all cells. Elements in yeast, *Drosophila,* and maize have been particularly well studied, as have retroviruses in mammalian cells.

2. Some transposable elements can be used as tools for cloning and gene manipulation. For instance, the *P* elements of *Drosophila* can be employed to transfer genes into the germline of a recipient fly. Another example — the T DNA segment of Ti plasmids (de- scribed in Chapter 15) — can be used to introduce cloned genes into certain plants.

3. One similarity between eukaryotic tranposable elements and their counterparts in prokaryotes is that transposition into a new site generates a short repeated sequence at the target site.

4. One difference between eukaryotic and prokaryotic transposable elements lies in the mechanism of transposition. Some eukaryotic transposable elements appear to transpose via an RNA intermediate; prokaryotic elements do not use an RNA intermediate.

Summary

■ Nature has devised many different ways of changing the genetic architecture of organisms. We are now beginning to understand the molecular processes behind some of these phenomena. Gene mutation, recombination between chromosomes (discussed in Chapters 17 and 18), and transposition can all be reasonably explained at the DNA level. Far from merely producing genetic waste, these processes undoubtedly all have important roles in evolution. This idea is strengthened through the knowledge that the processes themselves are to a large extent under genetic control: there are genes that affect the efficiency of mutation, recombination, and transposition.

Although different mechanisms of transposition are sometimes used, the analogies among the transposable elements of phages, bacteria, and eukaryotes are striking. At present, it is not known whether transposons are elements that normally play a role in the day-to-day transactions of the genome, as originally proposed by Barbara McClintock in the 1950s, or whether they are pieces of "selfish DNA" that exist for no purpose other than their own survival. Whatever the truth of this matter is, transposons certainly represent a completely un-

expected element of chaos in the genome, which geneticists have already harnessed into their team of analytical procedures. At the evolutionary level, transposons may be important in the sudden leaps that characterize the fossil record.

■ ■ ■

Solved Problems

1. The following table shows the reversion pattern of a series of mutations in the *gal* operon of *E. coli*.

   Here, a + indicates a high rate of reversion in the presence of each mutagen, a − depicts no reversion, and a "low" indicates a low rate of reversion. Which mutation is most likely to result from the insertion of a transposable element and why? Can you assign the other mutations to a category?

   Solution

   Transposable elements revert at low rates spontaneously, and this rate is not stimulated by base analogs, frameshift mutagens, alkylating agents, or UV. Based on these criteria, the *gal-3* mutation is most likely to result from an insertion, since it reverts at a low rate which is not stimulated by any of the mutagens. *gal-1* might be a frameshift, since it does not revert with 2AP and EMS, but does with ICR191, a frameshift mutagen, and UV. (Refer to Chapter 17 for details of each mutagen.) Likewise, *gal-2* is probably a frameshift, since it reverts only with ICR-191. The *gal-4* mutation is probably a deletion, since it is not stimulated to revert at all. The *gal-5* mutation appears to be a base substitution, since it reverts with 2AP, but not with ICR191.

2. Transposable elements have been referred to as "jumping genes," since they appeared to jump from one position to another, leaving the old locus and appearing at a new locus. In light of what we now know concerning the mechanism of transposition, how appropriate is the term "jumping genes" for bacterial transposable elements?

   Solution

   In bacteria, transposition occurs by two different modes. The conservative mode results in true jumping genes,

since in this case the transposable element excises from its original position and inserts at a new position. A second mode is termed the replicative mode. In this pathway, transposable elements move to a new location by replicating into the target DNA, leaving behind a copy of the transposable element at the original site. When operating by the replicative mode, transposable elements are not really jumping genes, since a copy does remain at the original site.

Problems

1. Suppose that you want to determine whether a new mutation in the *gal* region of *E. coli* is the result of an insertion of DNA. Describe two physical experiments that would allow you to demonstrate the presence of an insertion.

2. What is a polar mutation?

3. Explain the difference between the replicative and conservative modes of transposition. Briefly describe an experiment demonstrating each of these modes in prokaryotes.

4. Describe the generation of multiple drug-resistance plasmids.

5. Briefly describe the experiment that demonstrates that the transposition of the Ty1 element in yeast occurs via an RNA intermediate.

6. Explain how the properties of *P* elements in *Drosophila* make possible gene-transfer experiments in this organism.

7. When Rhoades took pollen from wholly pigmented anthers on plants of genotype a_1a_1DtDt and used this pollen to pollinate a_1a_1dtdt tester females, he found wholly pigmented kernels and, in addition, some dotted kernels. Explain the origin of *both* phenotypes.

8. In yeast, the *his4* region has three cistrons — A, B, and C, in that order — each of which mediates an enzymatic step of histidine synthesis. A certain spontaneous mutation is mapped in the cistron A, but these mutants are defective for all three functions (A, B, and C). The mutation is not suppressible by nonsense or frame-shift suppressors. Spontaneous reversion occurs quite frequently, but this rate is not enhanced by any mutagen. Discuss the possible nature of the mutation.

Mutation	Reversion				
	Spontaneous	2-aminopurine	ICR191	UV	EMS
gal-1	−	−	+	+	−
gal-2	−	−	+	−	−
gal-3	low	low	low	low	low
gal-4	−	−	−	−	−
gal-5	low	+	low	+	+

9. In *Drosophila*, M. Green found a *singed* allele *sn77-27* with some unusual characteristics. Females homozygous for this X-linked allele have singed bristles, but they have numerous patches of *sn*+ (wild-type) bristles on their heads, thoraxes, and abdomens. When these flies are bred to *sn* males, some females give only singed progeny, but others give both singed and wild-type progeny in variable proportions. Explain these results.

10. Crown-gall tumors are found in many dicotyledonous plants infected by the bacterium *Agrobacterium tumefaciens*. The tumors are caused by the insertion of DNA from a large plasmid carried by the bacterium into the plant DNA. A tobacco plant of type A (there are many types of tobacco plants) is infected, and it produces tumors. You remove tumor tissue and grow it on a synthetic medium. Some of these tumor cultures produce aerial shoots. You graft one of these shoots onto a normal tobacco plant of type B, and the graft grows to an apparently normal A-type shoot and flowers.

a. You remove cells from the graft and place them in synthetic medium, where they grow like tumor cells. Explain why the graft appears to be normal.

b. When seeds are produced by the graft, the resulting progeny are normal A-type plants. No trace of the inserted plasmid DNA remains. Propose a possible explanation for this "reversal."

11. Consider two maize plants:

a. Genotype $C/c^m Ac/+$, where c^m is an unstable allele caused by Ds insertion.

b. Genotype C/c^m, where c^m is an unstable allele caused by Ac insertion.

What phenotypes would be produced and in what proportions when (1) each plant is crossed to a base-pair substitution mutant c/c? (2) Plant (a) is crossed with plant (b)?

Assume that Ac and c are unlinked, that the chromosome breakage frequency is negligible, and that mutant c/c is Ac^+.

The Extranuclear Genome

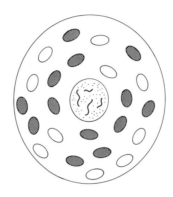

KEY CONCEPTS

Chloroplasts and mitochondria each contain multiple copies
of their own unique "chromosome" of genes.

Organelle DNA—and any variant phenotype coded therein—
is inherited generally through the maternal parent in a cross.

In mixtures of two genetically different mitochondrial
DNA's or chloroplast DNA's, it is commonly observed that a
sorting out process results in descendant cells of one type or
the other.

In "dihybrid" organelle mixtures, recombination can be detected.

Organelle genes code for organelle translation or
transcription components, or components of energy-
producing systems.

Most organelle-encoded polypeptides unite with nucleus-
encoded polypeptides to produce active proteins, and these
function in the organelle.

Eukaryotic plasmids show inheritance patterns that are in
some ways similar to those of organelles.

In the analysis of eukaryotic organisms, the notion that genes reside in the nucleus has become a firm part of the general dogma of genetics. Indeed, there is no need to question the fundamental truth of this notion. The mechanics of the nuclear processes of meiosis and mitosis, together with the understanding that genes are located on nuclear chromosomes, provide the basic set of operational rules for genetics. In fact, the great successes of genetics in this century have been in large part due to the success of these operational rules in predicting inheritance patterns — from simple eukaryotes to human beings.

However, like most fundamental truths, this one does have exceptions, and these exceptions are the subject of this chapter. In eukaryotes, the special inheritance patterns of some genes reveal that these genes must be located outside the nucleus. Such **extranuclear genes** ("extra-" means "outside") have been found in a variety of eukaryotes. At first, the very idea of such genes seemed heretical, but their existence has gradually been accepted over the years as more and more results have turned up to confirm their reality. The inheritance patterns shown by extranuclear genes are less commonly encountered than those of nuclear genes, and indeed there are fewer extranuclear genes than normal chromosomal genes. However, although they are exceptional in this sense, it is now clear that extranuclear genes play normal (but highly specialized) roles in the control of eukaryote phenotypes.

In the past decade or so, intense research activity has been directed toward the study of extranuclear inheritance. This activity has resulted from the discovery (and subsequent general availability) of new genetic and technological approaches. The great success of these analyses has led to a vastly improved understanding of the nature and function of extranuclear genes. We see in this chapter that the operational approaches in such research are often quite different from those employed to study nuclear genes (although there are similarities also). These unique analytical methods alone would merit special treatment, but their results also have revealed fascinating genetic systems that are novel and distinct enough in themselves to warrant major coverage in such a text as this.

A word of caution is in order here. A large proportion of the material presented in this chapter is based on frontier research. Although the experimental approaches to extranuclear inheritance are now well established and a wealth of indisputable information has been accumulated, there are still many unanswered questions. For one thing, only a rather small number of species has proved suitable for the specialized techniques developed thus far. Even in the case of the well-studied organisms, we must await further research in cell biology to fill in many details we do not yet understand.

Conventionally, the eukaryotic cell is divided into two domains: the nucleus and the cytoplasm. The cytoplasm itself is a highly organized heterogeneous array of organelles, membranes, and various molecules in solution (Figure 20-1). Extranuclear genes are found in the mitochondria and in the chloroplasts of green plants. These organelles serve the major functions of ATP synthesis and photosynthesis, respectively. The mitochondria and chloroplasts each contain their own set of unique and autonomous genes, linked together in an organellar chromosome. Extranuclear genes often are called extrachromosomal — a term that is potentially confusing because the genes found in the mitochondria and chloroplasts in fact do comprise extranuclear chromosomes. Nevertheless, it must be emphasized that the organellar chromosomes show organizational and inheritance patterns that are quite different from those of nuclear chromosomes. Extranuclear genes have also been called cytoplasmic genes based on their location in the cytoplasm.

A great deal is now known about the genetics and biochemistry of organellar genes. This chapter outlines some of the major steps in the evolution of our present understanding. It is a beautiful example of the power of genetic analysis combined with molecular biology.

> **Message** The genes that have been called cytoplasmic genes, extrachromosomal genes, or extranuclear genes are in fact located on a unique kind of chromosome inside cytoplasmic organelles.

Variegation in Leaves of Higher Plants

One of the first convincing examples of extranuclear inheritance was found in higher plants. In 1909, Carl Correns reported some surprising results from his studies on four-o'clock plants *(Mirabilis jalapa)*. He noted that the blotchy leaves of these variegated plants show patches of green and white tissue, but some branches carry only green leaves and others carry only white leaves (Figure 20-2).

Flowers appeared on all types of branches. Correns intercrossed a variety of different combinations by transferring pollen from one flower to another. Table 20-1 shows the results of such crosses. Two features of these results are surprising. First, there is a difference between reciprocal crosses: for example, white ♀ × green ♂ gives a different result from green ♀ × white ♂.

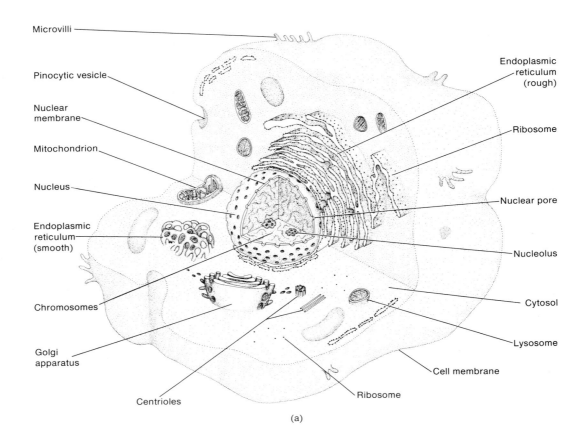

Microvilli

Pinocytic vesicle

Nuclear membrane

Mitochondrion

Nucleus

Endoplasmic reticulum (smooth)

Chromosomes

Golgi apparatus

Centrioles

Endoplasmic reticulum (rough)

Ribosome

Nuclear pore

Nucleolus

Cytosol

Lysosome

Cell membrane

Ribosome

(a)

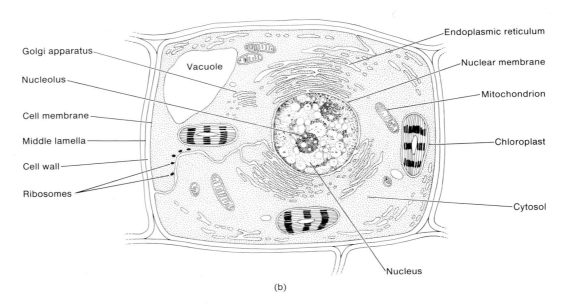

Golgi apparatus

Nucleolus

Cell membrane

Middle lamella

Cell wall

Ribosomes

Vacuole

Endoplasmic reticulum

Nuclear membrane

Mitochondrion

Chloroplast

Cytosol

Nucleus

(b)

Figure 20-1. The heterogeneity of the cytoplasm. The area called the cytoplasm (everything between the nuclear membrane and the cell membrane) embraces the cytosol (liquid phase) plus a rich assortment of organelles, membranes, and other structures. Extranuclear genes are found in the mitochondria and in the chloroplasts of green plants. (a) An animal cell. (b) A plant cell. (Part a from S. Singer and H. R. Hilgard, *The Biology of People.* Copyright © 1978 by W. H. Freeman and Company; part b from J. Janick et al., *Plant Science.* Copyright © 1974 by W. H. Freeman and Company)

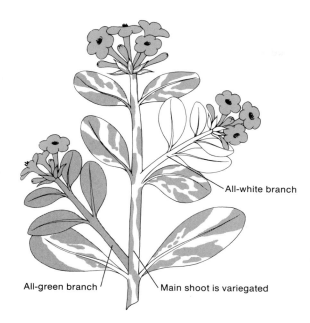

Figure 20-2. Leaf variegation in *Mirabilis jalapa,* the four-o'clock plant. Flowers may form on any branch (variegated, green, or white), and these flowers may be used in crosses.

Recall that Mendel never observed differences between reciprocal crosses. In fact, in conventional genetics, differences between reciprocal crosses are normally encountered only in the case of sex-linked genes. However, the results of the four-o'clock crosses cannot be explained by sex linkage.

The second surprising feature is that the phenotype of the maternal parent is solely responsible for determining the phenotype of all progeny. The phenotype of

TABLE 20-1. Results of crosses of variegated four-o'clock plants

Phenotype of branch bearing egg parent (♀)	Phenotype of branch bearing pollen parent (♂)	Phenotype of progeny
White	White	White
White	Green	White
White	Variegated	White
Green	White	Green
Green	Green	Green
Green	Variegated	Green
Variegated	White	Variegated, green, or white
Variegated	Green	Variegated, green, or white
Variegated	Variegated	Variegated, green, or white

the male parent appears to be irrelevant, and its contribution to the progeny appears to be zero! This phenomenon is known as **maternal inheritance.** Although the white progeny plants do not live long because they lack chlorophyll, the other progeny types do survive and can be used in further generations of crosses. In these subsequent generations, maternal inheritance always appears in the same patterns as the ones observed in the original crosses.

How can such curious results be explained? The differences in leaf color are known to be due to the presence of either green or colorless chloroplasts. The inheritance patterns might be explained if these cytoplasmic organelles are somehow genetically autonomous and furthermore are never transmitted via the pollen parent. For an organelle to be genetically autonomous, it must have its own genetic determinants that are responsible for its phenotype. In this case, the chloroplasts would carry their own genetic determinants responsible for chloroplast color. Thus, the suggestion arises that this organelle has its own genome. Its failure to be transmitted via the pollen parent is reasonable because the bulk of the cytoplasm of the zygote is known to come from the maternal parent via the cytoplasm of the egg. Figure 20-3 diagrams a model that formally accounts for all the inheritance patterns in Table 20-1.

Variegated branches apparently produce three kinds of eggs: some contain only white chloroplasts, some contain only green chloroplasts, and some contain both kinds of chloroplasts. The egg type containing both green and white chloroplasts produces a zygote that also contains both kinds of chloroplasts. In subsequent mitotic divisions, some form of cytoplasmic segregation occurs that segregates the chloroplast types in pure cell lines, thus producing the variegated phenotype in that progeny individual (Figure 20-4).

This process of sorting might be described as "mitotic segregation." However, this is an *extranuclear* phenomenon, not to be confused with the mitotic segregation of nuclear genes, so a new term is needed. This term — representing a hypothetical "black box" in which the segregation and (as we shall see) recombination of organelle genotypes occur — we shall call **cytoplasmic segregation and recombination,** referred to in this text by the convenient acronym CSAR. Throughout this chapter, CSAR crops up quite often; it appears to be a common behavior of extranuclear genomes. In one sense, CSAR is the cytoplasmic equivalent of meiosis, because meiosis is the process whereby the segregation and recombination of nuclear genes regularly occur. However, CSAR is purely hypothetical — no physical counterpart is known for this process.

We might suspect random chloroplast assortment to explain the green – white segregation in variegated plants, but this hypothesis has not been proved. Further-

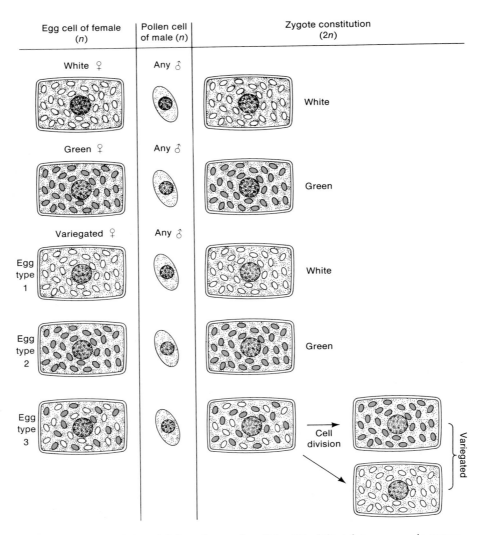

Egg cell of female (n)	Pollen cell of male (n)	Zygote constitution (2n)

Figure 20-3. A model explaining the results of the *Mirabilis jalapa* crosses in terms of autonomous chloroplast inheritance. The large, dark spheres are nuclei. The smaller bodies are chloroplasts, either green (color) or white. Each egg cell is assumed to contain many chloroplasts, and each pollen cell is assumed to contain no chloroplasts. The first two crosses exhibit strict maternal inheritance. If the maternal branch is variegated, three types of zygotes can result, depending on whether the egg cell contains only white, only green, or both green and white chloroplasts. In the last case, the resulting zygote can produce both green and white tissue, so a variegated plant results.

more, the large numbers of chloroplasts in cells make this an unlikely possibility. In a cell with, let's say, 40 chloroplasts, consisting of 20 green and 20 colorless chloroplasts, the production by random assortment of one daughter cell containing only green and one daughter cell containing only colorless chloroplasts would be a very rare event at best. To be on the safe side, we shall treat the CSAR process as hypothetical. This hypothetical CSAR process is diagrammed in Figure 20-5.

Poky *Neurospora*

In 1952, Mary Mitchell isolated a mutant strain of *Neurospora* that she called poky. This mutant differs from the wild-type in a number of ways: it is slow growing, it shows maternal inheritance of slow growth, and it has abnormal amounts of cytochromes. Cytochromes are mitochondrial electron-transport proteins necessary for the proper oxidation of foodstuffs to generate ATP en-

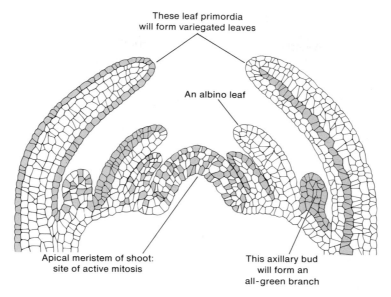

These leaf primordia
will form variegated leaves

An albino leaf

Apical meristem of shoot:
site of active mitosis

This axillary bud
will form an
all-green branch

Figure 20-4. In a plant containing both green and white chloroplasts, different green or white cell lines are established in the meristem (growing point). The egg from which such a plant is derived contains both types of chloroplasts. After fertilization and subsequent mitotic divisions, cytoplasmic segregation occurs, segregating the chloroplast types into pure cell lines. The color of any part of the plant thus depends on the particular cell types that happen to form the relevant leaf or shoot primordia.

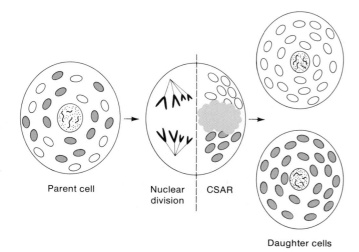

Parent cell

Nuclear
division

CSAR

Daughter cells

Figure 20-5. Two processes may be distinguished at cell division. Nuclear division (mitosis) is an observed physical process that parcels the genes on the nuclear chromosomes. The hypothetical CSAR process in the cytoplasm (for which no physical counterpart yet has been observed) would account for the segregation and recombination of organelle genes, producing the new organelle sets of the daughter cells. The diagrams show how CSAR can generate two different organelle sets from a mixed parental cell.

ergy. There are three main types of cytochrome: *a*, *b*, and *c*. In poky, there is no cytochrome *a* or *b*, and there is an excess of cytochrome *c*.

How can maternal inheritance be expressed in a haploid organism? It is possible to cross some fungi in such a way that one parent contributes the bulk to the cytoplasm to the progeny; this cytoplasm-contributing parent is called the female parent, even though no true sex is involved. Mitchell demonstrated maternal inheritance for the poky phenotype in the following crosses:

poky ♀ × wild-type ♂ ⟶ all poky

wild-type ♀ × poky ♂ ⟶ all wild-type

However, in such crosses, any nuclear genes that differ between the parental strains are observed to segregate in the normal Mendelian manner and to produce 1 : 1 ratios in the progeny (Figure 20-6). All poky progeny behave like the original poky strain, transmitting the poky phenotype down through many generations when crossed as females. Because poky does not behave like a nuclear mutation, it has been termed an **extranuclear mutation.** To hammer the point home, it also has been

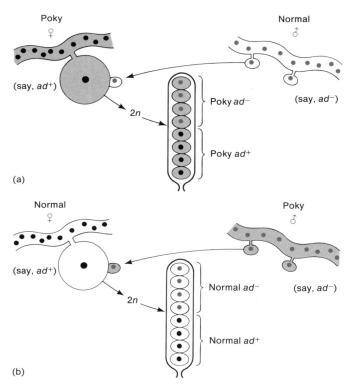

(a)

(b)

Figure 20-6. Explanation of the different results from reciprocal crosses of poky and normal *Neurospora*. Most of the cytoplasm of the progeny cells comes from the parent called female. Color shading represents cytoplasm with the poky determinants. The nuclear locus ad^+/ad^- is used to illustrate the segregation of the nuclear genes in the expected 1:1 Mendelian ratio.

called a **cytoplasmic mutation,** because it seems to be carried in the cytoplasm of the female parent.

But where in the cytoplasm is the mutation carried? The important clue is that several aspects of the mutant phenotype seem to involve mitochondria. For instance, the cells' slow growth suggests a lack of ATP energy, which is normally produced by mitochondria. Also, there are abnormal amounts of cytochromes in the mutants, and cytochromes are known to be located in the mitochondrial membranes. These indications have led geneticists to conclude that mitochondria are involved and have inspired researchers to design several interesting experiments designed to investigate mitochondrial autonomy.

David Luck labeled mitochondria with radioactive choline, a membrane component, and then followed their division autoradiographically in an unlabeled medium. He found that even after several doublings of mass, the radiation was distributed evenly among the mitochondria. Luck concluded that mitochondria can grow and divide in a fashion that seems to be autonomous rather than nucleus-directed. If mitochondria are

synthesized anew, or de novo, some of the resulting population of mitochondria would be expected to be unlabeled and the original mitochondria should remain heavily labeled.

In 1965, a research group led by Edward Tatum extracted purified mitochondria from a mutant called abnormal, which is similar to poky. With an ultrafine needle and syringe, the experimenters injected the mitochondria into wild-type recipient cells, using appropriate controls. The recipient cells were cultured. After several of these subcultures, the abnormal phenotype appeared! In transferring the mitochondria, the experimenters had transferred the hereditary determinants of the abnormal phenotype. Presumably, then, the extranuclear genes involved in this phenotype are located in the mitochondria. These inferences gained credibility with the discovery of DNA in the mitochondria of *Neurospora* and other species (but more of that story later).

The Heterokaryon Test

Maternal inheritance is one criterion for recognizing organelle-based inheritance. Another test has been applied in filamentous fungi such as *Neurospora* and *Aspergillus*. In principle, this test could be applied to other systems involving heterokaryons. A heterokaryon is made between the prospective extranuclear mutant (say, a slow-growth mutant) and a strain carrying a known nuclear mutation. If the phenotype of the nuclear mutation can be recovered from the heterokaryon in combi-

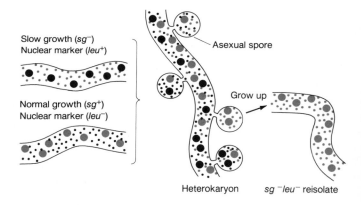

Figure 20-7. The heterokaryon test is used to detect extranuclear inheritance in filamentous (threadlike) fungi. A possible extranuclear mutation (here, sg^-, causing slow growth) is combined with a nuclear mutation (leu^-) to form a heterokaryon. Cultures of leu^- can be derived from the heterokaryon. If some of these cultures also are sg^- in phenotype, then sg is very likely to be an extranuclear gene, borne in an organelle. Since no nuclear recombination occurs in a heterokaryon, the sg^- phenotype must have been acquired by cytoplasmic contact.

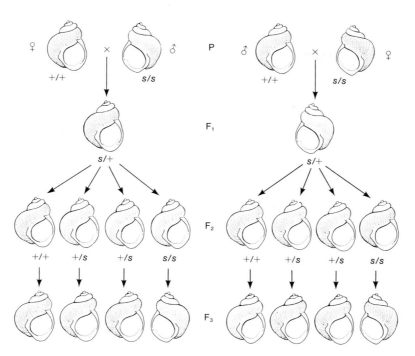

Figure 20-8. The inheritance of dextral (+) and sinistral (*s*) nuclear genes for shell coiling in a species of water snail. The direction of coiling is determined by the nuclear genotype of the mother, not the genotype of the individual involved. This explanation accounts for the initial difference between reciprocal crosses as well as the phenotypes of later generations. No organelle inheritance is involved (such a hypothesis would not explain the phenotypes of later generations).

nation with the phenotype of the slow-growth mutant being tested, then the growth mutant is a good candidate for an organelle-based mutation. This is because no diploidy occurs in a heterokaryon, so that there is no genetic exchange between nuclei. The slow-growth phenotype must have been transferred solely by cytoplasmic contact. The segregation of the pure extranuclear type from the mixed cytoplasm of the heterokaryon presumably involves the CSAR process (Figure 20-7).

> **Message** Extranuclear inheritance can be recognized by uniparental (usually maternal) transmission through a cross or by transmission via cytoplasmic contact.

Shell Coiling in Snails: A Red Herring

Does maternal inheritance always indicate extranuclear inheritance? Usually, but not always. It is possible for a maternal-inheritance pattern of reciprocal crosses to be generated by nuclear genes. In 1923, Alfred Sturtevant (who studied crossing-over in *Drosophila*) found a good example in the water snail *Limnaea*. Sturtevant analyzed the results of crosses between snails that differed in the direction of their shell coiling. On looking into the opening of the shell, it is seen that some snails coil to the right (dextral coiling) and that others coil to the left (sinistral coiling). All of the F_1 progeny of the cross dextral ♀ × sinistral ♂ are dextral, but all of the F_1 progeny of the cross sinistral ♀ × dextral ♂ are sinistral. Thus far, the situation seems similar to the one observed with poky or with chloroplast inheritance. However, the F_2 generation is all dextral from both pedigrees!

The F_3 generation in this case revealed that the inheritance of coiling direction involves nuclear genes rather than extranuclear genes. (The F_3 is produced by individually selfing the F_2 snails; this is possible because snails are hermaphroditic.) Sturtevant found that three-fourths of the F_3 snails were dextral and one-fourth were sinistral (Figure 20-8). This ratio reveals a Mendelian segregation in the F_2 generation. Apparently, dextral (+) is dominant to sinistral (*s*), but Sturtevant concluded that, strangely enough, the shell-coiling phenotype of any individual animal is determined by the genotype (not the phenotype) of its mother! We now know that this happens because the genotype of the mother's body determines the initial cleavage pattern of the developing embryo. Note that the segregation ratios shown in Figure 20-8 would never appear in the phenotypes of true organelle genes. This example shows that one generation of crosses is not enough to provide conclusive evidence that maternal inheritance is due to organelle-based inheritance. The term **maternal effect** can be used to describe the results in cases like the shell-coiling example to distinguish them from organelle-based inheritance.

Extranuclear Genes in *Chlamydomonas*

If an investigator could somehow design an ideal experimental organism with a simple life cycle, the result would be something like the unicellular freshwater alga *Chlamydomonas*. Like fungi, algae rarely have different sexes, but they do have mating types. In many algal and fungal species, two mating types are determined by alleles at one locus. A cross can occur only if the parents are of different mating types. The mating types are physically identical but physiologically different. Such species are called **heterothallic** (literally, "different-bodied"). In *Chlamydomonas*, the mating-type alleles are called mt^+ and mt^- (in *Neurospora* they are A and a; in yeast, a and α). Figure 20-9 diagrams the *Chlamydomonas* life cycle. As you might guess, tetrad analysis (see Chapter 6) is possible and in fact is routinely performed.

In 1954, Ruth Sager isolated a streptomycin-sensitive mutant of *Chlamydomonas* with a peculiar inheritance pattern. In the following crosses, *sm-r* and *sm-s* indicate streptomycin resistance and sensitivity, respectively, and *mt* is the mating-type gene discussed earlier:

$$sm\text{-}r\ mt^+ \times sm\text{-}s\ mt^- \longrightarrow \text{progeny all } sm\text{-}r$$

$$sm\text{-}s\ mt^+ \times sm\text{-}r\ mt^- \longrightarrow \text{progeny all } sm\text{-}s$$

Here we see a difference in reciprocal crosses; all progeny cells show the streptomycin phenotype of the mt^+ parent. Like the maternal inheritance phenomenon, this is a case of **uniparental inheritance.** In fact, Sager now refers to the mt^+ mating type as the female, using this analogy. However, there is no observable physical distinction between the mating types, as there would be if true sex were involved, nor is there a difference in the contribution of cytoplasm as seen in *Neurospora*. In these crosses, the conventional nuclear marker genes (such as *mt* itself) all behave normally and give 1 : 1 progeny ratios. For example, one-half of the progeny of the cross

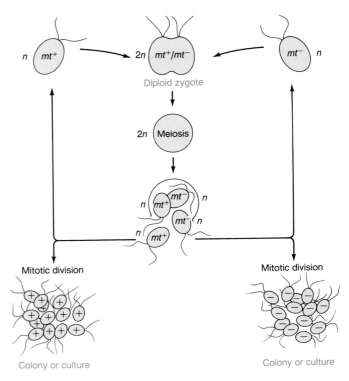

Figure 20-9. Diagrammatic representation of the life cycle of *Chlamydomonas*, a unicellular green alga. This organism has played a central role in research on organelle genetics. The nuclear mating-type alleles mt^+ and mt^- must be heterozygous in order for the sexual cycle to occur.

$sm\text{-}r\ mt^+ \times sm\text{-}s\ mt^-$ are $sm\text{-}r\ mt^+$ and one-half are $sm\text{-}r\ mt^-$.

To determine the sm-s or sm-r phenotypes of progeny, the cells are plated onto a medium containing streptomycin. While performing these experiments, Sager observed that the drug streptomycin itself acts as a mutagen. If a pure strain of *sm-s* cells is plated onto streptomycin, quite a significant proportion of streptomycin-resistant colonies will appear. In fact, treatment with streptomycin will produce a whole crop of different kinds of mutant phenotypes: all of them show uniparental inheritance (and hence presumably are extranuclear). Furthermore, streptomycin treatment does not produce nuclear mutants. Most of the mutants produced by streptomycin treatment show either resistance to one of several different drugs, or defective photosynthesis. The photosynthetic mutants are unable to make use of CO_2 from the air as a source of carbon, so they survive only if soluble carbon (in some form such as acetate) is added to their medium. (How would you go about selecting such mutants?) Table 20-2 lists several different extranuclear mutant types produced by the streptomycin treatment.

TABLE 20-2. Some of the *Chlamydomonas* mutants showing uniparental inheritance

Gene	Mutant phenotype
ac1 – ac4	Requires acetate
tm1 and *tm3 – tm9*	Cannot grow at 35°C
tm2	Conditional: grows at 35°C only in presence of streptomycin
ti1 – ti5	Forms tiny colonies on all media
ery1	Resistant to erythromycin at concentration of 50 μg/ml
kan1	Resistant to kanamycin at concentration of 100 μg/ml
spc1	Resistant to spectinomycin at concentration of 50 μg/ml
spi1 – spi5	Resistant to spiramycin at concentration of 100 μg/ml
ole1 – ole3	Resistant to oleandomycin at concentration of 50 μg/ml
car1	Resistant to carbamycin at concentration of 50 μg/ml
ele1	Resistant to eleosine at concentration of 50 μg/ml
ery3 and *ery11*	Resistant to erythromycin, carbamycin, oleandomycin, and spiramycin (at concentrations listed above for individual drugs)
sm2 and *sm5*	Resistant to streptomycin at concentration of 500 μg/ml
sm3	Resistant to streptomycin at concentration of 50 μg/ml
sm4	Requires streptomycin for survival

NOTE: all of these mutants were produced by treatment with streptomycin except for the *ti* mutants, which were produced by treatment with nitrosoguanidine.
SOURCE: R. Sager, *Cytoplasmic Genes and Organelles*, Academic Press.

These experiments reveal the existence of a mysterious "uniparental genome" in *Chlamydomonas* — that is, a group of genes that all show uniparental transmission in crosses. Where is this uniparental genome located? What is the mechanism of the uniparental transmission? When there seems to be no physical difference between mating types, why are these genes transmitted only by the mt^+ parent?

There is evidence to suggest that the uniparental genome in this case is in fact chloroplast DNA (cpDNA). About 15 percent of the DNA of a *Chlamydomonas* cell in rapidly growing culture is found in the single chloroplast of the cell. This DNA forms a band in a cesium chloride (CsCl) gradient that is distinct from the band of nuclear DNA (Figure 20-10). Furthermore, the precise position of the cpDNA band can be altered by adding the heavy isotope of nitrogen, ^{15}N, to the growth medium. The position of the band can be calibrated in terms of the buoyant density of the cpDNA. Using this technique, the cpDNA of the two parents in a cross can be labeled differently: one light (^{14}N), and one heavy (^{15}N). The buoyant densities of the cpDNA from these parental cells are 1.69 to 1.70, respectively. Although this difference seems small, it provides appreciably different band positions in a CsCl gradient. With differently labeled parents, the zygote DNA can be examined to see how it compares with the parents (Table 20-3). The results indicate that the cpDNA of the mt^- parent is in fact lost, inactivated, or destroyed in some way. This loss of cpDNA from the mt^- parent, of course, parallels the loss of uniparental genes (such as the *sm* genes) borne by the mt^- parent.

Other experiments have been performed using similar logic. For example, crosses can be made between

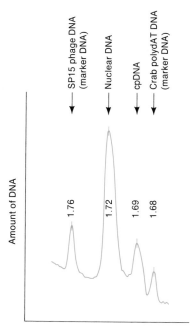

Figure 20-10. The chloroplast DNA (cpDNA) of *Chlamydomonas* can be detected in a CsCl density gradient as a band of DNA distinct from the nuclear DNA. In this particular experiment, two other DNA types were sedimented along with the *Chlamydomonas* DNA to act as known reference points in the gradient. The numbers represent buoyant densities (g/cm³) of the various DNA types. (Figures 20-10 and 20-11 after Ruth Sager, "Genetic Analysis of Chloroplast DNA in Chlamydomonas," in *Cytoplasmic Genes and Organelles*, Academic Press.)

Chlamydomonas strains having cpDNAs with distinctly different restriction-enzyme-digest patterns on electrophoretic gels. Again, the specific cpDNA digest pattern passed on to the progeny is that of the mt^+ parent only.

Message   The cpDNA of *Chlamydomonas* is inherited uniparentally. Because the behavior of this DNA parallels the behavior of the uniparentally transmitted genes, it can be inferred that these genes are located in the cpDNA.

Mapping Chloroplast Genes in *Chlamydomonas*

We have seen (Table 20-2) that *Chlamydomonas* has a large number of uniparentally inherited genes. Is there linkage among these genes? Are they all in one linkage group (on one chromosome), or are they arranged in several linkage groups? Or does each gene assort independently as if it were on its own separate piece of DNA? Of course, the way to approach this question in the true tradition of classical genetics is to perform a recombination analysis. But here we run into a problem. Both sets of parental DNA must be present if there is to be an opportunity for recombination, and we have seen that the cpDNA of the mt^- parent is eliminated in the zygote cell. Luckily, there is a way out of this dilemma.

In crosses of mt^+ *sm-r* $\times$ mt^- *sm-s,* about 0.1 percent of the progeny zygotes are found to contain both *sm-r* and *sm-s.* The presence of both alleles can be inferred from the fact that the products of meiosis of such cells show both sm-r and sm-s phenotypes among their number. Segregation of the two alleles presumably arises from a CSAR process in the zygote or at a later stage of cell division. Such zygotes are called **biparental zygotes,** for obvious regions, and their genetic condition is described as a **cytohet** ("*cyto*plasmically *het*erozygous"). It

TABLE 20-3. Buoyant densities of cpDNA in progeny from various *Chlamydomonas* crosses

Cross	Buoyant density of zygote cpDNA
^{14}N $mt^+ \times$ ^{14}N mt^-	1.69
^{15}N $mt^+ \times$ ^{15}N mt^-	1.70
^{15}N $mt^+ \times$ ^{14}N mt^-	1.70
^{14}N $mt^+ \times$ ^{15}N mt^-	1.69

SOURCE: Ruth Sager, *Cytoplasmic Genes and Organelles,* Academic Press.

appears as though the inactivation of the mt^- parent's cpDNA fails in these rare zygotes. However, whether this is what happens or not, the cytohets provide just the opportunity we need to study recombination: these cells contain *both* sets of parental cpDNA. The rarity of cytohet zygotes does pose a problem for research, but their frequency can be increased by treating the mt^+ parent with ultraviolet light before mating. After such treatment, 40–100 percent of the progeny zygotes are cytohets. (Note in passing that this observation implies that the m^+ cell plays a normal role in actively eliminating the mt^- cell's cpDNA. The ultraviolet irradiation of the mt^+ cell must inactivate such a function.)

Message   In *Chlamydomonas,* biparental zygotes (or cytohets) must be the starting point for all studies on the segregation and recombination of chloroplast genes.

A good map of the cpDNA has been obtained using the cytohets. One of the most profitable ways of studying linkage relationships is through the cosegregation patterns of two separate extrachromosomal genes in the same cross. For example, in a cross of the type mt^+ *sm-s* $ac^+ \times$ mt^- *sm-r* ac^-, biparental zygotes are first obtained. Cytoplasmic segregation can be detected in the two daughter cells arising from each product of meiosis, or in the daughter cells arising from the subsequent few mitotic divisions. It is reasonable to assume that if the two genes are linked closely, then when one gene undergoes cytoplasmic segregation, the other probably will too. More distantly linked genes should show proportionately lower frequencies of cosegregation.

Sager was able to qualify these procedures by using standard cell populations. She showed that a consistent additive-linkage map of the genes in the "uniparental genome" could be obtained. However, the map has one major inconsistency, which can be resolved only by assigning the shape of a circle to the cpDNA molecule! The map resulting from this and other mapping techniques is shown in Figure 20-11. Circular cpDNA has now been demonstrated in several plant species. It is worth recalling the history of classical genetic analysis. The genetic maps for nuclear chromosomes also were derived before their physical counterparts were demonstrated to exist. A great deal more detail is now available on the cpDNA maps in several organisms; we shall take another look at such maps after we have considered the extranuclear genome of yeast.

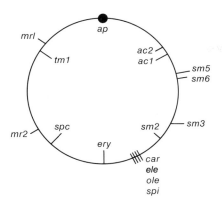

Figure 20-11. Circular map of *Chlamydomonas* cpDNA derived from genetic analysis. The map must be circular to accommodate all the linkage data.

Mitochondrial Genes in Yeast

Possibly the greatest success story in the clarification of extranuclear inheritance has been the development of the current view of the mitochondrial genome in baker's yeast (*Saccharomyces cerevisiae*). This achievement provides a classic example of the power of genetic analysis in combination with modern molecular techniques. Let's start with the contributions of genetic analysis.

Three kinds of mutants have been of particular importance: the petite, ant^R, and mit^- mutants. In the 1940s, Boris Ephrussi and his colleagues first described some curious mutants in yeast. The wild-type cells of yeast form relatively large colonies on the surface of a solidified culture medium. Among these large, or "grande," colonies an occasional small, or "petite," colony is found. When isolated, the **petite mutants** prove to be of three types on the basis of their inheritance patterns. The first type is called **segregational petites** because on crossing to a grande strain, half the ascospores give rise to grande colonies and the other half give rise to petite colonies. This 1:1 Mendelian segregation obviously indicates that the petite phenotype is due to a nuclear mutation in these cases. The second type is the **neutral petites,** which in crosses to a grande strain give ascospores that all grow into grande colonies—a clear case of uniparental inheritance. The third type of petite mutants is the **suppressive petites,** which give some ascospores that grow into grande colonies and some that grow into petite colonies. The ratio of grandes to petites is variable but strain-specific: some suppressive petites give exclusively petite offspring in such crosses. The suppressive petites obviously show a non-Mendelian inheritance pattern, and some also show uniparental inheritance.

In a yeast cross, the two parental cells fuse and apparently contribute equally to the cytoplasm of the resulting diploid cell (Figure 20-12). Furthermore, the inheritance of the neutral and suppressive petites is independent of mating type. In this sense, then, yeast is clearly quite different from *Chlamydomonas*. Nevertheless, since their inheritance is obviously extranuclear, the neutral and suppressive types have become known as the **cytoplasmic petites.**

Several properties of the cytoplasmic petites point to the involvement of mitochondria in their phenotype:

1. In cytoplasmic petites, the mitochondrial electron-transport chain is defective. Because this chain is responsible for ATP synthesis, petites must rely on the less efficient process of fermentation to provide their ATP. If they are placed in a medium containing a nonfermentable energy source such as glycerol, they will not grow.

2. These petites show no mitochondrial protein synthesis. Mitochondria normally possess their own unique protein-synthesizing apparatus, consisting of

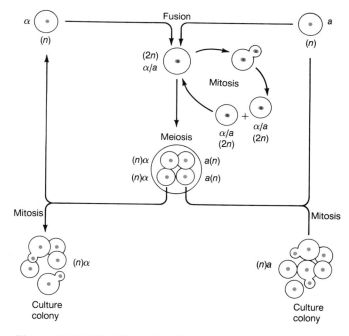

Figure 20-12. The life cycle of bakers' yeast (*Saccharomyces cerevisiae*). The alleles a and α determine mating type and are nuclear alleles. Cell fusion between haploid a and α cells produces a diploid cell. Normally, the cell then goes through a diploid mitotic cycle (budding). However, the cell can be induced (by plating on a special medium) to undergo meiosis (sporulation), producing haploid products. Note that budding involves the formation of a small growth on the side of the parent cell; this bud eventually enlarges and separates to become one of the daugther cells.

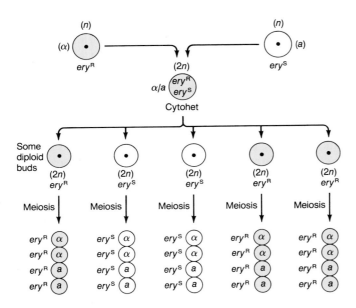

Figure 20-13. The special inheritance pattern shown by certain drug-resistant phenotypes in yeast. When diploid buds are induced to sporulate (undergo meiosis), the products of each meiosis show uniparental inheritance. (Only a representative sample of the diploid buds is shown here.) Note that the nuclear genes, represented here by the mating-type alleles *a* and α, segregate in a strictly Mendelian pattern. The alleles *ery*^R and *ery*^S determine erythromycin resistance and sensitivity, respectively.

a unique set of tRNA molecules and unique ribosomes, all of which are quite different from those operating outside the mitochondrion in the **cytosol,** or the nonorganellar phase of the cytoplasm.

3. Petites have massively altered **mitochondrial DNA (mtDNA).** Mitochondria in all organisms have their own unique mtDNA that, although smaller in amount, is quite different from nuclear DNA. Neutral petites are found to be totally lacking mtDNA, whereas suppressive petites show altered base ratios compared with the grandes from which they spring.

The second major class of yeast mutants, the ***ant*^R mutants,** were initially recognized at the phenotypic level by their resistance to antibiotics supplied in the medium. For example, strains have been obtained that are resistant to chloramphenicol (cap^R), erythromycin (ery^R), spiromycin (spi^R), paramomycin (par^R), and oligomycin (oli^R). These mutations each show a non-Mendelian inheritance pattern similar to that in the suppressive petites; that is, in a cross such as *ery*^R × *ery*^S, a strain-specific non-Mendelian ratio is seen among the random ascospore progeny. However, examination of specific meiosis by tetrad analysis reveals several now-familiar phenomena. Let's trace the cross through its se-

quential stages (Figure 20-13). When the parental cells fuse, the fusion product, as well as being diploid, is a cytohet. The diploid cells can then be allowed to bud mitotically. During this mitotic division, the CSAR process occurs and the daughter cells become either *ery*^R or *ery*^S. Therefore, when meiosis is induced by shifting the cells onto a special medium, all of the haploid products of any single meiosis are identical with respect to erythromycin sensitivity or resistance. At the level of single meioses, then, we see uniparental inheritance at work. (Note that the nuclear mating-type alleles *a* and α always segregate in a 1:1 ratio.) Similar results have been shown by all the *ant*^R mutations, pointing once again to their location in an "extrachromosomal genome."

The third important class of mutants, the ***mit*^− mutants,** were last to be discovered and required the development of special selective techniques. These mutants are similar to petite mutants in that they exhibit small colony size and abnormal electron-transport-chain functions, but they differ in that they have normal protein synthesis and are able to revert. In a way, *mit*^− mutants are like point-mutation petites. The inheritance of *mit*^− mutants is comparable to the patterns shown by *ant*^R types; that is, they show cytoplasmic segregation and also uniparental inheritance at meiosis.

Mapping the Mitochondrial Genome in Yeast

The demonstration of genetic determinants constituting an extranuclear genome immediately raises the question of the physical interrelationship of these determinants. Are they all linked together on one mitochondrial "chromosome," or are they located on separate structural units? Mapping the yeast mitochondrial genome has proceeded using many different approaches. A few representative analytical methods follow.

Recombination Mapping. We can set out to look for recombinants by using a cross between parents differing in two extranuclear gene pairs (a kind of "dihybrid cross"). For example, we can carry out the cross *ery*^R *spi*^R × *ery*^S *spi*^S, allow the resulting diploid cell to bud through several cell generations, and then induce the resulting cells to sporulate. We can then identify the genotype of each bud cell by observing the phenotype common to all its ascospores.

Four genotypes can result from such a cross (Figure 20-14), and all are in fact observed. An early cross yielded the following results:

ery^R *spi*^R	63 tetrads
ery^S *spi*^S	48 tetrads
ery^S *spi*^R	7 tetrads
ery^R *spi*^S	1 tetrad

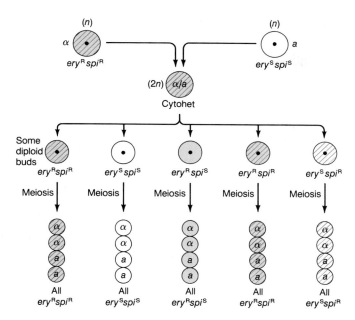

Figure 20-14. The study of inheritance in a cross between yeast cells differing with respect to two different drug-resistance alleles (*ery*^{R} = erythromycin resistance; *spi*^{R} = spiramycin resistance). Each diploid bud can be classified as parental or recombinant on the basis of these results. Note that the identity of the extranuclear genotype for all four products of meiosis confirms that the CSAR process must occur during the production of the diploid buds.

The genotypes *ery*^{S} *spi*^{R} and *ery*^{R} *spi*^{S}, of course, represent recombinants that have been produced in the CSAR process during the formation of diploid buds. (Note that these recombinants cannot have been produced by meiosis because the products of any given meiosis are identical with respect to the drug-resistance phenotypes.)

> **Message** In yeast, cytoplasmic segregation and recombination (CSAR) are achieved during bud formation in diploid cytohets. The segregation and/or recombination may be detected directly in cultures of the diploid buds or may be detected by observing the products of meiosis that result when the buds are induced to sporulate.

We seem to be just a short step away from the development of a complete map of all the drug-resistance genes in yeast. Unfortunately, the technique of recombination mapping has proved to be of only limited usefulness. For one thing, recombination involving mitochondria has proved to be a population phenomenon, similar to phage recombination. This is to say, so many rounds of recombination are occurring that most genes appeared to be unlinked. Linkage is detectable only for genes that are very close together. Furthermore, the process of recombination has been shown to be strongly influenced by a specific genetic factor, ω (omega), that is present in some mitochondrial genomes and not in others. The most useful mapping developments were still to come from less conventional analyses, examples of which are described in the following subsections.

Mapping by Petite Analysis. The petite mutations, the drug-resistance (*ant*^{R}) mutations, and the *mit*$^-$ mutations are apparently inherited on the mitochondrial genome of yeast. Some very effective techniques for mapping that genome have been developed through the combined study of these classes of mutants. Most of these approaches are based on the fact that petites represent deletions of the mtDNA. This fact opens up a new and different kind of genetic analysis that has been combined with new techniques of DNA manipulation to produce a rather complete genetic map of the mtDNA.

The pivotal observation came in studies where drug-resistant grande strains (such as *ery*^{R}) were used as the starting material for the induction of petite mutants. We have seen that petite mutants form spontaneously, but they can also be induced at high frequencies by the use of various specific mutagens, notably ethidium bromide. It is of interest to determine the drug resistance of petite cultures obtained from grande *ery*^{R} strains. Are they still *ery*^{R}, or has the petite mutation caused them to become *ery*^{S}? Unfortunately, the answer cannot be determined directly because drug resistance cannot be tested in petite cells!

However, genetic trickery comes to the rescue. The induced petites are combined with grande *ery*^{S} cells (which, of course, lack resistance to erythromycin) to form diploids, and the diploids are allowed to form diploid buds. Depending on the type of petite used, varying amounts of petite and grande diploid buds are formed. It is the grande diploid buds that are of interest here because they *can* be tested for drug resistance. If the original petites retain the drug resistance, then some of these diploid grande cells may be expected to have acquired that resistance through recombination during the CSAR process. In fact, for some petites, the derived grande diploid buds do prove to be of two phenotypes: *ery*^{S} (derived from the grande *ery*^{S} haploid) and *ery*^{R} (which must be derived from the petite haploid). This result indicates that only the genetic determinants for the grande phenotype are lost in the original mutation event that generates these particular petite mutants; the genetic determinants for drug resistance were not affected by the petite mutation.

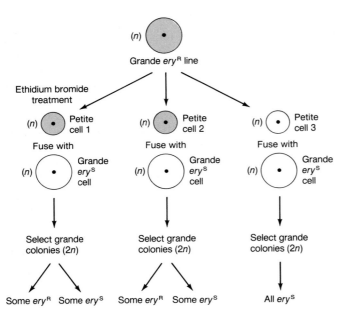

Figure 20-15. The retention or loss of drug resistance (ery^R) when petite mutants are induced from a drug-resistant grande culture. The petite cannot be tested directly for drug resistance, so any remaining drug resistance must be "rescued" through the mating of petite cells with drug-sensitive grande cells. Diploid buds can then be tested for drug resistance. In the examples shown here, the drug resistance is apparently retained in petite cells 1 and 2, but the drug resistance is apparently lost, together with the determinant for the grande phenotype, during the induction of the petite mutation in petite cell 3.

Of particular interest, however, are those petites with derived grande diploids that are all ery^S. In this case, the original mutation event apparently inactivates or destroys both the determinant for the grande phenotype *and* the determinant for drug resistance (Figure 20-15). The coincidental loss of several genetic determinants is characteristic of deletion mutations in general genetic analysis. In fact, it is now known that the petite phenotype is produced by large deletions of the mtDNA.

The discovery of marker loss in petites is interesting for two main reasons. First, because the petite mutation alters the mtDNA and causes the loss of a particular marker gene, the location of that marker can be assigned to the mtDNA beyond any reasonable doubt. This procedure can be used a routine test to detect a mitochondrial mutation.

Message Marker loss during petite induction is a good operational test for a gene located in mtDNA.

Second, the phenomenon of marker loss in petites has given rise to several mapping techniques, as follows. With one technique, it is possible to obtain a grande cell line that is resistant to several drugs; such a line might have the genotype $ery^R \, cap^R \, oli^R \, spi^R$. Petites can be induced in this line and then tested to see which of the resistance markers are retained in the petite lines. (Of course, this test must be performed by "rescuing" the markers in a cross with a drug-sensitive grande cell, as we have seen.) It then becomes a relatively simple matter to compare the frequencies with which various pairs of resistance genes are either retained or lost together, thus obtaining a good idea of which genes are closely linked. This idea can be extended in another mapping technique, which combines genetic analysis with physical techniques. Two different petite strains are derived from a grande line that shows multiple drug resistance. For example, suppose that strain A has retained cap^R, ery^R, and par^R, whereas strain B has retained cap^R, ery^R, oli^R, and ana^R. DNA is extracted from each strain and denatured. This DNA then is hybridized to a standard sample of grande DNA, both in individual experiments for DNA from each strain and with a mixture of DNA from A and B. Suppose that the amounts hybridized are a units in the case of A alone, b units in the case of B alone, and t units for the mixture of A and B.

Anthony Linnane used these values to measure the degree of overlap (h) of the petite deletions. From Figure 20-16, we see that if the retained (nondeleted) DNA

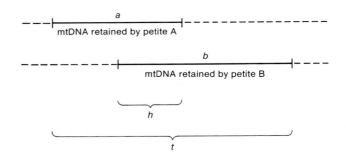

Figure 20-16. Measuring the extent of overlap between any two petite mtDNA's by hybridizing to a sample of grande mtDNA. In the case of petite A, a units of petite mtDNA hybridize to the grande mtDNA; in the case of petite B, b units hybridize. We assume that there is some overlap (homology) between the mtDNA's retained by the two petites. In that case, the total amount (t units) of a mixture of the petite mtDNA's that hybridizes to the grande mtDNA must be less than the sum of the individual totals by an amount h, so that $h = a + b - t$. This value h is proportional to the amount of overlap. When many pairs of different petites are studied in this way, the relative sizes and positions of these different mtDNA's can be fitted together with information about the genes retained in common by each pair to produce a map of the mitochondrial genome.

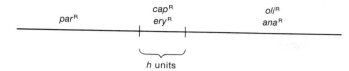

Figure 20-17. Genetic markers retained by specific petites can be correlated with regions of petite homology. Since petite induction causes the deletion of a part of the mitochondrial DNA, different petites will retain different genes. In the example, one petite retains cap^R, ery^R, and par^R and another retains cap^R, ery^R, oli^R, and ana^R. The *cap* and *ery* loci must be within the overlap region of size *h*, as determined by DNA hybridization techniques. The loci *par*, *oli*, and *ana* must lie somewhere outside this region, and their relative positions can be assigned through study of other pairs of petites. In this manner, a map of genetic markers can be built up.

in the two petites has a region in common, then the amount of DNA hybridized from the mixture will be less than the sum of the amounts bound when the petite DNA's are hybridized individually. In fact, a study of the diagram should convince you that this difference equals the amount of overlap, so that $h = a + b - t$. Because cap^R and ery^R are the only two alleles retained in both of these petites, we know that these genes must be located in the overlap region. Therefore, the value *h* is proportional to the maximum possible distance between *cap* and *ery*. Thus, we can insert genetic markers into map segments of defined size, and we have begun a comprehensive mapping process (Figure 20-17). This kind of

mapping has been extended to many petites, and a "library" of petites with well-defined retained regions has been established. These results have been combined with the results of marker-retention analysis to produce a complete genetic map of the mtDNA. This map is a circle. Electron-microscopic observation of mtDNA molecules also shows that they are circular.

We have stated that the petite mtDNA is a small retained piece from the grande mtDNA. However, we must reconcile this view with the fact that a typical petite cell contains more or less the same amount of mtDNA as does a typical grande cell, and petite mtDNA circles can be the same size as grande mtDNA circles. The answer to this puzzle is that the retained piece in the petite cell is present in as many tandem copies as it requires to make up an mtDNA of about the same length as the grande mtDNA (Figure 20-18). This model does not require any change in the interpretation of the petite-overlap experiments.

Once a defined set of petite deletions has been worked out, they can be very useful in subsequent mapping. For example, many *mit⁻* mutants have been mapped by fusing *mit⁻* with several specific petites. If the *mit⁻* mutation is in the region of mtDNA retained by the petite cell, then recombination can substitute petite DNA in place of the *mit⁻* DNA and the recombinant cell will be able to grow on a nonfermentable energy source.

Message   Marker loss or retention during petite induction and the overlap of petite deletions are novel genetic techniques used in mapping mitochondrial genomes.

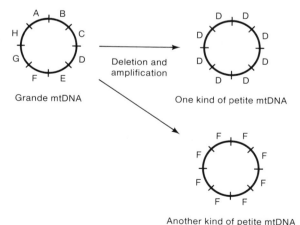

Figure 20-18. When a petite is produced from a grande cell, a large region of mtDNA may be deleted. Apparently, the DNA region retained by the petite (D or F in these examples) is amplified through tandem duplication to provide a chromosome of approximately the normal length.

Mapping by Restriction-Enzyme Analysis.   As we saw in Chapter 15, the availability of restriction enzymes provides a powerful new tool for the genetic analyst. This tool has been particularly useful in the analysis of mtDNA—not only in yeast (which is discussed here) but also in the analysis of any mtDNA or cpDNA that can be extracted in pure form from any organism.

A restriction enzyme attacks DNA and cuts it at a specific base sequence. These specific target sequences of the enzymes may, in a sense, be considered genetic markers. The positions of these target sequences may be mapped (Figure 20-19). Although this map is not particularly interesting in itself, it is an essential tool for the precise pinpointing of other genetic markers (such as drug-resistance genes) on the mtDNA.

One obvious way of using restriction analysis to map genes on mtDNA is in the case of the various unique mitochondrial RNA's, tRNA's, and rRNA's. A radioac-

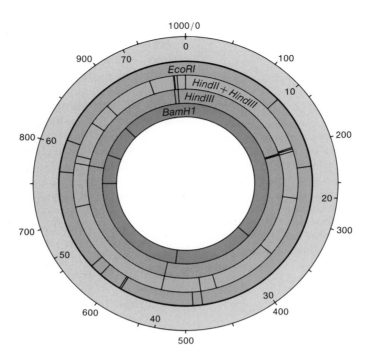

Figure 20-19. Map showing the target sites of various restriction enzymes on the yeast mtDNA. An arbitrary scale from 0 to 1000 is indicated on the outside of the outer circle; a scale in terms of thousands of DNA base pairs is indicated on the inside of the outer circle. The inner four bands show the fragments produced by treatment with particular restriction enzymes: *EcoRI, HindII + HindIII, HindIII,* and *BamHI.* The position for the zero point of this map is arbitrary. (After J. P. M. Sanders et al., "The Organization of Genes in Yeast Mitochondrial DNA, III." *Molecular and General Genetics* 157, 1977.)

tively labeled RNA can be used as a probe in a Southern hybridization (see page 407) to a restriction-enzyme digest. The band that "lights up" on autoradiography represents the approximate location of the RNA. Further similar experiments using different restriction enzymes will narrow the region of hybridization down to a precise locus on the mtDNA.

What about the *ant*[R] and *mit*[−] genes? One approach is to correlate the retained restriction fragments in a number of different petites with, say, the antibiotic-resistance markers also retained in those strains. In this way, specific markers may be associated with specific regions of the mtDNA, as defined by the restriction-enzyme target sites. Another approach is to take a petite that retains only a single drug-resistance gene — say, *ery*[R]. The mtDNA is extracted from the petite and put into an *E. coli* transcription system in vitro that makes RNA from DNA templates. The resulting RNA is complementary to the DNA being studied, so it is called complementary RNA (cRNA). This cRNA can be la-

beled through the use of radioactive uracil and then used as a probe. Its hybridization with the various restriction fragments can be tested, thus locating the portion of the mtDNA that is retained by this particular petite — and hence the locus of the *ery* gene. For example, if the cRNA from the *ery*[R] petite hybridizes with fragment X from one restriction enzyme and with fragment Y from another enzyme, then the *ery* locus must be located within the overlap region of these two fragments.

An Overview of the Mitochondrial Genome

Many specific areas of the yeast map have been subjected to more detailed dissection, such as sequencing and intron analysis. The present state of the yeast mtDNA map, based on the types of techniques we have consid-

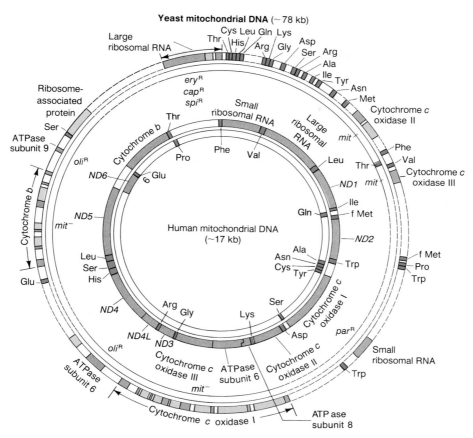

Figure 20-20. Maps of yeast and human mtDNA's. Each map is shown as two concentric circles corresponding to the two strands of the DNA helix. The human map has been produced exclusively by physical techniques. The yeast map has been produced by a combination of genetic and physical techniques, as discussed in the text. Note that the mutants used in the yeast genetic analysis are shown opposite their corresponding structural genes. Gray = exons and uninterrupted genes, red = tRNA genes, and pink = URF's (unassigned reading frames). tRNA genes are shown by their amino-acid abbreviations; ND genes code for subunits of NADH dehydrogenase. (NOTE: the human map is not drawn to the same scale as the yeast map.)

ered, is shown in Figure 20-20, together with the human mtDNA map for comparison. The various genes in yeast discussed in this chapter are shown together with their protein products. We can see that cap^R, ery^R, and spi^R are in fact alterations of the large mitochondrial rRNA genes, and that par^R is associated with changes in the small rRNA gene. Other ant^R mutations such as oli^R are associated with alterations in various subunits of the enzyme ATPase. On the other hand, mit^- mutations are in fact lesions in several subunits of cytochrome oxidase (I–III) or in the cytochrome-b gene. In addition, several other genes are indicated, such as those for the mitochondrial tRNA's and a gene for a ribosome-associated protein.

The map contains some surprises too. Most prominent is the occurrence of introns in several genes. Sub-

unit I of cytochrome oxidase contains nine introns! The discovery of introns in the mitochondrial genes is particularly surprising because they are relatively rare in yeast nuclear genes. Another surprise is the occurrence of long **unassigned reading frames (URF's).** These are sequences that have correct initiation codons and are uninterrupted by stop codons. These "genes in search of a function" are the subject of intense current research. Some URF's occurring within introns appear to be involved in specifying proteins important in the splicing out of the introns themselves at the RNA level. Notice that the human mtDNA is by comparison much smaller and more compact. There seems to be much less spacer DNA between the genes.

The overall view of the mitochondrial genome shows it to have two main functions: it codes for some proteins that are actually in or associated with the electron-transport chain, and it codes for some proteins, all the tRNA's, and both rRNA's necessary for mitochondrial protein synthesis. But in both of these processes, it is striking that the remaining necessary components are encoded by nuclear genes with mRNA that is translated on cytosolic ribosomes and that the products are transported to the mitochondrion (see Figure 20-21). Why this peculiar division of labor exists between nuclear and mitochondrial DNA is not known. Another curiosity is that some specific subunits are encoded by mtDNA in one organism but by nuclear DNA in another. Evidently, a transposition of information occurs between these organelles. There is also evidence of transposition between mitochondria and chloroplasts. Furthermore,

inactive "pseudogenes" are detectable in the nucleus, showing homology with mitochondrial genes.

Altogether, 25 yeast and 22 human mitochondrial tRNA's are shown on the maps in Figure 20-20. These tRNA's carry out all the translation that occurs in mitochondria. This is far less than the minimum of 32 required to translate nucleus-derived mRNA. The economy is achieved by a "more wobbly" wobble pairing (see page 339) of tRNA anticodons. The tRNA specificities in human mtDNA are shown in Figure 20-22. Notice that the codon assignments are in some cases different from the nuclear code. It is also known that there is variation between the mitochondria of different species. Hence, the genetic code is obviously not universal, as had been supposed for many years.

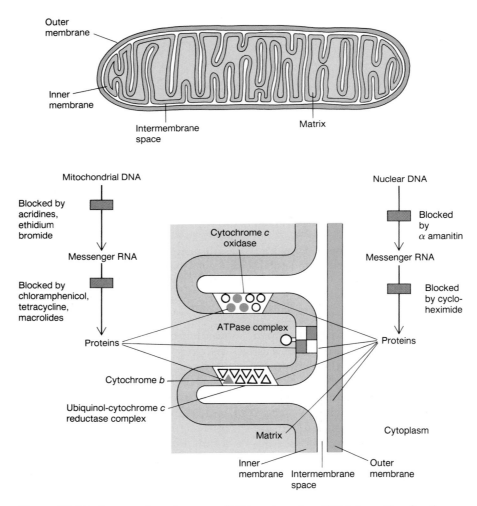

Figure 20-21. Cooperation of yeast mtDNA and nuclear DNA in coding for the protein components of the inner mitochondrial membrane. (a) Overview of mitochondrial structure. (b) Details of membrane constitution. Three functional units are involved in the cooperation. The mtDNA supplies six proteins, three subunits of cytochrome-*c* oxidase, two subunits of ATPase, and cytochrome *b*.

First letter	Second letter				Third letter
	U	C	A	G	
U	Phe	Ser	Tyr	Cys	U
	Phe	Ser	Tyr	Cys	C
	Leu	Ser	Stop	(Stop) Trp	A
	Leu	Ser	Stop	Trp	G
C	Leu	Pro	His	Arg	U
	Leu	Pro	His	Arg	C
	Leu	Pro	Gln	Arg	A
	Leu	Pro	Gln	Arg	G
A	Ile (Met)	Thr	Asn	Ser	U
	Ile	Thr	Asn	Ser	C
	(Ile) Met	Thr	Lys	(Arg) Stop	A
	Met	Thr	Lys	(Arg) Stop	G
G	Val	Ala	Asp	Gly	U
	Val	Ala	Asp	Gly	C
	Val	Ala	Glu	Gly	A
	Val	Ala	Glu	Gly	G

Figure 20-22. The genetic code of the human mitochondrion. The functions of the 22 tRNA types are shown by the 22 non-*stop* codon boxes.

Message Mitochondrial DNA has genes for mitochondrial translation components (mainly rRNA's and tRNA's) and for some subunits of the proteins associated with mitochondrial ATP production. Less understood regions include the introns, unassigned reading frames, and spacer DNA.

An Overview of the Chloroplast Genome

Although many of the concepts behind chloroplast inheritance have been derived from the kind of genetic studies we have followed in *Mirabilis* and *Chlamydomonas*, the current perspective on the overall organization of cpDNA has come mainly from molecular studies. In fact, at the time of this writing, the cpDNA of two species has been completely sequenced. One example (Figure 20-23) shows the organization of the cpDNA from the liverwort *Marchantia polymorpha*.

Typically, cpDNA molecules range from 120 to 200 kb in different plant species. In *Marchantia*, the molecular size is 121 kb. Figure 20-23 shows the functions of most of the genes in the *Marchantia* cpDNA. However, the sequencing of 121,000 nucleotides does not automatically tell us what genes are present. The presence of

genes is inferred by the detection of **open reading frames (ORF's)** — long sequences that begin with a start codon but are uninterrupted by stop codons except at the termini. Determining what these specific genes code for is more difficult; although most ORF's have been assigned, some of the ORF's are still unassigned reading frames (URF's). Most of the protein-encoding genes have been assigned by comparing the predicted amino-acid sequences with those of known chloroplast, mitochondrial, and bacterial proteins. Other genes, such as the rRNA's and tRNA's, have been assigned by hybridization experiments, using known genes from other organisms as probes.

There are about 136 genes in the *Marchantia* molecule, including four kinds of rRNA, 31 kinds of tRNA, and about 90 protein genes. Of the 90 protein genes, 20 code for photosynthesis and electron-transport functions. Genes coding for translation functions take up about one-half of the chloroplast genome and include the proteins and RNA types necessary for translation to occur in the organelle.

As is true of the mitochondrion, the cpDNA and nuclear DNA cooperate to provide subunits for functional proteins used inside the chloroplast. The nuclear components are translated outside in the cytosol and then transported into the chloroplast, where they are assembled together with the components synthesized in the organelle.

Also, notice the presence of a large inverted repeat in Figure 20-23. Such inverted repeats are found in the cpDNA of virtually all species of plants. However, there is some variation as to which genes are included in the inverted repeat region and therefore in the relative size of that region. One of the mysteries of the inverted repeat is that the duplicates are always identical within a species, yet to date no mechanism is known that ensures this complete identity.

Extragenomic plasmids in Eukaryotes

We saw in Chapter 10 that plasmids are common in bacteria. However, plasmids are also regularly, although less commonly, encountered in eukaryotes. Because plasmids are generally not associated with nuclear chromosomes, they can show a kind of non-Mendelian inheritance pattern similar in some ways to that of organelle DNA.

Most eukaryotic plasmids are silent at the phenotypic level and can only be detected using molecular techniques. The best known of these is the "two-micron circle" (2-μ) plasmid of yeast. This circular plasmid is probably located in the nucleoplasm. If a haploid strain

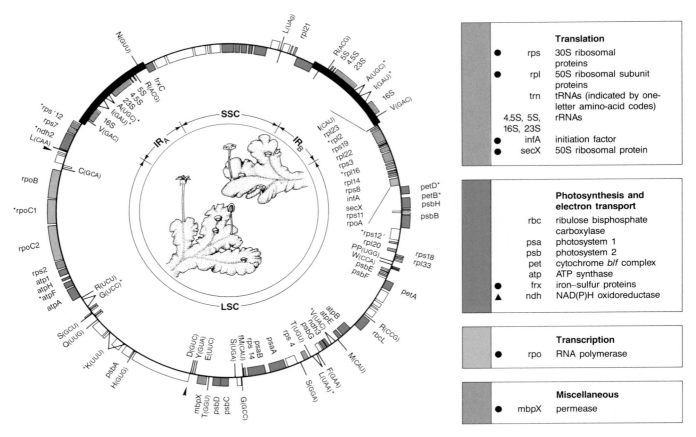

Figure 20-23. The chloroplast genome of the liverwort *Marchantia polymorpha.* IR$_A$ and IR$_B$, LSC and SSC on the inner circle indicate the inverted repeats, large single-copy and small single-copy regions, respectively. Genes shown inside the map are transcribed clockwise, and those outside are transcribed anticlockwise. A region flanked by two arrow heads is inverted in tobacco chloroplast DNA. Each of the tobacco IR regions carries *rps12, rps7, ndhB* (corresponds to *ndh2*), *trnL (CAA)*, *trnI (CAU)*, *rpl23, rpl2,* and several unidentified ORFs in addition to those found in the liverwort IR. Genes for rRNAs in the IR regions are represented by 16S, 23S, 4.5S and 5S, respectively. Genes for tRNAs are indicated by the one-letter amino acid code with the unmodified anticodon. Protein genes identified are indicated by gene symbols, and the remaining open boxes represent unidentified ORFs, approximately to scale. Genes containing introns are marked with asterisks. The boxes surrounding the gene map summarize the functions of the genes identified to date; groups with related functions are shown in different shades ORFs are shown in color. Genes identified on the basis of homology with genes from bacterial or mitochondrial genomes are marked ● and ▲, respectively. The central drawing depicts a male *(above)* and a female *(below) Marchantia* plant. The antheridia and archegonia are elevated on specialized stalks above the thallus, which contains the chloroplasts. *Marchantia* can also reproduce asexually: disks of green tissue (gemmae) grow from the bottom of cup-shaped structures on the thallus surface. When mature, the gemmae separate from the thallus and grow to produce new gametophyte plants. (From K. Umesono and H. Ozeki, *Trends in Genetics* 3, 1987.)

containing the plasmid is fused with another haploid strain lacking the plasmid, the asexual or sexual descendants all tend to have the plasmid. Although mysterious biologically, the 2-μ plasmid has assumed a very prominent role in the molecular biology of yeast because it has been genetically engineered to act as a gene vector for that organism (see Chapter 15). When nuclear DNA is spliced into the 2-μ vector and used to transform a recipient cell type, the insert is carried to the nucleus, where it may become incorporated into the chromosomes via a variety of recombination mechanisms.

Another interesting and well-studied plasmid type is associated with male sterility in corn. Male sterility in plants is of great importance in agriculture. Plants with the male-sterile trait produce no functional pollen. Interestingly, when male steriles are crossed as female parents using normal fertile plants as males, all of the progeny prove to be male-sterile and the inheritance pattern

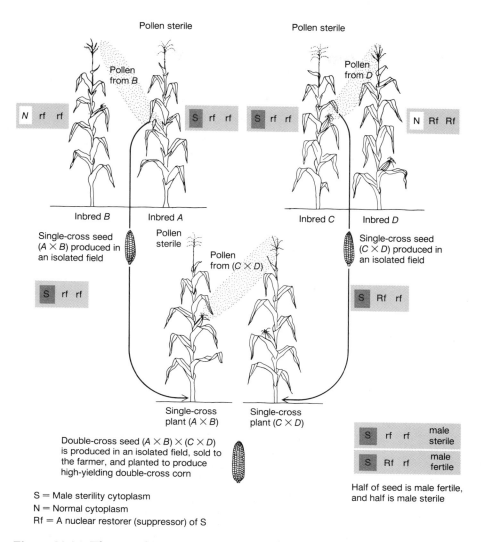

Pollen sterile

Pollen from B

Pollen sterile

Pollen from D

N rf rf

S rf rf

S rf rf

N Rf Rf

Inbred B

Inbred A

Inbred C

Inbred D

Single-cross seed (A × B) produced in an isolated field

Pollen sterile

Pollen from (C × D)

Single-cross seed (C × D) produced in an isolated field

S rf rf

S Rf rf

Single-cross plant (A × B)

Single-cross plant (C × D)

Double-cross seed (A × B) × (C × D) is produced in an isolated field, sold to the farmer, and planted to produce high-yielding double-cross corn

| S | rf | rf | male sterile |
| S | Rf | rf | male fertile |

Half of seed is male fertile, and half is male sterile

S = Male sterility cytoplasm
N = Normal cytoplasm
Rf = A nuclear restorer (suppressor) of S

Figure 20-24. The use of cytoplasmic male sterility to facilitate the production of hybrid corn. In this scheme, the hybrid corn is generated from four pure parental lines: *A, B, C,* and *D*. Such hybrids are called double-cross hybrids. At each step, selfing is prevented by appropriate combinations of cytoplasmic genes and nuclear restorer genes to ensure that the female parents will be male-sterile or pollen-sterile. (From J. Janick et al., *Plant Science.* Copyright © 1974 by W. H. Freeman and Company.)

is exclusively maternal. Therefore, we can conclude that the determinants of male sterility are located cytoplasmically. Indeed, the best-studied type of male sterility in corn is associated with two linear plasmids, S1 and S2, that are located within the mitochondria in addition to the mtDNA. The precise way in which the S1 and S2 plasmids are involved with male sterility is not yet fully understood, although much is known about the molecular properties of these plasmids. One intriguing property is that S1 and S2 can recombine with the mtDNA.

Male sterility in corn and other crop plants is used to facilitate the production of hybrid seed issuing from a cross between genetically different lines; such seeds usually result in larger, more vigorous plants. The big problem is how to prevent self-pollination, which interferes with the production of hybrid seed. Male sterility is useful here because selfing is impossible since there is no functional pollen. One breeding scheme is illustrated in Figure 20-24.

Our final example of eukaryotic plasmids is a type that determines senescence (aging) in the fungus *Neurospora.* Most *Neurospora* strains do not senesce and, given a large enough supply of medium, will keep growing forever. However, certain populations contain senescent

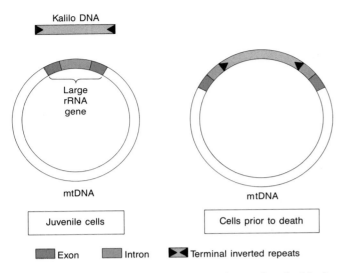

Figure 20-25. Senescence is *Neurospora* is associated with the insertion of a linear plasmid called kalilo DNA into the mtDNA. Here insertion is shown occurring in the intron of a gene for rRNA, but other insertion sites are known.

individuals. One such population is from Hawaii, where the senescent strains are called "kalilo," a Hawaiian word meaning "at death's door." Kalilo strains will grow only a certain distance through the culture medium before they die. However, crossing these strains before death occurs reveals that the ability to die is inherited in a strict maternal fashion. It has been found that the senescence determinant is a linear 9-kb plasmid called kalDNA. This plasmid can exist autonomously — and apparently innocuously — in the cytoplasm, but the onset of death is preceded by the physical insertion of kalDNA into the mtDNA. At death, most mtDNA molecules contain the full-length kalDNA insert (Figure 20-25). Presumably, death occurs because the kalDNA interferes with mitochondrial function. It can therefore be seen that kalDNA behaves like a molecular parasite.

Notice that the plasmids are associated with a definite phenotype in the examples from corn and *Neurospora*, whereas the yeast 2-μ plasmid has no detectable phenotypic manifestation. The existence and evolutionary position of eukaryotic plasmids is still a mystery, but the study of the structure and behavior of such elements will undoubtedly provide important clues about the fundamental properties of the genetic material.

> **Message** Extragenomic eukaryotic plasmids are inherited in a non-Mendelian manner that often mimics the inheritance patterns of organelle DNA.

How Many Copies?

For the genetic and the biochemical approaches to the study of organelle genomes that we have examined, it does not matter much how many copies of organelle genome are present per cell. (Even in the case of the nuclear genome, the number of genomes present per chromosome was in doubt until very recently, and many uncertainties still exist in that area.) However, it is interesting to ask how many copies of the organelle genome do exist, and an answer can be given.

The number of copies turns out to vary among species. More surprisingly, it also can vary within a single species. The leaf cells of the garden beet have about 40 chloroplasts per cell. The chloroplasts themselves contain specific areas that stain heavily with DNA stains; these areas are called **nucleoids,** and they are a feature commonly found in many organelles. Each beet chloroplast contains 4 to 18 nucleoids, and each nucleoid can contain from 4 to 8 cpDNA molecules. Thus, cells of a single beet leaf can contain as many as $40 \times 18 \times 8 = 5760$ copies of the chloroplast genome! Although *Chlamydomonas* has only one chloroplast per cell, the chloroplast contains 500 to 1500 cpDNA molecules, commonly observed to be packed in nucleoids.

A "typical" haploid yeast cell can contain 1 to 45 mitochondria, each having 10 to 30 nucleoids, with four or five molecules in each nucleoid.

How does this genome duplication relate to the CSAR process? How many genomes are present at the beginning of CSAR? Do all copies of the genome actually become involved in the CSAR process? Does CSAR occur when organelles fuse, and must the nucleoids fuse also? It seems that the number of organelles in a cell generally is too large to permit cytoplasmic segregation by the random assortment of organelles. How can the CSAR process segregate the many copies of the genome scattered through the cell? Few answers to such questions are available at present. The situation is rather like the one faced by the early geneticists, who knew about genes and linkage groups but knew nothing about their relation to the process of meiosis. As we have seen, such problems pose few limitations to progress in genetic research; they are problems for the cell biologist.

Summary

■ The extranuclear genome supplements the nuclear genome of eukaryotic organisms which we examined in earlier chapters. The existence of the extranuclear genome is recognized at the genetic level chiefly by uniparental transmission of the relevant mutant phenotypes. At the cellular level, a combination of genetic and biochemical techniques has demonstrated that the extranuclear genome is organelle DNA — either mitochondrial (mtDNA) or chloroplast (cpDNA).

The mtDNA of yeast is the best-understood organelle DNA. This DNA codes for unique mitochondrial translation components, and it also codes for some components of the respiratory enzymes normally found in the mitochondrial membranes. Mutations in the translational-component genes typically produce drug-resistant phenotypes; mutations in the respiratory-enzyme genes typically produce phenotypes involving respiratory insufficiency.

The chloroplast DNA is larger (about 2.5 times greater in diameter) and more complex than the yeast mtDNA. Mutations in cpDNA typically produce photosynthetic defects of drug resistance.

The organelle genes have been fully investigated in only a few organisms, but these studies have produced general analytical techniques that should have widespread application in the study of extranuclear genomes in many other organisms. Eukaryotic plasmids are extragenomic elements that are also inherited cytoplasmically but do not show the sorting-out processes of organelle genomes.

■ ■ ■

Solved Problems

1. In a mating type + strain of *Chlamydomonas,* a temperature-sensitive mutation arises that renders cells unable to grow at higher temperatures. This mutant is crossed to a wild-type stock, and all the progeny of both mating types, are temperature-sensitive. What can you conclude about the mutation?

Solution

We are told that the mutation arose in a mt^+ stock. Therefore, the cross must have been

$$mt^+, ts \times mt^-, ts^+$$

and the progeny must have been mt^+, ts and mt^-, ts. This is a clear-cut case of uniparental inheritance from the mt^+ parent to all the progeny. In *Chlamydomonas,* this type of inheritance pattern is diagnostic of genes in chloroplast DNA, so the mutation must have occurred in the chloroplast DNA.

2. Due to evolutionary conservation, organelle DNA shows homology across a wide range of organisms. Consequently, DNA probes derived from one organism often hybridize with the DNA of other species. Two probes derived from the cpDNA and mtDNA of a fir tree are hybridized to a Southern blot of the restriction digests of the cpDNA and mtDNA of two pine trees, R and S, that had been used as parents in a cross. The autoradiograms follow (numbers are in kb):

The cross R ♀ × S ♂ is made, and 20 progeny are isolated. They are all identical in regard to their hybridization to the two probes. The autoradiogram for each progeny is

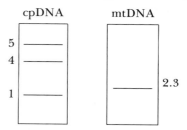

a. Explain the probe hybridizations of parents and progeny.

b. Explain the progeny results. Compare and contrast them with the results in this chapter.

c. What do you predict from the cross S ♀ × R ♂?

(NOTE: This question is based on results shown in several conifer species.)

Solution

a. For both cpDNA and mtDNA, the total amount of DNA hybridized by the probe is different in plants R and S. Hence, we can represent the DNA something like this (other fragment arrangements are possible):

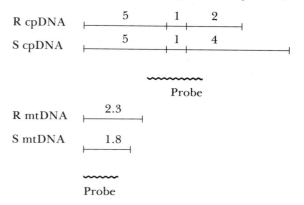

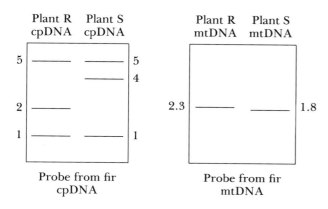

Thus, the probes reveal a (presumably neutral) restriction-fragment-length polymorphism of both the cpDNA and the mtDNA. These are useful organelle markers in the cross.

b. We can see that all of the progeny have inherited their mtDNA from the maternal parent R, because all show the same R mtDNA fragment hybridized by the probe. This is what we might have predicted, based on the predominantly maternal inheritance encountered in this chapter. However, cpDNA is apparently inherited exclusively paternally, because all progeny show the 5/4/1 pattern of the paternal plant S. This is surprising, but it is the only explanation of the data. In fact, all of the gymnosperms studied so far show paternal inheritance of cpDNA. The cause is unknown, but the phenomenon contrasts with that in angiosperms.

c. From this cross, we can predict that all progeny will show the paternal 5/2/1 pattern for cpDNA and the maternal 1.8-kb band for mtDNA.

Problems

1. How do the nuclear and organelle genomes cooperate at the protein level?

2. Name and describe two tests for cytoplasmic inheritance.

3. What is the basis for the green-white color variegation in the leaves of *Mirabilis*? If the cross is made

$$\text{variegated } ♀ \times \text{ green } ♂$$

what progeny types can be predicted? What about the reciprocal cross?

4. In *Neurospora* the mutant *stp* exhibits erratic stop-start growth. The mutant site is known to be in the mitochondrial DNA. If a *stp* strain is used as the female parent in a cross to a normal strain acting as the male, what type of progeny can be expected? What about the progeny from the reciprocal cross?

5. If a yeast cell carrying an mtDNA-based, antibiotic-resistance mutation is crossed to a normal cell and tetrads are produced, what ascus types can you expect with respect to resistance?

6. A new antibiotic mutation (ant^R) is discovered in a certain yeast. Cells of genotype ant^R are treated with ethidium bromide, and petite colonies are obtained. Some of these petites prove to have lost the ant^R determinant.

a. What can you conclude about the location of the ant^R gene?

b. Why didn't all of the petites lose the ant^R gene?

7. Two corn plants are studied. One is resistant (R) and the other is susceptible (S) to a certain pathogenic fungus. The following crosses are made, with the results shown:

$$S ♀ \times R ♂ \longrightarrow \text{progeny all R}$$
$$R ♀ \times S ♂ \longrightarrow \text{progeny all S}$$

What can you conclude about the location of the genetic determinants of R and S?

8. In *Chlamydomonas*, a certain probe picks up a restriction-fragment-length polymorphism in cpDNA. There are two morphs, as follows:

Morph 1: two bands, sizes 2 and 3 kb

Morph 2: two bands, sizes 3 and 5 kb

If the following crosses are made

$$mt^+ \text{ morph 1} \times mt^- \text{ morph 2}$$
$$mt^+ \text{ morph 2} \times mt^- \text{ morph 1}$$

what progeny types can be predicted from these crosses? Be sure to draw the DNA morphs with their restriction sites. Also draw a sketch of the autoradiogram.

9. In yeast, the following cross is dihybrid for two mitochondrial antibiotic genes:

$$MATa \; oli^R \; cap^R \times MAT\alpha \; oli^S \; cap^S$$

(*MATa* and *MATα* are the mating-type alleles in yeast.) What types of tetrads can be predicted from this cross?

10. In the genus *Antirrhinum*, a yellowish leaf phenotype called prazinizans (pr) is inherited as follows:

$$\text{normal} \times \text{pr} \longrightarrow 41,203 \text{ normal} + 13 \text{ variegated}$$
$$\text{pr} \times \text{normal} \longrightarrow 42,235 \text{ pr} + 8 \text{ variegated}$$

Explain these results based on a hypothesis involving cytoplasmic inheritance. (Explain both the majority *and* the minority classes of progeny.)

11. You are studying a plant with tissue comprised of both green and white sectors. You wish to decide whether this phenomenon is due to (1) a chloroplast mutation of the type discussed in this chapter, or (2) a dominant nuclear mutation that inhibits chlorophyll production and is present only in certain tissue layers of the plant as a mosaic. Outline the experimental approach you would use to resolve this problem.

12. A dwarf variant of tomato appears in a research line. The dwarf is crossed as female to normal plants, and all of the F_1 progeny are dwarfs. These F_1 individuals are selfed, and the F_2 progeny are all normal. Each of the F_2 individuals is selfed, and the resulting F_3 generation is $\frac{3}{4}$ normal and $\frac{1}{4}$ dwarf. Can these results be explained by: a. cytoplasmic inheritance? b. cytoplasmic inheritance plus nuclear suppressor gene(s)? c. maternal effect on the zygotes? Explain your answers.

13. Assume that diploid plant A has a cytoplasm genetically different from that of plant B. To study nuclear-cytoplasmic relations, you wish to obtain a plant with the cytoplasm of plant A and the nuclear genome predominantly of plant B. How would you go about producing such a plant?

14. Two species of *Epilobium* (fireweed) are intercrossed reciprocally as follows:

$$♀ \text{ } E. \text{ } luteum \times ♂ \text{ } E. \text{ } hirsutum \longrightarrow \text{all very tall}$$

$$♀ \text{ } E. \text{ } hirsutum \times ♂ \text{ } E. \text{ } luteum \longrightarrow \text{all very short}$$

The progeny from the first cross are backcrossed as females to *E. hirsutum* for 24 successive generations. At the end of this crossing program, the progeny still are all tall, like the initial hybrids.

a. Interpret the reciprocal crosses.

b. Explain why the program of backcrosses was performed.

15. One form of male sterility in corn is maternally transmitted. Plants of a male-sterile line crossed with normal pollen give male-sterile plants. In addition, some lines of corn are known to carry a dominant nuclear restorer gene (*Rf*) that restores pollen fertility in male-sterile lines.

a. Research shows that the introduction of restorer genes into male-sterile lines does not alter or affect the maintenance of the cytoplasmic factors for male sterility. What kind of research results would lead to such a conclusion?

b. A male-sterile plant is crossed with pollen from a plant homozygous for gene *Rf*. What is the genotype of the F_1? The phenotype?

c. The F_1 plants from (b) are used as females in a testcross with pollen from a normal plant (*rf rf*). What would be the result of this testcross? Give genotypes and phenotypes, and designate the kind of cytoplasm.

d. The restorer gene already described can be called *Rf-1*. Another dominant restorer, *Rf-2*, has been found. *Rf-1* and *Rf-2* are located on different chromosomes. Either or both of the restorer alleles will give pollen fertility. Using a male-sterile plant as a tester, what would be the result of a cross where the male parent was:

(i) Heterozygous at both restorer loci?

(ii) Homozygous dominant at one restorer locus and homozygous recessive at the other?

(iii) Heterozygous at one restorer locus and homozygous-recessive at the other?

(iv) Heterozygous at one restorer locus and homozygous-dominant at the other?

16. After treatment of *chlamydomonas* with streptomycin, streptomycin-resistant mutant cells are recovered. In the course of subsequent mitotic divisions, some of the daughter cells produced from some of these mutant cells show the normal phenotype. Suggest a possible explanation of this phenomenon.

17. Cosegregation mapping is performed in *Chlamydomonas* on four chloroplast markers: *m1*, *m2*, *m3*, and *m4*. The markers are considered pairwise in heterozygous condition, and cosegregation frequencies are obtained as follows:

	m1	*m2*	*m3*	*m4*
m1	——	29.0	18.0	18.4
m2		——	10.9	26.2
m3			——	8.8
m4				——

(For example, *m1* and *m2* cosegregate in 29 percent of the cell divisions followed.) Draw a rough genetic map based on these results.

18. In *Aspergillus*, a "red" mycelium arises in a haploid strain. You make a heterokaryon with a nonred haploid that requires *para*-aminobenzoic acid (PABA). From this heterokaryon, you obtain some PABA-requiring progeny cultures that are red, along with several other phenotypes. What does this information tell you about the gene determining the red phenotype?

19. On page 550, an experiment is described in which abnormal mitochondria are injected into normal *Neurospora*. The text mentions that "appropriate controls" are used. What controls would you use?

20. Adrian Srb crossed two closely related species, *Neurospora crassa* and *N. sitophila*. In the progeny of some of these crosses, a phenotype called aconidial (ac) appeared that involves a lack of conidia (asexual spores). The observed inheritance was

$$♀ \text{ } N. \text{ } sitophila \times ♂ \text{ } N. \text{ } crassa \longrightarrow \tfrac{1}{2} \text{ ac}, \tfrac{1}{2} \text{ normal}$$

$$♀ \text{ } N. \text{ } crassa \times ♂ \text{ } N. \text{ } sitophila \longrightarrow \text{all normal}$$

a. What is the explanation of this result? Explain all components of your model with symbols.

b. From which parent(s) did the genetic determinants for the ac phenotype originate?

c. Why were neither of the parental types ac?

21. Several crosses involving poky or nonpoky strains A, B, C, D, and E were made in *Neurospora*. Explain the results of the following crosses, and assign genetic symbols for each of the strains involved. (NOTE: poky strain D behaves just like poky strain A in all crosses.)

	Cross	Progeny
a.	nonpoky B ♀ × poky A ♂	all nonpoky
b.	nonpoky C ♀ × poky A ♂	all nonpoky
c.	poky A ♀ × nonpoky B ♂	all poky
d.	poky A ♀ × nonpoky C ♂	$\tfrac{1}{2}$ poky, all identical (e.g., D); $\tfrac{1}{2}$ nonpoky, all identical (e.g., E)
e.	nonpoky E ♀ × nonpoky C ♂	all nonpoky
f.	nonpoky E ♀ × nonpoky B ♂	$\tfrac{1}{2}$ poky; $\tfrac{1}{2}$ nonpoky

22. In yeast, an antibiotic-resistance haploid strain *ant*^{R} arises spontaneously. It is combined with a normal *ant*^{S} strain of opposite mating type to form a diploid culture that is then allowed to go through meiosis. Three tetrads are isolated:

Tetrad 1	Tetrad 2	Tetrad 3
α *ant*^{R}	α *ant*^{R}	a *ant*^{S}
α *ant*^{R}	a *ant*^{R}	a *ant*^{S}
a *ant*^{R}	a *ant*^{R}	α *ant*^{S}
a *ant*^{R}	α *ant*^{R}	α *ant*^{S}

a. Interpret these results.

b. Explain the origin of each ascus.

c. If an *ant*^{R} grande strain were used to generate petites, would you expect some of the petites to be *ant*^{S}? Explain your answer.

23. In yeast, two haploid strains are obtained that are both defective in their cytochromes; the mutants are designated *cyt1* and *cyt2*. The following crosses are made:

$$cyt1^- \times cyt1^+$$
$$cyt2^- \times cyt2^+$$

One tetrad is isolated from each cross:

cyt1$^-$	*cyt2*$^-$
cyt1$^-$	*cyt2*$^-$
cyt1$^+$	*cyt2*$^-$
cyt1$^+$	*cyt2*$^-$

a. Explain the difference in the natures of these two mutants.

b. What other ascus types could be expected from each cross?

c. How might the two genes involved in these mutants interact at the functional level?

24. In a marker-retention analysis in yeast, a multiply resistant strain *apt*^{R} *bar*^{R} *cob*^{R} is used to induce 500 petites. Table 20-4 shows, for different pairs of markers, the number of petites in which only one marker of the pair is lost. (For example, 87 petites were *apt*^{S} *bar*^{R}, and 120 were *apt*^{R} *bar*^{S}.) Use these results to draw a rough map of these mitochondrial genes.

TABLE 20-4.

Gene pair	Petites in which first marker is lost	Petites in which second marker is lost
apt bar	87	120
apt cob	27	18
bar cob	48	69

25. A grande yeast culture of genotype *cap*^{R} *ery*^{R} *oli*^{R} *par*^{R} is used to obtain petites by treatment with ethidium bromide. The petites are tested for (1) their drug resistance, and (2) the ability of their mtDNA to hybridize with various specific mtRNA types. In all, 12 petites are tested, and Table 20-5 shows the results. Use this information to plot a map of the mtDNA, showing the sequence of the 11 genetic loci involved in these phenotypes. Be sure to state your assumptions and draw a complete map.

26. The mtDNA is compared from two haploid strains of baker's yeast. Strain 1 (mating type α) is from North America, and strain 2 (mating type a) is from Europe. A single restriction enzyme is used to fragment the DNA's, and the fragments are separated on an electrophoretic gel. The sample from strain 1 produces two bands, corre-

TABLE 20-5.

Petite culture	Drug resistance (R) or sensitivity (S)				Ability (+ or −) of petite mtDNA to hybridize with mtRNA's						
	cap	ery	oli	par	rRNA$_{large}$	rRNA$_{small}$	tRNA$_1$	tRNA$_2$	tRNA$_3$	tRNA$_4$	tRNA$_5$
1	R	S	S	S	+	−	−	−	−	−	−
2	S	S	S	S	−	−	−	−	−	+	−
3	S	S	S	S	−	−	−	+	−	−	+
4	S	S	R	S	−	−	+	−	−	−	−
5	S	R	S	S	+	+	−	−	−	−	−
6	R	S	S	S	−	−	−	−	+	−	−
7	S	S	R	R	−	−	−	+	−	−	−
8	S	S	S	S	−	+	−	−	−	+	+
9	S	R	S	S	+	−	−	−	−	−	−
10	S	S	S	S	−	−	+	−	+	−	−
11	R	S	R	R	−	+	+	+	+	+	+
12	S	R	S	S	+	+	−	−	−	−	−

sponding to one very large and one very small fragment. Strain 2 also produces two bands, but they are of more intermediate sizes. If a standard diploid budding analysis is performed, what results do you expect to observe in the resulting cells and in the tetrads derived from them? In other words, what kinds of restriction-fragment patterns do you expect?

27. In yeast, some strains are found to have in their cytoplasm circular DNA molecules that are 2 micrometers (μm) in circumference. In some of these strains, this 2-μm DNA has a single EcoRI restriction site; in other strains, there are two such sites. A strain with one site is mated to a strain with two sites. All of the resulting diploid buds are found to contain both kinds of 2-μm DNA.

   a. Is the 2-μm DNA inherited in the same fashion as mtDNA?

   b. What do you think the 2-μm DNA is likely to represent functionally?

28. Circular mitochondrial DNA is cut with two restriction enzymes A and B, with the following results:

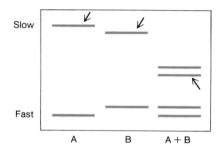

(The arrows indicate the bands that bound a radioactive mt rRNA-derived cDNA probe in a Southern blot.) Draw a rough map of the positions of the restriction site(s) of A and B, and also show approximately where the mt rRNA gene is located.

29. You are interested in the mitochondrial genome of a fungal species for which genetic analysis is very difficult but from which mtDNA can be extracted easily. How would you go about finding the positions of the major mitochondrially coded genes in this species? (Assume some evolutionary conservation for such genes.)

30. Rare senescent cultures of *Neurospora* have been found.

   a. When these cultures are crossed as male with nonsenescent partners, no senescent progeny are obtained. However, when they are crossed as female, some but not all progeny senesce; some senescent progeny die sooner than others and either sooner or later than the parental strain. Explain these findings.

   b. When mtDNA from senescent cultures is cut with EcoRI, the large normal fragment 1 is missing. In its place are four new Eco fragments B, C, E, and G, totaling 10 kb more than fragment 1. A map of the normal fragment 1 follows, where Eco represents EcoRI sites, Hi represents HindIII sites, and two segments cloned as probes are also shown:

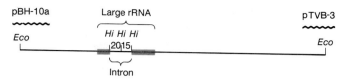

Probe pBH-10a binds to B but not to C, and probe pTVB-3 binds to C but not to B; neither probe binds to E or G. A HindIII digest shows the absence of fragment 20 and the appearance of two new large fragments, K1 and K2. K1 hybridizes to C, E, and G, and K2 hybridizes to B and E. Fragment C also hybridizes to HindIII fragment 15. What can you deduce from these data?

31. **a.** Early in the development of a plant, a mutation in cpDNA removes a specific BgIII restriction site (B) as follows:

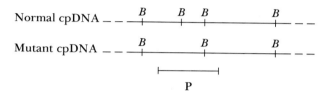

In this species, cpDNA is inherited maternally. Seeds from the plant are grown, and the resulting progeny plants are sampled for cpDNA. The cpDNA's are cut with BgIII and Southern blots are hybridized with the probe P shown. The autoradiograms show three patterns of hybridization:

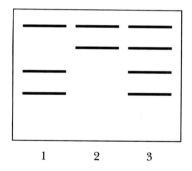

Explain the production of these three seed types.

   b. In *Gryllus* (a cricket) and *Drosophila*, rare females have been found that show mixtures of two mtDNA types, differing by the presence or absence of one specific restriction site. In the progeny of these females, each and every individual also proves to be a mixture of the parental mtDNA types. Contrast these results with the results from (a) and with other results in this chapter.

32. The mitochondrial genome of the turnip is a large, circular molecule 218 kb in size with a pair of 2-kb direct repeats located about 83-kb apart. However, when turnip mtDNA is examined carefully, three molecular types are seen: the 218-kb circle just described, a 135-kb circle bearing a single 2-kb repeat, and an 83-kb circle also bearing only one copy of the 2-kb repeat. Propose a model to explain the presence of the two smaller molecular types.

Genes and Differentiation

KEY CONCEPTS

In multicellular plants and animals as a fertilized egg divides,
its daughter cells undergo a very specific sequence of
movements and changes in form and function called
differentiation

•

The phenotype of a cell is a reflection of its genes. If the
nuclear DNA present at fertilization is faithfully replicated at
each division, then development must represent a
programmed sequence of gene activation and inactivation

•

It is possible to use molecular and cytological techniques to
demonstrate how specific loci produce products that are
characteristic of the cell's differentiated state

•

Mutations disrupt developmental events, thereby enabling
geneticists to study specific processes or cell types

•

Systems that represent unique cases of development merit
special attention: the mechanism whereby sex is determined,
how cells break out of their normal mode of activity to
become cancerous, and the way in which a limited number of
genes can generate a nearly endless array of different
products of the immune system called antibodies.

■ The fine structure of a single cell or the anatomical complexity of a multicellular organism offers an incredible challenge to the geneticist interested in how genes affect phenotype. For just as hair or eye color is a reflection of genotype, the proper location, size, function, and fine structure of an organ within a whole organism are also determined by genes. In humans, a single egg combined with a single spermatozoon constitutes the sole cellular legacy passed from parent to child; the resulting single-celled zygote will increase to about 10^{14} cells in an adult. As that seemingly nondescript cell proliferates, an anteroposterior and ventral-dorsal distinction will arise. From the original egg alone will be derived the amazing array of cell types present in a mature individual. The developmental fate of those cell types is so precisely controlled that two adjacent cells may have very different phenotypes and functions. Furthermore, in contrast to a new car, for example, which works only when all of its basic parts have been properly assembled, a developing zygote is—at every moment—a vibrant, functioning entity of integrated components. Each step of development is dependent on and builds from the preceding stage. Any disruption in the normal sequence of events results in an abnormal phenotype: from these we can learn about the normal sequence.

Once a cell has been committed to a specific fate, it is said to be **determined,** Henceforth, any derivations of that cell will be of the same type. In other words, once a cell is determined to be, say, a liver cell, regardless of the number of times it then divides, all its daughters will be liver cells. Thus, a state of determination is fixed and inherited from cell to cell. But does determination represent an actual change in the primary structure of DNA? If not, how is the determined state maintained? What is the mechanism of determination? Differentiation represents the phenotypic expression of a cell's determined fate. At what point is the definitive commitment made? What factor or factors are responsible for triggering it? What molecular mechanisms are involved? The phenomena being studied are extremely complex. In multicellular organisms, many layers of complexity —from cells to tissues to organs to systems and the whole organism—are involved as well as a time dimension. Nevertheless, these questions have become accessible experimentally.

We have come to understand the remarkable fidelity with which the genetic material is replicated and distributed at each cell division. Every complete normal genome contains the blueprint for the accurate reproduction of an organism in a given species. But how is the genetic program read to confer the complexity evident in a multicellar organism—or even within a single cell? It is clear from previous chapters that different genes are not transcribed all at once or at the same rate. Development and differentiation, therefore, must reflect the co-

ordinated regulation of genetic activity over time. Genetic analysis offers a way of probing the mechanism of control involved.

As we shall see, genetic analysis has yielded answers to basic questions about the events of development. What are the relative roles of the nucleus and the cytoplasm in differentiation? Does the genetic material remain completely intact after cellular differentiation has occurred? What is the evidence for differential gene activity as a basis for differentiation? How can mutations be used to probe developmental phenomena? What role does cellular regulation play in special phenomena such as sex determination, cancer, and the immune system? This chapter explores how genetic analysis has been used to answer these questions.

Nuclear and Cytoplasmic Factors in Development

For decades, classical embryologists argued that the nucleus plays a relatively passive role in directing the primary events that determine the developmental fates of cells and tissues. Different cells develop in different ways within the embryo, but each cell seems to inherit an identical set of nuclear genetic material. On the other hand, asymmetries are known to exist in the distribution of cytoplasmic material during cell divisions. Thus, it seems reasonable that the varying developmental patterns in different cells are under cytoplasmic rather than nuclear control.

Mitochondria and yolk, as examples, are not evenly distributed through the egg. The planes of division in early cleavages after fertilization result in the inclusion of different amounts of yolk and mitochondria in different daughter cells (Figure 21-1). Within a few divisions, differences in division rate and size become apparent among the cells of the embryo; such differences mean that environmental factors (oxygen, waste materials, temperature, and so on) are impinging on the various parts in different ways.

The importance of the cytoplasm can be demonstrated experimentally. A part of a cell cortex (outer layer) without nuclei is removed from one embryo and grafted onto another (Figure 21-2). The resulting embryo may develop an extra set of some structure (twinned spinal cords in the example illustrated), showing that the cytoplasm (without the nucleus) can exert a major influence on development.

Environmental Effects

Environmental effects are also important in determining the course of development. In the 1940s, Richard Goldschmidt showed that a variety of environmental

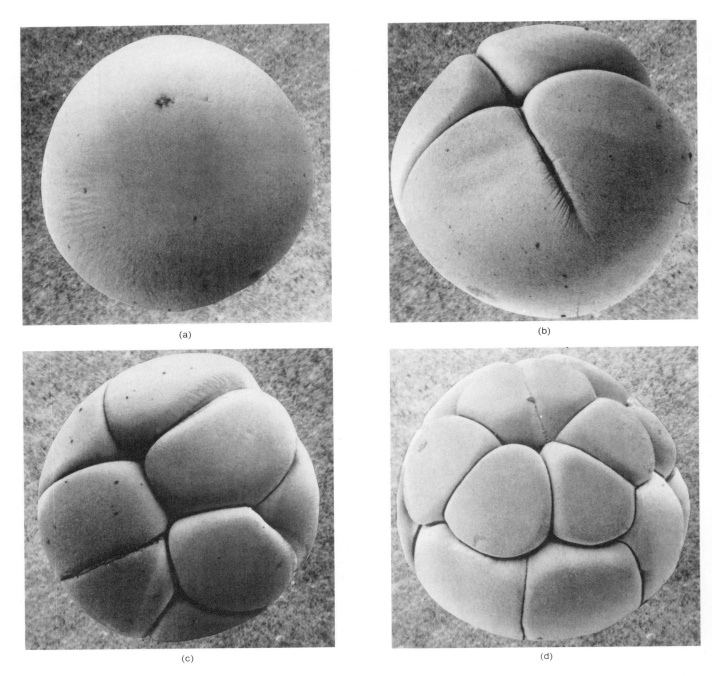

(a)

(b)

(c)

(d)

Figure 21-1. Early cleavages of an amphibian egg. It can be seen that asymmetries soon develop due to the presence of yolk at the lower (vegetal) pole of the zygote cell. (a) Undivided egg. (b) Second cleavage. (c) Eight-cell stage. (d) 16-cell stage. (Photograph courtesy of Lloyd M. Beidler, Florida State University.)

factors can produce phenotypic abnormalities in *Drosophila* that resemble the abnormal phenotypes of known genetic mutations. Such an environmentally induced phenotype is called a **phenocopy.** Phenocopies have subsequently been produced in a number of organisms by a variety of agents, including temperature (heat or cold), radiation, and an array of chemical compounds.

Goldschmidt showed that the agent must be applied at a specific critical period in development in order to obtain each particular phenocopy. In 1945, Walter Landauer demonstrated such **phenocritical periods** in the embryonic response of chickens to such compounds as insulin, boric acid, and pilocarpin. These studies revealed the existence of a nongenetic process that can produce

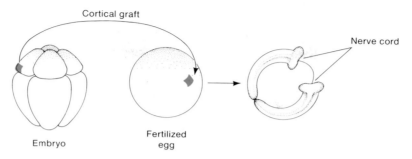

Figure 21-2. Experimental demonstration of extranuclear control of development. A graft of cell membrane and underlying cytoplasm is taken from the smaller cells of an embryo at the eight-cell stage and transplanted to the equatorial region of a fertilized egg. A second spinal cord develops in the resulting embryo. (From A. S. G. Curtis, *Journal of Embryology and Experimental Morphology* 10, 1972, 416.)

abnormalities—a process called **teratogenesis.** A variety of teratogenic agents (for example, the German measles virus and alcohol, thalidomide, and other drugs) has been proved to produce abnormalities in the human fetus.

> **Message** The egg cytoplasm is not homogeneous, and regional differences in the cytoplasm play a role in development. Furthermore, environmental agents can provoke specific defects during development.

Maternal Effects

While embryologists were searching in the cytoplasm for the key factors controlling development, geneticists were seeking evidence of nuclear control. There is no doubt that the egg cytoplasm does regulate the early events in embryonic development: for example, an enucleated frog egg will still go through the early cleavages. However, the architecture and molecular composition of the oocyte cytoplasm are themselves rigidly controlled by the nuclear genome of the mother! We have already considered an example of this: the *anucleolate (an)* mutation in *Xenopus*. The cross *an/+ × an/+* forms some *an/an* homozygotes that are incapable of forming ribosomes. Nonetheless, these embryos develop to twitching tadpoles before dying. Their survival is a result of the prefertilization amplification of ribosomal genes and the consequent store of ribosomes (which must reflect the maternal genotype). The most striking illustration of maternal genotypic influences on the phenotype of the offspring is seen in the coiling of snail shells (Figure 20-8).

Another dramatic illustration of maternal genotype on the oocyte is found in *Drosophila* eggs. In the *Drosoph-*

ila egg, the anterior end can be distinguished from the posterior end, where a special region of the cortex can be seen to have densely staining **polar granules.** The granules define a region containing pole cytoplasm, from which the cytoplasm of all gonadal cells is derived. An autosomal recessive mutation discovered in *D. pseudoobscura* is called grandchildless (gs), for reasons that will become apparent. In the cross *gs/+ × gs/+*, the *gs/gs* homozygotes formed are fertile. But all of the eggs produced by *gs/gs* females lack polar granules. Regardless of the genotype of their mates, *gs/gs* females produce only offspring that lack gonads (and thus are sterile and fail to produce grandchildren of the *gs/gs* females). Obviously, in this case, the maternal genotype plays a major role in determining the cytoplasmic makeup of eggs, thus affecting the phenotype of the offspring.

> **Message** Maternal genotype affects the architecture and composition of the egg cytoplasm and therefore can affect the phenotype of the offspring.

Is the cytoplasm or the nucleus the site for control of the developmental process? It should be obvious by now that this is rather like asking whether the chicken or the egg came first! In the context of development, the nucleus in isolation from the cytoplasm is as useless as a cell without its nucleus. Much of the immediate control of developmental processes seems to come from the cytoplasm, but nuclear genetic factors play a major role in determining the nature of the cytoplasm. Are the detectable changes during development a reflection of directed alterations of the genetic material of the nucleus, or does it remain unchanged from its state at fertilization?

Are Nuclei of Differentiated Cells Totipotent?

Classical studies of the embryology of organisms such as worms and insects have revealed that not all somatic cells carry the same number or amount of chromosomal elements. It has been suggested, therefore, that the developmental sequence may be controlled by a mechanism that specifies the order of genetic elimination. Mitosis maintains a constant chromosome number, but does each cell of a highly differentiated embryo contain all of the genetic material necessary for complete development? One way to confirm this would be to show that each cell's nucleus remains **totipotent,** or still capable of supporting complete development from egg to adult. Note that the answer may differ from species to species; throughout this chapter, each aspect of development will be considered in several systems—particularly mammalian, amphibian, and insect.

There are cases of multiple births in which "identical" siblings are derived from a single fertilized human egg (Figure 21-3). Thus, we can conclude that genetic information in people is faithfully reproduced through at least the first three cleavages after fertilization (two cleavages produce only four cells and identical quintuplets have been recorded). But it is known that early embryonic development is controlled primarily by maternal genotype. What of the nuclei in cells resulting from later divisions? Do they remain totipotent?

Many highly differentiated organisms can regenerate new organs and tissues. As examples, a starfish can regrow a lost "arm" (Figure 21-4), a reptile can re-form a lost tail, and the human body can repair a damaged liver. However, such regeneration is possible only in certain tissues. Regeneration of a complete organism from a single somatic cell is not observed among animals in nature.

Cloning in Plants.   In the 1950s, Frederick Steward demonstrated that highly differentiated phloem cells in the root of a carrot plant are totipotent. Steward was able to obtain an entire carrot plant from a single phloem cell (Figure 21-5). Using this method, he obtained a number of genetically identical plants from the somatic cells of a single plant. This asexual method of reproduction is now called **cloning** (in fact, the word "clone" is derived from a Greek word referring to a plant cutting). This work has important implications in terms of the economic potential for obtaining duplicates of any individual plant with a particularly useful phenotype. Almost all commercially available orchids are produced by cloning. One of the problems in breeding any organism that reproduces sexually is the constant

Figure 21-3. Identical quadruplets. (Photograph courtesy of Erika Stone, © Peter Arnold, Inc.)

Figure 21-4. Two "arms" (rays) of this starfish (*Asterias*) were severed. New rays have begun to form, and they will eventually grow into normal-sized rays. (Photograph courtesy of Grant Heilman.)

disruption of useful gene combinations by random assortment and crossing-over. Cloning retains genetic combinations intact, and this process is now used commercially to maintain lines of trees. Of course, gardeners and farmers have long used asexual reproduction in the grafting of a branch from one genotype onto another (permitting farmers to maintain and expand stocks of seedless oranges, for example) or in the generation of new plants from cuttings.

> **Message** The cloning of plants shows that some differentiated cells can remain totipotent.

The individual cells of multicellular organisms, when grown in cell cultures, can be treated and analyzed like microorganisms. Mutations can be induced and se-

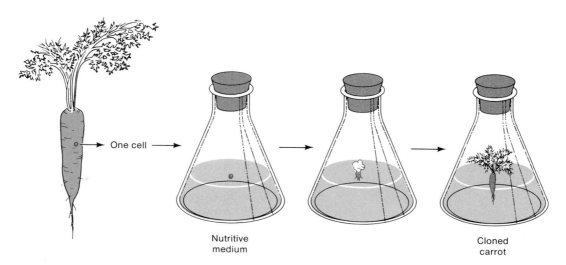

One cell

Nutritive medium

Cloned carrot

Figure 21-5. Cloning of a carrot plant from a single cell taken from the phloem tissue of a mature root. The nutritive media in the different flasks must contain appropriate plant hormones.

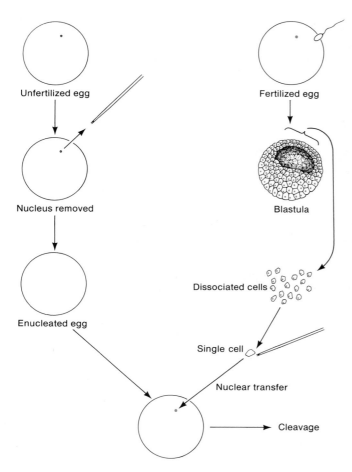

Figure 21-6. Transplantation of a nucleus from an embryonic amphibian cell into an enucleated egg. The nucleus of an unfertilized egg is removed mechanically by micropipette *(left).* Cells of a fertilized egg that has developed to the blastula stage are dissociated *(right),* so that the nucleus of a single cell can be removed and inserted into the enucleated egg *(bottom.)*

lected from these cells for such qualities as the increased content of certain amino acids. If such mutant cells can be stimulated to differentiate into mature plants, then the process of plant breeding can be tremendously simplified and accelerated. Such a procedure is possible in plants, because the totipotency of at least some kinds of differentiated plant cells has been demonstrated. However, are differentiated animal cells similarly totipotent?

Nuclear Transplantation in Amphibians. In the early 1950s, Robert Briggs and Thomas King developed techniques for manipulating nuclei in amphibian cells. They created **enucleated** frog eggs (eggs lacking nuclei) by

inserting fine glass pipettes into unfertilized eggs and sucking the nuclei out. This manipulation made it possible for them to study totipotency in animal cells. In order to test the developmental capacity of the nucleus from a differentiated cell within the cytoplasmic environment of an unfertilized egg, Briggs and King removed nuclei from differentiated cells and injected them into enucleated eggs (Figure 21-6). They found that an enucleated egg will divide when provided with a somatic nucleus. Furthermore, they found that a nucleus removed from a cell of the **blastula** (the developmental stage at which the embryo is still a hollow ball of cells) is totipotent. When the nucleus from a blastula cell is inserted into an enucleated egg, the egg divides normally and develops into an adult.

However, different results were obtained when the nucleus was taken from a cell at the **gastrula** stage of development (when folding of the layers of cells begins). A nucleus from a gastrula cell will not support normal development when inserted into an enucleated egg. It appears that some process of nuclear differentiation begins at gastrulation, destroying the totipotency of the nucleus from a frog somatic cell.

Even if each cell does not retain a complete complement of DNA, it is obvious that some regulation process must activate and inactivate specific parts of the genome as development occurs. Perhaps a certain amount of time is required for such activation and inactivation to take place. Thus, the renucleated egg may divide before the transplanted nucleus can return to a condition appropriate to the first division. Such an effect can lead to abnormal development even though the transplanted nucleus is in theory totipotent. To eliminate this possibility, Briggs and King performed another group of experiments. They kept nuclei in the cytoplasmic environment that encourages totipotency (cytoplasm from pregastrulation cells) by serially transplanting nuclei from gastrula embryos into enucleated eggs (Figure 21-7). Even after many serial transplantations, the nuclei remained incapable of supporting normal development.

These experiments show that somatic nuclei of the frog *Rana* differentiate irreversibly after gastrulation —at least under the conditions of the nucleus-transplantation experiments. However, John Gurdon repeated these experiments in 1964 with a different amphibian, the African clawed toad, *Xenopus laevis.* Gurdon took nuclei from highly differentiated cells of tadpole intestines and inserted them into eggs in which the nuclei had been destroyed by a beam of ultraviolet light (Figure 21-8). Although many of the eggs failed to develop or produced abnormal growth, a significant number did develop normally into adults with genetic markers that showed them to be identical clones of the nucleus-donor tadpole (not of the toad from which the unfertilized egg

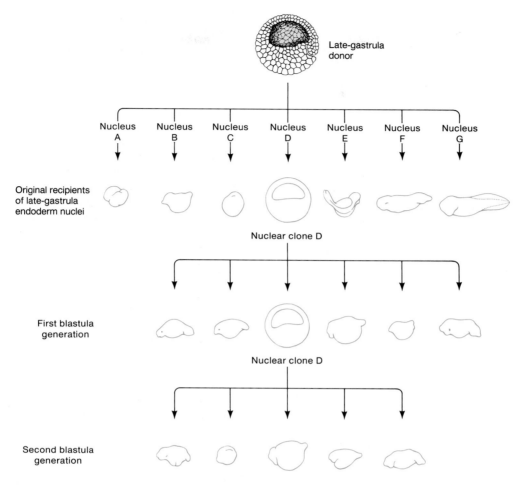

Late-gastrula
donor

Nucleus A   Nucleus B   Nucleus C   Nucleus D   Nucleus E   Nucleus F   Nucleus G

Original recipients
of late-gastrula
endoderm nuclei

Nuclear clone D

First blastula
generation

Nuclear clone D

Second blastula
generation

Figure 21-7. Serial transplants of gastrula nuclei in the frog *Rana*. All eggs except D are allowed to proceed to the end of their development (all develop abnormally). The D clone is allowed to proceed to the blastula stage. Then nuclei from the blastula cells are transplanted to enucleated eggs, and the same procedure is repeated. Normal development of any renucleated cells was never observed, even after many repetitions of this procedure. (From R. Briggs, *Cold Spring Harbor Symposium on Quantitative Biology* 21, 1956, 277.)

was taken). The results with the toad *Xenopus* clearly differ from those obtained with the frog *Rana*. Obviously, a nucleus can remain totipotent at a highly differentiated stage in at least one vertebrate organism.

It is not yet known why the nuclei of the frog *Rana* cease to be totipotent at gastrulation, whereas the nuclei of the toad *Xenopus* can remain totipotent in the tadpole stage. We must be very cautious about extrapolating from one species to another in this kind of research. Of course, the implications of techniques for cloning humans have received extensive popular attention. Aldous Huxley's *Brave New World* develops one scenario, now classic, based on this possibility. But is there

any reason to believe that nuclei remain totipotent in animals other than *Xenopus?*

Totipotency in Drosophila *and Mammals.* Karl Illmensee has obtained evidence to support the idea of totipotency in differentiated cells from both mice and *Drosophila.* In one set of experiments, Illmensee took cells from a teratocarcinoma, a type of tumor of the gonads, and injected them into the peritoneal cavities of normal mice. The ability of such cells to retain their phenotype is notable: injection of a single cell results in the growth of a teratocarcinoma in the mouse. Single cells from one genetically marked line were then injected into an em-

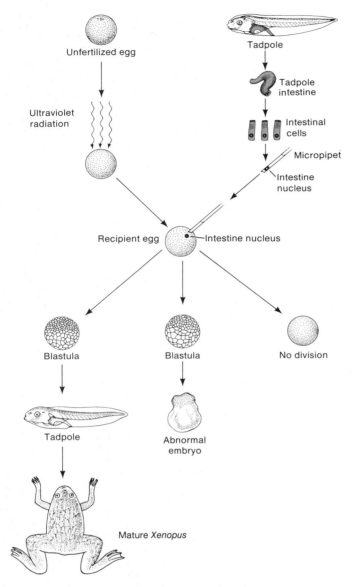

Unfertilized egg

Tadpole

Ultraviolet
radiation

Tadpole
intestine

Intestinal
cells

Micropipet

Intestine
nucleus

Recipient egg — Intestine nucleus

Blastula

Blastula

No division

Tadpole

Abnormal
embryo

Mature *Xenopus*

Figure 21-8. Cloning of the toad *Xenopus* by nuclear transplantation. The nucleus of an unfertilized egg is destroyed by a beam of ultraviolet light *(upper left)*, which specifically damages DNA. A nucleus taken from a gut cell of a tadpole *(upper right)* is transplanted into the enucleated egg *(center)*. Many of the resulting zygotes develop abnormally or fail to divide, but a few of the zygotes develop into mature individuals *(bottom left)*. (From J. B. Gurdon, "Transplanted Nuclei and Cell Differentiation." Copyright © 1968 by Scientific American, Inc. All rights reserved.)

bryo from another line (Figure 21-9). Next, the mosaic embryo was implanted into the uterus of a "pseudopregnant" female — a female whose uterus is receptive to embryo implantation following her mating to a sterile

male (Figure 21-10). The embryo developed normally, but the newborn mouse proved to be a **chimera** (or mosaic) containing a clone of cells derived from the injected tumor cell. This clone had been completely integrated into the organ or organs in which it developed. Illmensee found that if the clone of injected cells is included in the gonadal region, the animals can be mated and normal offspring can be derived from gametes formed from cells of the clone. Hence, the nuclei from the teratocarcinoma are indistinguishable from normal cells in their totipotency. Using a similar microinjection technique, Illmensee introduced single marked cells of *Drosophila* embryos into unfertilized eggs, which then developed into adult fruit flies. Again, the differentiated embryonic cells proved to be totipotent.

These results show that differentiated cells in many different organisms can be totipotent under certain experimental conditions. It seems reasonable to extrapolate from a small number of cases that most somatic cells in a highly differentiated organism may contain a complete copy of the DNA from the original zygote. In many such cells, the expression of parts of the genome is blocked in such a way that the nucleus from the differentiated cell cannot direct normal development when transplanted to an unfertilized egg. The nature of this blockage then becomes the subject for further research. How is the expression of DNA regulated during development?

> **Message** Studies in organisms as diverse as *Drosophila* and mammal reveal that mitosis accurately duplicates and distributes DNA sequences among the daughter cells at each division during development. Hence, each somatic cell contains the same genetic information as the newly fertilized egg from which the organism developed.

Turning Genes On and Off

We have seen that many cells carry a complete nuclear genome, but we know that certain genes exert their effects only at specific stages of development. A hypothesis involving the differential activation of various genes at different stages of development seems very attractive. Can such differential gene activation be demonstrated?

Early Embryonic Gene Regulation

We have seen that cell divisions just after fertilization are relatively rapid and are controlled by maternally derived information in the cytoplasm, with little genetic control from the zygotic nuclei. At what point in development

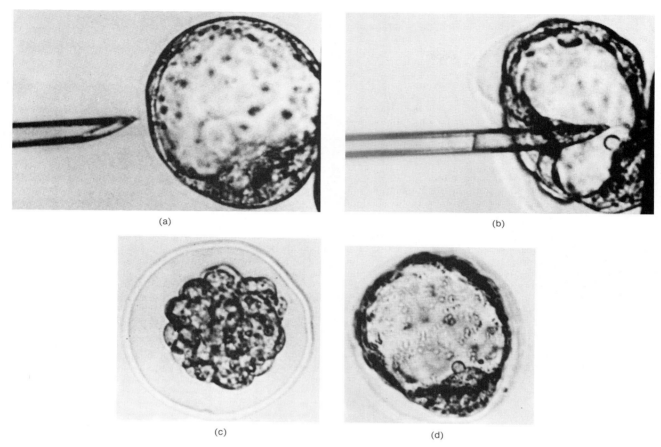

(a)

(b)

(c)

(d)

Figure 21-9. Formation of a mosaic embryo. (a) A mouse embryo is held on the end of a pipette while a teratocarcinoma cell is brought up to it. (b) The teratocarcinoma cell is injected into the center of the embryo. (c) The teratocarcinoma cell becomes integrated into the embryo. (d) All of the cells of the embryo (including the teratocarcinoma cell) multiply as development proceeds. The descendents of the injected cell will be integrated into tissues of the embryo; which tissues are involved will depend on the location of the injection. (From K. Illmensee, in *Genetic Mosaics and Chimeras in Mammals.* Edited by Liane B. Russell. Copyright © 1978 by Plenum Publishing Corp.)

do the embryonic nuclei begin to play a major role? (We know that they must play such a role because the zygote genotype does have a major effect on the phenotype.)

Nineteenth-century embryologists devised ingenious experimental approaches to answer this question. For example, Theodor Boveri studied species of sea urchin in which the embryos differ in morphology. He enucleated an egg from one species and fertilized it with sperm from another species. The sperm nucleus can promote development in such situations. Boveri showed that development of the embryo follows the pattern dictated by the genotype of the sperm, and he found the point in the developmental sequence at which this paternal influence becomes obvious. A similar approach can now be used to measure the time of appearance of some gene product (either RNA or protein) encoded by the paternal genes. Drugs such as actinomycin D that inhibit RNA synthesis can be tested to determine the develop-mental stage at which they produce lethal effects—showing when the products of the embryonic nuclei are required for survival. Such studies suggest that zygotic genes probably begin to act at approximately the time of gastrulation in amphibians. However, the lethal effects of actinomycin D on mouse embryos begin to appear around the eight-cell or 16-cell stage of development.

As we shall see, most studies of gene activation in development have concentrated on loci coding for specific cell products found in large quantities in highly differentiated cells. However, those genes acting early in development before differentiated tissues have appeared, can have a more profound effect on the organism as a whole. A method for cataloging them has been developed by Thomas Sargent and Igor Dawid. They studied *Xenopus* embryos, which are rich in maternally deposited mRNA's but in which no activity of zygotic genes occurs until the gastrula stage. The gastrula is the

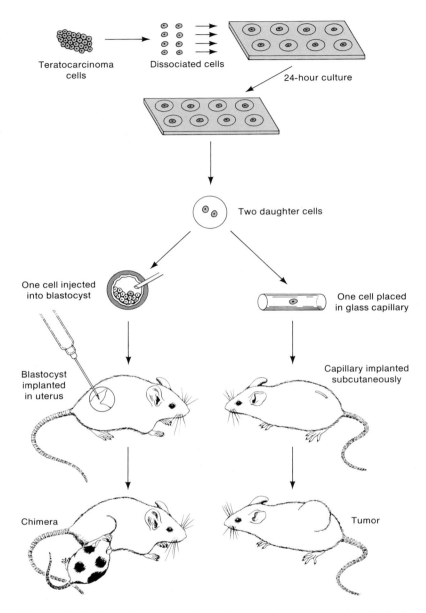

Figure 21-10. Experimental analysis of teratocarcinoma cells. Individual cells are allowed to divide once, and one of the daughter cells is injected into an embryo of the blastocyst stage *(left)*. The resulting mosaic embryo is then inserted into the uterus of a recipient female, where it develops normally. The other cell is placed in a glass capillary tube that is placed under the skin of another mouse, where it develops into a tumor *(right)*. Material from the tumor can be used for comparison with cells from the chimera. (From K. Illmensee and L. C. Stevens, "Teratomas and Chimeras." Copyright © 1979 by Scientific American, Inc. All rights reserved.)

first stage at which differentiation of the three primitive cell layers occurs. From gastrulae, Sargent and King isolated poly-A-containing RNA on poly-dT columns. From this gastrula mRNA (which contains all of the maternally inherited mRNA as well), cDNA was made. The cDNA was hybridized with mRNA taken from unferti-

lized eggs. All cDNA complementary to the maternally inherited transcripts formed double strands, but the gastrula-activated cDNA remained unpaired. The double strands were separated from the single strands. From the single strands, large cDNA molecules (400 to 2000 bases in length) were isolated and then inserted into

Labels in figure:
- Teratocarcinoma cells
- Dissociated cells
- 24-hour culture
- Two daughter cells
- One cell injected into blastocyst
- One cell placed in glass capillary
- Blastocyst implanted in uterus
- Capillary implanted subcutaneously
- Chimera
- Tumor

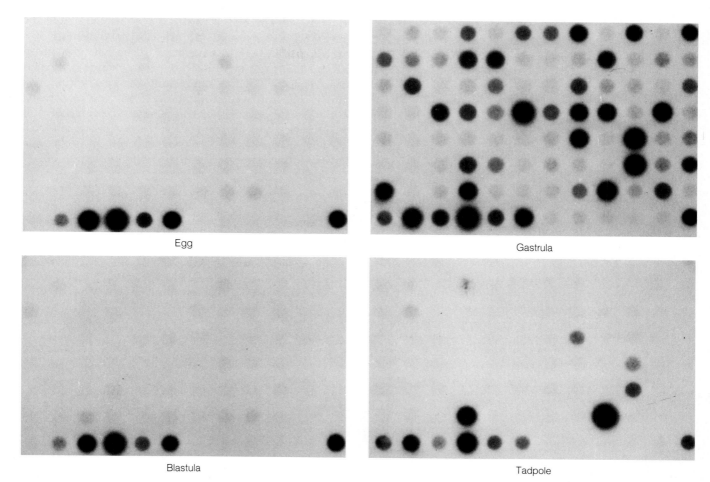

Egg

Gastrula

Blastula

Tadpole

Figure 21-11. Autoradiographs of DNA clones with radioactive mRNA taken from different developmental stages of *Xenopus*. The four panels contain an identical arrangement of the same set of 84 different clones. The bottom rows are reference DNAs. Radioactive mRNA obtained from egg, blastula, gastrula, and tadpole was hybridized to the filters; after the unbound material was removed, an autoradiograph was made. Note that none of the 84 DNAs is represented by pre-gastrula mRNAs and that the quantity of each mRNA can be measured by spot intensity. Finally, most of the "gastrula genes" are repressed in later stages, although the expression of one such gene is enhanced.

plasmids, until a library of 150,000 clones was derived. What these investigators obtained, in essence, was a large array of DNA's complementary to mRNA's transcribed at gastrulation. Of course, many of the sequences obtained in such procedures will be repeats of the same mRNA's, so the number of clones coding for a particular mRNA will reflect the abundance of those transcripts at gastrulation.

Sargent and Dawid bound their DNA to filters and then exposed them to radioactive mRNA from different stages. Autoradiographs showed that the cDNA's represent genes that are activated in gastrulation and that the level of transcription varies from gene to gene. Patterns of expression of such "early" genes can be seen in Figure 21-11. Most of these gastrula mRNA's declined in abundance in later stages, even though RNA was increased by six- to ten-fold in relative mass. One mRNA represented in the clones increased in the tadpole stage. Apparently, beginning with gastrulation, a specific group of genes is activated, with each gene being expressed in varying amounts; most are subsequently shut off by the tadpole stage. It is reasonable to assume that development proceeds according to the controlled activation and inactivation of specific genes.

Message Analysis of poly-A-containing transcripts shows that genes in amphibians are activated at gastrulation to produce varying numbers of mRNA molecules per cell. Most are shut off by the tadpole stage, at which time many new loci are activated. Development must proceed according to a sequence of turning genes on and off.

Visible Differentiation of Gene Activity

If differential gene activation is a mechanism of differentiation, is it possible to "see" this phenomenon occur? Let's take another look at the giant polytene chromosomes of Diptera (Chapter 8). Recall that certain stains show that these chromosomes contain regions rich in RNA and visibly swollen (Figure 8-4) suggesting that these regions may be cytologically visible sites of RNA synthesis. This has been verified by experiments in which polytene chromosomes are incubated in ³H-labeled uracil (which is incorporated into RNA, thereby identifying it). Autoradiographs of the chromosomes are then prepared, which reveal that certain parts of the chromosome do indeed incorporate the RNA precursor, especially in the swollen **puffs** and **Balbiani rings** (Figure 21-12).

In *Drosophila*, studies have shown that in a given polytene-chromosome-containing gland, puffing patterns do change predictably during larval development. Heat shock can also induce changes in puffing patterns that can be correlated with the appearance of specific mRNA's and proteins (Figure 21-13). Synthesis of a steroid hormone, ecdysone, near the end of larval life in *Drosophila* is accompanied by a major change in the pattern of puffing, which can be mimicked earlier in larval life by injection of the compound. These studies provide no evidence on whether the different patterns of puff information are the cause or the result of tissue differentiation. However, they do provide a very graphic demonstration of genes being turned on and turned off.

> **Message**   RNA-rich regions of polytene chromosomes—recognizable as cytologically visible puffs and Balbiani rings—correspond to RNA transcribed from chromosomal DNA. The puffing pattern is tissue- and developmental-stage-specific and provides a model for gene activation and inactivation in response to factors in the cytoplasm.

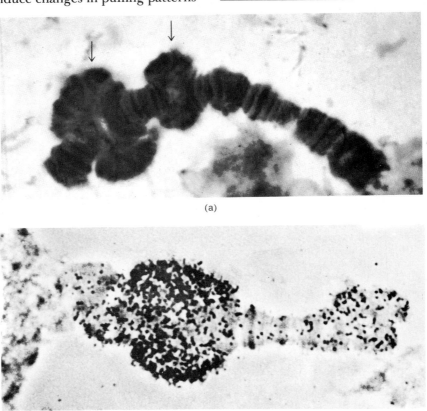

(a)

(b)

Figure 21-12. Balbiani rings in salivary gland chromosomes from the midge *Chironomus*. (a) Electron micrograph showing visible Balbiani rings (arrows). (b) Autoradiograph after incubation in a medium containing ³H-labeled uracil. The incorporation of the ³H-labeled uracil shows that heavy RNA synthesis is occurring in the Balbiani ring. (Photograph in (a) courtesy of Ulrich Clever; photograph in (b) courtesy of Pelling, Max Planck Institute for Biology.)

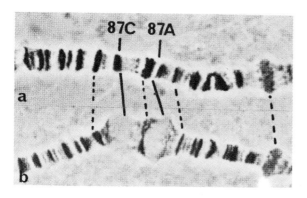

Figure 21-13. Induction of puffs in chromosome 3 of *Drosophila* by a 40-minute exposure to 37°C. (a) Control experiment with no heat shock. (b) Puffs formed after heat shock. (From M. Ashburner and J. J. Bonner, *Cell* 17, 241. Copyright © 1979 MIT Press.)

Turning Genes On and Off in Microorganisms

Development appears to reflect a coordinated sequence of gene activation and inactivation. What is the mechanism of control? It already appears that many mechanisms have evolved. However, the study of gene regulation in prokaryotes has provided the operon as a model for control in eukaryotes. By linking different operons, more complex control circuits develop. In the phage λ, for example, the very different alternate states of lysis and lysogeny are determined by two different repressors. The production of the λ repressor shuts off the *cro* gene, without which phage multiplication cannot proceed. Alternatively, production of the *cro* product shuts off the λ repressor. Similarly, the multiple activity of a single gene product as a repressor and an activator of different genes provides a cascading hierarchy of specific changes (Figure 21-14). Obviously, gene 1 at the apex of the pyramid has a major effect in contrast to the next levels of loci, which produce progressively restricted effects. This type of sequence, referred to as **combinatorial gene regulation,** serves to illustrate why differentiation is so often irreversible and requires a very specific sequence of reversals.

Message In microorganisms, a variety of mechanisms exist for turning genes on and off. The operon is the most refined mechanism, and by linking different operons together, a combinatorial sequence of gene control provides enormous variability.

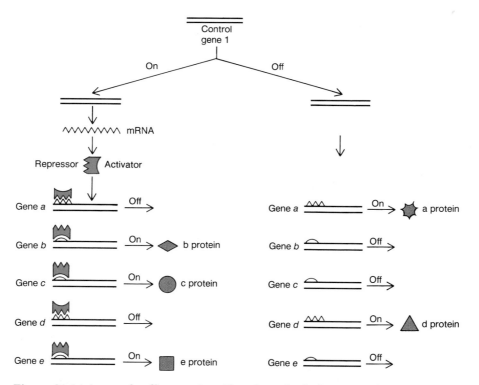

Figure 21-14 A cascade effect produced by a hypothetical gene 1, which behaves as a repressor or an activator for a series of five other operons. Enormous variability from cell to cell is possible.

Differentiated Tissue Reflects Different Gene Activity

Are the visible traits characteristic of a differentiated cell in multicellular organisms reflected in the array of gene products found in those cells? James Darnell and his group isolated a spectrum of moderately abundant mRNA's from mouse liver and then performed experiments to see whether these mRNA's were also present in mouse brain cells. These investigators separated poly-A-containing mRNA from total RNA on poly-dT columns. The liver mRNA was used as a template for reverse transcriptase to make double-stranded cDNA. The cDNA molecules were then inserted into plasmids, which were cloned in *E. coli*. Each cloned liver cDNA was tested for hybridization with labeled mRNA from mouse liver, brain, and hepatoma (a liver tumor). In this way, DNA complementary to mRNA only in liver or in all three cells was identified. These DNA's could be used to assay specific mRNA's from different tissues quantitatively. Darnell's group found no detectable liver-specific mRNA in brain or hepatoma cells. Even if an undetectable but low level of liver mRNA is made in brain cells, the maximum rate would be $\frac{1}{10}$ to $\frac{1}{50}$ of the rate in liver cells. Thus, at the molecular level, the differentiated phenotypes of specialized cells do indeed mirror the altered expression of genomic sequences. All of the cells tested express a common set of genes as well. It must be emphasized that the correlation of specific profiles of gene activity with cell or tissue type does not demonstrate a *causal* connection; that is, we can't say that differentiation is due to the gene products shown.

Message Isolation of mRNA from differentiated tissues shows that each tissue has its own set of expressed genes.

Inducible Systems in Multicellular Organisms

Mechanisms exist for altering gene activity in populations of individual microorganisms. But are there indications of how gene activity is controlled in different cells of the same individual?

Drosophila *Puffs.* Puffs are visible in highly specialized secretory glands of Diptera; both the pattern and size of specific puffs are determined by levels of the molting hormone, ecdysone. In principle, it should be possible to correlate a puff with a specific function in a secretory gland. Such evidence has come from studies on *Drosophila* salivary glands. The main function of these glands is

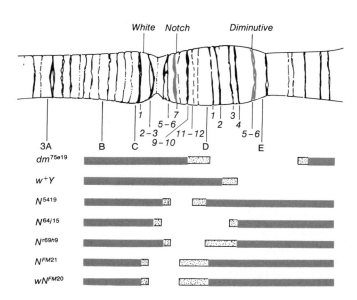

Figure 21-15. The 3A–D region near the tip of the X chromosome of a *Drosophila* salivary gland. The regions missing in the deficiencies are left as open spaces; the uncertain borders of the breakpoints are indicated by stippling. w^+Y is a duplication of a large segment of the X chromosome carried by a Y chromosome. The top three chromosomes carry the *Sgs-4* protein; the rest do not.

the production and extrusion of a thick saliva, which acts to "glue" the late third-instar larva to a solid surface when it pupates. The saliva begins to be found in the lumen of the gland in the middle of the third instar and is extruded about three hours before pupation. Günter Korge showed that the salivary secretion contains at least five proteins, one of which is called sgs-4 (salivary-gland secretion protein-4). Korge localized the *Sgs-4* locus to region 3C8–3D1 near the tip of the X chromosome, where a puff appears 5–6 hours before detectable levels of sgs-4. Steven Beckendorf and his group then more accurately localized *Sgs-4* by a series of overlapping deletions (Figure 21-15). DNA of the *Sgs-4* locus was recovered in two ways. Beckendorf's group used clones from a library of *Drosophila* DNA known to hybridize in situ to the 3C region. Other workers—Marc Muskavitch and David Hogness—isolated poly-A containing RNA's (presumed to be mRNA) of the glands at different developmental stages. From these types of mRNA, the experimenters made cDNA's to use as probes for mRNA's made at specific developmental stages. In this way, they recovered sequences that hybridize in the region 3C7–3D1 of the X chromosome. Once the DNA around the *Sgs-4* locus was identified, restriction maps were made in both laboratories, and nucleotide sequences flanking the ends of the part coding for RNA were determined. Muskavitch and Hogness then screened *Drosophila* strains of different geographical origin and found those that underproduced or overproduced sgs-4. Three underproducers from Japan carried a 52-base-pair (bp) deletion at a site 305 bp upstream from the 5' end of *Sgs-4;* a strain that produced no detectable sgs-4 had a 95-bp deletion at a site 392 bp upstream. In contrast, Beckendorf's group found two other types of underproducers of sgs-4. In one type, two mutants with single base changes 300 to 500 bp upstream from the *Sgs-4* structural gene made 50 percent less sgs-4 RNA. The other type was recovered when anomalous-sized DNA fragments, after restriction-enzyme treatment of a low-producing strain, were noted. A 1.3-kb insert was found 150 bp upstream from the *Sgs-4* structural gene. The insert, called hobo, reduces the expression of *Sgs-4* by 50- to 100-fold; of the transcripts produced, *four* different types (two starting in the hobo element) instead of only one are found. The ability to analyze the DNA of genetic variants of an inducible locus will provide a great deal of information about how gene expression is regulated.

Message   DNA in a specific puff on the *Drosophila* X chromosome codes for the production of the glue protein sgs-4. Strains of genetic variants that underproduce sgs-4 are found to carry deletions or insertions upstream from the 5' end of the structural gene. Thus, the regulatory controls of the structural gene are built into the DNA sequence ahead of it.

Chick Oviducts.   What factors are responsible for the control of a normal developmental process? In the 1960s, this question led Bert O'Malley to investigate the egg-producing system of female chickens. This complex system passes an egg through a series of chambers in the

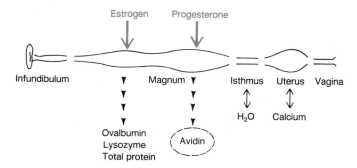

Figure 21-16. The functional segments of the chick oviduct. Eggs released by the ovary pass into the infundibulum and accumulate various components before being enclosed within a shell inside the uterus and extruded from the vagina.

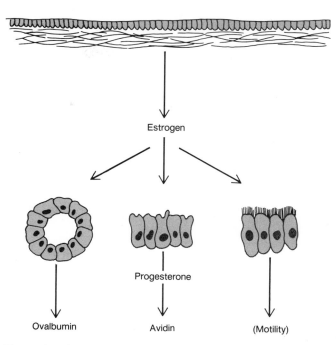

Figure 21-17. Effect of steroid hormones on the mucosal lining of the chick oviduct. Different cell types specialized for specific functions appear.

oviduct, where it is elaborated with the nutrients necessary to sustain a developing chick and finally enclosed in a protective shell (Figure 21-16). In newborn chicks, the oviduct is undeveloped and its walls are not yet differentiated for their egg-processing roles (Figure 21-17). Injection of newborn or older chicks with estrogen or with the synthetic steroid hormone diethylstilbestrol (DES) stimulates oviduct growth and the differentiation and production of various components of a mature egg. Thus, a normal developmental process can be controlled by the investigator. In the first six days of hormone treatment, the cells lining the oviduct undergo visible changes, but little ovalbumin (one of the main protein components of egg white) is detectable (Figure 21-18). In 6 to 15 days, ovalbumin synthesis reaches high levels. The increase in ovalbumin in the tubular cells is paralleled by a large increase in mRNA, which permitted O'Malley to purify the ovalbumin message. Ovalbumin mRNA is essentially undetectable before the administration of estrogen, but 18 days afterward, the amount increases until there are 48,000 molecules per cell, making up about one-half of all mRNA's in the cell. From ovalbumin mRNA, O'Malley made cDNA and used it as a probe to isolate the ovalbumin gene. The cDNA-DNA hybrids revealed the existence of introns in the structural gene.

But how is the ovalbumin gene activated? O'Malley showed that the steroid hormone alone does not act as an

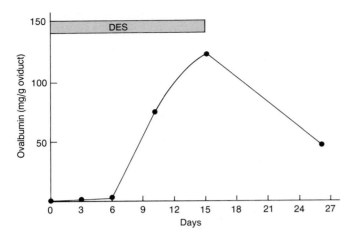

Figure 21-18. Effect of diethylstilbestrol (DES) on ovalbumin levels in immature chick oviducts. Chicks received 5 mg of DES daily for 15 days.

inducer. On entering the cell, the steroid is bound to two specific protein receptors, called A and B, to form an active complex. The hormone-receptor complex is then translocated to the nucleus, where it binds to DNA flanking the ovalbumin gene. By attaching the hormone-receptor complex to a column and passing DNA fragments through it, O'Malley was able to recover specifically bound sequences. He showed that a 114-bp segment, starting 135 bp upstream from the oviduct structural gene, is probably the binding site. Interestingly, within this segment is the very AT-rich sequence AAT TAAAAACTAATATTT that may be the actual binding site.

Other proteins of specific loci that are activated in specialized cells have been analyzed. These include hemoglobin in red blood cells, yolk proteins in amphibian eggs, and silk proteins in moths. The ability to isolate cloned DNA sequences of specific structural genes and flanking sequences provides a powerful method for de-

termining the importance of DNA sequence — and, ultimately, DNA structure — in gene regulation. The interesting question then becomes how does the regulating sequence actually affect transcription?

Message   Inducible genes in eukaryotic multicellular organisms can provide information about the specific sequences necessary for genetic activity. In the case of a chick-oviduct protein, steroid hormones induce the synthesis of specific mRNA's by forming a hormone-receptor protein complex, which then binds to a region flanking the structural gene.

The Genetics of Development

Classically, the study of development was a descriptive science wherein the process was carefully observed or inferred. Once a process has been documented, how can researchers probe the underlying mechanisms controlling it? An early method employed by embryologists was the mechanical disruption of normal sequences. Parts of an embryo were destroyed, removed, replaced, or shifted to determine the consequences and to infer the normal process from such results. What tools can a geneticist bring to the problem? A mutation is comparable to the embryologist's scalpel, providing a method of disrupting developmental sequences to produce a mutant phenotype. In a sense, every mutant phenotype reflects an alteration in development. But as with phenocopies, it is often difficult to infer the primary mode of gene action from a visibly abnormal phenotype.

Mutations Affecting Developmental Processes

Lethals.   Some examples of hereditary abnormalities in domesticated animals are economically important as

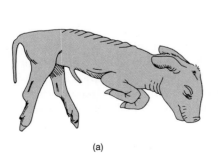

(a)

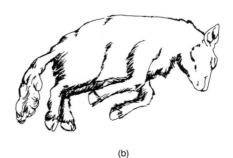

(b)

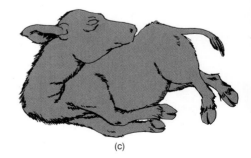

(c)

Figure 21-19. Lethal mutations with a similar phenotype in different domestic-animal species: (a) pig; (b) sheep; (c) cow.

Affected animals are born with muscles tightly flexed and often with the head pulled to one side.

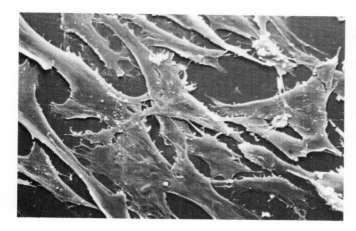

Figure 21-20. Scanning electron micrographs of connective tissue cells (fibroblasts) before *(left)* and after *(right)* infection with Rous sarcoma virus. The normal cells from a flat, extended sheet; transformed cells become round and clump together.

well as potentially useful in the study of the developmental defect. Figure 21-19 shows fetuses with extreme contraction of muscles in the legs and neck that probably is a result of a muscular or neurological defect. The mutations pictured are lethal in that the affected animal fails to survive to adulthood. There are many examples of lethal mutations in laboratory animals. Ernst Hadorn studied a number in *Drosophila* and defined the **effective lethal phase** (LP) as the developmental period during which death occurs. In surveying many lethals, he showed that the bulk of the LP's fell in the embryonic or pupal stages, the two periods of great change in a fly's life. Many of the mutations exhibit very specific patterns of the abnormality that is the ultimate cause of death. For example, the strain lethal giant larva (*lgl*) has a phenotype of a prolonged larval life and continued growth prior to death. This results from a defect in the production of hormones necessary for pupation. But can this phenotype be related to the primary action of the gene? The difficulty with as broad a phenotype as lethality, as with mutants with visible defects, is the pleiotropy of a gene. In Figure 4-10, we saw how a primary defect in cartilage formation in rats results in a number of characteristic anomalies, such as problems in breathing, eating, and heart beat, which are a secondary consequence of the cartilage lesion. So it is often difficult to determine the primary gene effect.

Temperature-sensitive Mutations. Can mutations that give other clues to their mode of actions be recovered? Lethal mutations that are conditionally expressed provide an extra dimension for analysis in a variety of organisms from viruses to mammals. We have already seen how the metabolic step affected by an auxotrophic mu-

tant can be inferred from the supplements necessary for survival. Another useful class is sensitivity to temperature. Recall that temperature sensitivity reflects the thermolability of mutant proteins. Studies of temperature-sensitive (ts) mutations under restrictive conditions reveal their mutant phenotype; the actual time of sensitivity can be inferred from "shift" studies. Let's look at an example involving a tumor virus called Rous sarcoma virus. Normally, mammalian connective-tissue cells grow in culture as a flat sheet (Figure 21-20); on infection with the virus, the cells become round and form clumps. This "transformation" is an in-vitro indication of the changes associated with tumor formation in vivo. A Rous sarcoma viral gene (*src*) codes for an enzyme that phosphorylates proteins. A temperature-sensitive allele of *src* permits cultures to shift between restrictive and permissive temperatures (Figure 21-21). This reveals that transformation is not irreversibly determined at a specific time but rather is maintained by the continued presence of wild-type enzymes.

The utility of imposing "pulses" of restrictive temperatures on a temperature-sensitive strain has been strikingly shown in the analysis of a mutation in *Drosophila* discovered by Thomas Grigliatti. The allele, *shibire* temperature-sensitive 1(*shi*^{ts1}), is named after the Japanese word for paralysis. Flies with the *shi*^{ts1} gene develop normally at 22°C, but when adults are transferred to a container and maintained at 29°C, they become paralyzed within minutes and eventually die. If shifted back to 22°C, however, the flies recover mobility within minutes. Cultures shifted from 22° to 29°C at different developmental stages revealed a succession of different lethal phases; Grigliatti found that larvae can also be paralyzed. When cultures were exposed to short

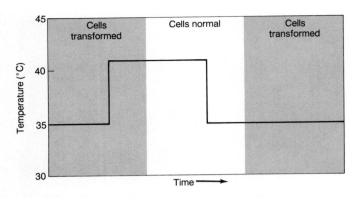

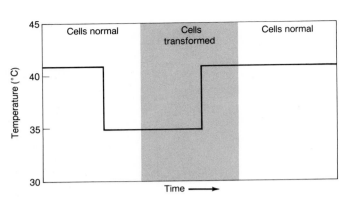

Figure 21-21. Shift studies of cells infected by Rous sarcoma virus carrying a temperature-sensitive allele of *src*. At 35°C, cells are transformed (*src* is active), but a few hours after a shift to 41°C, the cells revert to normal. On return to 35°C, the cells again become transformed after a few hours.

"pulses" of 29°C, a range of phenotypic defects in the bristles, hairs, and eyes of the surviving adults was found to be induced at different times (Figure 21-22). These pleiotropic effects of the *shi* locus are detectable only by varying the property of temperature sensitivity. The *shi*[ts1] gene specifies a product involved in the transmission of nerve impulses.

Reciprocal shifts give reciprocal results. These tests show that the *src* gene must be continuously active for the transformation phenotype to be sustained and that its product turns over in a few hours.

> **Message**   The conditional expression of a mutant phenotype permits a more precise determination of the time of gene activity and the nature of the primary defect. Temperature sensitivity is a particularly useful property.

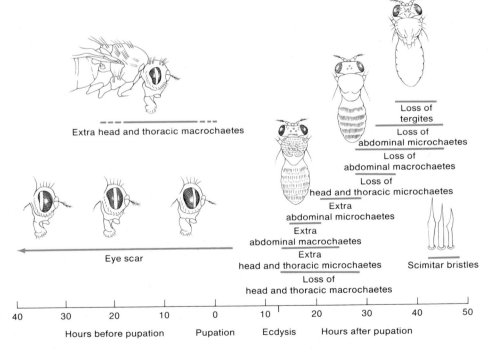

Figure 21-22 Temporal relationships of the temperature-sensitive periods for the various visible abnormalities of *shi*[ts1] flies, with diagrammatic representations of some of these abnormalities. The horizontal lines indicate the time during development when a heat pulse can produce the various abnormalities. Note the change in position of the eye scar with the changing time of the heat pulse (29°C). This corresponds to the location of mitotically active cells. (From Poodry et al., *Developmental Biology* 32, 1973, 381.)

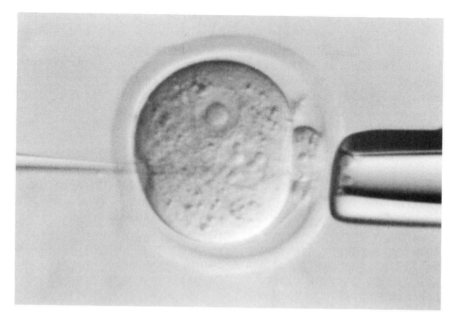

Figure 21-23 DNA introduced by injection into a one-cell fertilized rabbit ovum. The egg is held by a micropipette *(left)* while DNA is introduced into the pronucleus *(right)*. (From R. L. Brinster, *Cold Spring Harbor Symposium on Quantitative Biology* 50, 1985, 384.)

Transgenic Probes. Eventually, studies of development may permit the alteration of a phenotype through the control and modification of gene expression. The remarkable ability of cells to take up pieces of DNA and integrate them into the genome suggests a direct method of phenotypic modification. Can genes be mechanically injected into a cell and retain their biological integrity and function? What will be the fate of such molecules in subsequent generations?

John Gurdon developed a method of assaying the biological activity of both messenger RNA and cloned DNA by injecting either material directly into *Xenopus* eggs. This provided a model that was quickly copied by scientists using mouse eggs. In 1980, Frank Ruddle and his group reported the injection of DNA into mouse embryos and their subsequent genetic transformation. Pure DNA is injected into the pronuclei of newly fertilized mouse eggs (Figure 21-23). When linear pieces are used, about 25 percent of the mice that are born carry one or more copies of the introduced material; they are referred to as **transgenic** mice. The DNA is integrated into chromsomes although the mechanism of integration is far from clear.

Remarkably, DNA introduced in this way not only survives but, once integrated, can be expressed. Thus, DNA sequences that regulate gene expression can be assayed in vivo. A great deal of interest has been gener-ated by the reports of Ralph Brinster and Richard Palmiter that the introduction of rat, bovine, or human growth-hormone genes into mouse embryos results in up to a 100 percent increase in the growth of the mammals. There has resulted much intense activity to exploit the economic potential of such transgenic induction in rapid growth in domestic livestock. Although such studies in domestic animals are spectacular and presage useful applications, there are already indications that the resulting animals also exhibit undesirable characteristics. In the near future, the major value of transgenic experiments probably will continue to be in the study of the regulation of gene expression during development.

Special Cases of Differentiation

We have now seen how genetic analysis of development depends primarily on mutations that disrupt normal processes or mark certain groups of cells to be followed. In a sense, all final phenotypes are of interest to the developmental geneticist. In this section, we look at the determination of sex, the induction of cancer, and the regulation of immunity as specific instances of genetic regulation.

Sex Determination

Sex as a Developmental Phenotype.   The differentiation of individual organisms of a species into two sexes is a remarkable example of the role that genes play in development and merits special mention here. An extreme example of the importance of genes in this process is provided by the occurrence of XY human individuals who are chromosomally male but differentiate as well-developed females. This phenotype — **testicular feminization** — is apparently produced by a mutation on the X chromosome that acts only in XY zygotes by nullifying masculinization by the Y chromosome. In most higher organisms, the potential sex of a fertilized cell is determined at the time of gametic fusion by the combination of sex chromosomes or sex-determining genes in the zygote. Nevertheless, the actual differentiation of distinct sexual characteristics typically takes place much later in development and may even be altered by various sex-determined factors. The defect in testicular feminization syndrome is in a molecule that acts as a receptor for the male sex hormones, androgen, resulting in an inability of that hormone to carry out its normal function.

Hermaphrodites.   In many species of plants and animals, a single individual may function as both male and female. Such individuals are said to be **hermaphroditic** in animal species. Plants such as corn, in which the two sex organs appear at different locations on a plant, are termed **monoecious.** Plants with both female and male sexual organs in the same flower are called hermaphrodites. Earthworms and snails, which have both testes and ovaries in each individual, are examples of hermaphrodites in which the two kinds of sex organs develop from the same genotype. In such cases, some developmental mechanism must regulate the expression of specific sets of genes in the two regions of the organism where the different sex organs develop.

Environmental Effects.   In still other organisms, a single genotype can produce an individual of either sex. In such organisms, environmental or cytoplasmic elements play an important role in determining the sex of the individual. A few striking examples illustrate this phenotypic sex determination. The marine annelid *Ophryotrocha* differentiates into a sperm-producing male as a young animal and then changes into an egg-laying female when it gets older. If part of an older female is amputated, the worm reverts to the male form, indicating that size, rather than age, is the important factor controlling the sex of the individual. In the marine worm *Dinophilus*, on the other hand, sex appears to depend solely on the size of the egg produced by a female. Small eggs always produce males, whereas eggs 27 times as

large always develop into females. Recent research indicates that the sex of a turtle may depend on the temperature at which the individual develops. The mechanisms involved in environmental control of gene activation for the development of sexual organs obviously are analogous to those involved in the control of gene activation during differentiation of a zygote.

Sex Chromosomes.   In many organisms, the sex of an individual can be correlated with a specific genotype. The X – Y and W – Z sex-determining mechanisms were clearly established by the 1920s. In these cases, the heterogametic sexes produce two types of gametes, and the sex of a zygote is determined on fertilization by the genotypes of the fusing gametes. Among individuals from such species, mosaic individuals have been detected in which some parts of the organism are male and other parts are female.

Hormone Effects.   In many cases, hormones are important factors in the control of sex determination. Cases in which hormones can counteract chromosomal sex determination have been strikingly demonstrated in fish and frogs. Genetic markers can be used to distinguish the male and female eggs of the Japanese killifish called medaka. A male egg treated with a female hormone (estrogen) will develop into an individual with a male (XY) genotype and a female phenotype. These XY females can be mated with XY males to produce YY males. Treatment of a YY egg with estrogen also leads to the development of an individual with a female phenotype. A cross of YY females with YY males obviously must produce only male progeny. It is clear that the X and Y chromosomes in this species must be very similar in their genetic content; otherwise, the YY genotype would produce an abnormal phenotype due to a lack of genes carried on the X chromosome. Analogous results have been obtained in frogs, which have a W – Z sex-determining mechanism. A WZ female tadpole that is fed male hormones (androgens) will develop into an adult with a male phenotype. A cross of such WZ males with WZ females produces some WW females; these also can be modified by hormones to produce a male phenotype.

Ploidy.   In hymenopteran species (bees, wasps, and so on), sex is typically determined by the allelic condition of a sex locus. Thus, if any specific allele is either hemizygous or homozygous, the insect develops as a male; heterozygosity results in a female. Because there are many alleles of the locus, fertilized eggs are almost always females. Thus, a honeybee queen (diploid number is 32) can lay two types of eggs. By controlling the sphincter of her sperm receptable (which holds sperm previously obtained in matings with males), she produces a fertilized egg (a zygote having 32 chromosomes that will develop into a female) or an unfertilized egg (a zygote having 16 chromosomes that will develop into a male). In experimentally inbred lines, diploid males homozygous for a sex-locus allele can be recovered. The diploid female zygotes can differentiate into either workers or queens, depending on the diet they consume during development. This is a striking example of the chromosomal control of basic sexual constitution with environmental factors controlling subsequent differentiation.

Balance Theory. Recall C. B. Bridges's study of nondisjunction in *Drosophila* (Chapter 3), which showed that an XXY sex constitution forms phenotypically normal female flies and that an XO constitution forms sterile male flies. Obviously, the Y chromosome is necessary for male fertility but not for the development of the male phenotype. What does determine sex in *Drosophila?* Triploid flies carrying three X chromosomes and three sets of autosomes are normal females. Let's represent a complete set of autosomes as A, so that we can write this genotype as 3X:3A. When triploid females are crossed with normal males, some offspring having different combinations of X chromosomes and autosomes can survive. Bridges found that flies carrying two X chromosomes and three sets of autosomes (2X:3A) are **intersexes,** having intermediate phenotypic characteristics between those of the two sexes. He concluded that sex-determining genes are present on both the X chromosome and the autosomes and that the *balance* between these two kinds of chromosomes determines the phenotypic sex.

For example, suppose that the male-determining genes are on the autosomes and that the female-determining genes are on the X chromosome. In the normal flies, the XX genotype (with an X:A ratio of 2X:2A = 1.0) is female and the XY genotype (1X:2A = 0.5) is male. If the X:A ratio in fact determines the sex, it is not surprising to find that 3X:3A flies (ratio = 1.0) are female and that XO flies (1X:2A = 0.5) are male. The 2X:3A flies have an intermediate ratio of 0.67, and the corresponding phenotype is intermediate. This model is confirmed by the observation of 1X:3A (=0.33) individuals that have an extreme male phenotype and 3X:2A (=1.5) individuals that have an extreme female phenotype. (Both of these abnormal genotypes produce very weak individuals that would have a poor chance of survival in a normal environment.) This **balance model** for sex determination in *Drosophila* seems to explain the observations very well.

Message In *Drosophila*, sex is determined by the balance between male-determining genes on the autosomes and female-determining genes on the X chromosome.

The notion of a "balance" between sex chromosome and autosomal segments implies a weighing of blocks of chromatin. But are specific loci involved? The existence of major sex-influencing genes was suggested when Alfred Sturtevant discovered an autosomal gene called *transformer (tra).* Both *tra* and *transformer-2 (tra-2),* another locus on a different autosome, have no effect on XY males, but XX zygotes homozygous for either gene become phenotypic males. These transformed "males" are sterile, but have all of the external features of males, internal male genital duct systems, and male courtship behavior. An autosomal mutant makes both XX and XY zygotes intersexes, whereas yet another mutation, *intersex,* has no effect on XY males but makes XX flies intersexes. Clearly, normal alleles of important loci initiate the major events that direct a zygote to differentiate as one sex or the other.

Mating-Type Transposition in Homothallic Yeast. In the yeast *Saccharomyces cerevisiae,* transposition has been demonstrated as a normal part of the sexual cycle. Strains of *S. cerevisiae* can be either heterothallic or homothallic. A heterothallic strain is stable as either mating type, α or a, and the successful completion of the sexual cycle depends on the union of α and a strains. We have already seen that in such cases, the two phenotypes segregate 1:1 in the progeny, showing their determination by two nuclear alleles. The alleles are now called *MAT*α and *MAT*a. Homothallic strains, on the other hand, start out as either α or a mating type; however, any culture can go through the sexual cycle by itself without being paired with a strain of opposite mating type. In such cases, the products of meiosis (starting with, say, an α cell) are of two types: one half is α and (surprisingly) one half is a! Careful studies in which cell lineages were followed have shown that the way this remarkable process is achieved is through the switching of mating type by some cells, usually within two or three cell generations. Hence, a cell that starts out as, say, α quickly develops into a culture that is a mixture of α and a cells; these types then pair and go through the sexual cycle.

It became clear from genetic analyses that the process was under genetic control. Four important loci were initially shown to be involved. First, of course, there was the *MAT* locus itself. Second, an unlinked gene *HO* was essential for any kind of mating-type switching to occur, and its allele *ho* was inactive in this regard. Third and fourth, two loci both designated *HM* had to carry the appropriate alleles for switches of a specific mating type to occur. Linked 57 map units to the right of *MAT* was the locus initially designated *HM*α. Apparently, in order for α cells to switch to a, under the influence of *HO,* allele *HM*α was essential, and *hm*α was not adequate. Linked 65 map units to the left of *MAT* was the locus originally designated *HM*a. It appeared that in order for a cell to

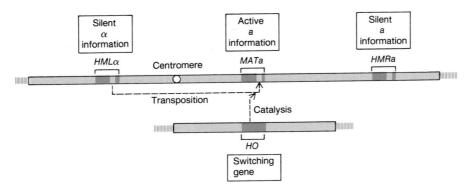

Figure 21-24. Mating-type switching in a yeast cell of genotype *HMLαMATaHMRa*. A silent copy of *MATα* information is present at the *HML* locus, designated *HMLα*, and a silent copy of *MATa* information is present at the *HMR* locus, designated *HMRa*. Under the influence of an unlinked gene *HO*, *HMLα* material is physically transposed to *MAT* with the exclusion of the *a* information residing there. At this point, the mating type of the cell switches from a to α.

switch to α, the allele *HMa* was a prerequisite, and *hma* would not work. However, there was a major paradox: the genotype *hmα*, *hma*, supposedly containing the inactive forms of both loci, would allow switching of either α or a cells!

Another important advance in the elucidation of transposition mechanics came from the observation that the *HMa–HMα* genotype will promote sequential switching, such as from *MATα* to *MATa* and then back to *MATα* again. This was not particularly surprising because the *HM* genes for both kinds of switching are present. What was surprising, however, was that if the culture was initially carrying a defective allele at the *MAT* locus, such as a defective *MATα*, then switching was first to *MATa* and then back, not to the defective *MATα*, but to a perfectly functional *MATα*. The defective *MATα* allelle seemed to have been removed and lost!

The model that was designed to explain all these results led to an exciting series of genetic and molecular tests that revealed a fascinating transposition system. The model, called the **cassette model,** was based on the idea that the *HM* loci were depots of "silent" *MAT* information that could be "played" like cassettes after their physical transposition to the *MAT* locus. At the *MAT* locus itself, the cassette that was inserted by transposition replaced the original cassette residing there, which was lost. However, at the silent *HM* depots, the cassettes that were transposed had to be copies of the cassettes residing there, because the *HM* genotype of the cells remained constant. According to the model, the *HMα* and *hmα* alleles were renamed *HMRa* and *HMRα*, respectively (the R in HMR stands for right). Thus, either silent a (*HMRa*) or silent α (*HMRα*) information can reside at this locus. To the left, *HMa* became *HMLα* while *hma* became *HMLa*, showing that either silent α

(*HMLα*) or silent a (*HMLa*) information could reside there too. The *hma hmα* paradox was explained because its new designation was *HMLa HMRα* and as such it was viewed as containing silent information of both mating types and hence could switch either α to a or a to α. Note that *HMLα HMRa* is a comparable genotype, whereas *HMLα HMRα* is capable only of switching a to α but not the reverse, and likewise *HMLa HMRa* can only switch α to a. The model is graphically depicted in Figure 21-24.

The cassette model was confirmed by a Southern blot analysis using a probe derived from a yeast plasmid carrying the *MAT* locus. The probe bound to three DNA segments, which were shown to be *HML*, *MAT*, and *HMR*. The three regions have now been sequenced; summary diagrams are shown in Figure 21-25. All three

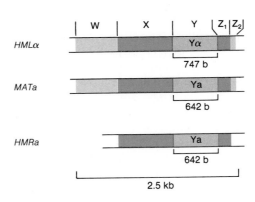

Figure 21-25. The structure of the *HML*, *MAT*, and *HMR* loci. *HML* and *MAT* always have W and Z₂, but *HMR* never does. Either Ya or Yα can exist at any of the three loci. Only the Y region is transposed.

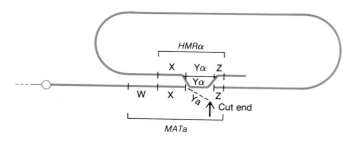

Figure 21-26. Transposition is effected by one of the *HM* genes (here, *HMR*) pairing with *MAT*, followed by an HO-enzyme-mediated cut at the Yα-Z boundary, degradation of Ya, and replacement with a copy of Yα DNA.

regions have similar sequences, but it turns out that only the Y region is unique to one mating type, and this is the part that is transposed.

How does transposition occur? One clue came from showing that homothallic cells regularly contained a double-stranded DNA break at the Y–Z junction of the *MAT* locus only. The break is produced there at a specific DNA target sequence by the product of the *HO* gene, which is a restriction-enzyme-like endonuclease. Either the *HML* or the *HMR* gene bends over the *MAT* gene, the *MAT* Y region is degraded, and Y information is inserted from *HMR* or *HML* by a gene conversion-like process (Figure 21-26). The precise involvement of the single strand break is currently under study.

> **Message** The ability of yeast to switch mating types is determined by the activation of either alternate state through the appropriate insertion into the site responsible for its expression.

Cancer

As we have seen, the study of abnormalities in mutants often provides insight into the normal process of development. In the case of cancer, there is a double payoff: knowledge about normal differentiation and control of cell division gained by studying this abnormal process and, as a bonus, the potential to cure the disease. Tumors are aggregates of cells derived from an initial aberrant founder cell that, although surrounded by normal tissue, is no longer integrated into that environment. Cancer cells often differ from their neighbors by a host of specific phenotypic changes, such as rapid division, invasion of new cellular territories, high metabolic rate, new membrane antigens, altered shapes, and so on. Clearly, the factors regulating normal cell differentiation have been altered, and the interest is in the primary causes of these alterations. An important question is

whether the multiplicity of changes is a reflection of several defects within the tumor cells or a pleiotropic result of a single lesion.

Environment versus Heredity. Is there an underlying basis for tumor indication common to all cancers? For decades, scientists have searched for such a mechanism. The observations that radiation and certain chemicals are carcinogenic point to one or more "targets" of such agents in cells. The known mutagenicity of many of the carcinogens also suggests that DNA may be one of the targets. Well-documented inherited types of cancer such as retinoblastoma and xeroderma pigmentosum do occur, but such types make up a small proportion of all known cancers. Indeed, most cancer experts now agree that up to 80 percent of all tumors are induced by environmental factors.

Tumor Viruses. How can researchers approach the question of the cause of tumor formation? This field of study is known as oncology. Back in 1910, Peyton Rous showed evidence for a "filterable agent" (which was how viruses were defined then) that induced the kind of tumor called sarcoma in chickens. For years, his colleagues didn't believe his claims, but eventually, the Rous sarcoma virus became an important tool in the study of tumors. A number of oncogenic (cancer-causing) viruses have been demonstrated in experimental animals, but in spite of an intensive search, comparable agents have proved elusive in humans. Nevertheless, the study of oncogenic viruses has led to concepts relevant to human disease.

In the early 1960s, Marguerite Vogt and Renato Dulbecco showed that hamster embryonic fibroblasts grown in vitro underwent a profound cellular transformation when infected with polyoma virus. As we have already seen, the transformed phenotype is a reflection of the cancerous state, since transformed cells form tumors on injection into an animal. (Most, but not all, oncogenic viruses induce transformation in vitro.) Vogt and Dulbecco demonstrated that transformation was a consequence of specific genes brought in by the virus. Mutation or deletion of those genes prevents transformation and tumor-forming capability. Thus, with oncogenic viruses at least, the tumor phenotype is the product of a limited number of genes.

Oncogenes. The Rous sarcoma virus belongs to a group called **retroviruses,** so named because, unlike polyoma, which has a DNA genome, their RNA is transcribed "backward" (hence, retro) by reverse transcriptase to make DNA. Transformation of cells by Rous sarcoma virus is found to result from the action of a single *src* gene at one end of the RNA molecule. Here, then, is a single gene that causes cancer. Such a gene is called an

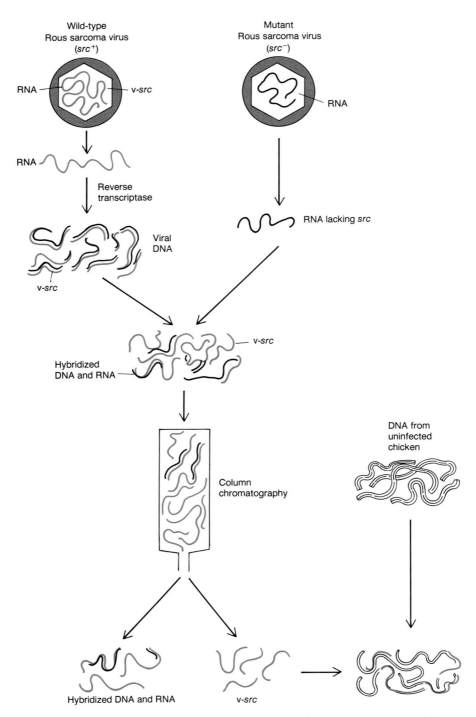

Figure 21-27 Demonstration that DNA that is homologous to oncogenes exists in normal cells. The oncogenic virus RNA is used to make DNA, which is hybridized to mutant viral RNA lacking the oncogene. The viral oncogene (V-*src*) can be separated by column chromatography as unhybridized material, which in turn can be used as a probe to hybridize with complementary sequences in normal chicken DNA. The latter sequences are called proto-oncogenes.

oncogene. What is the function of the *src* oncogene? The virus can multiply with no difficulty even when the *src* locus is deleted, so the oncogene is obviously not necessary for survival. By comparing proteins made by Rous sarcoma virus with and without *src,* Raymond Erikson and his colleagues identified a protein product: an enzyme, called a protein kinase, that phosphorylates (attaches phosphate groups to) the amino acid tyrosine. Phosphorylation affects the activity of a protein. Thus, the src protein alters protein activity and can be considered to be regulatory; it is found concentrated in cell membranes and appears to alter the adhesive properties of membrane proteins.

Origin of src.   But where did *src* come from if it plays no role in viral growth? Using radioactive *src* RNA (Figure 21-27), Michael Bishop's group found that complementary DNA exists in uninfected chicken cells. This shows that *src* DNA is a normal part of the chicken genome. Since this "normal" DNA sequence is assumed to be the precursor of the active oncogene, it is called a **proto-oncogene.** A comparison of viral DNA coding for the viral src protein with the homologous chicken DNA reveals an interesting difference (Figure 21-28). The chicken *src* DNA is typical of eukaryotes with a number of introns, whereas the virally made DNA lacks introns — suggesting that the virus somehow picks up processed mRNA from which the introns are cleaved. Nor is the existence of DNA related to *src* in normal cells unique. Almost every (15 out of 16) retrovirus oncogene tested so far has one or more closely related DNA sequences in uninfected host cells.

Is the attachment of an oncogene to the viral genome necessary for its oncogenic activity? George

Vande Wonde and Edward Scolnick attached isolated cellular oncogenes (one from a mouse and two from rats) to a viral promotor. This engineered DNA was found to transform cells in vitro, thereby suggesting that cancer-causing ability reflects the rate of DNA transcription.

> **Message**   Mammalian cell transformation provides an in-vitro assay for cancer induction. Viral induction of tumors is due to oncogenes that are not necessary for viral reproduction and that appear to be derived from the normal genome of uninfected cells. Tumor induction, therefore, is a reflection of abnormal regulation of gene activity.

The hope that oncogenic viruses would be the primary cause of cancer in humans has not been realized. Yet studies of viral oncogenes have converged with an exciting new approach.

Cell Transformation with DNA.   How could investigators test the involvement of DNA in tumor formation? In 1978, Chiaho Shih and Robert Weinberg devised a direct test based on the original method of Oswald Avery and his coworkers (page 258) to prove that DNA is the genetic material. Shih and Weinberg took DNA from tumors and used it as "transforming principle" to induce the cellular phenotypic changes that (unfortunately) are also called transformation. (In mammalian cells, this process of DNA uptake and expression is also referred to as *transfection.*) In other words, the researchers were asking whether or not the DNA of tumor viruses is oncogenic. They found that it is! Because the

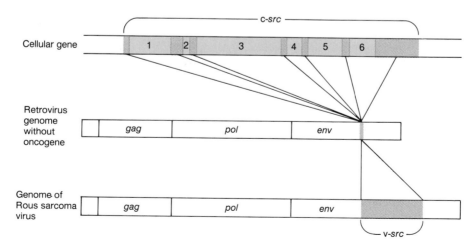

Figure 21-28 A comparison between the viral oncogene and the cellular proto-oncogene (*c-src*). The proto-oncogene consists of both exons and introns. Somehow a retrovirus *(center)* is able to pick up the RNA without the introns and attach it at the end of the virus genome.

DNA of tumor cells is oncogenic, cell transformation is a phenotypic indication of the uptake and incorporation of an oncogene. Then the question became whether or not large numbers of different genes present in the high concentration of DNA, to which the cells were exposed, were responsible. Shih and Weinberg ruled out this possibility by extracting DNA from transformed cells and retransforming normal cells. This could be repeated through three or four cycles. Since extraction of DNA is accompanied by its fragmentation into numerous pieces, only a few of which are actually taken up by the transformed cells, the factor(s) responsible for transformation must remain intact in a single molecule. This conclusion is based on the observation that mammalian DNA transfection results *not* in homologous recombination, as in bacteria or yeast, but in insertion at any of a number of nonhomologous sites. We can conclude, as in the case of oncogenic viruses, that the multiple phenotypic changes signaling cell transformation are pleiotropic effects of a small number of lesions (perhaps even one).

Message   DNA from tumor cells transforms cells in vitro; therefore, the basis for cancer formation rests in the DNA itself.

Recovery of Oncogenes from Normal Cells.   The retrovirus genome, being RNA, provides a convenient probe with which to search for complementary sequences in the host DNA. But without such a probe, how can an

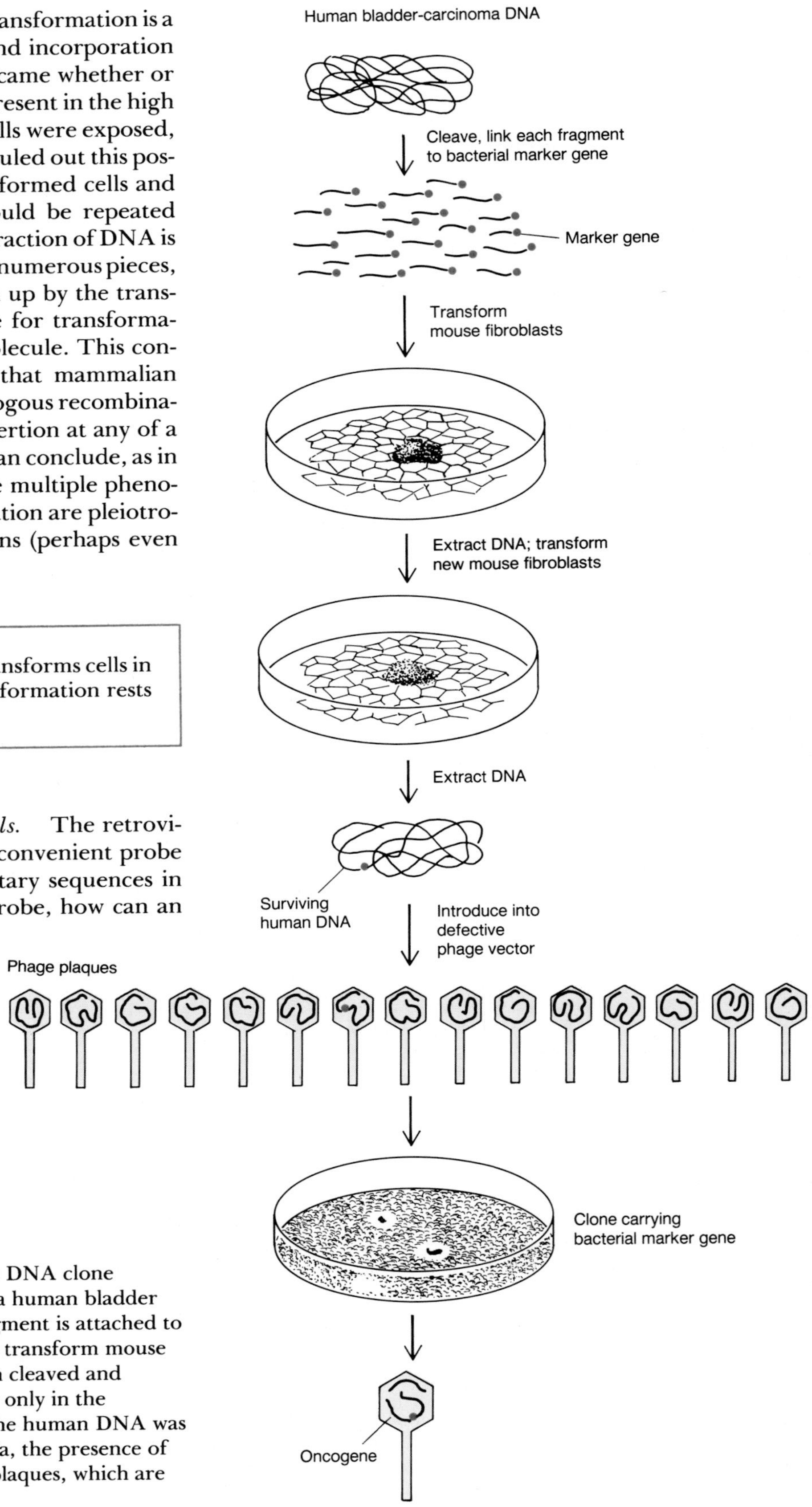

Figure 21-29. Method for identifying the DNA clone carrying a human oncogene. DNA from a human bladder tumor is extracted and cleaved; each fragment is attached to a bacterial marker gene and then used to transform mouse cells. DNA from transformed cells is then cleaved and cloned in a defective phage that can grow only in the presence of the bacterial gene to which the human DNA was originally linked. On plating onto bacteria, the presence of the bacterial gene is indicated by phage plaques, which are then tested for the oncogene.

oncogene be recovered from the vast number of sequences in a normal cell? Several different approaches have been successful.

Michael Wigler developed one of the techniques. He began by breaking DNA from a human bladder carcinoma into pieces and attaching a bacterial marker to each fragment (Figure 21-29). Cells were transformed after exposure to this DNA, and the DNA from these transformed cells was then extracted and cloned. Each clone was tested for the presence of the *bacterial gene* by infecting bacterial cells in which the phage vector was unable to grow unless it carried the bacterial marker. In this way, the bladder oncogene was readily detected.

> **Message** It is possible to isolate oncogenes from tumor-cell DNA using a variety of techniques. The identified oncogenes show that tumor formation results from a change in a small segment of the total genome.

This is just one of the current views. There are others about the number (≥ 2) and type of genes that must mutate to change from normal to malignant. The cell line used in Shih and Weinberg's experiments is already at least one step from normal; other cell lines give different results. Once oncogene DNA is isolated, it in turn can be used as a probe to hybridize with any complementary sequences that might exist in a normal genome. Such sequences have been found. Obviously, they cannot be identical to the oncogene sequences, or tumors would develop; but they are sufficiently closely related to form hybrids with the probe. Such oncogene-related sequences have been found in other mammals, in chickens, and even in *Drosophila*, showing that these sequences must have existed more than 600,000,000 years ago before these diverse groups separated.

Oncogene Fine Structure. Identification of a specific oncogene and a proto-oncogene allows researchers to determine the part of the oncogene DNA that is responsible for tumorigenicity. Weinberg's group found the answer using a classic method of genetic analysis—recombination. With restriction enzymes, oncogenes and proto-oncogenes were cleaved at identical points, and the segments were recombined (Figure 21-30). Each recombinant molecule was tested for its ability to transform cells. In this way, the active site was precisely pinpointed to a segment of 350 nucleotide pairs, which was then sequenced. Astonishingly, the difference between oncogene and proto-oncogene was in a single base pair of a 5000-bp sequence! As a result of a single amino-acid substitution, a normal protein apparently is sufficiently changed to produce an aberrant phenotype of cancer.

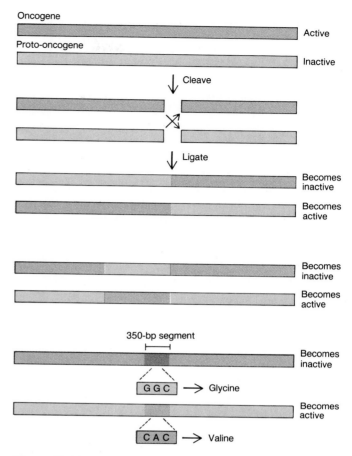

Figure 21-30. Location of the sequence responsible for oncogenicity. With restriction enzymes, the proto-oncogene and the oncogene are cleaved at the same site, the fragments are recombined, and the recombinants are tested for transforming activity. In this way, the exact position can be narrowed to a 350-bp segment, which is sequenced. The cause of oncogenic activity is single base change (G → A) that results in a valine substitution for glycine.

There are still many questions to be answered, but the identification and manipulation of oncogenes provides a powerful means of resolving the basic features of tumor formation. From an understanding of how a normal gene may be altered to shunt cell differentiation into a cancerous condition may come a clearer picture of the normal regulation of cell differentiation. Eventually, strategies for treating cancer may also result.

> **Message** Oncogenes may differ from related proto-oncogenes in normal cells by a single base change. The identification of and ability to manipulate oncogenic sequences provides a powerful tool for the study of cancer.

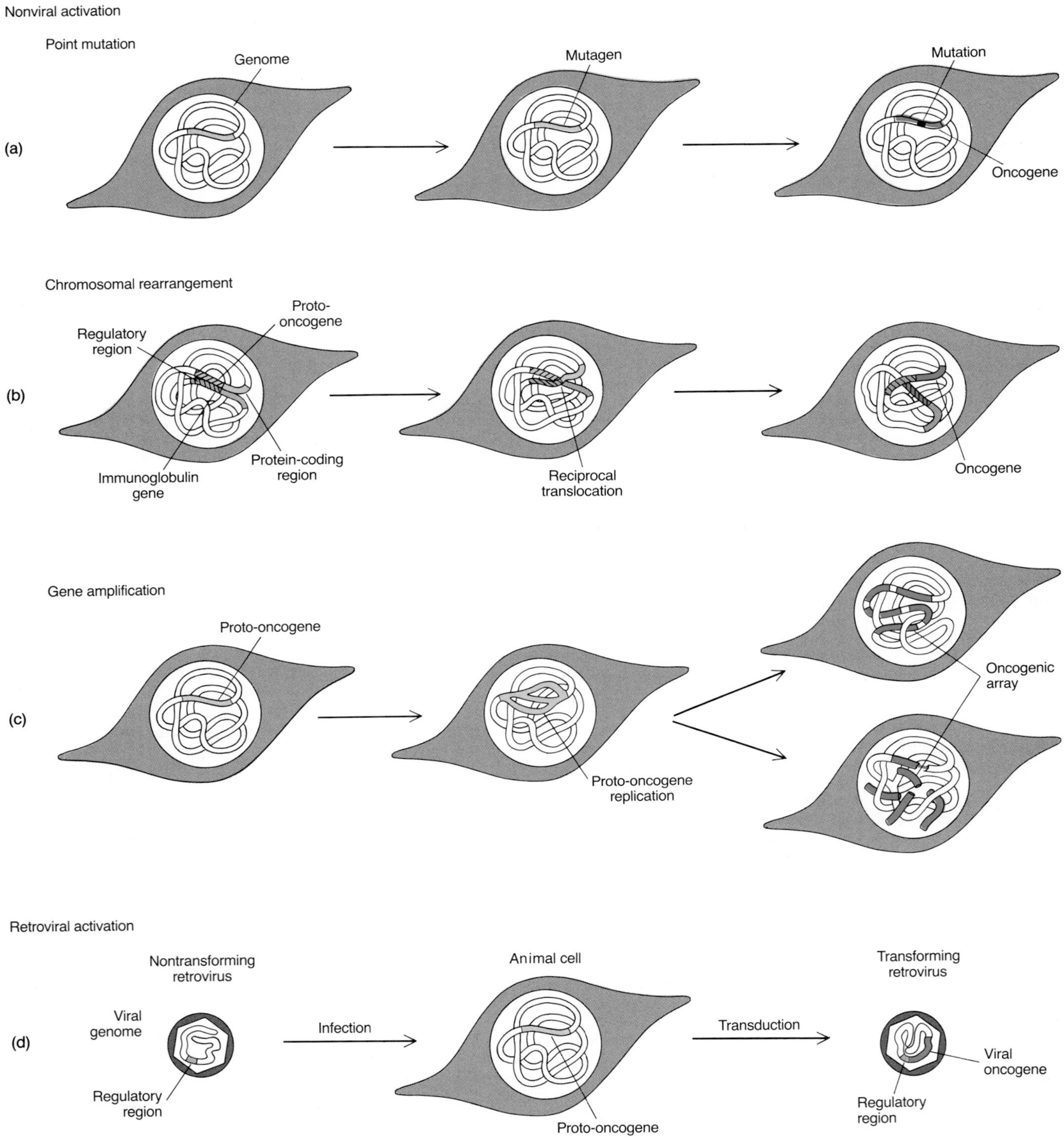

Figure 21-31. Proposed mechanisms of activation of proto-oncogenes. *From top down:* (a) A point mutation induced by a chemical or radiation carcinogen can cause a change in the protein, thereby initiating cancerous growth. (b) Another possibility is the induction of a chromosomal rearrangement that places the proto-oncogene next to the regulatory regions of an immunoglobulin gene. This will lead to the inappropriate expression of the proto-oncogene. (c) If the proto-oncogene is amplified either as repeat segments within the chromosome or extrachromosomally, then the gene product will be overproduced. (d) The final possibility is the retrovirus transfer of a proto-oncogene picked up in another animal cell.

The current model of how a proto-oncogene may be activated to produce a tumor is shown in Figure 21-31.

Immunogenetics

All aspects of a complex organism's phenotype ultimately can be traced to the differential expression of genes and eventually come under the umbrella of genetic regulation. Thus, the blood-producing system, liver, brain, and other organs and structures all represent specific examples of how differentiation occurs. In this sense, the immune system also may be seen as another example of specialized gene regulation. However, unique uses of the genetic material set this particular system apart from others. Immunogenetics has grown explosively in the past decade and is often taught today as an entire course. The immune system of a vertebrate organism provides the body's main line of defense against invasion by such disease-causing organisms as bacteria, viruses, and fungi. The system also attacks cancer cells produced by the organism itself. The complexity of the system can be readily seen in Figure 21-32.

We concentrate here on only one special aspect relating to the production of a specific molecule, the antibody. Antibody production invokes an exceptional method of generating genetic diversity.

Antibodies. When the body is invaded by a foreign agent — whether it is microbial, a chemical substance, or a larger structure such as dust or a pollen grain — one of the most powerful mechanisms for eliminating it is in the production of an **antibody**. The antibody is a protein that recognizes a specific stereochemical shape determined by the invading **antigen**. The antigenic portion may be an entire molecule or a part of a molecule or structure.

Humans seem to be able to manufacture an unlimited variety of antibodies against millions of potential antigens. The antibodies have been shown to be proteins belonging to a class called **immunoglobulins**. The genetic dilemma is that antibody specificity reflects its primary amino-acid sequence, yet the human genome has only perhaps a million genes. Furthermore, only a small number of these genes actually code for the immunoglobulins. The paradox of generating a large number of different products from a small number of genes is now being resolved at the molecular level.

Antibody Structure. Antibodies are made up of two pairs of identical subunits: light and heavy polypeptide chains bound together by disulfide bridges. The analysis of amino-acid sequences and the comparison of different antibodies have revealed that there are large areas that are similar in all antibodies (Figure 21-3); such areas are called **constant**. Within the **variable** domain, in which

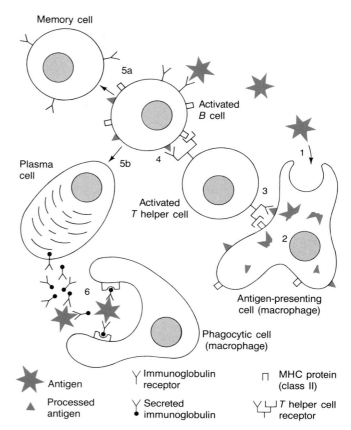

Figure 21-32. The cellular immune response to an infection. (1) The antigen is absorbed by a cell such as a macrophage. (2) The antigen is broken down into components which are displayed on the surface of the cell. The antigen is recognized by a helper T cell (3) which in turn activates B cells (4) which are also carrying pieces of the antigen. The activated B cells then differentiate into memory cells (5a) which respond in fugure infections with the same agent or become plasma cells (5b) which secrete antibody. The antibodies bind to the antigen (6) thereby forming a complex that is destroyed by macrophages. (From S. Tonegawa, "The Molecules of the Immune System." Copyright © 1985 by Scientific American, Inc. All rights reserved.)

amino-acid differences do occur from antibody to antibody, a large proportion of amino acids is nevertheless similar except for short **hypervariable** regions; in these, extreme variation is responsible for antibody specificity.

How does the variability come about? The ability to clone DNA coding for light and heavy chains has revealed a surprising mechanism. Separate DNA segments code for the constant and variable parts of the chains. In embryonic DNA, the constant and variable parts are not closely linked; in somatic cells, however, the two parts are brought into closer juxtaposition by some kind of

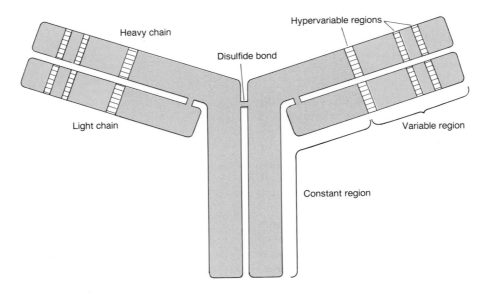

Figure 21-33 A diagrammatic representation of an antibody. It is made up of two pairs of identical polypeptides, each of which contains a light chain and a heavy chain linked by disulfide bridges. The constant regions are located in the stem region; the variable and hypervariable areas are located in the region of the branches.

somatic recombination event. There are families of variable sequences that, despite their differences, share certain amino acids that always differ between families. The kind of picture that emerges shows perhaps 100 to 200 different copies of the variable portion linked via a series of short, joining sequences to a single constant region (Figure 21-34). Somatic recombination links a specific variable segment to the constant gene; gene splicing maturation of RNA and protein removes excess coding portions.

From our standpoint, the important fact is that recombination between segments of homologous DNA molecules is the mechanism whereby enormous variety can be generated by shuffling linked sequences. Note

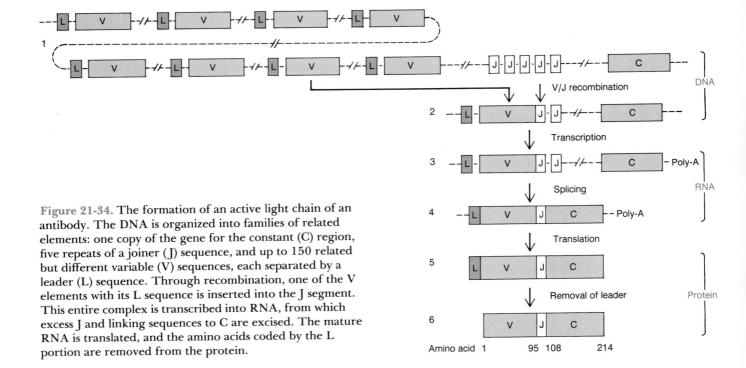

Figure 21-34. The formation of an active light chain of an antibody. The DNA is organized into families of related elements: one copy of the gene for the constant (C) region, five repeats of a joiner (J) sequence, and up to 150 related but different variable (V) sequences, each separated by a leader (L) sequence. Through recombination, one of the V elements with its L sequence is inserted into the J segment. This entire complex is transcribed into RNA, from which excess J and linking sequences to C are excised. The mature RNA is translated, and the amino acids coded by the L portion are removed from the protein.

that, just as Charles Yanofsky showed (Chapter 12), recombination between two different triplets can also generate still other new triplets in the hypervariable region (Figure 21-35).

> **Message** The enormous variability of antigens derives from a small number of genes. Two groups — variable and constant genes — can be linked together by somatic recombination. This provides a means of generating a vast array of different gene products.

Monoclonal Antibodies. The preceding discussion presents an extremely brief description of what is obviously a fascinating system. The tools of molecular biology are rapidly exposing the secrets of the immune system. When a foreign agent such as a virus, a bacterium, or some large molecule penetrates the physical barriers to its entry, cells of the immune system begin a search for antibodies to inactivate it. The agent may have several regions of antigenicity that an antibody may recognize. Thus, the immune response to an agent can be the production of a heterogeneous collection of antibodies, each specified by one antibody-producing cell called a **lymphocyte** or its clones.

Antibody-producing lymphocytes cannot be cultured long enough to provide a great deal of material. There are, on the other hand, tumors of the immune

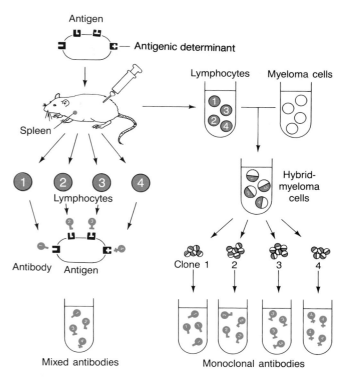

Figure 21-36. Recovery of monoclonal antibodies begins with the injection into a mouse of an antigen with several different sites capable of inducing specific antibodies. Each antibody is produced by a different lymphocyte. The antiserum contains a heterogeneous mixture of antibodies *(bottom left)*. Antibody-producing lymphocytes are recovered from the mouse and fused singly with myeloma cells. The hybrid cells that result are then isolated and grown separately to yield clones of identical daughter cells which generate a single antibody *(bottom right)*. (Figures 21-36 and 21-37 from C. Milstein, "Monoclonal Antibodies." Copyright © 1980 by Scientific American, Inc. All rights reserved.)

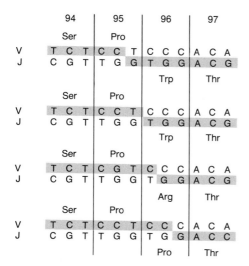

Figure 21-35. Recombination of certain points between V and J sequences can generate new codons. Amino acids 95 and 96 of an active light chain are at the position of exchanges between V and J. Exchanges in the codon for amino acid 96 can lead to arginine or proline instead of tryptophan.

system called **myelomas** that grow indefinitely in culture and produce large amounts of immunoglobulin. The problem with the myeloma globulins is that their target antigens cannot be determined *a priori*. In 1975, in the hope of surmounting this obstacle, Cesar Milstein fused myeloma cells with lymphocytes that had been induced to make antibodies against sheep red-blood cells (Figure 21-36; mice have since replaced sheep as the source of red blood cells). These **hybridomas** retained the proliferative properties of the tumor and the antibody specificity of the lymphocyte and thus manufactured large quantities of a specific antibody. Antibodies like this one that are recovered from isolated clones of specific hybridomas are termed **monoclonal antibodies.** Now large amounts of pure antibody can be readily obtained for specific antigens (Figure 21-37). For pharmaceutical use and in pure research and medical science monoclonal antibodies have quickly become an invaluable tool.

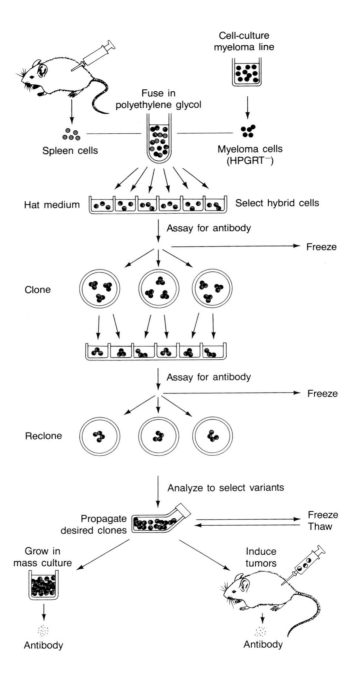

Cell-culture
myeloma line

Fuse in
polyethylene glycol

Spleen cells

Myeloma cells
(HPGRT⁻)

Hat medium   Select hybrid cells

Assay for antibody   Freeze

Clone

Assay for antibody   Freeze

Reclone

Analyze to select variants

Propagate   Freeze
desired clones   Thaw

Grow in   Induce
mass culture   tumors

Antibody   Antibody

Figure 21-37. Standard procedure for generating monoclonal antibodies. An antigen is injected into a mouse from which antibody-producing spleen cells are subsequently isolated. They are fused with myeloma cells deficient in enzymes TK and HPGRT, which are involved in minor metabolic pathways in the production of deoxynucleotide precursors, TMP and GMP, respectively. On a selective HAT medium containing hypoxanthine, aminopterin, and thymidine, aminopterin blocks the major nucleotide pathway while TK and HPGRT mutants cannot incorporate the precursors for the minor pathway. Thus, only hybrid cells can grow. Each clone is tested for antibody and preserved by freezing. Those lines selected for retention can be grown in culture to yield antibody *(bottom left)* or injected into hosts to induce tumors which yield antibody *(bottom right)*.

Summary

■ Developmental biologists seek an understanding of the mechanisms whereby a fertilized egg differentiates into the many cell types of a multicellular organism. In the early stages of development, each somatic cell receives a complete complement of DNA and, therefore, should be totipotent; this has been demonstrated in some plants and a few vertebrates and invertebrates. The cytoplasmic architecture and content of an egg is determined by the maternal genotype and controls the early developmental sequences.

Molecular models of regulated genes are best exemplified by the operon. By connecting products of an operon to other operons, as repressors or activators, a multitude of phenotypic fates is readily determined. From gastrulation onward, transcription sequences are stage-specific and reflect regulated gene activity. Specific phenotypes in highly differentiated cells such as chick oviduct or *Drosophila* salivary glands provide diagnostic criteria for molecular probes into the regulation of activity and the recovery of gene products. Mutations reveal that a great deal of regulation is effective in sequences upstream from the 5′ end of the structural gene.

All differentiated cells and tissues are the products of regulated gene activity, but certain systems within a whole animal are of special interest for different reasons: sex determination, cancer, and the immune system. Sex determination in *Drosophila* reflects a balance between male- and female-determining factors scattered throughout the genome. Nevertheless, mutations of major sex-determining elements can drastically affect sex differentiation. Cancer represents unregulated cell growth. Convergent studies of oncogenic viruses and the DNA of tumors reveal the existence of oncogenes, which are capable of inducing tumors. Oncogenes, in turn, appear to be derivatives of the DNA normally found in all cells. The immune system defends the body against invasion by foreign elements. Antibodies — proteins called immunoglobulins — are the mechanisms of destruction of foreign antigens. The enormous variety of antigens is determined by the somatic recombination of different genetic elements.

■　■　■

Solved Problems

1. You observe that, during larval development of *Drosophila*, an enzyme specifically found in the salivary glands appears, increases, and disappears in exactly the same pattern as a specific puff on the polytene X chromosome.

    a. Does this observation prove that the enzyme is specified by the puff? Explain.

    b. Design experiments to test the hypothesis that the puff specifies the enzyme.

    Solution

    a. The correlation of the puffing and the appearance of an enzyme is not proof that they are causally linked. It could be a fortuitous correlation of two unrelated phenomena or they could both happen to respond similarly to a common inducer. Of course, it is possible that there is a causal link between the state of the puff and the concentration of enzyme.

    b. There are a number of ways that one might test the relationship between the puff and the enzyme. The most direct way would be to determine the sequence of the DNA in the puff and the amino acid sequence of the enzyme and determine whether any segment of the DNA codes for the protein.

    A less direct method would be to select for mutations of the puffs and the enzyme. Thus, one might screen for mutations that result in a change in the time of appearance of the puff or of the enzyme so that in each mutant stock one could determine whether the appearance of the enzyme or puff, respectively, is also changed. Another class of mutations that can be selected is the absence of puff or enzyme; then the presence or absence of the enzyme or puff, respectively, can be checked. If the behavior of the puff and enzyme is similar in all these genetic combinations, then it can be concluded that the correlation between them is causally based.

2. Jonathan Jarvik and David Botstein developed the following method for determining the order of activity of different genes. Suppose that you obtain two different phage mutations, A and B, with an identical phenotype of incomplete head assembly. You recover a heat-sensitive mutation of A (the mutant dies at 40°C) and a cold-sensitive mutation of B (the mutant dies at 25°C); both mutants survive at 35°C. You construct a phage carrying both mutations (A^{hs} B^{CS}) and infect bacteria with it. You keep one culture of infected bacteria at 25°C for 10 minutes and then shift it to 40°C (call this the lo → hi culture). You keep another culture (hi → lo) at 40°C for 10 minutes and then shift it to 25°C. Interpret each of the following possible results of this experiment.

    a. Phages are released only in the hi → lo culture.

    b. Phages are released only in the lo → hi culture.

    c. No phages are released in either culture.

    d. Some phages are released in both cultures.

Solution

a. This suggests that the two genes act sequentially in the same pathway. The release of phages in the hi → lo culture suggests that B functions while the culture is at 40°C and subsequently at 25°C; there is sufficient product of the B-catalyzed reaction for A to act while B is knocked out.

b. This is interpreted in the same way as (a) to suggest that A acts first and is followed in the same process or pathway by B.

c. This suggests that there is no direct relationship between the two genes. They must both be vital at all times in the phage cycle so that at either temperature, one of the critical functions is lost and the phage cannot reach maturity.

d. One interpretation is that the products controlled by both genes are assembled into the final phage. At each temperature, either A or B products are made and incorporated into the forming phage particle. Upon shifting to the reciprocal temperature, the other product is manufactured and incorporated into the already partially formed particle. Thus, phages will be released in both shifts, but the number of particles produced by each infected bacterium should be considerably reduced. There are other ways the last two results can be interpreted. Perhaps you can think of others.

Problems

1. Review the preceding chapters and prepare a list of all the ways in which variegation (sectoring) of biological tissue can occur.

2. a. Of what significance are phenocopies in medical genetics?

    b. How can phenocopies be used to study gene action?

    c. How can an investigator determine whether an altered phenotype is due to a phenocopy or to a mutation?

3. Throughout history and mythology, there are stories of the birth of children to virgin mothers. Such parthenogenesis definitely does occur in many organisms. In theory, it could occur in mammals, including humans. Describe the mechanism that might be involved if any of the claims of the virgin birth of a son are true.

4. Cells of organisms that are phylogenetically distantly related can be fused to form single mononucleate hybrids. For example, human–mouse, human–fish, human–bird, human–insect, and human–plant hybrid cells have been created. These hybrids even grow and divide.

    a. What conclusions can be drawn from these observations?

    b. When a chicken erythrocyte (which has a nucleus and actively synthesizes hemoglobin) is fused to a mouse fibroblast, some of the inactive chicken DNA becomes transcriptionally active but hemoglobin synthesis is shut off. What conclusions can you draw from this observation?

5. In his book *In His Image: The Cloning of a Man*, David Rorvik claimed that a baby had been cloned from the cells of an elderly man. Few (if any) geneticists believe this story, but many scientists do feel that human cloning is a definite possibility in the future. Naturally occurring human clones (identical twins) do exist. How would the ability to produce human clones in the laboratory be useful in studying the relative importance of heredity and environment in determining the final phenotype of a human individual?

6. The figure on page 607 illustrates an experiment performed by J. Hämmerling in 1943 on two species of the unicellular alga *Acetabularia*. The "hat" of *A. mediterranea* is intact and umbrella-like, whereas the hat of *A. crenulata* is deeply indented. The nucleus is embedded in the base of the single cell in both species, and the hat is borne on a long cytoplasmic stem.

    a. The stem of an *A. mediterranea* individual is cut off just at the base, and the hat is removed by a cut at the top of the stem. The cut stem is grafted to the base of an *A. crenulata* individual, and a new hat forms (regenerates) on the top of the stem, as shown in the right side of the figure. A reciprocal transplant of an *A. crenulata* stem to an *A. mediterranea* base also leads to the regeneration of a new hat, as shown in the left side of the figure. In each case, the phenotype of the

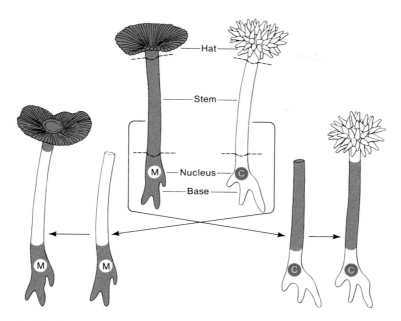

See Problem 6. Grafting experiments in *Acetabularia* have provided important information about the control of patterns of regeneration. When a cut stem from *A. crenulata* (C) is grafted to a base from *A. mediterranea* (M), the stem regenerates a hat of the *A. mediterranea* type (*left*). In the reciprocal experiment, the regenerated hat again matches the phenotype of the base (*right*). (After J. Hämmerling, *Z. Abstg. Vererb.* 81:114, 1943.)

new hat is normal for the base to which the stem is transplanted rather than for the stem from which the hat grows. On the basis of these results, discuss the cytoplasmic versus nuclear control of regeneration in *Acetabularia*.

b. If the nucleus of an *Acetabularia* cell is removed and the hat is then cut off, the enucleated cell immediately regenerates a new hat. However, if the upper part of the stem is removed along with the hat, the enucleated cell does not regenerate a new hat. Assume that some "hat-forming substance" is responsible for regeneration. Discuss the origin, qualitative control, and distribution of this substance in the intact cell.

c. If a stem with an intact hat is transplanted from one species of *Acetabularia* to the base of another species, the hat retains its original character. How does this observation modify your answer to (b)?

7. Boris Ephrussi and George Beadle have developed (and applied to good advantage) a technique for the study of genetic effects on hormone-like materials in *Drosophila*. The figure on page 608 diagrams this technique and some of the results of its application. A piece of the larval disk that would later have produced an adult eye is transplanted to a genetically different larva. The developmental interactions between host and transplant are then observed, particularly in terms of the color of pigment developed by the transplanted eye disk as the host larva matures.

a. The reciprocal transplants shown in part (a) of the figure are typical of a large majority of *Drosophila* transplantation experiments. The larval disk from a wild-type fly develops the pigment color characteristic of its own genotype, even when its differentiation and pigment development occur in a white-eyed host. Similarly, an eye disk from a white-eyed larva develops according to its own genotype and is not influenced noticeably by a wild-type host environment. Do these results allow you to conclude that the white-eyed gene affects the production of a circulating (hormone-like) material or that it acts directly in the developing eye tissue itself? Explain your answer.

b. The experiment shown in part (b) of the figure represents the kind of exception from which a good deal of information has been derived. Again, when wild-type eye disks are transplanted to either vermilion (*v*) or cinnabar (*cn*) host larvae, the transplants develop autonomously into wild-type eyes. However, when vermilion or cinnabar disks are transplanted to wild-type larvae, the disks do not develop according to their own genotypes; instead, they produce wild-type eyes! Explain these experimental results in detail, assuming that the body of the wild-type host is capable of providing the developing eye tissue with circulating or diffusing materials that compensate for the genetic blocks in vermilion and cinnabar flies.

c. Part (c) shows the results of reciprocal transplants between vermilion and cinnabar larvae. Cinnabar

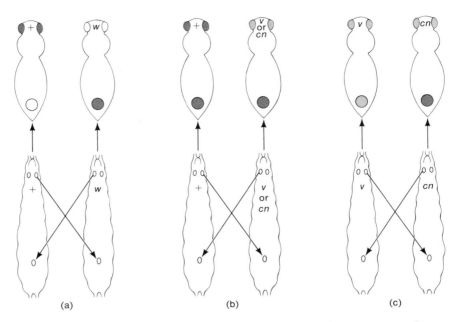

See Problem 7. Transplants of larval tissue have elucidated some aspects of gene action in *Drosophila*. (After B. Ephrussi, *Quarterly Review of Biology* 17, 1942, 329.)

disks developing in vermilion hosts maintain their cn phenotype, but vermilion disks develop as wild-type in cinnabar hosts. The genes are now known to be involved in the control of the sequential steps in a biochemical synthesis of hormone-like materials that is directly involved in production of eye pigment:

$$\text{tryptophane} \xrightarrow{v^+} \underset{(v^+ \text{ substance})}{\text{kynurenine}} \xrightarrow{cn^+} \underset{(cn^+ \text{ substance})}{\text{hydroxykynurenine}}$$

$$\downarrow$$

$$\downarrow$$

$$\text{pigment}$$

Using this information, give a detailed explanation of *all* the transplantation results shown in the figure.

8. One theory of aging (senescence) assumes that randomly occurring somatic mutations accumulate over time in different cells within an individual, gradually disrupting the normal cellular processes necessary for life. Another theory suggests that lifespan is a genetically determined character and that aging and death are simply the final stages of development and differentiation. These hypotheses are sufficiently specific to permit the design of direct tests. Design experiments to test each of these theories. (NOTE: choose your organism and the kinds of tests carefully.)

9. The following diagram shows a hypothetical operon circuit in which the structural genes *(G)* adjacent to the operators *(O)* of two operons specify enzymes (E) that act by converting substrates (S) into products (P). Each product

interacts with the repressor produced by the repressor gene *(RG)* to shut off the operator of the *other* operon. What would be the reaction of this system to variations in supplies of substrates S_1 and S_2?

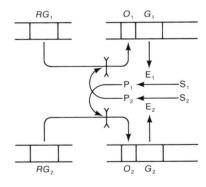

10. Different molecules in many complex structures (such as ribosomes and phage particles) can assemble spontaneously in a specific sequence. Design experiments to determine whether nucleosomal elements possess this capacity and whether the histones distinguish between DNA's.

11. The X-linked allele for glucose 6-phosphate deficiency *(Gd⁻)* also confers resistance to malaria. You make a microscopic examination of the blood of a woman suffering from malaria. You observe that about one-half of the cells contain parasites, whereas one-half of the cells are unaffected. Explain this result.

12. Women who are XO have a distinctive phenotype that includes short stature, webbed necks, and sterility. Men who are XXY can also be recognized phenotypically by

reduced axillary hair, enlarged breasts, and sterility. However, XYY males seem to be fertile, and phenotypically normal. What do these observations indicate about sex-chromosome function?

13. Through some aberrant circumstance, chromosome doubling occurs in certain somatic cells in a triploid *Drosophila* female. Will these clones of hexaploid cells differentiate as "male" or as "female" tissue?

14. The addition of duplication fragments of X chromosomes into nuclei from *Drosophila* intersexes (2X : 3A) shifts the balance of sexuality. Will this shift be in the direction of maleness or femaleness?

15. In *Drosophila*, an autosomal recessive *tra* acts only in females, transforming them into sterile males. What ratio of males and females would you expect among the progeny of a $+/tra \, ♀ \times +/tra \, ♂$ cross?

16. In many forms of cancer, immunologically detectable changes occur at the surface of cancer cells. In many cases, the new antigens made by a cancer cell are identical to those found on normal embryonic cells. Propose an explanation for these observations.

17. Methylcholanthrene is a potent tumor-causing chemical in mice. It is also a potent mutagen. If a methylcholanthrene-induced tumor (T_A) is removed from mouse A, the mouse is "cured." The T_A cells then can be kept alive in another mouse. When the T_A cells are reintroduced into the cured mouse, the mouse is seen to be immune (no tumors develop), although tumors can be induced by injecting T_A cells into mouse B, which has not previously been exposed to T_A cells. However, the mice surgically cured of tumor T_A are not immune to T_C cells removed from a methylcholanthrene-induced tumor in mouse C; such an injection induces a T_C-type tumor in mouse A. Propose an explanation for these observations in terms of gene–protein relations, bearing in mind that cell-surface proteins can act as immunological antigens.

18. You have two separate homothallic cultures of *Saccharomyces cerevisiae* with genotypes

$$MAT \; \alpha HO \; HMLa \; HMRa$$

$$MAT \; \alpha HO \; HML\alpha \; HMRa$$

Each culture is allowed to sporulate, and one tetrad is isolated from each. What mating behavior can be expected for the individual ascospores? (For each tetrad, categorize the mating behavior of each of the four ascospores as α, a, or homothallic. The order within the ascus is unimportant; also ignore recombination.)

19. In *Drosophila*, heterozygotes for dominant temperature-sensitive lethal mutations are viable and fertile at the permissive temperature, but they die at the restrictive temperature. What could be the molecular basis for such dominant lethality?

20. You are a molecular biologist studying the properties of muscle proteins in *Drosophila*. Design an experimental procedure for the detection and recovery of appropriate mutants. (This experiment has not been attempted, so you may be able to propose a clever scheme that is worth trying. NOTE: what would be the phenotype of a muscle mutant? Wouldn't it be lethal?)

21. Competition studies of nucleic acid hybridization provide a means of comparing sequences from different sources. In order to study mRNA in mice, suppose you attach a fixed amount of cold, single-stranded mouse DNA to nitrocellulose filters and, in the presence of a fixed amount of hot mRNA from liver, mix different amounts of cold mRNA from either lung or liver. You obtain the results shown in the following graph.

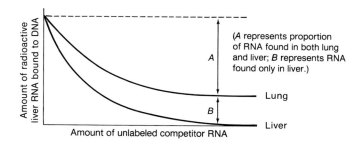

(A represents proportion of RNA found in both lung and liver; B represents RNA found only in liver.)

a. What conclusions do these results permit you to draw about differentiation?

b. How would you determine the mRNA sequences found only in liver and in no other body tissue?

22. Eukaryotic cells can be lysed without breaking down the nuclear membrane. This technique permits the separation of cytoplasmic from nuclear material. In completion studies, unlabled nuclear RNA is found to compete with the binding of labeled cytoplasmic RNA to the DNA. However, unlabled cytoplasmic RNA fails to interfere with the binding of a considerable amount of labeled nuclear RNA to the DNA.

a. What do these studies show?

b. Can you suggest a possible biological interpretation of these observations?

23. There is evidence to suggest that genes inherited from each parent are not activated simultaneously in the early embryo. Devise a molecular strategy to distinguish the time of activation of maternal and paternal genes.

Genetic Analysis of Development: Case Studies

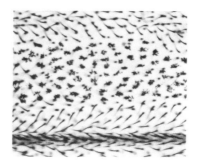

KEY CONCEPTS

Case studies show how genetic analysis can probe development.

∎

Using mutations as cell markers, the lineage of clonally
related cells can be traced in genetic mosaics.

∎

In mice, mosaics have revealed the cellular origins of fetal
skin and striated muscles, as well as insights into the basis of
hereditary disease.

∎

The unvarying position and movement of every cell during
the development of the embryo of the nematode worm
Caenorhabditis elegans permits the experimental disruption of
normal development surgically and genetically.

∎

In both *Caenorhabditis* and *Drosophila,* the range of cell types
that a specific embryonic cell can form is progressively
restricted over time.

∎

Factors designating the axes (anterior-posterior,
dorsal-ventral, lateral) and major body divisions (head,
thorax, and abdomen segments) are present in the egg cortex
of *Drosophila,* and these factors control which zygotic genes
are activated.

■ All living matter is organized on the same basic plan. The principal elements (carbon, hydrogen, oxygen, nitrogen, phosphorus, sulfur) are the primary constituents of all biological molecules; the main macromolecular components also are common to all cells. Interspecific comparisons of proteins and nucleic acids reveal a considerable sharing of common genes, even between organisms of considerable evolutionary separation. At the molecular level, the difference between people and apes is barely detectable. All of this suggests that the major determinative agents of species differences are timing, pattern, and degree of genetic activity rather than qualitative differences.

In Chapter 21, we saw how the activity of specific genes can be correlated with the appearance of products that are characteristic of differentiated cells. Thus, production of hemoglobin in the red blood cells of mammals, silk protein in the glands of silkworms, or glue in the salivary glands of *Drosophila* all reflect the activation of a specific gene or set of genes.

How and when does *determination*—the commitment of a cell to a specific developmental fate—occur? And how does an egg cell, from which all descendants receive an identical genome, partition the signals to the battery of genes that will be called into play at specific times in a designated group of cells? It is hard enough to decipher the three dimensions of a house from the two dimensions of a blueprint. The challenge for geneticists who wish to probe development is to determine how a linear sequence of instructions encoded in DNA can direct cells into tertiary and even higher levels of organization. In this chapter, we focus on three organisms—the mouse, a nematode worm, and *Drosophila*—to see how mutations and molecular tools can be used to trace basic events of development and to reveal underlying patterns in determination and differentiation.

Mutations in Aid of Embryology

In order to probe development, we need to be familiar with a great deal of descriptive biology of the developmental process so that we can formulate a question or develop an experimental procedure. If you have ever watched time-lapse movies of embryogenesis, you know that cell and tissue movements are extremely complex. In the dynamic process of embryogenesis, folding or migration can bring a group of cells to an area of the embryo very far from its original position in the blastula. How can we trace these cell movements during embryogenesis? The classical approach is physical: label specific cells of amphibian embryos with visible carbon particles, so that movements of the marked cells (or their descendants) can be traced during development. Using this marking technique, researchers can construct **fate maps** showing the destinies of the descendants of particular cells in the early stages of embryogenesis.

An alternative method is to mark cells *genetically*, so that they and their descendants can be distinguished phenotypically. This approach utilizes genetic mosaics, or individuals composed of several genetically distinct cell populations. Such mosaics do occur naturally. For example, the exchange of blood cells in the placentas of twin cattle can produce chimeras that carry two blood types. As discussed in Chapter 9, somatic nondisjunction in humans can produce a person whose cells carry different numbers of X and Y chromosomes. Furthermore, every female mammal is a mosaic in terms of the functional states of her two X chromosomes.

Tetraparental Mice

Recall Karl Illmensee's experimental generation of mosaics by injecting single cells into embryos of *Drosophila* and mice (Figure 21-9). Beatrice Mintz developed a very elegant technique for fusing the developing embryos of two different mouse genotypes (Figure 22-1). When implanted in a host female, such an embryo develops as a single mosaic individual. Because each of the original embryos had its own parents, the fused product can have *four* parents and, hence, is termed **tetraparental**.

The sex of the embryos being fused is not known at the time of fusion; by chance, about one-half of the fusions will be between a male embryo and a female em-

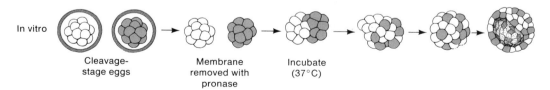

In vitro   Cleavage-stage eggs   Membrane removed with pronase   Incubate (37°C)

Figure 22-1. Formation of a genetic chimera by fusion of two mouse embryos of different genotypes. The embryos (at the eight-cell or 16-cell stage) are stripped of their surrounding membranes by treatment with the enzyme pronase, which breaks down proteins. The two cell clusters are placed together and incubated at 37°C. The cells adhere and mix together to form a single mosaic embryo. (Figures 22-1 and 22-2 from B. Mintz, *Proceedings of the National Academy of Sciences USA* 58, 1967, 345.)

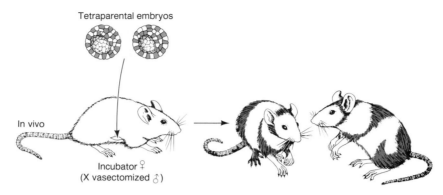

Tetraparental embryos

In vivo

Incubator ♀
(X vasectomized ♂)

Figure 22-2. Tetraparental mouse embryos are transplanted into a female whose womb is receptive for implantation after mating with a sterile male. The resulting offspring are mosaics for the fur genotypes of the two strains from which they are derived. The patterning of the fur phenotypes in dorsal-to-ventral stripes indicates the regions of skin that have been derived from a single embryonic cell.

bryo. Under the influence of the male hormones, most of these sexual chimeras differentiate as males, even though the presence of the XX cells can be demonstrated. All of the daughter cells of the cells introduced at the time of embryo fusion represent genetic clones, so the recovery of a genotypically identical cluster of cells indicates their common origin through division.

The Origin of Skin.   Mintz fused embryos from a mouse strain having black fur with embryos from a stock having white fur. If the cells from either genotype can become precursors of skin, then the pattern of fur color should reflect the clonal origins of the skin cells. Mintz found that the fur pattern of such mice always involves bands of black or white fur that circle the body to form a stripe from the stomach (ventral surface) to the back (dorsal surface) on each side (Figure 22-2). This information indicates that the cells from which descendants will produce fur pigment line up randomly in pairs along the midline of the embryo and that each then divides to form a clonal sheet extending halfway around the body to meet the other sheet. Clearly, mutations producing easily identifiable phenotypic features such as fur color are extremely useful as genetic markers in tracing cell lineage during development. We see later in this chapter that such genetic dissection of development has been honed to a fine edge in *Drosophila.*

The Origin of Striated Muscles.   Using tetraparental mice, Mintz set out to obtain a definitive solution to the problem of the origin of the multiple nuclei in a striated muscle cell. A muscle cell differentiates from precursor cells called **myoblasts,** each of which has a single nucleus. Does a muscle cell arise through the successive fusion of different myoblasts, or does a mononucleate myoblast undergo a series of nuclear divisions without

division of the cytoplasm? To distinguish between these two models, Mintz produced tetraparental mice from two lines that differed in the electrophoretic mobility of the enzyme isocitrate dehydrogenase-1. Electrophoretically distinct enzymes are produced by *Id-1*ᵃ and *Id-1*ᵇ homozygotes, whereas the *Id-1*ᵃ/*Id-1*ᵇ heterozygotes also exhibit an intermediate hybrid band indicative of a dimer containing one polypeptide unit from each phenotype (Figure 22-3). Formation of the hybrid dimer can occur only if both polypeptides are synthesized in the same cell, as shown by the fact that only the parental enzyme types are derived from the livers of mice that are chimeric for uninucleate cells of both homozygous lines.

The pattern of enzyme phenotypes in tetraparental mice permits researchers to test the two hypotheses for the origin of the muscle-cell nuclei (Figure 22-4). The

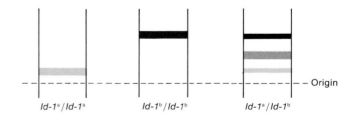

*Id-1*ᵃ/*Id-1*ᵃ   *Id-1*ᵇ/*Id-1*ᵇ   *Id-1*ᵃ/*Id-1*ᵇ   — Origin

Figure 22-3. Electrophoretic mobility of isocitrate dehydrogenase-1 of different genotypes. Each homozygote shows a single isocitrate dehydrogenase-1 band with a specific mobility in the electrical field. However, the heterozygote (*Id*ᵃ-*I*ᵃ/*Id*ᵇ-*I*ᵇ) shows three bands: one with the mobility of each parental strain and a band of intermediate mobility. The intermediate band stains approximately twice as intensely as either parental band. This pattern indicates that isocitrate dehydrogenase-1 is a dimer, and the intermediate band contains one polypeptide from each of the parental alleles.

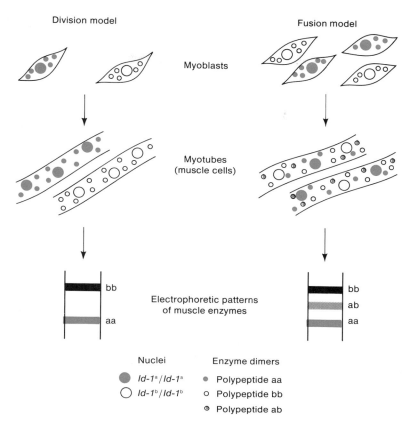

Figure 22-4. Predictions of differing results for the electrophoresis of enzymes based on two different models of the origin of multinucleate muscle cells, or myotubes. If a myotube is formed by successive divisions of the nucleus in a single myoblast, then the electrophoretic pattern should show only the parental enzyme bands. If a myotube is formed by the fusion of numerous myoblasts, a third band should appear as a result of the formation of hybrid enzyme dimers within the cell that contains the nuclei of both genotypes. The experimental results support the fusion model.

division model predicts that a single muscle cell will contain nuclei of only a single genotype; thus, any enzyme derived from the muscle cells of a tetraparental mouse should show only the parental patterns. The fusion model predicts that a single cell in a tetraparental mouse may contain the nuclei of both genotypes, so that the hybrid dimer can be formed. Therefore, Mintz's results (the formation of the hybrid dimer in the muscle cells of tetraparental mice) has provided definitive proof that a striated muscle cell is formed by myoblast fusion.

Muscular Dystrophy. Tetraparental mice have been used to provide precise solutions to other problems in developmental biology. For example, the sex-linked condition in humans called Duchenne's muscular dystrophy involves a muscular degeneration that develops progressively. It is difficult to determine biologically

whether this degeneration of muscle tissues is a direct result of the mutation or is a secondary effect due to genetically controlled changes in other tissues (such as nerve cells). A comparable hereditary condition exists in mice. Alan Peterson created tetraparental mice from normal lines and from lines carrying the gene for muscular dystrophy. These two lines also produce distinct forms of an enzyme. Peterson used the enzyme as a marker to determine the genetic origin of muscle cells in tetraparental mice showing the dystrophic phenotype. He found that such a mouse may have muscle cells derived from the normal line but that the nerve cells enervating the deteriorating muscles are always derived from the mutant line. Thus, it appears that the mutation has its direct effect on the nerve tissue; the muscular deterioration appears as a secondary effect due to the changes in nerve cells.

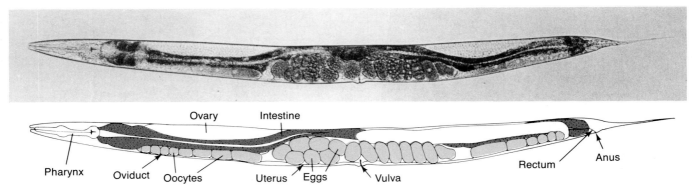

Figure 22-5. Adult *Caenorhabditis elegans*. (a) Drawing of an adult hermaphrodite, showing various organs and nuclei readily identified by their location. (b) A first-larval-stage (L1) hermaphrodite with about 500 nuclei, only a few of which can be seen on any focal plane. (Figures 22-5, 22-6, 22-9, and 22-10 from J. E. Sulston and H. R. Horvith, *Developmental Biology* 56, 1977, 111.)

> **Message**   Genetic mosaics can be generated in a variety of ways. Such mosaics provide an extremely useful tool for following the destinies of cells and their daughters through development.

Development in *Caenorhabditis elegans*

Another way of creating a developmental fate map is through the direct microscopic observation of cell division and movement. In the constantly shifting world of the geneticist's "favorite organism," a relative new-comer is an unobtrusive, nearly microscopic worm (the adult is just over 1 mm in length) called *Caenorhabditis elegans* (Figure 22-5). A free-living nematode, *C. elegans* possesses fully developed nervous and muscular systems that express an array of predictable behaviors. An adult is made up of only a few thousand cells and matures from an egg in just over two days (52 hours at 25°C). The lineages of every one of those cells has been traced by observing them under a microscope from egg to adult.

Genetics

Genetically, *C. elegans* has a number of attractive features. The small diploid number is 12 (a pair of X chromosomes and five pairs of autosomes). Most worms are XX and self-fertilizing hermaphrodites, but about one

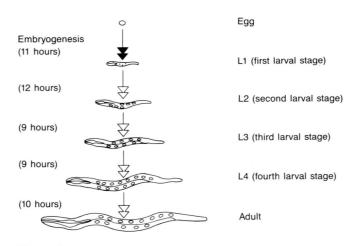

Figure 22-6. Life cycle of *C. elegans*. The relative sizes of each stage are indicated, as well as the number of hours of embryogenesis at 25°C. A lethargic period is indicated by the cross lines on the arrows at the end of each stage.

in every 500 worms is an XO male as a result of nondis-junction. As in plants, "selfing" by hermaphrodites gen-erates homozygous stocks of newly induced mutations more quickly:

P   +/+ (mutagen treated)

            Self-cross   Outcross

F₁   ⊗ +/m*   ×   +/+

F₂   m*/m*   +/m* × +/m*

F₃   m*/m*   m* (newly
                                   induced
                                   mutation)

The haploid genome of *C. elegans* consists of 8×10^7 base pairs coding for 2000 to 4000 genes, thereby mak-ing saturation of the genome with mutations possible.

Development

The life cycle of *C. elegans* is characterized by an embry-onic stage within the egg case, four successive larval stages, and the adult (Figure 22-6). A variety of mutant phenotypes can be recognized (Figure 22-7). Although *C. elegans* was introduced by Sydney Brenner as an ideal organism for genetic study, its simple nervous system and its invariant responses to specific stimuli (called stereotyped behavior) initially made it attractive as an organism for neurobiological study. However, the highly predictable pattern of cell movement and lineage during embryogenesis of the nematode has made its de-velopment accessible to genetic analysis.

Paul Laufer and his associates have shown that in the first four divisions after fertilization, the dorsal-ventral and anterior-posterior axes of the nematode are estab-lished. Six of the cells in the embryo can be shown to be the precursor cells for the six major cell lines in the larva (Figure 22-8). The pattern of cell division is unvarying from one worm to the next. Since the plane of cleavage determines cell position within the axes of the embryo, cell positions and movements are also very stereotyped. By the 28-cell stage, cells destined to form part of the intestine migrate into the cell mass. Cleavage ceases at about the 540-cell stage (some cells die at determined times).

As the multicellular embryo assumes a three-dimen-sional form, what determines the fate of each cell within it? It could be the cell's position, which responds to an underlying topological blueprint, or it could be an inter-nally preset ability to respond to neighbor cells. To an-swer this question, researchers have released the embryo from the egg case and destroyed one or more specific cells with a laser beam. Observing the surviving cells as

they move in the embryo then reveals that some cells continue to move normally to their proper positions, as if the embryo were still intact, while others behave aber-rantly and create novel phenotypes. Thus, there must be positional information to which any cell responds as well as predetermined cell-specific behaviors that are respon-sible for the precise movements of cells.

Cell Lineage

Determination of the entire cell lineage from fertilized egg to larva to mature adult in *C. elegans* is made possible by the predictably repetitious pattern of cell division and movement. From about the 50-cell stage, if one or more cells are killed, the remaining cells cannot fully replace them or fill their roles in some other way. In other words, development is invariant, and the potential fate of the cells appears to become progressively restricted as division proceeds.

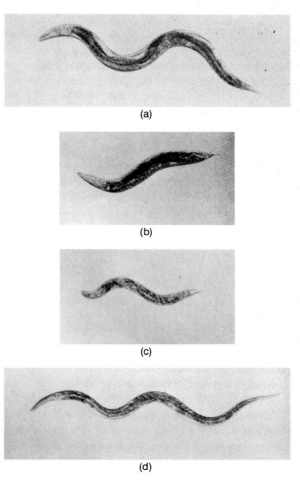

(a)

(b)

(c)

(d)

Figure 22-7. Some morphological mutants of *C. elegans*: (a) Wild-type. (b) A *dumpy* (*dpy*) mutant. (c) A *small* (*sma*) mutant. (d) A *blister* (*bli*) mutant. (From S. Brenner, *Genetics* 77, 1974, 26.)

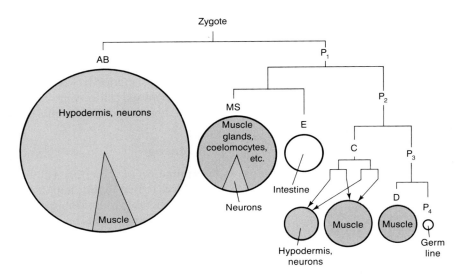

Figure 22-8. Cell origins of the major embryonic lineages in *C. elegans*. The columns at the left indicate the age of the animal in minutes and the number of nuclei at different intervals. Each division is indicated by a horizontal line; the letters represent the cell types. (From Hirsch et al., *Cold Spring Harbor Symposium on Quantitative Biology* 50, 1985, 1970.)

The L1 larva contains 558 nuclei while the hermaphrodite adult has 959 somatic nuclei. Unlike *Drosophila*, in which larval and adult cells are distinct, *C. elegans* simply builds adult organs from larval structures. The embryonic cells that will form adult structures are termed **blast cells** (Figure 22-9), and their derivatives are identified by a cell-lineage tree (Figure 22-10). You can see that the position of each cell and its ancestors can be read from the lineage.

Message The invariant pattern of cell division and movement in *C. elegans* yields a complete lineage of every cell from fertilized egg to larva to mature adult.

How does the description of a cell's lineage contribute to the analysis of the organism's development? For one thing, every cell at every stage can be recognized. By following each cell's position at different stages, a description of its movements can be reconstructed. Then by killing specific cells and observing the consequences, inferences can be made about the regulation of cell position and fate. The detection and analysis of mutations affecting specific cells is made even more informative by knowing the fate map of every cell.

The value of cell lineage information in the analysis of development is reflected in the studies of John E. Sulston and his group. Specific cells in the P-blast-cell line were **ablated** (killed with a laser beam); then the subsequent positions and behaviors of the surrounding

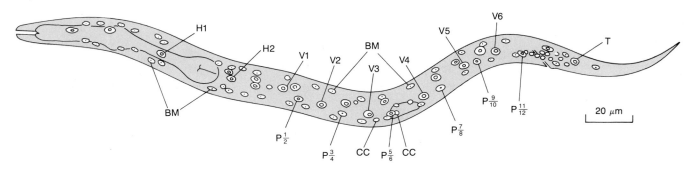

Figure 22-9. Positions of blast cells in an L1 larva. Each blast cell is a precursor to a specific and identifiable clone of differentiated cells in the adult.

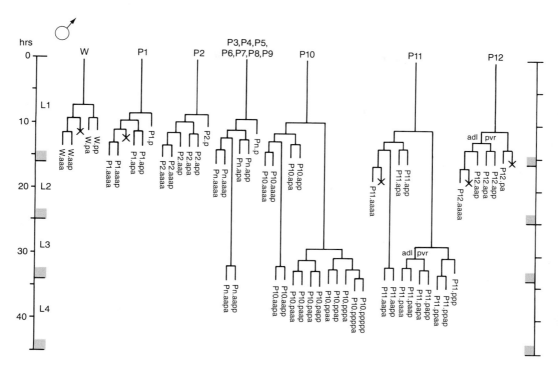

Figure 22-10. The lineage of 13 cell lines present in the L1 stage of *C. elegans*. The symbols identifying the parental cells appear on the top row. (Note that P3–P10 lineages are identical and therefore are represented as a single Pn family.) At each division, there is an anterior (a) and a posterior (p) cell. The history of each cell is indicated by its name. For example, P10.paap is a derivative of P10 that is the posterior cell in the first division, the anterior cell in the next two divisions, and the posterior cell in the last division.

TABLE 22-1. Cell lineage development after laser killing of particular cells in *Caenorhabditis elegans* larvae*

	Cell Pn and its fate				
Cell ablated	P8	P9	P10	P11	P12
—	P8	P9	P10	P11	P12
P8	—	P9	P10	P11	P12
P9	P8	—	P10	P11	P12
P10	P8	P10	—	P11	P12
P11	P8	P10	P11	—	P12
P12	P8	P9	P10	P11	—
P9, P10	P8	—	—	P11	P12
P10, P11	P8	P11	—	—	P12
P9, P10, P11	P8	—	—	—	P12

* The first row shows the fate of each cell in the intact animal. Subsequent rows show the cell fates after ablation of the specific cells listed in the first column. For example, when P10 is ablated, P8, P11, and P12 are unaffected, but P9 takes up the position and presumably the function of P10.

SOURCE: J. Kimble, J. Sulstan, and J. White, *Cell Lineage, Stem Cells, and Cell Determination*, p. 59. Edited by N. LeDouarin. Copyright © 1979 by Elsevier/North-Holland.

cells were observed. The results (Table 22-1) show that surviving cells are capable of compensating for the loss in some cases; for example, when P11 is ablated, P10 can fill in for P11 while P9 moves to P10. In other cases, the cells are already too rigidly determined to adjust and can only differentiate into their predetermined fate.

The Hermaphroditic Vulva

Cell lineage is useful in the analysis of the differentiation of a specific organ—the hermaphroditic vulva. This is an ideal organ for study because only three rounds of division within a five-hour interval generate all of the cells involved in its elaboration. Within the developing gonad of an L3-stage larva, seven cells are the prospective precursors of the adult vulva: an "anchor" cell induces three of the six hypodermis cells to become specific components of the vulva (Figure 22-11). Those three cells then undergo three rapid divisions apiece and differentiate into the characteristic phenotype in the L4 stage. By destroying each of the six hypodermal cells one at a time, John E. Sulston and John White found that any one of the other three cells may be recruited to make up for a lost cell. Thus, to begin with, all six cells are said to

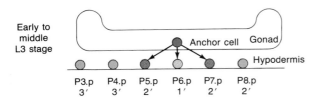

Early to middle L3 stage

Anchor cell   Gonad

Hypodermis

P3.p   P4.p   P5.p   P6.p   P7.p   P8.p
3′   3′   2′   1′   2′   2′

Figure 22-11. Cell components of the hermaphrodite vulva. The gonadal anchor cell acts on the six hypodermal cells to induce the three levels of differentiation. (Figures 22-11 and 22-12 from E. Ferguson, P. Sternberg, and R. Horvitz, *Nature* 326, 1987, 260.)

be **equipotent,** but once acted on by the anchor cells, their destinies are fixed.

Initially, all six hypodermal cells are equivalent and have the potential to differentiate into one of three cell types. Type 3′ is a single, multinucleate cell; the other two are recognizable, mononucleate, vulval derivatives called 2′ and 1′. How does the anchor cell affect the outcome? Paul Sternberg and Robert Horvitz isolated individual cells and showed that the fate of each of the six hypodermal cells is directly related to its proximity to the anchor cell. The cell closest to the anchor cell always becomes a 1′ cell, and the next closest cell always becomes a 2′ cell. The 3′ state can be regarded as a **ground state** because it develops when the triggering message from the anchor cell is absent or in low concentrations.

These results are most easily explained by postulating that an inducing substance is released by the anchor cell. As this hypothetical substance diffuses away from the anchor cell and through the embryo, a gradient of concentration occurs. So the closest hypodermal cell experiences the highest concentration and therefore becomes a 1′ cell; cells exposed to the next highest concentration become 2′ cells.

Horvitz and his group have recovered a number of mutations that fall into two phenotypic classes: *Multivulva (Muv)*, which have several vulva-like protrusions from their undersides, and *Vulvaless (Vul)* in which a vulva is absent. The researchers analyzed 23 genes for their effects on vulval development (Figure 22-12). The effect of the mutations can be arranged in a logical order, as if they act in succession during vulval formation. The reality of this inferred temporal sequence is shown by the fact that in double mutants, a gene acting on precursor cells is epistatic to (blots out) a gene acting after the cell's fate is fixed. After fertilization, obviously, the egg cell is totipotent, but the possible cell types that its daughter cells can become are progressively restricted with time. Sulston's study reveals that within cells already committed to becoming P cells, there is a temporally related narrowing of their potential fates. Once committed or determined, a cell's flexibility is limited. Thus, in *C. elegans*, we find a progressive constriction in developmental potential that reflects a hierarchy of determinative decisions.

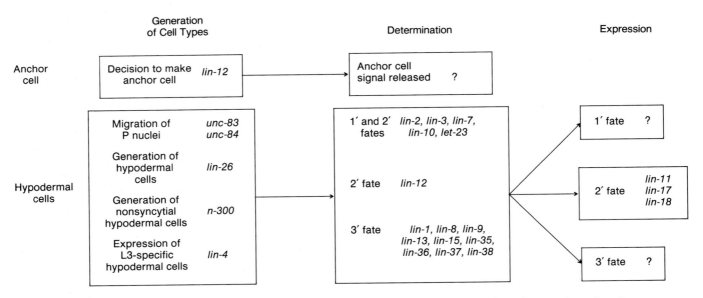

Figure 22-12. The inferred sequence of vulval development from the generation of cells that make up the organ, to the determination of the final cell phenotype, to the differentiation (expression) of the cells. The visible cell events at each stage are boxed. Mutations that disrupt different parts of vulval formation are listed. A question mark (?) denotes mutations that are expected but that have not been detected as yet.

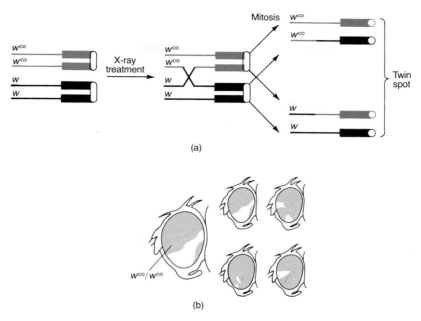

Figure 22-13. Production of twin spots of mutant tissue in a background of cells of a different phenotype. (a) *Drosophila* larvae heterozygous for the sex-linked alleles w and w^{co} are irradiated to induce a mitotic crossover in prospective eye cells. This crossover produces cells homozygous for w and w^{co} that are phenotypically distinct from the surrounding w/w^{co} tissue. (b) Some representative mosaic eyes in Drosophila. (Part b after H. J. Becker, *Verhandl. deutsch. zool. Ges.*, 1956.)

> **Message** Cell lineage in *C. elegans* provides the information that reveals a hierarchy of determinative events as the embryo proceeds through development. Cells proceed from totipotence through stages that become progressively more restrictive in their developmental potential.

Development in *Drosophila*

Drosophila has occupied the center stage of genetic research for almost a century. In addition to a detailed description of its anatomical and developmental processes, an extensive repertoire of mutations and genetic tricks has made *Drosophila* a favorite subject of molecular analysis. The sophisticated tools for tracking minute quantities of proteins, messenger RNA's, or DNA sequences to specific chromosome or cellular locations have led to a spectacular burst of new insights into the fruit fly.

Genetic Mosaics

We have already seen the value of genetic mosaicism as an analytical tool. However, the rare occurrence by chance of such mosaics makes any systematic use of this approach very difficult in many organisms. In *Drosophila*, a variety of procedures has been developed to make the production of numerous mosaics a relatively simple matter. Two general mechanisms are involved in the formation of the mosaics: mitotic crossing-over and chromosome loss.

Mitotic Crossing-over. We have already seen how mitotic pairing and crossing-over in *Drosophila* heterozygotes can produce "twin spots" of homozygous tissue (Figure 6-18). This provides a way of marking clonally related cells that can be presumed to come from a single crossover event. In 1957, Hans Becker used this method of marking in *Drosophila* larvae heterozygous for w^{co} and w (different alleles of the *white* locus). He chose these genotypes because the phenotypes of w^{co}/w^{co}, w/w, and w^{co}/w can be distinguished from one another. Becker irradiated young larvae to induce mitotic crossovers. Such crossovers in the w^{co}/w genotype lead to twin spots of homozygous w/w and w^{co}/w^{co} tissue (Figure 22-13). If the mitotic crossover is induced at an early stage of eye development, a large segment of the adult eye will derive from the daughter cells of the mitosis at which the crossover occurred and the twin spot will cover a large part of the eye. If the crossover is induced late in eye develop-

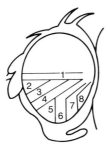

Figure 22-14. All of the cells in each sector of the lower eye of *Drosophila* are derived from one of eight original larval cells. (After H. J. Becker, *Verhandl. deutsch. zool. Ges.,* 1956.)

ment, then the twin spot will involve relatively few facets of the adult eye. Studying many such crossovers and mapping the twin spots, Becker was able to trace the **cell lineage** of the lower half of the eye to eight early larval cells. The daughter cells of each larval cell occupy one of the eight sectors shown in Figure 22-14. You can see how cell lineages are inferred from genetic mosaics.

In 1971, Antonio Garcia-Bellido and John Merriam used somatic crossing-over in an attempt to determine the time when a particular gene ceases to be active. They induced crossovers by irradiating females with the genotype *y Hw+/++sn³* at different developmental stages. (*Hw* is a dominant mutation that causes the growth of extra bristles along wing veins and is closely linked to *y*, the mutation for yellow body color; *sn³* produces short, gnarled bristles and is more distantly linked.) Induced crossovers between the *sn³* locus and the centromere produce twin spots; one spot of the genotype is *y Hw+/ y Hw+*, and the other is *++ sn³/++ sn³*. Garcia-Bellido and Merriam found that y Hw and sn³ mutant patches can be induced by irradiation at any time during the third-instar larval stage or during the first 24 hours of the pupal stage. However, if the irradiation occurs during the last 12 hours of the third-instar stage or the first 24 hours of the pupal stage, the sn³ patches also show the Hw phenotype. In other words, one spot of the twin spot is y Hw and the other spot is Hw sn³. No crossover pattern can explain this observation; the *Hw* alleles cannot be present in the *sn³* spot cells. Garcia-Bellido and Merriam concluded that the *Hw* allele somehow acts before the last 12 hours of larval life to imprint its phenotype irreversibly on the cell committed to form wing.

Chromosome Loss. In 1929, Alfred Sturtevant observed that eggs from females of *Drosophila simulans* that are homozygous for the autosomal recessive mutation *ca^nd* (claret-nondisjunction) commonly lose a chromosome during the first and second divisions after fertilization. If the lost chromosome is an X, then a mosaic of

XX (♀) and XO (♂) cells is formed. If the original fertilized egg is heterozygous (say, *wm/++*), then loss of the wild-type chromosome permits expression of the recessive phenotype in the hemizygous tissue (Figure 22-15). The resulting mosaic flies do *not* show a "salt-and-pepper" phenotype with male and female cells randomly arranged. The mosaic patterns observed indicate that the nuclei that are more closely related by cell division tend to stay together. This observation led Sturtevant to the realization that the distribution of mutant and nonmutant tissue in a mosaic can provide information about the spatial relationship in the *embryo* of prospective adult nuclei.

Since Sturtevant's time, several different mutants that cause chromosome loss have been studied. The

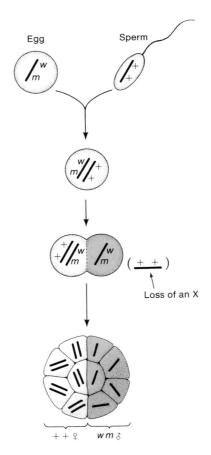

Figure 22-15. Production of a mosaic by the loss of an X chromosome shortly after fertilization in *Drosophila*. If the XO cells of the gynandromorph carry recessive markers on the X, then the mutant phenotype is visible. Loss of the X can be induced in a number of ways, including by irradiation or the presence of certain mutations. (From A. M. Srb, R. D. Owen, and R. S. Edgar, *general Genetics*, 2d ed. Copyright © 1965 by W. H. Freeman and Company)

most useful, described by Claude Hinton in 1955, is a ring X chromosome called $In(1)w^{vC}$ (we will call it w^{vC} for short) that has the completely mysterious property of great instability in the newly formed zygote nucleus. Thus, w^{vC} is lost at a very high frequency, and if the egg is initially w^{vC}/m, gynandromorphs of w^{vC}/m (wild-type ♀) and m/O (mutant ♂) cells are created. This is an extremely useful aberration.

In 1969, using data originally obtained by Sturtevant in the 1930s, Garcia-Bellido and Merriam set out to determine whether the pattern of mosaicism in adult flies can be used to infer relationships between cells in the embryo. But before we consider their research, we must understand the special nature of early embryonic development in *Drosophila*.

Early Embryonic Development

After fertilization the zygotic nucleus undergoes a rapid series of divisions without separating into cells. Thus, the newly fertilized egg develops as a syncytium (a single multinucleated cell). A region located in the posterior part of the egg is characterized by cytoplasmic inclusions called polar granules; as we have seen (page 575) these are maternally deposited elements that determine differentiation of the gonadal tissue. The nuclei divide synchronously. After the ninth division, when about 512 nuclei are present, they migrate to the periphery of the egg cytoplasm. After four more synchronous divisions, when there are 4000 to 8000 nuclei, cell membranes enclose them to form the mononucleated cells of the **blastoderm,** which is essentially a cell monolayer enclosing the yolk (Figure 22-16). All of these events occur within three hours after fertilization. An important feature of blastoderm formation is that the pattern of nuclear migration is very orderly. The nuclei most recently related by nuclear division remain nearer to one another than do more distantly related nuclei.

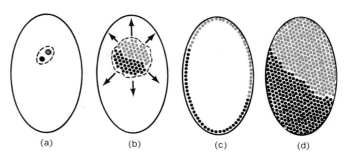

Figure 22-16. Early divisions leading to blastoderm formation in *Drosophila*. (a) The first division of the nucleus occurs shortly after fertilization. No division of the cytoplasm occurs. (b) After nine synchronous nuclear divisions without cell cleavage, the embryo is a syncitium containing about 500 nuclei. The nuclei now begin to migrate to the periphery of the egg cytoplasm. (c) After a few more nuclear divisions, the cell has some 4000–8000 nuclei in a layer near its surface. (d) Cell membranes form to enclose each nucleus in a separate cell, forming the blastoderm stage of the embryo. (From S. Benzer, "Genetic Dissection of Behavior." Copyright © 1973 by Scientific American, Inc. All rights reserved.)

After the blastoderm stage, cell movements and the folding of sheets of cells create the cell layers from which tissues will differentiate. Up to the blastoderm stage, the nuclei are totipotent, as Illmensee demonstrated by recovering adults after transplantation of blastoderm nuclei into enucleated eggs (Figures 21-9 and 21-10). However, after the blastoderm stage, each nucleus is restricted to a limited potential fate. Prospective cells for both various adult structures and larval structures are present in the surface of the blastoderm.

Fate Mapping. Now let's return to the analysis of Sturtevant's data by Garcia-Bellido and Merriam. Such an analysis should be possible because the related nuclei are

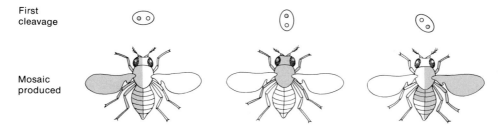

Figure 22-17. Orientation of the plane of the first nuclear division after fertilization in *Drosophila*. The spindle fibers can be tilted within the egg in any orientation with respect to the anterior–posterior poles of the cytoplasm. Because daughter nuclei tend to remain together and migrate to nearby locations at the egg surface, the orientation of the division plane determines the mosaic pattern observed in the adult (if the two nuclei produced from the first division are genetically different). (From S. Benzer, "Genetic Dissection of Behavior." Copyright © 1973 by Scientific American, Inc. All rights reserved.)

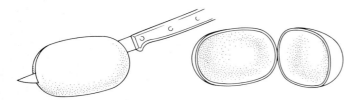

Figure 22-18. A hypothetical cut through the blastoderm of *Drosophila* to separate the right and left halves. The blastoderm fate maps are representations of the surface of one such half.

arranged in close proximity in the blastoderm. It appears that the orientation of the division plane for the first nuclear division is tilted randomly with respect to the anterior-posterior and dorsal-ventral axes of the egg, because all kinds of mosaic patterns are formed in the adult (Figure 22-17). We can also assume that cells destined to form adult structures and parts of the larvae are distributed throughout the surface of the blastoderm.

If cells destined to form two structures are located close to each other within the blastoderm (and hence are probably closely related), then the chance that they are genetically *different* in a mosaic is small. (This is because the dividing line between mutant and nonmutant cells determined at first cleavage is randomly distributed on the blastoderm.) The farther apart the positions of the cells are in the blastoderm, the greater the probability is that the dividing line will fall between them so that they will develop into genetically different adult structures. Thus, the "distance" between blastoderm cells destined to become different adult structures can be quantified as the percentage of mosaics in which the two adult structures are genotypically different:

$$
\text{Distance between blastoderm cells} = \frac{\text{Number of mosaics in which structures differ} \times 100}{\text{Total number of mosaics scored}}
$$

What these "distances" mean in real physical terms is not yet known—a situation similar to the circumstances under which linkage maps were first constructed. In this case, we can develop a two-dimensional map because the spatial distribution extends over the surface of a hollow spheroid. We can standardize the surface as that seen when the spheroid is cut along the axis of symmetry (Figure 22-18). Suppose that we map structures A and B 10 m.u. apart. We can introduce a second dimension by measuring the distance of A and B from a third structure (C). If A is 8 m.u. from C and B is 4

m.u. from C, then we obtain the map shown in Figure 22-19. This point to which an adult structure maps on the blastoderm surface can be regarded as a **focus** from which the growth and movement of the structure proceeds during development and differentiation.

Using this map distance between embryonic foci for adult structures, Garcia-Bellido and Merriam were able to construct an embryonic "fate map" like the one shown in Figure 22-20. These maps can be made as refined as we desire by studying ever more detailed features of the adult structures. Such maps are assumed to have some congruence with the actual spatial distribution of determined cells in the blastoderm.

Mapping Behavior. In 1972, Yoshiki Hotta and Seymour Benzer took this analysis one step further by analyzing the foci pertinent to a mutant behavioral phenotype. Consider a recessive mutation that causes the legs of a fly to twitch when the fly is anesthetized; we can map the cells responsible for the twitch in relation to the external markers. Let's call the behavioral mutant *kic* (for *kicker*) and make a cross to generate *y kic/+ +* flies. A female that has no yellow tissue but exhibits leg kicking must have internal cells that are *y kic/O*. We can calculate the distance from the focus for *kic* to the focus for the right front leg, for example, as

$$
\frac{\left(\begin{array}{c} \text{Number of } y \text{ nonkicking legs} \\ + \\ \text{Number of } y^{+} \text{ kicking legs} \end{array} \right)}{\text{Total mosaics}} \times 100
$$

where only the right front legs of mosaic flies are counted. Again, by triangulation with another focus, such as that for a thoracic bristle, we can map the focus for kicking.

Hotta and Benzer found that the mutants with presumed defects in leg muscles and nerves map near each

Figure 22-19. Embryonic map of cells destined to form adult structures A, B, and C of *Drosophila*. The map positions presumably represent the relative positions in the blastoderm of the cells from which these adult structures will eventually develop. The "distance" between blastoderm cells determined for different adult structures can be quantified as the percentage of mosaics in which two adult structures are genotypically different:

$$
\frac{\text{Number of mosaics in which structures are different}}{\text{Total number of mosaics recorded}} \times 100
$$

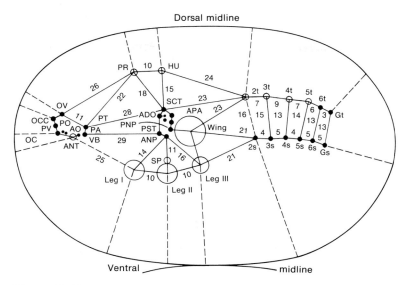

Figure 22-20. The embryonic fate map of adult *Drosophila* structures. The map shows the foci for various structures and individual bristles. The distances between them are indicated in map units called sturts. (From S. Benzer, "Genetic Dissection of Behavior." Copyright © 1973 by Scientific American, Inc. All rights reserved.)

other and near the foci of external leg structures. Their positions correspond very well with cytological studies of the neural and muscular tissues developing from blastoderm cells. Such cytological studies provide the reference points for orientation of the fate map to the poles of the blastoderm. With all of this evidence, we can construct a hypothetical map of the locations in the blastoderm of the ancestors of the adult structures. A revised fate map is presented in Figure 22-21. Of course, the mapping procedure becomes much more complex in the

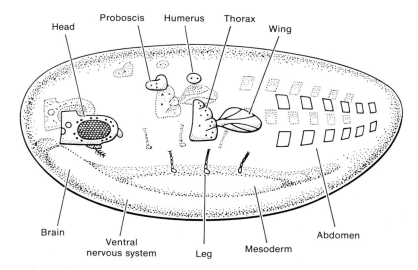

Figure 22-21. A redrawn version of the fate map in Figure 22-20, showing the adult fate of the cells that occupy specific regions of the embryo. Note that the anterior–posterior arrangement of adult structures is approximately retained in the arrangement of blastoderm cells. (From S. Benzer, "Genetic Dissection of Behavior." Copyright © 1973 by Scientific American, Inc. All rights reserved.)

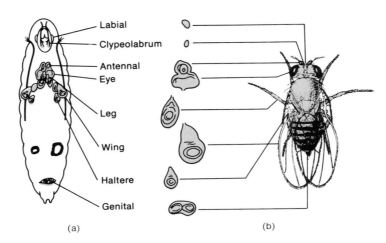

Figure 22-22. The imaginal disks of *Drosophila*. (a) The larva in the third-instar stage of development, showing the positions of the disks. The larva is significantly larger than the adult. (b) The adult fly, showing the part that is derived from each disk. Note that the disks can be identified by their differing sizes and shapes as well as by their positions in the larva. (From J. W. Fristrom, in *Problems in Biology*, University of Utah Press, 1969.)

case of the many neurological mutants that involve simultaneous defects in several anatomical positions.

Message The locations of cells destined to form particular adult structures can be mapped in relation to one another in the embryo by the use of mosaics.

Developmental Stages

In contrast to development in vertebrates, many organisms like insects go through two very distinct life forms. The mosquito and dragonfly, for example, live their early lives as nymphs underwater and emerge into the air as flying insects. *Drosophila* nuclei remain totipotent during the early stages of embryogenesis, but at some point they change. As the nuclei differentiate, they fall into two classes with very different prospects. One set of cells must develop into a larva that tunnels into its food and, although lacking vision, responds to temperature, gravity, light, and odors while crawling, eating, and excreting; it is a sophisticated organism with a central nervous system that coordinates its behavior in response to environmental stimuli. The other set of cells will shape a second organism that will emerge from the larval carcass as an adult fly bearing no resemblance to its larval predecessor and capable of seeing, walking, flying, and mating. This remarkable transformation is anticipated in the larva by the presence of packets of cells programmed to differentiate into external adult tissue on

exposure to the proper hormonal cues. These packets of cells are called **imaginal disks** (often known simply as disks). Although they can be distinguished from larval structures by their sizes, shapes, and locations in the larva (Figure 22-22), they nonetheless bear no resemblance to the adult structures they will form.

Disks. The cells in the disks are not yet recognizably adult, but they are already committed, or determined, to an adult fate. Indeed, within a disk mass, the cells are more than just a group that has been set aside to be directed to a more specific fate later. The cells in each disk are already preset to anticipate both their identity and location in the adult organ.

How do we know that the disks are already programmed for an adult fate? An eye disk, for example, can be removed and implanted in another larval host, which then pupates and emerges as an adult. This adult will have, somewhere within its body, extra adult eye structures derived from the implanted disk! Similar results are obtained with each of the other disk types. Furthermore, the parts of the adult structure that are formed depend on the specific parts of the disk that are injected. Thus, it is possible to obtain a fate map of the disk. The stimulus for the differentiation of the already determined disk is the molting hormone ecdysone, which is released in the late third-instar stage of the larva. This effect can be demonstrated by exposing disks isolated from late third-instar larvae to ecdysone and observing as they begin to differentiate.

On pupation, the larval carcass begins to break down. Its residues act as a thick medium, nourishing and embedding each of the disks. The disks now begin to differentiate into their specific adult structures, so that each part of the adult forms as a separate element. The separate elements then fuse with the correct neighbors to form a complete adult.

The imaginal disks play no functional role in the larva. A series of mutants has been recovered lacking all imaginal disks or specific sets of disks. These mutant larvae are completely viable until after much of pupation, when they die. The incredibly intricate and precise program whereby the disks are activated and differentiate into adult structures is acted out each time a fly is "born." Obviously, mutations that block various parts of this complex developmental process provide a great deal of information about the process itself—and evidence about the genetic system controlling this process.

> **Message** A *Drosophila* larva is composed of two distinct groups of cells: cells that contribute to the functions of the larva, and cells found in aggregates called imaginal disks that will, after pupation, form the adult. Within the disks, the cells are already determined in terms of both location and cell type in the adult.

Let's now go back and ask at what point in development are the nuclei of the embryo committed to becoming parts of specific imaginal disks? The answer has been provided by using techniques for culturing cells in vivo. Larval cells can be transplanted by injection into an adult host. The cells escape exposure to ecdysone, which triggers metamorphosis, so they simply proliferate as larval cells in the adult host. The clone of larval cells can then be removed from the adult host and used for experimentation. For example, the cell mass can be implanted in a larval host about to undergo metamorphosis, so that the differentiation of the cells on ecdysone stimulation can be studied. (Recall Illmensee's experiments, in which blastoderm nuclei were shown to be totipotent because they develop normally when transplanted to enucleated eggs, illustrated in Figures 21-9 and 21-10.)

In 1971, Lilian Chan and Walter Gehring performed an experiment with a *Drosophila* genotype distinct from the wild-type. They took a blastoderm and bisected it into anterior and posterior halves. They then dissociated the cells from each half, compacted them into pellets by centrifugation, and injected the pellets into adult hosts for culturing (Figure 22-23). Next they removed the cultures of larval cells from the adult host and injected them into wild-type larval hosts, where

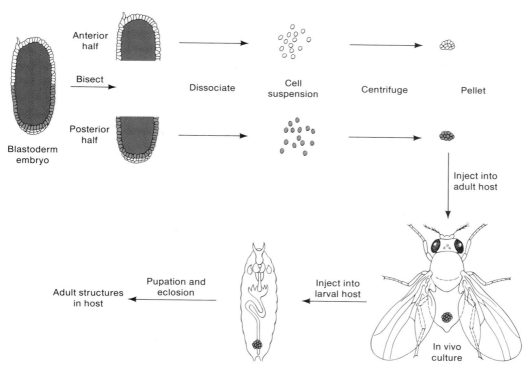

Figure 22-23. Experimental technique for testing the developmental fate of blastoderm cells of *Drosophila*. The cells multiply by mitotic division in the adult host but no not undergo differentiation. After a larger mass of cells is obtained from in vivo culture in the adult, their determination state is tested by implanting them into a larval host. These cells will now undergo differentiation, just as will the imaginal structures of the host larva. When an adult fly emerges, it can be dissected and the adult fate of the donor cells can be observed.

their differentiation could be observed after metamorphosis. (The distinctive genotype of the experimental cells distinguished them from the cells of the host.) Chan and Gehring found that the cells from the anterior half of the blastoderm developed into anterior adult structures (head and thorax), whereas the cells from the posterior half of the blastoderm developed into posterior adult structures (thorax and abdomen). These results indicate that the developmental fate of the blastoderm cells was already determined; the cells retained this programmed status through repeated rounds of division in the adult host.

Further evidence of early determination comes from studies on damaged embryos. If the blastoderm is punctured, burned, or subjected to ultraviolet irradiation in specific regions of the embryo, then the adult fly that develops from the embryo shows damage in the specific structures that correspond to the damaged part of the blastoderm. Because the cell nuclei remain totipotent until after the blastoderm stage, we can conclude that the early commitment is an effect of the cytoplasm. Further experiments indicate that this decision occurs at roughly the time when the nuclei are enclosed by membranes to form the blastoderm cells.

As we have seen in the example of snail-shell coiling (Figure 20-8), the egg cytoplasm contains developmental cues dictated by the genotype of the mother. Therefore, we would expect to find mutations that are expressed in females as an abnormal phenotype *in the offspring.* The *grandchildless* mutation of *Drosophila* mentioned earlier (page 575) is an example of just such a mutation.

> **Message**   The pattern of determination of embryonic cells is apparently established in the cytoplasm of the multinucleate egg and is under the control of the maternal genotype.

Maternal Mutants.   A *Drosophila* larva is made up of a series of similar but distinct segments arranged in order from head to abdomen (Figure 22-24). In fact, the larval segments correspond in arrangement to the order of adult parts. After the most anterior larval head segment, the next three segments correspond to the adult thorax. The posterior eight segments are congruent with the abdominal segments of the adult. Fate maps of the blastoderm nuclei reveal that cells that will form the three groups of body segments (head, thorax, abdomen) in the

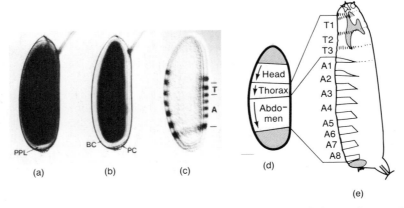

Figure 22-24. Embryonic development of *Drosophila*. The living embryo at ½ hour (a) and 3 hours (b) of development. A noncellular membrane, the chorion, surrounds the embryo. (PPL = posterior pole plasm that indicates the prospective gonads; BC = blastodermal cells; PC = pole cells that will form gonads. (c) Section through a 3-hour-old embryo stained for antibodies to the gene products of *fushi tarazu*. The seven regular stripes indicate segmentation patterns. (T = thorax; A = abdomen) (d) A blastoderm fate map, showing where the larval body regions will form. (e) The wild-type larva (lines connect to the blastodermal origin of each part. (Figures 22-24 and 22-25 from C. H. Nüsslein-Volhard, G. Frohnhöfer, and R. Lehman, *Science* 238, 1987, 1675 and 1678.)

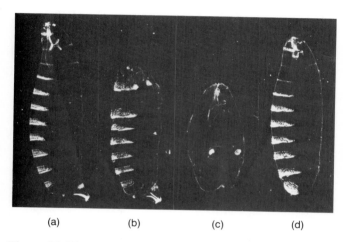

Figure 22-25. Externally visible patterns of wild-type and mutant *Drosophila* embryos. (a) Wild type. (b) *Bicoid* (lacking head and thoracic structures). (c) *Oscar* (lacking abdomen and pole plasm). (d) *Torsolike* (lacking the two unsegmented terminal regions—the anterior acron and posterior telson).

"ventralize" the embryo have more mesoderm and less ectoderm.

The maternal mutants not only affect larval segments and germ-line tissue, they also affect the disks that normally form in the affected blastodermal areas. Walter J. Gehring and Christiane Nüsslein-Volhard have cultured the anterior halves of wild-type and *bicaudal (bic)* embryos. The *bic/bic* eggs laid by females form two abdomens and no anterior structures. On metamorphosis, the bicaudal cells often form genital structures, whereas the wild-type cells from the same region never do. These mutations reveal the pattern of control over larval and adult topology by maternally acting genes. Presumably, factors deposited in the peripheral egg cytoplasm trigger or shut off different sets of genes, but the mechanism remains completely speculative at present.

> **Message** Maternally acting genes that cause embryonic lethality affect the factors that determine the anterior-posterior and dorsal-ventral axes in the egg. These effects span large areas and influence both larval and adult structures.

larvae also occupy a similar spatial distribution in the blastoderm (Figure 22-24). What could control the determination of nuclei in the blastoderm? An obvious possibility would be factors found in the egg cortex that are deposited in the ovaries under maternal genotypic control. Maternally acting mutations have been recovered that affect either the anterior-posterior or dorsal-ventral axes of the embryo and cause embryonic lethality regardless of the zygotic genotype. Females homozygous for such mutations are normal, and regardless of what genotype they are fertilized by, their eggs fail to hatch. From this group, mutations have been selected on the basis of observable defects in larval segmentation. After exhaustive searches for such mutations, it appears there may be fewer than 40 loci of this type.

Of the 30 maternal-effect genes detected, 18 control the development of one of the three main body sections (Figure 22-25). These 18 mutants affect the anterior, posterior, or terminal regions of the egg. Thus, single maternally active genes function within the already present anterior-posterior axis and determine whether or not certain segments will form, thereby affecting total segment number. It is possible that various loci are differentially sensitive to concentration gradients of macromolecules laid down in the egg.

The dorsal-ventral maternally acting mutants do not affect segment numbers but do affect the germ layers that lie along the anterior-posterior axis. Thus, mutants that "dorsalize" the embryo lack or have reduced mesodermal derivatives, whereas mutants that

Embryonic Genes Affecting Larval Segmentation. Genes acting in maternal ovaries establish the major axes of the embryo and therefore already constrain the developmental fate of zygotic nuclei. At what point do the zygotic genes begin to act and affect future differentiation? In other words, what are the earliest embryonic features to be affected by an embryo's own genes? We have already seen (Figure 22-26) that adult body segments have their origin in the comparable larval subdivisions.

The cuticular pattern of a newly hatched larva (Figure 22-26) is dominated by the three thoracic and eight abdominal segments. Each segment can be distinguished from the others, but they all share features such as a band of pinlike denticles (which allow the larva to cling to surfaces) along the anterior edge of the segment and a naked posterior portion. After a systematic search for early embryonic lethal mutations, Christiane Nüsslein-Volhard and Eric Wieschaus uncovered a remarkable group of defects. They found 15 different gene loci with effects that fell into three distinct categories (Figure 22-27): the **gap mutants,** which delete several continuous segments; the **pair-rule mutants,** which induce a repetitive deletion of specific parts of *alternating* segments; and the **segment-polarity mutants,** which are missing the same part of each segment with the deleted part replaced by a mirror-image duplication of the remaining part. (In many organisms, mirror-image dupli-

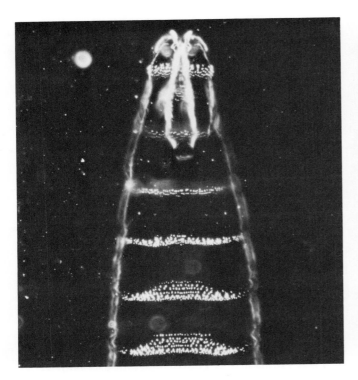

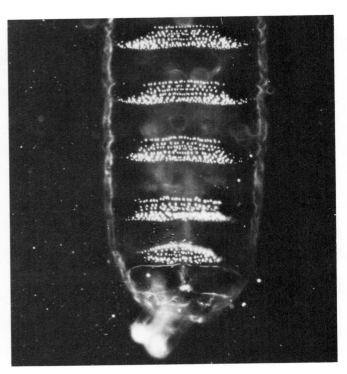

Figure 22-26. Segmentation pattern in a wild-type *Drosophila* larva. Note that the denticular pattern in each segment is recognizably different. The head is at the top. (Figures

22-26 and 22-27 from C. Nüsslein-Volhard and E. Wieschaus, *Nature* 287, 1980, 797 and 796.)

cations are frequently observed as a result of the accidental or induced loss of a body part.) These striking results clearly indicate that segmental arrangements are already established when these genes are activated. Their activation is required for the normal formation of those preset segments. With techniques for localizing gene transcripts or their protein products in situ, the actual pattern of gene activation can be specified (Figure 22-28).

Message   Within the maternally determined organization of the egg, segmentation genes are activated to affect gross segmental domains (gap), pairs of prospective segments (pair-rule), or repeat regions within segments (segment-polarity).

Homeotic Mutations.   Maternal and early embryonic mutations impose the major axes and segmentation patterns on both the larva and the adult. Within these broad patterns, specific genes act to coordinate further zones of distinction. The best illustration of this is the class of **homeotic mutations,** which transform one body or organ segment into a different but perfectly normal segment. Homeotic loci have long been assumed to repre-

sent critical switches that shunt groups of cells from one fate to another. An example is a mutation known as *ophthalmoptera*, which causes wing structures to develop from an eye disk. Another is *aristapedia*, which causes the feathery arista of the antennal complex to develop as a leg (Figure 22-29a and b).

In 1969, John Postlethwait and Howard Schneiderman studied the dominant mutation *Antennapedia*, which converts part of the antenna into leg structures. By inducing mitotic crossovers in a heterozygote, they were able to generate a mosaic of wild-type cells in an *Antennapedia* background. Postlethwait and Schneiderman showed that the replacement of parts is position-specific: a given antennal segment is always replaced by a specific part of the leg (Figure 22-30). These observations suggest that there is some overall developmental plan on which the specific details of leg or antennal development are overlaid.

Most of the homeotic mutations fall in two clusters on the right arm of chromosome 3: the *Antennapedia* complex (ANT-C) and the *bithorax* complex (BX-C). Genes in the ANT-C affect the head and thorax; genes in BX-C affect the posterior thoracic and abdominal segments. The first homeotic mutations in both ANT-C and BX-C were detected by their phenotypic effects in adults; the same mutations were subsequently found to affect the same segments in the larva.

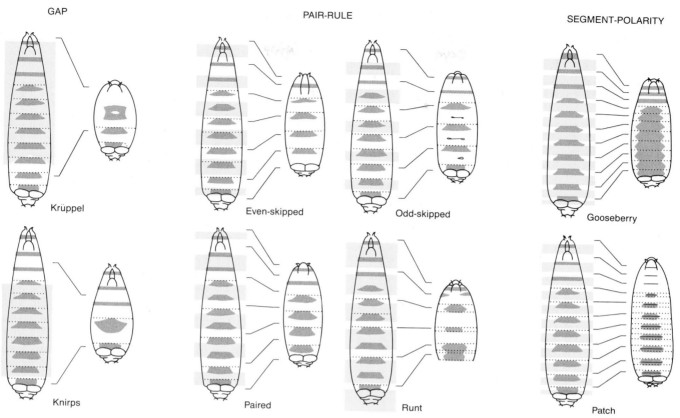

GAP PAIR-RULE SEGMENT-POLARITY

Krüppel Even-skipped Odd-skipped Gooseberry

Knirps Paired Runt Patch

Figure 22-27. Defects in the *Drosophila* larval-segment pattern, with representative mutants in each class. The denticular regions are indicated by the dotted areas; the boundary of each segment is the dotted line. The hatched regions are missing in each mutant, and the drawing to the right of each wild-type standard is the resulting mutant phenotype.

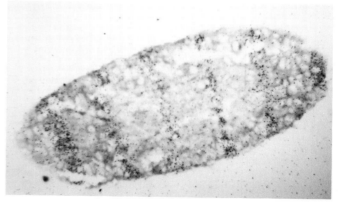

Figure 22-28. Location of RNA transcripts of the *enqrailed* gene in *Drosophila* embryos. A six-hour-old embryo is fixed, sectioned, and exposed to 3H-labeled DNA, which hybridizes to mRNA in the cells. The autoradiograph can be printed to show the silver grains as dark (left) or light (right) spots. The *enqrailed* gene specifies the anterior-posterior boundaries of compartments in both larval and adult segments. The bands of label correspond to the posterior parts of larval segments. (From A. Fjose, W. J. McGinnis, and W. J. Gehring. Reprinted by permission from *Nature* 313, 1985, 284. (Copyright © 1985 Macmillan Magazines Ltd.)

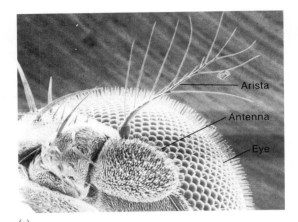

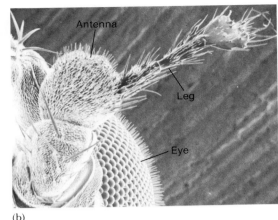

(a)   (b)

(c)

Figure 22-29. Scanning electron micrographs of the antennal complex of *Drosophila*, showing the effects of a homeotic mutation. (a) A wild-type fly. (b) A mutant fly in which the feathery arista is replaced by the distal part of a leg. This mutation is called *aristapedia*. (c) Four-winged *Drosophila* produced by combining three homeotic mutations of the *bithorax* gene complex. The haltere-bearing (3rd) thoracic segment closely resembles the wing-bearing (2nd) thoracic segment, as a result of reduction in function of three genes of the complex: *abx*, *bx³*, and *pbx*. (Part c provided by E. B. Lewis.)

Details of the Bithorax Complex. Calvin Bridges discovered *bx¹*, the first mutant in the BX-C, but it was Ed Lewis who analyzed the many additional mutants in the locus and doggedly persisted in its genetic analysis beginning in 1948. Now Lewis is participating in the molecular corroboration of all of his genetic inferences. The evolutionary history of the Diptera suggests that these two-winged flies came from four-winged flies. All that remains of the second pair of wings are vestigial balancing organs called halteres (Figure 22-31). In turn, the ancestral four-winged insects probably evolved from arthropod forms with multiple legs (perhaps from a mil-

lipede-like creature). Lewis reasoned that the evolutionary sequence must have occurred through an ancient mutation that suppressed leg formation in the abdominal segments and a later one that suppressed wing formation in one thoracic segment. Such changes do appear to have occurred and to have become a wild-type in the primordial dipteran; they are now clustered in the BX-C. Recent mutations within the gene locus can cause a loss of the suppressing function and a consequent reversal to a four-winged or multilegged insect (Figure 22-29c).

The adult fly consists of a series of segments; the cells in each segment come from a small number of adjacent founding cells in the embryo. Each segment has a unique set of external markers that permits its unambiguous

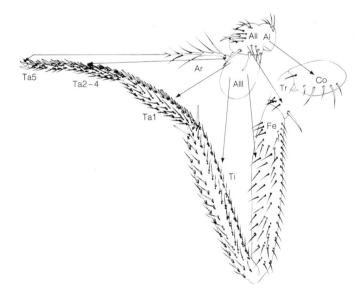

Figure 22-30. The correspondence between antennal and leg structures in *Drosophila*, based on position-specific transformations in homeotic antennae of *Antennapedia*. The symbols identify various corresponding parts of the structures. (From J. Postlethwait and H. Schneiderman, *Developmental Biology* 25:606, 1971.)

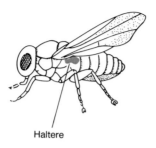

Haltere

Figure 22-31. The haltere of *Drosophila*. It is a vestigial wing that has evolved to serve as an organ of balance.

identification. Thus, the head is followed by the three thoracic segments (the prothorax, mesothorax, and metathorax, denoted by T1, T2, T3) and eight abdominals (A1 to A8), as shown in Figure 22-32. The segments visible in the adult are mirrored in those recognizable in the larvae, as seen at the left in the figure.

A strong *bx* mutation causes the anterior part of T3 to develop as the anterior part of T2. This is interpreted as showing that *bx*+ must act in a specific part of the fly to direct those cells to differentiate to anterior T3. The inactivation of *bx*+ by mutation leads to the automatic expression of those cells as anterior T2. We can conclude that *bx*+ does not act in the cells that normally form anterior T2, because T2 is unaffected by a *bx* mutation. So, it is assumed that *bx*+ acts like a binary switch that is turned on in anterior T3 and turned off in anterior T2. Flies homozygous for a deletion of the entire *bx* locus die as late embryos. Inspection of those embryos at death reveals that T3, as well as each abdominal segment, resembles T2 (which carries the wing and second leg). Thus, T2 can be considered to be the "developmental ground state" in the absence of BX-C activity. Compo-

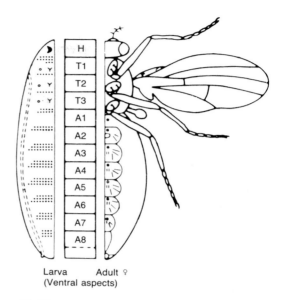

Larva   Adult ♀
(Ventral aspects)

Figure 22-32. Ventral view of body segments of *Drosophila* larva and adult stages. In the larva (left), each of the 12 segments can be identified. The larval segments correspond to those in the adult (right). (Courtesy of E. B. Lewis.)

nents of the BX-C are required to direct differentiation of the posterior segments along specialized lines. Recessive mutations that affect a specific segment transform it to a more anterior segment (that is, toward the ground state) and can be considered to be inactive alleles. There are also dominant mutations which transform structures away from the ground state.

The sequence of alleles on the genetic map has been correlated with DNA sequences that are known to overlap the BX-C locus (Figure 22-33). All the genetic data

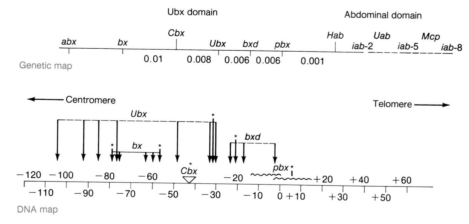

Figure 22-33. The genetic and DNA maps of the *bithorax* complex. The mutants appear to be located in two domains, the Ubx and Abdominal, which affect thoracic and abdominal development, respectively. In the DNA map, each arrow points to the position of an alteration. Those arrows marked with an asterisk represent mutants also mapped genetically and show complete congruence with the genetic map.

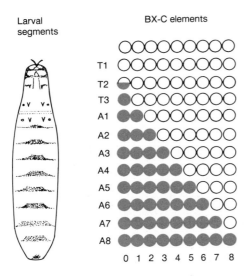

Figure 22-34. Model of function of the *bithorax* complex with larval segments. The BX-C is divided into nine elements (0 and 1 in the Ubx domain, the rest in the Abdominal). Active elements are filled in; inactive are open. The half-filled circle represents Ubx^+ in the posterior compartment of T2. Note that in each progressively posterior segment, one more element is active.

have been corroborated by sequence analysis. Lewis concludes, then, that there are spheres or domains of control by elements within the BX-C (Figure 22-34). In this model, Lewis proposes that each successive segment, proceeding from front to back, requires additional wild-type activity. We can now see how the loss of a function results in reversion toward the ground state (toward the more anterior segment). This model leads to predictions regarding the segments in which each of the genes in the BX-C will function; for example, bx^+ should act in the anterior part of T3 but not T2. Remarkably, the use of labeled probes for the gene products from different parts of the BX-C permits these predictions to be tested, and they have been confirmed by in situ hybridization (Figure 22-35). The BX-C story is a clear example of how genetic analysis, when coupled with molecular tools, can provide powerful probes into development.

> **Message** The phenotypic effects of various mutations within the *bithorax* complex reveal the genetic control by the BX-C locus over the normal differentiation of *Drosophila* body segments.

The Homeobox. In 1984, Walter J. Gehring and his associates hybridized the 3′ exon of the homeotic *Antp* gene to chromosomal DNA and noted there was also a small amount of hybridization with the adjacent gene *ftz*. The researchers found that the 3′ exon of *ftz* had a region of homology with the 3′ exon of *Antp*. Subsequently, the 3′ exon of *Ubx* in the BX-C was shown to share homology. The region of homology of these three homeotic genes spans a highly conserved segment of about 180 bp. Since a similar DNA sequence has been found in three different homeotic loci, the segment has been termed the **homeobox.** To date, more than 14 homeobox-containing genes have been isolated in *Drosophila*, all of which are homeotic or are involved in segmentation.

A great deal of excitement ensued when homeobox probes revealed sequences in higher organisms, including frogs, chickens, mice, and humans. The evolutionary span represented by these organisms covers some 500 million years, and the conservation of much of the homeobox sequence suggests that the gene plays a vital role in this process. To date, a great deal of descriptive information about the homeobox has been accumulated, but it is far too early to determine its significance.

The homeobox illustrates that as our tools for the recovery and comparison of DNA segments increase, the discoveries will exceed our understanding because we lack a theoretical framework in which to fit them. Now as scientists propose to develop the technology and commitment to sequence the entire human genome of 3 billion bp, the data will far exceed their ability to interpret them. These new molecular tools will take us into another purely descriptive phase of biology.

> **Message** The discovery of the homeobox has revealed the interesting fact that short sequences of highly conserved DNA may be shared by many genes with similar biological functions. These genes are also retained by many other organisms. Molecular tools now provide information for which scientists have yet to develop an adequate theoretical framework.

Determination within Disks

In the embryo, the imaginal disk originates as a small number of cells that are programmed to form specific parts of adult structures. The number of founding cells of a disk can be estimated by inducing genetic mosaics by chromosome loss or mitotic crossing-over, as already described. Suppose that we obtain XO/XX mosaics

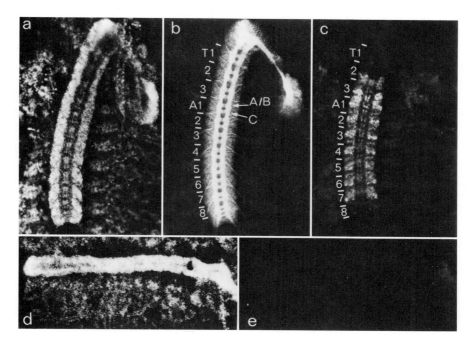

Figure 22-35. Visualization of *Ubx* gene product in the embryonic *Drosophila* nervous system by immunofluorescence. A bacterial clone containing the 5' *Ubx* exon was used to synthesize a partial *Ubx* protein sequence and antibodies to this were prepared. The antibodies were joined to a fluorescent dye and added to the embryo, where they were bound to the *Ubx* protein in situ. Fluorescence is seen from the posterior T2 through A8 segments as predicted by Lewis' model. (a) Wild-type embryo. (b) The same preparation as in (a) stained for neural tissue. (c) The same preparation as in (a) tagged with fluorescent antibodies for the *Ubx* protein. Note the total absence of product in T1 and the posterior localization in other segments. Thereby showing that *Ubx*-gene products are localized to the predicted areas. (d) An embryo, *Df bxd*[100], carrying a deletion for the *Ubx* region. (e) The same preparation as in (d) stained with fluorescent antibodies to the *Ubx* protein.

formed by a chromosome loss. If a disk originates from a single determined cell at the time the mosaic is induced, then no mosaic adult can have both male and female tissue in the adult derivative of that disk. On the other hand, if two cells begin a disk, then the adult structure they form can be a mosaic of 50 percent male and 50 percent female cells. Of course, the structure can be entirely male or female, depending on the location of the boundary between the X/O and X/X nuclei. Thus, the smallest proportion occupied by mutant tissue in a mosaic structure provides an estimate of the number of founding cells; among disks, this number ranges from 8 to 40. Determination must occur progressively as the cell number increases, since there are far more structures derived from a mature disk than there are cells in the founding group. It should be obvious that another way to derive a fate map of the cells within a disk is by cutting the disk into different fragments and implanting the fragments into larval hosts to see what adult structures they form.

Transdetermination. Ernst Hadorn and his students investigated various aspects of disk development, using the technique of adult culture and larval implantation that he had perfected. In the 1960s, Hadorn made an interesting observation in the course of these supposedly routine studies. One of the standard procedures in his laboratory involved injection of a disk or disk fragment into an adult host to increase the cell number in the research sample. Serial transplantation through successive adult hosts can provide a larger sample of material (Figure 22-36). Samples were tested after each transplantation to verify that the cells were unchanged by the culturing procedure. In some cases, an unexpected result was obtained. On successive transplants through adult hosts, the determined state of the disk sometimes changed. For

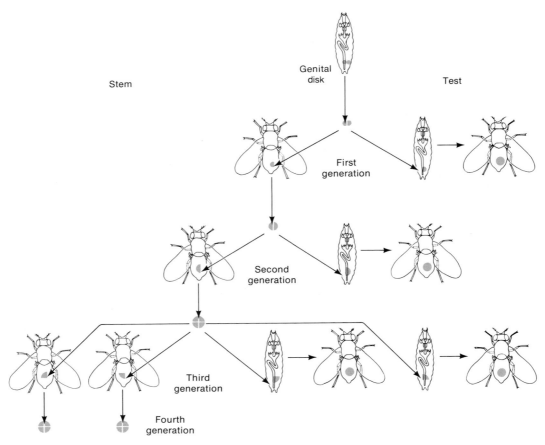

Figure 22-36. Serial transplants of a genital disk through adult *Drosophila*. At each generation, the population of disk cells is subdivided. Some of the cells are implanted in another adult for further growth, and the other cells are implanted into a larva where they will develop into adult structures. (From E. Hadorn, "Transdetermination in Cells." Copyright © 1968 by Scientific American, Inc. All rights reserved.)

example, a genital disk could eventually change so that it formed leg structures.

Hadorn referred to this change as a state of **transdetermination.** The nature of observed transdeterminations is not random. For example, a genital disk can become a leg disk, but it never changes directly to an eye disk (Figure 22-37). It has been proposed that transdetermination and homeotic pathways are controlled by a series of switches that can exist in stable, alternate states comparable to the control circuit of *cro* and *cI* for the λ repressor. Genetic or environmental triggers could then flip the switch from one state to another. We do not yet understand the molecular basis of transdetermination, but further study of the phenomenon should provide clues about the nature of the determination process. It is interesting to note that homeotic mutations cause alterations only in the same directions as known transdetermination pathways.

A great deal of research has been done on imaginal disks, but they remain a fertile source of new information about developmental mechanisms and processes.

Compartmentalization. In the early 1970s, Antonio Garcia-Bellido made interesting discoveries about disk development by inducing mitotic crossovers in the embryonic stage to produce a mosaic adult in which homozygous clones of cells are scattered in a heterozygous background. The clones are quite visible if the original embryo is a heterozygote for some mutation such as *multiple wing hairs (mwh),* which causes a cluster of hairs to appear where a single hair would normally grow on each cell of the wing. A clone of *mwh/mwh* derivatives is readily visible against the background of the wild phenotype (Figure 22-38). In different mosaic flies, these clones appear at positions all over the wing, and there is

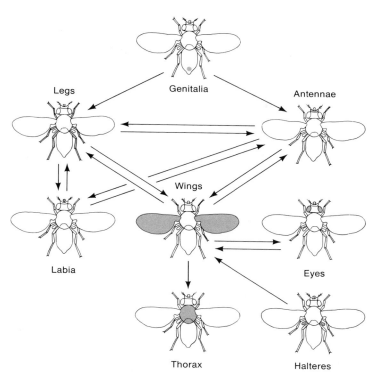

Figure 22-37. Types of transdetermination observed for disk cells. In each case, the colored areas indicate the parts of the adult that develop from the serially transplanted disk material. The arrows indicate the observed changes of fate. Note that cells of genital disks do transdetermine to form leg or antennal structures, but the reverse transdeterminations have not been observed. However, some transdeterminations do occur in either direction. (From E. Hadorn, "Transdetermination in Cells." Copyright © 1968 by Scientific American, Inc. All rights reserved.)

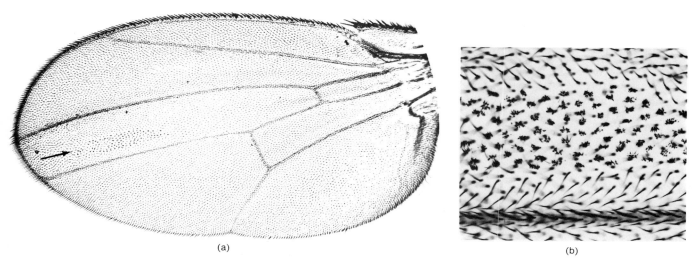

(a)   (b)

Figure 22-38. The wing of a *Drosophila* individual with clones of *mwh/mwh* cells among a background of wild-phenotype *mwh/+* cells. (a) The clones (arrow) are readily visible. (b) Higher magnification clearly shows the nature of the *mwh* phenotype. Such clones are produced by inducing somatic recombination with X rays. (From P. A. Lawrence, Medical Research Council, Cambridge, England.)

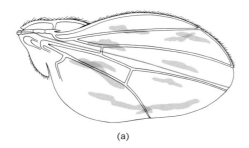

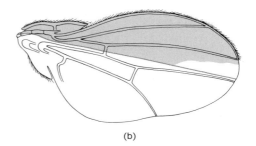

(a)   (b)

Figure 22-39. The fate of cells and their daughters in *Drosophila* disks is not rigidly fixed. (a) A plot of the positions of several mosaic patches (colored area) induced in different flies shows that they overlap and occupy many different parts of the wing. However, no single patch crosses the boundary between the third and fourth longitudinal wing veins. (b) Even when +/+ cells grow much more rapidly than the surrounding $M/+$ cells, the clone of +/+ cells never occupies an area in the adult greater than the compartment delineated by the boundary between the third and fourth veins. (From A. Garcia-Bellido, P. A. Lawrence, and G. Morata, "Compartments in Animal Development." Copyright © 1979 by Scientific American, Inc. All rights reserved.)

overlap of the boundaries of clones from different flies (Figure 22-39a).

Garcia-Bellido also produced mosaics with *Minute* (*M*) mutations, which retard development when hetero-zygous and are lethal when homozygous. Thus, when crossovers are induced in an $M/+$ embryo, the M/M cells die but the reciprocal +/+ cells divide much more quickly than the surrounding $M/+$ cells. The rates of

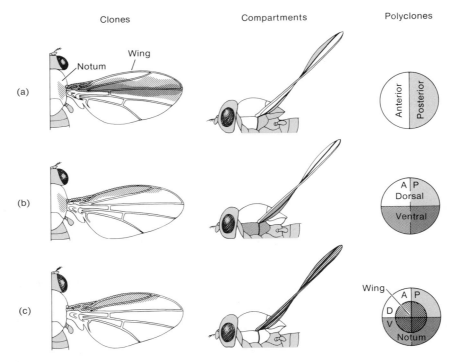

Figure 22-40. Changes in the compartment boundaries of the wing disk observed when genetically marked clones are induced at different developmental stages. (a) When the induction occurs in the embryo stage, a boundary separates the anterior from the posterior parts of the wing and notum. (b) When the induction occurs in young larvae, a second boundary separates the dorsal and ventral parts of the wing. (c) When the induction occurs in older larvae, a third boundary exists between wing and notum. (From A. Garcia-Bellido, P. A. Lawrence, and G. Morata, "Compartments in Animal Development." Copyright © 1979 by Scientific American, Inc. All rights reserved.)

division are so different that a single +/+ cell in an early stage of imaginal disk would be predicted to produce most of the cell population of the final disk. In fact, this does not happen; there seem to be rigid zones beyond which a clone will not expand. In the wing, the **boundary** lies between the third and fourth wing veins (Figure 22-39b). The areas separated by such boundaries are called **compartments.**

By noting the distribution and the boundaries of mutant tissue in mosaics induced at different times in development, Garcia-Bellido and his colleagues were able to map changes in cell determination during development. For example, consider the disk that forms the wing and the notum—the bit of thorax to which the wing is attached (Figure 22-40). If mutant clones are induced by irradiation of an embryo, this disk has two compartments that produce the anterior and posterior parts of the wing complex. If the mutant spots are induced in the early larval stage, the disk is divided into four compartments and a further compartmentalization between the wing and the notum occurs late in the larval stage. These results provide a clear picture of the successively more detailed determination of the cells in the wing disk through the process of development.

Soon after the blastoderm stage, the embryonic ectoderm can be shown to be made up of alternating groups of nuclei that represent more than one clonal family (a **polyclone**) (Figure 22-40). One polyclone will generate the anterior (A) compartment and another

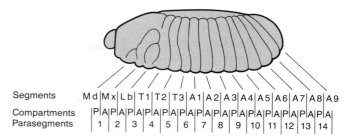

Figure 22-41. Subdivisions of the *Drosophila* embryo. The segments visible on the ectoderm of the embryo *(top)* are indicated by the segment symbols: Md (mandibular), Mx (maxillary), Lb (labial), T1–T3 (thoracic), and A1–A8 (abdominal). The anterior (A) and posterior (P) polyclones comprising each segment form AP compartments. Adjacent PA pairs are designated as parasegments. (From A. Martinez-Arias and P. Lawrence, *Nature* 313, 1985, 639.)

polyclone will generate the posterior (P) compartment of adult structures. One A polyclone and the P polyclone behind it will become an adult compartment (Figure 22-41). Alfonso Martinez-Arias and Peter Lawrence have pointed out another important pairing that is biologically meaningful: a P polyclone and the A polyclone posterior to it. This PA pair, in contrast to the AP compartment, is called a **parasegment.** Some genes act within the boundaries of parasegments.

Summary

■ Development and differentiation have become accessible to analysis by genetic techniques in a variety of organisms. We have selected three—the mouse, the nematode *C. elegans,* and the fruit fly *D. melanogaster*— as exemplary case studies.

Mutations are a vital tool for the developmental geneticist. In a genetic chimera, the cells of different genotypes permit the geneticist to trace cell lineages in various tissues. This technique has been exploited in mice by the formation of tetraparental individuals from the fusion of genotypically different embryos. These individuals have revealed the embryonic cells from which skin is derived, the formation of multinucleate striated muscle cells through the repeated fusion of precursors, and an apparent neurological basis for hereditary muscular dystrophy.

Development and behavior in *C. elegans* are invariant from individual to individual. Hence, the complete lineage of every cell in the adult has been traced from the egg by direct observation. By destroying specific cells or selecting organ-specific mutations, it can be demonstrated that determination—the commitment of a cell to a specific fate—is successively restricted to ever-narrowing possibilities. In other words, in the early embryo, a cell has the potential to differentiate into a number of cell types, but in later stages of development, the type of cell it can become is increasingly restricted.

The study of development in *Drosophila* now provides a theoretical framework within which observations are placed. By generating genetic mosaics, it is possible to determine the "distance" between various adult structures or behaviors on a "map" of the blastoderm. This map is based on the concept that the further apart two nuclei are in the blastoderm cortex, the greater the probability is that the adult structures they form will be genotypically different in a mosaic.

Maternally acting mutations indicate that the basic axes—the anterior-posterior axis and the dorsal-ventral axis—as well as the distinctive area of germ plasm are delineated by genes acting in the ovary. Both larval tissue and adult disk cells are affected by these basic body plans.

Early-acting embryonic genes function within the constraints of already spatially distinct parts of the egg. The larval and adult segments are congruent with each other, and homeotic genes that transform one body segment to another affect both larval and adult structures.

A detailed study of homeotic genes indicates that most of them are located in two clusters: the ANT-C, which affects the anterior body parts, and the BX-C, which affects the posterior body segments. So far, genetic predictions by Ed Lewis concerning the bithorax complex (BX-C) are being confirmed by molecular analysis of the locus.

Genetic mosaics produced by chromosome loss or somatic crossing-over have revealed the existence of developmental compartments within which cells are clonally related. Disk development demonstrates a pattern of anterior-posterior, dorsal-ventral, and dorsal-lateral axes that constrain the activity of genes. During development, the cells within compartments undergo progressive restrictions regarding their potential fates. Thus, at successive stages, gene activity is fine-tuned while the developmental potential of the cells is continuously reduced.

■ ■ ■

Solved Problems

1. After massive whole-body irradiation for cancer, the blood-forming capacity of the spleen is usually destroyed. If non-irradiated cells are injected into the blood system, they lodge in the spleen, form colonies or nodes from which the organ is eventually repopulated. How can you determine whether each focus of blood-producing cells develops from a single cell or a cluster of cells?

Solution

The study should be performed in a test animal such as a mouse. Take blood cells from two strains that differ detectably either in chromosome make-up or biochemical property such as enzyme activity. Mix the two genotypes and inject them into the irradiated animal. After allowing enough time for colony formation in the spleen, take cells from different colonies. Examine them separately for the chromosome or biochemical constitution. If each colony is a clone, they should exhibit only one or the other genotype, but never both. Colonies yielding both genotypes would indicate their origins from aggregates of injected cells.

Another way of answering the question would be to treat the cells to be injected into the irradiated animal with a chromosome breaking agent such as radiation or chemicals. Then inject the treated cells and allow them to repopulate the spleen. Inspect cells in each node for chromosome aberrations induced by the treatment. Again, if each node is a clone, where induced aberrations are detected, there should only be a single type per colony. In fact, this is the original experiment that was performed to show that each colony is a clonal derivative from a single cell.

2. In the embryogenesis of mammals, the inner cell mass or ICM (the prospective fetus) quickly separates from the cells that will serve as enclosing membranes and respiratory, nutritive and excretory channels between the mother and the fetus.

   a. Design experiments using mosaics in mice to determine when the two fates are decided.

   b. How would you trace the formation of different fetal membranes?

Solution

a. Just as we saw in the analysis of clonal derivatives in *Drosophila*, we must have markers that enable us to distinguish different cell lineages. This can be done with mice using strains that differ in chromosome or biochemical markers. (Other ways would be to use differences in sex chromosomes of XX and XY cells to induce chromosome loss or aberrations by irradiating embryos.)

   Decide on the marker difference (sex, chromosome, biochemical) to be used. One way to answer the question is to inject a single cell from one strain into embryos of the other at various developmental stages. Another approach is to fuse embryos of defined cell numbers from the two strains. In either case, inspect the embryos when the ICM and membranes are distinct and recognizable. When cell insertion or fusion results in membranes and ICM that are exclusively made up of one cell type and never a mosaic of the two, the two developmental fates have been set.

b. Carry out the same injection or fusion experiment on early embryos. Now look for the pattern of mosaicism. Correlate the occurrence of cells of similar genotype in different membranes. It should be possible to determine the cell lineage of cells in each set of membranes just as the cells of adult *Drosophila* are traced from embryonic mosaics.

Problems

1. There is a sex-linked gene in mice that causes muscular dystrophy. Alan Peterson fused embryos from the muscular-dystrophic strain with wild-type embryos. Because the two strains also differ in enzyme patterns, he could deter-

mine the parental origin of particular cells in the tetraparental mice. Peterson found that animals with muscle cells from the wild-type that are enervated by nerve cells from the dystrophic strain always have muscular dystrophy. However, in parabiosis studies (in which two animals are surgically bound together), attaching nerves from a dystrophic animal to the muscle of a wild-type mouse did not induce dystrophy in the muscles. Explain the significance of these results.

2. In the development of vulval cells in *C. elegans,* one anchor cell in the gonad interacts with six hypodermal cells (P*n*), which will become the parts of the vulva (Figure 22-11). The 22 vulval cells derived from the six P*n* cells have three distinct phenotypic fates: T_1, T_2, and T_3. If the anchor cell is ablated, all six P*n* lines differentiate into the T_3 state. The P*n* cell closest to the anchor cell develops the T_1 phenotype.

   a. Set up a model to explain these results.

   The anchor cell and the six P*n* cells can be isolated and grown in vitro.

   b. Design an experiment to test your model.

   c. If your model is correct, what do you predict?

3. There are two types of distinct myosin molecules in the muscle cells of *C. elegans:* pharyngeal muscles and body-wall muscles. They can be distinguished from each other and detected in single cells. Their cell lineage is indicated in the accompanying figure. James Priess and Nichol Thomson carried out the following experiments.

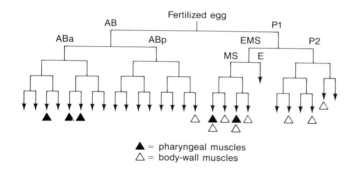

▲ = pharyngeal muscles
△ = body-wall muscles

(From J. Priess and N. Thomson, *Cell* 48, 1987, 241.)

   a. When ABa and ABp are physically interchanged, they develop according to their new position. What does that tell us?

   b. If P1a is ablated, AB derivatives do not make muscles. What does that suggest?

   c. If P1b is ablated, AB derivatives make myosin. What does that tell us?

   d. Design further experiments that exploit the differences in myosin.

4. The *Notch* locus near the tip of the X chromosome of *Drosophila* has been extensively studied genetically. Deletions of the entire locus produce a dominant visible phenotype of nicked wings and recessive lethality. In addition, several point mutations mapping throughout the locus have similar phenotypes. There are also alleles with recessive visible phenotypes affecting eyes, bristles, and wings. Design experiments to recover DNA spanning the *Notch* locus.

5. Curt Stern and Chiyoko Tokunaga looked for evidence of a **prepattern** — a pattern of organization within which developmental genes act. In other words, the researchers were trying to determine why a gene controlling, say, bristle formation, acts to produce a bristle at a specific position. To find out how a cell or group of cells knows where it is relative to the rest of the body, they induced mitotic crossing-over in *y ac*/+ + flies to generate *y ac/y ac* spots in a wild-type background. (*y* causes a yellow body color; *ac* deletes two bristles on each side of the thorax.) They obtained mosaics such as those shown in the accompanying figure. How do you interpret their results?

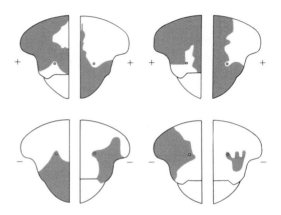

Mosaic thoraces from *y ac*/+ + flies. The dark area is mutant *y ac* tissue, while the white is wild type. The presence (+) or absence (−) of thoracic bristles affected by *ac* is always determined by the genotype of the cells and is unaffected by the extent of tissue of different genotype surrounding it.

(From C. Stern and C. Tokunaga, *Proceedings of the National Academy of Sciences* 57, 1967, 658–64.)

6. Mutations in *Drosophila* that affect behavior have been very useful in the analysis of the nervous system. Describe methods that could be developed to recover mutants with the following defects originating in the nervous system: a. blindness; b. inability to fly; c. abnormal feeding responses. (NOTE: flies taste with their feet. When they detect edible molecules, such as sugar, their proboscis is automatically lowered to feed.)

7. Let's assign numbers to each leg of *Drosophila* as follows: legs 1, 2, and 3 are in front, mid, and hind legs on the left side, respectively, and legs 4, 5, 6 are the front, mid, and hind legs on the right side. A fly normally walks by moving legs 1, 3, and 5 together and then moving legs 2, 4, and 6 together. A mutation called *wobbly* causes the fly to get its mid legs tangled up with its front or its hind legs. What could be wrong with the mutant, and how would you study it?

8. The sex-linked dominant mutation *Hyperkinetic-1* (Hk^1) causes a fly's legs to shake while it is etherized. Seymour

Benzer and Yoshiki Hotta generated mosaic flies with the tissue genotypes $y\ Hk^1/++$ and $y\ Hk^1/O$. They scored 600 fly sides for leg shaking and mutant cuticle tissue. Table 22-2 summarizes the results of this study. Draw a fate map of the foci for Hk^1-caused shaking, the three legs, the antenna, and the humeral bristle.

9. The evidence suggests that the major axes of a *Drosophila* egg have already been determined in the egg cortex by maternally acting genes. Design experiments to demonstrate that such cytoplasmic factors and patterns do, in fact, exist.

10. Nüsslein-Volhard and Wieschaus have shown that segmentation genes in *Drosophila* control the major division of the body. Many of the mutations within the BX-C and ANT-C have phenotypic effects with limits that coincide with segment, compartment, or parasegment boundaries. Molecular biologists have cloned DNA from many of the loci within the homeotic complexes.

   a. How would you determine the temporal pattern of gene action?

   b. How would you look for the spatial pattern of activity?

TABLE 22-2.

Structure	Color of cuticle tissue	Shaking of leg 1	
		Normal	Mutant
Leg 1	Wild-type	277	50.5
	Yellow	33	223.5
Leg 2	Wild-type	261.5	69
	Yellow	44	215.5
Leg 3	Wild-type	250.5	82.5
	Yellow	46.5	215.5
Antenna	Wild-type	253	82
	Yellow	84.5	179.5
Humeral bristle	Wild-type	241	94
	Yellow	66.5	197.5

Quantitative Genetics

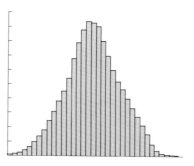

KEY CONCEPTS

In natural populations, variation in most characters takes the form of a continuous phenotypic range rather than discrete phenotypic classes. In other words, the variation is quantitative, not qualitative.

∎

Mendelian genetic analysis is extremely difficult to apply to such continuous phenotypic distributions, so statistical techniques are employed instead.

∎

A major task of quantitative genetics is to determine the ways in which genes interact with the environment to contribute to the formation of a given quantitative trait distribution.

∎

The genetic variation underlying a continuous character distribution can be the result of segregation at a single genetic locus or at numerous interacting loci which produce cumulative effects on the phenotype.

∎

The estimated ratio of genetic to environmental variation is *not* a measure of the relative contribution of genes and environment to phenotype.

∎

Estimates of genetic and environmental variance are specific to the single population and the particular set of environments in which the estimates were made.

■ Ultimately, the goal of genetics is the analysis of the genotypes of organisms. But the genotype can be identified—and therefore studied—only through its phenotypic effect. We recognize two genotypes as different from each other because the phenotypes of their carriers are different. There is a uniquely distinguishable phenotype for each genotype, and there is only a single genotype for each phenotype. At worst, when one allele is completely dominant, it may be necessary to perform a simple genetic cross to distinguish the heterozygote from a homozygote. Basic genetic experiments depend on the existence of a simple relationship between genotype and phenotype. For the most part, then, the study of genetics presented in the previous chapters is the study of allelic substitutions that cause *qualitative* differences in phenotype.

However, the actual variation among organisms is usually quantitative, not qualitative. Wheat plants in a cultivated field or wild asters at the side of the road are not neatly sorted into categories of "tall" and "short," any more than humans are neatly sorted into categories of "black" and "white." Height, weight, shape, color, metabolic activity, reproductive rate, and behavior are characteristics that vary more or less continuously over a range. Even when the character is intrinsically countable (such as eye-facet or bristle number in *Drosophila*), the number of distinguishable classes may be so large that the variation is nearly continuous. If we consider extreme individuals—say, a corn plant eight feet tall and another one three feet tall—a cross between them will not reproduce a Mendelian result. Such a corn cross will produce plants about six feet tall, with some clear variation among siblings. The F_2 from selfing the F_1 will not fall into two or three discrete height classes in ratios of $3:1$ or $1:2:1$. Instead, the F_2 will be continuously distributed in height from one parental extreme to the other. This behavior of crosses is not an exception, but is the rule, for most characters in most species. Mendel obtained his simple results because he worked with horticultural varieties of the garden pea that differed from each other by single allelic differences that had drastic phenotypic effects. Had Mendel conducted his experiments on the natural variation of the weeds in his garden, instead of abnormal pea varieties, he would never have discovered Mendel's laws. In general, size, shape, color, physiological activity, and behavior do not assort in a simple way in crosses.

The fact that most phenotypic characters vary continuously does not mean that their variation is the result of some genetic mechanisms different from the Mendelian genes we have been dealing with. The continuity of phenotype is a result of two phenomena. First, each genotype does not have a single phenotypic expression but a norm of reaction (see Chapter 1) that covers a wide phenotypic range. As a result, the phenotypic differ-ences between genotypic classes become blurred, and we are not able to assign a particular phenotype unambiguously to a particular genotype. Second, many segregating loci may have alleles that make a difference to the phenotype being observed. Suppose, for example, that five equally important loci affect some trait and that each locus has two alleles (call them $+$ and $-$). For simplicity, also suppose that there is no dominance and that a $+$ allele adds 1 unit to the trait whereas a $-$ allele adds nothing. Thus, there are $3^5 = 243$ different possible genotypes, ranging from

$$\frac{+\;+\;+\;+\;+}{+\;+\;+\;+\;+}$$

through

$$\frac{+\;+\;+\;+\;+}{-\;-\;-\;-\;-} \quad\text{to}\quad \frac{-\;-\;-\;-\;-}{-\;-\;-\;-\;-}$$

but there are only 11 phenotypic classes (10, 9, 8, . . . ,0) because many of the genotypes will have the same numbers of $+$ and $-$ alleles. For example, although there is only one genotype with $10 +$ alleles and therefore an average phenotypic value of 10, there are 51 different genotypes with $5 +$ alleles and $5 -$ alleles. These include a single pentuple heterozygote

$$\frac{+\;+\;+\;+\;+}{-\;-\;-\;-\;-},$$

20 different triple heterozygotes such as

$$\frac{+\;+\;+\;+\;-}{-\;-\;-\;+\;-} \quad\text{and}\quad \frac{+\;+\;-\;+\;+}{+\;-\;-\;-\;-}$$

and 30 different single heterozygotes such as

$$\frac{+\;+\;-\;+\;-}{+\;+\;-\;-\;-} \quad\text{and}\quad \frac{+\;+\;+\;-\;-}{+\;+\;-\;-\;-}$$

Thus, many different genotypes may have the same average phenotype. At the same time, because of environmental variation, two individuals of the same genotype may not have the same phenotype. This lack of a one-to-one correspondence between genotype and phenotype obscures the underlying Mendelian mechanism. If we cannot study the behavior of the Mendelian factors controlling such traits directly, then what can we learn about their genetics?

Using current experimental techniques, geneticists can answer the following questions about the genetics of a continuously varying character in a population (say, height in a human population). These questions constitute the study of *quantitative genetics*—the study of the genetics of continuously varying characters:

1. Is the observed variation in the character influenced *at all* by genetic variation? Are there alleles segregating in the population that produce some differential effect on the character, or is all the variation simply the result of environmental variation and developmental noise (page 10)?

2. If there is genetic variation, what are the norms of reaction of the various genotypes?

3. How important is genetic variation as a source of total phenotypic variation? Are the norms of reaction and the environments such that nearly all the variation is a consequence of environmental difference and developmental instabilities, or does genetic variation predominate?

4. Do many loci (or only a few) vary with respect to the character? How are they distributed over the genome?

5. How do the different loci interact with each other to influence the character? Is there dominance or epistasis (interaction among genes at different loci)?

6. Is there any nonnuclear inheritance (for example, any maternal effect)?

The precision with which these questions can be framed and answered varies greatly. In experimental organisms on the one hand, it is relatively simple to determine whether there is any genetic influence at all, but extremely laborious experiments are required to localize the genes (even approximately). In humans, on the other hand, it is extremely difficult to answer even the question of the presence of genetic influence for most traits because it is almost impossible to separate environmental from genetic effects in an organism that cannot be manipulated experimentally. As a consequence, we know a relatively large amount about the genetics of bristle number in *Drosophila* but essentially nothing about the genetics of complex human traits, except that a few (such as skin color) clearly are influenced by genes whereas others (such as the specific language spoken) clearly are not. It is the purpose of this chapter to develop the basic statistical and genetic concepts needed to answer these questions and to provide some examples of the applications of these concepts to particular characters in particular species.

Some Basic Statistical Notions

In order to consider the answers to these questions about the most common kinds of genetic variation, we must first examine a number of statistical tools that are essential in the study of quantitative genetics.

Distributions

The outcome of a cross for a Mendelian character can be described in terms of the proportions of the offspring that fall into several distinct phenotypic classes or often simply in terms of the presence or absence of a class. For example, a cross between a red-flowered plant and a white-flowered plant might be expected to yield all red-flowered plants or, if it were a backcross, $\frac{1}{2}$ red-flowered plants and $\frac{1}{2}$ white-flowered plants. However, we require a different mode of description for quantitative characters. The basic concept is that of the **statistical distribution.** If the heights of a large number of male undergraduates are measured to the nearest 5 centimeters (cm), they will vary (say, between 145 and 195 cm), but many more individuals will fall into the middle categories (say, between 170 and 180 cm) than at the extremes.

Representing each measurement class as a bar, with its height proportional to the number of individuals in each class, we can graph the result as shown in Figure 23-1a. Such a graph of numbers of individuals observed against measurement class is a **frequency histogram.** Now suppose that five times as many individuals are measured, each to the nearest centimeter. The classes in Figure 23-1a are now subdivided to produce a histogram like the one shown in Figure 23-1b. If we continue this process, refining the measurement but proportionately increasing the number of individuals measured, then the histogram eventually takes on the continuous appearance of Figure 23-1c, which is the **distribution function** of heights in the population.

Of course, this continuous curve is an idealization, because no measurement can be taken with infinite accuracy or on an unlimited number of individuals. Moreover, the measured variate itself may be intrinsically discontinuous because it is the count of some number of discrete objects such as eye facets or bristles. It is sometimes convenient, however, to develop concepts using this slightly idealized picture as a shorthand for the more cumbersome observed frequency histogram (Figure 23-1a). We should not forget, however, that the distribution function is indeed an idealization.

The Mode. Most distributions of phenotypes look roughly like those in Figure 23-1: a single-most frequent class, the **mode,** is located near the middle of the distribution, with frequencies decreasing on either side. There are exceptions to this pattern, however. Figure 23-2a shows the very asymmetric distribution of seed weights in the plant *Crinum longifolium.* Figure 23-2b shows a **bimodal** (two-mode) distribution of larval survival probabilities for different second-chromosome homozygotes in *Drosophila willistoni.*

A bimodal distribution may indicate that the population being studied could better be considered as a mixture of two populations, each with its own mode. For

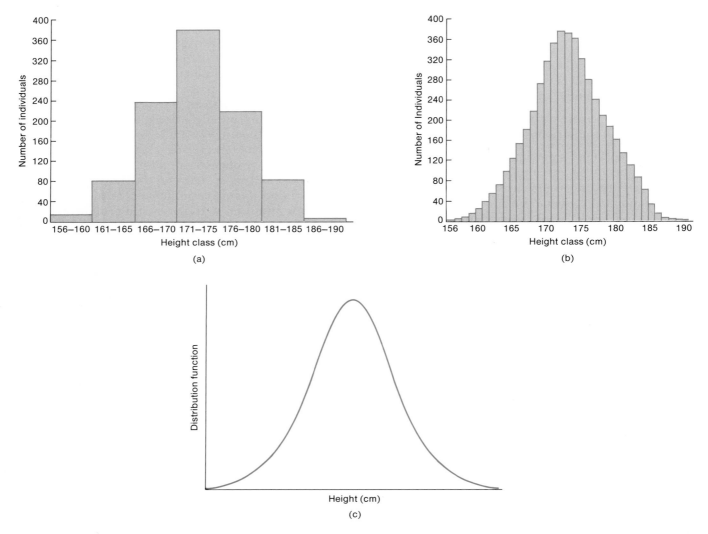

Figure 23-1. Frequency distributions for height of males. (a) A histogram with 5-cm class intervals. (b) A histogram with 1-cm class intervals. (c) The limiting continuous distribution.

example, suppose that we sample 100 undergraduates and measure their heights. These 100 individuals will include some males and some females. Thus, our sample is really a mixture of a sample from the population of male undergraduates and a sample from the population of female undergraduates, even though we have chosen the group in a single sampling operation. In Figure 23-2b, the left-hand mode probably represents severe one-locus mutations that are extremely deleterious when homozygous, whereas the right-hand mode is part of the distribution of "normal" viability. If the heights of both male and female undergraduates had been plotted in Figure 23-1, the distribution would have been bimodal, because the heights of females are distributed

around a mode at a value that is considerably smaller than the mode of the heights of males.

The Mean. Complete information about the distribution of a phenotype in a population can be given only by specifying the frequency of each measured class, but a great deal of information can be summarized in two statistics. First, we need some measure of the location of the distribution along the axis of measurement (for example, do the individual measurements tend to cluster around 100 cm or 200 cm)? One possibility is to give the measurement of the most common class, the mode. In Figure 23-1b, the mode is 172 cm; for females, the mode would be about 6 cm less. A more common measure of

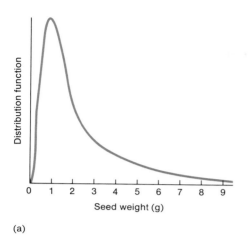

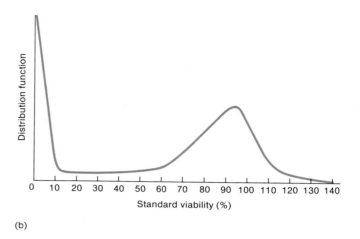

(a)

(b)

Figure 23-2. Asymmetric distribution functions. (a) Asymmetric distribution of seed weight in *Crinum longifolium*. (b) Bimodal distribution of survival of *Drosophila willistoni* expressed as a percentage of standard survival. (Adapted from S. Wright, *Evolution and the Genetics of Populations*, Vol. I. Copyright © 1968 by University of Chicago Press, 1968.)

location is the arithmetic average, or the **mean.** The mean of the measurement $(\bar{x})$ is simply the sum of all the measurements (x_i) divided by the number of measurements in the sample (N):

$$\text{Mean} = \bar{x} = \frac{x_1 + x_2 + x_3 + \cdots + x_N}{N} = \frac{1}{N}\sum x_i$$

where Σ represents summation and x_i is the ith measurement.

In a typical large sample, the same measured value will appear more than once, because several individuals will have the same value within the accuracy of the measuring instrument. For example, many individuals will be 170 cm tall. In such a case, $\bar{x}$ can be rewritten as the sum of all the measurement values, each weighted by how frequently it occurs in the population. From a total of N individuals measured, suppose that n_1 fall in the class with value x_1, that n_2 fall in the class with value x_2, and so on, so that $\Sigma n_i = N$. Then

$$\bar{x} = \frac{n_1}{N}x_1 + \frac{n_2}{N}x_2 + \cdots + \frac{n_k}{N}x_k$$

If we let f_i, be the **relative frequency** of the ith measurement class, so that

$$f_i = \frac{n_i}{N}$$

then we can rewrite the mean as

$$\bar{x} = f_1x_1 + f_2x_2 + \cdots + f_kx_k = \Sigma f_i x_i$$

where x_i equals the value of the ith measurement class. For the heights of male undergraduates in Figure 22-1, $\bar{x} = 173.5$ cm.

The Variance. A second characteristic of a distribution is the width of its spread around the central class. Two distributions with the same mean might differ very much in how closely the measurements are concentrated around the mean. The most common measure of variation around the center is the **variance,** which is defined as the average squared deviation of the observations from the mean, or

$$\begin{aligned}\text{Variance} &= s^2 \\ &= \frac{(x_1 - \bar{x})^2 + (x_2 - \bar{x})^2 + \cdots + (x_N - \bar{x})^2}{N} \\ &= \frac{1}{N}\sum (x_i - \bar{x})^2\end{aligned}$$

When more than one individual has the same measured value, the variance can be written

$$\begin{aligned}s^2 &= f_1(x_1 - \bar{x})^2 + f_2(x_2 - \bar{x})^2 + \cdots + f_k(x_k - \bar{x})^2 \\ &= \Sigma f_i(x_i - \bar{x})^2\end{aligned}$$

To avoid subtracting every value of x separately from the mean, we can use an alternative computing formula that is algebraically identical to the preceding equation:

$$s^2 = \frac{1}{N}\sum x_i^2 - \bar{x}^2$$

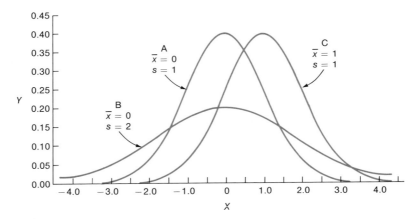

Figure 23-3. Three distribution functions with different means and standard deviations.

Because the variance is in squared units (square centimeters, for example), it is common to take the square root of the variance, which then has the same units as the measurement itself. This square-root measure of variation is called the **standard deviation** of the distribution:

$$\text{Standard deviation} = s = \sqrt{\text{variance}} = \sqrt{s^2}$$

Figure 23-3 shows two distributions having the same mean but different standard deviations (curves A and B) and two distributions having the same standard deviation but different means (curves A and C).

The mean and the variance of a distribution do not describe it completely, of course. They do not distinguish a symmetric distribution from an asymmetric one, for example. We can even construct symmetric distributions that have the same mean and variance but still have somewhat different shapes. Nevertheless, for the purposes of dealing with most quantitative genetic problems, the mean and variance suffice to characterize a distribution.

Correlation

Another statistical notion that is of use in the study of quantitative genetics is the association or **correlation** between variables. As a result of complex paths of causation, many variables in nature vary together but in an imperfect or approximate way. Figure 23-4a provides an example, showing the lengths of two particular teeth in several individual specimens of a fossil mammal, *Phenacodus primaevis*. The longer an individual's first lower molar is, the longer its second molar is, but the relationship between the two teeth is imprecise. Figure 22-4b shows that the total length and tail length in individual

snakes (*Lampropeltis polyzona*) are quite closely related to each other, whereas Figure 22-4c shows that the length and the number of caudal (tail) scales in these snakes seem to have no relation at all.

The usual measure of the precision of a relationship between two variables x and y is the **correlation coefficient**, (r_{xy}). It is calculated from the product of the deviation of each observation of x from the mean of the x values and the deviation of each observation of y from the mean of the y values—a quantity called the **covariance** of x and y (cov xy):

$$\text{cov } xy = \frac{(x_1 - \bar{x})(y_1 - \bar{y}) + (x_2 - \bar{x})(y_2 - \bar{y}) + \cdots}{N}$$
$$\frac{+ (x_N - \bar{x})(y_N - \bar{y})}{N}$$

$$= \frac{1}{N} \sum (x_i - \bar{x})(y_i - \bar{y})$$

$$\text{Correlation} = r_{xy} = \frac{\text{cov } xy}{s_x s_y}$$

This formula for the covariance is rather awkward computationally because it requires subtracting every value of x and y from the respective means $\bar{x}$ and $\bar{y}$. A formula that is exactly algebraically equivalent but that makes computation easier is

$$\text{cov } xy = \frac{1}{N} \sum x_i y_i - \overline{xy}$$

In the formula for correlation, the products of the deviations are divided by the product of the standard deviations of x and y (s_x and s_y). This normalization by the

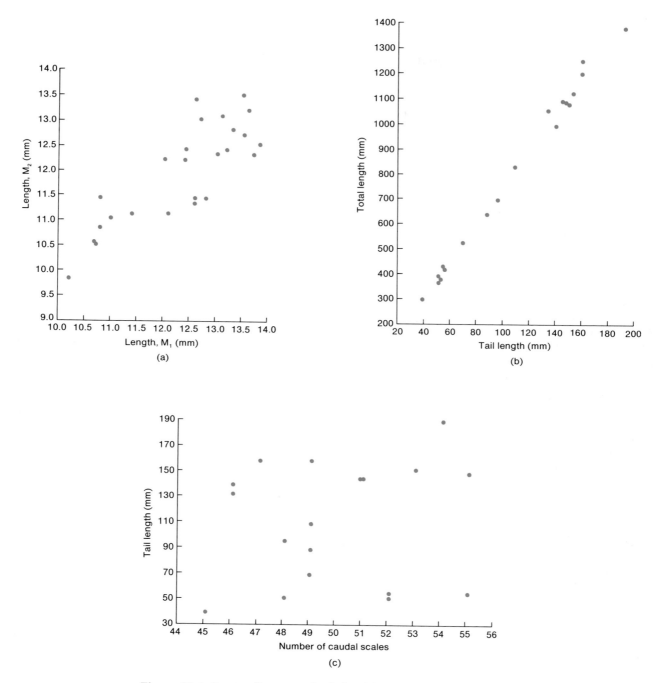

Figure 23-4. Scatter diagrams of relationships between pairs of variables. (a) Relationship between the lengths of the first and second lower molars (M_1 and M_2) in the extinct mammal *Phenacodus primaevis*. Each point gives the M_1 and M_2 measurements for one individual. (b) Tail length and body length of 19 individuals of the snake *Lampropeltis polyzona*. (c) Number of caudal scales and tail length related in the same 19 snakes in (b).

standard deviations has the effect of making r_{xy} a dimensionless number that is independent of the units in which x and y are measured. So defined, r_{xy} will vary from -1, which signifies a perfectly linear negative relation between x and y, to $+1$, which indicates a perfectly linear positive relation between x and y. If $r_{xy} = 0$, there is no linear relation between the variables. It is important to notice, however, that sometimes when there is no *linear* relation between two variables but there is a regular *nonlinear* relationship between them, one variable may be perfectly predicted from the other. Consider, for example, the parabola shown in Figure 23-5. The values of y are perfectly predictable from the values of x; yet $r_{xy} = 0$, because on average over the whole range of x values, larger x values are not associated with either larger or smaller y values. The data in parts (a), (b), and (c) in Figure 23-4 have r_{xy} values of 0.82, 0.99, and 0.10, respectively.

In these examples, correlations are described as the relation between a pair of measurements taken on the same individual, but a single measurement taken on pairs of individuals is also a subject for correlation analysis. Thus, we can determine the correlation between the height of a parent (x) and the height of an offspring (y) or the heights of an older sibling (x) and a younger sibling (y). This use of correlation is directly relevant to the problems encountered in quantitative genetics. Figure 23-6 shows the relation between the wing length of an offspring and the average wing length of its two parents (**midparent value**) in *Drosophila*.

Correlation and Identity. It is important to notice that correlation is not the same thing as identity. Values can be perfectly correlated without being equal. The variables x and y in the pairs

x	y
1	22
2	24
3	26

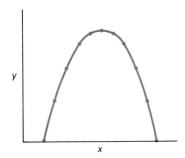

Figure 23-5. A parabola. Each value of y is perfectly predictable from the value of x, but there is no linear correlation.

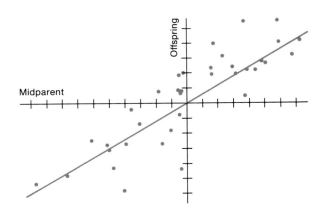

Figure 23-6. Relation between the wing lengths of individual *Drosophila* and the mean wing lengths of their two parents. (From D. Falconer, *Quantitative Genetics.* Copyright © 1981 by Longman Group Limited.)

are perfectly correlated ($r = +1.0$), although each value of y is about 20 units greater than the corresponding value of x. Two variables are perfectly correlated if, for a unit increase in one, there is a constant increase in the other (or a constant decrease if r is negative). The importance of the difference between correlation and identity arises when we consider the effect of environment on heritable characters. Parents and offspring can be perfectly correlated in some trait such as height, yet because of an environmental difference between generations, every child can be taller than the parents. This phenomenon appears in adoption studies, where children may be correlated with their biological parents but, on the average may be quite different from the parents as a result of a change in social situation.

Regression

The measurement of correlation provides us only with an estimate of the *precision* of relationship between two variables. A related problem is predicting the value of one variable given the value of the other. If x increases by two units, by how much will y increase? If the two variables are linearly related, then that relationship can be expressed as

$$y = bx + a$$

where b is the slope of the line relating y to x and a is the y-intercept of that line.

Figure 23-7 shows a scatter diagram of points for two variables, y and x, together with a straight line expressing the general linear trend of y with increasing x. This line, called the **regression line of y on x,** has been positioned so that the deviations of the points from the line are as small as possible. Specifically, if Δy is the dis-

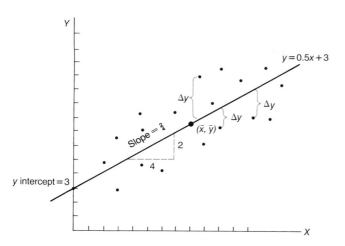

Figure 23-7. A scatter diagram showing the relation between two variables, x and y, with the regression line of y on x. This line, with a slope of $\frac{2}{4}$, minimizes the squares of the deviations. (Δy).

tance of any point from the line in the y direction, then the line has been chosen so that

$$\Sigma(\Delta y)^2 = \text{a minimum}$$

Any other straight line passed through the points on the scatter diagram will have a larger total squared deviation of the points from it.

Obviously, we cannot find this **least squares line** by trial and error. It turns out, however, that if the slope b of the line is calculated by

$$b = \frac{\text{cov } xy}{s_x^2}$$

and if a is then calculated from

$$a = \bar{y} - b\bar{x}$$

so that the line passes through the point $\bar{x}, \bar{y}$, then these values of b and a will yield the least-squares prediction equation.

Note that the prediction equation cannot predict y exactly for a given x, because there is scatter around the line. The equation predicts the *average y* for a given x, if large samples are taken.

Samples and Populations

The preceding sections have described the distributions and some statistics of particular assemblages of individuals that have been collected in some experiments or sets of observations. For some purposes, however, we are not really interested in the particular 100 undergraduates or

27 snakes that have been measured. Instead, we are interested in a wider world of phenomena, of which those particular individuals are representative. Thus, we might want to know the average height *in general* of undergraduates or the average seed weight *in general* of plants of the species *Crinum longifolium*. That is, we are interested in the characteristics of a **universe,** of which our small collection of observations is only a **sample.** The characteristics of any particular sample are, of course, not identical with those of the universe but vary from sample to sample. Two samples of undergraduates drawn from the universe of all undergraduates will not have exactly the same mean ($\bar{x}$) or variance (s^2), nor will these values for a sample typically be exactly equal to the mean and variance of the universe. We distinguish between the statistics of a sample and the values in the universe by using Roman letters, such as $\bar{x}$, s^2, and r, for sample values and Greek letters μ (mean), σ^2 (variance), and ρ (correlation) for the values in the universe.

The sample mean $\bar{x}$ is an approximation to the true mean μ of the universe. It is a statistical *estimate* of that true mean. So, too, s^2, s, and r are estimates of σ^2, σ, and ρ in the universe from which respective samples have been taken.

If we are interested in a particular collection of individuals not for its own sake but as a way of obtaining information about a universe, then we want the sample statistics such as $\bar{x}$, s^2, and ρ to be good estimates of the true values of μ, σ^2, and ρ, and so on. There are many criteria of what a "good" estimate is, but the one that seems clearly desirable is that if we take a very large number of samples, then the average value of the estimate over these samples should be the true value in the universe; that is, the estimate should be **unbiased.** It turns out that $\bar{x}$ is indeed an unbiased estimate of μ. If a very large number of samples are taken and $\bar{x}$ is calculated in each one, then the average of these $\bar{x}$ values will be μ. Unfortunately, s^2, as we have defined it, is not an unbiased estimate of σ^2. It tends to be a little too small, so that the average of many s^2 values is less than σ^2. The amount of bias is precisely related to the size N of each sample, and it can be shown that $[N/(N-1)]s^2$ is an unbiased estimate of σ^2. Thus, whenever we are interested in the variance of a set of measurements—not as a characteristic of the particular collection but as an estimate of a universe which the sample represents—then the appropriate quantity to use is $[N/(N-1)]s^2$ rather than s^2 itself. Note that this new quantity is equivalent to dividing the sum of squared deviations by $N-1$ instead of N in the first place, so that

$$\left(\frac{N}{N-1}\right)s^2 = \left(\frac{N}{N-1}\right)\frac{1}{N}\sum(x_i - \bar{x})^2$$
$$= \frac{1}{N-1}\sum(x_i - \bar{x})^2$$

This latter quantity is often called the "sample variance," but that term is really not correct. The sample variance is s^2, whereas $[N/(N-1)]s^2$ is an adjustment to the sample variance to make it an unbiased estimate of σ^2. Which of the two quantities used, depends on whether the primary interest is in the sample or in the universe.

All of these considerations about bias also apply to the sample covariance. In the formula for the correlation coefficient (page 646), however, the factor $N/(N-1)$ would appear in both the numerator and the denominator and therefore cancel out, so we can ignore it for the purposes of computation.

Genotypes and Phenotypic Distribution

Using the concepts of distribution, mean, and variance, we can understand the difference between quantitative and Mendelian genetic traits. Suppose that a population of plants contains three genotypes, each of which has some differential effect on growth rate. Further, assume that there is some environmental variation from plant to plant because of inhomogeneity in the soil in which the population is growing and that there is some developmental noise (see page 10) For each genotype, there will be a separate distribution of phenotypes with a mean and a standard deviation that depend on the genotype and the set of environments. Suppose that these distributions look like the three height distributions in Figure 23-8a. Finally, assume that the population consists of a mixture of the three genotypes but in the unequal proportions $1:2:3$. Then the phenotypic distribution of individuals in the population as a whole will look like the black line in

Figure 23-8b, which is the result of summing the three underlying separate genotypic distributions, weighted by their frequencies in the population. The mean of this total distribution is the average of the three genotypic means, again weighted by the frequencies of the genotypes in the population. The variance of the total distribution is produced partly by the environmental variation within each genotype and partly by the slightly different means of the three genotypes.

Two features of the total distribution are noteworthy. First, there is only a single mode. Despite the existence of three separate genotypic distributions underlying it, the population distribution as a whole does not reveal the separate modes. Second, any individual whose height lies between the two arrows could have come from any one of the three genotypes, because they overlap so much. The result is that we cannot carry out any simple Mendelian analysis to determine the genotype of an individual organism. For example, suppose that the three genotypes are the two homozygotes and the heterozygote for a pair of alleles at a locus. Let aa be the short homozygote and AA be the tall one, with the heterozygote being of intermediate height. Because there is so much overlap of the phenotypic distributions, we cannot know to which genotype a given individual belongs. Conversely, if we cross two individuals that happen to be a homozygote aa and a heterozygote Aa, the offspring will not fall into two discrete classes in a $1:1$ ratio but will cover almost the entire range of phenotypes smoothly. Thus, we cannot know that the cross is in fact $aa \times Aa$ and not $aa \times AA$ or $Aa \times Aa$.

If the hypothetical plants represented in Figure 23-8 are grown in a stressful environment but care is taken to provide each plant with exactly the same environment, then the picture changes. The phenotypic variance of each separate genotype is reduced because all of the

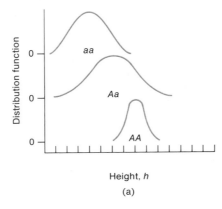

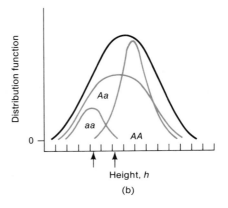

Figure 23-8. (a) Phenotypic distributions of three genotypes. (b) A population phenotypic distribution results from mixing individuals of the three genotypes in a proportion $1:2:3$.

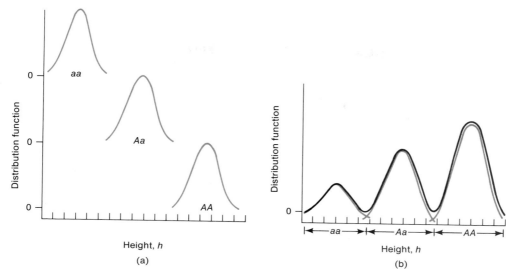

Figure 23-9. When the same genotypes as those in Figure 23-8 are grown in carefully controlled stress environments, the result is a smaller phenotypic variation in each genotype and a greater difference between genotypes.

plants are grown under identical conditions; at the same time, the differences between genotypes may become greater because the small differences in physiology may be exaggerated under stress. The result (Figure 23-9) would be a separation of the population as a whole into three nonoverlapping phenotypic distributions, each characteristic of one genotype. We could now carry out a perfectly conventional Mendelian analysis of plant height. A "quantitative" character has been converted into a "qualitative" one! This conversion has been accomplished by finding a way to make the differences between the means of the genotypes large compared to the variation within genotypes.

> **Message** A **quantitative** character is one for which the average phenotypic differences between genotypes are small compared to the variation between individuals within genotypes.

It is sometimes assumed that continuous variation in a character is a consequence of a large number of segregating genes that influence the measurement, so that continuous variation is to be taken as prima facie evidence for control by many genes. But, as we have just shown, this is not necessarily true. If the difference between genotypic means is small compared to the environmental variance, then the simple one-gene–two-allele case results in continuous variation.

Of course, if the range of a character is limited and if there are many segregating loci influencing it, then we expect the character to show continuous variation, because each allelic substitution must account for only a small difference in the trait. This **multiple-factor hypothesis** (that large numbers of genes, each with a small effect, are segregating to produce quantitative variation) has long been the basic model of quantitative genetics, although there is no convincing evidence that such groups of genes really exist. A special name, **polygenes,** has been coined for these hypothetical factors of small-but-equal effect as opposed to the genes of simple Mendelian analysis.

It is important to remember, however, that the *number* of segregating loci that influence a trait is not what separates quantitative and qualitative characters.

Even in the absence of large environmental variation, a very few genetically varying loci will produce variation that is indistinguishable from the effect of many loci of small effect. As an example, we can consider one of the earliest experiments in quantitative genetics, that of Wilhelm Johannsen on pure lines. By **inbreeding** (mating close relatives), Johannsen produced 19 homozygous lines of bean plants from an originally genetically heterogeneous population. Each line had a characteristic average seed weight ranging from 0.64 grams (g) per seed for the heaviest line down to 0.35 g per seed for the lightest line. It is by no means clear that all these lines were genetically different (for example, five of the lines had seed weights of 0.450, 0.453, 0.454, 0.454, and 0.455 g), but let's take the most extreme position — that the lines *were* all different. Obviously, these observations would be incompatible with a simple one-locus–two-allele model of gene action. In that case, if the original population were segregating for the two alleles, *A* and *a*, all inbred lines derived from that population would have to fall into one of two classes: *AA* or *aa*. If, in contrast, there were, say, 100 loci, each of small effect, segregating in the original population, then a vast number of different inbred lines could be produced, each with a different combination of homozygotes at different loci. One line might be

$$\frac{A \; B \; c \; D}{A \; B \; c \; D} \cdots \frac{Z}{Z} \cdots$$

and another might be

$$\frac{a \; b \; c \; D}{a \; b \; c \; D} \cdots \frac{z}{z} \cdots$$

and so on.

However, we do not need such a large number of loci to obtain the result observed by Johannsen. If there were only five loci, each with three alleles, then $3^5 = 243$ different kinds of homozygotes could be produced from the inbreeding process. If we make 19 inbred lines at random, there is a good chance (about 50 percent) that each of the 19 lines will belong to a different one of the 243 classes. So Johannsen's experimental results can easily be explained by a relatively small number of genes. Thus, there is no real dividing line between polygenic traits and other traits. It is safe to say that no phenotypic

trait above the level of the amino-acid sequence in a polypeptide is influenced by only one gene. Moreover, traits influenced by many genes are not equally influenced by them all. Some genes will have major effects on the trait; some minor effects.

Message The critical difference between Mendelian and quantitative traits is not the number of segregating loci but the size of phenotypic differences between genotypes compared to the individual variation within genotypic classes.

Norm of Reaction and Phenotypic Distribution

The phenotypic distribution of a trait, as we have seen, is a function of the average differences between genotypes and of the variation between genotypically identical individuals. But both of these are in turn functions of the environments in which the organisms develop and live. For a given genotype, each environment will result in a given phenotype (for the moment, ignoring developmental noise). Then a **distribution of environments** will be reflected biologically as a **distribution of phenotypes.** The way in which the environmental distribution is transformed into the phenotypic distribution is determined by the **norm of reaction,** as shown in Figure 23-10. The horizontal axis is environment (say, temper-

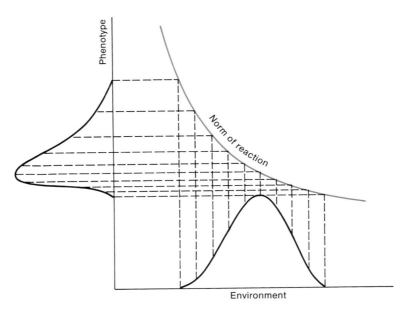

Figure 23-10. The distribution of environments on the horizontal axis is converted to the distribution of phenotypes on the vertical axis by the norm of reaction of a genotype.

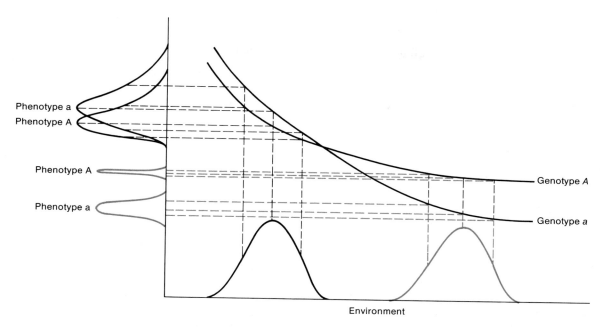

Phenotype a

Phenotype A

Phenotype A

Phenotype a

Genotype *A*

Genotype *a*

Environment

Figure 23-11. Two different environmental distributions are converted into different phenotypic distributions by two different genotypes.

ature) and the vertical axis is phenotype. The distribution on the horizontal axis represents the relative frequency of different environments. For example, most of these individuals developed at a temperature of 20°C, fewer at 18°C and 22°C, and so on. The norm of reaction for the genotype falls steeply at lower temperatures but flattens at high temperatures. Because each temperature produces a corresponding phenotype, as shown by the dashed lines, the distribution of temperatures results in the phenotypic distribution on the vertical axis. The norm of reaction is like a distorting mirror that reflects the environmental distribution onto the phenotypic axis.

By means of the same analysis, Figure 23-11 shows how a population consisting of two genotypes with different norms of reaction has a phenotypic distribution that depends on the distribution of environments. If the environments are distributed as shown by the black distribution curve, then the resulting population of plants will have a unimodal distribution, because the difference between genotypes is very small in this range of environments compared to the sensitivity of the norms of reaction to small changes in temperature. If the distribution of environments is shifted to the right, however, as shown by the colored distribution curve, a bimodal distribution of phenotypes results, because the norms of reaction are nearly flat in this environmental range but very different from each other.

Message A distribution of environments is reflected biologically as a distribution of phenotypes. The transformation of environmental distribution into phenotypic distribution is determined by the norm of reaction.

The Heritability of a Trait

The most basic question to be asked about a quantitative trait is whether or not the observed variation in the character is influenced by genes at all. It is important to note that this is not the same as asking whether or not genes play any role in the character's development. Gene-mediated developmental processes lie at the base of every character, but *variation* from individual to individual is not necessarily the result of *genetic variation*. Thus, the possibility of speaking any language at all depends critically on the structures of the central nervous system as well as of the vocal cords, tongue, mouth, and ears, which depend in turn on the nature of the human genome. There is no environment in which cows will speak. But although the particular language that is spoken by humans varies from nation to nation, that variation is totally nongenetic.

> **Message**   The question of whether or not a trait is heritable is a question about the role that differences in genes play in the phenotypic differences between individuals or groups.

Familiality and Heritability

In principle, it is easy to determine whether any genetic variation influences the phenotypic variation among organisms for a particular trait. If genes are involved, then (on average) biological relatives should resemble each other more than unrelated individuals do. This resemblance would be reflected as a positive correlation between parents and offspring or between siblings (offspring of the same parents). Parents who are larger than the average would have offspring who are larger than the average; the more seeds a plant produces, the more seeds its siblings would produce. Such correlations between relatives, however, are evidence for genetic variation *only if the relatives do not share common environments more than nonrelatives do.* It is absolutely fundamental to distinguish *familiality* from *heritability.* Traits are **familial** if members of the same family share them, for whatever reason. Traits are **heritable** only if the similarity arises from shared genotypes.

In experimental organisms, there is no problem in separating environmental from genetic similarities. The offspring of a cow producing milk at a high rate and the offspring of a cow producing milk at a low rate can be raised together in the same environment to see whether, despite the environmental similarity, each resembles its own parent. In natural populations, and especially in humans, this is difficult to do. Because of the nature of human societies, members of the same family not only share genes but also have similar environments. Thus, the observation of simple familiality of a trait is genetically uninterpretable. In general, people who speak Hungarian have Hungarian-speaking parents and people who speak Japanese have Japanese-speaking parents. Yet the massive experience of immigration to North America has demonstrated that these linguistic differences, although familial, are nongenetic. The highest correlations between parents and offspring for any social traits in the United States are those for political party and religious sect, but they are not heritable. The distinction between familiality and heredity is not always so obvious. The Public Health Commission, which originally studied the vitamin-deficiency disease pellagra in the southern United States in 1910 came to the conclusion that it was genetic because it ran in families!

To determine whether a trait is heritable in human populations, we must use adoption studies to avoid the usual environmental similarity between biological relatives. The ideal experimental subjects are identical twins raised apart, because they are genetically identical but environmentally different. Such adoption studies must be so contrived that there is no correlation between the social environment of the adopting family and that of the biological family. These requirements are exceedingly difficult to meet, so that in practice we know very little about whether human quantitative traits that are familial are also heritable. Skin color is clearly heritable, as is adult height—but even for these traits we must be very careful. We know that skin color is affected by genes from studies of crossracial adoptions and observations that the offspring of black African slaves were black even when they were born and raised in Canada. But are the differences in height between Japanese and Europeans affected by genes? The children of Japanese immigrants who are born and raised in North America are taller than their parents but shorter than the North American average, so we might conclude that there is some influence of genetic difference. However, second-generation Japanese-Americans are even taller than their Ameri-

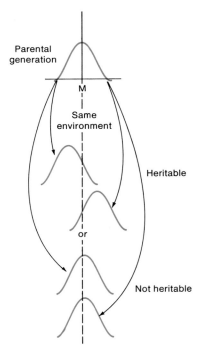

Figure 23-12. Standard method for testing heritability in experimental organisms. Crosses are performed within two populations of individuals selected from the extremes of the phenotypic distribution in the parental generation. If the phenotypic distributions of the two groups of offspring are significantly different from each other, then the trait is heritable. If the offspring distributions both resemble the distribution for the parental generation, then the trait is not heritable.

can-born parents! It appears that some environmental-cultural influence, or perhaps a maternal effect, is still felt in the first generation of births in North America. We cannot yet say whether genetic differences in height distinguish North Americans of, say, Japanese and Swedish ancestry.

Personality traits, temperament, and cognitive performance (including IQ scores) are all familial, but there are no well-designed adoption studies to show whether or not these traits are heritable. The only large-sample studies of human IQ performance claiming to involve randomized environments (Cyril Burt's reports on identical twins raised apart) have been shown to be completely fraudulent.

Figure 23-12 summarizes the usual method for testing heritability in experimental organisms. Individuals from both extremes of the distribution are mated with their own kind, and the offspring are raised in a common controlled environment. If there is an average difference between the two offspring groups, the trait is heritable. Most morphological traits in *Drosophila*, for example, turn out to be heritable — but not all of them. If flies with right wings that are slightly longer than their left wings are mated together, their offspring have no greater tendency to be "right-winged" than do the offspring of "left-winged" flies. As we shall see later (page 668), this method can also be used to obtain quantitative information about heritability.

> **Message** In experimental organisms, environmental similarity often can be readily distinguished from genetic similarity (or heritability). In humans, however, it is very difficult to determine whether a particular trait is heritable.

Determining Norms of Reaction

Remarkably little is known about the norms of reaction for any quantitative traits in any species. This is partly because it is difficult in most sexually reproducing species to replicate a genotype so that it can be tested in different environments. It is for this reason, for example, that we do not have a norm of reaction for any genotype for any human quantitative trait.

A few norm-of-reaction studies have been carried out with plants that can be clonally propagated. The results of one of these experiments are discussed on page 9. It is possible to replicate genotypes in sexually reproducing organisms by the technique of mating close relatives, or inbreeding. By selfing (where possible) or by mating brother and sister repeatedly generation after

generation, a **segregating line** (one that contains both homozygotes and heterozygotes at a locus) can be made homozygous.

In corn, for example, a single individual is chosen and self-pollinated. Then in the next generation, a single one of its offspring is chosen and self-pollinated. In the third generation, a single one of *its* offspring is chosen and self-pollinated, and so on. Suppose that the original individual in the first generation is already a homozygote at some locus. Then all of its offspring from self-pollination will also be homozygous and identical at the locus. Future generations of self-pollination will simply preserve the homozygosity. If, on the other hand, the original individual is a heterozygote, then the selfing $Aa \times Aa$ will produce $\frac{1}{4}$ AA homozygotes and $\frac{1}{4}$ aa homozygotes. If a single offspring is chosen in this subsequent generation to propagate the line, then there is a 50 percent chance that it is now a homozygote. If, by bad luck, the chosen individual should be still a heterozygote, there is another 50 percent chance that the selected individual in the third generation is homozygous, and so on. Of the ensemble of all heterozygous loci, then, after one generation of selfing, only $\frac{1}{2}$ will still be heterozygous; after two generations, $\frac{1}{4}$; after three, $\frac{1}{8}$. In the nth generation

$$\text{Het}_n = \frac{1}{2^n}\text{Het}_0$$

where Het_n is the proportion of heterozygous loci in the nth generation and Het_0 is the proportion in the 0th generation. When selfing is not possible, brother–sister mating will accomplish the same end, although more slowly. Table 23-1 is a comparison of the amounts of heterozygosity left after n generations of selfing and brother-sister mating.

TABLE 23-1. Heterozygosity remaining after various generations of inbreeding for two systems of mating

Generation	Remaining heterozygosity	
	Selfing	Brother-sister mating
0	1.000	1.000
1	0.500	0.750
2	0.250	0.625
3	0.125	0.500
4	0.0625	0.406
5	0.03125	0.338
10	0.000977	0.114
20	1.05×10^{-6}	0.014
n	$\text{Het}_n = \frac{1}{2}\text{Het}_{n-1}$	$\text{Het}_n = \frac{1}{2}\text{Het}_{n-1} + \frac{1}{4}\text{Het}_{n-2}$

To carry out a norm-of-reaction study of a natural population, a large number of lines are sampled from the population and inbred for a sufficient number of generations to guarantee that each line is virtually homozygous at all its loci. Each line is then homozygous at each locus for a randomly selected allele present in the original population. The inbred lines themselves cannot be used to characterize norms of reaction in the natural population, because such totally homozygous genotypes do not exist in the original population. Each inbred line can be crossed to every other inbred line to produce heterozygotes that reconstitute the original population, and an arbitrary number of individuals from each cross can be produced. If inbred line 1 has the genetic constitution *AA BB cc dd EE* · · · and inbred line 2 is *aa BB CC dd ee* · · ·, then a cross between them will produce a large number of offspring, all of whom are identically *Aa BB Cc ddEe* · · · and can be raised in different environments.

Inbreeding by mating of close relatives for many generations results in total homozygosity for the entire genome. In species like *Drosophila* in which the necessary dominant markers and crossover suppressors are available, it is possible to produce lines that are homozygous for only a single chromosome, rather than for the whole set, as shown for an autosome in Figure 23-13. A single male from the population to be sampled is crossed to a female carrying a chromosome with a crossover suppressor *C* (usually a complex inversion), a recessive lethal *l*, and a dominant visible marker M_1 heterozygous with a second dominant visible M_2. In the F_1, a *single* male carrying the ClM_1 chromosome is chosen. This male, which is also carrying a wild-type chromosome from the population, is again crossed to the marker stock. In the F_2, all flies showing the M_1 trait but not M_2 are necessarily all heterozygotes for copies of the original wild-type chromosome because ClM_1/ClM_1 is lethal, and no crossovers have taken place. In the F_3, all wild-type flies are identically homozygous for the wild-type chromosome and are now available to make a stock for norm-of-reaction studies and for crosses.

Figure 23-14 shows the norms of reaction of abdominal bristle number as a function of temperature for second-chromosome homozygotes of *Drosophila pseudoobscura*. Like the growth-rate norms of *Achillea* (Figure 1-8), the norms cross each other, with different genotypes having different temperature maxima. The heterozygotes (the natural genotypes) are more similar to each other than are the chromosomal homozygotes, which seldom, if ever, occur in a natural population.

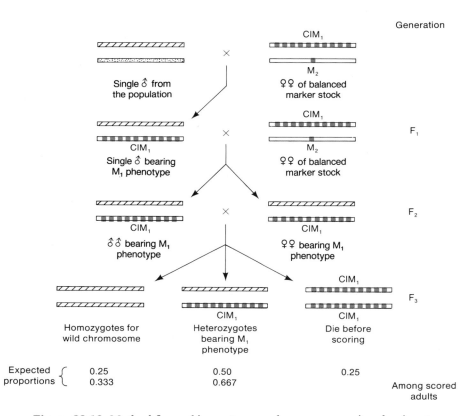

Figure 23-13. Method for making autosomes homozygous using dominant marker genes, M_1 and M_2, a crossover suppressor, C, and a lethal gene, l.

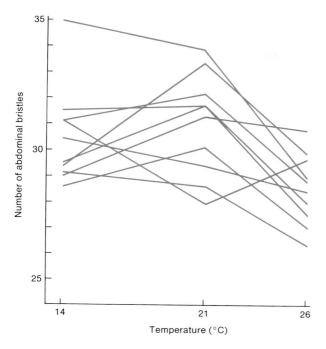

Figure 23-14. The number of abdominal bristles in different homozygous genotypes of *Drosophila pseudoobscura* at three different temperatures. (Data courtesy of A. P. Gupta.)

Results of Norm-of-Reaction Studies

Very few norm-of-reaction studies have been carried out for quantitative characters for the normally heterozygous genotypes found in natural populations. Only easily inbred species such as corn or *Drosophila* and clonal species such as strawberries have been studied to any degree. The outcomes of such studies resemble Figure 23-14. No genotype is consistently above or below other genotypes; instead, there are small differences among genotypes, and the direction of these differences is not consistent over a wide range of environments.

These facts have two important consequences. First, the selection of "superior" genotypes in domesticated animals and cultivated plants will result in very specifically adapted varieties that may not show their superior properties in other environments. To some extent, this problem is overcome by deliberately testing genotypes in a range of environments (for example, over several years and in several locations). It would be even better, however, if plant breeders could test their selections in a variety of controlled environments in which different environmental factors could be separately manipulated. The consequences of actual plant-breeding practices can be seen in Figure 23-15, where the yields of two varieties of corn are shown as a function of different farm envi-

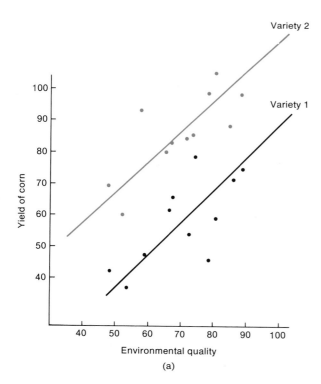

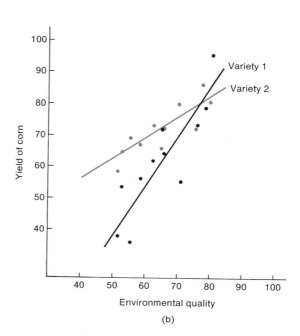

Figure 23-15. Yields of grain of two varieties of corn in different environments: (a) at a high planting density; (b) at a low planting density. (Data courtesy of W. A. Russell, *Proceedings of the 29th Annual Corn and Sorghum Research Conference,* 1974.)

ronments. Variety 1 is an older variety of hybrid corn; variety 2 is a later "improved" hybrid. These performances are compared at a low planting density, which prevailed when variety 1 was developed, and at a high planting density characteristic of farming practice when hybrid 2 was selected. At the high density, the new variety is clearly superior to the old variety in all environments (Figure 22-15a). At the low density, however, the situation is quite different. First, note that the new variety is less sensitive to environment than the older hybrid, as evidenced by its flatter norm of reaction. Second, the new "improved" variety is actually poorer under the best farm conditions. Third, the yield improvement of the new variety is not apparent under the low densities characteristic of earlier agricultural practice.

The second consequence of the nature of reaction norms is that even if it should turn out that there is genetic variation for various mental and emotional traits in the human species, which is by no means clear, this variation is unlikely to favor one genotype over another across a range of environments. We must beware of hypothetical norms of reaction for human cognitive traits that show one genotype unconditionally superior to another. Even putting aside all questions of moral and political judgment, there is simply no basis for describing different human genotypes as "better" or "worse" on any scale, unless the investigator is able to make a very exact specification of environment.

Message   Norm-of-reaction studies show only small differences among natural genotypes, and these differences are not consistent over a wide range of environments. Thus, "superior" genotypes in domesticated animals and cultivated plants may be superior only in certain environments. If it should turn out that humans exhibit genetic variation for various mental and emotional traits, this variation is unlikely to favor one genotype over another across a range of environments.

Quantifying Heritability

If a trait is shown to have some heritability in a population, then it is possible to quantify the degree of heritability. In Figure 23-8, we see that the variation among phenotypes in a population arises from two sources. First, there are average differences between the genotypes; second, each genotype exhibits phenotypic variance because of environmental variation. The total phenotypic variance of the population (s_p^2) can then be broken into two portions: the variance among genotypic

means (s_g^2), and the remaining variance (s_e^2). The former is called the **genetic variance,** and the latter is called the **environmental variance;** however, as we shall see, these names are quite misleading. The degree of heritability can then be defined as the proportion of the total variance that is due to the genetic variance:

$$H^2 = \frac{s_g^2}{s_p^2} = \frac{s_g^2}{s_g^2 + s_e^2}$$

H^2, so defined, is called the **broad heritability** of the character.

It must be stressed that this measure of "genetic influence" tells us what proportion of the population's *variation* in phenotype can be assigned to *variation* in genotype. It does not tell us what proportions of an *individual's* phenotype can be ascribed to its heredity and to its environment. This latter distinction is not a reasonable one. An individual's phenotype is a consequence of the interaction between its genes and its sequence of environments. It clearly would be silly to say that you owe 60 inches of your height to genes and 10 inches to environment. All measures of the "importance" of genes are framed in terms of the proportion of variance ascribable to their variation. This approach is a special application of the more general technique of the **analysis of variance** for apportioning relative weight to contributing causes. The method was, in fact, invented originally to deal with experiments in which different environmental and genetic factors were influencing the growth of plants. (For a sophisticated but accessible treatment of the analysis of variance written for biologists, see R. Sokal and J. Rohlf, *Biometry,* 2d ed., W. H. Freeman and Company, 1980.)

Methods of Estimating H²

Genetic variance and heritability can be estimated in several ways. Most directly, we can obtain an estimate of s_e^2 by making a number of homozygous lines from the population, crossing them in pairs to reconstitute individual heterozygotes, and measuring the phenotypic variance *within* each heterozygous genotype. Because there is no genetic variance within a genotypic class, these variances will (when averaged) provide an estimate of s_e^2. This value then can be subtracted from the value of s_p^2 in the original population to give s_g^2.

Other estimates of genetic variance can be obtained by considering the genetic similarities between relatives. Using simple Mendelian principles, we can see that one-half of the genes of full siblings will (on average) be identical. For identification purposes, we can label the alleles at a locus carried by the parents differently, so that they are, say, $A_1 A_2$ and $A_3 A_4$. Now the older sibling has a probability of $\frac{1}{2}$ of getting A_1 from its father, as does the younger sibling, so the two siblings have a chance of $\frac{1}{2} \times \frac{1}{2} = \frac{1}{4}$ of both carrying A_1. On the other hand, they might both have received an A_2 from their father, so again, they have a probability of $\frac{1}{4}$ of carrying a gene in common that they inherited from their father. Thus, the chance is $\frac{1}{4} + \frac{1}{4} = \frac{1}{2}$ that both siblings will carry an A_1 or that both siblings will carry an A_2. The other half of the time, one sibling will inherit an A_1 and the other will inherit an A_2. So, as far as paternally inherited genes are concerned, full siblings have a 50 percent chance of carrying the same allele. But the same reasoning applies to their maternally inherited gene. Averaging over their paternally and maternally inherited genes, one-half of the genes of full siblings are identical between them. Their **genetic correlation,** which is equal to the chance that they carry the same allele, is $\frac{1}{2}$.

If we apply this reasoning to half-siblings, say, with a common father but with different mothers, then we get a different result. Again, the two siblings have a 50 percent chance of inheriting an identical gene from their father, but this time they have no way of inheriting the same gene from their mothers because they have two different mothers. Averaging the maternally inherited and the paternally inherited genes thus gives a probability of $(\frac{1}{2} + 0)/2 = \frac{1}{4}$ that these half-siblings will carry the same gene.

The result is that the difference in genetic correlation between full and half-siblings is $\frac{1}{2} - \frac{1}{4} = \frac{1}{4}$. Let's contrast this with their **phenotypic correlations.** If the environmental similarity is the same for half- and full siblings — a very important condition for estimating heritability — then environmental similarities will cancel out if we take the difference in correlation between the two kinds of siblings. This difference in phenotypic correlation will then be proportional to how much of the variance is genetic. Thus

$$\begin{pmatrix} \text{Genetic correlation} \\ \text{of full siblings} \end{pmatrix} - \begin{pmatrix} \text{genetic correlation} \\ \text{of half-siblings} \end{pmatrix} = \frac{1}{4}$$

but

$$\begin{pmatrix} \text{Phenotypic} \\ \text{correlation} \\ \text{of full siblings} \end{pmatrix} - \begin{pmatrix} \text{phenotypic} \\ \text{correlation} \\ \text{of half-siblings} \end{pmatrix} = H^2 \times \frac{1}{4}$$

so an estimate of H^2 is

$$H^2 = 4 \begin{pmatrix} \text{correlation} \\ \text{of full siblings} \end{pmatrix} - \begin{pmatrix} \text{correlation} \\ \text{of half-siblings} \end{pmatrix}$$

where the correlation here is the *phenotypic* correlation.

We can use similar arguments about genetic similarities between parents and offspring and between twins to obtain two other estimates of H^2:

$$H^2 = 4 \begin{pmatrix} \text{correlation} \\ \text{of full siblings} \end{pmatrix} - 2 \begin{pmatrix} \text{parent–offspring} \\ \text{correlation} \end{pmatrix}$$

or

$$H^2 = 2 \begin{pmatrix} \text{correlation of} \\ \text{monozygotic twins} \end{pmatrix} - \begin{pmatrix} \text{correlation of} \\ \text{dizygotic twins} \end{pmatrix}$$

These formulas are derived from considering the genetic similarities between relatives. They are only approximate and depend on assumptions about the ways in which genes act. The first two formulas, for example, assume that genes at different loci add together in their effect on the character. The last formula is particularly inaccurate, because it also assumes that the alleles at each locus show no dominance (see the discussion of components of variance on page 665). If we ignore these problems of gene interaction, then the genetic correlation is $\frac{1}{2}$ between full siblings and $\frac{1}{4}$ between half-siblings. Substituting the values in the first formula gives $H^2 = 1$, which would be the case if all the variation were genetic. If there is nongenetic variation, it will be common to both half- and full siblings, making their correlations more similar and reducing the value of H^2. A similar argument applies, for example, to the twin formula; monozygotic twins have a genetic correlation of 1, whereas dizygotic twins are just full siblings and have a genetic correlation of $\frac{1}{2}$.

All of these estimates, as well as others based on correlations between relatives, depend *critically* on the assumption that environmental correlations between individuals are the same for all degress of relationship. If closer relatives have more similar environments, as they do in humans, the estimates of heritability are biased. It is reasonable to assume that most environmental correlations between relatives are positive, in which case the heritabilities would be overestimated. Negative environmental correlations can also exist. For example, if the members of a litter must compete for food that is in short supply, negative correlations in growth rates among siblings could occur.

In general, the presence of greater environmental correlation between close relatives makes heritability estimates uninterpretable. For this reason, there are no legitimate estimates of heritability for human quantitative traits. Despite their widespread use in human genetics, parent-offspring correlations are estimates of *familiality*, not of heritability. Even the difference between identical and fraternal twin correlations will not do, because there is the implicit assumption that identi-

660

cal twins are treated no more alike than fraternal twins —an assumption that is unwarranted by the facts. Volumes have been written on the heritability of human IQ, for example, and many modern genetics textbooks treat the numerical estimates of the heritability of IQ seriously (0.8 is the usual figure given). The absence of studies that estimate the environmental correlations between relatives, however, make all the numbers meaningless.

The Meaning of H^2

Attention to the problems of estimating broad heritability distracts from the deeper questions about the meaning of the ratio when it can be estimated. Despite its widespread use as a measure of how "important" genes are in influencing a trait, H^2 actually has a special and limited meaning.

First, H^2 is not a fixed characteristic of a trait but depends on the population in which it is measured and the set of environments in which that population has developed. In one population, alleles segregating at many loci may influence a trait. In another population, these loci may be homozygous. In such a homozygous population, the trait will show no heritability because $s_g^2 = 0$. That value does not mean that genes have no role in influencing the trait's development, but only that none of the variation between individuals within that population can be ascribed to genetic variation. Similarly, a population developing in a very homogeneous environment will have a smaller s_e^2 for a trait than a population developing in a varying environment. The lack of environmental heterogeneity will result in a high value of H^2, but that value does not mean the trait is insensitive to all environments.

> **Message** In general, the heritability of a trait is different in each population and in each set of environments; it cannot be extrapolated from one population and set of environments to another.

Second, the separation of variance into genetic and environmental components, s_g^2 and s_e^2, does not really separate the genetic and environmental causes of variation. Consider Figure 23-16, which shows two genotypes in a population with their two norms of reaction. In Figure 23-16a, the population is assumed to consist of the two genotypes in equal frequency, and the distribution of environments (shown on the horizontal axis) is centered toward the right. The phenotypic distribution (shown on the vertical axis) is composed to two underlying distributions that are very different in their means

and exhibit little environmental variation. H^2 has a high value. In Figure 23-16b, the environmental distribution has been shifted to the left. As a result, the average difference between genotypes is very much less. The important point here is that the so-called *genetic variance* has been changed by shifting the *environment*. In Figure 23-16c, we suppose that the environments are the same as in Figure 23-16b, but now genotype II has become extremely common in the population and genotype I is rare. Then the phenotypic distribution is almost entirely a reflection of norm-of-reaction II. The population has a smaller environmental variance (s_e^2) than does the population of Figure 23-16b because it has become enriched for the more developmentally stable genotype. In this case, the *environmental variance* has been changed by a change in *genotypes*. In general, genetic variance depends on the environments to which the population is exposed and environmental variance depends on the frequencies of the genotypes.

> **Message** Because genotype and environment interact to produce phenotype, no partition of variation can actually separate causes of variation.

Third, as a consequence of the argument just given, knowledge of the heritability of a trait does not permit us to predict how the distribution of that trait will change if either genotypic frequencies or environmental factors change markedly.

> **Message** A high heritability does not mean that a trait is unaffected by its environment.

Let's compare Figure 23-16a and b, for example. All that high heritability means is that for the particular population developing in the particular distribution of environments in which the heritability was measured, average differences between genotypes are large compared to environmental variation within genotypes. If the environment is changed, large differences in phenotype may occur.

Why All the Concern over H^2?

If we cannot predict the changeability of a trait by environmental or genetic manipulation when we know the broad heritability of that trait, then why have we spent so much effort developing and discussing H^2? There are two reasons. The first is a narrow, pedagogical one. Un-

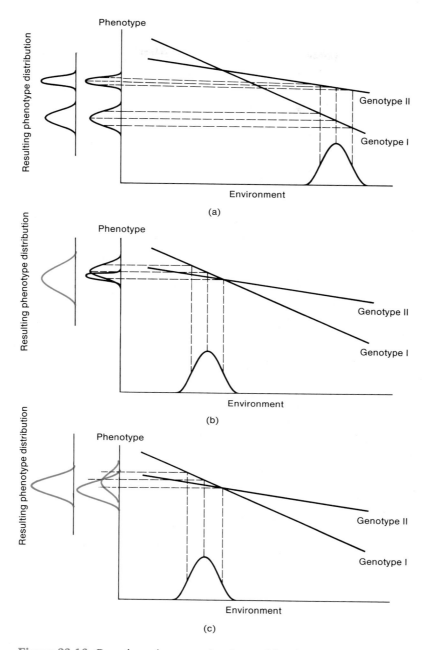

Figure 23-16. Genetic variance can be changed by changing the environment, and environmental variance can be changed by changing the relative frequencies of genotypes.

derstanding the broad heritability H^2 is a first step in developing the concept of **narrow heritability,** which is of considerable importance in plant and animal breeding (page 666).

The second reason for concerning ourselves with H^2 is that it is widely misused in the analysis of quantitative traits, especially human social and cognitive traits. Psychologists, social theorists, physicians, and others have

concerned themselves over and over again with the heritability of characteristics in the mistaken belief that the demonstration of heritability is equivalent to the demonstration of the unchangeability of characters by environmental or social means. Any intelligent and critical reading of the vast and growing popular, semi-popular, and semi-scientific literature on intelligence, criminality, aggression, alcoholism, and human nature requires a thorough understanding of the erroneous use of heritability arguments.

Perhaps the most famous example of the misuse of heritability arguments is the case of human IQ performance and social success. In 1969, an educational psychologist, A. R. Jensen, published a long paper in the *Harvard Educational Review,* asking the question (in its title) "How much can we boost IQ and scholastic achievement?" Jensen's conclusion was "not much." As an explanation and evidence of this unchangeability, he offered a claim of high heritability for IQ performance. A great deal of criticism has been made of the evidence offered by Jensen for the high heritability of IQ scores. But irrespective of the correct value of H^2 for IQ performance, the real error of Jensen's argument lies in his equation of high heritability with unchangeability. In fact, the heritability of IQ is *irrelevant* to the question raised in the title of his article.

To see why this is so, let's consider a hypothetical adoption study in which children are taken from their biological parents in infancy and raised by adoptive parents. The following table gives hypothetical IQ scores for the children, for their biological parents (average score of mother and father), and for their adoptive parents (average score of adoptive mother and father):

Children	Biological Parents	Adoptive Parents
110	90	118
112	92	114
114	94	110
116	96	120
118	98	112
120	100	116
Mean 115	95	115

Although these are hypothetical values, they do illustrate in an extreme form the features seen in actual adoption studies.

First, we can see that the children have a high correlation with their biological parents but a low correlation with their adoptive parents. In fact, in our hypothetical example, the correlation of children with biological parents is $r = 1.00$, but with adoptive parents it is $r = 0$. (Remember from page 648 that correlation between two sets of numbers does not mean that the two sets are identical but that for each unit increase in one set, there is a constant proportion increase in the other set.) This perfect correlation with biological parents and zero correlation with adoptive parents means that $H^2 = 1$, given the arguments developed on page 659. All of the variation in IQ score among the children is explained by the variation among the biological parents.

Second, however, we notice that each of the IQ scores of the children is 20 points higher than the IQ scores of their respective biological parents and that the mean IQ of the children is equal to the mean IQ of the adoptive parents. Thus, adoption has raised the average IQ of the children 20 points higher than the average IQ of their biological parents, so that as a *group* the children resemble their adoptive parents. So we have perfect heritability, yet high environmental plasticity.

An investigator who is seriously interested in knowing how genes might constrain or influence the course of development of any trait in any organism, must study directly the norms of reaction of the various genotypes in the population over the range of projected environments. No less detailed information will do. Summary measures such as H^2 are not first steps toward a more complete analysis and therefore are not valuable in themselves.

Message   Heritability is not the opposite of phenotypic plasticity. A character may have perfect heritability in a population and still be subject to great changes resulting from environmental variation.

Counting and Locating the Genes

It is not possible with purely genetic techniques to identify all of the genes that influence the development of a given trait. This is true even for simple qualitative traits —for example, the genes involved in determining the total antigenic configuration of the membrane of the human red blood cell. About 40 loci determining human blood groups are known at present; each has been discovered by finding at least one person with an immunological specificity that differs from the specificities of other people. Many other loci that determine red-cell membrane structure may remain undiscovered, because all of the individuals studied are genetically identical. *Genetic analysis* detects genes only when there is some allelic variation. In contrast, of course, *molecular analysis,* by dealing directly with DNA and its translated information, can identify genes even when they do not vary—provided the gene products can be identified.

Moreover, the power of classical genetic analysis to detect loci depends on the amount of genetic variation. Even in the case of qualitative human blood groups, only five were discovered in the first 40 years of research on the subject—these five being highly genetically variable in human populations. Then, during World War II, vast numbers of individuals were blood-typed in connection with the treatment of burns and other war-related injuries, so that loci for which most individuals are identically homozygous were discovered from the rare genetic variants.

The Method of Artificial Selection

The standard method of finding the loci that are segregating with respect to a given quantitative trait in a genetically well-marked species such as *Drosophila melanogaster* is to begin with two populations that are very divergent for the character. These may be two natural populations, but more frequently they are two subpopulations that have been created by artificial selection. For example, in one subpopulation, the largest individuals are chosen as parents in each generation; in the other subpopulation, the smallest individuals are chosen as parents. As generations pass, the upwardly selected line will become enriched for alleles that lead to larger size, whereas the alternative alleles will accumulate in the downwardly selected line. If selection is carried on for a long enough time, the two populations will become virtually homozygous for "high" and "low" alleles. Alternatively, many inbred lines could be made from the original population, and the most divergent lines then could be chosen as the "high" and "low" populations. Whichever method is used, the two divergent populations will probably differ only for those loci that were somewhat heterozygous in the original population.

Once the divergent lines have been established, each chromosome in one line can be separately substituted into the other line by using dominant marker stocks with crossover suppressors. An idealization of the method is shown in Figure 23-17, where A_I, A_{II}, A_{III}, B_I, $\cdots$ are dominant marker systems for chromosomes I, II, and III. Lines homozygous and heterozygous for various combinations of "+" and "−" chromosomes are measured for the trait; in this way, the contribution of each chromosome to the genetic differences between the lines is determined. An example of the application of this technique by J. Crow to the study of DDT resistance in *Drosophila melanogaster* is shown in Figure 23-18. Crow did not manufacture lines that were homozygous for "+" and "−" chromosomes, so we see only the effect of chromosomes when lines are heterozygous. The figure shows that every chromosome seems to have some genes that differ between the resistant and the susceptible line.

Linkage Analysis

The experiment illustrated in Figure 23-17 must be elaborated to find individual genes. A more detailed analysis can be carried out with recessive marker stocks that allow recombination. A cross between a multiply marked susceptible chromosome and a resistant chromosome, followed by a backcross to the marker stock, will result in a large number of recombinant types of varying resistance. In the simplest situation, all recombinants carrying a given short region will be resistant and all those not carrying such a region will be susceptible. This would provide strong evidence for a single locus of major effect in the region. In the worst case, all the recombinants will show partial resistance, indicating that no localization has been accomplished. Where local-

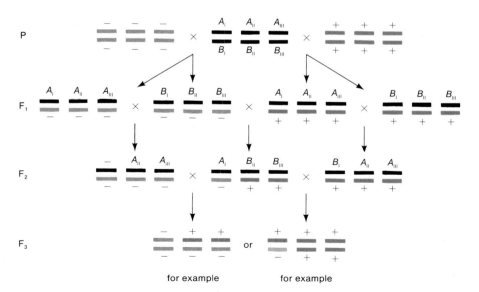

Figure 23-17. Scheme of mating to produce individuals with different combinations of chromosomes from a high-selection line (+) and a low-selection line (−). A_1, B_1, . . . are dominant marker crossover suppressors.

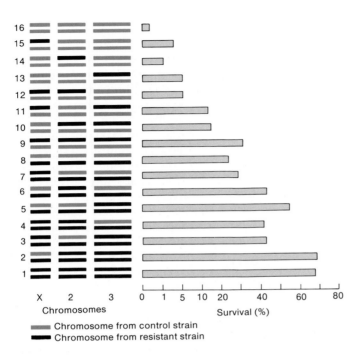

Figure 23-18. Survival of *Drosophila melanogaster* genotypes with different combinations of chromosomes from a selected line (black chromosomes) and a nonselected line (colored chromosomes). Black chromosomes are from strains that have been selected for resistance to DDT. (From J. Crow, *Annual Review of Entomology* 2, 1957, 228.)

ization experiments have been carried out — as, for example, for the sternopleural bristles on the side of the thorax of *Drosophila melanogaster* — most of the difference between selected lines has been localized to two or three genes with major effects plus a residue of genes with small effects.

The possibility exists that many quantitative characters (perhaps even most) will turn out to vary not as a consequence of the segregation of alleles at large numbers of loci with small effects (as the multiple-gene hypothesis supposes), but rather as a consequence of the segregation of very few loci with major effects. This question of the "genetic architecture" of quantitative traits is a leading problem in quantitative genetics. It could be solved by straightforward but excessively tedious linkage experiments in well-marked species such as *Drosophila melanogaster.* In *Homo sapiens,* such experiments have not been possible, and we have no information on the number or localization of loci contributing to quantitative traits. Even for a character with obvious genetic influence and with a relatively simple biochemical and developmental basis (such as skin color), we do not know the number of loci (except to say that there must be several) or their distribution over the human chromosome set.

Gene Action

The problem of detecting gene loci influencing quantitative characters is closely tied to the magnitude of the effects of different allelic substitutions, to the degree of dominance of alleles at each locus, and to the amount of epistasis among loci. The methods described in the preceding section are biased toward the detection of loci where allelic differences are of large effect. The process of selection by which the contrasting populations are established is relatively efficient only for allelic substitutions of large effect.

Even without being able to localize the genes, we can obtain some information about their actions and interactions. We can judge average dominance at the chromosomal level by comparing homozygotes and heterozygotes in chromosomal substitution experiments. In Figure 23-18, a comparison of lines 1 and 2 shows that the X chromosomes do not differ, whereas lines 15 and 16 show that they do. An explanation of this discrepancy would be the complete dominance of resistance alleles over susceptibility alleles. A comparison of lines 7, 9, and 10 supports the hypothesis of dominance because 7 and 9 are equally resistant, whereas 10 is much less so.

There is also evidence of specific epistatic interactions. Line 5 in the figure has both the substitutions of lines 2 and 3 and should be at least as susceptible as line 3; yet it is intermediate between 2 and 3. It would seem that heterozygosity for a susceptible X chromosome actually decreases the susceptibility when in combination with a susceptible second-chromosome heterozygote. Such evidences of dominance and epistasis of whole chromosomes throw only indirect light on individual gene action because with dominance or strong epistasis, a single gene of major effect would hide the effects of several other genes of small effect.

Evidence for non-nuclear effects can also be deduced from the chromosomal substitution experiments. For example, in Figure 23-18, lines 8 and 9, which differ in their resistance, have the same chromosomal constitution but are the result of reciprocal crosses and so have different cytoplasm.

More on Analyzing Variance

Knowledge of the broad heritability (H^2) of a trait in a population is not very useful in itself, but a finer subdivision of phenotypic variance can provide important information for plant and animal breeders. The genetic variation and the environmental variation can themselves each be further subdivided in a way that can provide information about gene action and the possibility of shaping the genetic composition of a population.

Additive and Dominance Variance

Our previous consideration of gene action suggests that the phenotypes of homozygotes and heterozygotes ought to have a simple relation. If one of the alleles codes for a less active gene product or one with no activity at all and if one unit of gene product is sufficient to allow full physiological activity of the organism, then we would expect complete dominance of one allele over the other, as Mendel observed for flower color in peas. If, on the other hand, physiological activity is proportional to the amount of active gene product, we would expect the heterozygote phenotype to be exactly intermediate between the homozygotes (show no dominance).

For many quantitative traits, however, neither of these simple cases is the rule. In general, heterozygotes are not exactly intermediate between the two homozygotes but are closer to one or the other (show partial dominance). The complexity of biochemical and developmental pathways is such that the phenotype of a heterozygote may not be intermediate between the two homozygotes, even though there is an equal mixture of the primary products of the two alleles in the heterozygote. Indeed, in some cases, the heterozygote phenotype may lie outside the phenotypic range of the homozygotes altogether—a feature termed **overdominance.** For example, newborn babies who are intermediate in size have a higher chance of survival than very large or very small newborns. Thus, if survival were the phenotype of interest, heterozygotes for genes influencing growth rate would show overdominance.

Suppose that two alleles, a and A, segregate at a locus influencing height. In the environments encountered by the population, the mean phenotypes (heights) and frequencies of the three genotypes might be:

	aa	Aa	AA
Phenotype	10	18	20
Frequency	0.36	0.48	0.16

There is genetic variance in the population; the phenotypic means of the three genotypic classes are different. Some of the variance arises because there is an average effect on phenotype of substituting an allele A for an allele a; that is, the average height of all individuals with A alleles is greater than that of all individuals with a alleles. By defining the average effect of an allele as the average phenotype of all individuals that carry it, we necessarily make the average effect of the allele depend on the frequencies of the genotypes.

The average effect is calculated by simply counting the a and A alleles and multiplying them by the heights of the individuals in which they appear. Thus, 0.36 of all the individuals are homozygous aa, each aa individual has two a alleles, and the average height of aa individuals is 10 cm. Heterozygotes make up 0.48 of the population, each has only one a allele, and the average phenotypic measurement of Aa individuals is 18 cm. The total "number" of a alleles is $2(0.36) + 1(0.48)$. Thus, the average effect of all the a alleles is

$$\bar{a} = \text{average effect of } a = \frac{2(0.36)(10) + 1(0.48)(18)}{2(0.36) + 1(0.48)}$$
$$= 13.20 \text{ cm}$$

and, by a similar argument

$$\bar{A} = \text{average effect of } A = \frac{2(0.16)(20) + 1(0.48)(18)}{2(0.16) + 1(0.48)}$$
$$= 18.80 \text{ cm}$$

This average difference in effect between A and a alleles of 5.60 cm accounts for some of the variance in phenotype—but not for all of it. The heterozygote is not exactly intermediate between the homozygotes; there is some dominance. The total genetic variance associated with this locus can then be partitioned between the **additive genetic variance** (s_a^2), the genetic variance associated with the average effect of substituting A for a, and the **dominance variance** (s_d^2), the genetic variance resulting from the partial dominance of A over a in heterozygotes. Thus

$$s_g^2 = s_a^2 + s_d^2$$

The components of variance in this example can be calculated using the definitions of mean and variance developed earlier in this chapter. Remembering that a mean is the sum of the values of a variable, each weighted by the frequency with which that value occurs (see page 645), we can calculate the mean phenotype to be

$$\bar{x} = \Sigma f_i x_i = (0.36)(10) + (0.48)(18) + (0.16)(20)$$
$$= 15.44 \text{ cm}$$

The total genetic variance that arises from the variation among the mean phenotypes of the three genotypes is

$$s_g^2 = \Sigma f_i(x_i - \bar{x})^2 = (0.36)(10 - 15.44)^2$$
$$+ (0.48)(18 - 15.44)^2$$
$$+ (0.16)(20 - 15.44)^2$$
$$= 17.13 \text{ cm}^2$$

The additive variance is calculated from the squared deviation of the average a effect from the mean, weighted by its frequency, plus the squared deviation of the average A effect from the mean, weighted by its frequency. The frequency of the a allele is (by counting alleles)

$$f_a = \frac{2(aa) + 1(Aa)}{2}$$

$$f_a = \frac{2(0.36) + 1(0.48)}{2} = 0.60$$

and the frequency of the A allele is

$$f_A = \frac{2(AA) + 1(Aa)}{2}$$

$$f_A = \frac{2(0.16) + 1(0.48)}{2} = 0.40$$

Thus, since there are two alleles present in diploid individuals

$$\begin{aligned}
s_a^2 &= 2[f_a(\bar{a} - \bar{x})^2 + f_A(\bar{A} - \bar{x})^2] \\
&= 2[(0.60)(13.20 - 15.44)^2 \\
&\quad + (0.40)(18.80 - 15.44)^2] \\
&= 15.05 \text{ cm}^2
\end{aligned}$$

and

$$s_d^2 = s_g^2 - s_a^2 = 17.13 - 15.05 = 2.08 \text{ cm}^2$$

The usefulness of this subdivision of genetic variance is in the prediction that can be made about the effect of selective breeding. This use becomes clear in an extreme case. Suppose that there is overdominance and the phenotypic means and frequencies of three genotypes are

	AA	Aa	aa
Phenotype	10	12	10
Frequency	0.25	0.50	0.25

It is apparent (and a calculation like the preceding one will confirm) that there is no average difference between the a and A alleles, because each has an effect of 11 units. So there is no additive genetic variance, although there is dominance variance. The largest individuals are heterozygotes. If a breeder attempts to increase height in this population by selecting breeding, mating these heterozygotes together will simply reconstitute the original population. Selection will be totally ineffective. This illustrates the general law that the effect of selection depends on the *additive* genetic variance and not on genetic variance in general.

The total phenotypic variance can now be written as

$$s_p^2 = s_g^2 + s_e^2 = s_a^2 + s_d^2 + s_e^2$$

We define a new kind of heritability, the **heritability in the narrow sense (h^2)**, as

$$h^2 = \frac{s_a^2}{s_p^2} = \frac{s_a^2}{s_a^2 + s_d^2 + s_e^2}$$

It is this heritability, not to be confused with H^2, that is useful in determining whether a program of selective breeding will succeed in changing the population. The greater the h^2 is, the greater the difference is between selected parents and the population as a whole that will be preserved in the offspring of the selected parents.

> **Message** The effect of selection depends on the amount of *additive* genetic variance and not on the genetic variance in general. Therefore, the narrow heritability h^2, not the broad heritability H^2, is relevant for a prediction of response to selection.

What has been described as the "dominance" variance is really more complicated. It is all the genetic variation that cannot be explained by the average effect of substituting A for a. If there is more than one locus affecting the character, then any epistatic interactions between loci will appear as variance not associated with the average effect of substituting alleles at the A locus. In principle, we can separate this **interaction variance (s_i^2)** from the dominance variance s_d^2. In practice, however, this cannot be done with any semblance of accuracy, so all the nonadditive variance appears as "dominance" variance.

Estimating Genetic Variance Components

Genetic components of variance are estimated from covariance between relatives. Ignoring the epistatic contributions to variance that are contained to some degree in all covariances between relatives, Table 23-2 shows some of the components of variance contained in the

TABLE 23-2. Proportion of the additive variance (s_a^2) and the dominance variance (s_d^2) contained in the genetic covariance between various related individuals

Relatives	Estimated proportion of	
	s_a^2	s_d^2
Cov (identical twins)	1	1
Cov (parent–offspring)	$\frac{1}{2}$	0
Cov (half-siblings)	$\frac{1}{4}$	0
Cov (full siblings)	$\frac{1}{2}$	$\frac{1}{4}$

covariances between relatives. These relations, together with the total phenotypic variance, can be used to estimate h^2. For example, we can see from Table 23-2 that the covariance between parent and offspring contains one-half of the additive variance. Thus

$$\text{cov (parent-offspring)} = \frac{s_a^2}{2}$$

Therefore

$$2 \text{ cov (parent-offspring)} = s_a^2$$

By the definition of correlation and taking into account that the phenotypic variance is the same in the parental and offspring generations, so that $s_p(\text{par}) = s_p(\text{off}) = s_p$, we can see that

$$\frac{2 \text{ cov (parent-offspring)}}{s_p^2} = \frac{2 \text{ correlation}}{\text{(parent-offspring)}}$$
$$= \frac{s_a^2}{s_p^2} = h^2$$

Thus, twice the parent-offspring correlation is an estimate of h^2.

There is yet another way to estimate h^2 that provides an insight into its real meaning. If we plot the phenotypes of the offspring against the average phenotypes of their two parents (the midparent value) we may observe a relationship like the one illustrated in Figure 23-19. The regression line will pass through the mean of all the

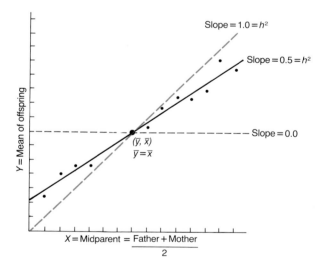

Figure 23-19. The regression (black line) of offspring measurements (y) on midparents (x) for a trait with narrow heritability (h^2) of 0.5. The colored line shows the regression slope if the trait were perfectly heritable.

parents and the mean of all the offspring, which will be equal to each other because no change has occurred in the population between generations. Moreover, taller parents have taller children and shorter parents have shorter children, so that the slope of the line is positive. But the slope is not unity; very short parents have children who are somewhat taller and very tall parents have children who are somewhat shorter than they themselves are. This slope of less than unity for the regression line arises because heritability is less than perfect. If the phenotype were additively inherited with complete fidelity, then the height of the offspring would be identical to the midparent value and the slope of the line would be 1. On the other hand, if the offspring have no heritable similarity to their parents, all parents will have offspring of the same average height and the slope of the line will be 0. This suggests that the slope of the regression line of the offspring value on the midparent value is an estimate of additive heritability. In fact, the relationship is precise. By definition (see page 649), the regression b, of the offspring value (y) on the midparent value (x) is

$$b = \frac{\text{cov } xy}{s_x^2}$$

But from Table 23-2, we know that

$$\text{cov (parent-offspring)} = \frac{s_a^2}{2}$$

and this covariance is not changed if we consider the covariance of the offspring with the average of their two parents. On the other hand, the variance of midparental values is the variance of an average of two measurements and is therefore only one-half of the variance of the measurements themselves. Obviously, average values vary less than their component numbers. So, letting $s_p^2 =$ phenotypic variance of the parents (which is the same as the phenotypic variance of the offspring generation) gives us

$$b = \frac{s_a^2/2}{s_p^2/2} = \frac{s_a^2}{s_p^2} = h^2$$

The fact that the slope equals the additive heritability now allows us to use h^2 to predict the effects of artificial selection. Suppose that we select parents for the next generation who are on the average 2 units above the general mean of the population from which they were chosen. If $h^2 = 0.5$, then the offspring who form the next, selected generation will lie $0.5(2.0) = 1.0$ unit above the mean of the present population, since the regression coefficient predicts how much increase in y will result from a unit increase in x. We can define the

selection differential as the difference between the selected parents and the unselected mean and the **selection response** as the difference between their offspring and the previous generation. Then

$$\text{Selection response} = h^2 \times \text{selection differential}$$

or

$$h^2 = \frac{\text{selection response}}{\text{selection differential}}$$

The second expression provides us with yet another way to estimate h^2: by selecting for one generation and comparing the response with the selection differential. Usually this is carried out for several generations, and the average response is used.

Remember that any estimate of h^2, just as for H^2, depends on the assumption of no greater environmental correlation between closer relatives. Moreover, h^2 in one population in one set of environments will not be the same as h^2 in a different population at a different time. Figure 23-20 shows the range of heritabilities reported in various studies for a number of traits in chickens. The very small ranges are generally close to zero. For most traits for which a substantial heritability has been reported in some population, there are big differences from study to study.

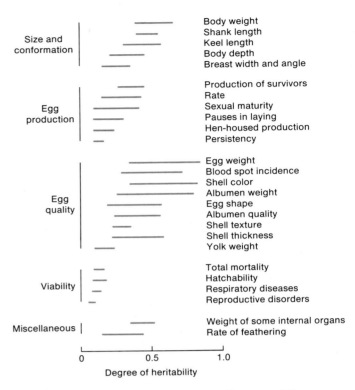

Figure 23-20. Ranges of heritabilities (h^2) reported for a variety of characters in chickens. (From I. M. Lenner and W. J. Libby. *Heredity, Evolution and Society.* Copyright © 1976 by W. H. Freeman and Company.)

Partitioning Environmental Variance

Environmental variance, like genetic variance, can also be further subdivided. In particular, developmental noise is usually confounded with environmental variance, but when a character can be measured on the left and right sides of an organism or over repeated body segments, it is possible to separate noise from environment. Table 23-3 shows the complete partitioning of variation for two characters in a population of *Drosophila melanogaster* raised under standard laboratory conditions. For each character, there is a substantial h^2 ($h^2 = s_a^2/s_p^2$), so we might expect selective breeding to increase or decrease bristle number and ovary size. Furthermore, nearly all the nongenetic variation is due to developmental noise. These values, however, are a con-

TABLE 23-3. Partition of the total phenotypic variance for two characters in a population of *Drosophila melanogaster*

Source of variation			Percentage of variance	
			Number of abdominal bristles	Ovary size
Additive genetic	s_a^2	$\Big\} s_g^2$	52	30
Dominance + epistatic variance	s_d^2		9	40
Environmental variance	s_e^2	$\Big\} s_c^2$	1	3
Developmental noise	s_n^2		38	27
Total	s_p^2		100	100

SOURCE: D. Falconer, *Quantitative Genetics,* Longman Group Limited. Copyright © 1981.

sequence of the relatively rigorously controlled environment of the laboratory. Presumably s_e^2 in nature would be considerably larger, with a consequent diminution in the relative sizes of h^2 and of s_a^2. As always, such studies of variation are applicable only to a particular population in a given distribution of environments.

The Use of h^2 in Breeding

Even though h^2 is a number that applies only to a particular population and a given set of environments, it is still of great practical importance to breeders (Figure 23-21). A poultry geneticist interested in increasing, say, growth rate, is not concerned with the genetic variance over all possible flocks and all environmental distributions. Given a particular flock (or a choice between a few particular flocks) under the environmental conditions approximating present husbandry practice, the question becomes can a selection scheme be devised to increase growth rate and, if so, how fast? If one flock has a lot of genetic variance and another only a little, the breeder will choose the former to carry out selection. If the heritability in the chosen flock is very high, then the mean of the population will respond quickly to the selection imposed, because most of the superiority of the selected parents will appear in the offspring. The higher h^2 is, the higher the parent–offspring correlation is. If, on the other hand, h^2 is low, then only a small fraction of the increased growth rate of the selected parents will be reflected in the next generation.

If h^2 is very low, some alternative scheme of selection or husbandry may be needed. In this case, H^2 together with h^2 can be of use to the breeder. Suppose that h^2 and H^2 are both low. This means that there is a lot of environmental variance compared to genetic variance. Some scheme of reducing s_e^2 must be used. One method is to change the husbandry conditions so that environmental variance is lowered. Another is to use **family selection.** Rather than choosing the best individuals, the breeder allows pairs to produce several progeny, and the *mating* is selected on the basis of the average performance of the progeny. By averaging over progeny, uncontrolled environmental and developmental noise variation is canceled out and a better estimate of the genotypic difference between pairs can be made so that the best pairs can be chosen as parents of the next generation.

If, on the other hand, h^2 is low but H^2 is high, then there is not much environmental variance. The low h^2 is the result of a small amount of additive genetic variance compared to dominance and interaction variance. Such a situation calls for special breeding schemes that make use of nonadditive variance. One such scheme is the **hybrid-inbred method,** which is used almost universally for corn. A large number of inbred lines are created by

Figure 23-21. Quantitative genetic theory has been extensively applied to poultry breeding. (Photograph courtesy of Welp, Inc., Bancroft, Iowa.)

selfing. These are then crossed in many different combinations (all possible combinations, if this is economically feasible), and the cross that gives the best hybrid is chosen. Then new inbred lines are developed from this best hybrid, and again crosses are made to find the best second-cycle hybrid. This scheme selects for dominance effects, because it takes the best heterozygotes; it has been the basis of major genetic advances in hybrid maize yield in North America since 1930. Yield in corn does not appear to have large amounts of nonadditive genetic variance, so it is debatable whether this technique *ultimately* produces higher-yielding varieties than those that would have resulted from years of simple selection techniques based on additive variance.

The hybrid method has been introduced into the breeding of all kinds of plants and animals. Tomatoes and chickens, as examples, are now almost exclusively hybrids. Attempts also have been made to breed hybrid wheat, but thus far the wheat hybrids obtained do not yield consistently better than the nonhybrid varieties now used.

Message The subdivision of genetic variation and environmental variation provides important information about gene action that can be used in plant and animal breeding.

Summary

■ Many — perhaps most — of the phenotypic traits we observe in organisms vary continuously. In many cases, the variation of the trait is determined by more than a single segregating locus. Each of these loci may contribute equally to a particular phenotype, but it is more likely that they contribute unequally. The measurement of these phenotypes and the determination of the contributions of specific alleles to the distribution must be made on a statistical basis in these cases. Some of these variations of phenotype (such as height in some plants) may show a normal distribution around a mean value; others (such as seed weight in some plants), will illustrate a skewed distribution around a mean value.

In other characters, the variation in one phenotype may be correlated with the variation in another. A correlation coefficient may be calculated for these two variables.

A quantitative character is one for which the average phenotypic differences between genotypes are small compared to the variation between the individuals within the genotypes. This situation may be true even for characters that are influenced by alleles at one locus. The distribution of environments is reflected biologically as a distribution of phenotypes. The transformation of environmental distribution into phenotypic distribution is determined by the norm of reaction. Norms of reaction can be characterized in organisms in which large numbers of genetically identical individuals can be produced.

By the use of genetically marked chromosomes, it is possible to determine the relative contributions of different chromosomes to variation in a quantitative trait, to observe dominance and epistasis from whole chromosomes, and, in some cases, to map genes that are segregating for a trait.

Traits are familial if members of the same family share them, for whatever reason. Traits are heritable, however, only if the similarity arises from shared genotypes. In experimental organisms, environmental similarities may be readily distinguished from genetic similarities, or heritability. In humans, however, it is very difficult to determine whether a particular trait is heritable. Norm-of-reaction studies show only small differences among genotypes, and these differences are not consistent over a wide range of environments. Thus, "superior" genotypes in domesticated animals and cultivated plants may be superior only in certain environments. If it should turn out that humans exhibit genetic variation for various mental and emotional traits, this variation is unlikely to favor one genotype over another across a range of environments.

The attempt to quantify the influence of genes on a particular trait has led to the determination of heritability in the broad sense (H^2). In general, the heritability of a trait is different in each population and each set of environments and cannot be extrapolated from one population and set of environments to another. Because H^2 characterizes present populations in present environments only, it is fundamentally flawed as a predictive device. Heritability in the narrow sense, h^2, measures the proportion of phenotypic variation that results from substituting one allele for another. This quantity, if large, predicts that selection for a trait will succeed rapidly. If h^2 is small, special forms of selection are required.

■　■　■

Solved Problems

1. In a study of body weight in a pig population, the total variance is calculated to be 3.90. The covariance between half-siblings is found to be 0.56. Estimate the narrow-sense heritability of body weight in this population.

Solution

From Table 23-2, we can see that the covariance for half-siblings is equal to $\frac{1}{4}$ of the additive genetic variance. Therefore

$$s_a^2 = 4 \times \text{cov} = 4 \times 0.56 = 2.24$$

and heritability in the narrow sense is

$$h^2 = \frac{s_a^2}{s_p^2}$$

so that

$$h^2 = \frac{2.24}{3.90} = 0.57 (57\%)$$

2. Two inbred lines of beans are intercrossed. In the F_1, the variance in bean weight is measured at 1.5. The F_1 is

selfed; in the F_2, the variance in bean weight is 6.1. Estimate the broad heritability of bean weight in this experiment.

Solution

The key here is to recognize that all the variance in the F_1 population must be environmental because all individuals must be of identical genotype. Furthermore, the F_2 variance must be a combination of environmental and genetic components, because all of the genes that are heterozygous in the F_1 will segregate in the F_2 to give an array of different genotypes that relate to bean weight. Hence, we can estimate

$$s_e^2 = 1.5$$
$$s_e^2 + s_g^2 = 6.1$$

Therefore

$$s_g^2 = 6.1 - 1.5 = 4.6$$

and broad heritability is

$$H^2 = \frac{4.6}{6.1} = 0.75 (75\%)$$

3. In an experimental population of *Tribolium* (flour beetles), the body length shows a continuous distribution with a mean of 6 mm. A group of males and females with body lengths of 9 mm are removed and interbred. The body lengths of their offspring average 7.2 mm. From these data, calculate the heritability in the narrow sense for body length in this population.

Solution

The selection differential is $9 - 6 = 3$ mm, and the selection response is $7.2 - 6 = 1.2$ mm. Therefore, the heritability in the narrow sense is

$$h^2 = \frac{1.2}{3} = 0.4 \ (40\%)$$

Problems

1. Distinguish between continuous and discontinuous variation in a population, and give some examples of each.

2. In a large herd of cattle, three different characters showing continuous distribution are measured and the variances in Table 23.4 are calculated:

   a. Calculate the broad- *and* narrow-sense heritabilities for each character.

   b. In the population of animals studied, which character would respond best to selection? Why?

   c. A project is undertaken to decrease mean fat content in the herd. The mean fat content is currently 10.5 percent. Animals of 6.5 percent fat content are interbred as parents of the next generation. What mean fat content can be expected in the descendants of these animals?

TABLE 23-4.

Variance	Characters		
	Shank length	Neck length	Fat content
Phenotypic variance	310.2	730.4	106.0
Environmental variance	248.1	292.2	53.0
Additive genetic variance	46.5	73.0	42.4
Dominance genetic variance	15.6	365.2	10.6

3. Suppose that two triple heterozygotes *Aa Bb Cc* are crossed. Assume that the three loci are in different chromosomes.

   a. What proportions of the offspring are homozygous at one, two, and three loci, respectively?

   b. What proportions of the offspring carry 0, 1, 2, 3, 4, 5, and 6 alleles (represented by capital letters), respectively?

4. In Problem 3, suppose that the average phenotypic effect of the three genotypes at the *A* locus is $AA = 4$, $Aa = 3$, $aa = 1$ and that similar effects exist for the *B* and *C* loci. Moreover, suppose that the effects of loci add to each other. Calculate and graph the distribution of phenotypes in the population (assuming no environmental variance).

5. In Problem 4, suppose that there is a threshold in the phenotypic character so that when the phenotypic value is above 9, the individual *Drosophila* has three bristles; when it is between 5 and 9, the individual has two bristles; and when the value is 4 or less, the individual has one bristle. Discuss the outcome of crosses within and between bristle classes. Given the result, could you infer the underlying genetic situation?

6. Suppose that the general form of a distribution of a trait for a given genotype is

$$f = 1 - \frac{(x - \bar{x})^2}{s_e^2}$$

over the range of x where f is positive.

   a. On the same scale, plot the distributions for three genotypes with the following means and environmental variances:

Genotype	$\bar{x}$	s_e^2	Approximate range of phenotype
	0.20	0.3	$x = 0.03$ to $x = 0.37$
2	0.22	0.1	$x = 0.12$ to $x = 0.24$
3	0.24	0.2	$x = 0.10$ to $x = 0.38$

   b. Plot the phenotypic distribution that would result if the three genotypes were equally frequent in a population. Can you see distinct modes? If so, what are they?

7. The following table shows a distribution of bristle number in *Drosophila*:

Bristle number	Number of individuals
1	1
2	4
3	7
4	31
5	56
6	17
7	4

Calculate the mean, variance, and standard deviation of this distribution.

8. The following sets of hypothetical data represent paired observations on two variables (x, y). Plot each set of data pairs as a scatter diagram. Look at the plot of the points, and make an intuitive guess about the correlation between x and y. Then calculate the correlation coefficient for each set of data pairs, and compare this value with your estimate.

   a. (1, 1); (2, 2); (3, 3); (4, 4); (5, 5); (6, 6).

   b. (1, 2); (2, 1); (3, 4); (4, 3); (5, 6); (6, 5).

   c. (1, 3); (2, 1); (3, 2); (4, 6); (5, 4); (6, 5).

   d. (1, 5); (2, 3); (3, 1); (4, 6); (5, 4); (6, 2).

9. A book on the problem of heritability of IQ makes the following three statements. Discuss the validity of each statement and its implications about the authors' understanding of h^2 and H^2.

   a. "The interesting question then is . . . 'How heritable'? The answer [0.01] has a very different theoretical and practical application from the answer [0.99]." [The authors are talking about H^2.]

   b. "As a rule of thumb, when education is at issue, H^2 is usually the more relevant coefficient, and when eugenics and dysgenics (reproduction of selected individuals) are being discussed, h^2 is ordinarily what is called for."

   c. "But whether the different ability patterns derive from differences in genes . . . is not relevant to assessing discrimination in hiring. Where it could be relevant is in deciding what, in the long run, might be done to change the situation."

   (From J. C. Loehlin, G. Lindzey, and J. N. Spuhler, *Race Differences in Intelligence.* Copyright © 1975 by W. H. Freeman and Company.

10. Using the concepts of norms of reaction, environmental distribution, genotypic distribution, and phenotypic distribution, try to restate the following statement in more exact terms: the "80 percent of the difference in IQ performance between the two groups is genetic." What would it mean to talk about the heritability of a difference between two groups?

11. Describe an experimental protocol involving studies of relatives that could estimate the broad heritability of alcoholism. Remember that you must make an adequate observational definition of the trait itself!

12. A line selected for high bristle number in *Drosophila* has a mean of 25 sternopleural bristles, whereas a low-selected line has a mean of only 2. Marker stocks involving the two large autosomes II and III are used to create stocks with various mixtures of chromosomes from the high (h) and low (l) lines. The mean number of bristles for each chromosomal combination is as follows:

$$\frac{h}{h}\frac{h}{h}\ 25.1 \qquad \frac{h}{l}\frac{h}{h}\ 22.2 \qquad \frac{l}{l}\frac{h}{h}\ 19.0$$

$$\frac{h}{h}\frac{h}{l}\ 23.0 \qquad \frac{h}{l}\frac{h}{l}\ 19.9 \qquad \frac{l}{l}\frac{h}{l}\ 14.7$$

$$\frac{h}{h}\frac{l}{l}\ 11.8 \qquad \frac{h}{l}\frac{l}{l}\ 9.1 \qquad \frac{l}{l}\frac{l}{l}\ 2.3$$

What conclusions can you reach about the distribution of genetic factors and their actions from these data?

13. Suppose that number of eye facets is measured in a population of *Drosophila* under various temperature conditions. Further suppose that it is possible to estimate total genetic variance s_g^2 as well as the phenotypic distribution. Finally, suppose that there are only two genotypes in the population. Draw pairs of norms of reaction that would lead to the following results:

   a. An increase in mean temperature decreases the phenotypic variance.

   b. An increase in mean temperature increases H^2.

   c. An increase in mean temperature increases s_g^2 but decreases H^2.

   d. An increase in temperature *variance* changes a unimodal into a bimodal phenotypic distribution (one norm of reaction is sufficient here).

14. The following variances and covariances between relatives have been found for egg weight in poultry: $s_p^2 = 14.8$; cov (mother-daughter) = 1.7; cov (sisters) = 2.7; cov (half-sisters) = 0.8. Calculate the broad and narrow heritabilities, H^2 and h^2.

15. Francis Galton compared the heights of male undergraduates to the heights of their fathers, with the results shown in the graph on page 673. The average height of all fathers is the same as the average height of all sons, but the individual height classes are not equal across generations. The very tallest fathers had somewhat shorter sons, whereas the very short fathers had somewhat taller sons. As a result, the best line that can be drawn through the points on the scatter diagram has a slope of about 0.67 (*solid line*) rather than 1.00 (*dashed line*). Galton used the term regression to describe this tendency for the phenotype of the sons to be closer than the phenotype of their fathers to the population mean.

   a. Propose an explanation for this regression.

   b. How are regression and heritability related here?

Scatter diagram for height data in father-son pairs (see Problem 15). The best line that can be drawn through the scatter points (black line) has a lesser slope than the line of identity (colored line). (After W. F. Bodmer and L. L. Cavalli-Sforza, *Genetics, Evolution, and Man.* Copyright © 1976 by W. H. Freeman and Company.)

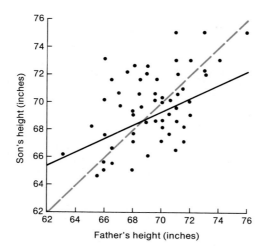

Population Genetics

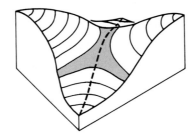

KEY CONCEPTS

The goal of population genetics is to understand the genetic composition of a population and the forces that determine and change that composition.

·

In any species, a great deal of genetic variation within and between populations arises from the existence of various alleles at different gene loci.

·

A fundamental measurement in population genetics is the frequency at which the alleles can occur at any gene locus of interest.

·

The frequency of a given allele in a population can be changed by recurrent mutation, selection, or migration, or by random sampling effects.

·

In an idealized population, in which no forces of change are acting, a randomly interbreeding population would show constant genotypic frequencies for a given locus.

Mendel's investigations of heredity—and indeed all of the interest in heredity in the nineteenth century—arose from two related problems: how to breed improved crops and how to understand the nature and origin of species. What is common to these problems (and what differentiates them from the problems of transmission and gene action) is that they are concerned with *populations* rather than with *individuals*. Studies of gene replication, protein synthesis, development, and chromosome movement all focus on processes that go on within the cells of individual organisms. But the transformation of a species, either in the natural course of evolution or by the deliberate intervention of human beings, is a change in the properties of a collectivity—of an entire population or a set of populations.

> **Message** Population genetics relates the heritable changes in populations of organisms to the underlying individual processes of inheritance and development.

Darwin's Revolution

The modern theory of evolution is so completely identified with the name of Charles Darwin (1809–1882) that many people think that the concept of organic evolution was first proposed by Darwin, but that is certainly not the case. Most scholars had abandoned the notion of fixed species, unchanged since their origin in a grand creation of life, long before publication of Darwin's *The Origin of Species* in 1859. By that time, most biologists agreed that new species arise through some process of evolution from older species; the problem was to explain *how* this evolution could occur. The theories that preceded Darwin's work were **transformational theories.** Such theories postulated that the species as a whole changed because each individual within the species changed in the same direction, just as the class reading this book will grow grayer and more wrinkled over the years because every individual in the class is aging. Jean-Baptiste Lamarck (1744–1829), for example, claimed that each individual changes slightly because of an inner striving or will to adapt itself to its environment. These small individual changes are passed on to the offspring, who in turn continue the process of change by striving to adapt to their environment.

Darwin broke fundamentally with these transformational hypotheses by creating a **variational theory.** This theory takes as its starting point the variation that exists among organisms within a species. Individuals of one generation are qualitatively different from one an-

other. Evolution of the species as a whole results from the differential rates of survival and reproduction of the various types, so that the relative frequencies of the types change over time. Evolution, in this view, is a sorting process rather than a transformational one. For Lamarck, evolution of the group was the consequence of environmentally directed changes in all individuals in the group, all of them in the same direction. For Darwin, evolution of the group resulted from the differential survival and reproduction of individual variants *already existing* in the group—variants arising in a way unrelated to the environment.

> **Message** Darwin proposed a new explanation to account for the accepted phenomenon of evolution. He argued that the population of a given species at a given time includes individuals of varying characteristics. The population of the next generation will contain a higher frequency of those types that most successfully survive and reproduce under the existing environmental conditions. Thus, the frequencies of various types within the species will change over time.

There is an obvious similarity between the process of evolution as Darwin described it and the process by which the plant or animal breeder improves a domestic stock. The plant breeder selects the highest-yielding plants from the current population and (as far as possible) uses them as the parents of the next generation. If the characteristics causing the higher yield are heritable, then the next generation should produce a higher yield. It was no accident that Darwin chose the term **natural selection** to describe his model of evolution through differential rates of reproduction of different variants in the population. As a model for this evolutionary process, he had in mind the selection that breeders exercise on successive generations of domestic plants and animals.

We can summarize Darwin's theory of evolution through natural selection in three principles:

1. **The principle of variation.** Among individuals within any population, there is variation in morphology, physiology, and behavior.

2. **The principle of heredity.** Offspring resemble their parents more than they resemble unrelated individuals.

3. **The principle of selection.** Some forms are more successful at surviving and reproducing than other forms in a given environment.

Clearly, a selective process can produce change in the population composition only if there are some variations to select among. If all individuals are identical, no amount of differential reproduction of individuals can affect the composition of the population. Furthermore, the variation must be in some part heritable if differential reproduction is to alter the population's genetic composition. If large animals within a population have more offspring than small ones but their offspring are no larger on average than those of small animals, then no change in population composition can occur from one generation to another. Finally, if all variant types leave, on average, the same number of offspring, then we can expect the population to remain unchanged.

> **Message** Darwin's principles of variation, heredity, and selection must hold true if there is to be evolution by a variational mechanism.

Variation and Its Modulation

Population genetics is the translation of Darwin's three principles into precise genetic terms. As such, it deals with the following problems:

1. The description of the genetic variation within and between populations, and the nature of the heredity of traits.

2. The study of the introduction of new variation into populations by the mutation and recombination of genes and chromosomes and the migration of individual organisms.

3. The study of the patterns of differential reproduction of genotypes as a result of variation in mating patterns, fertility, and survival of individuals of different genotypes.

4. The creation of a formal machinery of deduction (theoretical population genetics) that can be used to predict the effects of the introduction of variation and of the differential rate of reproduction of variants on the genetic composition of a population.

5. The observation of actual changes in the composition of populations over time, either in nature or in controlled culture, and the comparison of these changes with those predicted from the theory in order to check the adequacy of the entire theoretical structure.

6. The application of the theory to the controlled evolution of domesticated plants and animals, pests, pathogens, and other organisms relevant to human welfare.

> **Message** Population genetics is the study of inherited variation and its modulation in time and space.

Observations of Variation

Population genetics necessarily deals with genotypic variation, but by definition, only phenotypic variation can be observed. The relation between phenotype and genotype varies in simplicity from character to character. At one extreme, the phenotype may be the observed DNA sequence of a stretch of the genome. In this case, the distinction between genotype and phenotype disappears and we can say that we are, in fact, directly observing the genotype. At the other extreme lie the bulk of characters of interest to plant and animal breeders and to most evolutionists—the variations in yield, growth rate, body shape, metabolic ratio, and behavior that constitute the obvious differences between varieties and species. These characters have a very complex relation to genotype, and we must use the methods introduced in Chapter 23 to say anything at all about the genotypes. But as we have seen in Chapter 23, it is not possible to make very precise statements about the genotypic variation underlying quantitative characters. For that reason, most of the study of experimental population genetics has concentrated on characters with simple relations to the genotype, much like the characters studied by Mendel. A favorite object of study for human population geneticists, for example, has been the various human blood groups. The qualitatively distinct phenotypes of a given blood group—say, the MN group—are coded for by alternate alleles at a single locus, and the phenotypes are insensitive to environmental variations.

The study of variation, then, consists of two stages. The first is a description of the phenotypic variation. The second is a translation of these phenotypes into genetic terms and the redescription of the variation genetically. If there is a perfect one-to-one correspondence between genotype and phenotype, then these two steps merge into one, as in the case of the MN blood group. If the relation is more complex—for example, as the result of dominance, so that heterozygotes resemble homozygotes, it may be necessary to carry out experimental crosses or to observe pedigrees to translate phenotypes into genotypes. This is the case for the human ABO blood group (see page 74).

The simplest description of Mendelian variation is the frequency distribution of genotypes in a population. Table 24-1 shows the frequency distribution of the three genotypes at the MN blood-group locus in several human populations. Note that there is variation both within and between populations. More typically, instead of the frequencies of the diploid genotypes, the frequen-

TABLE 24-1. Frequencies of genotypes for alleles at the MN blood-group locus in various human populations

Population	Genotype			Allele frequencies	
	MM	MN	NN	$p(M)$	$q(N)$
Eskimo	0.835	0.156	0.009	0.913	0.087
Australian aborigine	0.024	0.304	0.672	0.176	0.824
Egyptian	0.278	0.489	0.233	0.523	0.477
German	0.297	0.507	0.196	0.550	0.450
Chinese	0.332	0.486	0.182	0.575	0.425
Nigerian	0.301	0.495	0.204	0.548	0.452

SOURCE: W. C. Boyd, *Genetics and the Races of Man.* D. C. Heath, 1950.

cies of the alternative alleles are used. If f_{AA}, f_{Aa}, and f_{aa} are the proportions of the three genotypes at a locus with two alleles, then the frequencies $p(A)$ and $q(a)$ of the alleles are obtained by counting alleles. Since each homozygote AA consists only of A alleles and only one-half of the alleles of each heterozygote Aa are type A, the total frequency (p) of A alleles in the population is

$$p = f_{AA} + \tfrac{1}{2} f_{Aa} = \text{frequency of } A$$

Similarly, the frequency q of a alleles is given by

$$q = f_{aa} + \tfrac{1}{2} f_{Aa} = \text{frequency of } a$$
$$p + q = f_{AA} + f_{aa} + f_{Aa} = 1.00$$

If there are multiple alleles, then the frequency for each allele is simply the frequency of its homozygote plus one-half the sum of the frequencies for all the heterozygotes in which it appears. Table 24-1 shows the values of p and q for each of the MN blood-group populations.

As an extension of p, which represents the **gene frequency** or **allele frequency**, we can describe variation at more than one locus simultaneously in terms of the **gametic frequencies.** Locus S (the secretor factor) is closely linked to the MN locus in humans. Table 24-2 shows the gametic frequencies of the four gametic types (MS, Ms, NS, and Ns) in various populations. The gametic frequency is obtained by summing up all the contributions of the different heterozygotes and homozygotes to the total pool of gametes. For example, the frequency of the MS gamete is given by

$$g(MS) = \text{frequency of } \frac{MS}{MS} + \tfrac{1}{2} \text{ frequency of } \frac{MS}{NS}$$
$$+ \tfrac{1}{2} \text{ frequency } \frac{MS}{Ms} + \tfrac{1}{2} \text{ frequency of } \frac{MS}{Ns}$$

Note that the last term in the sum involves the frequency of the double heterozygote MS/Ns. There is no contribution from the other double heterozygote Ms/NS, because it produces no MS gametes. If there were recombination, the Ms/NS heterozygote would produce MS gametes at a rate proportional to the recombination fraction. Thus, to give a gametic frequency description of a population at more than one locus simultaneously, we need to be able to distinguish coupling from repulsion double heterozygotes and to know the recombination fraction between the genes. For the human MNS system, there is essentially no recombination and the different types of heterozygotes can be distinguished by pedigree analysis.

A *measure* of genetic variation (as opposed to its *description* by gene frequencies) is the amount of **heterozygosity** at a locus in a population, which is given by the total frequency of heterozygotes at a locus. If one allele is in very high frequency and all others are near zero, then there will be very little heterozygosity because, by necessity, most individuals will be homozygous for the common allele. We expect heterozygosity to be greatest when there are many alleles at a locus, all at equal frequency. In Table 24-1, the heterozygosity is simply equal to the frequency of the MN genotype in each population. When more than one locus is considered, there are two possible ways of calculating heterozygosity. First, we can average the frequency of heterozygotes at each locus separately. Alternatively, we can take the gametic frequencies, as in Table 24-2, and calculate the proportion of all individuals who carry two different gametic forms. So, for example, a MS/Ms individual is a heterozygote, even though it is heterozygous only at one of the two loci. These two methods do not, in general, give the same value for heterozygosity. The heterozygosity based on gametic frequencies is always higher because it counts any individual as a heterozygote

TABLE 24-2. Frequencies of gametic types for the MNS system in various human populations

Population	Gametic type				Heterozygosity (H)	
	MS	Ms	NS	Ns	From gametes	From alleles
Ainu	0.024	0.381	0.247	0.348	0.672	0.438
Ugandan	0.134	0.357	0.071	0.438	0.658	0.412
Pakistan	0.177	0.405	0.127	0.291	0.704	0.455
English	0.247	0.283	0.080	0.390	0.700	0.469
Navaho	0.185	0.702	0.062	0.051	0.467	0.286

SOURCE: A. E. Mourant, *The Distribution of the Human Blood Groups,* Blackwell Scientific Pub., 1954.

who is heterozygous at *either* locus, even if the other locus is homozygous. The results of both calculations are given in Table 24-2.

Simple Mendelian variation can be observed within and between populations of any species at various levels of phenotype, from external morphology down to the amino-acid sequence of enzymes and other proteins. Indeed, with the new methods of DNA sequencing, variations in DNA sequence (such as third-position variants that are not differentially coded in amino-acid sequences and even variations in nontranslated intervening sequences) have been observed. Every species of organism ever examined has revealed considerable genetic variation, or **polymorphism,** reflected at one or more levels of phenotype, either within populations or between populations, or both. Genetic variation that might be the basis for evolutionary change is ubiquitous. The tasks for population geneticists are to describe that ubiquitous variation quantitatively in terms that allow evolutionary predictions and to build a theory of evolutionary change that can use these observations in prediction.

It is quite impossible in this text to provide an adequate picture of the immense richness of genetic variation that exists in species. We can consider only a few examples of the different kinds of Mendelian variation to gain a superficial sense of the genetic diversity within species. Each of these examples can be multiplied many times over in other species and with other traits.

Morphological Variation. The shell of the land snail *Cepaea nemoralis* may be pink or yellow, depending on two alleles at a single locus, with pink dominant to yellow. Also, the shell may be banded or unbanded (Figure 24-1) as a result of segregation at a second linked locus, with unbanded dominant to banded. Table 24-3 shows the variation of these two loci in several European colonies of the snail. The populations also show polymorphism for the number of bands and the height of the shells, but these characters have complex genetic bases.

Figure 24-2 shows another example of a naturally occurring morphological genetic polymorphism. The natural population of the illustrated mussel species in-

TABLE 24-3. Frequencies of snails (*Cepaea nemoralis*) with different shell colors and banding patterns in three French populations

Population	Yellow		Pink	
	Banded	Unbanded	Banded	Unbanded
Guyancourt	0.440	0.040	0.337	0.183
Lonchez	0.196	0.145	0.564	0.095
Peyresourde	0.175	0.662	0.100	0.062

SOURCE: Maxime Lamotte, *Bulletin Biologique de France et Belgique,* supplement 35, 1951.

cludes two forms: one blue, and the other brown. The difference is caused by an allelic difference at one locus.

Examples of naturally occurring morphological variation within plant species are *Plectritis* (page 5), *Collinsia* (blue-eyed Mary, page 35), and clover (Figure 4-5).

Figure 24-2. Dimorphism in the most common species of mussel found on the west coast of North America. Two morphological forms are found wherever the species occur: the blue form *(right),* and the brown form *(left).* Typically, the blue form is more frequent. The phenotypic difference is caused by an allelic difference at a single locus (B = brown and b = blue).

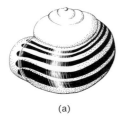   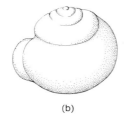

(a)   (b)

Figure 24-1. Shell patterns of the snail *Cepaea nemoralis.* (a) Banded. (b) Unbanded.

TABLE 24-4. Frequencies of third-chromosome inversions in *Drosophila pseudoobscura* from various North American populations

Population	Inversion				
	Standard	Arrowhead	Chiricahua	Pikes Peak	Others
Mount San Jacinto, CA	0.390	0.262	0.310	0.000	0.038
Prescott, AZ	0.110	0.790	0.090	0.010	0.000
South central Texas	0.002	0.117	0.000	0.703	0.178
Panamint Mountains, CA, 1937	0.138	0.674	0.184	0.000	0.004
1957	0.255	0.589	0.112	0.009	0.045
1963	0.243	0.409	0.053	0.129	0.166

SOURCE: T. Dobzhansky, W. W. Anderson, O. Pavlovsky, B. Spassky, and C. J. Wills, *Evolution* 18, 1964, 164–76.

TABLE 24-5. Frequencies of plants with supernumerary chromosomes and of translocation heterozygotes in a population of *Clarkia elegans* from California

No supernumeraries or translocations	Supernumeraries	Translocations	Both translocations and supernumeraries
0.560	0.265	0.133	0.042

SOURCE: H. Lewis, *Evolution* 5, 1951, 142–57.

Chromosomal Polymorphism. Although the karyotype is often regarded as a distinctive characteristic of a species, in fact, numerous species are polymorphic for chromosome number and morphology. Extra chromosomes (supernumeraries), reciprocal translocations, and inversions segregate in many populations of plants, insects, and even mammals. Table 24-4 gives the frequencies of various inversions in some populations of *Drosophila pseudoobscura* from western North America. Although one chromosomal arrangement is called "Standard," it is in fact impossible to say which of these types is "normal" and which are derived inversions. Indeed, in Mexico, there are no "Standard" chromosomes.

Variation within a population is not necessarily stable over time. Table 24-4 also shows the change in inversion frequencies in one population over a 25-year period.

Table 24-5 gives the frequencies of supernumerary chromosomes and of translocation heterozygosis in a population of the plant *Clarkia elegans* from California. The "typical" species karyotype would be hard to identify.

Immunological Polymorphism. A number of loci in vertebrates code for antigenic specificities such as the ABO blood types. Over 40 different specificities on human red cells are known, and several hundred are known in cattle. Another major polymorphism in humans is the HLA system of cellular antigens, which are implicated in tissue graft compatibility (Chapter 16). Table 24-6 gives the allelic frequencies for the ABO blood-group locus in some very different human populations. The polymorphism for the HLA system is vastly greater. There appear to be two main loci, each with five distinguishable alleles. Thus, there are $5^2 = 25$ different possible gametic types, making 25 different homozygous forms and $(25)(24)/2 = 300$ different heterozygotes. All genotypes are not phenotypically distinguishable, however, so only 121 phenotypic classes can be seen. L. L. Cavalli-Sforza and W. F. Bodmer report that, in a sample of only 100 Europeans, 53 of the 121 possible phenotypes were actually observed!

TABLE 24-6. Frequencies of the alleles I^A, I^B and i at the ABO blood-group locus in various human populations

Population	I^A	I^B	i
Eskimo	0.333	0.026	0.641
Sioux	0.035	0.010	0.955
Belgian	0.257	0.058	0.684
Japanese	0.279	0.172	0.549
Pygmy	0.227	0.219	0.554

SOURCE: W. C. Boyd, *Genetics and the Races of Man.* D. C. Heath, 1950.

Protein Polymorphism. In recent years, studies of genetic polymorphism have been carried down to the level of the polypeptides coded by the structural genes themselves. If there is a nonredundant codon change in a structural gene (say, GGU to GAU), this will result in an amino-acid substitution in the polypeptide produced at translation (in this case, glycine to aspartic acid). If a specific protein could be purified and sequenced from separate individuals, then it would be possible to detect genetic variation in a population at this level. In practice, this is tedious for large organisms and impossible for small ones unless a large mass of protein can be produced from a homozygous line.

There is, however, a practical substitute for sequencing that makes use of the change in the physical properties of a protein when an amino acid is substituted. Five amino acids (glutamic acid, aspartic acid, arginine, lysine, and histidine) have ionizable side chains that give a protein a characteristic net charge, depending on the pH of the surrounding medium. Amino-acid substitutions may directly replace one of these charged amino acids, or a noncharged substitution near one of them in the polypeptide chain may affect the degree of ionization of the charged amino acid, or a substitution at the joining between two α-helices may cause a slight shift in the three-dimensional packing of the folded polypeptide. In all of these cases, the net charge on the polypeptide will be altered.

To detect the change in net charge, protein can be subjected to the method of gel electrophoresis. Figure 24-3 shows the outcome of such an electrophoretic separation of variants of an esterase enzyme in *D. pseudoobscura*, where each track is the protein from a different individual. Figure 24-4 shows a similar gel for different variant human hemoglobins. In this case, each individual is heterozygous for the variant and normal hemoglobin A. Table 24-7 shows the frequencies of different alleles for three enzyme-coding loci in *D. pseudoobscura* in several populations: a nearly monomorphic locus (malic dehydrogenase), a moderately polymorphic locus (α-amylase), and a highly polymorphic locus (xanthine dehydrogenase).

The technique of gel electrophoresis (or sequencing) differs fundamentally from other methods of genetic analysis in allowing the study of loci that are not segregating, because the presence of a polypeptide is prima facie evidence of a structural gene. Thus, it has been possible to ask what proportion of all structural genes in the genome of a species is polymorphic and what the average heterozygosity is in a population. Very large numbers of species have been sampled by this method, including bacteria, fungi, higher plants, vertebrates, and invertebrates. The results are remarkably consistent over species. About one-third of structural-gene loci are polymorphic, and the average heterozygosity in a population over all loci sampled is about 10

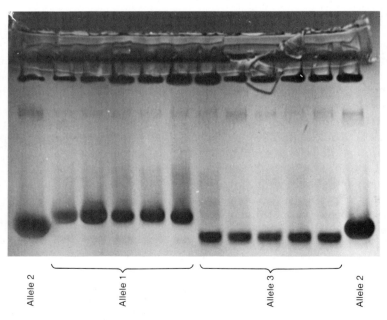

Allele 2 Allele 1 Allele 3 Allele 2

Figure 24-3. Electrophoretic gel showing homozygotes for three different alleles at the *esterase-5* locus in *Drosophila pseudoobscura*. Repeated samples of the same allele are identical, but there are repeatable differences between alleles.

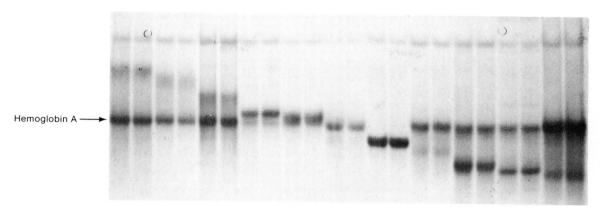

Figure 24-4. Electrophoretic gel showing heterozygotes of normal hemoglobin A and a number of different variant hemoglobin alleles. One of the dark-staining bands is marked as hemoglobin A. The second dark-staining band in each track represents the second protein derived from the second allele of the heterozygote; each of these proteins displays with a different electrophoretic mobility. Hemoglobin A is missing from tracks 13 and 14. (The dark-staining band with the same electrophoretic mobility in all tracks toward the top of the gel represents a protein other than hemoglobin.)

percent. This means that scanning the genome in virtually any species would show that about one in every ten loci is in heterozygous condition and that about one-third of all loci have two or more alleles segregating in any population. This represents an immense potential of variation for evolution. The disadvantage of the electrophoretic technique is that it detects variation only in structural genes. If most of the evolution of shape, physiology, and behavior rests on changes in regulatory genetic elements, then the observed variation in structural genes would be beside the point.

> **Message** Within species there exists great variation. The simplest *description* of variation is the frequency distribution of genotypes in a population. A *measure* of variation is the amount of heterozygosity in a population.

TABLE 24-7. Frequencies of various alleles at three enzyme-coding loci in four populations of *Drosophila pseudoobscura*

Locus (enzyme encoded)	Allele	Population			
		Berkeley	Mesa Verde	Austin	Bogotá
Malic dehydrogenase	A	0.969	0.948	0.957	1.00
	B	0.031	0.052	0.043	0.00
α-amylase	A	0.030	0.000	0.000	0.00
	B	0.290	0.211	0.125	1.00
	C	0.680	0.789	0.875	0.00
Xanthine dehydrogenase	A	0.053	0.016	0.018	0.00
	B	0.074	0.073	0.036	0.00
	C	0.263	0.300	0.232	0.00
	D	0.600	0.581	0.661	1.00
	E	0.010	0.030	0.053	0.00

SOURCE: R. C. Lewontin, *The Genetic Basis of Evolutionary Change,* Columbia University Press, 1974.

TABLE 24-8. Allelic frequencies at seven polymorphic loci in Europeans and black Africans

	Europeans			Africans		
Locus	Allele 1	Allele 2	Allele 3	Allele 1	Allele 2	Allele 3
Red-cell acid phosphatase	0.36	0.60	0.04	0.17	0.83	0.00
Phosphoglucomutase-1	0.77	0.23	0.00	0.79	0.21	0.00
Phosphoglucomutase-3	0.74	0.26	0.00	0.37	0.63	0.00
Adenylate kinase	0.95	0.05	0.00	1.00	0.00	0.00
Peptidase A	0.76	0.00	0.24	0.90	0.10	0.00
Peptidase D	0.99	0.01	0.00	0.95	0.03	0.02
Adenosine deaminase	0.94	0.06	0.00	0.97	0.03	0.00

SOURCE: R. C. Lewontin, *The Genetic Basis of Evolutionary Change,* Columbia University Press, 1974. Adapted from H. Harris, *The Principles of Human Biochemical Genetics.* North Holland, Amsterdam and London, 1970.

Variation within and among Populations

The various examples just given show that there are genetic differences among individuals within a population and also that the allelic frequencies differ among populations. The relative amounts of variation within and among populations vary from species to species, depending on history and environment. In humans, some gene frequencies (for example, those for skin color or hair form) obviously are well differentiated among populations and major geographical groups (so-called geographical "races"). If, however, we look at single structural genes identified immunologically or by electrophoresis rather than these outward phenotypic characters, the situation is rather different. Table 24-8 gives the allelic frequencies in a random sample of enzyme loci (chosen only for experimental convenience) in a European and an African population. Except for phosphoglucomutase-3, where the allelic frequencies are reversed between the populations, blacks and whites are very similar in their allelic distributions. In contrast to this random sample of loci, Table 24-9 shows the three loci for which Caucasians, Negroids, and Mongoloids are known to be most different from each other (Duffy and Rhesus blood groups and the P antigen) compared with the three polymorphic loci for which the races are most similar (Auberger blood group and Xg and secretor factors). Even for the most divergent loci, no race is homozygous for one allele that is absent in the other two races.

In general, different human populations show rather similar frequencies for polymorphic genes. Figure 24-5 is a **triallelic diagram** for the three main allelic classes I^A, I^B, and i of the ABO blood group. Each point represents the allelic composition of a population, where the three allelic frequencies can be read by taking the

lengths of the perpendiculars from each side to the point. The diagram shows that all human populations are bunched together in the region of high i, interme-

TABLE 24-9. Examples of extreme differentiation and close similarity in blood-group allelic frequencies in three racial groups

		Population		
Gene	Allele	Caucasoid	Negroid	Mongoloid
Duffy	Fy	0.0300	0.9393	0.0985
	Fy^d	0.4208	0.0607	0.9015
	Fy^b	0.5492	0.0000	0.0000
Rhesus	R_0	0.0186	0.7395	0.0409
	R_1	0.4036	0.0256	0.7591
	R_2	0.1670	0.0427	0.1951
	r	0.3820	0.1184	0.0049
	r'	0.0049	0.0707	0.0000
	others	0.0239	0.0021	0.0000
P	P_1	0.5161	0.8911	0.1677
	P_2	0.4839	0.1089	0.8323
Auberger	Au^a	0.6213	0.6419	no data
	Au	0.3787	0.3581	no data
Xg	Xg^a	0.67	0.55	0.54
	Xg	0.33	0.45	0.46
Secretor	Se	0.5233	0.5727	no data
	se	0.4767	0.4273	no data

SOURCE: A summary is provided in L. L. Cavalli-Sforza and W. F. Bodmer, *The Genetics of Human Populations,* W. H. Freeman and Company, 1971, pp. 724–31. See this source for loci and for data sources.

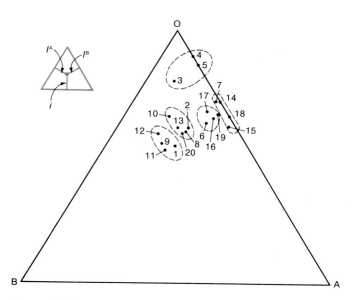

Figure 24-5. Triallelic diagram of the ABO blood-group allelic frequencies for human populations. Each point represents a population; the perpendicular distances from the point to the sides represent the allelic frequencies, as indicated in the small triangle. Populations 1 to 3 are African; 4 to 7 are American Indian; 8 to 13 are Asian; 14 to 15 are Australian aborigine; and 16 to 20 are European. Dashed lines enclose arbitrary classes with similar gene frequencies; these groupings do not correspond to "racial" classes. (After A. Jacquard, *Structures génétiques des populations.* Copyright © 1970 by Masson et Cie.)

diate I^A, and low I^B frequencies. Moreover, neighboring points (enclosed by dashed lines) do not correspond to geographical races, so that such races cannot be distinguished from each other by characteristic allelic frequencies for this gene. The study of polymorphic blood groups and enzyme loci in a variety of human populations has shown that about 85 percent of total human genetic diversity is found within local populations, about 8 percent is found among local populations within major geographical races, and the remaining 7 percent is found among the major geographical races. Clearly, the genes influencing skin color, hair form, and facial form that are well differentiated among races are not a random sample of structural gene loci.

> **Message** In general, the genetic variation among individuals within human races is much greater than the average variation between races.

Quantitative Variation

The variation in quantitative characters cannot be described in terms of allelic frequencies because individual loci and their alleles cannot be identified. Such variation can be characterized, however, by the amount of genetic variance (or the heritability of the trait) in the population. Figure 23-20 shows that many morphological and physiological traits in poultry have genetic variances of different amounts in different populations. A simple technique for estimating the additive genetic variance (page 665) of a character is to choose two groups of parents that are extremely different and then measure the difference between their offspring groups. The difference between offspring groups divided by the difference between parental groups is a measure of the heritability (h^2). Where this technique has been applied to morphological variation in *Drosophila*, for example, virtually every variable trait is found to have some genetic variance, so evolution of the trait can occur. Indeed, the method of estimating heritability just described is itself a kind of one-generation artificial-selection experiment.

It should not be supposed that all variable traits are heritable, however. Certain metabolic traits (such as resistance to high salt concentrations in *Drosophila*) show individual variation but no heritability. Left-right asymmetry is also nonheritable: "left-winged" flies (those with left wings slightly longer than their right wings) have no more left-winged offspring than do their right-winged companions. In general, behavioral traits have lower heritabilities than morphological traits, especially in organisms with more complex nervous systems that exhibit immense individual flexibility in central nervous states. Before any judgment can be made about the evolution of a particular quantitative trait, it is essential to determine if there is genetic variance for it in the population whose evolution is to be predicted. Thus, suggestions that such traits in the human species as performance on IQ tests, temperament, and social organization are in the process of evolving or have evolved at particular epochs in human history depend critically on evidence about the genetic variation for these traits. No such evidence is presently available.

One of the most important findings in evolutionary genetics has been the discovery of substantial genetic variation underlying characters that show no morphological variation! These are called **canalized characters,** because the final outcome of their development is held within narrow bounds despite disturbing forces. Development is such that all the different genotypes for canalized characters have the same constant phenotype over the range of environments that is usual for the species. The genetic differences are revealed if the organisms are put in a stressful environment or if a severe mutation stresses the developmental system. For example, all wild-type *Drosophila* have exactly four scutellar bristles (Figure 24-6). If the recessive mutant *scute* is present, the number of bristles is reduced, but, in addition, there is variation from fly to fly. This variation is heritable, and

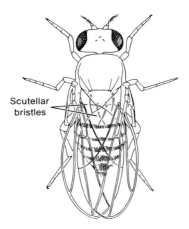

Figure 24-6. The scutellar bristles of the adult *Drosophila*. This is an example of a canalized character; all wild-type *Drosophila* will have four scutellar bristles in a very wide range of environments.

lines with zero or one bristles and lines with three or four bristles can be obtained by selection in the presence of the *scute* mutation. When the mutation is removed, these lines now have two and six bristles, respectively. Similar experiments have been performed using extremely stressful environments in place of mutants. The results demonstrate that genetic variation can affect bristle number but that this variation is not manifested phenotypically until the normal development of the fly is disturbed. A consequence of such hidden genetic variation is that a character that is phenotypically uniform in a species may nevertheless undergo rapid evolution if a stressful environment uncovers the genetic variation.

> **Message** Even characters with no apparent phenotypic variance may evolve when developmental conditions are changed drastically if genetic variation for the character is hidden by developmental canalization.

The Sources of Variation

The variational theory of evolution has a peculiar self-defeating property. If evolution occurs by the differential reproduction of different variants, we expect the variant with the highest rate of reproduction eventually to take over the population and all other genotypes to disappear. But then there is no longer any variation for further evolution. Genetic variation is the fuel for the evolutionary process, but differential reproduction consumes that fuel and so destroys the very condition necessary for further evolution. The possibility of continued evolution therefore is critically dependent on renewed variation. All species eventually become extinct, and it is possible that much of this extinction comes about because species fails to produce the variation on which the process of natural selection can operate. Unfortunately, we cannot study the genotypic variation in extinct species to know whether or not this is true.

For a given population, there are three sources of variation: mutation, recombination, and immigration of genes. However, recombination by itself does not produce variation unless alleles are segregating already at different loci; otherwise there is nothing to recombine, and immigration cannot provide variation if the entire species is homozygous for the same allele. Ultimately, the source of all variation must be mutation.

Variation from Mutations

Mutations are the *source* of variation, but the *process* of mutation does not itself drive evolution. The rate of change in gene frequency from the mutation process is very low because spontaneous mutation rates are low (Table 24-10). Let μ be the **mutation rate** from allele A to some other allele a (the probability that a gene copy A will become a during meiosis). If p_t is the frequency of the A allele in generation t, if $q_t = 1 - p_t$ is the frequency of the a allele, and if there are no other causes of gene-frequency change (no natural selection, for example), then the change in allelic frequency in one generation is

$$\Delta p = p_t - p_{t-1} = -\mu p_{t-1}$$

This tells us that the frequency of A decreases (and the frequency of a increases) by an amount that is proportional to the mutation rate μ and to the proportion p of all the genes that are still available to mutate. Thus Δp

TABLE 24-10. Some point-mutation rates in different organisms

Organism	Gene	Mutation rate per generation
Bacteriophage	Host range	2.5×10^{-9}
Escherichia coli	Phage resistance	2×10^{-8}
Zea mays (corn)	R (color factor)	2.9×10^{-4}
	Y (yellow seeds)	2×10^{-6}
Drosophila melanogaster	Average lethal	2.6×10^{-5}

SOURCE: T. Dobzhansky, *Genetics and the Origin of Species*, 3d ed., revised. Columbia University Press, 1951.

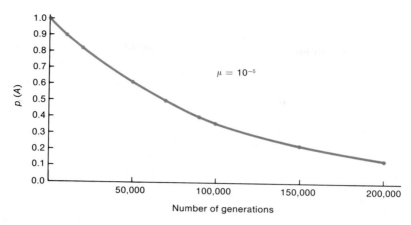

Figure 24-7. The change over generations in the frequency of a gene A due to mutation from A to a at a constant mutation rate (μ) of 10^{-5}.

gets smaller as the frequency of p itself decreases, because there are fewer and fewer A alleles to turn into a alleles. We can make the approximation that after n generations of mutation

$$p_n = p_0 e^{-n\mu}$$

where e is the base of the natural logarithms. This relation of gene frequency to number of generations is shown in Figure 24-7 for $\mu = 10^{-5}$. After 10,000 generations of continued mutation of A to a

$$p = p_0 e^{-(10^4)(10^{-5})} = p_0 e^{0.1} = 0.904\, p_0$$

So, if the population starts with only A alleles ($p_0 = 1.0$), it would still have only 10 percent a alleles after 10,000 generations at a rather high mutation rate and would require 60,000 additional generations to reduce p to 0.5. Even if mutation rates were doubled (say, by environmental mutagens), the rate of evolution would be very slow. For example, radiation levels of sufficient intensity to double the mutation rate over the reproductive lifetime of an individual human would be more than the amount necessary to cause death, so rapid genetic change in the species would not be one of the effects of increased radiation. Although we have many things to fear from environmental radiation pollution, turning into a species of monsters is not one of them.

If we look at the mutation process from the standpoint of the increase of a particular new allele rather than the decrease of the old form, the process is even slower. Most mutation rates that have been determined are the sum of all mutations of A to any mutant form with a detectable effect. Any *specific* base substitution is likely to be at least two orders of magnitude lower in frequency than the sum of all changes. So, precise reverse mutations ("back mutations") to the original allele A are unlikely, although many mutations may produce alleles that are *phenotypically* similar to the original.

> **Message**   Mutation rates are so low that mutation alone cannot account for the rapid evolution of populations and species.

It is not possible to measure locus-specific mutation rates for continuously varying characters, but the rate of accumulation of genetic variance can be determined. Beginning with a completely homozygous line of *Drosophila* derived from a natural population, between $\frac{1}{1000}$ and $\frac{1}{500}$ of the genetic variance in bristle number in the original population is restored each generation by spontaneous mutation. On the other hand, about $\frac{1}{60}$ of the genetic variance in the probability of survival present in a natural population is regenerated in an inbred line each generation. This difference reflects both the greater effect per mutation and the larger number of loci for genes that affect survival as opposed to those that affect bristle number.

Variation from Recombination

The creation of genetic variance by recombination can be a much faster process than its creation by mutation. When two chromosomes of *Drosophila* with "normal" survival are allowed to recombine for a single generation, they produce an array of chromosomes that have 25 to 75 percent as much genetic variance in survival as the original wild population from which the parent chromosomes were sampled. This is simply a consequence of the fact that a single homologous pair of chromosomes that is heterozygous at n loci (taking into account only single and double crossovers) can produce $n(n-1)/2$ new unique gametic types from one generation of recombination. If the heterozygous loci are well spread out on the chromosomes, these new gametic types will be frequent and a considerable variance will be generated. Asexual organisms or organisms like bacteria that very seldom undergo sexual recombination do not have this source of variation, so that new mutations are the only way in which a change in gene combinations can be achieved. As a result, asexual organisms may evolve more slowly under natural selection than sexual organisms.

Variation from Migration

A further source of variation is migration into a population from other populations with different gene frequencies. If p_t is the frequency of an allele in the recipient population in generation t, and P is the average allelic frequency over all the donor populations, and m is the proportion of migrants, then the gene frequency in the next generation is the result of mixing $(1-m)$ genes from the population with m genes from the donor populations. Thus

$$p_{t-1} = (1-m)p_t + mP = p_t + m(P-p_t)$$

and

$$\Delta p = p_{t-1} - p_t = m(P-p_t)$$

The change in gene frequency is proportional to the difference in frequency between the recipient population and the average of the donor populations. Unlike the mutation rate, the migration rate (m) can be large, so the change in frequency may be substantial.

We must understand *migration* as meaning any form of the introduction of genes from one population into another. So, for example, genes from Europeans have "migrated" into the population of African origin in North America steadily since the Africans were introduced as slaves. We can determine the amount of this migration by looking at the frequency of an allele that is found only in Europeans and not in Africans and comparing its frequency among blacks in North America.

We can use the formula for the change in gene frequency from migration if we modify it slightly to account for the fact that several generations of admixture have taken place. If the rate of admixture has not been too great, then (to a close order of approximation) the sum of the single-generation migration rates over several generations (let's call this M) will be related to the total change in the recipient population after these several generations by the same expression as the one used for changes due to migration. If, as before, P is the allelic frequency in the donor population and p_0 is the original frequency among the recipients, then

$$\Delta p_{total} = M(P-p_0)$$

so

$$M = \frac{\Delta p_{total}}{P-p_0}$$

For example, the Duffy blood-group allele Fy^a is absent in Africa but has a frequency of 0.42 in whites from the state of Georgia. Among blacks from Georgia, the Fy^a frequency is 0.046. Therefore, the total migration of genes from whites into the black population since the introduction of slaves in the eighteenth century is

$$M = \frac{\Delta p_{total}}{P-p} = \frac{0.046-0}{0.42-0} = 0.1095$$

When the same analysis is carried out on American blacks from Oakland (California) and Detroit, M turns out to be 0.22 and 0.26, respectively, showing either greater admixture rates in these cities than in Georgia or differential movement into these cities by American blacks who have more European ancestry. In any case, the genetic variation at the Fy locus has been increased by this admixture.

The Origin of New Functions

Point mutations or chromosomal rearrangements are themselves a limited source of variation for evolution because they can only alter a function or change one kind of function into another. To add quite new functions requires expansion in the total repertoire of genes through duplication and polyploidy, followed by a divergence between the duplicated genes, presumably by the usual process of mutation. Expansion of the genome by polyploidy has clearly been a frequent process, at least in plants. Figure 24-8 shows the frequency distribution of haploid chromosome numbers among dicotyledonous plant species. Note that even numbers are much more common than odd numbers—a consequence of frequent polyploidy.

Once an expansion in total DNA of the genome has occurred, it may require only a few base substitutions in a gene to provide it with a new function. For example, B. Hall has experimentally changed a gene to a new function in *Escherichia coli*. In addition to the *lac Z* genes specifying the usual lactose-fermenting β-galactosidase

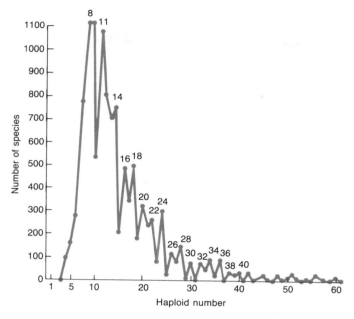

Figure 24-8. Frequency distribution of haploid chromosome numbers in dicotyledonous plants. (From Verne Grant, *The Origin of Adaptation.* Copyright © 1963 by Columbia University Press.)

activity in *E. coli*, another structural gene locus *ebg* specifies another β-galactosidase that does not ferment lactose, although it is induced by lactose. The natural function of this second enzyme is unknown. Hall was able to alter this gene into one specifying an enzyme that does ferment lactobionate. To do so, it was necessary to alter the regulatory element to a constitutive state and to produce three successive structural-gene mutations.

> **Message** Evolution would come to a stop by running out of variation if new genetic variation were not added to populations by mutation, recombination, and migration. Ultimately, all new variation is derived from gene and chromosome mutations.

The Effect of Sexual Reproduction on Variation

The evolutionary theorists of the nineteenth century encountered a fundamental difficulty in dealing with Darwin's theory of evolution through natural selection. The possibility of continued evolution by natural selection is limited by the amount of genetic variation.

But biologists of the nineteenth century, including Darwin, believed in one form or another of **blending inheritance,** a model postulating that the characteristics of each offspring are some intermediate mixture of the parental characters. Such a model of inheritance has fatal implications for a theory of evolution that depends on variation.

Suppose that some trait (say, height) has a distribution in the population and that individuals mate more or less at random. If intermediate individuals mated with each other, they would produce only intermediate offspring according to a blending model. The mating of a tall with a short individual also would produce only intermediate offspring. Only the mating of tall with tall individuals and short with short individuals would preserve extreme types (height variation). The net result of all matings would be an increase in intermediate types and a decrease in extreme types. The variance of the distribution would shrink, simply as a result of sexual reproduction. In fact, it can be shown that the variance is *cut in half* in each generation, so that the population would be essentially uniformly intermediate in height before very many generations had passed. There then would be no variation on which natural selection could operate. This was a very serious problem for the early Darwinists; it made it necessary for Darwin to assume that new variation is generated at a very rapid rate by the

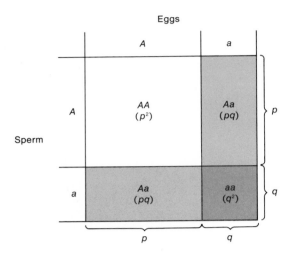

Eggs

	A	a	
A	AA (p^2)	Aa (pq)	p
a	Aa (pq)	aa (q^2)	q
	p	q	

Sperm

Figure 24-9. The Hardy-Weinberg equilibrium frequencies that result from random mating. The frequencies of A and a among both eggs and sperm are p and q ($= 1 - p$), respectively. The total frequencies of the zygote genotypes are p^2 for AA, $2pq$ for Aa, and q^2 for aa. The frequency of the allele A in the zygotes is the frequency of AA plus one-half of the frequency of Aa, or $p^2 + pq = p(p + q) = p$.

inheritance of characters acquired by individuals during their lifetimes.

The rediscovery of Mendelism changed this picture completely. Because of the discrete nature of the Mendelian genes and the segregation of alleles at meiosis, a cross of intermediate with intermediate individuals does *not* result in all intermediate offspring. On the contrary, extreme types (homozygotes) segregate out of the cross. To see the consequence of Mendelian inheritance for genetic variation, consider a population in which males and females mate with each other at random with respect to some gene locus A, a; that is, individuals do not choose their mates preferentially with respect to the partial genotype at the locus. Such random mating is equivalent to mixing all the sperm and all the eggs in the population together and then matching randomly drawn sperm with randomly drawn eggs.

If the frequency of allele A is p in both the sperm and the eggs and the frequency of allele a is $q = 1 - p$, then the consequences of random unions of sperm and eggs are shown in Figure 24-9. The probability that both the sperm and the egg will carry A is $p \times p = p^2$, so this will be the frequency of AA homozygotes in the next generation. In like manner, the chance of heterozygotes Aa will be $(p \times q) + (q \times p) = 2pq$, and the chance of homozygotes aa will be $q \times q = q^2$. The three genotypes, after a generation of random mating, will be in the frequencies $p^2 : 2pq : q^2$. As the figure shows, the allelic frequency of A has not changed and is still p. Therefore, in the second

generation, the frequencies of the three genotypes will again be $p^2 : 2pq : q^2$, and so on, forever.

Message   Mendelian segregation has the property that random mating results in an equilibrium distribution of genotypes after only one generation, so that genetic variation is maintained.

The equilibrium distribution

AA	Aa	aa
p^2	$2pq$	q^2

is called the **Hardy-Weinberg equilibrium** after those who independently discovered it. (A third independent discovery was made by the Russian geneticist Sergei Tschetverikov.)

The Hardy-Weinberg equilibrium means that sexual reproduction does not cause a constant reduction in genetic variation in each generation; on the contrary, the amount of variation remains constant generation after generation, in the absence of other disturbing forces. The equilibrium is the direct consequence of the segregation of alleles at meiosis in heterozygotes.

Numerically, the equilibrium shows that irrespective of the particular mixture of genotypes in the parental generation, the genotypic distribution after one round of mating is completely specified by the allelic frequency p. For example, consider three hypothetical populations, all having the same frequency of A ($p = 0.3$):

	AA	Aa	aa
I	0.3	0.0	0.7
II	0.2	0.2	0.6
III	0.1	0.4	0.5

After one generation of random mating, each of the three populations will have the same genotypic frequencies:

AA	Aa	aa
$(0.3)^2 = 0.09$	$2(0.3)(0.7) = 0.42$	$(0.7)^2 = 0.49$

and they will remain so indefinitely.

One consequence of the Hardy-Weinberg proportions is that rare alleles are virtually never in homozygous condition. An allele with a frequency of 0.001

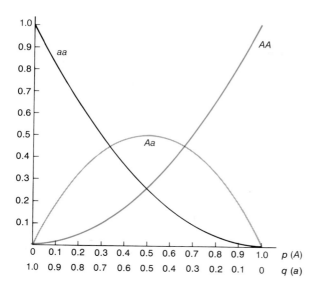

Figure 24-10. Curves showing the proportions of homozygotes *AA* (colored line), homozygotes *aa* (black line), and heterozygotes *Aa* (gray line) in populations of different allelic frequencies if the populations are in Hardy–Weinberg equilibrium.

occurs in homozygotes at a frequency of only one in a million; most copies of such rare alleles are found in heterozygotes. In general, since two copies of an allele are in homozygotes but only one copy of that allele is in each heterozygote, the relative frequency of the allele in heterozygotes (as opposed to homozygotes) is

$$\frac{2pq}{2q^2} = \frac{p}{q}$$

which for $q = 0.001$ is a ratio of $999:1$. Thus, the frequency of heterozygous carriers of rare genes that are deleterious in a homozygous condition is much greater than the frequency of the affected homozygotes. The general relation between homozygote and heterozygote frequencies is shown in Figure 24-10.

In our derivation of the equilibrium, we assumed that the allelic frequency p is the same in sperm and eggs. If p for males is not equal to p for females in the initial generation, then it takes one generation to equalize the frequencies between the sexes and a *second* generation to reach the Hardy-Weinberg equilibrium. As an extension of this effect, the Hardy-Weinberg equilibrium theorem does not apply to sex-linked genes even after two generations, if males and females start with unequal gene frequencies (see Problem 7 at the end of the chapter).

Random mating with respect to particular genes is quite common. Human beings, for example, do not choose mates with respect to their blood groups. Table 24-11 shows the result of sampling MN blood groups in various populations. Clearly, the populations are mating at random with respect to this locus.

Inbreeding and Assortative Mating

Random mating with respect to a locus is common, but it is not universal. Two kinds of deviation from random mating must be distinguished. First, individuals may mate with each other nonrandomly either because of their degree of common ancestry or their degree of genetic relationship. If mating between relatives occurs more commonly than would occur by pure chance, then the population is **inbreeding.** If mating between relatives is less common than would occur by chance, then the population is said to be undergoing **enforced outbreeding,** or **negative inbreeding.**

Second, individuals may tend to choose each other as mates, not because of their degree of genetic relationship but because of their degree of resemblance to each other at some locus. Bias toward mating of like with like is called **positive assortative mating.** Mating with unlike partners is called **negative assortative mating.** Assortative mating is never complete.

Inbreeding levels in natural populations are a consequence of geographical distribution, of the mechanism

TABLE 24-11. Comparison between observed frequencies of genotypes for the MN blood-group locus and the frequencies expected from random mating

Population	Observed			Expected		
	MM	*MN*	*NN*	*MM*	*MN*	*NN*
Eskimo	0.835	0.156	0.009	0.834	0.159	0.008
Egyptian	0.278	0.489	0.233	0.274	0.499	0.228
Chinese	0.332	0.486	0.182	0.331	0.488	0.181
Australian aborigine	0.024	0.304	0.672	0.031	0.290	0.679

NOTE: the expected frequencies are computed according to the Hardy-Weinberg equilibrium, using the values of p and q computed from the observed frequencies.

of reproduction, and of behavioral characteristics. If close relatives occupy adjacent areas, then simple proximity may result in inbreeding. The seeds of many plants, for example, fall very close to the parental source and the pollen is not widely spread, so a high frequency of sib mating occurs. Some plants (such as corn) can be self-pollinated as well as cross-pollinated, so that wind pollination results in some very close inbreeding. Yet other plants, like the peanut, are obligatorily selfed. Many small mammals (such as house mice) live and mate in restricted family groups that persist generation after generation. Humans, on the other hand, generally have complex mating taboos and proscriptions that reduce inbreeding.

Assortative mating for some traits is common. In humans, there is a positive assortative mating bias for skin color and height, for example. In plants and in many insects that produce only one generation per year, there is positive assortative mating for arrival at sexual maturity. An important difference between assortative mating and inbreeding is that the former is specific to a trait whereas the latter applies to the entire genome. Individuals may mate assortatively with respect to height but at random with respect to blood group. Cousins, on the other hand, resemble each other genetically on the average to the same degree at all loci.

For both positive assortative mating and inbreeding, the consequence to population structure is the same: there is an increase in homozygosity above the level predicted by the Hardy-Weinberg equilibrium. If two individuals are related, they have at least one common ancestor. Thus, there is some chance that an allele carried by one of them and an allele carried by the other are both descended from the identical DNA molecule. The result is that there is an extra chance of **homozygosity by descent,** to be added to the chance of homozygosity ($p^2 + q^2$) that arises from the random mating of unrelated individuals. The probability of homozygosity by descent is called the **inbreeding coefficient (F).** Figure 24-11 illustrates the calculation of the probability of homozygosity by descent. Individuals I and II are full sibs because they share both parents. We label each allele in the parents uniquely to keep track of them. Individuals I and II mate to produce individual III. If individual I is A_1A_3 and the gamete that it contributes to III contains the allele A_1, then we would like to calculate the probability that the gamete produced by II is also A_1. The chance is $\frac{1}{2}$ that II will receive A_1 from its father, and if it does, the chance is $\frac{1}{2}$ that II will pass A_1 on to the gamete in question. Thus, the probability that III will receive an A_1 from II is $\frac{1}{2} \times \frac{1}{2} = \frac{1}{4}$, and this is the chance that III — the product of a full-sib mating — will be homozygous by descent.

Such close inbreeding can have deleterious consequences. Let's consider a rare deleterious allele a, which,

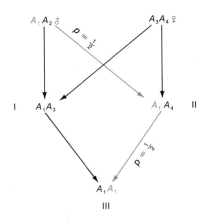

Figure 24-11. Calculation of homozygosity by descent for an offspring (III) of a brother-sister (I–II) mating. The probability that II will receive A_1 from its father is $\frac{1}{2}$; if it does, the probability that II will pass A_1 on to the generation producing III is $\frac{1}{2}$. Thus, the probability that III will receive an A_1 from II is $\frac{1}{2} \times \frac{1}{2} = \frac{1}{4}$.

when homozygous, causes a metabolic disorder. If the frequency of the allele in the population is p, then the probability that a random couple will produce a homozygous offspring is only p^2 (from the Hardy-Weinberg equilibrium). Thus, if p is, say, $\frac{1}{1000}$, then the frequency of homozygotes will be 1 in 1,000,000. Now suppose that the couple are brother and sister. If one of their common parents is a heterozygote for the disease, they may both receive it and may both pass it on to the offspring they produce. The probability of a homozygous aa offspring is

probability one or the other grandparent is Aa

$\times$ probability a is passed to male sib

$\times$ probability a is passed to female sib

$\times$ probability of a homozygous aa
$\qquad\qquad$ offspring from $Aa \times Aa$

$= (2pq + 2pq) \times \frac{1}{2} \times \frac{1}{2} \times \frac{1}{4}$

$= \dfrac{pq}{4}$

We assume that the chance that both grandparents are Aa is negligible. If p is very small, then q is nearly 1.0 and the chance of an affected offspring is close to $p/4$. For $p = \frac{1}{1000}$, there is one chance in 4000 of an affected child, compared to the one-in-a-million chance from a random mating. In general, for full sibs, the ratio of risks will be

$$\frac{p/4}{p^2} = \frac{1}{4p}$$

so the rarer the gene, the worse the *relative* risk of a defective offspring from inbreeding. For more distant relatives the chance of homozygosity by descent is, of course, less but still substantial. For first cousins, for example, the relative risk is $1/16p$ compared to random mating.

The population consequences of inbreeding depend on its intensity and form. Next we consider some examples.

Systematic Inbreeding. In experimental genetics (especially in plant and animal breeding), generation after generation of systematic selfing, full-sib, parent-offspring, or some other form of mating between relatives may be used to increase homozygosity. Such systematic mating schemes, when they are between close relatives, eventually lead to complete homozygosity of the population, but at different rates. Referring back to Table 22-1, we can see the amount of heterozygosity still left within lines after various numbers of generations of inbreeding. Which allele is fixed within a line is a matter of chance. If, in the original population from which the inbred lines are taken, allele A has frequency p and allele a has frequency $q = 1 - p$, then a proportion p of the homozygous lines established by inbreeding will be homozygous AA and a proportion q of the lines will be aa. What inbreeding does is to take the genetic variation present *within* the original population and convert it into variation *between* homozygous inbred lines sampled from the population (Figure 24-12).

Random Inbreeding. In a natural population, there will be some fraction of mating between relatives (or even selfing if that is possible) because of spatial proximity. However, there is no continuity of inbreeding within any specific family. If some proportion of wind-pollinated plants are selfed in a particular generation, these are not necessarily the progeny of selfed plants in the previous generation; they are distributed at random over selfed and outcrossed progeny. A consequence of such random inbreeding is that there is an equilibrium frequency of homozygotes and heterozygotes similar to the Hardy-Weinberg equilibrium, but with more homozygotes. Thus, genetic variation is still preserved, in contrast to the result of systematic experimental inbreeding.

Small Populations. Suppose that a population is founded by some small number of individuals who mate at random to produce the next generation. Also assume that no further immigration into the population ever occurs again. (For example, the rabbits now in Australia probably have descended from a single introduction of a few animals in the nineteenth century.) In later generations, then, everyone is related to everyone else, because their family trees have common ancestors here and there in their pedigrees. Such a population is then inbred, in the sense that there is some probability of a gene being homozygous by descent. Since the population is, of necessity, finite in size, some of the originally introduced family lines will become extinct in every generation, just as family names disappear in a closed human population because, by chance, no male offspring are left. As original family lines disappear, the population comes to be made up of descendants of fewer and fewer of the original founder individuals, and all the members of the population become more and more likely to carry the same alleles by descent. In other words, the inbreeding coefficient F increases and the heterozygosity decreases over time until finally F reaches 1.00 and heterozygosity reaches 0.

The rate of loss of heterozygosity per generation in such a closed, finite, randomly breeding population is inversely proportional to the total number ($2N$) of haploid genomes, where N is the number of diploid individuals in the population. In each generation, $1/2N$ of the remaining heterozygosity is lost, so that

$$H_t = H_0 \left(1 - \frac{1}{2N} \right)^t \cong H_0 e^{-t/2N}$$

where H_t and H_0 are the proportions of heterozygotes in the tth and original generations, respectively. As the

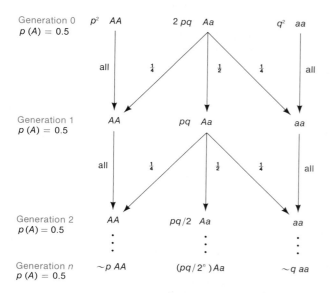

Figure 24-12. Repeated generations of self-fertilization (or inbreeding) will eventually split a heterozygous population into a series of completely homozygous lines. The frequency of AA lines among the homozygous lines will be equal to the frequency of allele A in the original heterozygous population.

number t of generations becomes very large, H_t approaches zero.

Which of the original alleles becomes fixed at each locus is a chance matter. If allele A_i has frequency p_i in the original population, then the probability is p_i that eventually the population will become homozygous A_iA_i. Suppose that a number of isolated island populations are founded from some large, heterozygous mainland population. Eventually, if these island populations remain completely isolated from one another, each will become homozygous for one of the alleles at each locus. Some will be homozygous A_1A_1, some A_2A_2, and so on. Thus, the result of this form of inbreeding is to cause genetic differentiation between populations.

> **Message**   Once again, we see that inbreeding is a process that converts genetic variation within a population into differences among populations by making each separate population homozygous for a randomly chosen allele.

Within each population, there is a change in allelic frequency from the original p_i to either 1 or 0, depending on which allele is fixed, but the average allelic frequency over all such populations remains p_i. Figure 24-13 shows the distribution of allelic frequencies among islands in successive generations, where $p(A_1) = 0.5$. In generation 0, all populations are identical. As time goes on, the gene frequencies among the populations diverge and some become fixed. After about $2N$ generations, every allelic frequency except the fixed classes ($p = 0$ and $p = 1$) is equally likely, and about one-half of the populations are totally homozygous. By the time $4N$ generations have gone by, 80 percent of the populations are fixed, one-half being homozygous AA and one-half being homozygous aa.

The process of differentiation by inbreeding in island populations is slow, but not on an evolutionary or geological time scale. If an island can support, say, 10,000 individuals of a rodent species, then after 20,000 generations (about 7000 years, assuming three generations per year), the population will be homozygous for about one-half of all the loci that were initially at the maximum of heterozygosity. Moreover, the island will be differentiated from other similar islands in two ways. For the loci that are fixed, many of the other islands will still be segregating and others will be fixed at a different allele. For the loci that are still segregating in all the islands, there will be a large variance in gene frequency from island to island, as shown in Figure 24-13.

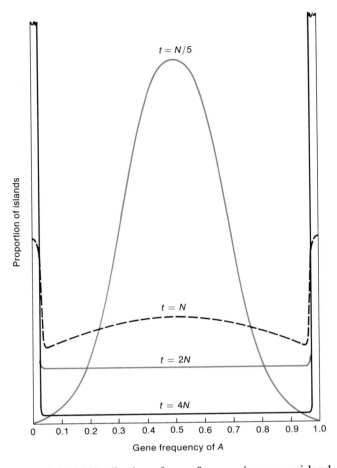

Figure 24-13. Distribution of gene frequencies among island populations after various numbers of generations of isolation, where the number of generations that have passed (t) is given in multiples of the population size (N).

The Balance between Inbreeding and New Variation

Any population of any species is finite in size, so all populations should eventually become homozygous and differentiated from one another as a result of inbreeding. Evolution would then cease. In nature, however, new variation is always being introduced into populations by mutation and by some migration between localities. Thus, the actual variation available for natural selection is a balance between the introduction of new variation and its loss through local inbreeding. The rate of loss of heterozygosity in a closed population is $1/2N$, so any effective differentiation between populations will be negated if new variation is introduced at this or a higher rate. If m is the migration rate into a given population and μ is the rate of mutation to new alleles, then roughly (to an order of magnitude) a population will retain most of its heterozygosity and will not differentiate much from other populations by local inbreeding if

$$m \geq \frac{1}{N} \quad \text{or} \quad \mu \geq \frac{1}{N}$$

or if

$$Nm \geq 1 \quad \text{or} \quad N\mu \geq 1$$

For populations of intermediate and even fairly large size, it is unlikely that $N\mu \geq 1$. For example, if the population size is 100,000, then the mutation rate must exceed 10^{-5}, which is somewhat on the high side for known mutation rates, although it is not an unknown rate. On the other hand, a migration rate of 10^{-5} per generation is not unreasonably large. In fact

$$m = \frac{\text{number of migrants}}{\text{total population size}} = \frac{\text{number of migrants}}{N}$$

Thus, the requirement that $Nm \geq 1$ is equivalent to the requirement that

$$Nm = N \times \frac{\text{number of migrants}}{N} \geq 1$$

or that

$$\text{number of migrant individuals} \geq 1$$

irrespective of population size! For many populations, more than a single migrant individual per generation is quite likely. Human populations (even isolated tribal populations) have a higher migration rate than this minimal value and, as a result, show remarkably little gene-frequency differentiation among populations. There is, for example, no locus known in humans for which one allele is fixed in some populations and an alternative allele is fixed in others (see Table 24-9).

Selection

Fitness and the Struggle for Existence

Darwin recognized that evolution consists of two processes, both of which must be explained. One is the origin of the *diversity* of organisms, and the second is the origin of the *adaptation* of these same organisms. Evolution is not simply the origin and extinction of different organic forms; rather, it is also a process that creates

some kind of match between the phenotypes of species and the environments in which they live. Darwin regarded "organs of extreme perfection" (such as the eye) as tests of his theory. His explanation of such organs was that there is a constant *struggle for existence.* Organisms with phenotypes that are better suited to the environment have a greater probability of surviving the struggle and will leave more offspring. Presumably, the better an organism can see, the better chance it has of locating food, defending itself, finding mates, and so on. Darwin called the process of differential survival and reproduction of different types **natural selection.** He chose that term by analogy with the **artificial selection** carried out by animal and plant breeders when they deliberately select some individuals of a preferred type to be the parents of the next generation, while rejecting others.

The relative probability of survival and rate of reproduction of a phenotype or genotype is now called its **Darwinian fitness.** Although geneticists sometimes speak loosely of the fitness of an individual, the concept of fitness really applies to classes of individuals and is a statement about the average survival and reproduction of the individuals in that class. Because of chance events in the life histories of individuals, even two organisms with identical genotypes and identical environments will differ in their survival and reproduction rates. No evolutionary prediction can be made from the unique life history of a single organism. It is the fitness of a genotype on average over all its possessors that matters.

Fitness is a consequence of the relationship between the phenotype of the organism and the environment in which the organism lives, so the same genotype will have different fitnesses in different environments. In part, this is because the exposure to different environments during development will result in different phenotypes for the same genotypes. But even if the phenotype is the same, the success of the organism depends on the environment. Having webbed feet is fine for paddling in water but a positive disadvantage for walking on land, as a few moments spent observing a duck walk will reveal. No genotype is unconditionally superior in fitness to all others in all environments.

Furthermore, the environment is not a fixed situation that is experienced passively by the organism. The environment of an organism is defined by the activities of the organism itself. Dry grass is part of the environment of a junco, so juncos that are more efficient at gathering it may waste less energy in nest building and thus have a higher reproductive fitness. But dry grass is part of a junco's environment *because juncos gather it to make nests.* The rocks among which the grass grows are not part of the junco's environment, although the rocks are physically present there. However, the rocks are part of the environment of thrushes; these birds use the rocks to break open snails. Moreover, the environment that is defined by the life activities of an organism evolves as a result of those activities. The structure of the soil that is in part determinative of the kinds of plants that will grow is altered by the growth of those very plants. Organisms define and alter the environment. Thus, as they evolve in response to the present environment, they find themselves in new environments that are direct consequences of their own evolution. As primitive plants evolved photosynthesis, they changed the earth's atmosphere from one that had had essentially no free oxygen and a high concentration of carbon dioxide to the atmosphere that we know today, which contains 21 percent oxygen and only 0.03 percent carbon dioxide. Plants that evolve today must do so in an environment created by the evolution of their own ancestors. Environment is both the cause and the result of the evolution of organisms. Organisms are both the cause and the result of changes in the environment. The human hand is at the same time the organ of human labor and the evolutionary product of that labor.

Darwinian or reproductive fitness is not to be confused with "physical fitness" in the everyday sense of the term, although they may be related. No matter how strong, healthy, and mentally alert the possessor of a genotype may be, that genotype has a fitness of zero if, for some reason, the possessor is sterile. Thus, such statements as "the unfit are outreproducing the fit so the species may become extinct" are meaningless. By definition, the unfit cannot outreproduce the fit, although some aspect of the phenotype of the more fit may be disadvantageous for some purpose. The fitness of a genotype is a consequence of all the phenotypic effects of the genes involved. Thus, an allele that doubles the fecundity of its carriers but at the same time reduces the average lifetime of its possessors by 10 percent will be more fit than its alternatives, despite its life-shortening property. The most common example is parental care. An adult bird that expends a great deal of its energy gathering food for its young will have a lower probability of survival than one that keeps all the food for itself. But a totally selfish bird will leave no offspring because its young cannot fend for themselves. As a consequence, parental care is favored by natural selection.

Two Forms of the Struggle for Existence

Darwin saw the "struggle for existence" as having two quite different forms, with different consequences for fitness. In one form, the organism "struggles" with the environment directly. Darwin's example was the plant that is struggling for water at the edge of a desert. The fitness of a genotype in such a case does not depend on whether it is frequent or rare in the population, because

fitness is not mediated through the interactions of individuals but is a direct consequence of the individual's physical relationship to the external environment. Fitness is then **frequency-independent.** Other examples are the differential probability that a seedling will survive freezing temperatures and the differential ability of ground squirrels to dig burrows for nesting.

The other form of struggle is between organisms competing for a resource in short supply or otherwise interacting so that their relative abundances determine fitness. If prey are in short supply but are relatively easy to catch, then the faster of two predators will have the higher fitness. Suppose that the faster lion (F) always wins out when it competes directly with the slower lion (S). When F types are rare, the S type that may have been in competition with an F for a particular prey animal will have another opportunity for a successful hunt because it will usually compete against another S the second time. However, if F types become more numerous, then the S type will have to compete directly with more F lions and will have fewer and fewer chances to acquire prey. Thus, its fitness will decrease more and more, compared with that of F, and will finally approach zero as S becomes rare. A more complex example is mimicry in butterflies. Some species of butterflies (such as the brightly colored orange and black monarchs) are distasteful to birds, who learn, after a few trials, to avoid attacking them (Figure 24-14). It is then advantageous for a palatable species (such as the viceroy butterfly) to evolve to look like the distasteful one, because birds will avoid the tasty mimics as well as the distasteful models. But as the frequency of the mimics increases, birds will increasingly have the experience that butterflies with this morphology are, in fact, good to eat. They will no longer avoid them, and the mimics will lose their fitness advantage. These are examples of **frequency-dependent fitness.**

For reasons of mathematical convenience, most models that have been developed to explain the mechanisms of natural selection have been constructed with frequency-independent fitness. In actual fact, however, a very large number of selective processes (perhaps most) are frequency-dependent. The kinetics of the evolutionary process depend on the exact form of frequency dependence, and, for that reason alone, it is difficult to make any generalizations. The result of *positive* frequency dependence (such as the competing predators, where fitness increases with increasing frequency) is quite different from the case of *negative* frequency dependence (such as the butterfly mimics, where fitness of a genotype declines with increasing frequency). For the sake of simplicity and to illustrate the main qualitative features of selection, we deal only with models of frequency-independent selection in this chapter, but convenience should not be confused with reality.

Figure 24-14. A blue jay eating a monarch butterfly *(a)*, which induces vomiting in the jay (b). Because of this experience, the jay later will refuse to eat a viceroy butterfly that is similar in appearance to the monarch, although jays that have never tried monarchs will eat the viceroys with no ill effects. (Photographs courtesy of Lincoln Brower.)

Measuring Fitness Differences

For the most part, the differential fitness of different genotypes can be most easily measured when the genotypes differ at many loci. In very few cases (except for laboratory mutants, horticultural varieties, and major metabolic disorders) does the effect of an allelic substitution at a single locus make enough difference to the phenotype to be reflected in measurable fitness differences. Figure 24-15 shows the probability of survival from egg to adult — that is, the **viability** — of a number of second-chromosome homozygotes of *Drosophila pseudoobscura* at three different temperatures. As is generally the case, the fitness (in this case, a component of the total fitness, viability) is different in different environments. A few homozygotes are lethal or nearly so at all three temperatures, whereas a few have consistently high viability. Most genotypes, however, are not consistent in viability between temperatures, and no genotype is unconditionally the most fit at all temperatures. The fitness of these chromosomal homozygotes was not measured in competition with each other; all are measured against a common standard, so we do not know whether they are frequency-dependent. An example of frequency-dependent fitness is shown in the estimates for inversion homozygotes and heterozygotes of *Drosophila pseudoobscura* in Table 24-12.

Examples of clear-cut fitness differences associated with single-gene substitutions are the many "inborn

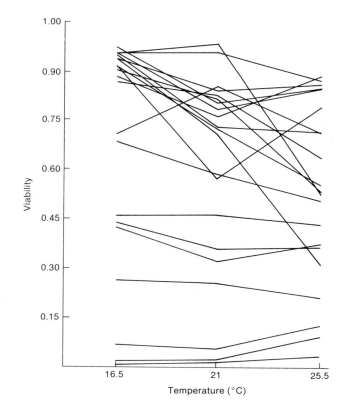

Figure 24-15. Viabilities of various chromosomal homozygotes of *Drosophila pseudoobscura* at three different temperatures.

TABLE 24-12. Comparison of fitnesses for inversion homozygotes and heterozygotes in laboratory populations of *Drosophila pseudoobscura* when measured in different competitive combinations

	Homozygotes			Heterozygotes		
Experiment	ST/ST	AR/AR	CH/CH	ST/AR	ST/CH	AR/CH
ST and AR alone	0.8	0.5	—	1.0	—	—
ST and CH alone	0.8	—	0.4	—	1.0	—
AR and CH alone	—	0.86	0.48	—	—	1.0
ST, AR, and CH together	0.83	0.15	0.36	1.0	0.77	0.62

errors of metabolism," where a recessive allele interferes with a metabolic pathway and causes lethality of the homozygotes. Two examples in humans are phenylketonuria (where tissue degeneration is the result of the accumulation of a toxic intermediate in the pathway of tyrosine metabolism) and Wilson's disease (where death results from copper poisoning because the pathway of copper detoxification is blocked). A case that illustrates the relation of fitness to environment is sickle-cell anemia. An allelic substitution at the structural-gene locus for the β chain of hemoglobin results in substitution of valine for the normal glutamic acid at chain position 6. The abnormal hemoglobin crystallizes at low oxygen pressure, and the red cells deform and hemolyze. Homozygotes $Hb^S Hb^S$ have a severe anemia, and survivorship is low. Heterozygotes have a mild anemia and under ordinary circumstances exhibit the same or only slightly lower fitness than normal homozygotes $Hb^A Hb^A$. However, in regions of Africa with a high incidence of falciparum malaria, heterozygotes ($Hb^A Hb^S$) have a *higher* fitness than normal homozygotes because the presence of some sickling hemoglobin apparently protects them from the malaria. Where malaria is absent, as in North America, the fitness advantage of heterozygosity is lost.

In contrast to chromosomal homozygotes, it has not been possible to measure fitness differences for most single-locus polymorphisms with the exception of some metabolic diseases. The evidence for differential net fitness for different ABO or MN blood types is shaky at best. The extensive enzyme polymorphism present in all sexually reproducing species has for the most part not been connected with measurable fitness differences, although in *Drosophila* clear-cut differences in the fitness of different genotypes have been demonstrated in the laboratory for a few loci such as α-amylase and alcohol dehydrogenase.

How Selection Works

Suppose that a population is mating at random with respect to a given locus with two alleles and that the population is so large that (for the moment) we can ignore

inbreeding. Just after the eggs have been fertilized, the zygotes will be in Hardy-Weinberg equilibrium:

Genotype	AA	Aa	aa
Frequency	p^2	$2pq$	q^2

and $p^2 + 2pq + q^2 = (p + q)^2 = 1.0$, where p is the frequency of A.

Further suppose that the three genotypes have probabilities of survival to adulthood (viabilities) of $W_{AA} : W_{Aa} : W_{aa}$. Then among the progeny once they have reached adulthood, the frequencies will be

Genotype	AA	Aa	aa
Frequency	$p^2 W_{AA}$	$2pq W_{Aa}$	$q^2 W_{aa}$

These adjusted frequencies do not add up to unity since the W's are all fractions smaller than 1. However, we can readjust them so that they do, without changing their relation to each other, by dividing each frequency by the sum of the frequencies after selection ($\overline{W}$):

$$\overline{W} = p^2 W_{AA} + 2pq W_{Aa} + q^2 W_{aa}$$

So defined, $\overline{W}$ is called the **mean fitness** of the population because it is, indeed, the mean of the fitnesses of all individuals in the population. After this adjustment, we have

Genotype	AA	Aa	aa
Frequency	$p^2 \dfrac{W_{AA}}{\overline{W}}$	$2pq \dfrac{W_{Aa}}{\overline{W}}$	$q^2 \dfrac{W_{aa}}{\overline{W}}$

We can now determine the frequency p' of the allele A in the next generation by summing up genes:

$$p' = AA + (\tfrac{1}{2})Aa = p^2 \frac{W_{AA}}{\overline{W}} + \frac{pq W_{Aa}}{\overline{W}} = p \frac{p W_{AA} + q W_{Aa}}{\overline{W}}$$

Finally, we note that the expression $pW_{AA} + qW_{Aa}$ is the mean fitness of A alleles, because A alleles occur with frequency p in homozygotes with another A and, in that condition, have a fitness of W_{AA}, whereas they occur with frequency q in heterozygotes with a and have a fitness of W_{Aa}. Using $\overline{W}_A$ to denote $pW_{AA} + qW_{Aa}$ yields the final new gene frequency:

$$p' = p\,\frac{\overline{W}_A}{\overline{W}}$$

In other words, after one generation of selection, the new value of the frequency of A is equal to the old value (p) multiplied by the ratio of the average fitness of A alleles to the fitness of the whole population. If the fitness of A alleles is greater than the average fitness of all alleles, then $\overline{W}_A/\overline{W}$ is greater than unity and p' is larger than p. Thus, the allele A increases in the population. Conversely, if $\overline{W}_A/\overline{W}$ is less than unity, A decreases. But the mean fitness of the population ($\overline{W}$) is the average fitness of the A alleles and of the a alleles. So if $\overline{W}_A$ is greater than the mean fitness of the population, it must be greater than $\overline{W}_a$, the mean fitness of a alleles.

> **Message** The allele with the higher average fitness increases in the population.

It should be noted that the fitnesses W_{AA}, W_{Aa}, and W_{aa} may be expressed as absolute probabilities of survival and absolute reproduction rates, or they may all be rescaled relative to one of the fitnesses, which is given the standard value of 1.0. This rescaling has absolutely no effect on the formula for p' because it cancels out in the numerator and denominator.

> **Message** The course of selection depends only on relative fitnesses.

An increase in the allele with the higher fitness means that the average fitness of the population as a whole increases, so that selection can also be described as a process that *increases mean fitness*. This rule is strictly true only for frequency-independent genotypic fitnesses, but it is close enough to a general rule to be used as a fruitful generalization. This maximization of fitness does not necessarily lead to any optimal property for the species as a whole because fitnesses are only defined relative to each other within a population. It is relative (not absolute) fitness that is increased by selection. The population does not necessarily become larger or grow faster, nor is it less likely to become extinct.

The Rate of Change of Gene Frequency

An alternative way to look at the process of selection is to solve for the *change* in allelic frequency in one generation:

$$\Delta p = p' - p = \frac{p\overline{W}_A}{\overline{W}} - p = \frac{p(\overline{W}_A - \overline{W})}{\overline{W}}$$

But $\overline{W}$, the mean fitness of the population, is the average of the allelic fitnesses $\overline{W}_A$ and $\overline{W}_a$, so that

$$\overline{W} = p\overline{W}_A + q\overline{W}_a$$

Substituting this expression for $\overline{W}$ in the formula for Δp and remembering that $q = 1 - p$, we obtain (after some algebraic manipulation)

$$\Delta p = \frac{pq(\overline{W}_A - \overline{W}_a)}{\overline{W}}$$

which is the general expression for a change in allelic frequency as a result of selection. This general expression is particularly illuminating. It says that Δp will be positive (A will increase) if the mean fitness of A alleles is greater than the mean fitness of a alleles, as we saw before. But it also shows that the speed of the change depends not only on the difference in fitness between the alleles but also on the factor pq, which is proportional to the frequency of heterozygotes ($2pq$). For a given difference in fitness of alleles, gene frequency will change most rapidly when the alleles A and a are in intermediate frequency, so that pq is large. If p is near 0 or 1 (that is, if A or a is nearly fixed), then pq is nearly 0 and selection will proceed very slowly. Figure 24-16 shows the S-shaped curve that represents the course of selection of a new favorable allele A that has recently entered a population of homozygotes aa. At first, the change in frequency is very small because p is still close to 0. Then it accelerates as A becomes more frequent, but it slows down again as A takes over and a becomes very rare. This is precisely what is expected from a selection process. When most of the population is of one type, there is nothing to select. For evolution by natural selection to occur, there must be genetic variance; the more variance, the faster the process.

An important consequence of the dependence of selection on genetic variance can be seen in a case of artificial selection. In the early part of this century, it became fashionable to advocate a program of **negative**

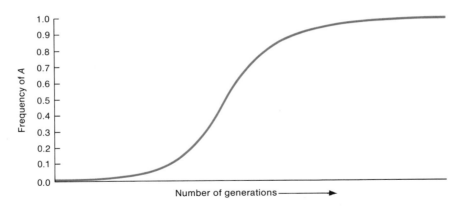

Figure 24-16. The time pattern of increasing frequency of a new favorable allele A that has entered a population of aa homozygotes.

eugenics. It was proposed that individuals with certain undesirable genetic traits (say, metabolic or nervous disorders) should be prevented from having any offspring. By this means, it was thought, the frequency of the trait in the population would be lowered, and the trait could eventually be eradicated. Suppose that such a program is completely efficient, so that every homozygote aa is prevented from reproducing. The fitnesses of the genotypes then are

Genotype	AA	Aa	aa
Fitness	1.0	1.0	0

If p is the frequency of A and q is the frequency of a, then

$$\overline{W}_A = pW_{AA} + qW_{Aa} = p(1) + q(1) = 1.0$$
$$\overline{W}_a = pW_{Aa} + qW_{aa} = p(1) + q(0) = p$$
$$\overline{W} = p^2 W_{AA} + 2pq W_{Aa} + q^2 W_{aa} = p\overline{W}_A + q\overline{W}_a$$
$$= p(1) + q(p) = p(1 + q)$$

so that

$$q' = q\,\frac{\overline{W}_a}{\overline{W}} = q\,\frac{p}{p(1+q)} = \frac{q}{1+q}$$

If we iterate this formula over generations, we obtain

$$q'' = \frac{q'}{1+q'} = \frac{q/(1+q)}{1+q/(1+q)} = \frac{q}{1+2q}$$

$$q''' = \frac{q}{1+3q}$$

$$q^{(n)} = \frac{q}{1+nq} = \frac{1}{n+(1/q)}$$

From this sequence, we can see what the fate of a negative eugenics program would be. A deleterious gene will already be rare in a population. Suppose that $q = \frac{1}{100}$, for example. Then after one generation, the frequency would be $\frac{1}{101}$; after two generations, $\frac{1}{102}$; and so on. It would take 100 generations to reduce the frequency to $\frac{1}{200}$ and another 200 generations to reduce it by half to $\frac{1}{400}$. But a human generation is 25 years, so it would require 2500 years (the time since the founding of the Roman republic) with perfectly efficient selection against the recessive just to reduce the frequency from $\frac{1}{100}$ to $\frac{1}{200}$. The negative eugenics plan clearly is impractical.

Of course, if the heterozygote for the deleterious genes (as, for example, in sickle-cell anemia) could be detected, then all copies of the gene could be removed from the population in a single generation if the heterozygotes were all prevented from having offspring. The only trouble with this suggestion is that every human being is heterozygous for several different deleterious recessive genes, so no one would be allowed to breed. Negative eugenics is no longer seriously proposed by geneticists as a population selection process, although individuals with particular inherited diseases in their families may be counseled against having children to spare the personal pain that would result for both the family and any child who is born defective.

When alternative alleles are not rare, selection can cause quite rapid changes in allelic frequency. Figure 24-17 shows the course of elimination of a malic dehydrogenase allele in a laboratory population of *Drosophila melanogaster*. The fitnesses in this case are

$$W_{AA} = 1.0 \qquad W_{Aa} = 0.75 \qquad W_{aa} = 0.40$$

Of course, the frequency of a is not reduced to 0, and further reduction in frequency will require longer and longer times, as shown in the negative eugenics case.

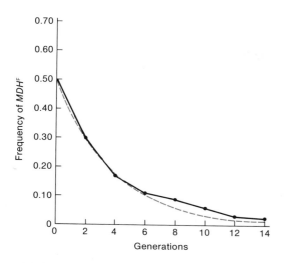

Figure 24-17. The loss of an allele of the malic dehydrogenase locus MDH^F due to selection in a laboratory population of *Drosophila melanogaster*. The colored dashed line shows the theoretical curve of change computed for fitnesses $W_{AA} = 1.0$, $W_{Aa} = 0.75$, and $W_{aa} = 0.4$. (From R. C. Lewontin, *The Genetic Basis of Evolutionary Change.* Copyright © 1974 by Columbia University Press. Data courtesy of E. Berger.)

Message   Unless alternative alleles are present in intermediate frequencies, selection (especially against recessives) is quite slow. Selection is dependent on genetic variation.

Balanced Polymorphism

Let's reexamine the general formula for allelic frequency change (page 697):

$$\Delta p = pq \frac{(\overline{W}_A - \overline{W}_a)}{\overline{W}}$$

Under what conditions will the process stop? When is $\Delta p = 0$? Two immediately obvious answers are when $p = 0$ or when $q = 0$ (that is, when either allele A or allele a, respectively, has been eliminated from the population). One of these events will eventually occur if $\overline{W}_A - \overline{W}_a$ is consistently positive or negative, so that Δp is always positive or negative irrespective of the value of p. The condition for such unidirectional selection is that the heterozygote fitness be somewhere between the fitnesses of the two homozygotes:

$(\overline{W}_A - \overline{W}_a)$ positive: $W_{AA} \geq W_{Aa} \geq W_{aa}$, so A is favored

$(\overline{W}_A - \overline{W}_a)$ negative: $W_{AA} \leq W_{Aa} \leq W_{aa}$, so a is favored

But there is another possibility for $\Delta p = 0$, even when p and q are not 0:

$$\overline{W}_A = \overline{W}_a$$

which can occur if the heterozygote is not intermediate between the homozygotes but has a fitness that is more extreme than either homozygote. Then $\overline{W}_A > \overline{W}_a$ for part of the range of values of p, whereas $\overline{W}_A < \overline{W}_a$ for the rest of the range. Just between those ranges is a value of p (denoted by $\hat{p}$) for which the mean fitnesses of the two alleles are equal. A little algebraic manipulation of

$$\overline{W}_A - \overline{W}_a = 0 = (\hat{p}W_{AA} + \hat{q}W_{Aa}) - (\hat{p}W_{Aa} + \hat{q}W_{aa})$$

gives us the solution for $\hat{p}$:

$$\hat{p} = \frac{(W_{aa} - W_{Aa})}{(W_{aa} - W_{Aa}) + (W_{AA} - W_{Aa})}$$

The equilibrium value is a simple ratio of the differences in fitness between the homozygotes and the heterozygote. As examples, let's consider three cases:

	W_{AA}	W_{Aa}	W_{aa}
Case 1	0.9	1.0	0.8
Case 2	0.8	1.0	0.6
Case 3	0.99	1.0	0.98

The fitness values in all three cases have exactly the same equilibrium value of $\hat{p} = \frac{2}{3}$, although the speed at which the population would reach the equilibrium differs very much in the three cases.

There are, in fact, two qualitatively different possibilities for $\hat{p}$. One possibility is that $\hat{p}$ is an *unstable* equilibrium. There will be no change in frequency if the population has exactly this value of p, but the frequency will move *away* from the equilibrium (toward $p = 0$ or $p = 1$) if the slightest perturbation of frequency occurs. This unstable case will exist when the heterozygote is *lower* in fitness than either homozygote; such a condition is an example of **underdominance.** The alternative possibility is a *stable* equilibrium, or **balanced polymorphism,** in which slight perturbations from the value of $\hat{p}$ will result in a return to $\hat{p}$. The condition for this balance is that the heterozygote be *greater* in fitness than either homozygote—a condition termed **overdominance.**

In nature, the chance that a gene frequency will remain balanced on the knife edge of an unstable equilibrium is negligible, so we should not expect to find naturally occurring polymorphisms in which heterozygotes fit less than homozygotes. On the contrary, the observation of a long-lasting polymorphism in nature might be taken as prima facie evidence of a superior heterozygote.

Unfortunately, life confounds theory. The *Rh* locus (rhesus blood group) in humans has a widespread polymorphism with Rh^+ and Rh^- alleles. In Europeans, the frequency of the Rh^- allele is about 0.4, whereas in Africans it is about 0.2. Thus, this must be a very old human polymorphism, antedating the origin of modern geographical races. But this polymorphism causes a maternal fetal incompatibility when an RH$^-$ mother (homozygous Rh^-/Rh^-) produces an RH$^+$ fetus (heterozygous Rh^-/Rh^+). This incompatability results in hemolytic anemia (from a destruction of red blood cells) and the death of the fetus in a moderate proportion of cases. Thus, there is selection against heterozygotes. This polymorphism is unstable and should have disappeared from the species, yet it exists in most human populations. There have been many hypotheses proposed to explain its apparent stability, but the mystery remains.

In contrast, no fitness difference at all can be demonstrated for many polymorphisms of blood groups (and for the ubiquitous polymorphism of enzymes revealed by electrophoresis). It has been suggested that such polymorphisms are not under selection at all but that

$$W_{AA} = W_{Aa} = W_{aa}$$

This situation of **selective neutrality** would, of course, also satisfy the requirement that $\overline{W}_A = \overline{W}_a$, but instead of a stable equilibrium, it gives rise to a **passive** (or **neutral**) **equilibrium** such that any allele frequency p is as good as any other. This leaves unanswered the problem of how the populations became highly polymorphic in the first place. Nevertheless, it is clear that the presence of a widespread polymorphism is not necessarily evidence for superior heterozygotes. The best case of overdominance for fitness at a single locus remains that of sickle-cell anemia, where the two homozygotes are at a disadvantage relative to the heterozygote for quite different reasons.

The best-studied cases of balanced polymorphism in nature and in the laboratory are the inversion polymorphisms in several species of *Drosophila*. Figure 24-18 shows the course of frequency change for the inversion *ST* against the alternative arrangement *CH* in a laboratory population of *Drosophila pseudoobscura*. The inversions *ST* and *CH* are part of a chromosomal polymorphism in natural populations of this species. The fitnesses estimated for the three genotypes in the laboratory are

$$W_{ST/ST} = 0.89 \qquad W_{ST/CH} = 1.0 \qquad W_{CH/CH} = 0.41$$

Applying the formula for the equilibrium value $\hat{p}$ we obtain $\hat{p} = 0.85$, which agrees quite well with the observations in Figure 24-18.

Multiple Adaptive Peaks

We must avoid taking an overly simplified view of the consequences of selection. At the level of the gene — or even at the level of the partial phenotype — the outcome of selection for a trait in a given environment is not

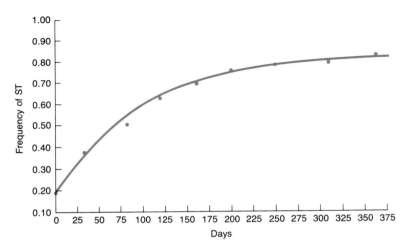

Figure 24-18. Changes in the frequency of the inversion *Standard (ST)* in competition with *Chiricahua (CH)* in a laboratory population of *Drosophila pseudoobscura*.

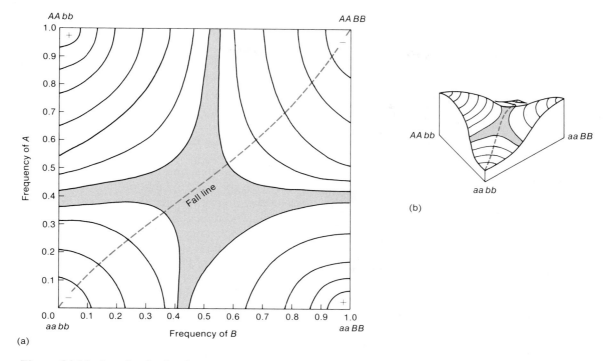

Figure 24-19. An adaptive landscape with two adaptive peaks (+), two adaptive valleys (−), and a topographic saddle in the center of the landscape. The topographic lines are lines of equal mean fitness. If the genetic composition of a population always changes in such a way as to move the population "uphill" in the landscape, then the final composition will depend on where the population began with respect to the fall (dashed) line. (a) Topographic map of the adaptive landscape. (b) A perspective sketch of the surface shown in the map.

unique. Selection to alter a trait (say, to increase size) may be successful in a number of ways. In 1952, F. Robertson and E. Reeve successfully selected to change wing size in *Drosophila* in two different populations. However, in one case the *number* of cells in the wing changed, whereas in the other case the *size* of the wing cells changed. Two different genotypes had been selected, both causing a change in wing size. The initial state of the population at the outset of selection determined which of these selections occurred.

The way in which the same selection can lead to different outcomes can most easily be illustrated by a simple hypothetical case. Suppose that the variation of two loci (there will usually be many more) influences a character and that (in a particular environment) intermediate phenotypes have the highest fitness. (For example, newborn babies have a higher chance of surviving birth if they are neither too big nor too small.) If the alleles act in a simple way in influencing the phenotype, then the three genetic constitutions *Aa Bb*, *AA bb*, and *aa BB* will produce a high fitness because they will all be intermediate in phenotype. On the other hand, very low fitness will characterize the double homozygotes *AA BB* and *aabb*. What will the result of selection be? We can predict the result by using the mean fitness $\overline{W}$ of a popu-

lation. As previously discussed, selection acts in most simple cases to increase $\overline{W}$. Therefore, if we calculate $\overline{W}$ for every possible combination of gene frequencies at the two loci, we can determine which combinations yield high values of $\overline{W}$. Then we should be able to predict the course of selection by following a curve of increasing $\overline{W}$.

The surface of mean fitness for all possible combinations of allelic frequency is called an **adaptive surface,** or an **adaptive landscape** (Figure 24-19). The figure is like a topographic map. The frequency of allele *A* at one locus is plotted on one axis, and the frequency of allele *B* at the other locus is plotted on the other axis. The height above the plane (represented by topographic lines) is the value of $\overline{W}$ that the population would have for a particular combination of frequencies of *A* and *B*. According to the rule of increasing fitness, selection should carry the population from a low-fitness "valley" to a high-fitness "peak." However, Figure 24-19 shows that there are two adaptive peaks, corresponding to a fixed population of *AA bb* and a fixed population of *aa BB*, with an adaptive valley between them. Which peak the population will ascend—and therefore what its final genetic composition will be—depends on whether the initial genetic composition of the population is on one side or the other of the dashed "fall line" shown in the figure.

The existence of multiple adaptive peaks for a selective process means that some differences between species are the result of history and not of environmental differences. For example, African rhinoceroses have two horns, and Indian rhinoceroses have one (Figure 24-20). We need not invent a special story to explain why it is better to have two horns on the African plains and one in India. It is much more plausible that the trait of having horns was selected, but that two long, slender horns and one short, stout horn are simply alternative adaptive features, and that historical accident differentiated the species. Explanations of adaptations by natural selection do not require that every difference between species be differentially adaptive.

It is important to note that nothing in the theory of selection requires that the different adaptive peaks be of the same height. The kinetics of selection is such that $\overline{W}$ increases, not that it necessarily reaches the highest possible peak in the field of gene frequencies. Suppose, for example, that a population is near the peak *AA bb* in Figure 24-19 and that this peak is lower than the *aa BB* peak. Selection alone cannot carry the population to *aa BB* because that would require a temporary decrease in $\overline{W}$ as the population descended the *AA bb* slope, crossed the saddle, and ascended the other slope. Thus, the force of selection is myopic. It drives the population to a *local* maximum of $\overline{W}$ in the field of gene frequencies —not to a *global* one.

Artificial Selection

In contrast to the difficulties of finding simple, well-behaved cases in nature that exemplify the simple formulas of natural selection, there is a vast record of the effectiveness of artificial selection in changing populations phenotypically. These changes have been produced by laboratory selection experiments and by selection of animals and plants in agriculture (as examples for increased milk production in cows and for rust resistance in wheat). No analysis of these experiments in terms of allelic frequencies is possible because individual loci have not been identified and followed. Nevertheless, it is clear that genetic changes have occurred in the populations and that some analysis of selected populations has been carried out according to the methods described in Chapter 23. Figure 24-21 shows, as an example, the large changes in average bristle number achieved in a selection experiment with *Drosophila melanogaster*. Figure 24-22 shows the changes in the number of eggs laid per chicken as a consequence of 30 years of selection.

For characters of high heritability, the usual method of selection is **truncation selection.** The individuals in a given generation are pooled (irrespective of their families), a sample is measured, and only those individuals

(a)

(b)

Figure 24-20. Differences in horn morphology in two geographically separated species of rhinoceroses: (a) the African rhinoceros. (b) The Indian rhinoceros. (Part a from Leonard Lee Rue, copyright © Tom Stack & Associates; part b copyright © Tom Stack & Associates.)

Message Under identical conditions of natural selection, two populations may arrive at two different genetic compositions as a direct result of natural selection.

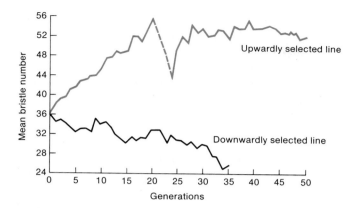

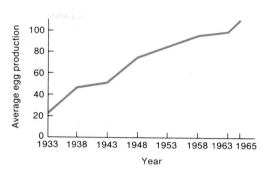

Figure 24-21. Changes in average bristle number obtained in two laboratory populations of *Drosophila melanogaster* through artificial selection for high bristle number in one population and for low bristle number in the other. The dashed segment in the curve for the upwardly selected line indicates a period of five generations during which no selections were performed. (From K. Mather and B. J. Harrison, "The Manifold Effects of Selection," *Heredity*, 3, 1949, 1–52.)

Figure 24-22. Changes in average egg production in a chicken population selected for its increase in egg-laying rate over a period of 30 years. (From I. M. Lerner and W. J. Libby, *Heredity, Evolution, and Society*, 2d. ed. Copyright © 1976 by W. H. Freeman and Company. Data courtesy of D. C. Lowry.)

above (or below) a given phenotypic value (the truncation point) are chosen as parents for the next generation. This phenotypic value may be a fixed value over successive generations; then selection is by **constant truncation.** More commonly, a fixed percent of the population representing the highest (or lowest) value of the selected character is chosen; then selection is by **proportional truncation.** With constant truncation, the intensity of selection decreases with time, as more and more of the population exceeds the fixed truncation point. With proportional truncation, the intensity of selection is constant, but the truncation point moves upward as the population distribution moves. Figure 24-23 illustrates these two types of truncation.

No matter which scheme of selection is used, narrow heritability will eventually decline as selected alleles be-

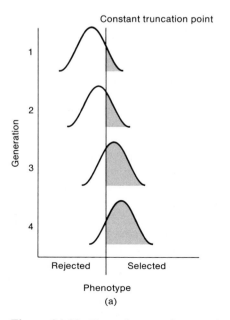

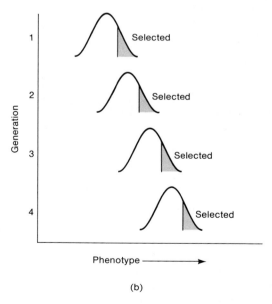

Figure 24-23. Two schemes of truncation selection for a continuously varying trait. (a) Constant truncation. (b) Proportional truncation.

come fixed; further progress becomes more and more difficult and eventually ceases when h^2 goes to 0. For characters of initially low heritability, some type of family selection is used (see page 669).

A common experience in artificial selection programs is that as the population becomes more and more extreme, its viability and fertility decrease. As a result, eventually no further progress under selection is possible, despite the presence of genetic variance for the character, because the selected individuals do not reproduce. The loss of fitness may be a direct phenotypic effect of the genes for the selected character, in which case nothing much can be done to improve the population further. Often, however, the loss of fitness is tied to linked sterility genes that are carried along with the selected loci. In such cases, a number of generations without selection allow recombinants to be formed, and selection can then be continued, as in the upwardly selected line in Figure 24-21.

We must be very careful in our interpretation of long-term agricultural selection programs. In the real world of agriculture, changes in cultivation methods, machinery, fertilizers, insecticides, herbicides, and so on are occurring along with the production of genetically improved varieties. Increases in average yields are consequences of all of these changes. For example, the average yield of corn in the United States increased from 40 bushels to 80 bushels per acre between 1940 and 1970. But experiments comparing old and new varieties of corn in common environments show that only about one-half of this increase is a direct result of new corn varieties (the other one-half being a result of improved farming techniques). Furthermore, the new varieties are more superior to the old ones at the high densities of modern planting for which they were selected.

Random Events

If a population is finite in size (as all populations are) and if a given pair of parents have only a small number of offspring, then even in the absence of all selective forces, the frequency of a gene will not be exactly reproduced in the next generation because of sampling error. If in a population of 1000 individuals, the frequency of a is 0.5 in one generation, then it may by chance be 0.493 or 0.505 in the next generation because of the chance production of a few more or a few less progeny of each genotype. In the second generation, there is another sampling error based on the new gene frequency, so the frequency of a may go from 0.505 to 0.511 or back to 0.498. This process of random fluctuation continues generation after generation, with no force pushing the frequency back to its initial state because the population

has no "genetic memory" of its state many generations ago. Each generation is an independent event. The final result of this random change in allelic frequency is that the population eventually drifts to $p = 1$ or $p = 0$. After this point, no further change is possible; the population has become homozygous. A different population, isolated from the first, also undergoes this **random genetic drift,** but it may become homozygous for allele A, whereas the first population has become homozygous for allele a. As time goes on, isolated populations diverge from each other, each losing heterozygosity. The variation originally present *within* populations now appears as variation *among* populations.

One form of genetic drift occurs when a small group breaks off from a larger population to found a new colony. This "acute drift," called the **founder effect,** results from a single generation of sampling, followed by several generations during which the population remains small. The founder effect is probably responsible for the virtually complete lack of blood group B in American Indians, whose ancestors arrived in very small numbers across the Bering Strait during the end of the last Ice Age, about 10,000 years ago. More recent examples are seen in religious isolates like the Dunkers and Old Order Amish of North America. These sects were founded by small numbers of migrants from their much larger congregations in central Europe. They have since remained nearly completely closed to immigration from the surrounding American population. As a result, their blood-group gene frequencies are quite different from those in the surrounding populations, both in Europe and in North America.

The process of genetic drift should sound familiar. It is, in fact, another way of looking at the inbreeding effect in small populations discussed earlier. Whether regarded as inbreeding or as random sampling of genes, the effect is the same. Populations do not exactly reproduce their genetic constitutions; there is a random component of gene-frequency change.

One result of random sampling is that most new mutations, even if they are not selected against, never succeed in entering the population. Suppose that a single individual is heterozygous for a new mutation. There is some chance that the individual in question will have no offspring at all. Even if it has one offspring, there is a chance of $\frac{1}{2}$ that the new mutation will not be transmitted. If the individual has two offspring, the probability that neither offspring will carry the new mutation is $\frac{1}{4}$, and so on. Suppose that the new mutation is successfully transmitted to an offspring. Then the lottery is repeated in the next generation, and again the allele may be lost. In fact, if a population is of size N, the chance that a new mutation is eventually lost by chance is $(2N - 1)/2N$. But if the new mutation is not lost, then the only thing

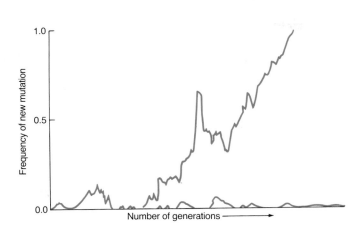

Figure 24-24. The appearance, loss, and eventual incorporation of new mutations during the life of a population. If random genetic drift does not cause the loss of a new mutation, then it must eventually cause the entire population to become homozygous for the mutation (in the absence of selection). (After J. Crow and M. Kimura, *An Introduction to the Population Genetics Theory.* Copyright © 1970 by Harper & Row.)

that can happen to it in a finite population is that eventually it will sweep through the population and become fixed! This event has the probability of $1/2N$. In the absence of selection, then, the history of a population looks like Figure 24-24. For some period of time, it is homozygous; then a new mutation appears. In most cases, the new mutant allele will be lost immediately or very soon after it appears. Occasionally, however, a new mutant allele drifts through the population, and the population becomes homozygous for the new allele. The process then begins again.

Even a new mutation that is slightly favorable selectively will usually be lost in the first few generations after it appears in the population, a victim of genetic drift. If a new mutation has a selective advantage of S in the heterozygote in which it appears, then the chance is only $2S$ that the mutation will ever succeed in taking over the population. So a mutation that is 1 percent better in fitness than the standard allele in the population will be lost 98 percent of the time by genetic drift.

> **Message** New mutations can become established in a population even though they are not favored by natural selection simply by a process of random genetic drift. Even new favorable mutations are often lost.

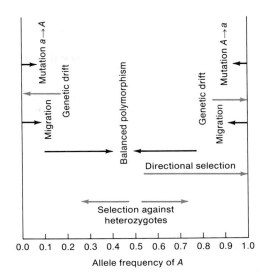

Figure 24-25. The effects on gene frequency of various forces of evolution.

A Synthesis of Forces

The genetic variation within and between populations is a result of the interplay of the various evolutionary forces (Figure 24-25). Generally, as Table 24-13 shows, forces that increase or maintain variation within populations prevent the differentiation of populations from each other, whereas the divergence of populations is a result of forces that make each population homozygous. Thus, random drift (inbreeding) produces homozygosity while causing different populations to diverge. This divergence and homozygosity are counteracted by the constant flux of mutation and the migration between localities, which introduce variation into the populations again and tend to make them more like each other.

TABLE 24-13. How the forces of evolution increase (+) or decrease (−) the variation within and among populations

Force	Variation within populations	Variation among populations
Inbreeding or genetic drift	−	+
Mutation	+	−
Migration	+	−
Selection		
directional	−	+/−
balancing	+	−
incompatible	−	+

The effects of selection are more variable. **Directional selection** pushes a population toward homozygosity, rejecting most new mutations as they are introduced but occasionally (if the mutation is advantageous) spreading a new allele through the population to create a new homozygous state. Whether or not such directional selection promotes differentiation of populations depends on the environment and on chance events. Two populations living in very similar environments may be kept genetically similar by directional selection, but, if there are environmental differences, selection may direct the populations toward different compositions. Advantageous new mutations are rare, so that a given mutation may occur in one population but not (for a very long time) in others. Directional selection will then, temporarily, cause divergence of the population in which the mutation has appeared. Given enough time, of course, the mutation should be incorporated in all populations, especially if there is migration between them. But populations and species do not last forever, so directional selection operating on rare mutants may in fact be a cause of much divergence.

A particular case of interest, especially in human populations, is the interaction between mutation and directional selection in a very large population. New deleterious mutations are constantly arising spontaneously or as the result of the action of mutagens. These mutations may be completely recessive or partly dominant. Selection removes them from the population, but there will be an equilibrium between their appearance and removal. If we let q be the frequency of the deleterious allele a and $p = 1 - q$ be the frequency of the normal allele, then the change in allelic frequency due to the mutation rate μ is

$$\Delta q_{mut} = \mu p$$

A simple way to express the fitnesses for a recessive deleterious gene is $W_{AA} = W_{Aa} = 1.0$ and $W_{aa} = 1 - s$, where s is the loss of fitness in homozygotes. We now can substitute these fitnesses in our general expression for allelic frequency change (see page 697) and obtain

$$\Delta q_{sel} = -\frac{pq(sq)}{1 - sq^2}$$

Equilibrium means that the increase in the allelic frequency due to mutation must exactly balance the decrease in the allelic frequency due to selection so that

$$\Delta \hat{q}_{mut} + \Delta \hat{q}_{sel} = 0$$

Remembering that $\hat{q}$ at equilibrium will be quite small, so that $1 - s\hat{q}^2 \simeq 1$, and substituting the terms for $\Delta \hat{q}_{mut}$ and $\Delta \hat{q}_{sel}$ in the formula above we have

$$\mu\hat{p} - \frac{s\hat{p}\hat{q}^2}{1 - s\hat{q}^2} \simeq \mu\hat{p} - s\hat{p}\hat{q}^2 = 0$$

or

$$\hat{q} = \sqrt{\frac{\mu}{s}}$$

So, for example, a recessive lethal ($s = 1$) mutating at the rate of $\mu = 10^{-6}$ will have an equilibrium frequency of 10^{-3}. Indeed, if we knew that a gene was a recessive lethal and had no heterozygous effects, we could estimate its mutation rate as the square of the allelic frequency. The basis for such calculations must be firm, however. Sickle-cell anemia was once thought to be a recessive lethal with no heterozygous effects, which led to an estimated mutation rate in Africa of 0.1 for this locus!

If we let the fitnesses be $W_{AA} = 1.0$, $W_{Aa} = 1 - hs$, and $W_{aa} = 1 - s$ for a partly dominant gene, then a similar calculation gives us

$$\hat{q} = \frac{\mu}{hs}$$

where h is the degree of dominance of the deleterious allele. Thus, if $\mu = 10^{-6}$ and the lethal is not totally recessive but has a 5 percent deleterious effect in heterozygotes ($s = 1.0$, $h = 0.05$), then

$$\hat{q} = \frac{10^{-6}}{5 \times 10^{-2}} = 2 \times 10^{-5}$$

which is smaller by two orders of magnitude than the equilibrium frequency for the purely recessive case. In general, then, we can expect deleterious, completely recessive genes to have much higher frequencies than partly dominant genes.

Selection favoring heterozygotes (balancing selection) will, for the most part, maintain more or less similar polymorphisms in different populations. However, again, if the environments are different enough between them, then the populations will show some divergence. The opposite of balancing selection is selection against heterozygotes, which produces unstable equilibria. Such selection will cause homozygosity and divergence among populations.

The Exploration of Adaptive Peaks

Random and selective forces should not be thought of as simple antagonists. Random drift may counteract the force of selection, but it can enhance it as well. The outcome of the evolutionary process is a result of the

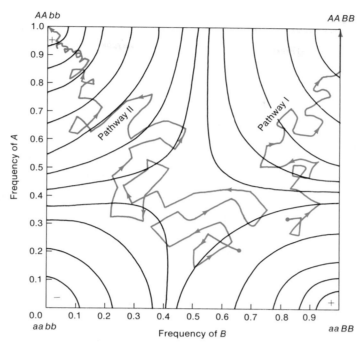

Figure 24-26. Selection and random drift can interact to produce different changes of gene frequency in an adaptive landscape. Without random drift, both populations would have moved toward *aaBB* as a result of selection alone.

simultaneous operation of these two forces. Figure 24-26 illustrates these possibilities. Note that there are multiple adaptive peaks in this landscape. Because of random drift, a population under selection does not ascend an adaptive peak smoothly. Instead, it takes an erratic course in the field of gene frequencies, like an oxygen-starved mountain climber. Pathway I shows a population history where adaptation has failed. The random fluctuations of gene frequency were sufficiently great that the population by chance became fixed at an unfit genotype. In any population, some proportion of loci are fixed at a selectively unfavorable allele because the intensity of selection is insufficient to overcome the random drift to fixation. Very great skepticism should be maintained toward naive theories about evolution that assume that populations always or nearly always reach an optimal constitution under selection. The existence of multiple adaptive peaks and the random fixation of less fit alleles are integral features of the evolutionary process. Natural selection cannot be relied on to produce the best of all possible worlds.

Pathway II in Figure 24-26, on the other hand, shows how random drift may improve adaptation. The population was originally in the sphere of influence of the lower adaptive peak; however, by random fluctuation in gene frequency, its composition passed over the adaptive saddle, and the population was captured by the higher, steeper adaptive peak. This passage from a lower to a higher adaptive stable state could never have occurred by selection in an infinite population, because by selection alone, $\overline{W}$ could never decrease temporarily in order to cross from one slope to another.

Message The interaction of selection and random drift makes possible the attainment of higher fitness states than are obtainable when natural selection is operating alone.

The Origin of Species

By a species (at least in sexually reproducing organisms), we mean a group of individuals that are biologically capable of interbreeding yet isolated genetically from other groups. **Speciation**—the origin of a new species —is the origin of a group of individuals capable of making a living in a new way and at the same time acquiring some barrier to genetic exchange with the species from which it arose. The genetic differentiation of a population by inbreeding, genetic drift, and differential selection is always threatened by the reintroduction of genes from other groups by migration. The reduction of gene migration to a very low value is therefore a prerequisite for speciation.

Generally, this reduction is the result of the geographical isolation of the population as a consequence of chance historical events: a few long-distance migrants may reach a new island; a part of the mainland may be cut off by a rise in sea level; an insect vector that formerly passed pollen from one population to another may become locally extinct; the grassy plain that connected the feeding grounds of two grazers may, by a slight change in rainfall pattern, become a desert. Once the population is isolated physically, the processes of genetic differentiation will go on unimpeded until the genetic constitution of the isolated population is so different from its parental group that there is real difficulty in interbreeding.

If migration is reestablished before this critical period, speciation will not occur and the divergent populations will once again converge. This has already happened in the human species, in which the genetic differentiation of geographical populations never has proceeded beyond some superficial physical traits and a mixed differentiation of frequencies at polymorphic loci. On the other hand, if populations are very divergent before they come back in contact with each other, hybrid offspring will have genotypes with such low fitness that they do not survive or are sterile. At this stage of differentiation, there is a definite selective advantage for the newly forming species to avoid mating with each other and so avoid the wastage of gametes. New (secondary) barriers to interbreeding may then be selected, thereby completing the speciation process. Evolutionists are not agreed on how frequent the selection of secondary barriers has been in speciation.

Beyond this generalized sketch of speciation, remarkably little can be said with certainty. Because species do not interbreed, it is difficult to analyze their differences genetically. A great deal must be made of the few cases in which some hybrid offspring can be produced in the laboratory or the garden. The methods of electrophoresis, immunology, and protein sequencing have made it possible to describe the differences in the proteins of species, but there are very few cases of species that have just recently separated. Thus, we do not know how much of the genome—or what part of it—is involved in the first divergence, nor do we know whether that divergence is often a consequence of selection in opposite directions or of random drift. A detailed genetic analysis of the process of speciation remains one of the most important tasks for population genetics.

Summary

■ Charles Darwin revolutionized the study of biology when he constructed a theory of evolution based on the principles that variation existed within populations, that variation was heritable, and that the phenotype of the individuals in the population changed through generations because of natural selection. These basic tenets of evolution—put forward by Darwin in *The Origin of Species* in 1859, prior to any knowledge of Mendelian genetics—have required only minor modification since that time. The study of changes within a population, or population genetics, relates the heritable changes in populations or organisms to the underlying individual processes of inheritance and development. Population genetics is the study of inherited variation and its modification in time and space.

Identifiable inherited variation within a population can be studied by observing morphological differences among individuals, examining the differences in specific amino-acid sequences of proteins, or even by examining, most recently, the differences in nucleotide sequences within the DNA. These kinds of observations have led to the conclusion that there is considerable polymorphism at many loci within a population. A measure of this variation is the amount of heterozygosity in a population. Population studies have shown that, in general, the genetic differences among individuals within human races are much greater than the average differences among races.

The ultimate source of all variation is mutation. However, within a population, the quantitative frequency of specific genotypes can be changed by recombination, immigration of genes, continued mutational events, and chance.

One property of Mendelian segregation is that random mating results in an equilibrium distribution of genotypes after one generation. However, inbreeding is one process that converts genetic variation within a population into differences between populations by making each separate population homozygous for a randomly chosen allele. On the other hand, for most populations, a balance is reached for any given environment between inbreeding, mutation from one allele to another, and immigration.

"Directed" changes of allelic frequencies within a population occur through the natural selection of a favored genotype. In many cases, such changes lead to homozygosity at a particular locus. On the other hand, the heterozygote may be more suited to a given environment than either of the homozygotes, leading to a balanced polymorphism.

Environmental selection of specific genotypes is rarely this simple, however. More often than not, phenotypes are determined by several interacting genes, and alleles at these different loci will be selected for at different rates. Furthermore, closely linked loci, unrelated to the phenotype in question, may have specific alleles carried along during the selection process. In general, genetic variation is the result of the interaction of evolutionary forces. For instance, a recessive, deleterious mutant may never be totally eliminated from a popula-

tion, because mutation will continue to resupply it to the population. Immigration can also reintroduce the undesirable allele into the population. And, indeed, a deleterious allele may, under environmental conditions of which we are unaware (including the remaining genetic makeup of the individual), be selected for.

Unless alternative alleles are in intermediate frequencies, selection (especially against recessives) is very slow, requiring many generations. In many populations, especially those of small size, new mutations can become established even though they are not favored by natural selection, simply by a process of random genetic drift. Such slow changes in different allelic frequencies throughout the genome can lead to the eventual formation of new races and new species.

■　■　■

Solved Problems

1. About 70 percent of all white North Americans can taste the chemical phenylthiocarbamide, and the remainder cannot. The ability to taste is determined by the dominant allele T, and the inability to taste is determined by the recessive allele t. If the population is assumed to be in Hardy-Weinberg equilibrium, what are the genotypic and allelic frequencies in this population?

Solution

Since 70 percent are tasters (TT) 30 percent must be nontasters (tt). This homozygous recessive frequency is equal to q^2, so to obtain q, we simply take the square root of 30:

$$q = \sqrt{0.30} = 0.55$$

Since $p + q = 1$, we can write

$$p = 1 - q = 1 - 0.55 = 0.45$$

Now we can calculate

$$p^2 = 0.45^2 = 0.20 \ (TT)$$
$$2pq = 2 \times 0.45 \times 0.55 = 0.50 \ (Tt)$$
$$q^2 = 0.3 \ (tt)$$

2. In a large natural population of *Mimulus guttatus* one leaf was sampled from each of a large number of plants. The leaves were crushed and run on an electrophoretic gel. The gel was then stained for a specific enzyme X. Six different banding patterns were observed as shown in the following frequencies:

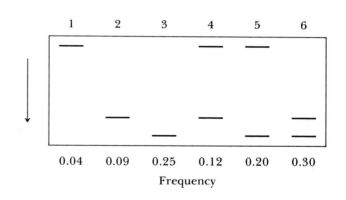

a. Assuming that these patterns are produced by a single locus, propose a genetic explanation for the six types.

b. How can you test your idea?

c. What are the allelic frequencies in this population?

d. Is the population in Hardy-Weinberg equilibrium?

Solution

a. Inspection of the gel reveals that there are only three band positions: we will call them slow, intermediate, and fast. Furthermore, any individual can show either one band or two. The simplest explanation of this is that there are three alleles of one locus (let's call them A^S, A^I, and A^F) and that the individuals with two bands are heterozygotes. Hence, $1 = SS$, $2 = II$, $3 = FF$, $4 = SI$, $5 = SF$, and $6 = IF$.

b. The hypothesis can be tested by making controlled crosses. For example, from a self of type 5, we can predict $\frac{1}{4}SS$, $\frac{1}{2}SF$, and $\frac{1}{4}FF$.

c. The frequencies can be calculated by a simple extension of the two-allele formulas. Hence

$$f(S) = 0.04 + \tfrac{1}{2}(0.12) + \tfrac{1}{2}(0.20) = 0.20$$
$$f(I) = 0.09 + \tfrac{1}{2}(0.12) + \tfrac{1}{2}(0.30) = 0.30$$
$$f(F) = 0.25 + \tfrac{1}{2}(0.20) + \tfrac{1}{2}(0.30) = 0.50$$

d. The Hardy-Weinberg genotypic frequencies are

$$(p + q + r)^2 = p^2 + q^2 + r^2 + 2pq + 2pr + 2qr$$
$$= 0.04 + 0.09 + 0.25 + 0.12 + 0.20 + 0.30$$

which are precisely the observed frequencies, so it appears that the population is in equilibrium.

3. In a large experimental *Drosophila* population, the fitness of a recessive phenotype is calculated to be 0.90 and the mutation rate to the recessive allele is 5×10^{-5}. If the population is allowed to come to equilibrium, what allelic frequencies can be predicted?

Solution

Here mutation and selection are working in opposite directions, so an equilibrium is predicted. Such an equilibrium is described by the formula

$$\hat{q} = \sqrt{\frac{\mu}{s}}$$

In the present question, $\mu = 5 \times 10^{-5}$ and $s = 1 - W = 1 - 0.9 = 0.1$. Hence

$$\hat{q} = \sqrt{\frac{5 \times 10^{-5}}{10^{-1}}} = 2.2 \times 10^{-2} = 0.022$$
$$\hat{p} = 1 - 0.022 = 0.978$$

4. In a population, q is currently 0.2. If the fitness of $aa = 0$, what will q be after 95 generations?

Solution

The formula needed here is

$$q^n = \frac{1}{n + (1/q)}$$

Thus

$$q^n = \frac{1}{95 + (1/0.2)} = \frac{1}{100} = 0.01$$

Problems

1. What are the forces that can change the frequency of an allele in a population?

2. In a population of mice, there are two alleles of the A locus, ($A1$ and $A2$). Tests showed that in this population there are 384 mice of genotype $A1A1$, 210 of $A1A2$, and 260 of $A2A2$. What are the frequencies of the two alleles in the population?

3. In a randomly mating laboratory population of *Drosophila*, 4 percent of the flies have black bodies (black is the autosomal recessive b) and 96 percent have brown bodies (the normal color B). If this population is assumed to be in Hardy-Weinberg equilibrium, what are the allelic frequencies of B and b and the genotypic frequencies of BB and Bb?

4. In a popualtion, the $D \rightarrow d$ mutation rate is 4×10^{-6}. If $p = 0.8$ today, what will p be after 50,000 generations?

5. You are studying protein polymorphism in a natural population of a certain species of a sexually reproducing haploid organism. You isolate many strains from various parts of the test area and run extracts from each strain on electrophoretic gels. You stain the gels with a reagent specific for enzyme X and find that in the population there is a total of, say, five electrophoretic variants of enzyme X. You speculate that these variants represent various alleles of the structural gene for enzyme X.

a. How could you demonstrate that this is so, both genetically and biochemically? (You can make crosses, make diploids, run gels, test enzyme activities, test amino-acid sequences, and so on.) Outline the steps and conclusions precisely.

b. Name at least one other possible way of generating the different electrophoretic variants, and say how you would distinguish this possibility from the one described here.

6. A study made in 1958 in the mining town of Ashibetsu in the Hokkaido province of Japan revealed the frequencies of MN blood-type genotypes (for individuals and for married couples) shown in the following table:

Genotype	Number of individuals or couples
Individuals	
$L^M L^M$	406
$L^M L^N$	744
$L^N L^N$	322
Total	1482
Couples	
$L^M L^M \times L^M L^M$	58
$L^M L^M \times L^M L^N$	202
$L^M L^N \times L^M L^N$	190
$L^M L^M \times L^N L^N$	88
$L^M L^N \times L^N L^N$	162
$L^N L^N \times L^N L^N$	41
Total	741

a. Show whether or not the population is in Hardy-Weinberg equilibrium with respect to the MN blood types.

b. Show whether mating is random with respect to MN blood types.

(Problem 2 is from J. Kuspira and G. W. Walker, *Genetics: Questions and Problems.* Copyright © 1973 by McGraw-Hill.)

7. For a sex-linked character in a species in which the male is the heterogametic sex, suppose that the allelic frequencies at a locus are different for males and females.

a. Letting the frequencies of A be 0.8 in males and 0.2 in females, show what happens to the allelic frequencies in successive generations in the two sexes.

b. Try to develop a general expression for the difference in the allelic frequencies in males (p) and females (P) in the nth generation, given that the initial values were p_0 and P_0, respectively.

8. Consider the populations that have the genotypes shown in the following table.

Population	AA	Aa	aa
1	1.0	0.0	0.0
2	0.0	1.0	0.0
3	0.0	0.0	1.0
4	0.50	0.25	0.25
5	0.25	0.25	0.50
6	0.25	0.50	0.25
7	0.33	0.33	0.33
8	0.04	0.32	0.64
9	0.64	0.32	0.04
10	0.986049	0.013902	0.000049

a. Which of the populations are in Hardy-Weinberg equilibrium?

b. What are p and q in each population?

c. In population 10, it is discovered that the $A \rightarrow a$ mutation rate is 5×10^{-6} and that reverse mutation is negligible. What must be the fitness of the aa phenotype?

d. In population 6, the a allele is deleterious; furthermore, the A allele is incompletely dominant, so that AA is perfectly fit, Aa has a fitness of 0.8, and aa has a fitness of 0.6. If there is no mutation, what will p and q be in the next generation?

9. Colorblindness results from a sex-linked recessive allele. One in every 10 males is colorblind.

a. What proportion of all women are colorblind?

b. By what factor is colorblindness more common in men (or, how many colorblind men are there for each colorblind woman)?

c. In what proportion of marriages would colorblindness affect one-half of the children of each sex?

d. In what proportion of marriages would all children be normal?

e. In a population that is not in equilibrium, the frequency of the allele for colorblindness is 0.2 in women and 0.6 in men. After one generation of random mating, what proportion of the female progeny will be colorblind? What proportion of the male progeny?

f. What will the allelic frequencies be in the male and in the female progeny in (e)?

(Problem 9 courtesy of Clayton Person.)

10. In a wild population of beetles of species X, you notice that there is a 3 : 1 ratio of shiny to dull wing covers. Does this prove that *shiny* is dominant? (Assume that the two states are caused by the alleles of one gene.) If not, what does it prove? How would you elucidate the situation?

11. It seems clear that most new mutations are deleterious. Why?

12. Most mutations are recessive to wild-type. Of those rare mutations that are dominant in *Drosophila*, for example, the majority turn out either to be chromosomal aberrations or to be inseparable from chromosomal aberrations. Can you explain why wild-type is usually dominant?

13. Ten percent of the males of a large and randomly mating population are colorblind. A representative group of 1000 from this population migrates to a South Pacific island, where there are already 1000 inhabitants and where 30 percent of the males are colorblind. Assuming that Hardy-Weinberg equilibrium applies throughout (in the two original populations before emigration and in the mixed population immediately following immigration), what fraction of males and females can be expected to be colorblind in the generation immediately following the arrival of the immigrants?

14. Using pedigree diagrams, find the probability of homozygosity by descent of the offspring of: a. parent-offspring matings; b. first-cousin matings; c. aunt-nephew or uncle-niece matings.

15. In a survey of Indian tribes in Arizona and New Mexico, it was found that albinos were completely absent or very rare in most groups (there is one albino per 20,000 North American Caucasians). However, in three Indian populations, albino frequencies are exceptionally high: one per 277 Indians in Arizona; one per 140 Jemez Indians in New Mexico; and one per 247 Zuni Indians in New Mexico. All three of these populations are culturally but not linguistically related. What possible factors might explain the high incidence of albinos in these three tribes?

16. In an animal population, 20 percent of the individuals are AA, 60 percent are Aa, and 20 percent are aa. What are the allelic frequencies? In this population, mating is always with *like phenotype* but is random within phenotype. What genotypic and allelic frequencies will prevail in the next generation? Such *assortative mating* is common in animal populations. Another type of assortative mating is that which occurs only between *unlike* phenotypes: answer the above question with this restriction imposed. What will the end result be after many generations of mating of both types?

17. In *Drosophila,* a stock isolated from nature has an average of 36 abdominal bristles. By selectively breeding only those flies with more bristles, the mean is raised to 56 in 20 generations! What is the source of this genetic flexibility? The 56-bristle stock is very infertile, so selection is relaxed for several generations and the bristle number drops to about 45. Why doesn't it drop back to 36? When selection is reapplied, 56 bristles are soon attained, but this time the stock is *not* sterile. How can this situation arise?

18. The fitnesses of three genotypes are $W_{AA} = 0.9$; $W_{Aa} = 1.0$; and $W_{aa} = 0.7$.

    a. If the population starts at the allelic frequency $p = 0.5$, what is the value of p in the next generation?

    b. What is the predicted equilibrium allelic frequency?

19. AA and Aa individuals are equally fertile. If 0.1 percent of the population is aa, what selection pressure exists against aa if the $A \rightarrow a$ mutation rate is 10^{-5}?

20. Gene B is a deleterious autosomal dominant. The frequency of affected individuals is 4.0×10^{-6}. The reproductive capacity of these individuals is about 30 percent that of normal individuals. Estimate μ, the rate at which b mutates to its deleterious allele B.

21. Of 31 children born of father-daughter matings, six died in infancy, 12 were very abnormal and died in childhood, and 13 were normal. From this information, calculate roughly how many recessive lethal genes we have, on average, in our human genomes. For example, if the answer is 1, then a daughter would stand a 50 percent chance of carrying the gene, and the probability of the union producing a lethal combination would be $\frac{1}{2} \times \frac{1}{4} = \frac{1}{8}$. (So obviously, 1 is not the answer.) Consider also the possibility of undetected fatalities in utero in such matings. How would they affect your result?

22. Let's define the **total selection cost** to a population of deleterious recessive genes as the loss of fitness per individual affected (s) multiplied by the frequency of affected individuals (q^2). Thus

$$\text{genetic cost} = sq^2$$

    a. Suppose that a population is at equilibrium between mutation and selection for a deleterious recessive gene, where $s = 0.5$ and $\mu = 10^{-5}$. What is the equilibrium frequency of the gene? What is the genetic cost?

    b. Suppose that we start irradiating individual members of the population, so that the mutation rate doubles. What is the new equilibrium frequency of the gene? What is the new genetic cost?

    c. If we do not change the mutation rate but we lower the selection intensity to $s = 0.3$ instead, what happens to the equilibrium frequency and the genetic cost?

Further Reading

Students interested in pursuing genetics further should start reading original research articles in scientific journals. Some important journals are *Cell, Current Genetics, Evolution, Gene, Genetic Research, Genetics, Heredity, Human Genetics, Journal of Medical Genetics, Journal of Molecular Biology, Molecular and General Genetics, Mutation Research, Nature, Plasmid, Proceedings of the National Academy of Sciences of the United States of America,* and *Science.* Useful review articles may be found in *Annual Review of Genetics, Advances in Genetics,* and *Trends in Genetics.*

Some particularly useful references, mostly general reviews, are listed below under the chapters to which they relate.

Chapter 1

Clausen, J., D. D. Keck, and W. W. Hiesey. 1940. *Experimental Studies on the Nature of Species,* Vol. 1: *The Effect of Varied Environments on Western North American Plants.* Carnegie Institute of Washington, Publ. No. 520, 1–452. This publication and the following one by the same authors are the classic studies of norms of reaction of plants from natural populations.

Clausen, J., D. D. Keck, and W. W. Hiesey. 1958. *Experimental Studies on the Nature of Species,* Vol 3: *Environmental Responses of Climatic Races of* Achillea. Carnegie Institute of Washington, Publ. No. 581, 1–129.

Milunsky, A., and G. J. Annas, eds. 1975. *Genetics and the Law.* New York: Plenum Press. Interesting accounts of the ramifications of genetics in the lives of individuals.

Schmalhausen, I. I. 1949. *Factors of Evolution: The Theory of Stabilizing Selection.* Philadelphia: Blakiston. The most general discussion of the relation of genotype and environment in the formation of phenotypic variation.

Chapter 2

Carlson, E. A. 1966. *The Gene: A Critical History.* Philadelphia: W. B. Saunders. A readable history of genetics.

Grant, V. 1975. *Genetics of Flowering Plants.* New York: Columbia University Press. One of the few texts on this subject.

Harpstead, D. 1971. "High-Lysine Corn." *Scientific American* (August). An account of the breeding of lines with increased amounts of normally limiting amino acids.

Hutt, F. B. 1964. *Animal Genetics.* New York: Ronald Press. A standard text on the subject with many interesting examples.

Jennings, P. R. 1976. "The Amplification of Agricultural Production." *Scientific American* (September). A discussion of genetics and the green revolution.

Olby, R. C. 1966. *Origins of Mendelism.* London: Constable. An enjoyable account of Mendel's work and the intellectual climate of his time.

Singer, S. 1978. *Human Genetics.* New York: W. H. Freeman and Co. A short and readable treatment of the subject.

Stern, C., and E. R. Sherwood. 1966. *The Origin of Genetics. A Mendel Source Book.* New York: W. H. Freeman and Co. A short collection of important early papers, including Mendel's papers and correspondence.

Sturtevant, A. H. 1965. *A History of Genetics.* New York: Harper & Row. Another useful historical text.

Todd, N. B. 1977. "Cats and Commerce." *Scientific American* (November). Includes some genetics of domestic cat coat colors and the use of this information to study cat migration throughout history.

Chapter 3

McLeish, J., and B. Snoad. 1958. *Looking at Chromosomes.* New York: Macmillan. A short classic book consisting of many superb photos of mitosis and meiosis.

Rick, C. M. 1978. "The Tomato." *Scientific American* (August). Includes an account of tomato genes and chromosomes and their role in breeding.

Stern, C. 1973. *Principles of Human Genetics,* 3d ed. New York: W. H. Freeman and Co. A standard text including many examples of the inheritance of human traits.

von Wettstein, D. et al. 1984. "The Synaptonemal Complex in Genetic Segregation." *Annual Review of Genetics* 18: 331–414. An up-to-date review of the structure and function of the complex.

Chapter 4

Bodmer, W. F., and L. L. Cavalli-Sforza. 1976. *Genetics, Evolution and Man.* New York: W. H. Freeman and Co. A very readable, well-illustrated book, including a clear account of HLA genetics.

Carlson, E. A. 1984. *Human Genetics.* Lexington: Heath. A readable and up-to-date account of human inheritance.

Day, P. R. 1974. *Genetics of Host-Parasite Interaction.* New York: W. H. Freeman and Co. A short technical account of the subject, which is of great importance to agriculture. It includes good examples of gene interaction.

Griffiths, A. J. F., and F. R. Ganders. 1984. *Wildflower Genetics.* Vancouver: Flight Press. A field guide to plant variation in natural populations and its genetic basis, including examples relevant to this chapter.

Searle, A. G. 1968. *Comparative Genetics of Coat Color in Mammals.* New York: Academic Press. A classic treatment of the subject with many examples relevant to this and other chapters

Silvers, W. K. 1979. *The Coat Colors of Mice.* New York: Springer-Verlag. A standard handbook on the subject, including many examples of gene interaction.

Chapter 5

O'Brien, S. J., ed. 1984. *Genetic Maps.* Cold Spring Harbor Press. A recent compendium of the detailed maps of 80 well-analyzed organisms.

Peters, J. A., ed. 1959. *Classic Papers in Genetics.* Englewood Cliffs, N.J.: Prentice-Hall. A collection of important papers in the history of genetics.

White, R., et al. 1985. "Construction of Linkage Maps with DNA Markers for Human Chromosomes." *Nature* 313: 101–104. An extension of the techniques of this chapter to DNA markers.

Chapter 6

Evans, D. A., W. R. Sharp, P. V. Ammirato, and Y. Yamada, eds. 1983. *Handbook of Plant Cell Culture,* Vol. 1. New York: Macmillan, Practical details of the techniques of propagation and breeding plants by cell, tissue, and organ culture.

Finchman, J. R. S., P. R. Day, and A. Radford. 1979. *Fungal Genetics,* 3d ed. London: Blackwell. A large, standard technical work. Good for tetrad analysis.

Kemp, R. 1970. *Cell Division and Heredity.* London: Edward Arnold. A short, clear introduction to genetics. Good for map functions and tetrad analysis.

Murray, A. W., and J. W. Szostak. 1983. "Construction of Artificial Chromosomes in Yeast." *Nature* 305:189–193. The first creation of new chromosomes by splicing together known telomere, centromere, replicator, and then DNA fragments by recombinant DNA technology. Includes tetrad analysis of markers on the new chromosomes.

Puck, T. T., and F-T. Kao. 1982. "Somatic Cell Genetics and Its Application to Medicine." *Annual Review of Genetics* 16: 225–272. A technical but readable review.

Ruddle, F. H., and R. S. Kucherlapati. 1974. "Hybrid Cells and Human Genes." *Scientific American* (July). A popular account of the use of cell hybridization in mapping human genes.

Stahl, F. W. 1969. *The Mechanics of Inheritance,* 2d ed. Englewood Cliffs, N.J.: Prentice-Hall. A short introduction to genetics, including some advanced material presented with a novel approach.

Chapter 7

Induced Mutations—A Tool in Plant Research. 1981. Vienna: International Atomic Energy Agency. A collection of papers by eminent workers in agricultural genetics illustrating the practical uses of mutations in plant breeding.

Lawrence, C. W. 1971. *Cellular Radiobiology.* London: Edward Arnold. A short standard text.

Neuffer, M. G., L. Jones, and M. S. Zuber. 1968. *The Mutants of Maize.* Madison: Crop Science Society of America. A color catalog of the many, and often bizarre, mutants used by corn geneticists.

Schull, W. J., et al. 1981. "Genetic Effect of the Atomic Bombs: A Reappraisal." *Science* 213: 1220–1227. A summary of all the indicators of potential genetic effects of the Hiroshima and Nagasaki explosions, concluding that "In no instance is there a statistically significant effect of parental exposure; but for all indicators the observed effect is in the direction suggested by the hypothesis that genetic damage resulted from the exposure."

Chapters 8 and 9

Epstein, C. J., et al. 1983. "Recent Developments in Prenatal Diagnosis of Genetic Diseases and Birth Defects." *Annual Review of Genetics* 17: 49–83. Includes amniocentesis.

Feldman, M. G., and E. R. Sears. 1981. "The Wild Gene Resources of Wheat." *Scientific American* (January). A general discussion of the genomes of wheat and its relatives, and how new genes can be introduced.

Friedmann, T. 1971. "Prenatal Diagnosis of Genetic Disease." *Scientific American* (November). An early article on amniocentesis and its uses.

Fuchs, F. 1980. "Genetic Amniocentesis." *Scientific American* (August).

German, J., ed. 1974. *Chromosomes and Cancer*. New York: Wiley. A large technical work, but readable, describing the relation of chromosome changes and cancer.

Hassold, T. J., and P. A. Jacobs, 1984. "Trisomy in Man." *Annual Review of Genetics* 18: 69–98. A comprehensive summary of trisomy, including a discussion of the maternal age effect.

Hulse, J. H., and D. Spurgeon. 1974. "Triticale." *Scientific American* (August). An account of the development and possible benefits of this wheat-rye amphidiploid.

Lawrence, W. J. C. 1968. *Plant Breeding*. London: Edward Arnold (*Studies in Biology*, No. 12). A short introduction to the subject.

Mangelsdorf, P. C. 1986. "The Origins of Corn." *Scientific American* (August).

Maniatis, T. E., et al. 1980. "The Molecular Genetics of Human Hemoglobins." *Annual Review of Genetics* 14: 145–178. A useful summary, which could be profitably read at this point in the course or after reading the material on molecular genetics.

Patterson, D. 1987 "The Causes of Down Syndrome." *Scientific American* (August).

Shepherd, J. F. 1982. "The Regeneration of Potato Plants from Protoplasts." *Scientific American* (May). A review by one of the leaders in this field.

Swanson, C. P., T. Mertz, and W. J. Young. 1967. *Cytogenetics*. Englewood Cliffs, N.J.: Prentice-Hall.

Chapter 10

Adelberg, E. A. 1966. *Papers on Bacterial Genetics*. Boston: Little, Brown.

Hayes, W. 1968. *The Genetics of Bacteria and Their Viruses*, 2d ed. New York: Wiley. The standard and classic text, written by a pioneer in the subject.

Lewin, B. 1977. *Gene Expression*, Vol. 1: *Bacterial Genomes*. New York: Wiley. An excellent set of volumes, all of which are relevant to various sections of this text.

Lewin, B. 1977. *Gene Expression*, Vol. 3: *Plasmids and Phages*. New York: Wiley.

Stent, G. S., and R. Calendar. 1978. *Molecular Genetics*, 2d ed. New York: W. H. Freeman and Co. A lucidly written account of the development of our present understanding of the subject, based mainly on experiments in bacteria and phage.

Chapter 11

Dickerson, R. E. 1983. "The DNA Helix and How It is Read." *Scientific American* (December). An article with some beautiful color models of DNA structures.

Kornberg, A. 1980. *DNA Replication*. New York: W. H. Freeman and Co. The definitive technical treatment of the subject, based on genetic and chemical analysis.

Wang, J. C. 1982. "DNA Topoisomerases." *Scientific American* (July). Diagrams different topological forms of DNA.

Watson, J. D. 1968. *The Double Helix*. New York: Atheneum. An enjoyable personal account of Watson and Crick's discovery, including the human dramas involved.

Chapter 12

Benzer, S. 1962. "The Fine Structure of the Gene." *Scientific American* (January). A popular version of the author's pioneer experiments.

Felsenfeld, G. 1985. "DNA." *Scientific American* (October).

Radman, M., and R. Wagner. 1988. "The High Fidelity of DNA Duplication." *Scientific American* (August).

Watson, J. D. 1976. *The Molecular Biology of the Gene*, 3d ed. Menlo Park, Calif.: Benjamin/Cummings. A superb development of the subject, written in a highly readable style and well illustrated.

Yanofsky, C. 1967. "Gene Structure and Protein Structure." *Scientific American* (May). This article gives the details of colinearity at the molecular level.

Chapter 13

Crick, F. H. C. 1962. "The Genetic Code." *Scientific American* (October). This article and the following one are popular accounts of code-cracking experiments.

Crick, F. H. C. 1966. "The Genetic Code: III." *Scientific American* (October).

Darnell, J. E., Jr. 1985. "RNA." *Scientific American* (October).

Doolittle, R. F. 1985. "Proteins." *Scientific American* (October).

Lake, J. A. 1981. "The Ribosome." *Scientific American* (August). Three-dimensional model of the ribosome.

Lane, C. 1976. "Rabbit Haemoglobin from Frog Eggs." *Scientific American* (August). This article describes experiments illustrating the universality of the genetic system.

Lawn, R. M., and G. A. Vehar. 1986. "The Molecular Genetics of Hemophilia." *Scientific American* (March).

Miller, O. L. 1973. "The Visualization of Genes in Action." *Scientific American* (March). A discussion of electron microscopy of transcription and translation.

Moore, P. B. 1976. "Neutron-Scattering Studies of the Ribosome." *Scientific American* (October). This article gives the details of ribosome substructure.

Nirenberg, M. W. 1963. "The Genetic Code: II." *Scientific American* (March). Another account of early code-cracking experiments.

Radman, M., and R. Wagner. 1988. "The High Fidelity of DNA Duplication." *Scientific American* (August).

Rich, A., and S. H. Kim. 1978. "The Three-Dimensional Structure of Transfer RNA." *Scientific American* (January). A presentation of the experimental evidence behind the structure described in this chapter.

Weinberg, R. A. 1985. "The Molecules of Life." *Scientific American* (October).

Chapter 14

Britten, R. J., and D. E. Kohne. 1970. "Repeated Segments of DNA." *Scientific American* (April).

Brown, S. W. 1966. "Heterochromatin." *Science* 151: 417–425. A nice review of the classic cytological observations.

Chambon, P. 1981. "Split Genes." *Scientific American* (May). The discovery of intervening sequences is described.

Davidson, E., and R. Britten. 1973. "Organization, Transcription and Regulation in the Animal Genome." *Quarterly Review of Biology* 48: 565–613. The analysis of renaturation kinetics of DNA fragments provides insights into chromosome structure.

Dupraw, E. J. 1970. *DNA and Chromosomes.* New York: Holt, Rinehart & Winston. A useful book on chromosome substructure, containing excellent photographs by the author.

Edgar, R. S., and R. H. Epstein. 1965. "The Genetics of a Bacterial Virus." *Scientific American* (February). In analyzing a large number of mutations, the clustering of functionally related genes became apparent.

Glover, D. M. et al. 1975. "Characterization of 6 Cloned DNAs from *Drosophila melanogaster* Including One That Contains the Genes for rRNA." *Cell* 5: 149–157. A technical paper showing how the isolation of random segments of *Drosophila* DNA in *E. coli* can be used to study chromosome structure.

Hayashi, S., et al. 1980. "Hybridization of tRNAs of *Drosophila melanogaster.*" *Chromosoma* 76: 65–84. A technical report showing how specific genes can be located cytologically by hybridization of labeled RNA to chromosomes in situ.

Judd, B. H., et al. 1972. "The Anatomy and Function of a Segment of the X Chromosome of *Drosophila melanogaster.*" *Genetics* 71: 139–156. A beautiful example of genetic analysis leading to a significant insight into chromosome structure.

Kavenoff, R., and B. H. Zimm. 1973. "Chromosome-Sized DNA Molecules from *Drosophila.*" *Chromosoma* 41. An elegant experiment involving the study of chromosome aberrations with physicochemical techniques.

Kornberg, R. D. 1974. "Chromatin Structure: A Repeating Unit of Histones and DNA." *Science* 184: 868–871. A technical review of the evidence for nucleosomes.

Kornberg, R. D., and A. Klug. 1981. "The Nucleosome." *Scientific American* (February). Detailed analysis of chromosome structure.

Lewin, B. 1977. *Gene Expression,* Vol. 2: *Eucaryotic Chromosomes.* New York: Wiley.

Lucchesi, J. C. 1973. "Dosage Compensation in *Drosophila.*" *Annual Review of Genetics* 7: 225–237. A technical survey of the phenomenon and its possible mechanisms.

Lyon, M. L. 1962. "Sex Chromatin and Gene Action in the Mammalian X Chromosome." *American Journal of Human Genetics* 14: 135–148. The first proposal that, in human females, one X is inactive, with supporting evidence of mosaic expression of sex-linked mutations.

Weinberg, R. A. 1985. "The Molecules of Life." *Scientific American* (October).

Chapter 15

Britten, R. J., and D. Kohne. 1968. "Repeated Sequences in DNA." *Science* 161: 529–540. One of the important summaries of the theoretical basis for distinguishing DNAs by renaturation.

Broda, P. 1979. *Plasmids.* New York: W. H. Freeman and Co. One of the few technical books on the subject.

Brown, D. D. 1973. "The Isolation of Genes." *Scientific American* (August). Illustrates the power of focusing molecular techniques on one specific locus with special properties.

Cohen, S. 1975. "The Manipulation of Genes." *Scientific American* (July). A summary of recombinant DNA techniques by one of the main innovators.

Fiddes, J. C. 1977. "The Nucleotide Sequence of a Viral DNA." *Scientific American* (December). This is a review of a landmark, the DNA sequence of an entire virus genome with an unexpected discovery.

Gilbert, W., and L. Villa-Komaroff. 1980. "Useful Proteins from Recombinant Bacteria." *Scientific American* (April). A description of the method of DNA sequencing used most extensively. Also discusses the potential application of recombinant DNA techniques.

Itakura, K., et al. 1977. "Expression in *E. coli* of a Chemically Synthesized Gene for the Hormone Somatostatin." *Science* 198. A technical paper well worth reading for its historical significance. It represents the start of bioengineering — using DNA manipulation to modify cells to produce a medically useful human protein.

Khorana, H. G., et al. 1972. "Studies on Polynucleotides. CIII. Total Synthesis of the Structural Gene for an Alanine Transfer Ribonucleic Acid from Yeast." *Journal of Molecular Biology* 72: 209–217. A classic technical paper.

Mertens, T. R. 1975. *Human Genetics: Readings on the Implications of Genetic Engineering.* New York: Wiley. A collection of popular articles.

Moses, P. B., and N.-H. Chua. 1988. "Light Switches for Plant Genes." *Scientific American* (April).

Murray, A. W., and J. W. Szostak. 1987. "Artificial Chromosomes." *Scientific American* (November).

Nathans, D., and H. O. Smith. 1975. "Restriction Endonucleases in the Analysis and Restructuring of DNA Molecules." *Annual Review of Biochemistry* 44: 273–293. A technical review of restriction enzymes by two pioneers in the field.

Sinsheimer, R. L. 1977. "Recombinant DNA." *Annual Review of Biochemistry* 46. A provocative article by a leading molecular biologist who has expressed concern about potential hazards of DNA manipulation.

Varmus, H. 1987. "Reverse Transcription." *Scientific American* (September).

Watson, J. D., J. Tooze, and D. T. Kurtz. 1983. *Recombinant DNA: A Short Course.* New York: W. H. Freeman and Co. An excellent introduction to the theory and applications of recombinant DNA.

White, R., and J. M. Lalouel. 1988. "Chromosome Mapping with DNA Markers." *Scientific American* (February).

Chapter 16

Herskowitz, I. 1973. "Control of Gene Expression in Bacteriophage Lambda." *Annual Review of Genetics* 7: 289–324. A technical review of the complex and well-analyzed regulation of lambda genes.

Jacob, F., and J. Monod. 1961. "Genetic Regulatory Mechanisms in the Synthesis of Proteins." *Journal of Molecular Biology* 3: 318–356. A classic paper setting forth the elements of an operon and the experimental evidence.

Maniatis, T., S. Goodbourn, and J. A. Fischer. 1987. "Regulation of Inducible and Tissue-Specific Gene Expression." *Science* 236: 1237–1244. A current review of regulatory elements in eukaryotes.

Maniatis, T., and M. Ptashne. 1976. "A DNA Operator-Repressor System." *Scientific American* (January). This article discusses the molecular structures of the components of the *lac* operon.

Miller, J. H., and W. S. Reznikoff, eds. 1978. *The Operon.* Cold Spring Harbor Laboratory. A valuable set of reviews on gene regulation in bacteria.

Moses, P. B., and N.-H. Chua. 1988. "Light Switches for Plant Genes." *Scientific American* (April).

Ptashne, M., and W. Gilbert. 1970. "Genetic Repressors." *Scientific American* (June). The exciting story of how repressors were identified and purified, thereby confirming the predictions of Jacob and Monod.

Steitz, J. A. 1988. "Snurps." *Scientific American* (June).

Weber, I. T., D. B. McKay, and T. A. Steitz. 1982. Two Helix DNA Binding Motif of CAP Found in *lac* Repressor and *gal* Repressor. *Nucleic Acids Research* 10: 5085–5102. Discussion of models for repressor-operator recognition.

Chapter 17

Auerbach, C. 1976. *Mutation Research.* London: Chapman & Hall. Standard text by a pioneer researcher.

Cairns, J. 1978. *Cancer: Science and Society.* New York: W. H. Freeman and Co. A fascinating discussion of the origins of human cancers, and of the role played by the environment.

Croce, C. M., and H. Koprowski. 1978. "The Genetics of Human Cancer." *Scientific American* (February). An excellent popular account.

Devoret, R. 1979. "Bacterial Tests for Potential Carcinogens." *Scientific American* (August). This article describes the use of mutation tests to screen for carcinogens and includes some details of DNA reactions.

Drake, J. W. 1970. *The Molecular Basis of Mutation.* San Francisco: Holden-Day. One of the few standard texts on the subject.

Miller, J. H. 1983. "Mutational Specificity in Bacteria." *Annual Review of Genetics* 17: 215–238. A discussion of the specificity of mutagens in bacteria.

Walker, G. C. 1984. "Mutagenesis and Inducible Responses to Deoxyribonucleic Acid Damage in *Escherichia coli*," *Microbiological Reviews* 48: 60–93. A review of mutagenesis and repair, with an excellent discussion of the SOS system.

Chapter 18

Alverts, B., D. Bray, J. Lewis, M. Raff, K. Roberts, and J. D. Watson. 1983. *Molecular Biology of the Cell.* New York: Garland. See pages 240 to 250 for an excellent description of recombination at the molecular level, with nice figures.

Stahl, F. W. 1979. *Genetic Recombination: Thinking About It in Phage and Fungi.* New York: W. H. Freeman and Co. A rather technical short book on recombination models.

Whitehouse, H. L. K. 1973. *Towards an Understanding of the Mechanism of Heredity,* 3d ed. London: Edward Arnold. An excellent general introduction to genetics stressing the historical approach and the pivotal experiments. It includes a good section on recombination models.

Chapter 19

Baltimore, D. 1985. "Retroviruses and Retrotransposons: The Role of Reverse Transcription in Shaping the Eukaryotic Genome." *Cell* 40: 481–482. A short review of recent work on RNA intermediates in transposition.

Bukhari, A. I., J. A. Shapiro, and S. L. Adhya, eds. 1977. *DNA Insertion Elements, Plasmids and Episomes.* Cold Spring Harbor Laboratory. An excellent large collection of short summary papers involving lower and higher life forms.

Cohen, S. N., and J. A. Shapiro. 1980. "Transposable Genetic Elements." *Scientific American* (February). A popular account stressing bacteria and phages.

Fedoroff, N. V. 1984. "Transposable Genetic Elements in Maize." *Scientific American* (June). An account of the early experiments in maize, with recent results on the molecular basis of transposition.

Shapiro, J. A., ed. 1983. *Mobile Genetic Elements.* New York: Academic Press. A comprehensive set of reviews covering the latest developments on transposition in later prokaryotes and eukaryotes.

Chapter 20

Borst, P., and L. A. Grivell. 1978. "The Mitochondrial Genome of Yeast." *Cell* 15: 705–723. A nontechnical review of the molecular biology of mtDNA.

Dujon, B. 1981. "Mitochondrial Genetics and Functions," in *The Molecular Biology of the Yeast* Saccharomyces. Edited by J. N. Strathern, E. W. Jones, and J. R. Broach. Cold Spring Harbor Laboratory. The most comprehensive review available. This book is useful for many other aspects of yeast genetics and molecular biology.

Gillham, N. W. 1978. *Organelle Heredity.* New York: Raven Press. A rather technical work on the subject, especially strong on genetic analysis.

Grivell, L. A. 1983. "Mitochondrial DNA." *Scientific American* (March). An excellent and up-to-date review including a discussion of the unique mitochondrial translation system and of splicing of mitochondrial introns.

Laughan, J. R., and S. Gabay-Laughan. 1983. "Cytoplasmic Male Sterility in Maize." *Annual Review of Genetics* 17: 27–48. A recent survey of the genetics and molecular biology of this phenotype.

Linnane, A. W., and P. Nagley. 1978. "Mitochondrial Genetics in Perspective: The Derivation of a Genetic and Physical Map of the Yeast Mitochondrial Genome." *Plasmid* 1: 324–345. An excellent review of the genetics and molecular biology of yeast mtDNA.

Osiewacz, H., and K. Esser. 1984. "The Mitochondrial Plasmid of *Podospora anserina:* A Mobile Intron of a Mitochondrial Gene." *Current Genetics* 8: 299–305. The paper that showed α senDNA is an intron.

Sager, R. 1965. "Genes Outside Chromosomes." *Scientific American* (January). A popular description of early *Chlamydomonas* experiments.

Sager, R. 1972. *Cytoplasmic Genes and Organelles.* New York: Academic Press.

Wright, R. M., et al. 1982. "Are Mitochondrial Structural Genes Selectively Amplified During Senescence in *Podospora anserina?*" *Cell* 29: 505–515. A key paper in the development of the fungal senescence story.

Chapter 21

Beerman, W., and U. Clever. 1964. "Chromosome Puffs." *Scientific American* (April). A description of the cytology of puffing.

Bishop, J. M. 1982. "Oncogenes." *Scientific American* (March). An exciting look at the first clues of a gene basis for cancers.

Davidson, E. H. 1986. *Gene Activity in Early Development,* 3d ed. New York: Academic Press. An update of one of the first books to try to develop a comprehensive molecular basis for understanding early development.

DeRobertis, E. M., and J. B. Gurdon. 1979. "Gene Transplantation and the Analysis of Development." *Scientific American* (December).

Gehring, W. J. 1985. "The Molecular Basis of Development." *Scientific American* (October).

Gurdon, J. B. 1968. "Transplanted Nuclei and Cell Differentiation." *Scientific American* (December). A review of nuclear transplantation and the successful cloning of a vertebrate.

Hayflick, L. 1980. "The Cell Biology of Human Aging." *Scientific American* (January). This intriguing report suggests that the aging process is a normal, regulated part of development—a point not treated in this textbook.

Hood, L. E., I. L. Weissman, W. B. Wood, and J. H. Wilson. 1984. *Immunology,* 2d ed. Menlo Park, Calif.: Benjamin Cummings. A fascinating and highly readable textbook on the immune system.

Illmensee, K., and L. C. Stevens. 1979. "Teratomas and Chimeras." *Scientific American* (April). A review of the use of genetic mosaics to analyze development in mammals.

Jonathan, P., et al. 1978. "The Assembly of a Virus." *Scientific American* (November). A study of the assembly of RNA and coat protein to form a functional tobacco mosaic virion.

Leder, P. 1982. "The Genetics of Antibody Diversity." *Scientific American* (May). An excellent review of a complex topic.

Loomis, W. F. 1986. *Developmental Biology.* New York: Macmillan. A contemporary overview of development.

Milstein, C. 1980. "Monoclonal Antibodies." *Scientific American* (October). This is a description of one of the most powerful techniques of cell biology by the Nobel-prize-winning originator.

Nomura, M. 1984. "The Control of Ribosome Synthesis." *Scientific American* (January).

Palmiter, R. D., and R. L. Brinster. 1985. "Transgenic Mice." *Cell* 41: 343–345.

Ptashne, M., et al. 1982. "Genetic Switch in a Bacterial Virus." *Scientific American* (November). This is the best-understood example of molecular control of gene activity.

Suzuki, D. T. 1970. "Temperature-Sensitive Mutations in *Drosophila melanogaster.*" *Science* 170: 695–706. A review of the usefulness of conditional mutations.

Tonegawa, S. 1985. "The Molecules of the Immune System." *Scientific American* (October). An up-to-date description of the complexity of the immune molecules by a recent Nobel prize winner.

Watson, J. D., N. H. Hopkins, J. W. Roberts, J. A. Steitz, and A. M. Weiner. 1987. *Molecular Biology of the Gene,* 4th ed. Reading, Mass.: Benjamin Cummings. A detailed, but highly readable updated edition of Watson's classic book.

Weinberg, R. A. 1983. "The Molecular Basis of Cancer." *Scientific American* (November). The discovery of oncogenes in normal human cells has signaled a radical new approach to cancer study.

Weinberg, R. A. 1988. "Finding the Anti-Oncogene." *Scientific American* (September).

Wood, W. B., and R. S. Edgar. 1967. "Building a Bacterial Virus." *Scientific American* (July). A fascinating review of the discovery that virus particles assemble in a specific sequence.

Chapter 22

Bender, W., et al. 1983. "Molecular Genetics of the Bithorax Complex in *Drosophila melanogaster.*" *Science* 221: 23. This is not an easy article to read, but it is well worth the effort. It illustrates the power of combining good genetic analysis with biochemistry.

Benzer, S. 1973. "The Genetic Dissection of Behavior." *Scientific American* (December). A beautiful illustration of the power of genetic analysis to probe development and behavior. Includes focus mapping.

Garcia-Bellido, A., et al. 1979. "Compartments in Animal Development." *Scientific American* (July). This article is not easy reading, but it points to an underlying principle in development that appears to be widespread.

Garcia-Bellido, A., and J. R. Merriam. 1969. "Cell Lineage of the Imaginal Discs of *Drosophila melanogaster.*" *Journal of Experimental Zoology* 170: 61–76. A clever genetic experiment to determine the end of gene activity in development.

Gardner, R. L. 1979. "The Relationship between Cell Lineage and Differentiation in the Early Mouse Embryo." In *Genetic Mosaics and Cell Differentiation,* W. J. Gehring, ed. New York: Springer-Verlag.

Gehring, W. J. 1985. "The Molecular Basis of Development." *Scientific American* (October). This is a comprehensive review of studies in *Drosophila.*

Hadorn, E. 1968. "Transdetermination in Cells." *Scientific American* (November). A fascinating look at a puzzling phenomenon.

McLaren, A. 1976. *Mammalian chimaeras.* Cambridge, England: Cambridge University Press. An old book that anticipated many of the uses to which mosaics would be put.

Molecular Biology of Development. Cold Spring Harbor Symposia on Quantitative Biology, 50, 1985. This is a collection of papers devoted to a broad look at development in a number of organisms. It is hard slogging, but gives an idea of the scope of work going on at the cutting edge.

Nesbitt, M. N., and S. M. Gartler. 1979. "The Applications of Genetic Mosaicism to Developmental Problems." *Annual Review of Genetics* 5: 143–162. An old paper that still conveys the potential of mosaics studies in mammal including humans.

Sternberg, P. W., and H. R. Horvitz. 1984. "Genetic Control of Cell Lineage during Nematode Development." *Annual Review of Genetics* 18: 489–524. A review and technical perspective on lineage studies.

Sulston, J. E., and H. R. Horvitz. 1977. "Post-embryonic Cell Lineages of the Nematode, *Caenorhabditis elegans.*" *Developmental Biology* 56: 110–156. The classic paper on the cell lineage descriptions; shows how the individual cells were traced.

Chapter 23

Bodmer, W. F., and L. L. Cavalli-Sforza. 1970. "Intelligence and Race." *Scientific American* (October). A popular treatment, including a discussion of heritability.

Falconer, D. S. 1981. *Introduction to Quantitative Genetics,* 2d ed. New York: Ronald Press. A widely read text with a strong mathematical emphasis.

Feldman, M. W., and R. C. Lewontin. 1975. "The Heritability Hangup." *Science* 190: 1163–1168. A discussion of the meaning of heritability and its limitations, especially in relation to human intelligence.

Lewontin, R. C. 1974. "The Analysis of Variance and the Analysis of Causes." *American Journal of Human Genetics.* 26: 400–411. A discussion of the meaning of the analysis of variance in genetics as a method for determining the roles of heredity and environment in determining phenotype.

Lewontin, R. C. 1982. *Human Diversity.* New York: Scientific American Books. Includes a discussion of quantitative variation in human populations.

Chapter 24

Beadle, G. W. 1980. "The Ancestry of Corn." *Scientific American* (January). A popular account of the various clues that led to the modern version.

Bodmer, W. F., and L. L. Cavalli-Sforza. 1976. *Genetics, Evolution, and Man.* New York: W. H. Freeman and Co.

Clarke, B. 1975. "The Causes of Biological Diversity." *Scientific American* (August). A popular account emphasizing genetic polymorphism.

Crow, J. F. 1979. "Genes That Violate Mendel's Rules." *Scientific American* (February). An interesting article on a topic called segregation distortion (not treated in this text) and its effects in populations.

Dobzhansky, T. 1951. *Genetics and the Origin of Species.* New York: Columbia University Press. The classic synthesis of population genetics and the processes of evolution. The most influential book on evolution since Darwin's *Origin of Species.*

Ford, E. B. 1971. *Ecological Genetics,* 3d ed. London: Chapman & Hamm. A nonmathematical treatment of the role of genetic variation in nature, stressing morphological variation.

Futuyma, D. J. 1979. *Evolutionary Biology.* Sunderland, Mass.: Sinauer Associates. The best modern discussion of population genetics, ecology, and evolution from both a theoretical and an experimental point of view.

Hartl, D. 1980. *Principles of Population Genetics.* Sunderland, Mass.: Sinauer Associates.

Lerner, I. M., and W. J. Libby. 1976. *Hereditary, Evolution and Society,* 2d ed. New York: W. H. Freeman and Co. A text meant for nonscience students.

Lewontin, R. C. 1974. *The Genetic Basis of Evolutionary Change.* New York: Columbia University Press. A discussion of the prevalence and role of genetic variation in natural populations. Both morphological and protein variations are considered.

Lewontin, R. C. 1982. *Human Diversity.* New York: Scientific American Books. Applies concepts of population genetics to human diversity and evolution.

Scientific American. September 1978. "Evolution." This volume contains several articles relevant to population genetics.

Wilson, A. C. 1985. "The Molecular Basis of Evolution." *Scientific American* (October).

Glossary

A Adenine, or adenosine.

abortive transduction The failure of a transducing DNA segment to be incorporated into the recipient chromosome.

acentric chromosome A chromosome having no centromere.

acrocentric chromosome A chromosome having the centromere located slightly nearer one end than the other.

active site The part of a protein that must be maintained in a specific shape if the protein is to be functional — for example, in an enzyme, the part to which the substrate binds.

adaptation In the evolutionary sense, some heritable feature of an individual's phenotype that improves its chances of survival and reproduction in the existing environment.

adaptive landscape The surface plotted in a three-dimensional graph, with all possible combinations of allele frequencies for different loci plotted in the plane, and mean fitness for each combination plotted in the third dimension.

adaptive peak A high point (perhaps one of several) on an adaptive landscape; selection tends to drive the genotype composition of the population toward a combination corresponding to an adaptive peak.

adaptive surface *See* **adaptive landscape.**

additive genetic variance Genetic variance associated with the average effects of substituting one allele for another.

adenine A purine base that pairs with thymine in the DNA double helix.

adenosine The nucleoside containing adenine as its base.

adenosine triphosphate *See* **ATP.**

adjacent segregation In a reciprocal translocation, the passage of a translocated and a normal chromosome to each of the poles.

ADP Adenosine diphosphate.

Ala Alanine (an amino acid).

alkylating agent A chemical agent that can add alkyl groups (for example, ethyl or methyl groups) to another molecule; many mutagens act through alkylation.

allele One of two or more forms that can exist at a single gene locus.

allele frequency A measure of the commonness of an allele in a population; the proportion of all alleles of that gene in the population that are of this specific type.

allopolyploid *See* **amphidiploid.**

allosteric transition A change from one conformation of a protein to another.

alternate segregation In a reciprocal translocation, the passage of both normal chromosomes to one pole and both translocated chromosomes to the other pole.

alternation of generations The alternation of gametophyte and sporophyte stages in the life cycle of a plant.

amber codon The codon UAG, a nonsense codon.

amber suppressor A mutant allele coding for tRNA whose anticodon is altered in such a way that the tRNA inserts an amino acid at an amber codon in translation.

Ames test A widely used test to detect possible chemical carcinogens; based on mutagenicity in the bacterium *Salmonella*.

amino acid A peptide; the basic building block of proteins (or polypeptides).

amniocentesis A technique for testing the genotype of an embryo or fetus in utero with minimal risk to the mother or the child.

AMP Adenosine monophosphate.

amphidiploid An allopolyploid; a polyploid formed from the union of two separate chromosome sets and their subsequent doubling.

amplification The production of many DNA copies from one master region of DNA.

anaphase An intermediate stage of nuclear division during which chromosomes are pulled to the poles of the cell.

aneuploid cell A cell having a chromosome number that differs from the normal chromosome number for the species by a small number of chromosomes.

angstrom (Å) A unit of length equal to 10^{-10} meter; many scientists now prefer to use the nanometer (1 nm = 10^{-9}m = 10 Å).

animal breeding The practical application of genetic analysis for development of lines of domestic animals suited to human purposes.

annealing Spontaneous alignment of two single DNA strands to form a double helix.

antibody A protein (immunoglobulin) molecule, produced by the immune system, that recognizes a particular foreign antigen and binds to it; if the antigen is on the surface of a cell, this binding leads to cell aggregation and subsequent destruction.

anticodon A nucleotide triplet in a tRNA molecule that aligns with a particular codon in mRNA under the influence of the ribosome, sot that the peptide carried by the tRNA is inserted in a growing protein chain.

antigen A molecule (typically found in the surface of a cell) whose shape triggers the production of antibodies that will bind to the antigen.

antiparallel A term used to describe the opposite orientations of the two strands of a DNA double helix; the 5′ end of one strand aligns with the 3′ end of the other strand.

AP sites Apurinic or apyrimidinic sites resulting from the loss of a purine or pyrimidine residue from the DNA.

Arg Arginine (an amino acid).

ascospore A sexual spore from certain fungus species in which spores are found in a sac called an ascus.

ascus In fungi, a sac that encloses a tetrad or an octad of ascospores.

asexual spores *See* **spore.**

Asn Asparagine (an amino acid).

Asp Aspartate (an amino acid).

ATP (adenosine triphosphate) The "energy molecule" of cells, synthesized mainly in mitochondria and chloroplasts; energy from the breakdown of ATP drives many important reactions in the cell.

attached X A pair of *Drosophila* X chromosomes joined at one end and inherited as a single unit.

attenuator A region adjacent to the structural genes of the *trp* operon; this region acts in the presence of tryptophan to reduce the rate of transcription from the structural genes.

autonomous controlling element A controlling element that apparently has both regulator and receptor functions combined in the single unit, which enters a gene and causes an unstable mutation.

autonomous replication sequence (ARS) A segment of a DNA molecule necessary for the initiation of its replication; generally a site recognized and bound by the proteins of the replication system.

autopolyploid A polyploid formed from the doubling of a single genome.

autoradiography A process in which radioactive materials are incorporated into cell structures, which are then placed next to a film of photographic emulsion, thus forming a pattern on the film corresponding to the location of the radioactive compounds within the cell.

autosome Any chromosome that is not a sex chromosome.

auxotroph A strain of microorganisms that will proliferate only when the medium is supplemented with some specific substance not required by wild-type organisms.

B chromosomes Small plant chromosomes of variable number between individuals of a species, having no known phenotypic role.

bacteriophage (phage) A virus that infects bacteria.

balanced polymorphism Stable genetic polymorphism maintained by natural selection.

Balbiani ring A large chromosome puff.

Barr body A densely staining mass that represents an X chromosome inactivated by dosage compensation.

base analog A chemical whose molecular structure mimics that of a DNA base; because of the mimicry, the analog may act as a mutagen.

bead theory The disproved hypothesis that genes are arranged on the chromosome like beads on a necklace, indivisible into smaller units of mutation and recombination.

bimodal distribution A statistical distribution having two modes.

binary fission The process in which a parent cell splits into two daughter cells of approximately equal size.

biparental zygote A *Chlamydomonas* zygote that contains cpDNA from both parents; such cells generally are rare.

blastoderm In an insect embryo, the layer of cells that completely surrounds an internal mass of yolk.

blending inheritance A discredited model of inheritance suggesting that the characteristics of an individual result from the smooth blending of fluidlike influences from its parents.

branch migration The process by which a single "invading" DNA strand extends its partial pairing with its complementary strand as it displaces the resident strand.

bridging cross A cross made to transfer alleles between two sexually isolated species by first transferring the alleles to an intermediate species that is sexually compatible with both.

broad heritability (H^2) The proportion of total phenotypic variance at the population level that is contributed by genetic variance.

bud A daughter cell formed by mitosis in yeast; one daughter cell retains the cell wall of the parent, and the other (the bud) forms a new cell wall.

buoyant density A measure of the tendency of a substance to float in some other substance; large molecules are distinguished by their differing buoyant densities in some standard fluid. Measured by density-gradient ultracentrifugation.

C Cytosine, or cytidine.

callus An undifferentiated clone of plant cells.

cAMP (cyclic adenosine monophosphate) A molecule that plays a key role in the regulation of various processes within the cell.

canalized character A character whose phenotype is kept within narrow boundaries even in the presence of disturbing environments or mutations.

cancer A syndrome that involves the uncontrolled and abnormal division of eukaryotic cells.

CAP (catabolite activator protein) A protein whose presence is necessary for the activation of the *lac* operon.

carbon source A nutrient (such as sugar) that provides carbon "skeletons" needed in the organism's synthesis of organic molecules.

carcinogen A substance that causes cancer.

carrier An individual who possesses a mutant allele but does not express it in the phenotype because of a dominant allelic partner; thus, an individual of genotype *Aa* is a carrier of *a* if there is complete dominance of *A* over *a*.

cassette model A model to explain mating-type interconversion in yeast. Information for both *a* and *α* mating types is assumed to be present as silent "cassettes"; a copy of either type of cassette may be transposed to the mating-type locus, where it is "played" (transcribed).

catabolite activator protein *See* **CAP.**

catabolite repression The inactivation of an operon caused by the presence of large amounts of the metabolic end product of the operon.

cation A positively charged ion (such as K⁺).

cDNA *See* **complementary DNA.**

cell division The process by which two cells are formed from one.

cell lineage A pedigree of cells related through asexual division.

centimorgan (cM) *See* **map unit.**

central dogma The hypothesis that information flows only from DNA to RNA to protein; although some exceptions are now known, the rule is generally valid.

centromere A kinetochore; the constricted region of a nuclear chromosome, to which the spindle fibers attach during division.

character Some attribute of individuals within a species for which various heritable differences can be defined.

character difference Alternative forms of the same attribute within a species.

chase *See* **pulse-chase experiment.**

chiasma (plural, **chiasmata**) A cross-shaped structure commonly observed between nonsister chromatids during meiosis; the site of crossing-over.

chimera *See* **mosaic.**

chi-square (χ^2) test A statistical procedure used to determine whether differences between sets of observed frequencies exceed those expected by chance if the sets were randomly selected from a single large population.

chloroplast A chlorophyll-containing organelle in plants that is the site of photosynthesis.

chromatid One of the two side-by-side replicas produced by chromosome division.

chromatid conversion A type of gene conversion that is inferred from the existence of identical sister-spore pairs in a fungal octad that shows a non-Mendelian allele ratio.

chromatid inference A situation in which the occurrence of a crossover between any two nonsister chromatids can be shown to affect the probability of those chromatids being involved in other crossovers in the same meiosis.

chromatin The substance of chromosomes; now known to include DNA, chromosomal proteins, and chromosomal RNA.

chromocenter The point at which the polytene chromosomes appear to be attached together.

chromomere A small beadlike structure visible on a chromosome during prophase of meiosis and mitosis.

chromosome A linear end-to-end arrangement of genes and other DNA, sometimes with associated protein and RNA.

chromosome aberration Any type of change in the chromosome structure or number.

chromosome loss Failure of a chromosome to become incorporated into a daughter nucleus at cell division.

chromosome map *See* **linkage map.**

chromosome mutation *See* **chromosome aberration.**

chromosome puff A swelling at a site along the length of a polytene chromosome; the site of active transcription.

chromosome rearrangement A chromosome aberration involving new juxtapositions of chromosome parts.

chromosome set The group of different chromosomes that carries the basic set of genetic information for a particular species.

chromosome theory of inheritance The unifying theory stating that inheritance patterns may be generally explained by assuming that genes are located in specific sites on chromosomes.

cis conformation In a heterozygote involving two mutant sites within a gene or gene cluster, the arrangement $a_1 a_2 / + +$.

cis dominance The ability of a gene to affect genes next to it on the same chromosome.

cis-trans test A test to determine whether two mutant sites of a gene are in the same functional unit or gene.

cistron Originally defined as a functional genetic unit within which two mutations cannot complement. Now equated with the term gene, as the region of DNA that encodes a single polypeptide (or functional RNA molecule such as tRNA or rRNA).

clone (1) A group of genetically identical cells or individuals derived by asexual division from a common ancestor. (2) *(colloquial)* An individual formed by some asexual process so that it is genetically identical to its "parent." (3) *See* **DNA clone.**

cM (centimorgan) *See* **map unit.**

code dictionary A listing of the 64 possible codons and their translational meanings (the corresponding amino acids).

codominance The situation in which a heterozygote shows the phenotypic effects of both alleles equally.

codon A section of DNA (three nucleotide pairs in length) that codes for a single amino acid.

coefficient of coincidence The ratio of the observed number of double recombinants to the expected number.

colinearity The correspondence between the location of a mutant site within a gene and the location of an amino-acid substitution within the polypeptide translated from that gene.

colony A visible clone of cells.

compartmentalization The existence of boundaries within the organism beyond which a specific clone of cells will never extend during development.

complementary DNA (cDNA) Synthetic DNA transcribed from a specific RNA through the action of the enzyme reverse transcriptase.

complementary gene action Intergenic complementation between mutant alleles at different loci to give wild-type phenotype.

complementary RNA (cRNA) Synthetic RNA produced by transcription from a specific DNA single-stranded template.

complementation The production of a wild-type phenotype when two different mutations are combined in a diploid or a heterokaryon.

complementation test *See* **cis-trans test.**

conditional mutation A mutation that has the wild-type phenotype under certain (permissive) environmental conditions and a mutant phenotype under other (restrictive) conditions.

conjugation The union of two bacterial cells, during which chromosomal material is transferred from the donor to the recipient cell.

conjugation tube *See* **pilus.**

conservative replication A disproved model of DNA synthesis suggesting that one-half of the daughter DNA molecules should have both strands composed of newly polymerized nucleotides.

constant region A region of an antibody molecule that is nearly identical with the corresponding regions of antibodies of different specificities.

constitutive heterochromatin Specific regions of heterochromatin always present and in both homologs of a chromosome.

controlling element A mobile genetic element capable of producing an unstable mutant target gene; two types exist, the regulator and the receptor elements.

copy-choice model A model of the mechanism for crossing-over, suggesting that crossing-over occurs during chromosome division and can occur only between two supposedly "new" nonsister chromatids; the experimental evidence does not support this model.

correction The production (possibly by excision and repair) of a properly paired nucleotide pair from a sequence of hybrid DNA that contains an illegitimate pair.

correlation coefficient A statistical measure of the extent to which variations in one variable are related to variations in another.

cosegregation In *Chlamydomonas,* parallel behavior of different chloroplast markers in a cross, due to their close linkage on cpDNA.

cotransduction The simultaneous transduction of two bacterial marker genes.

cotransformation The simultaneous transformation of two bacterial marker genes.

coupling conformation Linked heterozygous gene pairs in the arrangement $A B / a b$.

covariance A statistical measure used in computing the correlation coefficient between two variables; the covariance is the mean of $(x - \bar{x})(y - \bar{y})$ over all pairs of values for the variables x and y, where $\bar{x}$ is the mean of the x values and $\bar{y}$ is the mean of the y values.

cpDNA Chloroplast DNA.

cri-du-chat syndrome An abnormal human condition caused by deletion of part of one homolog of chromosome 5.

crisscross inheritance Transmission of a gene from male parent to female child to male grandchild — for example, X-linked inheritance.

cRNA *See* **complementary RNA.**

cross The deliberate mating of two parental types of organisms in genetic analysis.

crossing-over The exchange of corresponding chromosome parts between homologs by breakage and reunion.

crossover suppressor An inversion (usually complex) that makes pairing and crossing-over impossible.

cruciform configuration A region of DNA with palindromic sequences in both strands, so that each strand pairs with itself to form a helix extending sideways from the main helix.

CSAR (cytoplasmic segregation and recombination) An acronym used in this book to describe the process whereby organelle-based genes assort and recombine in a cytohet.

culture Tissue or cells multiplying by asexual division, grown for experimentation.

cyclic adenosine monophosphate *See* **cAMP.**

Cys Cysteine (an amino acid).

cytidine The nucleoside containing cytosine as its base.

cytochromes A class of proteins, found in mitochondrial membranes, whose main function is oxidative phosphorylation of ADP to form ATP.

cytogenetics The cytological approach to genetics, mainly involving microscopic studies of chromosomes.

cytohet A cell containing two genetically distinct types of a specific organelle.

cytoplasm The material between the nuclear and cell membranes; includes fluid (cytosol), organelles, and various membranes.

cytoplasmic inheritance Inheritance via genes found in cytoplasmic organelles.

cytosine A pyrimidine base that pairs with guanine.

cytosol The fluid portion of the cytoplasm, outside the organelles.

Darwinian fitness The relative probability of survival and reproduction for a genotype.

degenerate code A genetic code in which some amino acids may be encoded by more than one codon each.

deletion Removal of a chromosomal segment from a chromosome set.

denaturation The separation of the two strands of a DNA double helix, or the severe disruption of the structure of any complex molecule without breaking the major bonds of its chains.

denaturation map A map of a stretch of DNA showing the locations of local denaturation loops, which correspond to regions of high AT content.

deoxyribonuclease *See* **DNase.**

deoxyribonucleic acid *See* **DNA.**

development The process whereby a single cell becomes a differentiated organism.

developmental noise The influence of random molecular processes on development.

dicentric chromosome A chromosome with two centromeres.

dihybrid cross A cross between two individuals identically heterozygous at two loci—for example, *Aa Bb ×
Aa Bb.*

dimorphism A "polymorphism" involving only two forms.

dioecious plant A plant species in which male and female organs appear on separate individuals.

diploid A cell having two chromosome sets, or an individual having two chromosome sets in each of its cells.

directional selection Selection that changes the frequency of an allele in a constant direction, either toward or away from fixation for that allele.

dispersive replication Disproved model of DNA synthesis suggesting more-or-less random interspersion of parental and new segments in daughter DNA molecules.

distribution *See* **statistical distribution.**

distribution function A graph of some precise quantitative measure of a character against its frequency of occurrence.

DNA (deoxyribonucleic acid) A double chain of linked nucleotides (having deoxyribose as their sugars); the fundamental substance of which genes are composed.

DNA clone A section of DNA that has been inserted into a vector molecule, such as a plasmid or a phage chromosome, and then replicated to form many copies.

DNA polymerase An enzyme that can synthesize new DNA strands using a DNA template; several such enzymes exist.

DNase (deoxyribonuclease) An enzyme that degrades DNA to nucleotides.

dominance variance Genetic variance at a single locus attributable to dominance of one allele over another.

dominant allele An allele that expresses its phenotypic effect even when heterozygous with a recessive allele; thus if *A* is a dominant over *a*, then *AA* and *Aa* have the same phenotype.

dominant phenotype The phenotype of a genotype containing the dominant allele; the parental phenotype that is expressed in a heterozygote.

dosage compensation (1) Inactivation of X chromosomes in mammals so that no cell has more than one functioning X chromosome. (2) Regulation at some autosomal loci so that homozygous dominants do not produce twice as much product as the heterozygote.

dose *See* **gene dose.**

double crossover Two crossovers occurring in a chromosomal region under study.

double helix The structure of DNA first proposed by Watson and Crick, with two interlocking helices joined by hydrogen bonds between paired bases.

double infection Infection of a bacterium with two genetically different phages.

Down's syndrome An abnormal human phenotype including mental retardation, due to a trisomy of chromosome 21; more common in babies born to older mothers.

drift *See* **random genetic drift.**

duplicate genes Two identical allele pairs in one diploid individual.

duplication More than one copy of a particular chromosomal segment in a chromosome set.

dyad A pair of sister chromatids joined at the centromere, as in the first division of meiosis.

ecdysone A molting hormone in insects.

electrophoresis A technique for separating the components of a mixture of molecules (proteins, DNAs, or RNAs) in an electric field within a gel.

endogenote *See* **merozygote.**

endonuclease An enzyme that cleaves the phosphodiester bond within a nucleotide chain.

endopolyploidy An increase in the number of chromosome sets caused by replication without cell division.

endosperm Triploid tissue in a seed, formed from fusion of two haploid female and one haploid male nucleus.

enforced outbreeding Deliberate avoidance of mating between relatives.

enucleate cell A cell having no nucleus.

environment The combination of all the conditions external to the genome that potentially affect its expression and its structure.

environmental variance The variance due to environmental variation.

enzyme A protein that functions as a catalyst.

episome A genetic element in bacteria that can replicate free in the cytoplasm or can be inserted into the main bacterial chromosome and replicate with the chromosome.

epistasis A situation in which the differential phenotypic expression of genotypes at one locus depends upon the genotype at another locus.

equational division A nuclear division that maintains the same ploidy level of the cell.

ethidium A molecule that can intercalate into DNA double helices when the helix is under torsional stress.

euchromatin A chromosomal region that stains normally; thought to contain the normally functioning genes.

eugenics Controlled human breeding based on notions of desirable and undesirable genotypes.

eukaryote An organism having eukaryotic cells.

eukaryotic cell A cell containing a nucleus.

euploid A cell having any number of complete chromosome sets, or an individual composed of such cells.

excision repair The repair of a DNA lesion by removal of the faulty DNA segment and its replacement with a wild-type segment.

exconjugant A female bacterial cell that has just been in conjugation with a male and that contains a fragment of male DNA.

exogenote *See* **merozygote.**

exon Any non-intron section of the coding sequence of a gene; together, the exons constitute the mRNA and are translated into protein.

exonuclease An enzyme that cleaves nucleotides one at a time from an end of a polynucleotide chain.

expressivity The degree to which a particular genotype is expressed in the phenotype.

F$^-$ cell In *E. coli,* a cell having no fertility factor; a female cell.

F$^+$ cell In *E. coli,* a cell having a free fertility factor; a male cell.

F factor *See* **fertility factor.**

F$'$ factor A fertility factor into which a portion of the bacterial chromosome has been incorporated.

F$_1$ generation The first filial generation, produced by crossing two parental lines.

F$_2$ generation The second filial generation, produced by selfing or intercrossing the F$_1$.

facultative heterochromatin Heterochromatin located in positions that are composed of euchromatin in other individuals of the same species, or even in the other homolog of a chromosome pair.

familial trait A trait shared by members of a family.

family selection A breeding technique of selecting a pair on the basis of the average performance of their progeny.

fate map A map of an embryo showing areas that are destined to develop into specific adult tissues and organs.

fertility factor (F factor) A bacterial episome whose presence confers donor ability (maleness).

filial generations Successive generations of progeny in a controlled series of crosses, starting with two specific parents (the P generation) and selfing or intercrossing the progeny of each new (F$_1$, F$_2$, . . .) generation.

filter enrichment A technique for recovering auxotrophic mutants in filamentous fungi.

fingerprint The characteristic spot pattern produced by electrophoresis of the polypeptide fragments obtained through denaturation of a particular protein with a proteolytic enzyme.

first-division segregation pattern A linear pattern of spore phenotypes within the ascus for a particular allele pair, produced when the alleles go into separate nuclei at the first meiotic division, showing that no crossover has occurred between that allele pair and the centromere.

fitness *See* **Darwinian fitness.**

fixed allele An allele for which all members of the population under study are homozygous, so that no other alleles for this locus exist in the population.

fixed breakage point According to the heteroduplex DNA recombination model, the point from which unwinding of the DNA double helices begins, as a prelude to formation of heteroduplex DNA.

fluctuation test A test used in microbes to establish the random nature of mutation, or to measure mutation rates.

fMet *See* **formylmethionine.**

focus map A fate map of areas of the *Drosophila* blastoderm destined to become specific adult structures, based on the frequencies of specific kinds of mosaics.

formylmethionine (fMet) A specialized amino acid that is the very first one incorporated into the polypeptide chain in the synthesis of proteins.

forward mutation A mutation that converts a wild-type allele to a mutant allele.

frame-shift mutation The insertion or deletion of a nucleotide pair or pairs, causing a disruption of the translational reading frame.

frequency-dependent fitness Fitness differences whose intensity changes with changes in the relative frequency of genotypes in the population.

frequency-dependent selection Selection that involves frequency-dependent fitness.

frequency histogram A "step curve" in which the frequencies of various arbitrarily bounded classes are graphed.

frequency-interdependent fitness Fitness that is not dependent upon interactions with other individuals of the same species.

frequency-independent selection Selection in which the fitnesses of genotypes are independent of their relative frequency in the population.

fruiting body In fungi, the organ in which meiosis occurs and sexual spores are produced.

G Guanine, or guanosine.

gamete A specialized haploid cell that fuses with a gamete from the opposite sex or mating type to form a diploid zygote; in mammals, an egg or a sperm.

gametophyte The haploid gamete-producing stage in the life cycle of plants; prominent and independent in some species but reduced or parasitic in others.

gene The fundamental physical and functional unit of heredity, which carries information from one generation to the next; a segment of DNA, composed of a transcribed region and a regulatory sequence, that makes possible transcription.

gene conversion A meiotic process of directed change in which one allele directs the conversion of a partner allele to its own form.

gene dose The number of copies of a particular gene present in the genome.

gene family A set of genes descended from the same ancestral gene.

gene frequency *See* **allele frequency.**

gene interaction The collaboration of several different genes in the production of one phenotypic character (or related group of characters).

gene locus The specific place on a chromosome where a gene is located.

gene map A linear designation of mutant sites within a gene, based upon the various frequencies of interallelic (intragenic) recombination.

gene mutation A point mutation that results from changes within the structure of a gene.

gene pair The two copies of a particular type of gene present in a diploid cell (one in each chromosome set).

gene therapy The correction of a genetic deficiency in a cell by the addition of new DNA and its insertion into the genome

generalized transduction The ability of certain phages to transduce any gene in the bacterial chromosome.

genetic code The set of correspondences between nucleotide pair triplets in DNA and amino acids in protein.

genetic dissection The use of recombination and mutation to piece together the various components of a given biological function.

genetic markers Alleles used as experimental probes to keep track of an individual, a tissue, a cell, a nucleus, a chromosome, or a gene.

genetic variance Phenotypic variance resulting from the presence of different genotypes in the population.

genetics (1) The study of genes through their variation. (2) The study of inheritance.

genome The entire complement of genetic material in a chromosome set.

genotype The specific allelic composition of a cell— either of the entire cell or, more commonly, for a certain gene or a set of genes.

germinal mutations Mutations occurring in the cells that are destined to develop into gametes.

Gln Glutamine (an amino acid).
Glu Glutamate (an amino acid).
Gly Glycine (an amino acid).

gradient A gradual change in some quantitative property over a specific distance.

guanine A purine base that pairs with cytosine.

guanosine The nucleoside having guanine as its base.

gynandromorph A sexual mosaic.

half-chromatid conversion A type of gene conversion that is inferred from the existence of nonidentical sister spores in a fungal octad showing a non-Mendelian allele ratio.

haploid A cell having one chromosome set, or an organism composed of such cells.

haploidization Production of a haploid from a diploid by progressive chromosome loss.

Hardy-Weinberg equilibrium The stable frequency distribution of genotypes AA, Aa, and aa, in the proportions p^2, $2pq$, and q^2 respectively (where p and q are the frequencies of the alleles A and a), that is a consequence of random mating in the absence of mutation, migration, natural selection, or random drift.

harlequin chromosomes Sister chromatids that stain differently, so that one appears dark and the other light.

hemizygous gene A gene present in only one copy in a diploid organism—for example, X-linked genes in a male mammal.

hemoglobin (Hb) The oxygen-transporting blood protein in most animals.

heredity The biological similarity of offspring and parents.

heritability in the broad sense *See* **broad heritability.**

heritability in the narrow sense (h^2) The proportion of phenotypic variance that can be attributed to additive genetic variance.

hermaphrodite (1) A plant species in which male and female organs occur in the same flower of a single individual (*compare* **monoecious plant**). (2) An animal with both male and female sex organs.

heterochromatin Densely staining condensed chromosomal regions, believed to be for the most part genetically inert.

heteroduplex A DNA double helix formed by annealing single strands from different sources; if there is a structural difference between the strands, the heteroduplex may show such abnormalities as loops or buckles.

heteroduplex DNA model A model that explains both crossing-over and gene conversion by assuming the production of a short stretch of heteroduplex DNA (formed from both parental DNAs) in the vicinity of a chiasma.

heterogametic sex The sex that has heteromorphic sex chromosomes (for example, XY) and hence produces two different kinds of gametes with respect to the sex chromosomes.

heterogeneous nuclear RNA (HnRNA) A diverse assortment of RNA types found in the nucleus, including mRNA precursors and other types of RNA.

heterokaryon A culture of cells composed of two different nuclear types in a common cytoplasm.

heterokaryon test A test for cytoplasmic mutations, based on new associations of phenotypes in cells derived from specially marked heterokaryons.

heteromorphic chromosomes A chromosome pair with some homology but differing in size, shape, or staining properties.

heterothallic fungus A fungus species in which two different mating types must unite to complete the sexual cycle.

heterozygosity A measure of the genetic variation in a population; with respect to one locus, stated as the frequency of heterozygotes for that locus.

heterozygote An individual having a heterozygous gene pair.

heterozygous gene pair A gene pair having different alleles in the two chromosome sets of the diploid individual—for example, Aa or A^1A^2.

hexaploid A cell having six chromosome sets, or an organism composed of such cells.

high-frequency recombination (Hfr) cell In *E. coli*, a cell having its fertility factor integrated into the bacterial chromosome; a donor (male) cell.

His Histidine (an amino acid).

histocompatibility antigens Antigens that determine the acceptance or rejection of a tissue graft.

histocompatibility genes The genes that code for the histocompatibility antigens.

homeo box Short (~ 180 bp) homologous sequence found in the 3′ exon of homeotic genes in *Drosophila*. Similar sequences have been found in other organisms. Possibly involved in DNA binding during gene regulation.

homeologous chromosomes Partially homologous chromosomes, usually indicating some original ancestral homology.

homeotic mutations Mutations that can change the fate of an imaginal disk.

homogametic sex The sex with homologous sex chromosomes (for example, XX).

homolog A member of a pair of homologous chromosomes.

homologous chromosomes Chromosomes that pair with each other at meiosis or chromosomes in different species that have retained most of the same genes during their evolution from a common ancestor.

homothallic fungus A fungus species in which a single sexual spore can complete the entire sexual cycle (*compare* **heterothallic fungus**).

homozygote An individual having a homozygous gene pair.

homozygous gene pair A gene pair having identical alleles in both copies—for example, AA or A^1A^1.

host range The spectrum of strains of a given bacterial species that a given strain of phage can infect.

hot spot A part of a gene that shows a very high tendency to become a mutant site, either spontaneously or under the action of a particular mutagen.

hybrid (1) A heterozygote. (2) A progeny individual from any cross involving parents of differing genotypes.

hybrid dysgenesis A syndrome of effects including sterility, mutation, chromosome breakage, and male recombination in the hybrid progeny of crosses between certain laboratory and natural isolates of *Drosophila*.

hybridization in situ Finding the location of a gene by adding specific radioactive probes for the gene and detecting the location of the radioactivity on the chromosome after hybridization.

hybridize (1) To form a hybrid by performing a cross. (2) To anneal nucleic acid strands from different sources.

hydrogen bond A weak bond involving the sharing of an electron with a hydrogen atom; hydrogen bonds are important in the specificity of base pairing in nuclei acids and in determining protein shape.

hydroxyapatite A form of calcium phosphate that binds double-stranded DNA.

hyperploid Aneuploid containing a small number of extra chromosomes.

hypervariable region The part of a variable region that actually determines the specificity of an antibody.

hypha (plural, **hyphae**) A threadlike structure (composed of cells attached end to end) that forms the main tissue in many fungus species.

hypoploid Aneuploid with a small number of chromosomes missing.

Ig *See* **immunoglobulin.**

Ile Isoleucine (an amino acid).

imaginal disk A small group of cells in an insect larva that is destined to develop into an entire adult structure (for example, a leg).

immune system The animal cells and tissues involved in recognizing and attacking foreign substances within the body.

immunoglobulin (Ig) A general term for the kind of globular blood proteins that constitute antibodies.

in situ "In place"; *see* **hybridization in situ.**

in vitro In an experimental situation outside the organism (literally, "in glass").

in vivo In a living cell or organism.

inbreeding Mating between relatives.

inbreeding coefficient The probability of homozygosity that results because the zygote obtains copies of the *same* ancestral gene.

incomplete dominance The situation in which a heterozygote shows a phenotype quantitatively (but not exactly) intermediate between the corresponding homozygote phenotypes. (Exact intermediacy is no dominance.)

independent assortment *See* **Mendel's second law.**

inducer An environmental agent that triggers transcription from an operon.

infectious transfer The rapid transmission of free episomes (plus any chromosomal genes they may carry) from donor to recipient cells in a bacterial population.

inosine A rare base that is important at the wobble position of some tRNA anticodons.

insertion sequence (IS) A mobile piece of bacterial DNA (several hundred nucleotide pairs in length) that is capable of inactivating a gene into which it inserts.

insertional translocation The insertion of a segment from one chromosome into another nonhomologous one.

intercalating agent A chemical that can insert itself between the stacked bases at the center of the DNA double helix, possibly causing a frame-shift mutation.

interchromosomal recombination Recombination resulting from independent assortment.

interference A measure of the independence of crossovers from each other, calculated by subtracting the coefficient of coincidence from 1.

interphase The cell cycle stage between nuclear divisions, in which chromosomes are extended and functionally active.

interrupted mating A technique used to map bacterial genes by determining the sequence in which donor genes enter recipient cells.

interstitial region The chromosomal region between the centromere and the site of a rearrangement.

intervening sequence An intron; a segment of largely unknown function within a gene. This segment is initially transcribed, but the transcript is not found in the functional mRNA.

intrachromosomal recombination Recombination resulting from crossing-over between two gene pairs.

intron *See* **intervening sequence.**

inversion A chromosomal mutation involving the removal of a chromosome segment, its rotation through 180 degrees, and its reinsertion in the same location.

inverted repeat (IR) sequence A sequence found in identical (but inverted) form, for example, at the opposite ends of a transposon.

IR *See* **inverted repeat sequence.**

IS *See* **insertion sequence.**

isoaccepting tRNAs The various types of tRNA molecules that carry a specific amino acid.

isotope One of several forms of an atom having the same atomic number but differing atomic masses.

karyotype The entire chromosome complement of an individual or cell, as seen during mitotic metaphase.

kinetochore *See* **centromere.**

Klinefelter's syndrome An abnormal human male phenotype involving an extra X chromosome (XXY).

λ (lambda) phage One kind ("species") of temperate bacteriophage.

λdgal A λ phage carrying a *gal* bacterial gene and defective *(d)* for some phage function.

lampbrush chromosomes Large chromosomes found in amphibian eggs, with lateral DNA loops that produce a brushlike appearance under the microscope.

lawn A continuous layer of bacteria on the surface of an agar medium.

leader sequence The sequence at the 5′ end of an mRNA that is not translated into protein.

leaky mutant A mutant (typically, an auxotroph) that results from a partial rather than a complete inactivation of the wild-type function.

lesion A damaged area in a gene (a mutant site), a chromosome, or a protein.

lethal gene A gene whose expression results in the death of the individual expressing it.

Leu Leucine (an amino acid).

ligase An enzyme that can rejoin a broken phosphodiester bond in a nucleic acid.

line A group of identical pure-breeding diploid or polyploid organisms, distinguished from other individuals of the same species by some unique phenotype and genotype.

linear tetrad A tetrad that results from the occurrence of the meiotic and postmeiotic nuclear divisions in such a way that sister products remain adjacent to one another (with no passing of nuclei).

linkage The association of genes on the same chromosome.

linkage group A group of genes known to be linked; a chromosome.

linkage map A chromosome map; an abstract map of chromosomal loci, based on recombinant frequencies.

locus (plural, **loci**) *See* **gene locus.**

Lys Lysine (an amino acid).

lysis The rupture and death of a bacterial cell upon the release of phage progeny.

lysogen *See* **lysogenic bacterium.**

lysogenic bacterium A bacterial cell capable of spontaneous lysis due, for example, to the uncoupling of a prophage from the bacterial chromosome.

macromolecule A large polymer such as DNA, a protein, or a polysaccharide.

map unit (m.u.) The "distance" between two linked gene pairs where 1 percent of the products of meiosis are recombinant; a unit of distance in a linkage map.

mapping function A formula expressing the relationship between distance in a linkage map and recombinant frequency.

marker *See* **genetic markers.**

marker retention A technique used in yeast to test the degree of linkage between two mitochondrial mutations.

maternal effect The environmental influence of the mother's tissues on the phenotype of the offspring.

maternal inheritance A type of uniparental inheritance in which all progeny have the genotype and phenotype of the parent acting as the female.

mating types The equivalent in lower organisms of the sexes in higher organisms; the mating types typically differ only physiologically and not in physical form.

matroclinous inheritance Inheritance in which all offspring have the nucleus-determined phenotype of the mother.

mean The arithmetic average.

medium Any material on (or in) which experimental cultures are grown.

meiocyte Cell in which meiosis occurs.

meiosis Two successive nuclear divisions (with corresponding cell divisions) that produce gametes (in animals) or sexual spores (in plants and fungi) having one-half of the genetic material of the original cell.

meiospore Cell that is one of the products of meiosis in plants.

melting Denaturation of DNA.

Mendelian ratio A ratio of progeny phenotypes reflecting the operation of Mendel's laws.

Mendel's first law The two members of a gene pair segregate from each other during meiosis; each gamete has an equal probability of obtaining either member of the gene pair.

Mendel's second law The law of independent assortment; unlinked or distantly linked segregating gene pairs assort independently at meiosis.

merozygote A partially diploid *E. coli* cell formed from a complete chromosome (the endogenote) plus a fragment (the exogenote).

messenger RNA *See* **mRNA.**

Met Methionine (an amino acid).

metabolism The chemical reactions that occur in a living cell.

metacentric chromosome A chromosome having its centromere in the middle.

metaphase An intermediate stage of nuclear division when chromosomes align along the equatorial plane of the cell.

midparent value The mean of the values of a quantitative phenotype for two specific parents.

minimal medium A medium containing only inorganic salts, a carbon source, and water.

missense mutation A mutation that alters a codon so that it encodes a different amino acid.

mitochondrion A eukaryotic organelle that is the site of ATP synthesis and of the citric acid cycle.

mitosis A type of nuclear division (occurring at cell division) that produces two daughter nuclei identical to the parent nucleus.

mitotic crossover A crossover resulting from the pairing of homologs in a mitotic diploid.

mobile genetic element *See* **transposable genetic element.**

mode The single class in a statistical distribution having the greatest frequency.

modifier gene A gene that affects the phenotypic expression of another gene.

molecular genetics The study of the molecular processes underlying gene structure and function.

monocistronic mRNA An mRNA that codes for one protein.

monoecious plant A plant species in which male and female organs are found on the same plant but in different flowers (for example, corn).

monohybrid cross A cross between two individuals identically heterozygous at one gene pair — for example, *Aa* × *Aa.*

monoploid A cell having only one chromosome set (usually as an aberration), or an organism composed of such cells.

monosomic A cell or individual that is basically diploid but that has only one copy of one particular chromosome type and thus has chromosome number $2n - 1$.

mosaic A chimera; a tissue containing two or more genetically distinct cell types, or an individual composed of such tissues.

mRNA (messenger RNA) An RNA molecule transcribed from the DNA of a gene, and from which a protein is translated by the action of ribosomes.

mtDNA Mitochondrial DNA.

m.u. *See* **map unit.**

mu **phage** A kind ("species") of phage with properties similar to those of insertion sequences, being able to insert, transpose, inactivate, and cause rearrangements.

multimeric structure A structure composed of several identical or different subunits held together by weak bonds.

multiple allelism The existence of several known alleles of a gene.

multiple-factor hypothesis A hypothesis to explain quantitative variation by assuming the interaction of a large number of genes (polygenes), each with a small additive effect on the character.

multiplicity of infection The average number of phage particles that infect a single bacterial cell in a specific experiment.

mutagen An agent that is capable of increasing the mutation rate.

mutant An organism or cell carrying a mutation.

mutant allele An allele differing from the allele found in the standard, or wild-type.

mutant hunt The process of accumulating different mutants showing abnormalities in a certain structure or function, as a preparation for mutational dissection of that function.

mutant site The damaged or altered area within a mutated gene.

mutation (1) The process that produces a gene or a chromosome set differing from the wild-type. (2) The gene or chromosome set that results from such a process.

mutation breeding Use of mutagens to develop variants that can increase agricultural yield.

mutation event The actual occurrence of a mutation in time and space.

mutation frequency The frequency of mutants in a population.

mutation rate The number of mutation events per gene per unit of time (for example, per cell generation).

mutational dissection The study of the components of a biological function through a study of mutations affecting that function.

muton The smallest part of a gene that can be involved in a mutation event; now known to be a nucleotide pair.

myeloma A cancer of the bone marrow.

narrow heritability *See* **heritability in the narrow sense.**

negative assortative mating Preferential mating between phenotypically unlike partners.

negative control Regulation mediated by factors that block or turn off transcription.

Neurospora A pink mold, commonly found growing on old food.

neutral mutation (1) A mutation that has no effect on the Darwinian fitness of its carriers. (2) A mutation that has no phenotypic effect.

neutral petite A petite that produces all wild-type progeny when crossed with wild-type.

neutrality *See* **selective neutrality.**

nicking Nuclease action to sever the sugar-phosphate backbone in one DNA strand at one specific site.

nitrocellulose filter A type of filter used to hold DNA for hybridization.

nitrogen bases Types of molecules that form important parts of nucleic acids, composed of nitrogen-containing ring structures; hydrogen bonds between bases link the two strands of a DNA double helix.

nondisjunction The failure of homologs (at meiosis) or sister chromatids (at mitosis) to separate properly to opposite poles.

nonlinear tetrad A tetrad in which the meiotic products are in no particular order.

non-Mendelian ratio An unusual ratio of progeny phenotypes that does not reflect the simple operation of Mendel's laws; for example, mutant: wild ratios of 3:5, 5:3, 6:2, or 2:6 in tetrads indicate that gene conversion has occurred.

nonparental ditype (NPD) A tetrad type containing two different genotypes, both of which are recombinant.

nonsense codon A codon for which no normal tRNA molecule exists; the presence of a nonsense codon causes termination of translation (the end of the polypeptide chain). The three nonsense codons are called amber, ocher, and opal.

nonsense mutation A mutation that alters a gene so as to produce a nonsense codon.

nonsense suppressor A mutation that produces an altered tRNA that will insert an amino acid during translation in response to a nonsense codon.

norm of reaction The pattern of phenotypes produced by a given genotype under different environmental conditions.

NPD *See* **nonparental ditype.**

nu body *See* **nucleosome.**

nuclease An enzyme that can degrade DNA by breaking its phosphodiester bonds.

nucleoid A DNA mass within a chloroplast or mitochondrion.

nucleolar organizer A region (or regions) of the chromosome set physically associated with the nucleolus and containing rRNA genes.

nucleolus An organelle found in the nucleus, containing rRNA and amplified multiple copies of the genes coding for rRNA.

nucleoside A nitrogen base bound to a sugar molecule.

nucleosome A nu body; the basic unit of eukaryotic chromosome structure; a ball of eight histone molecules wrapped about by two coils of DNA.

nucleotide A molecule composed of a nitrogen base, a sugar, and a phosphate group; the basic building block of nucleic acids.

nucleotide pair A pair of nucleotides (one in each strand of DNA) that are joined by hydrogen bonds.

nucleotide-pair substitution The replacement of a specific nucleotide pair by a different pair; often mutagenic.

null allele An allele whose effect is either an absence of normal gene product at the molecular level or an absence of normal function at the phenotypic level.

nullisomic A cell or individual with one chromosomal type missing, with a chromosome number such as $n - 1$ or $2n - 2$.

ocher codon The codon UAA, a nonsense codon.

octad An ascus containing eight ascospores, produced in species in which the tetrad normally undergoes a post-meiotic mitotic division.

oncogene A gene that causes cancer.

opal codon The codon UGA, a nonsense codon.

open reading frame A nucleotide sequence with no stop codons, discovered by sequencing.

operator A DNA region at one end of an operon that acts as the binding site for repressor protein.

operon A set of adjacent structural genes whose mRNA is synthesized in one piece, plus the adjacent regulatory signals that affect transcription on the structural genes.

ORF (open reading frame) Generally synonymous with URF.

organelle A subcellular structure having a specialized function—for example, the mitochondrion, the chloroplast, or the spindle apparatus.

origin of replication The point of specific sequence at which DNA replication is initiated.

overdominance A phenotypic relation in which the phenotypic expression of the heterozygote is greater than that of either homozygote.

palindrome A sequence of DNA that is the same when each strand is read in the same direction.

paracentric inversion An inversion not involving the centromere.

parental ditype (PD) A tetrad type containing two different genotypes, both of which are parental.

partial diploid *See* **merozygote.**

particulate inheritance The model proposing that genetic information is transmitted from one generation to the next in discrete units ("particles"), so that the character of the offspring is not a smooth blend of essences from the parents (*compare* **blending inheritance**).

pathogen An organism that causes disease in another organism.

patroclinous inheritance Inheritance in which all offspring have the nucleus-based phenotype of the father.

PD *See* **parental ditype.**

pedigree A "family tree," drawn with standard genetic

symbols, showing inheritance patterns for specific phenotypic characters.

penetrance The proportion of individuals with a specific genotype who manifest that genotype at the phenotype level.

peptide *See* **amino acid.**

peptide bond A bond joining two amino acids.

pericentric inversion An inversion that involves the centromere.

permissive conditions Those environmental conditions under which a conditional mutant shows the wild-type phenotype.

petite A yeast mutation producing small colonies and altered mitochondrial functions. In cytoplasmic petites (neutral and suppressive petites), the mutation is a deletion in mitochondrial DNA; in segregational petites, the mutation occurs in nuclear DNA.

phage *See* **bacteriophage.**

Phe Phenylalanine (an amino acid).

phenocopy An environmentally induced phenotype that resembles the phenotype produced by a mutation.

phenotype (1) The form taken by some character (or group of characters) in a specific individual. (2) The detectable outward manifestations of a specific genotype.

phenotypic sex determination Sex determination by nongenetic means.

Philadelphia chromosome A translocation between the long arms of chromosomes 9 and 22, often found in the white blood cells of patients with chronic myeloid leukemia.

phosphodiester bond A bond between a sugar group and a phosphate group; such bonds form the sugar-phosphate backbone of DNA.

pilus (plural, **pili**) A conjugation tube; a hollow hairlike appendage of a donor *E. coli* cell that acts as a bridge for transmission of donor DNA to the recipient cell during conjugation.

plant breeding The application of genetic analysis to development of plant lines better suited for human purposes.

plaque A clear area on a bacterial lawn, left by lysis of the bacteria through progressive infections by a phage and its descendants.

plasmid Autonomously replicating extrachromosomal DNA molecule.

plate (1) A flat dish used to culture microbes. (2) To spread cells over the surface of solid medium in a plate.

pleiotropic mutation A mutation that has effects on several different characters.

point mutation A mutation that can be mapped to one specific locus.

Poisson distribution A mathematical expression giving the probability of observing various numbers of a particular event in a sample when the probability of an event on any one trial is very small, but on which large numbers of trials have been made.

poky A slow-growing mitochondrial mutant in *Neurospora.*

polar gene conversion A gradient of conversion frequency along the length of a gene.

polar granules Granules at the anterior end of the cytoplasm of a *Drosophila* egg.

polar mutation A mutation that affects the transcription or translation of the part of the gene or operon only on one side of the mutant site—for example, nonsense mutations, frame-shift mutations, and IS-induced mutations.

poly-A tail A string of adenine nucleotides added to mRNA after transcription.

polyacrilamide A material used to make electrophoretic gels for separation of mixtures of macromolecules.

polycistronic mRNA An mRNA that codes for more than one protein.

polygenes *See* **multiple-factor hypothesis.**

polymorphism The occurrence in a population (or among populations) of several phenotypic forms associated with alleles of one gene or homologs of one chromosome.

polypeptide A chain of linked amino acids; a protein.

polyploid A cell having three or more chromosome sets, or an organism composed of such cells.

polysaccharide A biological polymer composed of sugar subunits—for example, starch or cellulose.

polytene chromosome A giant chromosome produced by an endomitotic process in which the multiple DNA sets remain bound in a haploid number of chromosomes.

position effect Used to describe a situation where the phenotypic influence of a gene is altered by changes in the position of the gene within the genome.

position-effect variegation Variegation caused by the inactivation of a gene in some cells through its abnormal juxtaposition with heterochromatin.

positive assortative mating A situation in which like phenotypes mate more commonly than expected by chance.

positive control Regulation mediated by a protein that is required for the activation of a transcription unit.

primary structure of a protein The sequence of amino acids in the polypeptide chain.

Pro Proline (an amino acid).

probe Defined nucleic acid segment that can be used to identify specific DNA molecules bearing the complementary sequence, usually through autoradiography.

product of meiosis One of the (usually four) cells formed by the two meiotic divisions.

product rule The probability of two independent events occurring simultaneously is the product of the individual probabilities.

proflavin A mutagen that tends to produce frame-shift mutations.

prokaryote An organism composed of a prokaryotic cell, such as bacteria and blue-green algae.

prokaryotic cell A cell having no nuclear membrane and hence no separate nucleus.

promoter A regulatory region a short distance from the 5′ end of a gene that acts as the binding site for RNA polymerase.

prophage A phage "chromosome" inserted as part of the linear structure of the DNA chromosome of a bacterium.

prophase The early stage of nuclear division during which chromosomes condense and become visible.

propositus In a human pedigree, the individual who first came to the attention of the geneticist.

proto-oncogene A gene that, when mutated or otherwise affected, becomes an oncogene.

protoplast A plant cell whose wall has been removed.

prototroph A strain of organisms that will proliferate on minimal medium (*compare* **auxotroph**).

provirus A virus "chromosome" integrated into the DNA of the host cell.

pseudodominance The sudden appearance of a recessive phenotype in a pedigree, due to deletion of a masking dominant gene.

pseudogene An inactive gene derived from an ancestral active gene.

puff *See* **chromosome puff.**

pulse-chase experiment An experiment in which cells are grown in radioactive medium for a brief period (the pulse) and then transferred to nonradioactive medium for a longer period (the chase).

pulsed-field gel electrophoresis An electrophoretic technique in which the gel is subjected to electrical fields alternating between different angles, allowing very large DNA fragments to "snake" through the gel, and hence permit efficient separation of mixtures of such large fragments.

Punnett square A grid used as a graphic representation of the progeny zygotes resulting from different gamete fusions in a specific cross.

pure-breeding line or strain A group of identical individuals that always produce offspring of the same phenotype when intercrossed.

purine A type of nitrogen base; the purine bases in DNA are adenine and guanine.

pyrimidine A type of nitrogen base; the pyrimidine bases in DNA are cytosine and thymine.

quantitative variation The existence of a range of phenotypes for a specific character, differing by degree rather than by distinct qualitative differences.

quaternary structure of a protein The multimeric constitution of the protein.

R plasmid A plasmid containing one or several transposons that bear resistance genes.

random genetic drift Changes in allele frequency that result because the genes appearing in offspring do not represent a perfectly representative sampling of the parental genes.

random mating Mating between individuals where the choice of a partner is not influenced by the genotypes (with respect to specific genes under study).

reading frame The codon sequence that is determined by reading nucleotides in groups of three from some specific start codon.

realized heritability The ratio of the single-generation progress of selection to the selection differential of the parents.

reannealing Spontaneous realignment of two single DNA strands to re-form a DNA double helix that had been denatured.

receptor element A controlling element that can insert into a gene (making it a mutant) and can also exist (thus making the mutation unstable); both of these functions are nonautonomous, being under the influence of the regulator element.

recessive allele An allele whose phenotypic effect is not expressed in a heterozygote.

recessive phenotype The phenotype of a homozygote for the recessive allele; the parental phenotype that is not expressed in a heterozygote.

reciprocal crosses A pair of crosses of the type genotype A ♀ × genotype B ♂ and genotype B ♀ × genotype A ♂.

reciprocal translocation A translocation in which part of one chromosome is exchanged with a part of a separate nonhomologous chromosome.

recombinant An individual or cell with a genotype produced by recombination.

recombinant DNA A novel DNA sequence formed by the combination of two nonhomologous DNA molecules.

recombinant frequency (RF) The proportion (or percentage) of recombinant cells or individuals.

recombination (1) In general, any process in a diploid or partially diploid cell that generates new gene or chromosomal combinations not found in that cell or in its progenitors. (2) At meiosis, the process that generates a haploid product of meiosis whose genotype is different from either of the two haploid genotypes that constituted the meiotic diploid.

recombinational repair The repair of a DNA lesion through a process, similar to recombination, that uses recombination enzymes.

recon A region of a gene within which there can be no crossing-over; now known to be a nucleotide pair.

reduction division A nuclear division that produces daughter nuclei each having one-half as many centromeres as the parental nucleus.

redundant DNA *See* **repetitive DNA.**

regression A term coined by Galton for the tendency of the quantitative traits of offspring to be closer to the population mean than are their parents' traits. It arises from dominance, gene interaction, and nongenetic influences on traits.

regression coefficient The slope of the straight line that most closely relates two correlated variables.

regulator element *See* **receptor element.**

regulatory genes Genes that are involved in turning on or off the transcription of structural genes.

repetitive DNA Redundant DNA; DNA sequences that are present in many copies per chromosome set.

replication DNA synthesis.

replication fork The point at which the two strands of DNA are separated to allow replication of each strand.

replicon A chromosomal region under the influence of one adjacent replication-initiation locus.

reporter gene A gene whose phenotypic expression is easy to monitor; used to study promoter activity at different times or developmental stages, in recombinant DNA constructs in which the reporter is attached to a promoter region of particular interest.

repressor protein A molecule that binds to the operator and prevents transcription of an operon.

repulsion conformation Two linked heterozygous gene pairs in the arrangement *A b / a B*.

resolving power The ability of an experimental technique to distinguish between two genetic conditions (typically discussed when one condition is rare and of particular interest).

restriction enzyme An endonuclease that will recognize specific target nucleotide sequences in DNA and break the DNA chain at those points; a variety of these enzymes are known, and they are extensively used in genetic engineering.

restrictive conditions Environmental conditions under which a conditional mutant shows the mutant phenotype.

retrovirus An RNA virus that replicates by first being converted into double-stranded DNA.

reverse transcriptase An enzyme that catalyzes the synthesis of a DNA strand from an RNA template.

reversion The production of a wild-type gene from a mutant gene.

RF *See* **recombinant frequency.**

RFLP mapping A technique in which DNA restriction fragment length polymorphisms are used as reference loci for mapping in relation to known genes or other RFLP loci.

ribonucleic acid *See* **RNA.**

ribosomal RNA *See* **rRNA.**

ribosome A complex organelle that catalyzes translation of messenger RNA into an amino-acid sequence. Composed of proteins plus rRNA.

RNA (ribonucleic acid) A single-stranded nucleic acid similar to DNA but having ribose sugar rather than deoxyribose sugar and uracil rather than thymine as one of the bases.

RNA polymerase An enzyme that catalyzes the synthesis of an RNA strand from a DNA template.

rRNA (ribosomal RNA) A class of RNA molecules, coded in the nucleolar organizer, that have an integral (but poorly understood) role in ribosome structure and function.

S (Svedberg unit) A unit of sedimentation velocity, commonly used to describe molecular units of various sizes (because sedimentation velocity is related to size).

satellite A terminal section of a chromosome, separated from the main body of the chromosome by a narrow constriction.

satellite chromosomes Chromosomes that seem to be additions to the normal genome.

satellite DNA DNA that forms a separate band in a density gradient because of its different nucleotide composition.

scaffold The central core of a eukaryotic nuclear chromosome, from which DNA loops extend.

SCE *See* **sister-chromatid exchange.**

secondary structure of a protein A spiral or zigzag arrangement of the polypeptide chain.

second-division segregation pattern A pattern of ascospore genotypes for a gene pair showing that the two alleles separate into different nuclei only at the second meiotic division, as a result of a crossover between that gene pair and its centromere; can only be detected in a linear ascus.

second-site mutation The second mutation of a double mutation within a gene; in many cases, the second-site mutation suppresses the first mutation, so that the double mutant has the wild-type phenotype.

sector An area of tissue whose phenotype is detectably different from the surrounding tissue phenotype.

sedimentation The sinking of a molecule under the opposing forces of gravitation and buoyancy.

segregation (1) Cytologically, the separation of homologous structures. (2) Genetically, the production of two separate phenotypes, corresponding to two alleles of a gene, either in different individuals (meiotic segregation) or in different tissues (mitotic segregation).

segregational petite A petite that in a cross with wild-type produces $\frac{1}{2}$ petite and $\frac{1}{2}$ wild-type progeny; caused by a nuclear mutation.

selection coefficient (s) The proportional excess or deficiency of fitness of one genotype in relation to another genotype.

selection differential The difference between the mean of a population and the mean of the individuals selected to be parents of the next generation.

selection progress The difference between the mean of a population and the mean of the offspring in the next generation born to selected parents.

selective neutrality A situation in which different alleles of a certain gene confer equal fitness.

selective system An experimental technique that enhances the recovery of specific (usually rare) genotypes.

self To fertilize eggs with sperms from the same individual.

self-assembly The ability of certain multimeric biological structures to assemble from their component parts through random movements of the molecules and formation of weak chemical bonds between surfaces with complementary shapes.

semiconservative replication The established model of DNA replication in which each double-stranded molecule is composed of one parental strand and one newly polymerized strand.

semisterility The phenotype of individuals heterozygotic for certain types of chromosome aberration; expressed as a reduced number of viable gametes and hence reduced fertility.

senDNA An amplified section of mtDNA found in senescent *Podospora* cultures; circular and plasmid-like in nature.

Ser Serine (an amino acid).

sex chromosome A chromosome whose presence or absence is correlated with the sex of the bearer; a chromosome that plays a role in sex determination.

sex linkage The location of a gene on a sex chromosome.

sexduction Sexual transmission of donor *E. coli* chromosomal genes on the fertility factor.

sexual spore *See* **spore.**

shotgun technique Cloning a large number of different DNA fragments as a prelude to selecting one particular clone type for intensive study.

shuttle vector A vector (e.g. a plasmid) constructed in such a way that it can replicate in at least two different host species, allowing a DNA segment to be tested or manipulated in several cellular settings.

signal sequence The N-terminal sequence of a secreted protein, which is required for transport through the cell membrane.

silent mutation Mutation in which the function of the protein product of the gene is unaltered.

sister-chromatid exchange (SCE) An event similar to crossing-over that can occur between sister chromatids at mitosis or at meiosis; detected in harlequin chromosomes.

site-specific recombination Recombination occurring between two specific sequences that need not be homologous, and mediated by a specific recombination system.

S-9 mix A liver-derived supernatant used in the Ames test to activate or inactivate mutagens.

solenoid structure The supercoiled arrangement of DNA in eukaryotic nuclear chromosomes.

somatic cell A cell that is not destined to become a gamete; a "body cell," whose genes will not be passed on to future generations.

somatic mutation A mutation occurring in a somatic cell.

somatic-cell genetics Asexual genetics, involving study of somatic mutation, assortment, and crossing-over, and of cell fusion.

somatostatin A human growth hormone.

SOS repair The error-prone process whereby gross structural DNA damage is circumvented by allowing replication to proceed past the damage through imprecise polymerization.

spacer DNA Repetitive DNA found between genes; its function is unknown.

specialized (restricted) transduction The situation in which a particular phage will transduce only specific regions of the bacterial chromosome.

specific-locus test A system for detecting recessive mutations in diploids. Normal individuals treated with mutagen are mated to testers that are homozygous for the recessive alleles at a number of specific loci; the progeny are then screened for recessive phenotypes.

spindle The set of microtubular fibers that appear to move eukaryotic chromosomes during division.

splicing The reaction that removes introns and joins together exons in RNA.

spontaneous mutation A mutation occurring in the absence of mutagens, usually due to errors in the normal functioning of cellular enzymes.

spore (1) In plants and fungi, sexual spores are the haploid cells produced by meiosis. (2) In fungi, asexual spores are somatic cells that are cast off to act either as gametes or as the initial cells for new haploid individuals.

sporophyte The diploid sexual-spore-producing generation in the life cycle of plants—that is, the stage in which meiosis occurs.

stacking The packing of the flattish nitrogen bases at the center of the DNA double helix.

staggered cuts The cleavage of two opposite strands of duplex DNA at points near one another.

standard deviation The square root of the variance.

statistic A computed quantity characteristic of a population, such as the mean.

statistical distribution The array of frequencies of different quantitative or qualitative classes in a population.

strain A pure-breeding lineage, usually of haploid organisms, bacteria, or viruses.

structural gene A gene encoding the amino-acid sequence of a protein.

subvital gene A gene that causes the death of some proportion (but not all) of the individuals that express it.

sum rule The probability that one or the other of two mutually exclusive events will occur is the sum of their individual probabilities.

supercoil A closed double-stranded DNA molecule that is twisted on itself.

superinfection Phage infection of a cell that already harbors a prophage.

supersuppressor A mutation that can suppress a variety of other mutations; typically a nonsense suppressor.

suppressive petite A petite that in a cross with wild-type produces progeny of which variable non-Mendelian proportions are petite.

suppressor mutation A mutation that counteracts the effects of another mutation. A suppressor maps at a different site than the mutation it counteracts, either within the same gene or at a more distant locus. Different suppressors act in different ways.

synapsis Close pairing of homologs at meiosis.

synaptonemal complex A complex structure that unites homologs during the prophase of meiosis.

syncytium A single cell with many nuclei.

T (1) Thymine, or thymidine. (2) *See* **tetratype.**

tandem duplication Adjacent identical chromosome segments.

tautomeric shift The spontaneous isomerization of a nitrogen base to an alternative hydrogen-bonding condition, possibly resulting in a mutation.

T-DNA A portion of the Ti plasmid that is inserted into the genome of the host plant cell.

telocentric chromosome A chromosome having the centromere at one end.

telomere The tip (or end) of a chromosome.

telophase The late stage of nuclear division when daughter nuclei re-form.

temperate phage A phage that can become a prophage.

temperature-sensitive mutation A conditional mutation that produces the mutant phenotype in one temperature range and the wild-type phenotype in another temperature range.

template A molecular "mold" that shapes the structure or sequence of another molecule; for example, the nucleotide sequence of DNA acts as a template to control the nucleotide sequence of RNA during transcription.

teratogen An agent that interferes with normal development.

teratoma A tumor composed of a chaotic array of different tissue types.

terminal redundancy In phage, a linear DNA molecule with single-stranded ends that are longer than is necessary to close the DNA circle.

tertiary structure of a protein The folding or coiling of the secondary structure to form a globular molecule.

testcross A cross of an individual of unknown genotype or a heterozygote (or a multiple heterozygote) to a tester individual.

tester An individual homozygous for one or more recessive alleles; used in a testcross.

testicular feminization The creation of an apparent female phenotype in an XY individual as a result of an X-linked mutation.

tetrad (1) Four homologous chromatids in a bundle in the first meiotic prophase and metaphase. (2) The four haploid product cells from a single meiosis.

tetrad analysis The use of tetrads (definition 2) to study the behavior of chromosomes and genes during meiosis.

tetraparental mouse A mouse that develops from an embryo created by the experimental fusion of two separate blastulas.

tetraploid A cell having four chromosome sets; an organism composed of such cells.

tetratype (T) A tetrad type containing four different genotypes, two parental and two recombinant.

Thr Threonine (an amino acid).

three-point testcross A testcross involving one parent with three heterozygous gene pairs.

thymidine The nucleoside having thymine as its base.

thymine A pyrimidine base that pairs with adenine.

thymine dimer A pair of chemically bonded adjacent thymine bases in DNA; the cellular processes that repair this lesion often make errors that create mutations.

Ti plasmid A circular plasmid of *Agrobacterium tumifaciens* that enables the bacterium to infect plant cells and produce a tumor (crown gall tumor.)

totipotency The ability of a cell to proceed through all the stages of development and thus produce a normal adult.

trans conformation In a heterozygote involving two mutant sites within a gene or gene cluster, the arrangement $a_1 + / + a_2$.

transcription The synthesis of RNA using a DNA template.

transdetermination A specific change in the fate of an imaginal disk that can occur when the disk is cultured.

transduction The movement of genes from a bacterial donor to a bacterial recipient using a phage as the vector.

transfer RNA *See* **tRNA.**

transformation (1) The directed modification of a genome by the external application of DNA from a cell of different genotype. (2) Conversion of normal higher eukaryotic cells in tissue culture to a cancer-like state of uncontrolled division.

transgenic organism One whose genome has been modified by externally-applied new DNA.

transient diploid The stage of the life cycle of predominantly haploid fungi (and algae) during which meiosis occurs.

transition A type of nucleotide-pair substitution involving the replacement of a purine with another purine, or of a pyrimidine with another pyrimidine—for example, GC→AT.

translation The ribosome-mediated production of a polypeptide whose amino-acid sequence is derived from the codon sequence of an mRNA molecule.

translocation The relocation of a chromosomal segment in a different position in the genome.

transmission genetics The study of the mechanisms involved in the passage of a gene from one generation to the next.

transposable genetic element A general term for any genetic unit that can insert into a chromosome, exit, and relocate; includes insertion sequences, transposons, some phages, and controlling elements.

transposition *See* **translocation.**

transposon A mobile piece of DNA that is flanked by terminal repeat sequences and typically bears genes coding for transposition functions.

transversion A type of nucleotide-pair substitution involving the replacement of a purine with a pyrimidine, or vice versa—for example, GC→TA.

triplet The three nucleotide pairs that compose a codon.

triploid A cell having three chromosome sets, or an organism composed of such cells.

trisomic Basically a diploid with an extra chromosome of one type, producing a chromosome number of the form $2n + 1$.

tritium A radioactive isotope of hydrogen.

tRNA (transfer RNA) A class of small RNA molecules that bear specific amino acids to the ribosome during translation; the amino acid is inserted into the growing polypeptide chain when the anticodon of the tRNA pairs with a codon on the mRNA being translated.

Trp Tryptophan (an amino acid).

true-breeding line or strain *See* **pure-breeding line or strain.**

truncation selection A breeding technique in which individuals in whom quantitative expression of a phenotype is above or below a certain value (the truncation point) are selected as parents for the next generation.

Turner's syndrome An abnormal human female phenotype produced by the presence of only one X chromosome (XO).

twin spot A pair of mutant sectors within wild-type tissue, produced by a mitotic crossover in an individual of appropriate heterozygous genotype.

2μ (2 micron) plasmid A naturally-occurring extragenomic circular DNA molecule (Found in some yeast cells, with a circumference of 2μ). Engineered to form the basis for several types of gene vectors in yeast.

Tyr Tyrosine (an amino acid).

U Uracil, or uridine.

underdominance A phenotypic relation in which the phenotypic expression of the heterozygote is less than that of either homozygote.

unequal crossover A crossover between homologs that are not perfectly aligned.

uniparental inheritance The transmission of certain phenotypes from one parental type to all the progeny; such inheritance is generally produced by organelle genes.

unstable mutation A mutation that has a high frequency of reversion; a mutation caused by the insertion of a controlling element, whose subsequent exit produces a reversion.

uracil A pyrimidine base that appears in RNA in place of thymine found in DNA.

URF (unassigned reading frame) A DNA region, identified by nucleotide sequencing studies, that has an initiation and a termination codon, and therefore presumably represents a gene, but for which no phenotype or function is known.

uridine The nucleoside having uracil as its base.

Val Valine (an amino acid).

variable A property that may have different values in various cases.

variable region A region in an immunoglobin molecule that shows many sequence differences between antibodies of different specificities; the part of the antibody that binds to the antigen.

variance A measure of the variation around the central class of a distribution; the average squared deviation of the observations from their mean value.

variant An individual organism that is recognizably different from an arbitrary standard type in that species.

variate A specific numerical value of a variable.

variation The differences among parents and their offspring or among individuals in a population.

variegation The occurrence within a tissue of sectors with differing phenotypes.

vector In cloning, the plasmid or phage chromosome used to carry the cloned DNA segment.

viability The probability that a fertilized egg will survive and develop into an adult organism.

virulent phage A phage that cannot become a prophage; infection by such a phage always leads to lysis of the host cell.

wild-type The genotype or phenotype that is found in nature or in the standard laboratory stock for a given organism.

wobble The ability of certain bases at the third position of an anticodon in tRNA to form hydrogen bonds in various ways, causing alignment with several possible codons.

X linkage The presence of a gene on the X chromosome but not on the Y.

X-and-Y linkage The presence of a gene on both the X and Y chromosomes (rare).

X-ray crystallography A technique for deducing molecular structure by aiming a beam of X rays at a crystal of the test compound and measuring the scatter of rays.

Y linkage The presence of a gene on the Y chromosome but not on the X (rare).

zygote The cell formed by the fusion of an egg and a sperm; the unique diploid cell that will divide mitotically to create a differentiated diploid organism.

zygotic induction The sudden release of a lysogenic phage from an Hfr chromosome when the prophage enters the F⁻ cell, and the subsequent lysis of the recipient cell.

Answers to Selected Problems

Chapter 2

1. Mendel's first law states that alleles segregate during meiosis. Mendel's second law states that genes independently assort during meiosis.

3. The progeny ratio is 3 : 1, indicating a classical heterozygous by heterozygous mating. The parents must be Bb × Bb. Their black progeny must be BB and Bb in a 1 : 2 ratio and their white progeny must be bb.

5. a. $(\frac{1}{6})^3$

   b. $(\frac{1}{6})^3$

   c. $(\frac{1}{6})^3$

   d. $(\frac{5}{6})^3$

   e. $3(\frac{1}{6})^3$

   f. $2(\frac{1}{6})^3$

   g. $(\frac{1}{6})^2$

   h. $\frac{5}{9}$

7. Charlie or his mate, or both, obviously was not pure-breeding because his F2 progeny were of two phenotypes. Let A = black and white and a = red and white. If both parents were heterozygous, then red and white would have been expected in the F1 generation. It was not observed so only one of the parents was heterozygous. The cross is:

   P   Aa × AA
   F1 1 Aa : 1 AA

   Two F1 heterozygotes (Aa) when crossed would give

   1 AA (black and white) : 2 Aa (black and white) : 1 aa (red and white).

   If the red and white F2 progeny were from more than one mate of Charlie's, then the farmer acted correctly. However, if the F2 progeny came only from one mate, the farmer may have acted too quickly.

8. You are told that normal parents have affected offspring. This is the pattern with recessive disorders.

12. a. The man's father was probably a heterozygote. Therefore, the man has a 50 percent chance of inheriting the dominant allele.

b. $\frac{1}{4}$

13. a. Dominant

b. $Aa \times Aa \rightarrow 1\ AA : 2\ Aa : 1\ aa$

c. $\frac{1}{4}$ probability of normal, $\frac{3}{4}$ probability of dwarf

15. The plants are 3 spotted : 1 unspotted.

a. Let A = spotted and a = unspotted.

$$P \qquad Aa \text{ (spotted)} \times Aa \text{ (spotted)}$$
$$F1 \qquad 1\ AA : 2\ Aa : 1\ aa$$
$$3\ A\!-\!\text{(spotted)} : 1\ aa \text{ (unspotted)}$$

b. All unspotted plants should be true-breeding and $\frac{1}{3}$ of the spotted plants should be true-breeding.

20. Pedigree 1: recessive
Genotypes:
  generation 1: AA, aa
  generation 2: $Aa, Aa, Aa, A-, A-, Aa$
  generation 3: Aa, Aa
  generation 4: aa
Pedigree 2: dominant
Genotypes:
  generation 1: Aa, aa, Aa, aa
  generation 2: $aa, aa, Aa, Aa, aa, aa, Aa, Aa, aa$
  generation 3: $aa, aa, aa, aa, aa, A-, A-, A-, Aa, aa$
  generation 4: aa, aa, aa
Pedigree 3: dominant
Genotypes:
  generation 1: Aa, aa
  generation 2: Aa, aa, aa, Aa
  generation 3: aa, Aa, aa, aa, Aa, aa
  generation 4: aa, Aa, Aa, Aa, aa, aa
Pedigree 4: recessive
Genotypes:
  generation 1: $aa, A-, Aa, Aa$
  generation 2: $Aa, Aa, Aa, aa, A-, aa, A-, A-, A-, A-$
  generation 3: Aa, aa, Aa, Aa, aa, Aa

21.
$$\frac{1}{2}A \left\langle \begin{array}{l} \frac{1}{2}B = \frac{1}{4}\,AB \\ \frac{1}{2}b = \frac{1}{4}\,Ab \end{array} \right.$$
$$\frac{1}{2}a \left\langle \begin{array}{l} \frac{1}{2}B = \frac{1}{4}\,aB \\ \frac{1}{2}b = \frac{1}{4}\,ab \end{array} \right.$$

22. $\frac{1}{32}$

24. a. $Cc\ Ss \times Cc\ Ss$

b. $CC\ Ss \times CC\ ss$

c. $Cc\ SS \times cc\ SS$

d. $cc\ Ss \times cc\ Ss$

e. $Cc\ ss \times Cc\ ss$

f. $CC\ Ss \times CC\ Ss$

g. $Cc\ Ss \times Cc\ ss$

26. a. Two genes: shape and skin covering.

b. Bow and knock are alleles; hairy and smooth are alleles.

c. Bow is dominant (call it B). Hairy is dominant (call it H).

d. $Bb\ HH,\ Bb\ hh,\ Bb\ Hh,\ bb\ hh,\ BB\ Hh$

27. a. 1. $\frac{9}{128}$
   2. $\frac{9}{128}$
   3. $\frac{9}{64}$
   4. $\frac{55}{64}$

b. 1. $\frac{1}{32}$
   2. $\frac{1}{32}$
   3. $\frac{1}{16}$
   4. $\frac{15}{16}$

28. $\frac{4}{7}$ of the children will be affected.

Chapter 3

1. a. In mitosis the chromosome number remains unchanged, while in meiosis the chromosome number is halved.

b. In mitosis sister chromatids separate from each other, while in meiosis both homologous chromosomes and sister chromatids separate from each other.

c. Mitosis leads to two cells, while meiosis leads to four cells.

d. Homologous pairing occurs only in meiosis.

e. Recombination occurs at a much higher frequency during meiosis as compared to mitosis.

3.
$$P \qquad ad^-\ a \times ad^+\ \alpha$$
$$\text{Transient diploid} \qquad ad^+/ad^-\ a/\alpha$$
$$F1 \qquad 1\ ad^+\ a$$
$$1\ ad^-\ a$$
$$1\ ad^+\ \alpha$$
$$1\ ad^-\ \alpha$$

5.

	Mitosis	Meiosis
Fern	Sporophyte Gametophyte	Prothallus
Moss	Sporophyte Gametophyte	Archegonia Antheridia
Flowering plant	Sporophyte Gametophyte	Flowers
Pine tree	Sporophyte Gametophyte	Pinecones
Mushroom	Sporophyte Gametophyte	Hyphae
Frog	Somatic cells	Gonads
Butterfly	Somatic cells	Gonads
Snail	Somatic cells	Gonads

7. a. 46 chromosomes, each with 2 chromatids = 92 chromatids

   b. 46 chromosomes, each with 2 chromatids = 92 chromatids

   c. 46 chromosomes in each of 2 about-to-be-formed cells, each with 1 chromatid

   d. 23 chromosomes in each of 2 about-to-be-formed cells, each with 2 chromatids,

   e. 23 chromosomes in each of 2 about-to-be-formed cells, each with 1 chromatid

9. a. (4) 20 (or 10 pairs)

   b. The cell goes from 10 pairs of chromosomes, each with 2 chromatids, to 10 chromosomes, each with 1 chromatid.

10. $\frac{1}{8}$

11. $(\frac{1}{2})^{n-1}$

14. P GO (graceful female) $\times$ gg (gruesome male)

    F1   1 gO gruesome female

          1 Gg graceful male

The O can be a female-determining chromosome or no chromosome.

15. d. X-linked dominant

18. a. $\frac{1}{8}$

   b. $\frac{1}{4}$

   c. 0

21. a. Let B = black and Y = orange.

Females	Males
$X^B X^B$ = black	$X^B Y$ = black
$X^Y X^Y$ = orange	$X^Y Y$ = orange
$X^B X^Y$ = tortoise	

   b. P $X^Y X^Y$ (orange) $\times$ $X^B Y$ (black)

       F1 $X^B X^Y$ tortoise female

           $X^Y Y$ orange male

   c. P $X^B X^B$ (black) $\times$ $X^Y Y$ (orange)

       F1 $X^B X^Y$ tortoise female

           $X^B Y$ black male

   d. The mother had to have been tortoise-shell. The daughters are $\frac{1}{2}$ black, which means that their father was black.

   e. Males were orange or black, indicating that the mothers were tortoise-shell. Orange females mean that the father was orange.

22. e. $\frac{1}{4}$

26. a. Autosomal recessive: excluded

   b. Autosomal dominant: consistent

   c. X-linked recessive: excluded

   d. X-linked dominant: excluded

   e. Y-linked: excluded

27. a. Cross 6, bent $\times$ bent $\rightarrow$ normal, indicates that bent is dominant.

   b. X-linked

   c. Let B = bent and b = normal

	Parents		Progeny	
Cross	Female	Male	Female	Male
1	bb	BY	Bb	bY
2	Bb	bY	Bb, bb	BY, bY
3	BB	bY	Bb	BY
4	bb	bY	bb	bY
5	BB	BY	BB	BY
6	Bb	BY	BB, Bb	BY, bY

30. a. $\frac{1}{4} p^+ h^+$

         $\frac{1}{4} p^+ h$

         $\frac{1}{4} p \; h^+$

         $\frac{1}{4} p \; h$

   b. Because both genes are expressed in the sporophyte, they would not necessarily be expressed in the gametophyte.

   c. This is a dihybrid cross and the expected result is:

        9 $p^+ h^+$ green, hairy

        3 $p^+ h$ green, hairless

        3 $p \; h^+$ pink, hairy

        1 $p \; h$ pink, hairless

Chapter 4

1. The woman must be AO, so the mating is AO $\times$ AB. Their children will be:

Genotype	Phenotype
1 AA	A
1 AB	AB
1 AO	A
1 BO	B

3. The hairless dogs are heterozygous (Hh), and H is a recessive lethal. The cross is: $Hh \times Hh$. The progeny are:

     1 HH deformed, dead

     2 Hh hairless

     1 hh normal

To test this hypothesis, cross a hairless (Hh) with a normal (hh). The progeny should be $\frac{1}{2}$ hairless, $\frac{1}{2}$ normal.

5. To test the hypothesis that the erminette phenotype is a heterozygous phenotype, you could cross an erminette with either, or both, of the homozygotes. You should observe a 1:1 ratio in the progeny of both crosses.

7. $\frac{1}{4}$ fully fertile: $\frac{1}{2}$ partially fertile: $\frac{1}{4}$ fully sterile

9. e

12. Baby 1 from cross 4, baby 2 from cross 2, baby 3 from cross 1 and baby 4 from cross 3

14. Platinum must be dominant to normal color and heterozygous (Aa). An 82:38 ratio is very close to a 2:1. Because a 1:2:1 ratio is expected in a heterozygous cross, one genotype is nonviable. It must be the AA genotype, homozygous platinum, that is nonviable because the homozygous recessive genotype is normal color (aa). The platinum allele is a pleiotropic allele which governs coat color in the heterozygous state and is lethal in the recessive state.

16 A cross of the short-bristle female with a normal male results in two phenotypes with regard to bristles and an abnormal sex ratio of 2 females:1 male. All the males are normal, while the females are normal and short in equal numbers. The short bristle phenotype must be heterozygous and the allele must be a recessive lethal. The first cross was $Aa \times aY$.

Long-bristle females (aa) were crossed with long-bristle males (aY). All their progeny would be expected to be long-bristle (aa or aY).

Short-bristle females (Aa) were crossed with long-bristle males (aY). The progeny expected are:

1 Aa short females
1 aa long females
1 aY long males
1 AY nonviable

20. One parent could be $aaBB$ and the other parent could be $AAbb$. All offspring would be $AaBb$ (normal). The doubly heterozygous offspring have one copy of a functional allele for each gene, whereas each of the two parents is lacking a functional allele for one of the genes.

23. If c^{sr} = sun-red, c^o = orange, and c^p = pink, then the crosses and the results are:

Cross 1 $c^{sr}c^{sr} \times c^p c^p$ gives F1 $c^{sr}c^p$; F2 $3c^{sr}-:1 c^p c^p$

Cross 2 $c^o c^o \times c^{sr}c^{sr}$ gives F1 $c^{sr} c^o$; F2 $3c^{sr}-:1 c^o c^o$

Cross 3 $c^o c^o \times c^p c^p$ gives F1 $c^o c^p$; F2 $3c^o-:1 c^p c^p$

Cross 4 presents a new situation. Two genes are involved and there is epistasis. Let a stand for the scarlet gene and A for its colorless allele. Assume that there is a dominant allele, C, that blocks the expression of the gene being studied.

Cross 4   P $c^o c^o AA \times CCaa$

F1 $Cc^o Aa$

F2   9 $C-A-$ yellow

3 $C-aa$ scarlet

3 $c^o c^o A-$ orange

1 $c^o c^o aa$ orange (epistasis, with c^o blocking the expression of aa).

26. Let scarlet be represented by $aaBB$ and dark brown by $AAbb$. The F1 progeny must be $AaBb$. The F2 progeny are:

432 red $A-B-$

158 scarlet $aaB-$

139 brown $AAbb$

52 white $----$

This is a 9:3:3:1 ratio, indicating a dihybrid cross. The white phenotype must be $aabb$.

29. a. Two genes are involved. The mutants are recessive. The defect in line B is autosomal.

P   line A males $\times$ line B females

$d-EE \times DDee$

F1   200 wild-type males $D-Ee$
198 wild-type females $DdEe$

where d is the defective allele in line A and e is the defective allele in line B. The dash indicates that the location of the D/d gene is unknown.

b. The defect in line A is X-linked.

c.   P $dYEE \times DDee$

F1 $DYEe \times DdEe$

F2   Females

$\frac{1}{2} DD$ $\begin{cases} \frac{1}{4} EE = \frac{1}{8} \text{ DDEE wild-type} \\ \frac{1}{2} Ee = \frac{1}{4} \text{ DDEe wild-type} \\ \frac{1}{4} ee = \frac{1}{8} \text{ DDee scarlet} \end{cases}$

$\frac{1}{2} Dd$ $\begin{cases} \frac{1}{4} EE = \frac{1}{8} \text{ DdEE wild-type} \\ \frac{1}{2} Ee = \frac{1}{4} \text{ DdEe wild-type} \\ \frac{1}{4} ee = \frac{1}{8} \text{ Ddee scarlet} \end{cases}$

Wild-type: scarlet = 3:1

Males

$\frac{1}{2} DY$ $\begin{cases} \frac{1}{4} EE = \frac{1}{8} \text{ DYEE wild-type} \\ \frac{1}{2} Ee = \frac{1}{4} \text{ DYEe wild-type} \\ \frac{1}{4} ee = \frac{1}{8} \text{ DYee scarlet} \end{cases}$

$\frac{1}{2} dY$ $\begin{cases} \frac{1}{4} EE = \frac{1}{8} \text{ dYEE scarlet} \\ \frac{1}{2} Ee = \frac{1}{4} \text{ dYEe scarlet} \\ \frac{1}{2} ee = \frac{1}{8} \text{ dYee scarlet} \end{cases}$

Wild-type: scarlet = 3:5

32.

Cross	Results	Conclusion
$A-C-R- \times aaccRR$	50% colored	Colored or white will depend on the A and C genes. Because half the seeds are colored, one of the two genes is heterozygous.

32.

Cross	Results	Conclusion
$A-C-R- \times aaCCrr$	25% colored	Color depends on A and R here. If only one gene were heterozygous, 50% would be colored. Therefore, both A and R are heterozygous. The seed is $AaCCRr$.
$A-C-R- \times AAccrr$	50% colored	This confirms the above conclusion.

33. a.

1 td su	wild-type
1 td su^+	requires tryptophan
1 td^+ su^+	wild-type
1 td^+ su	wild-type

b. 1 tryptophan-dependent : 3 tryptophan-independent

34. Cross 1 is:

P $WwBbOo$ (white) $\times$ $WwBbOo$ (white)

F1 $\frac{3}{4}$ $W-----$white ($\frac{48}{64}$)
$\frac{1}{4}$ $WW----$lethal ($\frac{16}{64}$)
$\frac{1}{2}$ $Ww----$white ($\frac{32}{64}$)
$\frac{1}{4}$ $ww----$colored ($\frac{16}{64}$)

Among the colored animals, the following progeny will be found:

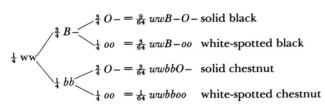

Cross 2 is:

P $WwBbOo$ (white) $\times$ $WwBboo$ (white)

F1 $\frac{3}{4}$ $W-----$white ($\frac{24}{32}$)
$\frac{1}{4}$ $WW----$lethal ($\frac{8}{32}$)
$\frac{1}{2}$ $Ww----$white ($\frac{16}{32}$)
$\frac{1}{4}$ $ww----$colored ($\frac{8}{32}$)

Among the colored animals, the following progeny will be found.

$\frac{3}{32}$ $wwB-O-$ solid black

$\frac{3}{32}$ $wwB-oo$ white-spotted black

$\frac{1}{32}$ $wwbbO-$ solid chestnut

$\frac{1}{32}$ $wwbboo$ white-spotted chestnut

36. Two genes are involved in disease resistance, resistance is dominant to susceptibility and the genes are autosomal. Let D = resistance to alpha and E = resistance to beta.

a.

P $ddEE \times DDee$

F1 $DdEe$

F2 9 $D-E-$ unaffected
3 $ddE-$ diseased
3 $D-ee$ diseased
1 $ddee$ diseased

b. If virulence is due to two separate genes:

let F = virulent in variety A

f = nonvirulent

G = virulent in variety B

g = nonvirulent

Then the cross is $Fg(\alpha) \times fG(\beta)$. Progeny would be of four types:

Fg = disease in A

fg = nonvirulent in A and B

FG = virulent in A and B

fG = virulent in B

In a test against the two varieties, the following results would be obtained, where + indicates infection:

	A	B
Fg	+	−
fg	−	−
FG	+	+
fG	−	+

If virulence is due to one gene, with two alleles:

let F = virulent in A

F' = virulent in B

Progeny from a cross would be F or F' in a 1 : 1 ratio. In a test against the two varieties, the following results would be obtained, where + indicates infection:

	A	B
F	+	−
F'	−	+

37. Pedigrees like this are quite common. They indicate lack of penetrance due to epistasis or environmental effects.

39.   Cross 1:

P $MM\ Dd\ ww \times mm\ dd\ ww$

F1 1 $Mm\ Dd\ ww$: 1 $Mm\ dd\ ww$

Cross 2:

P $mm\ Dd\ Ww \times MM\ dd\ ww$

F1 $\frac{1}{2}\ Mm\ Dd$ (or dd) Ww (or ww)

     $\frac{1}{4}\ Mm\ Dd\ ww$

     $\frac{1}{4}\ Mm\ dd\ ww$

40. a. Let line 1 be $AA\ BB$ and line 2 be $aa\ bb$. The F1 is $Aa\ Bb$. Assume that A blocks color in line 1 and bb blocks color in line 2. The F1 will be white because of the presence of A.

     The F2 are:

         $9\ A-B-$ white

         $3\ A-bb$ white

         $3\ aa\ B-$ red

         $1\ aa\ bb$ white

b.   Cross 1: $AA\ BB \times Aa\ Bb \rightarrow$ all $A-B-$, white

      Cross 2: $aa\ bb \times Aa\ Bb \rightarrow \frac{1}{4}\ Aa\ Bb$, white

                         $\frac{1}{4}\ Aa\ bb$, white

                         $\frac{1}{4}\ aa\ bb$, white

                         $\frac{1}{4}\ aa\ Bb$, red

Chapter 5

2.   P $Ad/Ad \times aD/aD$

           F1 Ad/aD

           F2 $1\ Ad/Ad$

                 $2\ Ad/aD$

                 $1\ aD/aD$

4. Ef and eF are recombinants equaling $\frac{1}{3}$ of the progeny. The two genes are 33.3 map units apart.

6. Because only parental types were recovered, the two genes must be quite close to each other, making recombination quite rare.

7. The female parent was AB/ab. The frequency of recombination between the two genes is 10 m.u.

9. b. 4%.

10. a. All four genes are linked.

b. and c. The map is

     B 10 m.u. A 30 m.u. C

    ——————+————————————+————————————+—

The parental chromosomes actually were:

$B\ (A,\ d)\ c/b\ (a,\ D)\ C$, where ( ) indicates that the order of the genes within is unknown.

d. Interference = .5

12. 20 m.u.

14. The genes are linked. The parental chromosomes were $ADH^F PGM^S/ADH^S PGM^F$. The recombination frequency is 16.88 m.u.

16.   P $abc/abc \times a^+b^+c^+/a^+b^+c^+$

           F1 $abc/a^+b^+c^+ \times abc/a^+b^+c^+$

           F2 $1364\ a^+-b^+-c^+-$

                $365\ abc/abc$

                $87\ aabbc^+-$

                $84\ a^+-b^+-cc$

                $47\ aab^+-c^+-$

                $44\ a^+-bb\ cc$

                 $5\ aa\ b^+-cc$

                 $4\ a^+-bb\ c^+-$

Recombination does not occur in the male *Drosophila*.

a. Use the frequency of abc/abc to estimate the frequency of $a^+b^+c^+$ gametes from the female.

Parentals	730 (2×365)
$a-b$ rec	91 ($a++,\ +bc = 47 + 44$)
$b-c$ rec	171 ($ab+,\ ++c = 87 + 84$)
DCO	$\underline{9}$ ($a+c,\ +b+ = 5 + 4$)
	1001

      $a-b = 10$ m.u.

      $b-c = 18$ m.u.

b. Coefficient of coincidence = .5.

18. The gene sequence is $v\ b\ lg$.

         $v-b = 18.0$ m.u.

         $b-lg = 28.0$ m.u.

         $CC = 0.79$.

20. Let: F = fat, L = long tail, Fl = flagella
The gene sequence is $F\ L\ Fl$.

        P $F\ L\ Fl/f\ l\ fl \times f\ l\ fl/f\ l\ fl$

          F1 $398\ F\ L\ Fl/f\ l\ fl$, parental

               $370\ f\ l\ fl/f\ l\ fl$, parental

               $72\ F\ L\ fl/\ f\ l\ fl$, c–o L–Fl

               $67\ f\ l\ Fl/\ f\ l\ fl$, c–o L–Fl

               $44\ f\ dL\ Fl/\ f\ l\ fl$, c–o F–L

               $35\ F\ l\ fl/\ f\ l\ fl$, c–o F–L

                $9\ f\ L\ fl/f\ l\ fl$, DCO

                $5\ F\ l\ Fl/f\ l\ fl$, DCO

        F   9.3 m.u. L 15.3 m.u. Fl

    ——————+————————————+————————————+—

23. a. $Rh^+ E$ $(R-E-)$

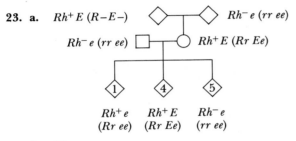

$Rh^- e$ $(rr\ ee)$

$Rh^- e$ $(rr\ ee)$ $Rh^+ E$ $(Rr\ Ee)$

      ①   ④   ⑤

  $Rh^+ e$ $Rh^+ E$ $Rh^- e$

  $(Rr\ ee)$ $(Rr\ Ee)$ $(rr\ ee)$

b. Yes

c. Dominant

d. As drawn, the pedigree indicates independent assortment. However, the data also support linkage, with the Re/re individual representing a crossover. The distance between the two genes would be 100% $(\frac{1}{10})$ = 10 m.u. There is no way to choose between the alternatives without more data.

24. The cross is:

    P $PAR/PAR \times par/par$

    F1 $PAR/par \times par/par$, a three-point testcross

a. 0.34.

b. 0.34.

c. 0.015.

d. 0.06.

26. a. $\chi^2 = 2.1266, P > .50$, nonsignificant. The hypothesis of no linkage cannot be rejected.

b. $\chi^2 = 6.6, P > .10$, nonsignificant. The hypothesis of no linkage cannot be rejected.

c. $\chi^2 = 66.0, P < .005$, significant. The hypothesis of no linkage must be rejected.

d. $\chi^2 = 11.60, P < .01$, significant. The hypothesis of no linkage must be rejected.

28. If h = hemophilia and b = color blindness, the genotypes for individuals in the pedigree can be written as:

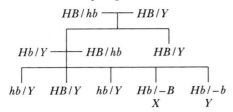

HB/hb ——— HB/Y

Hb/Y —— HB/hb HB/Y

hb/Y HB/Y hb/Y $Hb/-B$ $Hb/-b$

                         X   Y

The mother of the two women in question would produce the following gametes: 0.45 HB, 0.45 hb, 0.05 Hb, 0.05 hB. Woman X can be either Hb/HB (.45) or Hb/hB (.05), because she received B from her mother. If she is Hb/hB [.05/(.45 + .05) = .10 chance], she will produce the parental and recombinant gametes with the same probabilities as her mother. Thus, her child has a 45 percent chance of receiving hB, a 5 percent chance of receiving hb, and a 50 percent chance of receiving a Y from his father. The probability that her child will be a hemophiliac son is $(.1)(.50)(.5) = .025 = 2.5$ percent.

Woman Y can be either Hb/Hb (.05 chance) or Hb/hb (.45 chance), because she received b from her mother. If she is Hb/hb [.45/(.45 + .05) = .90 chance], she has a 50 percent chance of passing h to her child, and there is a 50 percent chance that the child will be male. The probability that she will have a son with hemophilia is: $(.9)(.5)(.5) = .225 = 22.5$ percent.

Chapter 6

1. a. $++, al-2\ al-2, ++, al-2\ al-2, al-2\ al-2, ++, al-2\ al-2, ++$

b. The 8 percent can be used to calculate the distance between the gene and the centromere.

3. a. $f(0) = e^{-2}2^0/0! = e^{-2} = .135$

b. $f(1) = e^{-2}2^1/1! = e^{-2}(2) = .27$

c. $f(2) = e^{-2}2^2/2! = e^{-2}(2) = .27$

5. a. The parents were

$$ad^-\ nic^+\ leu^+\ arg^- \times ad^+\ nic^-\ leu^-\ arg^+$$

b. Culture 16 resulted from a crossover between ad and nic. The reciprocal did not show up in the small sample.

7. a. 25.5 mu.

b. Hypothesis: no linkage, resulting in a $1:1:1:1$ ratio. $\chi^2 = 2.52$. With 3 degrees of freedom, the probability is greater than 10 percent that the genes are not linked. The hypothesis of no linkage can be accepted.

9. a. 0.704

b. 0

c. 0.176

d. 0.024

e. 0.096

f. 0

g. 0.024

11.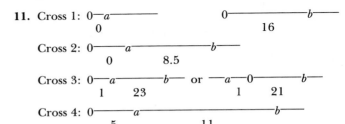

Cross 1: 0—a———————   0—————————b

            0   16

Cross 2: 0————a——————b——

            0   8.5

Cross 3: 0—a———b— or —a—0———b—

            1   23   1   21

Cross 4: 0———————a——————————b

            5   11

Cross 5: $PD = NPD$. The genes are not linked.

a-centromere = 100% $(\frac{1}{2})(22 + 8 + 10 + 20)/99 = 30.3$ m.u. and b-centromere = 100% $(\frac{1}{2})(24 + 8 + 10 + 20)/99 = 31.3$ m.u. For values this large, the genes are considered unlinked to their centromeres in tetrad analysis.

Cross 6: 0————————a—b or 0————————b—a

               >50   4   >50   4

Cross 7: —a—0———b—
    1   1.5

Cross 8: Same as cross 5.

Cross 9: —a————0————b—
        10.5   6.5

Cross 10: —0————a—   0——b—
         >50   5

or

—a————————b————0
      >50   5

or

—a—————————0———b
       >50   5

Cross 11: 0—a————   0——b———
         0   0

12.

a^+

	.9 M$_I$	.1 M$_{II}$
.8 M$_I$ b^+	.72 ⟨ .5 = .36 *PD* / .5 = .36 *NPD*	.08 *T*
.2 M$_{II}$	.18 *T*	.02 ⟨ .25 = .005 *PD* / .25 = .005 *NPD* / .50 = .01 *T*

a. 36.5% *PD*

b. 36.5% *NPD*

c. 27% *T*

d. 50% recombinants

e. 25% growth in minimal medium

14.   Cross 1: uncorrected RF = 26.5 m.u.
                    corrected RF: = 34.5 m.u.

            Cross 2: uncorrected RF = 19.5 m.u.
                    corrected RF: 23.5 m.u.

            Cross: uncorrected RF = 30 m.u.
                    corrected RF: 40.0 m.u.

17. Only *his-4* is located far enough from its centromere to result in 60 percent *T* asci. *His?* is *his-4*.

18. P $w\ t^+$ (white) $\times w^+\ t$ (tan)
Asci types:

4 black : 4 white = 4 $w^+\ t^+$: 4 $w\ t$ (*NPD*)

4 tan : 4 white = 4 $w^+\ t$: 4 $w\ t^+$ (*PD*)

4 white : 2 black : 2 tan = 2 $w\ t$ (white) : 2 $w\ t$ (white) :

2 $w^+\ t^+$(black) : 2 $w^+\ t$ (tan) (*T*)

There are two types of white, indicating that *w* is epistatic to *t*.

20. Let gg = green, Gg = yellowish, and GG = yellow. Mitotic crossing over can account for the observations.

The resulting daughter cells would be GG (yellow) and gg (green).

22. w/w^+ is not linked to either chromosome.

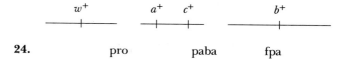

24.

	pro	paba	fpa
0————	—+—————	——+————	—+——

pro-paba = 100%(71)/154 = 71.4 relative m.u.
paba-fpa = 100%(35)/154 = 22.7 relative m.u.
pro-centromere = 100% (9)/154 = 5.8 relative m.u.

25. α 7

            β 1

            γ 5

            δ 6

            ϵ not on chromosomes 1–7

Chapter 7

1. The petal will now be Ww, or blue.

3. Plate the cells on medium lacking proline. Nearly all colonies will come from revertants. The remainder will be second-site suppressors.

5. Streak the yeast on minimal medium plus arginine. When colonies appear, replica-plate them onto minimal medium. The absence of growth in minimal medium will identify the arginine-requiring mutants.

7. Strain 2 carries a *mei* mutation (it results in 0 when crossed to strains 4 and 6). It must be recessive because some crosses result in full fertility (with 1, 3, and 5). The *mei* mutation was passed to strains 4 and 6 (0 when crossed with strain 2), and 7 and 8 (0 when crossed with either 4 or 6).

8. $e^{-\mu} \times 10^6 = \frac{37}{100}$ and $\mu = -\ln(.37) \times 10^6 = 1/10^{-6}$ cell divisions.

10. Assume that euchromatic chromosomes are genetically active and that heterochromatic chromosomes are genetically inactive. The female requires 10 active chromosomes while the male requires only 5. In order to have the normal sex ratio, the data suggest either that heterochromatic chromosomes normally must segregate from euchromatic chromosomes or that paternally derived chromosomes become heterochromatic in males.

When the female parent is irradiated, lethal mutations are induced. The mutations would be dominant in both sexes among the offspring. Thus, there are no progeny.

When the male parent is irradiated, lethal mutations again are induced. They would be dominant in the female offspring but recessive in male offspring, because the chromosomes donated by the male parent to male offspring normally are inactive.

12. Stain pollen grains, which are haploid, from a homozygous *Wx* parent. Look for red pollen grains, indicating mutations to *wx*, under microscope.

16. An X-linked disorder cannot be passed from father to son. Because the gene for hemophilia must have come from the mother, the nuclear power plant cannot be held responsible.

It is possible that the achondroplastic gene mutation was caused by exposure to radiation.

17. The mutation rate needs to be corrected for achondroplastic parents and put on a "per gamete" basis:

$$(10 - 2)/(2 \times 94,075) = 4.25 \times 10^{-5} \text{ gametes}$$

Revertants do not have to be considered.

21. **a.** reddish all over

b. reddish all over

c. many small, red spots

d. a few large, red spots

e. like c, but with fewer reddish patches

f. like d, but with fewer reddish patches

g. some large spots and many small spots

Chapter 8

3.   P $A-B-C-D-E-F- \times aa\ bb\ cc\ dd\ ee\ ff$

F1 $\frac{1}{2}\ A-B-C-D-E-F-$
          $\frac{1}{2}\ A-B-C-\ dd\ ee\ F-$

Because all progeny flies are $A-B-C-F-$, the wild-type must have been homozygous for these genes. This means that half the progeny received *DE* and half received *de* from the wild-type parent. No recombinants were seen. The best explanation is that the wildtype fly was heterozygous *DE/de* and that an inversion spanned these two genes.

5. From each of the statements concerning the rare cells, you should be able to draw the following conclusions:

Statement	Conclusion
a. Require leucine	*leu*+ lost
b. Do not mate	1 mating type lost
c. Will not grow at 37 degrees	*un*+ lost
d. Cross only with *a* type	*a* lost
e. Only nucleus 1 recovered	Deletion occurred in nucleus 2

Because *ad-3A*+ and *nic*+ function are required by the heterokaryon, these genes must have been retained. Therefore, the most reasonable explanation is that a deletion occurred in the left arm of chromosome 2 and that *leu*+, mating type *a* and *un*+ were lost.

7. The colonies that would not revert most likely had a deletion within the *ad-3B* gene.

9.

Mutant	Defect
1	Deletion of at least part of genes *h* and *i*
2	Deletion of at least part of genes *k* and *l*
3	Deletion of at least part of gene *m*
4	Deletion of at least part of genes *k*, *l*, and *m*
5	A deletion not within the *h−m* genes or a recessive point mutation

12. **a.** If the chromosome paired with itself, as diagrammed below, somatic crossing-over would yield the observed result.

b. Because the culture is haploid, only one copy of the chromosome is present. This eliminates the possibility of crossing-over between homologous chromosomes. Crossing-over could occur between *B/b* and the centromere of the chromosome diagrammed above, giving the result as below.

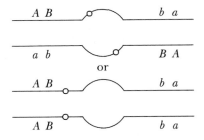

Unless nondisjunction first occurs, so that crossing-over could take place between homologous chromosomes, no other crossover products would be seen. If nondisjunction occurred, pairing would be as follows:

Single and double crossovers could give many different results, depending on where they occur.

16. **a.** The aberrant plant is semisterile, which suggests an inversion. Since the *d−f* and *y−p* frequencies of recombination in the aberrant plant are normal, the inversion must involve *b* through *x*.

b. In order to obtain recombinant progeny when an inversion is involved either a double crossover occurred within the inverted region or single crossovers occurred between *f* and the point of inversion that occurs someplace between *f* and *b*.

17. a. $0.1 = \frac{1}{2}(1 - e^{-m})$ or $m = -\ln(1 - .2) = 0.22$

b. The better general formula is $f(i) = e^{-m}m^i/i!$

    i. $f(0) = e^{-.22}(.22)^0/0! = 0.80$

    ii. $f(1) = e^{-.22}(.22)/1 = 0.176$

    iii. $f(2) = e^{-.22}(.22)^2/2 = 0.0042$

c.   i. Because no crossovers occur, the ratio of dark to light will be 8 : 0.

    ii. One crossover will result in a 4 : 4 ratio since a crossover within the inversion loop will result in all recombinant products being unbalanced.

    iii. Two crossovers will result in three different ratios, in a 1 : 2 : 1 overall ratio: 1 (8 : 0) : 2 (4 : 4) : 1 (0 : 8).

d.

Number of crossovers	Ratios among asci		
	8 : 0	4 : 4	8 : 0
0 (.8)	.8	0	0
1 (.176)	0	.176	0
2 (.0042)	.0011	.0021	.0011
Total	.8011	.1781	.0011

18. The F1 females are $y\ cv\ v\ f + +/+ + + + + +$. These are crossed with $y\ cv\ v\ f\ B\ car/Y$ males.

Class 1:   parental

Class 2:   parental

Class 3:   double crossover *y–cv* and *B–car*

Class 4:   reciprocal of class 3

Class 5:   double crossover *cv–v* and *v–f*

Class 6:   reciprocal of class 5

Class 7:   double crossover *cv–v* and *f–car*

Class 8:   reciprocal of class 7

Class 9:   double crossover *y–cv* and *v–f*

Class 10: reciprocal of class 9

Class 11: This class is identical to the male parent's X chromosome and could not have come from the female parent. Thus, the male sperm must have donated it to the offspring. In *Drosophila*, sex is determined by the ratio of X chromosomes to the number of sets of autosomes. The ratio in males is 1X : 2A, where *A* stands for the autosomes contributed by one parent (the ratio in females is 2X : 2A). Thus, this class of males must have arisen from the union of an X-bearing sperm with an egg that was the product of non-disjunction for X and contained only autosomes. This class should have only one sex

chromosome, which could be checked cytologically.

    Another possible explanation would be mutation to *Bar*.

21. If the *a* and *b* genes are on separate chromosomes, independent assortment should occur giving equal frequencies of *ab*, ++, *a*+, and +*b*. This was not observed; instead, the two genes are behaving as if they were linked with 10 m.u. between them. This is indicative of a reciprocal translocation in one of the parents, most likely the wild-type from nature.

    At meiosis prior to progeny formation, the chromosomes would look like the following (centromere not included since its position has no effect on the results):

Only alternate segregation avoids duplications and deletions for many genes. Therefore, the majority of the progeny would be parental *ab* and ++. The *a*+ and +*b* progeny would result from crossing-over between either a gene and the breakpoint locus.

25.

Cross	Meaning
1	Independent assortment of 2 genes.
2	2 genes linked at 1 m.u. distance. Therefore, a reciprocal translocation took place and both genes were very close to the breakpoint. The black spores resulted from alternate segregation, the white from adjacent segregation.
3	Half the spores were normal, nontranslocated, and half contained both translocated chromosomes.

27. The breakpoint can be treated as a gene which has two "alleles," one with normal fertility and one for semisterility. The problem thus becomes a two-point cross.

Parentals: 764 semisterile *Pr*

                727 normal *pr*

Recombinants: 145 semisterile *pr*

              $\underline{186 \text{ normal } Pr}$

              1822

100% (145 + 186)/1822 = 18.17 m.u.

28.

$$\begin{array}{ccc} leu & + & + & + \\ + & + & + & his \\ ad & + & + & + \end{array}$$

Because the short arm carries no essential genes, adjacent-1 segregation will yield progeny that are viable. Select for *leu*⁺, *his*⁺, and *ad*⁺ by omitting those components from a minimal medium.

31. Species B is probably the "parent" species. A paracentric inversion in this species would give rise to species D. Species E could then occur by a translocation of $z\ x\ y$ to $k\ l\ m$. Next, species A could result from a translocation of $a\ b\ c$ to $d\ e\ f$. Finally, species C could result from a pericentric inversion of $b\ c\ d\ e$.

Chapter 9

2. a. Obtain tetraploids with colchicine. Cross an $AAAA$ with an $A'A'A'A'$.

   b. Cross AA with $A'A'$, then double the chromosomes with colchicine treatment.

4. a. Cross a $6x$ with a $4x$.

   b. Cross AA with $aaaa$.

   c. The easiest way is to expose the Aa^* plant cells to colchicine for one cell division. This will result in a doubling of chromosomes to yield AAa^*a^*.

   d. Cross $6x$ ($aaaaaa$) with $2x$ (Aa).

   e. Obtain haploid cells from a plant and obtain resistant colonies by exposing them to the herbicide. Then expose the resistant colonies to colchicine to obtain diploids.

5. The polar nuclei can have the following combinations: AA BB, AA bb, $aaBB$ or $aabb$. Each can be fertilized by one of the following male gametes: AB, Ab, aB, or ab.

7. B can pair with B one-third of the time (leaving b to pair with b), and B can pair with b two-thirds of the time. If B pairs with B, the result is Bb, Bb, Bb, Bb, occurring one-third of the time. If B pairs with b, the tetrad is BB, BB, bb, bb, occurring one-third of the time, and Bb, Bb, Bb, Bb, occurring one-third of the time.

9. The data suggest that *G. thurberi* has 26 small chromosomes, *G. herbaceum* has 26 large chromosomes, and that *G. hirsutum* has 26 large and 26 small chromosomes. *G. hirsutum* is a polyploid derivative of a cross between the two Old World species. This could easily be checked by looking at the chromosomes.

10. a. The gametes for F/f would be: $\frac{1}{6}$ FF, $\frac{4}{6}$ Ff and $\frac{1}{6}ff$. When two genes are considered, the gametes are:

$$\frac{1}{6}FF\begin{cases}\frac{1}{6}\,GG=\frac{1}{36}\,FFGG\\ \frac{4}{6}\,Gg=\frac{4}{36}\,FFGg\\ \frac{1}{6}\,gg=\frac{1}{36}\,FFgg\end{cases}$$

$$\frac{4}{6}Ff\begin{cases}\frac{1}{6}\,GG=\frac{4}{36}\,FfGG\\ \frac{4}{6}\,Gg=\frac{16}{36}\,FfGg\\ \frac{1}{6}\,gg=\frac{4}{36}\,Ffgg\end{cases}$$

$$\frac{1}{6}ff\begin{cases}\frac{1}{6}\,GG=\frac{1}{36}\,ffGG\\ \frac{4}{6}\,Gg=\frac{4}{36}\,ffGg\\ \frac{1}{6}\,gg=\frac{1}{36}\,ffgg\end{cases}$$

   b. The cross is $FFffGGgg \times FFffGGgg$. For $FFFfGGgg$, consider each gene separately. The combination $FFFf$ can be achieved in two ways:

$$p(FFFf)=p(FF)\times p(Ff)+p(Ff)\times p(FF)$$
$$=\frac{1}{6}\times\frac{4}{6}+\frac{4}{6}\times\frac{1}{6}=\frac{2}{9}$$

The combination $GGgg$ can be achieved in three ways:

$$p(GGgg)=p(GG)\times p(gg)+p(gg)$$
$$\times p(GG)+p(Gg)\times p(Gg)$$
$$=\frac{1}{6}\times\frac{1}{6}+\frac{1}{6}\times\frac{1}{6}+\frac{4}{6}\times\frac{4}{6}=\frac{1}{2}$$

Therefore, the $p(FFFfGGgg)=\frac{2}{9}\times\frac{1}{2}=\frac{1}{9}$

The $p(ffffgggg)=p(ffff)\times p(gggg)=\frac{1}{6}\times\frac{1}{6}\times\frac{1}{6}\times\frac{1}{6}=\frac{1}{1296}$.

14. Assume that simple (nontranslocation) Down's syndrome is the case in both parents; that is, they have 47 chromosomes and are trisomy 21. Each parent has an equal probability of producing gametes with one and two chromosomes 21. The zygotes can have:

$p(2$ chromosomes $21)=\frac{1}{4}$ diploid (normal)

$p(3$ chromosomes $21)=\frac{1}{2}$ trisomy (Down's syndrome)

$p(4$ chromosomes $21)=\frac{1}{4}$ tetrasomy (lethal)

19. a. Loss of one X in the developing fetus after the two-celled stage.

   b. Nondisjunction leading to Klinefelter's syndrome (XXY) followed by a nondisjunctive event in one cell for the Y chromosome after the two-celled stage, leading to XX and $XXYY$.

   c. Nondisjunction for X at the one-celled stage.

   d. Either fused XX and XY zygotes or fertilization of an egg and polar body by one sperm bearing an X and another bearing a Y, followed by fusion.

   e. Nondisjunction of X at the two-celled stage or later.

21. The generalized cross is $AAA \times aa$, from which AAa progeny were selected. These progeny were crossed to aa individuals, yielding the results in the table. Assume for a moment that each allele can be distinguished from the other and let $1=A$, $2=A$, and $3=a$. The gametic combinations possible are:

$$1-2\ (AA)\ \text{and}\ 3\ (a)$$
$$1-3\ (Aa)\ \text{and}\ 2\ (A)$$
$$2-3\ (Aa)\ \text{and}\ 1\ (A).$$

Since diploid progeny were examined in the cross with aa, the haploid gametic ratio would be $2A:1a$, and the diploid ratio would also be 2 wild-type : 1 mutant. The table indicates that:

$$y\ \text{is on chromosome 1}$$
$$cot\ \text{is on chromosome 7}$$
$$h\ \text{is on chromosome 10}$$

23. P $a+c+e\times+b+d+$. Selection for $++++$. Since this rare colony gave rise to both parental types among asexual (haploid) spores, the best explanation is that the rare colony initially contained both marked chromosomes due to nondisjunction. That is, it was disomic. Subsequent mitotic nondisjunction yielded the two parental types, possibly because the disomic was unstable.

27. a. Mutation

b. Mitotic cross over between w and cnx

c. Nondisjunction

Chapter 10

1. The two techniques allow first for a localization of the mutant (interrupted-mating) and second for a precise location of the mutant (generalized transduction) within the general region.

2. M Z X W C N A L B R U

3. Strains 2, 3, and 7 are F⁻. Strains 1 and 8 are F⁺ and strains 4, 5 and 6 are Hfr.

6. a. The gene order is $arg\ bio\ leu$.

b. The $arg-bio$ distance is 12.76 m.u. The $bio-leu$ distance is 2.12 m.u.

7. To solve this problem, draw the Hfr and recipient chromosomes in both crosses and note the number of crossovers needed to get $Z_1^+ Z_2^+$ for the two possible gene orders.

> Order 1
>
> Hfr $\quad Z_1^- Z_2^+ ade^+ str^s$
>
> Recipient $\quad Z_1^+ Z_2^- ade^- str^r$
>
> Order 2
>
> Hfr $\quad Z_2^- Z_1^+ ade^+ str^s$
>
> Recipient $\quad Z_2^+ Z_1^- ade^- str^r$

From the number of crossovers required to get $Z^+ ade^- str^r$, the order must be $ade\ Z_2\ Z_1$.

11. The best explanation is that the integrated pro^+ was sexducted onto an F′ factor which was transferred into recipients early in the mating process. These cells now carry the F factor and are able to transmit F⁺ in the second cross as part of the F′ factor, which still carries pro^+.

16. a. Notice that each gene was transferred into about $\frac{1}{10}$ of the cells (single drugs tested). Also notice that pairwise testing gives low values whenever B is involved but fairly high rates when any drug but B is involved. This suggests that B resistance is not close to the other three genes and the low rates come from double crossovers.

b. A D C

18. a. $m-r$: The distance is 100% (1328)/10,342 = 12.8 m.u.
$r-tu$: The distance is 100% (2152)/10,342 = 20.8 m.u.
$m-tu$: The distance is 100% (2812)/10,342 = 27.2 m.u.

b. Because m and tu are farthest apart, the sequence is $m\ r\ tu$. At this point, the distance between m and tu can be corrected for double crossovers (classes $+ r +$ and $m + tu$). The final $m-tu$ distance is the sum of the two smaller distances, or 12.8 + 20.8 = 33.6.

c. Recall that $I = 1 - \text{C.C.} = 1 - \text{obs. DCO/exp. DCO}$. The observed DCO is $162 + 172 = 334$. The expected DCO would be $(.128)(.208)(10,342) = 275$. CC = 1.2. $I = 1 - 1.2 = -.2$. A negative value for I indicates that the occurrence of one crossover makes a second crossover more likely to occur than it would have without that first crossover. That is, more double crossovers occur than are expected.

21. a. Specialized transduction.

b. The prophage is located in the $cys-leu$ region.

24. The order is $d-a-e-c$. Notice that b is never cotransduced and is therefore distant from this group of genes.

Chapter 11

1. 35 percent.

2. Assume a diploid cell.

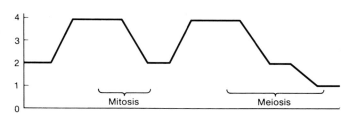

7. Chargaff's rule is that A = T and G = C. Since this is not observed, the most likely interpretation is that the DNA is single stranded. The phage would first have to synthesize a complimentary strand before it could begin to make multiple copies of itself.

12. The suggestion of this observation is that the mouse cancer cells have copies of the virus genome integrated into them. Thus, it may be that the viral genes somehow alter cell function, triggering malignancy. Alternatively, the viral genome may carry one or more genes which directly result in malignancy. In either case, viral infection may also be the mechanism by which human malignancy is triggered.

14. Let _____ indicate DNA that has incorporated bromodeoxyuridine and _____ indicate normal DNA.

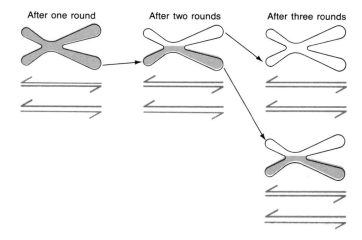

Chapter 12

2. A secondary cure would result if all lactose is removed from the diet. The disorder is recessive.

4. a. The main use is in detecting carrier parents and in diagnosing a disorder in the fetus.

    b. Because the values for normal individuals and carriers overlap for galactosemia, there is ambiguity if a person has 25 to 30 units as a result. That person could be either a carrier or normal.

    c. These genes are phenotypically dominant but are incompletely dominant at the molecular level. A minimal level of enzyme activity apparently is enough to ensure normal function and phenotype.

6. Assuming homozygosity, the children would be normal.

8.

Experiment	Result	Interpretation
v into v hosts	Scarlet	Defects in same gene
cn into cn hosts	Scarlet	Defects in same gene
v into w-t hosts	Wild-type	w-t provides v product
cn into w-t hosts	Wild-type	w-t provides cn product
cn into v hosts	Scarlet	v cannot provide cn product; cn later than v in metabolic pathway
v into cn hosts	Wild-type	cn provides v product; v defect earlier than cn

A simple test would be to grind up cn animals, inject v larvae with the material and look for wild-type development.

```
      controlled by              controlled by
        v locus                    cn locus
precursor ⤳ → v⁺ substance ⤳ → cn⁺ substance → wild-
           v                                      type
precursor ⤳ ‖                              cn    pigment
precursor → v⁺ substance ⤳ ‖
```

10. a. $E \rightarrow A \rightarrow C \rightarrow B \rightarrow D \rightarrow G$

    b. Mutant 1: grows on D and G; the block is at B → D
Mutant 2: grows on B, D, and G; the block is C → B
Mutant 3: grows on G, the block is D → G
Mutant 4: grows on B, C, D, and G; the block is A → C
Mutant 5: grows on all but E; the block is E → A

    c. 1,3 + 2,4 growth
1,2 + 2,4 no growth
1,2 + 2,4 + 1,4 growth

12. a. A defective enzyme B (from $m_2 m_2$) would yield red petals.

    b. Purple

    c. 9 $M_1 - M_2 -$ purple
3 $m_1 m_1 M_2 -$ blue
3 $M_1 - m_2 m_2$ red
1 $m_1 m_1\ m_2 m_2$ white

    d. Because they do not produce a functional enzyme.

14. The cis and trans burst size should be the same if the mutants are in different cistrons and, if they are in the same cistron, the trans burst size should be zero. Therefore, assuming rV is in A, rW also is in A and rU, rX, rY, and rZ are in B.

15. The cross is $pan\ 2x + \times + pan\ 2y$.

    a. If one centromere precociously divides, that will put three chromatids in one daughter cell and one in the other.

Daughter cell 1	Daughter cell 2	
$pan\ 2x +$	$+ pan\ 2y$	$pan\ 2x +$
————————0	————————0	
————————0	————————0	————————0

After meiosis II and mitosis, the first daughter cell would give rise to two pale ($pan\ 2x +$) and two white aborted (nullisomic) spores, while the second daughter cell would give rise to two black ($pan\ 2x +/+ pan\ 2y$) and two pale ($+ pan\ 2y$) spores. The same result (4 pale : 2 colorless : 2 black) would occur if only the other centromere divided early.

    b. If both centromeres divided precociously, each daughter cell would be $pan\ 2x +/+ pan\ 2y$. This would lead to 4 colorless (nullisomic) and 4 black (disomic) ascospores.

19. a. Here, + indicates nonoverlapping and − indicates overlapping. Therefore, 1 overlaps 3 and 5, 2 overlaps 5 only, 3 overlaps all but 2, 4 overlaps 3 only, and 5 overlaps all but 4. Putting all these pieces together yields:

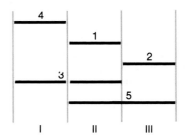

    b.

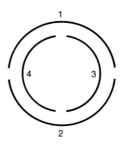

20. a.

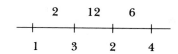

b. No

23. S^n will show dominance over s^f because there will be only 40 units of square factor in the heterozygote. Here, the functional allele is recessive. S^f may become dominant over time in two ways: (1) it could mutate slightly, so that it produces >50 units or (2) other modifying genes may mutate to increase the production of s^f.

28. a. Mutants *a* and *e* have point mutations within the same cistron. The other point mutations are all in different cistrons. There are at least four cistrons involved with leucine synthesis.

b. With the exception of two crosses ($a \times e$, $b \times d$), the frequency of prototrophic progeny is approximately 25 percent. This indicates independent assortment of $a + e$ with b, d, and c, and c with b and d. Cistrons b and d are linked:

$$RF = 100\% \ (4) \ 1/500 = 0.80 \text{ map units.}$$

29. a. There are three cistrons:

Cistron 1: mutants 1, 3, and 4

Cistron 2: mutants 2 and 5

Cistron 3: mutant 6

b. A/a 6 (1, 3, 4)(2, 5) B/b.

Chapter 13

2. a. The data do not indicate whether one or both strands are used for transcription in either case.

b. If the RNA is double stranded, the percentage purines ($A + G$) would equal the percentage pyrimidines ($U + C$), and the AG/UC ratio would be 1.0. This is clearly not the case for *E. coli*, which has a ratio of 0.80. Therefore, *E. coli* RNA is single stranded. The ratio for *B. subtilis* is 1.02. Either the RNA is double stranded, or there is an equal number of purines and pyrimidines in each strand.

3. A single nucleotide change should result in three adjacent amino acid changes in a protein. One and two adjacent amino acid changes would be expected to be much rarer than the three changes. This is directly opposite of what is observed in proteins.

6. a. Using Figure 12-19, there are eight cases in which knowing the first two nucleotides does not tell you the specific amino acid.

b. If you knew the amino acid, you would not know the first two nucleotides in the cases of *arg*, *ser*, and *leu* (Figure 12-19).

8. a. $\frac{1}{8}$

b. $\frac{1}{4}$

c. $\frac{1}{8}$

d. $\frac{1}{8}$

11. Mutant 1: A simple substitution of *arg* for *glu* exists, suggesting a nucleotide change. Two codons for *arg* are AGA and AGG, and one codon for *ser* is AGU. The final U for *ser* could have been replaced by either an A or a G.

Mutant 2: The *trp* codon (UGG) changed to a stop codon (UGA or UAG).

Mutant 3: Two frameshift mutations occurred:

5'GCN CCN (−U)GGA GUG AAA AA(+U OR C)
UGU/C CAU/C3'

Mutant 4: An inversion occurred after *trp* and before *cys*. The DNA original sequence was:

3'CGN GGN ACC TCA CTT TTT ACA/G
GTA/G5'

Therefore, the complementary RNA sequence was:

5'GCN CCN UGG AGU GAA AAA UGU/C
CAU/C3'

The DNA inverted sequence became:

3'CGN GGN ACC AAA AAG TGA ACA/G
GTA/G5'

Therefore, the complementary RNA sequence was:

5'GCN CCN UGG UUU UUC ACU UGU/C CAU/C3'

13. e

15.

3'	CGT	ACC	ACT	GCT	5'
5'	GCA	TGG	TGA	CGA	3'
5'	GCA	UGG	UGA	CGU	3'
3'	CGU	ACC	ACU	GCA	5'
	Ala	Trp	stop	nothing	

17. a. 120 nucleotides

b.

		I	II
Mutant 1: *Gln*	C	A	A/G
Mutant 2: *Ser*	A	G	U/C

The progeny from this cross are the two parental types plus:

arg	AGA/G	1 CO in II
arg	CGN	1 CO in I
his	CAC/U	1 CO in II
asn	AAC/U	DCO
lys	AAA/G	1 CO in I

The best data are twice the frequency of *his* or *lys*, which is 4×10^{-7}.

Chapter 14

1. c

6. The region of the heterochromatic knob contains a recognition point for whatever mechanism of migration that is in operation here.

9. **a.** The following factors will affect the resolving power of the technique:

1. The specific activity (amount of radioactivity per amount of RNA) of the RNA

2. The number of copies of DNA in each chromosome

3. The degree of endomitosis per cell

4. The degree of denaturation of the DNA

5. The type of radioactivity involved (determines both frequency of decay and size of the spot on the film)

6. The length of time of film exposure

7. The size of the silvergrain on the emission.

b. Excess cold RNA will compete with the radioactive RNA if it has the same sequence but excess cold RNA will not compete with the radioactive RNA if it has a different sequence.

c. You infer that the RNA is complementary to the DNA. This leads to the further inference that the region to which binding occurred is the region from which the RNA is transcribed.

10. Some of the suggestions that have been made are:

a. They are responsible for the homologous pairing seen in all *Drosophila* cells.

b. They facilitate synapsis.

c. They are part of the centromeres, with an unknown function.

Chapter 15

1. AAGCTT occurs, on average, every 4^6 bases. CCGG occurs, on average, every 4^4 bases.

4. Isolate the DNA and separate the two strands. Test each strand for the ability to hybridize with mRNA produced by the phage.

5. **a.**

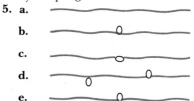

8. The structures result from palindromic DNA.

10.

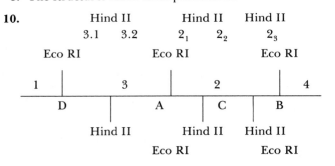

12. Reading from the bottom the sequence is:

Hind-Hae-Hae-Hind-Hae-Hae-Hae-Hind-Hind-Hae-
Hind-Hae-EcoR1

15. Serine would be substituted for glycine in the D gene (missense mutation). In the E gene, the AUG start codon would be eliminated and the E gene would not be translated.

17. This problem assumes a random distribution of nucleotides.
Alu 1 $(\frac{1}{4})^4$ = every 256 nucleotide pairs
*Eco*R1 $(\frac{1}{4})^6$ = every 4096 nucleotide pairs
*Acy*1 $(\frac{1}{4})^4(\frac{1}{2})^2$ = every 1024 nucleotide pairs

18. **a.**

1: *A*-1, *B*-2

2: *A*-2, *B*-1

3: *A*-1, *B*-1

4: *A*-2, *B*-2

b. Spores 3 and 4 are parental types, occuring in 70 percent of the population. Spores 1 and 2 are recombinants, occuring in 30 percent of the population.

c.

24.

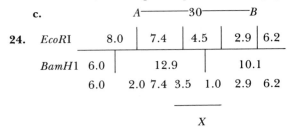

26. **a.** All individuals have the 5-kb band, indicating that the band sequence is not involved with the gene in question. All affected individuals, except the last son, have the 3-kb and 2-kb bands. These two bands are not seen in unaffected individuals. The suggestion is that the two bands are close to or part of the gene in question.

b. The last affected son indicates that the two bands are not part of the gene in question. He represents a crossover between the gene in question and the two bands.

c. The 2-kb and 3-kb bands are closely linked to the dominant allele. Their presence in an individual would indicate a high risk of developing the disorder, while their absence would indicate a low risk of developing the disorder. Exact risk cannot be stated until the map units between the two bands and the gene in question are determined. Although a rough estimate of map units can be made from the pedigree, the sample size is too small to make the estimate reliable.

Chapter 16

2. O^c mutants do not bind the repressor product of the I gene, and therefore the *lac* operon associated with the O^c operator cannot be turned off. Because an operator controls only the genes on the same DNA strand, it is *cis* (on the same strand) and dominant (cannot be turned off).

3. **a.** b is the Z gene, c is I, a is in the O region

b.
Line 1: I^c or O^c

Line 2: I^c or O^c

Line 3: I^c or I^s

Line 4: O^c and I^c or I^s

Line 5: O^c or O^o and I^c

Line 6: O^c or O^o and I^c

Line 7: O^c and I^c or I^s

5. **a.** A lack of only E_1 or only E_2 function indicates that both genes have enzyme products which are responsible for a conversion reaction. Since the two genes are in different linkage groups, they cannot be regulated by a single operator and promoter like the Z and Y genes of the *lac* operon. Type 3 mutants must be mutants of a site that produces a diffusable regulator of the E_1 and E_2 genes. The type 3 mutants identify a site that produces either a repressor (like I in the *lac* operon) or an activator (analogous to CAP) of the other two genes.

b. Separate operator and promotor mutants might be found for each gene.

7. Nonpolar Z^- mutants cannot convert lactose to allolactose, and thus the operon is never induced.

9. An operon is turned off by the mediator in negative control, and the mediator must be removed for transcription to occur. An operon is turned on by the mediator in positive control, and the mediator must be added for transcription to occur.

Chapter 17

3. The Streisinger model proposed that frame-shifts arise when loops in single-stranded regions are stabilized by slipped mispairing of repeated sequences. In the *lac* gene of *E. coli*, a four-base-pair sequence is repeated three times in tandem, and this is the site of a hotspot.

The sequence is:

5′ CTGG CTGG CTGG CTGG 3′

During replication the DNA must become single-stranded in short stretches for replication to occur. As the new strand is synthesized and becomes hydrogen-bonded to the template strand, it can pair out of register with that strand by a total of four bases. Depending on which strand, new or template, loops with regard to the other, there will be an addition or deletion of four bases.

5. **a.** Depurination results in the loss of adenine or guanine from the nucleotide. This apurinic site cannot specify a complementary base, which blocks replication. Under certain conditions, replication proceeds with a random insertion of a base opposite the apurinic site. In three-fourths of these insertions, a mutation will result.

b. Deamination of cytosine yields uracil. If left unrepaired, adenine is paired with uracil during replication, resulting in a transition mutation.

9. Leaky mutants are mutants with an altered protein product that retains a low level of function. Enzyme activity may, for instance, be reduced rather than abolished.

10. The wild-type contained a gene that increased the spontaneous mutation rate. This new gene seems to be unlinked to *ad-3*. Call the new gene B. Cross A(ad-3 B$^+$ × wild-type (ad-3$^+$ B). The progeny should reflect independent assortment.

Progeny: $\frac{1}{4}$ ad-3 B

$\frac{1}{4}$ ad-3 B$^+$

$\frac{1}{4}$ ad-3$^+$ B

$\frac{1}{4}$ ad-3$^+$ B$^+$

Further crosses should verify the above.

11. **a.** Because 5′UAA3′ does not contain G or C, a transition to a *GC* pair in the DNA cannot result in *UAA*. *UGA* and *UAG* have the DNA sense-strand sequence of ACT and ATC. A transition to either of these stop codons occurs from the nonmutant *ATT*. A DNA sequence of *ATT* results in an RNA sequence of UAA, itself a stop codon.

b. Yes. An example would be 3′UGG5′, which codes for *trp*, to 3′UAG5′.

c. No. In the three stop condons the only base that can be acted upon is G (in UAG, for instance). Replacing the G with an A would result in UAA, a stop condon.

13. To understand these data, recall that half of the progeny should come from the wild-type parent.

a. A lack of revertants suggests either a deletion or an inversion within the gene.

b. Prototroph A: since 100 percent of the progeny are prototrophic, this suggests a reversion at the original mutant site.

Prototroph B: Half of the progeny are parental prototrophs. The remaining prototrophs, 28 percent, are the result of the new mutation. Notice that 28 percent is approximately equal to the 22 percent auxotrophs. The suggestion is that an unlinked suppressor mutation occurred, yielding independent assortment with the *nic* mutant.

Prototroph C: There are 496 "revertant" prototrophs (the other 500 are parental prototrophs) and 4 auxotrophs. This suggests that a suppressor mutation occurred in a site very close (100% [4×2]/1000 = 0.8 m.u.) to the original mutation.

Chapter 18

2. Gene conversion results in a deviation from a 4 : 4 ratio, with the order unimportant. The following asci show gene conversion: 3, 6.

Ascus 4 is also produced by gene conversion. To recognize it as such, recall the sequence that gives rise to the eight meiotic products in *Neurospora*. The pattern generated could be produced only if the two DNA strands have a region of mismatch.

5. The actual mechanism of sister chromatid exchange induction by mutagens is unknown but would be expected to vary with mutagen effects. The endpoint must be a break in a DNA strand, very likely as part of post-replication repair.

7.

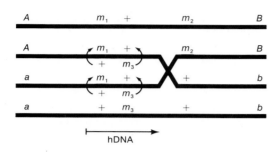

Order is $A/a - m_1 m_3 m_2 - B/b$ (or $m_3 m_1 m_2$)

Hybrid DNA entered m gene from left and ended between m_3 and m_2, and hence spanned m_1 and m_3.

One single excision-repair event corrected $+m_3$ to $m_1 +$ on both hybrid DNA molecules.

9. **a** and **b.** A heteroduplex that contains an unequal number of bases in the two strands has a larger distortion than a simple mismatch. Therefore, the former would be more likely to be repaired. For such a case, both heteroduplexes are repaired (leading to 6:2 and 2:6) more often than one (leading to 5:3 or 3:5) or none (leading to 3:1:1:3). The preference in direction, i.e., adding a base rather than subtracting, is analogous to TT dimer repair. In TT dimer repair, the unpaired, bulged nucleotides are treated as correct and the strand with the TT dimer is excised

A mismatch more often than not escapes repair, leading to a 3:1:1:3 ascus.

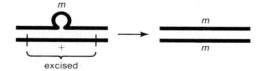

Transition mutations would not cause as large a distortion of the helix, and each strand of the heteroduplex should have an equal chance of repair. This would lead to 4:4 (two repairs each in the opposite direction), 5:3 (1 repair), 3:1:1:3 (no repairs or two repairs in opposite directions) and, less frequently, 6:2 (two repairs in the same direction).

c. Because excision repair excises the strand opposite the larger buckle, that is opposite the frameshift mutation, the *cis* transition mutation will also be retained. The nearby genes are converted because of the length of the excision repair.

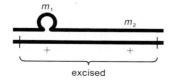

12. Rewrite the original cross:

$$\text{P} \quad A\, X\, y^+ \times a\, x^+\, y$$

The progeny of parental genotypes will be like either of the two parents. The backcrosses are as follows, with ' indicating progeny generation.

Cross 1: $a'\, x^+\, y \times A\, x\, y^+ \longrightarrow 10^{-5}$ prototrophs

Cross 2a: $A'\, x\, y^+ \times a\, x^+\, y \longrightarrow 10^{-5}$

Cross 2b: $A'\, x\, y^+ \times a\, x^+\, y \longrightarrow 10^{-2}$

Recombination is allowing for the higher rate of appearance of prototrophs. Cross 2 is obviously a backcross for some gene affecting the rate of recombination. Whatever that gene is, the allele in the A parent blocks recombination (cross 1 and cross 2a) dand the allele in the a parent allows recombination (cross 2b). It is unlinked to the his gene since cross 2 yields results in a 1:1 ratio. The allele that blocks recombination (in A) is dominant, while the allele that allows recombination is recessive. This is demonstrated by the original cross in which prototrophs occurred at the lower rate and by cross 1.

To test this interpretation, $\frac{1}{4}$ of the crosses between the $A\, x\, y^+$ and $a\, x^+\, y$ progeny should yield a high rate of recombination and therefore have a high frequency of prototrophs.

Chapter 19

2. Polar mutations affect the transcription or translation of the part of the gene or operon only on one side of the mutant site, usually described as downstream. Examples are nonsense mutations, frame-shift mutations, and IS-induced mutations.

4. R plasmids are the main carriers of drug resistence. They acquire these genes by transposition of drug resistance genes located between IR (inverted repeat) sequences. Once in a plasmid, the transposon carrying drug resistence can be transferred upon conjugation if it stays in the R plasmid, or it can insert into the host chromosome.

6. P elements are transposons (genes flanked by inverted repeats which allow for great mobility). Because they are transposons, they can insert into chromosomes. By inserting specific DNA between the inverted repeats of the P elements and injecting the altered transposons into cells, a high frequency of gene transfer will occur.

8. The best explanation is that the mutation is due to an insertion of a transposable element.

10. **a.** The expression of the tumor is blocked in plant B. This suggests that either plant B can suppress the functioning of the plasmid, which causes the tumor, or that plant A provides something to the tissue with the tumor-causing plasmid which plant B does not provide.

    b. Tissue carrying the plasmid, when grafted to plant B, appears normal, but the graft produces tumor cells in synthetic medium. This indicates that the plasmid sequences are present and capable of functioning in the right environment. However, the production of normal type A plants from seeds from the graft suggests a permanent loss of the plasmid during meiosis.

Chapter 20

1. Most organelle-encoded polypeptides unite with nucleus-encoded polypeptides to produce active proteins, and these active proteins function in the organelle.

3. Maternal inheritance of chloroplasts results in the green-white color variegation observed in *Mirabilis*.

   Cross 1: variegated ♀ × green ♂ ⟶ variegated progeny

   Cross 2: green ♀ × variegated ♂ ⟶ green progeny

6. **a.** The ant^R gene may be mitochondrial.

   b. Some of the petites must have been neutral petites, in which all mitochondrial DNA, including the ant^R gene, was lost. Other petites were suppressive, in which the ant^R gene was retained.

8. The cpDNA from mt^- *Chlamydomonas* is lost. The results of the crosses are:

   Cross 1: all morph 1; 2 kb and 3 kb bands

   Cross 2: all morph 2; 3 kb and 5 kb bands

10. Both crosses show maternal inheritance of a chloroplast gene. The rare variegated phenotype is probably due to a minor male contribution to the zygote. Variegation must result from a mixture of normal and prazinizan chloroplasts.

12. This pattern is observed when a maternal recessive nuclear gene determines phenotype. The crosses are:

    P   *dd* dwarf female × *DD* normal male

    F1   *Dd* dwarf (all dwarf because mother is *dd*)

    F2 $\frac{3}{4}$ *D*–: $\frac{1}{4}$ *dd* normal (all normal because mother is *Dd*)

    F3 $\frac{3}{4}$ normal (mother is *D*–): $\frac{1}{4}$ dwarf (mother is *dd*)

13. After the initial hybridization a series of backcrosses using pollen from B will result in the desired combination of cytoplasm A and nucleus B. With each cross the female contributes all of the cytoplasm and one half the nuclear contents, while the male contributes one half the nuclear contents.

17.
	m3		*m1*		*m2*		*m4*
		18.0		29.0		26.2	

    |---|---|
    | 10.9 | |

    | 8.8 |

    | 18.4 |

18. The red phenotype in the heterokaryon indicates that the red phenotype is caused by a cytoplasmic organelle allele.

21. Let poky be symbolized by (*c*). Let the nuclear suppresser of poky be symbolized by *n*. In order to do these problems, you cannot simply do the crosses in sequence. For instance, the parental genotypes in cross a must be written taking cross c into consideration.

Cross	Progeny
a. (+) + × (*c*) +	all (+) +
b. (+) *n* × (*c*) +	$\frac{1}{2}$ (+) *c* : $\frac{1}{2}$ (+) +
c. (*c*) + × (+) +	all (*c*) +
d. (*c*) + × (+) *n*	$\frac{1}{2}$ (*c*) + (= D) : $\frac{1}{2}$ (*c*) *n* (= E)
e. (*c*) *n* × (+) *n*	all (*c*) *n*
f. (*c*) *n* × (+) +	$\frac{1}{2}$ (*c*) *n* : $\frac{1}{2}$ (C) +

22. a and b. Each meiosis shows uniparental inheritance, suggesting cytoplasmic inheritance.

    c. Because ant^R is probably mitochondrial and because petites have been shown to result from deletions in the mitochondrial genome, ant^R may be lost in some petites.

24. The *apt-cob* pair had the lowest rate of loss (45 total), and the *apt-bar* pair had the highest rate of loss (207 total). This puts *cob* between *apt* and *bar*. The relative rate of loss is apt-*cob* : *cob-bar* or 45 : 117 = 1 : 2.6.

apt———*cob*————————*bar*
1   2.6

26. Some tetrads will show strain 1 type, some will show strain 2 type, and some will be recombinant.

30. **a.** The cytoplasm from senescent cultures is a mixture of normal and abnormal mitochondria. The mitochondrial types are distributed in different ratios to different spores. The abnormal mitochondria seem to have a replicative advantage over the normal since senescence seems, ultimately, to "win out" over normal nonsenescence. The rapidity of the onset of senescence seems to be related to the ratio of normal to abnormal mitochondria.

    b. The mutation is an insertion of about 10 kb. It carries bands E and G and splits the original fragment into two fragments, B and C.

Chapter 21

1. Somatic mutation, mitotic crossing over, mitotic nondisjunction, mitotic chromosome loss, position-effect variegation as a result of translocation and inversion, cytoplasmic mutation followed by segregation, unrepaired mismatch following recombination, fusion of two zygotes, X-chromosome inactivation, transposition.

3. Human females are XX; parthenogenesis cannot produce a normal XY male. However, a dominant mutation that causes a sex reversal could occur that would lead to a phenotypically male offspring through parthenogenesis. Such a male would be sterile because he would still be XX. The ovum carrying such a mutation would have to undergo nondisjunction at the second meiotic division to produce a diploid zygote.

6. a. Regeneration is controlled by the nucleus.

   b. Hat-forming substance, produced by the nucleus, is concentrated in the upper part of the stem. An enucleated cell cannot synthesize more of this substance when the upper part of the stem is removed.

   c. The hat-forming substance cannot function in the presence of a hat.

9. When an excess of S_1 exists, P_1 is produced, which shuts down operon 2 and allows operon 1 to stay active. When S_1 concentrations fall, the relative S_2 concentration rises, reversing the situation.

11. The woman was a heterozygote and X-inactivation randomly inactivated each X chromosome in roughly a 50:50 split. The cells containing the malarial parasite are those in which the normal allele is functioning and the mutant allele is inactivated, and parasite-free cells have a functional mutant allele and a nonfunctional normal allele.

14. Femaleness

18. Culture 1: two ascospores are homothallic, two are *a*.

    Culture 2: all ascospores are homothallic.

19. The rat GH gene inserted at an unknown location in the mouse chromosomes in 1 percent of the cases. When inserted, it was capable of being induced by heavy metals to produce larger mice. In the pedigree, the inserted gene is behaving as a dominant gene. The original mouse must have been heterozygous for the insertion because roughly 50 percent of its progeny respond to heavy metals.

    Potentially, this mechanism could be used in human gene therapy. Some of the problems are:

    1. The low efficiency of insertion.

    2. Random location of insertion.

    3. The mutant allele remains in the genome.

    4. Insertion could inactivate a gene, causing a new mutation.

22. a. Differences in tissue type are correlated with differences in mRNA.

    b. Do a similar competition study using labeled liver mRNA and increasing amounts of unlabeled mRNA from the whole body, minus the liver. The difference between them at maximum competition (minimum amount of label binding) would reflect the liver-specific mRNA.

23. The pool of cytoplasmic RNA is a subset of the nuclear pool, indicating that much RNA never leaves the nucleus. Some of this RNA which never leaves the nucleus is from intron transcripts and both leader and trailer transcripts.

Chapter 22

1. The studies suggest that muscular dystrophy is caused by defects in the nerves that subsequently cause muscular defects. The parabiosis studies show that the muscular defects cannot be induced in adult muscle. Therefore, the nerve effects on muscles must occur during development.

4. Compare restriction fragments of homozygous wild-type and *Notch*. You might also isolate labeled mRNA from homozygous wild-type cells and hybridize it to *Notch* cells.

6. a. The wild-type flies are attracted to light. Select flies that do not move toward light at the end of a tube.

   b. Select for flies that do not fly by placing a poisoned food source high in a fly cage to which flies must fly.

   c. Add a tasteless toxic compound to a sugar solution on filter paper in the bottom of the cage. Those flies that cannot taste sugar will not eat it, while those that can taste sugar will eat it and die.

7. The flies may have a structural defect in their legs that prevent normal walking, or they may have a neural defect that results in entanglement of the legs. Nerve and muscle action potentials could be compared in the mutant and wild-type. Electron microscopy could reveal structural abnormalities at that level. Biochemical analysis may reveal either abnormal structural proteins or abnormal enzymatic action.

8.

Leg 1: hyperkinetic: 14.3

Leg 2: hyperkinetic: 19.1

Leg 3: hyperkinetic: 21.7

Antenna: hyperkinetic: 27.8

Humeral bristle: hyperkinetic: 26.8

Chapter 23

3. a. Homozygotes at one locus can be homozygotic at A *or* at B *or* at C. The probability of being homozygotic is $\frac{1}{2}$ (for A/a: AA or aa) and the probability of being heterozygotic is $\frac{1}{2}$. Putting this all together:

   $$(\text{homozygotic at 1 locus}) = 3(\tfrac{1}{2})^3 = \tfrac{3}{8}$$
   $$(\text{homozygotic at 2 loci}) = 3(\tfrac{1}{2})^3 = \tfrac{3}{8}$$
   $$(\text{homozygotic at 3 loci}) = (\tfrac{1}{2})^3 = \tfrac{1}{8}$$

   b.

   $$pr(0 \text{ capital letters}) = \tfrac{1}{64}$$
   $$pr(1 \text{ capital letter}) = \tfrac{3}{32}$$
   $$pr(2 \text{ capital letters}) = \tfrac{15}{64}$$
   $$pr(3 \text{ capital letters}) = \tfrac{10}{32}$$
   $$pr(4 \text{ capital letters}) = \tfrac{15}{64}$$
   $$pr(5 \text{ capital letters}) = \tfrac{3}{32}$$
   $$pr(6 \text{ capital letters}) = \tfrac{1}{64}$$

4. For three genes there is a total of 27 genotypes which will occur in predictable proportions. For example, there are three genotypes which have two heterozygotes and a homozygote recessive ($AaBbcc$, $AabbCc$, $aaBbCc$). The frequency of this combination is $3(\tfrac{1}{2})(\tfrac{1}{2})(\tfrac{1}{4}) = \tfrac{3}{16}$ and the phenotypic score is $3 + 3 + 1 = 7$. The distribution of scores is:

Score	Proportion
3	$\frac{1}{64}$
5	$\frac{6}{64}$
6	$\frac{3}{64}$
7	$\frac{12}{64}$
8	$\frac{12}{64}$
9	$\frac{11}{64}$
10	$\frac{12}{64}$
11	$\frac{6}{64}$
12	$\frac{1}{64}$

7. Mean = 4.7
Variance = 0.2619
Standard deviation = 0.5117

8. a. 1.0

   b. 0.83

   c. 0.66

   d. −0.20

12. The effect of substituting a low for a high chromosome II can be seen within each row. In the first row, the differences are $25.1 - 22.2 = 2.9$ and $22.2 - 19.0 = 3.2$. In the second row the differences are 3.1 and 5.2. In the row, they are 2.7 and 6.8. The average difference is $23.9/6 = 3.98$, which actually tells you very little.

The effect of substituting one l chromosome for an h chromosome in chromosome II, and therefore going from homozygous hh to heterozygous hl, can be seen in a comparison in the differences along the rows in the first two columns. The average change is $(2.9 + 3.1 + 2.7)/3 = 2.9$. When chromosome II goes from heterozygous hl to homozygous ll, the average change is $(3.2 + 5.2 + 6.8)/3 = 5.1$.

The effect of substituting one l chromosome for an h chromosome in chromosome III, and therefore going from homozygous hh to heterozygous hl, can be seen in a comparison in the differences between rows ($25.1 - 23.0 = 2.1$, $22.2 - 19.9 = 2.3$, $19.0 - 14.7 = 4.3$, $23.0 - 11.8 = 11.2$, $19.9 - 9.1 = 10.8$, $14.7 - 12.4 = 12.4$). When chromosome III goes from homozygous hh to heterozygous hl, the average change is $(2.1 + 2.3 + 4.3)/3 = 2.9$. When it goes from heterozygous hl to homozygous ll, the average change is $(11.2 + 10.8 + 12.4)/3 = 11.5$.

A summary of these results appears below:

	Chromosome II	Chromosome III	Total
hh to hl	2.9	2.9	5.8
hl to ll	5.1	11.5	16.6

Now it should be clear that each set of alleles for both chromosomes is expressed in the phenotype, but that expression varies with the chromosome. Chromosome III appears to have a stronger affect on the phenotype than chromosome II (compare total amount of change). There

is some dominance of h over l for both chromosomes because the change from hh to hl is less than the change from hl to ll. Finally, there is definitely some epistasis occurring. Compare h/h h/h with both l/l h/h and h/h l/l. The difference in the first case is 6.0 and, in the second case, 13.3. The expected amount of change in going from h/h h/h to l/l l/l is therefore $6.0 + 13.3 = 19.3$. The l/l l/l phenotype should be $25.1 - 19.3 = 5.8$, but the observed value is 2.3.

13.

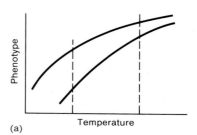

(a)

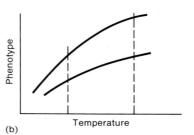

(b)

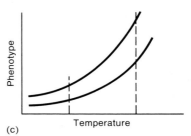

(c)

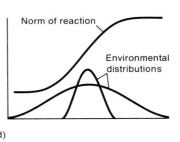

(d)

14. $H^2 = 0.5002$ or 0.5068, depending on which formula is used.

$h^2 = 0.2297$

Chapter 24

6. a. If the population is in equilibrium, $p^2 + 2pq + q^2 = 1$. Use p from the data to predict the frequency of q, then check the calculated values against the observed.

$$p = 0.5249$$

$$q = 1 - p = 0.4751 = \text{predicted value}$$

The phenotypes should be distributed as follows if the population is in equilibrium:

$$L^M L^M = p^2(1482) = 408$$

$$L^M L^M = 2pq(1482) = 739$$

$$L^N L^N = q^2(1482) = 334$$

The population is obviously in equilibrium.

b. If mating is random with respect to blood type, then the following frequency of matings should occur:

$$L^M L^M \times L^M L^M = p^2 \times p^2 \times 741 = 56.25$$

$$L^M L^M \times L^M L^N = 2p^2 \times 2pq \times 741 = 203.6$$

$$L^M L^M \times L^N L^N = 2p^2 \times q^2 \times 741 = 92$$

$$L^M L^N \times L^M L^N = 2pq \times 2pq \times 741 = 184.28$$

$$L^M L^N \times L^N L^N = 2 \times 2pq \times q^2 \times 741 = 166.8$$

$$L^N L^N \times L^N L^N = q^2 \times q^2 \times 741 = 37.75$$

The mating is obviously random with respect to blood type.

7. a.

Generation	p Male	p Female
0	0.8	0.2
1	0.2	0.5
2	0.5	0.35
3	0.35	0.425
.	.	.
.	.	.
.	.	.
n	$p_{f(n-1)}$	$p_{\frac{1}{2}[m(n-1)+f(n-1)]}$

where m = male and f = female.

b. Let p = frequency in males and P = frequency in females. For any generation

$$p_n = P_{n-1} \text{ and } P_n = \tfrac{1}{2}(p_{n-1} + P_{n-1}).$$

The difference between these two, d, is

$$d = \tfrac{1}{2}(p_{n-1} + P_{n-1}) - P_{n-1}$$

$$= \tfrac{1}{2}(p_{n-1} - P_{n-1})$$

Given the initial values of p_0 and P_0,

$$d_n = \tfrac{1}{2}^n (p_0 - P_0).$$

8.

Population	p	q	Equilibrium?
1	1.0	0.0	yes
2	0.5	0.5	no
3	0.0	1.0	yes
4	0.625	0.375	no
5	0.3775	0.625	no
6	0.5	0.5	yes
7	0.5	0.5	no
8	0.2	0.8	yes
9	0.8	0.2	yes
10	0.993	0.007	yes

c.

$$4.9 \times 10^{-6} = 5 \times 10^{-6}/s$$

where $s = 0.102$.

d.

Genotype	Frequency	Fitness	Gametes	A	a
AA	0.25	1.0	0.25	0.25	0.0
Aa	0.50	0.8	0.40	0.20	0.20
aa	0.25	0.6	0.15	0.0	0.15
				0.45	0.35

$$p = 0.45/(0.45 + 0.35) = 0.56$$

$$q = 0.35/(0.45 + 0.35) = 0.44$$

13. Prior to migration $q^A = 0.1$ and $q^B = 0.3$. Immediately after migration, $q^{A+B} = \tfrac{1}{2}(q^A + q^B) = \tfrac{1}{2}(0.1 + 0.3) = 0.2$. The frequency of affected males is 0.2 and the frequency of affected females is $(0.2)^2 = 0.04$.

16. The allele frequencies are:

$$A: 0.2 + \tfrac{1}{2}(0.60) = 50\%$$

$$a: \tfrac{1}{2}(0.60) + 0.2 = 50\%$$

The alleles will randomly unite within a phenotype. For $A-$, the mating population is $0.2\ AA + 0.6\ Aa$. The allele frequencies within this population are:

$$A: [0.2 + \tfrac{1}{2}(0.6)]/0.8 = 0.625$$

$$a: \tfrac{1}{2}(0.6)/0.8 = 0.375$$

The phenotypic frequencies that result are:

$$A-: p^2 + 2pq = (0.625)^2 + 2(0.625)(0.375)$$
$$= 0.3906 + 0.4688$$
$$= 0.8594$$

$$aa: q^2 = (0.375)^2 = 0.1406$$

However, because this subpopulation represents 0.8 of the total population, these figures must be adjusted to reflect that:

$$A-: (0.8)(0.8594) = 0.6875$$

$$aa: (0.8)(0.1406) = 0.1125.$$

The aa contribution from the other subpopulation will remain unchanged because there is only one genotype, aa. The contribution to the total phenotypic frequency is

0.20. Therefore, the final phenotypic frequencies are $A-$ $= 0.6875$ and $aa = 0.20 + 0.1125 = 0.3125$. These frequencies will remain unchanged over time, but the end result will be two separate populations, AA and aa, which will not interbreed.

If assortive mating is between unlike phenotypes, the two type of progeny will be Aa and aa. AA will not exist. The frequency of Aa will result from all $AA \times aa$ matings and one half of the $Aa \times aa$ matings. The matings will occur with the following frequency:

$$AA \times aa = (0.2)(0.2) = 0.04$$
$$Aa \times aa = (0.6)(0.2) = 0.12$$

Because these are the only matings that will occur, they must be put on a 100 percent basis by dividing by the total frequency of matings that occur:

$AA \times aa$: $0.04/0.16 = 0.25$, all of which will be Aa

$Aa \times aa$: $0.12/0.16 = 0.75$, half Aa and half aa

The phenotype frequencies in the next generation will be:

Aa: $0.25 + 0.75/2 = 0.625$

aa: $0.75/2 = 0.375$

In the second generation, the same method will result in a final ratio of $0.5\ Aa : 0.5\ aa$. These values will remain unchanged after the second generation of negative assortive mating.

18. **a.** $p' = 0.528$
    b. $\hat{p} = 0.75$

19. 0.01

20. Affected individuals $= Bb = 2pq = 4 \times 10^{-6}$. Because q is almost equal to 1.0, $2p = 4 \times 10^{-6}$. Therefore, $p = 2 \times 10^{-6}$.

$$u = hsp = (1.0)(0.7)(2 \times 10^{-6}) = 1.4 \times 10^{-6}$$

21. Therefore, the probability of not getting a recessive lethal genotype is $1 - \frac{1}{8} = \frac{7}{8}$. If there are n lethal genes, the probability of not being homozygous for any of them is $(\frac{7}{8})^n = \frac{13}{31}$. From log tables, $n = 6.5$, or an average of 6.5 recessive lethals in the human genome.

22. **a.** $\hat{q} = 4.47 \times 10^{-3}$
        $sq^2 = 10^{-5}$
    b. $\hat{q} = 6.32 \times 10^{-3}$
        $sq^2 = 2 \times 10^{-5}$
    c. $\hat{q} = 5.77 \times 10^{-3}$
        $sq^2 = 10^{-5}$

Index